HANDBOOK OF
POSITION LOCATION

SECOND EDITION

HANDBOOK OF POSITION LOCATION
Theory, Practice, and Advances

Edited by

S. A. (REZA) ZEKAVAT and R. MICHAEL BUEHRER

John B. Anderson, *Series Editor*

MATLAB and Simulink are registered trademarks of The MathWorks, Inc. See www.mathworks.com/ trademarks for a list of additional trade marks. **The MathWorks Publisher Logo identifies books that contain MATLAB® content. Used with permission. The MathWorks does not warrant the accuracy of the text or exercises in this book or in the software downloadable from** http://www .wiley.com/WileyCDA/WileyTitle/productCd-047064477X.html **and** http://www.mathworks.com/ matlabcentral/fileexchange/?term=authored%3A80973. **The book's or downloadable software's use or discussion of MATLAB® software or related products does not constitute endorsement or sponsorship by The MathWorks of a particular use of the MATLAB® software or related products.**

For MATLAB® and Simulink® product information, in information on other related products, please contact:

The MathWorks, Inc.
3 Apple Hill Drive
Natick, MA 01760-2098 USA
Tel: 508-647-7000
Fax: 508-647-7001
E-mail: info@mathworks.com
Web: www.mathworks.com

For general information on our other products and services or for technical support, please contact our Customer Care Department within the United States at (800) 762-2974, outside the United States at (317) 572-3993 or fax (317) 572-4002.

Wiley also publishes its books in a variety of electronic formats. Some content that appears in print may not be available in electronic formats. For more information about Wiley products, visit our web site at www.wiley.com.

Library of Congress Cataloging-in-Publication Data is available.

ISBN 978-1-119-43458-0

Printed in the United States of America.

Dedicated to our wives and children

S. A. (Reza) Zekavat: Fatemeh (wife), Maryam, Melica, Mona (children)
R. Michael Buehrer: Andrea (wife), Faith, JoHannah, Noah, Ellie, Ruthie and Judah (children)

CONTENTS

PART VI *Network Localization*

Part VII *Special Topics and Applications*

PREFACE

Radio systems capable of localization find applications in many areas including homeland security, law enforcement, emergency response, defense command and control, multi-robot coordination, and vehicle-to-vehicle and vehicle-to-pedestrian collision avoidance. In fact, high resolution localization is vital for many applications, including: traffic alerts, emergency services (e.g., indoor localization for firefighters), and battlefield command and control. These systems promise to dramatically reduce society's vulnerabilities to catastrophic events and improve its quality of life.

While work in the important areas of localization is progressing, limited resources are available to support graduate students and researchers: more specifically, a limited number of books have been published in this area, and the existing books don't have sufficient depth or breadth to be considered a handbook on positioning techniques.

This handbook presents a review of both classic and emerging techniques, and introduces advanced practical methods in localization and positioning. The chapters will allow any working engineer or graduate student to quickly come up to speed on a specific topic. The handbook will also help university professors to teach the fundamentals of wireless localization in graduate schools worldwide. The intended audience thus includes graduate students, researchers and industry.

Many chapters that explain the fundamentals of localization systems include MATLAB examples and their solutions. MATLAB examples have been designed to help researchers and graduate students learn the fundamental algorithms for positioning and get started with their research at a faster pace. All MATLAB codes are available online and the readers are provided with a unique user-id and password to access this code.

The handbook's second edition addresses shortcomings of the previous edition and includes additional emerging topics in localization. There has been considerable activity in the field since the first edition was published in 2011, including such applications as GPS, drones and autonomous vehicles. The new edition adds a complete review on techniques for localization error evaluation. A comprehensive chapter on Kalman filtering has also been added. A new Part V which focuses on Global Positioning (e.g., GPS), a subject of obvious importance, has been added to enhance a weakness in the first edition. In part VII, chapters on the new fields of autonomous driving and visible light localization have been added.

PART OPENINGS

PART I   FUNDAMENTALS OF POSITION LOCATION

The opening section includes five chapters, Chapter 1–5, which review the basic techniques, methods and research topics in the field of localization. The new Chapter 2 explains localization error assessment metrics. Chapters 3, 4, and 5, offer many

MATLAB examples and the associated code that is especially valuable for researchers and students who are new to the field.

Chapter 1 offers a short introduction to the localization techniques that will be discussed in the entire handbook. In addition, problems in the implementation of localization algorithms such as the availability of Line-of-Sight (LOS) are discussed. Techniques such as time-of-arrival (TOA), direction-of-arrival (DOA), Time-Difference-of-Arrival (TDOA) and received signal strength (RSS) are introduced in the chapter. In addition, the chapter reviews and compares examples of several localization systems such as Radar, Inertial Navigation System (INS) and Wireless Local Positioning Systems.

Chapter 2 offers an efficient review of the key measures for localization performance evaluation. It uses extensive tables to assess the error (and sources of error) in localization devices and methods. Moreover, it evaluates and compares different localization fusion methods in terms of complexity, convergence rate and accuracy. Furthermore, the chapter offers practical methods for error evaluation and correction that include calibration, data validation and partial error correction procedures.

Chapter 3 complements Chapters 1 and 2 by providing a theoretical analysis of algorithms for DOA, TOA and RSS localization. This Chapter introduces two categories of positioning algorithms based on TOA, TDOA, RSS and DOA measurements: (1) Algorithms that implement the nonlinear equations directly obtained from the relationships between the source and measurements. Corresponding examples, namely nonlinear least squares (NLS) and maximum likelihood (ML) estimators, are presented; and (2) Algorithms that attempt to convert the equations to simpler linear approximations. Examples of the second type are the linear least squares (LLS), weighted linear least squares (WLLS) and subspace approaches.

Chapter 4 presents fundamentals of wireless channel modeling. Specifically, Chapter 4 discusses channel parameters that impact different localization systems such as DOA, TOA and RSS. The spatial, temporal, and spectral correlation models for wireless channels are discussed. The chapter also discusses important channel models and the statistics of key parameters in those channel models.

Chapter 5 offers a nice and comprehensive overview of different Kalman filtering methods. The review includes details and issues in developing Kalman Filter code in MATLAB. It serves as a solid resource for engineers who need to develop MATLAB-based algorithms for Kalman filtering applications. The chapter provides many application-based examples including the application of localization to drone navigation, capsule endoscopy localization and satellite navigation/localization.

PART II   TOA AND DOA BASED POSITIONING

The second part of the handbook consists of five chapters, Chapters 6–10, which detail TOA and DOA localization methods. All chapters in this part include invaluable MATLAB examples and solutions and code that will help researchers as they begin their work in the relevant areas.

Chapter 6 presents principle methods of localization of a target based on the availability (and applicability) of a TOA estimation method. Although these techniques were introduced in Chapter 2, here we expand our analysis specifically on

TOA-based approaches. The chapter compares the performance of TOA and TDOA methods as well as their geometric interpretations. Linearization techniques are introduced to reduce the computational complexity of underlying nonlinear TOA and TDOA objective functions. The performance of the presented techniques is also investigated. Finally, we examine the impact of non-line-of-sight propagation on lateration-based positioning.

While the previous chapter relies on the existence of TOA measurements, **Chapter 7** complements Chapter 6 by presenting key TOA estimation methods. It introduces important measures that compare different TOA estimation techniques. Examples of these measures are sensitivity to signal-to-noise ratio, the number of reflections, resolution and complexity. Then the chapter introduces examples of popular TOA methods. Examples of these TOA methods are Maximum Likelihood, and sub-space methods such as MUSIC. The chapter compares the presented TOA techniques. In addition, the chapter introduces a novel TOA estimation method that is applied in the frequency domain and based on Independent Component Analysis.

Chapter 8 reviews TOA estimation methods via ultra-wide-band (UWB) systems. The chapter offers a short review of UWB signals/systems which are particularly relevant and useful for TOA. Moreover, it introduces recent research progress in UWB-based localization including fingerprinting and geometric techniques. In addition, the issues surrounding localization in indoor areas are highlighted in this chapter.

Chapter 9 presents DOA estimation methods. The chapter starts with a review of important antenna and antenna array parameters and their impact on DOA estimation. Next. it introduces important measures for comparing different types of DOA estimation techniques. Examples of these measures include calibration sensitivity, angular resolution and complexity. Popular DOA estimation methods such as "delay and sum," MUSIC and root MUSIC are introduced and compared. In addition, Chapter 9 studies new and practically implementable DOA estimation methods. Finally, this chapter investigates the impact of non-stationary signals. Specifically, non-stationary behavior creates low performance sample auto-correlation estimation which in turn reduces DOA performance in sub-space techniques such as MUSIC. The chapter presents a solution to this problem.

Chapter 10 investigates high resolution TOA estimation techniques for inhomogeneous media. It also studies straight line distance estimation within inhomogeneous media via a combined TOA and DOA estimation to calculate the true range between the transmitter and receiver as the propagation path of transmitted waveform is not straight-line.

PART III   RECEIVED SIGNAL STRENGTH BASED POSITIONING

One measurement that is ubiquitous in wireless systems is Received Signal Strength (RSS). Thus, while RSS-based localization is *typically* less accurate than TOA-based positioning, it is still a very important technique since it can be implemented with little to no modification to existing systems. Thus, the third part of the handbook which comprises five chapters, Chapters 11–15 study the fundamentals of RSS-based methods and their potential for indoor localization.

Specifically, three types of localization are presented: model-based techniques, kernel-based methods and fingerprinting techniques. This section reviews the applications and performance of RSS techniques in various environments including in buildings and in underground mining environments as well. All chapters in this part include MATLAB-based examples, solutions and the original code to help researchers quickly come up to sped in this area.

Chapter 11 addresses the fundamental aspects of location estimation based on received signal strength (RSS) measurements. Specifically, it provides an overview of RSS-based position location including major techniques and the primary sources of error. While the chapter introduces RF fingerprinting and proximity-based non-parametric methods, of particular interest in this chapter are range-based localization techniques that use a statistical radio propagation model. The chapter discusses geometric interpretations of both RSS and Differential RSS techniques, their solutions, achievable location accuracy, and optimal/practical estimators. Finally, it is shown via simulation how spatially correlated shadow fading, the number of anchor nodes, and path loss rate affect performance.

Chapter 12 compliments Chapter 11 by providing an overview of various indoor localization techniques employing RSS including lateration methods, machine learning classification, probabilistic approaches, and statistical supervised learning techniques and comparing their performance in terms of localization accuracy through measurement studies in real office building environments. The chapter further surveys emerging techniques and methods to support robust wireless localization and improve localization accuracy, including real-time RSS calibration via anchor verification, closely spaced multiple antennas, taking advantage of robust statistical methods to provide stability to contaminated measurements, utilizing linear regression to characterize the relationship between RSS and the distance to anchors, and exploiting RSS spatial correlation. Finally, the chapter concludes with a discussion of popular location-based applications.

Both the placement and the selection of anchors can affect the accuracy of location estimation significantly. Thus, in addition to developing methods to localize wireless devices, it is equally important to systematically investigate how the placement of the anchors specifically impacts localization performance.

Chapter 13 investigates the problem of anchor placement and anchor selection. In terms of anchor placement, how the geometric layout of the anchors affects the localization performance is studied. Different approaches to anchor placement including heuristic search methods, acute triangular based deployment, adaptive beacon placement, and optimal placement solutions via the maximum lambda and minimum error (maxL-minE) algorithm are discussed. Additionally, this chapter studies how smartly selected anchors can achieve comparable or better localization results as compared to using all the available anchors. A variety of effective strategies are investigated including a joint clustering technique, entropy-based information gain, convex hull selection, and smart selection in the presence of high density of anchors.

As opposed to model-based methods, **Chapter 14** explores the features and advantages of kernel-based localization. Kernel methods simplify received signal strength (RSS)-based localization by providing a means to learn the complicated relationship between RSS measurement vectors and position as opposed to assuming a specific model. The chapter discusses their use in self-calibrating indoor localization systems. The chapter reviews four kernel-based localization algorithms and presents a common framework for their comparison. Results from two simulations and an extensive measurement data set are presented which provide a quantitative comparison of the various techniques.

In **Chapter 15** the principles of radio-frequency (RF) fingerprinting methods, also known as database correlation methods (DCM), are presented. Although there is a wide variety of DCM implementations, they all share the same basic elements. These elements are identified and analyzed in this chapter. Different alternatives for building RF fingerprint correlation databases (CDBs) are also presented and the advantages and drawbacks of each of them are discussed, as well as their impact on the positioning accuracy. Different methods to numerically evaluate the similarity between each fingerprint within the CDB and the measured RF fingerprint are described, including an artificial neural network based approach. Two alternatives to reduce the correlation or search space are analyzed, including an approach based on genetic algorithms. Finally, field tests are used to evaluate the positioning accuracy of fingerprinting techniques in both GSM (Global System for Mobile Communications) and Wi-Fi (Wireless Fidelity) 802.11b networks.

PART IV LOS/NLOS LOCALIZATION – IDENTIFICATION – MITIGATION

Non-line-of-sight (NLOS) propagation is an important issue in localization systems since many techniques are based on lateration which requires range measurements and range measurements inherently assume LOS signal propagation. The fourth part of the handbook is composed of four chapters, Chapters 16–19, which examine NLOS identification, mitigation and localization methods. NLOS identification can avoid major positioning errors. Researchers have recently started working in this area and many novel techniques have been proposed. All three chapters come with critical MATLAB examples, and solutions.

Chapter 16 reviews many NLOS identification and localization techniques and compares them in terms of complexity and performance. NLOS identification techniques can be categorized into cooperative and non-cooperative. Cooperative localization techniques use multiple nodes to identify NLOS measurements. Non-cooperative NLOS identification are based on single-node channel measurements. These techniques rely on: (1) the range (TOA) statistics; (2) channel characteristics, such as received signal power, Rician K-factor and features extracted from the power delay profile; and (3) the consistency between the TOA measurement and path loss for LOS/NLOS, and the consistency between the direction of departure (DOD) and DOA. In the second group, suitable channel characteristics used in narrow/wide band systems, ultra-wide-band (UWB) systems and systems equipped with an antenna array are discussed. NLOS localization techniques can also be cooperative and non-cooperative. These techniques are categorized into: (1) NLOS localization using TOA, DOA measurements and an environment map; (2) localization using the measurements from reflectors. This chapter discusses the advantages and disadvantages of these techniques and the complexity and performance of each technique.

Chapter 17 introduces NLOS mitigation methods. Generally, the NLOS mitigation methods for geolocation can be grouped into four categories. Mitigation methods based on maximum likelihood, least squares and constrained optimization as well as robust statistics are discussed. These methods are then compared in terms of different performance measures. In addition, this chapter discusses a novel geolocation example using a single moving sensor. Numerical examples are then presented to demonstrate how position estimation can be achieved for the case of a single moving sensor with NLOS errors.

Chapter 18 investigates TOA estimation techniques for hybrid RSS-TOA localization in mixed LOS/NLOS environments. The chapter explains how to choose a sufficient number of TOA measurements so that the computational burden of the position estimation can be reduced without performance degradation. In addition, this chapter investigates how to determine a mobile's position using a number of TOAs while maintaining reasonable error performance. The results can help system designers to manage the tradeoff between accuracy and computational complexity.

Chapter 19 investigates the problem of mobile tracking in mixed line-of-sight (LOS)/non-line-of-sight (NLOS) conditions. The state-of-the-art methods in this field are first reviewed. Two types of Bayesian filters are studied: the Gaussian mixture filter (GMF) and the particle filter (PF). The modified extended Kalman filter (EKF) bank method, as one specific GMF, is described. The chapter also discusses the computation of a posterior Cramer-Rao lower bound (CRLB) for the mobile tracking problem.

PART V GLOBAL POSITIONING

With the rapid emergence of global positioning in autonomous vehicles, this part of the handbook includes four chapters, Chapters 20–23, that discuss the details of positioning based on Global Navigation Satellite Systems (GNSS).

Chapter 20 discusses the fundamentals of satellite navigation systems, i.e., GNSS. The chapter introduces the fundamentals of global position estimation using Time-of-Arrival estimation using signals received from a cluster of satellites. The impact of the satellite constellation geometry on the positioning accuracy is assessed. Further, an overview of current and future GNSS, signal formats and modulations is presented. This chapter also offers MATLAB examples as well as the corresponding code.

Chapter 21 introduces the general architecture of a GNSS receiver. The signal processing performed in the various blocks of the receiver is described in detail. The different functions of the receiver are explained including signal detection and acquisition for the satellites in view. The chapter also discusses low complexity signal detection methods such as the Fast Fourier Transform (FFT), and Bayesian detection theory as implemented in typical GNSS receivers. This chapter also includes MATLAB examples and their corresponding code.

Chapter 22 discusses the implementation of Kalman filtering for satellite navigation. In particular, the structure of the KF is exploited in a GNSS receiver to compute the position of the user and to integrate the standard GNSS receiver with an Inertial Navigation System (INS). It discusses the characterization of the inertial devices in terms of both deterministic and stochastic noise, calibration and alignment, as fundamental pre-requisites for any fruitful integration process. The chapter presents three typical GNSS-INS integration approaches: (a) loosely integrated, (2) tightly integrated and (3) ultra-tightly integrated architectures. Finally, the results of a live vehicular test campaign to compare two such implementations are shown and discussed.

Chapter 23 presents strategies and techniques that increase the sensitivity of GNSS receivers, to make them usable in difficult environments, such as urban canyons, indoor scenarios, deep forests, or even space. The chapter discusses assisted GNSS receiver. It also introduces approaches to increase the sensitivity at the acquisition stage, including delay and Doppler shift estimation/compensation, and the intrinsic limitations to coherent and non-coherent integration time extension.

PART VI   NETWORK LOCALIZATION

Classical position location is based on locating a target/mobile based on measurements to known anchors. In many applications what is needed are the positions of many nodes, some of which many not have connectivity to anchor nodes. In such cases, utilizing measurements between unlocalized nodes can be used to localize all of the nodes in the network. This is often termed *network localization, cooperative localization* or *collaborative localization* and is the subject of Part VI.

This part includes five chapters, Chapter 24–28, that detail several topics in the area of network-based or cooperative localization. The section also introduces techniques such as infrastructure-free local positioning system and wireless local positioning systems. Further, this part overviews the error characteristics of *ad hoc* positioning systems.

Chapter 24 describes the fundamentals of network (collaborative) position location, that is position location techniques where the nodes to be localized collaborate to determine their positions. This is sometimes also called cooperative position location or network position location to distinguish it from traditional position location techniques which localize a single node. After a brief introduction and definition of the problem, the chapter examines two main bounds for collaborative positioning: namely the Cramer-Rao Lower Bound (CRLB) and the weighted least squares solution. Although the nature of the bounds is somewhat different, they both provide guidance as to the achievable performance. Several sub-optimal approaches are described and compared in terms of their performance in various network configurations. Lastly, the chapter examines the impact of non-line-of-sight propagation and describes techniques to mitigate its effects.

In **Chapter 25** the authors review a series of results obtained in the field of localization that are based on polynomial optimization. First, they provide a review of a set of polynomial function optimization tools including the sum of squares (SOS) technique. Then they present several applications of these tools to various sensor network localization tasks. As a first application, they propose a method based on SOS relaxation for node localization using noisy measurements and describe the solution through semi-definite programming (SDP). Later, they examine the network localization problem of relative reference frame determination based on range and bearing measurements.

Chapter 26 considers cooperative localization using probabilistic inference. These techniques are able to obtain not only location estimates, but also a measure of the uncertainty of those estimates. Since these methods are computationally very expensive, message-passing methods, which are also known as belief propagation (BP) methods have been proposed. To reduce algorithmic complexity, the use of particle-based approximation via nonparametric belief propagation (NBP) can make BP acceptable for localization in sensor networks. In this chapter, after a brief introduction to cooperative localization, the authors describe BP/NBP techniques and generalizations (GBP) for loopy networks. Due to the poor performance of BP/NBP methods in loopy networks, they introduce three methods: GBP based on Kikuchi approximation (GBP-K), nonparametric GBP based on junction tree (NGBPJT), and NBP based on spanning trees (NBP-ST).

Chapter 27 examines network localization techniques using multi-hop techniques that are based on the idea of distance vector routing to find the positions of all nodes in the network. Specifically, this chapter examines the performance of four basic multi-hop algorithms: range/angle-free algorithms, range-based algorithms, angle-based algorithms and multimodal algorithms. The error characteristics of all four classes are examined under various conditions using both theoretical analysis and simulations. The authors discuss the trade-offs involved among the capabilities used in each node, the density of the network, the ratio of anchors (landmarks) and the quality of the position estimates obtained.

Chapter 28 examines the problem of localizing multiple nodes using bearing measurements. The chapter begins by treating the problem of localizing three agents moving in a plane when the inter-agent distances are known, and in addition, the angle subtended at each agent by lines drawn from two landmarks at known positions is also known. It is shown that there are in general a finite number of possible sets of positions for the three agents. Afterwards, the generalization of the result for more than three agents is presented. Examples are given to show the applicability of the methods.

Part VII   Special Topics and Applications

The final section of the book includes seven chapters, Chapters 29–35, that review several novel applications of position location systems. Many techniques and methods including GNSS, and RFID-based localization systems are discussed in detail. Moreover, an interesting application of localization to wild-life tracking is discussed. Next, an example of a remote positioning system, called Wireless Local Positioning System (WLPS) is introduced. Finally, the application of localization to autonomous driving is also discussed in two different chapters in this section.

Chapter 29 highlights localization technologies for the emerging field of autonomous driving. It reviews all localization technologies that are key to autonomous driving. It offers a detailed review of image-based localization methods. Localization using range sensors, vision sensors, and SLAM are highlighted. In addition, cooperative location estimation, filtering and sensor fusion are discussed. The chapter also presents localization techniques in use or in active research and briefly present future directions of localization in autonomous driving.

Chapter 30 introduces the use of RFID for navigation of robots. Mobile robots are becoming intelligent, autonomous systems, gradually becoming able to accomplish assigned tasks such as guidance, transportation, and human-robot cooperation. For these tasks, a robust and stable navigation system is a key requirement. The chapter investigates how RFID enables high precision navigation of such robots.

Chapter 31 offers an overview of the technologies that have paved the way for visible light communication and positioning systems. The chapter also focuses on visible light positioning (VLP) systems. It also discusses how arrangements of different light detectors can improve VLP system accuracy and usability. Many techniques to use visible light for communications are also discussed.

Chapter 32 examines fourth generation cellular systems and describes the positioning standard for the Long Term Evolution (LTE) cellular system. The chapter reviews the LTE architecture and air-interface, as well as the positioning requirements, positioning architectures and signaling. The simplest LTE-based techniques, the cell ID method, is discussed first. This method is often augmented with timing advance and angle of arrival information in what is called enhanced cell ID (E-CID) class of methods. Two approaches to fingerprinting-based positioning are described, including pattern matching and the self-learning adaptive enhanced cell ID (AECID) method. The LTE standard also supports a downlink observed time difference of arrival (OTDOA) approach which is discussed. Finally, the satellite navigation functionality provided by A-GNSS is described.

Chapter 33 describes the application of position location techniques to wildlife tracking. Specifically, the chapter begins with a discussion of currently available wildlife tracking devices (tags), and explains why tag mass is the primary design constraint. Current manual direction-finding methods are described, as are several automated implementations. The authors also discuss the need for generic asset (non-wildlife) tracking tags that are light and inexpensive, and review current asset tracking methods based on cellular and satellite platforms in this context. Two popular satellite-based location-tracking technologies, GPS and Argos, are described, as are several other tracking techniques. The shortcomings of existing systems motivate the need for a new approach that offers GPS-like accuracy, with vastly reduced energy consumption. Terrestrial TOA tracking methods are discussed as a lower energy cost solution, with specific sections dedicated to explaining the various concepts and their interplay in an integrated system. The chapter also introduces a few basic concepts needed for system analysis including spectrum utilization, auto-correlation, cross-correlation, signal to noise ratio, link budget and processing gain. Various positioning methods are compared using real and simulated data. The signal processing discussion details the computational requirements of real-time matched filtering, including the impact of Doppler shift. Several techniques used to implement real-time TOA receivers in embedded devices with limited computing resources, including the use of frequency domain processing via the FFT, and the intelligent reuse of data via time-shifting techniques are also described. The chapter concludes with a summary of the current performance of a TOA wildlife tracking system that the authors have implemented, its design limitations, and likely areas for improvement.

Chapter 34 introduces an active target remote positioning system called wireless local positioning system (WLPS) that has been recently developed and patented.

The structure and block diagram of WLPS is discussed in the chapter. The chapter studies the implementation of the WLPS receiver and its performance. In addition, the implementation of DOA and TOA estimation techniques in WLPS are presented. One of the features of WLPS signals, called cyclo-stationary, is investigated and its impact on DOA estimation and the detection process is discussed. The chapter also presents a variety of civilian and military applications of WLPS and its impact on safety and security.

Chapter 35 is a complementary chapter to autonomous driving. It offers near-ground channel path-loss modeling that is key to both autonomous driving sensors (such as those installed on vehicle bumpers), as well as the general concept of near ground sensor localization. An accurate channel pathloss is key to RSSI-based localization methods and this chapter details.

S. A. (REZA) ZEKAVAT
Michigan Tech
R. MICHAEL BUEHRER
Virginia Tech

CONTRIBUTORS

Ossama O. Abdelkhalik, Michigan Tech
Piyush Agrawal, University of Utah
Brian D. O. Anderson, The Australian National University and National ICT
Mojtaba Bahramgiri, Michigan Tech
Allert Bijleveld, NIOZ Royal Netherlands Institute for Sea Research
R. Michael Buehrer, Virginia Tech
Rafael Saraiva Campos, Universidade do Estado do Rio de Janeiro
Liang Chen, Wuhan University, Wuhan, China
Yinging Chen, Rutgers University, Piscataway, NJ
Kathryn Cortopassi, Cornell University
Fabrizio Dominici, Istituto Superiore Mario Boella
Fabio Dovis, Politecnico di Torino
Guillermo Enriquez, Waseda University
Gianluca Falco, Istituto Superiore Mario Boella, Torino
Emanuela Falletti, Istituto Superiore Mario Boella, Torino
Maurizio Fantino, Istituto Superiore Mario Boella
Baris Fidan, University of Waterloo
Rich Gabrielson, Cornell University
Shu Ting Goh, National University of Singapore
Marco Gruteser, Rutgers University
Shuji Hashimoto, Waseda University
Hatem Hmam, Defense Science & Technology Organization, Australia
Mohsen Jamalabdollahi, Michigan Tech
Tao Jia, FitBit
Thomas Kaiser, University of Duisburg Essen
Ari Kangas, Ericsson AB
Stuti Kansal, Michigan Tech
Jeong Heon Lee, Nokia
Alan Levesque, WPI
Joni Polili Lie, Nanyang Technological University, Singapore
Nicola Linty, Politecnico di Torino
Lisandro Lovisolo, Universidade do Estado do Rio de Janeiro
Simo Ali-Löytty, Tampere University of Technology
Robert MacCurdy, University of Colorado, Boulder
Davide Margaria, Istituto Superiore Mario Boella
Richard P. Martin, Rutgers University
William W. Melek, University of Waterloo, ON, Canada
Paolo Mulassano, Istituto Superiore Mario Boella
Dragoş Niculescu, University Politehnica of Bucharest
Sunhong Park, Waseda University

Neal Patwari, University of Utah
Robert Piche, Tampere University of Technology
Marco Pini, Istituto Superiore Mario Boella
Mohsen Pourkhaatoun, Microchip Technology
Letizia Lo Presti, Politecnico di Torino
Vladimir Savic, Polytechnic University of Madrid
Chong-Meng Samson See, DOS National Labs
Iman Shames, University of Melbourne
Bamrung Tau Sieskul, University of Vigo
Iana Siomina, Ericsson AB
H. C. So, City University of Hong Kong
Amir Torabi, Michigan Tech
Wade Trappe, Rutgers University
Zhonghai Wang, Michigan Tech
Torbjörn Wigren, Ericsson AB
Ami Woo, University of Waterloo, ON, Canada
Lenan Wu, Southeast University
Huilin Xu, Qualcomm
Wenjie Xu, Michigan Tech
Liuqing Yang, Colorado State University
Jie Yang, Florida State University, Tallahassee, FL
Swaroop Venkatesh, Qualcomm, Inc.
Santiago Zazo, Polytechnic University of Madrid
S. A. (Reza) Zekavat, Michigan Tech
Feng Zheng, University of Hannover

ABOUT THE COMPANION WEBSITE

This book is accompanied by a companion website:

www.wiley.com/go/zekavat/positionlocation2e

The website includes:
- MATLAB codes for various chapters in this book

PART 1

FUNDAMENTALS OF POSITION LOCATION

THE OPENING section includes five chapters, Chapter 1–5, which review the basic techniques, methods and research topics in the field of localization. The new Chapter 2 explains localization error assessment metrics. Chapters 3, 4, and 5, offer many MATLAB examples and the associated code that is especially valuable for researchers and students who are new to the field.

Chapter 1 offers a short introduction to the localization techniques that will be discussed in the entire handbook. In addition, problems in the implementation of localization algorithms such as the availability of Line-of-Sight (LOS) are discussed. Techniques such as time-of-arrival (TOA), direction-of-arrival (DOA), Time-Difference-of-Arrival (TDOA) and received signal strength (RSS) are introduced in the chapter. In addition, the chapter reviews and compares examples of several localization systems such as Radar, Inertial Navigation System (INS) and Wireless Local Positioning Systems.

Chapter 2 offers an efficient review of the key measures for localization performance evaluation. It uses extensive tables to assess the error (and sources of error) in localization devices and methods. Moreover, it evaluates and compares different localization fusion methods in terms of complexity, convergence rate and accuracy. Furthermore, the chapter offers practical methods for error evaluation and correction that include calibration, data validation and partial error correction procedures.

Chapter 3 complements Chapters 1 and 2 by providing a theoretical analysis of algorithms for DOA, TOA and RSS localization. This Chapter introduces two categories of positioning algorithms based on TOA, TDOA, RSS and DOA measurements: (1) Algorithms that implement the nonlinear equations directly obtained from the relationships between the source and measurements. Corresponding examples, namely nonlinear least squares (NLS) and maximum likelihood (ML) estimators, are presented; and (2) Algorithms that attempt to convert the equations to simpler linear approximations. Examples of the second type are the linear least squares (LLS), weighted linear least squares (WLLS) and subspace approaches.

Handbook of Position Location: Theory, Practice, and Advances, Second Edition.
Edited by S. A. (Reza) Zekavat and R. Michael Buehrer.
© 2019 by the Institute of Electrical and Electronics Engineers, Inc.
Published 2019 by John Wiley & Sons, Inc.
Companion Website: www.wiley.com/go/zekavat/positionlocation2e

Chapter 4 presents fundamentals of wireless channel modeling. Specifically, Chapter 4 discusses channel parameters that impact different localization techniques such as DOA, TOA and RSS. The spatial, temporal, and spectral correlation models for wireless channels are discussed. The chapter also discusses important channel models and the statistics of key parameters in those channel models.

Chapter 5 offers a nice and comprehensive overview of different Kalman filtering methods. The review includes details and issues in developing Kalman Filter code in MATLAB. It serves as a solid resource for engineers who need to develop MATLAB-based algorithms for Kalman filtering applications. The chapter provides many application-based examples including the application of localization to drone navigation, capsule endoscopy localization and satellite navigation/localization.

CHAPTER 1

WIRELESS POSITIONING SYSTEMS: OPERATION, APPLICATION, AND COMPARISON

S. A. (Reza) Zekavat,[1] Stuti Kansal,[1]
and Allen H. Levesque[2]

[1]Michigan Technological University, Houghton, MI
[2]Worcester Polytechnic Institute, Worcester, MA

RECENT YEARS have seen rapidly increasing demand for services and systems that depend upon accurate positioning of people and objects. This has led to the development and evolution of numerous positioning systems. This chapter provides an overview of the main positioning techniques: time of arrival (TOA), direction of arrival (DOA) and received signal strength indicator (RSSI). It then introduces positioning systems that are either in use or being developed for a variety of applications. Operations of these positioning systems are summarized using flowcharts and figures. In addition, the chapter compares positioning systems on the basis of system characteristics and performance parameters. Many of these positioning techniques and systems are introduced in greater details throughout different parts of this handbook. The chapter concludes by reviewing a number of emerging positioning systems and outlining some future applications.

1.1 INTRODUCTION

Positioning systems determine the location of a person or an object either relative to a known position or within a coordinate system [1]. In the last few decades, various positioning systems have been motivated by demand and developed.

Handbook of Position Location: Theory, Practice, and Advances, Second Edition.
Edited by S. A. (Reza) Zekavat and R. Michael Buehrer.
© 2019 by the Institute of Electrical and Electronics Engineers, Inc.
Published 2019 by John Wiley & Sons, Inc.
Companion Website: www.wiley.com/go/zekavat/positionlocation2e

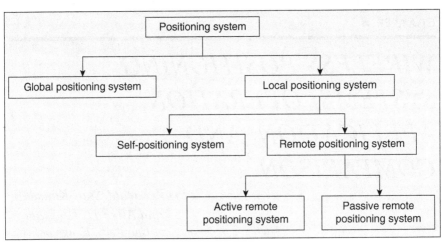

Figure 1.1   Range-based positioning system classification.

Some of the applications of positioning systems include (but are not limited to) law enforcement, security, road safety, tracking personnel, vehicles, and other assets, situation awareness, and mobile ad hoc networks.

In general, localization systems are divided into two categories: range-based and range-free. Range-based localization techniques utilize received signal features such as TOA [41, 42], DOA [43], and received signal strength (RSS). Range-free localization uses network connectivity and network localization.

As shown in Figure 1.1, range-based positioning systems can be classified into two categories:

1) Global positioning

2) Local positioning

Global positioning systems (GPS) allow each mobile to find its own position on the globe. A local positioning system (LPS) is a relative positioning system and can be classified into self- and remote positioning. Self-positioning systems allow each person or object to find its own position with respect to a static point at any given time and location. An example of these systems is the inertial navigation systems (INS).

A remote positioning system allows each node to find the relative position of other nodes located in its coverage area. Here, nodes can be static or dynamic. Remote positioning systems themselves are divided into:

a. Active target remote positioning

b. Passive target remote positioning

In the first cast, the target is active and cooperates in the process of positioning, while in the second the target is passive and noncooperative. Examples of active target positioning systems are radio-frequency identification (RFID), Wireless local positioning systems (WLPS) [2], and traffic alert and collision avoidance systems

(TCAS) [2]. Examples of passive target positioning systems are tracking radars and vision system. Figure 1.1 summarizes the classification of positioning systems.

This chapter reviews the operation of several key positioning systems and compares their operation, application, and pros and cons. Several key positioning parameters such as accuracy, capability in line-of-sight (LOS) versus non-LOS (NLOS) positioning, number of base stations required for positioning, and power consumption are considered as the benchmark for comparison. Moreover, this chapter uses tables to summarize information on the operating ranges of the positioning parameters for the positioning systems. This information will guide system designers in selecting a positioning system for a particular application based on requirements that may be specified using a combination of parameters discussed in this chapter.

Section 1.2 discusses the fundamentals of various techniques that form the basis of almost all positioning systems. Section 1.3 discusses the operation of several key positioning systems, while Section 1.4 compares the positioning systems and highlights their pros and cons. Section 1.5 outlines futuristic applications of several positioning systems.

1.2 BASIC METHODS USED IN POSITIONING SYSTEMS

Here, the fundamental techniques of positioning systems are explained. Different combinations of these techniques form the basis of various positioning systems.

Time-of-Arrival (TOA) Estimation

As detailed in Part II of this handbook, TOA estimation allows the measurement of range or distance; thus, enabling localization. Here, multiple base nodes collaborate to localize a target node via triangulation [3]. It is assumed that the positions of all base nodes are known. If these nodes are dynamic, a positioning technique such as GPS is used to allow base nodes to localize their positions (GPS-TOA positioning). In some circumstances, multiple base nodes may cooperate to find their own position before any attempt to localize a target node [4]. TOA estimation methods are discussed in Part II of this handbook (see Chapters 6–8). Specifically, the TOA estimation process would be complex in inhomogeneous media [39, 40]. Human body is an example of such media. Chapter 10 details TOA estimation in inhomogeneous media.

Assuming known positions of base nodes, and a coplanar scenario, three base nodes and three measurements of distances (TOA) are required to localize a target node (see Fig. 1.2a). In a non-coplanar case, four base nodes are required. Using the measurement of distance, the position of a target node is localized within a sphere of radius R_i with the receiver i at the center of the sphere (where R_i is directly proportional to the TOA τ_i as shown in Fig. 1.2a). The localization of the target node can be carried out either by base nodes using a master station or by the target node itself.

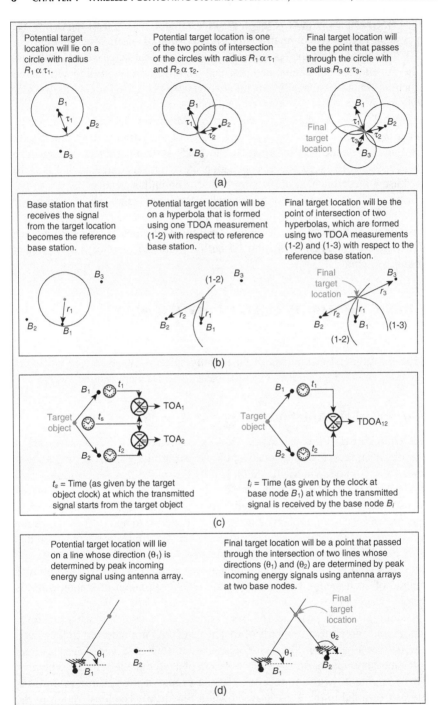

Figure 1.2 (a) Operation of TOA and RSSI, (b) operation of TDOA, (c) comparison of TOA and TDOA calculations, and (d) operation of DOA.

Although TOA seems to be a robust technique, it has a few drawbacks [5]:

a) It requires all nodes (base nodes and target nodes) to precisely synchronize: a small timing error may lead to a large error in the calculation of the distance R_i;

b) The transmitted signal must be labeled with a timestamp in order to allow the base node to determine the time at which the signal was initiated at the target node. This additional timestamp increases the complexity of the transmitted signal and may lead to additional source of error; and,

c) The positions of the base nodes should be known; thus, either static nodes or GPS-equipped dynamic nodes should be used.

Time-Difference-of-Arrival (TDOA) Estimation

As the name suggests, TDOA estimation requires the measurement of difference in time between the signals arriving at two base nodes. Similar to TOA estimation, this method assumes that the positions of base nodes are known [5]. The TOA difference at the base nodes can be represented by a hyperbola. A hyperbola is the locus of a point in a plane such that the difference of distances from two fixed points (called the foci) is a constant.

Assuming known positions of base nodes and a coplanar scenario, three base nodes and two TDOA measurements are required to localize a target node (see Fig. 1.2b). As shown in the figure, the base station that first receives the signal from the target node is considered the *reference base station*. The TDOA measurements are made with respect to the reference base station. For non-coplanar cases, the position of four base nodes and three TDOA measurements are required.

TDOA addresses the first drawback of TOA by removing the requirement of synchronizing the target node clock with the base node clocks. In TDOA, all base nodes receive the same signal transmitted by the target node. Therefore, as long as base node clocks are synchronized, the error in the arrival time at each base node due to unsynchronized clocks is the same.

As shown in Figure 1.2c, TOA is the time duration (or the relative time) between the start time (t_s) of signal at the transmitter (target node) and the end time (t_i) of the transmitted signal at the receiver (base node B_i). However, as shown in Figure 1.2c, TDOA is the time difference between the end times (t_i and t_j) of the transmitted signal at two receivers (base nodes B_i and B_j). Thus, in TDOA technique, only base nodes' clocks need to be synchronized to ensure minimum measurement error. In general, the complexity of target node clock synchronization is higher compared to base node clock synchronization. This is mainly due to the use of quartz clocks at target nodes, which are not as precise as the atomic clocks that are generally used for timing at base nodes [5]. Target node clock synchronization is further explained later in this chapter.

The base node clock can be synchronized externally using a backbone network or internally using timing standards provided at the nodes. The fact that synchronization of target nodes is not required enables many applications for TDOA-based systems. For example, in battlefield applications, a rescue team may localize the position of a soldier using its beacon signal without the need of synchronization of rescue team clocks with that of the soldier.

With respect to the second drawback of TOA, the transmitted signal from the target node in TDOA need not contain a timestamp, since a single TDOA measurement is the difference in the arrival time at the respective base nodes. This simplifies the structure of transmitted signals and removes potential sources of error. This advantage of TDOA is again exploited by many applications, such as emergency call localization on highways [6] and sound source localization by an artificially intelligent humanoid robot [7].

Direction-of-Arrival (DOA) Estimation

In DOA estimation, base nodes determine the angle of the arriving signal (see Fig. 1.2d). To allow base stations to estimate DOA, they should be equipped with antenna arrays, and each antenna array should be equipped with radio-frequency (RF) front-end components. However, this incurs higher cost, complexity, and power consumption. DOA estimation techniques are discussed in Chapter 9 (Part II) of this handbook.

Similar to TOA and TDOA estimation, in DOA estimation the positions of base nodes should be known. However, unlike TOA and TDOA, for the known position of a base node and a coplanar scenario, only two base nodes along with two DOA measurements are required. For a non-coplanar case, three base nodes are required. To determine the DOA, the main lobe of an antenna array is steered in the direction of peak incoming energy of the arriving signal [6].

Received Signal Strength Indicator (RSSI)

Similar to the TOA, in RSSI,multiple base nodes collaborate to localize a target node via triangulation (see Fig. 1.2a). However, instead of measuring TOA at base nodes, the estimation is carried out using the RSS [3]. In this method, the strength of the received signal indicates the distance travelled by the signal. For a coplanar case, assuming that the transmission strength and channel (or environment in which the signal is traveling) characteristics are known, three base nodes and three RSS measurements are required. Part III of this handbook studies RSS-based methods in detail.

Line of Sight (LOS) versus Non-LOS (NLOS)

Compared with RSSI, the performance characteristics of TOA, DOA, and TDOA techniques are very sensitive to the availability of LOS [36–38]. That is, in NLOS situations the computed TOA, DOA, and TDOA are subject to considerable error. However, the performance of the RSSI technique is altered only mildly by the lack of LOS: NLOS leads to a shadowing (random) effect in the power–distance relationship, which can be reduced using filtering techniques. Thus many NLOS identification, mitigation, and localization techniques have been designed. Part IV of this handbook introduces the details of these techniques.

Positioning, Mobility, and Tracking

The difficulty in achieving highly precise location estimates in many indoor and outdoor wireless environments has led a number of investigators to utilize parameter estimation techniques for positioning and tracking mobile targets. These techniques can be very beneficial, for example, in smoothing position tracks in mixed LOS/ NLOS situations. Kalman, Bayesian, or particle filters are widely used as state estimators. These state estimation methods can be applied with a variety of sensor technologies and positioning algorithms to improve positioning and tracking performance in many real-world environments. Part V of this handbook begins with a discussion of positioning as a state estimation problem and then discusses Kalman filtering and closely related techniques applicable in both indoor and outdoor applications.

Network Localization

Applications and services built upon wireless positioning can be implemented with different forms of infrastructure supporting the positioning function. GPS satellites, cellular base stations, and fixed wireless local area network (WLAN) access points are familiar infrastructures underlying many well-known applications and services, but for some applications they cannot be provided, for various economic and technical reasons. For some applications there is no supporting infrastructure at all, and methods must be devised to implement location-based services without infrastructure. In other cases, fixed infrastructure cannot provide a complete solution, and this has led to the development of network-based localization techniques. An important example of an application for wireless positioning systems is a wireless sensor network, comprising a number of geographically distributed autonomous sensors intended to cooperatively monitor some characteristics of their individual environments. Each sensor node is typically equipped with its application-specific sensors, a wireless transceiver, a microcontroller and a power source, usually a battery. Accurate positioning information for each sensor is essential for support of the network's application. Ideally, each sensor would have accurate knowledge of its own position, e.g., from GPS. However, size and cost constraints lead in turn to constraints on power and computational capabilities in the individual sensor nodes. Because of these constraints, a sensor network will typically be deployed with a small number of nodes, called *anchor or reference* nodes, having precise a priori location information, while a larger number of remaining nodes, called *unlocalized* nodes, will have no prior knowledge of their locations. An unlocalized node, due to power limitations or signal blockage, may not be able to communicate with anchor nodes. Thus, the unlocalized nodes will estimate their locations by communicating with each other, and schemes must be used to propagate the location information throughout the network. Techniques for accomplishing this are known as *collaborative position location, cooperative localization*, and *network localization*. Part VI of this handbook begins with a chapter on infrastructure-free tracking and then discusses several approaches to network localization.

1.3   OVERVIEW OF POSITIONING SYSTEMS

1.3.1   GPS

The GPS is based on a manmade constellation of 27 Earth-orbiting satellites (24 in operation and three extras in case one fails). Using these satellites, a person or object can localize their position in terms of latitude, longitude, and altitude [1]. These satellites orbit the Earth at an altitude of 12,000 miles and complete two rotations each 24 hours. The orbits of these satellites are arranged such that at any given time, anywhere on the Earth, at least four satellites are clearly visible. A GPS receiver placed on the Earth can localize its position using any set of four visible satellites.

While GPS can be effectively used for many navigational applications, it has limitations. It is not capable of positioning within buildings and mines due to signal attenuation. Its performance is also degraded in severe scattering environments, such as downtown urban areas. GPS is a self-positioning system. To enable this system for remote positioning, which is required for applications such as ad hoc networks, each node should be equipped with a communication system, which also allows it to transmit the self-localized data to other nodes. In addition, because GPS transmission features are known, these systems might be jammed by an adversary. This also limits its defense applications. Systems such as INS can be fused with GPS to enable localization in indoor areas and mines. In addition, WLPSs have been developed to enable localization in GPS-denied environments [45]. These systems are introduced in this chapter.

Two pieces of information are required to carry out the localization process via GPS:

1) The distance from the GPS receiver to satellites

2) The position of each satellite in terms of its latitude, longitude, and altitude (see Fig. 1.3a).

The receiver collects these pieces of information and analyzes and processes high-frequency, low-power radio signals received from the satellites. Mathematical details of localization using GPS are discussed in Part VII of this handbook (see Part V of this handbook).

Distance Measurement: Assuming that the clocks of a GPS receiver and a satellite are perfectly synchronized, the distance is measured using TOA estimation. Specifically, the lag between the signal transmitted by the satellite and the one generated at the GPS receiver is used to determine the distance (see Fig. 1.3a). Assuming that the satellite begins transmitting a long unique pattern (a pseudorandom code) at midnight and the GPS receiver also starts generating the same pattern at midnight, the lag is determined by comparing the two patterns.

As mentioned earlier, clock synchronization is required down to nanosecond precision for accurate calculations. Therefore, under ideal conditions, both the receiver and satellite should be equipped with high-precision clocks, for example, atomic clocks. However, since these clocks are expensive, the receiver manufacturers usually use ordinary quartz clocks. Because these clocks cannot be synchronized to

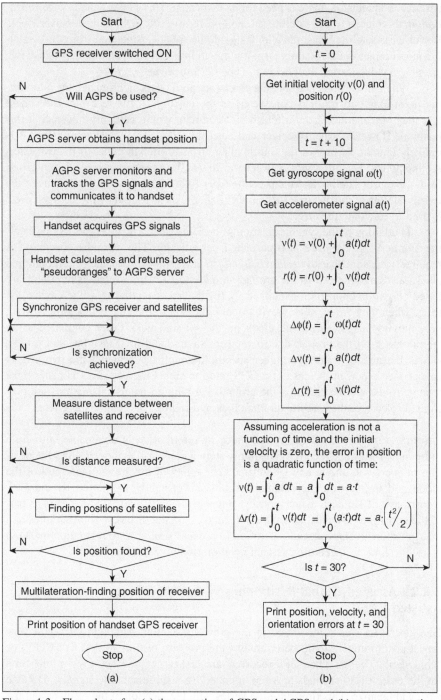

Figure 1.3   Flow charts for: (a) the operation of GPS and AGPS, and (b) error propagation in INS.

nanosecond precision, there is need for an extra step. This step is called *synchronization*. In this step, a fourth satellite is used to determine the error in the receiver clock. Because the satellite transmits a long signal, the spheres generated from three satellite measurements are certainly large enough to intersect each other and produce two possible candidates for the position of the GPS receiver.

When the receiver and satellite clocks are perfectly synchronized, the intersecting point closer to Earth is considered as the position of the receiver. The sphere that may be generated from a fourth measurement would certainly intersect at this position. However, when receiver and satellite clocks are not synchronized, it is unlikely that the surface of the fourth sphere passes through either of the two intersecting points. The difference between the distance of the estimated receiver position from the fourth satellite and the pseudorange of the fourth satellite (the radius of the fourth satellite or the distance to the fourth satellite as measured by the GPS receiver) is used to calculate the error.

In addition to the synchronization of ordinary quartz receiver clocks, the satellite atomic clocks [8] are also corrected periodically. This periodic correction is required to ensure that the *relativistic effects* are removed and the satellite atomic clocks are synchronized to the ground atomic clocks. These relativistic effects are based on two phenomenon explained by the theory of relativity: a) the clocks tick faster when they are in weak gravitational field, and b) the clocks tick slower when they moving. Thus, an atomic clock on the satellite ticks faster compared to an atomic clock on the ground due to weaker gravitational field in orbit; and it ticks slower because of relatively higher speed. Although theoretically the two effects cancel each other, in the case of a GPS satellite clock, the net effect is faster ticks relative to the atomic clock on the ground. Periodic on-board calculations are performed to correct the satellite atomic clock and remove the relativistic effects.

Satellite Positions: This second piece of information is obtainable with little difficulty, as the GPS receiver can simply store an almanac that determines the position of every satellite at any given time. The effect of the gravitational pull of the moon and the sun on the satellites' orbits is constantly monitored by the U.S. Department of Defense, which conveys any adjustments to all GPS receivers as part of the transmitted signals. When the information on the distance from satellites and their positions is known, multilateration (a process similar to triangulation in TOA) is used to find the three-dimensional position of a GPS receiver.

1.3.2   Assisted Global Positioning System (A-GPS or Assisted GPS)

GPS operation was summarized in the previous section. Although GPS is a very robust positioning system, there remains the problem of time to first fix (TTFF) or "cold start"; That is, when GPS receivers are first turned on, they need a long time (in the order of 30 seconds to a few minutes) to acquire satellite signals, navigate data, and localize. This time duration varies with the location of the receiver and the surrounding interference. In order to address this problem, assisted GPS (AGPS) has been developed.

AGPS consists of:

a) A wireless handset with a scaled-down version (with respect to the power requirements, computational capabilities, etc.) of a GPS receiver;

b) An AGPS server with a reference GPS receiver that can simultaneously monitor and track the same satellites as the wireless handset; and

c) A wireless network infrastructure consisting of base stations and a mobile switching center.

The AGPS server obtains handset position from the mobile switching center and can locate the cell of the handset, and even the sector of the handset, within a set if directional antennas are used at the cell base stations [1]. Because the AGPS server monitors and tracks the GPS satellites, it can predict the satellites that are sending the signals to the handset at any given point of time. Thus, the AGPS server can communicate the satellite information to the handset. This enables the handset to acquire GPS signals quickly when it is first turned on, reducing TTFF from minutes to less than a second. Once the satellite signals are acquired by the handset, it calculates the distances to satellites without clock synchronization. These satellite distances are sent back to the AGPS server for further computation, as can be seen in Figure 1.3a. Thus, the AGPS server also shares the computational load of the handset, reducing the handset battery power consumption.

1.3.3   INS

INS uses accelerometers and gyroscopes to track the position, velocity, and orientation of an object relative to a known starting point, velocity, and orientation. Gyroscopes and accelerometers are motion-sensing devices that measure the rate of rotation (angular velocity) and linear acceleration, respectively [9]. Assuming the initial position, velocity, and orientation are known for the object of interest, the updated position, velocity, and orientation are determined by integrating the information received from motion sensors. Thus, the object can continuously track its position, velocity, and orientation without the need for external information.

Actual spatial position and the movement of an object can be described by six parameters: three translational (linear acceleration in x, y, and z direction) and three rotational components (angular velocity in x, y, and z direction). In order to define the movement of the object, three orthogonal accelerometers and three orthogonal gyroscopes are mounted on the object. An orthogonal accelerometer is an instrument that measures acceleration along a single axis. The three orthogonal accelerometers are arranged so that they measure the linear acceleration in the north-south, east-west, and vertical directions. The orthogonal gyroscopes are also known as "integrating" gyroscopes, as their output is proportional to their angle of rotation about fixed axes.

Mathematical integration of the acceleration $a(t)$ yields the velocity $v(t)$, which in turn is integrated to determine the distance travelled from the starting point $r(t)$, as shown in Figure 3b. Orientation $\phi(t)$ can be found by integrating the angular velocity $\omega(t)$, also shown in Figure 3b. These calculations are performed periodically to trace the movement of the object with respect to the global reference frame. While undertaking the integration for the position of the object, acceleration due to gravity is subtracted from the vertical component of the acceleration.

The angular velocity and acceleration measurements made using motion sensors may have errors. When integrating these quantities, the errors in the measured values are propagated to the subsequently calculated position and orientation values. In addition, error is also introduced because the object numerically integrates the measurements at each time step. This error propagation in INS is called integration drift. The localization error can be adjusted to zero by a merger of the INS with other positioning systems such as GPS.

INS is used primarily by militaries to track submarines, warships, unmanned air vehicles, unmanned ground vehicles, missiles, airborne surveillance and navigation, search-and-rescue teams, artillery shells, etc. In addition, INS can be used for civilian applications such as the estimation of position and the orientation of a moving robot, law enforcement, underground tunnels/mines, and underwater vehicles.

INS Classification: There are two types of inertial navigation systems: a) stable platform systems and b) strapdown systems. The difference between the two types is the frame of reference in which the gyroscopes and accelerometers operate. The frame of reference can be the body of the object or the global reference frame.

Stable Platform System: In this system, the motion sensors are mounted on a platform that is held constant with respect to the global frame of reference. This is achieved by mounting the platform using gimbals, which allow the platform to rotate freely about all three axes. If the object rotates about any axis, the gyroscopes mounted on the platform send a feedback signal to the motor mounted on the appropriate gimbals. Based on the feedback signal, the appropriate motors rotate the gimbals in opposite direction and cancel the effect of object's rotation on the platform. This keeps the platform aligned to the global reference frame at all times. In order to track the orientation of the object, the angles between adjacent gimbals are measured and appropriate calculations are performed. To calculate the position of the object, the signals from the platform-mounted accelerometers are integrated as described above.

Strapdown System: In a strapdown system, the motion sensors are mounted rigidly on the object. Therefore, output quantities are measured in the body frame of reference. For orientation calculations, the signals from gyroscopes are directly integrated as described earlier. However, for position calculations, the acceleration signals from the three accelerometers are projected on the global axes. The projected accelerations are calculated by applying a 3×3 rotation matrix to the acceleration signals. The elements of the rotation matrix are generated using the orientation signals. These projected accelerations are then integrated to obtain the position of the object.

1.3.4 Integrated INS and GPS

GPS signals may not be available at all times and at all places. Thus, INS can be used for reliable navigation by filling the gaps in measurements between two GPS position computations. The INS can also be used in case of GPS outages resulting from jamming, obscuration caused by maneuvering, etc. In addition, GPS computations can also help in correcting the error propagation of the INS system. Many chapters of this handbook offer detailed information regarding GPS. Chapter 22 also details GPS and INS integration.

1.3.5   RFID

RFID is a wireless system that identifies tags attached to the object of interest. An RFID system consists of a reader and RFID tags. RFID systems are divided into two categories according to whether they use passive or active tags [10]. Passive tags do not contain a power source and thus are suitable for short-range applications. Passive RFID tags are equipped with an antenna that is excited by output signals at specific frequencies, and these tags are activated by the power of the received signal.

An active RFID system is in fact a full transceiver system including processors, antennas, and batteries. Thus, an active tag contains both a radio transponder and a power source for the transponder. An RFID reader constantly sends radio frequency electromagnetic waves, which are received by the RFID tag in its vicinity. The RFID tag modulates the wave, adding its identification information and sends it back to the reader. The reader converts the modulated signal into digital form to determine the tag identity. Active tags are ideally suitable for the identification of high-volume products moving through a processing unit.

RFID as a Positioning System: RFID can be used to localize the position of a target object. An active RFID tag can be attached to the object, which transmits a signal to the RFID reader. The concept of trilateration, as shown in Figure 2a, is used, along with the RSSI technique to localize the position of the tag. Because the objects to be positioned using RFID are usually in an enclosed environment, there are multipath effects, which decrease the accuracy of the system. In order to increase the accuracy of a RFID-based positioning system, the system utilizes additional readers and reference tags. However, these additional readers increase the cost of the system. In order to keep the costs down, Ni et al. [11] proposed an innovative approach that employs the idea of installing extra fixed reference tags. This approach is called the LANDMARC (location identification based on dynamic active RFID calibration). In a manner similar to the geographic landmarks we use in our daily lives, the fixed tags serve as reference points in the system.

1.3.6   WLPS

WLPS is a hybrid TOA and DOA positioning method with a variety of applications, including autonomous driving. Based on the classification shown in Figure 1.1, it can also be considered as an active remote positioning system. The system comprises a monitoring mobile unit (or dynamic base station) and a target mobile unit (or active target) [2, 12]. The active target contains a transponder and is assigned a unique identification (ID) code. As shown in Figure 1.4a, the dynamic base station (DBS) sends the ID request (IDR) signal to all active targets in its vicinity. The active targets respond by each transmitting a packet that includes its ID code back to the DBS. The DBS recognizes each target by its unique ID code. For positioning, TOA and DOA of the target are estimated by the DBS. As described earlier, DOA is estimated using antenna arrays mounted on the DBS. Using these measured values, the position of active targets can be localized relative to the known position of DBS.

WLPS can be considered as a node in a wireless ad hoc network, enabling all nodes (or specific nodes equipped with DBS) to localize all nodes located in their coverage area. The complexity of these systems lies mainly in the DBS, as they use antenna arrays for localization. The cost and complexity of transceiver (TRX) nodes is very low. In many applications, such as battlefield command and control, a small number of DBSs (expensive units carried by commanders) and a larger number of active targets (low-cost units carried by soldiers) are required. Thus, the overall cost of these systems across all nodes is minimal.

In the WLPS system, each node can independently find the location of the transceivers located in its coverage area. As discussed in Part II of this handbook, multipath effects reduce TOA and DOA estimation performance. Therefore, the localization performance of each DBS node could be low. Multiple DBS nodes can cooperate to reduce the estimation error of the TRX nodes in their coverage area [13, 14]. Chapter 34 details WLPS.

The WLPS can be used for space, outdoor, and indoor applications [15]. For outdoor and indoor applications, direct sequence code division multiple access (DS-CDMA) integrated with beamforming (supported by antenna arrays) provides a reasonable level of detection performance. WLPS enables many applications, such as road safety, security, defense, and robotic collaboration and coordination.

1.3.7 Traffic Alert and Collision Avoidance System (TCAS)

Traffic alert and collision avoidance system (TCAS) is used to detect and track target aircraft in the vicinity of the tracking aircraft [16]. It provides a warning signal to the pilot in the presence of another aircraft that can pose a danger of midair collision. This warning signal is provided to the pilot independent of the air traffic control (ATC) [16]. It consists of two components: interrogator and transponder. Each aircraft is equipped with both components. The interrogator in one aircraft interrogates transponders in other aircraft and analyzes the replies to determine range, bearing, and relative altitude (if reporting) of the intruder aircraft (see Fig. 1.4b). Range is determined by measuring the time elapsed from the interrogation signal to the receipt of the reply. A directional antenna is used to determine direction or bearing of the target aircraft. TCAS gets altitude information directly in the received reply from the transponder on the target aircraft. To determine the altitude, the time-frequency system is employed, which uses the synchronized time and frequency (via extremely accurate oscillators on board the aircraft) to transmit the encoded altitude information. Each aircraft is assigned a specific timeslot of few milliseconds during each one-second interval, used to transmit the encoded altitude signal.

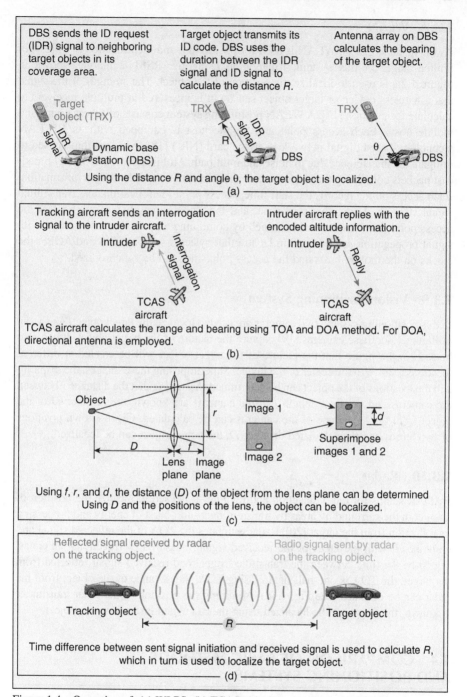

DBS sends the ID request (IDR) signal to neighboring target objects in its coverage area.

Target object transmits its ID code. DBS uses the duration between the IDR signal and ID signal to calculate the distance R.

Antenna array on DBS calculates the bearing of the target object.

Target object (TRX)

IDR signal

Dynamic base station (DBS)

TRX

IDR signal

R

DBS

TRX

θ

DBS

Using the distance R and angle θ, the target object is localized.

(a)

Tracking aircraft sends an interrogation signal to the intruder aircraft.

Intruder aircraft replies with the encoded altitude information.

Intruder

interrogation signal

TCAS aircraft

Intruder

Reply

TCAS aircraft

TCAS aircraft calculates the range and bearing using TOA and DOA method. For DOA, directional antenna is employed.

(b)

Object

r

D

f

Lens Image plane plane

Image 1

Image 2

Superimpose images 1 and 2

d

Using f, r, and d, the distance (D) of the object from the lens plane can be determined. Using D and the positions of the lens, the object can be localized.

(c)

Reflected signal received by radar on the tracking object.

Radio signal sent by radar on the tracking object.

Tracking object

R

Target object

Time difference between sent signal initiation and received signal is used to calculate R, which in turn is used to localize the target object.

(d)

Figure 1.4   Operation of: (a) WLPS, (b) TCAS, (c) vision system, and d) radar.

1.3.8 WLAN

As the name suggests, WLAN is used for positioning and identification of objects in limited-range. In this system, trilateration using the RSSI technique, shown in Figure 1.2a, is used to localize the position of the object. The strength of the signal that a wireless device or target object sends out is measured at multiple receivers to calculate the position [17]. A WLAN positioning system consists of access points and mobile hosts. Each access point and mobile host is equipped with an RF LAN technology-based digital network interface card (NIC) [18]. An algorithm is used to mitigate the interference due to noise and multipath. Mobile hosts periodically broadcast packets containing the start time and the transmitted signal strength information. Each access points records the start time, access point identification, and transmitted signal strength. Using this start time, and the signal strength measurement at the access point, the location is determined by combining empirical measurements with signal propagation modeling. Similar to other systems such as GPS and AGPS, the clocks on the mobile hosts and the access points need to be synchronized.

1.3.9 Vision Positioning System

In this positioning system, two cameras are used to localize the target object. As shown in Figure 1.4c, these cameras [19] capture the picture of the target object. It can also be seen in the figure that the picture of the target object will be created at different locations relative to the center of the image. Superimposing these images, the disparity d in the locations of the object can be determined. Assuming that the distance r between the cameras and the focal length f of the cameras are known, the distance D of the object from the lens plane of the cameras can be calculated. Given known positions of two lenses and the calculated distance D, the target object can be localized.

1.3.10 Radar

RADAR stands for radio detection and ranging. It is used to localize the position of a target in the surrounding areas by transmitting a short burst of energy and processing its reflection from the target [20]. Radar estimates the TOA of the reflected signal and combines it with the DOA of the received signal measured by directional antennas. Let Δt be the time between the transmitted signal and received signal reflected from the target, the TOA is one-half of Δt. Using TOA, the distance of the object from the radar can be obtained (see Fig. 1.4d). Assuming the position of the radar transmitter is known, the target can be localized using the calculated distance to the object.

1.4 COMPARISON OF BASIC METHODS AND POSITIONING SYSTEMS

This section compares basic localization methods and positioning systems in Tables 1.1 and 1.2. Several positioning parameters are used to compare different methods. Table 1.1 compares basic positioning methods previously discussed in the

TABLE 1.1.   Comparison of Basic Methods

	Accuracy (meter)[1]	LOS/NLOS	No. of Base Station(s)
TOA	M [3]	LOS	≥3
TDOA	M [3]	LOS	≥3
DOA	L [3]	LOS	≥2
RSSI	H to M [3]	Both	≥3

[1]Scale for accuracy of basic positioning methods: high (H), 0–50; medium (M), 50–100; low (L), >100.

chapter in terms of accuracy, a need for the availability of LOS, and the number of base station(s) required for localization. The table shows that on average the accuracy of the DOA estimate is poorer relative to TOA, TDOA, and RSSI estimates. This is mainly due to the fact that as the distance between the base station and the target increases, a small DOA error leads to higher localization error. The DOA error is very sensitive to the multipath environment and signal-to-noise ratio [20]. Complex algorithms could be used to improve the DOA performance. DOA performance in localization methods such as WLPS can be improved by incorporating its periodic transmission nature [21, 22]. RSSI is different from other methods when the availability of LOS is taken into account, since it operates in both LOS and NLOS environments. Other localization techniques are very sensitive to the availability of LOS.

Table 1.2 compares different positioning systems based on several parameters. The following observations can be made from the table:

- Most of the systems that operate based on the availability of LOS propagation have higher accuracy. Thus, NLOS introduces error in the calculation, decreasing the accuracy.
- Many existing positioning systems employ multiple base nodes to localize the target node. This may create a system cost problem for the designer.
- The power consumption for most positioning systems is medium to very low. Thus, using one of the low-power positioning systems does not impose additional constraint on the designer with respect to the required power consumption.
- All positioning systems except RFID-based systems can support mobility.
- Many positioning systems are well suited for both outdoor and indoor environments.

1.5   SUMMARY AND FUTURE APPLICATIONS

This chapter offers an overview of positioning techniques and systems that are currently in use for various applications. These positioning systems are usually based on one of the four basic methods: TOA, TDOA, DOA, and RSSI. We described the operation and compared their pros and cons. In addition, a comparison of the basic methods and positioning systems was provided.

TABLE 1.2. Comparison of Positioning Systems

				Accuracy (meter)[1]	LOS/NLOS	Environ.[2]	Power Consump. (W)[3]
Global Positioning System			GPS	H [17]	LOS	O	VL [24]
			AGPS	H to M [25]	LOS	O, I	VL [26]
Local Positioning System	Self-Positioning		INS	VH → VL[4] [9]	NLOS	O, I	M to H[5] [27]
	Remote Positioning	Active	RFID	H [11]	Both	I	VL [28]
			WLPS	varies with application	LOS	O, I	varies with application
			TCAS	L → VL[6] [29]	LOS	O	VH [30]
			WLAN	H to M [3]	Both	O, I	L [31]
		Passive	Vision	VH to H[7] [32]	LOS	O, I	M [33]
			Radar	VH [34]	LOS	O, I	H[8] [35]

				Multi- (M)/Single- (S) Node Positioning	No. of Base Station(s)	Dynamic(D)/ Static(S) Base Station	Absolute (A)/ Relative (R) Positioning
Global Positioning System			GPS	M	4[9]	D	A
			AGPS	M	4[9]	D	A
Local Positioning System	Self Positioning		INS	N/A	N/A	N/A	A
	Remote Positioning	Active	RFID	M	3	S	R
			WLPS	S	1	D	R
			TCAS	S	1	D	R
			WLAN	M	3	S	R
		Passive	Vision	S	1	D	R
			Radar	S	1	D	R

[1]Scale for accuracy of positioning systems: very high (VH): <1; high (H): 1–5; medium (M): 5–30; Low (L): 30–50; very low (VL): >50.

[2]O = outdoor, I = indoor.

[3]Scale for power consumption of positioning systems: very low (VL): <1; low (L): 1–10; moderate (M): 10–50; high (H): 50–200; very high (VH): > 200.

[4]The initial accuracy is high (few decimeters), which changes with time due to error propagation.

[5]Depends on INS classification (stable platform system consumes high power; strapdown system consumes moderate power).

[6]The initial accuracy is low (few degrees), which decreases with distance.

[7]The detection range of vision system is 1 m–95 m, which determines the accuracy.

[8]Within the high range, the power consumption depends on the peak power output.

[9]In GPS and AGPS, satellites are considered as base stations.

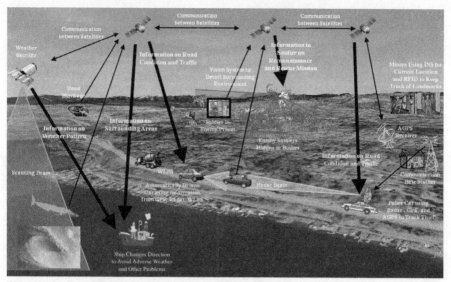

Figure 1.5   Futuristic applications of positioning systems.

The existing positioning systems are used in numerous applications that will likely be expanded in the future, in response to new demands. As shown in Figure 1.5, the future entails collaboration among various positioning systems, especially for applications related to situation awareness. For example, in the case of automatically driven car, the information on the road conditions and traffic (situation awareness) may be obtained using various existing position systems such as GPS, radar, and WLPS. As shown in the figure, the automatically driven car can derive information such as the distance of the surrounding traffic using radar and WLPS. GPS can help the automatically driven car recognize roadblocks caused by an accident or other factors, such as traffic. Thus, in this example, the automatically driven car can be aware of its situation using the positioning technologies.

In another example shown in the figure, a soldier on a reconnaissance or rescue mission can use the information from a vision system, radar, and GPS to be aware of its surroundings or situation. Also shown in the figure is the police car that keeps track of the thief and road conditions using signals from several positioning systems. Miners can also benefit from positioning systems such as INS and RFID to be aware of their current location and possible escape route in case of emergency. For the miners, it can be envisioned that RFID tags are installed along the mine tunnels as it is explored deeper in the earth crust. The information from these RFID tags can potentially include the depth of the mine, etc.

Finally, also shown in the figure is a ship that gathers information from a number of systems and selects its course accordingly. It should be noted that the figure provides only some of the applications where positioning systems collaborate among themselves or other systems (e.g., a weather forecast system) for situation awareness. This handbook reviews the details of many localization techniques that have been briefly introduced in this chapter.

REFERENCES

[1] G. M. Djuknic and R. E. Richton, "Geolocation and assisted GPS," *IEEE Comput.*, vol. 34, no. 2, pp. 123–125, 2001.

[2] H. Tong and S. A. Zekavat, "A novel wireless local positioning system via a merger of DS-CDMA and beamforming: Probability-of-detection performance analysis under array perturbations," *IEEE Trans. Veh. Technol.*, vol. 56, no. 3, pp. 1307–1320, 2007.

[3] M. Vossiek, L. Wiebking, P. Gulden, J. Wieghardt, C. Hoffmann, and P. Heide, "Wireless local positioning," *IEEE Microw. Mag.*, vol. 4, no. 4, pp. 77–86, Dec. 2003.

[4] S. Capkun, M. Hamdi, and J. P. Hubaux, "GPS-free positioning in mobile ad-hoc networks," in *Proc. 34th Annual Hawaii Int. Conf. on System Science*, Maui, HI, Jan. 3–6, 2001, p. 9008.

[5] T. S. Rappaport, J. H. Reed, and B. D. Woerner, "Position location using wireless communications on highways of the future," *IEEE Commun. Mag.*, pp. 33–41, Oct. 1996.

[6] M. Laoufi, M. Heddebaut, M. Cuvelier, J. Rioult, and J. M. Rouvaen, "Positioning emergency calls along roads and motorways using a GSM dedicated cellular radio network," in *Proc. of IEEE Vehicular Technology Conf.*, Boston, MA, Sep. 24–28, 2000, pp. 2039–2046.

[7] U.-H. Kim, J. Kim, D. Kim, H. Kim, and B.-J. You, "Speaker localization on a humanoid robot's head using the TDOA-based feature matrix," in *Proc. of IEEE 17th Int. Symp. on Robot and Human Interactive Communication*, Munich, Germany, Aug. 1–3, 2008, pp. 610–615.

[8] I. A. Getting, "The global position system," *IEEE Spectr.*, vol. 30, no. 12, pp. 36–47, Dec. 1993.

[9] M. Omerbashich, "Integrated INS/GPS navigation from a popular perspective," *J. Air Transp.*, vol. 7, no. 1, pp. 103–118, 2002.

[10] J. Werb and C. Lanzl, "Designing a positioning system for finding things and people indoors," *IEEE Spectr.*, vol. 35, no. 9, pp. 71–78, Sep. 1998.

[11] L. M. Ni, Y. Liu, Y. C. Lau, and A. P. Patil, "LANDMARC: Indoor location sensing using active RFID," *Wireless Netw.*, vol. 10, no. 6, pp. 701–710, 2004.

[12] H. Tong and S. A. Zekavat, "A novel wireless local positioning system via asynchronous DS-CDMA and beam-forming: Implementation and perturbation analysis," *IEEE Trans. Veh. Technol.*, vol. 56, no. 3, pp. 1307–1320, May 2007.

[13] W. Wang and S. A. Zekavat, "Comparison of semi-distributed multi-node TOA-DOA fusion localization and GPS-aided TOA (DOA) fusion localization for MANETs," *EURASIP J. Advances Signal Process.*, vol. 2008, Art. ID 439523, 2008.

[14] W. Wang and S. A. Zekavat, "A novel semi-distributed localization via multi-node TOA-DOA fusion," *IEEE Trans. Veh. Technol.*, vol. 58, no. 7, pp. 3426–3435, Sept. 2009.

[15] S. G. Ting, O. Abdelkhalik, and S. A. Zekavat, "Differential geometric estimation for spacecraft formations orbits via a novel wireless positioning," in *Proc. of IEEE Aerospace Conf.*, Big Sky, MT, Mar. 6–13, 2010.

[16] T. Williamson and N. A. Spencer, "Development and operation of the traffic alert and collision avoidance system (TCAS)," *IEEE Control Syst. Mag.*, vol. 77, no. 11, pp. 1735–1744, Nov. 1989.

[17] J. Hightower and G. Borriello, "Location systems for ubiquitous computing," *Computer*, vol. 34, no. 8, pp. 57–66, 2001.

[18] J. P. Ebert, B. Burns, and A. Wolisz, "A trace-based approach for determining the energy consumption of a WLAN network interface," in *Proc. European Wireless 2002*, Florence, Italy, Feb. 2002, pp. 230–236.

[19] K. Saneyoshi, "Drive assist system using stereo image recognition," in *Proc. IEEE Intelligent Vehicles Symp.*, Tokyo, Japan, Sept. 19–20, 1996, pp. 230–235.

[20] M. I. Skolink, *Radar Handbook*, New York: McGraw Hill, 2008

[21] J. Pourrostam, S. A. Zekavat, and H. Tong "Novel direction-of-arrival estimation techniques for periodic sense local positioning systems in wireless environment," in *Proc. of IEEE Radar '07*, Boston, MA, Apr. 17–20, 2007.

[22] H. Tong, J. Pourrostam, and S. A. Zekavat, "Optimum beam-forming for a novel wireless local positioning system: A stationary analysis and solution," *EURASIP J. Adv. Signal Process.*, vol. 2007, 2007, Art. ID 98243.

[23] M. Sekine, T. Senoo, I. Morita, and H. Endo, "Design method for an automotive laser radar system and future prospects for laser radar," in *Proc. of IEEE Intelligent Vehicles Symp.*, Detroit, MI, Jun. 29–Jul. 1, 1992, pp. 120–125.

[24] D. K. Shaeffer, A. R. Shahani, S. S. Mohan, H. Samavati, H. R. Rategh, M. del Mar Hershenson, M. Xue, C. P. Yue, D. J. Eddleman, and T. H. Lee, "A 115-mW, 0.5-µm CMOS GPS receiver with wide dynamic-range active filters," *IEEE J. Solid-State Circuits*, vol. 33, no. 12, pp. 2219–2231, Dec. 1998.

[25] V. Zeimpekis, G. Giaglis, and G. Lekakos, "A taxonomy for indoor and outdoor positioning techniques for mobile location services," *J. ACM SIGecom Exchanges*, vol. 3 no. 4, pp. 19–27, 2003.

[26] F. van Diggelen, "Indoor GPS theory and implementation," in *IEEE Position Location and Navigation Symp.*, Palms Springs, CA, Apr. 15–18, 2002, pp. 240–247.

[27] C. Piesch, T. Gulde, C. Sartorius, F. Friedl-Vallon, M. Seefeldner, M. Wölfel, C. E. Blom, and H. Fischer, "Design of a MIPAS instrument for high altitude aircraft," in *Proc. Second Int. Airborne Remote Sensing Conf. and Exhibition*, San Francisco, CA, Jun. 24–27, 1996.

[28] N. Cho, S.-J. Song, S. Kim, S. Kim, and H.-J. Yoo, "A 5.1-µW UHF RFID tag chip integrated with sensors for wireless environmental monitoring," in *IEEE European Solid-State Circuits Conf.*, Grenoble, France, Sep. 12–16, 2005, pp. 279–282.

[29] S. Ractliffe, "Air traffic control and mid-air collisions," *J. Electro. Commun. Eng.*, vol. 2, no. 5, pp. 202–208, Oct. 1990.

[30] H.-C. Lee, "Implementation of collision avoidance system using TCAS II to UAVs," *IEEE Aerosp. Electron. Syst. Mag.*, vol. 21, no. 7, pp. 8–13, Jul. 2006.

[31] J. Ebert, B. Burns, and A. Wolsiz, "A trace-based approach for determining the energy consumption of a WLAN network interface," *European Wireless Conf.*, Florence, Italy, Feb. 2002, pp. 230–236.

[32] S. Nedevschi, R. Schmidt, T. Graf, R. Danescu, D. Frentiu, T. Marita, F. Oniga, and C. Pocol, "High accuracy stereo vision system for far distance obstacle detection," in *IEEE Intelligent Vehicles Sym.*, Parma, Italy, Jun. 14–17, 2004, pp. 292–297.

[33] N. Sawasaki, M. Nakao, Y. Yamamoto, and K. Okabayashi, "An embedded feature-based stereo vision system for autonomous mobile robots" in *Proc. of IEEE Int. Conf. on Robotics and Automation*, Phoenix, AZ, Oct. 15–16, 2010, pp. 2693–2698.

[34] R. Hermann, M. Marc-Michel, K. Michael, and L. Urs, "Research activities in automotive radar," in *Proc. of MSMW Sym.*, Kharkov, Ukraine, Jun. 4–9, 2001, pp. 48–51.

[35] M. Edrich, "Ultra-lightweight synthetic aperture radar based on a 35 GHz FMCW sensor concept and online raw data transmission," *Proc. IEE Radar Sonar Navigation*, vol. 153, no. 2, pp. 129–134, Apr. 2006.

[36] N. Alsindi, Li Xinrong, and K. Pahlavan, "Analysis of time of arrival estimation using wideband measurements of indoor radio propagations," *IEEE Trans. Instru. Mea.*, vol. 56, no. 5, pp. 1537–1545, 2007.

[37] W. Xu and S. A. Zekavat, "Spatially correlated multi-user channels: LOS vs. NLOS," in *Proc. of IEEE DSP/SPE Workshop 2009*, Florida Marco Island, FL, Jan. 4–7, 2009.

[38] Z. Wang, W. Xu, and S. A. Zekavat, "A novel LOS and NLOS localization technique," *Proc. IEEE DSP/SPE Workshop 2009*, Florida Marco Island, FL, Jan. 4–7, 2009.

[39] M. Jamalabdollahi, and S. A. Zekavat, "High resolution ToA estimation via optimal waveform design," *IEEE Trans. Commun.*, vol. 65, no. 3, pp. 1207–1218, 2017.

[40] M. Jamalabdollahi and S. Zekavat, "Range measurements in non-homogenous, time and frequency dispersive channels via time and direction of arrival merger," *IEEE Trans. Geosci. Remote Sens.*, vol. 55, no. 2, pp. 742–752, 2017.

[41] M. Jamalabdollahi and S. A. Zekavat, "Joint neighbor discovery and time of arrival estimation in wireless sensor networks via OFDMA," *IEEE Sensors J.*, vol. 15, no. 10, pp. 5821–5833, 2015.

[42] M. Pourkhaatoun and S. A. Zekavat, "A novel high resolution ICA-based TOA estimation technique for multi-path environments," *IET Commun.*, vol. 5, no. 10, pp. 1440–1452, 2011.

[43] S. A. Zekavat, A. Kolbus, X. Yang, Z. Wang, J. Pourrostam, and M. Pourkhaatoon, "A novel implementation of doa estimation for node localization on software defined radios: Achieving high performance with low complexity," in *Proc. of IEEE ICSPC 2007*, Dubai, UAE, Nov. 26–27, 2007.

[44] R. Parker and S. Valaee, "Vehicular node localization using received-signal-strength indicator," *IEEE Trans. Veh. Technol.*, vol. 56, no. 6, pp. 3371–3380, Nov. 2007.

[45] S. A. Zekavat, Wireless Local Positioning System (WLPS), US Patent 7,489,935, 2004.

CHAPTER 2

LOCALIZATION SENSOR ERROR MEASURES AND ANALYSIS

Mojtaba Bahramgiri, S. A. (Reza) Zekavat
Michigan Technological University, MI, United States

THIS CHAPTER investigates error in the state-of-art localization *devices* and *methods*. Examples of localization *devices* include global positioning systems (GPS), inertial navigation systems (INS), radar and light detection and ranging (LiDAR). Examples of localization *methods* include time of arrival (TOA) and direction of arrival (DOA). A detailed review of localization devices and methods were discussed in Chapter 1. To evaluate localization devices and methods, this chapter presents two key localization error measures of probability of error and Cramer-Rao lower bound (CRLB). It uses extensive tables to assess the error (and sources of error) in localization devices and methods. Moreover, it evaluates and compares different localization fusion methods in terms of complexity, convergence rate, and accuracy. Furthermore, the chapter offers practical methods for error evaluation and correction that include calibration, data validation, and partial error correction procedures.

2.1 INTRODUCTION

Localization has diverse applications in navigation and robotics. Each application has different accuracy requirement. For example, smartphone routing and navigation needs relatively low localization accuracy, while high-precision localization is key to autonomous driving. In autonomous driving, an autonomous vehicle (AV) should locate itself and neighboring vehicles accurately to safely control driving speed and

Handbook of Position Location: Theory, Practice, and Advances, Second Edition.
Edited by S. A. (Reza) Zekavat and R. Michael Buehrer.
© 2019 by the Institute of Electrical and Electronics Engineers, Inc.
Published 2019 by John Wiley & Sons, Inc.
Companion Website: www.wiley.com/go/zekavat/positionlocation2e

lane changes. To assess the accuracy requirements of a localization method, a proper error measure should be defined.

This chapter studies two localization error measures, probability of error and CRLB. Probability of error is computed based on the error probability density function (PDF). It indicates the probability that the location measurements fall within a specific region and includes three categories of linear error probability (LEP), circular error probability (CEP) and spherical error probability (SEP). For a one-dimensional location measurement (e.g., range measurements made by a single node), the region is a line, and the probability-of-error measure refers to LEP. For a two- and three-dimensional (3-D) localization, the probability-of-error measure refers to CEP and SEP, respectively. CRLB is another measure, which represents the minimum possible variance of error in the estimation of a parameter such as TOA or DOA. Based on the nature of the localization method, probability-of-error, CRLB, or both might be used to evaluate the performance. For example, GPS offers the global location in space; thus, SEP can be used to evaluate GPS performance. Moreover, GPS is based on TOA estimation, and CRLB can be used to evaluate GPS TOA estimation error. Tracking radars may also use SEP or CEP (based on the nature of radar) to evaluate the localization performance.

The accuracy of localization methods, e.g., TOA and DOA, and technologies, e.g., GPS or radar, varies with channel parameters that are a function of availability of line of sight (LOS) or scattering nature. In many applications that require precise localization, such as autonomous driving, a set of sensors can be incorporated to enable high precision localization. Examples include *devices* such as cameras, LiDAR, radar, and INS. In addition, an AV may use network localization enabled via *gateways* such as vehicle-to-vehicle (V2V) and vehicle-to-infrastructure (V2I) communication and localization to allow high-precision localization. Chapter 29 details the localization methods for autonomous driving.

An AV fuses data attained across multiple entities to improve the localization performance. Measured localization error of each *device* or *gateway* impacts the fusion accuracy. In addition, the number of localization measurements impacts the complexity of the fusion method. Real-time implementation of fusion signal processing in futuristic smart cities infrastructures through the concept of Internet of Things (IoT) is key to the realization of the vision of safe and smart transportation. In addition, in the transition toward fully an autonomous world, we are facing a semiautonomous world in which autonomous and nonautonomous vehicles operate within the same environment. Thus, the accuracy of localization methods and eliminating erroneous localization data via implementation data reduction methods are critical to future localization systems.

This chapter investigates a variety of fusion methods, such as Kalman filter, particle filter, Bayesian filter, and deep fusion. Evaluating the specifications of fusion methods is key to efficient designing and analysis of localization systems. Thus, extensive tables compare these fusion methods in terms of accuracy, convergence rate, complexity, and sensitivity of measurement error. Finally, we will summarize data reduction methods to address erroneous data.

Section 2.2 presents two measures for error evaluation. We will use these measures to evaluate the performance of different methods and technologies in localization applications. Section 2.3 highlights error of different localization technologies. This section offers engineers a better view on different parameters that affect the error in localization technologies. This section compares the sources of error and the pros and cons for each method. Section 2.4 reviews a number of fusion methods and their applications. Different fusion methods have different speeds of convergence, sensitivity to erroneous observation, and complexity. Due to the change in conditions, performance of fusion methods may vary. Comparing fusion methods offers a good view for choosing the proper one based on the application. Finally, this section briefly reviews the practical error evaluation, partial error correction, and calibration methods.

2.2 SENSOR ERROR MEASURES

In smart transportation systems, sensors acquire data and feed them to a processor. Quality of acquired data affects the performance of whole system, and an error in data leads to system failure. Thus, it is important to study the parameters that affect the performance of sensors. Practically, error in sensor measurements is not zero in many applications, and we cannot neglect the effect of this error. Therefore, first, we should know the error and its origins, and next, minimize the error impact on system performance. In this section, we will introduce two measures for sensor error, error probability and Cramer-Rao bound.

2.2.1 Error Probability

Error probability refers to the probability that measurement error does not exceed a certain value. This measure indicates a region around the "true" value and a probability for the measured data to stay within that region [1–3] and corresponds to:

$$P\left(\left\|\hat{X} - X\right\| < \rho\right) = \alpha,\tag{2.1}$$

where X is the "true" vector (for an m-dimensional space), $\hat{X}$ is the measured vector, ρ is a radius around the "true" value that forms a region, and α is the probability that the measurement error is less than ρ. According to the nature of sensor output data, this region can be a line for one variable and a circle or a sphere for vectors. For an unbiased normal error model in localization, LEP, CEP, and SEP indicate the radius, ρ, in which the cumulative error is α (Figure 2.1).

Considering a zero mean Gaussian error model, the error probability density function corresponds to:

$$f(\mathcal{E}) = \frac{1}{(2\pi)^{\frac{n}{2}}|\Sigma|^{\frac{1}{2}}} \exp\left\{-\frac{1}{2}\mathcal{E}^T\Sigma^{-1}\mathcal{E}\right\},\tag{2.2}$$

where $\mathcal{E}$ is the error of measurement, and Σ is the error covariance matrix.

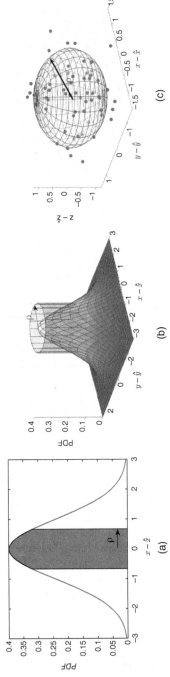

Figure 2.1 ρ is the distance from real position, which includes $\boldsymbol{\alpha}$ percent of measurements in (a) LEP, (b) CEP, and (c) SEP analysis.

Example 2.1   Sample Mean Estimator Probability of Error

Consider an m-dimensional localization system, which is based on sample mean estimator. We use N number of measurements modeled by:

$$Y_n = X + e_n, \quad n = 1, 2, \ldots, N. \tag{2.3}$$

Here, Y_n is the nth measurement vector of size m, X is a deterministic vector that represents true position of object and, e_n is a *iid* zero-mean Gaussian random variable with the variance of σ_n^2 for all m elements of the measurement vector. Investigate the error-probability measure.

Solution

The relationship between the estimated position and measurements corresponds to:

$$\hat{X} = \frac{1}{N} \sum_{n=1}^{N} Y_n \tag{2.4}$$

Where, $\hat{X}$ is the estimated position, N is the number of measurements, and Y_ns are the measurements. Assume that variance of e_n is constant and equals σ^2. The expected value of $\hat{X}$ is as follows:

$$E\left[\hat{X}\right] = E\left[\frac{1}{N} \sum_{n=1}^{N} X + e_n\right] = \frac{1}{N} \sum_{n=1}^{N} X + \frac{1}{N} \sum_{n=1}^{N} E[e_n] = X. \tag{2.5}$$

In (2.5), we observe that sample mean is an unbiased estimator for a random variable that is modeled by (2.3). Y_ns are *iid* random variables, so $Var(\hat{X})$ corresponds to:

$$Var\left(\hat{X}\right) = \frac{1}{N^2} \sum_{n=1}^{N} Var(Y_n) = \frac{1}{N^2} \sum_{n=1}^{N} \sigma^2 = \frac{\sigma^2}{N}. \tag{2.6}$$

Based on (2.3–2.6), the estimated value, $\hat{X}$, is a normally distributed random variable, that is,

$$\hat{X} \sim \mathcal{N}\left(X, \frac{\sigma^2}{N}\right). \tag{2.7}$$

Error of estimator, $\mathcal{E}$, is defined as the Euclidian distance between true value and estimated value of X, that is,

$$\mathcal{E} = \left\| X - \hat{X} \right\|. \tag{2.8}$$

Based on (2.7), the estimation error of (2.8) is a zero-mean Gaussian random variable, that is,

$$\mathcal{E} \sim \mathcal{N}\left(0, \frac{\sigma^2}{N}\right) \tag{2.9}$$

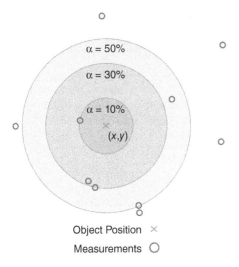

Object Position ×
Measurements ○

Figure 2.2   Example 2 simulation results via MATLAB where, $e_i \sim \mathcal{N}(0,1)$ in (2.3).

For instance, assume that m, the size of vector in (2.3), equals two, in other words, Y_n is a two-dimensional (2-D) variable. In this case, X represents the position of an object in 2-D plane, $X = (x, y)$. Using MATLAB, we can generate N independent Gaussian random vectors and calculate the CEP measure. Figure 2.2 depicts simulation results considering 10 samples and zero-mean Gaussian error model and unity variance. The radius of region, ρ, depends on the probability of error. Small ρ corresponds to small probability of error.

Here, only localization measurements with small error power fall within a region (e.g., a circle in a 2-D scenario) whose radius is characterized by the parameter ρ. As ρ increases, the region becomes bigger and samples with greater error power have higher chance to fall inside the region, which results in higher probability of error.

Figure 2.3 maintains a relationship across ρ and α. Assuming that all measurement errors have the same probability, we can simulate the CDF (cumulative distribution function) of error, and compare it with the theoretical CDF of error. Based on Figure 2.3, it is observed that lower probability of error leads to a smaller radius of the region around the true position of object, which contains a lower number of measurements.

Example 2.2   CEP Calculation of Measurements with Nonidentically Distributed Error

Consider a -dimensional 2-D localization system, which is based on sample mean estimator. We have two sets of measurements that are modeled by:

$$Y_n = X + e_n. \tag{2.10}$$

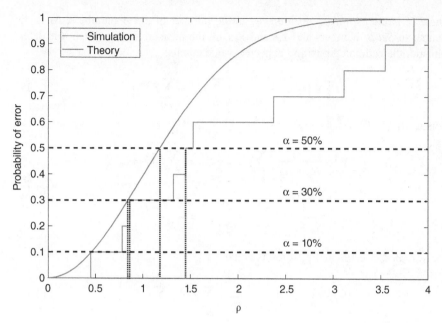

Figure 2.3 Theoretical versus simulated CDF of probability of error where $e_i \sim N(0,1)$ in Example 2.1.

Here, Y_n is the nth measurement vector of size 2, X is a deterministic vector that represents the true position of object and e_n is an independent nonidentically distributed zero-mean Gaussian random variable with a variance equal to σ_n^2 for all two elements of the measurement vector. *Set A* includes N_a number of measurements with zero-mean Gaussian noise and a variance equal to σ_a^2, and *Set B* includes N_b measurements with zero-mean Gaussian noise and a variance equal to σ_b^2.

(i) What is the statistics of error if we estimate the position only based on *Set A*?

(ii) Now let's apply the sample mean estimator to *Set A* and *Set B*. How will the variance of error in estimation be affected if we include *Set B* in the positioning process?

Solution

(i) Based on the solution of Example 2.1, we have $\mathcal{E} \sim \mathcal{N}\left(0, \dfrac{\sigma_a^2}{N_a}\right)$.

(ii) According to (2.4), the relationship of the estimated position and measurements corresponds to:

$$\hat{X} = \frac{1}{N_a + N_b}\left(\sum_{n=1}^{N_a} Y_n^{(A)} + \sum_{m=1}^{N_b} Y_m^{(B)}\right), \tag{2.11}$$

where $\hat{X}$ is the estimated position, N_a is the number of *Set A* members, N_b is the number of *Set B* members and Y_ns and Y_ms are the measurements. Given X is deterministic, the estimated position expected value equals:

$$E\left[\hat{X}\right] = E\left[\frac{1}{N_a + N_b}\left(\sum_{n=1}^{N_a}Y_n^{(A)} + \sum_{m=1}^{N_b}Y_m^{(B)}\right)\right]. \tag{2.12}$$

Now using (2.10):

$$E\left[\hat{X}\right] = \frac{1}{N_a + N_b}\left(\sum_{n=1}^{N_a}E\left[X + e_n^{(A)}\right] + \sum_{m=1}^{N_b}E\left[X + e_m^{(B)}\right]\right)$$

$$\tag{2.13}$$

$$= \frac{1}{N_a + N_b}\left(\sum_{n=1}^{N_a}X + \sum_{m=1}^{N_b}X\right) = \frac{N_aX + N_bX}{N_a + N_b} = X.$$

The variance of estimated position is equal to:

$$Var\left(\hat{X}\right) = Var\left(\frac{1}{N_a + N_b}\left(\sum_{n=1}^{N_a}Y_n^{(A)} + \sum_{m=1}^{N_b}Y_m^{(B)}\right)\right). \tag{2.14}$$

Because Y_ns and Y_ms are independent, Equation (2.14) can be rewritten as follows:

$$Var\left(\hat{X}\right) = \frac{1}{(N_a + N_b)^2}\left(\sum_{n=1}^{N_a}Var\left(Y_n^{(A)}\right) + \sum_{m=1}^{N_b}Var\left(Y_m^{(B)}\right)\right)$$

$$= \frac{1}{(N_a + N_b)^2}\left(\sum_{n=1}^{N_a}Var\left(e_n^{(A)}\right) + \sum_{m=1}^{N_b}Var\left(e_m^{(B)}\right)\right) \tag{2.15}$$

$$= \frac{N_a\sigma_a^2 + N_b\sigma_b^2}{(N_a + N_b)^2}.$$

According to (2.13) and (2.15), the distribution of estimated position corresponds to:

$$\hat{X} \sim \mathcal{N}\left(X, \frac{N_a\sigma_a^2 + N_b\sigma_b^2}{(N_a + N_b)^2}\right). \tag{2.16}$$

Let's define parameters β and γ as follows:

$$\beta = \frac{N_b}{N_a} \tag{2.17}$$

and

$$\gamma = \frac{\sigma_b}{\sigma_a} \tag{2.18}$$

Substituting (2.17) and (2.18) into (2.16), the variance of estimated position corresponds to

$$Var\left(\hat{X}\right)=\frac{1}{N_a}\left(\frac{\sigma_a^2+\beta\sigma_b^2}{(1+\beta)^2}\right)=\frac{\sigma_a^2}{N_a}\left(\frac{1+\beta\gamma^2}{(1+\beta)^2}\right). \qquad (2.19)$$

According to (2.18) and (2.19), estimation error is a zero-mean Gaussian random variable, that is,

$$\mathcal{E}\sim\mathcal{N}\left(0,\frac{\sigma_a^2}{N_a}\left(\frac{1+\beta\gamma^2}{(1+\beta)^2}\right)\right) \qquad (2.20)$$

Comparing (2.20) with the answer to part (*i*), $\frac{1+\beta\gamma^2}{(1+\beta)^2}$ appears as a scaling factor for error variance. It is deducted that:

- *If* $Var(\mathcal{E})>\frac{\sigma_a^2}{N_a}$, *including Set B in the estimation process increases error and decreases accuracy;*
- *If* $Var(\mathcal{E})=\frac{\sigma_a^2}{N_a}$, *including Set B in the estimation process would not affect the accuracy;*
- *If* $Var(\mathcal{E})>\frac{\sigma_a^2}{N_a}$, *including Set B in the estimation proces, decreases error and increases accuracy;*

Regardless of the effect of including *Set B* on accuracy, increasing the number of measurements results in more complexity in the process. Total number of measurements is equal to:

$$N_{Total}=N_a+N_b=N_a(1+\beta). \qquad (2.21)$$

Thus, complexity is proportional to $1+\beta$ *if we include Set B in estimation.*

Without losing generality, assume $\sigma_2>\sigma_1$, and γ is greater than unity (in this example, we take $\sigma_1\neq\sigma_2$). The normalized variance of error corresponds to:

$$Normalized\ Var(\mathcal{E})=\frac{Var(\mathcal{E})}{\sigma_a^2/N_a}=\frac{1+\beta\gamma^2}{(1+\beta)^2}. \qquad (2.22)$$

Figure 2.4 represents the relationship across the normalized variance (Eq. 2.22), β and γ. As was highlighted, including *Set B* in the estimation process improves or degrades accuracy. In other words, if the normalized variance is smaller than unity, then the accuracy improves; otherwise, accuracy decreases. Therefore, we are interested in finding a criterion that identifies the effects of including *Set B* in the estimation process. Setting the normalized variance of error to one, the criteria is:

$$\frac{1+\beta\gamma^2}{(1+\beta)^2}=1. \qquad (2.23)$$

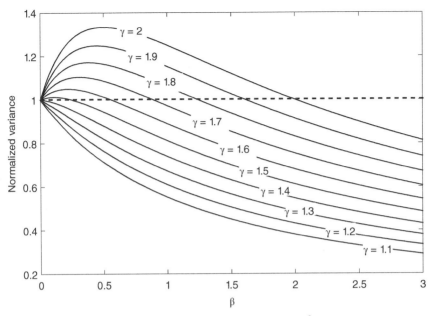

Figure 2.4 Normalized variance of error of estimation versus β and γ.

Applying simple mathematical manipulation leads to

$$\beta = \gamma^2 - 2. \tag{2.24}$$

Summary:

If $\beta > \gamma^2 - 2$, the normalized variance of error is smaller than unity and Set B *improves accuracy. β is a nonzero number, so if γ is less than $\sqrt{2}$, regardless of the β value, the normalized variance of error will be smaller than unity.*

Example 2.3   Positioning via Mixed LOS and NLOS Signals

There are several methods for finding the range between a base station and a user, and some of these methods are based on measuring the signal's time of flight. For instance, TOA estimation method estimates the distance accurately by sending a signal and measuring the time of flight. Non-line-of-sight (NLOS) signals compared to LOS signals have longer time of flight, which imposes error to range estimation (see Part IV of the handbook).

Now, consider two measurement sets, *Set A* and *Set B*, where *Set A* includes N_a measured range via LOS signals and *Set B* includes N_b measurements via NLOS signals. We aim to estimate location by sample mean estimator. Measurements of *Set A* are modeled as:

$$Y_n^{(A)} = X + e_n^{(A)} \text{ and, } e_n^{(A)} \sim \mathcal{N}(0, \sigma^2), \tag{2.25}$$

where X is the true position vector and $e_n^{(A)}$ represents the error. The measurement *Set B* is modeled as:

$$Y_m^{(B)} = X + e_m^{(B)} \text{ and, } e_m^{(B)} \sim \mathcal{N}(\alpha, \sigma^2),$$ (2.26)

where α is the error induced by the NLOS signal TOA measurement. Here, we aim to study the impact of NLOS signals on the position estimation if we employ sample mean estimator.

Solution

According to (2.4), the estimated position vector corresponds to:

$$\hat{X} = \frac{1}{N_a + N_b}\left(\sum_{n=1}^{N_a} Y_n^{(A)} + \sum_{m=1}^{N_b} Y_m^{(B)}\right).$$ (2.27)

The estimated position expected value equals:

$$E\left[\hat{X}\right] = \frac{1}{N_a + N_b}\left(\sum_{n=1}^{N_a} E\left[X + e_n^{(A)}\right] + \sum_{m=1}^{N_b} E\left[X + e_m^{(B)}\right]\right) = X + \frac{N_b \alpha}{N_a + N_b}.$$ (2.28)

Thus, NLOS signals lead to a biased estimation. Because the measurements are independent, the variance of estimated position corresponds to:

$$Var\left(\hat{X}\right) = \frac{1}{(N_a + N_b)^2}\left(\sum_{n=1}^{N_a} Var\left(X + e_n^{(A)}\right) + \sum_{m=1}^{N_b} Var\left(X + e_m^{(B)}\right)\right)$$

$$= \frac{\sigma^2(N_a + N_b)}{(N_a + N_b)^2} = \frac{\sigma^2}{(N_a + N_b)}.$$ (2.29)

Thus, according to (2.8), the error distribution is:

$$\mathcal{E} \sim \mathcal{N}\left(\frac{N_b \alpha}{N_a + N_b}, \frac{\sigma^2}{(N_a + N_b)}\right).$$ (2.30)

Because NLOS signals are available, the estimator is no longer *unbiased*. For a *biased* estimator, increasing the number of samples may not result in lower estimation error, thus, it is recommended to detect and discard NLOS signals. In real scenarios, the error mean and variance are stochastical parameters. To attain accurate localization, mean and variance statistics are modeled. The error in this modeling impacts the overall localization performance (see Part IV of handbook).

Using (2.1) and (2.2), the probability of error of Gaussian model is summarized in Table 2.1.

TABLE 2.1. LEP, CEP, and SEP Analysis for Unbiased Gaussian Error Model

Number of Dimensions	Measures	Relations
1	LEP	$\oint_{|x-\hat{x}|<\rho_{LEP}} \dfrac{1}{\sqrt{2\pi}\sigma} e^{\frac{-(x-\hat{x})^2}{2\sigma^2}}\, d\hat{x} = \alpha$
2	CEP	$\oiint_{\|\hat{X}-X\|<\rho_{CEP}} \dfrac{1}{2\pi|\Sigma|^{\frac{1}{2}}}\exp\left\{-\dfrac{1}{2}(X-\hat{X})^H \Sigma^{-1}(X-\hat{X})\right\}d\hat{X} = \alpha$
3	SEP	$\oiiint_{\|\hat{X}-X\|<\rho_{SEP}} \dfrac{1}{(2\pi)^{\frac{3}{2}}|\Sigma|^{\frac{1}{2}}}\exp\left\{-\dfrac{1}{2}(X-\hat{X})^H \Sigma^{-1}(X-\hat{X})\right\}d\hat{X} = \alpha$

2.2.2 Cramer–Rao Bound

In estimation theory, Cramer–Rao Bound (CRB) or Cramer–Rao Lower Bound (CRLB) is a lower limit for the variance of an estimation. Let $\theta = (\theta_1, \theta_2, ..., \theta_N)$ be a random sample from a distribution with $f_\theta(\theta\,|\,X)$, where X is a scalar parameter. CRLB is defined as:

$$Var\left(\hat{X}\right) \geq \frac{1}{I(X)}, \tag{2.31}$$

where $\hat{X}$ is the estimated value for X, and $I(X)$ is the Fisher information matrix [4]. The Fisher information matrix equals:

$$I(X) = E\left[\left(\frac{\partial l(\theta; X)}{\partial X}\right)^2\right] = -E\left[\frac{\partial^2 l(\theta; X)}{\partial X^2}\right]. \tag{2.32}$$

In (2.32), $l(\theta; X)$ is the natural logarithm of likelihood function of θ, that is, the measurement set. $l(\theta; X)$ is a function of the parameter X. In (2.32), $\frac{\partial}{\partial x}$ refers to partial derivative with respect to X.

Proof: Let $\theta = (\theta_1, \theta_2, ..., \theta_N)$ be a random sample of a distribution with $f_\theta(\theta\,|\,X)$, where X is a scalar parameter. For an *unbiased estimator* $\hat{\phi}(\theta)$, the expected value of estimator function corresponds to:

$$E\left[\hat{\phi}(\theta)\right] = \int \hat{\phi}(\theta)\, f_\theta(\theta|X)\, d\theta = X. \tag{2.33}$$

The second equality in (2.33) conveys the fact that the estimator is unbiased.

Applying partial derivative with respect to X:

$$\int \hat{\phi}(\theta) \frac{\partial f_\theta(\theta|X)}{\partial X} \, d\theta = \frac{\partial X}{\partial X} = 1. \tag{2.34}$$

Score function, $V(X)$, is defined as the partial derivative of likelihood function natural logarithm, which corresponds to:

$$V(X) = \frac{\partial l(\theta; X)}{\partial X} = \frac{\partial \log(f_\theta(\theta|X))}{\partial X} = \frac{1}{f_\theta(\theta|X)} \frac{\partial f_\theta(\theta|X)}{\partial X}. \tag{2.35}$$

Thus,

$$\frac{\partial f_\theta(\theta|X)}{\partial X} = V(X) f_\theta(\theta|X). \tag{2.36}$$

Substituting (2.36) into (2.34), we have:

$$\int \hat{\phi}(\theta) V(X) f_\theta(\theta|X) \, d\theta = E\left[\hat{\phi}(\theta) V(X)\right] = 1. \tag{2.37}$$

According to (2.35), the expected value of score function, $V(X)$, equals:

$$\begin{aligned}
E[V(X)] &= \int \frac{1}{f_\theta(\theta|X)} \frac{\partial f_\theta(\theta|X)}{\partial X} f_\theta(\theta|X) \, d\theta \\
&= \int \frac{\partial f_\theta(\theta|X)}{\partial X} d\theta \\
&= \frac{\partial}{\partial X} \int f_\theta(\theta|X) \, d\theta \\
&= \frac{\partial}{\partial X} 1 = 0 \cdot
\end{aligned} \tag{2.38}$$

The variance of score function, $V(X)$, is equal to Fisher information matrix, which is:

$$I(X) = Var(V(X)) = E[V^2] - E[V]^2 = E\left[\left(\frac{\partial \log(f_\theta(\theta|X))}{\partial X}\right)^2\right] = E\left[\left(\frac{\partial l(\theta; X)}{\partial X}\right)^2\right].$$

Moreover, based on (2.35):

$$I(X) = E\left[\left(\frac{\partial l(\theta; X)}{\partial X}\right)^2\right] = E\left[\frac{1}{f_\theta^2(\theta|X)} \left(\frac{\partial f_\theta(\theta|X)}{\partial X}\right)^2\right]. \tag{2.40}$$

Now, we show that the second equality of (2.32) is valid. The $Cov(V, \hat{\phi})$ corresponds to:

$$Cov(V, \hat{\phi}) = E[V\hat{\phi}] - E[V] \, E[\hat{\phi}]. \tag{2.41}$$

Based on (2.38), $E[V(X)] = 0$; thus, based on (2.37),

$$Cov(V, \hat{\phi}) = 1. \tag{2.42}$$

Since the absolute value of correlation coefficient for every two random variables is less than one, we have:

$$\rho^2(V, \hat{\phi}) \le 1, \tag{2.43}$$

that is,

$$\frac{Cov^2(V, \hat{\phi})}{Var(V)Var(\hat{\phi})} \le 1. \tag{2.44}$$

Using (2.41) and (2.39):

$$Var(\hat{\phi}) \ge \frac{1}{I(X)} \tag{2.45}$$

Based on (2.32), the Fisher information matrix can be calculated by second-order partial derivative of natural logarithm of likelihood function. Now, to evaluate $I(X)$ in (2.39), we calculate the secondorder partial differential equation of $l(\theta; X)$, that is:

$$
\begin{aligned}
\frac{\partial^2 l(\theta; X)}{\partial X^2} &= \frac{\partial}{\partial X}\left(\frac{1}{f_\theta(\theta|X)}\frac{\partial f_\theta(\theta|X)}{\partial X}\right) \\
&= \frac{1}{f_\theta(\theta|X)}\frac{\partial^2 f_\theta(\theta|X)}{\partial X^2} - \frac{1}{f_\theta^2(\theta|X)}\left(\frac{\partial f_\theta(\theta|X)}{\partial X}\right)^2.
\end{aligned}
\tag{2.46}
$$

Moreover, the expected value of (2.46) equals $I(X)$ (see Eq. 2.39), and corresponds to:

$$E\left[\frac{\partial^2 l(\theta; X)}{\partial X^2}\right] = E\left[\frac{1}{f_\theta(\theta|X)}\frac{\partial^2 f_\theta(\theta|X)}{\partial X^2}\right] - E\left[\frac{1}{f_\theta^2(\theta|X)}\left(\frac{\partial f_\theta(\theta|X)}{\partial X}\right)^2\right]. \tag{2.47}$$

Similar to (2.38), we have:

$$
\begin{aligned}
E\left[\frac{1}{f_\theta(\theta|X)}\frac{\partial^2 f_\theta(\theta|X)}{\partial X^2}\right] &= \int \frac{1}{f_\theta(\theta|X)}\frac{\partial^2 f_\theta(\theta|X)}{\partial X^2} f_\theta(\theta|X)\, d\theta \\
&= \frac{\partial^2}{\partial X^2}\int f_\theta(\theta|X)\, d\theta = \frac{\partial^2}{\partial X^2} 1 = 0.
\end{aligned}
\tag{2.48}
$$

Based on (2.40), the second term of (2.47) is $I(X)$. Thus, the second equality of (2.32) is valid; that is,

$$I(X) = E\left[\left(\frac{\partial l(\theta; X)}{\partial X}\right)^2\right] = -E\left[\frac{\partial^2 l(\theta; X)}{\partial X^2}\right]. \tag{2.49}$$

Note: For a *biased* estimator (see Eq. 2.31), CRLB corresponds to:

$$Var(\hat{X}) \geq \frac{\frac{\partial \phi(X)}{X}}{I(X)}, \tag{2.50}$$

where

$$\phi(X) = E\left[\hat{\phi}(\theta)\right]. \tag{2.51}$$

Because the estimator is *unbiased*, $\varphi(\Xi)$ is a function of (not equal to) X.

Example 2.4   Evaluating the Sample Mean Estimator Performance via CRLB

Let $\{Y_n\}_{n=1}^{N}$ be a measurement set from a distribution with $p\left(\{Y_n\}_{n=1}^{N}|X\right)$, where X is the "true" position vector. In this example, the goal is to find the CRLB of position estimation and evaluate the performance of the sample mean estimator.

Solution

Considering the proposed model for measurements in (2.3), the natural logarithm of likelihood function corresponds to:

$$l\left(\{Y_n\}_{n=1}^{N};X\right) = \log\left(p\left(\{Y_n\}_{n=1}^{N}|X\right)\right). \tag{2.52}$$

Here Y_ns are independent and identically distributed; hence, we can rewrite (2.52) as:

$$l\left(\{Y_n\}_{n=1}^{N};X\right) = \log\left[\prod_{n=1}^{N}p(Y_n|X)\right] = \sum_{n=1}^{N}\log(p(Y_n|X))$$

$$= \sum_{n=1}^{N}\left(-\frac{1}{2}\log(2\pi\sigma^2) - \frac{(Y_n-X)^2}{2\sigma^2}\right). \tag{2.53}$$

The partial derivative of (2.52) according to X is equal to:

$$\frac{\partial l\left(\{Y_n\}_{n=1}^{N};X\right)}{\partial X} = \sum_{n=1}^{N}\left(\frac{Y_n-X}{\sigma^2}\right). \tag{2.54}$$

The second-order partial derivative equation of natural logarithm of likelihood function is

$$\frac{\partial^2 l\left(\{Y_n\}_{n=1}^{N}|X\right)}{\partial X^2} = \sum_{n=1}^{N}\left(-\frac{1}{\sigma^2}\right) = -\frac{N}{\sigma^2}. \tag{2.55}$$

Thus,

$$I(X) = -E\left[\frac{\partial^2 l\left(\{Y_n\}_{n=1}^{N}|X\right)}{\partial X^2}\right] = \frac{N}{\sigma^2}. \tag{2.56}$$

Therefore, based on (2.56), variance of the estimated value of X, cannot be smaller than $\frac{\sigma^2}{N}$. Now, let's go back to the results of Example 2.1. In (2.7) the variance of sample mean estimator is presented. By comparing (2.7) and (2.56), we realize that if measurements are modeled according to (2.3), sample mean estimator has the best performance among the estimators.

As explained in first section, when the number of sensors are high, processing the data of all sensors applies a huge amount of processing to the system, which can reduce the performance or even cause system failure. Definitely, discarding the erroneous data leads to a better performance, but evaluation of the data before the processing will not be feasible. In this section, we introduced two measures for error evaluation of data, which can be employed in data evaluation, data fusion, data reduction, and many other applications.

2.3 ERROR EVALUATION OF INDIVIDUAL LOCALIZATION TECHNOLOGIES OF SMART TRANSPORTATION

2.3.1 GPS

GPS (global positioning system) is based on satellite communication and TOA-based range detection. The signals from at least four satellites are needed for a receiver to measure its absolute location (see Chapter 5). GPS localization updating rate is low, and its error is bounded due to system specifications and environmental conditions [5]. GPS-based localization is highly sensitive to obstruction, reflection, and time synchronization of receiver and satellites [6]. Sources of error in time synchronization are clock drift (in satellite and receiver), ionosphere and troposphere effects, the orbital error of satellite, and thermal noise in the receiver (Figure 2.5) [7]. The time offset due to all these effects is:

$$\rho_k^i(t) = R_k^i(t) + c\delta_k(t) + c\delta^i(t) + \epsilon^i(t) + \varsigma_k^i(t), \tag{2.57}$$

where t is the time, ρ_k^i is the code pseudorange between satellite i and receiver k, c is the speed of light, δ^i and δ_k are clock errors in satellite and receiver, respectively,

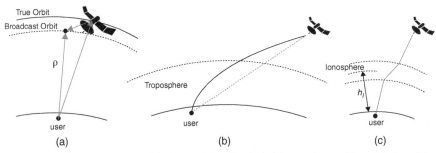

Figure 2.5   Sources of error in GPS-based localization: (a) error in satellite orbital position calculation caused by ephemeris error, (b) tropospheric effect on wave propagation, (c) ionosphere imposed error in delay estimation.

TABLE 2.2.   CRLB of Code-Based and Carrier-Based Pseudoranging Methods in GPS Receivers, Error Bound of Frequency Estimation in GPS Receivers [9]

ML Estimator	CRLB
Code-Based Pseudoranging	$\sigma_{\tau_{ML}} = \dfrac{3.444 \times 10^{-4}}{\sqrt{(C/N_0)WT}}$
Carrier-Based Pseudoranging	$\sigma_{\tau_{ML}} = \dfrac{1}{2\pi f_c \sqrt{2(C/N_0)T}}$
Frequency estimator	$\sigma_{f_{ML}} = \sqrt{\dfrac{3}{2\pi^2 (C/N_0)T^3}}$

ϵ^i is error due to ionosphere, troposphere, and orbit of satellite, and ς_k^i represents error due to the effect of thermal noise in receiver and multipath on TOA estimation [8]. In [9] the author provides the CRLB for ML (maximum likelihood) estimation in a GPS receiver, which has been summarized in Table 2.2. Part VI of this handbook details GPS TOA estimation. In Table 2.2, C/N_0 is the ratio of power in the code waveform to the one-sided power spectral density of the noise, W is the code bandwidth, T is signal observation time, and f_c is carrier frequency.

DGPS (Differential GPS) is an extension of GPS technology, which is based on satellite and terrestrial communication. A GPS receiver with a known location is considered the reference station, which communicates additional data to reduce the error of localization (Figure 2.6). Satellite clock, the combined effect of navigation message ephemeris error, and ionosphere and troposphere delay are the errors that are almost canceled in DGPS [9]. Satellite clock, ionosphere, troposphere, and satellite orbit errors are almost consistent in the tens-of-kilometers vicinity [7]. According to [8], the measured pseudorange of a receiver in DGPS technique under time synchronization assumption among the receiver and station is:

$$\rho_m^i = R_m^i + c\delta t_m + \epsilon_m, \tag{2.58}$$

where ρ_m^i is pseudorange measurement, R_m^i is real distance between satellite i and station k, c is the light propagation speed, δt_m represents drift in station clock with respect to satellite, and ϵ_m refers to pseudorange error. The GPS errors are highly correlated over time and space, as the DGPS is not capable of providing data to the end user simultaneously, so there are some bounded errors associated with different sources.

Table 2.3 summarizes ephemeris, tropospheric, and ionospheric errors in the DGPS technique. In Table 2.3, ϵ_s is error in satellite position estimation, d_m is the distance of reference station to satellite, p is distance between reference station and end user, φ_m is elevation angle of reference station, N_s is surface refractivity, ϕ_m' is

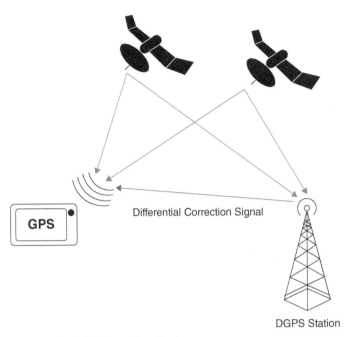

Figure 2.6.   DGPS-based localization.

TABLE 2.3.   Ephemeris, Tropospheric, and Ionospheric Errors in the DGPS Technique

Sources of error	Expected error
Ephemeris	$\dfrac{\epsilon_s}{d_m}\, p \sin^2 \phi_m$
Tropospheric	$\dfrac{p \csc \phi_m}{d_s} \times (1.4588 + .0029611 N_s)$
Ionospheric	$\dfrac{p}{d_m}\left\| \dfrac{p}{d_m} - \cos\phi'_m \right\| \dfrac{40.3}{f^2} \times TEC$

elevation angle at the reference station's ionospheric pierce point, f is carrier frequency, and TEC is total electron content (about 10^{16} to 10^{18} electron/m^2).

Assisted GPS (**AGPS**) has been developed to speed up the receivers for locating the user in weak signal conditions. GPS receivers demodulate 50 bps ephemeris data of satellite, which means they need 18–30 sec to estimate the position of each satellite [8]. This delays the GPS-based localization. Although DGPS and AGPS both employ a reference receiver to communicate additional data in real time, AGPS aims to improve robustness and coverage of localization services (Figure 2.7) [10]. The reference receiver enhances the start-up performance of the users' receiver, that

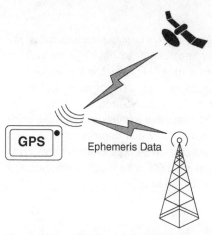

Figure 2.7   AGPS-based positioning system.

is, time to first fix (TTFF) by broadcasting position information of available satellites and time stamps. Furthermore, increasing the observation time results in better performance of the receiver for low C/N_0 and lower probability of acquisition failure. Table 2.4 offers an overview on GPS, DGPS, and AGPS, highlighting sources of error, latency, accuracy, and application of each technique.

TABLE 2.4.   Overview of GPS, DGPS, and AGPS Localization, Sources of Error, Latency, Accuracy, and Application of Each Technique

	Sources of Error	Latency	Accuracy	Application
GPS	• Clock drift in satellite and receiver • Ephemeris error • Ionospheric and tropospheric delay • Thermal noise and multipath error	High: 18 s up to 30 s for demodulating each satellite signal	Low: 5–20m	Under open sky
DGPS	• Thermal noise and multipath error	High (Same as GPS)	High ~ cm; up to < 10 m	In urban area
AGPS	• Clock drift in satellite and receiver • Ephemeris error • Ionospheric and tropospheric delay • Thermal noise and multipath error	Low: depends on cellular network BW; Could be less than a second	High	-In urban area -Indoor Localization

2.3.2 INS

INS consists of gyroscope and accelerometer sensors to enable inertial measurements. In INS, the velocity, position, and orientation of an object is tracked by integrating acceleration and angular velocity over time [11], which is known as dead reckoning. INS suffers error propagation in measurements. The accumulated error, which is known as drift, grows rapidly and makes the INS output unreliable for navigation purposes. Thus, usually INS is fused with other localization methods such as GPS. Here, we aim to evaluate the main parameters that contribute to the INS error propagation. The parameters include errors created by accelerometer and gyroscope sensors. In [12], authors provide output model for both accelerometer and gyroscope. The measurement of a gyro can be modeled as:

$$g_r = r + c_r + b_r + w_{gyro}, \tag{2.59}$$

where the g_r is gyro output, r is true object rotation, c_r is a constant offset, b_r is a moving bias or sensor drift, and w_{gyro} is a wideband sensor noise [13].

Accelerometer output is defined as zero for $g \approx 9.31 m/s^2$, which means in a case of falling toward the center of the Earth, accelerometer output would be zero [14]. Similar to gyro, the accelerometer output can be modeled as follows:

$$a_{\ddot{x}} = \ddot{x} + c_{\ddot{x}} + b_{\ddot{x}} + w_{accel}, \tag{2.60}$$

where $a_{\ddot{x}}$ is accelerometer output, $\ddot{x}$ is the true object acceleration, $c_{\ddot{x}}$ is a constant offset, $b_{\ddot{x}}$ is a moving bias, and w_{accel} is a white sensor noise. The moving bias or sensor shift is modeled as the first order of the Markov process. The noises are also modeled as zero-mean normal distribution. Table 2.5 represents the expected standard deviation of error in position and velocity measurements in a 2-D model of inertial measurement unit (IMU). In Table 2.5, $\sigma_{w_{accel}}$ and $\sigma_{w_{gyro}}$ are standard deviations of noises in gyroscope and accelerometer respectively, $T_s = \frac{1}{f_s}$ is the sampling period, and t shows the running time of the sensor.

Due to the unbounded error of gyroscope, derived orientation must be reinforced by the accelerometer and digital compass data to achieve long-term stability and information reliablity [15].

TABLE 2.5.   Standard Deviation of Position and Velocity Estimation Errors in 2-D IMU Model [12]

	North	East
Position	$\sigma_{P_{North}} = \sigma_{w_{accel}} \sqrt{\frac{1}{3} T_s t^3}$	$\sigma_{P_{East}} = \sigma_{w_{accel}} \sigma_{w_{gyro}} T_s t^2 \sqrt{\frac{1}{8}}$
Velocity	$\sigma_{V_{North}} = \sigma_{w_{accel}} \sqrt{T_s t}$	$\sigma_{V_{East}} = \sigma_{w_{accel}} \sigma_{w_{gyro}} T_s t \sqrt{\frac{1}{2}}$

TABLE 2 .6.   IMU Deterministic and Stochastic Sources of Error [12, 14]

	Deterministic Error Parameter	Stochastic Error Parameters	
Gyroscope	Constant offset	High frequency	Bias instability
	Scale factor		Scale factor instability
	Triad misalignment		
	Triad nonorthogonality	Low Frequency	White noise
Accelerometer	Constant offset	High frequency	Bias instability
	Scale factor		Scale factor instability
	Scale factor asymmetry		
	Scale factor nonlinearity	Low Frequency	White noise
	Triad misalignment		
	Triad Nonorthogonality		

INS uses IMU to calculate angular rate, linear velocity, and position. Table 2.6 presents the sources of error in IMU. In Table 2.6, deterministic errors are due to sensor fabrication. Scale factor error refers to the undesirable relationship between input and output signals. Misalignment and nonorthogonality refers to IMU axis error. Figure 2.8 demonstrates different types of errors in gyroscope and accelerometer sensor fabrication.

2.3.3   Camera

Detection and tracking based on vision data is one of the techniques that is widely used in autonomous vehicles (see Chapter 29). Relative object localization in autonomous driving offers useful data that can be directly used or fused with other sensors data to generate the required information for safe and efficient smart transportation. In addition to vehicle detection, driver monitoring [16, 17], lane tracking [18, 19], pedestrian and bicyclist detection are other applications of the camera in the new generation of vehicles [20].

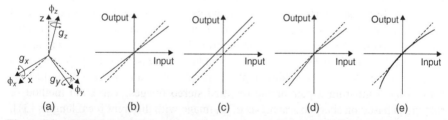

Figure 2.8   IMU deterministic errors: (a) nonorthogonality and misalignment error, (b) scale factor error, (c) constant offset, (d) error due to asymmetry, (e) error due to nonlinearity

Figure 2.9   Vision-based methods of detection and ranging.

In [21], a survey on lane estimation and tracking methods is presented. Detection and tracking are two important stages of using the camera for positioning. As it is shown in Figure 2.9, detection is based on feature extraction and pattern classification methods. Haar wavelet, gradient histogram (HOG), and adaptive local receptive field are popular methods for feature extraction. AdaBoost, neural network, and support vector machines (SVMs) are also the methods that can be applied for target classification [22]. Table 2.7 offers an overview on some of the feature extraction and classification methods, which are applicable for the autonomous driving.

The application of camera in smart transportation is not limited to detection. Researchers have offered techniques for tracking and range detecting of different objects, for example, vehicles [27, 28] and bicyclists [26, 29] to measure the dynamics and motion characteristics of objects and estimate their upcoming position. In general, monocular vision measures and estimates the dynamics in term of pixels, and stereo vision provides information in term of meters [27]. Stereo matching and DFD (depth from defocus) methods are popular for measuring the trajectory of a specific object. A stereo matching depth map, known as a disparity image [30], is formed by comparing a pair of the rectified stereo images. The DFD method of ranging is based on sharpness analysis of an image with different focal lengths [31]. Table 2.8, offers a comprehensive review of vision-based range detection methods and their imposed errors.

Table 2.7. An Overview of Feature Extraction and Classification Methods in Vision-Based Localization

Feature Extraction	Vision Type	Learning, Classification	Application	Comment
Haar Wavelet [20, 23, 24]	Monocular/ Stereo	AdaBoost/Active learning	−Vehicle detection	High TPR (True positive rate) while FPR (False positive rate) is < 1%
	Monocular	Support vector machine (SVM)	−Pedestrian detection	High FPR 80–90% TPR, while FPR is about 30%
SIFT (Scale Invariant Feature Transform) [25]	Monocular/ Stereo	GP (Gaussian process)/SVM)	−Vehicle detection −Motorbike detection	High recognition rate More than 95% recognition rate when active set size is more than 100.
HOG [26]	Monocular	AdaBoost	−Bicyclist detection	Original HOG has better performance than Integral HOG, but it is more complex to calculate.

TABLE 2.8. Overview of Vision-Based Range Detection Method for Vehicle Tracking Application

Methods	Vision Type	Tracking Method	Error Evaluation	Comments
Stereo Matching [30, 32]	Stereo	−Kalman filter −Extended Kalman filter (EKF) −Particle filter	Low (less than 1%)	High accuracy; high cost and complexity
SVM [28]	Stereo/ Monocular	−Kalman filter −Learning methods	Low translation error for various distance and speed.	Monocular SVM is cheaper than stereo SVM due to no baseline; monocular SVM has notably lower accuracy.
DFD [27, 31, 33]	Monocular	−Feature based −Kalman filter −Particle filter	Low (less than 1%)	Window choice impacts accuracy. Control over motor is required.

2.3.4   Network Localization

Positioning challenges in different types of networks, based on their topologies, physical layer characteristics, MAC (media access control) protocols and the routing protocol have been an interesting area for researchers for years [6]. Optimizing the latency, coverage, and power consumption in a network depends on the location of moving and fixed nodes. Hence, a large number of positioning systems are offered to address this challenge according to each network's requirements. Cellular networks, wireless LAN and ad hoc networks—as the most well-known networks—offer options for locating their members. Researchers designed various position aware services using location information.

TOA and RSS (received signal strength) are two ranging methods in networks that have been introduced in Chapter 1, and many chapters of this handbook offer in depth introduction of these techniques. TOA methods are divided into coarse (e.g., match filtering) and fine (e.g., the independent component analysis method detailed in Chapter 7). RSS methods are based on a selected path loss model. The accuracy of RSS estimation is highly sensitive to the proper selection of the path loss model. TDOA (time difference of arrival) and RTT (roundtrip time of arrival) are similar approaches with different measures. In TDOA, the distance is estimated based on the difference in time of arrival received in two points. Compared to TOA, TDOA does not need time synchronization across receivers and test node, but the receivers must be synchronized. RTT method is based on two way latency of signal propagation, so, like TDOA, the RTT method is independent of time synchronization (see Chapter 34). Wireless local positioning systems (WLPS) [34], introduced in detail in Chapter 34, incorporate RTT technique for ranging.

Table 2.9 presents the CRLB of TOA, TDOA, RTT, and RSS methods. In Table 2.9, B is signal bandwidth, T_s is signal duration, σ_{sh}^2 is noise power of

TABLE 2.9.   CRLB for TOA, TDOA, RTT and, RSS Methods of Ranging

Ranging Method	CRLB	Comment
TOA [35, 36]	$\sigma_{TOA} > \dfrac{1}{f_c \sqrt{8\pi^2} \sqrt{SNR \times B\,T_s}}$	Fine time synchronization required; sensitive to bandwidth, signal duration, and SNR.
TDOA [37]	$\sigma_{TDOA} \geq \sigma_{TOA}$	Time synchronization among the receivers required; the error of clock drift in the known node is omitted.
RTT	$\sigma_{RTT} \geq \sqrt{2}\sigma_{TOA}$	Large number of observations required for omitting the processing delay.
RSS [6]	$\sigma_{RSS} > \dfrac{ln10}{10}\dfrac{\sigma_{sh}}{n_p} d$	Sensitive to path loss model and distance (see [38], a survey on path loss models).

TABLE 2.10. Comparative Behavior of Delay Estimation Algorithms

	Accuracy	Comment
Correlation-Based	Low	Sensitive to signal duration, signal bank required, different signal design
Deconvolution-Based	Low	Sensitive to the channel model, channel estimation required
MLE (Maximum Likelihood Error)	Moderate	High complexity
Subspace-based	High	High complexity, sensitive to number of observations
ICA/PCA	High	High complexity for matrix decomposition

log-normal shadowing, n_p is path loss factor and f_c is frequency of carrier. Based on Section 2.3, a bound for error of each ranging method is presented in Table 2.9.

The accuracy of TOA, TDOA, and RTT depends on the employed TOA estimation method that is summarized in Table 2.10. In addition to accuracy, sensitivity to noise, the number of observation, and complexity are other considerations in designing a proper localization system that are detailed in Chapter 7 of this handbook. Based on the capability of cellular networks, positions of users can be estimated using different techniques and characterisitcs explained in Table 2.11 [6]. Chapter 3 introduces CRLB to evaluate different localization algorithms. Chapter 3 also presents mean and variance of the estimation's error as other performance measures for each algorithm. MSE (mean square error) is a common useful measure, which demonstrates precision of an estimation.

2.3.5 Range Detection Sensors

LiDAR (light detection and ranging), same as radar, is based on two-way propagation time estimation and measures distance to an object using laser range finder.

TABLE 2.11. Standard Cellular Network Positioning [6]

Network	Accuracy	Comment
GSM (with E-ODT) [39]	Low (50–120m)	Low accuracy with moderate latency
CDMA/GPRS (with AGPS)[40]	Good (less than 10 m)	GPS receiver is required
WCDMA (with IPDL, TA-IPDL, OTDOA-PE) [41]	Low (about 50 m)	Low coverage
Cellular ID	Low (depends on sector size)	Very low complexity

Today, LiDAR has a wide variety of applications in navigation, robotics, and autonomous driving. LiDAR systems have two key limitations: range and reflectivity. Due to multipath error and clutter, the performance of these sensors drops by increasing the distance of obstacle. Furthermore, the poor reflectivity of some materials limits the detectable objects [42]. LiDAR can provide a 3-D map of the vehicle surrounding using 32 or 64 lasers that measure the distance of objects within a 200 m range [43].

Millimeter Wave radars are a K_a band radar with a carrier frequency of 30–300 GHz. As observed in Table 2.2, high center frequency of these radars allows ultra-wideband signals, which guarantee high-resolution TOA estimation (see Chapter 7). Due to high attenuation caused by atmosphere, mm-wave radars have hard limitations over the range [44]. In addition, mm-wave radar performance depends on the material of objects and is not suitable for nonmetallic object detection due to the high reflection loss [42]. Antenna design considerations and harmful radiations created due to pencil beam signal propagation are other factors, which limit mm-wave radar applications.

The ultrasonic sensor provides proximity detection for low-speed moving objects. Opposed to LiDAR and mm-wave radars, ultrasonic sensors emit and receive acoustic waves, and therefore, Doppler effects impose error in case of high-speed moving object detection. Acoustic wave characteristics make an ultrasonic sensor suitable for detecting variety of objects located within short ranges [43]. Table 2.12, represents an overview of error evaluation and application of LiDAR, and ultrasonic and mm-mave radar.

TABLE 2.12.   Sources of Error and Accuracy of LiDAR and Ultrasonic and mm-Wave Radar in Autonomous Driving Application [42, 45]

Sensor	Sources of Error	Accuracy	Comments
LiDAR	−Error due to angles of laser −Divergence of the laser beam −Multipath in LiDAR −Poor reflectivity	Very high	Suitable for short range telemetry, high update rate, high accuracy, high cost
Ultrasonic	−Diffusion −Absorption −Doppler effect	Very high	Inexpensive, short-range sensor, which is highly sensitive to wavelength and it is proper for parking assistant systems.
mm-wave radar	−Misalignment −Multipath −Absorption	Very high	Compared to LiDAR, mm-wave radar has a shorter effective range and more sensitive to material reflectivity, not suitable for pedestrian or any nonmetallic objects detection.

2.4 DATA FUSION

Employing different sensors on autonomous vehicles leads researchers to study and implement different fusion methods and enable high-performance localization. Each sensor or technique, based on its characteristics and performance, provides some sort of information but cannot fulfill the requirements of safety in autonomous driving on its own. Today, different combinations of on-vehicle sensors are used to enable safe and reliable controlling systems. Examples include auto-break, auto-parking, and anti-collision distance management systems [46, 47], which facilitate the driving. In this section, we will review the fusion methods and some of their specifications.

When the covariance matrix of observed data increases or mean of error diverges away from zero, data fusion loses its stability; thus, researchers proposed self and cooperative calibration methods to cancel the biased error and bound the error power [48]. Sensitivity to error is another key factor for data fusion evaluation. The speed of convergence or sensitivity to the number of observations is also key to the fusion methods. Some sensors, such as INS and LiDAR, generate hundreds of samples per second. Thus, sensors with high latency and low updating frequency could cause low convergence speed or even data fusion instability when monitoring high-speed events. Practically, sensors' data are not necessarily synchronized, so taking the latency of each sensor into account increases both accuracy and complexity of localization system, but estimating the sensors parameters imposes error to the system. Some fusion methods, such as the interacting multiple model (IMM) are capable of handling asynchronous data [49–51]. Increasing the data sampling rate is another approach to overcome the asynchronous data fusion challenge [52], but some technologies such as GPS and cameras have hard limitation over their refreshment frequency [5]. Cloud assisted data fusion [53], cooperative positioning [54], and employing external source for data correction and sensor calibration [55] are samples of networks' application in smart transportation systems. A comparison across fusion methods and sensor combination is presented in Table 2.13.

2.5 PRACTICAL ERROR EVALUATION METHOD AND CORRECTION

In futuristic smart cities, autonomous vehicles (AVs) are envisioned to communicate big data with infrastructures, other AVs, and humans. Thus, in smart cities, real-time data processing and rigorous decision-making are key to safe and reliable smart transportation. Deployment of new generations of cellular networks, the emergence of the Internet of Things (IoT), and new methods of distributed data processing and data storing enables AVs to process and communicate data fast and efficiently. With the emergence of mm-wave communications, and small and well distributed access points, it is anticipated that some challenges, such as bandwidth limitations and coverage, will disappear and new challenges, such as real-time big data will emerge.

In addition, transitioning from the current nonautonomousto a fully autonomous world, we are facing the challenging period of a semiautonomous world. In

TABLE 2.13. Overview of Fusion Methods for Localization Application

Method	Sensors Combination	Sensitivity to Error	Sensitivity to Observation Rate	Asynchronous Observation Consideration	Comment
Kalman Filter	GPS, INS [5, 54, 56]	Low	High	No	Suitable cooperative positioning in Vehicular ad hoc networkVANET
				Yes	Cancels INS bias error, accurate approach to localization presented.
		Low	Low	No	A study on the effect of sensor calibration on data fusion is presented.
EKF	INS, GPS [48]	High	High	Yes	Based on weighted EKF, sensitive to biased error and noise power
	INS, GPS, LiDAR, camera [46]	Low	Moderate	Yes	Combining LiDAR and camera provides good detection.
	INS, camera, GPS [57]	High	Low	No	Good road tracking based on digital map correction and camera
Unscented Kalman Filter-IMM	GPS, camera, digital map [58]	Low	Low	No	SIFT offline image feature extraction has accurate positioning for 3D city model.
Dynamic Bayesian Network	GPS, camera, digital map [59]	High	Low	No	Compared to SIFT, Harris point match has less accuracy, sensitive to sensor error.
Deep fusion	Camera, LiDAR [60]	Low (accurate classification)	Low	No	CNN (Convolutional Neural Network)-based; good for LiDAR feature extraction

Table 2.1 Local Procedure and Data Correction position

Method	Sensors Combination	Sensitivity to Error	Sensitivity to Observation Rate	Asynchronous Observation Consideration	Comment
AROCCAM	GPS, camera [61]	High	Not mentioned	Yes	Good for asynchronous data fusion, sensitive to erroneous observation
Particle Filter (PF)	INS, GPS [50, 62]	Low	Low	Yes	Compatible with any posterior PDF. Based on entropy-weighted PF, information gain in data fusion is novel.
IF (Information Filter)	Camera, LiDAR, GPS [63]	Low	Low	No	Compared to unscented KF, UIF has better performance over nonlinear systems.
SCAAT [64] (Single Const. at a Time)	Not mentioned	Low	High	No	Proper for incomplete data fusion, based on Kalman filter.

TABLE 2.14.   Error Detection and Data Correction Methods

	Methods
Sensor Calibration	–DGPS [6]
	–INS correction by GPS [5, 56]
	–SOOP (signal of opportunity) as a supplement for any navigation systems [55]
Data Validation	–LiDAR scan can be used to correct the image-based detection methods [46]
	–GPS data correction by camera and digital map [58]
	–INS data evaluation by camera and digital map [59]
Measurement Error Detection	–Positioning in mixed LOS/NLOS conditions [65]
	–NLOS mitigation techniques [66–68]

the semiautonomous world, the confrontation between human and AI drivers, unreliability of localization systems and networks, big data processing and management, processing performance, and latency creates a state in which fulfilling safety requirements depends on employing high-performance data correction and data reduction methods.

A state vector as the input of data fusion can be either continuous, discrete, or hybrid. The accuracy of state vector estimation depends on the data error model and fusion method. Calibrating sensors, validating the sensor's output, and detecting erroneous measurements are different approaches for improving the performance of fusion methods (Table 2.14) [50]. Data reduction as another stage in localization systems is an important aspect of system design. Distributed processing methods and cloud and fog computation methods emerge in self-navigation systems to initiate a new era in transportation. High process capability of cloud-assisted computation systems enables employing big number of on-vehicle sensors on one hand, and on the other hand, the extension of external controlling and monitoring systems delivers a new level of safety in smart transportation.

2.6   SUMMARY

This chapter studied two measures for evaluating the performance of an estimator. It assessed the impact of increasing the number of measurements on accuracy. A criterion to evaluate the effect of increasing localization data was studied. Parameters that limit the accuracy for different localization technologies were introduced. To overcome these limitations, the output of different sensors can be fused; however, erroneous sensor data can induce error in the fusion process. Thus, efficient data fusion would be critical. The chapter presented diverse fusion methods and compared their performance using different sensor combinations. The chapter briefly reviewed some practical error evaluation and data reduction implementation.

REFERENCES

[1] S. Del Marco, "A series representation of the spherical error probability integral," *IEEE Transactions on Aerospace and Electronic Systems*, vol. 29, no. 4, pp. 1349–1356, 1993.

[2] S. Zhang, X. Duan, C. Li, X. Peng, and Q. Zhang, "CEP calculation based on weighted Bayesian mixture model," *Mathematical Problems in Engineering*, vol. 2017, 2017, doi.org/10.1155/2017/8070786.

[3] Z. Wang and S. Zekavat, "Comparison of semidistributed multinode TOA-DOA fusion localization and GPS-aided TOA (DOA) fusion localization for MANETs," *EURASIP Journal on Advances in Signal Processing*, vol. 2008, no. 1, pp. 439–523, 2008.

[4] H. Akaike, "Information theory and an extension of the maximum likelihood principle," *Selected Papers of Hirotugu Akaike*. Springer, New York, NY, 1998, pp. 199–213.

[5] S. Sukkarieh, E. M. Nebot, and H. F. Durrant-Whyte, "A high integrity IMU/GPS navigation loop for autonomous land vehicle applications," *IEEE Transactions on Robotics and Automation*, vol. 15, no. 3, pp. 572–578, 1999.

[6] G. Sun, J. Chen, W. Guo, and K. R. Liu, "Signal processing techniques in network-aided positioning: A survey of state-of-the-art positioning designs," *IEEE Signal Processing Magazine*, vol. 22, no. 4, pp. 12–23, 2005.

[7] N. Alam, A. T. Balaei, and A. G. Dempster, "Relative positioning enhancement in VANETs: A tight integration approach," *IEEE Transactions on Intelligent Transportation Systems*, vol. 14, no. 1, pp. 47–55, 2013.

[8] E. Kaplan and C. Hegarty, *Understanding GPS: Principles and Applications*. Boston, MA: Artech House, 2006.

[9] M. S. Grewal, L. R. Weill, and A. P. Andrews, *Global Positioning Systems, Inertial Navigation, and Integration*, 2nd edition. Hoboken: John Wiley & Sons, 2007.

[10] P. Misra and P. Enge, *Global Positioning System: Signals, Measurements, and Performance*, 2nd edition, Lincoln, MA: Ganga-Jamuna Press, 2006.

[11] K. Wen, C. K. Seow, and S. Y. Tan, "Inertial navigation system positioning error analysis and Cramér-Rao lower bound," *Position, Location and Navigation Symposium (PLANS), 2016 IEEE/ION*, pp. 213–218, 2016.

[12] W. Flenniken, J. Wall, and D. Bevly, "Characterization of various IMU error sources and the effect on navigation performance," *Ion GNSS*, 2005, pp. 967–978, 2005.

[13] N. El-Sheimy, H. Hou, and X. Niu, "Analysis and modeling of inertial sensors using Allan variance," *IEEE Transactions on Instrumentation and Measurement*, vol. 57, no. 1, pp. 140–149, 2008.

[14] D. Unsal and K. Demirbas, "Estimation of deterministic and stochastic IMU error parameters," *IEEE/ION Position Location and Navigation Symposium (PLANS)*, pp. 862–868, 2012.

[15] D. Jurman, M. Jankovec, R. Kamnik, and M. Topič, "Calibration and data fusion solution for the miniature attitude and heading reference system," *Sensors and Actuators A: Physical*, vol. 138, no. 2, pp. 411–420, 2007.

[16] J. C. McCall, D. P. Wipf, M. M. Trivedi, and B. D. Rao, "Lane change intent analysis using robust operators and sparse bayesian learning," *IEEE Transactions on Intelligent Transportation Systems*, vol. 8, no. 3, pp. 431–440, 2007.

[17] A. Doshi and M. M. Trivedi, "On the roles of eye gaze and head dynamics in predicting driver's intent to change lanes," *IEEE Transactions on Intelligent Transportation Systems*, vol. 10, no. 3, pp. 453–462, 2009.

[18] S. Sivaraman and M. M. Trivedi, "Improved vision-based lane tracker performance using vehicle localization," *IEEE Intelligent Vehicles Symposium (IV)*, pp. 676–681, 2010.

[19] S. Sivaraman and M. M. Trivedi, "Integrated lane and vehicle detection, localization, and tracking: A synergistic approach," *IEEE Transactions on Intelligent Transportation Systems*, vol. 14, no. 2, pp. 906–917, 2013.

[20] S. Sivaraman and M. M. Trivedi, "Combining monocular and stereo-vision for real-time vehicle ranging and tracking on multilane highways," *14th International IEEE Conference on Intelligent Transportation Systems (ITSC), 2011*, pp. 1249–1254, 2011.

[21] J. C. McCall and M. M. Trivedi, "Video-based lane estimation and tracking for driver assistance: Survey, system, and evaluation," *IEEE Transactions on Intelligent Transportation Systems*, vol. 7, no. 1, pp. 20–37, 2006.

[22] M. Enzweiler, A. Eigenstetter, B. Schiele, and D. M. Gavrila, "Multi-cue pedestrian classification with partial occlusion handling," *IEEE Conference on Computer Vision and Pattern Recognition (CVPR)*, , pp. 990–997, 2010.

[23] S. Sivaraman and M. M. Trivedi, "A general active-learning framework for on-road vehicle recognition and tracking," *IEEE Transactions on Intelligent Transportation Systems*, vol. 11, no. 2, pp. 267–276, 2010.

[24] M. Enzweiler and D. M. Gavrila, "A mixed generative-discriminative framework for pedestrian classification," *IEEE Conference on Computer Vision and Pattern Recognition, CVPR 2008*, pp. 1–8, 2008.

[25] A. Kapoor, K. Grauman, R. Urtasun, and T. Darrell, "Active learning with gaussian processes for object categorization," *IEEE 11th International Conference on Computer Vision, ICCV 2007*, pp. 1–8, 2007.

[26] H. Cho, P. E. Rybski, and W. Zhang, "Vision-based bicyclist detection and tracking for intelligent vehicles," *Intelligent Vehicles Symposium (IV), 2010 IEEE*, pp. 454–461, 2010.

[27] S. Sivaraman and M. M. Trivedi, "Looking at vehicles on the road: A survey of vision-based vehicle detection, tracking, and behavior analysis," *IEEE Transactions on Intelligent Transportation Systems*, vol. 14, no. 4, pp. 1773–1795, 2013.

[28] S. Song, M. Chandraker, and C. C. Guest, "High accuracy monocular SFM and scale correction for autonomous driving," *IEEE Transactions on Pattern Analysis and Machine Intelligence*, vol. 38, no. 4, pp. 730–743, 2016.

[29] C. Liu, R. Fujishiro, L. Christopher, and J. Zheng, "Vehicle–bicyclist dynamic position extracted from naturalistic driving videos," *IEEE Transactions on Intelligent Transportation Systems*, vol. 18, no. 4, pp. 734–742, 2017.

[30] H. Xue, S. Zhang, and D. Cai, "Depth image inpainting: Improving low rank matrix completion with low gradient regularization," *IEEE Transactions on Image Processing*, vol. 26, no. 9, pp. 4311–4320, Sept. 2017.

[31] E. Krotkov and R. Bajcsy, "Active vision for reliable ranging: Cooperating focus, stereo, and vergence," *International Journal of Computer Vision*, vol. 11, no. 2, pp. 187–203, 1993.

[32] D. Scharstein and R. Szeliski, "A taxonomy and evaluation of dense two-frame stereo correspondence algorithms," *International Journal of Computer Vision*, vol. 47, no. 1–3, pp. 7–42, 2002.

[33] Z. Chen, X. Guo, S. Li, X. Cao, and J. Yu, "A learning-based framework for hybrid depth-from-defocus and stereo matching," preprint arXiv:1708.00583, 2017.

[34] S. A. Zekavat, "Wireless local positioning system," Google Patents, 2009.

[35] N. Patwari, J. N. Ash, S. Kyperountas, A. O. Hero, R. L. Moses, and N. S. Correal, "Locating the nodes: cooperative localization in wireless sensor networks," *IEEE Signal Processing Magazine*, vol. 22, no. 4, pp. 54–69, 2005.

[36] A. Catovic and Z. Sahinoglu, "The Cramer-Rao bounds of hybrid TOA/RSS and TDOA/RSS location estimation schemes," *IEEE Communications Letters*, vol. 8, no. 10, pp. 626–628, 2004.

[37] Y. Qi and H. Kobayashi, "On relation among time delay and signal strength based geolocation methods," *Global Telecommunications Conference, 2003. GLOBECOM'03. IEEE*, vol. 7, pp. 4079–4083, 2003.

[38] C. Phillips, D. Sicker, and D. Grunwald, "A survey of wireless path loss prediction and coverage mapping methods," *IEEE Communications Surveys & Tutorials*, vol. 15, no. 1, pp. 255–270, 2013.

[39] J. H. Reed, K. J. Krizman, B. D. Woerner, and T. S. Rappaport, "An overview of the challenges and progress in meeting the E-911 requirement for location service," *IEEE Communications Magazine*, vol. 36, no. 4, pp. 30–37, 1998.

[40] H. Tong and S. Zekavat, "A novel wireless local positioning system via asynchronous DSCDMA and beamforming: implementation and perturbation analysis," *IEEE Transactions on Vehicular Technology*, vol. 56, no. 3, pp. 1307–1320, 2007.

[41] H. Tong, J. Pourrostam, and S. Zekavat, "Optimum beamforming for a novel wireless local positioning system: A stationarity analysis and solution," *EURASIP Journal on Advances in Signal Processing*, vol. 2007, p. 12, 2007.

[42] J. M. Anderson, K. Nidhi, K. D. Stanley, P. Sorensen, C. Samaras, and O. A. Oluwatola, *Autonomous Vehicle Technology: A Guide for Policymakers*. Rand Corporation, Santa Monica, CA, 2014.

[43] M. Nikowitz, *Fully Autonomous Vehicles: Visions of the Future or Still Reality?* Berlin: epubli, 2015.

[44] M. I. Skolnik, "Introduction to radar," *Radar Handbook*, vol. 2, McGraw-Hill, New York 1962.

[45] P. Chen, *Mechatronics and Manufacturing Technologies: Proceedings of the International Conference on Mechatronics and Manufacturing Technologies (MMT2016)*. World Scientific Wuhan, Hubai, China, Aug. 20–21 2016.

[46] A. K. Aijazi, P. Checchin, and L. Trassoudaine, "Multi sensorial data fusion for efficient detection and tracking of road obstacles for inter-distance and anti-colision safety management," *3rd International Conference on Control, Automation and Robotics (ICCAR), 2017*, pp. 617–621, 2017.

[47] N.-E. El Faouzi, H. Leung, and A. Kurian, "Data fusion in intelligent transportation systems: Progress and challenges–A survey," *Information Fusion*, vol. 12, no. 1, pp. 4–10, 2011.

[48] J. Sasiadek and Q. Wang, "Low cost automation using INS/GPS data fusion for accurate positioning," *Robotica*, vol. 21, no. 3, pp. 255–260, 2003.

[49] K. Jo, K. Chu, and M. Sunwoo, "Interacting multiple model filter-based sensor fusion of GPS with in-vehicle sensors for real-time vehicle positioning," *IEEE Transactions on Intelligent Transportation Systems*, vol. 13, no. 1, pp. 329–343, 2012.

[50] F. Caron, M. Davy, E. Duflos, and P. Vanheeghe, "Particle filtering for multisensor data fusion with switching observation models: Application to land vehicle positioning," *IEEE Transactions on Signal Processing*, vol. 55, no. 6, pp. 2703–2719, 2007.

[51] S. Maeyama, A. Ohya, and S. i. Yuta, "Non-stop outdoor navigation of a mobile robot-Retroactive positioning data fusion with a time consuming sensor system," *Proceedings. 1995 IEEE/RSJ International Conference on Intelligent Robots and Systems 95. 'Human Robot Interaction and Cooperative Robots,'* vol. 1, pp. 130–135, 1995.

[52] M.-M. Meinecke and M. A. Obojski, "Potentials and limitations of pre-crash systems for pedestrian protection," in *Proc. 2nd International Workshop Intelligent Transportation Systems*, 2005.

[53] F. H. Bijarbooneh, W. Du, E. C.-H. Ngai, X. Fu, and J. Liu, "Cloud-assisted data fusion and sensor selection for internet of things," *IEEE Internet of Things Journal*, vol. 3, no. 3, pp. 257–268, 2016.

[54] N. Alam, A. Kealy, and A. G. Dempster, "An INS-aided tight integration approach for relative positioning enhancement in VANETs," *IEEE Transactions on Intelligent Transportation Systems*, vol. 14, no. 4, pp. 1992–1996, 2013.

[55] C. Yang, T. Nguyen, and E. Blasch, "Mobile positioning via fusion of mixed signals of opportunity," *IEEE Aerospace and Electronic Systems Magazine*, vol. 29, no. 4, pp. 34–46, 2014.

[56] J. A. Rios and E. White, "Fusion filter algorithm enhancements for a MEMS GPS/IMU," *ION NTM*, pp. 28–30, 2002.

[57] Z. Hu and K. Uchimura, "Fusion of vision, GPS and 3D gyro data in solving camera registration problem for direct visual navigation," *International Journal of ITS*, vol. 4, no. 1, pp. 3–12, 2006.

[58] M. Dawood, C. Cappelle, M. E. El Najjar, M. Khalil, and D. Pomorski, "Vehicle geo-localization based on IMM-UKF data fusion using a GPS receiver, a video camera and a 3D city model," *IEEE Intelligent Vehicles Symposium (IV)*, pp. 510–515, 2011.

[59] C. Cappelle, M. E. El Najjar, D. Pomorski, and F. Charpillet, "Multi-sensors data fusion using dynamic Bayesian network for robotised vehicle geo-localisation," *International Conference on Information Fusion*, pp. 1–8, 2008.

[60] Y. Chen, C. Li, P. Ghamisi, X. Jia, and Y. Gu, "Deep fusion of remote sensing data for accurate classification," *IEEE Geoscience and Remote Sensing Letters*, 2017.

[61] C. Tessier, C. Cariou, C. Debain, F. Chausse, R. Chapuis, and C. Rousset, "A real-time, multi-sensor architecture for fusion of delayed observations: Application to vehicle localization," *IEEE Intelligent Transportation Systems Conference, 2006. ITSC'06*, pp. 1316–1321, 2006.

[62] H. Kim, B. Liu, C. Y. Goh, S. Lee, and H. Myung, "Robust vehicle localization using entropy-weighted particle filter-based data fusion of vertical and road intensity information for a large scale urban area," *IEEE Robotics and Automation Letters*, vol. 2, no. 3, pp. 1518–1524, 2017.

[63] L. Wei, C. Cappelle, and Y. Ruichek, "Camera/laser/GPS fusion method for vehicle positioning under extended NIS-based sensor validation," *IEEE Transactions on Instrumentation and Measurement*, vol. 62, no. 11, pp. 3110–3122, 2013.

[64] G. Welch and G. Bishop, "SCAAT: Incremental tracking with incomplete information," *Proceedings of the 24th Annual Conference on Computer Graphics and Interactive Techniques*, pp. 333–344, 1997.

[65] L. Chen and L. Wu, "Mobile positioning in mixed LOS/NLOS conditions using modified EKF banks and data fusion method," *IEICE Transactions on Communications*, vol. 92, no. 4, pp. 1318–1325, 2009.

[66] I. Guvenc and C.-C. Chong, "A survey on TOA based wireless localization and NLOS mitigation techniques," *IEEE Communications Surveys & Tutorials*, vol. 11, no. 3, 2009.

[67] C.-S. Chen, "A non-line-of-sight error mitigation method for location estimation," *International Journal of Distributed Sensor Networks*, vol. 13, no. 1, Jan. 19, 2017, doi: 10.1177/1550147716682739 2017.

[68] G. Mao, B. Fidan, and B. D. Anderson, "Wireless sensor network localization techniques," *Computer Networks*, vol. 51, no. 10, pp. 2529–2553, 2007.

CHAPTER 3

SOURCE LOCALIZATION: ALGORITHMS AND ANALYSIS

H. C. So

City University of Hong Kong,
Hong Kong

FINDING THE position of a source based on measurements from an array of spatially separated sensors has been an important problem in radar, sonar, global positioning systems (GPS), mobile communications, multimedia, and wireless sensor networks. Time of arrival (TOA), time difference of arrival (TDOA), time sum of arrival (TSOA), received signal strength (RSS), and direction of arrival (DOA) of the emitted signal are commonly used measurements for source localization. Basically, TOAs, TDOAs, TSOAs, and RSSs provide the distance information between the source and sensors, while DOAs are the source bearings relative to the receivers. However, positioning is not a trivial task because these measurements have nonlinear relationships with the source location.

This chapter introduces two categories of positioning algorithms based on TOA, TDOA, TSOA, RSS, and DOA measurements. The first class works on the nonlinear equations directly obtained from the nonlinear relationships between the source and measurements. Corresponding examples, namely, nonlinear least squares (NLS) and maximum likelihood (ML) estimators, will be presented. The second category attempts to convert the equations to linear. Here, we will discuss the linear least squares (LLS), weighted linear least squares (WLLS), and subspace approaches. In addition, under sufficiently small error conditions, we develop the mean and variance expressions for any positioning method which can be formulated as an unconstrained optimization problem. Assuming that the disturbances in the measurements are zero-mean Gaussian distributed, the Cramer-Rao lower bound

Handbook of Position Location: Theory, Practice, and Advances, Second Edition.
Edited by S. A. (Reza) Zekavat and R. Michael Buehrer.
© 2019 by the Institute of Electrical and Electronics Engineers, Inc.
Published 2019 by John Wiley & Sons, Inc.
Companion Website: www.wiley.com/go/zekavat/positionlocation2e

(CRLB), which is a lower bound on the variance attainable by any unbiased location estimator using the same data, will also be discussed.

The intended learning outcomes for this chapter include: (i) understanding the positioning algorithm development using TOA, TDOA, TSOA, RSS, and DOA measurements; and (ii) understanding the performance measures for position estimation.

3.1 INTRODUCTION

As discussed in Chapter 1, the position of a target of interest can be determined by utilizing its signal measured at an array of spatially separated receivers with a priori known locations. In fact, source localization has been one of the central problems in many fields, such as multiple-input multiple-output (MIMO) radar [1], sonar [2], telecommunications [3], mobile communications [4–6], wireless sensor networks [7, 8] as well as human–computer interaction [9]. For example, the position of an active talker can be tracked with the use of a microphone array in applications such as video conferencing, automatic scene analysis, and security monitoring. On the other hand, mobile terminal localization has been receiving considerable attention, especially after the Federal Communications Commission (FCC) in the U.S. adopted rules to improve 911 services by mandating the accuracy of locating an emergency caller to be within a specified range, even for a wireless phone user [10]. Apart from emergency assistance, mobile position information is also the key enabler for a large number of innovative applications, such as personal localization and monitoring, fleet management, asset tracking, travel services, location-based advertising, and billing. More recently, technological advances in wireless communications and microsystem integration have enabled the development of small, inexpensive, low-power sensor nodes that are able to collect surrounding data, perform small-scale computations, and communicate among their neighbors. These wirelessly connected nodes, when working in a collaborative manner, have great potential in numerous remote monitoring and control applications, such as habitat monitoring, healthcare, building automation, and battlefield surveillance, as well as environment observation and forecasting. Because sensor nodes are often arbitrarily placed, with their positions being unknown, node positioning is a fundamental and crucial issue for sensor network operation and management.

TOA, TDOA, TSOA, RSS, and DOA are commonly used measurements [1, 11] for source localization. Basically, TOAs, TDOAs, TSOAs, and RSSs provide the distance information between the source and sensors, while DOAs are the source bearings relative to the receivers. In two-dimensional (2-D) geometry, localization using TOAs or RSSs, TDOAs, TSOAs, and DOAs refer to determining the intersection of circles, hyperbolas, ellipses, and straight lines, respectively. Nevertheless, finding the source position is not an easy task because the location is nonlinear in these measurements. Given the TOA, TDOA, TSOA, RSS, or DOA information, the main focus in this chapter is on positioning algorithm development and analysis. Although 2-D source localization is considered, it is straightforward to extend the study to three-dimensional (3-D) space. We assume that there are no outliers in the measurements in order to achieve reliable location estimation. That is, the errors due to shadowing and multipath propagation in the RSSs are sufficiently small. On the other hand, line-of-sight (LOS) transmission [11] is assumed, so that there is a direct path between the source and each sensor, in estimating the TOAs, TDOAs, TSOAs, and DOAs. It is worth pointing out that non-line of sight (NLOS) occurs when there are obstructions between the source and transmitters/receivers, which can cause large positive biases in the corresponding distance information. For position estimation in the presence of NLOS propagation, the interested reader is referred to Part IV of this book.

The rest of this chapter is organized as follows. The measurement models of TOA, TDOA, TSOA, RSS, and DOA, and their positioning principles are first presented in Section 3.2. Positioning algorithms based on the location-bearing information, which are classified as nonlinear and linear approaches, are developed in Section 3.3. The first category deals with the nonlinear equations directly constructed from the TOA, TDOA, TSOA, RSS, or DOA measurements, which includes the NLS and ML estimators. On the other hand, the second approach converts the nonlinear equations to be linear, and LLS, WLLS and subspace methods will be presented. Note that the WLLS estimator is in fact a generalized version of the LLS technique, where a weighting function is involved. Section 3.4 contributes to the algorithm analysis and two important performance measures, namely, mean and variance, will be examined. Furthermore, the computation of CRLB, which is a lower bound on the variance attainable by any unbiased location estimator, will be presented. Finally, concluding remarks are given in Section 3.5. Symbols that are used in this chapter are listed in Table 3.1.

TABLE 3.1. List of Symbols

Symbol	Meaning
T	Transpose
-1	Inverse
$\dagger$	Pseudoinverse
E	Expectation operator
var	Variance
$\sim$	Distributed as
$\mathcal{N}(\mathbf{a}, \mathbf{C})$	Gaussian distribution with mean $\mathbf{a}$ and covariance $\mathbf{C}$
$|\mathbf{C}|$	Determinant of $\mathbf{C}$
$\mathbf{I}$	Identity matrix
$\mathbf{1}$	Vector with all elements equal 1
$\mathbf{0}$	Zero vector
$\tilde{\mathbf{a}}$	Variable for $\mathbf{a}$
$\hat{\mathbf{a}}$	Estimate of $\mathbf{a}$
$[\mathbf{a}]_m$	mth element of $\mathbf{a}$
$\|\mathbf{a}\|_2$	l_2-norm of $\mathbf{a}$
$[\mathbf{A}]_{m,n}$	(m, n) entry of $\mathbf{A}$
diag($\mathbf{a}$)	Diagonal matrix with vector $\mathbf{a}$ as main diagonal

3.2 MEASUREMENT MODELS AND PRINCIPLES FOR SOURCE LOCALIZATION

The TOA, TDOA, TSOA, RSS, and DOA measurement models and their basic positioning principles are presented in Sections 3.2.1 to 3.2.5, respectively. In fact, all the models can be generalized as:

$$\mathbf{r} = \mathbf{f}(x) + \mathbf{n}, \qquad (3.1)$$

where $\mathbf{r}$ is the measurement vector, $\mathbf{x}$ is the source position to be determined, $\mathbf{f}(\mathbf{x})$ is a known nonlinear function of $\mathbf{x}$, and $\mathbf{n}$ is an additive zero-mean noise vector. A comparison summary for the five measurement models is provided in Table 3.2.

3.2.1 TOA

TOA is the one-way propagation time of the signal traveling between a source and a receiver. This implies that the target and all receivers are required to be precisely synchronized to obtain the TOA information, although such synchronization is not needed if the round-trip or two-way TOA is measured. Multiplying TOA by the known propagation speed, denoted by c, provides the distance between source and receiver. For example, $c \approx 340 \text{ ms}^{-1}$ and $c \approx 3 \times 10^8 \text{ ms}^{-1}$ are the speeds of sound and light, respectively, in the in-air scenarios. In the absence of measurement error, each TOA corresponds to a circle centered at a receiver on which the source must lie in

TABLE 3.2. Comparison of Different Measurement Models

Model	Location Information	Advantages	Disadvantages
TOA	Range	- Accuracy is high.	- Time synchronization across source and all receivers is needed (Note that synchronization is not required for round-trip TOA). - LOS is assumed.
TDOA	Range difference	- Accuracy is high. - Time synchronization at source is not required.	- LOS is assumed.
TSOA	Range sum	- Accuracy is high. - Time synchronization at source is not required.	- Multiple transmitters and receivers are needed. - LOS is assumed.
RSS	Range	Simple and inexpensive Time synchronization is not required.	Accuracy is low.
DOA	Bearing	Only at least two receivers are needed. Time synchronization is not required.	Smart antennas are needed. LOS is assumed.

the 2-D space. As discussed in Chapter 1, geometrically, three or more circles deduced from the noise-free TOAs will result in a unique intersection, which is the source position, implying that at least three sensors are needed for 2-D positioning. Note that two TOA-circles generally have two intersection points, which correspond to two possible source locations. Nevertheless, these circles may not intersect or have multiple intersections, in the presence of disturbances, and hence it is not an effective way to solve the problem using the circles directly. In fact, with three or more receivers, it is more appropriate to convert the noisy TOAs into a set of circular equations from which the source position can be determined according to an optimization criterion, with the knowledge of the sensor array geometry.

Mathematically, the TOA measurement model is formulated as follows. Let $\mathbf{x} = \begin{bmatrix} x & y \end{bmatrix}^T$ be the unknown source position and $\mathbf{x}_l = \begin{bmatrix} x_l & y_l \end{bmatrix}^T$ be the known coordinates of the lth sensor, $l = 1, 2, \cdots, L$, where $L \geq 3$ is the number of receivers. The distance between the source and the lth sensor, denoted by d_l, is simply:

$$d_l = \|\mathbf{x} - \mathbf{x}_l\|_2 = \sqrt{(x - x_l)^2 + (y - y_l)^2}, \quad l = 1, 2, \cdots, L. \tag{3.2}$$

Without loss of generality, we assume that the source emits a signal at time 0 and the lth sensor receives it at time t_l. That is, $\{t_l\}$ are the TOAs, and we have a simple relationship between t_l and d_l:

$$t_l = \frac{d_l}{c}, \quad l = 1, 2, \cdots, L. \tag{3.3}$$

In practice, TOAs are subject to measurement errors. As a result, the range measurement based on multiplying t_l by c, denoted by $r_{\text{TOA},l}$, is modeled as:

$$r_{\text{TOA},l} = d_l + n_{\text{TOA},l} = \sqrt{(x - x_l)^2 + (y - y_l)^2} + n_{\text{TOA},l}, \quad l = 1, 2, \cdots, L, \tag{3.4}$$

where $n_{\text{TOA},l}$ is the range error in $r_{\text{TOA},l}$, which is resulted from the TOA disturbance.

Equation (3.4) can also be expressed in vector form as

$$\mathbf{r}_{\text{TOA}} = \mathbf{f}_{\text{TOA}}(\mathbf{x}) + \mathbf{n}_{\text{TOA}}, \tag{3.5}$$

where

$$\mathbf{r}_{\text{TOA}} = \begin{bmatrix} r_{\text{TOA},1} & r_{\text{TOA},2} & \cdots & r_{\text{TOA},L} \end{bmatrix}^T, \tag{3.6}$$

$$\mathbf{n}_{\text{TOA}} = \begin{bmatrix} n_{\text{TOA},1} & n_{\text{TOA},2} & \cdots & n_{\text{TOA},L} \end{bmatrix}^T, \tag{3.7}$$

and

$$\mathbf{f}_{\text{TOA}}(\mathbf{x}) = \mathbf{d} = \begin{bmatrix} \sqrt{(x - x_1)^2 + (y - y_1)^2} \\ \sqrt{(x - x_2)^2 + (y - y_2)^2} \\ \vdots \\ \sqrt{(x - x_L)^2 + (y - y_L)^2} \end{bmatrix}. \tag{3.8}$$

Here, $\mathbf{f}_{\text{TOA}}(\mathbf{x})$ represents the known function, which is parameterized by $\mathbf{x}$, and in fact it is the noise-free distance vector. The source localization problem based on TOA measurements is to estimate $\mathbf{x}$ given $\{r_{\text{TOA},l}\}$ or $\mathbf{r}_{\text{TOA}}$.

To facilitate the algorithm development and analysis, as well as CRLB computation, it is assumed that $\{n_{\text{TOA},l}\}$ are zero-mean uncorrelated Gaussian processes with variances $\{\sigma^2_{\text{TOA},l}\}$. It is worth mentioning that the zero-mean property indicates LOS transmission. The probability density function (PDF) for each scalar random variable $r_{\text{TOA},l}$, denoted by $p(r_{\text{TOA},l})$, has the form of

$$p(r_{\text{TOA},l}) = \frac{1}{\sqrt{2\pi\sigma^2_{\text{TOA},l}}} \exp\left(-\frac{1}{2\sigma^2_{\text{TOA},l}}(r_{\text{TOA},l} - d_l)^2\right), \tag{3.9}$$

which is characterized by its mean and variance, d_l and $\sigma^2_{\text{TOA},l}$, respectively. In other words, we can write $r_{\text{TOA},l} \sim \mathcal{N}(d_l, \sigma^2_{\text{TOA},l})$. While the PDF for $\mathbf{r}_{\text{TOA}}$, denoted by $p(\mathbf{r}_{\text{TOA}})$, is

$$p(\mathbf{r}_{\text{TOA}}) = \frac{1}{(2\pi)^{L/2}|\mathbf{C}_{\text{TOA}}|^{1/2}} \exp\left(-\frac{1}{2}(\mathbf{r}_{\text{TOA}} - \mathbf{d})^T \mathbf{C}_{\text{TOA}}^{-1}(\mathbf{r}_{\text{TOA}} - \mathbf{d})\right), \tag{3.10}$$

where $\mathbf{C}_{\text{TOA}}$ is the covariance matrix for $\mathbf{r}_{\text{TOA}}$ which corresponds to

$$\begin{aligned}
\mathbf{C}_{\text{TOA}} &= E\{(\mathbf{r}_{\text{TOA}} - \mathbf{d})(\mathbf{r}_{\text{TOA}} - \mathbf{d})^T\} \\
&= E\{\mathbf{n}_{\text{TOA}}\mathbf{n}_{\text{TOA}}^T\} \\
&= \text{diag}(\sigma^2_{\text{TOA},1}, \sigma^2_{\text{TOA},2}, \cdots, \sigma^2_{\text{TOA},L}).
\end{aligned} \tag{3.11}$$

In (3.11), the third equality is deduced using the assumption of uncorrelated $\{n_{\text{TOA},l}\}$. Using (3.11), (3.10) is simplified to:

$$p(\mathbf{r}_{\text{TOA}}) = \frac{1}{(2\pi)^{L/2} \prod_{l=1}^{L} \sigma_{\text{TOA},l}} \exp\left(-\frac{1}{2}\sum_{l=1}^{L} \frac{(r_{\text{TOA},l} - d_l)^2}{\sigma^2_{\text{TOA},l}}\right). \tag{3.12}$$

In other words, we can write $\mathbf{r}_{\text{TOA}} \sim \mathcal{N}(\mathbf{d}, \text{diag}(\sigma^2_{\text{TOA},1}, \sigma^2_{\text{TOA},2}, \cdots, \sigma^2_{\text{TOA},L}))$.

3.2.2 TDOA

TDOA is the difference in arrival times of the emitted signal received at a pair of sensors, implying that clock synchronization across all receivers is required. Nevertheless, the TDOA scheme is simpler than the TOA method because the latter needs source synchronization, which implies a higher hardware cost. Similar to TOA, multiplying TDOA by the known propagation speed leads to the range difference between a source and two receivers. As discussed in Chapter 1, geometrically, each noise-free TDOA defines a hyperbola on which the source must lie in the 2-D space, and the target location is given by the intersection of at least two hyperbolae. In the presence of disturbances, we estimate x from a set of hyperbolic equations converted from the TDOA measurements.

Mathematically, the TDOA measurement model is formulated as follows. We assume that the source emits a signal at the unknown time t_0, and the lth sensor receives it at time t_l, $l = 1, 2, \cdots, L$, with $L \geq 3$. There are $L(L - 1)/2$ distinct TDOAs from all possible sensor pairs, denoted by $t_{k,l} = (t_k - t_0) - (t_l - t_0) = t_k - t_l$, $k, l = 1, 2, \cdots, L$, with $k > l$. However, there are only $(L - 1)$ nonredundant TDOAs. Taking $L = 3$ as an example, the distinct TDOAs are $t_{2,1}$, $t_{3,1}$, and $t_{3,2}$, and we easily observe that $t_{3,2} = t_{3,1} - t_{2,1}$, which is redundant. In order to reduce complexity without sacrificing estimation performance, we should measure all $L(L - 1)/2$

TDOAs and then convert them to $(L - 1)$ nonredundant TDOAs for source localization [12]. Without loss of generality, we consider the first sensor as the reference, and the nonredundant TDOAs are $t_{l,1}$, $l = 2, 3, \cdots , L$.

Similar to (3.3) and (3.4), the range difference measurements deduced from the TDOAs are modeled as:

$$r_{\text{TDOA},l} = d_{l,1} + n_{\text{TDOA},l}, \quad l = 2, 3, \cdots, L, \tag{3.13}$$

where

$$d_{l,1} = d_l - d_1, \tag{3.14}$$

and $n_{\text{TDOA},l}$ is the range difference error in $r_{\text{TDOA},l}$ which is proportional to the disturbance in $t_{l,1}$. Following (3.5–3.8), the TDOA measurement model in vector form is

$$\mathbf{r}_{\text{TDOA}} = \mathbf{f}_{\text{TDOA}}(\mathbf{x}) + \mathbf{n}_{\text{TDOA}}, \tag{3.15}$$

where

$$\mathbf{r}_{\text{TDOA}} = \left[r_{\text{TDOA},2} \; r_{\text{TDOA},3} \cdots r_{\text{TDOA},L} \right]^T, \tag{3.16}$$

$$\mathbf{n}_{\text{TDOA}} = \left[n_{\text{TDOA},2} \; n_{\text{TDOA},3} \cdots n_{\text{TDOA},L} \right]^T, \tag{3.17}$$

and

$$\mathbf{f}_{\text{TDOA}}(\mathbf{x}) = \mathbf{d}_1 = \begin{bmatrix} \sqrt{(x - x_2)^2 + (y - y_2)^2} - \sqrt{(x - x_1)^2 + (y - y_1)^2} \\ \sqrt{(x - x_3)^2 + (y - y_3)^2} - \sqrt{(x - x_1)^2 + (y - y_1)^2} \\ \vdots \\ \sqrt{(x - x_L)^2 + (y - y_L)^2} - \sqrt{(x - x_1)^2 + (y - y_1)^2} \end{bmatrix}. \tag{3.18}$$

The source localization problem based on TDOA measurements is to estimate $\mathbf{x}$ given $\{r_{\text{TDOA},l}\}$ or $\mathbf{r}_{\text{TDOA}}$.

Assuming that $\mathbf{n}_{\text{TDOA}}$ is zero-mean and Gaussian distributed, the PDF for $\mathbf{r}_{\text{TDOA}}$, denoted by $p(\mathbf{r}_{\text{TDOA}})$, is

$$p(\mathbf{r}_{\text{TDOA}}) = \frac{1}{(2\pi)^{(L-1)/2} |\mathbf{C}_{\text{TDOA}}|^{1/2}} \exp\left(-\frac{1}{2}(\mathbf{r}_{\text{TDOA}} - \mathbf{d}_1)^T \mathbf{C}_{\text{TDOA}}^{-1}(\mathbf{r}_{\text{TDOA}} - \mathbf{d}_1) \right), \tag{3.19}$$

where $\mathbf{C}_{\text{TDOA}}$ is the covariance matrix for $\mathbf{r}_{\text{TDOA}}$. Alternatively, we can write $\mathbf{r}_{\text{TDOA}} \sim \mathcal{N}(\mathbf{d}_1, \mathbf{C}_{\text{TDOA}})$. Since all TDOAs are determined with respect to the first sensor, $n_{\text{TDOA},l}$, $l = 2, 3, \cdots, L$, are correlated. As a result, $\mathbf{C}_{\text{TDOA}}$ is not a diagonal matrix [12].

3.2.3 TSOA

TSOA arises in MIMO [1] and multistatic [13] systems consisting of two sets of sensors, namely, transmitters and receivers. A representative example is the distributed MIMO radar where there are widely separated transmit and receive antennas. As in the TDOA, clock synchronization across all transmitters and receivers is required. A transmitter emits a signal to the source, which is then reflected and collected at a receiver, and the total propagation time from the transmit antenna to the receive antenna is referred to as TSOA. Multiplying the TSOA by the signal propagation speed yields the transmitter-target-receiver distance or range sum, which is the sum of the distance between the transmitter and source and that between the source and receiver. Geometrically, each noise-free TSOA defines an ellipse on which the source must lie in the 2-D space, and the target location is given by intersection of at least three ellipses. In the presence of disturbances, we estimate $\mathbf{x}$ from a set of elliptic equations converted from the TSOA measurements. It is worth pointing out that the range, range difference, and range sum are the main distance-based location-bearing measurements, which correspond to circular, hyperbolic, and elliptic localization, respectively. Furthermore, each range, range difference, and range sum define a sphere, hyperboloid, and ellipsoid, respectively, in the 3-D case.

The TSOA measurement model is mathematically formulated as follows. To minimize the symbols used, we further introduce $\mathbf{x}_m^t = \left[x_m^t \ y_m^t \right]^T$ as the known position of the mth transmitter, $m = 1, 2, \cdots, M$, where $M \geq 1$ and employ $\{\mathbf{x}_l\}$ as the positions of the L receivers. Following (3.3) and (3.4), and noting that TSOA is the sum of two TOAs, the range sum measurements deduced from the TSOAs, denoted by $r_{\text{TSOA},m,l}$, are modeled as:

$$r_{\text{TSOA},m,l} = d_m^t + d_l + n_{\text{TSOA},m,l}, \quad m = 1, 2, \cdots, M, \quad l = 1, 2, \cdots, L, \quad (3.20)$$

where

$$d_m^t = \sqrt{(x - x_m^t)^2 + (y - y_m^t)^2}. \quad (3.21)$$

and $n_{\text{TSOA},m,l}$ is the range sum error in $r_{\text{TSOA},m,l}$, which is resulted from the TSOA disturbance.

Similar to (3.5–3.8), the TSOA measurement model in vector form is

$$\mathbf{r}_{\text{TSOA}} = \mathbf{f}_{\text{TSOA}}(\mathbf{x}) + \mathbf{n}_{\text{TSOA}}, \quad (3.22)$$

where

$$\mathbf{r}_{\text{TSOA}} = [r_{\text{TSOA},1,1} \ r_{\text{TSOA},2,1} \cdots r_{\text{TSOA},M,1} \ r_{\text{TSOA},1,2} \cdots r_{\text{TSOA},M,L}]^T, \quad (3.23)$$

$$\mathbf{n}_{\text{TSOA}} = [n_{\text{TSOA},1,1} \ n_{\text{TSOA},2,1} \cdots n_{\text{TSOA},M,1} \ n_{\text{TSOA},1,2} \cdots r_{\text{TSOA},M,L}]^T, \quad (3.24)$$

and

$$\mathbf{f}_{\text{TSOA}}(\mathbf{x}) = \mathbf{d}_2 = \begin{bmatrix} \sqrt{(x - x_1^t)^2 + (y - y_1^t)^2} + \sqrt{(x - x_1)^2 + (y - y_1)^2} \\ \sqrt{(x - x_2^t)^2 + (y - y_2^t)^2} + \sqrt{(x - x_1)^2 + (y - y_1)^2} \\ \vdots \\ \sqrt{(x - x_M^t)^2 + (y - y_M^t)^2} + \sqrt{(x - x_1)^2 + (y - y_1)^2} \\ \sqrt{(x - x_1^t)^2 + (y - y_1^t)^2} + \sqrt{(x - x_2)^2 + (y - y_2)^2} \\ \vdots \\ \sqrt{(x - x_M^t)^2 + (y - y_M^t)^2} + \sqrt{(x - x_L)^2 + (y - y_L)^2} \end{bmatrix}. \quad (3.25)$$

The TSOA-based localization problem is to estimate $\mathbf{x}$ given the ML measurements of $\{r_{\text{TSOA},m,l}\}$ or $\mathbf{r}_{\text{TSOA}}$.

Assuming that $\{n_{\text{TSOA},m,l}\}$ are zero-mean uncorrelated Gaussian processes with variances $\{\sigma^2_{\text{TSOA},m,l}\}$, the PDF for $\mathbf{r}_{\text{TSOA}}$, denoted by $p(\mathbf{r}_{\text{TSOA}})$, is:

$$p(\mathbf{r}_{\text{TSOA}}) = \frac{1}{(2\pi)^{ML/2}|\mathbf{C}_{\text{TSOA}}|^{1/2}} \exp\left(-\frac{1}{2}(\mathbf{r}_{\text{TSOA}} - \mathbf{d}_2)^T \mathbf{C}^{-1}_{\text{TSOA}}(\mathbf{r}_{\text{TSOA}} - \mathbf{d}_2)\right)$$

$$= \frac{1}{(2\pi)^{ML/2}\prod_{m=1}^{M}\prod_{l=1}^{L}\sigma_{\text{TSOA},m,l}} \exp\left(-\frac{1}{2}\sum_{m=1}^{M}\sum_{l=1}^{L}\frac{\left(r_{\text{TSOA},m,l} - d'_m - d_l\right)^2}{\sigma^2_{\text{TSOA},m,l}}\right),$$

$$\tag{3.26}$$

where $\mathbf{C}_{\text{TSOA}} = \text{diag}(\sigma^2_{\text{TSOA},1,1}, \sigma^2_{\text{TSOA},2,1}, \cdots, \sigma^2_{\text{TSOA},M,1}, \sigma^2_{\text{TSOA},1,2}, \cdots, \sigma^2_{\text{TSOA},M,L})$ represents the covariance matrix for $\mathbf{r}_{\text{TSOA}}$. That is, we can also write $\mathbf{r}_{\text{TSOA}} \sim \mathcal{N}(\mathbf{d}_2, \mathbf{C}_{\text{TSOA}})$.

3.2.4 RSS

RSS is the average power received at a sensor where the power is originated from the emitted source. It is commonly assumed that the received power follows an exponential decay model, which is a function of the transmitted power, path loss constant, and the distance between the source and sensor. This positioning scheme is simpler than those using TOA, TDOA, or TSOA measurements because synchronization among the source and/or sensors is not required. Once we are able to obtain the distances from the RSS measurements, the source location can be determined as in the TOA case with the use of at least three receivers.

The RSS measurement model is formulated as follows. Assuming that the source transmitted power is P_t and in the absence of disturbance, the average power received at the lth sensor, denoted by $P_{r,l}$, is modeled as [14]:

$$P_{r,l} = K_l P_t d_l^{-\alpha} = K_l P_t \|\mathbf{x} - \mathbf{x}_l\|_2^{-\alpha}, \quad l = 1, 2, \cdots, L, \tag{3.27}$$

where K_l accounts for all other factors that affect the received power, including the antenna height and antenna gain, while α is the path loss constant. Depending on the propagation environment, α can vary from 2 to 5. In particular, $\alpha = 2$ in free space. For detailed explanations of the channel models, the interested reader is referred to Chapter 4. It is assumed that P_t, K_l, $l = 1, 2, \cdots, L$ with $L \geq 3$ and α are known a priori. Field trials have validated that the disturbance in RSS is log-normal distributed. Accordingly, the log-normal path loss model can be expressed as:

$$\ln(P_{r,l}) = \ln(K_l) + \ln(P_t) - \alpha \ln(d_l) + n_{\text{RSS},l}, \, l = 1, 2, \cdots, L, \tag{3.28}$$

where the disturbance $n_{\text{RSS},l}$ is now Gaussian distributed. For simplicity, we assume that $n_{\text{RSS},l}$, $l = 1, 2, \cdots, L$, are zero-mean uncorrelated Gaussian processes with variances $\{\sigma^2_{\text{RSS},l}\}$. Let

$$r_{\text{RSS},l} = \ln(P_{r,l}) - \ln(K_l) - \ln(P_t). \tag{3.29}$$

The RSS measurement model is simplified to:

$$r_{\text{RSS},l} = -\alpha \ln(d_l) + n_{\text{RSS},l}, \quad l = 1, 2, \cdots, L. \tag{3.30}$$

The vector form of (3.30) is then:

$$\mathbf{r}_{\text{RSS}} = \mathbf{f}_{\text{RSS}}(\mathbf{x}) + \mathbf{n}_{\text{RSS}}, \tag{3.31}$$

where

$$\mathbf{r}_{\text{RSS}} = \left[r_{\text{RSS},1} \; r_{\text{RSS},2} \; \cdots \; r_{\text{RSS},L} \right]^T, \tag{3.32}$$

$$\mathbf{n}_{\text{RSS}} = \left[n_{\text{RSS},1} \; n_{\text{RSS},2} \; \cdots \; n_{\text{RSS},L} \right]^T, \tag{3.33}$$

and

$$\mathbf{f}_{\text{RSS}}(\mathbf{x}) = \mathbf{p} = -\alpha \begin{bmatrix} \ln\left(\sqrt{(x-x_1)^2+(y-y_1)^2}\right) \\ \ln\left(\sqrt{(x-x_2)^2+(y-y_2)^2}\right) \\ \vdots \\ \ln\left(\sqrt{(x-x_L)^2+(y-y_L)^2}\right) \end{bmatrix}. \tag{3.34}$$

Comparing with (3.8), we observe that $\mathbf{f}_{\text{RSS}}(\mathbf{x})$ also contains range information. The source localization problem based on RSS measurements is to estimate $\mathbf{x}$ given $\{r_{\text{RSS},l}\}$ or $\mathbf{r}_{\text{RSS}}$. Following the development in (3.9–3.12), the PDF for $\mathbf{r}_{\text{RSS}}$, denoted by $p(\mathbf{r}_{\text{RSS}})$, is determined as

$$
\begin{aligned}
p(\mathbf{r}_{\text{RSS}}) &= \frac{1}{(2\pi)^{L/2}|\mathbf{C}_{\text{RSS}}|^{1/2}} \exp\left(-\frac{1}{2}(\mathbf{r}_{\text{RSS}} - \mathbf{p})^T \mathbf{C}_{\text{RSS}}^{-1}(\mathbf{r}_{\text{RSS}} - \mathbf{p})\right) \\
&= \frac{1}{(2\pi)^{L/2}\, \Pi_{l=1}^{L}\sigma_{\text{RSS},l}} \exp\left(-\frac{1}{2}\sum_{l=1}^{L}\frac{(r_{\text{RSS},l} + \alpha\ln(d_l))^2}{\sigma_{\text{RSS},l}^2}\right),
\end{aligned}
\tag{3.35}
$$

where $\mathbf{C}_{\text{RSS}} = \text{diag}(\sigma_{\text{RSS},1}^2, \sigma_{\text{RSS},2}^2, \cdots, \sigma_{\text{RSS},L}^2)$. In other words, $\mathbf{r}_{\text{RSS}} \sim \mathcal{N}(\mathbf{p}, \mathbf{C}_{\text{RSS}})$.

It is worth mentioning that in case the source transmitted power is unknown, the differential RSS measurements [15], which correspond to range difference information as in the TDOA, can be employed for positioning.

3.2.5   DOA

DOA is the arrival angle of the emitted source signal observed at a receiver. As discussed in Chapter 1, from each DOA, we can draw a line of bearing (LOB) from the source to receiver and intersection of at least two LOBs will give the source location. Although this scheme does not require clock synchronization as in RSS-based positioning, an antenna array is needed to be installed at each receiver for DOA estimation. Let ϕ_l be the DOA between the source and the lth receiver, and we have:

$$\tan(\phi_l) = \frac{y - y_l}{x - x_l}, \quad l = 1, 2, \cdots, L, \tag{3.36}$$

with $L \geq 2$. Geometrically, ϕ_l is the angle between the LOB from the lth receiver to the target and the x-axis. The DOA measurements in the presence of angle errors, denoted by $\{r_{\text{DOA},l}\}$, are modeled as:

$$r_{\text{DOA},l} = \phi_l + n_{\text{DOA},l} = \tan^{-1}\left(\frac{y - y_l}{x - x_l}\right) + n_{\text{DOA},l}, \quad l = 1, 2, \cdots, L, \tag{3.37}$$

where $\{n_{\text{DOA},l}\}$ are the noises in $\{r_{\text{DOA},l}\}$, which are assumed zero-mean uncorrelated Gaussian processes with variances $\{\sigma_{\text{DOA},l}^2\}$. The vector form of (3.37) is:

$$\mathbf{r}_{\text{DOA}} = \mathbf{f}_{\text{DOA}}(\mathbf{x}) + \mathbf{n}_{\text{DOA}}, \tag{3.38}$$

where

$$\mathbf{r}_{\text{DOA}} = \begin{bmatrix} r_{\text{DOA},1} \ r_{\text{DOA},2} \cdots r_{\text{DOA},L} \end{bmatrix}^T, \tag{3.39}$$

$$\mathbf{n}_{\text{DOA}} = \begin{bmatrix} n_{\text{DOA},1} \ n_{\text{DOA},2} \cdots n_{\text{DOA},L} \end{bmatrix}^T, \tag{3.40}$$

and

$$\mathbf{f}_{\text{DOA}}(\mathbf{x}) = \phi = \begin{bmatrix} \tan^{-1}\left(\dfrac{y-y_1}{x-x_1}\right) \\ \tan^{-1}\left(\dfrac{y-y_2}{x-x_2}\right) \\ \vdots \\ \tan^{-1}\left(\dfrac{y-y_L}{x-x_L}\right) \end{bmatrix}. \tag{3.41}$$

The source localization problem based on DOA measurements is to estimate $\mathbf{x}$ given $r_{\text{DOA},l}$, $l = 1, 2, \cdots, L$, or $\mathbf{r}_{\text{DOA}}$. Following the development in (3.9–3.12), the PDF for $\mathbf{r}_{\text{DOA}}$, denoted by $p(\mathbf{r}_{\text{DOA}})$, is

$$\begin{aligned} p(\mathbf{r}_{\text{DOA}}) &= \frac{1}{(2\pi)^{L/2}|\mathbf{C}_{\text{DOA}}|^{1/2}} \exp\left(-\frac{1}{2}(\mathbf{r}_{\text{DOA}} - \phi)^T \mathbf{C}_{\text{DOA}}^{-1}(\mathbf{r}_{\text{DOA}} - \phi)\right) \\ &= \frac{1}{(2\pi)^{L/2} \prod_{l=1}^{L} \sigma_{\text{DOA},l}} \exp\left(-\frac{1}{2}\sum_{l=1}^{L}\frac{(r_{\text{DOA},l} - \phi_l)^2}{\sigma_{\text{DOA},l}^2}\right), \end{aligned} \tag{3.42}$$

where $\mathbf{C}_{\text{DOA}} = \text{diag}\left(\sigma_{\text{DOA},1}^2, \sigma_{\text{DOA},2}^2, \cdots, \sigma_{\text{DOA},L}^2\right)$. In other words, we can write $\mathbf{r}_{\text{DOA}} \sim \mathcal{N}(\phi, \mathbf{C}_{\text{DOA}})$.

3.3   ALGORITHMS FOR SOURCE LOCALIZATION

Two approaches for source localization, namely, nonlinear and linear, are presented in Sections 3.3.1 and 3.3.2, respectively. Generally speaking, the nonlinear methodology [16–19] directly employs (3.1) to solve for $\mathbf{x}$ by minimizing the least squares (LS) or weighted least squares (WLS) cost function constructed from the following error function:

$$\mathbf{e}_{\text{nonlinear}} = \mathbf{r} - \mathbf{f}(\tilde{\mathbf{x}}), \tag{3.43}$$

where $\tilde{\mathbf{x}} = \begin{bmatrix} \tilde{x} \ \tilde{y} \end{bmatrix}^T$ is the optimization variable for $\mathbf{x}$, which corresponds to the NLS or ML estimator, respectively. On the other hand, the linear techniques convert (3.1) into a set of linear equations in $\mathbf{x}$:

$$\mathbf{b} = \mathbf{Ax} + \mathbf{q}, \tag{3.44}$$

TABLE 3.3. Comparison of Different Position Estimators

Estimator	Advantages	Disadvantages
NLS	- Accuracy is generally high. - Noise statistics are not needed.	- Global solution may not be guaranteed. - Complexity is high if grid or random search is involved.
ML	- Accuracy is highest.	- Global solution may not be guaranteed. - Complexity is high if grid or random search is involved. - Noise statistics are needed.
LLS or subspace	- Global solution is guaranteed. - Simple and computationally efficient - Noise statistics are not needed.	- Accuracy is generally low.
WLLS	- Global solution is guaranteed. - Highest accuracy can be achieved with constraints.	- Noise statistics are needed. - Iterations may be required.

where $\mathbf{b}$ and $\mathbf{A}$ are known, while $\mathbf{q}$ is the transformed noise vector. Based on (3.44), we construct:

$$\mathbf{e}_{\text{linear}} = \mathbf{b} - \mathbf{A}\tilde{\mathbf{x}}. \tag{3.45}$$

Applying the LS or WLS techniques to (3.45) results in the LLS [20, 21], WLLS [22–25] and subspace [30–35] estimators. A comparison summary for the position estimators examined in this chapter is shown in Table 3.3.

3.3.1 Nonlinear Methods

The nonlinear methodology attempts to find the source location directly from (3.4), (3.13), (3.20), (3.30), and (3.37), which includes the NLS and ML estimators. Global convergence of these schemes may not be guaranteed because their optimization cost functions are multi-modal. Moreover, the NLS method is simpler and is a practical choice when the noise information is unavailable. On the other hand, the ML estimator can be considered as a weighted version of the NLS method by utilizing the noise covariance, and it is optimum in the sense that its estimation performance can attain CRLB.

3.3.1.1 NLS: The NLS approach minimizes the LS cost functions directly constructed from (3.4), (3.13), (3.20), and (3.37), which is presented one by one as follows. Detailed discussion is provided for TOA-based positioning, while the results can be straightforwardly applied to the remaining four cases.

A. TOA-Based Positioning: Based on (3.4) and (3.5), the NLS cost function, denoted by $J_{\text{NLS,TOA}}(\tilde{\mathbf{x}})$, is:

$$J_{\text{NLS,TOA}}(\tilde{\mathbf{x}}) = \sum_{l=1}^{L} \left(r_{\text{TOA},l} - \sqrt{(\tilde{x} - x_l)^2 + (\tilde{y} - y_l)^2} \right)^2$$

$$= \left(\mathbf{r}_{\text{TOA}} - \mathbf{f}_{\text{TOA}}(\tilde{\mathbf{x}}) \right)^T \left(\mathbf{r}_{\text{TOA}} - \mathbf{f}_{\text{TOA}}(\tilde{\mathbf{x}}) \right). \tag{3.46}$$

The NLS position estimate is equal to $\tilde{\mathbf{x}}$ which corresponds to the smallest value of $J_{\text{NLS,TOA}}(\tilde{\mathbf{x}})$; that is:

$$\hat{\mathbf{x}} = \arg\min_{\tilde{\mathbf{x}}} J_{\text{NLS,TOA}}(\tilde{\mathbf{x}}). \tag{3.47}$$

Finding $\hat{\mathbf{x}}$ is not a simple task, as there are local minima apart from the global minimum in the 2-D surface of $J_{\text{NLS,TOA}}(\tilde{\mathbf{x}})$. Basically, there are two directions for solving (3.47). The first one attempts to perform a global exploration using grid search or random search techniques such as genetic algorithm [17] and particle swarm optimization [18]. On the other hand, the second direction corresponds to a local search, which is an iterative algorithm requiring an initial position estimate, denoted by $\hat{\mathbf{x}}^0$. If $\hat{\mathbf{x}}^0$ is sufficiently close to $\mathbf{x}$, it is expected that $\hat{\mathbf{x}}$ can be obtained in the iterative procedure. In this chapter, three commonly used local search schemes, namely, Newton–Raphson, Gauss–Newton, and steepest descent methods, are presented. For more advanced local search techniques of source localization, the interested reader is referred to [19].

The iterative Newton–Raphson procedure for $\hat{\mathbf{x}}$ is

$$\hat{\mathbf{x}}^{k+1} = \hat{\mathbf{x}}^k - \mathbf{H}^{-1}(J_{\text{NLS,TOA}}(\hat{\mathbf{x}}^k))\nabla(J_{\text{NLS,TOA}}(\hat{\mathbf{x}}^k)), \tag{3.48}$$

where $\mathbf{H}(J_{\text{NLS,TOA}}(\hat{\mathbf{x}}^k))$ and $\nabla(J_{\text{NLS,TOA}}(\hat{\mathbf{x}}^k))$ are the corresponding Hessian matrix and gradient vector computed at the kth iteration estimate, namely, $\hat{\mathbf{x}}^k$, and they have the forms of:

$$\mathbf{H}\left(J_{\text{NLS,TOA}}(\mathbf{x})\right) = \frac{\partial^2 J_{\text{NLS,TOA}}(\mathbf{x})}{\partial \mathbf{x} \partial \mathbf{x}^T}$$

$$= \begin{bmatrix} \dfrac{\partial^2 J_{\text{NLS,TOA}}(\mathbf{x})}{\partial x^2} & \dfrac{\partial^2 J_{\text{NLS,TOA}}(\mathbf{x})}{\partial x \partial y} \\[3mm] \dfrac{\partial^2 J_{\text{NLS,TOA}}(\mathbf{x})}{\partial y \partial x} & \dfrac{\partial^2 J_{\text{NLS,TOA}}(\mathbf{x})}{\partial y^2} \end{bmatrix}, \tag{3.49}$$

with

$$\frac{\partial^2 J_{\text{NLS,TOA}}(\mathbf{x})}{\partial x^2}$$

$$= \sum_{l=1}^{L} 2 \left(\frac{(x - x_l)^2}{(x - x_l)^2 + (y - y_l)^2} - \frac{\left(r_{\text{TOA},l} - \sqrt{(x - x_l)^2 + (y - y_l)^2} \right)(y - y_l)^2}{\left[(x - x_l)^2 + (y - y_l)^2 \right]^{3/2}} \right), \tag{3.50}$$

$$\frac{\partial^2 J_{\text{NLS,TOA}}(\mathbf{x})}{\partial x \partial y} = \frac{\partial^2 J_{\text{NLS,TOA}}(\mathbf{x})}{\partial y \partial x} = \sum_{l=1}^{L} \frac{2 r_{\text{TOA},l}(x - x_l)(y - y_l)}{\left[(x - x_l)^2 + (y - y_l)^2 \right]^{3/2}}, \tag{3.51}$$

and

$$
\frac{\partial^2 J_{\text{NLS,TOA}}(\mathbf{x})}{\partial y^2}
$$

$$
= \sum_{l=1}^{L} 2 \left(\frac{(y - y_l)^2}{(x - x_l)^2 + (y - y_l)^2} - \frac{\left(r_{\text{TOA},l} - \sqrt{(x - x_l)^2 + (y - y_l)^2} \right)(x - x_l)^2}{\left[(x - x_l)^2 + (y - y_l)^2 \right]^{3/2}} \right), \qquad (3.52)
$$

while

$$
\nabla \left(J_{\text{NLS,TOA}}(\mathbf{x}) \right) =
\begin{bmatrix}
\dfrac{\partial J_{\text{NLS,TOA}}(\mathbf{x})}{\partial x} \\[2mm]
\dfrac{\partial J_{\text{NLS,TOA}}(\mathbf{x})}{\partial y}
\end{bmatrix}
\qquad (3.53)
$$

$$
= -2
\begin{bmatrix}
\displaystyle\sum_{l=1}^{L} \dfrac{\left(r_{\text{TOA},l} - \sqrt{(x - x_l)^2 + (y - y_l)^2} \right)(x - x_l)}{\sqrt{(x - x_l)^2 + (y - y_l)^2}} \\[4mm]
\displaystyle\sum_{l=1}^{L} \dfrac{\left(r_{\text{TOA},l} - \sqrt{(x - x_l)^2 + (y - y_l)^2} \right)(y - y_l)}{\sqrt{(x - x_l)^2 + (y - y_l)^2}}
\end{bmatrix}.
$$

For the Gauss–Newton method, the updating rule is:

$$
\hat{\mathbf{x}}^{k+1} = \hat{\mathbf{x}}^{k} + \left(\mathbf{G}^{T} \left(\mathbf{f}_{\text{TOA}}(\hat{\mathbf{x}}^{k}) \right) \mathbf{G} \left(\mathbf{f}_{\text{TOA}}(\hat{\mathbf{x}}^{k}) \right) \right)^{-1} \mathbf{G}^{T} \left(\mathbf{f}_{\text{TOA}}(\hat{\mathbf{x}}^{k}) \right) \left(\mathbf{r}_{\text{TOA}} - \mathbf{f}_{\text{TOA}}(\hat{\mathbf{x}}^{k}) \right), \qquad (3.54)
$$

where $\mathbf{G}\left(\mathbf{f}_{\text{TOA}}(\hat{\mathbf{x}}^{k}) \right)$ is the Jacobian matrix of $\mathbf{f}_{\text{TOA}}(\hat{\mathbf{x}}^{k})$ computed at $\hat{\mathbf{x}}^{k}$ and has the following expression:

$$
\mathbf{G}\left(\mathbf{f}_{\text{TOA}}(\mathbf{x}) \right) =
\begin{bmatrix}
\dfrac{\partial \sqrt{(x - x_1)^2 + (y - y_1)^2}}{\partial x} & \dfrac{\partial \sqrt{(x - x_1)^2 + (y - y_1)^2}}{\partial y} \\[4mm]
\dfrac{\partial \sqrt{(x - x_2)^2 + (y - y_2)^2}}{\partial x} & \dfrac{\partial \sqrt{(x - x_2)^2 + (y - y_2)^2}}{\partial y} \\[4mm]
\vdots & \vdots \\[2mm]
\dfrac{\partial \sqrt{(x - x_L)^2 + (y - y_L)^2}}{\partial x} & \dfrac{\partial \sqrt{(x - x_L)^2 + (y - y_L)^2}}{\partial y}
\end{bmatrix}
$$

$$
=
\begin{bmatrix}
\dfrac{x - x_1}{\sqrt{(x - x_1)^2 + (y - y_1)^2}} & \dfrac{y - y_1}{\sqrt{(x - x_1)^2 + (y - y_1)^2}} \\[4mm]
\dfrac{x - x_2}{\sqrt{(x - x_2)^2 + (y - y_2)^2}} & \dfrac{y - y_2}{\sqrt{(x - x_2)^2 + (y - y_2)^2}} \\[4mm]
\vdots & \vdots \\[2mm]
\dfrac{x - x_L}{\sqrt{(x - x_L)^2 + (y - y_L)^2}} & \dfrac{y - y_L}{\sqrt{(x - x_L)^2 + (y - y_L)^2}}
\end{bmatrix}.
\qquad (3.55)
$$

Finally, the iterative procedure for the steepest descent method is:

$$\hat{\mathbf{x}}^{k+1} = \hat{\mathbf{x}}^k - \mu \nabla \left(J_{\text{NLS,TOA}} \left(\hat{\mathbf{x}}^k \right) \right), \tag{3.56}$$

where μ is a positive constant, which controls the convergence rate and stability. Generally speaking, a larger value of μ increases the convergence speed and vice versa. In practice, we should choose a sufficiently small μ to ensure algorithm stability.

Starting with $\hat{\mathbf{x}}^0$, the iterative procedure of (3.48), (3.54), or (3.56) is terminated according to a stopping criterion, which indicates convergence. Typical choices of stopping criteria include number of iterations and $\left\| \hat{\mathbf{x}}^{k+1} - \hat{\mathbf{x}}^k \right\|_2 < \epsilon$, where ϵ is a sufficiently small positive constant. As a brief comparison, both Newton–Raphson and Gauss–Newton methods provide fast convergence, but matrix inverse is required, and the latter is simpler in the sense that second-order differentiation of $J_{\text{NLS,TOA}}(\tilde{\mathbf{x}})$ is not involved. On the other hand, the steepest descent method is stable, but its convergence rate is slow and can be considered as an approximation form of (3.48), where the Hessian matrix is omitted.

Example 3.1

Consider a 2-D geometry of $L = 4$ receivers with known coordinates at (0, 0), (0, 10), (10, 0), and (10, 10), while the unknown source position is $(x, y) = (2, 3)$. Note that the source is located inside the square bounded by the four receivers. For presentation simplicity, the range error variance, $\sigma_{\text{TOA},l}^2$, is assigned proportional to d_l^2, and we define the signal-to-noise ratio (SNR) as $d_l^2 / \sigma_{\text{TOA},l}^2$. Examine the convergence rates of the Newton–Raphson, Gauss–Newton, and steepest descent methods for the NLS approach in a single trial at SNR = 30 dB.

Solution

The MATLAB program for this example is called Example3_1.m. There are three main parts in the program, namely, generating range measurements, position estimation using the NLS estimator, which is realized by the Newton–Raphson, Gauss–Newton, and steepest descent methods, and displaying results. Figures 3.1 and 3.2 show the estimates of x and y, respectively, versus number of iterations at SNR = 30 dB. The initial guess is chosen as $\hat{\mathbf{x}}^0 = \begin{bmatrix} 3 & 2 \end{bmatrix}^T$, and the step size of the steepest descent method is $\mu = 0.1$. Although all schemes provide the same position estimates upon convergence, it is observed that the Newton–Raphson and Gauss–Newton methods converge in around three iterations, while the steepest descent algorithm needs approximately 15 iterations to converge. Note that in the presence of noise, $\hat{\mathbf{x}} \neq \mathbf{x}$ in a single trial, although $E\{\hat{\mathbf{x}}\} = \mathbf{x}$ for small error conditions.

B. TDOA-Based Positioning: Similarly, with the use of (3.13–3.15), the NLS cost function for TDOA-based positioning, denoted by $J_{\text{NLS,TDOA}}(\tilde{\mathbf{x}})$, is:

$$J_{\text{NLS,TDOA}}(\tilde{\mathbf{x}}) = \sum_{l=2}^{L} \left(r_{\text{TDOA},l} - \sqrt{(\tilde{x} - x_l)^2 + (\tilde{y} - y_l)^2} + \sqrt{(\tilde{x} - x_1)^2 + (\tilde{y} - y_1)^2} \right)^2$$

$$= \left(\mathbf{r}_{\text{TDOA}} - \mathbf{f}_{\text{TDOA}}(\tilde{x}) \right)^T \left(\mathbf{r}_{\text{TDOA}} - \mathbf{f}_{\text{TDOA}}(\tilde{x}) \right), \tag{3.57}$$

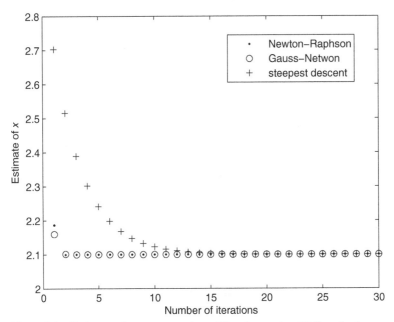

Figure 3.1 Estimate of *x* versus number of iterations using NLS method.

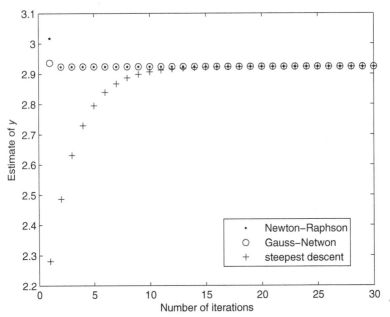

Figure 3.2 Estimate of *y* versus number of iterations using NLS method.

and the NLS position estimate is:

$$\hat{\mathbf{x}} = \arg \min_{\tilde{\mathbf{x}}} J_{\text{NLS,TDOA}}(\tilde{\mathbf{x}}). \tag{3.58}$$

C. TSOA-Based Positioning: Based on (3.20–3.22), the NLS cost function for TSOA-based positioning, denoted by $J_{\text{NLS,TSOA}}(\tilde{\mathbf{x}})$, is:

$$J_{\text{NLS,RSS}}(\tilde{\mathbf{x}}) = \sum_{m=1}^{M} \sum_{l=1}^{L} \left(r_{\text{TSOA},m,l} - \sqrt{(\tilde{x} - x_m^t)^2 + (\tilde{y} - y_m^t)^2} - \sqrt{(\tilde{x} - x_l)^2 + (\tilde{y} - y_l)^2} \right)^2$$

$$= \left(\mathbf{r}_{\text{TSOA}} - \mathbf{f}_{\text{TSOA}}(\tilde{\mathbf{x}}) \right)^T \left(\mathbf{r}_{\text{TSOA}} - \mathbf{f}_{\text{TSOA}}(\tilde{\mathbf{x}}) \right), \tag{3.59}$$

and the NLS position estimate is:

$$\hat{\mathbf{x}} = \arg \min_{\tilde{\mathbf{x}}} J_{\text{NLS,TSOA}}(\tilde{\mathbf{x}}). \tag{3.60}$$

D. RSS-Based Positioning: Employing (3.30) and (3.31), the NLS cost function for RSS-based positioning, denoted by $J_{\text{NLS,RSS}}(\tilde{\mathbf{x}})$, is:

$$J_{\text{NLS,RSS}}(\tilde{\mathbf{x}}) = \sum_{l=1}^{L} \left(r_{\text{RSS},l} + \alpha \ln \left(\sqrt{(\tilde{x} - x_l)^2 + (\tilde{y} - y_l)^2} \right) \right)^2$$

$$= \left(\mathbf{r}_{\text{RSS}} - \mathbf{f}_{\text{RSS}}(\tilde{\mathbf{x}}) \right)^T \left(\mathbf{r}_{\text{RSS}} - \mathbf{f}_{\text{RSS}}(\tilde{\mathbf{x}}) \right), \tag{3.61}$$

and the NLS position estimate is:

$$\hat{\mathbf{x}} = \arg \min_{\tilde{\mathbf{x}}} J_{\text{NLS,RSS}}(\tilde{\mathbf{x}}). \tag{3.62}$$

E. DOA-Based Positioning: With the use of (3.37) and (3.38), the NLS cost function for DOA-based positioning, denoted by $J_{\text{NLS,DOA}}(\tilde{\mathbf{x}})$, is then:

$$J_{\text{NLS,DOA}}(\tilde{\mathbf{x}}) = \sum_{l=1}^{L} \left(r_{\text{RSS},l} - \tan^{-1} \left(\frac{y - y_l}{x - x_l} \right) \right)^2$$

$$= \left(\mathbf{r}_{\text{DOA}} - \mathbf{f}_{\text{DOA}}(\tilde{\mathbf{x}}) \right)^T \left(\mathbf{r}_{\text{DOA}} - \mathbf{f}_{\text{DOA}}(\tilde{\mathbf{x}}) \right), \tag{3.63}$$

and the NLS position estimate is

$$\hat{\mathbf{x}} = \arg \min_{\tilde{\mathbf{x}}} J_{\text{NLS,DOA}}(\tilde{\mathbf{x}}). \tag{3.64}$$

3.3.1.2 ML: Assuming that the error distribution is known, the ML approach maximizes the PDFs of TOA, TDOA, TSOA, RSS, and DOA measurements to obtain the source location. When the disturbances in the measurements are zero-mean Gaussian distributed, it is shown in the following that maximization of (3.10), (3.19), (3.26), (3.35), or (3.42) will correspond to a weighted version of the NLS scheme.

A. TOA-Based Positioning: To facilitate the maximization of (3.10), we consider its logarithmic version:

$$\ln(p(\mathbf{r}_{\text{TOA}})) = \ln \left(\frac{1}{(2\pi)^{L/2} |\mathbf{C}_{\text{TOA}}|^{1/2}} \right) - \ln \left(\frac{1}{2} (\mathbf{r}_{\text{TOA}} - \mathbf{d})^T \mathbf{C}_{\text{TOA}}^{-1} (\mathbf{r}_{\text{TOA}} - \mathbf{d}) \right). \tag{3.65}$$

As the first term is independent of $\mathbf{x}$, maximizing (3.65) is in fact equivalent to minimizing the second term, and the ML estimate is derived as:

$$\hat{\mathbf{x}} = \arg\min_{\tilde{\mathbf{x}}} \ln\left(\left(\mathbf{r}_{\text{TOA}} - \mathbf{f}_{\text{TOA}}(\tilde{\mathbf{x}})\right)^T \mathbf{C}_{\text{TOA}}^{-1}\left(\mathbf{r}_{\text{TOA}} - \mathbf{f}_{\text{TOA}}(\tilde{\mathbf{x}})\right)\right)$$

$$= \arg\min_{\tilde{\mathbf{x}}} \left(\mathbf{r}_{\text{TOA}} - \mathbf{f}_{\text{TOA}}(\tilde{\mathbf{x}})\right)^T \mathbf{C}_{\text{TOA}}^{-1}\left(\mathbf{r}_{\text{TOA}} - \mathbf{f}_{\text{TOA}}(\tilde{\mathbf{x}})\right), \tag{3.66}$$

Or we can write:

$$\hat{\mathbf{x}} = \arg\min_{\tilde{\mathbf{x}}} J_{\text{ML,TOA}}(\tilde{\mathbf{x}}), \tag{3.67}$$

where $J_{\text{ML,TOA}}(\tilde{\mathbf{x}})$ denotes the ML cost function for TOA-based positioning, which has the form of:

$$J_{\text{ML,TOA}}(\tilde{\mathbf{x}}) = \left(\mathbf{r}_{\text{TOA}} - \mathbf{f}_{\text{TOA}}(\tilde{\mathbf{x}})\right)^T \mathbf{C}_{\text{TOA}}^{-1}\left(\mathbf{r}_{\text{TOA}} - \mathbf{f}_{\text{TOA}}(\tilde{\mathbf{x}})\right)$$

$$= \sum_{l=1}^{L} \frac{\left(r_{\text{TOA},l} - \sqrt{(\tilde{x} - x_l)^2 + (\tilde{y} - y_l)^2}\right)^2}{\sigma_{\text{TOA},l}^2}. \tag{3.68}$$

Comparing (3.67) and (3.68) and (3.57) and (3.58), it is observed that in the presence of zero-mean Gaussian noise, the ML estimator generalizes the NLS method because the former is a weighted version of the latter. Intuitively speaking, when $\sigma_{\text{TOA},l}^2$ is large, which corresponds to a large noise in $r_{\text{TOA},l}$, a small weight of $1/\sigma_{\text{TOA},l}^2$ is employed in $\left(r_{\text{TOA},l} - \sqrt{(\tilde{x} - x_l)^2 + (\tilde{y} - y_l)^2}\right)^2$ in (3.68), and vice versa. When $\mathbf{C}_{\text{TOA}}^{-1}$ is proportional to the identity matrix or $\sigma_{\text{TOA},l}^2$, $l = 1, 2, \cdots, L$ are identical, the ML estimator is reduced to the NLS method. To compute (3.67), we can follow the numerical methods discussed in the NLS approach. In particular, the Newton–Raphson procedure for (3.67) is:

$$\hat{\mathbf{x}}^{k+1} = \hat{\mathbf{x}}^k - \mathbf{H}^{-1}\left(J_{\text{ML,TOA}}(\hat{\mathbf{x}}^k)\right)\nabla\left(J_{\text{ML,TOA}}(\hat{\mathbf{x}}^k)\right), \tag{3.69}$$

where

$$\mathbf{H}\left(J_{\text{ML,TOA}}(\mathbf{x})\right) = \frac{\partial^2 J_{\text{ML,TOA}}(\mathbf{x})}{\partial \mathbf{x} \partial \mathbf{x}^T}. \tag{3.70}$$

with

$$\left[\mathbf{H}\left(J_{\text{ML,TOA}}(\mathbf{x})\right)\right]_{1,1}$$

$$= \sum_{l=1}^{L} \frac{2}{\sigma_{\text{TOA},l}^2}\left(\frac{(x - x_l)^2}{(x - x_l)^2 + (y - y_l)^2} - \frac{\left(r_{\text{TOA},l} - \sqrt{(x - x_l)^2 + (y - y_l)^2}\right)(y - y_l)^2}{\left[(x - x_l)^2 + (y - y_l)^2\right]^{3/2}}\right), \tag{3.71}$$

$$\left[\mathbf{H}\left(J_{\text{ML,TOA}}(\mathbf{x})\right)\right]_{1,2} = \left[\mathbf{H}\left(J_{\text{ML,TOA}}(\mathbf{x})\right)\right]_{2,1} = \sum_{l=1}^{L} \frac{2r_{\text{TOA},l}(x - x_l)(y - y_l)}{\sigma_{\text{TOA},l}^2\left[(x - x_l)^2 + (y - y_l)^2\right]^{3/2}}, \tag{3.72}$$

and

$$\left[\mathbf{H}\big(J_{\mathrm{ML,TOA}}(\mathbf{x})\big)\right]_{2,2}$$

$$=\sum_{l=1}^{L}\frac{2}{\sigma_{\mathrm{TOA},l}^{2}}\left(\frac{(y-y_{l})^{2}}{(x-x_{l})^{2}+(y-y_{l})^{2}}-\frac{\big(r_{\mathrm{TOA},l}-\sqrt{(x-x_{l})^{2}+(y-y_{l})^{2}}\,\big)(x-x_{l})^{2}}{\big[(x-x_{l})^{2}+(y-y_{l})^{2}\big]^{3/2}}\right),$$

$$(3.73)$$

while

$$\nabla\big(J_{\mathrm{ML,TOA}}(\mathbf{x})\big)=-2\left[\begin{array}{c}\displaystyle\sum_{l=1}^{L}\frac{\big(r_{\mathrm{TOA},l}-\sqrt{(x-x_{l})^{2}+(y-y_{l})^{2}}\,\big)(x-x_{l})}{\sigma_{\mathrm{TOA},l}^{2}\sqrt{(x-x_{l})^{2}+(y-y_{l})^{2}}}\\[4mm]\displaystyle\sum_{l=1}^{L}\frac{\big(r_{\mathrm{TOA},l}-\sqrt{(x-x_{l})^{2}+(y-y_{l})^{2}}\,\big)(y-y_{l})}{\sigma_{\mathrm{TOA},l}^{2}\sqrt{(x-x_{l})^{2}+(y-y_{l})^{2}}}\end{array}\right]\qquad(3.74)$$

On the other hand, the corresponding Gauss–Newton and steepest descent algorithms are, respectively:

$$\hat{\mathbf{x}}^{k+1}=\hat{\mathbf{x}}^{k}+\big(\mathbf{G}^{T}\big(\mathbf{f}_{\mathrm{TOA}}(\hat{\mathbf{x}}^{k})\big)\mathbf{C}_{\mathrm{TOA}}^{-1}\mathbf{G}\big(\mathbf{f}_{\mathrm{TOA}}(\hat{\mathbf{x}}^{k})\big)\big)^{-1}\mathbf{G}^{T}\big(\mathbf{f}_{\mathrm{TOA}}(\hat{\mathbf{x}}^{k})\big)\mathbf{C}_{\mathrm{TOA}}^{-1}\big(\mathbf{r}_{\mathrm{TOA}}-\mathbf{f}_{\mathrm{TOA}}(\hat{\mathbf{x}}^{k})\big)$$

$$(3.75)$$

and

$$\hat{\mathbf{x}}^{k+1}=\hat{\mathbf{x}}^{k}-\mu\nabla\big(J_{\mathrm{ML,TOA}}(\hat{\mathbf{x}}^{k})\big).\qquad(3.76)$$

Example 3.2

Repeat the test of Example 3.1 using the ML approach.

Solution

The MATLAB program for this example is called Example3_2.m, and its structure is identical to that of Example3_1.m. Figures 3.3 and 3.4 show the estimates of x and y, respectively, versus number of iterations at SNR = 30 dB with $\hat{\mathbf{x}}^{0}=[3\ 2]^{T}$. Similar to the NLS approach, all schemes provide the same position estimates upon convergence, but the Newton–Raphson and Gauss–Newton methods converge faster than the steepest descent algorithm. Nevertheless, it is difficult to see that the ML estimator is superior to the NLS approach in terms of positioning accuracy based on a single run.

Example 3.3

Consider a 2-D geometry of $L=4$ receivers with known coordinates at (0, 0), (0, 10), (10, 0), and (10, 10), while the unknown source position is $(x, y) = (2, 3)$. The range error variance, $\sigma_{\mathrm{TOA},l}^{2}$, is assigned proportional to d_{l}^{2} with SNR $= d_{l}^{2}/\sigma_{\mathrm{TOA},l}^{2}$. Compare the mean square position error (MSPE) performance of the NLS and ML approaches for SNR $\in [-10, 60]$ dB. The MSPE is defined as $E\big\{(\hat{x}-x)^{2}+(\hat{y}-y)^{2}\big\}$.

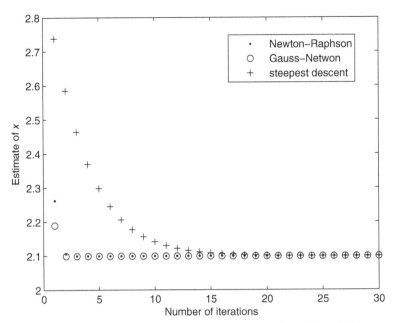

Figure 3.3   Estimate of x versus number of iterations using ML method.

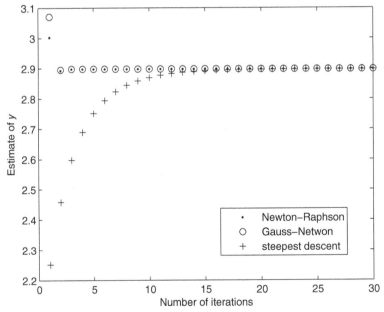

Figure 3.4   Estimate of y versus number of iterations using ML method.

Solution

The MATLAB program for this example is provided as Example3_3.m, and its structure is similar to those of Example3_1.m and Example3_2.m. We compute the empirical MSPE based on 1000 independent runs, which is given as $\sum_{i=1}^{1000}\left[(\hat{x}_i-x)^2+(\hat{y}_i-y)^2\right]/1000$, where $(\hat{x}_i, \hat{y}_i)$ denotes the position estimate of the ith run. Without loss of generality, the NLS and ML estimators are realized by the Newton–Raphson scheme. Figure 3.5 shows the MSPEs of the two schemes at SNR $\in [-10, 60]$ dB. Note that the dB scale is employed in both axes to facilitate the presentation. As expected, the ML estimator is superior to the NLS method, and the former MSPE is smaller by around 1 dB at SNR $\in [10, 60]$ dB.

B. TDOA-Based Positioning: Similarly, incorporating (3.19), the ML cost function for TDOA-based positioning, denoted by $J_{\mathrm{ML,TDOA}}(\tilde{\mathbf{x}})$, is:

$$J_{\mathrm{ML,TDOA}}(\tilde{\mathbf{x}}) = \left(\mathbf{r}_{\mathrm{TDOA}} - \mathbf{f}_{\mathrm{TDOA}}(\tilde{\mathbf{x}})\right)^T \mathbf{C}_{\mathrm{TDOA}}^{-1}\left(\mathbf{r}_{\mathrm{TDOA}} - \mathbf{f}_{\mathrm{TDOA}}(\tilde{\mathbf{x}})\right), \qquad (3.77)$$

and the ML position estimate is:

$$\hat{\mathbf{x}} = \arg\min_{\tilde{\mathbf{x}}} J_{\mathrm{ML,TDOA}}(\tilde{\mathbf{x}}). \qquad (3.78)$$

C. TSOA-Based Positioning: Based on (3.26), the ML cost function for TSOA-based positioning, denoted by $J_{\mathrm{ML,TSOA}}(\tilde{\mathbf{x}})$, is:

$$J_{\mathrm{ML,TSOA}}(\tilde{\mathbf{x}}) = \left(\mathbf{r}_{\mathrm{TSOA}} - \mathbf{f}_{\mathrm{TSOA}}(\tilde{\mathbf{x}})\right)^T \mathbf{C}_{\mathrm{TSOA}}^{-1}\left(\mathbf{r}_{\mathrm{TSOA}} - \mathbf{f}_{\mathrm{TSOA}}(\tilde{\mathbf{x}})\right), \qquad (3.79)$$

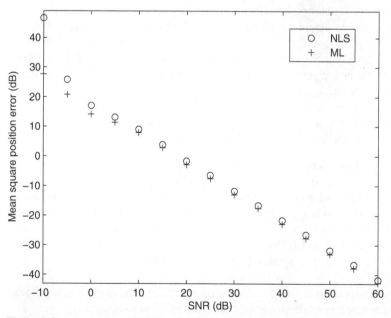

Figure 3.5   Mean square position error comparison for nonlinear methods.

and the ML position estimate is:

$$\hat{\mathbf{x}} = \arg\min_{\tilde{\mathbf{x}}} J_{\mathrm{ML,TSOA}}(\tilde{\mathbf{x}}). \tag{3.80}$$

D. RSS-Based Positioning: Employing (3.35), the ML cost function for RSS-based positioning, denoted by $J_{\mathrm{ML,RSS}}(\tilde{\mathbf{x}})$, is:

$$J_{\mathrm{ML,RSS}}(\tilde{\mathbf{x}}) = \left(\mathbf{r}_{\mathrm{RSS}} - \mathbf{f}_{\mathrm{RSS}}(\tilde{\mathbf{x}})\right)^T \mathbf{C}_{\mathrm{RSS}}^{-1}\left(\mathbf{r}_{\mathrm{RSS}} - \mathbf{f}_{\mathrm{RSS}}(\tilde{\mathbf{x}})\right), \tag{3.81}$$

and the ML position estimate is:

$$\hat{\mathbf{x}} = \arg\min_{\tilde{\mathbf{x}}} J_{\mathrm{ML,RSS}}(\tilde{\mathbf{x}}). \tag{3.82}$$

E. DOA-Based Positioning: Using (3.42), the ML cost function for DOA-based positioning, denoted by $J_{\mathrm{ML,DOA}}(\tilde{\mathbf{x}})$, is then:

$$J_{\mathrm{ML,DOA}}(\tilde{\mathbf{x}}) = \left(\mathbf{r}_{\mathrm{DOA}} - \mathbf{f}_{\mathrm{DOA}}(\tilde{\mathbf{x}})\right)^T \mathbf{C}_{\mathrm{DOA}}^{-1}\left(\mathbf{r}_{\mathrm{DOA}} - \mathbf{f}_{\mathrm{DOA}}(\tilde{\mathbf{x}})\right), \tag{3.83}$$

and the ML position estimate is:

$$\hat{\mathbf{x}} = \arg\min_{\tilde{\mathbf{x}}} J_{\mathrm{ML,DOA}}(\tilde{\mathbf{x}}). \tag{3.84}$$

3.3.2 Linear Methods

The basic idea of the linear localization methodology is to convert the nonlinear expressions of (3.4), (3.13), (3.20), (3.30) or (3.37), into a set of linear equations with zero-mean disturbances, assuming that the measurement errors are sufficiently small. As the corresponding optimization cost functions are now unimodal, it is always guaranteed to obtain the global solution. Three linear positioning methods, namely, LLS, WLLS, and subspace estimators, will be presented as follows. Analogous to NLS and ML estimators, the WLLS method is a weighted version of the LLS scheme and it provides higher localization accuracy, although the mean and covariance of the errors in the linear equations are required for the weight computation. On the other hand, the subspace technique first relates $\mathbf{x}$ with the squared pair-wise distances among the source and receivers. Then, source localization is achieved using an eigenvalue decomposition (EVD) procedure.

3.3.2.1 Linear Least Squares: The LLS approach attempts to reorganize (3.4), (3.13), (3.20), (3.30), and (3.37) into linear equations in $\mathbf{x}$, and the position is then estimated using the ordinary LS technique. For TOA, TDOA, and RSS measurements, we have to introduce an intermediate variable that is a function of the source position in the linearization process. While in TSOA-based positioning, M or L extra variables are needed. The LLS location estimators based on TOA, TDOA, TSOA, RSS, and DOA information are developed one by one as follows.

A. TOA-Based Positioning: To convert the TOA measurements into linear models in $\mathbf{x}$, we first consider squaring both sides of (3.4) to obtain:

$$r_{\mathrm{TOA},l}^2 = (x - x_l)^2 + (y - y_l)^2 + n_{\mathrm{TOA},l}^2 + 2n_{\mathrm{TOA},l}\sqrt{(x - x_l)^2 + (y - y_l)^2}, \quad l = 1, 2, \cdots, L. \tag{3.85}$$

Let

$$m_{\text{TOA},l} = n_{\text{TOA},l}^2 + 2n_{\text{TOA},l}\sqrt{(x-x_l)^2 + (y-y_l)^2} \tag{3.86}$$

be the noise component in (3.85) and introduce a dummy variable R of the form:

$$R = x^2 + y^2. \tag{3.87}$$

Substituting (3.86) and (3.87) into (3.85) yields:

$$
\begin{aligned}
r_{\text{TOA},l}^2 &= (x-x_l)^2 + (y-y_l)^2 + m_{\text{TOA},l} \\
\Rightarrow r_{\text{TOA},l}^2 &= x^2 - 2x_l x + x_l^2 + y^2 - 2y_l y + y_l^2 + m_{\text{TOA},l} \\
\Rightarrow -2x_l x - 2y_l y + R + m_{\text{TOA},l} &= r_{\text{TOA},l}^2 - x_l^2 - y_l^2, \quad l = 1, 2, \cdots, L.
\end{aligned}
\tag{3.88}
$$

Let

$$
\mathbf{A} =
\begin{bmatrix}
-2x_1 & -2y_1 & 1 \\
-2x_2 & -2y_2 & 1 \\
\vdots & \vdots & \vdots \\
-2x_L & -2y_L & 1
\end{bmatrix},
\tag{3.89}
$$

$$\boldsymbol{\theta} = \begin{bmatrix} x & y & R \end{bmatrix}^T, \tag{3.90}$$

$$\mathbf{q} = \begin{bmatrix} m_{\text{TOA},1} & m_{\text{TOA},2} & \cdots & m_{\text{TOA},L} \end{bmatrix}^T, \tag{3.91}$$

and

$$
\mathbf{b} =
\begin{bmatrix}
r_{\text{TOA},1}^2 - x_1^2 - y_1^2 \\
r_{\text{TOA},2}^2 - x_2^2 - y_2^2 \\
\vdots \\
r_{\text{TOA},L}^2 - x_L^2 - y_L^2
\end{bmatrix}.
\tag{3.92}
$$

The matrix form for (3.88) is then:

$$\mathbf{A}\boldsymbol{\theta} + \mathbf{q} = \mathbf{b}, \tag{3.93}$$

where the observed $\mathbf{r}_{\text{TOA}}$ of (3.5) is now transformed to $\mathbf{b}$, $\mathbf{A}$ is constructed from the known receiver positions, and $\boldsymbol{\theta}$ contains the source location to be determined. When $\{n_{\text{TOA},l}\}$ are sufficiently small such that

$$
\mathbf{q} \approx
\begin{bmatrix}
2n_{\text{TOA},1}\sqrt{(x-x_1)^2 + (y-y_1)^2} \\
2n_{\text{TOA},2}\sqrt{(x-x_2)^2 + (y-y_2)^2} \\
\vdots \\
2n_{\text{TOA},L}\sqrt{(x-x_L)^2 + (y-y_L)^2}
\end{bmatrix}
\tag{3.94}
$$

can be considered a zero-mean vector as $E\{n_{\text{TOA},l}\} = 0$; that is, $E\{\mathbf{q}\} \approx \mathbf{0}$, we can approximate (3.93) as:

$$\mathbf{A}\boldsymbol{\theta} \approx \mathbf{b}. \tag{3.95}$$

Analogous to (3.46), the LS cost function based on (3.95), denoted by $J_{\text{LLS,TOA}}(\tilde{\theta})$, is:

$$
\begin{aligned}
J_{\text{LLS,TOA}}(\tilde{\theta}) &= \left(\mathbf{A}\tilde{\theta} - \mathbf{b}\right)^T \left(\mathbf{A}\tilde{\theta} - \mathbf{b}\right) \\
&= \tilde{\theta}^T \mathbf{A}^T \mathbf{A} \tilde{\theta} - 2\tilde{\theta}^T \mathbf{A}^T \mathbf{b} + \mathbf{b}^T \mathbf{b},
\end{aligned}
\tag{3.96}
$$

which is a quadratic function of $\tilde{\theta}$, indicating that there is a unique minimum in $J_{\text{LLS,TOA}}(\tilde{\theta})$. The LLS estimate corresponds to:

$$
\hat{\theta} = \arg\min_{\tilde{\theta}} J_{\text{LLS,TOA}}(\tilde{\theta}),
\tag{3.97}
$$

which can be easily computed by differentiating (3.96) with respect to $\tilde{\theta}$ and setting the resultant expression to zero:

$$
\left. \frac{\partial J_{\text{LLS,TOA}}(\tilde{\theta})}{\partial \tilde{\theta}} \right|_{\tilde{\theta}=\hat{\theta}} = \mathbf{0}
$$

$$
\Rightarrow 2\mathbf{A}^T \mathbf{A} \hat{\theta} - 2\mathbf{A}^T \mathbf{b} = \mathbf{0}
\tag{3.98}
$$

$$
\Rightarrow \mathbf{A}^T \mathbf{A} \hat{\theta} = \mathbf{A}^T \mathbf{b}
$$

$$
\Rightarrow \hat{\theta} = \left(\mathbf{A}^T \mathbf{A}\right)^{-1} \mathbf{A}^T \mathbf{b}.
$$

The LLS position estimate is simply extracted from the first and second entries of $\hat{\theta}$; that is,

$$
\hat{\mathbf{x}} = \left[[\hat{\theta}]_1 [\hat{\theta}]_2 \right]^T.
\tag{3.99}
$$

In the literature, Equation (3.99) is also referred to as the LS calibration method [20].

An alternative way for LLS TOA-based positioning is to eliminate R in (3.88) by employing the differences between any two equations [21]. For simplicity, and without loss of generality, subtracting the first equation of (3.88) from the remaining $(L-1)$ equations, R is removed and we have:

$$
\begin{aligned}
&-2(x_l - x_1)x - 2(y_l - y_1)y + m_{\text{TOA},l} - m_{\text{TOA},1} \\
&= r_{\text{TOA},l}^2 - x_l^2 - y_l^2 - r_{\text{TOA},1}^2 + x_1^2 + y_1^2, \, l = 2, 3, \cdots, L,
\end{aligned}
\tag{3.100}
$$

or in matrix form:

$$
\mathbf{A}\mathbf{x} + \mathbf{q} = \mathbf{b},
\tag{3.101}
$$

where $\mathbf{A}$, $\mathbf{q}$, and $\mathbf{b}$ are now modified to:

$$
\mathbf{A} = \begin{bmatrix}
-2(x_2 - x_1) & -2(y_2 - y_1) \\
-2(x_3 - x_1) & -2(y_3 - y_1) \\
\vdots & \vdots \\
-2(x_L - x_1) & -2(y_L - y_1)
\end{bmatrix},
\tag{3.102}
$$

$$
\mathbf{q} = \left[m_{\text{TOA},2} - m_{\text{TOA},1} \; m_{\text{TOA},3} - m_{\text{TOA},1} \; \cdots \; m_{\text{TOA},L} - m_{\text{TOA},1} \right]^T,
\tag{3.103}
$$

and

$$\mathbf{b} = \begin{bmatrix} r_{\text{TOA},2}^2 - x_2^2 - y_2^2 - r_{\text{TOA},1}^2 + x_1^2 + y_1^2 \\ r_{\text{TOA},3}^2 - x_3^2 - y_3^2 - r_{\text{TOA},1}^2 + x_1^2 + y_1^2 \\ \vdots \\ r_{\text{TOA},L}^2 - x_L^2 - y_L^2 - r_{\text{TOA},1}^2 + x_1^2 + y_1^2 \end{bmatrix}. \tag{3.104}$$

Assuming sufficiently small noise conditions and following (3.94–3.98), a variant of the LLS method using TOA measurements is:

$$\hat{\mathbf{x}} = (\mathbf{A}^T \mathbf{A})^{-1} \mathbf{A}^T \mathbf{b}. \tag{3.105}$$

   B. TDOA-Based Positioning: In a similar manner, we first rewrite (3.13) with the use of (3.14) as

$$r_{\text{TDOA},l} + \sqrt{(x-x_1)^2 + (y-y_1)^2} = \sqrt{(x-x_l)^2 + (y-y_l)^2} + n_{\text{TDOA},l}, \quad l = 2,3,\cdots,L. \tag{3.106}$$

Let

$$m_{\text{TDOA},l} = n_{\text{TDOA},l}^2 + 2n_{\text{TDOA},l}\sqrt{(x-x_l)^2 + (y-y_l)^2} \tag{3.107}$$

be the modified noise component and introduce a dummy variable R_1 of the form:

$$R_1 = \sqrt{(x-x_1)^2 + (y-y_1)^2}. \tag{3.108}$$

Squaring both sides of (3.106) and employing (3.107) and (3.108), we get:

$$2(x_1-x_l)(x-x_1) + 2(y_1-y_l)(y-y_1) - 2r_{\text{TDOA},l}R_1 + m_{\text{TDOA},l}$$
$$= r_{\text{TDOA},l}^2 - (x_1-x_l)^2 - (y_1-y_l)^2, l = 2,3,\cdots,L. \tag{3.109}$$

Let

$$\mathbf{A} = 2 \begin{bmatrix} x_1 - x_2 & y_1 - y_2 & -r_{\text{TDOA},2} \\ x_1 - x_3 & y_1 - y_3 & -r_{\text{TDOA},3} \\ \vdots & \vdots & \vdots \\ x_1 - x_L & y_1 - y_L & -r_{\text{TDOA},L} \end{bmatrix}, \tag{3.110}$$

$$\boldsymbol{\theta} = \begin{bmatrix} x - x_1 \ y - y_1 \ R_1 \end{bmatrix}^T, \tag{3.111}$$

$$\mathbf{q} = \begin{bmatrix} m_{\text{TDOA},2} \ m_{\text{TDOA},3} \ \cdots \ m_{\text{TDOA},L} \end{bmatrix}^T, \tag{3.112}$$

and

$$\mathbf{b} = \begin{bmatrix} r_{\text{TDOA},2}^2 - (x_1-x_2)^2 - (y_1-y_2)^2 \\ r_{\text{TDOA},3}^2 - (x_1-x_3)^2 - (y_1-y_3)^2 \\ \vdots \\ r_{\text{TDOA},L}^2 - (x_1-x_L)^2 - (y_1-y_L)^2 \end{bmatrix}. \tag{3.113}$$

The matrix form for (3.109) is then identical to (3.93), and the LLS estimate of $\boldsymbol{\theta}$ corresponds to (3.98), provided that the noises in $\mathbf{q}$ are sufficiently small such that $\{n_{\text{TDOA},l}^2\}$ can be ignored. Note that in order to solve (3.109) in a unique manner, the minimum number of receivers is now $L = 4$ because of the introduction of the additional variable of R_1. Finally, the LLS position estimate is obtained from the first and second entries of $\hat{\theta}$; that is,

$$\hat{\mathbf{x}} = \left[[\hat{\theta}]_1 + x_1 \ [\hat{\theta}]_2 + y_1 \right]^T. \tag{3.114}$$

It is worth noting that if (3.109) is divided by $r_{\text{TDOA},l}$, we yield

$$\frac{(x_1 - x_l)(x - x_1)}{r_{\text{TDOA},l}} + \frac{(y_1 - y_l)(y - y_1)}{r_{\text{TDOA},l}} - R_1 + \frac{m_{\text{TDOA},l}}{r_{\text{TDOA},l}}$$
$$= r_{\text{TDOA},l} - \frac{(x_1 - x_l)^2}{r_{\text{TDOA},l}} - \frac{(y_1 - y_l)^2}{r_{\text{TDOA},l}}, l = 2, 3, \cdots, L. \tag{3.115}$$

Based on (3.115), we can obtain a variant of LLS–TDOA positioning algorithm by straightforwardly following the development in (3.100–3.105). Nevertheless, the corresponding algorithm analysis will be extremely difficult, if not impossible, because most of the terms in (3.115) are inversely proportional to the random variables of $\{r_{\text{TDOA},l}\}$.

B. TSOA-Based Positioning: We first reorganize (3.20) as:

$$r_{\text{TSOA},m,l} - d_l = d_m^t + n_{\text{TSOA},m,l}, \quad m = 1, 2, \cdots, M, \quad l = 1, 2, \cdots, L. \tag{3.116}$$

Assign

$$m_{\text{TSOA},m,l} = n_{\text{TSOA},m,l}^2 + 2d_m^t n_{\text{TSOA},m,l} \tag{3.117}$$

as the modified noise component, which is similar to (3.107). With the use of (3.3) and (3.21), squaring both sides of (3.116) yields:

$$2(x_l - x_m^t)x + 2(y_l - y_m^t)y + 2r_{\text{TSOA},m,l}d_l + m_{\text{TSOA},m,l}$$
$$= r_{\text{TSOA},m,l}^2 + x_l^2 + y_l^2 - (x_m^t)^2 - (y_m^t)^2,$$
$$m = 1, 2, \cdots, M, \quad l = 1, 2, \cdots, L. \tag{3.118}$$

Let

$$\mathbf{A} = 2 \begin{bmatrix} x_1 - x_1^t & y_1 - y_1^t & r_{\text{TSOA},1,1} & 0 & \cdots & 0 \\ x_1 - x_2^t & y_1 - y_2^t & r_{\text{TSOA},2,1} & 0 & \cdots & 0 \\ \vdots & \vdots & \vdots & \vdots & \ddots & \vdots \\ x_1 - x_M^t & y_1 - y_M^t & r_{\text{TSOA},M,1} & 0 & \cdots & 0 \\ x_2 - x_1^t & y_2 - y_2^t & 0 & r_{\text{TSOA},1,2} & \cdots & 0 \\ \vdots & \vdots & \vdots & \vdots & \ddots & \vdots \\ x_L - x_M^t & y_L - y_M^t & 0 & \cdots & 0 & r_{\text{TSOA},M,L} \end{bmatrix}, \tag{3.119}$$

$$\theta = \begin{bmatrix} x \ y \ d_1 \ d_2 \cdots d_L \end{bmatrix}^T, \tag{3.120}$$

$$\mathbf{q} = \begin{bmatrix} m_{\text{TSOA},1,1} \ m_{\text{TSOA},2,1} \cdots m_{\text{TSOA},M,1} \ m_{\text{TSOA},1,2} \cdots m_{\text{TSOA},M,L} \end{bmatrix}^T, \tag{3.121}$$

and

$$
\mathbf{b} = \begin{bmatrix}
r_{\text{TSOA},1,1}^2 + x_1^2 + y_1^2 - \left(x_1^t\right)^2 - \left(y_1^t\right)^2 \\[4pt]
r_{\text{TSOA},2,1}^2 + x_1^2 + y_1^2 - \left(x_2^t\right)^2 - \left(y_2^t\right)^2 \\[4pt]
\vdots \\[4pt]
r_{\text{TSOA},M,1}^2 + x_1^2 + y_1^2 - \left(x_M^t\right)^2 - \left(y_M^t\right)^2 \\[4pt]
r_{\text{TSOA},1,2}^2 + x_2^2 + y_2^2 - \left(x_1^t\right)^2 - \left(y_1^t\right)^2 \\[4pt]
\vdots \\[4pt]
r_{\text{TSOA},M,L}^2 + x_L^2 + y_L^2 - \left(x_M^t\right)^2 - \left(y_M^t\right)^2
\end{bmatrix}.
\tag{3.122}
$$

The matrix form for (3.118) is then identical to (3.93), and the LLS estimate of $\boldsymbol{\theta}$ corresponds to (3.98) when the noises in $\mathbf{q}$ are sufficiently small such that $\{n_{\text{TDOA},l}^2\}$ can be ignored. It is clear from (3.120) that there are L introduced variables, which are the L distances between the source and receivers. Note that (3.20) can also be reorganized as another set of linear equations via including $\{d_m^t\}$ instead of $\{d_l\}$ in (3.120). Finally, the LLS position estimate is obtained from the first and second entries of $\hat{\boldsymbol{\theta}}$:

$$
\hat{\mathbf{x}} = [[\hat{\boldsymbol{\theta}}]_1 \ [\hat{\boldsymbol{\theta}}]_2]^T.
\tag{3.123}
$$

D. RSS-Based Positioning: To convert (3.30) to linear equations in $\mathbf{x}$, we need a few steps as follows. First, taking the inverse logarithm of (3.30) yields:

$$
e^{-2r_{\text{RSS},l}/\alpha} = \left[(x - x_l)^2 + (y - y_l)^2\right]e^{-2n_{\text{RSS},l}/\alpha}, \quad l = 1, 2, \cdots, L.
\tag{3.124}
$$

It is well known [36] that if q is a Gaussian variable with mean μ and variance σ^2, the mean and variance of e^q would be $E\{e^q\} = e^{\mu + \sigma^2/2}$ and $\text{var}(e^q) = \left(e^{\sigma^2} - 1\right)e^{2\mu + \sigma^2}$, respectively. Applying the transformation results, the mean and variance of $e^{-2r_{\text{RSS},l}/\alpha}$ are calculated as:

$$
E\left\{e^{-2r_{\text{RSS},l}/\alpha}\right\} = d_l^2 e^{2\sigma_{\text{RSS},l}^2/\alpha^2}, \quad l = 1, 2, \cdots, L
\tag{3.125}
$$

and

$$
\text{var}\left(e^{-2r_{\text{RSS},l}/\alpha}\right) = d_l^4 e^{4\sigma_{\text{RSS},l}^2/\alpha^2}\left(e^{4\sigma_{\text{RSS},l}^2/\alpha^2} - 1\right).
\tag{3.126}
$$

Although $e^{-2r_{\text{RSS},l}/\alpha}$ is a biased estimate of d_l^2, its bias can be removed via dividing it by $e^{2\sigma_{\text{RSS},l}^2/\alpha^2}$. That is, an unbiased estimate of d_l^2 is

$$
\hat{d}_l^2 = e^{-2r_{\text{RSS},l}/\alpha - 2\sigma_{\text{RSS},l}^2/\alpha^2}.
\tag{3.127}
$$

Following the development in (3.85–3.94), we can obtain the approximate linear equations as in (3.95), but now $\mathbf{b}$ has the form of:

$$\mathbf{b} = \begin{bmatrix} e^{-2r_{RSS,1}/\alpha - 2\sigma^2_{RSS,1}/\alpha^2} - x_1^2 - y_1^2 \\ e^{-2r_{RSS,2}/\alpha - 2\sigma^2_{RSS,2}/\alpha^2} - x_2^2 - y_2^2 \\ \vdots \\ e^{-2r_{RSS,L}/\alpha - 2\sigma^2_{RSS,L}/\alpha^2} - x_L^2 - y_L^2 \end{bmatrix}. \tag{3.128}$$

The LLS estimates of $\boldsymbol{\theta}$ and $\mathbf{x}$ follow (3.98) and (3.99), respectively.

E. DOA-Based Positioning: Moving $n_{DOA,l}$ to the left-hand side and taking tangent on both sides of (3.37), we obtain:

$$\tan(r_{DOA,l} - n_{DOA,l}) = \frac{\sin(r_{DOA,l} - n_{DOA,l})}{\cos(r_{DOA,l} - n_{DOA,l})} = \frac{y - y_l}{x - x_l}, \quad l = 1, 2, \cdots, L. \tag{3.129}$$

Performing cross-multiplication in (3.129) yields:

$$\sin(r_{DOA,l} - n_{DOA,l})(x - x_l) = \cos(r_{DOA,l} - n_{DOA,l})(y - y_l), \quad l = 1, 2, \cdots, L. \tag{3.130}$$

When $\{n_{DOA,l}\}$ are sufficiently small such that $\cos(n_{DOA,l}) \approx 1$ and $\sin(n_{DOA,l}) \approx n_{DOA,l}$, we have the following approximations:

$$\begin{aligned} \sin(r_{DOA,l} - n_{DOA,l}) &= \sin(r_{DOA,l})\cos(n_{DOA,l}) - \cos(r_{DOA,l})\sin(n_{DOA,l}) \\ &\approx \sin(r_{DOA,l}) - \cos(r_{DOA,l})n_{DOA,l}, \end{aligned} \tag{3.131}$$

and

$$\begin{aligned} \cos(r_{DOA,l} - n_{DOA,l}) &= \cos(r_{DOA,l})\cos(n_{DOA,l}) + \sin(r_{DOA,l})\sin(n_{DOA,l}) \\ &\approx \cos(r_{DOA,l}) + \sin(r_{DOA,l})n_{DOA,l}. \end{aligned} \tag{3.132}$$

Substituting (3.131) and (3.132) into (3.130), we obtain:

$$\sin(r_{DOA,l})x - \cos(r_{DOA,l})y + m_{DOA,l} = \sin(r_{DOA,l})x_l - \cos(r_{DOA,l})y_l, \quad l = 1, 2, \cdots, L, \tag{3.133}$$

where $m_{DOA,l}$ is the transformed noise component which is expressed as:

$$m_{DOA,l} = n_{DOA,l}[(x_l - x)\cos(r_{DOA,l}) + (y_l - y)\sin(r_{DOA,l})], \quad l = 1, 2, \cdots, L. \tag{3.134}$$

As a result, we can construct (3.101), but now $\mathbf{A}$, $\mathbf{q}$, and $\mathbf{b}$ are:

$$\mathbf{A} = \begin{bmatrix} \sin(r_{DOA,1}) & -\cos(r_{DOA,1}) \\ \sin(r_{DOA,2}) & -\cos(r_{DOA,2}) \\ \vdots & \vdots \\ \sin(r_{DOA,L}) & -\cos(r_{DOA,L}) \end{bmatrix}, \tag{3.135}$$

$$\mathbf{q} = \begin{bmatrix} m_{DOA,1} & m_{DOA,2} & \cdots & m_{DOA,L} \end{bmatrix}^T, \tag{3.136}$$

and

$$
\mathbf{b} = \begin{bmatrix} \sin(r_{\text{DOA},1})x_1 - \cos(r_{\text{DOA},1})y_1 \\ \sin(r_{\text{DOA},2})x_2 - \cos(r_{\text{DOA},2})y_2 \\ \vdots \\ \sin(r_{\text{DOA},L})x_L - \cos(r_{\text{DOA},L})y_L \end{bmatrix}. \tag{3.137}
$$

Noting that $E\{\mathbf{q}\} = \mathbf{0}$, the LLS position estimator based on DOA measurements is also given by (3.105).

3.3.2.2 WLLS

A. TOA-Based Positioning: Although the LLS approach is simple, it offers optimum estimation performance only when the disturbances in the linear equations are independent and identically distributed. From (3.94) and (3.103), it is obvious that the LLS–TOA-based positioning algorithms are suboptimal. Taking (3.96) as an illustration, the localization accuracy can be improved if we include a symmetric weighting matrix, say, $\mathbf{W}$, in the cost function, denoted by $J_{\text{WLLS,TOA}}(\tilde{\boldsymbol{\theta}})$. The resultant expression is referred to as the WLS cost function, which has the form of:

$$
\begin{aligned} J_{\text{WLLS,TOA}}(\tilde{\boldsymbol{\theta}}) &= (\mathbf{A}\tilde{\boldsymbol{\theta}} - \mathbf{b})^T \mathbf{W}(\mathbf{A}\tilde{\boldsymbol{\theta}} - \mathbf{b}) \\ &= \tilde{\boldsymbol{\theta}}^T \mathbf{A}^T \mathbf{W} \mathbf{A} \tilde{\boldsymbol{\theta}} - 2\tilde{\boldsymbol{\theta}}^T \mathbf{A}^T \mathbf{W} \mathbf{b} + \mathbf{b}^T \mathbf{W} \mathbf{b}. \end{aligned} \tag{3.138}
$$

According to (3.93) and (3.94), we have $E\{\mathbf{b}\} = \mathbf{A}\boldsymbol{\theta}$, which corresponds to the linear unbiased data model. As a result, we can follow the best linear unbiased estimator (BLUE) [22], [37] to determine the optimum $\mathbf{W}$, which is equal to the inverse of the covariance of $\mathbf{q}$. That is, the weighting matrix is similar to that of the ML methodology. Employing (3.94), we obtain:

$$
\begin{aligned} \mathbf{W} &= \left[E\{\mathbf{q}\mathbf{q}^T\} \right]^{-1} \\ &\approx \left[\text{diag}\left(4\sigma_{\text{TOA},1}^2 d_1^2, 4\sigma_{\text{TOA},2}^2 d_2^2, \cdots, 4\sigma_{\text{TOA},L}^2 d_L^2 \right) \right]^{-1} \\ &= \frac{1}{4} \text{diag}\left(\frac{1}{\sigma_{\text{TOA},1}^2 d_1^2}, \frac{1}{\sigma_{\text{TOA},2}^2 d_2^2}, \cdots, \frac{1}{\sigma_{\text{TOA},L}^2 d_L^2} \right). \end{aligned} \tag{3.139}
$$

As $\{d_l\}$ are not available, a practical choice of $\mathbf{W}$ is to replace d_l with $r_{\text{TOA},l}$ which is valid for sufficiently small error condition:

$$
\mathbf{W} = \frac{1}{4} \text{diag}\left(\frac{1}{\sigma_{\text{TOA},1}^2 r_{\text{TOA},1}^2}, \frac{1}{\sigma_{\text{TOA},2}^2 r_{\text{TOA},2}^2}, \cdots, \frac{1}{\sigma_{\text{TOA},L}^2 r_{\text{TOA},L}^2} \right). \tag{3.140}
$$

Following (3.97) and (3.98), the WLLS estimate of $\boldsymbol{\theta}$ is:

$$
\begin{aligned} \hat{\boldsymbol{\theta}} &= \arg\min_{\boldsymbol{\theta}} J_{\text{WLLS,TOA}}(\hat{\boldsymbol{\theta}}) \\ &= (\mathbf{A}^T \mathbf{W} \mathbf{A})^{-1} \mathbf{A}^T \mathbf{W} \mathbf{b}. \end{aligned} \tag{3.141}
$$

The WLLS position estimate is then given as (3.99).

With only a moderate increase of computational complexity [22], Equation (3.141) is superior to (3.98) in terms of estimation performance. Nevertheless, the localization accuracy can be further enhanced by making use of $[\hat{\boldsymbol{\theta}}]_3$ according to the relation of (3.87) as follows. When $\hat{\mathbf{x}}$ of (3.141) is sufficiently close to $\mathbf{x}$, we have:

$$
\begin{aligned}
[\hat{\boldsymbol{\theta}}]_1^2 - x^2 &= \left([\hat{\boldsymbol{\theta}}]_1 + x\right)\left([\hat{\boldsymbol{\theta}}]_1 - x\right) \\
&\approx 2x\left([\hat{\boldsymbol{\theta}}]_1 - x\right).
\end{aligned}
\tag{3.142}
$$

Similarly, for $[\hat{\boldsymbol{\theta}}]_2$,

$$
[\hat{\boldsymbol{\theta}}]_2^2 - y^2 \approx 2y\left([\hat{\boldsymbol{\theta}}]_2 - y\right).
\tag{3.143}
$$

Based on (3.87) and with the use of (3.142) and (3.143), we construct:

$$
\mathbf{h} = \mathbf{Gz} + \mathbf{w},
\tag{3.144}
$$

where

$$
\mathbf{h} = \left[[\hat{\boldsymbol{\theta}}]_1^2 \quad [\hat{\boldsymbol{\theta}}]_2^2 \quad [\hat{\boldsymbol{\theta}}]_3\right]^T,
\tag{3.145}
$$

$$
\mathbf{G} = \begin{bmatrix} 1 & 0 \\ 0 & 1 \\ 1 & 1 \end{bmatrix},
\tag{3.146}
$$

$$
\mathbf{z} = \left[x^2 y^2\right]^T,
\tag{3.147}
$$

and

$$
\mathbf{w} = \left[2x\left([\hat{\boldsymbol{\theta}}]_1 - x\right) 2y\left([\hat{\boldsymbol{\theta}}]_2 - y\right)[\hat{\boldsymbol{\theta}}] - R\right]^T.
\tag{3.148}
$$

Note that $\mathbf{z}$ is the parameter vector to be determined. To find the covariance of $\mathbf{w}$, we utilize the result of BLUE that the covariance of $\hat{\mathbf{x}}$ is of the form of [37]:

$$
E\left\{\left[[\hat{\boldsymbol{\theta}}]_1 - x \quad [\hat{\boldsymbol{\theta}}]_2 - y \quad [\hat{\boldsymbol{\theta}}]_3 - R\right]\left[[\hat{\boldsymbol{\theta}}]_1 - x \quad [\hat{\boldsymbol{\theta}}]_2 - y \quad [\hat{\boldsymbol{\theta}}]_3 - R\right]^T\right\} = \left(\mathbf{A}^T \mathbf{W} \mathbf{A}\right)^{-1}.
\tag{3.149}
$$

Employing (3.148) and (3.149), the optimal weighting matrix for (3.144), denoted by $\boldsymbol{\Phi}$, is then:

$$
\boldsymbol{\Phi} = \left[\text{diag}(2x, 2y, 1)\left(\mathbf{A}^T \mathbf{W} \mathbf{A}\right)^{-1} \text{diag}(2x, 2y, 1)\right]^{-1}.
\tag{3.150}
$$

As a result, the WLLS estimate of $\mathbf{z}$ is

$$\hat{\mathbf{z}} = \left(\mathbf{G}^T \Phi \mathbf{G}\right)^{-1} \mathbf{G}^T \Phi \mathbf{h}. \tag{3.151}$$

As there is no sign information for $\mathbf{x}$ in $\mathbf{z}$, the final position estimate is determined as:

$$\hat{\mathbf{x}} = \left[\operatorname{sgn}\left(\left[\hat{\theta}\right]_1\right)\sqrt{[\hat{z}]_1} \quad \operatorname{sgn}\left(\left[\hat{\theta}\right]_2\right)\sqrt{[\hat{z}]_2} \right]^T, \tag{3.152}$$

where sgn represents the signum function. In the literature, this approach is called the two-step WLS estimator [23] where (3.87) is exploited in an implicit manner. Alternatively, an explicit way is to minimize (3.138) subject to the constraint of (3.87) which can be solved by the method of Lagrange multipliers [24–25].

B. TDOA-Based Positioning: Similarly, the WLLS version for (3.106) is also computed using (3.141), where the weighting matrix is now

$$
\begin{aligned}
\mathbf{W} &= \left[4\operatorname{diag}(d_2, d_3, \cdots, d_L)\mathbf{C}_{\text{TDOA}} \operatorname{diag}(d_2, d_3, \cdots, d_L) \right]^{-1} \\
&\approx \left[4\operatorname{diag}(r_{\text{TDOA},2} + R_1, r_{\text{TDOA},3} + R_1, \cdots, r_{\text{TDOA},L} + R_1)\mathbf{C}_{\text{TDOA}} \right. \\
&\quad \left. \times \operatorname{diag}(r_{\text{TDOA},2} + R_1, r_{\text{TDOA},3} + R_1, \cdots, r_{\text{TDOA},L} + R_1) \right]^{-1} \\
&\approx \left[4\operatorname{diag}(r_{\text{TDOA},2} + [\hat{\theta}]_3, r_{\text{TDOA},3} + [\hat{\theta}]_3, \cdots, r_{\text{TDOA},L} + [\hat{\theta}]_3)\mathbf{C}_{\text{TDOA}} \right. \\
&\quad \left. \times \operatorname{diag}(r_{\text{TDOA},2} + [\hat{\theta}]_3, r_{\text{TDOA},3} + [\hat{\theta}]_3, \cdots, r_{\text{TDOA},L} + [\hat{\theta}]_3) \right]^{-1},
\end{aligned}
\tag{3.153}
$$

while $\mathbf{A}$, θ, and $\mathbf{b}$ are defined as in (3.110), (3.111), and (3.113), respectively. As the estimate of R_1 is not available at the beginning, we first use the LLS estimator to obtain $[\hat{\theta}]_3$. The WLLS position estimate is given by (3.114). The two-step WLS estimator for TDOA-based positioning can be derived by following (3.142–3.152) [23].

C. TSOA-based Positioning: Ignoring $\{n^2_{\text{TSOA},m,l}\}$ in (3.117) and following (3.139), the weighting matrix for the WLLS estimator using TSOA measurements is obtained as:

$$
\begin{aligned}
\mathbf{W} &= \left[E\{\mathbf{q}\mathbf{q}^T\} \right]^{-1} \\
&\approx \frac{1}{4} \operatorname{diag}\left(\frac{1}{\sigma^2_{\text{TSOA},1,1}(d_1^t)^2}, \frac{1}{\sigma^2_{\text{TSOA},2,1}(d_2^t)^2}, \cdots, \frac{1}{\sigma^2_{\text{TSOA},M,1}(d_M^t)^2}, \frac{1}{\sigma^2_{\text{TSOA},1,2}(d_1^t)^2}, \cdots, \right. \\
&\quad \left. \frac{1}{\sigma^2_{\text{TSOA},M,L}(d_M^t)^2} \right).
\end{aligned}
\tag{3.154}
$$

while $\mathbf{A}$, $\boldsymbol{\theta}$ and $\mathbf{b}$ are defined as in (3.119), (3.120), and (3.122), respectively. As the estimates of $\{d'_m\}$ are not available at the beginning, we first use the last L elements in the LLS estimate of (3.120) and (3.116) for their determination. The WLLS position estimate is given by (3.99). Via exploiting the relationship between x and $\{d_l\}$ in (3.120), several two-step WLS estimators for TSOA-based positioning have been derived [26–28].

   D. *RSS-Based Positioning:*   With the use of (3.126) and (3.127), the variance of $\hat{d}_l^2$ in (3.127) is derived as:

$$
\begin{aligned}
\operatorname{var}\left(\hat{d}_l^2\right) &= \operatorname{var}\left(e^{-2r_{\text{RSS},l}/\alpha}\right)\left(e^{-2\sigma_{\text{RSS},l}^2/\alpha^2}\right)^2 \\
&= d_l^4\left(e^{4\sigma_{\text{RSS},l}^2/\alpha^2} - 1\right) \\
&\approx \left(\hat{d}_l^2\right)^2\left(e^{4\sigma_{\text{RSS},l}^2/\alpha^2} - 1\right) \\
&= e^{-4r_{\text{RSS},l}/\alpha}\left(1 - e^{-4\sigma_{\text{RSS},l}^2/\alpha^2}\right), \quad l = 1, 2, \cdots, L.
\end{aligned}
\tag{3.155}
$$

As a result, the WLLS estimate of $\boldsymbol{\theta}$ is also given by (3.141), where $\mathbf{b}$ is defined in (3.128) and the optimum weighting matrix is now:

$$
W = \operatorname{diag}\left(\frac{e^{4r_{\text{RSS},1}/\alpha}}{1 - e^{-4\sigma_{\text{RSS},1}^2/\alpha^2}}, \frac{e^{4r_{\text{RSS},2}/\alpha}}{1 - e^{-4\sigma_{\text{RSS},2}^2/\alpha^2}}, \cdots, \frac{e^{4r_{\text{RSS},L}/\alpha}}{1 - e^{-4\sigma_{\text{RSS},L}^2/\alpha^2}}\right).
\tag{3.156}
$$

The development of the two-step WLS estimator for RSS-based positioning then follows (3.142–3.152) straightforwardly [29].

   E. *DOA-Based Positioning:*   To derive the weighting matrix for WLLS DOA-based positioning, we first assume sufficiently small error conditions such that $r_{\text{DOA},l} \approx \phi_l$, $l = 1, 2, \cdots, L$. Furthermore, with the use of

$$
\cos(\phi_l) = \frac{x - x_l}{d_l}
\tag{3.157}
$$

and

$$
\sin(\phi_l) = \frac{y - y_l}{d_l}
\tag{3.158}
$$

$m_{\text{DOA},l}$ in (3.134) can then be approximated as:

$$
\begin{aligned}
m_{\text{DOA},l} &\approx n_{\text{DOA},l}[(x_l - x)\cos(\phi_l) + (y_l - y)\sin(\phi_l)] \\
&= -n_{\text{DOA},l}d_l, \quad l = 1, 2, \cdots, L.
\end{aligned}
\tag{3.159}
$$

Similar to (3.139), the corresponding optimum weight is:

$$
\begin{aligned}
W &= \left[E\{\mathbf{qq}^T\} \right]^{-1} \\
&= \left[\mathrm{diag}\left(\sigma^2_{\mathrm{DOA},1} d_1^2, \sigma^2_{\mathrm{DOA},2} d_2^2, \cdots, \sigma^2_{\mathrm{DOA},L} d_L^2 \right) \right]^{-1} \\
&= \mathrm{diag}\left(\frac{1}{\sigma^2_{\mathrm{DOA},1} d_1^2}, \frac{1}{\sigma^2_{\mathrm{DOA},2} d_2^2}, \cdots, \frac{1}{\sigma^2_{\mathrm{DOA},L} d_L^2} \right).
\end{aligned}
\tag{3.160}
$$

As d_l, $l = 1, 2, \cdots, L$, are unavailable and they are not directly related to $\{r_{\mathrm{DOA},l}\}$, we use the following iterative procedure to find $\mathbf{x}$:

(i) Find $\hat{\mathbf{x}}$ using the LLS estimator.

(ii) Construct $\mathbf{W}$ based on (3.160) using the computed $\hat{\mathbf{x}}$.

(iii) Use (3.141) with $\mathbf{A}$ and $\mathbf{b}$ defined in (3.135) and (3.137) to determine an updated $\mathbf{x}$.

(iv) Repeat steps (ii) and (iii) until a stopping criterion is reached.

3.3.2.3 Subspace Approach:

The subspace positioning approach using TOA measurements is presented as follows. We first define a $L \times 2$ matrix $\mathbf{X}$:

$$
\mathbf{X} = \begin{bmatrix}
x_1 - x & y_1 - y \\
x_2 - x & y_2 - y \\
\vdots & \vdots \\
x_L - x & y_L - y
\end{bmatrix},
\tag{3.161}
$$

which is parameterized by $\mathbf{x}$. With the use of $\mathbf{X}$, the multidimensional similarity matrix [30], denoted by $\mathbf{D}$, is constructed as:

$$
\mathbf{D} = \mathbf{X}\mathbf{X}^T,
\tag{3.162}
$$

whose (m, n) entry can be shown to be

$$
\begin{aligned}
[\mathbf{D}]_{m,n} &= (x_m - x)(x_n - x) + (y_m - y)(y_n - y) \\
&= 0.5 \left[(x_m - x)^2 + (y_m - y)^2 + (x_n - x)^2 + (y_n - y)^2 - (x_m - x_n)^2 - (y_m - y_n)^2 \right] \\
&= 0.5 \left(d_m^2 + d_n^2 - d_{mn}^2 \right),
\end{aligned}
\tag{3.163}
$$

where $d_{mn} = d_{nm} = \sqrt{(x_m - x_n)^2 + (y_m - y_n)^2}$ is of known value because it represents the distance between the mth and nth receivers. We then represent $\mathbf{D}$ using EVD:

$$
\mathbf{D} = \mathbf{U}\Lambda\mathbf{U}^T,
\tag{3.164}
$$

where $\Lambda = \text{diag}(\lambda_1, \lambda_2, \ldots, \lambda_L)$ is the diagonal matrix of eigenvalues of $\mathbf{D}$ with $\lambda_1 \geq \lambda_2 \geq \cdots \geq \lambda_L \geq 0$, and $\mathbf{U} = [\mathbf{u}_1 \, \mathbf{u}_2 \, \cdots \, \mathbf{u}_L]$ is an orthonormal matrix whose columns are the corresponding eigenvectors. Noting that the rank of $\mathbf{D}$ is 2, we have $\lambda_3 = \lambda_4 = \cdots \lambda_L = 0$. As a result, Equation (3.164) can also be written as:

$$
\begin{aligned}
\mathbf{D} &= \mathbf{U}_s \Lambda_s \mathbf{U}_s^T \\
&= \mathbf{U}_s \Lambda_s^{1/2} \left(\mathbf{U}_s \Lambda_s^{1/2} \right)^T \\
&= \mathbf{U}_s \Lambda_s^{1/2} \Omega \left(\mathbf{U}_s \Lambda_s^{1/2} \Omega \right)^T,
\end{aligned}
\tag{3.165}
$$

where $\mathbf{U}_s = [\mathbf{u}_1 \, \mathbf{u}_2]$, $\Lambda_s = \text{diag}(\lambda_1, \lambda_2)$, and $\Lambda_s^{\frac{1}{2}} = \text{diag}\left(\lambda_1^{\frac{1}{2}}, \lambda_2^{\frac{1}{2}}\right)$ denote the signal subspace components, while Ω is a rotation matrix, such that $\Omega \Omega^T = \mathbf{I}$. Comparing (3.162) and (3.165) yields:

$$
\mathbf{X} = \mathbf{U}_s \Lambda_s^{1/2} \Omega.
\tag{3.166}
$$

We then determine the unknown rotation matrix via pseudoinverse:

$$
\begin{aligned}
\Omega &= \left(\mathbf{U}_s \Lambda_s^{1/2} \right)^{\dagger} \mathbf{X} \\
&= \left[\left(\mathbf{U}_s \Lambda_s^{1/2} \right)^T \left(\mathbf{U}_s \Lambda_s^{1/2} \right) \right]^{-1} \left(\mathbf{U}_s \Lambda_s^{1/2} \right)^T \mathbf{X} \\
&= \Lambda_s^{-1/2} \mathbf{U}_s^T \mathbf{X}.
\end{aligned}
\tag{3.167}
$$

Substituting (3.167) in (3.166) results in

$$
\mathbf{X} = \mathbf{U}_s \mathbf{U}_s^T \mathbf{X},
\tag{3.168}
$$

which implies that $\mathbf{x}$ can be extracted from the signal subspace eigenvectors. As d_l, $l = 1, 2, \cdots, L$, are not available, we construct a practical $\mathbf{D}$ according to:

$$
[\mathbf{D}]_{m,n} = 0.5 \left(r_{\text{TOA},m}^2 + r_{\text{TOA},n}^2 - d_{mn}^2 \right).
\tag{3.169}
$$

In the presence of measurement errors, it is shown [31] that $\mathbf{U}_s \Lambda_s^{1/2}$ is the LS estimate of $\mathbf{X}$ up to a rotation, and thus now (3.168) becomes an approximate relation. To derive the position estimate, we first rewrite $\mathbf{X}$ as

$$
\mathbf{X} = \mathbf{Y} - \mathbf{1}\mathbf{x}^T,
\tag{3.170}
$$

where

$$
\mathbf{Y} = \begin{bmatrix} x_1 & y_1 \\ x_2 & y_2 \\ \vdots & \vdots \\ x_L & y_L \end{bmatrix}.
\tag{3.171}
$$

Substituting (3.170) into (3.168), and with the use of the subspace relation of $\mathbf{U}_s\mathbf{U}_s^T = \mathbf{I} - \mathbf{U}_n\mathbf{U}_n^T$, where $\mathbf{U}_n = [\mathbf{u}_3 \ \mathbf{u}_4 \cdots \mathbf{u}_L]$, which corresponds to the noise subspace, we get:

$$\mathbf{U}_n\mathbf{U}_n^T\mathbf{1}\mathbf{x}^T \approx \mathbf{U}_n\mathbf{U}_n^T\mathbf{Y}. \tag{3.172}$$

Following the LLS procedure in (3.95–3.98), the subspace-based estimate of $\mathbf{x}$ using TOA measurements is computed as:

$$\begin{aligned}
\hat{\mathbf{x}} &= \left(\left(\mathbf{U}_n\mathbf{U}_n^T\mathbf{1}\right)^{\dagger}\mathbf{U}_n\mathbf{U}_n^T\mathbf{Y}\right)^T \\
&= \frac{\mathbf{Y}^T\mathbf{U}_n\mathbf{U}_n^T\mathbf{1}}{\mathbf{1}^T\mathbf{U}_n\mathbf{U}_n^T\mathbf{1}}.
\end{aligned} \tag{3.173}$$

It is worth pointing out that the classical multidimensional scaling approach [32] is a variant of the subspace technique. For subspace-based positioning with TDOA and RSS measurements, the interested reader is referred to [33, 34] and [35], respectively.

Example 3.4

Repeat the test of Example 3.3 using the linear approaches. That is, compare the MSPE performance of LLS, WLLS, two-step WLS, and subspace methods for SNR $\in [-10, 60]$ dB.

Solution

The MATLAB program for this example is called Example3_4.m, and its structure is identical to that of Example3_3.m. Figure 3.6 shows the MSPEs of different linear schemes at SNR $\in [-10, 60]$ dB. It is observed that the two-step WLS scheme, which exploits the constraint of (3.87), offers the highest localization performance, followed by WLLS, LLS, and subspace estimators.

3.4 PERFORMANCE ANALYSIS FOR LOCALIZATION ALGORITHMS

The CRLB is a lower bound on variance attainable by any unbiased estimators using the same data, and thus it serves as a key benchmark to compare with the mean square error (MSE) of positioning algorithms. Nevertheless, it is possible that the MSE of a biased estimator is less than the CRLB. The procedure for CRLB computation using the TOA, TDOA, TSOA, RSS, and DOA measurements in the presence of Gaussian noises is discussed in Section 3.4.1. In Section 3.4.2, we present the theoretical mean and MSE expressions for positioning estimators whose derivation is based on minimization or maximization of a cost function.

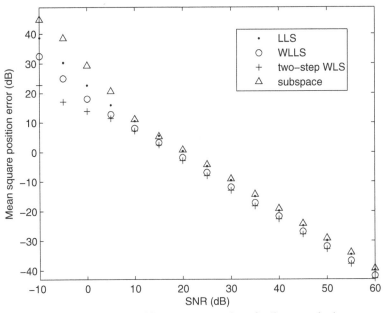

Figure 3.6   Mean square position error comparison for linear methods.

3.4.1   CRLB Computation

The key in producing the CRLB is to construct the corresponding Fisher information matrix (FIM). The diagonal elements of the FIM inverse are the minimum achievable variance values. Considering the general measurement model of (3.1), the standard procedure to compute the CRLB is summarized using the following steps:

(i) Compute the second-order derivatives of the logarithm of the measurement PDF with respect to $\mathbf{x}$; that is, $\partial^2 \ln p(\mathbf{r})/(\partial\mathbf{x}\partial\mathbf{x}^T)$.

(ii) Take the expected value of $\partial^2 \ln p(\mathbf{r})/(\partial\mathbf{x}\partial\mathbf{x}^T)$ to yield $\mathbf{I}(\mathbf{x}) = -E\{\partial^2 \ln p(\mathbf{r})/(\partial\mathbf{x}\partial\mathbf{x}^T)\}$, where $\mathbf{I}(\mathbf{x})$ denotes the FIM.

(iii) The CRLBs for x and y are given by $[\mathbf{I}^{-1}(\mathbf{x})]_{1,1}$ and $[\mathbf{I}^{-1}(\mathbf{x})]_{2,2}$, respectively.

Alternatively, when the measurement errors are zero-mean Gaussian distributed, $\mathbf{I}(\mathbf{x})$ can also be computed as [23]:

$$\mathbf{I}(\mathbf{x}) = \left[\frac{\partial \mathbf{f}(\mathbf{x})}{\partial \mathbf{x}}\right]^T \mathbf{C}^{-1}\left[\frac{\partial \mathbf{f}(\mathbf{x})}{\partial \mathbf{x}}\right], \qquad (3.174)$$

where $\mathbf{C}$ denotes the covariance matrix for $\mathbf{n}$. We now utilize (3.173) to determine the FIMs for positioning with TOA, TDOA, TSOA, RSS, and DOA measurements based on (3.8), (3.18), (3.25), (3.34), and (3.41), and their associated noise covariance matrices, respectively. The FIM based on TOA measurements of (3.5), denoted by $\mathbf{I}_{\text{TOA}}(\mathbf{x})$, is:

$$\mathbf{I}_{\text{TOA}}(\mathbf{x}) = \left[\frac{\partial \mathbf{f}_{\text{TOA}}(\mathbf{x})}{\partial \mathbf{x}}\right]^{T} \mathbf{C}_{\text{TOA}}^{-1} \left[\frac{\partial \mathbf{f}_{\text{TOA}}(\mathbf{x})}{\partial \mathbf{x}}\right]. \tag{3.175}$$

It is straightforward to show that

$$\left[\frac{\partial \mathbf{f}_{\text{TOA}}(\mathbf{x})}{\partial \mathbf{x}}\right] = \begin{bmatrix} \dfrac{x-x_1}{\sqrt{(x-x_1)^2+(y-y_1)^2}} & \dfrac{y-y_1}{\sqrt{(x-x_1)^2+(y-y_1)^2}} \\[3ex] \dfrac{x-x_2}{\sqrt{(x-x_2)^2+(y-y_2)^2}} & \dfrac{y-y_2}{\sqrt{(x-x_2)^2+(y-y_2)^2}} \\[1ex] \vdots & \vdots \\[1ex] \dfrac{x-x_L}{\sqrt{(x-x_L)^2+(y-y_L)^2}} & \dfrac{y-y_L}{\sqrt{(x-x_L)^2+(y-y_L)^2}} \end{bmatrix} \tag{3.176}$$

$$= \begin{bmatrix} \dfrac{x-x_1}{d_1} & \dfrac{y-y_1}{d_1} \\[2ex] \dfrac{x-x_2}{d_2} & \dfrac{y-y_2}{d_2} \\[1ex] \vdots & \vdots \\[1ex] \dfrac{x-x_L}{d_L} & \dfrac{y-y_L}{d_L} \end{bmatrix}.$$

Employing (3.176) and (3.11), (3.175) becomes:

$$\mathbf{I}_{\text{TOA}}(\mathbf{x}) = \begin{bmatrix} \displaystyle\sum_{l=1}^{L} \dfrac{(x-x_l)^2}{\sigma_{\text{TOA},l}^2 d_l^2} & \displaystyle\sum_{l=1}^{L} \dfrac{(x-x_l)(y-y_l)}{\sigma_{\text{TOA},l}^2 d_l^2} \\[3ex] \displaystyle\sum_{l=1}^{L} \dfrac{(x-x_l)(y-y_l)}{\sigma_{\text{TOA},l}^2 d_l^2} & \displaystyle\sum_{l=1}^{L} \dfrac{(y-y_l)^2}{\sigma_{\text{TOA},l}^2 d_l^2} \end{bmatrix}. \tag{3.177}$$

The CRLBs for x and y are given by $\left[\mathbf{I}_{\text{TOA}}^{-1}(\mathbf{x})\right]_{1,1}$ and $\left[\mathbf{I}_{\text{TOA}}^{-1}(\mathbf{x})\right]_{2,2}$, respectively. That is, the corresponding CRLB for $\mathbf{x}$, denoted by $\text{CRLB}_{\text{TOA}}(\mathbf{x})$, is

$$\text{CRLB}_{\text{TOA}}(\mathbf{x}) = \left[\mathbf{I}_{\text{TOA}}^{-1}(\mathbf{x})\right]_{1,1} + \left[\mathbf{I}_{\text{TOA}}^{-1}(\mathbf{x})\right]_{2,2}. \tag{3.178}$$

In a similar manner, the FIMs for TDOA, TSOA, RSS, and DOA measurements are:

$$\mathbf{I}_{\text{TDOA}}(\mathbf{x}) = \left[\frac{\partial \mathbf{f}_{\text{TDOA}}(\mathbf{x})}{\partial \mathbf{x}}\right]^{T} \mathbf{C}_{\text{TDOA}}^{-1} \left[\frac{\partial \mathbf{f}_{\text{TDOA}}(\mathbf{x})}{\partial \mathbf{x}}\right]; \tag{3.179}$$

$$\mathbf{I}_{\text{TSOA}}(\mathbf{x}) = \left[\frac{\partial \mathbf{f}_{\text{TSOA}}(\mathbf{x})}{\partial \mathbf{x}}\right]^{T} \mathbf{C}_{\text{TSOA}}^{-1} \left[\frac{\partial \mathbf{f}_{\text{TSOA}}(\mathbf{x})}{\partial \mathbf{x}}\right]; \tag{3.180}$$

$$\mathbf{I}_{\text{RSS}}(\mathbf{x}) = \left[\frac{\partial \mathbf{f}_{\text{RSS}}(\mathbf{x})}{\partial \mathbf{x}}\right]^{T} \mathbf{C}_{\text{RSS}}^{-1} \left[\frac{\partial \mathbf{f}_{\text{RSS}}(\mathbf{x})}{\partial \mathbf{x}}\right]; \tag{3.181}$$

and

$$\mathbf{I}_{\text{DOA}}(\mathbf{x}) = \left[\frac{\partial \mathbf{f}_{\text{DOA}}(\mathbf{x})}{\partial \mathbf{x}}\right]^{T} \mathbf{C}_{\text{DOA}}^{-1} \left[\frac{\partial \mathbf{f}_{\text{DOA}}(\mathbf{x})}{\partial \mathbf{x}}\right], \tag{3.182}$$

where

$$\left[\frac{\partial \mathbf{f}_{\text{TDOA}}(\mathbf{x})}{\partial \mathbf{x}}\right] = \begin{bmatrix} \dfrac{x - x_2}{d_2} - \dfrac{x - x_1}{d_1} & \dfrac{y - y_2}{d_2} - \dfrac{y - y_1}{d_1} \\[2ex] \dfrac{x - x_3}{d_3} - \dfrac{x - x_1}{d_1} & \dfrac{y - y_3}{d_3} - \dfrac{y - y_1}{d_1} \\[2ex] \vdots & \vdots \\[2ex] \dfrac{x - x_L}{d_L} - \dfrac{x - x_1}{d_1} & \dfrac{y - y_L}{d_L} - \dfrac{y - y_1}{d_1} \end{bmatrix}; \tag{3.183}$$

$$\left[\frac{\partial \mathbf{f}_{\text{TSOA}}(\mathbf{x})}{\partial \mathbf{x}}\right] = \begin{bmatrix} \dfrac{x - x_1^t}{d_2^t} + \dfrac{x - x_1}{d_1} & \dfrac{y - y_1^t}{d_1^t} + \dfrac{y - y_1}{d_1} \\[2ex] \dfrac{x - x_2^t}{d_2^t} + \dfrac{x - x_1}{d_1} & \dfrac{y - y_2^t}{d_2^t} + \dfrac{y - y_1}{d_1} \\[2ex] \vdots & \vdots \\[2ex] \dfrac{x - x_M^t}{d_M^t} + \dfrac{x - x_1}{d_1} & \dfrac{y - y_M^t}{d_M^t} + \dfrac{y - y_1}{d_1} \\[2ex] \dfrac{x - x_1^t}{d_1^t} + \dfrac{x - x_2}{d_2} & \dfrac{y - y_1^t}{d_1^t} + \dfrac{y - y_2}{d_2} \\[2ex] \vdots & \vdots \\[2ex] \dfrac{x - x_M^t}{d_M^t} + \dfrac{x - x_L}{d_L} & \dfrac{y - y_M^t}{d_M^t} + \dfrac{y - y_L}{d_L} \end{bmatrix}; \tag{3.184}$$

$$\left[\frac{\partial \mathbf{f}_{\text{RSS}}(\mathbf{x})}{\partial \mathbf{x}}\right] = -\alpha \begin{bmatrix} \dfrac{x - x_1}{d_1^2} & \dfrac{y - y_1}{d_1^2} \\[2ex] \dfrac{x - x_2}{d_2^2} & \dfrac{y - y_2}{d_2^2} \\[2ex] \vdots & \vdots \\[2ex] \dfrac{x - x_L}{d_L^2} & \dfrac{y - y_L}{d_L^2} \end{bmatrix}; \tag{3.185}$$

and

$$\left[\frac{\partial \mathbf{f}_{\text{DOA}}(\mathbf{x})}{\partial \mathbf{x}}\right] = \begin{bmatrix} -\dfrac{y - y_1}{d_1^2} & \dfrac{x - x_1}{d_1^2} \\[2ex] -\dfrac{y - y_2}{d_2^2} & \dfrac{x - x_2}{d_2^2} \\[2ex] \vdots & \vdots \\[2ex] -\dfrac{y - y_L}{d_L^2} & \dfrac{x - x_L}{d_L^2} \end{bmatrix}. \tag{3.186}$$

Their corresponding CRLBs are then determined according to (3.178).

3.4.2   Mean and Variance Analysis

When the position estimator corresponds to minimizing or maximizing a continuous cost function, the mean and MSE expressions of $\hat{\mathbf{x}}$, can be produced with the use of Taylor's series expansion as follows [32, 38]. Let $J(\tilde{\mathbf{x}})$ be a differentiable function of $\tilde{\mathbf{x}}$, and the estimate $\hat{\mathbf{x}}$ is given by its minimum or maximum. This implies that

$$\left.\frac{\partial J(\tilde{\mathbf{x}})}{\partial \tilde{\mathbf{x}}}\right|_{\tilde{\mathbf{x}}=\hat{\mathbf{x}}} = \mathbf{0}. \tag{3.187}$$

At small estimation error conditions, such that $\hat{\mathbf{x}}$ is located at a reasonable proximity of the ideal solution of $\mathbf{x}$, using Taylor's series to expand (3.187) around $\mathbf{x}$ up to the first-order terms, we have

$$
\begin{aligned}
\mathbf{0} = \left.\frac{\partial J(\tilde{\mathbf{x}})}{\partial \tilde{\mathbf{x}}}\right|_{\tilde{\mathbf{x}}=\hat{\mathbf{x}}} &\approx \left.\frac{\partial J(\tilde{\mathbf{x}})}{\partial \tilde{\mathbf{x}}}\right|_{\tilde{\mathbf{x}}=\mathbf{x}} + \left.\frac{\partial^2 J(\tilde{\mathbf{x}})}{\partial \tilde{\mathbf{x}} \partial \tilde{\mathbf{x}}^T}\right|_{\tilde{\mathbf{x}}=\mathbf{x}} (\hat{\mathbf{x}}-\mathbf{x}) \\
\Rightarrow -\left.\frac{\partial J(\tilde{\mathbf{x}})}{\partial \tilde{\mathbf{x}}}\right|_{\tilde{\mathbf{x}}=\mathbf{x}} &\approx \left.\frac{\partial^2 J(\tilde{\mathbf{x}})}{\partial \tilde{\mathbf{x}} \partial \tilde{\mathbf{x}}^T}\right|_{\tilde{\mathbf{x}}=\mathbf{x}} (\hat{\mathbf{x}}-\mathbf{x}) \\
\Rightarrow -\nabla(J(\mathbf{x})) &\approx \mathbf{H}(J(\mathbf{x}))(\hat{\mathbf{x}}-\mathbf{x}),
\end{aligned} \tag{3.188}
$$

where $\mathbf{H}(J(\mathbf{x}))$ and $\nabla(J(\mathbf{x}))$ are the corresponding Hessian matrix and gradient vector evaluated at the true location. When the second-order derivatives inside the Hessian matrix are smooth enough around $\mathbf{x}$, we have [39]:

$$\mathbf{H}(J(\mathbf{x})) \approx E\{\mathbf{H}(J(\mathbf{x}))\}. \tag{3.189}$$

Employing (3.189) and taking the expected value of (3.188) yields the mean of $\hat{\mathbf{x}}$:

$$E\{\hat{\mathbf{x}}\} \approx \mathbf{x} - [E\{H(J(\mathbf{x}))\}]^{-1} E\{\nabla(J(\mathbf{x}))\}. \tag{3.190}$$

When $\hat{\mathbf{x}}$ is an unbiased estimate of $\mathbf{x}$, $E\{\hat{\mathbf{x}}\} = \mathbf{x}$, indicating that the last term in (3.190) is a zero vector. Utilizing (3.188) and (3.189) again and the symmetric property of the Hessian matrix, we obtain the covariance for $\hat{\mathbf{x}}$, denoted by $\mathbf{C}_x$:

$$
\begin{aligned}
C_x &= E\left\{\left(\mathbf{x} - E\{\hat{\mathbf{x}}\}\right)\left(\mathbf{x} - E\{\hat{\mathbf{x}}\}\right)^T\right\} \\
&\approx E\left\{\left(\mathbf{x} - \hat{\mathbf{x}}\right)\left(\mathbf{x} - \hat{\mathbf{x}}\right)^T\right\} \\
&\approx [E\{H(J(\mathbf{x}))\}]^{-1} E\{\nabla(J(\mathbf{x})) \nabla^T(J(\mathbf{x}))\}[E\{H(J(\mathbf{x}))\}]^{-1}.
\end{aligned} \tag{3.191}
$$

That is, the variances of the estimates of x and y are given by $[\mathbf{C}_x]_{1,1}$ and $[\mathbf{C}_x]_{2,2}$, respectively.

We first take the ML cost function for TOA-based positioning in (3.68), namely, $J_{\text{ML,TOA}}(\tilde{\mathbf{x}})$, as an illustration. As $E\{r_{\text{TOA},l}\} = d_l$, $l = 1, 2, \cdots, L$, the expected value of $\mathbf{H}(J_{\text{ML,TOA}}(\mathbf{x}))$ can be easily determined with the use of (3.70–3.73) as:

$$
E\{\mathbf{H}(J_{\text{ML,TOA}}(\mathbf{x}))\} = 2 \begin{bmatrix} \displaystyle\sum_{l=1}^{L} \frac{(x-x_l)^2}{\sigma_{\text{TOA},l}^2 d_l^2} & \displaystyle\sum_{l=1}^{L} \frac{(x-x_l)(y-y_l)}{\sigma_{\text{TOA},l}^2 d_l^2} \\ \displaystyle\sum_{l=1}^{L} \frac{(x-x_l)(y-y_l)}{\sigma_{\text{TOA},l}^2 d_l^2} & \displaystyle\sum_{l=1}^{L} \frac{(y-y_l)^2}{\sigma_{\text{TOA},l}^2 d_l^2} \end{bmatrix}. \quad (3.192)
$$

On the other hand, the expected value of $\nabla(J_{\text{ML,TOA}}(\mathbf{x}))$ is

$$
E\{\nabla(J_{\text{ML,TOA}}(\mathbf{x}))\} = \mathbf{0}. \quad (3.193)
$$

As $\mathbf{H}(J_{\text{ML,TOA}}(\mathbf{x}))$ is nonsingular, it is clear from (3.190) that the ML position estimate is unbiased. Furthermore, noting that $E\{(r_{\text{TOA},l} - d_l)^2\} = \sigma_{\text{TOA},l}^2$, $l = 1, 2, \cdots, L$, and employing (3.74), we have:

$$
E\{J_{\text{ML,TOA}}(\mathbf{x}) J_{\text{ML,TOA}}^T(\mathbf{x})\} = 4 \begin{bmatrix} \displaystyle\sum_{l=1}^{L} \frac{(x-x_l)^2}{\sigma_{\text{TOA},l}^2 d_l^2} & \displaystyle\sum_{l=1}^{L} \frac{(x-x_l)(y-y_l)}{\sigma_{\text{TOA},l}^2 d_l^2} \\ \displaystyle\sum_{l=1}^{L} \frac{(x-x_l)(y-y_l)}{\sigma_{\text{TOA},l}^2 d_l^2} & \displaystyle\sum_{l=1}^{L} \frac{(y-y_l)^2}{\sigma_{\text{TOA},l}^2 d_l^2} \end{bmatrix}.
$$
$$ (3.194) $$

Substituting (3.192) and (3.194) into (3.191) yields:

$$
\mathbf{C}_x = \begin{bmatrix} \displaystyle\sum_{l=1}^{L} \frac{(x-x_l)^2}{\sigma_{\text{TOA},l}^2 d_l^2} & \displaystyle\sum_{l=1}^{L} \frac{(x-x_l)(y-y_l)}{\sigma_{\text{TOA},l}^2 d_l^2} \\ \displaystyle\sum_{l=1}^{L} \frac{(x-x_l)(y-y_l)}{\sigma_{\text{TOA},l}^2 d_l^2} & \displaystyle\sum_{l=1}^{L} \frac{(y-y_l)^2}{\sigma_{\text{TOA},l}^2 d_l^2} \end{bmatrix}^{-1}
$$

$$ (3.195) $$

$$
= \mathbf{I}_{\text{TOA}}^{-1}(\mathbf{x}),
$$

which implies that the ML estimator is optimum in the sense that its variance attains the CRLB in (3.178).

The mean and variance expressions can also be applied in the linear approach, which corresponds to minimizing a quadratic cost function. Considering the WLLS cost function in (3.138), the corresponding Hessian matrix and gradient vector are determined as:

$$
\left. \frac{\partial J_{\text{WLLS,TOA}}^2(\tilde{\theta})}{\partial \tilde{\theta} \partial \tilde{\theta}^T} \right|_{\tilde{\theta} = \theta} = \mathbf{H}(J_{\text{WLLS,TOA}}(\theta))
$$

$$ (3.196) $$

$$
= 2\mathbf{A}^T \mathbf{W} \mathbf{A},
$$

and

$$\frac{\partial J_{\mathrm{WLLS,TOA}}(\tilde{\theta})}{\partial \tilde{\theta}}\bigg|_{\hat{\theta}=\theta} \quad \nabla\big(J_{\mathrm{WLLS,TOA}}(\theta)\big) \tag{3.197}$$
$$= 2\mathbf{A}^T \mathbf{W}(\mathbf{A}\theta - \mathbf{b}).$$

With the use of (3.91) and (3.93), the expected value of $\nabla\big(J_{\mathrm{WLLS,TOA}}(\theta)\big)$ is:

$$E\big\{\nabla\big(J_{\mathrm{WLLS,TOA}}(\theta)\big)\big\} = 2\mathbf{A}^T \mathbf{W} E\{\mathbf{q}\} \tag{3.198}$$
$$= \mathbf{0}.$$

From (3.190), (3.196), and (3.198), it is observed that the WLLS estimator offers an unbiased estimate of $\mathbf{x}$. Furthermore, with the use of 3.139, the expected value of $\nabla\big(J_{\mathrm{WLLS,TOA}}(\theta)\big)\nabla^T\big(J_{\mathrm{WLLS,TOA}}(\theta)\big)$ is computed as:

$$E\big\{\nabla\big(J_{\mathrm{WLLS,TOA}}(\theta)\big)\nabla^T\big(J_{\mathrm{WLLS,TOA}}(\theta)\big)\big\} = 4\mathbf{A}^T \mathbf{W} E\{\mathbf{q}\mathbf{q}^T\}\mathbf{W}\mathbf{A} \tag{3.199}$$
$$= 4\mathbf{A}^T \mathbf{W}\mathbf{A}.$$

Substituting (3.196) and (3.199) into (3.191), the covariance matrix of $\hat{\theta}$ denoted by $\mathbf{C}_\theta$, is

$$\mathbf{C}_\theta = \big(\mathbf{A}^T \mathbf{W}\mathbf{A}\big)^{-1}, \tag{3.200}$$

which is consistent with the BLUE analysis [37] in (3.149). Finally, the variances of the estimates of x and y are given by $[\mathbf{C}_\theta]_{1,1}$ and $[\mathbf{C}_\theta]_{2,2}$, respectively. Nevertheless, it has been proved [22] that the MSE of the WLLS estimator is larger than the corresponding CRLB.

To study the theoretical performance of the two-step WLS estimate of (3.152), we first note that this approach implicitly solves the following constrained optimization problem:

$$\min J_{\mathrm{WLLS,TOA}}(\tilde{\theta}), \text{ subject to } [\hat{\theta}]_1^2 + [\hat{\theta}]_2^2 = [\hat{\theta}]_3. \tag{3.201}$$

Substituting the constraint into $J_{\mathrm{WLLS,TOA}}(\tilde{\theta})$, the two-step WLS estimate is equivalent to:

$$\hat{\mathbf{x}} = \arg\min_{\tilde{x}} J_{\mathrm{CWLS,TOA}}(\tilde{\mathbf{x}}), \tag{3.202}$$

where

$$J_{\mathrm{CWLS,TOA}}(\tilde{\mathbf{x}}) = \sum_{l=1}^{L} \frac{1}{\sigma_{\mathrm{TOA},l}^2 d_l^2}\big(-2x_l\tilde{x} - 2y_l\tilde{y} + \tilde{x}^2 + \tilde{y}^2 - r_{\mathrm{TOA},l}^2 + \tilde{x}^2 + \tilde{y}^2\big)^2. \tag{3.203}$$

Applying (3.190) and (3.191) on (3.203), it is shown [24] that the two-step WLS estimate is unbiased and its covariance is identical to (3.195). In other words,

Equation (3.152) is also an optimal position estimate. Similarly, the optimality of the two-step WLS estimators using TDOA, TSOA, and RSS measurements have been proved [25, 27]. Furthermore, it can be shown that the WLLS estimator for DOA-based positioning is unbiased and its MSE equals the CRLB. Finally, it is worth pointing out that by considering the subspace techniques as LS cost function optimization problems [31, 32], their means and variances can be produced using (3.190) and (3.191).

Example 3.5

Consider a 2-D geometry of $L = 4$ receivers with known coordinates at $(0, 0)$, $(0, 10)$, $(10, 0)$, and $(10, 10)$, while the unknown source position is $(x, y) = (2, -3)$. Note that the source is located outside the square bounded by the four receivers. The range error variance, $\sigma_{TOA,l}^2$, is assigned proportional to d_l^2 with SNR $= d_l^2/\sigma_{TOA,l}^2$. Compare the MSPE performance of the nonlinear and linear approaches with CRLB for SNR $\in [-10, 60]$ dB.

Solution

The MATLAB program for this example is called Example3_5.m, which can be considered as the combined version of Example3_3.m and Example3_4.m. The function for CRLB computation is also included. We compute the MSPE based on 1000 independent runs. The NLS and ML estimators are realized by the Newton–Raphson scheme, and their initial guesses are created by the LLS and WLLS algorithms, respectively. The MSPEs of the nonlinear and linear approaches are shown in Figures 3.7 and 3.8, respectively. In Figure 3.7, we observe that the ML estimator is superior to the NLS method, and its MSPE can attain the CRLB for sufficiently high SNR conditions, namely, SNR ≥ 35 dB, which agrees with (3.195). In Figure 3.8, it is seen that the two-step WLS estimator achieves the optimal estimation performance at SNR ≥ 25 dB, while the LLS, WLLS, and subspace methods can only offer suboptimal accuracy. Note that the MSPEs of the ML and two-step WLS estimators can be less than CRLB when SNR ≤ 0 dB because their estimates become biased for sufficiently large noise conditions. It is worth mentioning that the results of Figures 3.5 and 3.6 can also be produced from Example3_5.m by modifying the source location. In doing so, we will again observe the optimality of the ML and two-step WLS estimators and the suboptimality of the NLS, LLS, WLLS, and subspace schemes. Note that the CRLB for $(x, y) = (2, 3)$ is smaller than that of $(x, y) = (2, -3)$. This aligns with the conventional wisdom [40, 41] that better estimation performance can be achieved when the source location falls within the convex hull of the receivers.

Example 3.6

Study the effect of the number of receivers on the MSPE performance of the nonlinear and linear approaches at SNR $= 30$ dB. Start with the minimum number of sensors,

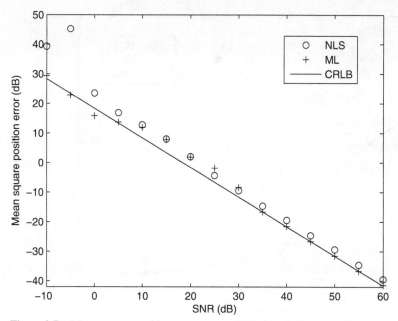

Figure 3.7   Mean square position error versus SNR for nonlinear methods.

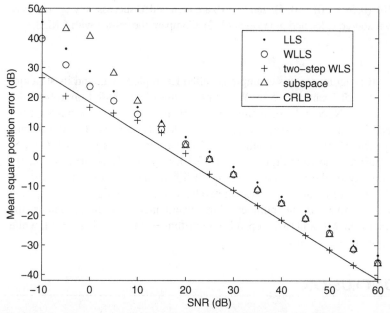

Figure 3.8   Mean square position error versus SNR for linear methods.

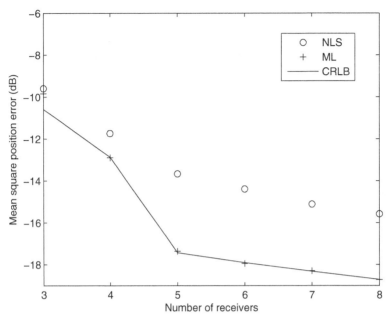

Figure 3.9   Mean square position error versus receiver number for nonlinear methods.

namely, $L = 3$, and their positions are $(0, 0), (10, 0),$ and $(10, 10)$. The receivers with coordinates $(0, 10), (0, 5), (5, 0), (10, 5),$ and $(5, 10)$ are then added successively up to $L = 8$. The unknown source is located at $(x, y) = (2, 3)$. Compare the results with CRLB.

Solution

The MATLAB program for this example is called Example3_6.m, and its structure is similar to that of Example3_5.m. We compute the MSPE based on 1000 independent runs. The NLS and ML estimators are realized by the Newton–Raphson scheme, and their initial guesses are created by the LLS and WLLS algorithms, respectively. The MSPEs of the nonlinear and linear approaches versus $L \in [3, 8]$ at SNR = 30 dB are shown in Figures 3.9 and 3.10, respectively. Apart from verifying the optimality of the ML and two-step WLS estimators and suboptimality of the NLS, WLLS, LLS, and subspace methods, we also see that the estimation performance improves as the number of receivers increases. In particular, the improvement of the ML and two-step WLS algorithms is large for $L \in [3, 5]$, while the gain is not significant at $L \geq 5$.

3.5   CONCLUSION

Source localization using TOA, TDOA, TSOA, RSS, and DOA measurements has been studied. After introducing the measurement models and their positioning principles, the nonlinear and linear approaches for determining the source location, which include the NLS, ML, LLS, WLLS, and subspace estimators, are presented.

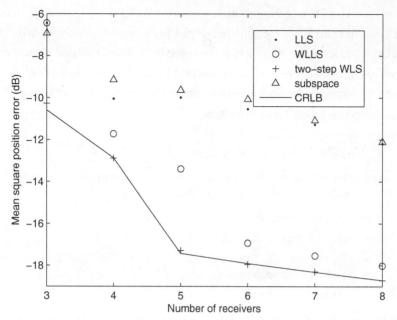

Figure 3.10   Mean square position error versus receiver number for linear methods

The nonlinear approach solves the nonlinear equations directly constructed from the TOA, TDOA, TSOA, RSS, and DOA measurements while the linear methodology converts the nonlinear equations to be linear. Moreover, mean and variance analysis for a class of position estimators that correspond to an unconstrained optimization problem as well as CRLB computation are discussed. In the presence of zero-mean Gaussian measurement errors, the estimation performance of the ML and constrained WLLS methods can achieve the CRLB, while the remaining estimators can only offer suboptimal localization accuracy.

APPENDIX

Apart from the example programs, we have also provided the MATLAB functions for TOA-based positioning as follows:

- `crlb.m`—function for CRLB computation
- `gn_ml.m`—function for ML estimator realized by Gauss–Newton method
- `gn_nls.m`—function for NLS estimator realized by Gauss–Newton method
- `grad_ml.m`—function for gradient computation in ML estimator
- `grad_nls.m`—function for gradient computation in NLS estimator
- `hessian_ml.m`—function for Hessian matrix computation in ML estimator
- `hessian_nls.m`—function for Hessian matrix computation in NLS estimator
- `jacob.m`—function for Jacobian matrix computation in Gauss–Newton method

- `lls.m`—function for LLS algorithm
- `nr_ml.m`—function for ML estimator realized by Newton–Raphson method
- `nr_nls.m`—function for NLS estimator realized by Newton–Raphson method
- `sd_ml.m`—function for ML estimator realized by steepest descent method
- `sd_nls.m`—function for NLS estimator realized by steepest descent method
- `sub.m`—function for subspace algorithm
- `wlls.m`—function for WLLS algorithm
- `wls2.m`—function for two-step WLS algorithm

Furthermore, brief descriptions for their usages can be obtained with the use of `help` command. For example, typing `help crlb` in MATLAB gives:

```
v = crlb(X, sigma2)
v = sum of the CRLB diagonal elements
X = first row is the source position; second to last
    rows are receiver positions
sigma2 = noise variance vector
```

That is, the inputs to the function are X and `sigma2` and the output is v, which is the sum of the CRLBs for x and y. The X is a $(L + 1) \times 2$ matrix whose first row contains the source position and remaining rows correspond to the receiver positions, while `sigma2` is a vector of the form $\left[\sigma^2_{\text{TOA},1} \ \sigma^2_{\text{TOA},2} \cdots \sigma^2_{\text{TOA},L} \right]^T$.

ACKNOWLEDGEMENT

The author thanks Dr. Lanxin Lin for her help in developing the MATLAB programs in this chapter.

REFERENCES

[1] J. Li and P. Stoica, *MIMO Radar Signal Processing*. Hoboken, NJ: John Wiley & Sons, 2009.

[2] G. C. Carter, Ed., *Coherence and Time Delay Estimation: An Applied Tutorial for Research, Development, Test, and Evaluation Engineers*. New York: IEEE, 1993.

[3] Y. Huang and J. Benesty, Eds., *Audio Signal Processing for Next-Generation Multimedia Communication Systems*. Kluwer Academic Publishers: Dordrecht, Boston, 2004.

[4] J. C. Liberti and T. S. Rappaport, *Smart Antennas for Wireless Communications: IS-95 and Third Generation CDMA Applications*. Upper Saddle River, NJ: Prentice-Hall, 1999.

[5] J. J. Caffery, Jr., *Wireless Location in CDMA Cellular Radio Systems*. Kluwer Academic Publishers: Dordrecht, Boston, 2000.

[6] A. Jagoe, *Mobile Location Services: The Definitive Guide*. Upper Saddle River, NJ: Prentice Hall, 2003.

[7] M. Ilyas and I. Mahgoub, *Handbook of Sensor Networks: Compact Wireless and Wired Sensing Systems*. London: CRC Press, 2005.

[8] I. Stojmenovic, *Handbook of Sensor Networks: Algorithms and Architectures*. New York: Wiley, 2005.

[9] V. G. Reju, A. W. H. Khong, and A. B. Sulaiman, "Localization of taps on solid surfaces for human-computer touch interfaces," *IEEE Trans. Multimedia*, vol. 15, no. 6, pp. 1365–1376, October 2013.

[10] Revision of the Commission's Rules to Ensure Compatibility With Enhanced 911 Emergency Calling, RM-8143, CC Docket No.94-102, July 26, 1996. Available at: http://www.fcc.gov/pshs/ services/911-services/enhanced911/Welcome.html.

[11] H. Wymeersch, J. Lien, and M. Z. Win, "Cooperative localization in wireless networks," *Proc. IEEE*, vol. 97, no. 2, pp. 427–450, February 2009.

[12] H. C. So, Y. T. Chan, and F. K. W. Chan, "Closed-form formulae for optimum time difference of arrival based localization," *IEEE Trans. Signal Process.*, vol. 56, no. 6, pp. 2614–2620, June 2008.

[13] L. Rui and K. C. Ho, "Efficient closed-form estimators for multistatic sonar localization," *IEEE Trans. Aerosp. Electron. Syst.*, vol. 51, no. 1, pp. 600–614, January 2015.

[14] H. L. Song, "Automatic vehicle location in cellular communications systems," *IEEE Trans. Veh. Technol.*, vol. 43, pp. 902–908, November 1994.

[15] L. Lin, H. C. So, and Y. T. Chan, "Accurate and simple source localization using differential received signal strength," *Digit. Signal Process.*, vol. 23, no. 3, pp. 736–743, May 2013.

[16] D. J. Torrieri, "Statistical theory of passive location systems," *IEEE Trans. Aerosp. Electron. Syst.*, vol. 20, no. 2, pp. 183–198, March 1984.

[17] M. Marks and E. Niewiadomska-Szynkiewicz, "Two-phase stochastic optimization to sensor network localization," *Proc. IEEE International Conference on Sensor Technologies and Applications*, Valencia, Spain, pp. 134–139, October 2007.

[18] P.-J. Chuang and C.-P. Wu, "An effective PSO-based node localization scheme for wireless sensor networks," *Proc. IEEE International Conference on Parallel and Distributed Computing, Applications and Technologies*, Dunedin, New Zealand, pp. 187–194, December 2008.

[19] C. Mensing and S. Plass, "Positioning algorithms for cellular networks using TDOA," *Proc. IEEE International Conference on Acoustics, Speech and Signal Processing*, Toulouse, France, vol. 4, pp. 513–516, May 2006.

[20] J. C. Chen, R. E. Hudson, and K. Yao, "Maximum-likelihood source localization and unknown sensor location estimation for wideband signals in the near field," *IEEE Trans. Signal Process.*, vol. 50, no. 8, pp. 1843–1854, August 2002.

[21] A. J. Fenwick, "Algorithms for position fixing using pulse arrival times," *IEE Proc. – Radar, Sonar and Navigation*, vol. 146, no. 4, pp. 208–212, August 1999.

[22] F. K. W. Chan, H. C. So, J. Zheng, and K. W. K. Lui, "Best linear unbiased estimator approach for time-of-arrival based localization," *IET Signal Process.*, vol. 2, no. 2, pp. 156–162, June 2008.

[23] Y. T. Chan and K. C. Ho, "A simple and efficient estimator for hyperbolic location," *IEEE Trans. Signal Process.*, vol. 42, no. 8, pp. 1905–1915, August 1994.

[24] K. W. Cheung, H. C. So, W.-K. Ma, and Y. T.Chan, "Least squares algorithms for time-of-arrival based mobile location," *IEEE Trans. Signal Process.*, vol. 52, no. 4, pp. 1121–1128, April 2004.

[25] K. W. Cheung, H. C. So, W.-K. Ma, and Y. T. Chan, "A constrained least squares approach to mobile positioning: Algorithms and optimality," *EURASIP J. Adv. Signal Process.*, vol. 2006, Article ID 20858, pp. 1–23, 2006.

[26] Y. Du and P. Wei, "An explicit solution for target localization in noncoherent distributed MIMO radar systems," *IEEE Signal Process. Lett.*, vol. 21, no. 9, pp. 1093–1097, September 2014.

[27] M. Einemo and H. C. So, "Weighted least squares algorithm for target localization in distributed MIMO radar," *Signal Process.*, vol. 115, pp. 144-150, October 2015.

[28] R. Amiri, F. Behnia, and H. Zamani, "Asymptotically efficient target localization from bistatic range measurements in distributed MIMO radars," *IEEE Signal Process. Lett.*, vol. 24, no. 3, pp. 299–303, March 2017.

[29] H. C. So and L. Lin, "Linear least squares approach for accurate received signal strength based source localization," *IEEE Trans. Signal Process.*, vol. 59, no. 8, pp. 4035–4040, August 2011.

[30] Q. Wan, Y.-J. Luo, W.-L. Yang, J. Xu, J. Tang, and Y.-N. Peng, "Mobile localization method based on multidimensional similarity analysis," *Proc. IEEE International Conference on Acoustics, Speech and Signal Processing*, vol. 4, Philadelphia, USA, pp. 1081–1084, March 2005.

[31] H. C. So and K. W. Chan, "A generalized subspace approach for mobile positioning with time-of-arrival measurements," *IEEE Trans. Signal Process.*, vol. 55, no. 10, pp. 5103–5107, October 2007.

[32] K. W. Cheung and H. C. So, "A multidimensional scaling framework for mobile location using time-of-arrival measurements," *IEEE Trans. Signal Process.*, vol. 53, no. 2, pp. 460–470, February 2005.

[33] H.-W. Wei, Q. Wan, Z.-X. Chen, and S.-F. Ye, "Multidimensional scaling-based passive emitter localisation from range-difference measurements," *IET Signal Process.*, vol. 2, no. 4, pp. 415–423, December 2008.

[34] W. Jiang, C. Xu, L. Pei, and W. Yu, "Multidimensional scaling-based TDOA localization scheme using an auxiliary line," *IEEE Signal Process. Lett.*, vol. 23, no. 4, pp. 546–550, April 2016.

[35] L. Lin, H. C. So, and F. K. W. Chan, "Multidimensional scaling approach for node localization using received signal strength measurements," *Digit. Signal Process.*, vol. 34, pp. 39–47, Boston, November 2014.

[36] Y. Viniotis, *Probability and Random Processes for Electrical Engineers*. New York: McGraw-Hill, 1998.

[37] S. M. Kay, *Fundamentals of Statistical Signal Processing: Estimation Theory*. Englewood Cliffs, NJ: Prentice-Hall, 1993.

[38] H. C. So, Y. T. Chan, K. C. Ho, and Y. Chen, "Simple formulas for bias and mean square error computation," *IEEE Signal Process. Mag.*, vol. 30, no. 4, pp. 162–165, July 2013.

[39] V. H. MacDonald and P. M. Schultheiss, "Optimum passive bearing estimation in a spatially incoherent noise environment," *J. Acoust. Soc. Am.*, vol. 46, no. 1, pt.1, pp. 37–43, July 1969.

[40] J. N. Ash and R. L. Moses, "On optimal anchor node placement in sensor localization by optimization of subspace principal angles," *Proc. IEEE International Conference on Acoustics, Speech and Signal Processing*, Las Vegas, NV, USA, pp. 2289–2292, March 2008.

[41] K. W. K. Lui and H. C. So, "A study of two-dimensional sensor placement using time-difference-of-arrival measurements," *Digit. Signal Process.*, vol. 19, no. 4, pp. 650–659, July 2009.

CHAPTER 4

CHANNEL MODELING AND ITS IMPACT ON LOCALIZATION

S. A. (Reza) Zekavat
Michigan Technological University

T HIS CHAPTER reviews concepts and models for wireless channels that are critical to localization systems. Channel modeling impact on the design of radio system is discussed. In addition, key channel parameters for different localization techniques are investigated. Channel parameters essential to position location methods such as time-of-arrival (TOA) and direction-of-arrival (DOA), and line-of-sight (LOS) and non-line-of-sight (NLOS) localization are discussed.

4.1 INTRODUCTION

Recent advances in wireless sensor networks (WSN) and wireless localization have enabled their applications in a range of systems from autonomous driving and collision avoidance to multirobot collaboration in support of search and rescue operations [1–5]. In collision avoidance, a wireless system implemented on the vehicle bumper is used to locate nearby vehicles. Multiple unmanned vehicles equipped with wireless localization systems may collaborate to accomplish tasks, such as finding earthquake victims in building remains. Firefighters may use localization systems to locate themselves in a burning building. In these systems, a target may communicate with a set of base nodes and use TOA and/or DOA estimation to allow localization [1, 2].

The communication system is a key component in all WSNs and wireless localization systems [6–9]. On the other hand, optimal design of a communications system demands the knowledge of various *channel characteristics* [10–12]. For example, the symbol duration should be selected larger than channel delay spread to avoid intersymbol interference (ISI), which negatively impacts communication system performance. Equalizers implemented at the receiver mitigate ISI effects; however, equalizers increase complexity and energy consumption, thus, reducing the

Handbook of Position Location: Theory, Practice, and Advances, Second Edition.
Edited by S. A. (Reza) Zekavat and R. Michael Buehrer.
© 2019 by the Institute of Electrical and Electronics Engineers, Inc.
Published 2019 by John Wiley & Sons, Inc.
Companion Website: www.wiley.com/go/zekavat/positionlocation2e

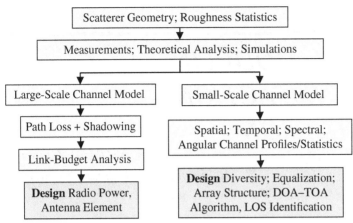

Figure 4.1   Impact of channel on radio design.

battery life of the mobile platform. Another example is proper design of antenna elements based on the channel coherence distance in multi-input–multi-output (MIMO) systems: if coherence distance is small compared to interantenna element spacing, high-performance space-time coding is ensured, while high DOA estimation requires coherent channel across all antenna elements. Figure 4.1 shows the impact of channel modeling on radio system design.

Due to their low cost, small size, ease of installation, and benefits for personnel safety, localization and WSN systems have been found applications in many areas, such as military surveillance, environmental monitoring, disaster area search and rescue operations, intrusion detection, inventory tracking, or supervisory control and data acquisition (SCADA) of remote sensors [9, 13]. For example, sensors distributed on the forest floor collect environmental data to accurately identify extended area fluxes of gases and energy. Another example of autonomous collaborative robots can be found in NASA's plan to set up a lunar colony in which a near-ground WSN is installed on the moon's surface.

In general, radio system parameters should be optimally designed to reflect the propagation environment. The parameters (e.g., frequency and bandwidth); signal processing techniques (e.g., TOA estimation, diversity method, equalization); and the design (e.g., antenna structure) of current communication systems should be revamped to address the challenges of very near-ground channels.

As shown in Figure 4.1, channel parameters are vital for the design of wireless receiver and transmitters in different near-ground WSN and localization applications. Examples of these parameters include:

a) Channel power loss: to design receiver radio frequency (RF) components;

b) Channel delay spread: to select transmission bandwidth, digital modulation, equalization methods, TOA estimation, and frequency diversity);

c) Channel angular spread: to properly implement DOA estimation, beamforming techniques, and angular diversity methods;

d) Channel Doppler spread and channel coherence time: to plan signal processing techniques, the required processing capacity, and time diversity;

e) Channel coherence distance: to properly adjust the antenna spacing for space diversity and/or beamforming;

f) Channel space-time, TOA statistics, and amplitude correlation: to design and evaluate the performance of MIMO systems, and LOS and NLOS identification in localization applications; and

g) Channel power-delay profile (PDP) and power-angle-profile (PAP) information: important for NLOS target localization and environment recognition.

Propagation models will account for several system variables, including, but not limited to:

(1) antenna parameters: (*i*) mutual coupling, (*ii*) antenna spacing, (*iii*) number of antenna elements in MIMO systems, (*iv*) polarization, and (*v*) bandwidth;

(2) frequency and bandwidth,

(3) transmitter and receiver antenna heights,

(4) environment structure (e.g., outdoor versus indoor, urban versus rural), and

(5) terrain profile, including grass, sand, gently/rapidly sloping ground, dense/sparse woods, marshes, water, urban/suburban/rural areas, diurnal water vapor, and temperature effects.

Channel models are widely discussed in the literature [6, 14–16]. Here, we offer a short summary of channel models and parameters critical to localization systems.

4.2   CHANNEL MODEL

As shown in Figure 4.2, a simple model for a channel consists of an impulse response and an additive noise. The channel impulse response is in general time varying. Note that the channel is always subject to change due to the movements. However, within a short time period represented by a parameter called channel coherence time, the channel can be considered time invariant. Accordingly, wireless channels are considered quasistatic. Thus, channel impulse response is represented by a time-invariant model within a short time period.

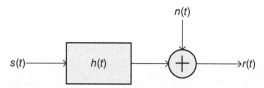

Figure 4.2   Linear time invariant model well represents a channel for a short time period (quasistatic model).

Figure 4.2 represents a model of linear time invariant (LTI) channel. In Figure 4.2, the received signal can be represented in terms of the transmitted signal as:

$$r(t) = s(t) * h(t) + n(t), \tag{4.1}$$

where "$*$" refers to convolution operation.

The receiver of localization systems that estimate DOA is equipped with an antenna array. An antenna array is represented by the array vector $AV(\phi_i)$ (ϕ_i is the DOA of path $i \in \{1, 2, \ldots, N$, N refers to the number of paths) that corresponds to:

$$\overline{AV}(\phi_i) = \left[1, e^{-j2\pi \frac{d \cdot \sin \phi_i}{\lambda}}, \ldots, e^{-j2\pi \frac{(M-1)d \cdot \sin \phi_i}{\lambda}} \right]. \tag{4.2}$$

Here, M is the number of antennas in the array, λ is the wavelength, and d is the distance between antenna elements. The array factor has been discussed in detail in Chapter 9.

Assuming an antenna array at the receiver, the LTI channel impulse response $\vec{h}(t)$ is represented by:

$$\vec{h}(t) = \sum_{i=1}^{N} \overline{AV}(\phi_i) \cdot a_i e^{-j\theta_i} \delta(t - \tau_i). \tag{4.3}$$

In (4.3), a_i refers to the amplitude, θ_i refers to phase, and τ_i refers to the TOA of the signal. The statistics of a_i are very important in many localization systems. If the receiver is not equipped with an antenna array, Equation (4.3) can be simplified to:

$$h(t) = \sum_{i=1}^{N} a_i e^{-j\theta_i} \delta(t - \tau_i). \tag{4.4}$$

A channel is represented by different profiles defined in time, frequency, and angle domains that include: power delay profile (PDP), power spectrum profile (PSP), and power angle profile (PAP). Here, power refers to $E(a_i^2)$; and $E(.)$ refers to expectation. For example, PDP shows how the power changes with time, PSP represents how the power varies with (Doppler) frequency (e.g., due to channel movement). PDP and PSP are defined for any type of antennas. However, PAP is only defined for directional antenna elements. Based on the definition of power in different domains, it is clear that a_i would not only have components in time domain but also in frequency and angle domains. An example of power profile has been sketched in Figure 4.3. The y-axis is the power, while the x-axis could be time (τ), Doppler frequency (v), and azimuth angle (ϕ).

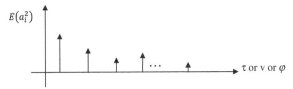

Figure 4.3   Power profile.

Based on the model presented by Fleury in [16], the received signal can be comprehensively represented by a model that well represents the movement of the scatterers (while considering the receiver static) and the DOA of the signal, which corresponds to:

$$r(x;t) = \iiint h(\Omega,\tau,v) e^{j2\pi\Omega \cdot x/\lambda} e^{j2\pi vt} s(t-\tau) d\tau dv d\Omega. \tag{4.5}$$

In (4.5), x represents the position vector of the receiver with respect to the coordinate origin, $s(t)$ is the transmitted signal, v represents the Doppler frequency, and Ω represents the DOA vector in space (a unit vector); that is,

$$\Omega(\varphi,\vartheta) = \langle \cos\varphi \sin\vartheta, \sin\varphi \sin\vartheta, \cos\vartheta \rangle. \tag{4.6}$$

In (4.6), (φ,ϑ) are the azimuth and elevation angle pairs. Moreover, in (4.5):

$$h(\Omega,\tau,v) = \sum_{i=1}^{N} \alpha_i \delta(\tau-\tau_i) \delta(\Omega-\Omega_i) \delta(v-v_i) \tag{4.7}$$

Note that the channel model in (4.7) represents the motion of the scatterers; however, it assumes receiver is static. Here, α_i is the complex amplitude, and N is the number of scatterers. The model in (4.7) is regarded as a comprehensive model for the channel. In Section 4.4, we will use this model to calculate important channel statistics.

4.3   IMPORTANT STATISTICS FOR RECEIVED SIGNAL STRENGTH (RSS)

RSS methods rely on the statistics of the power of impulse response. Considering a fully random arrival at any given time, that is, θ_i is uniform in $[0,2\pi)$, and θ_i is independent of θ_j for $i \neq j$, and θ_i independent of a_j for all values of i and j, using (4.4) the average power of impulse response corresponds to:

$$E(|h(t)|^2) = \sum_{i=1}^{N} E(a_i^2). \tag{4.8}$$

In reality, measurements are used to calculate PDP, PSP, or PAP. Next, PDP, PSP and/or PAP are used to calculate the average power. For example, given PDP, that is, $P_D(\tau)$ over $[0, \tau_{max}]$, where 0 represents the first time of arrival and τ_{max} represents the maximum time over which an arrival is detected, the average power corresponds to:

$$\bar{P}_r = \frac{1}{\tau_{max}} \int_0^{\tau_{max}} P_D(\tau) d\tau. \tag{4.9}$$

Note that $\bar{P}_r$ is also equal to the average power in spectrum $P_s(v)$ and angle $P_A(\varphi)$ profiles, that is:

$$\bar{P}_r = \frac{1}{2v_{max}} \int_{f_o-v_{max}}^{f_o+v_{max}} P_S(v) dv \tag{4.10}$$

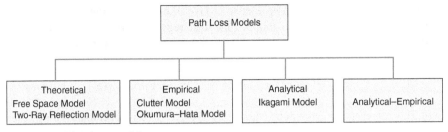

Figure 4.4 Path loss models.

and

$$\bar{P}_r = \frac{1}{2\pi} \int_{-\pi}^{\pi} P_A(\varphi) d\varphi \tag{4.11}$$

In (4.10), f_0 is the center frequency and v_{max} is the maximum Doppler frequency. Models developed for the average received power are also called path loss models. Assuming the average transmitted power corresponds to $\bar{P}_t$, path loss corresponds to:

$$L_{path-loss} = -10 log \bar{P}_r / \bar{P}_t. \tag{4.12}$$

Many path loss models have been proposed, which can be categorized as shown in Figure 4.4. Theoretical models are only based on some theoretical calculations. Examples of theoretical models are:

a. **Free-space model:** a function of the distance of transmitter and receiver d (in km) and the frequency f (in MHz):

$$L_{path-loss}(dB) = 32.45 + 20 log(d_{km}) + 20 log(f_{MHz}). \tag{4.13}$$

This model is an appropriate model for the points close to high-altitude antennas.

b. **Two-ray model:** a function of the distance of transmitter and receiver d (in km), and the height of the receiver h_r (in meters) and the transmitter h_t (in meters):

$$L_{path-loss}(dB) = 40 log(d_{km}) - 20 log(h_{r,meters}) - 20 log(h_{t,meters}) + 120. \tag{4.14}$$

This model assumes that there are only two ways between the transmitter and the receiver. One is the direct way, and the other one is the reflected way from the ground. Therefore, it is also called the ground reflection model. It should be noted that neither the model in (4.13) nor the model represented in (4.14) represent good models for many wireless environments. (4.14) shows that the signal power attenuates with an order of four in the two-ray model compared to free-space model of (4.13). This higher attenuation occurs mostly when the transmitter and receiver are closer to the ground and can be explained via the Fresnel zone. Section 4.6.4 briefly highlights how the Fresnel zone can be used to explain the impact of near-ground propagation on the large-scale channel model.

TABLE 4.1.   Values for the Parameter n in Different Environments

Environment	n
Free-pace Model	2.0
Urban	2.7–3.5
Suburban	3–5
In building	1.6–1.8
Obstructed in building	4–6
Obstructed factories	2–3

Empirical models take into account the measurements in the model. An example of an empirical model is the clutter model, which is the addition of the two-ray model in (4.14) and a constant k that accounts for clutter. The constant is calculated for any given environment based on the measurements made. The constant is calculated to maintain the minimum mean square error (MMSE) between the measured valued and the model. The clutter model corresponds to:

$$L_{path-loss}(dB) = 40\log(d_{Km}) - 20\log(h_{r,meters}) - 20\log(h_{t,meters}) + 120 + k. \quad (4.15)$$

Analytical models consider many parameters in the environment and the geometry of the environment to create a path loss model. An example of an analytical model is Ikagami model.

Finally, the most famous model used for path loss is the analytical-empirical model. In (4.13) we observed the coefficient of 20 (i.e., 2 × 10) was used for the $log(d_{km})$, and in (4.14) we observed the coefficient of 40 (i.e., 4 × 10) was used for $log(d_{km})$. Analytical-empirical models incorporate the general coefficient of $n \times 10$ for the distance. Here, the parameter n is a function of many variables such as frequency, the height of receiver and transmitter and the environment. The relevant model is called log-distance model and corresponds to:

$$L_{dB}(d) = L(d_o) + 10 \cdot n \cdot \log\left(\frac{d}{d_0}\right). \quad (4.16)$$

In (4.16) the distance d_o is a distance in which a priori information of path loss can be created. Typically, it can be considered close enough to the transmitter such that the free space model of (4.13) is used for the computation of this loss.

Note that the value of n in (4.16) is mainly calculated via measurements. Table 4.1 presents typical values for the parameter n. The value of n is computed based on the measurements and using MMSE estimation, that is, minimizing the squared error between the measurements and the model.

Example 4.1   Determination of n in the Log-Distance Model

Assuming log-distance model of (4.16), provide an equation determining the value of n that minimizes the mean square error (MSE) between the predicted loss from log-distance model and measurements. The measurements of the attenuation are summarized in the following table:

Distance	d_0	d_1	...	d_N
Path Loss	L_0	L_1	...	L_N

Solution

The calculated path loss corresponds to:

$$L_{dB}(d) = \underbrace{L_{dB}(d_0) + 10 \cdot n \cdot \log\left(\frac{d}{d_0}\right)}_{Path\ Loss}.$$

Hence, the MSE is:

$$\text{MSE} = \frac{1}{N} \sum_{i=1}^{N} (L_{dB}(d_i) - L_i)^2. \tag{4.17}$$

Replacing $L_{dB}(d)$ with (4.16) and taking $L_{dB}(d_0) = L_0$,

$$MSE = \frac{1}{N} \sum_{i=1}^{N} \left(L_0 - L_i + 10 \cdot n \cdot \log\frac{d_i}{d_0} \right)^2. \tag{4.18}$$

To find the value of n that minimizes the MSE, we differentiate (4.18) and equate it to zero, that is,

$$\frac{dMSE}{dn} = \frac{1}{N} \sum_{i=1}^{N} 10 \cdot \log\frac{d_i}{d_0} \left(L_0 - L_i + 10 \cdot n \cdot \log\frac{d_i}{d_0} \right) = 0.$$

This leads to:

$$n = \frac{\sum_{i=1}^{N} \log\frac{d_i}{d_o}(L_i - L_o)}{10 \sum_{i=1}^{N} \left(\log\left(\frac{d_i}{d_o}\right) \right)^2}. \tag{4.19}$$

Replacing n in (4.18) with the value of n in (4.19) we find the minimum MSE. This minimum MSE indeed represents the variance of error in path loss. This, error is usually due to the shadowing effects. Usually, a random term is added to the deterministic term of path loss defined in (4.14) and (4.16) that takes into account shadowing effects. For example, considering the log-distance path loss model of (4.16), we have:

$$L_{dB}(d) = \underbrace{L_{dB}(d_0) + 10 \cdot n \cdot \log\left(\frac{d}{d_0}\right)}_{Path\ Loss} + \underbrace{\chi(\sigma)}_{Shadowing}. \tag{4.20}$$

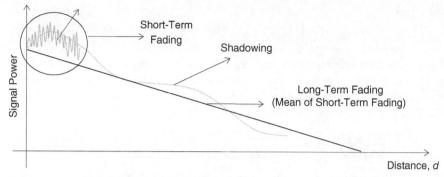

Figure 4.5   Short-term and long-term fading.

In (4.20), σ corresponds to the squared root of MMSE calculated by (4.18). A log-normal statistic is considered for shadowing effects $\chi(\sigma)$.

Path loss plus shadowing is also called *large scale* or *long-term fading* in the context of wireless channel models. Figure 4.5 represents how the amplitude a_i changes with distance. As is observed, the amplitude a_i varies fast. The local average (e.g., over small distances such as 10 m) of a_i is represented by path loss and shadowing. Moreover, the global average of channel models (e.g., over 100 m) is the path loss only. Channel models required for other localization methods are based on the small-scale variations of the amplitude a_i which is also called *small scale* or *short-term fading*.

4.4   IMPORTANT STATISTICS FOR TOA, TDOA, AND DOA

In many TOA estimation methods, a priori information or assumption on the time, spectrum, or spatial statistics of the channel are critical. The first assumption in the process of data acquisition in many signal processing techniques is the independency of samples taken in the time domain. For example, in the multiple signal classification (MUSIC) algorithm used for TOA and DOA estimation (see Chapters 7 and 9) the data can be collected in time, frequency, or space domains should be considered independent. Important statistics of channel can be extracted from the PDP, PSP, or PAP introduced earlier in this chapter.

Important statistics of channels include channel coherence time, coherence frequency, and coherence distance. All these features can be extracted using the comprehensive channel model introduced in (4.5–4.7). The model in (4.5) can be represented by its Fourier transform [16]:

$$r(x;t) = \int S(f)H(x,f,t)e^{j2\pi ft}\,df. \tag{4.21}$$

Here,

$$S(f) = \int s(t)e^{-j2\pi ft}\,dt \tag{4.22}$$

and

$$H(x,f,t) = \iiint h(\Omega,\tau,v) e^{j2\pi\Omega \cdot x/\lambda} e^{j2\pi vt} e^{-j2\pi ft} \, d\Omega \, d\tau \, dv \qquad (4.23)$$

Fourier transform offers a duality between time and frequency. Comparing (4.23) with the definition of Fourier transform, we observe space-angle (x,Ω), frequency-delay (f,τ), and time-Doppler frequency (t,v) duality. Based on this duality:

> **coherence distance** is calculated based on **angular spread,**
>
> **coherence frequency** is calculated based on **delay spread,**
>
> **coherence time** is calculated based on **Doppler spread.**

Assuming wide sense stationary (WSS) in space, frequency, and time domains, the associated correlation corresponds to:

$$R(\Delta x, \Delta f, \Delta t) = E\left[H(x + \Delta x, f + \Delta f, t + \Delta t) H^*(x,f,t)\right]. \qquad (4.24)$$

Here, $H(x,f,t)$ is defined in (4.23). The function in (4.24) is called spatial-spectral-temporal correlation function. From the basics of random process, we also know that the correlation function in (4.24) is the inverse Fourier transform of the joint angular-delay-Doppler density function, $P(\Omega,\tau,v)$, that is:

$$R(\Delta x, \Delta f, \Delta t) = \int e^{j2\pi\Omega \cdot \Delta x/\lambda} e^{j2\pi v \Delta t} e^{-j2\pi \Delta f \tau} P(\Omega,\tau,v) \, d\Omega \, d\tau \, dv. \qquad (4.25)$$

Here, we assume $\int P(\Omega,\tau,v) \, d\Omega \, d\tau \, dv =$. In addition,

$$\int P(\Omega) \, d\Omega = 1, \quad \int P(\tau) \, d\tau = 1, \quad \int P(v) \, dv = 1.$$

Note that $P(\Omega)$, $P(\tau)$, and $P(v)$ correspond to the normalized power PAP, PDP, and PSP, respectively. The nonnormalized power version of these profiles are $P_D(\tau)$, $P_S(v)$, $P_A(\varphi)$ introduced in (4.9–4.11), respectively.

Equation (4.25) is used in the rest of this section to configure the statistics of the channel based on the angle, delay, and Doppler frequency profiles.

4.4.1 PDP Statistics and Impact on Localization and Radio Design

PDP represents the average power and delay relationship, $P_D(\tau)$. It is mainly due to the effects of *multipath* in the environment. Accordingly, the signal is received through multiple paths, each with a different length. This is equivalent to the arrival of the signal at different times. Multipath may lead to destructive addition of the signals which creates *fading effect*. Diversity techniques should be implemented in order to cope with multipath effects. Thus, in general, multipath makes the detection process of signals in wireless environments complicated. Many receivers are forced to implement signal processing techniques to extract the signal through multipath.

Multipath effects are not desirable for localization techniques as well. As discussed in Chapters 7 and 9 of this handbook, in general, as the number of multipath increases, both TOA and DOA estimation techniques will be subject to lower performance. Multipath effects in a wireless environment are characterized by the spread in the time delay created due to the availability of scatterers (buildings, cars, street signs, etc).

Using PDP, we can find the channel delay spread that corresponds to:

$$\tau_{rms} = \sqrt{\frac{\int_0^{\tau_{max}} (\tau - \overline{\tau})^2 P_D(\tau) d\tau}{\int_0^{\tau_{max}} P_D(\tau) d\tau}}. \tag{4.26}$$

Here, τ_{max} refers to the maximum expected TOA. $\overline{\tau}$ is the average TOA, and is:

$$\overline{\tau} = \frac{\int_0^{\tau_{max}} \tau P_D(\tau) d\tau}{\int_0^{\tau_{max}} P_D(\tau) d\tau}. \tag{4.27}$$

Recall that from (4.9):

$$\overline{P}_r = \frac{1}{\tau_{max}} \int_0^{\tau_{max}} P_D(\tau) d\tau. $$

The channel delay spread calculated in (4.26) impacts the receiver and transmitter design. Typically, if the symbol duration of is higher than the channel spread, the *channel is called flat*. In this case, the channel model of (4.8) can be simplified to one single arrival, that is,

$$h(t) = a_o e^{-j\theta_o} \delta(t - \tau_o), \tag{4.28}$$

and a simple receiver can be designed. When delay spread is in the order of or bigger than the transmitted symbol duration, the channel is called *frequency selective*. Frequency selective channels need the implementation of equalizers at the receiver. This makes the receiver structure complex. Therefore, for a given channel with a given delay spread, the symbol duration or the modulation scheme can be properly adjusted to ensure a flat channel. Orthogonal frequency division multiplexing (OFDM) has also been proposed in order to maintain a flat channel across transmissions.

Based on (4.25), coherence frequency is represented by:

$$R(\Delta x, \Delta f, \Delta t)\big|_{\Delta x=0, \Delta t=0} = R(\Delta f) = \int e^{-j2\pi\Delta f\tau} P(\tau) d\tau. \tag{4.29}$$

The coherence frequency is defined as:

$$f_c = \{\min \Delta f > 0 \mid R(\Delta f) = c\}. \tag{4.30}$$

Here, c represents a value that can vary based on the model of PDP, $P_D(\tau)$. The inverse of channel spread represents *coherence frequency*. A rule of thumb equation for coherence frequency in terms of the channel delay spread in (4.26) is:

$$f_c \cong \frac{1}{5\tau_{rms}}. \tag{4.31}$$

The expression of (4.29) in terms of $P_D(\tau)$ ($P(\tau)$ is the normalized power version of $P_D(\tau)$) corresponds to

$$R(\Delta f) = \frac{\displaystyle\int_0^{\tau_{max}} e^{-j2\pi\Delta f\tau} P_D(\tau) d\tau}{\displaystyle\int_0^{\tau_{max}} P_D(\tau) d\tau}.$$

Coherence frequency has important applications in the design of receivers. Some examples are:

1. It is a representation of the frequency band over which the channel is highly correlated. In the design of frequency diversity receivers, the separation of the transmission channels should be kept higher than f_c to ensure independent channels across frequency bands used for transmission.

2. It is a representation of the channel bandwidth. If the transmission bandwidth is higher than f_c, the channel distorts the transmitted signal. In this case, the channel is called frequency selective. That is, the channel allows some frequency components to cross and avoids others. If the transmission bandwidth is lower than f_c, the channel frequency band would be flat over the signal bandwidth and allows the signal to follow without distortion. If the channel behaves as frequency selective over the transmission bandwidth, equalizers should be developed in order to undo the effect of the channel. Equalizers increase the complexity, power consumption, and cost of the receiver. As mentioned, OFDM is a solution for frequency selective channels.

Figure 4.6 represents the impact of the channel on the transmitted signal. As observed in this figure, a flat fading channel does not impact the received signal, while a frequency selective channel renders the received signal distorted. It should be noted that the channel behavior is not an absolute phenomenon. It depends on the transmission bandwidth. If the transmission bandwidth is higher than the channel coherence

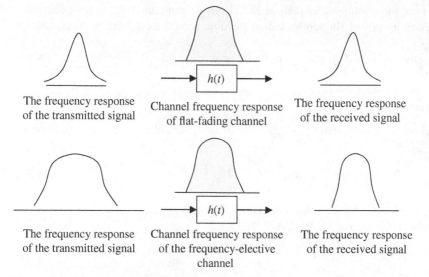

Figure 4.6   Flat (top) channel versus frequency-selective (bottom) one.

frequency, then the channel acts as a frequency selective channel. As shown in Figure 4.6, the channel remains unchanged. But for one signal bandwidth it is flat and for another, it is frequency selective.

Example 4.2   Generation of Equal Power-Correlated Gaussian and Rayleigh Random Variables

In many simulations for wireless communication and localization systems, it is required to generate colored noise and equal power correlated fade amplitudes. An example is in multicarrier systems such as OFDM when fade amplitudes across multiple carriers are correlated. Another example is the correlation across fade amplitudes of arrivals at different time frames (time correlation) based on the coherence time discussed. In this example, we offer a simple technique to generate correlated fades.

First use the command `randn` in MATLAB to generate 1000 sets of four equal power-correlated zero mean Gaussian random vector $X = [x_1, x_2, \ldots, x_n]^T$ (here, T refers to transpose), where each variable has unity power, that is, $E[x_i x_i^*] = 1$. In this case, the correlation coefficient matrix is $K_{XX} = E[XX^H]$; H refers to the Hermitian. Assume that K_{XX} is positive semidefinite, that is,

$$a^H K_{XX} a \geq 0, a \in \mathbb{C}^n. \tag{4.32}$$

Assume the matrix elements of K_{XX} have the parameters $(\rho_g)_{ij}$ that are created based on the Rayleigh autocorrelation function of:

$$(\rho_r)_{ij} = \frac{1}{1 + \left(\dfrac{i-j}{8}\right)^2}. \tag{4.33}$$

Next, use the command `randn` in MATLAB to generate 1000 sets of four correlated unity power Rayleigh random variable with the correlation coefficient of (4.33).

Draw the histograms (and their envelopes) of all of generated random variables.

Solution

Based on the assumption of (4.32), we can apply singular value decomposition to K_{XX}:

$$K_{XX} = \left(U\Lambda^{1/2}\right)\left(U\Lambda^{1/2}\right)^H. \tag{4.34}$$

Here, $UU^H = I_n$, I_n is an identity matrix. Cholesky factorization can be used to decompose the matrix K_{XX} to $U\Lambda^{1/2}$. Here,

$$R = U\Lambda^{1/2} \tag{4.35}$$

is called the Cholesk factorization of K_{XX}. In MATLAB, the command `chol` can be used to decompose the autocorrelation created by (4.33). The creation of autocorrelation of Gaussian using the autocorrelation of Rayleigh is discussed in this solution.

Now, if we use `randn` in MATLAB to generate $W = [w_1, w_2, ..., w_n]^T$ with zero-mean equal power-independent Gaussian random elements, that is, $E(WW^H) = \sigma^2 I_n; \sigma^2 = E(w_i w_i^H)$, we can create correlated Gaussian noise by applying (4.35), that is,

$$Z = RW. \tag{4.36}$$

In this case,

$$E(ZZ^H) = E(RW(RW)^H) = RE(WW^H)R^H = \sigma^2 RI_n R^H = \sigma^2 K_{XX}.$$

Now, to create colored noise, we use an available Gaussian autocorrelation (as discussed later in this solution, the autocorrelation of Gaussian can be generated from the autocorrelation of Rayleigh) and apply Cholesky factorization to create R in (4.35). Next, we generate four independent Gaussian random variables using `randn` to create W in (4.36). Finally, we use (4.36) to create the colored noise Z. The flow chart for creating colored noise is shown in Figure 4.7. The results are sketched in Figure 4.8a. We generate $N = 1000$ samples in order to create the histogram (probability density function (PDF)) of each random variable. See the MATLAB codes in the file Chapter_4_Example_2.m.

Now, we discuss the generation of correlated Rayleigh. Here, we assume that $G = [g_1, g_2, ..., g_n]^T$, $g_i, i \in \{1, 2, ..., n\}$, are correlated unity power complex Gaussian random variables, that is,

$$g_i = x_i + jy_i.$$

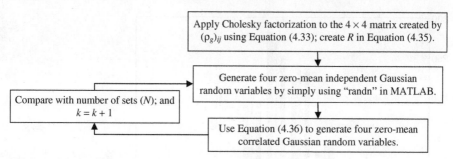

Figure 4.7   Flow chart of the generation of N sets of correlated Gaussian (colored noise) generation.

Here, x_i and y_i are two independent zero-mean Gaussian random variables, and:

$$E(x_i)^2 = E(y_i)^2 = 1/2$$

Now, the amplitude of g_i,

$$r_i = |g_i| = \sqrt{x_i^2 + y_i^2},$$ (4.37)

is a unity power Rayleigh random variable, that is,

$$E(r_i)^2 = 1.$$

The authors in [17] have maintained a mathematical relationship between the correlation coefficient of Rayleigh amplitudes $(\rho_r)_{ij}$ and the correlation coefficient of Gaussian random variable $(\rho_g)_{ij}$. However, in many applications in wireless systems, the autocorrelation of Rayleigh is given, and we need to generate correlated Rayleigh random variables. For example, in [18], a relationship similar to (4.33) is maintained between the fade amplitude, a, at two different frequencies f_i, f_j. That is,

$$(\rho_r)_{ij} = E\left(a(f_i)a(f_j)\right) = \frac{1}{1 + \left(\dfrac{f_i - f_j}{f_c}\right)^2}.$$ (4.38)

Here, f_c is to the coherence frequency introduced in (4.31). Now, if $f_i - f_j = (i - j)\Delta f$, Δf is the frequency deviation of two subcarriers in OFDM systems, and $\dfrac{f_c}{\Delta f} = 8$, (i.e., the channel is highly correlated over eight consecutive subcarriers), then (4.33) is created.

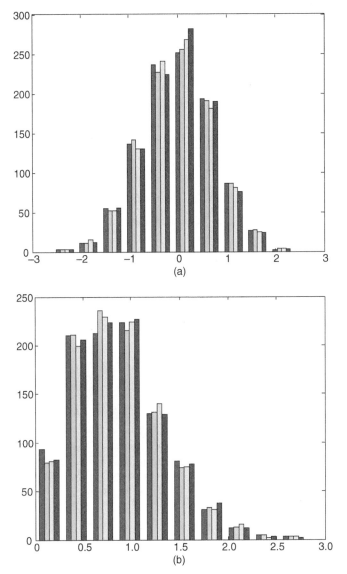

Figure 4.8   Generated correlated (a) Gaussian and (b) Rayleigh random variables.

In other words, given a relationship between the Rayleigh amplitudes, such as the one given in (4.33) (see [18]), we intend to generate the corresponding Gaussian variables. Then we use correlated Gaussian random variables to create correlated Rayleigh random variable. Using the equation in [17], we have generated the following equation to maintain the relationship of correlated Rayleigh and correlated Gaussian:

$$(\rho_g)_{ij} = 1.615(\rho_r)_{ij}^3 - 3.171(\rho_r)_{ij}^2 + 2.494(\rho_r)_{ij} + 0.062 . \tag{4.39}$$

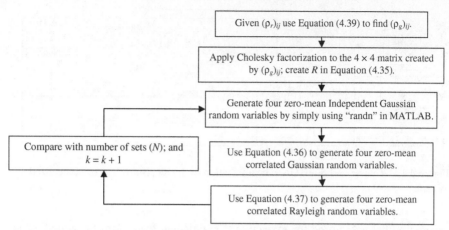

Figure 4.9   The flowchart of correlated Rayleigh random variables.

Therefore, the flowchart for the generation of correlated Rayleigh corresponds to Figure 4.9. The results are shown in Figure 4.8b. See Chapter _4_Example_2.m file for the MATLAB codes.

4.4.2 PSP Statistics and Impact on Localization and Radio Design

PSP represents the relationship of average power and frequency. This is mainly due to the movement and variation of the channel. The variation of the channel leads to Doppler effects. Accordingly, the central frequency of the received signal is deviated. Doppler effects complicate the process of synchronization and phase tracking. As discussed in Part II, synchronization is an essential component of TOA and DOA methods. Figure 4.10 represents a typical receiver used in wireless local positioning systems (WLPS; see Chapter 34). As shown in Figure 4.10, synchronization is applied in two stages: coarse synchronization in the RF front-end, and fine synchronization in the baseband.

PSP statistics are very critical for the data acquisition process for DOA and TOA techniques. Channel PSP represents how fast the channel varies. In many TOA and DOA estimation methods, it is essential to obtain independent observations. If a channel varies slowly, the data acquisition process may last for longer time periods. Thus, while channel variation is not desirable for receiver design and synchronization, it is quite supportive for the data acquisition process.

Another example is in LOS identification techniques that is detailed in Chapter 16. Here, it is very important to extract the desired statistics, such as the statistics of signal amplitude. In order to compute the signal statistical mean via sample mean, samples should be independent as well. Thus, the channel should vary fast enough to ensure the independence assumption.

It should also be noted that in the process of sampling, it is always assumed that the channel stays stationary. Stationarity is an important assumption for

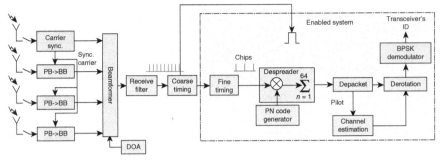

Figure 4.10 Coarse and fine TOA synchronization in the receiver of a typical localization system.

ergodicity that is required for extracting the statistical features of the signal using its samples. The assumption of stationarity is very critical for many DOA and TOA techniques (e.g., MUSIC) when it is required to compute the sample autocovariance matrix of the observed signal. As discussed in Chapter 9, if stationary property does not hold, the error in autocovariance matrix estimation created via samples would be high. This error reduces the performance of DOA and TOA estimation.

The transmission scheme of the signal may violate the assumption of stationarity across multiple symbols arrived in one transmission. For example, in WLPS (see Chapters 1, 9 and 34), the periodic transmissions of signals do not support the stationarity. Therefore, as discussed in Chapter 9, MUSIC-based DOA estimation would involve serious errors. However, in Chapters 9 and 34 we have shown that the periodic nature of WLPS signals creates a cyclostationary process. This cyclostationary process maintains stationarity across the same symbol of multiple transmissions. Thus, in place of calculating the sample autocorrelation based on the symbols within one transmission (package), the same symbol across multiple transmissions (packages) is used.

PSP represents the variation of average power with frequency, $P_B(v)$, that is used to calculate the frequency spread (also called Doppler spread), B_{rms}; that is:

$$B_{rms} = \sqrt{\frac{\int_{f_o - v_{max}}^{f_o + v_{max}} (v - \bar{B})^2 P_B(v) dv}{\int_{f_o - v_{max}}^{f_o + v_{max}} P_B(v) dv}}. \tag{4.40}$$

Here, f_o is the center frequency and v_{max} is the maximum expected Doppler frequency calculated based on the maximum speed in the environment. In addition, in (4.40)

$$\bar{B} = \frac{\int_{f_o - f_{max}}^{f_o + v_{max}} v P_B(v) dv}{\int_{f_o - v_{max}}^{f_o + v_{max}} P_B(v) dv}. \tag{4.41}$$

Based on (4.25), coherence time is represented by:

$$R(\Delta x, \Delta f, \Delta t)\big|_{\Delta x=0, \Delta f=0} = R(\Delta t) = \int e^{j2\pi v \Delta t} P(v) dv. \tag{4.42}$$

The coherence time is defined as:

$$T_c = \{\min \Delta t > 0 \mid R(\Delta t) = c\}. \tag{4.43}$$

Here, c represents a value that can vary based on the model of $P(v)$.

The coherence time refers to the time over which the channel parameters are highly correlated and corresponds to:

$$T_c \cong \frac{0.423}{B_{rms}}. \tag{4.44}$$

Equation (4.42) can also be written in terms of $P_B(v)$, that is,

$$R(\Delta t) = \frac{\displaystyle\int_{f_o - f_{max}}^{f_o + v_{max}} e^{j2\pi v \Delta t} P_B(v) dv}{\displaystyle\int_{f_o - f_{max}}^{f_o + v_{max}} P_B(v) dv}.$$

Recall that $P(v)$ is the normalized version of $P_B(v)$.

If the channel symbol duration is much less than the channel coherence time, channel fading remains unchanged over symbol duration. In that case, the channel is called *slow fading*. Usually, the channel stays unchanged over multiple symbol durations. In that case, channel specifications can be tracked and estimated. This is specifically very critical for the synchronization process discussed in Figure 4.10.

If symbol duration is in the order of (or smaller than) channel coherence time, channel is called *fast fading*. The channel model of time varying channels corresponds to:

$$h(t;\tau) = \sum_{i=1}^{N(t)} a_i(t) e^{-j\theta_i(t)} \delta(\tau - \tau_i(t)). \tag{4.45}$$

Based on the model presented by Fleury in (4.6–4.7), a more comprehensive time-varying channel model that well represents the movement of both the scatterers and the receiver, and the DOA of the signal, can be well represented by:

$$h(t;\Omega,\tau,v) = \sum_{i=1}^{N(t)} a_i(t) e^{-j\theta_i(t)} \delta(\tau - \tau_i(t)) \delta(\Omega - \Omega_i(t)) \delta(v - v_i(t)). \tag{4.46}$$

Based on (4.46) and consistent with (4.5), the received signal corresponds to:

$$r(x;t) = \iiint h(t;\Omega,\tau,v)e^{j2\pi\Omega \cdot x(t)/\lambda}e^{j2\pi vt}s(t-\tau)\,d\tau\,dv\,d\Omega. \qquad (4.47)$$

In a fast-fading channel, the channel parameters should be estimated with a faster pace. In other words, a faster synchronization rate is required. This makes the process of frequency and phase tracking complex. It also increases the power consumption.

As discussed above, there are cons associated to fast-fading channels. The pros of fast-fading channels follow. In order to implement time diversity methods, a signal should be retransmitted after a duration that is larger than T_c in (4.44). Now, if T_c is small enough, then time diversity can be implemented. If channel coherence time is smaller than the symbol duration, time diversity can be implemented even within the duration of symbol. However, as mentioned, in wireless channels coherence time is usually much larger than the symbol time. Techniques that enable time diversity within symbol duration by creating a virtual fast-fading channel have been proposed. An example, beam pattern scanning or beam pattern oscillation, has been proposed in [19–25].

In beam pattern oscillation technique, a base station antenna array sweeps the beam pattern directed to the mobile such that at all times, the intended mobile lies within the half-power beam width (HPBW) of the antenna pattern; the beam pattern moves to create L independent fades within each symbol duration T_S; this leads to large performance benefits due to L-fold diversity gains; after each T_S the antenna beam returns to its initial position and sweeps the same area of space over T_S (leading to an oscillating antenna pattern and easing parameter estimation); the movement of the beam pattern, as a percentage of HPBW, is small, thereby allowing the beam pattern to maintain directionality. Figure 4.11 represents the proposed technique. Antennas, antenna arrays, and their parameters such as beam pattern are discussed in detail in Chapter 9.

In the antenna beam pattern oscillation technique, two categories of channel coherence time are defined: (1) the real channel coherence time, T_C, when the proposed technique is not in place; and (2) the virtual channel coherence time, T_{CB}, when the proposed method is in place. Note that $T_{CB} \ll T_C$; thus, T_{CB} is dominant when beam pattern oscillation is in place. Now, the question is: How can we estimate channel parameters?

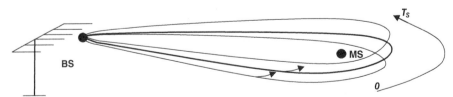

Figure 4.11   Antenna beam pattern scanning allows time diversity within each symbol duration T_S.

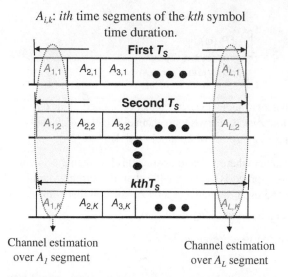

Figure 4.12   Channel estimation in beam pattern scanning method.

Usually, in wireless systems channel parameters are estimated assuming a slow-fading channel, that is, no change in the channel behavior within T_C. A similar concept is applied to estimate a channel in the proposed beam pattern scanning system. Figure 4.12 explains the channel estimation process. Because $T_C \gg T_s$, and due to the cyclic nature of beam pattern motion, the channel fades are identical in the consecutive segments $A_{l,k}$ and $A_{l,k+1}$, $l \in \{1, 2, ..., L\}$, $k \in \{1, 2, ..., K\}$. The number of symbols, K, used for channel estimation should meet: $K < T_C / T_s$. This enables fading parameters (e.g., phases), in each segment of T_s to be tracked, permitting coherent detection.

4.4.3   PAP Statistics and Impact on Localization and Radio Design

The PAP is mainly used to represent the effect of directional antennas on the localization. As the angular spread increases, the directional antennas and particularly antenna arrays would be able to receive signals from multiple angles. This specification can be used to create angular diversity at the receiver. In addition, angular dispersion impacts spatial correlation across antenna arrays. Spatial correlation is required to design antenna arrays and their spacing in order to ensure coherent channel across antenna arrays whenever antennas are used to create beamforming and independent channels whenever antennas are used to create space diversity.

The PAP is represented by $P_A(\varphi)$ and used to determine the angular spread, which corresponds to:

$$\varphi_{rms} = \sqrt{\frac{\int_{-\pi}^{\pi} (\varphi - \overline{\varphi})^2 P_A(\varphi) d\varphi}{\int_{-\pi}^{\pi} P_A(\varphi) d\varphi}}. \tag{4.48}$$

Here,

$$\bar{\varphi} = \frac{\int_{-\pi}^{\pi} \varphi P_A(\varphi) d\varphi}{\int_{-\pi}^{\pi} P_A(\varphi) d\varphi}. \tag{4.49}$$

If φ_{rms} is less than HPBW of the antenna, angle diversity cannot be applied. If the channel is also flat, the channel is called *low rank*. Otherwise the channel is called *high rank*.

Based on PAP, two different categories of channels are defined [26]:

a. **Low-Rank Channel:** a channel where its angular spread is less than the antenna HPBW or the channel delay spread is much smaller than the symbol duration. Low-rank channels cannot be used to create angular or spectral diversity.

b. **High Rank Channel:** a channel that is not low rank.

Based on (4.25), coherence distance is represented by:

$$R(\Delta x, \Delta f, \Delta t)\big|_{\Delta f=0, \Delta t=0} = R(\Delta x) = \int e^{j2\pi\Omega\cdot\Delta x/\lambda} P(\Omega) d\Omega. \tag{4.50}$$

Here, assuming the power profile is only defined in azimuth angle, we can rewrite (4.29) as:

$$R(\Delta x) = \frac{\int e^{j2\pi\Delta x\cos\varphi/\lambda} P_A(\varphi) d\varphi}{\int_{-\pi}^{\pi} P_A(\varphi) d\varphi}. \tag{4.51}$$

The coherence distance is defined as:

$$D_c = \{\min \Delta x > 0 \mid R(\Delta x) = c\}. \tag{4.52}$$

Here, c represents a value that can vary based on the model of $P_A(\varphi)$.

Different PAP models have been proposed that include uniform, Gaussian [27], Laplacian [28], Secan square [29], and Von Mises [30]. Each of these models is appropriate for different urban and indoor areas. Using each of these models, different values for the coherence distance is obtained.

4.5   SUMMARY OF DIFFERENT CHANNEL CATEGORIES

Based on PDP and PSP, four different channel categories can be defined. Defining:

HPBW:	Half-power beam width of Antenna Array
τ_{rms} :	Delay spread (see (4.26))
T_c :	Coherence time (see (4.44))
T_s :	Symbol duration
f_c :	Coherence bandwidth (see (4.31))
B_{rms} :	Doppler spread (see (4.40))
B_s :	Signal bandwidth

TABLE 4.2. **Definition of Different Categories of Channels**

Channel Category	Definition in Time Domain	Definition in Frequency Domain
Flat slow-fading channel	$\tau_{rms} < T_s < T_c$	$B_{rms} < B_s < f_c$
Flat fast-fading channel	$\tau_{rms} < T_s$ and $T_c < T_s$	$B_s < f_c$ and $B_{rms} > B_s$
Frequency-selective slow-fading channel	$\tau_{rms} > T_s$ and $T_c > T_s$	$B_s > f_c$ and $B_{rms} < B_s$
Frequency-selective fast-fading channel	$\tau_{rms} > T_s > T_c$	$B_{rms} < B_s < f_c$

Table 4.2 summarizes the definition of different channel categories. In addition, based on PDP and PAP, two categories of channels are defined:

1. Low Rank: $\tau_{rms} < T_s$ and $\varphi_{rms} < HPBW$
2. High Rank: $\tau_{rms} > T_s$ and/or $\varphi_{rms} > HPBW$

4.6   STATISTICS OF AMPLITUDE, PHASE, AND TOA

Here, we discuss the statistics of important parameters in the channel model of (4.3).

Fade Amplitude

One of the most common statistics in a NLOS transmission for the fade amplitude is the Rayleigh distribution. The received signal is an addition of a number of signals from different paths. This addition leads to a Rayleigh distribution. The PDF for the NLOS Rayleigh function corresponds to:

$$f(a) = \frac{a}{\sigma^2} exp\left\{-\frac{a^2}{2\sigma^2}\right\}, \tag{4.53}$$

where a is the Rayleigh random variable. If LOS is available between the transmitter and the receiver; the fade would be a combination of deterministic amplitude due to LOS signal and random amplitude due to NLOS reception. The PDF of this combination follows Rician distribution. The PDF for the Rician function corresponds to:

$$f(a) = \frac{q}{\sigma^2} exp\left\{-\frac{a^2 + \gamma^2}{2\sigma^2}\right\} I_o\left(\frac{a\gamma}{\sigma^2}\right), \tag{4.54}$$

where:

a = Rician random variable
γ = LOS amplitude (deterministic)
I_o = Zero-order modified Bessel function of the first kind, defined as

$$I_o(x) = \int_{-\pi}^{\pi} e^{-x cos\theta}\, d\theta.$$

The power of LOS corresponds to $\gamma^2/2$ and the power NLOS is σ^2. The total average (including LOS and NLOS) power is:

$$\bar{P} = \sigma^2 + \gamma^2/2. \tag{4.55}$$

Here, the k-factor corresponds to:

$$k_{dB} = 10 \, log \frac{\gamma^2}{2\sigma^2}. \tag{4.56}$$

It can be observed that when the LOS amplitude γ approaches 0, Equation (4.54) reduces to (4.53). In other words, the Rician distribution is converted into a Rayleigh distribution. The k-factor has applications in LOS identification methods (see Chapter 16 for details). Many other statistical models for the received signal amplitude, such as Nakagami [15], have been introduced in the literature as well.

Fade Phase Statistics

The statistics of the phase is simply considered uniform within $[0,2\pi)$. A simple way to represent the reason for uniform statistics is as follows: if we transmit a signal $A cos \omega t$ and the distance of the transmitter and receiver is L, and assuming free space only, the received signal would correspond to:

$$A cos(\omega(t - \tau)).$$

Here, $\tau = \dfrac{L}{C}$ refers to the delay in the reception of the signal at the receiver due to the distance L. Thus, the received signal corresponds to:

$$A cos\left(\omega\left(t - \frac{L}{C}\right)\right) = A cos\left(\omega t - \frac{2\pi L}{\lambda}\right), \tag{4.57}$$

Where λ refers to the wavelength of the transmitted signal. Now, considering a frequency of 3 GHz, the wavelength would be in the order of 10 cm. Note that even small variations in the order of 10 cm in the distance L leads to the phase rotation in the order of 2π. Therefore, intuitively speaking, the phase cannot be determined, and a uniform PDF can be a good statistic for the phase.

The uniform phase statistics can also be evaluated mathematically using the fact that the received signal is usually an addition of multipath signals, each with independent phases and amplitudes. In other words, assuming $A cos \omega t$ is the transmitted signal, the received signal corresponds to:

$$Re\left(\sum_{i=1}^{N} B_{1i} cos \omega_i t + j B_{2i} sin \omega_i t\right).$$

Assuming:

$$\sum_{i=1}^{N} B_{1i} \cos\omega_i t + j B_{2i} \sin\omega_i t = X + jY .$$

It can be depicted that the real part X and the imaginary part Y are two independent Gaussian random variables. Accordingly, it can be mathematically shown that the amplitude of $X + jY$ would be Rayleigh (or Rician) and the phase would be uniform.

TOA

Signal TOA generally follows the Poisson process. In this case, the probability of making n observations within a time range of Δt is:

$$P(n \text{ observations within } \Delta t) = e^{-\lambda \cdot \Delta t} \cdot \frac{(\lambda \cdot \Delta t)^n}{n!}. \tag{4.58}$$

However, in some environments, it might be explained via the modified or group Poisson process. In a modified Poisson process, we assume that the incoming signal arrives at the mobile in groups at different time delays. Examples of these environments are clustered housing complexes in suburban areas. This clustered delay results in the PDP shown in Figure 4.13.

As shown in Figure 4.13, in the clustered Poisson, if there is an observation within a time period, the probability of observing more arrivals is higher. In addition, if we do not observe an arrival within a period of time, the probability of observing more arrivals in the next period is lower.

For a modified Poisson process, the parameter λ is replaced with $\lambda(t)$. In this case, if an observation is made within a given time period, $\lambda(t) = \lambda_2$ and if an observation is not made, $\lambda(t)$ is changed to $\lambda_1 < \lambda_2$. For a typical modified Poisson, the reader is referred to Suzuki [31] and Hashemi [15].

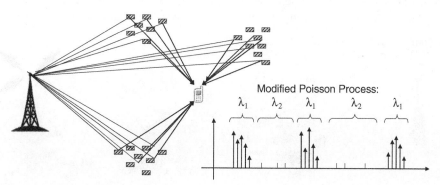

Figure 4.13   Modified Poisson process.

4.7 OTHER CHANNEL MODELS

4.7.1 Geometric-Based Single Bounce Statistical Channel Modeling

This is an important channel modeling technique method widely used in the literature to find statistical representations of channels including DOA, and TOA, and PDF. In this model, it is assumed that there is only one reflection (single bounce) between a transmitter and receiver. It is also usually assumed that the position of reflectors follows a statistical distribution. An example is uniform distribution of scatterers in an environment.

As detailed in Chapter 16, this model is also considered for NLOS localization and identification of mobiles. In this model, it is assumed that there is only one reflector between the transmitter and receiver. The geometry and the position of reflectors or the statistics of the position of reflectors is considered in this channel modeling technique. An example of this channel modeling technique has been used in [23] to model the channel coherence time considering the oscillating beam technique explained in Section 4.4.2.

4.7.2 Circular and Elliptical Geometrical Models

Two well-known geometrical models that have been used in the literature are circular and elliptical. As shown in Figure 4.14, a circular channel model is used for the scenario that the height of the transmitter is much higher than that of the receiver. This model is appropriate for macrocells (e.g., in rural areas) when the height of base station is very high. In this case, we assume the receiver is located in the center of circle, and the scatterers are located on the edge of the circle. This modeling has been used by Ertel [32] to calculate DOA and TOA distributions.

As shown in Figure 4.14, elliptical channel models are used for the scenario that the height of both transmitter and receiver are equivalently low. This model is very useful in downtown areas and for wireless sensor networks, and in microcells when the height of base stations is low. In this case, it is assumed that both transmitter and receiver are located at the focal points of an ellipse, and the scatterers are located at the boundary of the ellipse.

4.7.3 Rough Surface Channel Modeling

This modeling is very critical for measuring the channel coherence distance. The authors in [33] have already investigated the impact of channel rough surface. In

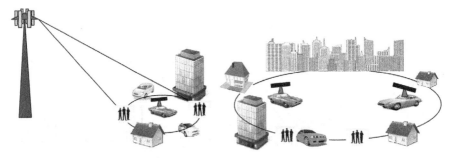

Figure 4.14   Circular (left) and elliptical (right) channel models.

general, a surface is called rough if the variation of the height of roughness, Δh, is in the order of the transmission wavelength, λ. More specifically, the ground is considered rough if:

$$\Delta h > \frac{\lambda}{32}. \tag{4.59}$$

Usually, irregularity height is considered a random variable. A zero-mean Gaussian random with the variance of σ_G^2 well represents the statistics of this random variable, which corresponds to:

$$f_G(h) = \frac{1}{\sqrt{2\pi\sigma_G^2}} e^{-h^2/2\sigma_G^2}. \tag{4.60}$$

Usually, the height of the ground in different positions are correlated with the correlation distance Lc is assumed, based on which,

$$C_h(l) = e^{l^2/L_c^2}. \tag{4.61}$$

Here, L_c is called the correlation length. It represents the distance of two points on the ground, for which the heights of antennas are highly correlated. If the variations of the roughness are very high, L_c would be very small (or $L_c \to 0$). In this case, it is also expected that the roughness is very high as well. If the variations of the roughness are very low, then L_c would be very large (or $L_c \to \infty$). In this case, it is also expected that the roughness is very low.

Therefore, we can conclude that the roughness can be represented by two measures:

a. the variations in the roughness that is measured by (4.59) and

b. the coherence distance L_c.

Typically, we may compare L_c with the wavelength λ, and conclude that the ground surface is rough if:

$$\Delta h > \frac{\lambda}{32} \text{ and } L_c < 5\lambda. \tag{4.62}$$

4.7.4 Near-Ground Channel Modeling

Comparing (4.14) and (4.13), we highlighted the impact of ground on the attenuation. We also explained that the attenuation due to ground effect can be explained via Fresnel zone. Near-ground channels are key components of many applications, such as autonomous driving when localization sensors are installed on the vehicle (specifically their bumpers). See Chapter 35 for details.

The conditions that should be met to call a channel near ground have not been clearly defined in the literature. One definition is based on the Fresnel zone. The Fresnel zones are represented by ellipsoids with their foci at transmitter and receiver. The main axis of this ellipsoid is represented by the line connecting the transmitter and the receiver, which is the LOS. As shown in Figure 4.15, the intersections of these Fresnel zones with a plane perpendicular to the LOS are concentric circles. The radius of each circle is approximated by [34, 59]:

$$r_n = \sqrt{\frac{n\lambda d_1 d_2}{d_1 + d_2}}, \tag{4.63}$$

where d_1 and d_2 are the distances of the plane from transmitter and receiver, respectively, λ is the wavelength of the transmitted wave, and r_n is the radius of nth circle. The approximation is valid for $d_1, d_2 \gg r_n$. The radii of the circles depend on the location of the plane and reach their maximum of $r_{n,max} = \sqrt{n\lambda d}/2$ when the plane is midway between the terminals where $d = d_1 + d_2$. The radius of the first circle is shown in Figure 4.15. A considerable portion of wave energy is concentrated in the first Fresnel zone. A lower percentage of energy is concentrated between zone 1 and zone 2 and and an even lower one between 2 and 3, etc. Fresnel zones have been frequently used in order to investigate the effect of tall buildings and the power received by radios located at the shadow of tall buildings.

Given that the first Fresnel zone includes most of the energy, sometimes it is used to divide the LOS into two regions of near and far, represented by the break distance (BD) denoted here with the parameter d_B. In the near region, mean signal attenuation is almost equivalent to free-space wave-front spreading loss, whereas beyond d_B obstruction of the first Fresnel zone also contributes to attenuation loss, which results in a steeper falloff rate of the signal strength [59]. Assuming $h_t, h_r \gg \lambda$, we have [59]:

$$d_B \approx \frac{4h_t h_r}{\lambda}. \tag{4.64}$$

We also define a critical distance, d_C, at which 57% of the first Fresnel circle is obstructed by ground. In this case, the diffraction loss should be included. Assuming $h_t, h_r \gg \lambda$, we have:

$$d_C \approx \frac{12.5h_t h_r}{\lambda}. \tag{4.65}$$

Considering these distances, the near-ground path loss model in dB is summarized as [59]:

$$L_{NG}(dB) = \begin{cases} L_{fs}, & d \leq d_B \\ L_{fs} + L_{ex}, & d_B \leq d \leq d_C, \\ L_{fs} + L_{ex} + L_{ke}, & d \geq d_C \end{cases} \tag{4.66}$$

where

$$L_{fs}(dB) = -27.56 + 20\log_{10}(f) + 20\log_{10}(d). \tag{4.67}$$

$$L_{ex}(dB) = 20\log_{10}\left|1 + \frac{l_d}{l_r} R_\alpha^{eff} e^{(-j\Delta\varphi)}\right| \tag{4.68}$$

$$L_{ke}(dB) = 20\log_{10}\left(0.5 + \frac{0.877(h_t + h_r)}{\sqrt{\lambda d}}\right). \tag{4.69}$$

Here, f is in MHz and d, h_t, h_r, and λ are all in meters. In addition, in (4.68), $l_r = \frac{25h_t h_r}{2\lambda} + 0.08\lambda$, $l_d = \frac{25h_t h_r}{2\lambda} - 0.08\lambda$, and R_α^{eff} is the effective reflection coefficient where $\alpha = v, h$ denotes the vertical or horizontal incident polarization, respectively [62, 63]. Moreover,

$$\Delta\varphi = \frac{\pi}{2}v^2.$$ (4.70)

Here, v is the dimensionless Fresnel–Kirchhoff diffraction parameter that corresponds to [65]:

$$v = h\sqrt{\frac{2d}{\lambda d_1 d_2}} = \frac{h}{r_n}\sqrt{2n}.$$ (4.71)

In (4.71), h is the obstruction height and h / r_n is the Fresnel zone clearance. Chapter 35 details the derivation of these parameters.

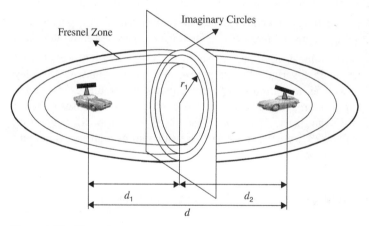

Figure 4.15   Fresnel zone.

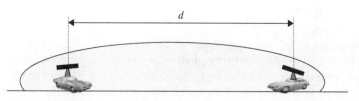

Figure 4.16   Obstructed Fresnel zone by the ground.

4.7.5   Foliage Effects

Loss of the waveform via absorption by trees and vegetation can be substantial. This loss, commonly referred to as the *foliage factor,* should be accounted for in deploying communications systems. The models introduced to calculate foliage factor focus mainly on the path loss created by foliage. These models provide an estimation of the additional attenuation due to foliage that is within the LOS path. There are a variety of different models and a wide variation in foliage types. In general, foliage loss is mathematically modeled as:

$$L(dB) = Af^{\alpha}d^{\beta}. \tag{4.72}$$

Here, f is the frequency and d is the foliage depth along LOS in meters. Parameters A, α, and β are dependent on foliage type and density. These parameters can be evaluated empirically. Table 4.3 summarizes different foliage models based on previous measurements taken by various researchers.

Investigation of the impact of foliage on near-ground propagation has depicted that [39], [40]:

  (i) The path loss exponent increases with increasing forest density as a result of the increased number of obstructions between transmitter and receiver;

 (ii) use of omnidirectional antennas resulted in a lower path-loss exponent, and was able to capture more energy, but suffered from reduced overall range; and

(iii) a significant amount of multipath exists in forest environments, evidenced by as much as 35:1 variation in the minimum and maximum RMS delay spreads as well as the minimum and maximum maximum excess delay spreads. Furthermore, fading analysis demonstrated that small-scale fading followed a Ricean distribution, with k-factors of at least 8 dB, suggesting that even in NLOS scenarios the receiver still captures a strong, dominant multipath signal.

4.8   INHOMOGENEOUS MEDIA CHANNEL MODELING

Inhomogenous refers to dispersive media with spatially or temporally varying electromagnetic (EM) properties. In inhomogenous channels, each node might be placed in an electromagnetically diverse medium. There are many challenging tasks in inhomogenous media, including localization, characterization, layer thickness estimation, irregularity detection, high-resolution TOA measurements and straight-line range estimation [47, 48]. Chapter 9 discusses TOA estimation techniques in inhomogeneous environments.

Emerging applications of inhomogeneous media include endoscopy capsule localization, cancer detection, drug delivery, underwater monitoring/surveillance, and underground or mine safety and monitoring. It is critical to many of these applications to find the location information of measured data or observed events.

TABLE 4.3. Summary of Foliage Models

Model	Expression	Notes
Weissberger's [41]	$L_{fo}(dB) = \begin{cases} 0.45 f^{0.284} d_{fo}^{1}, & 0 < d_{fo} \leq 14\,m \\ 1.33 f^{0.284} d_{fo}^{0.588}, & 14\,m < d_{fo} \leq 1400\,m \end{cases}$	f in GHZ; d_{fo} in meters; $0.230\,GHz \leq f \leq 95\,GHz$ for dense, dry, in leaf trees
COST 235 [45]	$L_{fo}(dB) = \begin{cases} 26.6 f^{-0.2} d_{fo}^{0.5}, & out\ of\ leaf \\ 15.6 f^{-0.009} d_{fo}^{0.26}, & in\ leaf \end{cases}$	f in MHZ; d_{fo} in meters; $9.6\,GHz \leq f \leq 57.6\,GHz$; $d_{fo} \leq 200\,m$
ITU-R	$L_{fo}(dB) = 0.2 f^{0.3} d_{fo}^{0.6}$	f in MHZ; d_{fo} in meters; for UHF $d_{fo} \leq 400\,m$
Fitted ITU-R [46]	$L_{fo}(dB) = \begin{cases} 0.37 f^{0.18} d_{fo}^{0.59}, & out\ of\ leaf \\ 0.39 f^{0.39} d_{fo}^{0.25}, & in\ leaf \end{cases}$	f in MHZ; d_{fo} in meters; for SHF
Lateral ITU-R [64]	$L_{fo}(dB) = 0.48 f^{0.43} d_{fo}^{0.13}$	f in MHZ; d_{fo} in meters; for VHF $d_{fo} \leq 5\,km$; for UHF $d_{fo} \leq 1.2\,km$

Modeling inhomogeneous channels via direct measurements is complex and time consuming. Computational methods can properly model inhomogeneous media. Inhomogeneous layered and volumetric media propagation problems can be tackled by techniques such as the extended boundary condition method (EBCM) [49–53], distorted-wave Born approximation (DWBA) [54], mean field theory (MFT) [56], small perturbation method (SPM) [57], and Kirchhoff approximation (KA) [58]. For smaller computational domains (e.g., the human body), full-wave EM simulators such as the finite element method (FEM) and finite difference time domain (FDTD) solution are suitable candidates [60, 61].

4.9   CONCLUSIONS

This chapter presented a summary of important topics on channel modeling that support localization methods. The chapter supports the material in many chapters presented throughout this handbook. Examples include channel models required for RSS indicator techniques, DOA and TOA estimation methods, and LOS identification techniques.

ACKNOWLEDGMENT

Some of my students, including Dr. Wenjie Xu, created figures in this chapter; in addition, Table 4.3 was created with the support of Drs. Amir Torabi and Imran Aslam. Dr. Amir Torabi also contributed to Section 4.6.4.

REFERENCES

[1] H. Tong and S. A. Zekavat, "A novel wireless local positioning system via asynchronous DS-CDMA and beam-forming: Implementation and perturbation analysis," *IEEE Trans. Veh. Technol.*, vol. 56, no. 3, pp. 1307–1320, May 2007.

[2] H. Tong, J. Pourrostam, and S. A. Zekavat, "Optimum Beam-forming for a novel wireless local positioning system: A stationarity analysis and solution," *EURASIP J. Adv. Signal Process.*, vol. 2007, Article ID 98243, 12 pages, 2007.

[3] A. I. Mourikis and S. I. Roumeliotis, "Performance analysis of multirobot cooperative localization," *IEEE Trans. Robot.*, vol. 22, no. 4, pp. 666–681, Aug. 2006.

[4] N. Patwari, J. N. Ash, S. Kyperountas, A. O. Hero, R. L. Moses, and N. S. Correal, "Locating the nodes: Cooperative localization in wireless sensor networks," *IEEE Signal Process. Mag.*, vol. 22, no. 4, pp. 54–69, July 2005.

[5] H. Wymeersch, J. Lien, and M. Z. Win, "Cooperative localization in wireless networks," *Proc. IEEE*, vol. 97, no. 2, pp. 427–450, Feb. 2009.

[6] T. S. Rappaport, *Wireless Communications: Principles and Practice*. Upper Saddle River, NJ: Prentice Hall, 1996.

[7] W. Hawkins, B. L. F. Daku, and A. F. Prugger, "Vehicle localization in underground mines using a particle filter," *IEEE Canadian Conference on Electrical and Computer Engineering*, pp. 2159–2162, 1–4 May 2005.

[8] K. Pahlavan and P. Krishnamurthy, *Networking Fundamentals*, John Wiley, West Sussex, UK, 2009.

[9] I. F. Akyildiz, S. Weilian, Y. Sankarasubramaniam, and E. Cayirci, "A survey on sensor networks," *IEEE Commun. Mag.*, vol. 40, no. 8, pp. 102–114, 2002.

[10] D. W. Matolak, I. Sen, W. Xiong, and R. D. Apaza, "Channel measurement/modeling for airport surface communications: Mobile and fixed platform results," *IEEE Aerosp. Electron. Syst. Mag.*, vol. 22, no. 10, pp. 25–30, Oct. 2007.

[11] O. Norklit, P. C. F. Eggers, and J. B. Andersen, "Jitter diversity in multipath environments," *Proc. Vehicular Technology Conference, VTC'95*, Chicago, IL, USA, vol. 2, pp. 853–857, 1995.

[12] W. C. Wong, R. Steele, B. Glance and D. Horn, "Time diversity with adaptive error detection to combat Rayleigh fading in digital mobile radio," *IEEE Trans. Commun.*, vol. COM-31, no. 3, pp. 378–387, 1983.

[13] A. Hugine, H. I. Volos, J. Gaeddert, and R. M. Buehrer, "Measurement and characterization of the near-ground indoor ultra wideband channel," *Wireless Communications and Networking Conference*, vol. 2, no., pp. 1062–1067, 2006.

[14] T. Rappaport, *Wireless Communications: Principles and Practice*, 2nd ed., Upper Saddle River, NJ: Prentice Hall, 2001.

[15] H. Hashemi, The indoor radio propagation channel," *Proc. IEEE*, vol. 81, no. 7, pp. 943–968, 1993.

[16] B. H. Fleury, "First- and second-order characterization of direction dispersion and space selectivity in the radio channel," *IEEE Trans. Inf. Theory*, vol. 46, no. 6, pp. 2027–2044, 2000.

[17] R. B. Ertel and J. H. Reed, "Generation of two equal power correlated Rayleigh fading envelopes," *IEEE Commun. Lett.*, vol. 2, pp. 276–278, Oct. 1998.

[18] W. C. Jakes, *Microwave Mobile Communication*. New York: IEEE Press, 1974.

[19] S. A. Zekavat and C. R. Nassar, "High performance wireless via the merger of CI chip shaped DS-CDMA and oscillating-beam smart antenna arrays," *EURASIP J. Appl. Signal Process.*, vol. 2004, no. 9, pp. 1376–1383, Aug. 2004.

[20] S. A. Zekavat and C. R. Nassar, "Transmit diversity via oscillating beam pattern adaptive antennas: An evaluation using geometric-based stochastic circular-scenario channel modeling," *IEEE Trans. Wireless Commun.*, vol. 4, no. 3, pp. 1134–1141, July 2004.

[21] S. A. Zekavat, C. R. Nassar, and S. Shattil, "Merging multi-carrier CDMA and oscillating-beam smart antenna arrays: Exploiting directionality, transmit diversity and frequency diversity," *IEEE Trans. Commun.*, vol. 52, no. 1, pp. 110–119, Jan. 2004.

[22] S. A. Zekavat and C. R. Nassar, " Achieving high capacity wireless by merging multi-carrier CDMA systems and oscillating-beam smart antenna arrays," *IEEE Trans. Veh. Technol.*, vol. 52, no. 2, pp. 772–778, July 2003.

[23] S. A. Zekavat and C. R. Nassar, "Smart antenna arrays with oscillating beam patterns: Characterization of transmit diversity using semi-elliptic-coverage geometric-based stochastic channel modeling," *IEEE Trans. Commun.*, vol. 50, no. 10, pp. 1549–1556, Oct. 2002.

[24] S. A. Zekavat, C. R. Nassar, and S. Shattil, "Oscillating beam adaptive antennas and multi-carrier systems: Achieving transmit diversity, frequency diversity and directionality," *IEEE Trans. Veh. Technol.*, vol. 51, no. 5, pp. 1030–1039, Sept. 2002.

[25] S. A. Zekavat and P. T. Keong, "Beam-pattern-scanning dynamic-time block coding: Achieving high performance" *IEEE Trans. Wireless Commun.*, vol. 5, no. 9, pp. 2334–2337, Sept. 2006.

[26] J. Fuhl, A. F. Molisch, E. Bonek, "Unified *channel* model for mobile radio systems with smart antennas," *IEE Proc. Radar, Sonar and Navigation*, vol. 145, no. 1, pp. 32–41, 1998.

[27] K. I. Pedersen, P. E. Mogensen, and B. Fleury, "Power azimuth spectrum in outdoor environments," *Electron. Lett.*, vol. 33, no. 18, pp. 1583–1584, Aug. 1997.

[28] R. Vaughan, "Pattern translation and rotation in uncorrelated source distributions for multiple beam antenna design," *IEEE Trans. Antennas Propag.*, vol. 46, pp. 982–990, July 1998.

[29] S. A. Zekavat and C. R. Nassar, "Power-azimuth-spectrum modeling for antenna array systems: A geometric-based approach," *IEEE Trans. Antennas Propag.*, vol. 51, no. 12, Dec. 2003.

[30] G. S. Watson, *Statistics on Spheres*. New York: Wiley, 1983.

[31] H. Suzuki, "A statistical model for urban channel propagation," *IEEE Trans. Commun.*, vol. 25, pp. 673–680, July 1977.

[32] R. B. Ertel and J. H. Reed, "Angle and time of arrival statistics for circular and elliptical scattering models," *IEEE J. Sel. Areas Commun.*, vol. 17, no. 11, pp. 1829–1840, Nov. 1999.

[33] W. Xu, S. A. Zekavat, and H. Tong, "A novel spatially correlated multi-user MIMO channel modeling: Impact of surface roughness," *IEEE Trans. Antennas Propag.*, vol. 57, no. 8, pp. 2429–2438, Aug. 2009.

[34] N. Blaunstein, *Radio Propagation in Cellular Networks*, Norwood, MA: Artech House, 2000, Chaps. 3–4.

[35] A. Alvira and I. T. Corbell, "Microwave system parameters for reliable communications," *Trans. Am. Inst. Electr. Eng. Part III. Power Apparatus and Systems,* vol. 74, no. 3, pp. 1100–1107, Jan. 1955.

[36] H. R. Anderson, *Fixed Broadband Wireless System Design*. West Sussex, UK: John Willey & Sons Inc., 2003, Ch. 2.

[37] J. S. Seybold, *Introduction to RF Propagation*. Hoboken, NJ: John Willey & Sons Inc., 2005, pp. 174–178.

[38] I. Aslam, and S. A. Zekavat, "Channel path loss modeling for near ground antennas," *IET Sensor Syst.*, vol. 2, no. 2, pp. 103–107, 2012.

[39] C. R. Anderson, H. I. Volos, W. C. Headley, F. C. B. F. Muller, and R. M. Buehrer, "Low antenna ultra wideband propagation measurements and modeling in a forest environment," *IEEE Wireless Communications and Networking Conference*, Las Vegas, NV, pp. 1229–1234, April 2008.

[40] C. R. Anderson, H. I. Volos, W. C. Headley, F. C. B. F. Muller, and R. M. Buehrer, "Low-antenna ultra wideband spatial correlation analysis in a forest environment," *IEEE Vehicular Technology Conference*, Barcelona, Spain, April 2009.

[41] M. A. Weissberger, "An initial critical summary of models for predicting the attenuation of radio waves by foliage," Electromagnetic Compatibility Analysis Center, Annapolis, MD, ECAC-TR-81-101, 1981.

[42] European Commission: CORDIS, "Radio propagation effects on next-generation, fixed-service terrestrial telecommunication systems," Final Report, Luxembourg, 1996.

[43] J. D. Parsons, *The Mobile Radio Propagation Channel*, 2nd ed., West Sussex, England: Wiley, pp. 52–53, 2000.

[44] M.O. Nuaimi and R.B.L. Stephen, "Measurement and prediction model optimization for signal attenuation in vegetation media at centimeter wave frequencies," *Proc. Institute of Electronics, Engineering and Microwave Antenna Propagation*, vol. 145, pp. 201–206, June 1998.

[45] M. Jamalabdollahi and S. A. Zekavat, "High resolution TOA estimation via optimal waveform design," *IEEE Trans. Commun.*, vol. 65, no. 3, pp. 1207–1218, 2017.

[46] M. Jamalabdollahi, S. Zekavat, "Range measurements in non-homogenous, time and frequency dispersive channels via time and direction of arrival merger," *IEEE Trans. Geosci. Remote Sens.*, vol. 55, no. 2, pp. 742–752, 2017.

[47] K. Sayrafian-Pour, W. Yang, J. Hagedorn, J. Terrill, K. Yazdandoost, et al. "A statistical path loss model for medical implant communication channels," *2009 IEEE 20th International Symposium on Personal, Indoor and Mobile Radio Communications*, 2009.

[48] D. B. Smith, D. Minniutti, T. Lamahewa, L. Hanlen, et al., "Propagation models for body-area networks: A survey and new outlook," *IEEE Antennas Propag. Mag.*, vol. 55 no. 5, p. 97–117, 2013.

[49] W. Chao, W. and Z. Xiaojuan, "Second-order perturbative solutions for 3-D electromagnetic radiation and propagation in a layered structure with multilayer rough interfaces," *IEEE J. Sel. Topics Appl. Earth Observ. Remote Sens.*, vol. 8, no. 1, pp. 180–194, 2015.

[50] W. Chao, Z. Xiaojuan, and F. Guangyou, "Bistatic scattering from three-dimensional layered structures with multilayer rough interfaces," *IEEE Geosci. Remote Sens. Lett.*, vol. 11, no. 3, pp. 676–680.

[51] K. Chih-hao, and M. Moghaddam, "Scattering from multilayer rough surfaces based on the extended boundary condition method and truncated singular value decomposition," *IEEE Trans. Antennas Propag.*, vol. 54, no. 10, pp. 2917–2929.

[52] M. Yu Song M., L. Yee Hui, and N. Boon Chong, "Empirical near ground path loss modeling in a forest at VHF and UHF bands, *IEEE Trans. Antennas Propag.*, vol. 57, no. 5, pp. 1461–1468, 2009.

[53] R. Janaswamy, *Radiowave Propagation and Smart Antennas for Wireless Communications.*, New York, Boston, London, Moscow: Kluwer, 2002.

[54] N. Alsindi, L. Xinrong, and K. Pahlavan, "Analysis of time of arrival estimation using wideband measurements of indoor radio propagations," *IEEE Trans. Instrum. Meas.*, vol. 56 no. 5, pp. 1537–1545, 2007.

[55] A. Torabi, S. A. Zekavat, and Kamal Sarabandi, "Wideband wireless channel characterization for multi-antenna systems over a random rough dielectric ground," *IEEE Trans. Wireless Commun.*, vol. 15, no. 5, pp. 3103–3113, 2016.

[56] A. Torabi and S. A. Zekavat, "Millimeter wave directional channel modeling," *Proc. IEEE WiSEE' 15*, Orlando, FL, Dec. 14–16, 2015.

[57] A. Torabi and S. A. Zekavat, "Near-ground channel modeling for distributed cooperative communications," *IEEE Trans. Antennas Propag.*, vol. 64, no. 6, pp. 2494–2502, 2016.

[58] Shen, B., R. Yang, S. Ullah, K. Kwak, et al., "Linear quadrature optimisation-based non-coherent time of arrival estimation scheme for impulse radio ultra-wideband systems," *IET Commun.*, vol. 4, no. 12, p. 1471–1483, 2010.

[59] P. J. Voltz, and D. Hernandez, "Maximum likelihood time of arrival estimation for real-time physical location tracking of 802.11a/g mobile stations in indoor environments," *Position Location and Navigation Symposium, 2004. PLANS*, 2004.

[60] K. Sarabandi and C. Tsenchieh, "Electromagnetic scattering from slightly rough surfaces with inhomogeneous dielectric profiles," *IEEE Trans. Antennas Propag.*, vol. 45, pp. 1419–1430, 1997.

[61] L. DaHan and K. Sarabandi, "On the effective low-grazing reflection coefficient of random terrain roughness for modeling near-earth radiowave propagation," *IEEE Trans. Antennas Propag.*, vol. 58, pp. 1315–1324, 2010.

[62] L. DaHan and K. Sarabandi, "Near-Earth wave propagation characteristics of electric dipole in presence of vegetation or snow layer," *IEEE Trans. Antennas Propag.*, vol. 53, pp. 3747–3756, 2005.

[63] A. Goldsmith, *Wireless Communications*. Cambridge: Cambridge University Press, 2005.

CHAPTER 5

AN INTRODUCTION TO KALMAN FILTERING IMPLEMENTATION FOR LOCALIZATION AND TRACKING APPLICATIONS

Shu Ting Goh, S. A. (Reza) Zekavat, Ossama Abdelkhalik

THIS CHAPTER investigates the implementation of linear and nonlinear Kalman filters for localization, target tracking, and navigation. The chapter first formulates the positioning problem in the estimation context. Next, a deterministic derivation for Kalman filters is presented. This chapter introduces several types of Kalman filters used for localization, which include extended Kalman filter (EKF), unscented Kalman filter (UKF), ensemble Kalman filter (EnKF), and constrained Kalman filter (CKF). Implementation examples for localization, target tracking, and navigation of these Kalman filters are offered, and their associated MATLAB codes are presented. The "Further Reading" section offers relevant and more advanced topics to the reader.

5.1   INTRODUCTION

In general, an estimation algorithm predicts the quantities of interest via direct or indirect observations. However, observations contain error, which leads to uncertainty in estimation. Thus, estimation algorithms include two categories: unbiased and biased. Unbiased estimation algorithms, such as least square algorithms, are generally formulated based on perfect measurement or no observation error. Biased estimation algorithms such as Kalman filters are formulated by considering various sources of system error. Almost all localization problems include a channel and a transducer, while the error can be induced by the channel, the transducer, or both. In radio frequency (RF) communication, the channel is multipath and the transducer

Handbook of Position Location: Theory, Practice, and Advances, Second Edition.
Edited by S. A. (Reza) Zekavat and R. Michael Buehrer.
© 2019 by the Institute of Electrical and Electronics Engineers, Inc.
Published 2019 by John Wiley & Sons, Inc.
Companion Website: www.wiley.com/go/zekavat/positionlocation2e

Figure 5.1 Travelled distance of a train.

is an RF receiver. In most of these scenarios, compared to other filtering methods, Kalman filtering works more efficiently.

Estimation algorithms have been implemented in numerous applications, such as aircraft and spacecraft tracking, spacecraft attitude determination and control, orbit determination, wildfire and weather prediction, plant state estimation and stochastic control, and identification of vibratory systems [1–8]. This chapter mainly presents the estimation algorithm for both target tracking and navigation applications. Such applications include vehicular navigation, aircraft tracking and navigation, satellite orbit and attitude determination, etc.

To better understand the concept of localization-based estimation, consider a train traveling from Chicago, IL to Detroit, MI. For illustrative purposes (see Figure 5.1), suppose that the train is moving in a straight line and pulled by a locomotive with a constant force, F. When the train passes each station, its arrival time, t_k, at the particular station is recorded. If we aim to calculate the travelled distance (from Chicago) of the train within a time period, t_k, we can use the following simple procedure. Acceleration of the train corresponds to:

$$a = \frac{F}{m}. \tag{5.1}$$

Velocity of the train at time t_k is:

$$v_k = v_{k-1} + \int_{t_{k-1}}^{t_k} a\,dt = a\Delta t + v_{k-1}. \tag{5.2}$$

Distance from Chicago (or any station) at time t_k is:

$$x_k = x_{k-1} + \int_{t_{k-1}}^{t_k} v_k\,dt = x_{k-1} + v_{k-1}\Delta t + a\frac{\Delta t^2}{2}. \tag{5.3}$$

If we know the engine force, F, and the train velocity at Chicago, v_0, we can compute the position and velocity of the train at time t_k. Let us denote this distance (or position) by x_k and velocity by v_k. Now, assume you aim to measure the distance x_2 after the train arrives at the Battle Creek station at time t_2 and denote the measured quantity as $\bar{x}_2$. What if x_2 is not equal to $\bar{x}_2$? Which one should we believe? Why are they different?

The measurement $\bar{x}_k$ might be different from the calculated x_k due to error. The mathematical model that we use in our calculations may have uncertainties; for example, we may have uncertainty in the velocity v_k, and/or in the engine force, F.

Moreover, friction force might not be modelled in the above mathematical model, and this friction force affects the motion of the train and will definitely cause error in the calculated value. Furthermore, the time measurement t_k may not be accurate! In addition, the Battle Creek station's time may have an error of half a second. Thus, there is indeed a true value for the distance from Chicago at time t_k; this true value is not equal to x_k or $\overline{x}_k$. This true value is unknown due to (a) uncertainties in the mathematical model and/or initial conditions, and (b) measurement errors. What is the solution? If the computed quantity x_k along with the measured one $\overline{x}_k$ leads to a better guess for the true distance, then the true value is approached through estimation.

In an estimation problem, true quantities are estimated based on the information collected from the measurements and based on our knowledge about the mathematical model. There are several ways to do this combination. For instance, in the train example, the average of $\overline{x}_k$ and x_k is computed. This average is a guess for the true distance of the train at time t_k. We can also apply a weighted average for the two quantities depending on the quality of each quantity. So, if we know that our measurement is poor, we assign less weight to the measurement compared to the computed quantity. This may lead to a better estimate for the true quantity, compared to the average with equal weights. There are many ways for estimating the true quantity from the measurements and mathematical model; thus, it is preferred to apply this estimation step optimally. In other words, we may construct a penalty function and compute the estimate for true value that will minimize this penalty function. The estimate will still use the same set of information: the measurements and the mathematical model. Now, another question: If you were to choose the penalty function, how would you pick it?

Our objective is to get the best estimate for the true quantity. We may choose the penalty function as the variance between the true quantity and the estimated quantity.

Curve fitting is a very simple form of estimation that uses a least squares technique to fit the data points to a model equation. Here, the model equation has unknown parameters that are estimated to best fit the data. There is one difference between the train example and the curve-fitting problem. In the curve-fitting problem, we estimate parameters, while in the train example, we estimate the distance from Chicago to a current train position, which is a variable.

5.2   THE ESTIMATION PROBLEM

Two major elements in the estimation problem include the desired state vector (what you want to know), $\mathbf{x}$ and the measurement vector (the available information), $\tilde{y}$. The variable $\mathbf{x}$ often consists of the parameters or variables that need to be estimated. The parameters and variables can be stacked together to become a single state vector. The underlying physics that describe the system behaviour and evolution over time is called the system's mathematical model (these are the distance and velocity equations in the train example described above). The system's mathematical model is usually written in the form

$$\dot{\mathbf{x}}(t) = \mathbf{F}(t)\mathbf{x}(t) + \mathbf{B}(t)\mathbf{u}(t) + \mathbf{w}(t),$$

where $\mathbf{u}(t)$ is the input to the system and $\mathbf{w}(t)$ is the process noise. Moreover, functions $\mathbf{F}(t)$, and $\mathbf{B}(t)$ are the mathematical representation for the system's dynamics, and input model, respectively. The process noise is due to the uncertainties in the system's model; for example, the clock error of the timing system in each train station leads to a system noise. The process noise is generally assumed Gaussian.

Measurements are collected and used in the estimator to update the system's state vector. All measurements are gathered in a vector called the measurement vector, $\tilde{\mathbf{y}}$. The mathematical model for the measurement vector may be written as

$$\tilde{\mathbf{y}}(t) = \mathbf{C}(t)\mathbf{x}(t) + \mathbf{v}(t),$$

where $\mathbf{v}(t)$ is the measurement noise. The matrix $\mathbf{C}(t)$ is known as the observation matrix, which is often the linear representation of $\tilde{\mathbf{y}}(t)$ by $\mathbf{x}(t)$. An estimator computes the best estimate, $\hat{\mathbf{x}}$, for the system's state vector, $\mathbf{x}$, given the system's mathematical model and the measurement vector.

5.2.1 Estimation Problem Classifications

5.2.1.1 Linear versus Nonlinear

Linear problems are those in which both the system's model and the measurement vector are linear functions of the state vector $\mathbf{x}$. In this linear case, we can implement a linear filter to estimate the state vector $\mathbf{x}$. One of the popular linear filters is the Kalman filter. Other examples are the H_∞ filter and the least square estimator. On the other hand, we will have a nonlinear problem in hand if:

(a) the system model is represented by a nonlinear function with respect to $\mathbf{x}$,

(b) the measurement model is nonlinear with respect to $\mathbf{x}$, or

(c) both the system and measurement models are nonlinear with respect to $\mathbf{x}$.

The estimation algorithm for a nonlinear system could be formulated through the approximation methods via Taylor series expansion that is the case in EKF and nonlinear least square algorithms. The estimation algorithm could also utilize a large estimation size to compute the mean and variance, such as the particle filter (PF) and UKF for nonlinear problem. However, performance trade-off is often considered for nonlinear estimation, such as the trade-off between stability and convergence rate in EKF and computational cost for both PF and UKF.

5.2.1.2 Sequential versus Batch

A batch filter does not work in real time. It collects measurements over time and then processes them all together to produce the system's state estimation. A sequential estimator, on the other hand, works in real time. It updates the system's states every time a measurement is received. The update utilizes the current measurement and the estimate of the state at the previous time. Many applications, including localization, do need to implement sequential estimators for real-time processing. Kalman filter is an example of a sequential filter.

In general, the nonlinear least square algorithm is similar to linear least square algorithm, with the exception that the first-order Taylor series expansion is utilized in nonlinear least square algorithms. All Kalman filter variations utilize similar state vector update algorithms, as in linear Kalman filter with different Kalman gain estimation methods. Therefore, it can be observed that a linear filter is a fundamental component of nonlinear estimation.

5.3 FORMULATION OF LOCALIZATION AS AN ESTIMATION PROBLEM

Based on our discussion, a summary of the estimation problem corresponds to:

> *given a set of noisy measurements for a dynamic system, and the mathematical model that describes the dynamics of the system, with some uncertainty, it is desired to estimate the system states in an optimal sense.*

The localization problem can be viewed as an estimation problem. A typical localization system provides measurements for the ranges and the directions of some targets. For example, if a system similar to radar is used for localization, the measurements would be the time of arrival (TOA) of a signal and/or the direction of arrival (DOA) of the signal. If we know the mathematical model that describes target motions, we can implement an estimation method to compute the position and velocity of moving targets. It is expected that the estimated positions and velocities will have better accuracies compared with the measured positions and velocities.

5.4 DISCRETE KALMAN FILTER

A Kalman filter is the optimal estimator in a minimum variance sense. In other words, a Kalman filter is the optimal filter that minimizes the difference between the estimated states and the true states. A mathematical derivation for the Kalman filter is first presented. A discussion on how to implement the linear Kalman filter and some insight into the filter equations follows. The mathematical derivation follows the derivations presented in [9] and is offered here for completeness.

In this chapter, unless otherwise specified, the overhead notation $\hat{.}$ denotes the estimated value and $\tilde{.}$ denotes the measurement.

The Kalman filter generally consists of four processes:

1. The initialization process starts with an initial guess for the state vector $\hat{\mathbf{x}}_0$ and state error covariance $\hat{\mathbf{P}}_0$.

2. The system is propagated in forward time step to predict $\hat{\mathbf{x}}_k^-$ and $\hat{\mathbf{P}}_k^-$ at the time k. Here, "$-$" denotes a priori estimate.

3. Kalman gain $\mathbf{K}_k$ is estimated based on the $\hat{\mathbf{P}}_k^-$ and given measurement noise covariance.

4. At the current time t_k, $\hat{\mathbf{x}}_k^-$ is updated to $\hat{\mathbf{x}}_k^+$, and $\hat{\mathbf{P}}_k^-$ is updated to $\hat{\mathbf{P}}_k^+$ based on given $\tilde{y}_k$ and estimated $\mathbf{K}_k$. Here, "+" denotes the posteriori estimate.

The Kalman filter process is generally similar for all Kalman filter variations such as EKF, UKF, EnKF, CKF, etc. However, each Kalman filter variation may have its own method in Kalman gain and update computation. The next few sections present Kalman filter variation algorithms in detail.

5.4.1 Kalman Filter Derivation

This section presents a deterministic derivation for the Kalman filter equations. A probabilistic derivation can be found in [9]. Let the mathematical model for the system dynamics (sometimes called the truth model) be expressed as

$$\mathbf{x}_{k+1} = \mathbf{A}_k \mathbf{x}_k + \mathbf{B}_k \mathbf{u}_k + \mathbf{D}_k \mathbf{w}_k, \tag{5.4}$$

where $\mathbf{x}_k$ is the state vector, $\mathbf{A}_k$, $\mathbf{B}_k$ are the dynamics matrices, $\mathbf{u}_k$ is the input vector, $\mathbf{w}_k$ is a vector representing the uncertainties in the dynamics, and $\mathbf{D}_k$ is a mapping matrix that maps $\mathbf{w}_k$ into the dynamic model $\mathbf{A}_k \mathbf{x}_k + \mathbf{B}_k \mathbf{u}_k$. The subscript k is the time index. The measurement model is

$$\tilde{\mathbf{y}}_k = \mathbf{C}_k \mathbf{x}_k + \upsilon_k, \tag{5.5}$$

where $\tilde{\mathbf{y}}_k$ is the measurement vector, $\mathbf{C}_k$ maintains the relation between the measured vector and the system states, which is often known as observation matrix, and υ_k is the measurement noise. We assume that the measurement noise and the model uncertainties are both white noises. Let $E\{.\}$ denote the expectation operator, and define $\mathbf{R}_k$ and $\mathbf{Q}_k$ as

$$\mathbf{R}_k = E\left\{\upsilon_k \upsilon_k^T\right\} \tag{5.6}$$

$$\mathbf{Q}_k = E\left\{\mathbf{w}_k \mathbf{w}_k^T\right\} \tag{5.7}$$

$$E\left\{\upsilon_k \mathbf{w}_k^T\right\} = E\left\{\mathbf{w}_k \upsilon_k^T\right\} = \mathbf{0} \tag{5.8}$$

The truth dynamic model is not known because of the uncertainty term. We assume an estimated state vector, $\hat{\mathbf{x}}_k$, and an estimated measurement vector, $\hat{\mathbf{y}}_k$. The estimated state vector is propagated according to the dynamic model:

$$\hat{\mathbf{x}}_{k+1}^- = \mathbf{A}_k \hat{\mathbf{x}}_k^+ + \mathbf{B}_k \mathbf{u}_k. \tag{5.9}$$

Then, the propagated state estimate is updated according to the relationship

$$\hat{\mathbf{x}}_k^+ = \hat{\mathbf{x}}_k^- + \mathbf{K}_k \left[\tilde{\mathbf{y}}_k - \mathbf{C}_k \hat{\mathbf{x}}_k^- \right], \tag{5.10}$$

where $\mathbf{K}_k$ is a gain matrix. The superscript (−) denotes the vector after propagation and before update. The superscript (+) denotes the vector after update. The measurement estimate is computed as

$$\hat{\mathbf{y}}_k = \mathbf{C}_k \hat{\mathbf{x}}_k^-. \tag{5.11}$$

Now, we define the following errors, $\tilde{\mathbf{x}}_k$, and their covariance $\mathbf{P}_k$:

$$\tilde{\mathbf{x}}_k^- = \hat{\mathbf{x}}_k^- - \mathbf{x}_k \qquad\qquad \mathbf{P}_k^- = E\left\{ \tilde{\mathbf{x}}_k^- \tilde{\mathbf{x}}_k^{-T} \right\}$$

$$\tilde{\mathbf{x}}_k^+ = \hat{\mathbf{x}}_k^+ - \mathbf{x}_k \qquad\qquad \mathbf{P}_k^+ = E\left\{ \tilde{\mathbf{x}}_k^+ \tilde{\mathbf{x}}_k^{+T} \right\} \tag{5.12}$$

$$\tilde{\mathbf{x}}_{k+1}^- = \hat{\mathbf{x}}_{k+1}^- - \mathbf{x}_k \qquad\qquad \mathbf{P}_{k+1}^- = E\left\{ \hat{\mathbf{x}}_{k+1}^- \hat{\mathbf{x}}_{k+1}^{-T} \right\}.$$

Substituting the definitions ((5.11) and (5.12)) into the propagation and update equations in (5.9) and (5.10), we obtain the following covariance propagation and update equations, respectively:

$$\mathbf{P}_{k+1}^- = \mathbf{A}_k \mathbf{P}_k^+ \mathbf{A}_k^T + \mathbf{D}_k \mathbf{Q}_k \mathbf{D}_k^T \tag{5.13}$$

$$\mathbf{P}_k^+ = \mathbf{P}_k^- - \mathbf{P}_k^- \mathbf{C}_k^T \mathbf{K}_k^T - \mathbf{K}_k \mathbf{C}_k \mathbf{P}_k^- + \mathbf{K}_k \left(\mathbf{C}_k \mathbf{P}_k^- \mathbf{C}_k^T + \mathbf{R} \right) \mathbf{K}_k^T. \tag{5.14}$$

Here, we assume there is no error correlation between the state estimation error and measurement noise, such that $E\left\{ \tilde{\mathbf{x}}_k^- \upsilon_k^T \right\} = E\left\{ \upsilon_k \tilde{\mathbf{x}}_k^{-T} \right\} = \mathbf{0}$. The gain $\mathbf{K}_k$ is computed so as to minimize an error function. The error function, J, quantifies the error between the estimated states and the true states:

$$J(\mathbf{K}_k) = Tr(\mathbf{P}_k^+), \tag{5.15}$$

where the $Tr(.)$ operator returns the trace of the matrix. So,

$$\frac{\partial J}{\partial \mathbf{K}_k} \equiv -2(\mathbf{I} - \mathbf{K}_k \mathbf{C}_k) \mathbf{P}_k^- \mathbf{C}_k^T + 2\mathbf{K}_k \mathbf{R} = \mathbf{0}. \tag{5.16}$$

Here, $\mathbf{I}$ is identity matrix. Solving for the gain $\mathbf{K}_k$, we get

$$\mathbf{K}_k = \mathbf{P}_k^- \mathbf{C}_k^T \left(\mathbf{C}_k \mathbf{P}_k^- \mathbf{C}_k^T + \mathbf{R}_k \right)^{-1}. \tag{5.17}$$

Using the definition in (5.17), the covariance update in (5.14) may be simplified to:

$$\mathbf{P}_k^+ = (\mathbf{I} - \mathbf{K}_k \mathbf{C}_k) \mathbf{P}_k^-. \tag{5.18}$$

Substituting (5.10) into (5.9), and (5.18) into (5.13), we get

$$\hat{\mathbf{x}}_{k+1} = \mathbf{A}_k \hat{\mathbf{x}}_k + \mathbf{B}_k \mathbf{u}_k + \mathbf{A}_k (\tilde{\mathbf{y}}_k - \mathbf{C}_k \hat{\mathbf{x}}_k)$$

$$\mathbf{K}_k = \mathbf{P}_k \mathbf{C}_k^T (\mathbf{C}_k \mathbf{P}_k \mathbf{C}_k^T + \mathbf{R}_k)^{-1} \tag{5.19}$$

$$\mathbf{P}_{k+1} = \mathbf{A}_k \mathbf{P}_k \mathbf{A}_k^T - \mathbf{A}_k \mathbf{K}_k \mathbf{C}_k \mathbf{P}_k \mathbf{A}_k^T + \mathbf{D}_k \mathbf{Q}_k \mathbf{D}_k^T.$$

Equations (5.19) are the Kalman filter equations without the a priori and posterior notations.

5.4.2 Kalman Filter Implementation and Practical Considerations

The implementation of Kalman filter is straight forward. The process is shown in Figure 5.2. The linear Kalman filter can be implemented directly without much trouble if the dynamic and measurement models are known; however, there are always some challenges for beginners. Before implementing the Kalman filter (including any variation of Kalman filter), some key consideration must be taken into account:

- How do we initialize state vector $\hat{x}_0$?
- How do we choose our initial state error covariance $\mathbf{P}_0$?
- How do we choose our process noise covariance $\mathbf{Q}_k$?
- How do we model process noise in estimated model?

How Do We Initialize $\hat{\mathbf{x}}_0$?
Usually in the literature, $\hat{\mathbf{x}}_0$ has always been assumed similar to the true value. For research and study purposes, the assumption is acceptable. For application purposes, Kalman filter requires collaboration between multiple sensors to offer a good initial estimation. For localization applications, Kalman filter could be initialized via global positioning system (GPS) data. The dilution of precision from GPS data could be used as a reference to compute the initial $\mathbf{P}_0$ for Kalman filter. For attitude estimation application, the quaternion vector is highly suggested to be the state vector, as the state vector can be initialized using $\hat{\mathbf{x}}_0 = \begin{bmatrix} 0 & 0 & 0 & 1 \end{bmatrix}^T$, with $\mathbf{P}_0 = 1$ for each quaternion axis. For other applications where initialization accuracy is quite crucial but no sensor provides a good initial guess, deterministic methods may offer initial estimates. If no deterministic methods are available, other variations of Kalman filter such as UKF or EnKF can be used.

How Do We Select $\mathbf{P}_0$?
In fact, there isn't any standard formulation to initialize $\mathbf{P}_0$. For many simulation studies, $\mathbf{P}_0$ is often initialized either based on the tuning process or predicted error from sensor measurements, such as GPS data. But for some applications such as

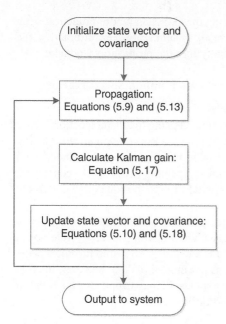

Figure 5.2   Kalman filter flow diagram.

attitude estimation, which uses either Euler angles or quaternion vectors, the possible maximum error can be used for $\mathbf{P}_0$. For example, the 180°could be used if Euler angle is used because the range of Euler angle is between $-180°$ to $180°$ (or 0 to 360° and -360 to 0 degrees). Also, 1 could be used for quaternion vector because the magnitude of quaternion always equals 1. In general, $\mathbf{P}_0$ should not be a zero matrix even if we assume $\hat{\mathbf{x}}_0$ is perfect.

How to Choose the $\mathbf{Q}_k$?
A similar case can also be considered to determine $\mathbf{Q}_k$. In theory, $\mathbf{Q}_k$ is equivalent to the variance of sensor's random walk error (if there is an input function), which is usually available in sensor's datasheet. However, in practice, this method would not guarantee the stability of Kalman filter. In fact, $\mathbf{Q}_k$ is indirectly correlated with the measurement noise covariance, $\mathbf{R}_k$. If Kalman filter is unstable for a given set of $\mathbf{Q}_k$ and $\mathbf{R}_k$, then $\mathbf{Q}_k$ could be tuned (or scaled) to a higher value.

Moreover, if a dynamic system does not include an input function, such as the spacecraft orbit propagator, then $\mathbf{Q}_k$ needs to be tuned to obtain the near optimal solution for Kalman filter. The impact of tuning process on $\mathbf{Q}_k$ will be discussed in the next section.

How Do We Model Process Noise in Propagation Model?
For application implementation, the process noise is never modelled in the propagation model. The process noise is the noise or error that contains in the input data provided by sensors (e.g., accelerometer or gyroscope errors), or the integration error (e.g., eight-bit microcontroller's floating point error). For simulation purposes, the

process noise can be artificially generated using the Gaussian random function, such as MATLAB's *randn()* and added into the input system. However, in practice, we should avoid adding the process noise manually into the propagation model.

Example 5.1 Permanent Magnet Wind Turbine Rotation Speed Tracking

We consider a permanent magnet wind turbine's blade tracking scenario such that the rotation speed of wind turbine blade is estimated using the blade location sensor and without using the gyroscope. The wind turbine blade's rotation speed tracking is a key application. It could cause a catastrophic failure (such as Nordtank wind system accident in February 2008) if the rotation speed is not properly controlled. The failure of gear box and brake causes the wind turbine to rotate at an extremely high speed and eventually collapse.

In this wind turbine tracking scenario, a sensor provides the location of a particular wind blade at 1 Hz sampling rate with an error of 1 degree white noise. The particular wind blade initial true location is 2 degrees. The wind turbine's blade rotation speed, $\hat{\omega}$ is predicted at 18 revolutions per minute (RPM); however, the true rotation speed, ω, is 19 RPM. Thus, a prediction bias, β, may happen for the rotation speed. Then, the dynamic system of the wind turbine blade is given as

$$\theta_{k+1} = \theta_k + \omega\Delta t + \beta\Delta t, \tag{5.20}$$

where Δt is the sampling time.

The blade location θ_k and rotation speed bias are our desired states; thus, the state vector is $\mathbf{x} = [\theta \quad \beta]^T$, and the predicted rotation speed will become the input, $\mathbf{u} = \hat{\omega}\Delta t$. The process noise of the system model is assumed 0.01 degrees, and the state space model corresponds to (5.4) and (5.5), as follows:

Figure 5.3 Wind turbine rotation.

$$\mathbf{A} = \begin{bmatrix} 1 & \Delta t \\ 0 & 1 \end{bmatrix}$$

$$\mathbf{B} = \begin{bmatrix} 1 \\ 0 \end{bmatrix}$$

$$\mathbf{C} = \begin{bmatrix} 1 & 0 \end{bmatrix}$$ (5.21)

$$\mathbf{D} = \begin{bmatrix} 1 \\ 0 \end{bmatrix}.$$

Matrix **C** shows that only the location of the wind turbine blade is measured. The initial state vector and state error covariance are

$$\hat{\mathbf{x}}_0 = \begin{bmatrix} 0 \\ 0 \end{bmatrix}$$ (5.22)

$$\mathbf{P}_0 = \begin{bmatrix} 100 & 0 \\ 0 & 25 \end{bmatrix}.$$

The simulation is run for 15 minutes with the measurement noise randomly generated. Figure 5.4 presents the estimated rotation speed bias and the true speed bias. It shows that the Kalman filter successfully estimates the rotation speed bias within 30 seconds (or 30 sampling points based on 1-Hz sampling rate). On the other hand, Figure 5.5 presents the estimated rotation speed bias error with three sigma boundaries. The three sigma boundaries play an important role in Kalman filter performance evaluation such that 99.7% of estimation error should fall within the three sigma boundaries (positive and negative boundaries). Furthermore, it helps verify the stability of the Kalman filter based on given set of process noise. If error is mainly out of the three sigma boundaries, it could result in error divergence!

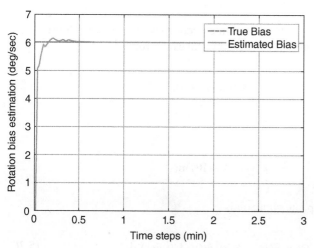

Figure 5.4   Estimated wind turbine blade rotation speed bias versus true rotation speed bias.

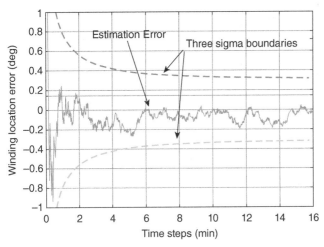

Figure 5.5 Estimation error of wind turbine blade rotation speed bias.

5.5 CONTINUOUS-TIME KALMAN FILTER

If the system is a continuous-time process and the measurements are also continuous functions, then a continuous-time Kalman filter can be developed for estimation. The detailed derivation of the continuous-time Kalman filter follows similar approach as that of the discrete Kalman filter. The reader can find detailed derivation in [9–11]. Here, we present the filter equations and an example for a continuous-time system. The model for a continuous-time system takes the form

$$\dot{\mathbf{x}}(t) = \mathbf{F}(t)\mathbf{x}(t) + \mathbf{B}(t)\mathbf{u}(t) + \mathbf{D}(t)\mathbf{w}(t).$$
(5.23)

Here, $\mathbf{F}(t)$ is the continuous-time dynamic model ($\mathbf{A}_k$ in (5.4) is discrete-time dynamic model), and the measurement model takes the form

$$\tilde{\mathbf{y}}(t) = \mathbf{C}(t)\mathbf{x}(t) + \upsilon(t).$$
(5.24)

The estimated state propagation equation is

$$\dot{\hat{\mathbf{x}}} = \mathbf{F}(t)\hat{\mathbf{x}}(t)^{+} + \mathbf{B}(t)\mathbf{u}(t),$$
(5.25)

and the estimated state update equation is

$$\hat{\mathbf{x}}(t)^{+} = \hat{\mathbf{x}}(t)^{-} + \mathbf{K}(t)\big[\tilde{\mathbf{y}}(t) - \mathbf{C}(t)\hat{\mathbf{x}}(t)^{-}\big],$$
(5.26)

where the Kalman gain is

$$\mathbf{K}(t) = \mathbf{P}(t)\mathbf{C}(t)\mathbf{R}(t)^{-1}. \tag{5.27}$$

Equation (5.27) shows that the Kalman gain for a continuous-time Kalman filter is different than the linear Kalman filter, where $\mathbf{K}(t)$ is a function of the inverse of measurement noise covariance only while $\mathbf{K}_k$ is expressed in terms of the inverse of $\mathbf{C}_k\mathbf{P}_k\mathbf{C}_k^T + \mathbf{R}_k$. The difference is due to the reason that $\mathbf{K}(t)$ is derived based on $\Delta t \to 0$ in a continuous-time system, with details that can be found in [10]. In addition, the state error covariance matrix propagation is:

$$\dot{\mathbf{P}}(t) = \mathbf{F}(t)\mathbf{P}(t) + \mathbf{P}(t)\mathbf{F}(t)^T + \mathbf{D}(t)\mathbf{Q}(t)\mathbf{D}(t)^T - \mathbf{P}(t)\mathbf{C}(t)\mathbf{R}(t)^{-1}\mathbf{C}(t)^T\mathbf{P}(t). \tag{5.28}$$

Note that the update process of state error covariance $\mathbf{P}(t)$ has been directly integrated into the $\dot{\mathbf{P}}(t)$ in (5.28). Furthermore, both $\mathbf{Q}(t)$ and $\mathbf{R}(t)$ are the instantaneous covariance matrices of the dynamics uncertainties and measurement noises for a continuous-time case:

$$E\{\upsilon(t)\upsilon(t)^T\} = \mathbf{R}(t)\delta(t-\tau) \tag{5.29}$$

$$E\{\mathbf{w}(t)\mathbf{w}(t)^T\} = \mathbf{Q}(t)\delta(t-\tau) \tag{5.30}$$

As compared to the propagation of $\mathbf{P}_k$ in (5.19), (5.28) requires the time integration of matrix $\dot{\mathbf{P}}(t)$ in order to compute $\mathbf{P}(t)$ at the next time step. One method is to convert $\dot{\mathbf{P}}(t)$ into a $n^2 \times 1$ vector. After the integration, transform the $n^2 \times 1$ vector into $n \times n$ matrix $\mathbf{P}(t)$. However, it should be noted that if the processor used for Kalman filter is a low-cost and low-performance processor, then the processing time for $\dot{\mathbf{P}}(t)$ might be significant if n is very large. An alternative method is to propagate $\mathbf{P}(t)$ via (5.13) [12]; however, this requires $\mathbf{D}(t)\mathbf{Q}(t)\mathbf{D}(t)^T$ to be expressed using time-discrete format. One possible method is the transformation of time continuous into discrete time using the exponential matrix function:

$$\mathcal{A} = \begin{bmatrix} -\mathbf{F}(t) & \mathbf{D}(t)\mathbf{Q}(t)\mathbf{D}(t)^T \\ \mathbf{0} & \mathbf{F}(t)^T \end{bmatrix} \Delta t \tag{5.31}$$

$$\mathcal{B} = e^{\mathcal{A}} \equiv \begin{bmatrix} \mathcal{B}_{11} & \mathcal{B}_{12} \\ \mathbf{0} & \mathcal{B}_{22} \end{bmatrix} = \begin{bmatrix} \mathcal{B}_{11} & \phi^{-1}\mathcal{Q} \\ \mathbf{0} & \phi^T \end{bmatrix}. \tag{5.32}$$

Here, $\mathbf{0}$ is a square matrix in which all elements equal to zero, $\mathcal{Q}$ is the discrete-time process noise, and ϕ is state transition matrix. Then, both $\mathcal{Q}$ and ϕ can be used to replace $\mathbf{D}_k\mathbf{Q}_k\mathbf{D}_k^T$ and $\mathbf{A}_k$, respectively in (5.13) to enable the discrete-time propagation of $\mathbf{P}(t)$.

The implementation of the continuous Kalman filter is shown by the flowchart of Figure 5.6. Given system dynamics and initial conditions for the system and the

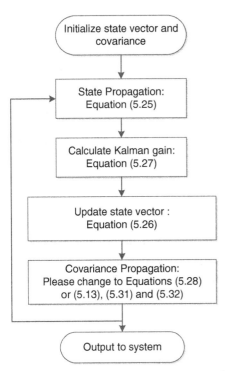

Figure 5.6   Continuous-time Kalman filter flow diagram.

confidence level in the initial conditions (initial covariance), we compute the Kalman gain, $\mathbf{K}(t)$, and update the states using (5.26) and (5.27). The states are then propagated using (5.25). The covariance is propagated using (5.28) (or (5.13)).

5.6   EKF

If the system dynamics and/or the measurements model are nonlinear, then the extended Kalman filter (EKF) can be implemented. The basic idea of the EKF is to linearize the nonlinear equations around the current estimate and then apply a linear Kalman filter to the linearized model. In general, there are two types of EKF, continuous-time EKF and continuous-discrete EKF. This section presents continuous-discrete EKF. More detailed derivation of continuous-time EKF can be found in [10].

Assume the truth model and measurements for the nonlinear system have the form

$$\dot{\mathbf{x}}(t) = \mathbf{f}(\mathbf{x}(t), \mathbf{u}(t), t) + \mathbf{D}(t)\mathbf{w}(t) \tag{5.33}$$

$$\tilde{\mathbf{y}}(t) = \mathbf{c}(\mathbf{x}(t), t) + \upsilon(t). \tag{5.34}$$

The error between the estimates and the truth model for the nonlinear system can be represented in terms of the first-order Taylor series expansion, which is given as

$$f(\mathbf{x}(t),\mathbf{u}(t),t) \cong f(\hat{\mathbf{x}}(t),\mathbf{u}(t),t) + \frac{\partial f}{\partial \mathbf{x}}\bigg|_{\hat{x}(t)} [\mathbf{x}(t) - \hat{\mathbf{x}}(t)] \tag{5.35}$$

$$c(\mathbf{x}(t),t) = c(\hat{\mathbf{x}}(t),t) + \frac{\partial c}{\partial \mathbf{x}}\bigg|_{\hat{x}(t)} [\mathbf{x}(t) - \hat{\mathbf{x}}(t)]. \tag{5.36}$$

The $\hat{\mathbf{x}}(t)$ in (5.35) and (5.36) is expressed without either a priori or posterior notation for generalization purposes. The EKF is generally structured as a similar fashion as in continuous-time Kalman filter (continuous-time Kalman filter derivation see [10]), where the estimated state obeys the following equation:

$$\dot{\hat{\mathbf{x}}}(t) = f(\hat{\mathbf{x}}(t),\mathbf{u}(t),t) + \mathbf{K}(t)[\tilde{\mathbf{y}}(t) - c(\hat{\mathbf{x}}(t),t)]. \tag{5.37}$$

Similar to (5.12), we define $\tilde{\mathbf{x}}(t) = \hat{\mathbf{x}}(t) - \mathbf{x}(t)$ (also for $\dot{\tilde{\mathbf{x}}}(t)$). Then, substituting (5.33), (5.35) and (5.36) into (5.37) and leads to

$$\dot{\tilde{\mathbf{x}}}(t) = [F(\hat{\mathbf{x}}(t),t) - \mathbf{K}(t)\mathbf{C}(\hat{\mathbf{x}}(t),t)]\tilde{\mathbf{x}}(t) - \mathbf{D}(t)\mathbf{w}(t) + \mathbf{K}(t)\upsilon(t), \tag{5.38}$$

with

$$\mathbf{F}(\hat{\mathbf{x}}(t),t) \equiv \frac{\partial f}{\partial \mathbf{x}}\bigg|_{\hat{x}(t)}, \ \mathbf{C}(\hat{\mathbf{x}}(t),t) \equiv \frac{\partial c}{\partial \mathbf{x}}\bigg|_{\hat{x}(t)}. \tag{5.39}$$

Although the EKF could be updated using (5.37), the EKF update process could also be formulated into a continuous-discrete process, which follows the linear Kalman filter update structure. To avoid any confusion for the reader, here we define $\hat{\mathbf{x}}_k^- \equiv \hat{\mathbf{x}}(t_{k-1} \mid t_k)$ and $\hat{\mathbf{x}}_k^+ \equiv \hat{\mathbf{x}}(t_k \mid t_k)$ such that $\hat{\mathbf{x}}_k^-$ is a prior estimate of $\hat{\mathbf{x}}(t)$ and $\hat{\mathbf{x}}_k^+$ is the posterior estimation of $\hat{\mathbf{x}}(t)$; then the EKF update is:

$$\hat{\mathbf{x}}_k^+ = \hat{\mathbf{x}}_k^- + \mathbf{K}_k[\tilde{\mathbf{y}}_k - c(\hat{\mathbf{x}}_k^-)]. \tag{5.40}$$

Also, the covariance of (5.40) can be obtained in a similar way as in (5.14):

$$\mathbf{P}_k^+ = \mathbf{P}_k^- - \mathbf{P}_k^- \mathbf{C}(\hat{\mathbf{x}}_k^-)^T \mathbf{K}_k^T - \mathbf{K}_k \mathbf{C}(\hat{\mathbf{x}}_k^-)\mathbf{P}_k^- + \mathbf{K}_k\left(\mathbf{C}(\hat{\mathbf{x}}_k^-)\mathbf{P}_k^-\mathbf{C}(\hat{\mathbf{x}}_k^-)^T + \mathbf{R}\right)\mathbf{K}_k^T \tag{5.41}$$

By following the similar minimization process of the trace of the covariance as in (5.15) to (5.17), the Kalman gain and the covariance updates are given as

$$\mathbf{K}_k = \mathbf{P}_k^- \mathbf{C}(\hat{\mathbf{x}}_k^-)^T \left(\mathbf{C}(\hat{\mathbf{x}}_k^-)\mathbf{P}_k^-\mathbf{C}(\hat{\mathbf{x}}_k^-)^T + \mathbf{R}_k\right)^{-1} \tag{5.42}$$

$$\mathbf{P}_k^+ = (\mathbf{I} - \mathbf{K}_k\mathbf{C}(\hat{\mathbf{x}}_k^-))\mathbf{P}_k^-. \tag{5.43}$$

The covariance propagation is derived similar to the continuous-time Kalman filter:

$$\dot{\mathbf{P}} = \mathbf{F}(\hat{\mathbf{x}}_k^+)\mathbf{P}_k^+ + \mathbf{P}_k^+\left(\hat{\mathbf{x}}_k^+\right)^T + \mathbf{D}_k\mathbf{Q}_k\mathbf{D}_k^T, \tag{5.44}$$

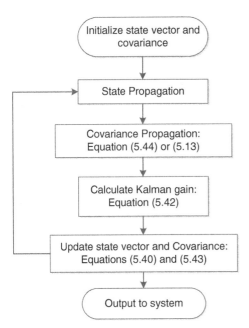

Figure 5.7 EKF flow diagram.

where the $\mathbf{P}_k^-$ is computed through the time integration of $\dot{\mathbf{P}}$. Similar to the continuous-time Kalman filter, we could also compute the discrete-time process noise covariance $\mathcal{Q}$ and the state transition matrix as shown in (5.32); then the discrete-time propagation of $\mathbf{P}_{k+1}^-$ in (5.13) can be used.

While EKF is easy implement, not all the Jabocian matrixes in (5.39) could be easily computed. If the analytical model of Jacobian matrix is not possible to be derived, then the finite element method such as shown in [7] and [13] could be considered.

The implementation of EKF is summarized in Figure 5.7. The nonlinear dynamic system and its respective state error covariance are first propagated. Then, the respective estimated measurement is computed using (5.34), followed by the state and covariance update using (5.40) and (5.43).

Example 5.2 Localization of Spacecraft Flying in Formation

We consider a two-spacecraft formation flying scenario as shown in Figure 5.8 (or the scenario in [14]), where spacecraft-1 (S/C-1) is being tracked by radar on the ground. For simplicity, the output of radar is preprocessed and the range, azimuth angle, λ, and elevation angle, ξ, of S/C-1 are expressed with respect to the Earth center. S/C-1 is equipped with a wireless local positioning system (WLPS; detailed in Chapter 34), a dynamic base station (DBS), and spacecraft-2 (S/C-2) is equipped with a WLPS transponder (TRX) [15], such that the relative range, azimuth angle, φ, and elevation angle, ϕ, of S/C-2 is measured by S/C-1.

Figure 5.8   Two spacecraft flying in formation.

Given the following spacecraft parameters with the WLPS measurement noises are 10 meters in range and 0.01° for azimuth and elevations angles, and radar measurement noises are 1 meter in range and 0.001° for azimuth and elevation angles, write an EKF simulation example such that both spacecraft positions and velocities are continuously estimated via radar and WLPS measurement. The suggested sampling rate is 10 seconds.

In this example, we assume the WLPS has infinite tracking range and the signal occultation by Earth is temporarily ignored. The Earth gravitational constant is given as $\mu = 38600.4\,\text{km}^3\text{s}^{-2}$ and Earth radius is 6378.145 km. The Keplerian parameters for two spacecraft at same epoch are shown in Table 5.1.

We desire to estimate the position $\mathbf{r}$ and velocity $\mathbf{v}$ of two spacecraft in Figure 5.8; therefore, the desired state vector is:

$$\mathbf{x} = \begin{bmatrix} \mathbf{r}_1^T & \mathbf{v}_1^T & \mathbf{r}_2^T & \mathbf{v}_2^T \end{bmatrix}^T. \tag{5.45}$$

The conversion of satellite Keplerian parameters into position and velocity vectors in an inertial reference frame can be found in [16] and [17]. The dynamic model of the ith satellite in term of position and velocity is given as:

$$\dot{\mathbf{r}}_i = \mathbf{v}_i$$

$$\dot{\mathbf{v}}_i = -\frac{\mu \mathbf{r}_i}{\|\mathbf{r}_i\|^3}. \tag{5.46}$$

TABLE 5.1.   Keplerian Parameters for Two Spacecraft in Formation

	Semimajor Axis	Eccentricity	Inclination	Right Ascension of Ascending Node	Argument of Perigee	True Anomaly
S/C 1	18378.145 km	0.3	25°	30.0°	20°	30°
S/C 2	18378.145 km	0.3	25°	30.1°	20°	30°

The notation $\|\cdot\|$ denotes the magnitude of a vector. Equation (5.46) shows that it is a nonlinear dynamic model; therefore linearization is required. Thus, $F(\hat{x}_k^+)$ in (5.39) corresponds to

$$F(\hat{x}_k^+) = \begin{bmatrix} \frac{\partial f}{\partial r_1} & \mathbf{0}_{6\times6} & \mathbf{0}_{6\times6} & \mathbf{0}_{6\times6} \\ \mathbf{0}_{6\times6} & \mathbf{0}_{6\times6} & \frac{\partial f}{\partial r_2} & \mathbf{0}_{6\times6} \end{bmatrix},$$
(5.47)

where

$$\frac{\partial f}{\partial r_i} = \frac{\mu}{\|r_i\|^5} \begin{bmatrix} \mathbf{0}_{3\times1} & \mathbf{0}_{3\times1} & \mathbf{0}_{3\times1} & \mathbf{I}_{3\times3} \\ 2r_{i,x}^2 - r_{i,y}^2 - r_{i,z}^2 & 3r_{i,x}r_{i,y} & 3r_{i,x}r_{i,z} & \mathbf{0}_{1\times3} \\ 3r_{i,x}r_{i,y} & 2r_{i,y}^2 - r_{i,x}^2 - r_{i,z}^2 & 3r_{i,y}r_{i,z} & \mathbf{0}_{1\times3} \\ 3r_{i,x}r_{i,z} & 3r_{i,y}r_{i,z} & 2r_{i,z}^2 - r_{i,x}^2 - r_{i,y}^2 & \mathbf{0}_{1\times3} \end{bmatrix}$$
(5.48)

$$r_i \equiv \begin{bmatrix} r_{i,x} & r_{i,y} & r_{i,z} \end{bmatrix}^{\mathrm{T}}.$$
(5.49)

Here, $\mathbf{0}_{m\times n}$ is an m by n matrix with all elements equal to zero. Next, we formulate the measurement model for both radar and WLPS measurements. The estimated measurement vector is given as

$$c(\hat{x}_k^-) \equiv \begin{bmatrix} h_{\text{radar}}^{\mathrm{T}} & h_{\text{WLPS}}^{\mathrm{T}} \end{bmatrix}^{\mathrm{T}},$$
(5.50)

where the subscript radar and WLPS represent the measurement model for the radar and WLPS, respectively. The detailed radar and WLPS measurement models are:

$$h_{\text{radar}}^{\mathrm{T}} \equiv \begin{bmatrix} \|r_1\| \\ \lambda \\ \xi \end{bmatrix} = \begin{bmatrix} \|r_1\| \\ \tan^{-1}\dfrac{r_{1,y}}{r_{1,x}} \\ \tan^{-1}\dfrac{r_{1,z}}{\sqrt{r_{1,x}^2 + r_{1,y}^2}} \end{bmatrix}$$
(5.51)

$$h_{\text{WLPS}}^{\mathrm{T}} \equiv \begin{bmatrix} \|r_{12}\| \\ \varphi \\ \phi \end{bmatrix} = \begin{bmatrix} \|r_{12}\| \\ \tan^{-1}\dfrac{r_{12,y}}{r_{12,x}} \\ \tan^{-1}\dfrac{r_{12,z}}{\sqrt{r_{12,x}^2 + r_{12,y}^2}} \end{bmatrix},$$
(5.52)

with $r_{12} = r_2 - r_1 \equiv \begin{bmatrix} r_{12,x} & r_{12,y} & r_{12,z} \end{bmatrix}^{\mathrm{T}}.$

$$C(\hat{\mathbf{x}}_k^-) \equiv \begin{bmatrix} \dfrac{\partial \mathbf{h}_{\text{radar}}}{\partial \mathbf{r}_1} & \mathbf{0}_{3\times3} & \mathbf{0}_{3\times3} & \mathbf{0}_{3\times3} \\ -\dfrac{\partial \mathbf{h}_{\text{WLPS}}}{\partial \mathbf{x}} & \mathbf{0}_{3\times3} & \dfrac{\partial \mathbf{h}_{\text{WLPS}}}{\partial \mathbf{x}} & \mathbf{0}_{3\times3} \end{bmatrix}$$ (5.53)

$$\frac{\partial \mathbf{h}_{\text{radar}}}{\partial \mathbf{r}_1} = \begin{bmatrix} \dfrac{r_{1,x}}{\|\mathbf{r}_1\|} & \dfrac{r_{1,y}}{\|\mathbf{r}_1\|} & \dfrac{r_{1,z}}{\|\mathbf{r}_1\|} \\ \dfrac{-\sin\lambda}{\|\mathbf{r}_1\|\cos\xi} & \dfrac{\cos\lambda}{\|\mathbf{r}_1\|\cos\xi} & 0 \\ \dfrac{-\cos\lambda\sin\xi}{\|\mathbf{r}_1\|} & \dfrac{-\sin\lambda\sin\xi}{\|\mathbf{r}_1\|} & \dfrac{\cos\xi}{\|\mathbf{r}_1\|} \end{bmatrix}$$ (5.54)

$$\frac{\partial \mathbf{h}_{\text{WLPS}}}{\partial \mathbf{x}} = \begin{bmatrix} \dfrac{r_{2,x}-r_{1,x}}{\|\mathbf{r}_{12}\|} & \dfrac{r_{2,y}-r_{1,y}}{\|\mathbf{r}_{12}\|} & \dfrac{r_{2,z}-r_{1,z}}{\|\mathbf{r}_{12}\|} \\ \dfrac{-\sin\varphi}{\|\mathbf{r}_{12}\|\cos\phi} & \dfrac{\cos\varphi}{\|\mathbf{r}_{12}\|\cos\phi} & 0 \\ \dfrac{-\cos\varphi\sin\phi}{\|\mathbf{r}_{12}\|} & \dfrac{-\sin\varphi\sin\phi}{\|\mathbf{r}_{12}\|} & \dfrac{\cos\phi}{\|\mathbf{r}_{12}\|} \end{bmatrix}.$$ (5.55)

Finally, the process noise matrix and its respective mapping matrix, $\mathbf{D}(t)$, are given as:

$$Q = \begin{bmatrix} 10^{-6} \times \mathbf{I}_{3\times3} & \mathbf{0}_{3\times3} \\ \mathbf{0}_{3\times3} & 10^{-6} \times \mathbf{I}_{3\times3} \end{bmatrix}$$ (5.56)

$$\mathbf{D}(t) = \begin{bmatrix} \mathbf{0}_{3\times3} & \mathbf{0}_{3\times3} \\ \mathbf{I}_{3\times3} & \mathbf{0}_{3\times3} \\ \mathbf{0}_{3\times3} & \mathbf{0}_{3\times3} \\ \mathbf{0}_{3\times3} & \mathbf{I}_{3\times3} \end{bmatrix}.$$ (5.57)

Figure 5.9 shows the root mean square error (RMSE) of the estimated position of both S/C-1 and S/C-2. The simulation is run for 12 hours. It shows that the RMSE is approximately 50 meters even though the radar range error is only 1 meter. This is due to the reason that the azimuth and elevation error of radar contributed higher error for such a high-altitude satellite (12,000 km from Earth surface).

The three-axis position estimation errors of S/C-1 together with three-sigma boundaries are plotted in Figure 5.10. The figure shows that with a proper tuning of process noise, the position error always falls within the three-sigma boundaries. For additional study, what would happen if we further lower the process noise parameter? Try to lower the process noise in (5.56) by a magnitude order each time the simulation is conducted. Observe the spacecraft position's RMSE and also the three-axis position error.

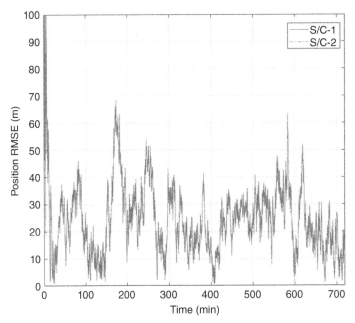

Figure 5.9   RMSE of S/C-1 and S/C-2.

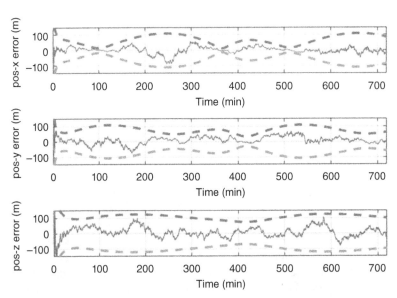

Figure 5.10   S/C-1 position errors (solid line) with three-sigma boundaries (dotted line).

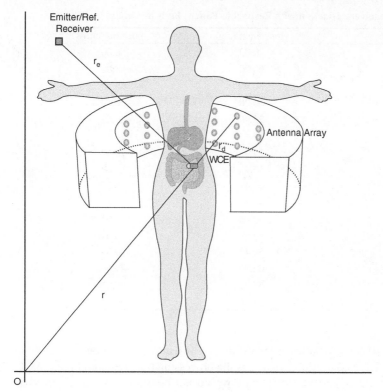

Figure 5.11   The WCE tracking process.

Example 5.3   Wireless Capsule Endoscopy Tracking

The wireless capsule endoscopy (WCE) is a pill-size medical device equipped with a camera and RF transmitter. The WCE is able to travel through human intestines, capture images of intestine interior, and wirelessly transmit images to a receiver located outside of the body. However, precise location of the WCE is required to be known in order to determine the location of each received image.

In this example, we use the EKF algorithm to track the WCE via DOA and TOA of the WCE signal (for details, see [18] and [19]). Similar to WPLS principles highlighted in Example 5.2, the system consists of multiple units that function similar to the DBS in the WLPS structure, although here DBS units are not moving and are installed on a cylinder that surrounds a patient (see Figure 5.11). DBS units periodically transmit an identification (ID) signal to the WCE. WCE consists of a simple TRX unit of WLPS. Once the WCE receives the signal, it will transmit the ID signal together with the image to the receivers surrounding the patient. The receivers' location is shown in Table 5.2. The TOA of the ID signal travels from emitter to WCE and from WCE to ith receiver is:

$$c_{i,\text{TOA}}(\mathbf{x}) = \frac{1}{c}\| r_i - r_d \| + \frac{1}{c}\| r_e - r_i \| + t_p, \tag{5.58}$$

TABLE 5.2. Receivers' Location with Respect to Patient Body as Origin

x-position (cm)	y-position (cm)	z-position (cm)
60.00	0.00	−60.00
0.00	60.00	−60.00
−60.00	0.00	−60.00
0.00	−60.00	−60.00
55.43	22.96	−26.67
−22.96	55.43	−26.67
−55.43	−22.96	−26.67
22.96	−55.43	−26.67
42.43	42.43	6.67
−42.43	42.43	6.67
−42.43	−42.43	6.67
42.43	−42.43	6.67
22.96	55.43	40.00
−55.43	22.96	40.00
−22.96	−55.43	40.00
55.43	−22.96	40.00

where $c = 3 \times 10^{10}$ cm/s denotes speed of light, r_d is receiver's location, r_e is emitter location, r_i is the WCE position, and t_p is the processing time required by the WCE to process the ID signal. For simplicity, we do not consider any signal delay or refraction due to the change to travelling medium as presented in [20].

In additional, each receiver consists of a set of antenna array that is capable of providing azimuth angle, φ, and elevation angle, ϕ, measurements:

$$c_{i,\mathrm{DOA}}(\mathbf{x}) = \begin{bmatrix} \varphi \\ \phi \end{bmatrix} = \begin{bmatrix} \tan^{-1} \dfrac{r_{i,y} - r_{d,y}}{r_{i,x} - r_{d,x}} \\ \tan^{-1} \dfrac{r_{i,z} - r_{d,z}}{\sqrt{(r_{i,x} - r_{d,x})^2 + (r_{i,y} - r_{d,y})^2}} \end{bmatrix}. \tag{5.59}$$

Here, $r_{i,x}$, $r_{i,y}$, and $r_{i,z}$ are x, y, and z components of $\mathbf{r}_i$. A similar case is applied for $r_{d,x}$, $r_{d,y}$, and $r_{d,z}$. For simplicity, the Jacobian matrix for (5.58) and (5.59) are not shown here (in fact, the Jacobian matrix of (5.59) is similar to the one presented in Example 5.2). The movement of the WCE is shown in Figure 5.12 (detail location data available in the provided MATLAB code). We assume that the WCE motion can be expressed in terms of the linear motion as follows:

$$\hat{\mathbf{x}}_{k+1}^- = \mathbf{A}_k \hat{\mathbf{x}}_k^+ + \mathbf{B}_k \mathbf{u}_k, \tag{5.60}$$

with

$$\mathbf{A}_k = \begin{bmatrix} \mathbf{I}_{3\times3} & \Delta t \mathbf{I}_{3\times3} \\ \mathbf{0}_{3\times3} & \mathbf{I}_{3\times3} \end{bmatrix} \tag{5.61}$$

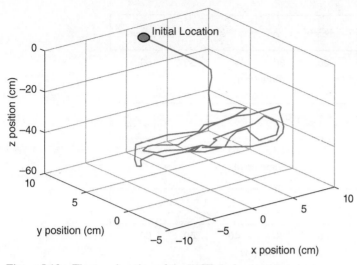

Figure 5.12 The true location of the WCE during tracking process.

$$\mathbf{B}_k = \mathbf{D}_k = \begin{bmatrix} 0.5\Delta t^2 \mathbf{I}_{3\times3} \\ \Delta t \mathbf{I}_{3\times3} \end{bmatrix}. \tag{5.62}$$

The WCE is equipped with an accelerometer, which is presented as the input, $\mathbf{u}_k$. Because the output of accelerometer contains noise, the input vector can be written as

$$\mathbf{u}_k = \mathbf{a}_k + \mathcal{N}(\mathbf{0}, \eta_u^2), \tag{5.63}$$

where $\mathbf{a}_k$ is the accelerometer reading, and $\mathcal{N}(\mathbf{0}, \eta_u^2)$ denotes the zero-mean Gaussian noise with the standard deviation of η_u.

Simulations assume the sampling period of 1 s, the processing time of 1 ms, the emitter location of [200, 100, 50], the TOA error of 0.1 cm, and the DOA error of 0.1°. The accelerometer noise is $\eta_u = 10$ mg in each axis. The discretized process noise matrix is given as:

$$\mathcal{Q} = \eta_u^2 \begin{bmatrix} \frac{1}{3}\Delta t^3 \mathbf{I}_{3\times3} & \frac{1}{2}\Delta t^2 \mathbf{I}_{3\times3} \\ \frac{1}{2}\Delta t^2 \mathbf{I}_{3\times3} & \Delta t \mathbf{I}_{3\times3} \end{bmatrix}. \tag{5.64}$$

The WCE estimation error is shown in Figure 5.13. The results depict that the EKF can estimate the WCE location, with its estimation error falling within the three-sigma boundaries. What will happen if we increase the TOA and DOA error? What happens if new process noise is required instead of the one provided by the accelerometer?

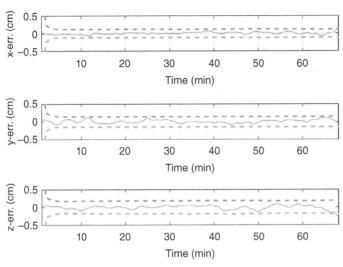

Figure 5.13   Three-axis WCE position estimation error.

5.7 UKF

In the previous section, the algorithm of EKF has been presented. The EKF algorithm is a low computational cost estimator, but requires performing first-order differentiation (or linearization) on the nonlinear dynamic and/or measurement model to compute the Kalman gain and covariance propagation. However, the linearization may result in a singularity matrix problem. For example, if the azimuth angle is 90°, it could result in a singularity issue in (5.54), such as division by zero. An alternative method for Kalman gain estimation is required to avoid such issues.

The UKF is introduced in [21]. UKF maps its state vector onto a specific boundary, which is normally formulated through its state error covariance. The mapped vectors are generally known as the *sigma points*. The mean and standard

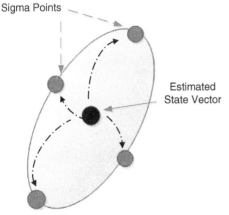

Figure 5.14   Example of sigma points in UKF.

deviation of the estimation performance of these sigma points can be computed and used for Kalman gain and covariance estimation.

Let the state vector, $\hat{\mathbf{x}}_k$, have the dimension of $n \times 1$, and the state error covariance $\hat{\mathbf{P}}_k$ have the dimension of $n \times n$. A matrix S_k, which has the dimension of $n \times n$, is computed from the square root matrix of $\hat{\mathbf{P}}_k$ such that

$$S_k \leftarrow n \text{ columns from } \sqrt{(n+\lambda)\left(\hat{\mathbf{P}}_k + \mathcal{Q}\right)}. \tag{5.65}$$

S_k contains a series of column vectors, $\sigma_k(i)$, that correspond to:

$$S_k \equiv \begin{bmatrix} \sigma_k(1) & \sigma_k(2) & \cdots & \sigma_k(n) \end{bmatrix}. \tag{5.66}$$

The square root matrix in (5.65) is typically obtained through Cholesky decomposition. It is noted that the parameter λ in (5.65) needs to be carefully tuned such that

$$\lambda = \alpha^2(n+\kappa) - n \tag{5.67}$$

$$\kappa = 3 - n, \quad \text{for } n \neq 3. \tag{5.68}$$

Both α and κ are constant values. The parameter α determines the spread of the sigma points, with $1 \times 10^{-4} \leq \alpha \leq 1$. Although κ is typically defined in (5.68), if $n = 3$, then κ is required to be carefully selected. The UKF *sigma points* correspond to:

$$\Xi(+) = \begin{bmatrix} \hat{\mathbf{x}}_k + \sigma_k(1) & \hat{\mathbf{x}}_k + \sigma_k(2) & \cdots & \hat{\mathbf{x}}_k + \sigma_k(n) \end{bmatrix} \tag{5.69}$$

$$\Xi(-) = \begin{bmatrix} \hat{\mathbf{x}}_k - \sigma_k(1) & \hat{\mathbf{x}}_k - \sigma_k(2) & \cdots & \hat{\mathbf{x}}_k - \sigma_k(n) \end{bmatrix} \tag{5.70}$$

$$\chi = \begin{bmatrix} \hat{\mathbf{x}}_k & \Xi(+) & \Xi(-) \end{bmatrix}. \tag{5.71}$$

There are a total of $2n+1$ column vectors in χ. The UKF requires all column vectors of χ to be evaluated (or propagated), that is,

$$\chi_{k+1}(i) \leftarrow \dot{\chi}(i) = f(\chi_k(i), \mathbf{u}(t), t), \quad i = 0, 1, 2, \ldots, 2n \tag{5.72}$$

$$\gamma_{k+1}(i) = c(\chi_{k+1}(i), t), \quad i = 0, 1, 2, \ldots, 2n. \tag{5.73}$$

This represents a total of $2n+1$ propagations $f(\chi_k(i), \mathbf{u}(t), t)$ and estimated measurement computation $c(\chi_{k+1}(i), t)$ are required. After all the sigma points are propagated and the respected estimated measurement are computed, the mean estimate and output are computed based on the sigma points:

$$\hat{\mathbf{x}}_{k+1}^- = \sum_{i=0}^{2n} W_i^{mean} \chi_{k+1}(i) \tag{5.74}$$

$$\hat{\mathbf{y}}_{k+1}^{-} = \sum_{i=0}^{2n} W_i^{mean} \Upsilon_{k+1}(i),$$ (5.75)

where

$$W_0^{mean} = \frac{\lambda}{n+\lambda}$$ (5.76)

$$W_i^{mean} = \frac{1}{2(n+\lambda)}, \quad i = 1, 2, \ldots, 2n.$$ (5.77)

Instead of propagating $\hat{\mathbf{P}}_k^+$, the UKF calculates the state error covariance, $\hat{\mathbf{P}}_{k+1}^-$, output covariance, $\mathbf{P}_{k+1}^{yy}$, and cross-correlation matrix, $\mathbf{P}_{k+1}^{xy}$, directly from the sigma points, which respectively correspond to:

$$\hat{\mathbf{P}}_{k+1}^{-} = \sum_{i=0}^{2n} W_i^{cov} \left[\boldsymbol{\chi}_{k+1}(i) - \hat{\mathbf{x}}_{k+1}^{-} \right] \left[\boldsymbol{\chi}_{k+1}(i) - \hat{\mathbf{x}}_{k+1}^{-} \right]^{\mathrm{T}}$$ (5.78)

$$\mathbf{P}_{k+1}^{yy} = \sum_{i=0}^{2n} W_i^{cov} \left[\Upsilon_{k+1}(i) - \hat{\mathbf{y}}_{k+1}^{-} \right] \left[\Upsilon_{k+1}(i) - \hat{\mathbf{y}}_{k+1}^{-} \right]^{\mathrm{T}}$$ (5.79)

$$\mathbf{P}_{k+1}^{xy} = \sum_{i=0}^{2n} W_i^{cov} \left[\boldsymbol{\chi}_{k+1}(i) - \hat{\mathbf{x}}_{k+1}^{-} \right] \left[\Upsilon_{k+1}(i) - \hat{\mathbf{y}}_{k+1}^{-} \right]^{\mathrm{T}},$$ (5.80)

where

$$W_0^{cov} = \frac{\lambda}{n+\lambda} + \left(1 - \alpha^2 + \beta\right)$$ (5.81)

$$W_i^{cov} = \frac{1}{2(n+\lambda)}, \quad i = 1, 2, \ldots, 2n.$$ (5.82)

Finally, the mean state and covariance are updated, similar to the standard Kalman filter:

$$\hat{\mathbf{x}}_{k+1}^{+} = \hat{\mathbf{x}}_{k+1}^{-} + \mathbf{K} \left[\hat{\mathbf{y}}_{k+1} - \hat{\mathbf{y}}_{k+1}^{-} \right]$$ (5.83)

$$\hat{\mathbf{P}}_{k+1}^{+} = \hat{\mathbf{P}}_{k+1}^{-} - \mathbf{K} \left(\mathbf{P}_{k+1}^{yy} + \mathbf{R} \right) \mathbf{K}^{\mathrm{T}}$$ (5.84)

$$\mathbf{K} = \mathbf{P}_{k+1}^{xy} \left(\mathbf{P}_{k+1}^{yy} + \mathbf{R} \right)^{-1}.$$ (5.85)

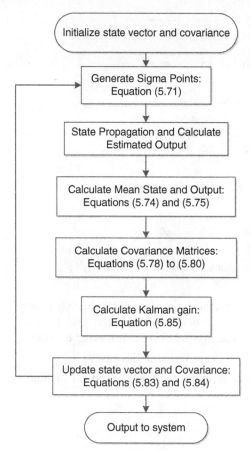

Figure 5.15   The UKF flow diagram.

Note that the Kalman gain is computed using the output covariance and cross-correlation matrix. The summary of the UKF flowchart is shown in Figure 5.15. First, sigma points are generated using the state error covariance. There are a total of $2n+1$ sigma point vectors. Then, all sigma point vectors are propagated, and the respective output is computed. Next, the mean of propagated sigma points and respective mean output are computed. Then, the new state and output error covariance and cross-correlation matrix are computed with respect to the mean sigma points and output. Finally, the Kalman gain is computed and both mean state vector and state error covariance are updated.

Example 5.4   Wireless Capsule Endoscopy Tracking via UKF

The same WCE problem as in Example 5.3 is considered for this UKF example. The process/measurement noise variances and initial state conditions remain the same. However, additional parameters are needed for the UKF algorithm. These parameters are $\alpha = 0.1$ and $\beta = 0.5$.

The UKF's three-axis position estimation error is shown in Figure 5.16. In addition, Figure 5.17 compares the average RMSE of UKF and EKF algorithms. As

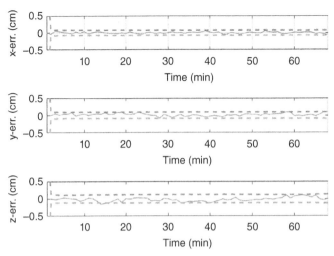

Figure 5.16   UKF three-axis WCE position estimation error.

compared to the results in EKF, the UKF overall estimation error is approximately 10% lower than the EKF. This shows that UKF performs better than EKF for nonlinear systems. For more in-depth study, you may address the following questions:

Question: What will happen if both α and β are varied in this case?

Question: What will be the performance difference if UKF is applied for linear systems?

5.8   CKF

Previous sections presented the commonly used Kalman filter variations. While the Kalman filter has shown its applications in localization problem since the Apollo

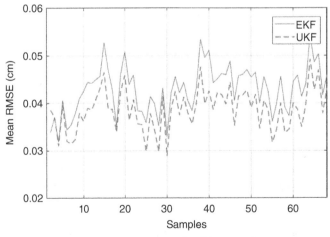

Figure 5.17   UKF and EKF average RMSE comparison.

mission era, it still has its limitations. The Kalman filter does not include any constrain condition for the purpose of estimation. If an element in the state vector is required to fulfill a certain constraint condition, or if the Kalman filter needs to be implemented on a more complex problem, a modification on the Kalman filter algorithm is needed.

From this section onward, we will present several variations of Kalman filter that are specifically designed for applications that possess constrained estimation, low computational cost (fixed gain) and highly nonlinear system models.

In constrained estimation in the CKF, the state vector is required to fulfill the following constraint condition:

$$\mathbf{b} = \mathbf{d}(\mathbf{x}), \tag{5.86}$$

where $\mathbf{d}(\mathbf{x})$ could be either linear or nonlinear model. Authors in [22] have detailed various CKF algorithms. These CKF algorithms are:

- Model reduction method

- Perfect measurement method

- Projection method

The implementation of each method will be briefly presented in this section.

5.8.1 Model Reduction Method

In the model reduction method, the size of the state vector is reduced. This method is achievable if a given constrained condition is linear, and one element of the state vector is constrained with respect to other elements of the state vector. For example, let us consider a 3×1 state vector, $\mathbf{x} = \begin{bmatrix} x_1 & x_2 & x_3 \end{bmatrix}^T$, with:

$$\mathbf{A}_k = \begin{bmatrix} 1 & 0 & 2 \\ -2 & 1 & 3 \\ 0 & 1 & 1 \end{bmatrix}. \tag{5.87}$$

If the constrained condition is given as

$$x_3 = x_1 - 2x_2, \tag{5.88}$$

then we can substitute the expression of (5.88) into (5.87) to achieve model reduction, such that the new $\mathbf{A}_k$ matrix (with the size of 2×2) equals

$$\mathbf{A}_k = \begin{bmatrix} 3 & -4 \\ 1 & -5 \end{bmatrix}. \tag{5.89}$$

In addition, the new state vector has the size of 2×1, which is $\mathbf{x} = \begin{bmatrix} x_1 & x_2 \end{bmatrix}^T$. In general, the model reduction is simple and useful for any linear system with a linear constrained condition.

5.8.2   Perfect Measurement Method

For the perfect measurement method, the constraint function in (5.86) is appended into the measurement vector in (5.34). Then, the new measurement vector becomes:

$$\tilde{\mathbf{y}} = \begin{bmatrix} \mathbf{c}(\mathbf{x}) \\ \mathbf{d}(\mathbf{x}) \end{bmatrix} + \begin{bmatrix} \upsilon \\ \mathbf{0} \end{bmatrix}. \tag{5.90}$$

In addition, $\mathbf{C}(\hat{\mathbf{x}})$ in (5.42) and (5.43) includes the expression of both $\mathbf{c}(\mathbf{x})$ and $\mathbf{d}(\mathbf{x})$. The perfect measurement method is then processed similar to the standard Kalman filter in Sections 5.4 and 5.6. It should be noted that for the perfect measurement case, there is a possibility where the term $\left(\mathbf{CP^-C^T} + \mathbf{R}\right)$ in (5.17) is not invertible due to insufficient matrix rank, as

$$\mathbf{R} = \begin{bmatrix} \mathrm{E}\{\upsilon\upsilon^T\} & \mathbf{0} \\ \mathbf{0} & \mathbf{0} \end{bmatrix}. \tag{5.91}$$

If matrix rank is an issue, then we can assume the virtual perturbation σ_b in the constrained equation, such that $\mathbf{R}$ becomes:

$$R = \begin{bmatrix} \mathrm{E}\{\upsilon\upsilon^T\} & \mathbf{0} \\ \mathbf{0} & \mathrm{E}\{\sigma_b\sigma_b^T\} \end{bmatrix}. \tag{5.92}$$

While virtual perturbation is included in the measurement covariance matrix, it does not mean that noise is added into the constraint.

5.8.3   Projection Method

In the projection method, the constraint condition is applied after the Kalman filter update process such that

$$\hat{\mathbf{x}}_k^{++} = \hat{\mathbf{x}}_k^+ + \mathbf{L}\left(\mathbf{b} - \mathbf{d}\left(\hat{\mathbf{x}}_k^+\right)\right). \tag{5.93}$$

The superscript "++" presents the update of $\hat{\mathbf{x}}_k^+$. The matrix $\mathbf{L}$ is the constraint gain. There are several methods to compute the $\mathbf{L}$ matrix, such as Kalman gain-like and the least square (or weighted least square) approaches. For the Kalman gain-like approach, $\mathbf{L}$ corresponds to:

$$\mathbf{L} = -\mathbf{P}_k^+\mathbf{D}^T\left(\mathbf{DP}_k^+\mathbf{D}^T\right)^{-1}, \tag{5.94}$$

where $\mathbf{D} \equiv \dfrac{\partial \mathbf{d}\left(\hat{\mathbf{x}}_k^+\right)}{\partial \hat{\mathbf{x}}}$. For the weighted least square approach, $\mathbf{L}$ is given as:

$$\mathbf{L} = -\mathbf{W}^{-1}\mathbf{D}^T\left(\mathbf{DW}^{-1}\mathbf{D}^T\right)^{-1}. \tag{5.95}$$

If $\mathbf{W} = \mathbf{I}$, Equation (5.95) becomes standard least square approach. For projection method, the $\mathbf{P}_k^+$ does not need updates. However, if $\mathbf{P}_k^+$ requires updates along with $\hat{\mathbf{x}}_k^+$, then additional steps are considered to avoid the discontinuity caused by updated $\mathbf{P}_k^+$ at the constraint application [23]. Therefore, the constraint update in (5.93) and respective $\mathbf{P}_k^+$ update are modified to:

$$\hat{\mathbf{x}}_k^{++} = \hat{\mathbf{x}}_k^+ + \alpha\mathbf{L}\left(\mathbf{b} - \mathbf{d}\left(\hat{\mathbf{x}}_k^+\right)\right) \tag{5.96}$$

$$\hat{\mathbf{P}}_k^{++} = \left(\mathbf{I} - \beta\mathbf{L}\mathbf{D}\right)\hat{\mathbf{P}}_k^+,$$

where $0 < \alpha < 1$ and $0 < \beta < 1$.

Both α and β are controlled parameters, which control the ratio of constrain update being applied to $\hat{\mathbf{x}}_k^+$ and $\hat{\mathbf{P}}_k^+$. The parameter β controls the speed of covariance convergence, and the parameter α controls the amplitude of updates in the state due to the application of constraint. It is important to note that α and β in (5.96) are different than the α and β in UKF.

CKF Summary: In general, if the model reduction method or perfect measurement is used for constrained estimation, then the standard Kalman filter flow shown in Figure 5.2 can be used. If the project method is considered for constrained estimation, then the CKF flow diagram is shown in Figure 5.18. The flow diagram of projection method is similar to that of CKF, with the exception that additional state updates are applied after the standard Kalman update and before the state vector and error covariance are propagated.

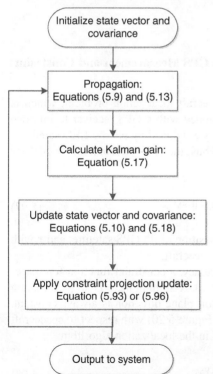

Figure 5.18   CKF's projection method flow diagram.

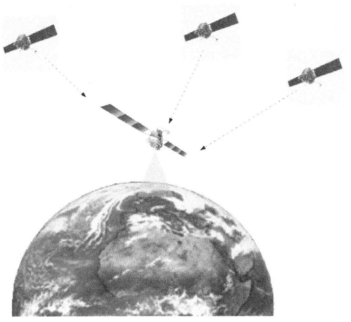

Figure 5.19   Illustration of a spacecraft receiving GPS satellites' signal.

Example 5.5   Spacecraft Localization via GPS Measurement and Constraint Function

Let's consider a low earth orbit (LEO) spacecraft localization problem, which is shown in Figure 5.19. The spacecraft is equipped with a GPS receiver to improve its positioning accuracy. Due to the interference induced by nearby LEO satellites, the GPS measurement error is 100 meters. Thus, the GPS measurement model for spacecraft is:

$$\tilde{\mathbf{y}}_{GPS} = \| \mathbf{r}_{i,GPS} - \mathbf{r}_{S/C} \| + v_{GPS}. \tag{5.97}$$

The $\mathbf{r}$ with subscript "i, GPS" indicates the position of ith GPS satellite, and $\mathbf{r}$ with subscript "S/C" denotes the position of the spacecraft.

To further improve the positioning accuracy, a constraint function for spacecraft localization algorithm has been considered. Assuming there is no perturbation and Earth-flattening effect, the spacecraft orbit plane direction is always constant. Thus, an orthogonal vector ("h" direction in Figure 5.20) with respect to spacecraft orbit plane is used as the constraint function in the localization algorithm:

$$\mathbf{b} = \frac{[\mathbf{r}_{S/C} \times] \mathbf{v}_{S/C}}{\|[\mathbf{r}_{S/C} \times] \mathbf{v}_{S/C}\|}, \tag{5.98}$$

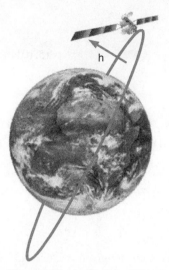

Figure 5.20 Indication of the angular moment direction, "h," in a spacecraft orbit.

where $[a \times]$ is cross-product matrix $\mathbf{v}$ with subscript "S/C" that denotes the velocity of the spacecraft. Here, the perfect measurement method is used; thus, the measurement vector for CKF is:

$$\mathbf{y}_{CKF} = \begin{bmatrix} \mathbf{y}_{GPS} \\ \mathbf{b} \end{bmatrix} \tag{5.99}$$

In this example, it is assumed that four (and only four) GPS satellites are always visible by the spacecraft. The corresponding Jacobian matrix for (5.97), (5.98), and (5.99) are:

$$\mathbf{C}(\hat{\mathbf{x}}_k^-) = \begin{bmatrix} \dfrac{\partial c_{1,GPS}(\hat{\mathbf{x}})}{\partial \mathbf{r}} & \mathbf{0}_{3\times3} \\ \dfrac{\partial c_{2,GPS}(\hat{\mathbf{x}})}{\partial \mathbf{r}} & \mathbf{0}_{3\times3} \\ \dfrac{\partial c_{3,GPS}(\hat{\mathbf{x}})}{\partial \mathbf{r}} & \mathbf{0}_{3\times3} \\ \dfrac{\partial c_{4,GPS}(\hat{\mathbf{x}})}{\partial \mathbf{r}} & \mathbf{0}_{3\times3} \\ \dfrac{\partial \mathbf{b}(\hat{\mathbf{x}})}{\partial \mathbf{r}} & \dfrac{\partial \mathbf{b}(\hat{\mathbf{x}})}{\partial \mathbf{v}} = \end{bmatrix}, \tag{5.100}$$

with

$$\frac{\partial c_{i,\text{GPS}}(\hat{\mathbf{x}})}{\partial \mathbf{r}} = \frac{\mathbf{r}_{\text{s/c}} - \mathbf{r}_{i,\text{GPS}}}{\|\mathbf{r}_{i,\text{GPS}} - \mathbf{r}_{\text{s/c}}\|} \tag{5.101}$$

$$\frac{\partial \mathbf{b}(\hat{\mathbf{x}})}{\partial \mathbf{r}} = \frac{-[\mathbf{v}_{\text{s/c}} \times]}{\|[\mathbf{r}_{\text{s/c}} \times]\mathbf{v}_{\text{s/c}}\|} + \frac{1}{2} \frac{[\mathbf{r}_{\text{s/c}} \times]\mathbf{v}_{\text{s/c}}([[\mathbf{r}_{\text{s/c}} \times]\mathbf{v}_{\text{s/c}} \times]\mathbf{v}_{\text{s/c}})^{\text{T}}}{\|[\mathbf{r}_{\text{s/c}} \times]\mathbf{v}_{\text{s/c}}\|^{3/2}}$$

$$\frac{\partial \mathbf{b}(\hat{\mathbf{x}})}{\partial \mathbf{v}} = \frac{[\mathbf{r}_{\text{s/c}} \times]}{\|[\mathbf{r}_{\text{s/c}} \times]\mathbf{v}_{\text{s/c}}\|} - \frac{1}{2} \frac{[\mathbf{r}_{\text{s/c}} \times]\mathbf{v}_{\text{s/c}}([[\mathbf{r}_{\text{s/c}} \times]\mathbf{v}_{\text{s/c}} \times]\mathbf{r}_{\text{s/c}})^{\text{T}}}{\|[\mathbf{r}_{\text{s/c}} \times]\mathbf{v}_{\text{s/c}}\|^{3/2}}.$$

The spacecraft true and initial estimate Keplerian parameters are given in Table 5.3. A total of 40 Monte Carlo simulations are conducted at a 1-second sampling rate, for 6000 seconds. The positioning accuracy of CKF is compared with standard EKF without the constraint function in this example.

Simulations are conducted assuming the process noise standard deviation is 1 meter and the standard deviation of virtual perturbation is 0.001. The conversion and propagation of a GPS's almanac into position and velocity are provided in the MATLAB code. The propagation model and partial derivative of spacecraft orbit model are provided in Example 5.2.

Figure 5.21 presents the comparison of positioning RMSE between CKF and EKF. The result clearly shows that CKF has a lower positioning RMSE than the EKF. **Question:** Did reader notice MATLAB warning message on the near singularity matrix for $(\mathbf{CP}^-\mathbf{C}^{\text{T}} + \mathbf{R})$?

5.9 MEASUREMENT FUSION KALMAN FILTER

The Kalman filter inherently is a multisensor fusion algorithm. This could be easily proven by considering that the measurement vector of Kalman filter contains M number of sensor measurements:

TABLE 5.3. Keplerian Parameters

Item	True	Initial Estimates
Semimajor axis (km)	6978.145	6978.145
Eccentricity	0.0	0.0
Inclination (°)	98	97.8
Argument of perigee (°)	0	0
Right ascension of ascending node (°)	100	100
True anomaly (°)	15	15.2

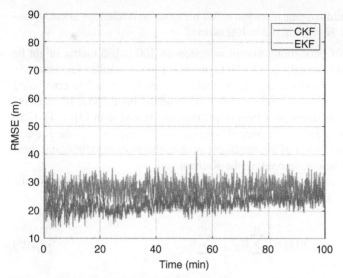

Figure 5.21   Positioning RMSE comparison between CKF and EKF.

$$\mathcal{Y}(t) = \left[\left(\tilde{\mathbf{y}}^1(t) \right)^T \quad \left(\tilde{\mathbf{y}}^2(t) \right)^T \quad \cdots \quad \left(\tilde{\mathbf{y}}^M(t) \right)^T \right]^T \tag{5.102}$$

The respective observation matrix and measurement covariance from (5.5) and (5.6) correspond to:

$$\mathcal{C}(t) = \begin{bmatrix} \mathbf{C}^1(t) \\ \mathbf{C}^2(t) \\ \vdots \\ \mathbf{C}^M(t) \end{bmatrix} \tag{5.103}$$

$$\mathcal{R}(t) = \begin{bmatrix} \mathbf{R}_1(t) & 0 & \cdots & 0 \\ 0 & \mathbf{R}_2(t) & 0 & \vdots \\ \vdots & 0 & \ddots & 0 \\ 0 & \cdots & 0 & \mathbf{R}_M(t) \end{bmatrix}. \tag{5.104}$$

Each of the sensor measurements may differ from each other, such that the measurement vector in (5.102) consists of GPS, magnetometer, sun tracker, star tracker, and other measurement readings. Refer to the Kalman gain and update in (5.19); without any additional steps or algorithms, the Kalman filter automatically fuses all measurement readings and updates the state vector.

Question: What would happen if we have 100 measurement readings at each sampling time, such that $\mathcal{R}(t)$ is a 100×100 matrix?

Question: The cost of performing matrix inversion on 100×100 matrix might be very high. Is it possible to split $\mathcal{Y}(t)$, $\mathcal{C}(t)$, and $\mathcal{R}(t)$ into a smaller vector/matrix size, such as 5×5 for $\mathcal{R}(t)$, and perform measurement fusion before computing the Kalman gain to avoid performing matrix inverse on the huge 100×100 matrix?

The alternative measurement fusion methods are available in [3], [24], [25]. These techniques aim to reduce the overall computational complexity of the Kalman filter. They require the fusion of measurement vector, observation matrix, and measurement covariance before computing the Kalman gain.

Here, let the overhead notation "$\overline{\cdot}$" denote the fused vector/matrix. For the first alternative method, the measurement fusion algorithm corresponds to:

$$\bar{\mathbf{y}}(t) = \left[\sum_{j}^{M} \mathbf{R}_j^{-1} \right]^{-1} \sum_{j}^{M} \mathbf{R}_j^{-1} \mathbf{y}^j(t) \tag{5.105}$$

$$\bar{\mathbf{C}}(t) = \left[\sum_{j}^{M} \mathbf{R}_j^{-1} \right]^{-1} \sum_{j}^{M} \mathbf{R}_j^{-1} \mathbf{C}^j(t) \tag{5.106}$$

$$\bar{\mathbf{R}}(t) = \left[\sum_{j}^{M} \mathbf{R}_j^{-1} \right]^{-1} . \tag{5.107}$$

The fused covariance $\bar{\mathbf{R}}(t)$ in (5.107) indicates that a more accurate measurement vector can be obtained through the measurement fusion process. However, it may result in overconfidence in the error estimation of the fused measurements, and accordingly underestimation of covariance matrix. Therefore, another measurement fusion algorithm is introduced in [3] to avoid covariance to get underestimated. This algorithm corresponds to:

$$\bar{\mathbf{y}}(t) = \left[\sum_{j}^{M} w_j \mathbf{R}_j^{-1} \right]^{-1} \sum_{j}^{M} w_j \mathbf{R}_j^{-1} \mathbf{y}^j(t) \tag{5.108}$$

$$\bar{\mathbf{C}}(t) = \left[\sum_{j}^{M} w_j \mathbf{R}_j^{-1} \right]^{-1} \sum_{j}^{M} w_j \mathbf{R}_j^{-1} \mathbf{C}^j(t) \tag{5.109}$$

$$\bar{\mathbf{R}}(t) = M \left[\sum_{j}^{M} w_j \mathbf{R}_j^{-1} \right]^{-1} , \tag{5.110}$$

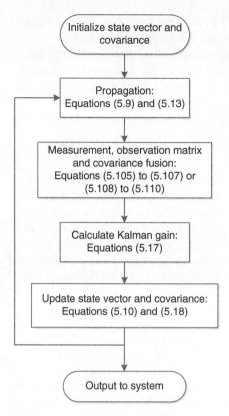

Figure 5.22   Flow chart of MFKF.

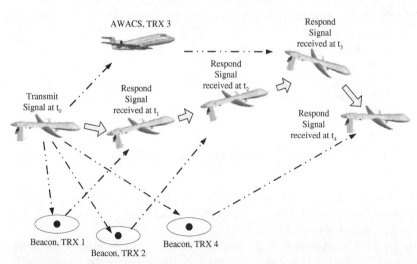

Figure 5.23   Illustration of transmission between UAV (DBS) and TRXs.

where w_j is a weighting factor, such that $\sum_j^m w_j = 1$. Introducing w_j allows accurate estimation via applying a weighting factor to each measurement vector.

The overall measurement fusion Kalman filter (MFKF) process still remains similar to Sections 5.4 and 5.6. However, the measurement fusion is introduced before the Kalman gain computation, which is shown by the flow diagram in Figure 5.22.

Example 5.6   Unmanned Aerial Vehicle Localization via Time-Delayed Measurement

In this example, we use the WLPS (see Example 5.2 and Chapter 34) principles for unmanned aerial vehicle (UAV) localization. Here, each UAV is equipped with DBS and simple TRX units are distributed between the initial point and the destination to allow UAVs to travel safely to the destination via GPS-free localization. It should be noted that GPS can be subject to jamming or error due to scattering or non-line-of-sight (NLOS) in urban areas. We consider the simplified localization problem in [3] where a UAV travels from an airbase to a destination at a straight-line constant speed direction, such that:

$$\mathbf{A}_k = \begin{bmatrix} \mathbf{I}_{3\times3} & \mathbf{0}_{3\times3} \\ \mathbf{0}_{3\times3} & \mathbf{0}_{3\times3} \end{bmatrix}, \quad \mathbf{B}_k = \mathbf{0}. \tag{5.111}$$

There are a number of beacon transponders (or TRX in Figure 5.23 and locations are shown in Figure 5.24) along the airbase to the destination. The UAV is equipped with a transceiver (or DBS), which periodically broadcasts a signal. TRXs located within the communication range receive the DBS ID signal, and reply to the DBS. The TRX locations are assumed known. Thus, the signal submitted by TRX includes the TRX location. When the DBS receives the signal from a particular TRX, it computes the round-trip TOA and the DOA of the signal. However, due to the signal's travelling time delay, the UAV receives the responded signal at different time instances, t_1, t_2, t_3, etc. Here, it is assumed that the last responded signal received by the UAV is always within a sampling time period. The total TOA for signal travel from UAV to beacon and from beacon back to UAV is:

$$c(\mathbf{x})_{j,\text{TOA}} = \frac{1}{c}\|\mathbf{r} - \mathbf{b}_j\| + \delta t_{r/j} \frac{1}{c}\|\mathbf{b}_j - \mathbf{r}\| + \delta t_{j/r}, \tag{5.112}$$

where $\mathbf{r}$ is UAV position, $\mathbf{b}_j$ is jth beacon position, c is speed of light, $\delta t_{r/j}$ denotes the signal travelling time delay from UAV to jth beacon, and $\delta t_{j/r}$ denotes the signal travelling time delay from jth beacon to UAV.

On the other hand, the DOA of a beacon with respect to the UAV remains the same as in (5.52); thus, it is not shown in this example. Overall, the measurement vector in terms of polar coordinates (expressed in term of TOA and DOA) is:

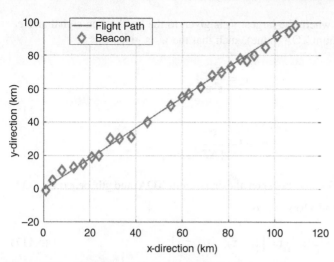

Figure 5.24   UAV flight path and beacons location.

$$\tilde{\mathbf{y}}(\mathbf{t}) = \left[\left(\tilde{\mathbf{y}}_1^p \right)^{\mathrm{T}} \quad \left(\tilde{\mathbf{y}}_2^p \right)^{\mathrm{T}} \quad \cdots \quad \left(\tilde{\mathbf{y}}_M^p \right)^{\mathrm{T}} \right]^{\mathrm{T}}$$

(5.113)

$$\tilde{\mathbf{y}}_j^p \equiv \begin{bmatrix} \mathrm{c}(\mathbf{x})_{j,\mathrm{TOA}} \\ \mathrm{c}(\mathbf{x})_{j,\mathrm{DOA}} \end{bmatrix} + \mathbf{v}_j(t),$$

where $\mathbf{v}_j(t)$ is measurement noise vector.

To conduct the measurement fusion process, we first convert the measurements from polar coordinates into Cartesian coordinates, such as

$$\tilde{\mathbf{y}}_j^c = polar2cartesian \left(\mathrm{c}(\mathbf{x})_{TOA,j}, \mathrm{c}(\mathbf{x})_{DOA,j} \right).$$

(5.114)

The details of (5.114), including the measurement noise covariance in polar coordinate $\mathbf{R}_j$ into Cartesian coordinate $\mathbf{R}_j^c$ are shown in [3] and [26]. It should be noted that we ignore the signal travelling time delay error during the polar-to-Cartesian coordinate conversion and fusion process. That is the distance between ith beacon and UAV is assumed to be:

$$\| \hat{r} - \mathbf{b}_j \| \cong \frac{c}{2} \mathrm{c}(\mathbf{x})_{j,\mathrm{TOA}}.$$

(5.115)

The signal travelling time delay is not estimated in MKFK because the computational cost of signal travelling time delay estimation is high. Instead, the additional error due to the signal travelling time delay could be compensated for through the fusion algorithm. Next, the weighting factor, w_j in (5.108) is computed based on the

TOA of each beacon, where the largest w_j is given to the last beacon signal arrived at the UAV (or maximum TOA value), such that the w_j is given as:

$$w_j = \frac{\tau_j}{\tau^T \tau}$$

(5.116)

$$\tau_j = 1 - \frac{\Delta T_j}{\| \Delta T \|},$$

where ΔT_j is the difference between the maximum TOA and jth beacon's TOA, $\Delta T_j = \max(c(\mathbf{x})_{\text{TOA}}) - c(\mathbf{x})_{j,\text{TOA}}$. Also,

$$\tau = \begin{bmatrix} \tau_1 & \tau_2 & \cdots & \tau_M \end{bmatrix}$$

(5.117)

$$\Delta T = \begin{bmatrix} \Delta T_1 & \Delta T_2 & \cdots & \Delta T_M \end{bmatrix}.$$

Next, the fusion of $\mathbf{R}_j^c$ and $\tilde{y}_j^c$ is given as follows:

$$\bar{\mathbf{R}}^c(t) = M \left[\sum_j^M w_j \left(\mathbf{R}_j^c \right)^{-1} \right]^{-1}$$

(5.118)

$$\bar{y}(t) = \left[\sum_j^M w_j \left(\mathbf{R}_j^c \right)^{-1} \right]^{-1} \sum_j^M w_j \left(\mathbf{R}_j^c \right)^{-1} \left[\mathbf{b}_j + \tilde{y}_j^c \right].$$

(5.119)

Note that $\tilde{y}_j^c$ is the UAV position with respect to the jth beacon location. Therefore, the jth beacon location, $\mathbf{b}_j$, with respect to the airbase location, is needed to maintain the origin of all $\tilde{y}_j^c$, the airbase location. On the other hand, it could be easily shown that the resultant observation matrix of (5.109) corresponds to:

$$\bar{\mathbf{C}}(t) = [-\mathbf{I}_{3\times3} \quad \mathbf{0}_{3\times3}].$$

(5.120)

For simulation purposes, the initial position and velocity of the UAV, with respect to the origin, that is, the airbase location, is

$$\mathbf{r}(t) = \begin{bmatrix} 0.1 & -0.1 & 9.2 \end{bmatrix}^T \text{ km}$$

(5.121)

$$\mathbf{v}(t) = \begin{bmatrix} 0.6 & 0.055 & -0.0048 \end{bmatrix}^T \text{ km/s}.$$

We assume our initial estimated position and velocity of the UAV contains 10 km and 0.1 km/s random error. Accordingly, the initial state error covariance is

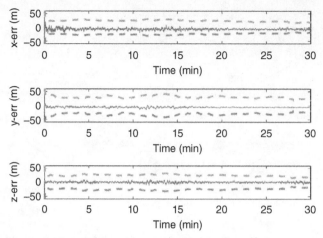

Figure 5.25   MFKF position estimation error.

$$\mathbf{P}(t_0) = \begin{bmatrix} 225\mathbf{I}_{3\times3} & 0_{3\times3} \\ 0_{3\times3} & 0.01\mathbf{I}_{3\times3} \end{bmatrix}. \tag{5.122}$$

Furthermore, the location of each beacon with respect to airbase position is provided in the MATLAB code. In addition, the discretized process noise covariance is given as

$$\mathcal{Q} = \eta_u^2 \begin{bmatrix} \dfrac{1}{3}\Delta t^3 \mathbf{I}_{3\times3} & 0.5\Delta t^2 \mathbf{I}_{3\times3} \\ 0.5\Delta t^2 \mathbf{I}_{3\times3} & \Delta t \mathbf{I}_{3\times3} \end{bmatrix}, \tag{5.123}$$

where $\eta_u = 0.001$ is the process noise parameter, which is obtained through tuning process.

The TOA error is 100 m, DOA error is $0.1°$, the sampling time is 0.1 seconds, and the beacon range limit is 30 km. The simulation is conducted for 30 minutes. The position estimation error of MFKF is shown in Figure 5.25. It shows that the measurement fusion method allows the algorithm to have a fast convergence rate, and the errors are always within the predicted three-sigma boundary.

5.10   GAIN-SCHEDULED KALMAN FILTER

Not every Kalman filter variation requires the state error covariance matrix to compute the Kalman gain matrix. In other words, some Kalman filter may compute the Kalman gain off-line and store it in a memory, such as the gain-scheduled Kalman filter (GSKF). The GSKF is often applied for the case where the dynamic system model is time invariant, so that the Kalman gain is a constant matrix and can be computed off-line. As often, the Kalman gain matrix in GSKF is expressed in term of both process noise and measurement noise covariance.

As shown in Figure 5.26, an appropriate stored Kalman gain is chosen at each sampling time step, based on a given operation condition, and before it proceeds to Kalman update process.

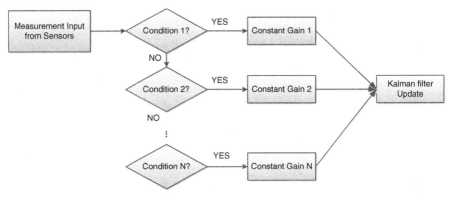

Figure 5.26 Selection of scheduled gain in Kalman update process.

Computing Kalman gain off-line would greatly reduce the computational requirement. This is a significant advantage for applications where computational cost is crucial. For example, the microcontroller in the Cubesat [27] has very limited memory (e.g., 64 kB), and it is required to process many tasks, such as attitude determination and control, power system control, telemetry and telecommand, etc. Therefore, implementing the GSKF would allow the microcontroller to achieve a faster attitude determination process and allocate additional processing time for other tasks.

The overall GSKF flow diagram is shown in Figure 5.27. The GSKF equations are similar to the standard Kalman filter, with the exception that the computation of Kalman gain and state error covariance are waived.

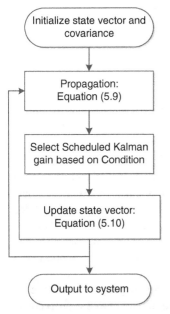

Figure 5.27 Flow chart of GSKF.

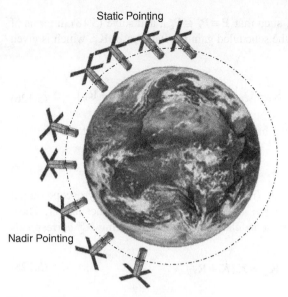

Figure 5.28   Illustration of satellite opration.

Example 5.7   Satellite Attitude Estimation Using GSKF

We consider a simplified version of a attitude determination problem for a nanosatellite, namely the VELOX-I operation scenario [27]. The nanosatellite has two pointing modes, which are static pointing (without any rotation) and nadir pointing (the z-axis of satellite always points to Earth center). Due to the limited computational resource in the nanosatellite's microcontroller, the GSKF is chosen for the attitude determination process. In this example, the GSKF equations are briefly presented. For detailed derivations, please see [27].

In this example, the attitude of VELOX-I is estimated in terms of *quaternion vector*. Quaternion is a 4×1 vector, which equals $\mathbf{q} = \begin{bmatrix} q_{13}^{\mathrm{T}} & q_4 \end{bmatrix}^{\mathrm{T}}$. However, because the magnitude of $\mathbf{q}$ always equals to one, and the scalar q_4 is merely a function of q_{13}, the state estimated vector only includes q_{13}, that is, $\hat{\mathbf{x}} = q_{13}$.

The attitude sensors of the VELOX-I outputs its measured quaternion through its sun sensors and magnetometer measurements. Thus, the observation matrix in (5.11) corresponds to:

$$\mathbf{C}_k = \mathbf{I}_{3 \times 3}. \tag{5.124}$$

For static case where satellites do not rotate, we have $\mathbf{A}_k = \mathbf{I}_{3 \times 3}$. Substituting (5.13) into (5.17), with the condition that $\mathbf{C}_k = \mathbf{I}_{3 \times 3}$, we have

$$\mathbf{P}_k^+ = \left(\mathbf{I} - \left(\mathbf{P}_{k-1}^+ + \mathbf{Q}_k \right) \left(\mathbf{P}_{k-1}^+ + \mathbf{Q}_k + \mathbf{R}_k \right)^{-1} \right) \left(\mathbf{P}_{k-1}^+ + \mathbf{Q}_k \right), \tag{5.125}$$

If we assume Δt is very small, such that $\mathbf{P} \equiv \mathbf{P}_k^+ \cong \mathbf{P}_{k-1}^+$, and solve (5.18) in terms of matrix $\mathbf{P}$, then we could have the scheduled gain for static case, $\mathbf{K}_k^{st}$, which is given as:

$$\mathbf{K}_k^{st} = \left(\sqrt{\mathbf{Q}_k\mathbf{Q}_k + 4\mathbf{Q}_k\mathbf{R}_k} + \mathbf{Q}_k\right)\left(\sqrt{\mathbf{Q}_k\mathbf{Q}_k + 4\mathbf{Q}_k\mathbf{R}_k} + \mathbf{Q}_k + \mathbf{R}_k\right)^{-1}. \quad (5.126)$$

For the nadir pointing case, such that the satellite rotates at a constant rate, we have the system dynamic model as

$$\overline{\mathbf{A}}_k = \mathbf{I}_{3\times3} + [\hat{\boldsymbol{\omega}}\times]\Delta t, \quad (5.127)$$

where $\hat{\boldsymbol{\omega}}$ is the expected body rate (it could be different than the gyroscope output due to noise and bias in gyroscope) and $[\boldsymbol{a}\times]$ is the cross product matrix. Then, applying a similar procedure, the scheduled gain for nadir pointing $\mathbf{K}_k^{np}$ corresponds to:

$$\mathbf{K}_k^{np} = \mathcal{K}(\mathcal{K} + \mathbf{R}_k)^{-1}, \quad (5.128)$$

with

$$\mathcal{K} = \frac{\sqrt{\left(\mathbf{Q}_k - \mathbf{R}_k\left(\overline{\mathbf{A}}_k\overline{\mathbf{A}}_k - \mathbf{I}_{3\times3}\right)\right)^2 + 4\overline{\mathbf{A}}_k\overline{\mathbf{A}}_k\mathbf{Q}_k\mathbf{R}_k} - \mathbf{Q}_k - \mathbf{R}_k\left(\overline{\mathbf{A}}_k - \mathbf{I}_{3\times3}\right)}{2} + \mathbf{Q}_k. \quad (5.129)$$

Before we proceed to simulation, several precautions have to be taken into consideration. Due to the nature of quaternion vector, where its magnitude always equals one, the addition between two quaternion vectors $\mathbf{q}_A$ and $\mathbf{q}_B$ (or multiplication) is not a straightforward method. Instead, it is given as:

$$\mathbf{q}_A \otimes \mathbf{q}_B = [\boldsymbol{\Psi}(\mathbf{q}_A) \quad \mathbf{q}_A]\mathbf{q}_B, \quad (5.130)$$

with

$$\boldsymbol{\Psi}(\mathbf{q}) = \begin{bmatrix} q_4\mathbf{I}_{3\times3} - [\boldsymbol{q}_{13}\times] \\ -\boldsymbol{q}_{13}^T \end{bmatrix}. \quad (5.131)$$

Similarly, for subtraction between $\mathbf{q}_A$ and $\mathbf{q}_B$ (such that $\mathbf{q}_A - \mathbf{q}_B$), it is given as $\mathbf{q}_A \otimes \mathbf{q}_B^{-1}$, with

$$\mathbf{q}^{-1} = \begin{bmatrix} -\boldsymbol{q}_{13} \\ q_4 \end{bmatrix}. \quad (5.132)$$

The simulation assumptions are as follows: The output quaternion error by the VELOX-I attitude sensor is $0.1°$, the gyroscope noise is $1°$, there is no bias error in the gyroscope, the sampling time is 1 second, and the total simulation time is 1 hour. The operation mode switching is provided in Figure 5.29, and the initial true quaternion and estimated quaternion are given as follows:

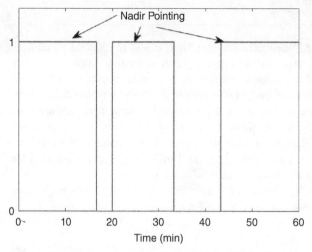

Figure 5.29   Operation mode and interval for VELOX-I.

$$\mathbf{q}_{\text{true}} = \begin{bmatrix} -0.3244 & -0.6786 & -0.6411 & -0.1522 \end{bmatrix}^{\mathrm{T}} \qquad (5.133)$$

$$\hat{\mathbf{q}} = \begin{bmatrix} 0 & 0 & 0 & 1 \end{bmatrix}^{\mathrm{T}}.$$

Figure 5.30 presents the RMSE of the estimated quaternion. It shows that the RMSE is lower than 0.4° for most of the time. Furthermore, it shows a very fast convergence rate due to the fact that the GSKF works in linear conditions.

Question: One could replace the gain-scheduled part with the standard Kalman filter gain computation process and compare the difference. What will be the overall RMSE difference?

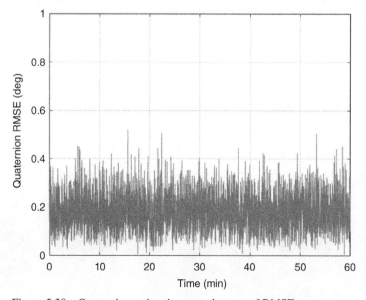

Figure 5.30   Quaternion estimation error in term of RMSE.

5.11   EnKF

The EnKF is a sequential Monte Carlo Kalman filter. It was introduced by authors in [28] for the purpose of large-size nonlinear system modelling, such as weather prediction. The EnKF utilizes a large number of ensemble members, which are generally randomly distributed around a given region, as shown in Figure 5.31. Then, the error statistic is computed based on the estimated output and all ensemble members are updated. This process allows all ensemble members to be distributed around the ensemble with the lowest estimation error. As compared to the standard Kalman filter, which is shown in Figure 5.32, the EnKF is able to achieve a faster error convergence rate, especially for highly nonlinear system.

The ensemble matrix $\hat{\mathbf{X}}_k^-$ consists of N number of ensemble members (or state vectors), $\hat{\mathbf{x}}_k^-(N)$. The ensemble matrix $\hat{\mathbf{X}}_k^-$ and respective estimated output $\hat{\mathbf{Y}}_k$ are given as:

$$\hat{\mathbf{X}}_k^- = \begin{bmatrix} \hat{\mathbf{x}}_k^-(1) & \hat{\mathbf{x}}_k^-(2) & \cdots & \hat{\mathbf{x}}_k^-(N) \end{bmatrix} \tag{5.134}$$

$$\hat{\mathbf{Y}}_k = \begin{bmatrix} \hat{\mathbf{y}}_k(1) & \hat{\mathbf{y}}_k(2) & \cdots & \hat{\mathbf{y}}_k(N) \end{bmatrix}. \tag{5.135}$$

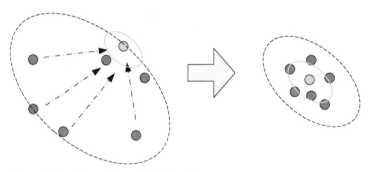

Figure 5.31   Illustration of EnKF update process.

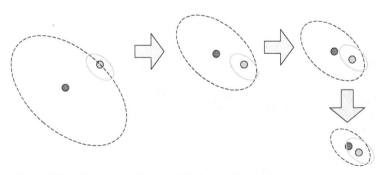

Figure 5.32   Illustration of standard Kalman filter update process.

A few advantages of the EnKF include: (1) it does not require the initial state estimated vector to be close to the true values as in EKF, especially for highly nonlinear systems; (2) similar to the UKF, the EnKF does not require any linearization; (3) it does not require covariance estimation in many applications (some applications still require the covariance estimation for covariance inflation purposes to avoid divergence issues in EnKF).

The only major drawback of EnKF is that it requires much higher computational cost than other Kalman filter algorithms.

The EnKF uses a similar Kalman update as in UKF. However, UKF only updates the mean estimated state vector, while EnKF updates the entire ensemble matrix, $\hat{\mathbf{X}}_k^-$, that is:

$$\hat{\mathbf{X}}_k^+ = \hat{\mathbf{X}}_k^- + \mathbf{K}\left(\tilde{\mathbf{Y}}_k - \hat{\mathbf{Y}}_k\right). \tag{5.136}$$

Here, $\tilde{\mathbf{Y}}_k$ denotes the observation matrix. The $\tilde{\mathbf{Y}}_k$ matrix generally consists of a repeated copy of the measurement vector, $\tilde{\mathbf{y}}_k$, such that:

$$\tilde{\mathbf{Y}}_k = \begin{bmatrix} \tilde{\mathbf{y}}_k + \vartheta_1 & \tilde{\mathbf{y}}_k + \vartheta_2 & \cdots & \tilde{\mathbf{y}}_k + \vartheta_N \end{bmatrix}, \tag{5.137}$$

where ϑ denotes the simulated random measurement error. It is noted that all column vectors, $\tilde{\mathbf{y}}_k$ in $\hat{\mathbf{Y}}_k$, (defined in (5.34)) are the same, with ϑ_i is added as oscillation. The Kalman gain is computed using the following equation:

$$\mathbf{K} = \frac{1}{N-1}\Delta\mathbf{X}_k\Delta\mathbf{Y}_k^{\mathrm{T}}\left(\frac{1}{N-1}\Delta\mathbf{Y}_k\Delta\mathbf{Y}_k^{\mathrm{T}} + \mathbf{R}\right)^{-1}, \tag{5.138}$$

where $\mathbf{R}$ is the measurement noise covariance as defined in (5.6).

Equation (5.138) shows that no linearization is required for Kalman gain, and it is similar to the Kalman gain computation method in UKF (see (5.85)). The difference of Kalman gain computation between EnKF and UKF is that EnKF assumes equal weight for all the ensemble members. Both $\Delta\mathbf{X}_k$ and $\Delta\mathbf{Y}_k$ are defined as

$$\Delta\mathbf{X}_k = \hat{\mathbf{X}}_k^-(\mathbf{I}_{N\times N} - \mathbf{1}_N)$$

$$\tag{5.139}$$

$$\Delta\mathbf{Y}_k = \hat{\mathbf{Y}}_k(\mathbf{I}_{N\times N} - \mathbf{1}_N),$$

where the $\mathbf{1}_N$ denotes $N \times N$ matrix with all elements equal to $1/N$.

The EnKF implementation is straightforward, as shown in Figure 5.33. However, one should notice that the EnKF may experience divergence if all state vectors converge into one optimal solution, such that $\Delta\mathbf{X}_k = \Delta\mathbf{Y}_k = \mathbf{0}$. Furthermore, the EnKF ensembles may drift away from the truth solution due to the chaotic forecast dynamic modeling, while the predicted state error covariance is being underestimated. Several precaution steps have been introduced in the literature, such as covariance inflation or utilizing the resampling method similar to particle filters [29–31]. These details are not within the scope of this handbook; thus, they are not covered.

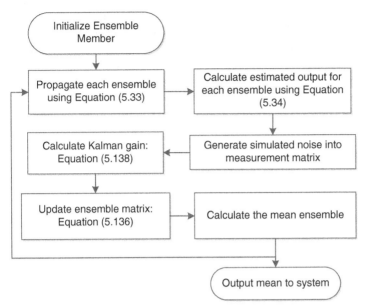

Figure 5.33   The EnKF flow diagram.

Example 5.8   Aircraft Tracking by Communication Tower

A two-dimensional aircraft tracking problem is considered in this example [32]. Here, the pilot is consistently communicating with two communication towers (as illustrated in Figure 5.34). The communication tower is able to measure the DOA of the signal transmitted by aircraft and the Doppler shift of the aircraft signal:

$$\tilde{\mathbf{y}} = \begin{bmatrix} \Delta f \\ \phi_1 \\ \phi_2 \end{bmatrix} + \nu, \tag{5.139}$$

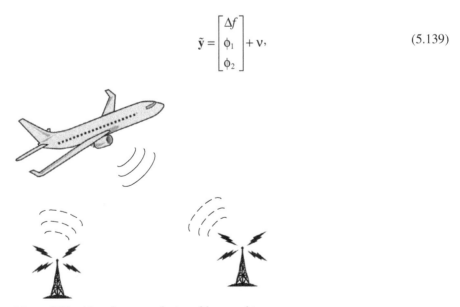

Figure 5.34   Aircraft communicates with ground tower.

where ϕ_1 is the DOA angle measured by tower 1, ϕ_2 is the DOA angle measured by tower 2, and $\Delta f = \dfrac{\|\mathbf{v}(t)\|}{c} f$ is the Doppler shift frequency, with $c = 3 \times 10^5$ km/s. The communication frequency between pilot and communication tower is 4 GHz (C-band frequency), and the aircraft is flying at a constant acceleration based on the following dynamic system model:

$$\mathbf{A}_k = \begin{bmatrix} \mathbf{I}_{2\times2} & \mathbf{I}_{2\times2} \\ \mathbf{0}_{2\times2} & \mathbf{I}_{2\times2} \end{bmatrix} \qquad (5.140)$$

$$\mathbf{B}_k = \begin{bmatrix} 0.5\Delta t^2 \mathbf{I}_{2\times2} \\ \Delta t \mathbf{I}_{2\times2} \end{bmatrix}. \qquad (5.141)$$

The initial position, velocity, and constant acceleration of the vehicle is

$$\mathbf{r}(t) = \begin{bmatrix} 30 & 50 \end{bmatrix}^T \text{ km}$$

$$\mathbf{v}(t) = \begin{bmatrix} -0.1 & -0.12 \end{bmatrix}^T \text{ km/s} \qquad (5.142)$$

$$\mathbf{a}(t) = \begin{bmatrix} -0.00025 & 0 \end{bmatrix}^T \text{ km/s}^2.$$

The locations of two communication towers are given as:

$$\mathbf{p}_{tower1} = \begin{bmatrix} 20 & 25 \end{bmatrix}^T \text{ km}$$

$$\qquad (5.143)$$

$$\mathbf{p}_{tower2} = \begin{bmatrix} -5 & 25 \end{bmatrix}^T \text{ km} \cdot$$

The Doppler shift measurement noise is 10 Hz, the one-dimensional DOA noise is $1°$, the accelerometer output noise is 10 cm/s, the sampling period is 1 second, and the number of ensembles is 64. The initial ensembles are assumed to be randomly distributed around the true position and velocity, with the standard deviation of 3 km in position and 0.1 km/s in velocity, and the simulated noise is 1% of the sensor's measurement noise. The simulation is conducted for 300 seconds and compared with the performance of EKF.

The estimated flight path by EnKF and EKF with respect to the ground truth is shown in Figure 5.35. The initial error of EKF is assumed to be 1-km random error. The results show that the EnKF can achieve better accuracy than EKF in nonlinear system estimation.

Question: What will happen if we lower the number of ensembles?

Question: What is the lowest number before the EnKF starts to fail?

Question: What happens if we increase the initial error of EKF to be same as EnKF, which is 3-km random error?

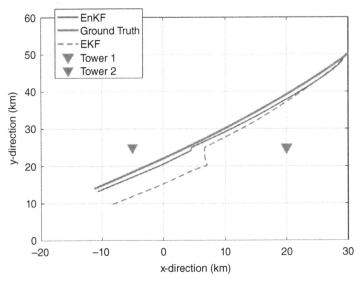

Figure 5.35   Estimated flight path by EnKF and EKF.

5.12   SUMMARY: CHOOSING THE KALMAN FILTER TYPE

The fundamentals of Kalman filters are based on a linear dynamic system. However, its variations, such as EKF and UKF, are all derived based on the standard Kalman filter theory, such as the Kalman gain matrix, state update, and propagation processes.

There is always a question of choosing the "right" Kalman filter. Which Kalman filter should we use for a given scenario? In fact, for many tracking applications, EKF is more than sufficient. However, as the complexity of tracking application increases, EKF may not be a good choice for highly nonlinear applications. In this case, a higher-complexity Kalman filter algorithm such as UKF should be considered.

Table 5.4 offers a simple summary of each Kalman filter variation that has been introduced in this chapter. It can be observed that the CKF, MFKF, and GSKF are for special applications. CKF, MFKF, and GSKF for different tracking applications may require the algorithm to be specifically designed. If computational cost is not the issue, either EnKF or UKF can be considered. In fact, EnKF can be used as the initialized algorithm such that once the error tolerance reaches a lower threshold, then either the EKF or UKF could be used to replace the EnKF, with the mean ensemble to be the initial estimates of EKF (or UKF).

5.13   FURTHER READINGS

Kalman filter implementation for localization problems is not restricted to the techniques presented in this chapter. Several other techniques have been published in the literature:

- extended particle filter, which is a mixed of EKF and particle filter [33, 34];
- ensemble particle filter, which is a merger between EnKF and PF [35];

TABLE 5.4. Comparison of Each Kalman Filter Variation

Kalman Filter Type	Advantages	Disadvantages
Discrete Kalman filter	Simple and fast	Cannot work in nonlinear systems
Continuous-time Kalman filter	Simple and fast	Cannot work in nonlinear systems
EKF	Low computational complexity Easy to be implemented	Requires good initial estimation Not applicable for highly nonlinear systems
UKF	No linearization required	High computational cost Requirse more parameters tuning
CKF	Considers constraint for state estimate vector	Only applicable for certain applications
MFKF	Lower computational cost	Only applicable for certain applications
GSKF	No need compute Kalman gain online	Only applicable for certain application
EnKF	More flexible on initial error Usable for highly nonlinear systems	Higher computational cost than UKF

- cubature Kalman filter is a variation of UKF, where the sigma point transformation is based on the spherical-radial rules [36–38];
- square-root spherical UKF, in which the sigma points are transformed based on the unit hyper sphere project to achieve lower computational cost than the standard UKF [39–41].

Furthermore, not all localization problems involve time variation in the state vector. The state vector in some localization problems may always be constant even though the tracked object is moving at all times. For example, we can estimate the spacecraft's two-line elements at a specific epoch instead of the spacecraft position and velocity [7]. Then, the dynamic model will always be a time-invariant matrix!

This chapter does not cover all variation of Kalman filters and its derivation in detail; thus, we suggest the reader explore publications listed in the references for further study on the Kalman filter. Authors also suggest reviewing [9–11] for a more detailed understanding of the Kalman filter derivation.

REFERENCES

[1] E. J. Lefferts, F. L. Markley, and M. D. Shuster, "Kalman filtering for spacecraft attitude estimation," *J. Guid. Control Dynam.*, vol. 5, pp. 417–429, 1982.

[2] J. L. Crassidis and F. L. Markley, "Unscented filtering for spacecraft attitude estimation," *J. Guid. Control Dynam.*, vol. 26, pp. 536–542, 2003.

[3] S. T. Goh, O. Abdelkhalik, and S. A. Zekavat, "A weighted measurement fusion Kalman filter implementation for UAV navigation," *Aerosp. Sci. Technol.*, vol. 28, pp. 315–323, 2013.

[4] S. T. Goh, O. Abdelkhalik, and S. A. Zekavat, "Spacecraft formation orbit estimation using WLPS-based localization," *Int. J. Navig. Obs.*, vol. 2011, p. 12, 2011.

[5] T. Srivas, T. Artés, R. A. de Callafon, and I. Altintas, "Wildfire spread prediction and assimilation for FARSITE using ensemble Kalman filtering," *Procedia Comput. Sci.*, vol. 80, pp. 897–908, 2016.

[6] A. T. Nair, T. K. Radhakrishnan, K. Srinivasan, and S. R. Valsalam, "Kalman filter based state estimation of a thermal power plant," presented at the Int. Conf. on Process Automation, Control and Computing, Coimbatore, India, Jul. 2011.

[7] S. T. Goh and K. S. Low, "Real-time estimation of satellite's two-line elements via positioning data," presented at the IEEE Aerospace Conf., Big Sky, MT, Mar. 2018.

[8] A. Suarez, H. D. Reeves, D. Wheatley, and M. Coniglio, "Comparison of ensemble Kalman filter–based forecasts to traditional ensemble and deterministic forecasts for a case study of banded snow," *Weather Forecast.*, vol. 27, pp. 85–105, 2012.

[9] Y. Bar-Shalom, X. Li, and T. Kirubarajan, *Estimation with Applications to Tracking and Navigation, Theory Algorithms and Software*. New York: John Wiley & Sons, 2001.

[10] J. L. Crassidis and J. L. Junkins, *Optimal Estimation of Dynamic Systems*. Boca Raton, FL: Chapman & Hall/CRC, 2004.

[11] A. Gelb, *Applied Optimal Estimation*. Cambridge, MA: MIT Press, 1974.

[12] S.-G. Kim, J. L. Crassidis, Y. Cheng, A. M. Fosbury, and J. L. Junkins, "Kalman filtering for relative spacecraft attitude and position estimation," *J. Guid. Control Dynam.*, vol. 30, pp. 133–143, 2007.

[13] D. Vallado and P. Crawford, "SGP4 orbit determination," presented at the AIAA/AAS Astrodynamics Specialist Conf. and Exhibit, Honolulu, Hawaii, 2008.

[14] S. T. Goh, O. Abdelkhalik, and S. A. Zekavat, "Spacecraft constellation orbit estimation via a novel wireless positioning system," presented at the 2009 AAS/AIAA Space Flight Mechanics Meeting, Savannah, Georgia, Feb. 2009.

[15] H. Tong and S. A. Zekavat, "A novel wireless local positioning system via a merger of DS-CDMA and beamforming: Probability-of-detection performance analysis under array perturbations," *IEEE Trans. Veh. Technol.*, vol. 56, pp. 1307–1320, 2007.

[16] D. D. Mueller, R. R. Bate, and J. E. White, *Fundamentals of Astrodynamics*. New York: Dover, 1971.

[17] D. Vallado, *Fundamentals of Astrodynamics and Applications*, 3rd ed. Hawthorne, CA: Microcosm Press/Springer, 2007.

[18] A. R. Nafchi, S. T. Goh, and S. A. Zekavat, "Circular arrays and inertial measurement unit for DOA/TOA-based endoscopy capsule localization: Performance and complexity investigation," *IEEE Sensors J.*, vol. 14, pp. 3791–3799, 2014.

[19] A. R. Nafchi, S. T. Goh, and S. A. Zekavat, "High performance DOA/TOA-based endoscopy capsule localization and tracking via 2D circular arrays and inertial measurement unit," presented at the Wireless Sensor System Workshop 2013, Baltimore, MD, Nov. 2013.

[20] S. T. Goh, S. A. Zekavat, and K. Pahlavan, "DOA-based endoscopy capsule localization and orientation estimation via unscented Kalman filter," *IEEE Sensors J.*, vol. 14, pp. 3819–3829, 2014.

[21] E. A. Wan and R. V. D. Merwe, "The unscented Kalman filter for nonlinear estimation," presented at the Symp. on Adaptive Systems for Signal Processing, Communications, and Control Symp., Lake Louise, Alberta, Canada, 2000.

[22] D. Simon, *Optimal State Estimation: Kalman, H∞, and Nonlinear Approaches*. New York: John Wiley & Sons, 2006.

[23] S. T. Goh, O. Abdelkhalik, and S. A. Zekavat, "Constraint estimation of spacecraft position," *J. Guid. Control Dynam.*, vol. 35, pp. 387–397, 2012.

[24] J. A. Roecker and C. D. McGillem, "Comparison of two-sensor tracking methods based on state vector fusion and measurement fusion," *IEEE Trans. Aerosp. Electron. Syst.*, vol. 24, pp. 447–449, 1988.

[25] Q. Gan and C. J. Harris, "Comparison of two measurement fusion methods for Kalman-filter-based multisensor data fusion," *IEEE Trans. Aerosp. Electron. Syst.*, vol. 37, pp. 273–279, 2001.

[26] S.-T. Park and J. G. Lee, "Improved Kalman filter design for three-dimensional radar tracking," *IEEE Trans. Aerosp. Electron. Syst.*, vol. 37, pp. 727–739, 2001.

[27] M. D. Pham, K. S. Low, S. T. Goh, and S. Chen, "Scheduled gain EKF for nano-satellite attitude determination system," *IEEE Trans. Aerosp. Electron. Syst.*, vol. 51, pp. 1017–1028, 2015.

[28] G. Evensen, "Sequential data assimilation with a nonlinear quasi-geostrophic model using Monte Carlo methods to forecast error statistics," *J. Geophysical Res.,* vol. 99, pp. 10143–10162, 1994.

[29] S. T. Goh, J. J. Soon, and K. S. Low, "An inequality constrained ensemble Kalman filter for parameter estimation application," presented at the IEEE Aerospace Conf., Big Sky, MT, 2018.

[30] T. M. Hamill, J. S. Whitaker, and C. Snyder, "Distance-dependent filtering of background error covariance estimates in an ensemble Kalman filter," *Mon. Weather Rev.,* vol. 129, pp. 2776–2790, 2001.

[31] J. L. Anderson and S. L. Anderson, "A Monte Carlo implementation of the nonlinear filtering problem to produce ensemble assimilations and forecasts," *Mon. Weather Rev.,* vol. 127, pp. 2741–2758, 1999.

[32] B. Cui and J. Zhang, "The improved ensemble Kalman filter for multisensor target tracking," presented at the Int. Symp. on Information Science and Engineering, Shanghai, China, 2008.

[33] A. Levy, S. Gannot, and E. A. P. Habets, "Multiple-hypothesis extended particle filter for acoustic source localization in reverberant environments," *IEEE Trans. Audio Speech Lang. Process.,* vol. 19, pp. 1540–1555, 2011.

[34] P. Aggarwal, D. Gu, S. Nassar, Z. Syed, and N. El-Sheimy, "Extended particle filter (EPF) for INS/GPS land vehicle navigation application," presented at the Institute of Navigation Satellite Division Technical Meeting, Fort Worth, TX, 2007.

[35] S. Robert and H. R. Künsch, "Localizing the ensemble Kalman particle filter," *Tellus A,* vol. 69, pp. 1–14, 2017.

[36] I. Arasaratnam and S. Haykin, "Cubature Kalman filters," *IEEE Trans. Autom. Control,* vol. 54, pp. 1254–1269, 2009.

[37] A. Roy and D. Mitra, "Multi-target trackers using cubature Kalman filter for Doppler radar tracking in clutter," *IET Signal Proc.,* vol. 10, pp. 888–901, 2016.

[38] Z. Ding and B. Balaji, "Comparison of the unscented and cubature Kalman filters for radar tracking applications," presented at the IET Int. Conf. on Radar Systems, Glasgow, UK, 2012.

[39] H. Aung, K. S. Low, and S. T. Goh, "State-of-charge estimation of lithium-ion battery using square root spherical unscented Kalman filter (Sqrt-UKFST) in nanosatellite," *IEEE Trans. Power Electron.,* vol. 30, pp. 4774–4783, 2015.

[40] K. Zhao and Z. You, "Square root spherical simplex unscented Kalman filter for micro satellite attitude measurement," presented at the Int. Conf. on Information Engineering and Computer Science, Wuhan, China, 2010.

[41] X. Tang, X. Zhao, and X. Zhang, "The square-root spherical simplex unscented Kalman filter for state and parameter estimation," presented at the Int. Conf. on Signal Processing, Beijing, China, 2008.

TOA AND DOA BASED POSITIONING

THE SECOND part of the handbook consists of five chapters, Chapters 6–10, which detail TOA and DOA localization methods. All chapters in this part include invaluable MATLAB examples and solutions and code that will help researchers as they begin their work in the relevant areas.

Chapter 6 presents principle methods of localization of a target based on the availability (and applicability) of a TOA estimation method. Although these techniques were introduced in Chapter 3, here we expand our analysis specifically on TOA-based approaches. The chapter compares the performance of TOA and TDOA methods as well as their geometric interpretations. Linearization techniques are introduced to reduce the computational complexity of underlying nonlinear TOA and TDOA objective functions. The performance of the presented techniques is also investigated. Finally, we examine the impact of non-line-of-sight propagation on lateration-based positioning.

While the previous chapter relies on the existence of TOA measurements, Chapter 7 complements Chapter 6 by presenting key TOA estimation methods. It introduces important measures that compare different TOA estimation techniques. Examples of these measures are sensitivity to signal-to-noise ratio, the number of reflections, resolution and complexity. Then the chapter introduces examples of popular TOA methods. Examples of these TOA methods are Maximum Likelihood, and sub-space methods such as MUSIC. The chapter compares the presented TOA techniques. In addition, the chapter introduces a novel TOA estimation method that is applied in the frequency domain and based on Independent Component Analysis.

Chapter 8 reviews TOA estimation methods via ultra-wide-band (UWB) systems. The chapter offers a short review of UWB signals/systems which are particularly relevant and useful for TOA. Moreover, it introduces recent research progress in UWB-based localization including fingerprinting and geometric techniques. In addition, the issues surrounding localization in indoor areas are highlighted in this chapter.

Handbook of Position Location: Theory, Practice, and Advances, Second Edition.
Edited by S. A. (Reza) Zekavat and R. Michael Buehrer.
© 2019 by the Institute of Electrical and Electronics Engineers, Inc.
Published 2019 by John Wiley & Sons, Inc.
Companion Website: www.wiley.com/go/zekavat/positionlocation2e

Chapter 9 presents DOA estimation methods. The chapter starts with a review of important antenna and antenna array parameters and their impact on DOA estimation. Next. it introduces important measures for comparing different types of DOA estimation techniques. Examples of these measures include calibration sensitivity, angular resolution and complexity. Popular DOA estimation methods such as "delay and sum," MUSIC and root MUSIC are introduced and compared. In addition, Chapter 9 studies new and practically implementable DOA estimation methods. Finally, this chapter investigates the impact of non-stationary signals. Specifically, non-stationary behavior creates low performance sample auto-correlation estimation which in turn reduces DOA performance in sub-space techniques such as MUSIC. The chapter presents a solution to this problem.

Chapter 10 investigates high resolution TOA estimation techniques for inhomogeneous media. It also studies straight line distance estimation within inhomogeneous media via a combined TOA and DOA estimation to calculate the true range between the transmitter and receiver as the propagation path of transmitted waveform is not straight-line.

CHAPTER 6

FUNDAMENTALS OF TIME-OF-ARRIVAL-BASED POSITION LOCATION

R. Michael Buehrer and Swaroop Venkatesh
Virginia Tech, Blacksburg, VA

THIS CHAPTER presents an overview of time-based positioning. In particular, it focuses on lateration approaches, which use time-of-arrival (TOA) or time-difference-of-arrival (TDOA) measurements to estimate range or pseudorange. We present the basic approaches to each as well as geometric interpretations of the approaches. Also, due to the fact that lateration involves inherently nonlinear functions, we discuss linearization techniques, which ease the computational requirements at the expense of accuracy. Finally, we examine the impact of non-line-of-sight (NLOS) propagation, a major impediment in lateration-based positioning.

6.1 INTRODUCTION

The problem of position location can be described as the following: Given a set of transceivers at known locations (hereafter termed anchors), how can we determine the position of a transceiver at an unknown location (hereafter termed the mobile) based on wireless (i.e., radio frequency [RF]) communications between the anchors and mobile? Classic approaches to solving this problem are based on a relatively small set of measurements on the wireless signals. Specifically, position location is done using a measurement of the angle of arrival (AOA), TOA, or received signal strength (RSS) of individual signals at either the mobile, the anchors, or both. In this document, we are concerned with determining position based on measurements of the TOA.

TOA measurements can be used to determine the position of the mobile based on the fact that the time a signal takes to travel from point A to point B (often termed

Handbook of Position Location: Theory, Practice, and Advances, Second Edition.
Edited by S. A. (Reza) Zekavat and R. Michael Buehrer.
© 2019 by the Institute of Electrical and Electronics Engineers, Inc.
Published 2019 by John Wiley & Sons, Inc.
Companion Website: www.wiley.com/go/zekavat/positionlocation2e

the time of flight) is directly related to the distance between the two points, $d = c\tau$, where c is the speed of light, 2.98×10^8 m/s, and τ is the time of flight. There are three basic means of using TOA to perform position location, typically termed in the literature as TOA positioning, TDOA positioning, and time-sum-of-arrival (TSOA) positioning. TSOA is similar to TDOA and is not particularly common, so we will limit our description to TOA and TDOA positioning techniques.

6.2 TDOA POSITIONING

Let us assume that we can measure the TOA of a signal transmitted by the mobile at each of N anchors (base stations) whose positions are known a priori. This measurement is typically done by correlating the received signal with an expected signal (e.g., a known synchronization word) and by comparing the correlator output with an accurate clock. TOA estimation techniques are discussed in detail in Chapter 7. Note that we will assume in this discussion that all of the receivers have access to the same clock or can be synchronized to the same clock. Further, we will label these times as t_i, which represents the TOA at the ith anchor.*

Assuming that the signal was transmitted at time t_M, these received times can be related to the distance between the mobile and the ith anchor. Let τ_i be the time of flight between the mobile and the ith anchor, (x_i, y_i) the two-dimensional position of the ith anchor, and d_i the distance between the mobile and the ith anchor. Using these definitions, we can write the following equation relating the measurement t_i with the position of the mobile (x, y):

$$t_i = \tau_i + t_M \tag{6.1a}$$

$$= \frac{d_i}{c} + t_M \tag{6.1b}$$

$$= \frac{\sqrt{(x_i - x)^2 + (y_i - y)^2}}{c} + t_M. \tag{6.1c}$$

Unfortunately, with only this information, we cannot determine the position of the mobile since we have only one equation with three unknowns (x, y, and t_M). However, we can eliminate one of these variables (t_M) by taking the difference between two arrival time measurements taken at two anchors ($i \neq j$):

$$t_i - t_j = \tau_i + t_M - (\tau_j - t_M) \tag{6.2a}$$

$$= \tau_i - \tau_j \tag{6.2b}$$

$$= \frac{d_i}{c} - \frac{d_j}{c} \tag{6.2c}$$

$$= \frac{\sqrt{(x_i - x)^2 + (y_i - y)^2} - \sqrt{(x_j - x)^2 + (y_j - y)^2}}{c}. \tag{6.2d}$$

---

*Note that in the preliminary discussion of these techniques, we will assume the noiseless case. We will discuss the impact of noise in Section 6.5.

We now have one equation and two unknowns, which still does not allow us to determine the position. In other words, there are an infinite number of (x, y) pairs that would make (6.2d) true. Thus, we must have more information. If we take a third measurement at a third anchor, we can now define two equations by taking two differences:

$$t_i - t_j = \frac{\sqrt{(x_i - x)^2 + (y_i - y)^2} - \sqrt{(x_j - x)^2 + (y_j - y)^2}}{c} \tag{6.3a}$$

and

$$t_i - t_k = \frac{\sqrt{(x_i - x)^2 + (y_i - y)^2} - \sqrt{(x_k - x)^2 + (y_k - y)^2}}{c}. \tag{6.3b}$$

We now have two nonlinear equations and two unknowns. It should be noted that a third equation using $t_j - t_k$ could also be created, but it is not independent of the previous two equations in (6.3b) and thus does not provide any new information. The two nonlinear equations in two unknowns may have one or two solutions. In practice, there may typically be only one, but in order to guarantee that there are no multiple solutions, we require an additional measurement.

6.2.1   Geometric Interpretation

TDOA positioning is also commonly known as hyperbolic positioning. This is based on the fact that the values of (x, y) that solve the equation in (6.1c) when plotted form a hyperbola. The intersection of two hyperbolae (as in (6.3b)) represents the solution and the estimate location of the mobile. To see this, consider two anchors located on a relative coordinate system at $(-\delta, 0)$ and $(\delta, 0)$. If $\Delta\tau$ is the measured TDOA, we can write

$$\Delta\tau c = \sqrt{(x + \delta)^2 + y^2} - \sqrt{(x - \delta)^2 + y^2}. \tag{6.4}$$

Defining $m = \Delta\tau c$ and rearranging, we have

$$m + \sqrt{(x - \delta)^2 + y^2} = \sqrt{(x + \delta)^2 + y^2}. \tag{6.5}$$

Rearranging, we can obtain

$$\frac{x^2}{\left(\delta^2 - \dfrac{m^2}{4}\right) \Big/ \left(\dfrac{4\delta^2}{m^2} - 1\right)} - \frac{y^2}{\left(\delta^2 - \dfrac{m^2}{4}\right)} = 1, \tag{6.6}$$

which is the equation of a hyperbola centered at the origin. Note that there are two branches to the hyperbola defined in the previous equation. However, in obtaining (6.6), we squared the value of m, hiding the fact that $\Delta\tau$ is either positive or negative. The sign of $\Delta\tau$ determines which branch of the hyperbola is valid.

Thus, TDOA positioning is classically known as *hyperbolic* positioning since each time difference measurement defines a hyperbola. The intersection of these hyperbolae defines the location of the mobile in the absence of noise.

Example 6.1

As an example, consider two anchors located at $(x_0, y_0) = (0, 0)$ and $(x_1, y_1) = (125, 150)$ with a mobile station located at $(x, y) = (325, 150)$. Plot the hyperbola defined by the TDOA seen by the two anchors along with the base stations and mobile position. Assume that the anchor (base station) located at (125, 150) observes a shorter distance (earlier time).

Solution

In the absence of noise, the difference in distance measured at the two base stations is

$$m = \sqrt{(325-0)^2 + (150-0)^2} - \sqrt{(325-125)^2 + (150-150)^2} = 158. \quad (6.7)$$

From (6.6), we can determine the parameters of the hyperbola centered at the origin. Specifically,

$$\delta = \frac{1}{2}\sqrt{(125-0)^2 + (150-0)^2} = 98. \quad (6.8)$$

The resulting equation for a hyperbola centered at the origin (along the x-axis) is then

$$\frac{x^2}{79} - \frac{y^2}{98} = 1. \quad (6.9)$$

To obtain the true hyperbola, we must translate the above hyperbola to the midpoint between the two anchors (62.5, 75.0) and rotate it by $\theta = \mathrm{atan}(-150/-125) = -130°$. The two base stations, the mobile, and the resulting hyperbola, which defines the possible locations of the mobile, are given in Figure 6.1. We can see that there are an infinite number of possible points for the location of the mobile based on the

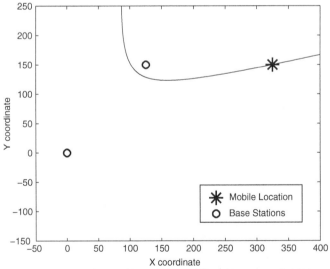

Figure 6.1   Possible locations of the mobile with two received time measurements from Example 6.1 (the line represents the possible locations of the mobile based on a single TDOA measurement).

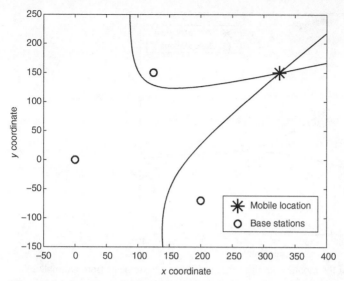

Figure 6.2   Location of the mobile with three received time measurements (two TDOA measurements) from Example 6.2.

observed time difference, including the true location. Note that since the base station at (125,150) observes the shorter distance, we only plot the top portion of the hyperbola. Had the other base station observed the closer distance, we would have plotted the bottom portion of the hyperbola.

Example 6.2

Repeat Example 6.1, adding a third base station at (200, −70).

Solution

Adding a third base station defines two additional hyperbolae, although one is redundant. Defining a second hyperbola using the difference in distances seen by the base station at the origin and the new base station, we now have two hyperbolae, which (in this case) intersect at a single point as shown in Figure 6.2. However, as we will see in the next example, three base stations does not always guarantee a unique solution.

Example 6.3

Repeat example 6.2 with three base stations located at (0, −300), (−100, 300), and (100, 0).

Solution

Using the same approach as in Examples 6.1 and 6.2, we obtain three hyperbolae, which are plotted in Figure 6.3. In this case, we plot all three possible hyperbolic equations, although one is redundant as mentioned earlier. As we can see, due to the geometry of the situation, there are two possible mobile locations due to the

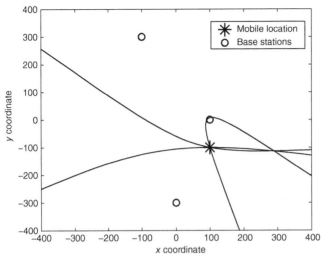

Figure 6.3 Location of the mobile with three received time measurements from Example 6.3 when an ambiguity arises.

nonlinear nature of the defining equations, one of which is clearly the true solution. However, since all three hyperbolae intersect in two locations, we cannot differentiate between the two. Thus, TDOA requires four base station measurements to guarantee a unique solution, although in many cases three is sufficient.

6.2.2 Uplink versus Downlink Measurements

TDOA positioning can be done using either uplink (mobile-to-base) signals or downlink (base-to-mobile) signals. When the measurements are on uplink signals, the base stations must be synchronized and must record the time at which they observe a specified mobile signal. The different measurements must then be communicated to a central location for TDOA calculation (and presumably position location estimation). The main requirements are that the mobile transmits a known signal for timing estimation and that all base stations have a common time.

If TDOA positioning is done via downlink signals, the base stations must transmit a known signal at a synchronized time, which can be measured by the mobile. The mobile can compare the arrival times of the various base station signals to obtain a time difference measurement. The base station transmit time is not needed, and the mobile can compute its own position given a sufficient number of time difference measurements.

6.3 TOA POSITIONING

Returning to our original problem, if, in addition to TOA measurements at the base station receivers, we also have knowledge of the time of transmission at the mobile (t_M), the situation changes. Again, our equation can be written as

$$t_i = \tau_i + t_M \tag{6.10a}$$

$$t_i - t_M = \frac{d_i}{c} \tag{6.10b}$$

$$= \frac{\sqrt{(x_i - x)^2 + (y_i - y)^2}}{c}. \tag{6.10c}$$

As before, with only a single base station, we cannot solve for the position of the mobile. However, it is a fundamentally different situation than that with TDOA, as we will see shortly. With two base stations, we have two nonlinear equations and two unknowns:

$$t_i - t_M = \frac{\sqrt{(x_i - x)^2 + (y_i - y)^2}}{c} \tag{6.11a}$$

and

$$t_j - t_M = \frac{\sqrt{(x_j - x)^2 + (y_j - y)^2}}{c}. \tag{6.11b}$$

In general, these two equations will have two solutions, although in some scenarios (e.g., if the mobile is equidistant from the two base stations) there will be a single solution. If we have three base stations, we are guaranteed to have a single solution.

6.3.1   Geometric Interpretation

If we return to the equation that defines the relationship between the received times and the mobile position, we have

$$\tau_i = \frac{\sqrt{(x_i - x)^2 + (y_i - y)^2}}{c}. \tag{6.12}$$

Squaring both sides and rearranging, we have

$$\frac{(x_i - x)^2}{(c\tau_i)^2} + \frac{(y_i - y)^2}{(c\tau_i)^2} = 1. \tag{6.13}$$

We can clearly see that this equation defines a circle of points upon which the mobile must lie with the center at the ith base station and the radius being the distance between the mobile and the base station. If we have measurements at two base stations, we define two circles that can intersect at one or two points. This can be seen in the following example.

Example 6.4

Let us repeat Example 6.2 for TOA positioning. Specifically, consider two anchors located at (0, 0) and (125, 150) with a mobile station located (325, 150). Plot the circles defined by the noise-free distance measured by the two anchors, along with their known positions.

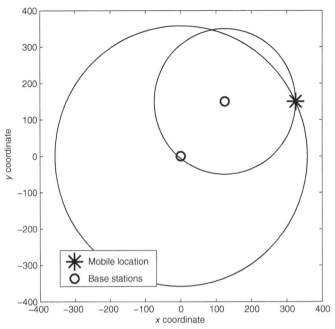

Figure 6.4   Possible locations of the mobile with two received time measurements and the mobile transmit time (t_M) from Example 6.4.

Solution

The two circles have centers at the locations of the anchors and radii of

$$d_1 = \sqrt{(325-0)^2 + (150-0)^2} = 358 \tag{6.14}$$

and

$$d_2 = \sqrt{(325-125)^2 + (150-150)^2} = 200, \tag{6.15}$$

respectively. The two resulting circles are plotted in Figure 6.4. We can see that there are two possible locations for the mobile based on the two intersections, one of which is the true position. Thus, we cannot unambiguously determine the mobile's position with only two TOA measurements.

Example 6.5

Repeat Example 6.4 if we also have the time measurement from a third base station located at (200, −70).

Solution

Now we have three circles, with the third circle centered at (200, −70) and a radius of 253. With three such measurements, there is no ambiguity, as can be seen in Figure 6.5. Unlike the case of TDOA, with three measurements in TOA, we are guaranteed a unique solution, unless the three anchors are colinear.

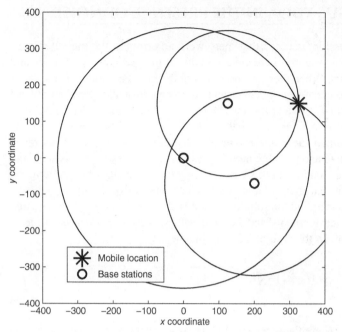

Figure 6.5   Possible locations of the mobile with three received time measurements and the mobile transmit time (t_M) from Example 6.5.

6.4   TDOA VERSUS TOA

It should be clear from the previous discussion that the two approaches (TOA and TDOA) have some fundamental differences. Although they both use time measurements, TOA exploits additional information, and thus TDOA has more ambiguity than TOA. This can be clearly seen in the geometric interpretations. When the two techniques have only a single base station measurement, TDOA cannot limit the mobile's position in any way. It could potentially be anywhere in two-dimensional space. In the case of TOA, on the other hand, because it utilizes the time of transmission from the mobile, it can say something about the mobile's position. The extra information allows the mobile's position to be limited to a circle with the base station at its center. If we increase the number of base stations to two, the mobile's position can now be located on a hyperbola with TDOA. However, with TOA, the extra information allows us to limit the location to either one or two points defined by the intersection of two circles. Thus, while TDOA has an infinite number of possible locations, TOA provides only two possible locations. Increasing the number of base stations to three, TOA is now guaranteed to provide a single unique solution equal to the true location of the mobile. TDOA, on the other hand, can result in either one or two solutions depending on the geometry of the situation. Only when there are four base stations or more are both TOA and TDOA guaranteed to provide a single (and the same) solution. Of course, this assumes that there is no noise in the system. When there is noise in the system, the two approaches will nearly always provide different solutions. That will be discussed next.

6.5   TOA VERSUS TDOA IN THE PRESENCE OF NOISE

The preceding discussion assumed that there were no errors in the measurements. Of course, in practical systems, there are errors due to thermal noise, multipath, and possibly system errors. When there are no errors in the measurements, the measurements can be used to create sets of consistent equations. Provided there are at least three measurements in TOA* and four in TDOA, we are guaranteed to get one solution from this set of equations, and that solution corresponds to the true mobile position. However, when there is error in the measurements, the equations made from the measurements and possible mobile positions are typically inconsistent and thus do not provide a solution. In this case, we require a means for determining the "best" solution given that there is none that agrees with all of the measurements. The typical approach is to find the solution (i.e., position) that minimizes the squared error between the measurements and implied values. This is known as the least squares (LS) solution. In the case of TDOA, this means

$$[\hat{x}, \hat{y}] = \arg\min_{x,y} \sum_{i=1}^{N} \sum_{j \neq i} \left\{ c(\tilde{t}_i - \tilde{t}_j) - \sqrt{(x_i - x)^2 + (y_i - y)^2} + \sqrt{(x_j - x)^2 + (y_j - y)^2} \right\}^2,$$

(6.16)

where $\tilde{t}_i$ for $i = 1, \ldots N$ are the received TOA measurements ($\tilde{t}_i = t_i + n_i$, where n_i is the measurement noise) at the N base stations located at (x_i, y_i). In the case of TOA, the LS solution is also possible, but the equations are somewhat different:

$$[\hat{x}, \hat{y}] = \arg\min_{x,y} \sum_{i=1}^{N} \left[c(\tilde{t}_i - t_M) - \sqrt{(x_i - x)^2 + (y_i - y)^2} \right]^2.$$

(6.17)

It should be clear that even given the same measurements, $\tilde{t}_i$, the solutions of these two formulations will not be the same in general. As an example, Figure 6.6 shows error contours for the two techniques using the same noisy measurements of arrival time $\tilde{t}_i$.

In Figure 6.6, the error is plotted using a contour plot where each contour represents a constant level of error between the noisy measurements and the distances implied by the corresponding (x, y) coordinates. Although an exact interpretation of the error contours is beyond the scope of this chapter, it is clear from the plot that even though both plots rely on LS, the presence of additional information in the form of the mobile transmit time results in a very different error surface and a different solution. In particular, the shape of the hyperbola as compared to the circle results in a more skewed error surface. This is particularly true in the case where the mobile location is outside the convex hull defined by the base station locations.

Consider the case demonstrated by Figures 6.2 and 6.5. Now consider that each of the three time measurements has a small amount of error ranging from 2 to 25 ns (roughly corresponding to 0.5–8.0 m). The resulting circles and hyperbola are

---

* It should be noted that the use of three TOA measurements also assumes that the transmit time is known. If this must be measured, then the two techniques use the same number of time measurements.

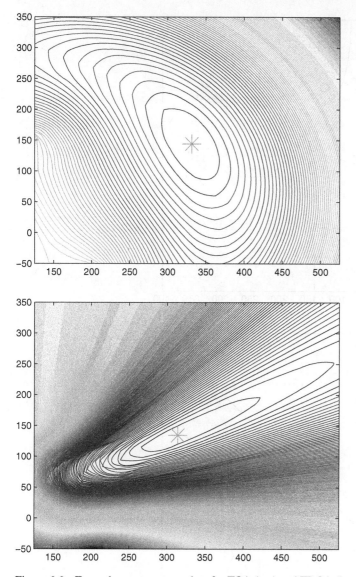

Figure 6.6   Example error contour plots for TOA (top) and TDOA (bottom) for the scenarios depicted in Figures 6.2 and 6.5. The asterisk denotes the true node location. See color insert.

plotted in Figure 6.7, along with the true mobile position and the two LS solutions. First, it can be seen that in the case of TOA, the circles do not intersect at a single point, meaning that there is no solution to the set of nonlinear equations. This is not true for TDOA since there are only two independent curves and there will generally be a solution (albeit not necessarily the true location). The LS solution can be thought of as the point in the two-dimensional plane that is closest to the three curves. We

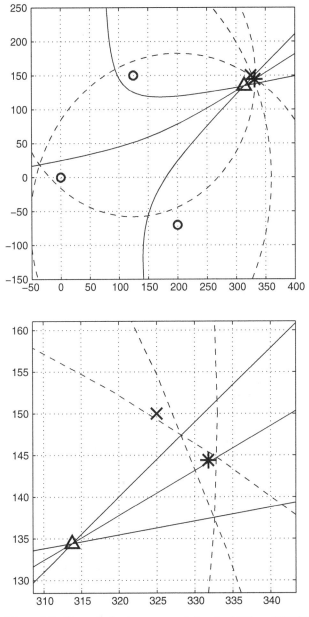

Figure 6.7   Example least squares solution for TOA and TDOA (top) and close-in view (bottom) ("x" represents true position; the triangle represents the TDOA solution; the asterisk represents the TOA solution; the solid lines represent hyperbolae; and the dashed lines represent the circles).

can see that the two solutions are clearly different. And in this case, due to the geometry of hyperbolas, where a small error can result in large changes in the curve, especially far from the focus, TDOA results in a worse solution (i.e., further from the true location) than TOA. Specifically, the TOA solution provided an error of approximately 9 m, whereas the TDOA solution provided an error of approximately 19 m, or an error twice as big as TOA. This is not unexpected since TOA uses more information than TDOA, and we have assumed that t_M is known.

6.6   LINEARIZATION

Let

$$\theta = \begin{bmatrix} x \\ y \end{bmatrix} \qquad (6.18)$$

and $\hat{\theta}$ be the estimate of θ. Then, the vector of observed distances using TOA measurement **r** can be written as

$$\mathbf{r} = \mathbf{f}(\theta) + \mathbf{n}, \qquad (6.19)$$

where $\mathbf{f}(\theta)$ is a function $\mathbb{R}^2 \to \mathbb{R}^N$, which maps the two-dimensional position vector θ to an N-dimensional vector of distances to each of the N anchors:

$$\mathbf{f}(\theta) = \begin{bmatrix} \sqrt{(x_1 - x)^2 + (y_1 - y)^2} \\ \sqrt{(x_2 - x)^2 + (y_2 - y)^2} \\ \dots \\ \sqrt{(x_N - x)^2 + (y_N - y)^2} \end{bmatrix}. \qquad (6.20)$$

Using this notation, the LS estimate can be written as

$$\hat{\theta} = \arg\min_{\theta} [\mathbf{r} - \mathbf{f}(\theta)]^T [\mathbf{r} - \mathbf{f}(\theta)] \qquad (6.21a)$$

and

$$\hat{\theta} = \arg\min_{\theta} \phi(\theta), \qquad (6.21b)$$

where $\phi(\theta)$ is termed the cost function. Clearly, this is a nonlinear estimator that is difficult to solve. There are many techniques to find the global minimum of the cost function, which can be very complex and prone to getting trapped in local minima. Thus, it is desirable to find linear approximations of this estimate. We discuss two such techniques in the following.

6.6.1   Taylor Series Approximation

One approach to simplifying the nonlinear least squares (NLS) estimator of (6.21a) is to employ a Taylor series approximation to the function $\mathbf{f}(\theta)$. Specifically, we create a linear function by using the first two terms of a Taylor series expansion about an initial solution θ_o. For the ith component of $\mathbf{f}(\theta)$, we have

$$f_i(x, y) \approx f_i(x_o, y_o) + \left.\frac{\partial f_i(\theta)}{\partial x}\right|_{\theta=\theta_o} \Delta x + \left.\frac{\partial f_i(\theta)}{\partial y}\right|_{\theta=\theta_o} \Delta y, \qquad (6.22)$$

where $\theta = \theta_o + \Delta\theta$ or $x = x_o + \Delta x$, $y = y_o + \Delta y$. Putting together the N equations, we have

$$\mathbf{f}(\theta) \approx \mathbf{f}(\theta_o) + \mathbf{H}(\theta - \theta_o), \qquad (6.23)$$

where $\mathbf{H}$ is an $N \times 2$ matrix defined as

$$\mathbf{H} = \begin{bmatrix} \left.\dfrac{\partial f_1(\theta)}{\partial x}\right|_{\theta=\theta_o} & \left.\dfrac{\partial f_1(\theta)}{\partial y}\right|_{\theta=\theta_o} \\ \vdots & \vdots \\ \left.\dfrac{\partial f_N(\theta)}{\partial x}\right|_{\theta=\theta_o} & \left.\dfrac{\partial f_N(\theta)}{\partial y}\right|_{\theta=\theta_o} \end{bmatrix}. \qquad (6.24)$$

For the functions of interest, we have

$$\frac{\partial f_i(\theta)}{\partial x} = \frac{x_i - x}{\sqrt{(x_i - x)^2 + (y_i - y)^2}} \qquad (6.25)$$

and

$$\frac{\partial f_i(\theta)}{\partial y} = \frac{y_i - y}{\sqrt{(x_i - x)^2 + (y_i - y)^2}}. \qquad (6.26)$$

Now, returning to our cost function $\varphi(\theta)$, we have

$$\phi(\theta) \approx [\mathbf{r} - \{\mathbf{f}(\theta_o) + \mathbf{H}(\theta - \theta_o)\}]^T [\mathbf{r} - \{\mathbf{f}(\theta_o) + \mathbf{H}(\theta - \theta_o)\}] \qquad (6.27a)$$

$$= [\mathbf{r} - \mathbf{f}(\theta_o) + \mathbf{H}\theta_o - \mathbf{H}\theta]^T [\mathbf{r} - \mathbf{f}(\theta_o) + \mathbf{H}\theta_o - \mathbf{H}\theta]. \qquad (6.27b)$$

Now, combining terms that are independent of θ, we can define $\tilde{\mathbf{r}} = \mathbf{r} - \mathbf{f}(\theta_o) + \mathbf{H}\theta_o$, which gives

$$\phi_L(\theta) = [\tilde{\mathbf{r}} - \mathbf{H}\theta]^T [\tilde{\mathbf{r}} - \mathbf{H}\theta] \qquad (6.28)$$

and

$$\hat{\theta} = \arg\min_{\theta} \phi_L(\theta). \qquad (6.29)$$

This is now a linear least squares (LLS) problem. The solution to this problem is well-known and can be written as

$$\hat{\theta} = (\mathbf{H}^T\mathbf{H})^{-1} \mathbf{H}^T \tilde{\mathbf{r}}. \qquad (6.30)$$

Substituting for $\tilde{\mathbf{r}}$ in (6.30) gives

$$\hat{\theta} = \left(H^T H\right)^{-1} H^T \left(r - f(\theta_o) + H\theta_o\right) \tag{6.31a}$$

$$= \theta_o + \left(H^T H\right)^{-1} H^T \left(r - f(\theta_o)\right). \tag{6.31b}$$

Thus, given an initial estimate of the position $\hat{\theta}_0$, we can create a new estimate, which is closer to the true global minimizer. If we use this new estimate as the initial estimate in (6.31b), we can obtain an updated estimate. This suggests an iterative procedure defined by

$$\hat{\theta}[n+1] = \hat{\theta}[n] + \left(H^T H\right)^{-1} H^T \left(r - f(\hat{\theta}[n])\right), \tag{6.32}$$

where it should be noted that the matrix H must also be updated at each step as indicated in (6.24). The benefit of this approach is a significant reduction in computational complexity, while the negative aspect is that it requires an initial solution that is reasonably close to the true solution.

Example 6.6

Consider a TOA system with anchors at (1, 6), (4, 4), and (5, 8). The reported measurements based on TOA are 3.0, 2.0, and 2.23 m. Using several initial estimates, find the mobile position using the Taylor series linearization approach.

Solution

The convergence of the position estimate for 20 different initial estimates is plotted in Figure 6.8 (top). In this case, we can see that all of the initial estimates converge to the same location, (4, 6), which is indeed the true mobile location. Note that the anchors surround the node location. In this case, virtually any initial starting point converges to the true location.

Example 6.7

Repeat Example 6.6 for anchor locations of (1, 6), (4, 7), and (5, 8) reported measurements based on the TOA of 4.12, 5.0, and 5.39 m.

Solution

The results are shown in Figure 6.8 (bottom) for 20 different initial solutions. In this case, since the anchors are nearly colinear and the node to be located is outside the convex hull of the anchors, there are multiple local minima in the cost function, and the solution depends on the initial solution. More specifically, there are two minima located at (0, 10), which is the true location, and at (3.32, 2.49), which is a local minimum.

6.6.2 Differencing

A second method for linearizing the cost function given in (6.21a) is to take differences between squared range functions [7]. More specifically, if we square the observations r_i, we have the following set of nonlinear equations:

(a)

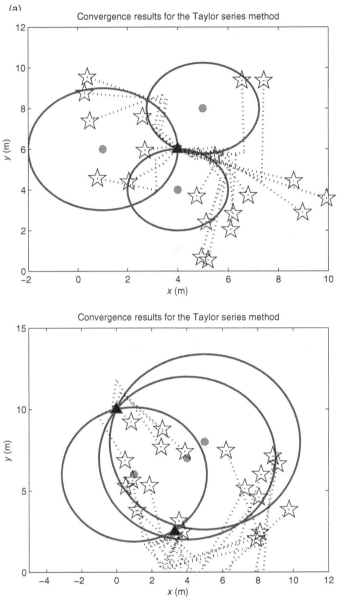

Figure 6.8 Example convergence plots for Taylor series iteration from Examples 6.6 (top) and 6.7 (bottom). Convergence is shown for 20 random starting points for good (top) and bad (bottom) geometry. Disks represent base station (anchor) locations; stars represent initial estimates; dark triangles represent the final solutions.

$$r_1^2 = (x_1 - x)^2 + (y_1 - y)^2 \tag{6.33a}$$

$$r_2^2 = (x_2 - x)^2 + (y_2 - y)^2 \tag{6.33b}$$

$$\vdots$$

$$r_N^2 = (x_N - x)^2 + (y_N - y)^2. \tag{6.33c}$$

Now, if we take the difference between two equations, we obtain

$$r_2^2 - r_1^2 = (x_2 - x)^2 + (y_2 - y)^2 - (x_1 - x)^2 - (y_1 - y)^2 \tag{6.34a}$$

$$= x_2^2 - 2x_2 x + x^2 + y_2^2 - 2y_2 y + y^2 - \ldots \tag{6.34b}$$

$$x_1^2 + 2x_1 x - x^2 - y_1^2 + 2y_1 y - y^2$$

$$= 2(x_1 - x_2)x + 2(y_1 - y_2)y + x_2^2 - x_1^2 + y_2^2 - y_1^2. \tag{6.34c}$$

Rearranging, we have

$$2(x_1 - x_2)x + 2(y_1 - y_2)y = r_2^2 - r_1^2 - x_2^2 + x_1^2 - y_2^2 + y_1^2. \tag{6.35}$$

Taking

$$M = \binom{N}{2} \text{ differences,}$$

we obtain a system of equations,

$$\mathbf{A}\boldsymbol{\theta} = \mathbf{c}, \tag{6.36}$$

where

$$\mathbf{A} = \begin{bmatrix} 2(x_1 - x_2) & 2(y_1 - y_2) \\ 2(x_1 - x_3) & 2(y_1 - y_3) \\ \vdots & \vdots \\ 2(x_{N-1} - x_N) & 2(y_{N-1} - y_N) \end{bmatrix} \tag{6.37}$$

and

$$\mathbf{c} = \begin{bmatrix} r_2^2 - r_1^2 - x_2^2 + x_1^2 - y_2^2 + y_1^2 \\ r_3^2 - r_1^2 - x_3^2 + x_1^2 - y_3^2 + y_1^2 \\ \vdots \\ r_N^2 - r_{N-1}^2 - x_N^2 + x_{N-1}^2 - y_N^2 + y_{N-1}^2 \end{bmatrix} \tag{6.38}$$

The LLS solution to (6.36) is

$$\hat{\boldsymbol{\theta}} = (\mathbf{A}^T \mathbf{A})^{-1} \mathbf{A}^T \mathbf{c}. \tag{6.39}$$

To understand the impact of linearizing the circular constraints in this way, it is useful to consider the geometric interpretation of this solution. Clearly, any solution to a pair of circular equations will also satisfy the linear equations. This can be seen by considering any two range constraints:

$$0 = (x_i - x)^2 + (y_i - y)^2 - r_i^2 \tag{6.40a}$$

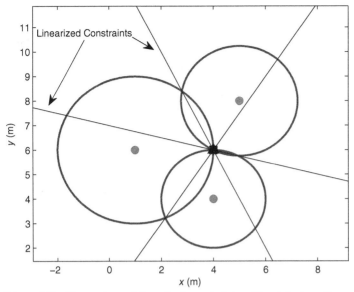

Figure 6.9   Comparison of linearized constraints with original nonlinear constraints from Example 6.8.

and

$$0 = (x_j - x)^2 + (y_j - y)^2 - r_j^2. \tag{6.40b}$$

In general, there are two values of (x, y) that satisfy these two constraints (assuming that they are consistent—i.e., there is no noise* in the range measurement r). The two solutions will also satisfy the linear equation found by subtracting one equation from the other. In other words, the two solutions should fall on the line defined by

$$y = \frac{x_i - x_j}{y_j - y_i} x + \frac{r_i^2 - r_j^2 + x_j^2 - x_i^2 + y_j^2 - y_i^2}{2(y_j - y_i)}. \tag{6.41}$$

This will be seen in the following example.

Example 6.8

Consider three anchors located at $(x_1, y_1) = (1, 6)$, $(x_2, y_2) = (4, 4)$, and $(x_3, y_3) = (5, 8)$, and a node located at $(x, y) = (4, 6)$ as shown in Figure 6.9. Find and plot both the circular equations and the linear equations resulting from the differences.

Solution

There are three nonlinear constraints on the solution given by

$$9 = (x-1)^2 + (y-6)^2, \tag{6.42}$$

---

*If there is noise in the range equations, there may not be a solution since they may be inconsistent.

$$4 = (x-4)^2 + (y-4)^2, \tag{6.43}$$

and

$$5 = (x-5)^2 + (y-8)^2. \tag{6.44}$$

The three corresponding circles are plotted in Figure 6.8. The anchors are designated by disks, while the node location is designated by a solid square. Also on the plot are the

$$M = \frac{3!}{2!} = 3 \text{ corresponding linear plots:}$$

$$y = 1.5x, \tag{6.45}$$

$$y = -2x + 14, \tag{6.46}$$

and

$$y = -0.25x + 7. \tag{6.47}$$

Two things can be noticed: (1) Each line intersects the two intersection points of the corresponding circles, and (2) the three lines intersect at the true node location, which is also where the three circles intersect.

Thus, in the absence of noise, there is no loss in solving the linear equations as opposed to the original nonlinear equations; that is, both sets of equations have a single solution and it is the same. However, this is not true in the presence of noise. When noise is present, the three original equations (in general) will not be consistent and will thus not have a solution. This can be seen in the following example.

Example 6.9

Consider the same scenario as in Example 6.8. However, due to measurement noise, the three range measurements are 2.75, 1.9, and 2.4. In other words, the three nonlinear constraints are given by

$$7.56 = (x-1)^2 + (y-6)^2, \tag{6.48}$$

$$3.61 = (x-4)^2 + (y-4)^2, \tag{6.49}$$

and

$$5.76 = (x-5)^2 + (y-8)^2. \tag{6.50}$$

Plot the nonlinear and linearized constraints.

Solution

The three corresponding circles are plotted in Figure 6.10. Also plotted are the resulting linear constraints. Again, a few things can be seen. First, the nonlinear equations are indeed inconsistent, as seen by the fact that there is no intersection between the three circles. Second, the three linear equations do intersect at one point. This is due to the fact that although

$$\frac{3!}{2!} = 3$$

distinct equations can be created, only $M - 1 = 2$ of them are independent. This can be seen from the linear equations given in the preceding example. Specifically, $(6.47) = 0.5 ((6.45) + (6.46))$. Also plotted on the figure are the NLS solution and

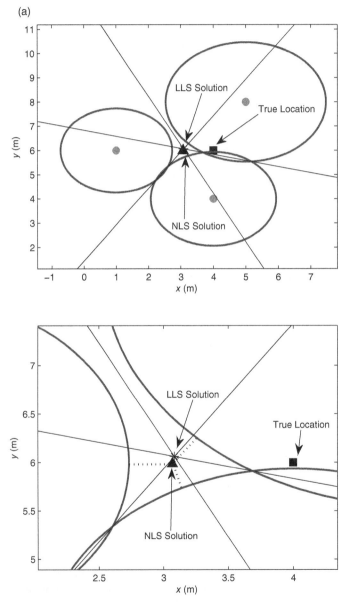

Figure 6.10   Example of LLS and NLS solutions for good geometry with noise (top) and close-up plot (bottom) from Example 6.9.

the LLS solution. It can be seen that the two give nearly identical results. This is due to the fact that the geometry of the anchors relative to the true node location is good. The NLS solution can be viewed as the point that is closest in Euclidean distance to the three circles as shown in Figure 6.10 (bottom). The LLS solution is the point closest in Euclidean distance to the lines defined by the linear constraints. In this case, since there are only two independent lines, the LLS solution corresponds to the intersection. If there were additional anchors, the lines would not necessarily all intersect at one point since there would be three independent lines.

Example 6.10

Consider three anchors located at $(x_1, y_1) = (1, 6)$, $(x_2, y_2) = (4, 7)$, and $(x_3, y_3) = (5, 8)$, and a node located at $(x, y) = (0, 10)$ as shown in Figure 6.11. Due to measurement noise, the three three nonlinear constraints are given by

$$24.5 = (x-1)^2 + (y-6)^2, \tag{6.51}$$

$$26.0 = (x-5)^2 + (y-8)^2, \tag{6.52}$$

and

$$28.1 = (x-4)^2 + (y-7)^2. \tag{6.53}$$

The three corresponding circles are also plotted in Figure 6.11. The corresponding linear constraints are plotted, along with the NLS and LLS solutions. In this case, we can see that a small amount of measurement noise does not significantly impact the NLS solution, but it does dramatically affect the LLS solution. This highlights the fact that in bad geometric conditions, the LLS approach, although much simpler, will generally perform poorly.

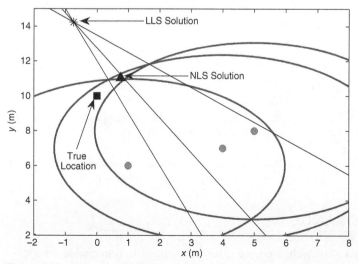

Figure 6.11   Example of LLS and NLS solutions for bad geometry from Example 6.10.

6.6.3 Linearization of TDOA

Like TOA, the TDOA cost function can also be linearized. More specifically, if we replace $\mathbf{f}(\theta)$ with $\mathbf{g}(\theta)$ and $\mathbf{r}$ with $\Delta\mathbf{r}$, where

$$\mathbf{g}(\theta) = \begin{bmatrix} \sqrt{(x_1-x)^2+(y_1-y)^2} - \sqrt{(x_2-x)^2-(y_2-y)^2} \\ \sqrt{(x_1-x)^2+(y_1-y)^2} - \sqrt{(x_3-x)^2-(y_3-y)^2} \\ \vdots \\ \sqrt{(x_{N-1}-x)^2+(y_{N-1}-y)^2} - \sqrt{(x_N-x)^2-(y_N-y)^2} \end{bmatrix} \tag{6.54}$$

and

$$\Delta\mathbf{r} = \begin{bmatrix} r_1 - r_2 \\ r_1 - r_3 \\ \vdots \\ r_{N-1} - r_N \end{bmatrix} \tag{6.55}$$

in the development from the previous section, the final iteration is then

$$\hat{\theta}[n+1] = \hat{\theta}[n] + (\mathbf{D}^T\mathbf{D})^{-1}\mathbf{D}^T\left(\Delta\mathbf{r} - g\left(\hat{\theta}[n]\right)\right), \tag{6.56}$$

where

$$\mathbf{D} = \begin{bmatrix} \dfrac{\partial g_1(\theta)}{\partial x}\bigg|_{\theta=\theta_o} & \dfrac{\partial g_1(\theta)}{\partial y}\bigg|_{\theta=\theta_o} \\ \vdots & \vdots \\ \dfrac{\partial g_M(\theta)}{\partial x}\bigg|_{\theta=\theta_o} & \dfrac{\partial g_M(\theta)}{\partial y}\bigg|_{\theta=\theta_o} \end{bmatrix} \tag{6.57}$$

(note that $\mathbf{D}$ must be updated at each iteration),

$$M = \binom{N}{2},$$

and

$$\frac{\partial g_1(\theta)}{\partial x} = \frac{x_1-x}{\sqrt{(x_1-x)^2+(y_1-y)^2}} - \frac{x_2-x}{\sqrt{(x_2-x)^2+(y_2-y)^2}} \tag{6.58}$$

$$\frac{\partial g_1(\theta)}{\partial y} = \frac{y_1-y}{\sqrt{(x_1-x)^2+(y_1-y)^2}} - \frac{y_2-y}{\sqrt{(x_2-x)^2+(y_2-y)^2}}. \tag{6.59}$$

6.7 PSEUDORANGE

When employing a TOA technique, we assume that there is systemwide time synchronization; that is, for each TOA measurement, we require that the transmitter and receiver both have the same interpretation of time (i.e., their clocks are synchronized). In most systems, it is possible to closely synchronize the transmitters to one another or the receivers to one another, but synchronizing transmitters to receivers is difficult at best. In such a case, there remains a system bias that must be included in the localization equations.

The best known example of this is the global positioning system (GPS). Specifically, in GPS (which is discussed in more detail in Part V), the transmitters (satellites) can be closely synchronized. The receiver, which is relatively inexpensive, is also synchronized to the system, but only loosely. As a result, there exists a bias in the range measurements, and we term these estimates as *pseudoranges*.

To see this, let us return to (6.1c). Specifically, rewriting the equation for the *i*th link,

$$t_i = \frac{\sqrt{(x_i - x)^2 + (y_i - y)^2}}{c} + t_{M_i}, \tag{6.60}$$

where t_{M_i} is the transmission time at the *i*th transmitter, and t_i is the received time at the receiver of interest. Now, if the clocks at each transmitter and the receiver are not perfectly synchronized, we have

$$t_i - \eta = \frac{\sqrt{(x_i - x)^2 + (y_i - y)^2}}{c} + t_{M_i} - \varepsilon_i, \tag{6.61}$$

where η is the error between the receiver clock and system time and ε_i is the error between the *i*th transmitter's clock at system time. Now, in a system like GPS where the transmitters have known locations (i.e., are anchors) and are part of the infrastructure, regular calibration allows for the essential elimination of clock error at the transmitters. Thus, synchronized transmitters result in

$$\begin{aligned} \varepsilon_i &= 0 \quad \forall i \\ t_{M_i} &= t_M \quad \forall i \end{aligned}. \tag{6.62}$$

Thus, after multiplying by the speed of light, we have a set of equations of the form

$$\rho_i = \sqrt{(x_i - x)^2 + (y_i - y)^2} - b, \tag{6.63}$$

where $b = c\eta$ is a common range bias and ρ_i are termed pseudoranges. Thus, we need to solve the following system of equations for the position $\{x, y\}$ and the range bias (due to the clock bias) b:

$$\rho_1 = \sqrt{(x_1 - x)^2 + (y_1 - y)^2} - b, \tag{6.64}$$

$$\rho_2 = \sqrt{(x_2 - x)^2 + (y_2 - y)^2} - b, \tag{6.65}$$

$$\rho_N = \sqrt{(x_N - x)^2 + (y_N - y)^2} - b. \tag{6.66}$$

Now, similar to range equations and TDOA equations, this system of equations can be solved using nonlinear optimization or using linear approximations, where we must now solve for $\tilde{\theta} = \left[\tilde{\theta}^T, b\right]^T$. It should be noted that in the absence of error, this equates to hyperbolic positioning. This can be seen by considering two simultaneous equations in the absence of noise. Specifically, if we solve (6.64) for b and substitute it into (6.65), we have

$$\rho_2 = \sqrt{(x_2 - x)^2 + (y_2 - y)^2} + \rho_1 - \sqrt{(x_1 - x)^2 + (y_1 - y)^2}. \tag{6.67}$$

Rearranging, we have

$$\rho_2 - \rho_1 = \sqrt{(x_2 - x)^2 + (y_2 - y)^2} - \sqrt{(x_1 - x)^2 + (y_1 - y)^2}, \tag{6.68}$$

which is clearly equivalent to a TDOA equation, and we have shown that the solutions to this equation fall on a hyperbola. This will also be seen in the following example.

Example 6.11

Consider three anchors located at $(x_1, y_1) = (1, 6)$, $(x_2, y_2) = (4, 4)$, and $(x_3, y_3) = (5, 8)$, and a node located at $(x, y) = (4, 6)$. Further, assume that the observed values for ρ are $\rho_1 = 2.5000$, $\rho_2 = 1.5000$, and $\rho_3 = 1.7361$. For anchors 1 and 2, plot the circular equations for several values of b. Show that the intersections of these circles fall on a hyperbola.

Circles for anchors 1 and 2 are plotted in Figure 6.12 for values of $b = \{0.25, 0.5, 0.75, 1.0\}$. We can see that the intersections of circles with the same value of b lie on a hyperbola. In other words, the hyperbola defines the locus of points, which

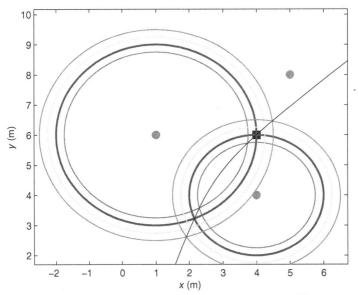

Figure 6.12   Example of pseudoranges from Example 6.11. Circles represent candidate ranges for a specific bias value. Intersections of common circles represent potential solutions and lie on a hyperbola.

satisfy the pair of equations for different values of bias, b. It should be noted that the true solution does indeed lie on the hyperbola, specifically the pair of circles corresponding to a bias of 0.5 m. If this were repeated for the first and third equations, another hyperbola would be defined, which would intersect the first at the true node location. Note that depending on the node geometry, we are not guaranteed that there would be only one intersection. A fourth observation would be required to resolve any ambiguity.

Note that when there is noise in the system, we must resort to LS-type solutions in an analogous manner as shown previously for TOA and TDOA systems.

6.8   THE IMPACT OF NLOS PROPAGATION

In TOA-based position location systems deployed in dense multipath propagation environments, especially indoors or in urban scenarios, the line-of-sight (LOS) path between nodes may be obstructed. As a result, in NLOS conditions, TOA-based range estimates are positively biased with high probability since the first arriving multipath component travels a distance that is in excess of the true LOS distance. A similar effect is seen in the case of RSS-based range estimates, where the received signal power is reduced due to the obstruction of the LOS path. These effects result in range estimates that are often much larger than the true distances, and therefore, in NLOS scenarios, the accuracy of node location estimates can be adversely affected.

6.8.1   Impact of NLOS Bias Errors

Again, suppose that an "unlocalized" node's (unknown) location is $\boldsymbol{\theta} = [x\ y]^T$. Let $\mathcal{A}$ denote the set of anchor locations, and $\mathcal{L} \in \mathcal{A}$ denote the set of locations of anchors, which provide LOS range estimates, with cardinality $m_L = |\mathcal{L}|$. The known locations of the LOS anchors are denoted by $\{\boldsymbol{\theta}_{Lj}\}$, $j = 1, 2, \ldots, m_L$. Similarly, $\mathcal{N} \in \mathcal{A}$ represents the set of locations of anchors that provide NLOS range estimates, with $m_N = |\mathcal{N}|$, and the known set of locations of the NLOS anchors is represented by $\{\boldsymbol{\theta}_{Nj}\}$, $j = 1, 2, \ldots, m_N$.

The LOS range estimates $\{r_{Lj}\}$, $j = 1, 2, \ldots, m_L$ are modeled as unbiased Gaussian [1] estimates of the actual internode distances $R_{Lj} = \|x - x_{Lj}\|$:

$$r_{Lj} = d_{Lj} + n_{Lj}, j = 1, 2, \cdots, m_L, \tag{6.69}$$

where n_{Lj} represents zero-mean Gaussian range measurement noise in the jth LOS range estimate $n_{Lj} \sim \mathcal{N}\left(0, \sigma_{Lj}^2\right)$. The range measurement noise variance σ_{Lj}^2 can be modeled as [2]

$$\sigma_{Lj}^2 = K_E d_{Lj}^{\beta_L}, \tag{6.70}$$

where β_L is the LOS path loss exponent and K_E is a proportionality constant (governed by the transmit power and the receiver noise floor) that determines the accuracy of range estimation. This model arises due to the fact that the accuracy of TOA-based range estimates can be shown [3, 4] to be inversely proportional to the received signal-to-noise ratio (or, more generally, the signal-to-interference-and-noise ratio) assuming matched-filter detection. The above model for the accuracy of range estimates [2] applies to both TOA- and RSS-based range estimates when $\beta = 2$. The vector of LOS range estimates is denoted by $r_L = \begin{bmatrix} r_{L1} & r_{L2} & \cdots & r_{Lm_L} \end{bmatrix}^T$.

The NLOS range estimates are assumed to be positively biased Gaussian estimates [5] of the true distances:

$$r_{Nj} = d_{Nj} + n_{Nj} + b_{Nj}, \, j = 1, 2, \cdots, m_N, \tag{6.71}$$

where $d_{Nj} = \|x - x_{Nj}\|$, $n_{Nj} \sim \mathcal{N}(0, \sigma_{Nj}^2)$, and $\sigma_{Nj}^2 = K_E d_{Nj}^{\beta_N}$. b_{Nj} are the NLOS bias errors, and β_N is the effective path loss exponent under NLOS conditions. We assume that the bias errors are always positive: $b_{Nj} > 0$, $\forall j$. The bias errors b_{Nj} and the range measurement noise n_{Nj} are assumed to be independent random variables. Although we make no assumptions about the statistical distribution [6] of the NLOS bias errors in the following development, for the purpose of simulation we assume that bias errors are uniformly distributed: $b_{Nj} \sim \mathcal{U}(0, B_{max})$, where B_{max} represents the maximum possible bias.* Additionally, we assume that the bias errors are, with high probability, much larger than the range measurement noise $B_{max} \gg \sigma_{Nj}$, $j = 1, 2, \cdots, m_N$. Finally, without loss of generality, we assume that the coordinate axes are selected such that $\theta \geq 0$.

6.8.2   Discarding NLOS Range Estimates

As we saw previously, when we have at least three range estimates, the LS estimator [7] can be used to compute an estimate $\hat{\theta}$ of the unlocalized node's location θ. We define the *localization error*, a measure of the accuracy of the location estimate $\hat{\theta}$, as

$$\Omega = \|\theta - \hat{\theta}\|^2 \, (\text{m}^2). \tag{6.72}$$

It must be noted that Ω is a random variable, with different instances corresponding to different realizations of the range measurement noise, bias errors, and anchor locations. Therefore, we characterize the accuracy of location estimates through the mean μ_Ω and standard deviation σ_Ω of the localization error defined in (6.72); smaller values of both μ_Ω and σ_Ω indicate more-accurate node location estimates.

When $m_N = 0$ and $m_L \geq 3$, the LS estimator provides accurate estimates of a node's location [7]. However, when $m_N > 0$, we need effective ways to incorporate NLOS information into the estimation procedure. The Cramer–Rao lower bound (CRLB) analysis presented in [6] and [8] showed that, in the absence of prior statistical information on the NLOS range estimates, the minimum variance unbiased estimator (MVUE) discards the biased NLOS range estimates and utilizes only LOS range information while computing location estimates. However, this approach may not be optimal when using practical estimators, such as the LS estimator, that do not achieve the CRLB. Figure 6.13a, b shows the impact of directly (without mitigation of the bias errors) incorporating NLOS range estimates into the LS solution for two

---

*Practically speaking, B_{max} would depend on the propagation environment and physical layer parameters such as the transmit power.

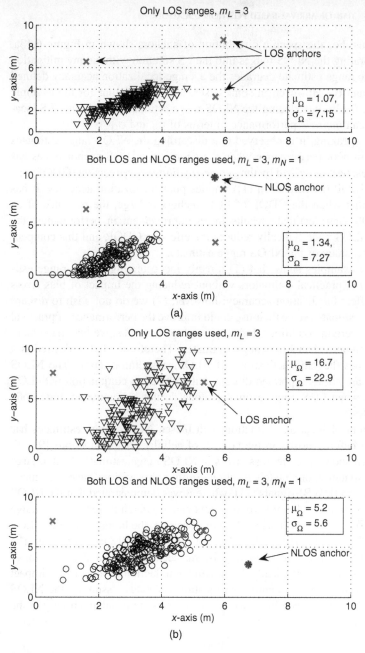

Figure 6.13 (a) This example shows several instances of the LS location estimate $\hat{\theta}$, one for each realization of the range estimates: for (i) (top), only $m_L = 3$ LOS estimates, and (ii) (bottom) also including $m_N = 1$ NLOS range estimate. The NLOS range estimate is treated exactly like a LOS range estimate and is directly incorporated into the LS solution in the bottom figure. In this case, the addition of the biased NLOS range estimate *degrades* localization accuracy with respect to μ_Ω and σ_Ω. (b) In this case, the addition of the biased NLOS range estimate *improves* localization accuracy with respect to μ_Ω and σ_Ω. In both cases, $\boldsymbol{x} = [3\ 3]^T$, $\beta_L = 2$, $\beta_L = 2.5$, $K_E = 0.1$, and $B_{max} = 4$ m. Reprinted with permission from [24], copyright 2007 IEEE.

specific scenarios. For the specific distribution of anchors shown in Figure 6.13a, directly incorporating the NLOS ranges into the LS solution without any mitigation of the bias in the range estimate degrades the average localization accuracy defined in terms of μ_Ω and σ_Ω. However, in some cases, and in particular for the example shown in Figure 6.13b, introducing the NLOS range estimate directly into LS location estimation can *improve* performance in terms of μ_Ω and σ_Ω.

Generally speaking, it is observed that discarding the NLOS range estimates does not result in poor performance when the geometry of LOS anchor nodes has certain properties, best described by the *geometric dilution of precision* (GDOP) [1], where a large GDOP (as defined in [1]) implies poor localization accuracy. It has been observed that when the GDOP of LOS anchors is large, the presence of an additional NLOS range estimate results in an improvement in performance: the addition of an NLOS node typically reduces the effective GDOP, and this compensates for the inaccuracy of the NLOS range estimate.

These two examples show that (1) directly incorporating NLOS range estimates into existing practical estimators without reducing the impact of bias errors can adversely affect localization accuracy; however, (2) we do not wish to discard the NLOS range estimates since their use could improve the performance of practical estimators under certain conditions. Indeed, in indoor networks, we may have more NLOS range estimates than LOS range estimates. Therefore, what is desired is a method that allows the "soft activation" of NLOS range information: The NLOS range estimates are not incorporated directly but are used in conjunction with LOS range estimates when LOS range estimates alone do not guarantee accurate node location estimates.

The problem of location estimation with biased NLOS range estimates has been considered before, mostly in the context of cellular communications [9, 10], where it has been shown that the bias errors in NLOS range estimates lead to large errors in the computation of a node's location. Similar observations have been made with ultra-wideband (UWB) signals [11], where it was demonstrated that the NLOS bias errors can be on the order of several meters and are much larger than the range measurement errors in LOS scenarios. Broadly speaking, the literature on the NLOS problem falls in two categories: NLOS identification and NLOS mitigation. The former deals with the problem of distinguishing between LOS and NLOS range estimates, whereas the latter typically deals with the reduction of the adverse impact of NLOS range errors on the accuracy of location estimates, assuming the NLOS range estimates have been identified. We will examine both problems briefly in the next two sections.

6.8.3   NLOS Identification

Statistical NLOS identification techniques for cellular systems that rely on a time series of range measurements have been examined [5, 12]. A comprehensive discussion of a decision-theoretic framework for NLOS identification based on the statistical distributions of range estimates in LOS and NLOS scenarios was presented in [12]. Such approaches will not be applicable to systems with low or no mobility, and specific features of multipath channels can sometimes be exploited to enhance the accuracy of NLOS identification. For example, UWB signals possess higher temporal resolution and robustness to multipath fading than narrowband and wide-band signals. These characteristics create considerable differences in the behavior of (1) the TOA, (2) RSS, and (3) the temporal energy dispersion of the received signal in different propagation scenarios. The temporal dispersion of received signal energy in indoor UWB propagation channels has been extensively characterized (e.g., see [13]) and is quantified through the delay spread statistics, such as the root-mean-squared delay spread (RDS) [13]. The variations of the TOA, RSS, and RDS estimates from UWB signals can be utilized to distinguish between LOS and NLOS propagation scenarios.

A major issue concerning the practical application of the statistical decision-theoretic method is the availability of complete statistical information corresponding to received signal parameters, and particularly their variation with distance. When the conditional distributions of the received signal parameters are unknown, we may resort to "conventional" hypothesis testing, where the received signal parameter is compared with a threshold, followed by a decision on the channel state depending on whether the estimate is smaller or larger than the threshold. Due to the strong dependence of the statistics of the TOA and RSS estimates on the true distance between transmit and receive antennas, the threshold used for hypothesis testing these estimates must be dependent on the true distance.

On the other hand, it can be shown [14] that significantly lower channel state estimation error probabilities can be obtained with RDS estimates, even with noisy and biased distance estimates. Figure 6.14 shows a histogram of the RDS observed

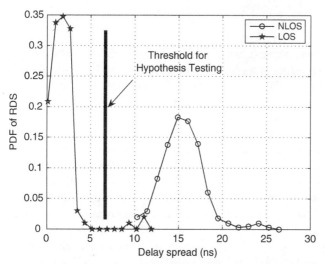

Figure 6.14 Hypothesis testing of the RDS for the nature of the UWB channel: the histogram of the RDS in LOS and NLOS scenarios for distances 1–30 m. We see that since the empirical probability density function (PDF) of the RDS for the LOS and NLOS cases are well separated, the performance of NLOS identification using this approach is not sensitive to the value of the threshold selected. Reprinted with permission from [14], copyright 2007 IEEE.

TABLE 6.1. RDS Statistics from Measurement Campaigns [14]

Indoor LOS				
Measurement Campaign	Mean μ_L (ns)	Standard Deviation σ_L (ns)	$\mathcal{M}$	P_e
AT&T [26, 27]	4.71	2.31	0.50	0.16
	3.55	1.65	0.63	0.20
Samsung/SAIT [28]	14.00	1.53	0.14	3.5×10^{-3}
	12.87	1.87	0.23	0.024
Time domain [29]	5.27	3.37	0.36	0.087
MPRG, VT [13]	3.34	2.17	0.18	3.3×10^{-3}
IEEE CM1 [30]	6	1.1		

Indoor NLOS				
Measurement Campaign	Mean μ_N (ns)	Standard Deviation σ_N (ns)	$\mathcal{M}$	P_e
AT&T [26, 27]	8.20	3.30	0.50	0.16
	7.35	3.45	0.63	0.20
Samsung/SAIT [28]	38.61	8.03	0.14	3.5×10^{-3}
	26.51	5.22	0.23	0.024
Time domain [29]	14.59	3.41	0.36	0.087
MPRG, VT [13]	16.08	2.41	0.18	3.3×10^{-3}
IEEE CM2 [30]	8	0.75	0.43	0.13
IEEE CM3 [30]	14.5	2.26	0.19	5.5×10^{-3}
IEEE CM4 [30]	25	3.7	0.10	3.4×10^{-5}

in a number of indoor measurements taken in LOS and NLOS scenarios, *for all measurement distances* (1–30 m) in an indoor propagation study [13, 14]. We see that the regions of support for the density functions of the RDS estimates remain sufficiently separated to use conventional hypothesis testing, without requiring a distance-dependent threshold. The statistics of the RDS estimates observed in several measurement campaigns [15, 16], summarized in Table 6.1, indicate that the distinction between the RDS estimates in LOS and NLOS propagation scenarios is general and is not restricted to a given set of measurements. The statistics of the RDS estimates corresponding to the channel model adopted by the IEEE 802.15.3a subcommittee [17] are also shown. It is important to emphasize that for each of the measurement scenarios listed in Table 6.1, differences in the propagation environments, deconvolution algorithms and their associated parameters, types and directionality of antennas, and so on, result in the variation of the observed values of the RDS across the measurement campaigns.

Assuming that the RDS estimate is positive and Gaussian [15], the minimum probability of error P_e^* and the optimal threshold T^* for hypothesis testing can be computed in a straightforward manner, as detailed in [18]. In general, the values of P_e^* and T^*, respectively, depend on the means $\{\mu_L, \mu_N\}$ and the standard deviations

$\{\sigma_L, \sigma_N\}$ of the RDS estimates. In order to allow for a unified comparison of performance using different sets of measurements, we define a metric:

$$\mathcal{M} = \frac{\sqrt{\sigma_L \sigma_N}}{|\mu_N - \mu_L|}. \tag{6.73}$$

Intuitively, the metric $\mathcal{M}$ is a measure of the extent of the overlap of the probability density functions of the RDS estimates in LOS and NLOS scenarios: As $\mathcal{M}$ decreases, P_e^* decreases. As the parameters $\{\mu_L, \mu_N, \sigma_L, \sigma_N\}$ vary, the values of the metric vary as well. However, there is a strong correlation between the value of $\mathcal{M}$ defined earlier and the value of P_e^*. Further, P_e^* largely depends on the value of $\mathcal{M}$ rather than on the specific values of the parameters $\{\mu_L, \mu_N, \sigma_L, \sigma_N\}$.

Table 6.1 lists the values of $\mathcal{M}$ and P_e^* computed using (6.73) [14] from the RDS statistics in the same table. Note that the probability of classification error, P_e, assumes (truncated) Gaussian statistics. We see that for the sets of measurements, the probability of channel state estimation error can vary from 10^{-5} to 0.2. It must be emphasized that these values of P_e are obtained under the assumption that the distribution of the RDS estimate under LOS and NLOS conditions is positive and (truncated) Gaussian distributed. The direct application of the hypothesis testing to received signal measurements over all distances [13] results in a probability of error P_e of approximately 0.02. No knowledge of the distance or statistical distributions was assumed.

6.8.4   NLOS Mitigation

Equipped with the ability to distinguish between LOS and NLOS range estimates, the CRLB analysis presented in [6] and [8] characterizes the performance of the MVUE [19] of a node's location, given a mixture of (unbiased) LOS and (biased) NLOS range estimates. This analysis showed that, in the absence of a priori statistical information on the NLOS range estimates, the MVUE discards the biased NLOS range estimates and utilizes only LOS range information while estimating sensor locations. However, as was demonstrated, in the case of practical nonefficient [19] estimators such as the commonly used LS estimator [7], discarding NLOS range information does not necessarily improve performance.

A semidefinite programming (SDP) approach to node localization based on connectivity information was investigated in [20], and a quadratic programming approach with NLOS range estimates was discussed in [21], but these approaches result in high computational complexity [22]. The residual weighting algorithm (Rwgh) was proposed in [23], whose advantage is that NLOS identification is not required a priori. However, this algorithm implicitly assumes that the range measurement noise is much smaller than the NLOS bias introduced, in order to inherently distinguish between the residuals from LOS and NLOS range estimates. More importantly, it relies on the availability of a large number of range estimates, several of which are LOS, so that the set of range estimates finally selected to compute a node's location results in the smallest residual error. However, in indoor networks, situations may arise where *only NLOS* range estimates are available while estimating a node's location.

6.9 HANDLING NLOS ERRORS: A LINEAR PROGRAMMING APPROACH

In this section, we show that the problem of node localization given LOS range information can be cast into the form of a linear program [24]. We then modify the linear program to utilize additional NLOS range information, resulting in a method that utilizes a mixture of LOS and NLOS range estimates to estimate a node's location.

6.9.1 LOS Range Estimates

The LOS range estimates, which are modeled as unbiased estimates of the true ranges, can be used to define conditions satisfied by the unknown node location $\boldsymbol{\theta}$. Specifically, if we linearize the the LOS range equations using differencing, we have from (6.39) the LS solution:

$$\hat{\boldsymbol{\theta}} = \left(A^T A\right)^{-1} A^T c. \tag{6.74}$$

This linear system of equations can be converted to a linear program [25] if the objective function is a linear function of the unknowns. Specifically, if we define

$$Z \triangleq \sum_{i} \sum_{j,j>i} |e_{ij}|$$

and then replace the unconstrained variable e_{ij} by $e_{ij}^+ - e_{ij}^-, e_{ij}^+ \geq 0, e_{ij}^- \geq 0$, we can write an alternative *linearized* objective function that is to be minimized as

$$Z \triangleq \sum_{i} \sum_{j,j>i} \left(e_{ij}^+ + e_{ij}^-\right). \tag{6.75}$$

It must be noted that in the optimal solution that minimizes Z, only one term among $\{e_{ij}^+, e_{ij}^-\}$ will be equal to $|e_{ij}|$, with the other being zero [25]. The constraints are then given by

$$a_{ij}x + b_{ij}y - e_{ij}^+ + e_{ij}^- = c_{ij}, \quad i, j = 1, 2, 3, \cdots, m_L, \ i < j. \tag{6.76}$$

Since there are now $2M$ nonnegative slack variables, the vector z of $(2M + 2)$ variables can be written as $z = \begin{bmatrix} x \ y \ \boldsymbol{\varepsilon}^T \end{bmatrix}^T$, where

$$\boldsymbol{\varepsilon} = \begin{bmatrix} e_{12}^+ \ e_{12}^- \ e_{13}^+ \ e_{13}^- \cdots e_{(m_L-1)m_L}^+ \ e_{(m_L-1)m_L}^- \end{bmatrix}_{1 \times 2M}^T. \tag{6.77}$$

Thus, the linear program can be formulated in *standard form* [25] as $\min Z = f_L^T z$, such that

$$\begin{bmatrix} A | J \end{bmatrix} z = c, z \geq \mathbf{0}, \tag{6.78}$$

where A and c were respectively defined in (6.37) and (6.38),

$$J = \begin{bmatrix} -1 & 1 & 0 & 0 & \cdots & \cdots & 0 & 0 \\ 0 & 0 & -1 & 1 & \cdots & \cdots & 0 & 0 \\ \vdots & \vdots & & & & & & \\ 0 & & 0 & 0 & 0 & 0 & -1 & 1 \end{bmatrix}_{M \times 2M}, \tag{6.79}$$

and $f_L = \begin{bmatrix} 0_{2 \times 1}^T & 1_{2M \times 1}^T \end{bmatrix}^T$. Here, $0_{k \times l}$ represents a $k \times l$ matrix of zeros and $1_{k \times l}$ represents a $k \times l$ matrix of ones.

We now have a linear program that can be used to solve for a node's location given LOS range estimates. In this linear program, the objective function Z defined in (6.75) is a function of the distances of a point θ to the straight lines resulting from the differences between two ranging circles. If we use NLOS range estimates in a similar manner, by incorporating them into the objective function, we could potentially degrade the accuracy of the location estimate. Instead, as described in the following section, we can use the NLOS range estimates to constrain the feasible region for θ without affecting the objective function defined using LOS range estimates, thereby limiting the possibility of large errors, particularly when the number of LOS range estimates is small.

6.9.2   NLOS Range Estimates

As the bias errors in the NLOS range estimates are always positive and are assumed to be much larger than the range measurement noise, we know each NLOS range estimate r_{Ni} defined in (6.71) is, with high probability, larger than the true range d_{Ni}, $i = 1, 2, \ldots, m_N$. Based on this observation, we can convert the NLOS range estimates into inequalities for $i = 1, 2, \ldots, m_N$:

$$\|\theta - \theta_{Ni}\| \le r_{Ni} \Rightarrow (x - x_{Ni})^2 + (y - y_{Ni})^2 \le r_{Ni}^2. \tag{6.80}$$

These inequalities imply that the feasible region for θ lies *in the interior* of each of the circular constraints defined by (6.80). Note that this assumption cannot be made if the standard deviation of the zero-mean measurement noise and the positive bias in (6.71) are comparable. Once again, these are nonlinear constraints on x and y. However, these constraints can be relaxed to the following linear constraints, as suggested in [22]:

$$x - x_{Ni} \le r_{Ni}, -x + x_{Ni} \le r_{Ni}, y - y_{Ni} \le r_{Ni}, -y + y_{Ni} \le r_{Ni}, \quad (i = 1, 2, \cdots, m_N). \tag{6.81}$$

This essentially relaxes the circular constraints to rectangular constraints. It is readily seen that the new rectangular feasible region contains the original (convex) feasible region formed by the intersection of the original circular regions. We can now write the above four constraints for the ith NLOS range estimate in standard form [25]:

$$x - x_{Ni} + u_{1i} = r_{Ni}, -x + x_{Ni} + u_{2i} = r_{Ni}, y - y_{Ni} + v_{1i}$$
$$= r_{Ni}, -y + y_{Ni} + v_{2i} = r_{Ni} \tag{6.82}$$
$$u_{1i}, u_{2i}, v_{1i}, v_{2i} \ge 0, \quad i = 1, 2, \cdots, m_N,$$

where u_{1i}, u_{2i}, v_{1i}, and v_{2i} are the slack variables corresponding to the ith NLOS range estimate. Defining $w_i = \left[u_{1i}\ u_{2i}\ v_{i1}\ v_{2i} \right]^T_{1\times 4}$ and $z_i = \left[x\ y\ w_i^T \right]^T$ as the vectors of variables corresponding to the ith NLOS range estimate, we can express the above equations in matrix form as

$$\left[B_1 | I_{4\times 4} \right] z_i = r_i,\ z_i \geq 0,$$

where

$$B_1 = \begin{bmatrix} 1 & 0 \\ -1 & 0 \\ 0 & 1 \\ 0 & -1 \end{bmatrix},\ r_i = \begin{bmatrix} r_{Ni} + x_{Ni} \\ r_{Ni} - x_{Ni} \\ r_{Ni} + y_{Ni} \\ r_{Ni} - y_{Ni} \end{bmatrix}, \tag{6.83}$$

and $I_{n\times n}$ denotes an $n \times n$ identity matrix. We can now stack the constraints corresponding to each of the m_N NLOS range estimates to form a system of $N = 4m_N$ equations as follows:

$$\left[B\ I_{N\times N} \right] z = r,\ z \geq 0, \tag{6.84}$$

where

$$B = \left[B_1^T\ B_1^T\ \cdots\ B_1^T \right]^T_{(2\times N)},\ r = \left[r_1^T\ r_2^T\ \cdots\ r_{m_N}^T \right]^T_{(1\times N)}, \tag{6.85}$$

and

$$w = \left[w_1^T\ w_2^T\ \cdots\ w_{m_N}^T \right]^T_{(1\times N)}, \tag{6.86}$$

with the vector of variables being defined as

$$z = \left[x\ y\ w^T \right]^T_{(1\times(N+2))}. \tag{6.87}$$

It is important to note that in the above analysis, no objective function was defined based on the NLOS range estimates, and only a feasible region for x was derived. The feasible region can further be constrained by including the tangents at the intersection points of the circular constraints defined in (6.80) to reduce the size of the feasible region (we shall call this linear program [LP]–extended). In the following subsection, we integrate the constraints and objective function obtained using LOS range estimates with the NLOS constraints defined earlier, for the problem of node location estimation given any mixture of LOS and NLOS range estimates, such that $m_L \geq 3$, $m_N \geq 0$.

6.9.3   Combining the LOS and NLOS Range Information

Based on the previous subsections, given $m_L \geq 3$ LOS range estimates and $m_N \geq 0$ NLOS range estimates, we can combine them into a single linear program. We define the vector of variables as

$$z = \left[x\ y\ \varepsilon\ w \right]^T_{(1\times(2M+N+2))}, \tag{6.88}$$

where $\boldsymbol{\varepsilon}$ and $\boldsymbol{w}$ are respectively defined in (6.77) and (6.86). The objective function Z is defined as

$$Z = \boldsymbol{f}^T \boldsymbol{z}, \qquad (6.89)$$

where $\boldsymbol{f}^T = \begin{bmatrix} 0 & 0 & \boldsymbol{1}_{2M \times 1} & \boldsymbol{0}_{N \times 1} \end{bmatrix}_{1 \times (2M+N+2)}$. The complete linear program is then formulated as

$$\min Z = \boldsymbol{f}^T \boldsymbol{z}, \quad \text{such that}$$

$$\boldsymbol{Dz} = \boldsymbol{g}, \boldsymbol{z} \geq \boldsymbol{0},$$

where

$$\boldsymbol{D} = \begin{bmatrix} \boldsymbol{A}| & \boldsymbol{J} & \boldsymbol{0}_{M \times N} \\ \boldsymbol{B}| & \boldsymbol{0}_{2M \times N} & \boldsymbol{I}_{N \times N} \end{bmatrix}_{(M+N) \times (2+2M+N)}, \boldsymbol{g} = \begin{bmatrix} \boldsymbol{c} \\ \boldsymbol{r} \end{bmatrix}_{(2+2M+N) \times 1}.$$

In the above equations, the matrices $\boldsymbol{A}$, $\boldsymbol{J}$, and $\boldsymbol{B}$ are respectively defined in (6.37), (6.79), and (6.85), and the vectors $\boldsymbol{c}$ and $\boldsymbol{r}$ are defined in (6.38) and (6.85), respectively.

It must be pointed out that in the above linear program, LOS range information is used to define both the objective function and the feasible region, whereas the NLOS range information is used only to define the feasible region. This allows the NLOS range estimates to "assist" in improving the accuracy of location estimates by limiting the size of the feasible region but does not allow the NLOS bias errors to adversely affect node localization accuracy, since the NLOS range information plays no part in defining the objective function. The above approach works for any mixture of LOS and NLOS range estimates, provided $m_L \geq 3$, $m_N \geq 0$.

To examine the relative performance of this method with techniques that discard NLOS measurements, consider a $W \times W \, \text{m}^2$ area of interest with $W = 100 \, \text{m}$, containing $N_A = 80$ randomly (uniformly) distributed anchors. Several obstructions are randomly dispersed over the area of interest. At time t, if the path between the mobile's true location $\boldsymbol{x}(t)$ and a given anchor contains any portion of the obstructions, then the range estimates from that anchor are assumed to be given by (6.71); otherwise, the range estimates are given by (6.69). The density of nodes in this simulation was selected such that the total number of range estimates $(m_L + m_N)$ received at each instant of time lies between 3 and 8. The mobile is assumed to move at a speed $v = 2.5 \, \text{m/s}$ in a direction

$$\boldsymbol{e}(0) = \frac{1}{\sqrt{2}} \begin{bmatrix} 1 & 1 \end{bmatrix}^T$$

at $t = 0$ and subsequently reflects off the boundary of any obstacle encountered.

Figure 6.15a compares the true location of a mobile node with the location estimates computed using the LS and LP approaches. The mobile node estimates its location based on range information from anchors every $T_s = 1$ second between $t = 0$ and $t = 50$ seconds. We see that the LP-based NLOS mitigation schemes outperform the two LS-based schemes. This is more evident in Figure 6.15b, which compares the localization error achieved by the estimation approaches versus time. We see that

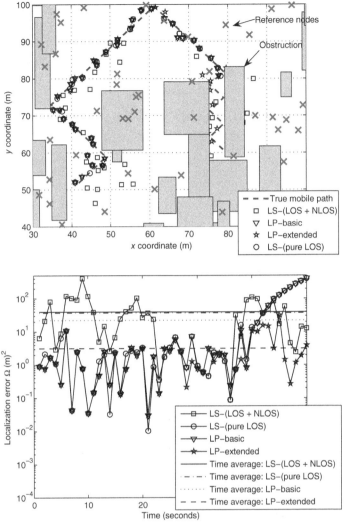

Figure 6.15   Simulation of a 2-D location-aware network in a NLOS environment: the values of the simulation parameters used are: $P_T = 1$ mW, $K_p = 1 \times 10^5$, $\xi_T = 15$ dB, $\beta_L = 2$, $\beta_N = 3$, $K_E = 0.1$, and $B_{max} = 10$ m. The total number of anchors is $N_A = 80$, and the mobile node moves at a speed of $v = 2.5$ m/s through a $W \times W$ m^2 area, where $W = 100$ m. (top) The true mobile trajectory and the mobile location estimates obtained using the LS and the LP-based location estimators. (bottom) A comparison of the different location estimators in terms of the localization error Ω versus time. Reprinted with permission from [24], copyright 2007 IEEE.

on average, the LP-based NLOS mitigation schemes achieve much higher localization accuracy than the LS-based location estimators.

6.10   CONCLUSIONS

In this chapter, we have reviewed the basic theory behind TOA-based position location, including TOA and TDOA techniques. We discussed linearization approaches and the geometric interpretations of the two approaches. Finally, we examined the impact of NLOS propagation and how it can be mitigated.

REFERENCES

[1] N. Patwari, A. O. Hero III, M. Perkins, N. S. Correal, and R. J. O'Dea, "Relative location estimation in wireless sensor networks," *IEEE Trans. Signal Process.*, vol. 51, no. 8, pp. 2137–2148, 2003.

[2] Y. Qi and H. Kobayashi, "On relation among time delay and signal strength based geolocation methods," in *Proc. of IEEE Global Telecommunications Conf. (GLOBECOM '03)*, Dec. 2003, pp. 4079–4083.

[3] Y. Zhang, L. Ackerson, D. Duff, C. Eldershaw, and M. Yim, "STAM: A system of tracking and mapping in real environments," *IEEE Wireless Comm. Mag.*, vol. 11, pp. 87–96, 2004.

[4] S. Gezici, Z. Tian, G. B. Giannakis, H. Kobayashi, A. F. Molisch, H. V. Poor, and Z. Sahinoglu, "Localization via ultra-wideband radios," *IEEE Signal Process. Mag.*, vol. 22, pp. 70–84, 2005.

[5] M. P. Wylie and J. Holtzman, "The non-line of sight problem in mobile location estimation," in *5th IEEE Int. Conf. on Universal Personal Communications*, Sep. 1996, pp. 827–831.

[6] Y. Qi, H. Kobayashi, and H. Suda, "Analysis of wireless geolocation in a non-line-of-sight environment," *IEEE Trans. Wireless Comm.*, vol. 5, pp. 672–681, 2006.

[7] J. J. Caffery, "A new approach to the geometry of TOA location," in *Proc. of IEEE Vehicular Technology Conf.*, vol. 4, Sep. 2000, pp. 1943–1949.

[8] Y. Qi and H. Kobayashi, "Cramer–Rao lower bound for geolocation in nonline-of-sight environment," in *Proc. of the Int. Conf. on Acoustics, Speech and Signal Processing (ICASSP)*, vol. 3, May 2002, pp. 2473–2476.

[9] M. I. Silventoinen and T. Rantalainen, "Mobile station emergency locating in GSM," in *IEEE Int. Conf. on Personal Wireless Communications*, Feb. 1995, pp. 232–238.

[10] J. Caffery and G. Stuber, "Overview of radiolocation in CDMA cellular systems," *IEEE Commun. Mag.*, vol. 36, pp. 38–45, 1998.

[11] B. Denis, J. Keignart, and N. Daniele, "Impact of NLOS propagation upon ranging precision in UWB systems," in *Proc. of 2003 IEEE Conf. on Ultra Wideband Systems and Technologies*, Nov. 2003, pp. 379–383.

[12] J. Borras, P. Hatrack, and N. B. Mandayam, "Decision theoretic framework for NLOS identification," in *IEEE 48th Vehicular Technology Conf.*, May 1998, pp. 1583–1587.

[13] R. Buehrer, W. Davis, A. Safaai-Jazi, and D. Sweeney, "Ultrawideband propagation measurements and modelling—DARPA netex final report," Tech. Rep., Jan. 2004. Available: http://www.mprg.org/people/buehrer/ultra/darpa_netex.shtml

[14] S. Venkatesh and R. M. Buehrer, "NLOS identification in ultra-wideband systems based on received signal statistics," in *IEEE Proc. on Microwaves, Antennas and Propagation: Special Issue on Antenna Systems and Propagation for Future Wireless Communications*, vol. 1, Dec. 2007, pp. 1120–1130.

[15] Z. Irahhauten, H. Nikookar, and G. J. Janssen, "An overview of ultrawide band indoor channel measurements and modeling," *IEEE Microw. Wireless Compon. Lett.*, vol. 14, pp. 386–388, 2004.

[16] C.-C. Chong, Y. Kim, and S. Lee, "Channel model parameterization of the indoor residential environment," Tech. Rep., Jul. 2004. Submitted to IEEE P802.15 Working Group for Wireless Personal Area Networks (WPANs). Available: http://grouper.ieee.org/groups/802/15/pub/2004/July04/

[17] J. Foerster, "Channel modeling sub-committee report final (Doc: IEEE 802-15-02/490r1-SG3a)," Tech. Rep., Feb. 2002. Available: http://grouper.ieee.org/groups/802/15/pub/2002/Nov02/

[18] S. Venkatesh, "The design and modeling of ultra-wideband position location networks," Ph.D. Thesis, Virginia Tech, 2007.

[19] S. M. Kay, *Fundamentals of Statistical Processing, Volume I: Estimation Theory.* Upper Saddle River, NJ: Prentice Hall, 1993.

[20] L. Doherty, K. Pister, and L. E. Ghaoui, "Convex position estimation in wireless sensor networks," in *Proc. of IEEE Conf. on Computer Communications (INFOCOM)*, Apr. 2001, pp. 1655–1663.

[21] X. Wang, Z. Wang, and R. O'Dea, "A TOA-based location algorithm reducing the errors due to non-line-of-sight (NLOS) propagation," *IEEE Trans. Veh. Technol.*, vol. 52, pp. 112–116, 2003.

[22] E. G. Larsson, "Cramer–Rao bound analysis of distributed positioning in sensor networks," *IEEE Signal Process. Lett.*, vol. 11, pp. 334–337, 2004.

[23] P. C. Chen, "A non-line of sight error mitigation algorithm in location estimation," in *IEEE Conf. on Wireless Communications and Networking*, Sep. 1999, pp. 316–320.

[24] S. Venkatesh and R. M. Buehrer, "NLOS mitigation in UWB location-aware networks using linear programming," *IEEE Trans. Veh. Technol.*, vol. 56, pp. 3182–3198, 2007.

[25] M. S. Bazaraa, J. J. Jarvis, and H. D. Sherali, *Linear Programming and Network Flows.* New York: John Wiley & Sons, 1995.

[26] S. S. Ghassemzadeh, L. J. Greenstein, T. Sveinsson, A. Kavcic, and V. Tarokh, "UWB delay profile models for residential and commercial indoor environments," *IEEE Trans. Veh. Technol.*, vol. 54, pp. 1235–1244, 2005.

[27] S. S. Ghassemzadeh, L. J. Greenstein, A. Kavcic, T. Sveinsson, and V. Tarokh, "UWB indoor delay profile model for residential and commercial environments," in *2003 IEEE Conf. on Vehicular Technology (VTC 2003-Fall)*, Oct. 2003, pp. 3120–3125.

[28] C.-C. Chong, Y.-E. Kim, S. K. Yong, and S.-S. Lee, "Statistical characterization of the UWB propagation channel in indoor residential environment," *IEEE Trans. Commun.*, vol. 5, pp. 503–512, 2005.

[29] S. M. Yano, "Investigating the ultra-wideband indoor wireless channel," in *55th IEEE VTS Conf. on Vehicular Technology*, Oct. 2002, pp. 1200–1204.

[30] A. F. Molisch, "IEEE 802.15.4a channel model-final report," Tech. Rep. IEEE 802.15-04-0662-01-04a, Sep. 2004.

TOA ESTIMATION TECHNIQUES: A COMPARISON

Mohsen Pourkhaatoun and S. A. (Reza) Zekavat
Michigan Technological University, Houghton, MI

T**HIS CHAPTER** introduces time-of-arrival (TOA) techniques used in many localization systems. The focus of this chapter is on wideband TOA estimation techniques. Chapter 7 introduces ultra-wideband localization methods. Different categories of TOA techniques that include correlation-based approaches, deconvolution methods, maximum likelihood (ML)-based methods, subspace-based methods, and blind signal separation (BSS)-based approaches are discussed. Mathematical details of the introduced techniques are presented. These techniques are compared in terms of different performance measures. In addition, the chapter investigates blind TOA techniques and introduces a new technique based on independent component analysis (ICA). Finally, the chapter discusses multiband range estimation that enables cognitive radios (CRs) to attain higher localization performance.

7.1 INTRODUCTION

Estimating the distance is the core of many positioning systems such as radar [1], sonar [2], and wireless local positioning systems (WLPSs) [3, 4]. In addition, localization is required in mobile ad hoc networks (MANET) for routing and resource allocation [5]. TOA-based positioning systems measure the distance between a transmitter and a receiver based on the estimation of the signal propagation delay, which is highly affected by multipath channels and their associated delay profiles [6]. In these techniques, it is assumed that the line-of-sight (LOS) signal is available. LOS and non-line-of-sight (NLOS) identification techniques are discussed in Part III of this handbook. Figure 7.1 presents the simple structure of TOA-based positioning systems [7].

Handbook of Position Location: Theory, Practice, and Advances, Second Edition.
Edited by S. A. (Reza) Zekavat and R. Michael Buehrer.
© 2019 by the Institute of Electrical and Electronics Engineers, Inc.
Published 2019 by John Wiley & Sons, Inc.
Companion Website: www.wiley.com/go/zekavat/positionlocation2e

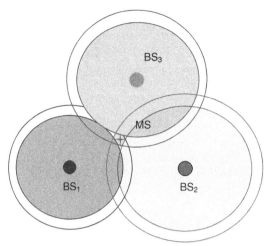

Figure 7.1 TOA-based positioning system.

In this scenario, three or more base stations (BSs) measure the TOA of the transmission from each mobile station (MS). Knowing the speed of light, the distance between each BS and the MS is calculated. The locus of points at the estimated range of MS from each BS is a circle, and the intersections of circles give the MS location. Due to errors in TOA measurements, the circles do not intersect at a unique point. Thus, it is necessary to find a location that best fits the measurements [7]. Accordingly, the accuracy of the estimated range in these systems is a function of TOA estimation error. The basis of time-based techniques is the receiver's ability to accurately estimate the TOA of the LOS signal. Generally, different measures are used to evaluate a TOA estimation algorithm in wireless channels.

1. *Sensitivity to Bandwidth.* It is also called the ability to separate closely spaced multipaths. A multipath is the major source of error in TOA estimation algorithms [8, 9]. In wireless environments, due to the availability of scatterers such as hills, buildings, walls, people, cars, and street signs, a transmitted signal arrives at the receiver through multiple paths. The overlap of the TOA of these paths is a source of error in the estimation of the LOS signal. In particular, this error is significant when the time-delay difference between the arrived multipaths is less than the minimum pulse duration (inverse of bandwidth) of the transmitted signal. Figure 7.2 illustrates two scenarios of multipath components in a received signal. Here, many multipath signals arrive promptly after the LOS signal. This creates error in the estimated time delay. It is observed that lower duration of the signal allows the separation of the arrived signals through multiple paths. Lower duration of the signal is equivalent to higher bandwidths. In other words, the TOA estimation error is a function of signal bandwidth [10, 11]. In the case of a closely spaced multipath, increasing the bandwidth improves the performance of a closely spaced multipath mitigation algorithm. It is obvious that algorithms with the ability to separate closely spaced multipaths possess less sensitivity to increasing bandwidths.

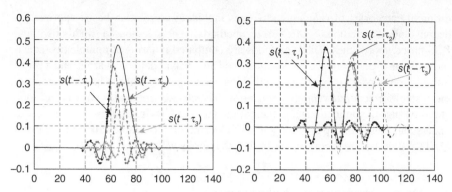

Figure 7.2   Effect of bandwidths of received multipath signal.

2. *Sensitivity to Noise.* Another major source of error is additive noise. Even in the absence of multipath signals, the accuracy of the arrival time is limited by noise. For a given signal-to-noise ratio (SNR), the time-delay estimation can only attain a certain accuracy [8]. This accuracy varies for different delay estimation methods. Sensitivity to noise is also explained by robustness. Lower sensitivity to noise is equivalent to higher robustness of the technique. Some techniques are robust as long as the SNR is higher than certain thresholds, and they may lose their robustness after that threshold.

3. *Complexity.* In the majority of these applications, the complexity of the TOA estimation algorithm has a crucial role in implementation. From the design point of view, according to the application, the best trade-off between accuracy and complexity should be selected.

4. *A Priori Information.* In some algorithms, a priori information is needed as an input to the system, which is not always easy to attain. This information includes but is not limited to channel profile parameters such as the number of paths and coarse estimation of the first path delay. As a part of the ranging system, the priori information should be estimated and applied to the TOA estimation algorithm. In addition to increasing the complexity, the error in estimation of these parameters may be introduced uncertainly in the TOA estimation algorithm, which degrades the performance of a full system.

In the literature, many high-resolution time-delay estimation algorithms have been proposed in order to accurately estimate TOA by determining the delay of the first incoming signal path. These methods can be performed either in the *time domain* by estimating TOA based on the impulse response of the channel, or in the *frequency domain* by detecting the channel frequency response (CFR) [11]. Spectral estimation methods, namely, super-resolution time-delay estimation algorithms, have been used in the literature for different applications [12, 13]. Specifically, they have been employed in the frequency domain to estimate multipath time delays and channel coefficients.

In this chapter, we discuss TOA estimation techniques, which are applicable for wireless multipath channels. After introducing the signal model, we introduce different categories of ranging techniques including correlation-based approaches, deconvolution methods, ML-based methods, subspace-based methods, and BSS-based approaches. These methods are compared in terms of complexity, sensitivity to noise, performance in closely spaced and distant paths scenarios, and a priori information needed as input. The following sections represent the block diagram of the system for accurate ranging application. The chapter concludes with a brief look at radio ranging applications and their application in CR. In order to help the reader understand these concepts better, some TOA estimation examples and their relevant MATLAB codes are incorporated in this chapter as well.

7.2 TOA ESTIMATION METHODS

The multipath wireless channel can be modeled via a channel impulse response (CIR); that is,

$$h(t) = \sum_{m=1}^{M} \alpha_m \delta(t - \tau_m). \tag{7.1}$$

In Equation 7.1, M represents the number of multipath components, and α_m and τ_m are the complex attenuation and propagation delays of the mth path, respectively. In general, due to the mobility of the transmitter, the receiver, and objects in the environment, the parameters α_m and τ_m are time-varying random variables. However, if the measurement time interval is selected less than the channel coherence time, these parameters do not vary significantly within the observation interval. Without loss of generality, we assume $\tau_1 < \tau_2 < \ldots < \tau_M$. Therefore, τ_1 represents the delay of the first received path, which is considered to be the desired TOA (assuming the availability of LOS).

Accordingly, the received signal $x(t)$ in a multipath channel corresponds to

$$x(t) = \sum_{m=1}^{M} \alpha_m s(t - \tau_m) + v(t). \tag{7.2}$$

In Equation 7.2, $s(t)$, $0 < t < T_0$ represents a known transmitted signal, with the bandwidth B, center frequency f_c, and duration T_0, and $v(t)$ is a zero-mean white Gaussian noise, and it is independent of the transmitted signal $s(t)$.

Examples 7.1 and 7.2 represent how a multipath channel is realized and how transmitted and received signals in a wireless system are modeled via MATLAB.

Example 7.1 Indoor Channel Model Generation

Different channel models have been proposed for wireless communication. Implement Saleh–Valenzuela (S-V) channel model [14], which is suitable for indoor multipath propagation, using a tap delay line structure. Assume that multipath components arrive in clusters. The cluster arrival rate is described by a Poisson process. In other words, two Poisson models are employed to model the arrival time. The first Poisson model is for the arrival of clusters, and the second Poisson model is for paths (or rays) within each cluster. For simplicity in this example, assume the number of clusters, and the number of rays in each cluster, is known.

Solution

Considering the maximum length of the channel and the maximum duration of each cluster, we generate the arrival time of each cluster using a uniform random generator (use command "max_channel_length*rand(1,cluster_nom)"). The same procedure is applied for generating the arrival time of each ray within one cluster. MATLAB codes can be found online at ftp://ftp.wiley.com/public/sci_tech_med/matlab_codes.

Based on the S-V model, the power of each cluster is a function of two components. First, the average power of subsequent clusters is assumed to decay exponentially by time. In addition, the fading in each cluster is modeled using a lognormal random variable. In the simulation, the lognormal random variable can be generated by a Gaussian random variable. Given a random variable $x \sim N(0, 1)$, the variable $z = e^{\mu + \sigma x}$ has a lognormal distribution with parameters μ and σ. As mentioned before, each cluster is composed of many multipath rays. We can consider the same model for generating the power of each ray in any given cluster. Figure 7.3 presents an illustration of exponential decay of mean cluster power and ray power within clusters.

In this step, we should model the phase rotation of each path in our channel model. Here, we assume the phase of each ray is uniformly distributed over $[0, 2\pi]$. In MATLAB, this phase can be generated by the "rand" command, such as "exp(-sqrt(-1)* 2 * pi * rand(size(h)))." Figure 7.4 presents a sample continuous time channel model. The outputs of the above statistical model are two vectors. The

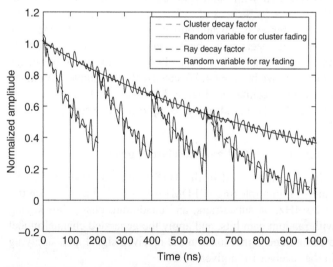

Figure 7.3   Decay of mean cluster power and ray power within clusters.

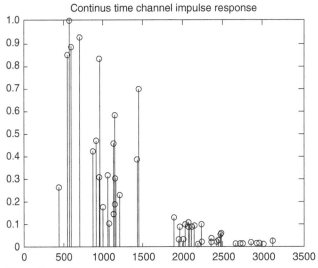

Figure 7.4   Channel impulse response.

first one includes the time delays of different paths, and the second one presents the corresponding complex amplitudes of those paths. As a result, we will have a continuous model of our indoor multipath wireless channel.

Now, we should convert this model to a discrete model in a consistent manner. At the first step, we quantize time in the continuous-time model's (time, value) pairs to t_s/N, where ts is the desired resolution and N is a reasonably large integer as upsample rate. In our simulation, we assume $t_s = 50\,ns$ and $N = 128$. If multiple (time, value) pairs get quantized into the same time bin, their values should be added. In this case, we have a discrete CIR with bandwidth equal N time greater than our required bandwidth (N/t_s).

To obtain our desired complex baseband equivalent channel model sampled at 20 MHz, we first shift the whole spectrum by the desired center frequency (multiply CIR by $e^{-j2\pi f_c t}$ in time domain), apply low-pass filtering (with cutoff frequency $\pm 10\,MHz$), and then decimate down by $N = 64$. We can use the "resample" command in MATLAB to implement this decimation. Figure 7.5 presents the result of these operations. MATLAB codes can be found online at ftp://ftp.wiley.com/public/sci_tech_med/matlab_codes.

Example 7.2   Generating Transmitted and Received Signals

Different transmitted signals can be used in the TOA estimation system. Use an orthogonal frequency division multiplexing (OFDM)-based system as a transmitter with the bandwidth of 20 MHz, 64 subcarriers, and quadrature phase-shift keying (QPSK) modulation with random data bits, and apply the generated channel model in Example 7.1 to the transmitted signal to form the received signal after applying white additive noise in the receiver for a given SNR.

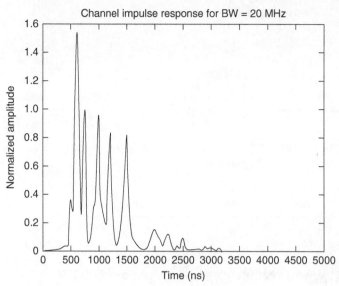

Figure 7.5   CIR of baseband equivalent channel model.

Solution

Based on the Nyquist rate, the sampling rate is considered 20 MHz for the complex baseband model. In our simulation, first, we generate 64 complex numbers with unit amplitude and a random phase, which is randomly selected from these values: $\{-3\pi/4, -\pi/4, \pi/4, 3\pi/4\}$. Using the inverse fast Fourier transform (IFFT) module ("ifft" command in MATLAB), we convert these data to time domain. For some high-resolution approaches such as multiple signal classification (MUSIC), we assume multiple symbols to estimate TOA. In this case, the distance between two consecutive symbols is selected properly to avoid intersymbol interference (ISI) due to the multipath effect. Figure 7.6a represents the real part of a baseband transmitted signal. MATLAB codes can be found online at ftp://ftp.wiley.com/public/sci_tech_med/matlab_codes.

Now, we model the channel and the additive noise in the receiver. Figure 7.7 presents the model block diagram.

The multipath channel is applied by convolving the transmitted signal with the complex baseband channel model (use the "conv" command for convolution operation). Consequently, additive noise is modeled by a white complex Gaussian random variable. The power of this noise is extracted from the power of the received signal and the desired SNR. In practice, power is considered as a known parameter, which can be measured by the received signal. Figure 7.6b presents the real part of the baseband received signal.

Next, we present algorithms that estimate τ_1, that is, the first time delay that corresponds to the LOS from the received signal $x(t)$ defined in (7.2).

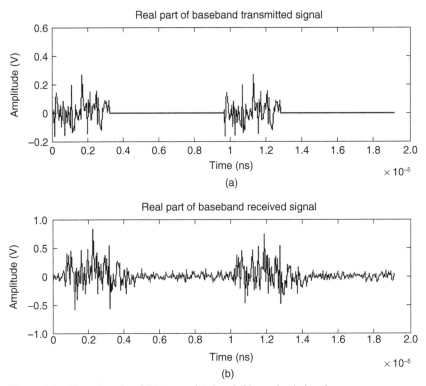

Figure 7.6   Time domain of (a) transmitted and (b) received signals.

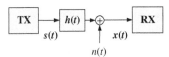

Figure 7.7   Block diagram of transceiver model.

7.2.1   Conventional Correlation-Based Techniques

The TOA is the earliest arrival time that maximizes the cross correlation between the received signals and the known template signal [15, 16]. TOA estimation based on the cross correlation operation is called a cross correlator. In general, an optimal estimation of TOA can be obtained using a correlation receiver with the received waveform as the template signal (or, equivalently, a matched filter that is matched to the received waveform), as shown in Figure 7.8, and choosing the time shift of the template signal that produces the maximum correlation with the received signal [17, 18]. However, due to the multipath channel, the received waveform has many unknown parameters to be estimated. Hence, the optimal correlation-based TOA estimation is unfeasible.

In practice, in a conventional correlation-based receiver, the transmitted waveform can be used as the template signal. However, this technique would be still

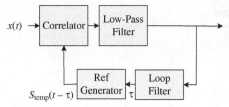

Figure 7.8   The correlation-based TOA estimation receiver.

suboptimal in a multipath environment [19, 20]. In general, TOA-based range errors due to multipath channels are many times greater than those caused by additive noise alone. Essentially, all late-arriving multipath components are self-interferences that effectively decrease the SNR of the desired LOS signal. Rather than finding the highest peak of the cross correlation, in the multipath channel, the receiver must find the first-arriving peak because there is no guarantee that the LOS signal will be the strongest of the arriving signals [21].

Pros and Cons: In general, the main advantage of correlation-based methods is their low *complexity*, which makes them applicable to many online systems. Moreover, this method does not need a *priori information*. However, in addition to high *sensitivity to noise*, these methods seem to be appropriate only for single-path channels. In other words, *closely spaced multipath signals* cannot be resolved via these methods. It has been shown that the resolution of TOA estimation in the correlation-based methods is roughly determined by the minimum pulse width of the transmitted signal, or, equivalently, the signal bandwidth [15]. Due to the scarcity of the available bandwidth, in practice, correlation-based methods cannot provide adequate accuracy for many applications. For example, if a 200-MHz bandwidth is used in an indoor communication system, the absolute distance estimation errors would be about 1.5 m if the LOS signal is detectable, which is not enough for ranging application. On the other hand, it is always desirable to achieve higher ranging accuracy using a given bandwidth. This motivates using more complex TOA estimation techniques. Here, we present an example of a correlation-based TOA estimation method.

Example 7.3   Simple Correlation-Based TOA Estimation

Calculate the correlation between transmitted signal and received signal generated in Example 7.2 and find the TOA from the output of this correlator.

Solution

In this example, we present a simple realization of the correlation-based method. Cross correlation between the transmitted signal and the received signal can be implemented by the "xcorr" command in MATLAB. Here, we calculate the absolute value of the output, which is a complex number. The first-arriving peak is estimated by comparing the absolute value of the cross-correlator output with a certain threshold, which is set based on the measured power of noise. MATLAB codes can be found

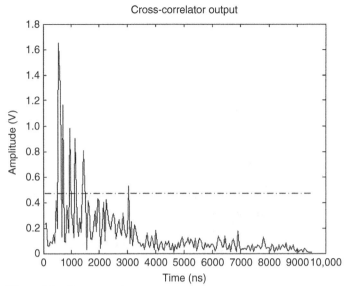

Figure 7.9   Cross-correlator output.

Figure 7.10   Block diagram of the inverse filtering method.

online at ftp://ftp.wiley.com/public/sci_tech_med/matlab_codes. Figure 7.9 presents the output of the correlator and the defined threshold in this method.

7.2.2   Deconvolution Methods

Deconvolution methods are essentially inverse filters [22]. Based on (7.2), the received signal $x(t)$ can be represented as a convolution of the transmitted signal and the CIR; that is, $x(t) = s(t) \otimes h(t) + v(t)$. This relationship in the frequency domain is expressed as

$$X(f) = S(f)H_f(f) + V(f). \tag{7.3}$$

In Equation 7.3, $X(f)$, $S(f)$, and $V(f)$ are Fourier transforms of $x(t)$, $s(t)$, and $v(t)$, respectively, and $H_f(f)$ is the frequency response of $h(t)$ defined in (7.1).

Therefore, by dividing $X(f)$ by $S(f)$, the channel response and, among other channel parameters, the TOA of the first path can be estimated. The dual of this operation (which is in the frequency domain) is called deconvolution, which is applied to the time domain. Figure 7.10 illustrates this procedure.

In practice, discrete samples of the frequency domain can be obtained by sweeping the signal at different frequencies [23]. Here, we take the samples over the frequency domain of the transmitted signal. Based on the bandwidth (B), and the center frequency (f_c) of the transmitted signal, we define $X(n) = X(f)|_{f=f_n}$,

$S(n) = S(f)|_{f=f_n}$, $H(n) = H_f(f)|_{f=f_n}$, and $V(n) = V(f)|_{f=f_n}$, where $f_n = f_o + n \cdot \Delta f$, $n = \{1, \ldots, N\}$. Here, f_o (i.e., the lowest frequency component within the signal bandwidth) corresponds to $f_o = f_c - B/2$. Δf is the frequency sample spacing, which corresponds to the frequency resolution of sampling in the frequency domain and $N = B/\Delta f$ represents the number of samples. This signal model can be written in the vector format as

$$X = SH + V. \tag{7.4}$$

In Equation 7.4, $X = [X(1), \ldots, X(N)] \in C^N$, $S = \text{diag}(S(1), \ldots, S(N)) \in C^{N \times N}$, $V = [V(1), \ldots, V(N)] \in C^N$, and $H = [H(1), \ldots, H(N)] \in C^N$. Now, if we define $Y = S^{-1}X$ for the frequency range within the bandwidth B, the transmitted signal S can be removed from X. Therefore, we have

$$Y = H + W, \tag{7.5}$$

where $W = S^{-1}V$ represents the additive noise in vector Y. Accordingly, one of the adverse effects of inverse filtering, when noise is present, is noise enhancement. Therefore, while dividing $X(f)$ by $S(f)$, great care should be taken as very low signal levels can cause noisy results in the inverse filtering process.

The noise enhancement effect can be reduced by using the so-called constrained inverse filtering methods. These methods are constrained in the sense that they do not allow the output values to lie outside some predefined set. Among the constrained inverse filtering methods, the best-known ones are the least squares (LS) techniques [24, 25], which can be applied in both time and frequency domains [26]. Here, we present this method in the frequency domain.

In this approach, the TOA is estimated from the estimated CIR defined in (7.1). First, we recover the CIR with a simple model, which includes L equally spaced taps distributed within $[0, LT_s]$:

$$\tilde{h}(t) = \sum_{l=1}^{L} \tilde{h}_l \delta(t - \tilde{\tau}_l). \tag{7.6}$$

In Equation 7.6, $\tilde{\tau}_l = (l-1)T_s$, where T_s represents the tap interval, which is set equal to the inverse of the sampling rate and $\tilde{h}_l$ is the channel coefficient of the lth tap. Here, we assume the multipath channel length is less than LT_s. Accordingly, we estimate the CIR via the Fourier coefficients of $\tilde{h}(t)$ on the N samples of CFR that is presented in (7.5); that is,

$$H(n) = \sum_{l=1}^{L} \tilde{h}_l e^{-j2\pi f_n \tilde{\tau}_l}, \quad n \in \{0, \ldots, N-1\}, \tag{7.7}$$

where $f_n = f_o + n \cdot \Delta f$, $n = \{1, \ldots, N\}$. Here, f_o (the lowest frequency component within the signal bandwidth) corresponds to $f_o = f_c - B/2$. Δf is the frequency sample spacing, which corresponds to the frequency resolution of sampling in frequency domain, and $N = B/\Delta f$ represents the number of samples.

In the matrix form, (7.7) corresponds to $H = F_L\tilde{h}$, where $H = [H(1), \ldots, H(N)]$ is the frequency response of channel, which is defined in (7.4); $\tilde{h} = \left[\tilde{h}_1, \ldots, \tilde{h}_L\right]$ and $F_L \in C^{N \times L}$ are the Fourier transform matrix with elements $F_{nl} = \dfrac{1}{\sqrt{N}} e^{-j2\pi kl/N}$ for $n \in \{0, \ldots, N-1\}$ and $l \in \{0, \ldots, L-1\}$. Now, we can rewrite (7.5) as

$$Y = F_L\tilde{h} + W, \tag{7.8}$$

where Y and W are defined in (7.5). Here, we define R_W as the covariance matrix of noise W. Based on the Gauss–Markov theorem [27], the minimum-variance unbiased linear estimator of $\tilde{h}$ observing Y corresponds to

$$\hat{h} = \left(F_L^H R_W^{-1} F_L\right)^{-1} F_L^H R_W^{-1} Y. \tag{7.9}$$

Now, using $\hat{h}$ as an estimation of the CIR presented in (7.1), the TOA can be estimated. Based on the definition of CIR, there is a sharp change of amplitude in the recovered channel around the first time delay. The location of this sharp change can be detected and used to estimate the TOA by comparing the amplitude of the CIR with a certain threshold.

Pros and Cons: In general, although deconvolution methods can better resolve *closely spaced multipaths* compared to the correlation-based method, their offered resolution is not enough for many high-precision ranging systems. In addition, these methods suffer from the same limitation as correlation-based approaches. As mentioned before, the noise enhancement is a known drawback of this category of methods. Although implementing LS techniques for deconvolution methods reduces the sensitivity to noise, they still suffer from lack of robustness at high noise levels. In addition, as a frequency domain approach, a coarse estimation of TOA should be known to feed a proper window of data to the fast Fourier transform (FFT) engine. Therefore, this approach needs a priori information of channel. Another drawback associated with the deconvolution methods is their increased complexity and memory requirements (mainly due to matrix inversions), which increase the implementation cost of this method in a real system. The next example presents a realization of this method.

Example 7.4   IFFT-Based TOA Estimation

Implement the deconvolution method in the frequency domain using the FFT/IFFT module. Convert the received signal generated in Example 7.2 to the frequency domain. Remove the transmitted signal and then estimate the CIR (presented in Eq. 7.9) by applying inverse Fourier transform. Consequently, find the TOA from the estimated CIR.

Solution

In this example, we implement IFFT-based TOA estimation as a simplified version of convolution-based methods. First, we convert the received data to the frequency

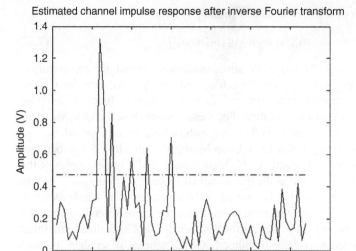

Figure 7.11   Channel impulse response estimated by the deconvolution method.

domain (use the "fft" command with length 64). Then, the CFR is estimated by removing the transmitted signal from the received signal in the frequency domain. Here, we assume the length of channel is less than the length of the transmit signal. Therefore, the CIR (presented in Eq. 7.9) can be calculated using IFFT from the estimated CFR (use the "ifft" command with length 64). Figure 7.11 depicts the CIR estimated by IFFT as a realization of the deconvolution method. Similar to the correlation-based example, we use a threshold to estimate the TOA, which is defined based on the measured noise power. This threshold can be selected based on the measured noise power in the estimated CIR. It is obvious that the performance of this algorithm is degraded in low SNR regimes. The MATLAB code of this algorithm can be found online at ftp://ftp.wiley.com/public/sci_tech_med/matlab_codes.

7.2.3   ML-Based Methods

The algorithms grouped here under the generic name of Maximum Likelihood (ML) ML-based algorithms refer to those that estimate the multipath time delays and coefficients via the ML estimation method [28, 29]. From (7.2), we define the unknown channel coefficients and delays into a $\theta \in C^{2M \times 1}$ as the channel parameter vector:

$$\theta = [\alpha_1, \ldots, \alpha_M, \tau_1, \ldots, \tau_M].$$ (7.10)

The log-likelihood function of θ takes the form

$$\ln[\Lambda(\theta)] = -\int_0^{T_0} \left| x(t) - \sum_{m=1}^{M} \alpha_m s(t - \tau_m) \right|^2 dt.$$ (7.11)

The ML estimation of θ can be obtained from (7.11) via

$$\hat{\theta}_{ML} = \arg\max_{\theta} \{\ln[\Lambda(\theta)]\}. \tag{7.12}$$

Due to the large number of paths in a realistic multipath channel, the complexity of this approach is very high [30]. On the other hand, the ML criterion for channel estimation requires a multidimensional optimization of an error function, implying a huge, complex computational solution. For these reasons, based on the ML criterion, some suboptimal, less-complex TOA algorithms have been proposed.

Multipath Estimating Delay Lock Loop Multipath estimating delay lock loop (MEDLL) [31] and pulse subtraction (PS) algorithms [32, 33] are examples of this category of methods. The same as the traditional DLL method, presented in Section 2.1, a correlator (match filter) has been used in these systems. The whole idea is to cancel multipath interference via subtraction from the correlation function of a certain reference pulse, which is the transmitted signal. The Generalized Maximum Likelihood generalized maximum likelihood (GML) scheme is another approach specifically for the detection of the first path in a multipath channel [30, 34–36]. In this method, the arrival time and the amplitude of the strongest path are determined by correlation. Subsequently, the ML-based approach tries to find the signal components that arrived earlier than the peak component. This step reduces the number of unknown parameters. In addition to reducing computational complexity, an iterative nonlinear programming technique is employed by which the unknown parameters are estimated in a sequential manner [36]. Specifically, the arrival time of each component signal is estimated individually while all other parameters are fixed. The advantage of the GML-based algorithm is that it is a recursive algorithm and can accurately estimate TOA [34, 35]. However, the main drawback is that it requires a very high sampling rate, which cannot be practically used in many applications.

Pros and Cons: ML-based methods separate the paths serially one by one. This approach mitigates the multipath effect in a ranging system. However, it is not effective for *closely spaced multipath* components. In addition, based on the estimation theory [28], the ML estimator is an optimum solution for a single-path scenario in the presence of additive noise. This means that the variance of ML estimator error is less than other approaches, when only a single path is available in the received signal. Although in multipath scenarios the ML-based method is not the best solution, the sensitivity to noise for this category of methods is less than correlation- and deconvolution-based approaches. In addition, no *prior information* is needed for the implementation of ML-based methods. Nevertheless, due to multidimensional search in the presented optimization problem, the complexity of these methods is much higher than other proposed approaches.

7.2.4 Subspace-Based Techniques

The subspace-based algorithms involve the decomposition of the space spanned by the observation vector (i.e., the vector formed by the received signal samples) into several subspaces, usually a noise subspace and a signal subspace [13, 37]. These algorithms use the orthogonality property between noise and signal subspaces to estimate channel parameters. The subspace approaches involve eigenvector (EV)

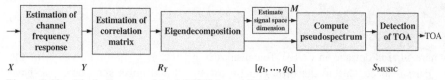

Figure 7.12   Block diagram of the MUSIC-based TOA estimation method.

decomposition of high-order matrices, and they remain usually very complex for practical applications. The main advantage of subspace-based methods is their increased resolution in parameter estimation. The most known subspace-based methods that have been employed in delay estimation applications are the MUSIC [38–40] and estimation of signal parameters via rotational invariance technique (ESPRIT) [41–43]. In [37], it is depicted that these two methods are quite similar, both from the point of view of their performance and complexity. Accordingly, here, we consider and explain only the MUSIC algorithm.

Figure 7.12 presents the block diagram of MUSIC-based TOA estimation. The MUSIC technique is based on eigenvalue decomposition (EVD) of the covariance matrix of the observation data.

Here, we represent the frequency domain channel vector $\boldsymbol{H}$ defined in (7.5) as $\boldsymbol{H} = \boldsymbol{UA}$, where $\boldsymbol{U} = \left[\boldsymbol{u}(\tau_1), \ldots, \boldsymbol{u}(\tau_M)\right] \in C^{N \times M}$, $\boldsymbol{u}(\tau_m) = \left[1, e^{-j2\pi\Delta f\tau_m}, \ldots, e^{-j\pi\Delta f(N-1)\tau_m}\right]^T \in C^N$, $\boldsymbol{A} = \left[\alpha_1, \ldots, \alpha_M\right]$, $\alpha_m = \alpha_m e^{-j2\pi f_o\tau_m}$ for $m \in \left\{1, \ldots, M\right\}$, f_o, and Δf are defined in (7.6), and N and M represent the number of samples and reflections, respectively. Therefore, we rewrite (7.6) as

$$\boldsymbol{Y} = \boldsymbol{UA} + \boldsymbol{W}. \tag{7.13}$$

The covariance matrix of $\boldsymbol{Y}$ corresponds to

$$\boldsymbol{R}_Y = E\left\{\boldsymbol{YY}^H\right\} = \boldsymbol{UR}_A\boldsymbol{U}^H + \sigma_v^2\boldsymbol{I}, \tag{7.14}$$

where $\boldsymbol{R}_A = E\{\boldsymbol{AA}^H\}$. In practice, we need Q observations of $\boldsymbol{Y}$ to form the covariance matrix $\boldsymbol{R}_Y$. It is essential to mention that these observations should have the same statistics. In other words, time domain observation should be made within channel coherence time [14].

Since the propagation delays $\{\tau_m\}_{m=1}^M$ in (7.1) can be assumed independent, and $\boldsymbol{U}$ is a full rank matrix, the column vectors of $\boldsymbol{U}$ are linearly independent. If we assume the magnitude of the parameters $\{\alpha_m\}_{m=1}^M$ is constant and the phase is a uniform random variable in $[0, 2\pi]$, the covariance matrix $\boldsymbol{R}_A$ would be nonsingular. Therefore, based on the theory of linear algebra, the rank of the matrix $\boldsymbol{UR}_A\boldsymbol{U}^H$ is M, where $M < Q$. Consequently, the $Q - M$ smallest eigenvalues of $\boldsymbol{R}_Y$ are all equal to noise power. The EVs corresponding to $Q - M$ smallest eigenvalues of $\boldsymbol{R}_Y$ are called noise EVs, while the EVs corresponding to M largest eigenvalues are called signal EVs. Thus, the Q-dimensional subspace can be split into two orthogonal subspaces, known as signal subspace and noise subspace, by the signal EVs and noise EVs, respectively. The noise subspace is defined by $\boldsymbol{P} = [\boldsymbol{q}_{M+1}, \ldots, \boldsymbol{q}_Q]$, where $\{\boldsymbol{q}_k\}_{k=M+1}^Q$ are noise EVs. Here, we define MUSIC pseudospectrum as

$$S_{\text{MUSIC}} = \frac{1}{|\boldsymbol{Pu}(\tau)|^2},\tag{7.15}$$

where $\boldsymbol{u}(\tau)$ is defined in (7.13). Because vector $\{\boldsymbol{u}(\tau_m)\}_{m=1}^{M}$ is a component of the signal subspace, due to the orthogonality of signal and noise subspaces, we have $\boldsymbol{Pu}(\tau_m) = 0$. Multipath delays $\{\tau_m\}_{m=1}^{M}$ can be determined by finding the delay values at which S_{MUSIC} achieves its maximum value. Therefore, the TOA is estimated by detecting the first peak of the pseudospectrum in the delay axis.

Pros and Cons: Because of utilizing the orthogonality between different multipaths and noise, subspace methods possess good performance in the *separation of closely spaced* multipaths. In addition, due to separating signal and noise subspaces, the effect of noise is reduced in these methods. Therefore, high resolution of time estimates and low sensitivity to noise are the advantages of subspace-based delay estimation [37, 38]. However, the complexity associated with matrix eigendecomposition makes these algorithms quite difficult to use in practical applications. Another drawback of these methods is that the number of paths M should be known as a priori information of channel. An error in the estimation of M reduces the estimation performance.

Example 7.5   Pseudospectrum of MUSIC Algorithm

Implement a realization of the MUSIC-based TOA estimation algorithm in the frequency domain. First, convert multiple observations of the received signal generated in Example 7.2 to the frequency domain. Then, form the MUSIC pseudospectrum of these data and based on that, estimate the TOA as the time delay corresponding to the first peak of this spectrum.

Solution

As mentioned before, multiple observations are needed to form the subspace-based methods. In general, these observations could be collected in time, space, or frequency domains. Independent of the nature of observations, within the simulations, these observations would be independent. Here, we repeat the transmit symbol in the transmitter to generate these observations.

At the first step of the algorithm, we convert data to frequency domain (use the "fft" command with length 64). Consequently, after removing the transmitted signal and estimating the CFR for all observations, we apply MUSIC to form the pseudospectrum of the received signal. In this example, we assume the number of paths equals the assumed maximum channel length. Using this assumption, we use the "S=pmusic(X,N)" command in MATLAB to implement the MUSIC method. This command returns the pseudospectrum of a discrete time signal vector X in the vector S. N is the number of paths with complex amplitude in the signal X. In data matrix X, each row is interpreted as a separate observation. Figure 7.13 illustrates the result of this algorithm. Similar to the previous examples, we use a threshold to estimate the TOA from this pseudospectrum. This threshold is defined based on the measured

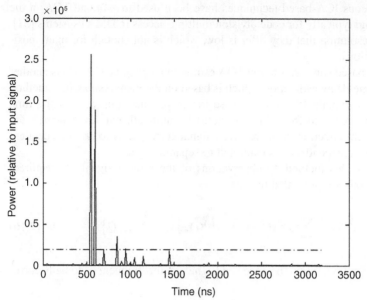

Figure 7.13   MUSIC pesudospectrum.

noise power. MATLAB codes can be found online at ftp://ftp.wiley.com/public/
sci_tech_med/matlab_codes.

7.2.5   BSS-Based Algorithms

BSS, also known as blind source separation, is the separation of a set of signals from
a set of mixed signals, without the aid of information (or with very little information)
about the source signals or the mixing process [44].

In the literature [45, 46], channel parameter estimation has been studied in the
context of convolutive BSS in both time and frequency domains. Here, the observed
signal consists of an unknown source signal mixed with itself at different time
delays. The task is to estimate the source signal and channel parameters by observing
the incoming signal without a priori knowledge of the system. Blind separation of
convolutive mixtures of signals is composed of *blind source deconvolution* methods
and *instantaneous blind source separation* techniques. In these techniques, channel
parameters such as the convolving system, the time delays, and mixing coefficients
should be estimated. These operations are not necessary for the range estimation
problems. In addition, these methods suffer from high complexity and low time
resolution. Therefore, they are not suitable for range estimation.

ICA is an instantaneous BSS technique that has a variety of applications in
statistical signal processing [47–50]. Based on the central limit theorem, the sum of
multiple independent components has a Gaussian probability density function (PDF).
ICA incorporates a cost function to develop a filter that increases the non-Gaussianity
of the observed signal. This signal is formed by the addition of multiple independent
components. This filter ultimately separates those independent components (signals).

Instantaneous ICA-based techniques have been used to estimate TOA in the time domain, and typically for code division multiple access (CDMA) systems [51]. However, the resolution that they offer is low, which is not enough for many positioning applications.

The ICA-based super-resolution TOA estimation technique [52, 53] is another approach for time-delay estimation, which is based on the complex FastICA method in the frequency domain [54–56]. Because the amplitudes, phases, and the time delays of signals received from different paths in multipath wireless channels are independent, each reflection in the received signal corresponds to one independent component. This independency is exploited to separate signals.

To perform this method, Q observations of the arrived signal are required. These observations are presented by

$$x^{(q)}(t) = \sum_{m=1}^{M} \alpha_m^{(q)} s(t - \tau_m) + v^{(q)}(t), q \in \{1, 2, \dots, Q\}. \tag{7.16}$$

Applying Fourier transform, the received signal in the frequency domain corresponds to

$$X^{(q)}(f) = S(f) \sum_{m=1}^{M} \alpha_m^{(q)} e^{-j2\pi f \tau_m} + V^{(q)}(f). \tag{7.17}$$

In Equation 7.17, $X^{(q)}(f)$, $S(f)$, and $N_m^{(q)}(f)$ are Fourier transforms of $x^{(q)}(t)$, $s(t)$, and $v^{(q)}(t)$, respectively. Here, parameters $\alpha_m^{(q)}$ and τ_m are frequency independent [14]. Similar to (7.4), we sample Equation 7.17 in frequencies $f_n = f_o + n \cdot \Delta f$, $n = \{1, \dots, N\}$. Here, f_o (the lowest frequency component within the signal bandwidth) corresponds to $f_o = f_c - B/2$. Δf is the frequency sample spacing, which corresponds to the frequency resolution of sampling in the frequency domain, and $N = B/\Delta f$ represents the number of samples. This signal model can be written in the vector format as

$$X = AV^T S + V, \tag{7.18}$$

where $A \in C^{Q \times M}$ is the coefficient matrix with elements $A_{mq} = \alpha_m^{(q)} e^{-j2\pi f_o \tau_m}$, $m \in \{1, \dots, M\}$, $q \in \{1, \dots, Q\}$, $V = [v(\tau_1), \dots, v(\tau_M)] \in C^{N \times M}$, $v(\tau_m) = [e^{-j2\pi \Delta f \tau_m}, \dots, e^{-j2\pi N \Delta f \tau_m}]^T \in C^N$ for $m \in \{1, \dots, M\}$ and $S = \text{diag}[S(f_1), \dots, S(f_N)] \in C^{M \times N}$ is the transmitted signal, and $V \in C^{Q \times N}$ presents the frequency domain noise matrix. M, N, and Q represent the number of paths, samples, and observations, respectively. Here, the source signals $Y \in C^{M \times N}$ is defined as

$$Y = V^T S + N'. \tag{7.19}$$

In Equation 7.19, matrix $N' \in C^{M \times N}$ represents the effective additive noise. Comparing (7.19) and (7.18), the relationship of V and $N' = [N_1', \dots, N_M']^T$ corresponds to

$$V = AN'. \tag{7.20}$$

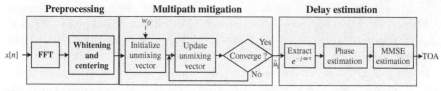

Figure 7.14   Block diagram of the super-resolution TOA estimation method.

Based on (7.19), each row of Y represents one signal reflection out of M reflections. In other words, for $Y = [y_1, \dots, y_M]^T$ and $\omega = [2\pi\Delta f, \dots, 2\pi N\Delta f]^T$, we have

$$y_m = Se^{-j\cdot\omega\cdot\tau_m} + N'_m, m \in \{1, \dots, M\}. \tag{7.21}$$

In Equation 7.21, y_m, ω, and N'_m are $N \times 1$ vectors.

Here, we present an algorithm that estimates τ_1, that is, the first time delay that corresponds to the LOS signal, by observing matrix $X = AY$ and estimating matrix Y.

Figure 7.14 presents the block diagram of the ICA-based TOA estimation method. In the proposed ICA algorithm, we assume zero mean for both the observation and source process. Therefore, the preprocessing module applies centering and whitening to the complex data in the frequency domain. Centering and whitening transforms (7.18) to

$$Z = \hat{C}^{-1/2}(X - \bar{X}), \tag{7.22}$$

where $Z \in C^{M \times N}$ $\hat{C} = (X - \bar{X})(X - \bar{X})^H/N$, and $\bar{X} = X\mathbf{1}_N\mathbf{1}_N^T/N$. In Equation 7.22, the whitening matrix $\hat{C}^{-1/2} \in C^{M \times Q}$ has been extracted from EVs and eigenvalues of the covariance matrix of X [57].

Observing X, the proposed ICA algorithm computes the unmixing matrix $W = [w_1, \dots, w_M] \in C^{M \times M}$ to optimally estimate the source signals (independent components) from their linear transformation [47].

ICA calculates the unmixing matrix W by increasing the non-Gaussianity of the mixed signals [47]. FastICA, a category of ICA methods, minimizes differential negentropy (a measure of non-Gaussianity) [48–50]. Generally, the estimation of negentropy is difficult; hence, in many practical situations, approximations of negentropy are incorporated [47, 48]. Here, we represent negentropy as $J_G(w^H Z)$, where Z has been defined in (7.22), and $w \in \{w_1, \dots, w_M\}$ is the unmixing vector. The proposed cost function is $J(Y) = E[|G(Y)|^2]$, where $G:C \to C$. The nonlinear function G for noncircular sources should not be symmetric [54]. By definition, a symmetric function is a function that does not change by any permutation of its variable. In addition, the function G should be second-order differentiable [54, 55]. A number of functions that have these properties include $\cosh^{-1}(y)$, e^y, $e^{-\cosh y}$, $e^{-\sin y}$, and e^{-y}. However, the optimum function G should be selected to optimize a cost function such that it minimizes the interference of other components with the extracted one.

In the literature, there are different approaches for solving this optimization problem [47]. In the proposed FastICA [48–50], a fixed-point iteration scheme is utilized to solve this problem. The iterative algorithm finds the weight vector by maximizing the non-Gaussianity of the projection $w^H Z$ for the data matrix Z. In order to attain faster convergence [54], a quasi-Newton version of complex maximization of non-Gaussianinty (CMN) is used. It has been shown that the optimization problem can be implemented iteratively with the constraint $\|w\| = 1$ using [54]

$$
\begin{cases}
w_{(r)}^+ = -E\left\{ Z \left(G^* \left(w_{(r)}^H Z \right) \circ g \left(w_{(r)}^H Z \right) \right)^T \right\} + E\left\{ g \left(w_{(r)}^H Z \right) \circ g^* \left(w_{(r)}^H Z \right) \right\} w_{(r)} \\
+ E\left\{ Z Z^T \right\} E\left\{ G^* \left(w_{(r)}^H Z \right) \circ g' \left(w_{(r)}^H Z \right) \right\} w_{(r)}^*, \quad w_{(r+1)} = w_{(r)}^+ / \left\| w_{(r)}^+ \right\|.
\end{cases}
\tag{7.23}
$$

In Equation 7.23, $\circ$ denotes the element-wise product, Z is defined in (7.22), $G(\cdot)$ is the cost function of the FastICA algorithm, $g(\cdot)$ and $g'(\cdot)$ are the first and second differentiations of $G(\cdot)$, respectively, and $w_{(r)}$ represents the normalized unmixing vector in the rth iteration.

The algorithm starts after whitening the input data and creating the matrix Z and using the initial vector w_0. Usually, in the ICA algorithm, the unmixing vector is randomly initialized. Depending on this initialization, the algorithm converges randomly to one of the independent components. Here, applying a constraint to this initial vector forces the algorithm to converge to the independent components corresponding to the first received path, that is, the LOS signal, in order to extract the TOA corresponding to the LOS in the first iteration [53]:

$$
w_0 = ZS * 1_N.
\tag{7.24}
$$

In Equation 7.24, 1_N is an $N \times 1$ vector composed of 1's. Hence, selecting vector w_0 defined in (7.24) as the initial vector in (7.23) would force the algorithm to converge to the independent component corresponding to the smallest delay (τ_1), if $J_G\left(Se^{-j \cdot \omega \cdot \tau} \right)$ is a monotonically increasing function of τ. It is observed that the cost function $J_G(\cdot)$, defined based on $G_1 = \cosh^{-1}(y)$ and $G_2 = e^{-y}$, is a good candidate for our complex ICA algorithm.

After applying FastICA to reduce the effect of multipath, we should find τ_1, as the first delay presented in (7.1), from the extracted vector $\hat{u}_1 \in C^N$, that is, the estimation of the elements of the normalized source signal matrix U in (7.10). Therefore, $\hat{u}_1$, a scaled version of y_1 in (7.21), is equal to

$$
\hat{u}_1 = Se^{-j \cdot \omega \cdot \tau_1} + P,
\tag{7.25}
$$

where $S = \text{diag}[S(f_1), \ldots, S(f_N)]$ is a known transmitted signal defined in (7.18), ω is introduced in (7.21), τ_1 is the desired TOA that should be estimated, and vector $P \in C^N$ models the total noise in the estimated vector.

For the frequency range within the bandwidth B, the transmitted signal S can be removed from $\hat{u}_1$. Now, our problem is defined as the estimation of τ_1 from the noisy samples of $e^{-j\cdot\omega\cdot\tau_1}$, which is equivalent to the frequency estimation of complex sinusoids from signal $x(t) = e^{-j\omega_0 t} + n(t)$ [58–60].

Based on the presented block diagram in Figure 7.14, an algorithm is developed to estimate time delay τ_1 based on the phase of estimated data. This calculated phase corresponds to

$$\hat{\phi}_n = 2\pi n \Delta f \tau_1 + n_{\omega_n}. \tag{7.26}$$

Here, Δf is the frequency sample spacing defined in (7.6).

Equation 7.26 is a linear combination of the variable τ_1, which allows this variable to be extracted using the minimum mean square error (MMSE) criterion, such that

$$\hat{\tau}_1 = \frac{6 \sum_{n=1}^{N} k\hat{\phi}_n}{2\pi \Delta f N (N+1)(ZN+1)}. \tag{7.27}$$

In Equation 7.27, $\hat{\phi}_n, n \in \{1, \dots, N\}$ is defined in (7.26).

Pros and Cons: It has been shown that the proposed technique has a good TOA estimation performance in terms of estimation error standard deviation, compared to other high-resolution methods. Due to the whitening process, this technique has lower sensitivity to noise. In addition, this technique uses the principles of the independency between multipath delays, which is another property of multipath components that allows us to *separate closely spaced reflections*. However, the computational complexity of these methods is high. In addition, similar to subspace-based methods, the number of paths should be known in these algorithms. Therefore, a priori information is essential for implementation.

7.3   COMPARISON OF TOA ESTIMATION TECHNIQUES

In the previous section, we presented different categories of TOA estimation methods. We have already detailed the pros and cons of the discussed TOA estimation techniques. Here, we compare and summarize the advantages of these algorithms in terms of different criteria:

1. *Ability to Separate Closely Spaced Multipaths.* As discussed in the previous sections, correlation-based methods are appropriate only for single-path channels, as these methods cannot resolve closely spaced multipath signals. Although this problem has been improved using deconvolution methods, the offered resolution by deconvolution does not satisfy the requirements of many high-precision ranging systems, such as the WLPS system [33]. In ML-based

methods, because paths are separated one by one serially, the effect of multipath components is mitigated. However, this is not an effective method for *closely spaced multipaths*. Therefore, this category of methods does not still provide enough resolution for high-precision ranging systems. High-resolution techniques such as subspace- and BSS-based methods have been proposed to provide the ability to separate closely spaced multipaths. These methods utilize the orthogonality or independency of multipath reflections to separate them. Therefore, the sensitivity of these methods to closely arrived signals is minimal.

2. *Sensitivity to Noise.* As discussed in the literature [15, 16], the performance of correlation-based methods degrades by increasing the power of additive noise. Although LS techniques reduce the sensitivity to noise, they still suffer from lack of robustness at high noise levels. The ML-based approach has better performance in the presence of additive noise. However, in low SNR regimes, these methods do not perform properly. High-resolution methods such as subspace- and BSS-based methods reduce the effect of noise by utilizing different observations of received signals. Therefore, the sensitivity to noise in these approaches is less than other presented methods.

3. *Complexity.* Generally, there is always a trade-off between accuracy and complexity. This trade-off varies with application. The main advantage of correlation-based methods is their low complexity, which makes them applicable to many online systems. Although the complexity of deconvolution methods is higher than that of correlation methods, this category of method is still a good candidate for a real-time ranging system. Other proposed methods such as ML-, subspace-, and BSS-based approaches suffer from high complexity, which increases the cost of implementation.

4. *A Priori Information.* As discussed in previous sections, only in correlation-based and ML-based methods priori information is not required. The implementation of all other TOA estimation techniques needs priori information, such as coarse estimation of TOA, and the number of paths. For example, for all frequency domain TOA estimation methods such as the BSS-based approach, the coarse estimation of TOA should be known to select the proper window of data and to implement the algorithm. Generally, a low-resolution method such as the correlation-based method is applied to find a coarse estimation of TOA. In high-resolution algorithms such as subspace- and BSS-based methods, the number of paths should be known. In these methods, EVD is used to estimate the number of paths. However, EVD is part of techniques such as MUSIC, and therefore this part of the algorithm does not add extra complexity in the full system.

Table 7.1 summarizes different discussed TOA estimation algorithms in terms of presented criteria.

TABLE 7.1.   Comparative Behavior of Delay Estimation Algorithms

	Sensitivity to Noise	Ability to Separate Closely Spaced Paths	Complexity	A Priori Information
Correlation-based approaches	Low	Low	Low	None
Deconvolution methods	Low to moderate	Low	Moderate	Coarse estimation of TOA
ML-based method	Moderate	Moderate	Moderate to high	None
Subspace-based methods	Good	Good	High	Number of paths
BSS-based method	Good	Good	High	Number of paths

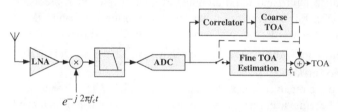

Figure 7.15   Block diagram of a single-band TOA estimation system.

7.4 RANGE ESTIMATION SYSTEM DESIGN

In this section, we discuss the architecture of a range estimation system [61, 62]. It is obvious that this structure can be combined with any range estimation techniques. At the first step, we consider a single-band scenario.

7.4.1 Single-Band Range Estimation Architecture

Figure 7.15 presents the block diagram of a conventional receiver with range estimation. Here, we assume a known single-band transmitted signal with the bandwidth B and center frequency f_c.

In this system, the received signal is amplified via a low-noise amplifier (LNA) and is down-converted to the baseband signal. A low-pass filter is applied to eliminate an aliasing effect. Next, the baseband signal is sampled by an analog-to-digital converter (ADC) with the sampling rate of f_s.

Assuming an ideal low-pass filter, and based on the Nyquist rate, the sampling rate is selected equal to the signal bandwidth; that is, $f_s = B$. Accordingly, the additive noise in the receiver, which is a zero-mean white Gaussian noise, is filtered via the ideal low-pass filter such that the spectrum of filtered noise is limited to the frequency range of $[-B/2, B/2]$. According to sampling theorem [63, 64], after applying the Nyquist sampling rate ($f_s = B$), the power spectrum of additive noise in the sampled baseband signal becomes flat [65]. Therefore, the noise in the sampled baseband signal is modeled as white Gaussian noise.

Here, the proposed TOA estimation technique is composed of two steps: coarse TOA acquisition and fine TOA estimation (high resolution) [66]. In the coarse TOA estimation module, a correlator is used to detect the received signal. Comparing the output of this correlator to a threshold, the starting point of the TOA is estimated. This estimation is used to select and apply the optimum window of data to the fine TOA estimation module. It is essential to mention that these block diagrams can be combined with any TOA estimation, which was presented in the previous section.

7.4.2 Multiband Range Estimation: General Architecture

In this part, we introduce multiband radio architecture that enables high-precision TOA estimation [67, 68]. First, we assume that the neighboring sub-bands K are used in this system. In other words, the total bandwidth B is divided into a set of K smaller sub-bands with bandwidths $\{B_k\}_{k=1}^K$, such that $B = \sum_{k=1}^K B_k$. Figure 7.16b represents this scenario for $K = 3$.

Here, we can use the receiver structure presented in Figure 7.15 for each sub-band. In other words, the same accuracy as the single-band scenario can be achieved with the same parallel structure as the one presented in Figure 7.17 (for $K = 3$ sub-bands).

Based on this structure, all K signals are transmitted and received simultaneously, hence requiring K separate receivers. As shown in Figure 7.17, we can share and utilize a single wideband LNA, but the rest of the transceiver has to be parallelized. Here, $f_{c_k}, k \in \{1, \ldots, K\}$ is the center frequency of sub-band k. We can show that

$$f_{c_k} = f_c - B/2 + \sum_{i=1}^{k-1} B_i + B_k/2, \quad k \in \{1, \ldots, K\}. \tag{7.28}$$

In Equation 7.28, f_c and B represent the center frequency and the bandwidth of single-band signal $s(t)$ defined in (7.2), respectively, and $\{B_k\}_{k=1}^K$ are bandwidths of sub-bands.

Note that here, the wideband signal with the bandwidth B is replaced by K different signals of bandwidths $\{B_k\}_{k=1}^K$. Therefore, although the cost and the complexity of the parallel structure are higher than the single-band scenario, it relaxes the design requirements on the bandwidth of the filters and the sampling rate of ADC.

There is another option for the architecture that eliminates the parallel structure of Figure 7.17. With the assumption that the channel is static during the transmission period of all sub-band signals, we can estimate the channel

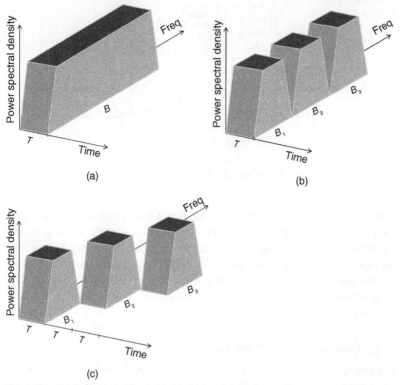

Figure 7.16   Spectrum of a transmitted signal: (a) single band, (b) parallel multiband, and (c) sequential multiband.

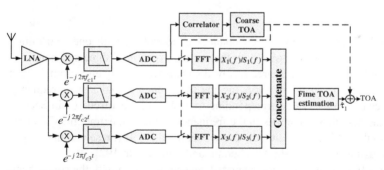

Figure 7.17   Parallel structure for TOA estimation.

response of each band in different time slots (Fig. 7.16c for $K = 3$). Figure 7.18 presents the proposed architecture of this approach for a three-sub-band scenario ($K = 3$).

The new structure reduces the complexity of parallel methods by changing the parallel radio architecture to a single radio that sequentially hops over the K sub-bands in the time domain. Comparing this structure with the single-band

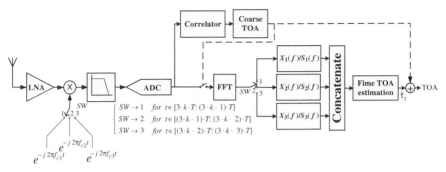

Figure 7.18   Proposed architecture for high-resolution TOA estimation.

scenario presented in Figure 7.15, we reduce the signal bandwidth at each time slot, and thus we decrease the sampling rate and the filtering bandwidth in our system.

The bandwidth used in the above scheme does not need to be contiguous [69, 70]. This is especially helpful when the entire band is not contiguously available. The next section discusses how this noncontiguous spectrum is incorporated in TOA estimation algorithms. Here, we present an example of a contiguous multiband concatenation scenario.

Example 7.6   Contiguous Multiband Concatenation

Implement a contiguous multiband scenario with three equal sub-bands. Each band has a 20-MHz bandwidth. Corresponding to each center frequency, there is one CIR in the generated channel model. Three consecutive center frequencies, which are separated by 20 MHz, are selected.

Solution

To generate the received signal, each channel is convolved with one transmitted signal. Next, we add additive noise with the same power for all three sub-bands. In the receiver, we first convert all three received signals to the frequency domain (use "fft" with length 64). Then, by removing the transmitted signal in the frequency domain, we calculate the CFR corresponding to each band. As discussed in Section 4.2, we concatenate these channels in the frequency domain to extract the CFR corresponding to the 60-MHz bandwidth. Consequently, the CIR (presented in Eq. 7.9) is calculated using IFFT (use the "ifft" command with length 192).

Figure 7.19 compares the concatenated channel response with the single-band scenario. It can be observed that the achieved resolution in a multiband scenario is higher than that in a single-band scenario. MATLAB codes can be found online at ftp://ftp.wiley.com/public/sci_tech_med/matlab_codes.

7.4.3   Noncontiguous Multiband Scenario

With the spectrum becoming a more limited resource, utilizing the radio spectrum effectively becomes crucial for communication systems. As a solution to this

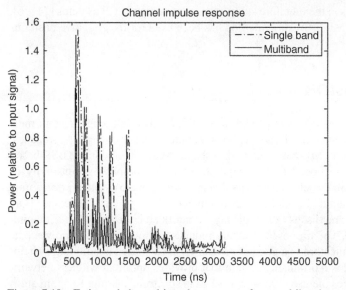

Figure 7.19   Estimated channel impulse response for a multiband scenario.

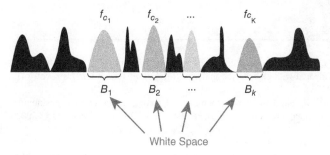

Figure 7.20   White space band in spectrum.

problem, CR systems [71, 72] provide the capability to improve the use of available radio frequency (RF) by sensing the environment and optimally selecting bands for communication that reduce interference effects. All locations in a spectrum where a CR can operate are named as "white space" (Fig. 7.20).

This available spectrum can be used in both communication and positioning (ranging) systems [73]. Utilizing white spaces as noncontiguous bands, a ranging system can achieve the resolution corresponding to the total available bandwidth. Similar to a contiguous scenario (Fig. 7.16b, c), the proposed parallel and sequential multiband architectures presented in the previous section can be used for noncontiguous sub-bands. In this case, instead of using the center frequency presented in (7.28), the new center frequency, based on the availability of the spectrum, should be applied in the system.

Limited works in the literature study range estimation via CR systems [74–76]. In these works, the ML estimation method has been used for the cognitive positioning system (CPS).

7.5 CONCLUSION

This chapter introduces different categories of TOA estimation techniques in multipath wireless environments. These methods include correlation-based approaches, deconvolution methods, ML-based methods, subspace-based methods and BSS-based approaches. These methods are compared in terms of complexity, sensitivity to noise, performance in closely spaced and distant paths scenarios, and a priori information needed as input. In addition, different architectures for a practical TOA estimation system are discussed. Typically, a multiband scenario for a ranging system is presented. The proposed structure reduces complexity and improves performance. This architecture is a good candidate for a CR-based positioning system. CR technology [71] is a promising system that provides optimum wireless network solutions. It optimizes the use of available RFs via spectrum sharing. Ranging algorithms, which provide distance information of the users, play a crucial role in improving the quality of service of CR networks.

REFERENCES

[1] E. Bosse, R. M. Turner, and D. Brookes, "Improved radar tracking using a multipath model: maximum likelihood compared with eigenvector analysis," *IEEE Proc. Radar Son. Nav.*, vol. 141, no. 4, pp. 213–222, 1994.

[2] F. V. F. de Lima and C. M. Furukawa, "Development and testing of an acoustic positioning system: Description and signal processing," in *IEEE Proc. of Ultrasonic Symp.*, Oct. 2002, pp. 849–852.

[3] H. Tong and S. A. Zekavat, "A novel wireless local positioning system via a merger of DS-CDMA and beamforming: Probability-of-detection performance analysis under array perturbations," *IEEE Trans. Veh. Technol.*, vol. 56, no. 3, pp. 1307–1320, 2007.

[4] H. Tong, J. Pourrostam, and S. A. Zekavat, "Optimum beam-forming for a novel wireless local positioning system: A stationarity analysis and solution," *EURASIP J. Adv. Signal Process.*, vol. 2007, pp. 1–12, 2007.

[5] Z. Zhao, B. Zhang, J. Zheng, Y. Yan, and J. Ma, "Providing scalable location service in wireless sensor networks with mobile sinks," *IET Commun.*, vol. 3, no. 10, pp. 1628–1637, 2009.

[6] I. Guvenc and C. C. Chong, "A survey on TOA based wireless localization and NLOS mitigation techniques," *IEEE Commun. Surv. Tutorials*, vol. 11, no. 3, pp. 107–124, 2009.

[7] Y. Gu, A. Lo, and I. Niemegeers, "A survey of indoor positioning systems for wireless personal networks," *IEEE Commun. Surv. Tutorials*, vol. 11, no. 1, pp. 13–32, 2009.

[8] N. Patwari, J. Ash, S. Kyperountas, A. Hero, R. Moses, and N. Correal, "Locating the nodes," *IEEE Signal Process. Mag.*, vol. 22, no. 4, pp. 54–69, 2005.

[9] K. Pahlavan, X. Li, and J. Makela, "Indoor geolocation science and technology," *IEEE Commun. Mag.*, vol. 40, pp. 112–118, 2002.

[10] B. Alavi and K. Pahlavan, "Modeling of the TOA based distance measurement error using UWB indoor radio measurements," *IEEE Commun. Lett.*, vol. 10, no. 4, pp. 275–277, 2006.

[11] K. Pahlavan and P. Krishnamurthy, *Principles of Wireless Networks: A Unified Approach*. Englewood Cliffs, NJ: Prentice Hall, 2002.

[12] P. Stoica, "List of references on spectral line analysis," *Signal Process.*, vol. 31, no. 3, pp. 329–340, 1993.

[13] P. Stoica and R. Moses, *Spectral Analysis of Signals*. Upper Saddle River, NJ: Prentice Hall, 2005.

[14] A. Saleh and R. Valenzuela, "A statistical model for indoor multipath propagation," *IEEE J. Sel. Area. Comm.*, vol. 5, no. 2, pp. 128–137, 1987.

[15] C. H. Knapp and C. G. Carter, "The generalized correlation method for estimation of time delay," *IEEE Trans. Acoust. Speech Signal Process.*, vol. ASSP-21, pp. 320–327, 1976.

[16] D. Hertz, "Time delay estimation by combining efficient algorithms and generalized cross-correlation methods," *IEEE Trans. Acoust. Speech Signal Process.*, vol. 34, no. 1, pp. 1–7, 1986.

[17] G. Fock, J. Baltersee, P. Schulz-Rittich, and H. Meyr, "Channel tracking for rake receivers in closely spaced multipath environments," *IEEE J. Sel. Area. Comm.*, vol. 19, pp. 2420–2431, 2001.

[18] S. Glisic and B. Vucetic, *Spread Spectrum CDMA Systems for Wireless Communications*. Norwood, MA: Artech House, 1997.

[19] H. V. Poor, *An Introduction to Signal Detection and Estimation*. New York: Springer, 1994.

[20] C. E. Cook and M. Bernfeld, *Radar Signals: An Introduction to Theory and Applications*. New York: Academic Press, 1970.

[21] G. L. Turin, "An introduction to matched filters," *IRE Trans. Inf. Theory*, vol. IT-6, no. 3, pp. 311–329, 1960.

[22] M. D. Hahm, Z. I. Mitrovski, and E. L. Titlebaum, "Deconvolution in the presence of Doppler with application to specular multipath parameter estimation," *IEEE Trans. Signal Process.*, vol. 45, pp. 2203–2219, 1997.

[23] K. Pahlavan and A. Levesque, *Wireless Information Networks*. Hoboken, NJ: John Wiley & Sons, 1995.

[24] N. R. Yousef, A. H. Sayed, and N. Khajehnouri, "Detection of fading overlapping multipath components," *IEEE Trans. Signal Process.*, vol. 86, no. 9, pp. 2407–2425, 2006.

[25] A. Sayed, A. Tarighat, and N. Khajehnouri, "Network-based wireless location," *IEEE Signal Process. Mag.*, vol. 22, no. 4, pp. 24–40, 2005.

[26] T. G. Manickam, R. J. Vaccaro, and D. W. Tufts, "A least squares algorithm for multipath time-delay estimation," *IEEE Trans. Signal Process.*, vol. 42, pp. 3229–3233, 1994.

[27] A. H. Sayed, *Fundamentals of Adaptive Filtering*. New York: Wiley, 2003.

[28] H. Saarnisaari, "ML time delay estimation in a multipath channel," in *Proc. of IEEE Int. Symp. on Software Test. Anal. (ISSTA)*, Sep. 1996, pp. 1007–1011.

[29] J. P. Iannello, "Large and small error performance limits for multipath time delay estimation," *IEEE Trans. Acoust. Speech Signal Process.*, vol. ASSP-34, pp. 245–251, 1986.

[30] C. Falsi, D. Dardari, L. Mucchi, and M. Z. Win, "Time of arrival estimation for UWB localizers in realistic environments," *EURASIP J. Appl. Signal Process.*, vol. 2006, pp. 1–13, 2006.

[31] R. D. J. V. Nee, J. Siereveld, P. C. Fenton, and B. R. Townsend, "The multipath estimating delay locked loop: Approaching theoretical accuracy limits," in *Proc. IEEE Position Location Navigation Symp.*, 1994, pp. 246–251.

[32] E. S. Lohan, R. Hamila, A. Lakhzouri, and M. Renfors, "Highly efficient techniques for mitigating the effects of multipath propagation in DS-CDMA delay estimation," *IEEE Trans. Wirel. Commun.*, vol. 4, no. 1, pp. 149–162, 2005.

[33] R. Hamila, E. S. Lohan, and M. Renfors, "Multipath delay estimation in GPS receivers," in *Proc. Nordic Signal Processing Symp. (NORSIG)*, Jun. 2000, pp. 417–420.

[34] C. D. Wann and L. K. Chang, "Modified GML algorithm with simulated annealing for estimation of signal arrival time," in *IEEE TENCON*, 2007, pp. 1–4.

[35] C. D. Wann and L. K. Chang, "Modified GML algorithm for estimation of signal arrival time in UWB systems," in *IEEE GLOBCOM*, 2006, pp. 1–5.

[36] J. Y. Lee and R. A. Scholtz, "Ranging in a dense multipath environment using an UWB radio link," *IEEE J. Sel. Area. Comm.*, vol. 20, no. 9, pp. 1677–1683, 2002.

[37] A. Jakobsson, A. L. Swindlehurst, and P. Stoica, "Subspace-based estimation of time delays and Doppler shifts," *IEEE Trans. Signal Process.*, vol. 46, pp. 2472–2483, 1998.

[38] X. Li and K. Pahlavan, "Super-resolution TOA estimation with diversity for indoor geolocation," *IEEE Trans. Wirel. Commun.*, vol. 3, no. 1, pp. 224–234, 2004.

[39] Y. Y. Wang, J. T. Chen, and W. H. Fang, "TST-MUSIC for joint DOA-delay estimation," *IEEE Trans. Signal Process.*, vol. 49, no. 4, pp. 721–729, 2001.

[40] P. Stoica and A. Nehorai, "MUSIC, maximum likelihood and Cramer–Rao bound," *IEEE Trans. Acoust. Speech Signal Process.*, vol. 37, no. 5, pp. 720–741, 1989.

[41] A.-J. Van der Veen, M. C. Vanderveen, and A. J. Paulraj, "Joint angle and delay estimation using shift-invariance properties," *IEEE Signal Process. Lett.*, vol. 4, no. 5, pp. 142–145, 1997.

[42] H. Saarnisaari, "TLS-ESPRIT in a time delay estimation," in *Proc. IEEE Vehicular Technology Conf.*, May 1997, pp. 1619–1623.

[43] S. Al-Jazzar and J. Caffery Jr., "ESPRIT-based joint AOA/delay estimation for CDMA systems," in *IEEE Wireless Communications and Networking Conference (WCNC '04)*, Mar. 2004, pp. 2244–2249.

[44] J. F. Cardoso, "Blind signal separation: statistical principles," *Proc. IEEE*, vol. 86, no. 10, pp. 2009–2025, 1998.

[45] H. Sawada, S. Araki, R. Mukai, and S. Makino, "Grouping separated frequency components by estimating propagation model parameters in frequency-domain blind source separation," *IEEE T. Audio Speech Lang. Process.*, vol. 15, no. 5, pp. 1592–1604, 2007.

[46] Y. G. Won and S. Y. Lee, "Convolutive blind signal separation by estimating mixing channels in time domain," *Electron. Lett.*, vol. 44, no. 21, pp. 1277–1278, 2008.

[47] A. Hyvarinen, J. Karhunen, and E. Oja, *Independent Component Analysis*. New York: John Wiley & Sons, 2001.

[48] A. Hyvärinen, "Fast and robust fixed-point algorithms for independent component analysis," *IEEE Trans. Neural Netw.*, vol. 10, no. 3, pp. 626–634, 1999.

[49] E. Bingham and A. Hyvarinen, "A fast fix-point algorithm for independent component analysis of complex valued signal," *Int. J. Neural Syst.*, vol. 10, no. 1, pp. 1–8, 2000.

[50] P. Z. Tichavský, P. Koldovský, and E. Oja, "Performance analysis of the fast ICA algorithm and Cramér–Rao bounds for linear independent component analysis," *IEEE Trans. Signal Process.*, vol. 54, pp. 1189–1203, 2006.

[51] R. Cristescu, T. Ristaniemi, J. Joutsensalo, and J. Karhunen, "CDMA delay estimation using a fastica algorithm," in *Proc. IEEE Int. Symp. on Personal, Indoor, and Mobile Radio Communications (PIMRC 2000)*, pp. 1117–1120, Sep. 2000.

[52] M. Pourkhaatoun, S. A. Zekavat, and J. Pourrostam, "A high resolution ICA based time delays estimation technique," in *Proc. of IEEE RADAR Conf.*, Apr. 2007, pp. 320–324.

[53] M. Pourkhaatoun and S. A. Zekavat, "A novel ICA-based TOA estimation technique: Achieving high resolution, high reliability, and, low cost," *Proc. IEEE International Workshop on Signal Processing and its Applications, WOSPA '08*, pp. 18–20, Mar. 2008.

[54] M. Novey and T. Adali, "Complex ICA by negentropy maximization," *IEEE Trans. Neural Netw.*, vol. 19, no. 4, pp. 596–609, 2008.

[55] M. Novey and T. Adali, "Extending the complex fast ICA algorithm to noncircular sources," *IEEE Trans. Signal Process.*, vol. 56, no. 5, pp. 2148–2154, 2008.

[56] T. Adali, M. Novey, and J. F. Cardoso, "Complex ICA using nonlinear functions," *IEEE Trans. Signal Process.*, vol. 56, no. 9, pp. 4536–4544, 2008.

[57] T. Chen and Q. Lin, "Dynamic behavior of the whitening process," *IEEE Signal Process. Lett.*, vol. 5, no. 1, pp. 25–26, 1998.

[58] R. J. Kenefic and A. H. Nuttall, "Maximum likelihood estimation of the parameters of tone using real discrete data," *IEEE J. Oceanic Eng.*, vol. OE-12, no. 1, pp. 279–280, 1987.

[59] D. Tufts and R. Kumaresan, "Estimation of frequencies of multiple sinusoids: Making linear prediction perform like maximum likelihood," *Proc. IEEE*, vol. 70, pp. 975–989, 1982.

[60] H. C. So and K. W. Chan, "Reformulation of Pisarenko harmonic decomposition method for single-tone frequency estimation," *IEEE Trans. Signal Process.*, vol. 52, no. 4, pp. 1128–1135, 2004.

[61] S. Mirabbasi and K. Martin, "Classical and modern receiver architectures," *IEEE Commun. Mag.*, vol. 38, no. 11, pp. 132–139, 2000.

[62] W. Namgoong and T. H. Meng, "Direct-conversion RF receiver design," *IEEE Trans. Commun.*, vol. 49, pp. 518–529, 2001.

[63] B. Porat, *Digital Processing of Random Signals: Theory and Methods*. Englewood Cliffs, NJ: Prentice Hall, 1993.

[64] A. V. Oppenheim, R. W. Schafer, and J. R. Buck, *Discrete-Time Signal Processing*. Upper Saddle River, NJ: Prentice Hall, 1999.

[65] D. Manolakis, V. Ingle, and S. Kogon, *Statistical and Adaptive Signal Processing*. New York: McGraw-Hill, 2000.

[66] S. Wu, Q. Zhang, and N. Zhang, "A two-step TOA estimation method for IR-UWB ranging systems," in *Proc. of IEEE CNSR*, May 2007, pp. 302–310.

[67] E. Saberinia and A. H. Tewfik, "Ranging in multiband ultrawideband communication systems," *IEEE Trans. Veh. Technol.*, vol. 57, no. 4, pp. 2523–2530, 2008.

[68] H. Xu, C.-C. Chong, I. Guvenc, and L. Yang, "High-resolution TOA estimation with multi-band OFDM UWB signals," in *Proc. of ICC*, May 2008, pp. 4191–4196.

[69] H. Celebi and H. Arslan, "Ranging accuracy in dynamic spectrum access networks," *IEEE Commun. Lett.*, vol. 11, no. 5, pp. 405–407, 2007.

[70] H. Celebi and H. Arslan, "Cognitive positioning systems," *IEEE T. Wirel. Commun.*, vol. 6, no. 12, pp. 4475–4483, 2007.

[71] A. B. Fette, *Cognitive Radio Technology (Communications Engineering)*. Burlington, MA: Newnes, 2006.

[72] J. Mitola III and G. Q. Maguire Jr., "Cognitive radio: Making software radios more personal," *IEEE Personal Commun.*, vol. 6, no. 4, pp. 13–18, 1999.

[73] H. Celebi and H. Arslan, "Utilization of location information in cognitive wireless networks," *IEEE Wireless Commun.*, vol. 14, no. 4, pp. 6–13, 2007.

[74] S. Gezici, H. Celebi, H. V. Poor, and H. Arslan, "Fundamental limits on time delay estimation in dispersed spectrum cognitive radio systems," *IEEE Trans. Wireless Comm.*, vol. 8, no. 1, pp. 78–83, 2009.

[75] H. Celebi, K. A. Qaraqe, and H. Arslan, "Performance comparison of time delay estimation for whole and dispersed spectrum utilization in cognitive radio systems," in *Int. Conf. on Cognitive Radio Oriented Wireless Networks and Communications (CROWNCOM '09)*, Jun. 2009, pp. 1–6.

[76] F. Kocak, H. Celebi, S. Gezici, K. A. Qaraqe, H. Arslan, and H. V. Poor, "Time delay estimation in dispersed spectrum cognitive radio systems," *EURASIP J. Adv. Signal Process.*, vol. 2010, pp. 1–20, 2010, Art. ID 675959.

CHAPTER **8**

WIRELESS LOCALIZATION USING ULTRA-WIDEBAND SIGNALS

Liuqing Yang[1] and Huilin Xu[2]
[1]Colorado State University, Fort Collins, CO
[2]QUALCOMM Incorporated, San Diego, CA

REVIEWING THIS chapter, the reader will have an overall understanding of ultra-wideband (UWB)-related localization and ranging techniques including systems, algorithms, performance analyses, and comparisons. In Section 8.1, we briefly review the background knowledge of UWB. Section 8.2 introduces the recent research progress on UWB localization including fingerprinting and geometric techniques. Since time-based localization, is most widely adopted for UWB, Sections 8.3 and 8.4 are dedicated to time-of-arrival (TOA) estimation with UWB signals.

8.1 INTRODUCTION TO UWB

According to the U.S. Federal Communications Commission (FCC), UWB refers to the wireless technology that can access the frequency spectrum larger than 500 MHz or a fractional bandwidth exceeding 20%. Since the U.S. permitted the operation of UWB devices in shared or nongovernment frequency bands [1], UWB has attracted broad interests in areas of wireless communications and localization. Due to the unique advantages, including wall penetration capability, low transmission power, simple transceiver structure, and high temporal as well as spatial resolutions, UWB can potentially provide huge capacity and centimeter-level localization accuracy.

8.1.1 Regularization

In 2002, the United States permitted the unlicensed use of UWB technology in shared or nongovernment frequency bands [1]. The FCC order specifies operational limitations for applications including imaging systems and vehicular radar, indoor,

Handbook of Position Location: Theory, Practice, and Advances, Second Edition.
Edited by S. A. (Reza) Zekavat and R. Michael Buehrer.
© 2019 by the Institute of Electrical and Electronics Engineers, Inc.
Published 2019 by John Wiley & Sons, Inc.
Companion Website: www.wiley.com/go/zekavat/positionlocation2e

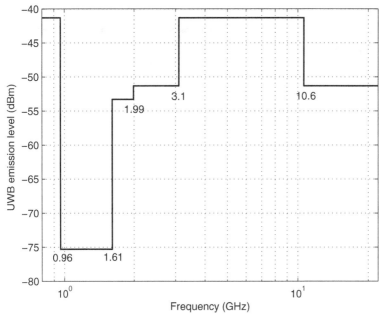

Figure 8.1 FCC emission limits for indoor UWB communications.

and handheld UWB systems. Depending on the implementation scenario, a UWB localization device can be categorized into any of these systems. In order to avoid harmful interference from UWB devices to existing systems, FCC has set the emission limit for each type of UWB implementation, which is called the emission mask. For example, the highest power spectral density (PSD) emission level for indoor UWB devices is −41.3 dBm/MHz in the range of 3.1–10.6 GHz (see Fig. 8.1).

In Europe, the European Commission's (EC) Radio Spectrum Committee approved the operation of UWB in 2006 [2]. The EC identified the frequency bands of 3.4–5.0 GHz and 6.0–8.5 GHz with potential extension to 9 GHz for UWB devices. In Singapore, the Info-communications Development Authority of Singapore (IDA) was established to support the development of UWB in 2003. In Japan, the UWB spectral mask was approved in 2006 [3].

8.1.2 Transmission Approaches

A UWB spectrum can be accessed either by generating a series of extremely short duration pulses or the aggregation of a number of narrowband subcarriers having a bandwidth over 500 MHz. This leads to two transmission approaches for UWB technology, namely, impulse radio (IR) and multiband orthogonal frequency division multiplexing (MB-OFDM). IR transmits pulses that occupy the entire allowable frequency band. In MB-OFDM, the UWB spectrum is exploited by transmitting orthogonal frequency division multiplexing (OFDM) symbols on several sub-bands in a frequency-hopping manner.

IR transmits information by altering the positioning, amplitude, polarity, or shape of the UWB pulse, which corresponds to pulse position modulation (PPM), pulse amplitude modulation (PAM), on–off keying (OOK), or pulse shape modulation (PSM). The continuous transmission of uniformly spaced pulses gives rise to strong spectrum spikes occurring at integer multiples of the pulse repetition frequency. Therefore, time hopping (TH) and direct sequence (DS) techniques are adopted to mitigate the frequency spikes by randomizing the pulse train. TH and DS can also enable multiaccess in UWB systems. TH is often applied to low data rate UWBs with low pulse duty cycles. DS is suitable for medium to high data rate transmissions with high duty cycles.

MB-OFDM accesses the entire UWB spectrum by altering the carrier frequency according to some time–frequency (TF) hopping pattern. The advantages of MB-OFDM include high spectral efficiency, resilience against narrowband interference, and tolerance to multipath effects. Therefore, MB-OFDM is very suitable for high data rate short-range applications. In addition, due to the multiband operation, MB-OFDM can achieve higher-frequency diversity than single-band OFDM.

8.1.3 Standards

After the FCC's authorization of UWB, attempts have been made by the Institute of Electrical and Electronics Engineers (IEEE) to incorporate UWB into technical standards as the PHY technology. The standard groups include the IEEE 802.15.3a for high-rate wireless personal area networks (WPANs) in short range less than 10 m and the IEEE 802.15.4a for low-rate WPANs.

The IEEE 802.15.3a task group (TG3a) was established to identify a higher data rate amendment to IEEE 802.15.3 for an alternative physical layer (PHY). Two UWB-based proposals used to compete for the final approval. One is the DS-UWB and the other is the MB-OFDM. MB-OFDM standards of the European Computer Manufacturers Association (ECMA) have been accepted by the International Organization for Standardization (ISO) [4, 5]. In ECMA MB-OFDM, the entire UWB spectrum of 3.1–10.6 GHz is divided into 14 bands, each with a bandwidth of 528 MHz. The 14 bands are further grouped into six groups, as shown in Figure 8.2. OFDM symbols are transmitted within a group by switching the frequency band according to the TF code. TG3a was officially disbanded in 2006 due to regulatory and market uncertainty.

The IEEE standard 802.15.4a is an improved version of IEEE 802.15.4 with an alternative UWB-based PHY. It provides data communications and high-accuracy

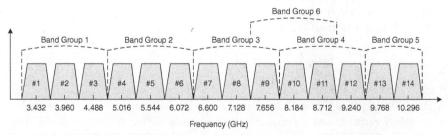

Figure 8.2   ECMA MB-OFDM bands labeled with the corresponding center frequencies.

positioning for low data rate networking with ultralow complexity, ultralow power consumption, and scalable data rates of 110 and 851 kbps and 6.81 and 27.24 Mbps. The PHY can operate on three different bands: the subgigahertz band (250–750 MHz), the low band (3.1–5.0 GHz), and the high band (6.0–10.6 GHz). IEEE 802.15.4a is the first UWB standard that incorporates wireless localization. It supports both the two-way and one-way ranging protocols. In 2007, the IEEE 802.15.4a standard was approved by the IEEE Standards Association (SA) [6].

8.1.4   UWB Channels

Due to the huge bandwidth of UWB signals, the UWB channel is significantly different from narrowband wireless channels. UWB channels are characterized by the dense multipaths that come in clusters. Since a deterministic channel model cannot describe the wireless propagation environment very well, the UWB channel is usually studied by extracting a statistical model from measurements. Parameters of the model reflect the statistical properties of various propagation environments.

As discussed in Chapter 4, the wireless channel is usually expressed by the following model:

$$h(t) = \sum_{l=0}^{\infty} h_l \delta(t - \tau_l), \tag{8.1}$$

where h_l and τ_l are the amplitude and delay of the lth channel path. In UWB channels, multipath components come in clusters, each of which is otherwise resolved as a single path by narrowband signals. Modeling of UWB channels is often based on the Saleh-Valenzuela (S-V) model with impulse response [7]:

$$h(t) = \sum_{m=0}^{+\infty} \sum_{n=0}^{+\infty} \alpha_{m,n} \exp(j\theta_{m,n}) \delta(t - T_m - \tau_{m,n}), \tag{8.2}$$

where $\alpha_{m,n}$ and $\theta_{m,n}$ are the multipath gain and phase, respectively. T_m is the arrival time of the first path of the mth cluster, and $\tau_{m,n}$ is the delay of the nth ray inside the mth cluster relative to T_m. Theoretically, each channel realization can contain an infinite number of multipath components, but in practice, the number of multipaths is usually considered to be finite by ignoring the very weak trailing paths.

In the S-V model, the phases of multipath component $\theta_{m,n}$ are modeled as independent uniform random variables in $[0, 2\pi)$, and the amplitude $\alpha_{m,n}$ is an independent Rayleigh random variable with the power delay profile (PDP)

$$E\{\alpha_{m,n}^2\} = E\{\alpha_{0,0}^2\} e^{-T_m/\Gamma} e^{-\tau_{m,n}/\gamma}, \tag{8.3}$$

where Γ and γ are the constant decay rates for clusters and rays, respectively. The cluster and ray arrival times are represented by Poisson distributed random variables with arrival rates Λ and λ, respectively, according to the following probability density functions (PDFs):

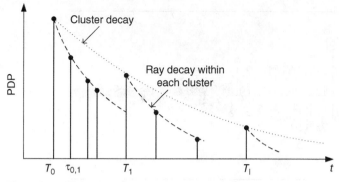

Figure 8.3   PDP of the Saleh–Valenzuela channel model.

$$p(T_m | T_{m-1}) = \Lambda \exp[-\Lambda(T_m - T_{m-1})], m > 0 \qquad (8.4)$$

and

$$p(\tau_{m,n} | \tau_{m,(n-1)}) = \lambda \exp[-\lambda(\tau_{m,n} - \tau_{m,(n-1)})], n > 0. \qquad (8.5)$$

An example of the PDP of the S-V model is illustrated in Figure 8.3.

Example 8.1

Generate PDP realizations for the S-V channel model in MATLAB with the following parameters: $\Lambda = 1/200\,\text{ns}^{-1}$, $\lambda = 1/20\,\text{ns}^{-1}$, $\Gamma = 60\,\text{ns}$, and $\gamma = 20\,\text{ns}$. Count the number of clusters contained in each PDP realization.

Solution

The MATLAB codes in Chapter_8_Example_1.m generate PDP realizations for the S-V channel model. MATLAB codes can be found online at ftp://ftp.wiley.com/ public/sci_tech_med/matlab_codes. In the first while loop, clusters are generated with their arrival times determined by PDF (8.4). The loop terminates if the newly generated cluster arrival time is larger than the maximum channel delay, and this cluster arrival time is discarded. In the for loop, rays are generated for each cluster with their arrival times determined by PDF (8.5). Then, the power of each ray is calculated according to (8.3). For each cluster, the while loop terminates if the newly generated ray arrival time is larger than the maximum channel delay. When log scale is used for the y-axis, it is clear that both clusters and rays decay exponentially.

The IEEE 802.15.3a and 4a UWB channels are modified versions of the S-V model with the main differences summarized as follows [8, 9]. Both UWB channels have incorporated the path loss and the lognormally distributed shadowing. The phase $\theta_{m,n}$ is constrained to 0 or π with equal probability in the IEEE 802.15.3a channel model, which results in a real valued baseband channel model (see Fig. 8.4). Distributions of the amplitude $\alpha_{m,n}$ are modified to better fit the UWB channel parameters. The lognormal distribution and the Nakagami-m distribution with the

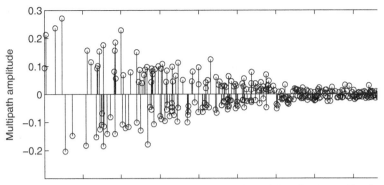

Figure 8.4   A typical channel realization of the IEEE 802.15.3a channel in the line-of-sight environment with a propagation distance less than 4 m.

lognormally distributed m-parameter are used for the IEEE 802.15.3a and 4a channel models, respectively. Variable ray decay rates are adopted for the IEEE 802.15.4a channel model to account for the linear dependence on the cluster decay rate. For the IEEE 802.15.4a channel model, distributions of ray arrival times are replaced by the mixture of two Poisson processes.

8.2   UWB LOCALIZATION TECHNIQUES

As discussed in Chapters 1 and 2, a wireless localization system consists of multiple anchor nodes with their positions known to the system and target nodes with unknown positions. Inference of the unknown target position relies on the measurement of certain parameters of signals transmitted between the target and anchors. These parameters can be the received signal strength (RSS), direction of arrival (DOA), TOA, time difference of arrival (TDOA), channel PDP, or the channel impulse response (CIR). Based on the measurement, the target position can be estimated by either fingerprinting or the geometric technique, which will be introduced in the rest of this chapter. The relation among these techniques is illustrated in Figure 8.5.

8.2.1   Fingerprinting Localization

As discussed in Chapter 15, fingerprinting estimates the target position by finding the best-matched pattern or fingerprint for the measurement within a map. Fingerprinting is a good substitute to geometric localization in harsh environments where the geometric relationship between measurements and node positions cannot be established. Fingerprinting localization consists of the off-line and in-line phases. In the off-line phase, the map is generated for the environment with each entry recording the measured fingerprint and the corresponding target position. In the in-line phase, the location fingerprint is extracted for the target with unknown position and the position is estimated by pattern matching.

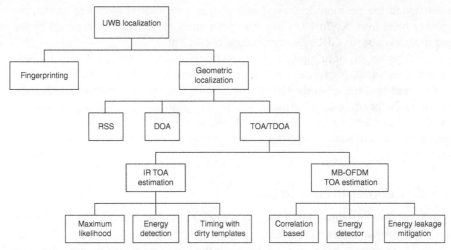

Figure 8.5   Wireless localization with UWB signals.

Suppose that, in the off-line phase, pattern measurements $\{p_i\}_{i=1}^K$ are observed at K positions $\{r_i\}_{i=1}^K$ over the environment and stored in the map. In the in-line localization phase, the target position r can be estimated by the nearest neighbor method as $\hat{r} = r_k$ with

$$k = \underset{i \in [1,\ldots,K]}{\arg\min} \|p - p_i\|. \tag{8.6}$$

Besides this straightforward method, many more sophisticated and better-performing algorithms have been reported in the literature, such as the weighted nearest k neighbors, support vector machine, neural network, and maximum likelihood (ML) [10–14].

RSS is the most common parameter used in narrowband fingerprinting localization. Since UWB signals resolve the channel much better than narrowband signals, UWB fingerprints can provide more position-dependent information so that the localization accuracy is improved. In [13], a vector of seven parameters extracted from the CIR is used as the fingerprint, which can be expressed as

$$p = \left[\tau_m, \tau_{\mathrm{rms}}, \tau_{\max}, P, N, P_1, \tau_1\right]. \tag{8.7}$$

In (8.7), the channel parameters include the mean excess delay τ_m, the root mean square (RMS) delay spread τ_{rms}, the maximum excess delay $\tau_{\max}$, the total received power P, the number of multipath components N, the first path power P_1, and the first path delay τ_1.

In [15], the PDF of the CIR is extracted as the fingerprint based on measurements obtained in the frequency band from 3 to 6 GHz. In the localization phase, the target position is estimated by finding the best-matched fingerprint from the map based on the ML criterion. Malik and Allen [16] used the CIR as fingerprint, which contains all the position-dependent information. The CIR is measured in the frequency band from 3.1 to 10.6 GHz. The position of the target is determined by the maximization of the CIR cross-correlation coefficient.

UWB fingerprinting results in a small ambiguity region even with a single anchor node so that high localization accuracy is guaranteed for both the line-of-sight (LOS) and non-line-of-sight (NLOS) scenarios. The main drawback is that the process of generating the fingerprint map is time-consuming, and a large amount of storage is needed to record the channel parameters. Therefore, in contrast to the geometric localization, fingerprinting is more suitable for small-size indoor environments.

8.2.2   Geometric Localization

Similar to the conventional narrowband system, UWB-based geometric localization uses channel parameter measurements inlcuding RSS, DOA, TOA, and TDOA (see Fig. 8.5). Among these, DOA measurement is the basis for triangulation localization. RSS and TOA give the range information that is used by the trilateration technique (see Fig. 8.6). For the time-based localization, in the absence of the common time base, TDOA is used instead of TOA and leads to hyperbolic localization (Fig. 8.7).

RSS and DOA are not very suitable for UWB localization. This is because the RSS approach does not benefit from the huge bandwidth of UWB [17], and

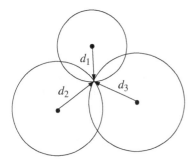

Figure 8.6   Trilateration for RSS- and TOA-based localizations with three anchor nodes.

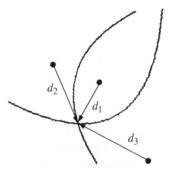

Figure 8.7   TDOA-based hyperbolic localization with three anchor nodes.

measuring the DOA often requires costly antenna arrays that are especially unaffordable to UWB-based wireless sensor networks. For these reasons, time-based localization is the most widely adopted solution for UWB systems such as the IEEE 802.15.4a low-rate WPAN [6].

TOA Estimation: As mentioned in Section 8.1.4, the wireless channel has an impulse response, $h(t) = \sum_{l=0}^{\infty} h_l \delta(t - \tau_l)$, with h_l and τ_l being the amplitude and delay of the lth multipath component. The first path delay τ_0 is the TOA of the transmission link, which can be converted to the separation distance between the transmitter and the receiver:*

$$d = c \cdot \tau_0, \tag{8.8}$$

where c is the speed of light.

For a narrowband system, if signal $s(t)$ is transmitted, the received signal can be expressed as

$$r(t) = As(t - T) + n(t), \tag{8.9}$$

where A and T capture the joint effects of all channel multipath components and $n(t)$ is the additive noise. Given (8.9), the matched filtering estimator can only provide the optimal estimate of T, which is not the first path delay. For a UWB system, due to the multipath resolution capability of UWB signals, the received signal can be expressed as

$$r(t) = \sum_{l=0}^{\infty} h_l s(t - \tau_l) + n(t). \tag{8.10}$$

Given (8.10), the optimal TOA estimate can be obtained based on the ML or Bayesian criterion depending on the level of a priori knowledge. However, the required complexity could be unaffordable due to the huge number of multipath components in typical indoor or urban UWB channels. In order to avoid this problem, various low-complexity and suboptimal TOA estimators have been developed for both IR and MB-OFDM. The rest of this chapter will focus on the introduction of these techniques.

Position Estimation: Let us consider a time-based two-dimensional localization system with M anchors located at positions $\{(x_i, y_i)\}_{i=1}^{M}$ and a target at the unknown position (x, y). TOAs are measured between the target and each anchor node, which can be expressed as

$$\hat{\tau}_i = c\sqrt{(x_i - x)^2 + (y_i - y)^2} + \eta_i, \quad i = 1, \dots, M. \tag{8.11}$$

In (8.11), c is the speed of light and η_i is the TOA error between the ith anchor node and the target. For the hyperbolic localization, TDOA is obtained by subtracting $\hat{\tau}_1$ from $\hat{\tau}_j$, $j = 2, \dots, M$, which can be expressed as

---

*Here, a LOS propagation is assumed such that the first path is the direct path. NLOS issues will be discussed later.

$$\hat{\tau}_{1,j} = c\left(\sqrt{(x_i - x)^2 + (y_i - y)^2} - \sqrt{(x_1 - x)^2 + (y_1 - y)^2}\right) + \eta_{1,j}. \qquad (8.12)$$

As discussed in Parts I and II of this book, various techniques have been developed to estimate the target position from the TOA and TDOA measurements. The simplest way is to directly solve (8.11) and (8.12) by assuming measurements are noise free [18]. The limitation of this method is that due to the measurement error, the geometric equations may not give a unique solution of the target position. This implies that the three circles in Figure 8.6 may not intersect at one point. An alternative is to search for the target position that minimizes some cost functions based on (8.11) and (8.12). This is similar to the pattern matching algorithms used by the fingerprinting localization. If a priori knowledge of measurement errors is available, the localization accuracy can be further improved by incorporating statistical criteria such as ML and Bayesian [19]. For detailed information on related techniques, please refer to Parts I and II of this book.

8.2.3   NLOS Issues

When signals propagate through the channel, an NLOS effect may emerge if the LOS signal component is obstructed by objects across the direct path between the transmitter and the receiver. NLOS propagation is often encountered in harsh indoor and urban environments. For time-based localization, NLOS propagation induces extra delay to the TOA measurement, which results in location estimation errors.

In the presence of NLOS, the TOA measurement relationship (8.11) needs to be modified accordingly. Suppose that links between the target and anchor nodes with indices $i = 1, ..., M_1$ experience LOS propagation, and the other $M_2 = M - M_1$ links undergo NLOS propagation. Then, the TOA measurement $\hat{\tau}_i$ can be expressed as

$$\hat{\tau}_i = \begin{cases} c\sqrt{(x_i - x)^2 + (y_i - y)^2} + \eta_i, & i = 1, ..., M_1 \\ c\sqrt{(x_i - x)^2 + (y_i - y)^2} + \eta_i + n_i, i = M_1 + 1, ..., M \end{cases}. \qquad (8.13)$$

In (8.13), η_i is the measurement noise and n_i is the positive NLOS error of the TOA estimate.

For time-based localization, Cramer–Rao bound (CRB) analysis indicates that if the statistical information of NLOS errors is unknown, then only the LOS TOA estimates contribute to the location estimation of the target; if the knowledge of $\{n_i\}_{M_1+1}^{M_2}$ is available, then both the LOS and NLOS TOA estimates contribute to thelocation estimation [20]. These results suggest that when the statistical information of NLOS error is unknown, the NLOS TOA estimates need to be identified and discarded before position estimation. However, with the knowledge of NLOS statistics, the NLOS TOA estimates also provide useful information for position estimation. Based on these, various NLOS identification and mitigation algorithms have been developed in the literature. For further details on NLOS-related techniques, please refer to Part IV of this book and survey papers [21, 22].

8.3   TOA ESTIMATION FOR IR UWB

In this section, we will introduce TOA estimation techniques for IR UWB that have been reported in the literature (see Table 8.1). As shown in Figure 8.5, three typical TOA estimators will be covered in detail, including the ML TOA estimator, the energy detection-based TOA estimator, and the timing with dirty templates (TDT) technique. All techniques in Table 8.1 will be reviewed and compared at the end of this section.

8.3.1   System Model

Before the discussion of TOA estimation techniques, we will first briefly introduce the IR system model in the following. In an IR system, each information symbol is transmitted over a T_s period that consists of N_f frames. During each frame of duration T_f, a data modulated pulse $p(t)$ with duration $T_p \ll T_f$ is transmitted from the antenna. The transmitted signal is

$$v(t) = \sqrt{\varepsilon} \sum_{n=0}^{\infty} s(n) p_T(t - nT_s), \qquad (8.14)$$

where ε is the energy per pulse, $s(n)$ is the PAM information symbol, and $p_T(t)$ denotes the symbol-long waveform

$$p_T(t) = \sum_{n=0}^{N_f-1} c_{ds}(n) p(t - nT_f - c_{th}(n)T_c). \qquad (8.15)$$

In (8.15), $c_{th}(n)$ and $c_{ds}(n)$ are the TH and DS codes, respectively.

An example of the transmitted signal with TH and DS codes is shown in Figure 8.8. The TH and DS codes randomize the spectrum of the transmitted signal to avoid strong spectral spikes. In addition, they can be used to separate users in a multiaccess UWB system. For low-rate applications, the frame duration T_f and TH codes are usually set to avoid both interframe interference (IFI) and intersymbol interference (ISI).

The transmitted signal propagates through the multipath channel with impulse response $h(t) = \sum_{l=0}^{L-1} h_l \delta(t - \tau_l)$, with $h = [h_0, h_1, \ldots, h_{L-1}]^T$ and $\tau = [\tau_0, \tau_1, \ldots, \tau_{L-1}]^T$ being amplitudes and delays of the L multipath components, respectively. The received waveform is thus given by

$$r(t) = \sqrt{\varepsilon} \sum_{n=0}^{\infty} s(n) \sum_{l=0}^{L-1} h_l p_T(t - nT_s - \tau_l) + \eta(t), \qquad (8.16)$$

where $\eta(t)$ is the additive noise.

TABLE 8.1. TOA Estimation Algorithms for IR UWB

Reference	Method	Type	Comments
23	Estimate delays and amplitudes of all multipath components. Use the estimated delay of the first path as TOA estimate.	Maximum likelihood (ML)	Optimal estimation can be potentially achieved. Complexity can be reduced by decoupling the joint estimation to individual paths.
24	Estimate the TOA from Nyquist rate samples of the signal. Reduce the search range of each path delay by ignoring detected paths.	Generalized maximum likelihood (GML)	Reduce the complexity of the ML TOA estimator. Complexity is still high due to the Nyquist rate sampling of the received signal.
25	Estimate all taps of the sampled channel. Use the first tap delay as TOA estimate.	ML	The computational complexity of the ML estimator is reduced at the price of limited accuracy due to sub-Nyquist sampling.
28	Use the energy samples of the received signal. Estimate TOA with an adaptively selected threshold.	Energy detection (ED)	The complexity of the receiver is reduced with ED. The threshold is adaptively selected. Accuracy is limited by the sampling rate.
27	Threshold-based TOA estimator with matched filtering and squaring operations on the received signal	Threshold crossing	TOA estimation performance is limited by the sampling rate. Compared to the ED-based estimator, this TOA estimator is more effective in mitigating noise.
50	Estimate the TOA in two steps: coarse timing with low-rate energy samples and fine timing with higher-rate sampling.	ED, correlation	Low complexity is maintained by the low sampling rate in coarse timing. High timing accuracy is obtained by fine timing with higher-rate sampling.
29	Correlate adjacent symbol-long signal segments. Estimate TOA at the peak of the cross correlation.	Timing with dirty templates (TDT)	TOA estimation can be carried out in both data-aided (DA) and non-data-aided (NDA) modes. The required analog delay line can be avoided by the digital receiver with low-resolution analog-to-digital converters (ADCs). Timing ambiguity exists due to the plateau of objective function.
30	TOA estimation with a pair of orthogonal UWB pulses	TDT	NDA timing performance is improved with the increased complexity of transceivers.

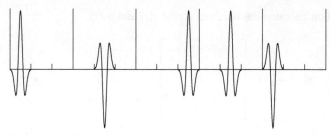

Figure 8.8   Transmitted signals for IR with TH and DS codes. In this example, the TH codes are {0, 1, 2, 1, 0} and the DS codes are {1, −1, 1, 1, −1}.

8.3.2   ML TOA Estimation

The ML algorithm estimates not only the TOA τ_0 but also the nuisance parameters $\{\tau_l\}_{l=1}^{l=L-1}$ and $\{h_l\}_{l=0}^{l=L-1}$. Suppose that the TH and DS codes have been synchronized and $s(n)$s are training symbols known at the receiver. The received signal segment captured in duration $[0, T]$ is used for TOA estimation. The ML criterion finds the estimates of the channel parameters $\hat{h} = \left[\hat{h}_0, \hat{h}_1, \ldots, \hat{h}_{L-1}\right]^T$ and $\hat{\tau} = \left[\hat{\tau}_0, \hat{\tau}_1, \ldots, \hat{\tau}_{L-1}\right]^T$ such that the squared error

$$E = \int_0^T |r(t) - \hat{r}(t)|^2 \, dt \tag{8.17}$$

is minimized between the received signal $r(t)$ and its reconstructed version,

$$\hat{r}(t) = \sqrt{\mathcal{E}} \sum_{n=0}^{\infty} s(n) \sum_{l=0}^{L-1} \hat{h}_l \, p_T(t - nT_s - \hat{\tau}_l), \tag{8.18}$$

where $p_T(t)$ is the symbol level transmit pulse shaper defined in (8.15).

Based on this criterion, the ML estimates of τ and h can be obtained as (see [23])

$$\hat{\tau} = \arg\max_{\bar{\tau}} \left\{ \chi^{\mathcal{H}}(\bar{\tau}) R^{-1}(\bar{\tau}) \chi(\bar{\tau}) \right\} \tag{8.19}$$

and

$$\hat{h} = R^{-1}(\hat{\tau}) \chi(\hat{\tau}), \tag{8.20}$$

where $R(\tau)$ is the matrix with each element being the correlation of $p_T(t)$ at different lags of τ:

$$R(\tau) = \int_0^T \begin{bmatrix} p_T(t - \tau_0) \\ p_T(t - \tau_1) \\ \vdots \\ p_T(t - \tau_{L-1}) \end{bmatrix} \left[p_T(t - \tau_0), p_T(t - \tau_1), \ldots, p_T(t - \tau_{L-1}) \right] dt, \tag{8.21}$$

and $\chi(\tau)$ is the correlation between the received signal $r(t)$ and $p_T(t)$:

$$\chi(\tau) = \int_0^T r(t) \begin{bmatrix} p_T(t-\tau_0) \\ p_T(t-\tau_1) \\ \vdots \\ p_T(t-\tau_{L-1}) \end{bmatrix} dt. \tag{8.22}$$

The ML estimator is too computationally intensive to be feasible due to the huge number of multipath components of the UWB channel. In [23], a simplified method is adopted by assuming that multipath components are resolvable so that the joint estimation of all parameters is decoupled to individual paths. In [24], the generalized maximum likelihood (GML) estimator reduces the searching region of path delays by ignoring already detected paths. The complexity of the ML criterion is further reduced if the estimation is based on the sampled channel [25].

8.3.3 Energy Detection-Based TOA Estimation

The energy detection-based TOA estimator adopts sub-Nyquist rate sampling of the received signal [26]. Compared to the ML TOA estimator, which usually requires Nyquist rate sampling, this estimator can considerably reduce the computational complexity. This estimator works for the low duty cycle modulation where no IFI is present in the received waveform. It also requires synchronized TH codes at the receiver to allow for TOA estimation.

Suppose that a signal segment in the duration $[0, T]$ is captured at the receiver, which contains the received waveform for a single transmitted pulse. If the received waveforms of multiple transmitted pulses are captured for a larger duration, samples from all received pulses are combined according to their TH codes to alleviate noise [26]. The signal segment is first passed through a band-pass filter to remove the out-of-band noise (see Fig. 8.9). The filtered waveform $r_{BP}(t)$ is processed by a square-law device. The output is sampled every T_{int} seconds by the integrate-and-dump operation to generate:

$$v_k = \int_{kT_{int}}^{(k+1)T_{int}} |r_{BP}(t)|^2 dt, \tag{8.23}$$

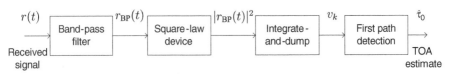

Figure 8.9 Diagram of the energy detection-based TOA estimator.

where $k = 0, \ldots, K - 1$ with $K = \lfloor T/T_{\text{int}} \rfloor$. The true TOA τ_0 is contained in the $N_{\tau_0} = \lfloor \tau_0/T_{\text{int}} \rfloor th$ sample, and samples v_k, $0 \leq k \leq N_{\tau_0} - 1$ only contain noise.

Sampling of the received signal can also be carried out by the matched filter receiver [27]. If the received UWB pulse $p_r(t)$ is known, the sampled signal sequence can be expressed as

$$v_k = \left(\int r_{BP}(t) p_r(t - kT_{\text{int}}) dt \right)^2.$$ (8.24)

In (8.24), the squaring operation is used to remove the random polarity of the signal. Compared to (8.23), the matched filter receiver is more effective in mitigating the noise contained in $r_{BP}(t)$ but less effective in collecting energy, especially when the sampling rate is low [19].

These samples of the signal are fed to the first path detection module. The simplest way is to estimate the TOA by picking out the sample that contains the strongest energy (see Fig. 8.10):

$$\hat{n}_p = \underset{k \in [0, \ldots, K-1]}{\arg \max} \{ v_k \}.$$ (8.25)

As a result, the TOA estimate is given by

$$\hat{\tau}_0 = \hat{n}_p T_{\text{int}} + \frac{1}{2} T_{\text{int}}.$$ (8.26)

TOA estimation by finding the peak as (8.25) is not accurate because the first path is not necessarily the strongest one. A more accurate method is the threshold-based estimator where the TOA estimate is determined by finding the first path that exceeds the threshold [26, 28]. The resolution of the energy detection-based TOA estimator is T_{int}, which resulted from the sub-Nyquist sampling. In order to improve resolution, Gezici and Poor [17] proposed a two-step TOA estimator wherein a rough TOA estimate is first obtained by the energy detection estimator, and then the rough estimate is refined with higher-rate sampling.

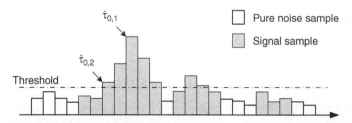

Figure 8.10   Illustration of the energy detection-based TOA estimator for IR. $\hat{\tau}_{0,1}$ and $\hat{\tau}_{0,2}$ are TOA estimates based on the detection of the sample with the strongest energy and the first crossing of the threshold.

8.3.4 TDT

The ML and energy detection-based TOA estimators require Nyquist or sub-Nyquist rate sampling. The TDT technique is a low-complexity TOA estimation algorithm that only relies on symbol rate sampling. TDT provides a rough TOA estimate that can be refined by estimators based on higher-rate sampling.

TDT was first proposed for IR with PAM. In this case, the received waveform can be expressed as (cf. (8.16)):

$$r(t) = \sqrt{\mathcal{E}} \sum_{n=0}^{+\infty} s(n) \, p_R(t - nT_s - \tau_0) + \eta(t), \tag{8.27}$$

where $p_R(t)$ is the received symbol level pulse shaper. TDT can accommodate IFI but requires ISI to be absent. Without loss of generality, the TH and DS codes are assumed to have a period of N_f, which is the number of frames in a symbol.

A pair of successive waveform segments is correlated to generate the symbol rate samples (see Fig. 8.11):

$$x(k;\tau) = \int_0^{T_s} r_{2k}(t;\tau) r_{2k-1}(t;\tau) dt, \tau \in [0, T_s), k = 1, 2, \ldots, \tag{8.28}$$

where the symbol-long segments are given by

$$r_k(t;\tau) = r(t + kT_s + \tau), t \in [0, T_s). \tag{8.29}$$

With $\eta(t)$ in (8.27) being the band-pass-filtered zero-mean additive white Gaussian noise (AWGN), it is proved in [29] that the symbol rate sample is Gaussian with the mean square of

$$E\{x^2(k;\tau)\} = \frac{1}{2}\mathcal{E}_R^2 + \frac{1}{2}\mathcal{E}_D^2(\tau) + \sigma_\zeta^2, \tag{8.30}$$

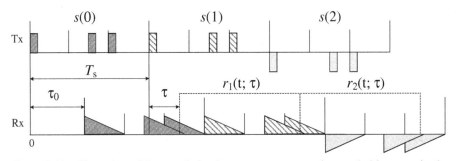

Figure 8.11 Illustration of the correlation between two consecutive symbol-long received segments $r_1(t; \tau)$ and $r_2(t; \tau)$ for TDT algorithms. Tx, transmitter; Rx, receiver.

where $\varepsilon_D(\tau) = \varepsilon_A(\tau) - \varepsilon_B(\tau)$ and ε_R is the constant energy of $p_R(t)$. $\varepsilon_D(\tau)$ reaches its maximum at the correct timing point where $\varepsilon_A(\tau) = 0$ and $\varepsilon_B(\tau) = \varepsilon_R$. This proves that, in the presence of the AWGN, the mean square of the symbol rate samples $x(k; \tau)$ is maximized when correct timing is achieved.

The ensemble mean (8.30) is replaced in practice by its sample mean obtained from K pairs of symbol-long received segments $K^{-1} \sum_{k=1}^{K} x^2(k; \tau)$. This results in the TDT algorithm that the TOA estimate can be obtained as

$$\hat{\tau}_{0,nda} = \underset{t \in [0, T_s)}{\arg\max} \left\{ y_{nda}(K; \tau) \right\}, \tag{8.31}$$

where $y_{nda}(K; \tau)$ is the objective function

$$y_{nda}(K; \tau) = \frac{1}{K} \sum_{k=1}^{K} \left(\int_0^{T_s} r_{2k}(t; \tau) r_{2k-1}(t; \tau) dt \right)^2, \tag{8.32}$$

which is the sample mean form of the ensemble expression in (8.30). This TOA estimator is non-data-aided (NDA) since no training sequence is required.

The number of samples K required for reliable TOA estimation can be reduced considerably in the data-aided (DA) mode with the training sequence $s(n)$ comprising a repeated pattern $[1, 1, -1, -1]$. Using this sequence, the objective function can still be formulated using (8.32). With the training sequence, orders of the averaging and squaring operations can be exchanged to obtain better estimates of the mean square of $x(k; \tau)$ in (8.30). Based on these, the following objective functions can be formed for the DA TDT:

$$y_{da1}(K; \tau) = \frac{1}{K} \sum_{k=1}^{K} \left(\int_0^{T_s} r_{2k}(t; \tau) r_{2k-1}(t; \tau) dt \right)^2,$$

$$y_{da2}(K; \tau) = \left(\frac{1}{K} \sum_{k=1}^{K} \int_0^{T_s} r_{2k}(t; \tau) r_{2k-1}(t; \tau) dt \right)^2, \tag{8.33}$$

$$y_{da3}(K; \tau) = \left(\int_0^{T_s} \overline{r}(t + \tau) \overline{r}(t + \tau - T_s) dt \right)^2,$$

where

$$\overline{r}(t) = \frac{1}{K} \sum_{k=1}^{K} (-1)^k r(t + 2kT_s).$$

The TOA estimate can be obtained with these objective functions, respectively, by

$$\hat{\tau}_{0,dai} = \underset{\tau \in [0, T_s)}{\arg\max} \left\{ y_{dai}(K; \tau) \right\}, \quad i = 1, 2, 3. \tag{8.34}$$

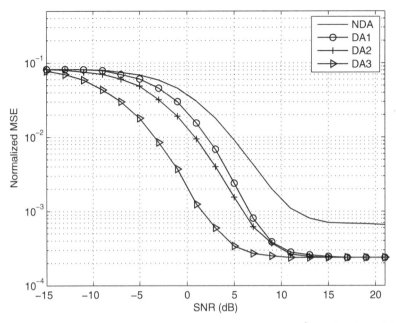

Figure 8.12   The MSE of the TOA estimate normalized by T_s^2 as a function of the SNR.

Analysis in [29] shows that the DA TDT outperforms the NDA TDT. In addition, even with the same training pattern, the DA estimators can have a different performance with slightly different objective functions. In Figure 8.12, the NDA and three DA TDT algorithms are simulated to obtain the mean square error (MSE) of the TOA estimate normalized by T_s^2. Simulation is performed in the IEEE 802.15.3a LOS channel with the transmission distance less than 4 m. The UWB pulse is the second derivative of the Gaussian function with unit energy and duration $T_p \approx 1$ ns. One symbol contains $N_f = 32$ frames each with duration $T_f = 32$ ns. A random TH code uniformly distributed over $[0, N_c - 1]$ is adopted, with $N_c = 35$ and $T_c = 1$ ns. Frame-level coarse timing is performed; that is, $N_i = N_f$. $K = 8$ symbol rate samples are used to obtain one TOA estimate. Simulation resolutions show that DA TDTs outperform the NDA TDT. Among the three DA TDT algorithms, the one based on $y_{da3}(K; \tau)$ in (8.33) achieves the best performance.

8.3.5   Discussions on IR-Based TOA Estimation

The ML TOA estimator [23] estimates the amplitudes and delays of all multipath components. The estimate of the first path delay τ_0 automatically becomes the TOA estimate. It is the optimal one when no channel statistical information is known at the receiver. When channel statistics are known, the Bayesian estimator provides the optimal estimate. Complexity of these optimal estimators is too high because a large number of unknown parameters need to be jointly estimated. A simplified version of the ML TOA estimator [23] is the GML TOA estimator [24]. The GML estimator searches the direct path in a narrowed range by ignoring the already detected paths so that the complexity of TOA estimation is reduced. The disadvantage of the GML estimator is that the required Nyquist rate sampling is still unaffordable for many UWB applications.

In [25], the ML TOA estimator is developed based on the discrete time sampled channel. The TOA estimation process is simplified because each tap delay can only come from a finite number of values. In [28], the low-complexity threshold-based TOA estimator is investigated based on the energy detection receiver structure. The threshold is adaptively selected according to the minimum and maximum of the energy samples of the received signal. An alternative of [28] is [27], where the received signal is sampled via matched filtering. Then, the direct path is detected when the output signal crosses the threshold. For detailed discussions on various threshold selection techniques, please refer to [19] and [26].

Performance of these low-complexity TOA estimators ([25, 27, 28]) is essentially limited by the sampling rate of the receiver. In order to simultaneously maintain low complexity and high timing accuracy, Gezici and Poor [17] proposed a two-step TOA estimation procedure. In the coarse timing step, a rough estimate of the TOA is obtained based on the energy detection scheme with low-rate sampling of the received signal. In the fine-timing step, the TOA estimate is refined with a higher-rate sampling of signals in a small ambiguity region.

The TDT is a low-complexity TOA estimation algorithm that relies on the symbol rate samples of the received signal [29, 31]. At the receiver, adjacent symbol-long signal segments serve as the correlation template to each other. Then, the TOA estimate is determined as the instant when the correlation reaches its maximum. The advantage is that timing estimation can be achieved without any knowledge of the CIR. These estimators are suitable for obtaining coarse timing, which can be further refined by applying fine TOA estimation techniques in a much reduced ambiguity region.

The application of TDT is not limited to the single PAM UWB with analog receivers. It has been proved in [31] that TDT is operational for IR with PPM. In [32], TDT is applied to the digital IR receiver with extremely low-resolution analog-to-digital converters (ADCs). In [30], the TDT TOA estimator is established for IR with the PSM. By adopting a pair of orthogonal pulses, the timing performance can be improved even if no training symbols are transmitted.

8.4 TOA ESTIMATION FOR MB-OFDM UWB

In this section, various TOA estimation techniques will be reviewed for the MB-OFDM UWB system (see Table 8.2). As shown in Figure 8.5, three typical methods will be covered in detail, including the correlation-based TOA estimator, energy detection-based TOA estimation, and TOA estimation by mitigating energy leakage. All techniques in Table 8.2 will be introduced and compared at the end of this section.

TABLE 8.2. TOA Estimation Algorithms for MB-OFDM UWB

Reference	Method	Type	Precision	Comments
35	Calculate the correlation between two identical half-symbol segments. TOA is determined by peak picking the cross correlation.	Correlation	Coarse	The correlation plateau causes timing ambiguity in the ISI free part of the CP.
34	Correlation of multiple symbols with optimal training sequence	Correlation	Coarse	The sharper correlation function enables better timing performance.
36 and 45	Correlation based on CP	Correlation	Coarse	Performance degrades when the channel becomes more dispersive.
46 and 51	Estimate TOA by peak picking the correlation between the received signal and the transmitted training signal.	Matched filtering	Coarse	TOA error arises when the strongest path is not the direct path.
46	Detect TOA when the matched filter output exceeds the threshold.	Matched filtering	Fine	Performance depends on the selection of threshold.
41 and 52	Estimate TOA by finding the maximum energy of L consecutive taps of the sample spaced channel.	Energy detection	Fine	Prior knowledge of the sample spaced channel length L is required at the receiver.
48	Estimate CIR. Use the estimated delay of the first path as the TOA estimate.	Modified GML	Fine	Computational complexity is reduced by narrowing down the search range of each path.
39 and 42	Estimate the TOA by finding the tap prior to which energy leakage is minimized.	Energy leakage minimization	Fine	Computational complexity is low with equally sampled channel. Mistiming is mitigated by minimizing the energy leakage effect.

8.4.1 System Model

MB-OFDM combines the basic OFDM technique with the frequency-hopping technique. Within each symbol duration, a signal block is transmitted in two steps. First, the baseband signal is generated in the same manner as the traditional single-band OFDM system. Then, the baseband signal is upconverted to the center frequency of a sub-band according to some predefined TF hopping pattern and is transmitted from the antenna. In the ECMA MB-OFDM UWB, each user can access up to three sub-bands, each having a bandwidth of 528 MHz [4].

Figure 8.13 shows the transmitter and receiver diagrams of the MB-OFDM system. For an MB-OFDM system with B frequency sub-bands, a block of information symbols $s_b = [s_{b,1}, \ldots, s_{b,K}]^T$ is multicarrier modulated to the bth frequency band, $b \in [1, B]$ on K orthogonal digital subcarriers to form the signal block

$$x_b = F^{\mathcal{H}} s_b, \tag{8.35}$$

where F is the $K \times K$ discrete Fourier transform (DFT) matrix. A cyclic prefix (CP) is added to each block to mitigate the interblock interference (IBI) induced by the multipath channel. After digital-to-analog conversion (DAC), the signal is carrier modulated and transmitted from the antenna. The transmitted signal then propagates through the channel.

At the receiver, the arriving waveform is carrier demodulated, sampled, and digitalized by the ADC to obtain baseband discrete time samples. After the symbol

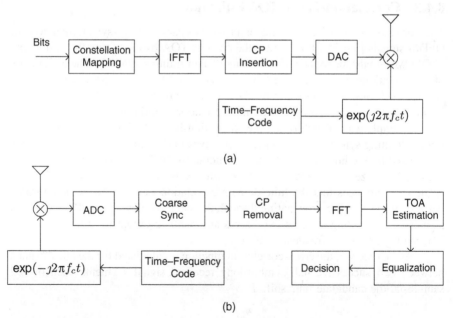

(a)

(b)

Figure 8.13   Illustration of the modulation and demodulation of MB-OFDM signals. IFFT, inverse fast Fourier transform.

level coarse timing, the CP is removed and the baseband signal is multicarrier demodulated with fast Fourier transform (FFT) to generate frequency domain signals $\{v_{b,k}\}_{k=1}^{K}$. It can be readily shown that

$$v_{b,k} = s_{b,k} H_{b,k} + \xi_{b,k}, \quad k = 1, \ldots, K, b = 1, \ldots, B, \qquad (8.36)$$

where $\xi_{b,k}$ is the noise and $H_{b,k}$ is the Fourier transform (FT) coefficient of the CIR:

$$H_{b,k} = \sum_{l=0}^{L-1} h_l \cdot \exp(-j\omega_{b,k}\tau_l), \qquad (8.37)$$

where $\omega_{b,k}$ is the frequency of the kth subcarrier on the bth frequency band.

Based on (8.36), the estimate of the frequency domain channel $H_{b,k}$ can be easily formed as

$$\hat{H}_{b,k} = \frac{v_{b,k}}{s_{b,k}} = H_{b,k} + \eta_{b,k}, \quad k = 1, \ldots, K, b = 1, \ldots, B, \qquad (8.38)$$

where $\eta_{b,k} = \xi_{b,k}/s_{b,k}$ is the noise term on the (b, k)th subcarrier (see [33]). For the channel estimation-based TOA estimator, the CIR will be reconstructed from $H_{b,k}$ by various algorithms based on the basic idea of the inverse Fourier transform.

8.4.2   Correlation-Based TOA Estimator

The correlation-based TOA estimator provides coarse timing synchronization for OFDM signals [34, 35]. It estimates the channel TOA by exploiting the periodicity of OFDM signals due to the transmission of CP or training symbols. The complexity of this technique is low because CIR does not need to be reconstructed. After the coarse synchronization provided by these algorithms, accuracy of the rough TOA estimate can be improved by fine timing estimation algorithms.

A simple way of performing the correlation-based TOA estimation is to transmit the training symbol with two identical halves in the time domain [35]. This can be realized by modulating an $N/2$-long pseudonoise (PN) sequence on even subcarriers and $N/2$ zeros on odd subcarriers, with N being an even integer. The PN sequence guarantees that the time domain correlation function of the half symbol has a sharp peak. After propagating through the same wireless channel, the received signals corresponding to these two halves are identical except for a phase rotation induced by the carrier frequency offset.

At the receiver, the cross-correlation function is calculated for the bth sub-band between two successive half-symbol-long received signal segments each of $N/2$ samples at the candidate time shift d:

$$P_{b,d} = \sum_{n=0}^{\frac{N}{2}-1} r_{b,d+n}^* r_{b,d+n+\frac{N}{2}}, \qquad (8.39)$$

where $r_{b,n}$ is the nth sample of the received signal in the bth sub-band. The timing metric is then defined as

$$M_{b,d} = \frac{|P_{b,d}|^2}{(R_{b,d})^2},$$
(8.40)

where $R_{b,d}$ is the received energy for the second half symbol involved in the TOA estimation

$$R_{b,d} = \sum_{n=0}^{\frac{N}{2}-1} \left| r_{d+n+\frac{N}{2}} \right|^2.$$
(8.41)

This timing metric shows a plateau with the width being the length of the CP minus the length of the CIR. Any candidate time shift selected inside the plateau can be used as the rough TOA estimate, which will induce no IBI and therefore cause no degradation of the symbol error rate (SER) performance.

Based on the correlation obtained for each sub-band, the TOA can be estimated at the kth sample instant:

$$k = \arg\max_d \sum_{b=1}^{B} P_{b,d}.$$
(8.42)

Timing accuracy of the basic correlation-based TOA estimator can be improved using multiple training symbols so that the timing metric will show a sharp trajectory. In [36], the periodicity due to CP is also exploited for TOA estimation to reduce the overhead by transmitting training symbols.

8.4.3   Energy Detection-Based TOA Estimator

After coarse timing synchronization, the TOA estimation accuracy can be improved by fine timing algorithms that require reconstruction of the CIR. Basically, the same fine-timing algorithms for IR can be applied to MB-OFDM with slight modifications, after CIR is reconstructed. For example, the ML criterion can be used to jointly estimate the delays and amplitudes of all multipaths as well as other nuisance parameters. Then, the delay estimate of the first path becomes the TOA estimate [37]. For the purpose of low-complexity estimation, the CIR can be sampled as the equally spaced sequence in a similar manner as the energy detection-based TOA estimator and the matched filtering TOA estimator for IR.

Let $r_{b,n}$ denote the baseband received signal from the bth sub-band and $\{s_{b,n}\}_{n=0}^{N-1}$ the transmitted synchronization sequence. The sampled channel is obtained by the matched filtering operation [38]

$$h_{b,d} = \sum_{n=0}^{N-1} r_{b,d+n}^* s_{b,n}.$$
(8.43)

Sampling of the channel can also be carried out by fitting the frequency domain channel information to an equally spaced sequence [39]. In order to distinguish the estimated channel from the physical channel, in the rest of this section, the term "tap" is used for samples in the estimated CIR and "path" for arriving rays in the original physical channel.

In [34], fine timing is achieved by finding the first segment of K samples with energy exceeding a threshold. Both K and the threshold depend on the channel environment. A similar method exists for IR, which is known as the p–max technique [26]. In [40], a joint channel order and TOA estimation technique based on the Akaike information, which is proposed for single-band OFDM, can be readily extended to the multiband case. Next, the energy detection-based TOA estimator will be presented as an example to show how the single-band algorithms can be extended to the multiband system.

For the energy detection-based TOA estimator, it is assumed that the CIR has been sampled into discrete taps for all B sub-bands. L taps of the sampled channel contain channel information. The nth tap of the sampled channel for the bth sub-band can be expressed as follows:

$$\bar{h}_{b,n} = \begin{cases} h_{b,n-L_1} + \eta_{b,n}, n \in [L_1, L_1 + L - 1] \\ \eta_{b,n}, n \in [0, L_1 - 1] \quad \text{and} \quad n \in [L_1 + L, L_1 + L + L_2 - 1] \end{cases}, \quad b \in [1, B], \quad (8.44)$$

where $h_{b,n}$ are noise-free samples of the CIR; L_1 and L_2 represent the ambiguity of coarse timing.

After the channel is sampled, the energy detection criterion is adopted for TOA estimation. In particular, for a single band, this estimator detects the starting point of the channel by seeking the maximum total energy of a length L segment in the channel estimate sequence given by (8.44). With the availability of multiple sub-bands, this TOA estimator simply combines the energy from all sub-bands via noncoherent combining. Then, the index of the first channel tap can be estimated as

$$\bar{k} = \arg\max_p \left(\sum_{b=1}^{B} \sum_{n=p}^{p+L-1} |\bar{h}_{b,n}|^2 \right). \quad (8.45)$$

Figure 8.14 shows the probability of mistiming curves for the energy detection-based TOA estimator. In the simulation, the sampled channel has $L = 12$ independent Nakagami-m distributed channel taps and $L_1 = L_2 = 5$ pure noise terms. The channel has an exponentially decaying PDP, with the last tap being 20 dB weaker than the first tap. It is proved in [41] that the energy detection-based TOA estimation can achieve a diversity gain of mB. This implies that by the noncoherent combining, the energy detection-based TOA algorithm can achieve a higher diversity gain, which is proportional to the number of sub-bands and the channel diversity.

Example 8.2

Use the codes in Chapter_8_Example_2.m to simulate the mistiming probability for the energy detection-based TOA estimator. MATLAB codes can be found online at ftp://ftp.wiley.com/public/sci_tech_med/matlab_codes. The channel has independent

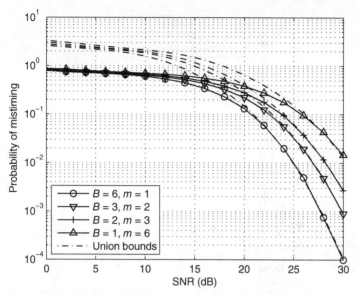

Figure 8.14   Probability of mistiming as a function of SNR for the energy detection-based TOA estimator.

Nakagami-m distributed samples. Try different values of the Nakagami-m parameter and the number of sub-bands. Confirm that the same diversity gain can be acquired for a fixed mB.

Solution

The channel has taps with Nakagami-m distributed envelopes and uniformly distributed phases. Note that Nakagami-m random variables are generated by MATLAB function gamrnd() by using the relationship between the Nakagami-m distribution and the gamma distribution. In particular, if the distribution of random variable X is Nakagami(m, Ω) with $m = E^2[X^2]/\text{Var}[X^2]$ and $\Omega = E[X^2]$, $Y = X^2$ can be generated by the distribution $Y \sim \text{gamma}(m, \Omega/m)$.

8.4.4   TOA Estimation by Suppressing Energy Leakage

Channel estimation by sampling the CIR results in the energy leakage phenomenon that has been ignored in many TOA estimation algorithms for OFDM and MB-OFDM. Energy leakage means that the energy of one channel path disperses into all taps in the sampled CIR when this path is missampled (see Fig. 8.15). The counterpart of energy leakage in IR is the interpath interference (IPI). Energy leakage needs to be mitigated because it induces TOA estimation error. Based on this, a TOA estimation criterion that simultaneously suppresses the energy leakage is proposed in [39] and [42].

The channel can be estimated by sampling the CIR at intervals of T_p by $\bar{L} = \lceil T_h/T_p \rceil$ taps, each with amplitude and delay $\bar{h}_n$ and $\bar{\tau}_n = \bar{\tau}_0 + nT_p$, $0 \le n \le \bar{L}-1$, respectively, with T_h being the maximum channel delay spread. For every possible

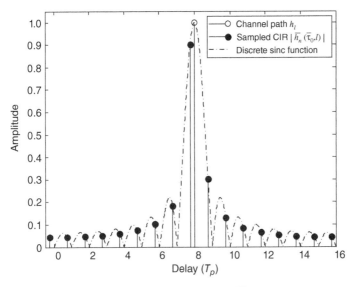

Figure 8.15 Illustration of the energy leakage $|\bar{h}(\bar{\tau}_0, l|$ from the lth channel path when missampling occurs.

first tap delay $\bar{\tau}_0$, the sampling is performed by assuming that the tap amplitudes $\bar{h}_l(\bar{\tau}_1)$ satisfy (cf. (8.37))

$$\sum_{n=0}^{\bar{L}-1} \bar{h}_n(\bar{\tau}_0)\exp(-j\omega_{b,k}\bar{\tau}_n) = \hat{H}_{b,k}, \quad k=1,\ldots,K, b=1,\ldots,B, \qquad (8.46)$$

where the dependence of $\bar{h}_n s$ on $\bar{\tau}_0$ is explicitly shown. The tap interval T_p is set as the inverse of the signal bandwidth, which is known as the time domain resolution of the system [43]. When T_p is smaller than the system resolution, the sampling equation tends to be ill-conditioned and unsolvable [39].

When training data are constant modulus over subcarriers, channel estimation of (8.46) is equivalent to the time domain sampling based on matched filtering (see (8.43)). The difference is that (8.46) can adopt the coherent combining of multiband channel information that does not directly apply to the time domain matched filtering. Besides, (8.46) is a more flexible channel estimator due to the free parameter $\bar{\tau}_0$.

Let $\bar{h}_n(\bar{\tau}_0, l)$ denote the contribution from the lth channel path at the nth tap so that $\bar{h}_n(\bar{\tau}_0) = \sum_{l=0}^{L-1} \bar{h}_n(\bar{\tau}_0, l) + \bar{\eta}_n(\bar{\tau}_0)$, with $\bar{\eta}_n(\bar{\tau}_0)$ being the noise. It turns out that

$$\bar{h}_n(\bar{\tau}_0, l) = h_l \exp\left(\frac{j\pi(N-1)(\bar{\tau}_n - \tau_l)}{NT_p}\right) \frac{\sin\left(\pi(\bar{\tau}_n - \tau_l)/T_p\right)}{N\sin\left(\pi(\bar{\tau}_n - \tau_l)/(NT_p)\right)} \qquad (8.47)$$

has the form of the discrete sinc function $\sin(\pi t/T_p)/(N\sin(\pi t/(NT_p)))$ sampled at $t = (\bar{\tau}_n - \tau_l)$, $n \in [0, \bar{L}-1]$ (see Fig. 8.15). For multipath channels, energy leakage is

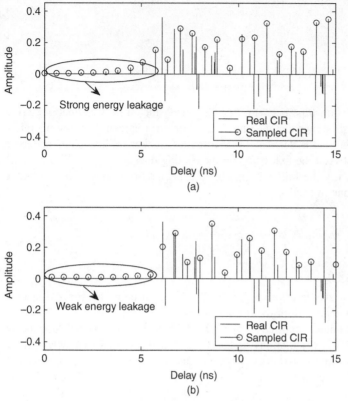

Figure 8.16   Illustration of the energy leakage prior to the first arriving path with two different values of the first tap delay $\bar{\tau}_0$.

always present because the equally spaced sequence cannot exactly sample all paths. In this case, the IPI due to energy leakage becomes the main source of error for TOA estimation.

TOA can be estimated when energy leakage prior to TOA is minimized in the sampled channel so that a sharp jump of amplitude shows up around the leading path (Fig. 8.16b). This is in contrast to Figure 8.16a, where strong energy leakage disperses before the channel TOA. The sharp jump of amplitude can be detected by searching the value of $\bar{\tau}_0$ that maximizes the following energy ratio between two adjacent taps of the sampled CIR:

$$\gamma_n(\bar{\tau}_0) = \frac{\left|\bar{h}_n(\bar{\tau}_0)\right|^2}{\left|\bar{h}_{n-1}(\bar{\tau}_0)\right|^2}, \quad n \in [L_1, L_2],\tag{8.48}$$

where $[L_1, L_2]$ represents the ambiguity region of the first channel path after coarse timing synchronization. The TOA estimate τ_0 is then obtained as the delay of the tap where (8.48) is maximized:

$$\hat{\tau}_0 = \hat{\bar{\tau}}_0 + \hat{n}T_p \tag{8.49}$$

with

$$\left(\hat{\bar{\tau}}_0, \hat{n}\right) = \underset{0 \le \bar{\tau}_0 < T_h, L_1 \le n \le L_2}{\arg\max} \gamma_n(\bar{\tau}_0). \tag{8.50}$$

In [39], the relationship between $\gamma_n(\bar{\tau}_0)$ and the TOA estimation error is analyzed. Analysis shows that the time estimation error is reduced when the interference from trailing paths decreases due to either weaker trailing paths or larger time separation between the first path and the trailing ones.

In [39] and [42], the following modified energy ratio is shown to be a better measure than the one in (8.48):

$$\gamma_n(\bar{\tau}_0, M) = \frac{\left|\bar{h}_n(\bar{\tau}_0)\right|^2}{\dfrac{1}{M} \displaystyle\sum_{i=n-1}^{n-M} \left|\bar{h}_i(\bar{\tau}_0)\right|^2}, \quad n \in [L_1, L_2]. \tag{8.51}$$

This modified form avoids the energy ratio $\gamma_n(\bar{\tau}_0)$ from being maximized at taps that contain strong energy from the trailing paths.

The TOA estimator is tested in all eight IEEE 802.15.4a channel models, as shown in Figure 8.17. Simulations are performed using the ECMA MB-OFDM system parameters with the first three sub-bands and $M = 5$. Performance in the LOS channel is generally better than that in the corresponding NLOS channel. This is

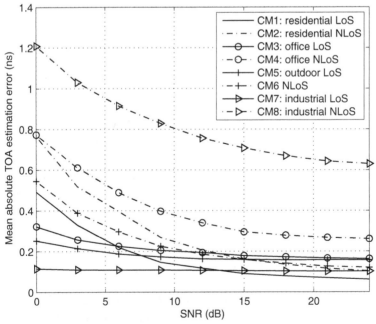

Figure 8.17 The mean absolute TOA estimation error as a function of the SNR.

because in the NLOS channel, the first path may not be strong and the estimator is more affected by the trailing paths. However, there is an exception for the outdoor environment at high signal-to-noise ratio (SNR) values. From their channel realizations, it can be found that channel paths in the NLOS outdoor channel are weaker but sparser than the LOS outdoor channel. As a result, the NLOS outdoor channel can be even better resolved than the LOS outdoor channel, and therefore the determination of its first path will be less interfered by the trailing paths.

8.4.5 Discussions on MB-OFDM-Based TOA Estimation

In [35], the timing estimation of OFDM signals is carried out by correlating the received signal of the transmitted training sequence with two identical halves. The TOA estimate is obtained at the time instant when the correlation coefficient reaches its maximum. Due to the protection of CP, the signal component of the correlation shows a plateau, which corresponds to the ISI-free part of CP. This implies that any timing synchronization point selected inside this range induces no ISI. This correlation-based method provides low-complexity coarse TOA estimation for single-band OFDM. For MB-OFDM, it can be readily implemented if the overall correlation is obtained by summing up sub-band correlations [44].

In [34], an enhanced version of [35] is presented, which uses multiple training OFDM symbols with their polarities determined by the training symbol pattern. If the training symbol pattern is properly chosen, the correlation has a sharp trajectory so that the resilience against noise is improved. Simulation results in [34] show that this estimator outperforms those in [35] at low SNR values. Timing performance of [34] is comparable to that of [35]. In [36] and [45], timing of OFDM signals is achieved by exploiting the redundancy introduced by CP. An advantage of this CP-based technique is high spectral efficiency. The drawback is that, since the CP part of the received symbol also contains ISI, performance of this estimator degrades drastically when the multipath channel becomes more dispersive.

The correlation-based TOA estimators are low complexity because the correlation function can be iteratively calculated. However, these estimators cannot provide fine TOA estimates due to the timing ambiguity induced by the correlation plateau. In order to obtain a fine TOA estimate, the estimator needs to resolve the multipath components from the received signal, which results in the channel estimation-based TOA estimator. For OFDM, channel estimation can be carried out by matched filtering [38, 46] or by converting the frequency domain channel information into the time domain [39, 47]. After channel estimation, the TOA estimate can be determined by detecting the leading edge.

In [38] and [46], matched filtering is adopted to sample the received signal with the transmitted training sequence. Output of the matched filter is the sample spaced discrete time channel. This is the same as the matched filtering TOA estimation for IR UWB (e.g., see [26]). Therefore, various first path detection algorithms developed for IR can be applied to MB-OFDM with slight modification (e.g., see [26] and [28]). Depending on the level of a priori knowledge, TOA can be estimated by several means, such as using the strongest sample of the sampled channel [46], finding the first significant path energy that exceeds a threshold, searching for the segment of L samples with the strongest total energy with L being the length of the sample spaced channel [41], and finding the strongest total energy of the segment weighted by the channel PDP [19]. Similar to the IR system, performance of these MB-OFDM TOA estimators is limited by the sampling rate. In addition, matched filtering TOA estimators are more computationally intensive than correlation-based TOA estimators.

An alternative means of estimating the CIR is to transform the frequency domain channel information to time domain after subcarrier demodulation. Due to the huge number of multipath components of UWB channels, the optimal ML and Bayesian channel estimators are too computationally intensive to be feasible. Therefore, research has mainly focused on the development of low-complexity channel estimators such as the GML and modified GML ones, which reduce the path searching region [24, 48], and the model-based estimator [42] by sampling the CIR with an equally spaced sequence. One major advantage of these estimators is that the multiband channel information can be coherently combined, which is not the case for the correlation-based and matched filtering estimators.

For coherent combining, the multiple sub-bands are treated as a single huge band, which allows the estimator to achieve the full resolution facilitated by the entire bandwidth [42, 49]. In comparison, the noncoherent combining for correlation-based and matched filtering estimators can only obtain the resolution facilitated by the sub-band bandwidth. Advantages of the noncoherent combining are the lower computational complexity than the coherent combining and no requirement for the estimation of the random phase rotation after carrier demodulation [39, 42].

8.5 CONCLUSIONS

In this chapter, we introduced localization and TOA estimation techniques for IR and MB-OFDM UWB systems. Background knowledge of UWB was reviewed at the beginning of this chapter. Due to their huge bandwith, UWB signals can provide orders of magnitude improvement on the timing resolution compared to conventional narrowband signals. For this reason, this chapter has focused on the time-based localizations for UWB. In particular, various TOA estimation algorithms were reviewed, including both coarse and fine-timing techniques.

Depending on the way of accessing the UWB bandwidth, UWB systems can be categorized into two types, namely, IR and the MB-OFDM UWB. For IR UWB, the ML criterion was first revisited, which is the optimal yet computationally demanding TOA estimation approach. In order to achieve low-complexity TOA estimation, two practical approaches were then introduced, including the energy detection-based TOA estimator and the TDT algorithms. These two approaches rely on sub-Nyquist rate and symbol rate sampling of the received signal, respectively. For the MB-OFDM UWB, the correlation-based TOA estimator was first reviewed, which provides coarse timing by exploiting the periodicity of OFDM signals. After coarse timing, the rough TOA estimate can be refined by the channel estimation-based

TOA estimator. To this end, two fine-timing techniques were introduced based on energy detection and suppression of energy leakage criteria, respectively.

Some issues important for UWB and also for general wireless localization, such as the NLOS mitigation and multiuser localization, were not discussed in this chapter. For details on these topics, please refer to corresponding chapters of this book.

REFERENCES

[1] Federal Communications Commission, "FCC First Report and Order: In the matter of revision of Part 15 of the Commission's rules regarding ultra-wideband transmission systems," FCC 02-48, Apr. 2002.

[2] European Commission, "Draft commission decision on the harmonised use of radio spectrum by equipment using ultra-wideband technology in the Community," RSCOM06-90 Final, Dec. 2006.

[3] Ministry of Internal Affairs and Communications, Japan, "Technical conditions on UWB radio systems in Japan," AWF3/14, 2006.

[4] ECMA International, "ECMA-368: High rate ultra wideband PHY and MAC standard," 1st ed., Dec. 2005.

[5] ECMA International, "ECMA-369: MAC-PHY interface for ECMA-368," 1st ed., Dec. 2006.

[6] IEEE 802.15.4 Working Group, "Part 15.4: Wireless medium access control (MAC) and physical layer (PHY) specifications for low-rate wireless personal area networks (WPANs), amendment 1: Add alternate PHYs," Approved: Mar. 2007, IEEE-SA Standards Board.

[7] A. A. M. Saleh and R. A. Valenzuela, "A statistical model for indoor multipath propagation," *IEEE J. Sel. Area. Comm.*, vol. 5, no. 2, pp. 128–137, 1987.

[8] A. F. Molisch, K. Balakrishnan, D. Cassioli, C.-C. Chong, S. Emami, A. Fort, J. Karedal, J. Kunisch, H. Schantz, U. Schuster, and K. Siwiak, "IEEE 802.15.4a channel model—Final report," Nov. 2004.

[9] A. F. Molisch, J. R. Foerster, and M. Pendergrass, "Channel models for ultrawideband personal area networks," *IEEE Wireless Commun.*, vol. 10, no. 6, pp. 14–21, 2003.

[10] R. Battiti, M. Brunato, and A. Villani, "Statistical learning theory for location fingerprinting in wireless LANs," Tech. Rep. DIT-02-0086, 2002.

[11] S. Gezici, H. Kobayashi, and H. Poor, "A new approach to mobile position tracking," in *Proc. of IEEE Sarnoff Symp. Advances in Wired and Wireless Communications*, 2003, pp. 204–207.

[12] M. McGuire, K. N. Plataniotis, and A. N. Venetsanopoulos, "Location of mobile terminals using time measurements and survey points," *IEEE Trans. Veh. Technol.*, vol. 52, no. 4, pp. 999–1011, 2003.

[13] C. Nerguizian, C. Despins, and S. Affès, "Geolocation in mines with an impulse response fingerprinting technique and neural networks," in *Proc. of IEEE 60th Vehicular Technology Conf. (VTC2004-Fall)*, vol. 5, Sep. 2004, pp. 3589–3594.

[14] C. Steiner and A. Wittneben, "Low complexity location fingerprinting with generalized UWB energy detection receivers," *IEEE Trans. Signal Process.*, vol. 58, no. 3, pp. 1756–1767, 2010.

[15] C. Steiner, F. Althaus, F. Troesch, and A. Wittneben, "Ultra-wideband geo-regioning: A novel clustering and localization technique," *EURASIP J. Adv. Signal Process.*, vol. 2008, 2008, Article ID 296937.

[16] W. Q. Malik and B. Allen, "Wireless sensor positioning with ultrawideband fingerprinting," in *Proc. of First European Conf. on Antennas and Propagation (EuCAP)*, 2006, pp. 1–5.

[17] S. Gezici and H. V. Poor, "Position estimation via ultra-wide-band signals," *Proc. IEEE*, vol. 97, no. 2, pp. 3864–3403, 2009.

[18] A. H. Sayed, A. Tarighat, and N. Khajehnouri, "Network-based wireless location," *IEEE Signal Process. Mag.*, vol. 22, no. 4, pp. 24–40, 2005.

[19] Z. Sahinoglu, S. Gezici, and I. Guvenc, *Ultra-Wideband Positioning Systems: Theoretical Limits, Ranging Algorithms, and Protocols*. New York: Cambridge University Press, 2008.

[20] Y. Qi, H. Kobayashi, and H. Suda, "Analysis of wireless geolocation in a non-line-of-sight environment," *IEEE Trans. Wireless Commun.*, vol. 5, no. 3, pp. 672–681, 2006.

[21] I. Guvenc, C.-C. Chong, and F. Watanabe, "NLOS identification and mitigation for UWB localization systems," in *Proc. of Wireless Communications and Networking Conf.*, 2007, pp. 1571–1576.

[22] J. Khodjaev, Y. Park, and A. S. Malik, "Survey of NLOS identification and error mitigation problems in UWB-based positioning algorithms for dense environments," *Ann. Telecommun.*, vol. 65, no. 5–6, pp. 301–311, 2010.

[23] M. Z. Win and R. A. Scholtz, "Characterization of ultra-wide bandwidth wireless indoor channels: A communication-theoretic view," *IEEE J. Sel. Area. Comm.*, vol. 20, no. 9, pp. 1613–1627, 2002.

[24] J. Y. Lee and R. A. Scholtz, "Ranging in a dense multipath environment using an UWB radio link," *IEEE J. Sel. Area. Comm.*, vol. 20, no. 9, pp. 1677–1683, 2002.

[25] Z. Tian and G. B. Giannakis, "Data-aided ML timing acquisition in ultra-wideband radios," in *Proc. of IEEE Conf. on Ultra-Wideband Systems and Technologies*, Reston, VA, Nov. 16–19, 2003, pp. 142–146.

[26] D. Dardari, A. Conti, U. Ferner, A. Giorgetti, and M. Z. Win, "Ranging with ultrawide bandwidth signals in multipath environments," *Proc. IEEE*, vol. 97, no. 2, pp. 404–426, 2009.

[27] D. Dardari, C.-C. Chong, and M. Z. Win, "Threshold-based time-of-arrival estimators in UWB dense multipath channels," *IEEE Trans. Commun.*, vol. 56, no. 8, pp. 1366–1378, 2008.

[28] I. Guvenc and Z. Sahinoglu, "Threshold-based TOA estimation for impulse radio UWB systems," in *IEEE Int. Conf. on Ultra-Wideband*, 2005, pp. 420–425.

[29] L. Yang and G. B. Giannakis, "Timing ultra-wideband signals with dirty templates," *IEEE Trans. Commun.*, vol. 53, no. 11, pp. 1952–1963, 2005.

[30] M. Ouertani, H. Xu, L. Yang, H. Besbes, and A. Bouallegue, "Orthogonal bi-pulse UWB: Timing and (de)modulation," *Elsevier Phys. Commun.*, vol. 1, no. 4, pp. 237–247, 2008.

[31] L. Yang, "Timing PPM-UWB signals in ad hoc multi-access," *IEEE J. Sel. Area. Comm.*, vol. 24, no. 4, pp. 794–800, 2006.

[32] H. Xu and L. Yang, "Timing with dirty templates for low-resolution digital UWB receivers," *IEEE Trans. Wireless Commun.*, vol. 7, no. 1, pp. 54–59, 2008.

[33] Z. Wang, G. Mathew, Y. Xin, and M. Tomisawa, "An iterative channel estimator for indoor wireless OFDM systems," in *Proc. of IEEE Int. Conf. on Communication Systems (ICCS)*, Singapore, Oct. 30–Nov. 1, 2006, pp. 1–5.

[34] H. Minn, V. K. Bhargava, and K. B. Letaief, "A robust timing and frequency synchronization for OFDM systems," *IEEE Trans. Wireless Commun.*, vol. 2, no. 4, pp. 799–807, 2003.

[35] T. M. Schmidl and D. C. Cox, "Robust frequency and timing synchronization for OFDM," *IEEE Trans. Commun.*, vol. 45, no. 12, pp. 1613–1621, 1997.

[36] B. Yang, K. B. Letaief, R. S. Cheng, and Z. Cao, "Timing recovery for OFDM transmission," *IEEE J. Sel. Area. Comm.*, vol. 18, pp. 2278–2291, 2000.

[37] B. H. Fleury, M. Tschudin, R. Heddergou, D. Dahlhaus, and K. I. Pedersen, "Channel parameters estimation in mobile radio environments using SAGE algorithm," *IEEE J. Sel. Area. Comm.*, vol. 17, no. 3, pp. 434–449, 1999.

[38] Z. Ye, C. Duan, P. Orlik, and J. Zhang, "A low-complexity synchronization design for MB-OFDM ultra-wideband systems," in *Proc. of Int. Conf. on Communications*, Beijing, China, May 19–23, 2008, pp. 3807–3813.

[39] H. Xu and L. Yang, "TOA estimation for MB-OFDM UWB by suppressing energy leakage," *Proc. of MILCOM Conf.*, San Jose, CA, Oct. 31–Nov. 3, 2010, pp. 718–723.

[40] E. G. Larsson, G. Liu, J. Li, and G. B. Giannakis, "Joint symbol timing and channel estimation for OFDM based WLAN," *IEEE Commun. Lett.*, vol. 5, no. 8, pp. 325–327, 2001.

[41] H. Xu, L. Yang, Y. T. Morton, and M. Miller, "Mistiming performance analysis of the energy detection based TOA estimator for MB-OFDM," *IEEE Trans. Wireless Commun.*, vol. 8, no. 8, pp. 3980–3984, 2009.

[42] H. Xu, C.-C. Chong, I. Guvenc, and L. Yang, "High-resolution TOA estimation with multi-band OFDM UWB signals," in *Proc. of Int. Conf. on Communications*, Beijing, China, May 19–23, 2008, pp. 4191–4196.

[43] O. Simeone, Y. Bar-Ness, and U. Spagnolini, "Pilot-based channel estimation for OFDM systems by tracking the delay subspace," *IEEE Trans. Wireless Commun.*, vol. 3, no. 1, pp. 315–325, 2004.

[44] Y. Li, H. Minn, and R. M. A. Rajatheva, "Synchronization, channel estimation, and equalization in MB-OFDM systems," *IEEE Trans. Wireless Commun.*, vol. 7, no. 11, pp. 4341–4352, 2008.

[45] J.-J. van de Beek, M. Sandell, and P. O. Borjesson, "ML estimation of time and frequency offset in OFDM systems," *IEEE Trans. Signal Process.*, vol. 45, no. 7, pp. 1800–1805, 1997.

[46] C. W. Yak, Z. Lei, S. Chattong, and T. T. Tjhung, "Timing synchronization for ultra-wideband (UWB) multi-band OFDM systems," in *Proc. of Vehicular Technology Conf.*, vol. 3, Dallas, TX, Sep. 25–28, 2005, pp. 1599–1603.

[47] S. Zhang and J. Zhu, "SAGE based channel estimation and delay tracking scheme in OFDM systems," in *Proc. of Vehicular Technology Conf.*, Stockholm, Sweden, May 30–Jun. 1, 2005, pp. 788–791.

[48] C.-D. Wann and S.-W. Yang, "Modified GML algorithm for estimation of signal arrival time in UWB systems," *presented at Proc. of Global Telecommunications Conf.*, San Francisco, CA, Nov. 27–Dec. 1, 2006.

[49] E. Saberinia and A. H. Tewfik, "Ranging in multiband ultrawideband communication systems," *IEEE Trans. Veh. Technol.*, vol. 57, no. 4, pp. 2523–2530, 2008.

[50] S. Gezici, Z. Sahinoglu, A. F. Molisch, H. Kobayashi, and H. V. Poor, "Two-step time of arrival estimation for pulse-based ultra-wideband systems," *EURASIP J. Adv. Signal Process.*, vol. 2008, Article ID 529134.

[51] R. A. Saeed, S. Khatun, B. M. Ali, and M. A. Khazani, "Performance of ultra-wideband time-of-arrival estimation enhanced with synchronization scheme," *ECTI Trans. Electr. Eng. Eletron Commun.*, vol. 4, no. 1, pp. 78–84, 2006.

[52] C. R. Berger, S. Zhou, Z. Tian, and P. Willett, "Performance analysis on an MAP fine timing algorithm in UWB multiband OFDM," *IEEE Trans. Commun.*, vol. 56, no. 10, pp. 1606–1611, 2008.

CHAPTER 9

AN INTRODUCTION TO DIRECTION-OF-ARRIVAL ESTIMATION TECHNIQUES

S. A. (Reza) Zekavat
Michigan Technological University, Houghton, MI

T HIS CHAPTER introduces the fundamentals of direction-of-arrival (DOA) techniques for localization systems. The main goal is the introduction of DOA methods via antenna arrays. In order to better introduce the principles of DOA, first, we briefly introduce antenna systems and antenna arrays. The reader will learn important parameters of an antenna and antenna arrays and the notion of beamforming (BF). The chapter introduces different measures for comparing DOA estimation methods and examples of DOA estimation techniques, and discusses their pros and cons. Moreover, a new DOA estimation method that improves the performance and complexity for implementation on software-defined radios is introduced. Finally, we discuss a DOA estimation method that is suitable for periodic sense transmissions in localization techniques such as wireless local positioning systems (WLPSs) (see Chapters 1 and 34).

9.1 INTRODUCTION

DOA estimation techniques have applications in BF, detection, and localization. In general, DOA estimation techniques can be divided into two categories: (1) real-time, those that possess lower complexity and can be easily implemented for real-time applications, and (2) non-real-time, those that are very complex and can hardly be implemented for real-time applications. Beamforming and detection processes need real-time DOA estimation, while achieving high precision is not vital: Coarse

Handbook of Position Location: Theory, Practice, and Advances, Second Edition.
Edited by S. A. (Reza) Zekavat and R. Michael Buehrer.
© 2019 by the Institute of Electrical and Electronics Engineers, Inc.
Published 2019 by John Wiley & Sons, Inc.
Companion Website: www.wiley.com/go/zekavat/positionlocation2e

DOA estimation is sufficient for this process. Localization applications require high-performance DOA estimation; however, real-time DOA estimation is not essential in many localization applications.

Compared to time-of-arrival (TOA) estimation techniques, DOA estimation needs the implementation of antenna arrays. Each antenna element maintains connection through a radio frequency (RF) component. RF components are one of the expensive components of radio systems. In addition, the power consumption of these components is relatively high. Thus, it is expected that, compared to TOA estimation techniques, DOA estimation needs higher complexity and power consumption. However, as discussed in Chapter 1, in DOA estimation techniques, only two base nodes (equipped with antenna arrays) are sufficient to maintain full localization of a target node. This adds a higher flexibility to DOA estimation techniques compared with TOA or received signal strength indicator (RSSI) estimation methods.

In addition, TOA and RSSI estimation methods are suitable methods for homogeneous mediums with known specifications. Air and space are examples of those mediums. However, TOA and RSSI methods may need major modification (see Chapter 10 with a focus on TOA method) for inhomogeneous mediums such as water. The temperature and the percentage of salt in different layers of water are different. Accordingly, the speed of propagation of signals would be different. This creates error in TOA estimation. In addition, the loss of different layers of water is not the same. Accordingly, RSSI methods are involved with considerable error in inhomogeneous mediums. DOA estimation is also affected by the nonhomogeneity of environments; however, the impact is milder specifically if the DOA is getting closer to the antenna boresight. This makes DOA-based localization mildly better for inhomogeneous media, such as the human body and the localization of sensors (e.g., endoscopy capsules) inside the body.

Section 9.2 offers an overview on antennas. This chapter reviews DOA estimation techniques that are based on antenna arrays; thus, Section 9.3 introduces antenna arrays. Driven by the demand, many DOA estimation techniques have been proposed [1, 2]; examples are spatial spectral estimation methods, such as delay and sum (DAS) [3], and eigenstructure methods, such as multiple signal classification (MUSIC) [4], root MUSIC [5], and estimation of signal parameters via rotational invariance technique (ESPRIT) [6]. DOA estimation is possible via multiple antennas (separated by half wavelength) installed at the receiver. Section 9.4 discusses the details of many important DOA estimation techniques. It has been depicted that performance of DOA estimation in many systems such as WLPSs that use signals wherein periodic nature is low. Section 9.5 discusses this problem and introduces a solution to this problem. Section 9.6 concludes the chapter.

9.2   ANTENNAS AND THEIR PARAMETERS

This section briefly introduces antenna systems. Antenna is the key component in the transmission and reception of electromagnetic (EM) waves. Antennas radiate EM waves because of time-varying electric fields created by a time-varying signal (e.g., sinusoidal waveform). Antennas come in all shapes and sizes but are essentially metallic structures used for the radiation and reception of radio waves. At high frequencies, even a short wire can act as an antenna. Antennas are divided into two main categories: (1) directional antennas and (2) omnidirectional antennas.

Directional antennas propagate energy only in some specific directions. Examples are dish antennas that are used for satellite communication or in radars. Omnidirectional antennas emit the EM energy in all directions. Practically, it is hard to build an omnidirectional antenna; however, some antennas, such as dipole (or monopole) antennas, are considered omnidirectional at least in one plane such as horizon or azimuth angle. However, these antennas may not propagate energy in the elevation angle. These antennas are used in applications such as broadcasting; for example, by TV or radio stations. In general, antennas possess many important parameters, which include the following:

1. antenna beam pattern,
2. antenna half-power beamwidth (HPBW),
3. main lobe power to the first side lobe power ratio,
4. main lobe power to non-main lobe power (all side lobe power) ratio,
5. antenna impedance,
6. antenna return loss,
7. antenna bandwidth,
8. antenna gain, and
9. antenna polarization.

The *antenna beam pattern* represents the variation of the power or amplitude of the signal with angle. Here, angles are represented in terms of azimuth angle and elevation angle. The azimuth angle is the angle that is defined by setting a reference direction (e.g., north) in the horizontal plane (e.g., a plane parallel to the earth), and the elevation angle is the angle that is defined with respect to the line perpendicular to this horizontal plane. These angles are shown in Figure 9.1. The antenna beam pattern is generally a function of the antenna shape, dimension, and frequency.

As shown in Figure 9.2, in general, the beam pattern consists of a main lobe, some side lobes, and some nulls. It is usually desired to transmit the signal in the direction of the main lobe. Therefore, side lobes are not desirable and should be

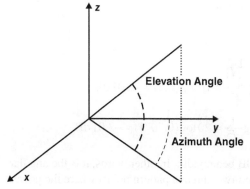

Figure 9.1   Azimuth and elevation angles.

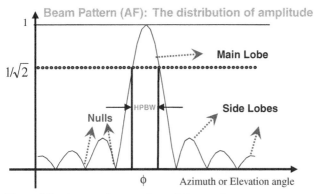

Figure 9.2   Antenna beam pattern. AF, array factor.

mitigated. Figure 9.2 shows the beam pattern of a directional antenna. The beam pattern of an ideal omnidirectional antenna is represented by a sphere.

9.2.1   Antenna HPBW

Considering that the y-axis in Figure 9.1 is the signal amplitude normalized to the main lobe maximum value, the HPBW can be found if we cross the $1/\sqrt{2}$ line and the beam pattern. The term "half power" comes from the fact that the output power has fallen to half of its midband level. This can be understood by examining the equations below:

$$\text{Power ratio (dB)} = 100\log\frac{P_2}{P_1},$$

where P_2 is the output power and P_1 is the input power, and

$$\text{Voltage ratio (dB)} = 20\log\frac{V_2}{V_1},$$

where V_2 is the output voltage and V_1 is the input voltage.

When $P_2 = \frac{1}{2}P_1$,

$$\text{Power ratio (dB)} = 10\log\frac{\frac{1}{2}P_1}{P_1} = 10\log(0.5) = -3.01\,\text{dB}$$

$$\text{Voltage ratio (dB)} = 20\log\frac{1}{\sqrt{2}} = 20\log(0.7071) = -3.01\,\text{dB}.$$

As a result, HPBW is also called a 3-dB beamwidth. In other words, it is the angular distance (azimuth or elevation) between two antenna pattern points where the power

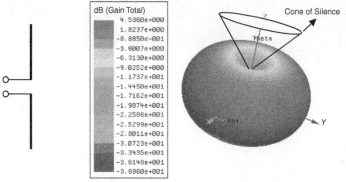

Figure 9.3   Dipole antenna structure and its beam pattern. See color insert.

becomes half of its maximum value. The elevation angle is measured with respect to its projection on the *xy*-plane.

The structure of a dipole antenna and its 3-D beam pattern are shown in Figure 9.3. As seen in this figure, its pattern is not fully isotropic or omnidirectional. Typically, the power at the top and bottom of the antenna is almost zero. This area is called the cone of silence.

9.2.2   First Side Lobe to the Main Lobe Power Ratio

This parameter refers to the maximum power of the largest (usually first) side lobe to the maximum power of the main lobe. To avoid interference effects in many wireless communications systems, it is desirable to reduce this ratio as much as possible. For example, in cellular systems, to avoid interuser interference effects, it is desirable to use directional antennas and to direct the main lobe of these antennas toward the desired users. However, the side lobe of these antennas may create interference. Adaptive antennas are those that direct their main lobe toward the desired users and adjust their nulls toward the interfering users.

9.2.3   Non-Main Lobe Power (All Side Lobe Power) to Main Lobe Power Ratio

This is another characteristic of a directional antenna and represents how well a directional antenna can suppress the side lobes. It could be considered as a better measure for an antenna compared with the *first side lobe to the main lobe power ratio*, as it considers the power of all side lobes and, accordingly, the maximum possible interference level.

9.2.4   Antenna Impedance

This parameter represents the equivalent impedance of the antenna structure. Similar to other transmission lines (cable, waveguides, etc.), or any circuit element, an antenna can be well modeled by a combination of capacitors, inductors, and resistors.

Thevenin Equivalent

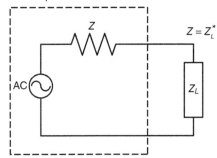

Figure 9.4 Thevenin circuit.

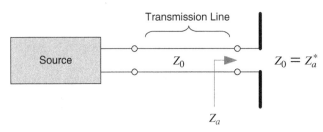

Figure 9.5 Antenna impedance.

Then, an antenna can be represented by its impedance. In general, the impedance of any element is not constant and varies with frequency. In addition, in circuit theory, we have learned that a given Thevenin circuit has the maximum energy transfer when the load impedance Z_L matches the circuit impedance Z, as shown in Figure 9.4. Because Z and Z_L are both functions of frequency, we can match the antenna load only for certain frequencies.

At the transmitter end, the antenna is connected to the source, as shown in Figure 9.5. It is important to match the antenna impedance to the source to allow for maximum power transfer from the source to the antenna. The impedance is normally seen through the antenna terminals (Z_a). In Figure 9.5, Z_0 refers to the transmission line (e.g., connecting cable) characteristic impedance.

If the impedance is not matched properly, the maximum transfer of energy does not occur. Under these circumstances, the reflections of energy can occur, which leads to interference and unwanted standing waves in the transmission line. This reduces the power transform efficiency. As we mentioned, the impedance of the antenna varies with frequency. Therefore, the antenna would be properly matched only at certain frequencies.

9.2.5 Antenna Return Loss

Antenna return loss refers to the amount of power reflected by the antenna divided by the amount of power transmitted to the antenna due to the mismatch between the antenna and the course and transmission line, as shown in Figure 9.5; that is,

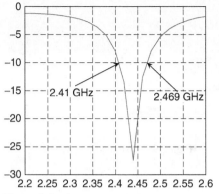 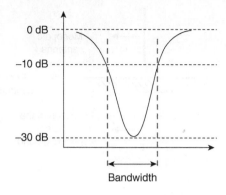

Figure 9.6   Antenna return loss.

$$\text{Return loss} = 10\log\frac{P_{\text{Reflected}}}{P_{\text{Transmitted}}}. \tag{9.1}$$

Here, $P_{\text{Reflected}}$ refers to the reflected power by the antenna, and $P_{\text{Transmitted}}$ refers to the transmitted power to the antenna. As $P_{\text{Reflected}}$ reduces to zero, the return loss reduces. The return loss is measured with respect to the frequency. As frequency changes, the antenna impedance would vary, and accordingly, the mismatch increases. As shown in Figure 9.6, in some frequencies, the return loss is minimum.

9.2.6   Antenna Bandwidth

Antenna bandwidth is defined based on the antenna return loss. Typically, the frequencies where the return loss is less than $-10°$dB determine the antenna bandwidth. Higher antenna bandwidth allows higher transmission bandwidth. Higher transmission bandwidth allows higher throughput, transmission rate, or capacity. Later, we will observe that the transmission rate of a transmitter is not only a function of the antenna but also of many other transmitter components.

9.2.7   Antenna Gain

Antenna gain measures the change of the antenna radiation intensity with respect to the azimuth and elevation direction. The following expressions illustrate how the antenna gain is determined:

$$\text{Antenna gain} = G(\theta, \phi) = \eta D(\theta, \phi). \tag{9.2}$$

Here,

η = antenna efficiency = $\dfrac{R_r}{R_r + R_d}$,

R_r = radiation resistance, and

R_d = antenna resistance.

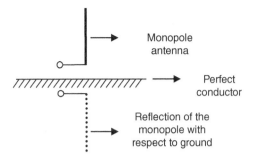

Figure 9.7   Monopole antenna.

In addition,

$$D(\theta, \phi) = \text{Antenna directivity} = \frac{\text{Radiation power in space angle unit}}{\text{Average radiation power in space angle unit}}$$

and

$$\text{Space angle unit} = d\Omega = \sin\theta \cdot d\theta \cdot d\phi.$$

It is clear that if the antenna loss is zero, R_d would be equal to zero. In this case, antenna efficiency would be unity; however, in general, $R_d > 0$. In addition, the radiation resistance R_r is a function of antenna length and/or dimensions. If the antenna dimension (d) increases with respect to the antenna wavelength λ, for example, if d approaches infinity, the radiation resistance R_r increases. As a result, as the antenna length becomes longer, its efficiency η tends toward its maximum value of 1. Hence, theoretically, the length of an antenna should be infinitely large to achieve full efficiency. In other words,

$$\boxed{\text{If } d \to \infty, \text{ then } R_r \to \infty \text{ and then } \eta \to 1.}$$

However, in practice, it is not acceptable to infinitely increase the antenna dimension. The length of $\lambda/2$ leads to a reasonably high antenna efficiency for dipole antennas. As shown in Figure 9.7, monopole antenna has a length that is half of a dipole antenna. Monopole antennas are usually installed on (perfect) conductors, such as the ground surface or the top of vehicles. The monopole antenna, along with its projection with respect to the ground surface, creates a beam pattern similar to a dipole antenna. An example of monopole antennas is your car radio antenna. The length of a monopole antenna is usually selected $\lambda/4$.

Antenna Gain Is Usually Measured in dBi: dBi refers to the gain of antenna, when it is measured with respect to an omnidirectional antenna. An omnidirectional antenna (also called isotropic antenna) is an antenna that emits EM energy uniformly in all directions. The term "i" in dBi refers to *isotropic*. As the antenna gain increases, the energy transferred to certain points would increase compared to an isotropic antenna. For example, if two antennas both transmit 1 W of power, and one is an omnidirectional antenna and the other is directional, a 5-dBi gain of the directional

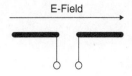

Figure 9.8   Horizontal polarization.

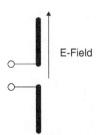

Figure 9.9   Vertical polarization.

antenna represents that the gain of that antenna is 5 dB more than the omnidirectional antenna that transmits the same amount of power. In other words, there are points in the space in the direction of maximum power that receive $10^{0.5} = 3$ times more power with the directional antenna compared with an omnidirectional one.

9.2.8   Antenna Polarization

Antenna polarization refers to the orientation of an EM wave's electric field. The three most common antenna polarizations are

1. horizontal polarization,
2. vertical polarization, and
3. circular polarization.

A horizontal polarization antenna's electric field is parallel to the ground (Fig. 9.8). "Fishbone"-type television antennas are examples of horizontal polarized antennas. A vertical polarization antenna's electric field is perpendicular to the ground (Fig. 9.9). Examples of vertical polarized antennas include AM radio towers and car radio antennas.

Horizontal and vertical polarized antennas radiate EM waves in a linear plane only. A circular polarized antenna radiates waves in the horizontal, vertical, and all planes in between both linear planes. Circular polarization is commonly used for satellite communications. A helix antenna is an example of circular polarized antennas.

9.3   ANTENNA ARRAYS

An antenna array is an array of antenna elements located at a distance, d, from each other. Figure 9.10 represents a patch antenna element, and Figure 9.11 is an antenna array made of patch antenna elements. Antenna arrays are also called multi-input–multioutput (MIMO) systems. MIMO systems that experience a coherent channel across all antenna elements can be used for BF and DOA estimation. If

Figure 9.10   Patch antenna element.

Figure 9.11   Patch antenna array.

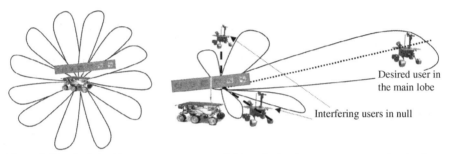

Figure 9.12   Switched beam smart antennas (left) and adaptive smart antennas (optimal beamformers) (right).

the channel is not coherent across antenna elements, MIMO systems are mainly used for space diversity in order to improve the performance of detection.

Antenna arrays that are used for BF and DOA estimation are examples of directional antennas. As shown in Figure 9.12, these antennas can direct the main part of the produced energy toward a specific direction in the main lobe. Antenna arrays that are used for BF and DOA estimation are also called *smart antennas*.

9.3.1   Smart Antennas

These antennas are called smart not because of the antenna, as it is usually a metallic structure, but because of signal processing algorithms that enable them to find the direction of the desired users and to form a beam optimally toward that user. Two important categories of smart antennas are *switched beam smart antennas* and *adaptive antennas*. Figure 9.12 represents these categories of smart antennas.

Switched beam antennas are formed by fixed beam patterns. As the source moves from one direction to another, the antenna is switched from one beam to another one. In this category of smart antennas, the interference effects of other sources are not optimally removed. In switched beam smart antennas, a simple BF technique such as conventional BF is applied to the antennas.

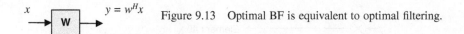

$x \longrightarrow \boxed{\mathbf{W}} \longrightarrow y = w^H x$

Figure 9.13   Optimal BF is equivalent to optimal filtering.

Adaptive smart antennas are the second category of antennas in which, based on the direction of the desired and interfering sources, an adaptive beam is formed. The goal is to optimally receive the signal of a desired source and to remove the signal of interfering sources. Adaptive smart antennas are equipped with optimal BF techniques. The goal of an optimal beamformer is to direct the signal toward a specific user or source and to create nulls toward undesired users or sources.

An optimal beamformer is indeed a filter that incorporates the signals observed over each antenna element (space domain) and each time (time domain) to extract the desired signal and null interfering signals. As shown in Figure 9.13, the input to the beamformer x is a matrix that consists of all observed signals (in time, space, or frequency domains). The optimal beamformer finds the weight matrix w such that the output y represents the best estimate of the desired signal.

As shown in Figure 9.12, an optimal beamformer forms a beam toward desired users and steers nulls toward interfering users. Different BF techniques have been introduced, which include capon and linear constrained minimum variance (LCMV) [7], and maximum noise fraction (MNF) [8].

The conventional or Fourier BF technique is the simplest BF method in which a fixed phase consistent (based on known DOA) is applied to each antenna element to form a beam toward the desired users. In conventional BF, the DOA of a desired user should be known a priori. In addition, this technique does not optimally remove the impact of interfering users. In conventional BF, w of Figure 9.13 corresponds to the array factor, v defined in (9.3).

An important factor in all DOA estimation techniques is that the phase and frequency across all antenna elements should be properly synchronized. In other words, the same carrier signal reference should be applied to all antenna elements.

9.3.2   Important Parameters of Antenna Arrays

Besides the parameters defined for antenna elements, antenna arrays possess other parameters, which include the following:

1. array vector,
2. array factor, and
3. mutual impedance.

Array Vector: This parameter of an antenna array corresponds to

$$v = \left[1, \ e^{-j2\pi \frac{d \cdot \sin \phi}{\lambda}}, \ldots, e^{-j2\pi \frac{(M-1)d \cdot \sin \phi}{\lambda}} \right]. \tag{9.3}$$

In this equation, M is the number of antenna elements, ϕ corresponds to the DOA, and λ is the wavelength. The array vector indeed represents the relative signal

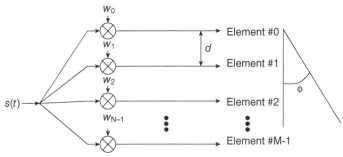

Figure 9.14   Computing the array vector of an antenna array.

current (or voltage) of each antenna element with respect to the other antenna elements. Based on Figure 9.14, it is observed that if the signal arrived in the first antenna element can be represented by the phasor v, the signal of the second antenna with respect to the first one would have the phase difference of $e^{-j2\pi\frac{d\cdot\sin\phi}{\lambda}}$, and the third antenna with respect to the first one corresponds to

$$e^{-j2\pi\frac{2d\cdot\sin\phi}{\lambda}},$$

and the Mth antenna with respect to the first one is,

$$e^{-j2\pi\frac{(M-1)d\cdot\sin\phi}{\lambda}}.$$

Array Factor: Considering Figure 9.14 and the array vector, it is observed that the signal received at the antenna receiver after application of the weights $w = [w_1, w_2, \ldots, w_M]$, and the addition process corresponds to $w \cdot v^T$; that is,

$$AF = \sum_{m=1}^{M} w_m e^{-j2\pi\frac{(m-1)d\cdot\sin\phi}{\lambda}}. \tag{9.4}$$

The normalized array vector assuming $w_m = 1$, $\forall m$ corresponds to

$$AF_n = \frac{1}{M}\sum_{m=1}^{M} e^{-j2\pi\frac{(m-1)d\cdot\sin\phi}{\lambda}} = \frac{1}{M}\cdot\frac{1-e^{-j2\pi\frac{Md\cdot\sin\phi}{\lambda}}}{1-e^{-j2\pi\frac{d\cdot\sin\phi}{\lambda}}}.$$

The magnitude of the array factor represents the antenna beam pattern and corresponds to

$$|AF_n| = \frac{1}{M}\cdot\frac{\sin\left(\frac{M}{2}\cdot\frac{2\pi d\cdot\sin\phi}{\lambda}\right)}{\sin\left(\frac{1}{2}\cdot\frac{2\pi d\cdot\sin\phi}{\lambda}\right)}. \tag{9.5}$$

Essentially, Equation (9.5) represents the antenna gain corresponding to a conventional BF if the antenna elements are omni directional antennas.

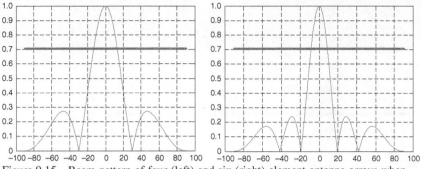

Figure 9.15   Beam pattern of four (left) and six (right) element antenna arrays when $d = \lambda/2$.

Example 9.1   Antenna Array Pattern

Use the magnitude of the normalized array factor in (9.5), assume $d = \lambda/2$, and write a MATLAB code and sketch the beam pattern of an antenna array composed of four and six antenna elements. Compare their HPBW.

In addition, sketch the beam pattern assuming higher antenna element distancing, for example, $d = 2\lambda$, for six element antenna arrays. What happens if the number of antenna elements increases beyond six elements?

Solution

The beam pattern MATLAB code is in the file Chapter_9_Example1.m. MATLAB codes can be found online at ftp://ftp.wiley.com/public/sci_tech_med/matlab_codes. As it is observed, the code does not need any specific MATLAB toolbox. The corresponding beam pattern for four and six element antenna arrays and $d = \lambda/2$ is sketched in Figure 9.15. Based on this figure and the MATLAB code results, the HPBW of four-element antenna arrays corresponds to $26.3°$ and that for six-element antenna arrays corresponds to $17.18°$. As the number of antenna elements increases from 4 to 6 elements, the HPBW decreases. Therefore, it is expected that as we increase the number of antenna elements, using any DOA algorithm, we will be able to better resolve sources. In other words, we expect that the resolution of the DOA estimation technique increases.

It is observed that as the number of antenna elements increases, HPBW decreases. In addition, the side lobe level of six-element antenna arrays is lower than the four-element antennas. Figure 9.16 sketches the beam pattern for $d = 2\lambda$ and six-element antenna arrays. It is shown that as the number of antenna elements increases, a higher number of grating lobes are generated. As a result, for DOA estimation techniques to avoid any confusion, the spacing of antenna elements should be selected to be $d = \lambda/2$.

Mutual Coupling: As shown in Figure 9.17, the signal incident on one antenna element is usually reflected from that antenna element and impacts the signal received by neighboring antenna elements. In other words, each antenna acts as a

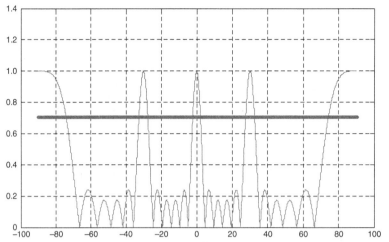

Figure 9.16   Antenna pattern when $d = 2\lambda$. Many grating lobes avoid proper DOA estimation.

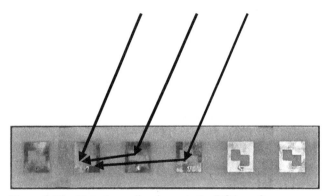

Figure 9.17   Mutual coupling impacts the DOA performance.

parasitic element for neighboring antenna elements. This situation creates mutual coupling between antenna elements.

In this case, the array vector in (9.3) does not any more represent the relationship between the signals received at different antennas. Indeed, the array vector represents the relative current or voltage of each antenna element with respect to another antenna element assuming no mutual effect between these elements. Now, if the mutual coupling exists, the received signal (voltage or current) over one antenna element is an addition of the received signal that is coming directly from the far-field source and the reflected signals from neighboring antenna elements; that is,

$$
\begin{bmatrix} W_1 \\ \vdots \\ W_M \end{bmatrix} = \begin{bmatrix} u_{11} & \cdots & u_{1M} \\ \vdots & \ddots & \vdots \\ u_{M1} & \cdots & u_{MM} \end{bmatrix} \begin{bmatrix} v_1 \\ \vdots \\ v_M \end{bmatrix}.
\tag{9.6}
$$

In (9.6), W_i refers to the received signal (voltage or current) at the antenna element $i \in \{1, 2, \ldots, M\}$ and $u_{ii} = 1$ for $i \in \{1, 2, \ldots, M\}$, and u_{ij} represents the mutual coupling of antenna element j on antenna element i, and v_i is the ith element of the array vector in (9.3). It is expected that $|u_{ij}| < 1$. In addition, usually, $|u_{ij}| > |u_{im}|$, if $|i - j| < |i - m|$. In addition, it is expected that $|u_{ij}| \ll 1$, if $|i - j| > 1$. Thus, the effect of mutual coupling can be ignored for non neighboring antenna elements in most scenarios.

Mutual coupling has a negative impact on DOA estimation techniques. This problem is usually mitigated by antenna calibration techniques. Many methods have addressed the calibration issue in smart antennas in the literature. Some works emphasize calibrating mutual coupling between antenna elements [9–11], while others emphasize calibrating a receiver mismatch [12, 13].

In a smart antenna systems, the mutual coupling and the receiver mismatch exist simultaneously. The mutual coupling is altered by the antenna structure, distances between antenna elements, and so on. The receiver mismatch is due to the independence of receivers and depends on receiver configuration (amplifier gain or attenuator attenuation).

If we compensate antenna element mutual coupling and receiver mismatch simultaneously, the best performance would be achieved. If we cannot compensate them simultaneously, we should compensate the one generating the largest error. The proposed techniques in [10] and [11] all need a far-field fixed source implemented in a lab environment, that is, off line calibration. The methods presented in [12] and [13] apply extra hardware in each receiver, which increases system cost and complexity.

9.4   DOA ESTIMATION METHODS*

Different DOA estimation methods have been proposed and discussed in the literature (see [1]). In this section, we introduce the fundamentals of some techniques and compare them in terms of different measures. The section offers some MATLAB-based examples on the implementation of MUSIC and root MUSIC methods.

DOA Estimation Requirements

The first assumption in many DOA estimation techniques is the fact that the source is located in the far field. Usually, the far field is considered a distance of beyond about 10λ. Thus, for a 3-GHz frequency, the far field is beyond 1 m. In many applications, the usual distance of the antenna array and the target might be small. Thus, the implementation of DOA estimation for near-field situations might be considered. Many papers have already discussed DOA estimation techniques for near-field conditions [14, 15].

In general, as shown in Figure 9.18, DOA estimation needs an array of antennas located at a distance (d) from each other. Usually, d is selected to be half a wavelength to avoid any grating lobe and confusion in DOA (see Example 9.1). It

---

*Some of the materials in this section are partially published in [3].

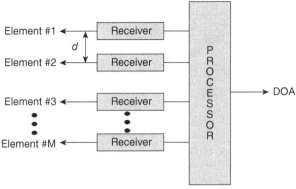

Figure 9.18   The general block diagram of a DOA estimator.

should be noted that *receiver synchronization* is very critical in the DOA estimation problem, as DOA estimation is computed by measuring the relative phase of signals received across a number of antenna elements.

Figure 9.19 represents the detailed block diagram of the WLPS (a single-node localization system introduced in Chapters 1 and 34), which needs DOA and TOA estimation. In this figure, DDC refers to digital down converter. As shown in Figure 9.19, each antenna element is connected to an RF front end and a receiver. It is observed that both DOA estimation and BF are applied in the baseband, and after full amplitude, frequency and phase synchronization were applied across the signals received from antenna elements. This synchronization is required to extract the estimated DOA with minimal error.

Another important requirement of DOA estimation techniques is *full calibration* of the system. *Calibration is needed because*

1. the RF front end of each antenna is not exactly similar to the other one and each may apply a different phase to the signal; and

2. even if the RF front end of antenna elements are fully similar, the interaction of propagation effects of antenna elements are not consistent. The latest effect is due to the antenna array mutual coupling. The impact of mutual coupling on the received phase is a function of DOA; that is, as the DOA varies, the induced phase due to mutual coupling changes.

To address the calibration problem, two general approaches have been proposed in the literature:

1. *Antenna Calibration Techniques*: These techniques are usually incorporated offline. They use lab experiments to find error in DOA estimation due to the mutual coupling and receiver mismatch in order to fix the DOA estimation problem. The receiver RF front and the specifications of antennas may change with time and temperature. Thus, offline calibration techniques may not perform over a long run and signal processing technique should also be used.

2. *Signal Processing Methods*: These techniques estimate the error in DOA estimation and apply corrections to the measured DOA estimation [16–18].

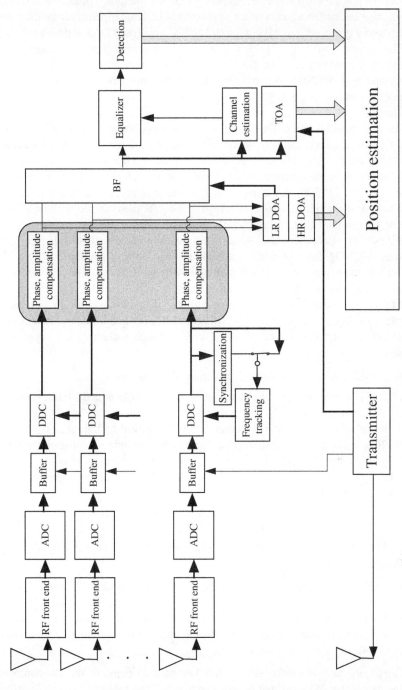

Figure 9.19 The detailed structure of a positioning system. ADC, analog-to-digital converter. LR, low resolution; HR, high resolution.

Examples of DOA estimation techniques proposed include spectral estimation methods (also called DAS), maximum likelihood (ML), and eigenstructure methods (includes many types, such as MUSIC, root MUSIC, and fused DAS and root MUSIC [3]). The authors in [1], [7], and [19] have reviewed these techniques. In this chapter, we introduce some popular DOA estimation methods such as DAS, MUSIC, and root MUSIC. We will also introduce some new extensions of these techniques, including fused DAS and root MUSIC and cyclostationary-based DOA estimation.

As mentioned earlier, DOA estimation is possible via multiple antennas installed at the receiver. DAS applies several sets of complex weights to antenna elements for each angle. The delayed signals are summed and the power is computed. The DOA is determined by analyzing the output power from all sets of complex weights. DAS DOA estimation performance is low; however, its low complexity would allow online coarse DOA estimation. It uses simple arithmetic and can be easily implemented on field programmable gate array (FPGA) systems.

MUSIC computes a spatial spectrum by estimating the noise subspace and determines the DOA from its dominant peak. Root MUSIC is similar to MUSIC in many aspects except that the DOA is computed by roots closer to the unit circle of a polynomial formed by the noise subspace.

In this section, first, some criteria to compare different DOA estimation techniques are introduced. Next, DAS, MUSIC, root MUSIC, and a fusion of DAS and root MUSIC is introduced. The pros and cons of each technique are separately discussed.

Important Measures in DOA Estimation Techniques

- *DOA Estimation Error.* DOA estimation error is determined by the variance of the error in the DOA estimation process. This is the most important measure when comparing different DOA estimation techniques. This error itself is a function of different parameters that include signal-to-noise ratio (SNR), multipath, and antenna calibration. The estimation error is usually measured in terms of root mean square error (RMSE) and corresponds to

$$\text{RMSE} = \sqrt{\frac{1}{N}\sum_{n=1}^{N}\left(\hat{\theta}_n - \theta_{\text{DOA}}\right)^2}.$$

Here, $\hat{\theta}_n, n \in \{1, 2, \ldots, N\}$, is the estimated DOA in the nth snap shot and θ_{DOA} is the real DOA.

- *Resolution.* This represents the minimum angular separation of two incoming signals, which the DOA estimation method can resolve. This parameter itself is a function of SNR as well. As the SNR decreases, the resolution decreases as well. A technique that can resolve two sources with lower angular separation has a priority over other techniques.

- *Sensitivity to Multipaths.* As the number of multipaths in the environment increases, the DOA estimation error increases. The rate of change of performance with the number of multipaths is considered as a measure for the goodness of a technique.

- *Sensitivity Sensor Errors Such as Calibration.* We refer to the fact that calibration is an important factor in creating a good DOA performance. Some DOA estimation techniques are heavily altered by antenna and RF component calibration, while other techniques are mildly altered.

- *Sensitivity to SNR.* SNR plays an important role in the performance of DOA estimation techniques. Specifically, performance results show that as SNR reduces, the sensitivity of all techniques increases. Lower SNR leads to higher sensitivity to SNR.

9.4.1 DAS

In DAS, an array of complex weights, $\vec{w} = [w_1, w_2, \ldots, w_M]$ (M is the number of antennas), is applied to the set of incoming signals over all antennas. These complex weights delay the signal by changing its phase. The delayed signals are then summed, and the output power is measured. If the complex weights are properly selected, the signals constructively interfere, resulting in a high output power. The relative phase delay $\phi_i(\theta)$ of the signal is a function of the DOA, θ_{DOA}, wavelength λ, antenna spacing $d = \lambda/2$, and the antenna number i; that is,

$$\phi_i(\theta_{DOA}) = -2\pi(i-1)d \cdot \sin\theta_{DOA}/\lambda. \tag{9.7}$$

This phase delay is relative to that of antenna 1. Ignoring noise effects, the received signal at antenna i is

$$r_i = Re\left[Ae^{j(\omega_0 t + \theta_0 + \phi_i(\theta_{DOA}))} \right].$$

Here, $\omega_0 = 2\pi f_0$, f_0 is the center frequency, $\phi_i(\theta_{DOA})$ has been introduced in (9.7), and θ_0 refers to the phase shift in the signal due to the propagation time delay and the modulation phase. Considering a far-field condition, the phase θ_0 would be the same for all received signals over all antennas.

The applied weight w_i to the ith antenna element is

$$w_i(\theta) = e^{j\phi_i^w(\theta)}, \phi_i^w(\theta) = 2\pi(i-1)d \times \sin\theta/\lambda. \tag{9.8}$$

The delayed (over all M antennas) and summed signal is

$$S(\theta, \theta_{DOA}) = \sum_{i=1}^{M} s_i(\theta, \theta_{DOA}) = Ae^{(j\omega_0 t + \theta_0)} \sum_{i=1}^{M} e^{j\left(\phi_i(\theta_{DOA}) + \phi_i^w(\theta)\right)}. \tag{9.9}$$

In (9.9), ϕ_i and ϕ_i^w are defined in (9.7) and (9.8), respectively. If $\theta = \theta_{DOA}$, then $\phi_i^w = -\phi_i$; hence, the relative phase would be zero and the power maximizes. The algorithm calculates the associated power, $P(\theta_l, \theta_{DOA})$, of $L = 1 + (180/\Delta\theta)$ angles $\theta_l \in [-90°, +90°]$, $l \in \{1, 2, \ldots, L\}$ and finds the angle that maximizes the power that is called the DOA. These angles are equally spaced with the resolution of $\Delta\theta$. Hence, a power profile for each DOA (normalized to the power of the first antenna) is created that corresponds to

$$P(\theta_l, \theta_{\text{DOA}}) = \frac{1}{A^2}|S(\theta_l, \theta_{\text{DOA}})|^2, \, l \in \{1, 2, \dots, L\}. \tag{9.10}$$

Here, $|\cdot|$ refers to the absolute value. If $\phi_i + \phi_i^w = 0$, then $S = M \cdot A e^{(j\omega t + \theta_0)}$, and the associated maximum power corresponds to $P_{\max} = \frac{1}{A^2}|S|^2 = M^2$. Mathematically,

$$\hat{\theta}_{\text{DOA}} = \arg\max_{\theta_l} P(\theta, \theta_{\text{DOA}}). \tag{9.11}$$

The received signal r_i experiences zero-mean Gaussian noise. Hence, in (9.11), $P(\theta_l, \theta_{\text{DOA}})$ estimation involves an (exponentially distributed) error. Here, we consider that N snapshots (e.g., $N \geq 25$) of $P(\theta_l, \theta_{\text{DOA}})$ are measured in the time domain. Hence, $\hat{P}_i(\theta_l, \theta_{\text{DOA}}) = P(\theta_l, \theta_{\text{DOA}}) + n_i$, $i \in \{1, 2, \dots, N\}$, n_i is the power measurement error. Assuming independent measurements, ML estimation leads to temporal averaging over N snapshots; that is,

$$P_{\text{out}}(\theta_l, \theta_{\text{DOA}}) = \hat{P}(\theta_l, \theta_{\text{DOA}}) = \frac{1}{N}\sum_{i=1}^{N} P_i(\theta_l, \theta_{\text{DOA}}). \tag{9.12}$$

The peak power in $P_{\text{out}}(\theta_l, \theta_{\text{DOA}})$ should correspond to the DOA, but there are two problems associated with this approach: (1) Noise can easily throw off a simple peak search, and (2) this only leads to a resolution as fine as $\Delta\theta$.

A polynomial interpolator improves the performance of the peak finder of (9.11). Considering (θ_m, P_m) as the point with the maximum sampled power, the points $\{(\theta_{m-n}, P_{m-n}), \dots, (\theta_m, P_m), \dots, (\theta_{m+n}, P_{m+n})\}$ are used for the polynomial fit. The polynomial coefficients are found using a least squares fit. DOA is the angle that maximizes the polynomial. We call this DOA estimator DAS1.

The second DAS algorithm (DAS2) fits all of the sampled data points to a theoretical set of data. For each given DOA, based on (9.10), a lookup table is created off-line for the power profile, $\left\{P_{\text{true}}\left(\theta_l, \theta_{\text{DOA}}^{(z^{(k)})}\right)\right\}_{l=1}^{L}$, over all $\theta_l \in [-90°, +90°]$ and all true DOA, $\theta_{\text{DOA}}^{(z^{(k)})}$, $z^{(k)} \in \{1, 2, \dots, Z^{(k)}\}$, $Z^{(k)} = Q^{(k)}/\Delta\theta_{\text{DOA}}^{(k)}$, $Q^{(0)} = 180$, $Q^{(k+1)} = 2\Delta\theta_{\text{DOA}}^{(k)}$, $\Delta\theta_{\text{DOA}}^{(k+1)} \ll \Delta\theta_{\text{DOA}}^{(k)}$. Here, $k \in \{0, 1, 2, \dots, K\}$ refers to the iteration number (an integer); K is selected large enough to achieve the desired performance. In each iteration $k \geq 1$, a search is conducted around the priori estimated angle $\hat{\theta}_{\text{DOA}}^{(k-1)}$. The angle $\Delta\theta_{\text{DOA}}^{(k)}$ is decreased stepwise to increase angle resolution with minimal computation cost. Based on minimum mean square error (MMSE) criteria,

$$\hat{\theta}_{\text{DOA}}^{(k)} = \arg\min_{\theta_{\text{DOA}}^{(z^{(k)})}} \frac{1}{L}\sum_{l=1}^{L}\left[P_{\text{out}}(\theta_l, \theta_{\text{DOA}}) - P_{\text{true}}\left(\theta_l, \theta_{\text{DOA}}^{(z^{(k)})}\right)\right]^2. \tag{9.13}$$

The value of $\Delta\theta$ specifies the DOA estimation performance. Accordingly, DAS2 corresponds to the following steps: (1) The received power profiles are generated based on (9.10), over N snapshots; (2) $P_{\text{out}}(\theta_l, \theta_{\text{DOA}})$ is calculated using (9.12);

(3) $\hat{\theta}_{DOA}^{(k)}$ is estimated via $\Delta\theta_{DOA}^{(k)}$, $k = 0$ by (9.13); (4) $k = k + 1$, and go to (3) for a new search around $\hat{\theta}_{DOA}^{(k)}$; that is, from $\hat{\theta}_{DOA}^{(k)} - \Delta\theta_{DOA}^{(k-1)}$ to $\hat{\theta}_{DOA}^{(k)} + \Delta\theta_{DOA}^{(k-1)}$. The performance of DAS2 is higher than that of DAS1. However, DAS2 suffers from higher computation cost.

The number of multiplications (NOM) is the complexity measure. For DAS1 with quadratic polynomial, NOM is

$$NOM = 4 \cdot N \cdot M \cdot L + 6 \cdot P + 360. \tag{9.14}$$

Using $N = 5$ snapshots, $M = 6$ antennas, $\Delta\theta = 5°$, and $P = 5$ sample points used in the quadratic fit, NOM = 4830. For DAS2, NOM corresponds to

$$NOM = 4 \cdot N \cdot M \cdot L + L \cdot \sum_{k=0}^{K} \left(\frac{2\Delta\theta_{DOA}^{(k-1)}}{\Delta\theta_{DOA}^{(k)}} + 1 \right). \tag{9.15}$$

Using $N = 5$ snapshots, $M = 6$ antennas, $\Delta\theta = 5°$, and $K = 2$, and $\Delta\theta_{DOA}^{(k)} = 10°, 1°, 0.1°$ for $k = 0, 1, 2$, respectively, NOM = 6697.

In general, DAS is a simple and low-complexity DOA estimation technique. However, its sensitivity with respect to SNR, calibration errors, and number of reflections is high. Moreover, its resolution is low. These points will be presented and discussed via simulations in this section.

9.4.2 MUSIC and Root MUSIC

In these techniques, it is assumed that (1) a priori estimation of the number of sources is available and (2) the number of antenna elements is higher than the number of sources.

MUSIC: The received signal at each antenna element $I \in \{1, 2, \ldots, M\}$ is

$$x_i(t) = w_i(\theta) \cdot y_i(t) + n_i(t). \tag{9.16}$$

Here, $w_i(\theta)$ is defined in (9.8); y_i is the ith element of incident signal vector; and N_i is the ith element of noise. We define $\vec{w}(\theta)$, $\vec{y}$, $\vec{n}$, and X as arrays corresponding to all M antenna elements.

By applying temporal averaging over $n \in \{1, 2, \ldots, N\}$ snapshots, the received signal sample covariance matrix R_X is calculated. Each row of the eigenvector matrix V_N of R_X consists of the eigenvectors corresponding to the set of smallest eigenvalues. These eigenvectors are called noise subspaces. The noise subspaces are orthogonal to the array vector $\vec{w}$; that is, $\vec{w}^H(\theta) \cdot V_N = 0$. MUSIC estimates the DOA of incident signals by locating the peaks of the MUSIC spectrum defined as

$$H(\theta) = \frac{1}{\left(\vec{w}^H(\theta) \cdot V_N \cdot V_N^H \cdot w(\theta) \right)}. \tag{9.17}$$

The orthogonality of $\vec{w}(\theta)$ and V_N minimizes the denominator of $H(\theta)$ and generates the peaks of the MUSIC spatial spectrum. Searching for the directions that maximizes $H(\theta)$ requires a substantial computational complexity.

Example 9.2   MUSIC Spectrum

The MUSIC spectrum has been defined in (9.17). Use MATLAB and sketch this spectrum assuming two sources at $10°$ and $50°$ and six element antennas. Repeat the problem when the two sources are at $10°$ and $14°$. Can this method resolve two sources when they are as close as $2°$ (e.g., 10 and 12)? Change the number of antenna elements to 12 and study the effect of increasing the number of antenna elements. Does it impact the resolution?

Solution

To generate the spectrum, we first generate (9.16). Next, the spectrum is generated. The spectrum has been sketched in Figure 9.20a. It is observed that even if the two sources are as close as $4°$, they can be resolved; however, if the sources are not as close as $3°$, they cannot be resolved.

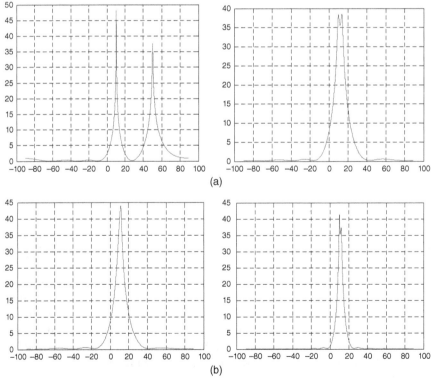

Figure 9.20   (a) Six-element antenna MUSIC spectrum for $10°$ and $50°$ (left) and for $10°$ and $14°$ (right). (b) MUSIC Spectrum for sources at $10°$ and $12°$ for 6-element (left) and for 12-element (right) antennas.

Figure 9.20b sketches the scenario that the sources are as close as 2°. It is observed that the MUSIC algorithm is not able to resolve the two sources any more. Thus, the resolution of this algorithm for a six-element antenna is in the order of 4°. Now, if the number of antenna elements is increased to 12, higher resolution can be attained. Typically, it is observed that a resolution of 2° is achievable. This conclusion is consistent with our discussions in Example 9.1: As the number of antenna elements increases, lower HPBW is attained. However, it should be noted that increasing the number of antenna elements from 6 to 12 improves the resolution from 4° to 2°, while the complexity of the antenna element system as well as the MUSIC algorithm highly increases. It should be noted that increasing the number of antenna elements and the complexity associated with MUSIC leads to higher power consumption as well. Thus, the number of antenna elements should be wisely selected. The associated codes have been provided in the file Chapter_9_Example_2.m. MATLAB codes can be found online at ftp://ftp.wiley.com/public/sci_tech_med/matlab_codes.

The block diagram of the MUSIC has been sketched in Figure 9.21a. In general, the MUSIC algorithm is a complex algorithm, as it needs to find the eigenvalues of a matrix. In addition, it usually needs the prior information of the number of sources (or reflections). However, its sensitivity with respect to the number of reflections, calibration error, and SNR is relatively lower compared to almost all DOA estimation techniques. As will be discussed later in this section, root MUSIC performs better than MUSIC, and its complexity is relatively lower.

Root MUSIC: MUSIC calculates the spectrum for a finite number of directions based on the selection of the angle increment. Such a searching method has large computational and storage requirements [20]. To refine DOA estimation and to alleviate the algorithm complexity, root MUSIC has been proposed by Barabell [5]. As summarized in Figure 9.21b, root MUSIC finds the roots of the polynomial represented by

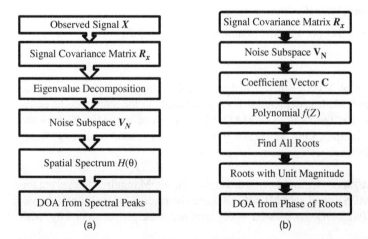

(a)   (b)

Figure 9.21   Block diagram of the (a) MUSIC and (b) root MUSIC algorithms.

$$f(z) = H^{-1}(z). \tag{9.18}$$

Its complexity is lower compared with MUSIC [7]. In (9.18), z refers to $w_i(\theta)$, defined in (9.8). Equation (9.18) can be rewritten as

$$H^{-1}(z) = \sum_{k=-L+1}^{L-1} \left(C_k \cdot z^k \right), \tag{9.19}$$

where C_k is the sum of the elements of the matrix $C = V_N V_N^H$ along the kth diagonal

$$C_k = \sum_{m-n=k} C_{mn}. \tag{9.20}$$

Note that the magnitude of z corresponds to one, and its phase contains information of DOA.

Root MUSIC evaluates all the roots of $H^{-1}(z)$ and assigns those roots that are closer to the unit circle and the remaining roots that are farther from the unit circle to noise. Among all roots, the closer a root is to the unit circle, the more likely it stands for DOA, which can be found from

$$\theta = \sin^{-1}\left(\frac{-\lambda \cdot \arg(z)}{2\pi \cdot d} \right). \tag{9.21}$$

Hence, to estimate DOA using the root MUSIC algorithm, it calculates $2N - 2$ roots of the polynomial $H^{-1}(z)$, where N represents the number of antennas, then it finds the root, which lies on or is closest to the unit circle. Based on the definition of z, it should be noted that unlike MUSIC, root MUSIC is only applicable in the case of uniform linear arrays (ULAs). This is considered a limitation of root MUSIC algorithm.

Example 9.3   Root MUSIC

Repeat Example 9.2 for root MUSIC assuming a source at 30°.

Solution

The roots of the root MUSIC in the z-plane and the amplitude of the roots are sketched in Figure 9.22. It is depicted that only 30° leads to the amplitude of unity, and it is considered as the DOA. The MATLAB program of root MUSIC is in the file Chapter_9_Example_3.m. MATLAB codes can be found online at ftp://ftp.wiley.com/public/sci_tech_med/matlab_codes.

Complexity Analysis: Complexity is defined as the NOM required to execute the algorithm. Both MUSIC and root MUSIC compute the received signal sample covariance matrix R_X (with NOM_A), as well as eigenvalues and eigenvectors of R_X (with NOM_B):

$$\text{NOM}_A = 4 \cdot M \cdot N^2 + N^2 \tag{9.22}$$

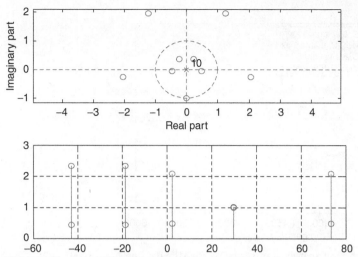

Figure 9.22 Root MUSIC roots in z-plane and the amplitude.

and

$$\text{NOM}_B = \frac{26}{3} N^3, \tag{9.23}$$

where M is the number of snapshots and N is the number of antennas. Specifically, MUSIC computes spatial spectrum $H^{-1}(z)$ over a certain range of angles, which leads to the following NOM:

$$\text{NOM}_C = (4 \cdot M^3 - 2 \cdot M^2 + M) \cdot f(\theta_1, \theta_2, stp). \tag{9.24}$$

In (9.24),

$$f(\theta_1, \theta_2, stp) = \pi \cdot \left[\sin\left(\frac{\pi \cdot \theta_2}{180}\right) - \sin\left(\frac{\pi \cdot \theta_1}{180}\right) \right] \div stp, \tag{9.25}$$

where $[\theta_1, \theta_2]$ is the angular range to compute the MUSIC spectrum and stp is the number of angular steps used to compute the MUSIC spectrum. Hence, the total NOM of MUSIC is

$$\begin{aligned}
\text{NOM}_1 &= \text{NOM}_A + \text{NOM}_B + \text{NOM}_C \\
&= 4 \cdot M^2 \cdot N + \frac{26}{3} \cdot M^3 + M^2 + (4 \cdot M^3 - 2 \cdot M^2 + M) \cdot f(\theta_1, \theta_2, stp).
\end{aligned} \tag{9.26}$$

On the other hand, root MUSIC specifically computes the roots of a polynomial, which costs

$$\text{NOM}_D = \frac{220}{3} N^3 - 212N^2 + 206N - 78. \tag{9.27}$$

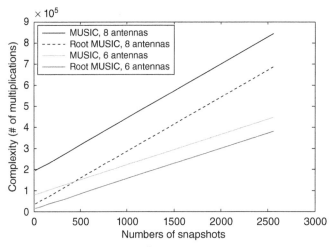

Figure 9.23   Numbers of snapshots versus NOM for MUSIC; step size = 0.001.

Hence, the total NOM of root MUSIC is

$$\text{NOM}_2 = \text{NOM}_A + \text{NOM}_B + \text{NOM}_D$$
$$= 4 \cdot M^2 \cdot N + 82 \cdot M^3 - 211 \cdot M^2 + 206 \cdot M - 78. \quad (9.28)$$

Figure 9.23 compares the complexity of MUSIC and root MUSIC algorithms. As it is observed, the complexity of MUSIC is higher than root MUSIC. This is due to the fact that MUSIC requires an exhaustive search through all possible steering vectors to estimate DOA.

Comparison of MUSIC and Root MUSIC: *Sensitivity to SNR:* Figure 9.24 compares the RMSE in the DOA estimation of MUSIC and root MUSIC. RMSE is defined as

$$\text{RMSE} = \sqrt{\frac{1}{N} \sum_{i=1}^{N} \left(\hat{\theta}_i - \theta_{\text{DOA}} \right)^2}, \quad (9.29)$$

where N is the number of DOA estimations, $\hat{\theta}_i$ is the ith estimated DOA, and θ_{DOA} is the real DOA.

Figure 9.24 is sketched assuming one signal source, six antennas, 50 snapshots, 50 estimations, and DOA = 30°.

It is observed that as SNR increases, the DOA RMSE of both MUSIC and root MUSIC algorithms improves. Using the same NOM, root MUSIC is more robust to SNR. MUSIC costs four times more NOM to attain comparable performance with root MUSIC.

Figure 9.25 illustrates the accuracy of MUSIC and root MUSIC within a DOA range of −80° to 80°. Here, it is assumed that one signal source, six antennas, 50 snapshots, and 50 estimations are available. In addition, SNR = 20 dB. The RMSE

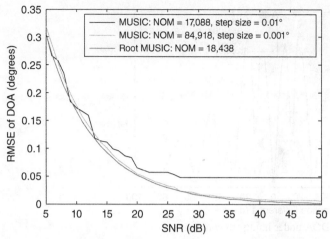

Figure 9.24   RMSE of DOA versus SNR.

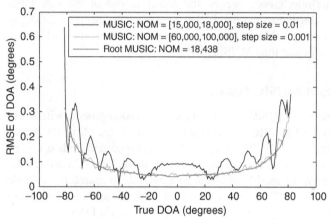

Figure 9.25   RMSE of DOA versus true DOA.

of MUSIC and root MUSIC are significantly low within a DOA range of −50° to 50°, beyond which the RMSE increases drastically. With comparable NOM, root MUSIC has lower RMSE and stable performance over a wide range of angles.

In general, the sensitivity of root MUSIC with respect to SNR, number of reflections, and calibration error (see [21]) is lower than that of MUSIC.

Sensitivity to Multipaths: Figure 9.26 shows the impact of a multipath on RMSE. In this figure, we have considered a line-of-sight (LOS) signal and single reflection that is considered a non-line-of-sight (NLOS) signal, and the power of NLOS to LOS is considered 0.1. In addition, six antennas, 50 snapshots, and 50 estimations are assumed. It is observed that the errors take on a fringe pattern with respect to the angular difference between LOS and NLOS signals. This occurs

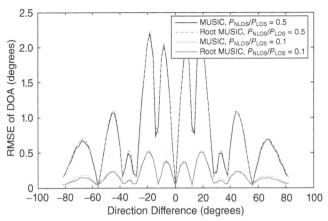

Figure 9.26   RMSE of DOA under multipath conditions.

because the main lobe of the reflected signal interferes with the side lobes of the LOS signal. When this interference occurs, the power in one of the side lobes increases enough to be considered the main lobe, generating an incorrect DOA.

In general, for a comparable DOA estimation performance, root MUSIC is more computationally efficient than MUSIC.

9.4.3   DAS and Root MUSIC Fusion

We can merge DAS and root MUSIC to support the localization process with lower complexity: We modify root MUSIC to accept a priori knowledge of the DOA provided by DAS. Classical root MUSIC computes a polynomial and finds all of its roots. The DOA is a function of the phase of one of these roots. To reduce the computational complexity, we modify classical root MUSIC to search for a single root using Newton's method [9]. The magnitude of the initial root z_0 is 1; its phase is $-(2\pi d/\lambda)\cdot\sin\theta_{\text{coarse}}$. Here, θ_{coarse} refers to the DOA estimated via DAS.

First, the tangent line of the polynomial is found, and then the root of that tangent line is computed. The zero of the tangent is a better approximation of the polynomial root. The iteration equation is

$$z_{i+1} = \left[f(z_i) / f'(z_i) \right] - z_i. \tag{9.30}$$

Here, $f(z)$ is defined in (9.18). Once the iterative method converges to a final root, z_{Final}, the phase of that root would be the fine DOA estimation: $\hat{\theta}_{\text{Fine}} = \sin^{-1}(-\lambda \cdot \arg(z_{\text{Final}})/2\pi \cdot d)$. Here, $\arg(z_{\text{Final}})$ corresponds to the phase of z_{Final}.

If the error in the coarse DOA estimation is large, and the seed root for Newton's method is beyond some threshold, the algorithm may converge to an incorrect root. This problem can be easily detected. Classical root MUSIC operates by finding the roots close to 1. This is used to check the validity of Newton's method: If the root found has a magnitude near 1, it would be valid. The roots far from 1 (outside of some threshold) would be invalid.

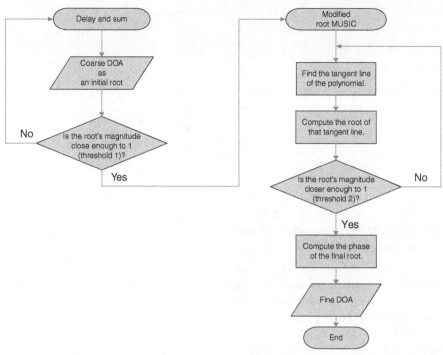

Figure 9.27   Flowchart of delay and sum + root MUSIC.

There are several ways to proceed from here: (1) Discard the result and wait for a new estimate from the first stage; (2) discard the result, modify the phase of the seed value, and try again; or (3) discard the result and use the coarse DOA approximation as the final DOA. The choice made depends on the importance of a fine DOA approximation, how often a DOA approximation must be generated, and the power constraints of the system. Figure 9.27 summarizes the fusion algorithm.

By merging Newton's method with classical root MUSIC, only one root has to be calculated for DOA estimation with comparable accuracy, while classical root MUSIC requires the calculation of $2M-2$ roots. Thus, the computational burden of the polynomial rooting step is reduced. For modified root MUSIC, NOM is reduced to

$$\text{NOM} = 38/3 \cdot M^3 + 4 \cdot N \cdot M^2 - 3 \cdot M^2 + 148 \cdot M + 26 \qquad (9.31)$$

For improved performance in the second stage, $N = 50$ snapshots are used with modified root MUSIC. Considering $M = 6$ antennas, NOM = 10,742 multiplications is needed, while for the same performance, classical root MUSIC requires 18,522 multiplications. Hence, fine estimation with modified root MUSIC is more cost-effective than with classical root MUSIC. It should be considered that the offline DOA estimation does not need an updating rate as high as online DOA estimation (required for detection). This will reduce the overall computation cost as well.

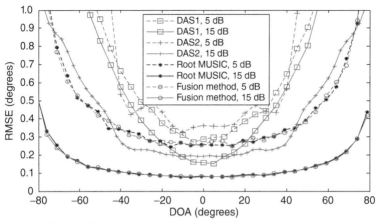

Figure 9.28   RMSE of estimations versus true DOA: one signal source, six antennas, 50 snapshots, and 500 estimations.

Simulations and Performance Analysis: Monte Carlo simulations are conducted to evaluate the proposed algorithms assuming ULA, a single source, and Gaussian noise. For DAS1, quadratic polynomial, and for DAS2, $K = 2$, and $\Delta\theta_{DOA}^{(k)} = 10^{o}, 1^{o}, 0.1^{o}$ for $k = 0$, 1, 2, respectively, are considered. For the fusion method, the error of the coarse DOA estimation is taken uniform over $[-2^{\circ}, 2^{\circ}]$. Figure 9.28 illustrates that the RMSE increases as the DOA moves away from the boresight (0°). DAS has an error between 0.2° and 0.5° when the DOA is between -40° and $+40^{\circ}$. In addition, the stability of DAS1 is lower than DAS2. The RMSE of root MUSIC and the modified root MUSIC are practically identical. Their RMSE is less than half of that of DAS methods. In addition, their stability over DOA is higher than DAS. Finally, Figure 9.28 shows that increasing SNR improves the performance.

Figure 9.29 shows the accuracy of DOA estimation methods as a function of complexity. For all techniques, the RMSE decreases drastically with complexity, until a certain threshold. This threshold is unique for each method. Beyond this threshold, the impact of adding more complexity on the RMSE is negligible. Considering six antennas, this threshold is about 20,000, 12,000, 5000, and 4000 for DAS2, root MUSIC, DAS1, and fusion methods, respectively. Considering the cost of online (coarse) DOA (DAS1), the combined cost of both coarse and fine estimations is still less than that of a single estimation with root MUSIC. The figure also shows that for DAS (root MUSIC), as the number of antennas increases, for the same performance, the complexity would decrease (increase).

The performance of DOA estimation techniques and their ability to discriminate signals coming from different directions is very important. In Figure 9.30, the corresponding performance of modified root MUSIC is illustrated. The incoming signal is simulated as the sum of LOS and NLOS signals. The error is measured as a function of $\Delta\Phi$ (the angular difference between LOS and NLOS signals) and the ratio of the power of the NLOS and LOS signals (P_{NLOS}/P_{LOS}). The error takes on a fringe pattern with respect to $\Delta\Phi$. This occurs because the main lobe of the reflected

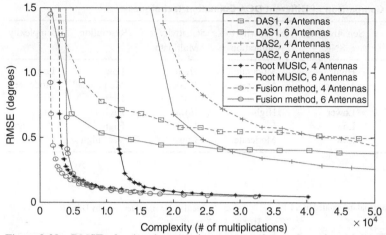

Figure 9.29   RMSE of estimation techniques versus the number of multiplications, DOA = 30°, 500 estimations, and SNR = 15°dB.

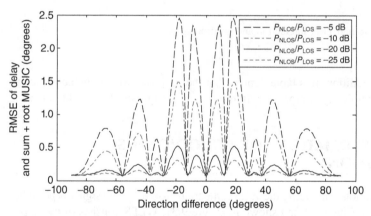

Figure 9.30   RMSE of the fusion method under multipath conditions LOS signal + single reflection DOA = 30°, six antennas, 50 snapshots, and 500 estimations.

signal interferes with the side lobes of the LOS signal. When this interference occurs, the power in one of the side lobes increases enough to be considered the main lobe, generating an incorrect DOA.

9.4.4   Comparison

Table 9.1 compares different DOA estimation methods in terms of different measures. In general, it is depicted that root MUSIC has much better properties compared

TABLE 9.1.   Comparison of Different DOA Estimation Techniques

	Sensitivity to SNR	Sensitivity to Calibration	Sensitivity to Multipath	Resolution	Complexity
DAS	High	Moderate	High	Low	Low
Max entropy	Moderate	Moderate	Moderate	Moderate	Moderate
MUSIC	Low	Higher	Low	High	High
Root MUSIC	Lower	High	Lower	Higher	High
ESPRIT	Low	Low	Low	High	Very high
Fusion	Low	Moderate	Lowest	Very high	Moderate

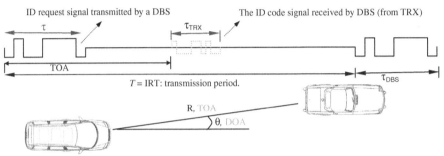

Figure 9.31   DBS and TRX interaction and positioning process.

to MUSIC. In addition, this table depicts that the proposed fusion method leads to much lower complexity compared with other methods proposed in the literature.

9.5   DOA ESTIMATION FOR PERIODIC SENSE TRANSMISSION*

In many applications, including the WLPS, the transmission of the signal is periodic. As mentioned in Chapter 1 and shown in Figure 9.31, WLPS consists of two types of nodes: (1) a multiantenna radio called dynamic base station (DBS), and (2) a single-antenna radio called transceiver (TRX). The localization process starts at the DBS (see Fig. 9.31). It transmits a signal and requests the availability of the TRX in its coverage area. Each TRX transmits a unique ID code back to the DBS as soon as it detects a request sent by the DBS and informs its availability.

   The DBS receiver estimates the round-trip time and thus the distance between the TRX and DBS. The round trip is calculated by estimating the TOA of the TRX's response signal (ID code) with reference to the starting time of the DBS's ID response (IDR) signal (see Fig. 9.31). To calculate the range, the TRX response time delay T_d should be taken into account. DBS uses an antenna array to find the DOA

---

*The materials in this section have been partially published in [24].

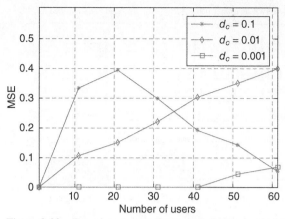

Figure 9.32   Error in autocovariance calculation due to lack of stationary assumption for different duty cycles.

of the TRX. Thus, the DBS is capable of localizing the TRX independently via DOA–TOA estimation (see Fig. 9.31). When multiple DBS nodes perform DOA–TOA positioning, a fusion across those DBS nodes improves the positioning performance [22, 23].

As shown in Figure 9.31, the DBS transmits the IDR signal periodically with the period of ID request repetition time (IRT) to all mobiles in its coverage area. This periodic transmission will create a cyclostationary property at the received signals at both the TRX and DBS receivers, which can be exploited to improve TOA, DOA, channel estimation, and, accordingly, receiver performance.

This periodic sense transmission may avoid stationarity of received signals, which in turn avoids a good estimation of sample autocorrelation of observed signals. Autocorrelation of observed signals should be calculated for applications such as optimal BF, TOA, and DOA estimation (e.g., in MUSIC algorithm). Typically, Figure 9.32 represents the error in the estimation of the sample covariance matrix defined as

$$\text{MSE} = \sum_{m=1}^{M} \sum_{u=1}^{M} \left| \hat{R}(m, u) - R(m, u) \right|.$$

Here, M is the number of antenna array elements, $R(m, u)$ is the (m, u) element of the real (estimated) covariance matrix. It is observed that this error is a function of the number of users (nodes) and the duty cycle of the periodic sense signals that is denoted by $d_c = \tau / T$. The parameters τ and T have been depicted in Figure 9.31.

Periodic sense transmission of the signals creates a cyclostationary process in the received signal over each band. As shown in Figure 9.33, the cyclostationary period can be interpreted as the time over which the interference experienced by each transmitted symbol remains unchanged. Cyclostationarity enables high-performance covariance matrix estimation, which leads to optimum BF, detection, and localization. The cyclostationary process is available for a time period, T_{cy}, called cyclostationary coherence time, which is a function of the bandwidth (B) over which

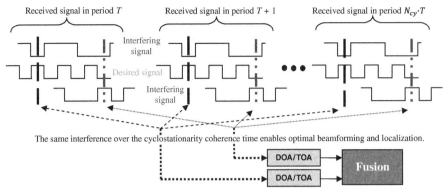

Figure 9.33 Representation of the cyclostationary process and its role in improving localization performance.

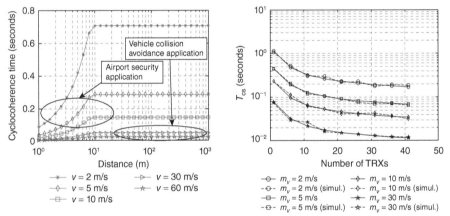

Figure 9.34 Cyclostationarity duration in terms of distance (left) and number of TRXs and their average speed (right).

the signal is transmitted, the beamwidth of antenna array (θ_B), the distance between two nodes that are involved in the localization of each other (d), the speed of nodes (v), and the speed of light (c) [24]. T_{cy} is approximated by the inverse of *cyclostationary Doppler spread*, $B_{d,cy}$, and

$$B_{d,cy} = \max\left(\frac{Bv}{\sqrt{2}c}, \frac{\sqrt{2}v}{\theta_B d}\right). \tag{9.32}$$

Theoretical and simulation T_{cy} results have been sketched in Figure 9.34 [24]. These results are computed by defining and allocating a probability to the cyclostationary time (see [24] for details). As expected, by increasing the speed of vehicle or the number of TRX sets, the cyclostationary time decreases.

An important issue of the cyclostationary-based autocorrelation estimator is the maximum possible value of the number of periods T (see Fig. 9.33) over which the cyclostationarity holds (i.e., the ratio of T_{cy} and T). A larger value $N_{cy} = T_{cy}/T$ leads to better covariance matrix estimation. Typically, T_{cy} is in the order of 1 second for a speed of 2 m/s and the distance of 10 m between nodes, 300-MHz bandwidth, and 27° of HPBW. Now, in Figure 9.33, if the period of transmission over each band is 1 ms, then within $N_{cy} = 1000$ periods, the channel would remain unchanged over each transmitted symbol.

Therefore, as depicted in Figure 9.33, based on the cyclostationary principles, the interference across a symbol remains unchanged over multiple periods determined by T_{cy}. Hence, the estimated autocorrelation of the observed signal $\bar{y}(n, \omega)$ (considering all antenna elements) for each symbol, n, and across multiple periods, ω, corresponds to

$$\hat{R}(n) = \frac{1}{N_{cy}} \sum_{\omega=1}^{N_{cy}} \bar{y}(n, \omega) \bar{y}^H(n, \omega). \tag{9.33}$$

As shown in Figure 9.33, using (9.33), DOA and TOA can be estimated across each symbol, n. Next, these DOA and TOA are fused to improve the positioning performance. Moreover, in optimal BF, Equation (9.33) is used to compute the weight matrix applied to the antenna elements.

We incorporate cyclostationarity and estimate DOA using each bit of the received signal in all frames via the MUSIC method and then find the best estimate of DOA based on those observations (see Fig. 9.34). ML estimator is applied across DOA estimations, which correspond to

$$\hat{\omega}_{ML} = \frac{\displaystyle\sum_{k=1}^{K} \frac{\hat{\omega}_k}{\sigma_k^2}}{\displaystyle\sum_{k=1}^{K} \frac{1}{\sigma_k^2}}. \tag{9.34}$$

The optimal DOA is computed from the optimal spatial frequency $\hat{\omega}_{ML}$ via

$$\hat{\theta} = \sin^{-1}\left(\frac{\lambda}{2\pi d} \hat{\omega}_{ML}\right). \tag{9.35}$$

Here, primary DOA estimates are obtained from different bits of consecutive frames. In other words, after receiving signals within all IRT frames within T_{cs}, we apply MUSIC to the bth, $b \in \{1, 2, \ldots, N_b\}$ bit of all frames and calculate DOA (N_b is the number of bits in an ID frame); that is,

$$\hat{\omega}_b = \omega + \Delta\omega_b; b = 1, 2, \ldots, N_b..$$

Similarly, for a statistically large number of samples, the estimation errors are zero-mean Gaussian variables; that is, $\Delta\omega_b \sim N(0, \sigma_b^2), b = 1, 2, \ldots, N_b$.

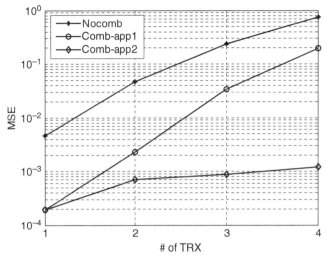

Figure 9.35 Performance comparison of proposed approaches in terms of SNR for two TRXs.

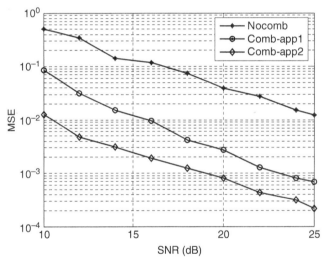

Figure 9.36 Performance comparison of proposed approaches in terms of number of TRXs at SNR = 20°dB.

Figures 9.35 and 9.36 represent the MSE performance simulation results of the estimated DOA. The assumptions include the following: (1) Two TRXs are present at 15° and 25°, respectively; (2) a DBS receiver equipped with a ULA with six elements and half-wavelength element spacing; (3) binary phase-shift keying (BPSK) modulation; (4) 20 bits for each ID frame; (5) the difference between the TOAs of the two TRXs is $5T_b$; and (6) $T_{cs} = 20IRT$; this assumption would be reasonable selecting $IRT \approx 50\,\mu s$ [6], as the T_{cs} calculated in Figure 9.34 is in the order of 100 ms. Figure 9.35 shows that at $MSE \approx 10^{-2}$; about 10-dB (15-dB) improvement is achievable via the first (second) estimation approach compared to a system without combining.

Figure 9.36 represents the variation of MSE with the number of TRXs, assuming SNR = 20°dB and the difference between TOAs of the desired TRX and j^{th} TRX is $5T_b$, $10T_b$ and $-5T_b$ for $j = 2, 3, 4$, respectively. It is observed that (1) when only one TRX is available, the performance of both combining approaches is identical (as expected); (2) considering two TRXs, the second approach leads to better MSE performance; and (3) as the number of TRXs increases, the difference between MSE performances increases. It should be pointed out that better performance of the second combining approach is achievable with a minimal expense of computing the estimation error variance.

The results in Figures 9.35 and 9.36 represent three scenarios. The worst performance is associated with the scenario that fusion is not available and DOA is estimated by applying MUSIC across all bits of one IRT period (no combining). In this case, it is clear that there is a major error in the covariance matrix estimation of MUSIC that reduces the performance of DOA estimation. In the second scenario (comb-app1), fusion is applied across multiple estimated DOAs (in a number of IRT periods within cyclostationary time). Here, each DOA is estimated by applying MUSIC to all bits of the same IRT period. In this case, the performance improves; however, since covariance estimation involves error, the performance is ideal.

The third scenario (comb-app2) is consistent with Figure 9.33 where covariance matrix estimation is applied across the same bit of multiple ITR periods. Next, fusion is applied across all bits. In this case, covariance is ideally estimated and the best performance is achievable. The results represent a major difference in performance for a higher number of TRXs. It is due to the fact that as the number of TRXs increases, the MSE in the autocorrelation matrix increases (see Fig. 9.32).

9.6 CONCLUSION

The chapter presented the basics of antenna arrays and DOA estimation techniques. In addition, it presented a DOA fusion method that merges a simple DOA estimation method, which is DAS, and a complex method, which is root MUSIC. It is observed that this merger leads to higher performance and lower complexity. This makes this DOA estimation technique a good candidate for the implementation on software-defined radios. In addition, we proposed a DOA estimation method for periodic sense transmissions that is inherent in localization systems such as radar and WLPS. It is depicted that by exploiting cyclostationarity, we can improve the performance of DOA estimation methods.

ACKNOWLEDGMENTS

Many former students in my group have contributed in the preparation of pictures and MATLAB programs for this chapter. Specifically, I would like to thank Dr. Zhonghai Wang and Ms. Xiaofeng Yang.

REFERENCES

[1] L. C. Godara, "Limitations and capabilities of directions-of-arrival estimation techniques using an array of antennas: A mobile communications perspective," in *Proc. IEEE Int. Symp. on Phased Array Systems and Technology*, 1996, pp. 327–333.

[2] D. H. Johnson, "The application of spectral estimation methods to bearing estimation problems," *Proc. IEEE*, vol. 70, pp. 1018–1028, 1982.

[3] S. A. Zekavat, A. Kolbus, X. Yang, Z. Wang, J. Pourrostam, and M. Pourkhaatoon, "A novel implementation of DOA estimation for node localization on software defined radios: Achieving high performance with low complexity," *presented at IEEE ICSPC 2007*, Dubai, UAE, Nov. 26–27, 2007.

[4] R. O. Schmidt, "Multiple emitter location and signal parameter estimation," *IEEE Trans. Antenn. Propag.*, vol. AP-34, pp. 276–280, 1986.

[5] A. J. Barabell, "Improving the resolution performance of eigenstructure-based direction-finding algorithms," presented at IEEE ICASSP, Boston, MA, 1983.

[6] R. Roy and T. Kailath, "ESPRIT—Estimation of signal parameters via rotational invariance techniques," *IEEE Trans. Acoust. Speech Signal Process.*, vol. ASSP-37, pp. 984–995, 1989.

[7] L. C. Godara, "Application of antenna arrays to mobile communications. II. Beam-forming and direction-of-arrival considerations," *Proc. IEEE*, vol. 85, no. 8, pp. 1195–1245, 1997.

[8] S. A. Zekavat and F. Emdad, "Achieving higher uplink performance and capacity via non overlapping window adaptive maximum noise fraction beamforming technique," *IET Commun.*, vol. 5, no. 5, pp. 598–605, 2011.

[9] T. Su, K. Dandekar, and H. Ling, "Simulation of mutual coupling effect in circular arrays for direction-finding applications," *Microw. Opt. Tech. Lett.*, vol. 26, no. 5, pp. 331–336, 2000.

[10] A. Manikas and N. Fistas, "Modeling and estimation of mutual coupling between array elements," in *IEEE Int. Conf. on Acoustics, Speech, and Signal Processing*, Adelaide, SA, Australia, Apr. 19–22, 1994, pp. 553–556.

[11] G. Sommerkorn, D. Hampicke, R. Klukas, A. Richter, A. Schneider, and R. S. Thoma, "Uniform rectangular antenna array calibration issues for 2-D ESPRIT application," presented at the *4th European Personal Mobile Communications Conf., EPMCC 2001*, Vienna, Austria, Feb. 20–22, 2001.

[12] L. Lengier and R. Farrell, "Amplitude and phase mismatch calibration testbed for 2 × 2 tower-top antenna array system," in *2007 China–Ireland Int. Conf. on Information and Communications Technologies*, Dublin, Ireland, Aug. 28–29, 2007, pp. 165–172.

[13] Y. Yasui, S. Kobayakawa, and T. Nakamura, "Adaptive array antenna for W-CDMA systems," *Fujitsu Sci. Tech. J.*, vol. 38, no. 2, pp. 192–200, 2002.

[14] T. Huang and A. S. Mohan, "Effects of array mutual coupling on near-field DOA estimation," in *IEEE CCECE 2003 Canadian Conf. on Electrical and Computer Engineering*, vol. 3, 2003, pp. 1881–1884.

[15] D.-S. Yang, J. Shi, and B.-S. Liu, "Adaptive target tracking for wideband sources in near field," in *12th Int. Conf. on Information Fusion, FUSION '09*, 2009, pp. 649–655.

[16] M. Lin and L. Yang, "Blind calibration and DOA estimation with uniform circular arrays in the presence of mutual coupling," *IEEE Antennas Wirel. Propag. Lett.*, vol. 5, no. 1, pp. 315–318, 2006.

[17] N. B. Poh, J. P. Lie, H. E. Meng, and A. Feng, "A practical simple geometry and gain/phase calibration technique for antenna array processing," *IEEE Trans. Antenn. Propag.*, vol. 57, no. 7, pp. 1963–1972, 2009.

[18] Y. Horiki and E. H. Newman, "A self-calibration technique for a DOA array with near-zone scatterers," *IEEE Trans. Antenn. Propag.*, vol. 54, no. 4, pp. 1162–1166, 2006.

[19] F. Li and R. J. Vaccaro, "Sensitivity analysis of DOA estimation algorithms to sensor errors," *IEEE Trans. Aerosp. Electron. Syst.*, vol. 28, no. 3, pp. 708–717, 1992.

[20] E. Santos and M. Zoltowski, "Power spectrum estimation with low rank beamforming," in *Proc. GLOBECOM*, 2005, pp. 2184–2188.

[21] S. Nemirovsky and M. A. Doron, "Sensitivity of MUSIC and root-MUSIC to gain calibration errors of 2D arbitrary array configuration," in *Sensor Array and Multichannel Signal Processing Workshop Proc.*, 2004, pp. 594–598.

[22] W. Wang and S. A. Zekavat, "A novel semi-distributed localization via multi-node TOA-DOA fusion," *IEEE Trans. Veh. Technol.*, vol. 58, no. 7, pp. 3426–3435, 2009.

[23] W. Wang and S. A. Zekavat, "Comparison of semi-distributed multi-node TOA-DOA fusion localization and GPS-aided TOA (DOA) fusion localization for manets," *EURASIP J. Adv. Signal Process.*, vol. 2008, Dec. 2008, Article ID 439523.

[24] H. Tong, J. Pourrostam, and S. A. Zekavat, "Optimum beam-forming for a novel wireless local positioning system: A stationarity analysis and solution," *EURASIP J. Adv. Signal Process.*, vol. 2007, Dec. 2007, Article ID 98243.

CHAPTER 10

POSITIONING IN INHOMOGENEOUS MEDIA

Mohsen Jamalabdollahi and S. A. (Reza) Zekavat, Michigan Tech

IN CHAPTERS 6–9 of this handbook, the details of time-of-arrival (TOA) and direction-of-arrival (DOA) estimation techniques were reviewed. This chapter studies techniques of measuring the straight-line range between sensor nodes distributed within Inhomogeneous media (IHM), which are critical to localization and body scanning applications. This chapter proposes high-resolution TOA estimation in IHM. In IHM, TOA cannot be mapped into the actual range between the transmitter and receiver, as the propagation path of transmitted waveform is not a straightline. This chapter discusses how the range in IHM is computed via a combination of TOA and DOA estimation.

10.1 INTRODUCTION

Range measurement is desired in the variety of applications that are performed in free space, such as digital communication, wireless sensor networks (WSNs), radar, and wireless indoor and outdoor localization. WSNs have emerging applications in homogeneous media (e.g., free space) as well as IHM. Examples of homogeneous media include environmental monitoring, search and rescue, health, road traffic monitoring, and pollution sensing. Examples of IHM include endoscopy capsule localization, cancer detection, drug delivery, underwater monitoring and surveillance, and mine safety and underground monitoring. Each of these applications needs location information of measured data or observed events.

Range-based localization is possible via received signal strength (RSS) [1], signal TOA [2], time difference of arrival (TDOA) [3], DOA, [4] or their combinations [5, 6]. Range-free methods utilize connectivity information [7]. Although some range-free approaches offer simple methods with acceptable performance, they have limitations such as network topology and ranging accuracy. This justifies the employment of range-based approaches.

Handbook of Position Location: Theory, Practice, and Advances, Second Edition.
Edited by S. A. (Reza) Zekavat and R. Michael Buehrer.
© 2019 by the Institute of Electrical and Electronics Engineers, Inc.
Published 2019 by John Wiley & Sons, Inc.
Companion Website: www.wiley.com/go/zekavat/positionlocation2e

Within all proposed range-based approaches, TOA estimation has received considerable attention because of its high precision and low complexity [8]. TOA estimation techniques are divided into two different categories, the traditional matched filter-based and the super-resolution techniques. The former incorporates a predesigned waveform with autocorrelation properties close to the delta function at the output of the matched filter. In the latter, TOA is calculated via maximizing the pseudospectrum of the corresponding signal subspace achievable via decomposition of the matched filter output in frequency domain [9]. Examples of super-resolution techniques are maximum likelihood (ML) [10], multiple signal classification (MUSIC) [11], and estimation of signal parameter via rotational invariance technique (ESPRIT) [9].

The resolution of all proposed techniques depends on the transmitted waveform bandwidth. Incorporating wideband waveforms in free space enables high TOA resolution; however, increasing the bandwidth of transmitted waveform in dispersive channels intensifies signal dispersion in time domain [12, 13]. Here, signal dispersion refers to the principle that different frequency components of a transmitted electromagnetic (EM) waveform propagate with different velocities due to the frequency dependency of a medium's relative permeability and/or permittivity. Therefore, increasing a transmitted waveform's bandwidth leads to higher dispersion, which alleviates the nonlinear distortion of transmitted waveform. The nonlinearity nature of the imposed distortion by waveform dispersion leads to a received signal that is barely detectable by the matched filter of a transmitted waveform, which dramatically degrades the TOA estimator performance [13].

Acoustic range measurement is a reliable approach for underwater localization; however, it cannot be applied to IHM such as underwater-airborne channels due to strong reflections and attenuation in the water/air boundary. Authors of [14–19] propose TOA estimation in a dispersive medium for specific scenarios, which cannot be extended to IHM. In [14–16], time delay estimation of buried target echoes for ground penetration radar (GPR) is addressed; meanwhile, the received echoes do not represent frequency dispersion assuming low conductivity and small layer thickness of submedia.

A low conductivity assumption is feasible for dry media; however, when the water content of media exceeds 10% by weight, the frequency dispersion must be considered due to the dielectric relaxation of water [12]. In [17], the transmitter location is estimated by combining the measured TOA data with the knowledge about the shape and position of the medium, which is not a feasible assumption in IHM. Authors in [18] propose sensor node localization via TOA measurements, meanwhile the procedure of TOA estimation in soil (i.e., a frequency dispersive medium) has not been discussed. In [19], TOA estimation of short-range seismic signals in dispersive environments is addressed; however, the system model does not represent the frequency dispersion due to low bandwidth of applied seismic signals.

In this chapter, applications of localization in IHM are introduced in Section 10.2. As explained in previous chapters, TOA estimation performance is improved as bandwidth increases. However, higher bandwidth in IHM leads to high dispersion. Section 10.3 discusses the impact of exploiting wideband waveform in IHM consisting of dispersive submedia. High-resolution TOA estimation via bandlimited signals is proposed in Section 10.4. Exploiting bandlimited signals mitigates waveform dispersion in IHM. In addition, the low resolution of TOA estimation imposed by signal limited bandwidth is addressed via the maximum rising level detector (MRLD) technique. Section 10.5 proposes TOA estimation in IHM via wideband signals. The proposed technique exploits preallocated orthogonal frequency division multiplexing (OFDM) subcarriers to construct a wideband ranging signal. The TOA measurement technique (via bandlimited or wideband waveforms) discussed in this chapter is combined with DOA measurements to construct a system of linear equations as a function of the thickness of available submedia. Once the thickness of available submedia is estimated, the straight-line range between transmitter and receiver is computed via the method introduced in Section 10.6. IHM channel and layer thickness estimation applications for underground sensing and tomography or wireless nanosensors (WNS) is discussed in Section 10.7.

10.2 IMPACT OF WAVEFORM DISPERSION ON TOA ESTIMATION

Traditional TOA estimation approaches exploit a waveform with good autocorrelation properties and estimate TOA by maximizing the output of the matched filter. Moreover, the channel impulse response (CIR) can be estimated at the output of matched filter [20, 21]. Matched filter represents low complexity and outstanding performance in Multi-Path Frequency Selective (MPFS) channels. However, its resolution is limited by available bandwidth.

Super-resolution methods such as MUSIC, minimum variance distortionless response (MVDR), and ESPRIT [9] offer higher resolution. However, their computational complexity (imposed by autocorrelation matrix singular value decomposition) is high. Moreover, super-resolution methods require the transmission of a waveform with good autocorrelation properties in order to estimate CIR at the output of matched filter [22]. Here, the waveform with good autocorrelation properties refers to any waveform with a low integrated side lobe level (ISL), defined as [20, 21]:

$$ISL = \sum_{m=1}^{N-1} \left| \sum_{n=m}^{N-1} s_n s_{N+n-m}^* \right|^2 , \tag{10.1}$$

where S_n is the nth sample of a transmitted waveform with length N. Designing a sequence with good correlation properties has significant theoretical and practical interest [20, 21]. Here, designing a low-ISL waveform for TOA estimation in IHM entails extra limitations. In homogeneous media (e.g., free space), the resolution of TOA methods increases as the transmitted waveform bandwidth increases. However, wideband signals disperse within IHM consisting of multiple submedia and cannot offer low-ISL properties at the output of matched filter.

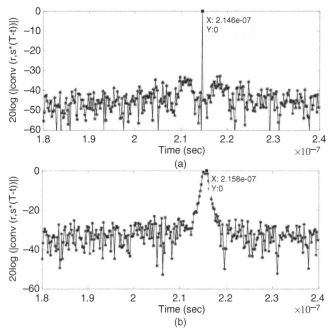

Figure 10.1 Impact of media on detection of waveform in (a) free space, (b) the human body (dispersive media).

Figure 10.1 depicts an example of absolute value of matched filter output in (a) free space and (b) the human body exploiting the Golomb sequence s_n, such that [23]:

$$s_n = e^{j\pi \frac{n(n-1)}{N}}, \qquad \text{for } 1 \leq n \leq N. \tag{10.2}$$

Figure 10.1a is the absolute value of the convolution of the Golomb sequence proposed in (10.2) with $N = 128$, propagated within free space. However, Figure 10.1b depicts the absolute value of convolution of the same Golomb sequence propagated within the human body consisting of five body organs. Here, the waveform is up-converted to carrier frequency of 5 GHz, considering the sampling time of 1 ns. It is observed that the output corresponding to the free-space channel represents sharp autocorrelation levels with very low ISL, while the autocorrelation levels corresponding to the propagated waveform within the human body contains high ISL due to dependency of signal velocity to permittivity, which changes with frequency.

Table 10.1 shows the details of permittivities of selected organs at simulations of Figure 10.1b at maximum (5.5 HHz) and minimum (4.5 GHz) frequencies of transmitted waveform achievable via measurements by [24] (here, permeabilities are considered as unity for simplicity). This phenomena is known as waveform dispersion due to various velocities of frequency components of the propagated signal. Theoretically, the propagation velocity of an EM wave such as $e^{j\pi f_0 t}$ in dispersive media is:

$$\vartheta(f_0) = \frac{c}{\sqrt{\mu(f_0)\varepsilon(f_0)}}, \tag{10.3}$$

TABLE 10.1.   Relative Permittivities of Human Body Organs at Various Frequencies

Frequency\Organ	Body Fluid	Liver	Fat	Muscle	Stomach
4.5 GHz	66.392	39.972	5.076	50.186	58.758
5.5 GHz	65.182	38.553	4.980	48.883	57.009

where c is the universal speed of light, and $\mu(f_0)$ and $\varepsilon(f_0)$ are media's relative permeability and permittivity and frequency f_0. Hence, exploiting higher bandwidth leads to more difference in frequency component velocity, which leads to higher dispersion. In order to mitigate dispersion issues, we need to consider:

1. low-ISL waveforms to improve TOA performance; and
2. bandwidth limitations imposed by the dispersive channels.

In [20] and [21], it is depicted that the optimum waveform for TOA with low ISL has a near-white power spectrum; that is, it includes all frequencies within $[-f_s/2 + f_s/2]$, f_s is the sampling frequency. This condition is not consistent with the required bandwidth limitations discussed earlier. In the following sections, we propose two possible approaches to address TOA estimation in IHM consisting of multiple dispersive submedia.

10.3   TOA ESTIMATION IN IHM VIA BANDLIMITED SIGNALS

The traditional matched filter-based TOA estimation approach can be applied to IHM if the bandwidth of transmitted waveform is chosen such that the waveform dispersion is negligible. However, avoiding dispersion entails very limited bandwidth, which leads to poor resolution. For instance, the permittivity change over 100 MHz at the center frequency of 5 GHz is negligible considering the human body as propagation media. However, exploiting waveform with 100-MHz bandwidth offers ranging resolution that is limited to $\frac{3 \times 10^8}{\sqrt{50}} \times \frac{1}{10^8} \cong 42 \ cm$ (considering average relative permittivity of the human body equal to 50 at 5 GHz), which is not feasible.

In this section, first the mathematical approach for waveforms with very low side-lobes at the output of matched filter was proposed. Next, the MRLD technique is discussed to attain higher TOA resolution.

10.3.1   Waveform Design

The desired waveform sequence $s_n = e^{j\varphi_n}$ for $n = 1, 2, \ldots, N$, is achievable by minimizing the sidelobes of autocorrelation over its center value. Different criteria for optimum waveform design, such as cyclic algorithms (CAs), CA-pruned (CAP), CA-new (CAN), or its weighted version, weighted-CAN (WECAN) [20, 21, 23], can be applied for unimodular sequences (any sequence in the form of $s_n = e^{j\varphi_n}$). Moreover, various techniques such as iterative twisted approximation (ITROX) [25] or monotonically error-bound improving technique (MERIT) [26] are proposed to design real value sequences with very low autocorrelation sidelobe levels. In all of these techniques, the optimum sequences are derived by minimizing the ISL (10.1) or weighted ISL (WISL) functions. Here, we aim discuss minimization of WISL that corresponds to [21, 23]:

$$\hat{\varphi} = \underset{\varphi}{\mathrm{argmin}} \; WISL = \underset{\varphi}{\mathrm{argmin}} \sum_{n=k}^{N-1} w_k |y_k|^2 \; s.t. \; y_k = \sum_{n=1}^{N-1} s_n s_{N+n-k}^*, \qquad (10.4)$$

where $\hat{\varphi} = [\varphi_1, \varphi_2, \ldots, \varphi_N]^T$, and N, s_n and w_k is the length of desired waveform, the nth sample of the desired waveform and corresponding weight value, respectively. In addition, $(.)^*$ indicates the complex conjugate notation. Considering the Parseval energy equality in time-frequency domain and defining the $2N$ point discrete Fourier transform of weighted sidelobes $w_n y_n$, $n = 1, 2, \ldots, N-1$ as S_p, $p = 1, 2, \ldots, 2N$, it can be shown that [23]:

$$WISL = \frac{1}{4N} \sum_{p=1}^{2N} [S_p - N]^2, \qquad (10.5)$$

in which,

$$S_P = \frac{1}{\sqrt{2N}} \sum_{n=-(N-1)}^{N-1} w_k^2 y_k e^{-j\frac{2\pi k p}{2N}}. \qquad (10.6)$$

Substituting for y_k from (10.4) leads to:

$$S_P = \sum_{l=1}^{N} \sum_{k=1}^{N} e^{-j\varphi k} e^{j\frac{\pi k p}{N}} \Gamma_{k,l} e^{j\varphi l} e^{-j\frac{\pi l p}{N}}. \qquad (10.7)$$

Here, $\Gamma_{k,l}$ represents the elements at the kth row and lth column of squared weights matrix Γ, as its elements represent the squared values of ISL weights, that is, w_k^2, as shown in (10.6). This leads to the final format of waveform design objective function as follows:

$$\hat{\varphi} = \underset{\varphi}{\mathrm{argmin}} \sum_{p=1}^{2N} [S_p - N]^2 s.t.: S_p = \sum_{l=1}^{N} \sum_{k=1}^{N} e^{-j\varphi k} e^{j\frac{\pi k p}{N}} \Gamma_{k,l} e^{j\varphi l} e^{-j\frac{\pi l p}{N}}. \qquad (10.8)$$

In [27] the proposed optimization problem is addressed via the trust-region algorithm by solving $\nabla_\varphi \sum_{p=1}^{2N} [S_p - N]^2 = 0$. The trust-region technique aims to find the optimum step size without exceeding the trust-region radius, which leads to quick convergence to a more interesting area [31], and therefore much better performance compared to Levenberg–Marquardt approach.

Figures 10.2a and b depicts the autocorrelation levels of the designed waveforms exploiting the proposed waveform design technique for $N = 128$ and 512, respectively, using the following squared weights values:

$$\Gamma = \begin{bmatrix} 1 & \gamma_1 & \cdots & \gamma_{N-1} \\ \gamma_1 & 1 & \cdots & \gamma_{N-2} \\ \vdots & \vdots & \ddots & \vdots \\ \gamma_{N-1} & \gamma_{N-2} & \cdots & 1 \end{bmatrix} \; for: \gamma_n = \begin{cases} 8 & n = N \\ 0.6 & N - N_0 \le n \le N-1, \\ 0 & elsewhere \end{cases} \qquad (10.9)$$

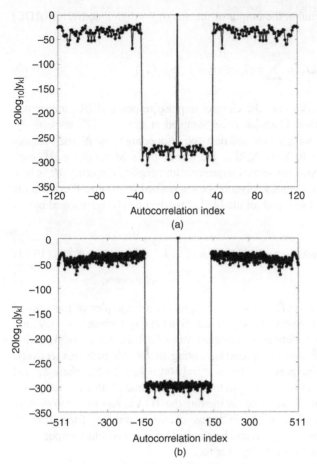

Figure 10.2   Autocorrelation level of designed waveform, (a) $N = 128$, (b) $N = 512$.

where $N_0 = 35$ and 140 for simulations presented in Figures 10.2a and b, respectively. Moreover, the waveform $s_n = e^{j\varphi_n}$ for $n = 1, 2, \ldots, N$, (or its phase sequence φ in (10.8)) is initialized exploiting the Golomb sequence defined in (10.2). In order to ease the implementation of the trust-region algorithm, the useful MATLAB function *fsolve* can be used. Examples of designed waveform are available online at [28].

10.3.2 High-Resolution TOA Estimation via Bandlimited Signals

Figure 10.3 shows the block diagrams corresponding to the proposed MRLD technique. This section aims to explain the MRLD technique by following the signal path in the depicted block diagrams in Figure 10.3. Considering transmitted waveform with very low sidelobes at the autocorrelation output proposed at Section 10.3.1, the baseband received signal at the output of the analog-to-digital convertor (ADC) corresponds to:

$$r(kT_s) = \sum_{l=0}^{L-1} h(lT_s)s(kT_s - \tau_l) + v(kT_s), \tag{10.10}$$

where $h(kT_s)$, $s(kT_s)$, and $v(kT_s)$ are the channel impulse response (CIR), transmitted waveform, and additive white Gaussian noise sampled at time $t = kT_s$, respectively. Moreover, T_s and τ_l are the sample time and delay corresponding to the lth path, respectively. As shown in Figure 10.3, the MRLD receiver employs M parallel paths, each incorporating a delayed version of sampled sequence with sampling frequency Mf_s, where $f_s = \frac{1}{T_s}$ and T_s represents the sample interval of the transmitted waveform. Therefore, the kth sample at the mth correlation path of the MRLD receiver can be represented by:

$$r_m(kT_s)$$
$$= \begin{cases} \sum_{l=0}^{L-1} h(l\tilde{T}_s)s((Mk+m)\tilde{T}_s - \tau_l) + v((Mk+m)\tilde{T}_s), & k_{\tau_0} \le k \le k_{\tau_{L-1}} + N \\ v((Mk+m)\tilde{T}_s), & elsewhere, \end{cases} \tag{10.11}$$

where $\tilde{T}_s = \frac{1}{MT_s}$, and $s(k\tilde{T}_s)$, $r(k\tilde{T}_s)$, and $v(k\tilde{T}_s)$ represent the samples of transmitted and received signals, and zero-mean complex additive Gaussian noise at $t = k\tilde{T}_s$, and $h(l\tilde{T}_s)$, $l = 0, 1, \dots, ML - 1$, denote the complex value CIR. Furthermore, M and τ_l are the oversampling rate and delay corresponding to the lth path, respectively. Defining the coarse and fine part of delay as $\tau_l = (Mk_{\tau_l} + k_{\delta_l})\tilde{T}_s + \tilde{\delta}_l$, where k_{τ_l} and k_{δ_l} are the sample indices corresponding to the coarse and fine of lth path delay, and $\tilde{\delta}_l$ is the remnant of the fine TOA. Here, estimating the TOA is equivalent to discovering the index corresponding to τ_0 in terms of new sampling time, that is, $(Mk_{\tau_0} + k_{\delta_0})\tilde{T}_s$. Considering (10.11), it can be shown that matched filter output at the mth correlation path of receiver corresponds to:

$$g_m(kT_s)$$
$$= \begin{cases} \sum_{l=0}^{L-1} h(lT_s) \sum_{n=k}^{N+k-1} s((n-k_{\tau_0})T_s)s^*((n-k_{\tau_0})T_s) + \tilde{v}(kT_s), & m = k_{\delta_0} \\ \\ \sum_{l=0}^{L-1} h(lT_s) \sum_{n=k}^{N+k-1} s\left((n-k_{\tau_0} - \frac{m-k_{\delta_0}}{M})T_s\right)s^*((n-k_{\tau_0})T_s) + \tilde{v}(kT_s), & m \ne k_{\delta_0}. \end{cases} \tag{10.12}$$

Here, it is observed that the matched filter output corresponding to the correlation path including the original waveform samples, that is, $m = k_{\delta_0}$, can be simplified to:

$$g_m(kT_s) = \sum_{l=0}^{L-1} h_l y_{k-k_{\tau_0}-l} + \tilde{v}_{m,k} \qquad \text{for: } m = k_{\delta_0}, \tag{10.13}$$

where y_k represents the autocorrelation function of the transmitted waveform defined in (10.4). Considering transmission of waveform with very low ISL obtained in Section 10.3, Equation (10.13) can be written as:

$$g_m(kT_s) = \begin{cases} h_l + \tilde{v}_{m,k} & k_{\tau_0} \le k \le k_{\tau_0} + L - 1 \\ \tilde{v}_{m,k} & elsewhere \end{cases}. \tag{10.14}$$

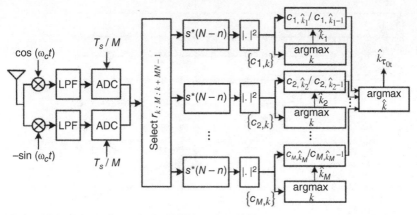

Figure 10.3   Block diagram of MRLD receiver [19].

Here, it is observed that the matched filter output for $m = k_{\delta_0}$ contains the noisy version of channel impulse response for $k_{\tau_0} \leq k \leq k_{\tau_0} + L - 1$, and colored noise elsewhere. However, for $m \neq k_{\delta_0}$ the sidelobe corresponding to oversampled version of waveform $s\left(\left(n - k_{\tau_0} - \frac{m - k_{\delta_0}}{M}\right)T_s\right)$ is available. The absence of sidelobes in the correlation path corresponding to $m = k_{\delta_0}$, is key to reveal higher TOA resolution exploiting the MRLD technique. Therefore, the MRLD receiver finds the maximum absolute values of the matched filters output at each correlation path, that is, $\hat{k}_m$, $m = 1, 2, \ldots, M$. Then, it estimates the TOA via finding the correlation path (m^*) by maximizing the rising levels $\left(\frac{c_{m,\hat{k}_m}}{c_{m,\hat{k}_m - 1}}\right)$, among the obtained maximum values of matched filter outputs. Thus, the MRLD technique is efficiently implementable based on three simple steps:

1. searching for the peak of correlation levels at each path using the transmitted waveform matched filter ($s^*(N - n)$);

2. measuring the rising level corresponding to the revealed peaks at each path; and

3. searching for the maximum among the measured rising levels.

Therefore, the path containing the maximum rising level (m^*), among M available correlation paths is achievable via:

$$m^* = \underset{m}{\operatorname{argmax}}\left(\frac{c_{m,\hat{k}_m}}{c_{m,\hat{k}_m - 1}}\right), \ s.t. \ c_{m,\hat{k}_m} = \underset{k}{\operatorname{argmax}}\left|\sum_{n=k}^{k+N-1} r_m(nT_s)s^*_{N+n-k}\right|^2. \quad (10.15)$$

Using (10.15), the estimated TOA can be represented by:

$$\hat{\tau}_0 = \left(M\hat{k}_{m^*} + m^*\right)\tilde{T}_s, \quad (10.16)$$

where m^* and $\hat{k}_{m^*}$ denote the correlation path number containing the maximum rising level and its corresponding index of maximum correlation level, respectively, achievable via (10.15).

Figure 10.4 depicts the MSE of the estimated TOA exploiting the designed waveform in Section 10.3 for $N = 512$, and the MRLD receiver in the presence of unknown MPFS channels in (a) indoor (b) outdoor environments. Here, different

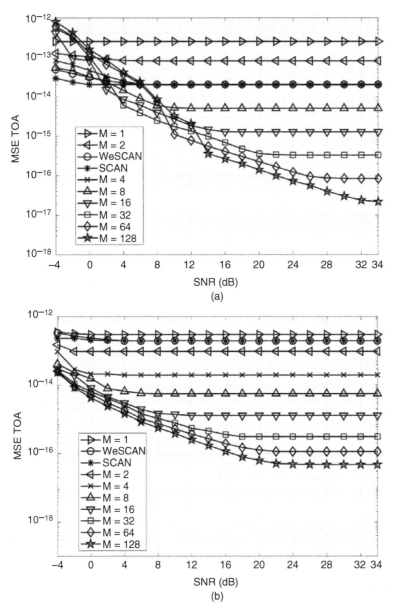

Figure 10.4 The MSE of TOA exploiting different numbers of correlation paths M versus SCAN and WeSCAN [23] with the same bandwidth (1 MHz); (a) indoor MPFS; (b) outdoor MPFS channels.

numbers of correlation paths are applied to the MRLD receiver to investigate its impact on the proposed TOA estimator and compare them with the stop-band CA-pruned (SCAN) and weighted SCAN (WeSCAN) [23] approaches.

As shown in Figure 10.4, the resolution of TOA estimator improves by increasing the number of correlation paths in the MRLD receiver specifically at higher signal-to-noise (SNR) regimes. This is due to the impact of extra resolvable paths created by increasing the oversampling rate. Therefore, according to the simulation results shown in Figure 10.4a, it is concluded that for indoor channels, a lower oversampling rate should be selected at low SNR regimes, while for higher SNR regimes, a higher oversampling rate is feasible. For outdoor scenarios, resolution of the TOA estimator improves by increasing the number of correlation paths in the MRLD receiver as shown in Figure 10.4b; however, it is observed that the proposed MRLD technique performance is more reliable even at lower SNR regimes.

10.4 TOA ESTIMATION IN IHM VIA WIDEBAND SIGNALS

TOA estimation in IHM via wideband signals needs to deal with waveform dispersion described in Section 10.3. Here, we present the new method proposed in [13], which exploits preallocated orthogonal frequency division multiplexing (OFDM) subcarriers to construct a wideband ranging signal. Exploiting OFDM subcarriers to construct ranging waveform has been leveraged in many works [29, 30]. Here, we show that each frequency component of the propagated waveform is received with different time delays and phases, which dramatically increases the number of unknowns in the received signal system model. Then, we propose a novel approach, which reduces the number of unknowns by linear approximation of imposed delays and phases to the allocated OFDM subcarriers in frequency domain. These justified approximations replace the imposed delays with the delay corresponding to the waveform spectrum center. This delay is the desired unknown, which is estimated by exploiting the ML estimator. Then, the proposed TOA measurement technique exploits different carrier frequencies combined with DOA measurements to construct a system of linear equations as a function of the thickness of available submedia.

10.4.1 System Model of Propagated Signal

The process of TOA estimation initiates with the transmission of a ranging waveform. Here, 2K OFDM subcarriers are utilized to construct the ranging waveform corresponding to:

$$s_n = \frac{1}{\sqrt{N}} \sum_{k=1}^{2K} S_k e^{j2\pi m_k \Delta f n T_s}, \tag{10.17}$$

where N and T_s are the total number of available subcarriers or the number of waveform samples, and sampling interval, respectively, which leads to a ranging waveform with length NT_s. Moreover, m_k and S_k are the frequency index of the kth subcarrier, the OFDM subcarrier spacing, and the kth frequency domain symbol known at both transmitter and receiver, respectively. To maintain orthogonality across subcarriers, subcarrier spacing must satisfy $\Delta f = \frac{1}{NT_s}$, which leads to:

$$s_n = \frac{1}{\sqrt{N}} \sum_{k=1}^{2K} S_k e^{j\frac{2\pi n m_k}{N}}. \tag{10.18}$$

Exploiting symmetric subcarriers constructs a uniform waveform in frequency domain, which enables the linear approximation of imposed delay to each subcarrier, which entails:

$$m_k = \begin{cases} K, K-1, \ldots, 1, & k = 1, 2, \ldots, K \\ N-K, N-K+1, \ldots, N-1, & k = K+1, \ldots, 2K \end{cases}. \qquad (10.19)$$

Due to the dependence of propagation velocity on the relative permittivity of submedia (as discussed in Section 10.3 and (10.3)), frequency dispersion of the transmitted waveform at receiver is inevitable. Thus, waveforms with higher-frequency components arrive with different delay compared to lower-frequency components [27]. Therefore, for the dispersed version of the ranging waveform, we can write:

$$\tilde{s}_l(t) = \frac{1}{2|\omega_K|} \int_{-\omega_K}^{\omega_K} S(\omega) e^{j\omega(t - \tau(\omega))} d\omega, \qquad (10.20)$$

where $S(\omega)$ is the Fourier transform of the transmitted waveform $s(t)$, and $\tau(\omega)$ and $2|\omega_K|$ are the time delay imposed by the line-of-sight (LOS) path into the waveform component with the frequency of ω and the bandwidth of transmitted waveform, respectively. Applying the received baseband signal to the dual channel ADC with sampling interval T_s, leads to:

$$r_n = \sum_{l=0}^{L-1} h_l \sum_{k=1}^{2K} S_k e^{-j \frac{2\pi m_k (nT_s - \tau_{k,l})}{NT_s}} + v_n, \qquad (10.21)$$

where N, K and S_k and T_s are defined in (10.17) and (10.18), respectively. Moreover, r_n and v_n represent the nth sample of the base-band received signal and additive, white zero-mean Gaussian noise, respectively, and $\tau_{k,l}$ represents the time delay of the kth subcarrier propagated at lth path of IHM with channel impulse response h_l, $l = 0, 1, \ldots, L-1$.

10.4.2   Very High-Resolution TOA Estimation via OFDM Subcarriers

According to TOA definition, it is desired to estimate the delay corresponding to the first arrived component ($m_k = K$) of the transmitted signal propagated from the shortest path, which is equivalent to the value of $\tau_{0,K}$. However, here, a novel technique is proposed, which exploits frequency domain analysis to estimate $\tau_{0,0}$ that represents the delay corresponding to the spectrum center of the transmitted waveform $m_k = 0$, propagated from the LOS path.

Considering N possible samples corresponding to propagation delay and signal expansion due to frequency dispersion, a receiver selects $N_r = 3N$ samples from the time origin and converts them into frequency domain applying discrete Fourier transform (DFT). Therefore, 2K samples of the received signal in frequency domain are achievable via:

$$R_k = S_k \sum_{l=0}^{L-1} h_l e^{-j \frac{2\pi m_k \tau_{l,k}}{NT_s}} + V_k. \qquad (10.22)$$

The proposed system model in (10.22) represents 2K nonlinear equations corresponding to 2K allocated subcarriers forming a system of nonlinear equations with $L(2K + 1)$ unknowns (L and $2KL$ correspond to h_l and $\tau_{k,l}$, respectively). Such a underdetermined system leads to infinitely many solutions (or no feasible solution), and demands reducing the number of unknowns.

Considering (10.22) as the system model for the received samples in frequency domain and known transmitted OFDM symbols (S_k), a receiver calculates $X_k = Y_k / Y_{k+K}$ for $Y_k = R_k / S_k$ which leads to (see [13] for details):

$$X_k = x_k e^{-j\frac{2\pi m_k(\tau_{0,k}+\tau_{0,k+K})}{NT_s}} e^{j\frac{2\pi m_k \tau_{0,k+K}}{NT_s}} + \tilde{V}_k. \qquad (10.23)$$

It is shown that (see remark 1 in [13]) the overall imposed delays to transmitted subcarriers corresponding to each path can be approximated as a line such that $\tau_{l,k} \cong a_l(k-1)+b_l$, given $\left|\tau_{l,k} - \hat{\tau}_{l,k}\right| < \eta T_s$, $k = 1,2,\ldots,2K$, where:

$$a_l = \frac{1}{c}\sum_{m=1}^{M} d_{l,m}\left(\sqrt{\varepsilon_{r,m,K+1}} - \sqrt{\varepsilon_{r,m,K}}\right), \qquad (10.24)$$

$$b_l = \frac{1}{c}\sum_{m=1}^{M} d_{l,m}\sqrt{\varepsilon_{r,m,1}}, \qquad (10.25)$$

where c denotes the universal speed of light and η is and arbitrary small number $\eta < 1$. Moreover, $\varepsilon_{r,m,k}$ and $d_{l,m}$ are the relative permittivity of the mth media at the kth subcarrier frequency and its corresponding propagation range, respectively.

Assuming all imposed delays ($\tau_{0,k}$) are placed on the approximated line, the average of all equally spaced delays from the line center, that is, $\tau_{0,k}$ and $\tau_{0,k+K}$, are the same and equal to $\tau_{0,0}$, which is defined as the delay corresponding to the spectrum center of the transmitted waveform or the desired TOA. Therefore, 2K unknown delays in (10.13) can be replaced by applying $(\tau_{0,k} + \tau_{0,k+K}) = 2\tau_{0,0}$, which leads to:

$$X_k = x_k e^{-j\frac{4\pi m_k \tau_{0,0}}{NT_s}} e^{j\frac{2\pi\tau_{0,k+K}}{T_s}} + \tilde{V}_k. \qquad (10.26)$$

Substituting $\tau_{0,k+K}$ with its corresponding coarse and fine delays such that $\tau_{0,k+K} = (n_{0,k+K} + \delta_{0,k+K})T_s$ gives:

$$X_k = x_k e^{-j\frac{4\pi m_k \tau_{0,0}}{NT_s}} e^{j2\pi\delta_{0,k+K}} + \tilde{V}_k, \quad k = 1,2,\ldots,K. \qquad (10.27)$$

The proposed system model in (10.27) forms K nonlinear and noisy equations containing one desired unknown, $\tau_{0,0}$, and 2K undesired unknowns (x_k and $\delta_{0,k+K}$ for $k = 1,2,\ldots,K$). Although the values of fine delays ($\delta_{0,k}$) are negligible in terms of delay, ignoring them dramatically degrades the system model accuracy due to the impact of the exponential term $e^{j2\pi\delta_{0,k+K}}$. The number of unknowns in (10.27) can be reduced by exploiting a linear approximation of the fine delays exploiting $\delta_{0,k+K} = -\beta(m_k - 1) + \delta_{2K}$, such that $\left|\delta_{0,k+K} - \hat{\delta}_{0,k+K}\right| < \eta T_s$, (see remark 2 in [13]), where:

$$\acute{\beta} = \frac{1}{c} \sum_{m=1}^{M} d_{0,m} \left(\sqrt{\varepsilon_{r,m,K+2}} - \sqrt{\varepsilon_{r,m,K+1}} \right). \tag{10.28}$$

According to the fine delay approximation, the term $e^{j2\pi\delta_{0,k+K}}$ in (10.27), can be replaced by $x_0 e^{-j\frac{4\pi mk}{N}\beta}$, for $x_0 = e^{j2\pi(\beta+\delta_{0,2K})}$ and $\beta = \frac{N\acute{\beta}}{2}$. Substituting $x_0 e^{-j\frac{4\pi mk}{N}\beta}$ into (10.27), the final approximated form of the proposed system model for the received signal in frequency domain corresponds to:

$$X_k = \tilde{x}_k e^{-j\frac{4\pi mk}{N}[2(\alpha+\beta)]} + \tilde{V}_k, \quad k = 1,2,\ldots,K, \tag{10.29}$$

where $\tilde{x}_k = x_0 x_k$, $\alpha = \frac{\tau_{0,0}}{T_s}$, and β denote two unknowns needing to be estimated to compute $\tau_{0,0}$. Defining $n^{(1)} = 2(\alpha+\beta)$, it can be shown that (see appendix A in [13]), the ML estimator of $n^{(1)}$ can be written as:

$$n^{(1)} = \underset{n}{\operatorname{argmax}} \left| \mathbf{w}_n^T \mathbf{x} \right|, \tag{10.30}$$

where, $\mathbf{w}_n = \left[e^{-j\frac{2\pi Kn}{N}}, e^{-j\frac{2\pi(K-1)n}{N}}, \ldots, e^{-j\frac{2\pi n}{N}} \right]^T$ and $\mathbf{x} = [X_1, X_2, \ldots, X_K]^T$, for X_k defined in (10.29). Exploiting (10.30), one can estimate $(\alpha+\beta)$; however, at least one linearly independent equation in terms of α and β is required to estimate the desired TOA via $\tau_{0,0} = \alpha T_s$. To address this issue, [13] proposes repeating the entire measurement procedure, applying the same number of allocated subcarriers (2K) and waveform samples (N), while the sampling interval is changed, for instance to $T_s^{(p)} = pT_s$, where $T_s^{(p)}$ represents the sampling interval at the pth ($p \geq 2$) measurement.

The performance of the proposed TOA estimation technique depends on the accuracy of the frequency domain detectors proposed in (10.30). However, the returned value by (10.30) is the time index and hence, $0 \leq \hat{n}^{(1)} \leq N-1$; meanwhile, the value of $2(\alpha+\beta)$ can be a negative number or larger than N. In order to resolve this issue, the estimated TOA via $\hat{\tau}_{0,0} = \alpha T_s$ needs to be inspected regarding its acceptable range as described in [13].

Figure 10.5a and b depict the average error of estimated TOA for an underground medium considering multipath channels. In Figure 10.5a–d the average TOA error is evaluated versus SNR for $N = 512, 1024, 2048,$ and 4096, respectively, where N denotes the length of the transmitted waveform.

As shown in Figure 10.5, increasing the number of allocated subcarriers improves the average TOA estimation error at medium-to-high SNR regimes (see Fig. 5a–d). This improvement is acquired, as allocating a larger number of subcarriers increases the frequency domain samples in (10.19). Moreover, it is observed that increasing N with applying the same number of allocated subcarriers K decreases the performance of the proposed technique as shown in Figure 10.5a and b. This is due to the very fact that increasing N decreases the subcarrier spacing at OFDM subcarrier, which leads to lower bandwidth of transmitted signal. However, the TOA estimation limit is achievable by increasing the number of allocated subcarriers K, as shown in Figure 10.5c and d. Moreover, in Figure 10.5a and b it can be observed that the average TOA estimation error increases at higher allocated subcarriers and low SNR regimes for shorter waveform length. In addition

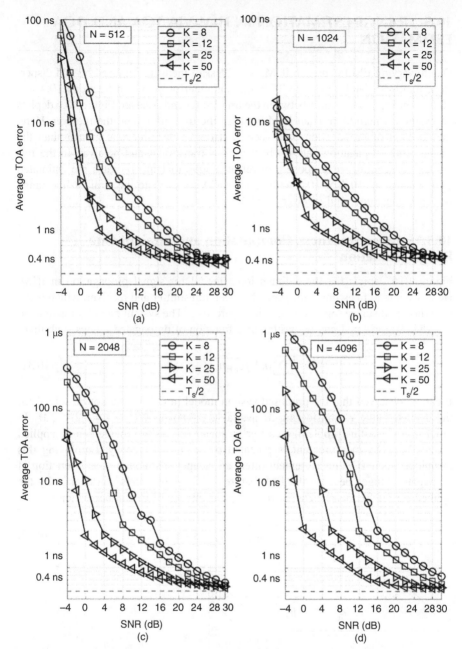

Figure 10.5   Impact of the number of allocated subcarriers and transmitted signal length on the estimated TOA average error in underground channels; (a) (b), (c), and (d).

to the positive impact of increasing the number of allocated subcarriers at medium-to-high SNR regimes, exploiting a larger number of subcarriers extends the in-band noise power, which affects the TOA estimator performance at low SNR regimes specifically

10.5   RANGE ESTIMATION IN HIM VIA TOA AND DOA ESTIMATION

Due to varying EM features in IHM, regardless of the error imposed by EM dispersion, multipath, and noise, the TOA measurement proposed in Sections 10.4 and 10.5 cannot reveal the range between the transmitter and receiver. Figure 10.6 depicts an example where there is an IHM between the transmitter and the receiver. Here, the addition of propagation path vectors at each submedia r_i is greater than the straight-line propagation path (r) between the transmitter and receiver. In the following subsection, we propose a novel technique that incorporates the estimated TOA in Section 10.4 or 10.5 with signal DOA to estimate the straight-line range between transmitter and receiver.

10.5.1   Layer Thickness Computation and Straight-Line Range Estimation

Figure 10.6 depicts the three-dimensional model for signal propagation in IHM. Here, we intend to estimate the range between transmitter and receiver, which is the straight-line distance referred to as r in Figure 10.6. The system model for estimated TOA via Algorithm 1 can be written as a function of propagated ranges such that:

$$\hat{\tau} = \frac{1}{c} \sum_{m=1}^{M} |r_m| \sqrt{\varepsilon_{r,m} \mu_{r,m}} + v_\tau. \tag{10.31}$$

It can be observed that the proposed system model in (10.31) cannot be solved for the desired range $\mathbf{r}$, considering M available unknowns of r_m, $m = 1, 2, \ldots, M$. In order to address this problem, a carrier frequency diversity scheme can be applied where the TOA measurement is repeated per carrier frequency. Employing this technique, each TOA measurement offers an independent equation as a function of propagated ranges because the EM features of available submedia change by the applied carrier frequency. Therefore, the qth measured TOA can be written as:

$$\hat{\tau}^{(q)} = \frac{1}{c} \sum_{m=1}^{M} |r_m^{(q)}| \sqrt{\varepsilon_{r,m}^{(q)} \mu_{r,m}^{(q)}} + v_\tau^{(q)}, \quad q = 1, 2, \ldots, Q, \tag{10.32}$$

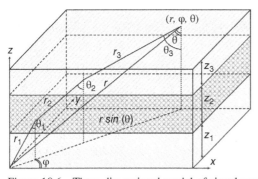

Figure 10.6   Three-dimensional model of signal propagation in IHM.

where $\hat{\tau}^{(q)}$, $r_m^{(q)}$ and $v_\tau^{(q)}$ are the estimated TOA, the range of propagation path, and TOA measurement error, all at the qth measurement, respectively. Moreover, $\varepsilon_{r,m}^{(q)}$, $\mu_{r,m}^{(q)}$ represent relative permittivity and permeability in the mth sub-media at the qth measurement, respectively. It can be observed that each measurement $q = 1, 2, \ldots, Q$ in (10.32) leads to a different set of M unknowns, $r_m^{(q)}$, $m = 1, 2, \ldots, M$. This creates a set of Q independent equations containing MQ unknowns. We propose to replace these unknowns by the thickness of available submedia to construct a set of Q independent equations as a function of M available unknowns z_m, $m = 1, 2, \ldots, M$, such that $r_m^{(q)} = z_m cos\left(\theta_m^{(q)}\right)$. Here, z_m represents the layer thickness. Applying $r_m^{(q)} = z_m cos\left(\theta_m^{(q)}\right)$, Snell's law, and some mathematical manipulations, Equation (10.32) can be written as (see [13] for more details):

$$\hat{\tau}^{(q)} = \frac{1}{c} \sum_{m=1}^{M} \frac{z_m \sqrt{\varepsilon_{r,m}^{(q)} \varepsilon_{r,M}^{(q)} \mu_{r,m}^{(q)} \mu_{r,M}^{(q)}}}{\sqrt{\varepsilon_{r,m}^{(q)} \varepsilon_{r,M}^{(q)} - sin^2\left(\theta_M^{(q)}\right) \varepsilon_{r,m}^{(q)} \mu_{r,m}^{(q)}}} + v_\tau^{(q)}, \tag{10.33}$$

where $\theta_M^{(q)}$ denotes the DOA corresponding to the qth measurement at the Mth submedium (including receiver medium) as shown in Figure 10.6. Here, it is assumed that receiver enjoys array antenna, which allows DOA estimation, recruiting state-of-the art techniques such as Root-MUSIC [31] and ESPRIT [32, 33]. Considering Gaussian distribution with zero mean and variance σ_v^2 for TOA measurements' noise, the ML estimation of thickness of available submedia can be derived by solving the following optimization problem:

$$\underset{z_1, \ldots z_M}{argmin} \sum_{q=1}^{Q} \frac{1}{\sigma_v^2} \left| \hat{\tau}^{(q)} - \frac{1}{c} \sum_{m=1}^{M} z_m \gamma_m^{(q)} \right|^2 \quad for: \gamma_m^{(q)} = \frac{\sqrt{\varepsilon_{r,m}^{(q)} \varepsilon_{r,M}^{(q)} \mu_{r,m}^{(q)} \mu_{r,M}^{(q)}}}{\sqrt{\varepsilon_{r,M}^{(q)} \mu_{r,M}^{(q)} - sin^2\left(\theta_M^{(q)}\right) \varepsilon_{r,m}^{(q)} \mu_{r,m}^{(q)}}}. \tag{10.34}$$

The least square solution for the proposed ML estimator in (10.30) can be represented by:

$$\hat{z} = \left(\Upsilon^T \Upsilon\right)^{-1} \Upsilon^T \tau, \tag{10.35}$$

where $\hat{z} = [\hat{z}_1, \hat{z}_2, \ldots, \hat{z}_M]^T$, $\tau = [\hat{\tau}^{(1)}, \hat{\tau}^{(2)}, \ldots, \hat{\tau}^{(Q)}]^T$ and $\Upsilon = [\Upsilon_1, \Upsilon_2, \ldots, \Upsilon_M]^T$ for

$$\Upsilon_m = \left[\gamma_m^{(1)}, \gamma_m^{(2)}, \ldots, \gamma_m^{(Q)}\right]^T. \tag{10.36}$$

Given the estimated thickness of available submedia using (10.32), it can be shown that (see appendix B in [26]), the straight-line propagation range between transmitter and receiver is achievable via $\hat{d} = \frac{1}{Q} \sum_{q=1}^{Q} \hat{d}^{(q)}$, where:

$$\hat{d}^{(q)} = D \sqrt{1 + \left[\frac{1}{D} \sum_{m=1}^{M} \frac{\hat{z}_m sin\left(\theta_M^{(q)}\right) \sqrt{\varepsilon_{r,M}^{(q)} \mu_{r,M}^{(q)}}}{\sqrt{\varepsilon_{r,M}^{(q)} \mu_{r,M}^{(q)} - sin^2\left(\theta_M^{(q)}\right) \varepsilon_{r,M}^{(q)} \mu_{r,M}^{(q)}}}\right]^2}, \tag{10.33}$$

where $D = \sum_{m=1}^{M} z_m$.

Figure 10.7a and b depicts the impact of the number of applied measurements and (a) average TOA measurements error given perfect DOA measurements, (b) average DOA measurements error given perfect TOA measurements on the estimated range underground channels, respectively, considering $N = 4096$ and $k = 50$. It is observed that the proposed technique offers the average range error of less than 1 m for

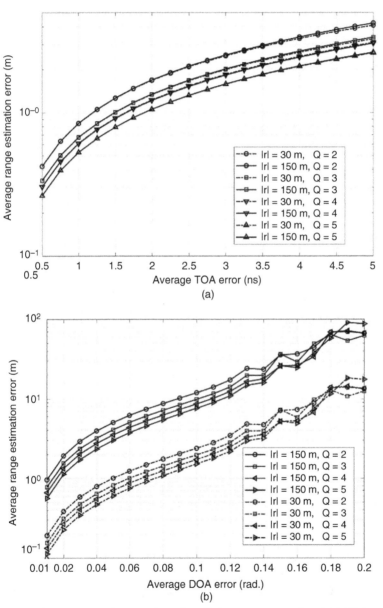

Figure 10.7 Impact of the number of applied measurements Q and (a) TOA measurements error, (b) DOA measurements error.

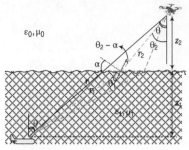

Figure 10.8   Impact of rough surface on range measurements in IHM.

underwater-airborne and underground channels at TOA measurements error around or less than 2 ns, which is achievable at $SNR \geq 4 \ dB$. Moreover, it is observed that the average ranging error decreases as the number of applied measurements is increased, as expected. Moreover, it is observed that the proposed technique offers proper range estimation even at imperfect DOA measurements, as depicted in Figure 10.7b.

10.5.2   Impact of Rough Surface at the Sublayer Boundaries

Figure 10.8 depicts the airborne-underwater scenario, another example of IHM assuming a rough surface. A rough surface entails two modifications to the proposed system model of straight-line range estimation technique. First, the angle between horizontal line and tangential line of the incident angle (α) must be added to the proposed system model in (10.33). In (10.33) it is assumed that the tangential line of the incident angle corresponding to the shortest path is horizontal ($\alpha = 0$) as shown in Figure 10.6. A rough surface at the boundaries of submedia adds one extra unknown per boundaries to the system linear equations proposed in (10.33), which entails extra TOA/DOA measurements. Second, different paths corresponding to various incident angles must be considered which leads to an MPFS channel as described in (10.21).

10.6   FURTHER APPLICATIONS OF INHOMOGENEOUS LOCALIZATION

10.6.1   Underground Layer Detection

In Section 10.6 straight-line range estimation in IHM was proposed. As discussed, the proposed technique estimates the thickness of available submedia by exploiting multiple TOA/DOA measurements applying carrier frequency diversity.

Figure 10.9 shows an example of the proposed technique for underground tomography via multiple TOA/DOA measurements. In the proposed straight-line range estimation technique, the EM properties of available submedia such as permittivity or permeability are assumed known; meanwhile the thickness is calculated exploiting the proposed system of linear equations proposed in (10.35). The same

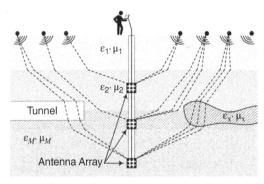

Figure 10.9   Underground tomography via TOA/DOA measurements.

criteria can be used to verify the existence of specific submedia or bulk, as shown in Figure 10.9. Here, the proposed technique can be exploited to verify the existence of underground tunnel and its coordinates. Some prior works proposed radio frequency (RF) tomography for underground tunnel detection [34–36]; however, the proposed techniques are not capable of offering precise information about the coordinates of tunnels. Moreover, the proposed techniques do not investigate the case of frequency dispersion at echo signal, which is highly probable by increasing the soil water content.

Furthermore, exploiting the proposed technique, one can verify the existence of specific bulk among a set of possible candidates. Exploiting this technique, the number of mine exploration sampling is decreased. Considering cost, time, and geological impacts of available sampling techniques such as core drilling [37], the proposed technique offers an efficient alternative for underground exploration.

10.6.2   Nanosensor Node Localization within the Human Body

Applications of WSNs in healthcare are extremely wide. Here, we consider applications that either employ exterior WSNs or body area networks (BANs) for localizing active sensors, such as endoscopy capsules, or WSNs within the human body formed by Nanosensors (NS) and Nanosensor networks (NSNs). Localization of endoscopy capsules in the human body is a practical application of three-dimensional localization of a sensor node within IHM [4, 6].

In [38], the authors discuss the problem of wideband localization in homogenous human tissues using a fast finite difference time domain (FDTD) technique. In [39] ultra-wideband signals are used for time-of-flight (ToF) and range estimation considering the average human body's permeability. However, the detection of transmitted wideband signals in frequency dispersive channels is not discussed. Moreover, the proposed technique does not consider the refraction effect among different human body organs and tissues, which tremendously increases the propagated distance and its corresponding ToF. In [40], the propagation time of wideband waveform is calculated, exploiting the correlator output via adaptive template synthesis method [41] in human body. Then, the position of transmitter is estimated by minimizing the objective function corresponding to the estimated propagation time and its system model for all available sensors at the first stage. However, the system model proposed for detection of wideband signals in the human body does not consider the frequency dispersion. Moreover, the percentage of each tissue assumed known is only a feasible assumption if prior information about the implanted node location is available. In [13] and [30], the authors propose range estimation in IHM considering frequency-dispersive submedia. The proposed techniques are evaluated for airborne to underwater and underground with layers containing different water contents [13] environments and the human body [30]. However, the proposed technique considers flat-fading channel in media due to a high attenuation rate of reflected signals.

10.7 CONCLUSION

In this chapter, the problem of TOA-based straight-line range measurement in IHM containing dispersive submedia was proposed. Here, we show that traditional TOA estimation techniques are not feasible due to transmitted waveform dispersion. Therefore, we introduced using bandlimited waveform. Here, the low resolution of TOA estimation imposed by limited bandwidth is compensated for using the MRLD receiver. Although the proposed technique offers promising results in terms of increasing TOA resolution, the proposed resolution does not fulfill the requirements of some IHM environments, such the as human body.

To address the resolution issue, a novel technique for TOA estimation in IHM was proposed, incorporating OFDM subcarriers. In the proposed technique, a waveform constructed of symmetric OFDM subcarrier is transmitted through the IHM channel. Here, we show that the delay imposed by frequency dispersion of media can be approximated by a line, which allows the reduction of unknowns (i.e., delays imposed to transmitted subcarriers). Exploiting these approximations, we proposed a precise method that estimates the delay of the subcarrier corresponding to the center of waveform spectrum. Simulation results in three different IHM indicate that the applied approximations are thoroughly feasible. This leads to perfect TOA estimation from medium to high-SNR regimes.

The introduced TOA estimation offers enough resolution; however, TOA cannot be mapped into the actual range between the transmitter and receiver, as the propagation path of transmitted waveform is not straight-line. In order to resolve this issue, a novel technique that enjoys TOA/DOA estimation was proposed. Here, the proposed TOA measurement technique exploits different carrier frequencies combined with DOA measurements to construct a system of linear equations as a function of the thickness of available submedia. Once the thicknesses of available submedia are estimated, the straight-line range between transmitter and receiver is estimated. Simulation results confirm that the proposed technique allows a good range resolution given noisy TOA and DOA measurements for different IHM.

Moreover, this chapter discussed state-of-the-art techniques for high-resolution TOA and straight-line range estimation in IHM, which can be applied to the human body. The proposed technique opens a new way forward in localization of implanted sensor nodes such as capsule endoscopy or nanosensor tracking within the human body. Comparing sensor movement models and velocity with the entire range measurement time period, it is concluded that the media between implanted sensor nodes and receiver sensor array remains static, which enables tracking of moving sensor nodes within the human body. Moreover, incorporating the proposed technique for detection, localization, and measurements of target tissues within the human body by exploiting the proposed sublayer thickness estimation method opens a new research area on cancer detection methodology, which remains as an open problem for future studies.

REFERENCES

[1] X. Tian, R. Shen, D. Liu, Y. Wen, and X. Wang, "Performance analysis of RSS fingerprinting based indoor localization," *IEEE Trans. Mobile Comput.*, vol. 16, no. 10, pp. 2847–2861, 2017.

[2] Z. Abu-Shaban, X. Zhou, and T. D. Abhayapala, "A novel TOA-based mobile localization technique under mixed LOS/NLOS conditions for cellular networks," *IEEE Trans. Veh. Technol.*, vol. 65, no. 11, pp. 8841–8853, 2016.

[3] X. Qu, L. Xie, and W. Tan, "Iterative constrained weighted least squares source localization using TDOA and FDOA measurements," *IEEE Trans. Signal Process.*, vol. 65, no. 15, pp. 3990–4003, 2017.

[4] G. Shu Ting, S. A. Reza Zekavat, and K. Pahlavan, "DOA-based endoscopy capsule localization and orientation estimation via unscented Kalman filter," *IEEE Sensors J.*, vol. 14, no. 11, pp. 3819–3829, 2014.

[5] S. Jeong, J. Kang, K. Pahlavan, and V. Tarokh, "Fundamental limits of TOA/DOA and inertial measurement unit-based wireless capsule endoscopy hybrid localization," *Int. J. Wirel. Inf. Netw.*, vol. 24, no. 2, pp. 169–179, 2017.

[6] A. R. Nafchi, G. Shu Ting, and S. A. Reza Zekavat, "Circular arrays and inertial measurement unit for DOA/TOA/TDOA-based endoscopy capsule localization: Performance and complexity investigation," *IEEE Sensors J.*, vol. 14, no. 11, pp. 3791–3799, 2014.

[7] T. He, C. Huang, B. M. Blum, J. A. Stankovic, and T. Abdelzaher, "Range-free localization schemes for large scale sensor networks,", in *Proc. of the 9th Annual Int. Conf. on Mobile Computing and Networking*, San Diego, CA, 2003, pp. 81–95.

[8] X. Enyang, D. Zhi, and S. Dasgupta, "Source localization in wireless sensor networks from signal time-of-arrival measurements," *IEEE Trans. Signal Process.*, vol. 59, no. 6, pp. 2887–2897, 2011.

[9] R. Roy, A. Paulraj, and T. Kailath, "Estimation of signal parameters via rotational invariance techniques-ESPRIT," presented at the 30th Annual Technical Symp., International Society for Optics and Photonics, San Diego, CA, 1986.

[10] L. Joon-Yong and R. A. Scholtz, "Ranging in a dense multipath environment using an UWB radio link," *IEEE J. Sel. Areas Commun.*, vol. 20, no. 9, pp. 1677–1683, 2002.

[11] J. Wang and Z. Shen, "An improved MUSIC TOA estimator for RFID positioning," in *2002 Int. Radar Conf.*, Oct. 2002, pp. 478–482.

[12] D. J. Daniels, *Ground Penetrating Radar*, vol. 1. IET, 2004.

[13] M. Jamalabdollahi and S. Zekavat, "ToA ranging and layer thickness computation in nonhomogeneous media," *IEEE Trans. Geosci. Remote Sens.*, vol. 55, no. 2, pp. 742–752, Feb. 2017.

[14] L. Qu, Q. Sun, T. Yang, L. Zhang, and Y. Sun, "Time-delay estimation for ground penetrating radar using ESPRIT with improved spatial smoothing technique," *IEEE Geosci. Remote Sens. Lett.*, vol. 11, no. 8, pp. 1315–1319, 2014.

[15] S. Zheng, X. Pan, A. Zhang, Y. Jiang, and W. Wang, "Estimation of echo amplitude and time delay for OFDM-based ground-penetrating radar," *IEEE Geosci. Remote Sens. Lett.*, vol. 12, no. 12, pp. 2384–2388, 2015.

[16] C. Le Bastard, Y. Wang, V. Baltazart, and X. Derobert, "Time delay and permittivity estimation by ground-penetrating radar with support vector regression," *IEEE Geosci. Remote Sens. Lett.*, vol. 11, no. 4, pp. 873–877, 2014.

[17] M. Mirbach and W. Menzel, "Time of arrival based localization of UWB transmitters buried in lossy dielectric media," presented at the 2012 IEEE Int. Conf. on Ultra-Wideband (ICUWB), Syracuse, NY, Sep. 2012.

[18] H. Sahota and R. Kumar, "Network based sensor localization in multi-media application of precision agriculture, Part 2: Time of arrival," presented at the 2014 IEEE 11th Int. Conf. on *Networking, Sensing and Control* (ICNSC), Miami, FL, Apr. 2014.

[19] I. Thanasopoulos and J. Avaritsiotis, "Wavelet analysis of short range seismic signals for accurate time of arrival estimation in dispersive environments," *IET Sci. Meas. Technol.*, vol. 5, no. 4, pp. 125–133, 2011.

[20] H. Hao, P. Stoica, and L. Jian, "Designing unimodular sequence sets with good correlations— Including an application to MIMO radar," *IEEE Trans. Signal Process.*, vol. 57, no. 11, pp. 4391–4405, 2009.

[21] P. Stoica, H. Hao, and L. Jian, "New algorithms for designing unimodular sequences with good correlation properties," *IEEE Trans. Signal Process.*, vol. 57, no. 4, pp. 1415–1425, 2009.

[22] H. Xu, C.-C. Chong, I. Guvenc, F. Watanabe, and L. Yang, "High-resolution TOA estimation with multi-band OFDM UWB signals," presented at the IEEE Int. Conf. on Communications, Beijing, China, May 2008.

[23] H. He, J. Li, and P. Stoica, *Waveform Design for Active Sensing Systems: A Computational Approach*. Cambridge University Press, 2012.

[24] C. Gabriel, "Compilation of the dielectric properties of body tissues at RF and microwave frequencies," DTIC Document, 1996.

[25] M. Soltanalian and P. Stoica, "Computational design of sequences with good correlation properties," *IEEE Trans. Signal Process.*, vol. 60, no. 5, pp. 2180–2193, 2012.

[26] M. Soltanalian and P. Stoica, "Designing unimodular codes via quadratic optimization," *IEEE Trans. Signal Process.*, vol. 62, no. 5, pp. 1221–1234, 2014.

[27] M. Jamalabdollahi and R. Zekavat, "High-resolution ToA estimation via optimal waveform design," *IEEE Trans. Commun.*, vol. 65, no. 3, pp. 1207–1218, 2017.

[28] M. Jamalabdollahi, 2016. Available: http://www.ece.mtu.edu/ee/faculty/rezaz/wlps/research.html

[29] M. Jamalabdollahi and S. A. R. Zekavat, "Joint neighbor discovery and time of arrival estimation in wireless sensor networks via OFDMA," *IEEE Sensors J.*, vol. 15, no. 10, pp. 5821–5833, 2015.

[30] M. Jamalabdollahi and S. R. Zekavat, "OFDMA-based high resolution sensor node ToA estimation in non-homogenous medium of human body," presented at the 2016 10th Int. Symp. on Medical Information and Communication Technology (ISMICT), Worcester, MA, Mar. 2016.

[31] A. J. Barabell, "Improving the resolution performance of eigenstructure-based direction-finding algorithms," in *IEEE Int. Conf. on ICASSP'83 Acoustics, Speech, and Signal Processing*, vol. 8, Boston, MA, Apr. 1983, pp. 336–339.

[32] V. U. Prabhu and D. Jalihal, "An improved ESPRIT based time-of-arrival estimation algorithm for vehicular OFDM systems," in *VTC Spring 2009—IEEE 69th Vehicular Technology Conf.*, Barcelona, Spain, 2009, pp. 1–4.

[33] H. Saarnisaari, "TLS-ESPRIT in a time delay estimation," in *1997 IEEE 47th Vehicular Technology Conf.*, vol. 3, Phoenix, AZ, May 1997, pp. 1619–1623.

[34] H. K. Choi and J. W. Ra, "Detection and identification of a tunnel by iterative inversion from cross-borehole CW measurements," *Microw. Opt. Technol. Lett.*, vol. 21, no. 6, pp. 458–465, 1999.

[35] J. A. Martinez-Lorenzo, C. M. Rappaport, and F. Quivira, "Physical limitations on detecting tunnels using underground-focusing spotlight synthetic aperture radar," *IEEE Trans. Geosci. Remote Sens.*, vol. 49, no. 1, pp. 65–70, 2011.

[36] L. L. Monte, D. Erricolo, F. Soldovieri, and M. C. Wicks, "Radio frequency tomography for tunnel detection," *IEEE Trans. Geosci. Remote Sens.*, vol. 48, no. 3, pp. 1128–1137, 2010.

[37] W. P. Gundy, "Core drill," 1980, Google Patents.

[38] P. Swar, K. Pahlavan, and U. Khan, "Accuracy of localization system inside human body using a fast FDTD simulation technique," presented at the 2012 6th Int. Symp. on Medical Information and Communication Technology (ISMICT), La Jolla, CA, Mar. 2012.

[39] A. Camlıca, B. Fidan, and M. Yavuz, "Implant localization in human body using adaptive least-squares based algorithm," in *ASME 2013 Int. Mechanical Engineering Congress and Exposition*, Volume 3B: American Society of Mechanical Engineers, doi:10.1115/IMECE2013-66039.

[40] M. Kawasaki and R. Kohno, "A TOA-based positioning technique of medical implanted devices," presented at the Third Int. Symp. on Medical Information and Communication Technology (ISMCIT09), Montreal, Canada, 2009.

[41] K. Taniguchi and R. Kohno, "Design and analysis of synthesized template waveform for receiving UWB signals," *IEICE Trans. Fundam. Electron. Commun. Comp. Sci.*, vol. 88, no. 9, pp. 2299–2309, 2005.

PART *III*

RECEIVED SIGNAL STRENGTH BASED POSITIONING

ONE MEASUREMENT that is ubiquitous in wireless systems is Received Signal Strength (RSS). Thus, while RSS-based localization is *typically* less accurate than TOA-based positioning, it is still a very important technique since it can be implemented with little to no modification to existing systems. Thus, the third part of the handbook which comprises five chapters, Chapters 11–15, studies the fundamentals of RSS-based methods and their potential for indoor localization. Specifically, three types of localization are presented: model-based techniques, kernel-based methods and fingerprinting techniques. This section reviews the applications and performance of RSS techniques in various environments including in buildings and in underground mining environments as well. All chapters in this part include MATLAB-based examples, solutions and the original code to help researchers quickly come up to sped in this area.

Chapter 11 addresses the fundamental aspects of location estimation based on received signal strength (RSS) measurements. Specifically, it provides an overview of RSS-based position location including major techniques and the primary sources of error. While the chapter introduces RF fingerprinting and proximity-based non-parametric methods, of particular interest in this chapter are range-based localization techniques that use a statistical radio propagation model. The chapter discusses geometric interpretations of both RSS and Differential RSS techniques, their solutions, achievable location accuracy, and optimal/practical estimators. Finally, it is shown via simulation how spatially correlated shadow fading, the number of anchor nodes, and path loss rate affect performance.

Chapter 12 compliments Chapter 11 by providing an overview of various indoor localization techniques employing RSS including lateration methods, machine learning classification, probabilistic approaches, and statistical supervised learning

Handbook of Position Location: Theory, Practice, and Advances, Second Edition.
Edited by S. A. (Reza) Zekavat and R. Michael Buehrer.
© 2019 by the Institute of Electrical and Electronics Engineers, Inc.
Published 2019 by John Wiley & Sons, Inc.
Companion Website: www.wiley.com/go/zekavat/positionlocation2e

techniques and comparing their performance in terms of localization accuracy through measurement studies in real office building environments. The chapter further surveys emerging techniques and methods to support robust wireless localization and improve localization accuracy, including real-time RSS calibration via anchor verification, closely spaced multiple antennas, taking advantage of robust statistical methods to provide stability to contaminated measurements, utilizing linear regression to characterize the relationship between RSS and the distance to anchors, and exploiting RSS spatial correlation. Finally, the chapter concludes with a discussion of popular location-based applications.

Both the placement and the selection of anchors can affect the accuracy of location estimation significantly. Thus, in addition to developing methods to localize wireless devices, it is equally important to systematically investigate how the placement of the anchors specifically impacts localization performance.

Chapter 13 investigates the problem of anchor placement and anchor selection. In terms of anchor placement, how the geometric layout of the anchors affects the localization performance is studied. Different approaches to anchor placement including heuristic search methods, acute triangular based deployment, adaptive beacon placement, and optimal placement solutions via the maximum lambda and minimum error (maxL-minE) algorithm are discussed. Additionally, this chapter studies how smartly selected anchors can achieve comparable or better localization results as compared to using all the available anchors. A variety of effective strategies are investigated including a joint clustering technique, entropy-based information gain, convex hull selection, and smart selection in the presence of high density of anchors.

As opposed to model-based methods, **Chapter 14** explores the features and advantages of kernel-based localization. Kernel methods simplify received signal strength (RSS)-based localization by providing a means to learn the complicated relationship between RSS measurement vectors and position as opposed to assuming a specific model. The chapter discusses their use in self-calibrating indoor localization systems. The chapter reviews four kernel-based localization algorithms and presents a common framework for their comparison. Results from two simulations and an extensive measurement data set are presented which provide a quantitative comparison of the various techniques.

In **Chapter 15** the principles of radio-frequency (RF) fingerprinting methods, also known as database correlation methods (DCM), are presented. Although there is a wide variety of DCM implementations, they all share the same basic elements. These elements are identified and analyzed in this chapter. Different alternatives for building RF fingerprint correlation databases (CDBs) are also presented and the advantages and drawbacks of each of them are discussed, as well as their impact on the positioning accuracy. Different methods to numerically evaluate the similarity between each fingerprint within the CDB and the measured RF fingerprint are described, including an artificial neural network based approach. Two alternatives to reduce the correlation or search space are analyzed, including an approach based on genetic algorithms. Finally, field tests are used to evaluate the positioning accuracy of fingerprinting techniques in both GSM (Global System for Mobile Communications) and Wi-Fi (Wireless Fidelity) 802.11b networks.

CHAPTER 11

FUNDAMENTALS OF RECEIVED SIGNAL STRENGTH-BASED POSITION LOCATION

Jeong Heon Lee and R. Michael Buehrer

Virginia Tech, Blacksburg, VA

THIS CHAPTER addresses the fundamental aspects of location estimation based on received signal strength (RSS) measurements. Specifically, we give an overview of RSS-based position location, including major techniques and primary sources of error. While we introduce radio frequency (RF) fingerprinting and proximity-based nonparametric methods, of particular interest in this chapter are range-based localization techniques using a statistical radio propagation model. We discuss their geometric interpretations, solutions, achievable location accuracy, and optimal/practical estimators. We show via simulation results the effect of spatially correlated shadow fading, the number of anchor nodes, and path loss (PL) rate.

11.1 INTRODUCTION AND MOTIVATION

Position location has recently been of significant interest in wireless networks due to its crucial role in an increasing number of applications. In cellular networks and wireless local area networks (WLANs), position information is valuable not only to fulfill the wireless Enhanced 911 (E-911) mandate for locating wireless 911 callers but also to meet the increasing demand for navigation/tracking and location-based services (LBSs) [1–3]. In ad hoc wireless technologies such as wireless sensor networks (WSNs) and cognitive radio networks (CRNs), the task of localizing sensors or radios with unknown positions is important for the operation and configuration of the network. The operational scenarios include data collection, surveillance, and inventory tracking, among many others. Also, the knowledge of node position can

Handbook of Position Location: Theory, Practice, and Advances, Second Edition.
Edited by S. A. (Reza) Zekavat and R. Michael Buehrer.
© 2019 by the Institute of Electrical and Electronics Engineers, Inc.
Published 2019 by John Wiley & Sons, Inc.
Companion Website: www.wiley.com/go/zekavat/positionlocation2e

be useful for routing, interference avoidance, and other basic functions [4]. This significance has stimulated research into a variety of localization techniques based on either RSS, differential received signal strength (DRSS), time of arrival (TOA), time difference of arrival (TDOA), angle of arrival (AOA), or their hybrids [1, 5–8].

11.1.1 Why Is RSS Attractive for Localization?

The task of location estimation has attracted a great deal of attention from academia, industry, and the military for over 60 years. In the early years, most research activities were driven by the military demand for target detection and tracking (e.g., radar and sonar). The most stringent requirement of military applications is the accuracy of location estimation, while the system cost and complexity are typically not a major concern. On the other hand, the aforementioned civilian location applications, which have led to many recent location studies from the academia and industry, are very concerned about the cost, complexity, and feasibility of a location system. For many practical location applications, the goal of a location system designer is to minimize the system requirements despite reasonable degradation in location accuracy. To this end, an RSS-based approach is an attractive candidate for position location in wireless networks.

It should be noted that in the early years, location applications employed only a small number of sophisticated receivers or sensor/antenna arrays, typically as few as two or three, particularly for long-distance positioning. Consequently, it was typically preferred to use TDOA/TOA or AOA measurements over RSS to achieve high location accuracy [9–11]. However, the recent proliferation of wireless devices and networks has enabled a larger number of observation points that are thus relatively close to the mobile target. Further, new wireless applications and services for which the RSS approach is suitable have been developed, particularly in indoor and urban non-line-of-sight (NLOS) environments. In these scenarios, an RSS-based approach is a viable, cost-effective solution that can be applied to a broad range of applications while providing comparable location accuracy.

Despite lower location accuracy with a small number of nodes in general, RSS-based localization is a simple, low-complexity method that can be integrated into another type of location system as a hybrid approach. Particularly, RSS values are readily available in (nearly) every wireless system without additional hardware or system modifications. In fact, RSS information is required by many wireless standards and specifications for the purpose of basic radio functions such as clear channel assessment, link quality estimation, handover, and resource management. Further, it may be the only ranging information available, for example, in severe multipath environments or for surveillance/security applications.

11.1.2 Problem Statement and Outline

The problem of position location based on signal measurements can be described as a series of three major phases: (1) signal observation, (2) extraction of position-related signal parameters, and (3) estimation of location coordinates. Specifically, upon the arrival of a signal, one or more signal parameters or features, among which are signal power, arrival time, and direction of arrival, are extracted depending on

TABLE 11.1.   A List of Radio Location Systems Developed

Signal Parameter	Examples of Radio Location Systems
RSS	Radar [12], QRSS [13], SpotOn [14], Nibble [15]
TOA	GPS [16], A-GPS [17], UWB PAL [18], and many UWB/CDMA systems [19–21]
TDOA	LORAN [22], AHLoS [23], Cricket [24], standard cellular positioning systems including E-OTD (for GSM/GPRS), U-TDOA (for GSM/4G), A-FLT (for cdmaOne/CDMA2000), and OTDOA (for WCDMA) [25]
AOA	APS [26], SDP [27], and nonstandard cellular positioning systems using an (adaptive) array antenna [19, 28, 29]

the type of location system, as exemplified in Table 11.1. The parameters are then used to estimate the target position by exploiting prior (statistical) knowledge of the observed data. Since a location estimator is subject to various environmental and systematic errors during any localization phase, it is vital to understand the source of these errors to develop a robust location system (see Section 11.2).

In this chapter, we deal with the problem of locating an RF energy-emitting device or node based on the power levels (i.e., RSS) measured at a set of reference points, referred to as *anchors*. Localization techniques using RSS measurements are classified mainly into (1) range-based positioning, (2) RF fingerprinting, and (3) proximity-based positioning (see Section 11.3). While we introduce all three techniques, of specific interest is the method of range-based positioning (see Sections 11.4–11.6).

Let us consider a typical source localization problem as follows. Suppose that given m anchor nodes with known coordinates $x_i = [x_i, y_i]^T$, their RSS measurements are translated into the distance $\{\hat{d}_i\}_{i=1}^{m}$ to the signal source using a radio propagation model. Then, denoting the unknown target position as $\theta = [x, y]^T$, we can formulate a basic mathematical problem for position location, referred to as *lateration*, as

$$\hat{d}_i^2 = \|\theta - x_i\|^2$$
$$= (x - x_i)^2 + (y - y_i)^2, \quad i = 1, \dots, m. \tag{11.1}$$

By solving this set of equations ($m \geq 3$), the position (x, y) can be determined. The solution can be interpreted geometrically, thus giving another (more intuitive) perspective of the problem (see Section 11.4). This two-dimensional (2-D) single source localization problem, which is assumed in this chapter, can readily be extended to three-dimensional (3-D) and to multiple sources by incorporating necessary node coordinates into the problem.

There are two primary factors that pose a challenge in solving the location problem: (1) sources of error and (2) optimality of a solution. Specifically, due to noise and other error sources, additional anchor measurements are valuable for better location accuracy. However, there exists no unique solution that satisfies with equality the resulting overdetermined system. Also, the problem formulation as in (11.1) does not fully exploit all the available data and statistical information. Thus, it is often preferred to find an optimal or practical location estimator by solving the problem:

$$\text{Minimize} \quad \phi(\theta) \tag{11.2}$$

$$\text{subject to} \quad g_i(\theta) \leq 0, \quad i = 1, \ldots, l_g \tag{11.3}$$

$$h_j(\theta) = 0, \quad j = 1, \ldots, l_h, \tag{11.4}$$

where θ may include nuisance parameters in addition to the parameters (x, y) of interest. $\phi(\theta)$ is a linear or nonlinear objective function incorporating our knowledge of RSS observations, a signal/system model, and error statistics (if known a priori). The constraints $g_i(\theta)$ and $h_j(\theta)$ are optional but should be exploited if possible to reduce a feasible region and to avoid excessive location errors. For instance, knowledge of NLOS measurements or radio coverage can be used to impose the constraints. The objective in solving the problem given in (11.2) is to find the optimal solution θ^*; that is, the best location estimate, such that $\phi(\theta^*) \leq \phi(\theta)$ for each feasible point θ. In Section 11.5, we discuss various estimators that are widely used for location estimation. The lower bound on estimator covariance can be used to measure the achievable location accuracy of an unbiased estimator, which is usually employed as a benchmark. In Section 11.6, we present simulation results on the performance of location techniques using RSS and DRSS.

11.2 SOURCES OF LOCATION ERROR AND MITIGATION

Localization performance is fundamentally limited by various estimation biases and errors. However, RSS measurements are so unpredictable that it is important to understand their sources of error to design a robust location system with a desired accuracy. We next discuss major sources of error as well as key techniques developed to mitigate the errors.

11.2.1 Multipath Fading and NLOS Propagation

Multipath fading is the major concern in wireless environments, which degrades the reliability and accuracy of a location system considerably. This error source, which causes frequency-selective fading, is random and unpredictable by nature [30]. It varies with node geometry, mobile position, and the surrounding environment. The effects of multipath fading are of great concern for many envisioned location applications that aim for operation in indoor and urban areas, where a line-of-sight (LOS) path is typically blocked and substantial scattering exists.

A great deal of research has been devoted to mitigating NLOS bias effects in location estimation [31]. A primary observation concerning NLOS signals exploited in many studies is that NLOS-corrupted measurements are positively biased so that a constrained optimization problem as in (11.2) can be formed. Also, researchers have proposed algorithms that selectively remove or scale NLOS measurements by examining the least squares (LS) residual error of an estimator for TOA [32], TDOA [33], and AOA [34]. Recently, ultra-wideband (UWB)-based ranging techniques have been of considerable interest as a promising indoor location solution [18, 20, 21].

11.2.2 Shadow Fading

Assuming multipath fading is averaged out or diminished, a received signal power envelope fluctuates slowly over distance (refer to (11.5)). Then, the dominant error source is large-scale shadow fading. As discussed in Section 11.3, shadow fading values at different locations are spatially correlated [35, 36]. In the study of cellular phone systems, a primary effort to mitigate this effect on link quality is made through macrodiversity. However, few effective techniques have been known for position location. If some measurements are known to be reliable, the LS residuals of the other measurements can be exploited to selectively mitigate their effects. Note that RSS-based range or location estimators are biased over typical wireless channels.

11.2.3 Systematic Bias or Error

In practice, radio location systems usually encounter systematic errors mainly due to imperfect receiver measurements, radio miscalibration, and hardware/software accuracy. While most location studies ignore this type of error because of its unpredictable and complex nature, it can be detrimental to location estimation. Systematic errors often bias location estimators, thereby making the mean of the estimator differ from the true value. When the errors are constant to mobile target position or static over time for stationary targets, they cannot be eliminated by repeating measurements or by averaging over a number of observations. Nevertheless, recent advancements in radio and digital signal processing technologies have considerably reduced their impact. Also, some techniques such as clock offset correction and hybrid localization methods have been proposed to tackle this issue. If channel/noise states are time varying, a recursive Bayesian approach such as Kalman filters and particle filters can be employed.

11.2.4 Geometric Node Configuration

The geometry of anchors relative to the mobile is a crucial factor that impacts the localization performance. This error source has been studied particularly for the global positioning system (GPS) in early location research and is often termed the geometric dilution of precision (GDOP) [16, 37]. To minimize the impact of node geometry, a location system designer can increase node density or seek optimal anchor positions [38–40]. It is important to note that such a geometric solution has recently become more attractive as wireless connectivity has been rapidly increasing, along with a growing interest in WSNs/CRNs.

11.3 TECHNIQUES USING RSS FOR POSITION LOCATION

We next describe major RSS-based localization approaches: (1) range-based positioning, (2) RF fingerprinting, and (3) proximity-based positioning.

11.3.1 Range-Based Positioning

The signal power or RSS observed over wireless channels changes in a random and unpredictable manner, and thus can only be characterized statistically. Therefore, a statistical model for RSS is employed to estimate a transmitter–receiver distance or "range" d_i, which is then used to infer location coordinates via lateration as described in Section 11.4. On the other hand, noting that this method is suboptimal, we can incorporate the model directly into an optimization framework as in Section 11.5.

Statistical Model for RSS: In wireless transmission media, a signal transmitted by a mobile device travels along a number of different paths of varying lengths, referred to as *multipaths*. This radio propagation causes signal distortions and fades, which are attributed to reflection, scattering, diffraction, and/or refraction from buildings, trees, furniture, and other obstructions in the environment [30]. The overall loss in signal strength is typically characterized as a product of three factors, namely, local mean propagation loss, long-term or slow fading, and short-term or fast fading. The first two factors are considered large-scale effects, whereas short-term fading is a small-scale effect.

Let us consider a location system where m anchor nodes estimate their distances d_i to the mobile of interest using the observed signal strength P_r. The received power or RSS (dBm) at a transmitter–receiver distance d_i for the ith anchor is characterized as

$$P_r(d_i) = P_t - \underbrace{\left(\overline{PL}(d_i) + \mathcal{M}_{F_i} + X_{\sigma_i}\right)}_{\text{Total propagation loss on the }i\text{th link}}, \tag{11.5}$$

where P_t (dBm) is the mobile's transmit power and $\overline{PL}(d_i)$ (dB) is the local mean propagation or PL as a function of distance d_i. The small-scale fading $\mathcal{M}_F$ (dB) generally varies abruptly (as much as 30 or 40 dB) over a distance of only a fraction of a wavelength. On the other hand, X_σ (dB) is the slow-term fading due to shadowing effects, which is the local mean of the fast-fading signal power. Thus, by relating the received power $P_r(d_i)$ to a PL model for $\overline{PL}(d_i)$, we can estimate d_i.

The PL is typically modeled as a function of $PL(d_0)$ (dB) measured at a close-in reference distance d_0 ($<d_i$) or predicted by an empirical model (e.g., (11.11) with $d_i = d_0$). As many measurement campaigns [30, 35] and analytical results [41] have shown, the relationship between the PL and distance can be captured in a log-distance equation:

$$PL(d_i)(\text{dB}) = \overline{PL}(d_0) + 10 n_p \log_{10}\left(\frac{d_i}{d_0}\right) + X_{\sigma_i}, \tag{11.6}$$

where n_p is termed the PL exponent or gradient, indicating that the transmitted signal power P_t decays with d^{n_p} on average. The value of n_p typically ranges from two (in the free space or clear LOS channels) to five, which tends to increase with more NLOS paths [30]. Assuming that the effect of small-scale fading is reduced by averaging it out over a range of frequencies, space, or some time period, location estimation using (11.6) is mainly subject to large-scale shadow fading $X_{\sigma_i} \sim \mathcal{N}\left(0, \sigma_S^2\right)$;

it is empirically modeled as a lognormal random variable with zero mean and variance σ_S^2 (dB). This environment-dependent variability is one of the most influential yet unavoidable factors in RSS-based localization. Hence, the received power P_r at d_i is also lognormally distributed with mean $\overline{P}_r$ as

$$P_r(d_i)(\text{dBm}) \sim \mathcal{N}\left(\overline{P}_r(d_i), \sigma_S^2\right), \tag{11.7}$$

where the ensemble mean received power

$$\overline{P}_r(d_i)(\text{dBm}) = P_r(d_0) - 10n_p(\log_{10} d_i - \log_{10} d_0). \tag{11.8}$$

Equivalently, the simplified RSS observation υ_i at anchor i is defined as

$$
\begin{aligned}
\upsilon_i(\text{dB}) &= P_r(d_0) - P_r(d_i) \\
&= L(d_i) + X_{\sigma_i,}
\end{aligned} \tag{11.9}
$$

where $L(d_i) = 10n_p(\log_{10} d_i - \log_{10} d_0)$. Noting that d_i is a function of the unknown target position (x, y), we will use the notation $L_i(\boldsymbol{\theta})$ interchangeably depending on the context. Due to the lognormal shadowing term, the observation υ_i is also lognormally distributed with mean L_i and the probability of density function (PDF).

$$f_V(\upsilon_i; \boldsymbol{\theta}) = \frac{1}{\sqrt{2\pi}\sigma_s} \exp\left(\frac{-(\upsilon_i - L_i(\boldsymbol{\theta}))^2}{2\sigma_S^2}\right). \tag{11.10}$$

Basics of Differential RSS: The received signal power or $P_r(d_0)$ in (11.9) is a function of two types of parameters; that is, transceiver and environmental parameters, as reflected in the Friis free-space equation [30]:

$$P_r(d_i) = P_t \frac{G_t \mathcal{L}_t^{-1} G_r \mathcal{L}_r^{-1} \lambda^2}{(4\pi)^2 d_i^2}, \tag{11.11}$$

where G_t and $\mathcal{L}_t$ (or G_r and $\mathcal{L}_r$) denote antenna gain and system loss factors of a transmitter (or a receiver), respectively. λ is the wavelength of the transmitted signal. In this unobstructed LOS channel model, the signal power decays according to the inverse square law (i.e., $n_p = 2$). To minimize location error, we should know or estimate the system parameters as precisely as possible, thus requiring offline calibrations. However, in many practical situations, this manual effort may be too costly or infeasible. Even if the environmental parameters can be accurately determined or known a priori, the transmitter parameter values may not be readily available at the anchors, or could be erroneously reported or even falsified. In most studies, the values are assumed to be perfectly known a priori. This assumption is made by relying on some form of cooperation from reliable signal sources (e.g., via a *predefined* beacon/pilot signal) and an accurate radio calibration of the transmitter. This passive dependency raises security concerns for location systems subject to various attacks.

One means of eliminating or reducing the need for knowledge of the system parameters is to change the observation to DRSS:

$$
\begin{aligned}
\upsilon_{ij}(\text{dB}) &= \upsilon_j - \upsilon_i \\
&= L(d_i, d_j) + \Delta X_{\sigma_{ij}},
\end{aligned} \tag{11.12}
$$

where

$$L(d_i, d_j) = 10n_p(\log_{10} d_j - \log_{10} d_i), \tag{11.13}$$

and $L(d_i, d_j)$ (or $L_{ij}(\boldsymbol{\theta})$) is a differential log-distance PL model with $i, j \in \{1, \ldots, m\}$, $i < j$. As a general rule of notation in this chapter, the subscripts fall in this range unless indicated otherwise. It can be noticed that all or most of the transmitter uncertainties in $P_r(d_0)$ are removed. Also, note that RSS measurements at m anchor nodes yield an unordered set of M distinct DRSS measurements and corresponding PL equations where

$$M = \binom{m}{2} = \frac{m(m-1)}{2} = \underbrace{m-1}_{\text{basic}} + \underbrace{\frac{(m-1)(m-2)}{2}}_{\text{redundant}}, \tag{11.14}$$

in which an ij-pair and a ji-pair are counted only once. This means that the whole set of size M can be determined by a linear combination of the $m-1$ basic (or nonredundant) measurements. The geometric interpretation of these equations for localization is presented in Section 11.4.

Example 11.1 Statistical Features for DRSS

We have discussed that shadow fading X_σ and the RSS observation can be modeled as lognormal.

(a) Find the joint PDF of two shadowing components $X_{\sigma i}$ and $X_{\sigma y}$ at anchor positions i and j.

(b) Derive the mean and variance for the random variable $\Delta X_{\sigma ij} = X_{\sigma j} - X_{\sigma i}$.

(c) Find the PDF for the DRSS observation υ_{ij} in (11.12).

Solution

(a) Because $P_r(d_i)$ is a Gaussian random variable (conditioned on the distance d_i), the random variables $P_r(d_i)$ and $P_r(d_j)$ from the same signal source are jointly Gaussian in the log domain. Specifically, the two random variables $X_{\sigma i}$ and $X_{\sigma j}$ can be related by a bivariate Gaussian distribution so that the joint probability density of $X_{\sigma i}$ and $X_{\sigma j}$ is

$$f_{X_{\sigma i}, X_{\sigma j}}(\eta_i, \eta_j) = \frac{1}{2\pi\sigma_s^2\sqrt{1-\rho_{s_{ij}}^2}} \exp\left\{\frac{-\left(\eta_i^2 - 2\rho_{s_{ij}}\eta_i\eta_j + \eta_j^2\right)}{2\sigma_s^2\left(1-\rho_{s_{ij}}^2\right)}\right\}, \tag{11.15}$$

where $\rho_{s_{ij}}$ is the correlation coefficient reflecting the degree of spatial correlation between the shadow fading components at any two locations, i and j.

(b) The PDF for $\Delta X_{\sigma ij}$, which represents the difference of the correlated Gaussian random variables, can be derived using (11.15) as

$$f_{\Delta X_\sigma}(\Delta\eta_{ij}) = \frac{1}{2\pi\sigma_s^2\sqrt{1-\rho_{s_{ij}}^2}}$$
$$\cdot \int_{-\infty}^{+\infty} \exp\frac{-\left(\eta_i^2 - 2\rho_{s_{ij}}\eta_i(\eta_i - \Delta\eta_{ij}) + (\eta_i - \Delta\eta_{ij})^2\right)}{2\sigma_s^2\left(1-\rho_{s_{ij}}^2\right)} d\eta_i, \tag{11.16}$$

so that

$$f_{\Delta X_\sigma}(\Delta\eta_{ij}) = \frac{1}{2\sigma_s\sqrt{\pi(1-\rho_{s_{ij}})}}\exp\left(\frac{-\Delta\eta_{ij}^2}{4\sigma_s^2(1-\rho_{s_{ij}})}\right). \tag{11.17}$$

It is clear that ΔX_σ is Gaussian with zero mean and variance $\hat{\sigma}_S^2 = 2(1-\rho_{S_{ij}})\sigma_S^2$. The variance $\hat{\sigma}_S^2$ can also be derived as $\hat{\sigma}_S^2 = 2\sigma_S^2 - 2C(X_{\sigma_i}, X_{\sigma_j})$, where the covariance $C(X_{\sigma_j}, X_{\sigma_j}) = \rho_{S_{ij}}\sigma_S^2$.

(c) From (11.12) and (11.17), we can see that the observation υ_{ij} is lognormally distributed with mean L_{ij} and variance $\hat{\sigma}_S^2 = 2(1-\rho_{S_{ij}})\sigma_S^2$. Thus, its PDF is

$$f_V(\upsilon_{ij};\theta) = \frac{1}{2\sigma_s\sqrt{\pi(1-\rho_{s_{ij}})}}\exp\left(\frac{-\upsilon_{ij}-L_{ij}(\theta)^2}{4\sigma_s^2(1-\rho_{s_{ij}})}\right). \tag{11.18}$$

Spatial Correlation of Shadow Fading: Although most existing studies in RSS-based localization simply assume that shadowing noise components at two locations are independent (i.e., $\rho_s = 0$), in reality the spatial correlation of shadow fading is often substantial due to similar terrain or obstacles on the signal propagation paths between the source and anchor nodes. It has been found in previous empirical studies that a typical value for the correlation coefficient ranges from 0.2 to 0.8 in indoor [35] and outdoor [36] channels, and as the angle and distance between a pair of reference locations decrease, the correlation tends to increase. Therefore, when performing a computer simulation, it is important to take the correlation into account for accurate analysis of signal strength-based location estimation [42].

A correlated shadow fading vector $X_\sigma = [X_{\sigma_1}, \ldots, X_{\sigma_m}]^T$ can be generated by the following procedure. First, we obtain an $m \times m$ covariance matrix K whose ij-element is computed as

$$K_{ij}(d_{ij}) = \sigma_S^2 \exp\left(-\frac{d_{ij}}{D_c}\ln 2\right) = \begin{cases} \sigma_S^2, & \text{if } d_{ij} = 0 \\ 0, & \text{if } d_{ij} = +\infty \end{cases}, \tag{11.19}$$

where D_c is termed the correlation distance. The above equation indicates that the spatial correlation between any two locations separated by a distance d_{ij} exponentially decays with a constant correlation distance D_c. Previous empirical studies have shown that this location-dependent shadowing model reflects well the real wireless environment [43, 44]. The symmetric and nonnegative definite matrix K is then decomposed into $K = LL^T$ by means of Cholesky factorization [45], which can be used to create $X_\sigma = Lw$. Here, $w = [w_1, \ldots, w_m]^T$ is a vector of m zero-mean, unit-variance, uncorrelated random variables. The resulting "shadowing map" generated by one simulation instance is shown in Figure 11.1.

In position network simulations with large data sizes (e.g., large-scale WSNs), the above procedure via a matrix operation with (11.19) may be too computationally intensive and memory constrained. An alternative means for approximating the effect of correlated shadowing is to use a distance-independent covariance matrix (assuming $\rho_{S_{ij}} = \rho_{S_{ji}}$),

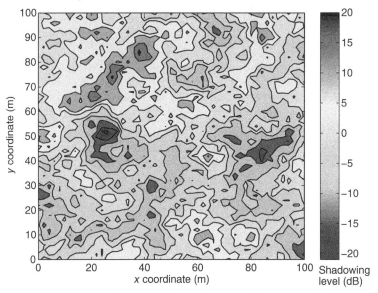

Figure 11.1 A shadowing map with spatial correlation generated using (11.19) with $D_c = 10$ m and $\sigma_S = 6$ dB. See color insert.

$$K_{ij} = \begin{cases} \sigma_S^2 & \text{if } i = j \\ \rho_{S_{ij}} \sigma_S^2 & \text{if } i \neq j \end{cases}, \tag{11.20}$$

instead of (11.19). Note in this model that as the correlation ρ_S increases, the overall shadowing variance of ΔX_σ in (11.12) tends to decrease. This relationship can be seen from (11.17).

11.3.2 RF Fingerprinting

We have discussed earlier that multipath fading degrades the performance of a location system significantly. Especially in indoor and urban environments where many recent location applications are targeted, the effect of fading is often so severe that other positioning techniques based on TOA/TDOA and AOA encounter difficulties. Also, range-based positioning using RSS will experience higher shadowing variance under greater multipath fading.

In the multipath location problem, we can exploit the high variability of multipath signals to relate each position $z_j = [x_j, y_j]^T$ to its unique signal signature $\mathcal{F}_j$, which is referred to as an *RF fingerprint* [12]. The more the signal strength varies over different locations, the more selective fingerprints can be collected, thus leading to better location accuracy. The simplest form of the signal signature $\mathcal{F}_j$ is a vector of RSS readings at each anchor position, thereby avoiding sophisticated hardware. When each receiver can provide its measured multipath delay profile, the fingerprinting resolution can be improved. However, since the acquisition of the multipath profile usually requires wide signal bandwidth and sophisticated hardware, it may be too costly unless the existing infrastructure provides that capability.

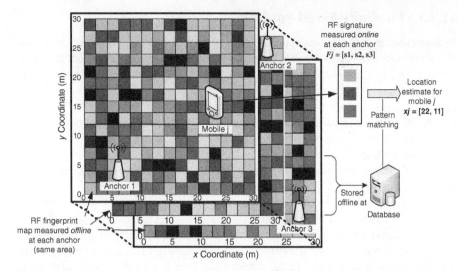

Figure 11.2   An illustration of RF fingerprinting for locating a mobile device. See color insert.

As illustrated in Figure 11.2, the process of RF fingerprinting can be divided into the following two primary phases as similarly implemented in [12]:

1. **Offline Training Phase.** For each of N positions $\{z_j\}_{j=1}^{N}$ (on a grid) in the area of interest, signal measurements are taken at m observation (or receiver) points to produce the associated RF fingerprints $\{\mathcal{F}_j\}_{j=1}^{N}$, where $\mathcal{F}_j = [s_{j1}, \ldots, s_{jm}]^T$. The observed data s_j can be an RSS value, a multipath delay profile, and/or an orientation of the operator's body. Then, the fingerprints are collected to construct a "radio map," which is stored in a database prior to the localization process.

2. **Online Localization Phase.** Upon arrival of a signal, the signal signature of interest $\overline{\mathcal{F}_i} = \left[\overline{S}_{i1}, \ldots, \overline{S}_{im} \right]^T$ for mobile i is extracted and compared to the radio map using one or more pattern matching techniques (e.g., k-nearest neighbors, naive Bayes classifiers). Then, the mobile position is estimated as $\hat{z}_i = \left[\hat{x}_i, \hat{y}_i \right]^T$ by selecting the best match or by interpolating among the best matches on the radio map.

Despite attractive aspects of RF fingerprinting, there are practical issues that hinder widespread adoption of the approach. In particular, the construction of accurate RF fingerprints or a radio map requires considerable offline effort and cost due to offline measurements, manual calibration, thorough site planning, and more. Also, since wireless channels and networks are inherently time varying, the radio map needs to be updated regularly (e.g., when large furniture or a wall is removed). Therefore, the success of this RSS approach will depend on how to address unpredictable wireless environments, interference, user behaviors (e.g., antenna/body orientation), costly hardware, and other practical issues. Fingerprinting approaches are discussed in detail in the following chapters.

11.3.3 Proximity-Based Positioning

In distributed ad hoc wireless applications such as WSNs, a majority of sensors with unknown positions try to localize themselves through cooperation—what is sometimes referred to as *relative positioning* [6, 46]. Their absolute positions are then found with the aid of a small number of reference nodes with *global* coordinate knowledge. Relative position location can be accomplished using range-based techniques. However, in large-scale networks, iterative optimization algorithms generally create challenging issues of computational inefficiency, nonguaranteed global optimality, and convergence to construct a global map of node locations. Also, for some WSN applications, just coarse-grain sensor locations are sufficient. Relative or collaborative techniques are described in detail in Chapters 24–28.

In such location scenarios, range-free localization techniques based on *proximity* or *connectivity* information are attractive. By "connectivity," we mean whether or not unlocalized sensors/nodes are within communications range (or radio range) with others in the network. The principle of this approach is that the relative positions of (neighboring) nodes in the network are found according to the proximity constraints. As a descriptive example, a pair of unlocalized sensor position vectors "push" each other to be outside the radio range when the sensors are not connected. On the other hand, when connected, the position vectors "pull" each other to be within the radio range, but this effort may be restricted by other neighbors that do the same to them. When RSS readings are available, this simple proximity information can be enhanced by taking into account approximate distances (or ranges) among the nodes. This principle can be implemented by various means among which we choose a dimensionality reduction approach next to demonstrate the proximity-based concept. Other approaches are described with references in [7].

Dimensionality Reduction Using Geographical Proximity: In recent years, the problem of *dimensionality reduction* has arisen in various fields of research dealing with large amounts of high-dimensional data, including psychology, computer vision, artificial intelligence, and cognitive sciences [47, 48]. The popularity is a consequence of the growing awareness that the underlying structure of relations among the observed data can be revealed and their similarity (or proximity geometrically) can be observed through the concept of dimensionality reduction. Specifically, in this process, the fundamental information of high-dimensional data can be embedded and visualized as a set of points in a low-dimensional space. For example, a number of images of a person's face depicting different appearance variations such as illuminations and poses are explored and then visualized as a set of 2-D or 3-D dimensional vectors in a Euclidean space, where axes are associated with observed modes of variability [47].

Recently, many attempts have been made to develop efficient algorithms that discover a geometric embedding structure of input data, while integrating classical algorithmic features of dimensionality reduction, such as guaranteed global optimality and convergence. Unlike traditional parametric approaches such as LS and maximum likelihood (ML) estimation, through which unknown parameters of some data model are optimized iteratively, they estimate the parameters based on the functional dependence of the observed data on distance measures. The type of dependence, whether linear or nonlinear, classifies dimensionality reduction algorithms. Multidimensional scaling (MDS) and principal component analysis (PCA) are classical linear dimensionality reduction algorithms [49], whereas Isomap [47], locally linear embedding (LLE) [48], Hessian-based locally linear embedding (HLLE) [50], and Laplacian eigenmap (LE) [51] are those developed for nonlinear dimensionality reduction.

Among many types of MDS methods, classical MDS is the simplest one for quantitative data (i.e., RSS measurements or proximity data), engaging one similarity matrix, and the proximity between the data is treated as a Euclidean distance [52, 53]. The key of the algorithm is to find interpoint Euclidean distances d_{ij} between a pair of data points $z_i = [x_{i1}, \ldots, x_{i\mathcal{D}}]^T$ and $z_j = [x_{j1}, \ldots, x_{j\mathcal{D}}]^T$. These distances are then related to the proximity measures p_{ij} of the data of size n and dimension $\mathcal{D}$. To reveal the *linear* relationship, a linear transformation is applied to relate d_{ij} to p_{ij} such that $d_{ij} = \alpha + \beta p_{ij}$, where p_{ij} are elements of an $n \times n$ proximity matrix **P**. It can be shown that the matrix **P** is double centered so that the elements of **P** satisfy the relation [53]

$$-\frac{1}{2}\left(p_{ij}^2 - \frac{1}{n}\sum_{i=1}^{n} p_{ij}^2 - \frac{1}{n}\sum_{j=1}^{n} p_{ij}^2 + \frac{1}{n^2}\sum_{i=1}^{n}\sum_{j=1}^{n} p_{ij}^2 \right) = \sum_{k=1}^{\mathcal{D}} x_{ik}x_{jk}. \qquad (11.21)$$

Using singular value decomposition (SVD), the double-centered matrix on the left side, say, **G**, can be found as $\mathbf{G} = \mathbf{L}\mathbf{\Sigma}\mathbf{L}^T$ so that we have a coordinate matrix $X = \mathbf{L}\mathbf{\Sigma}^{1/2}$. By computing the first ∂ largest eigenvalues and associated eigenvectors, we can determine the coordinates of the sensors in the space of lower dimension ∂ ($\partial < \mathcal{D}$) (e.g., $\partial = 2$ in 2-D localization).

Despite some good features of the MDS algorithm, such as the closed-form solution, global optimality, and convergence, the assumed linear relationship between the proximity measure p_{ij} and the Euclidean distance d_{ij} should hold for reconstructing the original geometric map. However, when exploiting useful proximity-related information such as the degree of proximity using a nonlinear PL model in (11.6) or spatial correlation of the data [54], the linearity assumption may be an oversimplification. Also, the classical MDS technique needs global knowledge of the network, thus hindering its adoption in many application scenarios. To address this issue, nonlinear manifold algorithms can be employed for dimensionality reduction.

The main idea of manifold algorithms is that despite the nonlinearity of the intrinsic embedding structure of the data in a high-dimensional space, a local neighborhood region (where K points are grouped) is approximately linear. Thus, interpoint Euclidean distances or weights in the small region can be used to represent the data while preserving the original nonlinear manifold. Isomap is similar to classical MDS except for the first stage that finds the geodesic shortest path for a pair of all non neighboring nodes. Most manifold algorithms try to find the eigenvectors associated with the $\partial + 1$ largest eigenvalues from the ∂-dimensional embedding coordinates. The detailed description of the algorithms can be found in [47–51].

Note that these relative positioning algorithms produce the embedded relative map of sensor coordinates. To obtain a global (or absolute) map of sensor coordinates, we need to exploit knowledge of the anchors' positions (if known) to transform the relative map into the global map via a geometric transformation, including translation, rotation, scaling, and/or reflection as similarly done in (11.30).

Example 11.2   Simulating Dimensionality Reduction Algorithms

We now use dimensionality reduction algorithms implemented in MATLAB for sensor localization. Download the MATLAB toolbox of the algorithms, named toolbox_dimreduc, available in [55]. Consider a simple 2-D location scenario as shown in Figure 11.3a, where 45 unlocalized sensors (denoted by hollow circles) are deployed on a 30×30 m grid with placement error $e_p \sim \mathcal{N}(0, 0.5^2)$ over the channel with $\sigma_S = 5$ dB and $n_p = 3$. Four anchor nodes (denoted by filled circles), one at each corner, are located at known coordinates. Generate RSS data using (11.7) and (11.8) so that each algorithm takes as input Euclidean distances between the data.

(a) Find the intrinsic dimensionality of the data.

(b) Construct the 2-D embedding for each algorithm and reconstruct the original location map. Compare the results with MDS and other nonlinear dimensionality reduction algorithms.

Solution

(a) In general, the original dimension embedded in the data can be found by examining major principal components and the eigenvalues, which correspond to the variance explained by the principal components. For example, as shown in Figure 11.3b, the intrinsic dimensionality can be determined using Isomap by seeking the "knee" point on the curve from which the variance begins to exhibit little change over dimensionality after dropping abruptly.

(b) In Figure 11.4, we show 2-D embeddings (left column) and reconstructed location maps (right column) using classical MDS (Figs. 11.4a, b) and Isomap with different neighborhood sizes ($K = 5, 10$) (Figs. 11.4c–f). From the embedded relative map of sensor coordinates (which is the output of the algorithm), the knowledge of the anchors' coordinates is used to transform the relative map into the global map. From the simulation results, we can see that the nonlinear algorithm, Isomap, reconstructs the original location map better than classical MDS. However, not all the nonlinear algorithms perform as desired.

11.4   GEOMETRIC INTERPRETATIONS OF RSS/DRSS POSITIONING

The problem of wireless localization is to find the unknown mobile position, which is equivalent to a point in the 2-D or 3-D Cartesian coordinate system. Thus, it is natural to represent the point in a geometric form so that the problem can be described

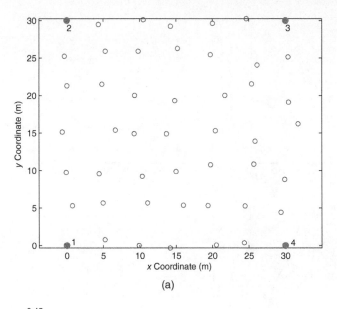

(a)

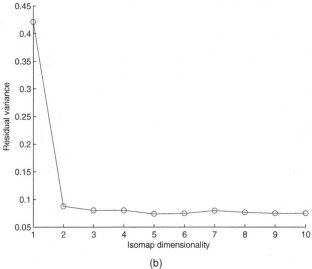

(b)

Figure 11.3   (a) Grid node placement with four anchors (at each corner) and 45 unlocalized sensors placed erroneously. (b) A typical residual variance curve to measure the embedded original dimension ($\partial = 2$, $\sigma_S = 5\,dB$, $n_p = 3$).

using basic geometric equations. The geometric interpretation not only sheds light on the problem but also facilitates the development of new location algorithms and theories. This section provides geometric interpretations of range-based positioning using RSS and DRSS, along with the mathematical representations of RSS/DRSS lateration. We show that although the two approaches exhibit the same geometric shape (i.e., circular), the geometry and redundancy of the circles are different.

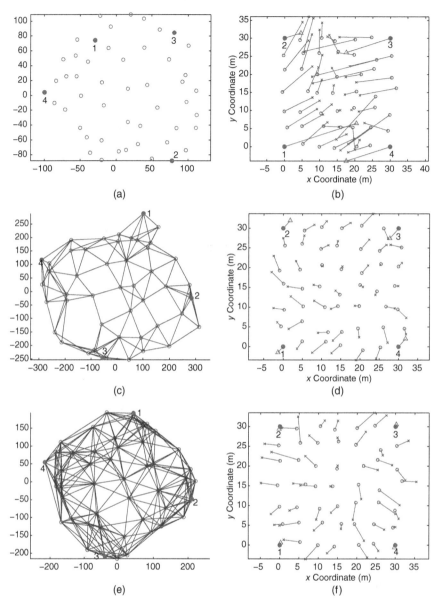

Figure 11.4 (a, b) MDS: 2-D embedding (left) and reconstructed map of sensor coordinates (right). (c, d) Isomap: 2-D embedding (left) and reconstructed map (right) using a neighborhood graph with $K = 5$. (e, f) Isomap: 2-D embedding (left) and reconstructed map (right) with $K = 10$. The unfilled circles and crosses indicate true and estimated positions, respectively.

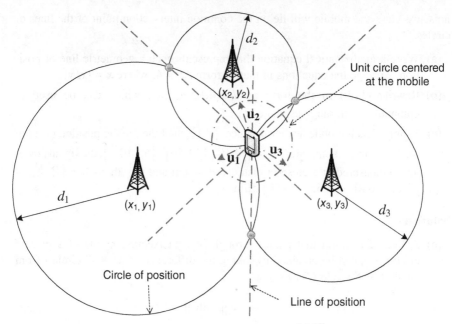

Figure 11.5 Trilateration using RSS range measurements $\left\{\hat{d}_i\right\}_{i=1}^{m}$ ($m = 3$) in a noiseless scenario (i.e., $\hat{d}_i = d_i$).

11.4.1 RSS-Based Lateration

We have described that the mobile position can be determined using a set of m range estimates. In RSS-based positioning, the range estimates can be represented by m circles with radii $\left\{\hat{d}_i\right\}_{i=1}^{m}$ in the 2-D space (or spheres in the 3-D space) centered at each anchor position $\{x_i\}_{i=1}^{m}$ as described by (11.1). The circumference of the circle defines an uncertainty of mobile position (x, y). This means that, in a noiseless case, the position can be found at the common intersection point of m circles as illustrated in Figure 11.5. Due to the nonlinearity of the equations, it is necessary for 2-D source localization to have $m \geq 3$ to avoid an ambiguous solution. Even when this condition is satisfied, in practice there exists no unique intersection point due to various sources of error (including shadow fading), which perturb the circles or range estimates. Thus, it is usually better to have $m > 3$.

Example 11.3 Geometric Solution of RSS Trilateration

Consider a range-based location network of three anchor nodes with known coordinates $x_i = [x_i, y_i]$. Based on RSS measurements subject to noise, suppose that each anchor i estimates its distance d_i to some mobile device. Then, we can estimate the mobile position (x, y) by solving the system of three nonlinear equations in (11.1) (i.e., $m = 3$) in a geometric approach. Assuming the circles all intersect each other, as described in Figure 11.5, we can connect two intersection points for each pair of circles to generate a line of position on which the mobile is supposed to lie. In a noiseless case, the mobile will lie on the common intersection point of the lines or circles.

(a) Write a mathematical equation that represents such a geometric line of position. Arrange the equations in matrix form $Ax = b$, where $x = [x, y]^T$.

(b) Rewrite (11.9) in the form of $d_i = f(d_0, n_p, \upsilon_i)$, which can be used to compute the ith range estimate $\hat{d}_i$.

(c) Estimate each mobile anchor distance $\{d_i\}_{i=1}^3$ and the mobile position (x, y) in the following example: $\{x_i\}_{i=1}^3 = \{[0, 0]^T, [30, 50]^T, [50, 10]^T\}$, the signal power levels measured at a close-in distance $d_0 = 1$ m and the ith anchor $\{P_{r,i}\}_{i=0}^3 = \{-10, -56.81, -43.55, -55.06\}$ dBm, $n_p = 3$.

Solution

(a) A line of position that passes through two intersection points of a pair of circles, i, j, can be obtained by taking the difference of an RSS circle i from another RSS circle j for $i < j$ as

$$(x_j - x_i)x + (y_j - y_i)y = \frac{1}{2}\left\{\|x_j\|^2 - \|x_i\|^2 - \left(\hat{d}_j^2 - \hat{d}_i^2\right)\right\} \quad i < j \quad (11.22)$$

or

$$y = -\left(\frac{x_j - x_i}{y_j - y_i}\right)x + \frac{\|x_j\|^2 - \|x_i\|^2 - \left(\hat{d}_j^2 - \hat{d}_i^2\right)}{2(y_j - y_i)}, \quad i < j. \quad (11.23)$$

For m circles (or anchors), we can only use $m - 1$ basic (or nonredundant) lines to form an original system of linear equations. For trilateration (i.e., $m = 3$), two basic lines of position are used to produce the linear system $Ax = b$, where

$$A = \begin{bmatrix} x_2 - x_1 & y_2 - y_1 \\ x_3 - x_1 & y_3 - y_1 \end{bmatrix}, b = \frac{1}{2}\begin{bmatrix} \|x_2\|^2 - \|x_1\|^2 - \left(\hat{d}_2^2 - \hat{d}_1^2\right) \\ \|x_3\|^2 - \|x_1\|^2 - \left(\hat{d}_3^2 - \hat{d}_1^2\right) \end{bmatrix}. \quad (11.24)$$

Then, the mobile position (x, y) can be estimated as $x = A^{-1}b$.

(b) Using (11.9), it is not difficult to obtain a range estimator as

$$\hat{d}_i = d_0 \cdot 10^{\frac{\upsilon_i}{10 n_p}}, \quad (11.25)$$

which is indeed an ML estimator for the distance d_i due to the lognormality of shadow fading or υ_i.

(c) By substituting the given parameter values into (11.25), we have the distance estimates as $\{\hat{d}_i\}_{i=1}^3 = \{36.34, 13.13, 31.77\}$ meters. The mobile position is then estimated as

$$x = A^{-1}b = \begin{bmatrix} 30 & 50 \\ 50 & 10 \end{bmatrix}^{-1}\begin{bmatrix} 2274.10 \\ 1455.63 \end{bmatrix} = \begin{bmatrix} 22.75 \\ 31.84 \end{bmatrix}. \quad (11.26)$$

In this example with the true mobile position $(x, y) = (26, 42)$, the location error is 10.67 m. We can improve the location accuracy by (1) exploiting additional anchor measurements to increase the robustness to shadow fading (e.g., via an LS estimator) and (2) incorporating the underlying statistical model in (11.9) directly into an optimization framework in the form of (11.2) (see Example 11.4). Before discussing such a more effective (or optimal) approach in the next section, we now explore geometric aspects of DRSS-based positioning.

11.4.2   DRSS-Based Lateration

Recall that DRSS and TDOA data are generated in the same manner (i.e., differences in RSS or TOA data, which both define a circle centered at each anchor). While TDOA defines a hyperbolic curve of position (see Chapter 6), the geometric interpretation of DRSS position is circular, as we will show.

Geometry of Relative DRSS Positioning: Suppose that the relative distance $D_{ij} = \|x_i - x_j\|$ of a pair of anchors or sensors i, j is known a priori. Here, $\{x_i\}_{i=1}^{m}$ denotes the ith *absolute* coordinate vector $x_i = [x_i, y_i]^T$, which may be unknown. Next, the middle of the link connecting the pair is translated to the origin of the local coordinate system (X, Y) for the pair. Thus, the two nodes are located at $\left(-\dfrac{D_{ij}}{2}, 0\right)$ and $\left(\dfrac{D_{ij}}{2}, 0\right)$ on the local x-axis, respectively. In detecting a signal from some unknown mobile, the local coordinate system produces a geometric function of $L_{ij}(\theta)$ using (11.13) as

$$L_{ij}(\theta) = 5n_p\left[\log_{10}\left(\left(X - \frac{D_{ij}}{2}\right)^2 + Y^2\right) - \log_{10}\left(X + \frac{D_{ij}}{2}\right)^2 + Y^2\right]$$

$$= 5n_p \log_{10}\left(1 - \frac{2D_{ij}X}{\left(X + \dfrac{D_{ij}}{2}\right)^2 + Y^2}\right). \tag{11.27}$$

Then, a local y coordinate of the source location is estimated as

$$Y = \pm\sqrt{\frac{2D_{ij}X}{1 - h_{ij}} - \left(X + \frac{D_{ij}}{2}\right)^2}, \tag{11.28}$$

where $h_{ij} = 10^{\upsilon_{ij}/5n_p}$. Thus, a pair of nodes form a *local* DRSS circle,

$$\left(X + \frac{D_{ij}}{2}\left(\frac{h_{ij} + 1}{h_{ij} - 1}\right)\right)^2 + Y^2 = \frac{h_{ij}D_{ij}^2}{\left(h_{ij} - 1\right)^2}, \tag{11.29}$$

on which a target node is assumed to (or will in a noiseless case) be located. As noticed in Figure 11.6a ($D_{ij} = 10$ m), the focus of the circle (marked as "×") is located

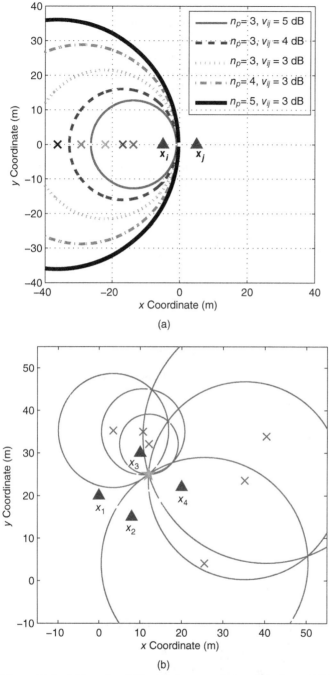

Figure 11.6 (a) Local and (b) global geometric interpretations of DRSS-based positioning in a noiseless case [8]. The source, anchor, and the center of each DRSS circle are indicated by "★," "▲," and "×", respectively.

at $\left(-\dfrac{D_{ij}}{2}\left(\dfrac{h_{ij}+1}{h_{ij}-1}\right),0\right)$ on the x-axis in the local coordinate system. Clearly, the geometry (focus and radius) of the circles is different from that in RSS/TOA-based lateration shown in Figure 11.5, where the centers of the circles are located at the node/anchor positions. It is also noted that the circles are impacted by the PL gradient n_p and observation υ_{ij}, as well as the relative distance D_{ij}. As shown in Figure 11.6a, the circle grows as either υ_{ij} decreases or n_p increases. This is because, for fixed n_p and decreasing υ_{ij} (i.e., DRSS $P(d_i, d_j)$ in (11.12) is decreasing), the source is further from the ith node *relative to* the distance to the jth node. On the other hand, for fixed υ_{ij} and smaller n_p, h_{ij} becomes larger for the same DRSS value so that the pattern is reversed. Note that unlike RSS/TOA-based lateration, where each measurement results in a circle, a pair of sensor/anchor measurements corresponds to a single circle.

With three or more distinct local circles ($m \geq 4$), the target position can be estimated. It can be noticed from (11.29) that the observed υ_{ij} and the relative distance D_{ij} are the only information necessary for relative positioning. Such a framework for relative positioning is particularly useful for developing a distributed location algorithm without requiring prior knowledge of absolute anchor positions [56, 57]. Further, as discussed earlier, the use of DRSS does not necessitate any cooperation from the signal source to obtain transmitter-related parameter values. Consequently, scarce wireless resources such as bandwidth and energy can be saved.

Geometry of Absolute DRSS Positioning: Although simply knowing the relative location will suffice for some scenarios such as WSNs/CRNs, global knowledge of the source position is essential in many location applications. To this end, we can transform local geometric systems $X = [X, Y]^T$ in (11.29) built by pairs of neighboring nodes into a global geometric system $x = [x, y]^T$ via a linear transformation. This is a linear mapping in the form of $x_s = \mathbf{T}^{(k)} X_s$, $k = 1, \dots, M$, where, in the problem of single source location,*

$$x_s = \begin{pmatrix} x \\ 1 \end{pmatrix}_{3\times 1}, \ \mathbf{T}^{(k)} = \begin{pmatrix} c & -s & t_{13} \\ s & c & t_{23} \\ 0 & 0 & 1 \end{pmatrix}_{3\times 3}, \ X_s = \begin{pmatrix} X \\ 1 \end{pmatrix}_{3\times 1}. \qquad (11.30)$$

Here, the rotation elements c and s are $\cos(\varphi_{ij})$ and $\sin(\varphi_{ij})$, where $\varphi_{ij} = \arctan\left(\dfrac{y_j - y_i}{x_j - x_i}\right)$ is the angle of the vector pointing from the ith anchor to the jth anchor relative to the x-axis measured counterclockwise. The translation elements $t_{13} = \dfrac{1}{2}(x_i + x_j)$ and $t_{23} = \dfrac{1}{2}(y_i + y_j)$ denote the x and y coordinates of the center of the link connecting an ij-pair of anchors, respectively. If υ_{ij} is negative, the circles are reflected.

---

*For multiple source localization, a unique solution usually cannot be found so that an optimization scheme (e.g., LS) needs to be employed.

Also, a set of *global* or absolute DRSS circles centered at $c_{d_k} = [x_{d_k}, y_{d_k}]^T$ can be obtained directly using a system of nonlinear equations ($M > m \geq 4$):

$$\left(x - x_{d_k}\right)^2 + \left(y - y_{d_k}\right)^2 = r_{d_k}^2, \quad k = 1, 2, \ldots, M, \tag{11.31}$$

where

$$x_{d_k} = \frac{h_{ij} x_i - x_j}{h_{ij} - 1}, \quad y_{d_k} = \frac{h_{ij} y_i - y_j}{h_{ij} - 1}, \quad r_{d_k} = \frac{\sqrt{h_{ij}} \cdot D_{ij}}{|h_{ij} - 1|} \tag{11.32}$$

and $|\cdot|$ denotes the absolute value. In Figure 11.6b, by using different $M = \binom{5}{2} = 10$ pairs of nodes, 10 global circles are formed, which all intersect at the source position (x, y). Unlike the local circle in (11.29), the center of the global circle is a function of absolute node coordinates. Note that both the circle's radius and focus will be affected by shadow fading unlike in other range-based circular positioning methods (i.e., RSS and TOA).

Let us now look into the geometric implication of the DRSS redundancy discussed in Section 11.3.1. In the simple noiseless example with three anchors ($m = 3$), we discussed earlier that two basic and one redundant measurements exist. Accordingly, using (11.31), we can create three distinct geometric circles, yet only the two associated with the basic measurements are independent. This is reflected by the fact that the circle associated with the redundant measurement intersects both of the other circles at the same two points at which the two independent circles meet. Thus, we need a fourth anchor in order to have an unambiguous solution. Despite the need for an additional independent measurement as compared to other circular positioning methods, in the presence of noise, the redundant DRSS measurements increase robustness against noise as demonstrated in Section 11.6.

Geometric Solution of DRSS Location: As was similarly done for the RSS case, a geometric solution of DRSS positioning can be obtained from the difference of a DRSS circle k and the other circles l for $k < l$. Then, we have

$$\left(x_{d_l} - x_{d_k}\right)x + \left(y_{d_l} - y_{d_k}\right)y = \frac{1}{2}\left\{\|c_{d_l}\|^2 - \|c_{d_k}\|^2 - \left(r_{d_l}^2 - r_{d_k}^2\right)\right\} \quad k < l, \tag{11.33}$$

which leads us to estimate the DRSS solution $x = A^{-1}b$ ($m = 4$), where

$$A = \begin{bmatrix} x_{d_2} - x_{d_1} & y_{d_2} - y_{d_1} \\ x_{d_3} - x_{d_1} & y_{d_3} - y_{d_1} \end{bmatrix}, b = \frac{1}{2}\begin{bmatrix} \|c_{d_2}\|^2 - \|c_{d_1}\|^2 - \left(r_{d_2}^2 - r_{d_1}^2\right) \\ \|c_{d_3}\|^2 - \|c_{d_1}\|^2 - \left(r_{d_3}^2 - r_{d_1}^2\right) \end{bmatrix}.$$

When additional DRSS measurements are available, the overdetermined system can be solved by an LS estimator as presented in the following section.

11.5   LOCATION ESTIMATORS

We now discuss two fundamental issues in location estimation: (1) determining achievable location accuracy and (2) searching for optimal and practical estimators.

11.5.1   Theoretical Limits for Location Estimation

Optimality Criterion: In location estimation, an optimality criterion needs to be defined not only to find an optimal location estimator but also to measure the performance of any estimator under consideration. In signal processing, the most widely used criterion is the *mean square error* (MSE), defined as

$$
\begin{aligned}
\mathrm{MSE}\left(\hat{\theta}\right) &= E\left\{\left\|\hat{\theta}-\theta\right\|^{2}\right\} \\
&= \mathrm{Tr}\left(C\left(\hat{\theta}\right)\right)+\left\|E\left(\hat{\theta}\right)-\theta\right\|^{2},
\end{aligned}
\tag{11.34}
$$

where $\mathrm{Tr}(\cdot)$ and $C(\cdot)$ indicate the matrix trace and the covariance. The first and second terms represent the total *variance* and *bias*—the sum of the variances and biases in estimating each position coordinate—respectively. Note that the MSE is a natural, direct measure of location estimation error, as it measures the average squared norm error of the estimator $\hat{\theta}$.

Cramer–Rao Lower Bound (CRLB):   We have discussed in Section 11.2 that a location system is subject to various errors, which may not be predictable or measured individually. Therefore, in designing a location system, one of the most important tasks is to determine the achievable location accuracy of the system. The theoretical limit on an *unbiased* location estimator can be measured by the CRLB. The CRLB, as the name indicates, is a lower bound on the covariance of an unbiased location estimator $\hat{\theta}$, which must satisfy

$$
C\left(\hat{\theta}\right)-F^{-1}(\theta)\geq 0.
\tag{11.35}
$$

Here, $F(\theta) = -E[\nabla_{\theta}(\nabla_{\theta}\ln f_V(v;\theta))^T]$ is the Fisher information matrix (FIM) [58], given by

$$
[F(\theta)]_{kl} = -E\left[\frac{\partial^2 \ln f_V\left(v;\theta\right)}{\partial\theta_k\,\partial\theta_l}\right],
\tag{11.36}
$$

where $f_V(v;\theta)$ is the joint PDF of the observation vector v. If some estimator is unbiased and attains the CRLB, it must be the minimum variance unbiased (MVU) estimator whose covariance matrix is $F^{-1}(\theta)$ of dimension $\eth\times\eth$ for $\eth$-dimensional θ [58].

We now obtain the CRLB for an RSS location estimator. For mathematical simplicity, let us assume that the elements of the observation vector v in (11.9) are independent and identically distributed.* Then, using (11.10), we have the *log-likelihood function* $\ln f_V(v;\theta)$ in (11.36) as

$$
\ell(\theta) = -\frac{1}{2\sigma_S^2}\sum_{i=1}^{m}\left(v_i - L_i(\theta)\right)^2.
\tag{11.37}
$$

By substituting (11.37) into (11.36), we can derive the FIM for RSS-based location estimation as

---

*In practice, the observations v at different anchors are spatially correlated, as discussed in Section 11.3.1. In this case, the FIM will include a correlation factor since the joint PDF $f_V(v;\theta)$ is a function of a non-diagonal covariance matrix for the shadow fading term.

$$[F(\theta)]_{kl} = \begin{cases} \dfrac{1}{\sigma_S^2} \displaystyle\sum_{i=1}^{m} \left(\dfrac{\partial L_i(\theta)}{\partial \theta_k} \right)^2, & k = l \\[3mm] \dfrac{1}{\sigma_S^2} \displaystyle\sum_{i=1}^{m} \left(\dfrac{\partial L_i(\theta)}{\partial \theta_k} \dfrac{\partial L_i(\theta)}{\partial \theta_l} \right), & k \neq l \end{cases} \tag{11.38}$$

where the gradient of $L_i(\theta)$ is derived in Section 11.5.3 (see (11.47)). The FIM and CRLB for DRSS-based location estimation can be derived in a similar manner by considering spatial correlation.

Since the CRLB assumes unbiasedness, it can measure the achievable location accuracy in terms of the MSE. Specifically, from (11.35), the MSE of any *unbiased* estimator is lower bounded as

$$\begin{aligned} \text{MSE}(\hat{\theta}) &= \text{Tr}\left(C(\hat{\theta})\right) \\ &\geq \left[F^{-1}(\theta)\right]_{11} + \left[F^{-1}(\theta)\right]_{22} \\ &= \frac{F_{11} + F_{22}}{F_{11}F_{22} - F_{12}^2}. \end{aligned} \tag{11.39}$$

In the above derivation, the key assumption was the *unbiasedness* of the estimator. However, many practical estimators, including ML/LS estimators and many "optimal" estimators are biased. In this case, the CRLB will not provide a lower bound on location accuracy [59]. For biased estimators, the uniform CRLB can be used as the lower bound on the MSE instead [60].

11.5.2   ML Estimator

In signal processing, it is customary to seek an unbiased estimator first and then to find the one that exhibits the least variability, that is, the MVU estimator [58]. However, many practical issues in location estimation not only challenge us to find the MVU estimator but also make it suboptimal in terms of the MSE (compare (11.34) and (11.39)). Specifically, since in practice we often encounter small data sizes (e.g., a small number of anchor nodes) and/or correlated location errors, many practical estimators are inefficient and biased. Although an MVU estimator exists, it may be outperformed by biased estimators [59–61]. In the rest of the section, we describe three popular approaches to obtain practical estimators, which are biased in many cases.

Given a set of data and an underlying statistical model, perhaps the most popular approach to parameter estimation is based on the ML principle. This approach is attractive due to the fact that ML estimators can be found even for complex estimation problems and are efficient and unbiased *asymptotically* (i.e., for large data sizes) [58]. The ML estimator $\hat{\theta}_{\text{ML}}$ is defined as one that maximizes the likelihood function or, equivalently, the log-likelihood function $\ell(\theta)$. Thus, the ML location estimator using RSS measurements can be found by solving $\partial \ell(\theta)/\partial \theta = 0$, where $\ell(\theta)$ is from (11.37). The solution is the maximizer of $\ell(\theta)$. Since this maximization problem is nonlinear and nonconvex, we need either to employ an iterative algorithm to find the globally optimal solution or to linearize the equation to have a closed-form approximate solution. This is the same procedure used to solve the problem of nonlinear LS estimation as presented next.

11.5.3   Nonlinear LS Estimator

The ML estimator can be derived only if the statistical properties of the observed data are known. Even when a statistical distribution of the data is known, we still need to know or estimate the underlying statistical parameters of the distribution. Specifically, for RSS-based location estimation using the statistical model given in Section 11.3.1, RSS measurements are jointly Gaussian but are not independent with unknown covariances. In particular, when the spatial correlation factor of shadow fading (which cannot be measured at every position in the feasible region) is involved, the corresponding covariance matrix is nondiagonal with unknown parameters.

The LS approach is widely used in practice due to its applicability to various location problems without resorting to statistical assumptions about the observed data. Its ease of implementation due to the special structure of the estimator also makes it attractive. Specifically, regardless of the complexity of the problem, we can easily form an LS location estimation problem. For example, in DRSS-based localization, we can incorporate the redundant DRSS data in an LS framework without knowledge of the statistical properties of the correlated shadowing. Note that it is difficult to exploit the redundancy using many well-known statistical methods. Particularly, a covariance matrix of all the measurements is typically nonpositive definite or singular due to the data colinearity [62]. Also, as noticed by comparing (11.37) and (11.46), an LS estimator is indeed an ML estimator when the residuals $r_i = v_i - L_i(\theta)$ are jointly normally distributed; $r \sim \mathcal{N}(0, \sigma_S^2 I)$.

Despite the attractive aspects of the LS approach, it has some limitations. One of the most notable issues is its nonrobustness to data *outliers*. Hence, the LS location problem should be constrained in the form of (11.2). The constraints can be set by exploiting the inherent characteristics of environmental and system constraints associated with network connectivity (or radio range), spatial correlation, node geometry, and so on.

LS Optimization Framework: Since an LS-RSS optimization framework can be handily derived from its DRSS counterparts, let us first consider the LS-DRSS formulation. Given the DRSS observations v_{ij} in (11.12), an LS-DRSS location estimator determines a parameter vector $\theta = [x, y, n_p]^T$ of source coordinates and PL gradient n_p, possibly subject to some constraints $[l_\theta, u_\theta]$ on θ as

$$\hat{\theta}_D = \underset{\theta}{\text{argmin}} \left\{ \phi_D(\theta) = \frac{1}{2} \sum_{\substack{i,j \in \{1, \dots, m\} \\ i<j}} r_{ij}^2(\theta) \right\} \tag{11.40}$$

$$\text{subject to } l_\theta \le \theta \le u_\theta,$$

where the residual $r_{ij}(\theta) = v_{ij} - L_{ij}(\theta)$, and $L_{ij}(\theta)$ is the PL model for DRSS given in (11.13). With a residual vector $r_D = [r_{12}, \dots, r_{1m}, r_{23}, \dots, r_{(m-1)m}]^T$, we have

$\phi_D(\theta) = (1/2)\|r_D(\theta)\|^2$. The subscripts D and R indicate the DRSS-based location estimator (DLE) and RSS-based location estimator (RLE), respectively. When the equations are common to both DLE and RLE, we omit their subscripts. The first derivative of ϕ_D with respect to θ, that is, the gradient vector field, is

$$\nabla\phi_D(\theta) = \sum_{i=1}^{m-1}\sum_{j=i+1}^{m} r_{ij}(\theta)\nabla r_{ij}(\theta) \tag{11.41}$$

$$= -J_D(\theta)^T r_D(\theta),$$

where J_D is the $M \times 3$ Jacobian matrix of the vector $L_D(\theta)$:

$$J_D(\theta) = \left[\frac{\partial L_{ij}(\theta)}{\partial\theta_k}\right]_{\substack{i,j\in\{1,\ldots,m\},i<j \\ k=1,2,3}}. \tag{11.42}$$

The ijth row vector J_{ij} for DLE is found to be

$$J_{D.ij} = -\frac{10n_p}{\ln 10}\left[z_{ij,x}, z_{ij,y}, -\frac{1}{n_p}\ln\left(\frac{d_j}{d_i}\right)\right]_{\substack{i,j\in\{1,\ldots,m\} \\ i<j}}, \tag{11.43}$$

in which, geometrically, $z_{ij,x}$ and $z_{ij,y}$ are x- and y-elements of the *difference* vector $z_{ij,x} = z_{ij,x}e_x + z_{ij,y}e_y$, given by

$$z_{ij,x} = \frac{u_{j,x}}{d_j} - \frac{u_{i,x}}{d_i} \quad\text{and}\quad z_{ij,y} = \frac{u_{j,y}}{d_j} - \frac{u_{i,y}}{d_i}, \tag{11.44}$$

where

$$u_{i,x} = \frac{x_i - x}{d_i}, u_{j,x} = \frac{x_j - x}{d_j}, u_{i,y} = \frac{y_i - y}{d_i}, u_{j,y} = \frac{y_j - y}{d_j}. \tag{11.45}$$

The basis vectors e_x and e_y are unit vectors in the direction of x- and y-axes, respectively. $u_{i,x}$ and $u_{i,y}$ are x- and y-elements of the unit position vector u_i of the ith anchor with respect to the target position, as expressed by $u_i = u_{i,x}e_x + u_{i,y}e_y$. This geometric unit vector represents the direction from the source to each anchor node as illustrated in Figure 11.5. It should be noted that the difference vector z_{ij} is formed by a pair of the ith and jth unit position vectors *inversely* scaled by their respective position vector lengths d_i and d_j, respectively, unlike the TDOA case [37].

We now turn to the LS-RSS formulation. Given the simplified RSS observations $\{v_j\}_{j=1}^{m}$ in (11.9), the problem of LS-RSS optimization can be formed as

$$\hat{\theta}_R = \underset{\theta}{\arg\min}\left\{\phi_R(\theta) = \frac{1}{2}\sum_{j=1}^{m} r_j^2(\theta)\right\} \tag{11.46}$$

$$\text{subject to } l_\theta \leq \theta \leq u_\theta,$$

where the residual $r_j(\theta) = v_j - L_j(\theta)$, and $L_j(\theta)$ is the PL model for RSS given in (11.9). By denoting the residual vector r for RLE as $r_R = [r_1, \ldots, r_m]^T$, we have $\phi_R(\theta) = \frac{1}{2}\|r_R(\theta)\|^2$. By setting $i = 0$ in (11.41–11.45), where the terms associated with i are thus removed, we can obtain the RSS counterparts. This modification replaces the vector J_{ij} in (11.43) by J_j for RLE as

$$J_{R,j} = -\frac{10n_p}{\ln 10}\left[\frac{u_{j,x}}{d_j}, \frac{u_{j,y}}{d_j}, -\frac{\ln d_j}{n_p}\right]_{j=1,\dots,m}, \tag{11.47}$$

which forms the $m \times 3$ Jacobian matrix J_R of the log-distance vector $L_R(\theta)$ for RLE. Thus, we can obtain the gradient vector field $\nabla\phi_R(\theta)$ for RLE as

$$\nabla\phi_R(\theta) = -J_R(\theta)^T r_R(\theta). \tag{11.48}$$

Example 11.4 Improving Location Accuracy in Example 11.3

Repeat Example 11.3(c) by estimating the mobile position $\theta = [x, y]^T$ using the above LS-RSS framework in (11.46) (assuming $n_p = 3$ known a priori). In this case, the distance estimates are not needed, yet a numerical optimization algorithm is necessary to deal with the nonconvex objective function. Using one of the algorithms implemented by the MATLAB optimization toolbox function lsqnonlin, the solution (i.e., minimizer) is found to be $\hat{\theta} = [27.07, 36.03]^T$. The previous location error of 10.67 m is then reduced to 6.07 m, which will be further improved with more RSS measurements.

11.5.4 Linear LS Estimator

The LS objective or merit function $\phi(\theta)$ formulated in (11.40) or (11.46) is inherently nonlinear and nonconvex (i.e., multimodal) with respect to the unknown parameters θ. Consequently, there exists no closed-form solution, and thus it is imperative to solve the problem numerically; a numerical algorithm tries to find an optimal solution iteratively, starting from some initial guess. Due to the unknown curvature and complexities of the objective function, not only can the number of required iterations be large, but the algorithm may also converge to a local minimum.

In the case where the computational complexity and convergence rate are of concern, one may seek to find a closed-form solution through the *linear LS approach*, converted from the original nonlinear LS function ϕ. There are two popular methods for the conversion. The first method is to linearize the original LS function directly to form a system of linear LS equations. In general, the LS equations are a function of either a statistical RSS/DRSS model considered here or a simpler distance/range model in (11.1). Second, as discussed in Section 11.4, a geometric system of nonlinear equations of position (i.e., circles) can be linearized to form a system of linear LS equations, as in (11.23) or (11.33).

Let us first look into the first linearization method to obtain a closed-form approximate solution. In order to apply the linear LS approach by modifying $\phi(\theta)$ directly, we first need to linearize the PL model $L(\theta)$ at some point $\overline{\theta}$ so that

$$L(\theta) \approx L(\overline{\theta}) + J(\overline{\theta})(\theta - \overline{\theta}). \tag{11.49}$$

Then, we can solve the normal equations $J^T J \overline{h} = J^T r$, where the step size $\overline{h} = \hat{\theta} - \overline{\theta}$, to obtain the linear LS solution [45, 58]

$$\hat{\theta} = \overline{\theta} + \left(J(\theta)^T J(\theta) \right)^{-1} J(\theta)^T r(\theta) \Big|_{\theta = \overline{\theta}}. \tag{11.50}$$

As noted, this linear LS solution is very simple and computationally attractive. Nevertheless, if the evaluated point $\overline{\theta}$ is far from the global minimum, the first-order approximation does not accurately represent the actual function, and no further improvement can be made. Note that this solution is equivalent to the first iteration of a Gauss–Newton method with a starting point $\theta_0 = \overline{\theta}$. Thus, the appropriate determination of the linearization point $\overline{\theta}$ is crucial to have the desired performance of the linear LS estimator.

An alternative linear LS approach is to exploit the geometric relationship between the range estimates described in Section 11.4. Specifically, we formulate a linear LS framework with the residual $r = b - A\theta$, such that

$$\hat{\theta} = \underset{\theta}{\operatorname{argmin}} \frac{1}{2} (b - A\theta)^T (b - A\theta). \tag{11.51}$$

For the RSS case ($m \geq 3$), from (11.23), we have

$$A = \begin{bmatrix} x_{12} & x_{13} & \cdots & x_{1m} \\ y_{12} & y_{13} & \cdots & y_{1m} \end{bmatrix}_{2 \times (m-1)}^T, b = \begin{bmatrix} b_{12} & b_{13} & \cdots & b_{1m} \end{bmatrix}_{1 \times (m-1)}^T,$$

where $x_{ij} = x_j - x_i$, $y_{ij} = y_j - y_i$, and $b_{ij} = \frac{1}{2} \left\{ \|x_j\|^2 - \|x_i\|^2 - (d_j^2 - d_i^2) \right\}$. Similarly, for the DRSS case ($m \geq 4$), from (11.33) we have

$$A = \begin{bmatrix} x_{d12} & x_{d13} & \cdots & x_{d1M} & x_{d23} & \cdots & x_{d(M-1)M} \\ y_{d12} & y_{d13} & \cdots & y_{d1M} & y_{d23} & \cdots & y_{d(M-1)M} \end{bmatrix}_{2 \times_M C_2}^T$$

$$b = \begin{bmatrix} b_{12} b_{13} & \cdots & b_{1M} & b_{23} & \cdots & b_{(M-1)M} \end{bmatrix}_{1 \times_M C_2}^T,$$

where $x_{dij} = x_{dj} - x_{di}$, $y_{dij} = y_{dj} - y_{di}$, and $b_{ij} = \frac{1}{2} \left\{ \|c_{dj}\|^2 - \|c_{di}\|^2 - (r_{dj}^2 - r_{di}^2) \right\}$.

It is then straightforward to find a linear LS estimator using the closed-form LS solution [45, 58]:

$$\hat{\theta} = \left(A^T A \right)^{-1} A^T b. \tag{11.52}$$

We note that in the DRSS case, the size of **A** increases rapidly with additional measurements. The matrix inversion operation is usually not a problem with a small- or medium-scale network. However, for large-scale networks, the linear LS approach may be erroneous and computationally challenging due to computational and ill-conditioning issues [45].

Since the second linear LS approach does not require an initial solution in the noniterative procedure, for RSS-based positioning it generally performs better than the first linearization method if we have no knowledge of a good linearization point. This can be inferred from Figure 11.5, since shadow fading only perturbs the circles' circumferences. However, in the case of DRSS, the shadowing noise affects both the circumferences and centers of DRSS circles, which will result in further degradation of the estimate, as can be seen in Figure 11.7. This also implies that DRSS

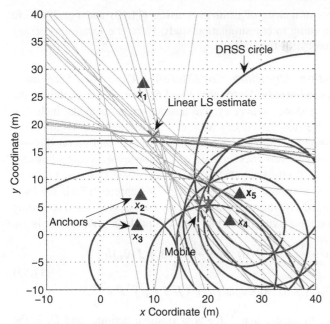

Figure 11.7   A geometric linear LS approach, which linearizes a system of nonlinear geometric DRSS equations (i.e., circles).

positioning is more sensitive to the geometry of anchors or nodes than its RSS counterpart. Note that in both localization cases, a linear LS approach is also not robust to data outliers.

11.6   PERFORMANCE EVALUATION

11.6.1   Simulation Settings

In this section, the performance of range-based location techniques based on RSS/ DRSS measurements (i.e., RLE/DLE) is evaluated using the nonlinear LS approach presented in Section 11.5.3. For simplicity, we only estimate the position parameters given prior knowledge* of n_p (i.e., $\theta = [x, y]^T$). Regarding the geometry of nodes, $m + 1$ anchor and unknown source locations are randomly placed in a $30 \times 30\,\text{m}^2$ area for every simulation iteration to reflect the effect of node geometry [37]. The results are parameterized by several primary factors affecting location estimation—specifically, the number of anchor nodes, spatial correlation of shadowing, shadowing variance, and PL gradient. To evaluate the impact of spatial correlation of shadow fading, we adopt the model in (11.20) for the generation of a correlated shadowing map. Despite its simpler form as compared to (11.19), in

---

*When n_p is jointly estimated, the estimator performances are comparable to those presented here. The value may be determined during an offline training phase.

general, the average performances of estimators using (11.19) and (11.20) were found to be similar according to our simulation study. Here, the location accuracy is shown in terms of the RMSE (in meters).

Numerical Optimization Algorithm Considered: There have been many iterative algorithms proposed specifically for nonlinear LS optimization, exploiting the special structure of φ to enhance efficiency. We tested various methods, including steepest descent, Gauss–Newton, Levenberg–Marquardt (LM), and trust region (TR), and found that the method of TR using a subspace technique and preconditioned conjugate gradients [45, 63] performs best in terms of location accuracy and robustness to random initial solutions. It effectively tackles scenarios where LM may not work properly, such as negative curvature and poor scaling. This approach finds the local minimizer in the constrained TR whose size at the kth iteration Δ_k is adjusted according to its performance during the previous iteration. The TR quadratic subproblem is

$$\theta_{k+1} = \min_{\theta_{k+1} \leftarrow h_k} \left\{ r(\theta_k)^T J(\theta_k) h_k + \frac{1}{2} h_k^T J(\theta_k)^T J(\theta_k) h_k : \right.$$
$$\left. \|D_k h_k\| \leq \Delta_k, h_k \in \mathrm{span}\left[J^T r, (J^T J + \gamma I)^{-1} J^T r \right] \right\}, \tag{11.53}$$

where γ is chosen so as to ensure that $J^T J + \gamma I$ is positive definite, and D_k is the diagonal scaling matrix. This algorithm is implemented by the MATLAB optimization toolbox function lsqnonlin. All the results presented next are obtained by this approach. Since a single start for the optimization can bias the comparison due to the multimodal LS error functions, we evaluate the functions multiple times using random starting points and then select the minimizer.

11.6.2   Simulation Results

Impact of Number of Anchor Nodes and Spatial Correlation: In Figure 11.8, the performance of RLE and DLE with respect to the number of anchor nodes is presented for $\sigma_s = 5\,\text{dB}$ and $n_p = 3$. The results are parameterized by the shadowing correlation to demonstrate its impact. Note that an ideal assumption was made for RLE in that perfect information about the radio/propagation parameters in (11.9) was available (i.e., $P_r(d_0)$ was known perfectly).

We summarize the performance of the two techniques as follows. First, higher correlation actually improves the accuracy of RLE when the number of anchors is small ($m < 9$) but deteriorates its performance when the number is increased beyond $m = 9$. When the number of anchors is small (i.e., small data size), the RLE is biased, and the correlation tends to help location estimation. On the other hand, as more anchors are added (i.e., more RSS measurements), RLE becomes unbiased and approaches the CRLB asymptotically. In this case, it is observed that the error increases with higher correlation values. The second key observation is that the performance of DLE with higher correlation is always better for different numbers of anchors. When the correlation is high, in general DLE outperforms RLE. Note that other range-based positioning techniques, by contrast, perform worse in highly correlated shadowing environments such as indoor/urban wireless networks. Third,

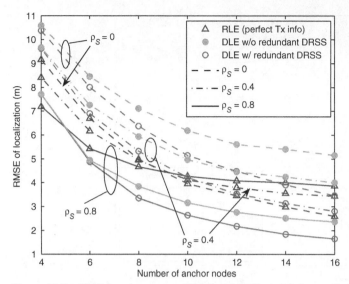

Figure 11.8   RMS location errors of RLE and DLE versus the number of anchor nodes m ($\sigma_S = 5\,\mathrm{dB}$ and $n_p = 3$) [8]. Tx, transmitter.

incorporating the redundant DRSS measurements is beneficial for localization. Specifically, a performance advantage can be found with more anchors for both DLE and RLE, but the rate of improvement with additional anchors is higher for DLE due to the redundancy. Thus, when the network size m is large, DLE becomes comparable to RLE even with lower correlation. In other words, the larger the number of anchors involved, the larger the rate of increase in DRSS redundancy as noted in (11.14). As a result, its susceptibility to fading and measurement noise can be mitigated, as similarly found for TDOA [64].

Impact of Correlated Shadow Fading: In Figure 11.9, the impact of lognormal shadow fading on RLE and DLE is shown by varying σ_S. Due to the two-sided localization behavior of RLE observed above, two cases of $m = 6$ and $m = 14$ are considered in Figure 11.9a, b, respectively. From the results, similar observations can be made. Specifically, for $m = 6$, the correlation improves the performance of both RLE and DLE, but the improvement rate of DLE is higher. When the correlation is low ($\rho_S < 0.4$), RLE outperforms DLE, whereas higher correlation makes DLE superior to RLE. On the other hand, for $m = 14$, the correlation degrades the performance of RLE, whereas the accuracy of DLE is noticeably improved. We also see that the redundant DRSS information improves the performance of DLE.

Impact of PL and Spatial Correlation: We now examine the effect of PL gradient n_p, which indicates the rate of PL over distance. In addition to the above study of shadow fading, this investigation is especially valuable as more location applications have considered indoor and urban environments. For instance, one of the greatest concerns for the wireless E-911 mandate is the unavailability or large error of the emergency caller's location information in indoor areas.

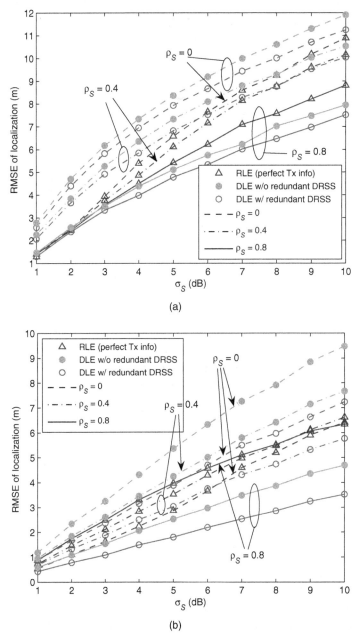

Figure 11.9 RMS location errors of RLE and DLE versus standard deviation of shadow fading σ_S ($n_p = 3$) for (a) $m = 6$ and (b) $m = 14$ [8].

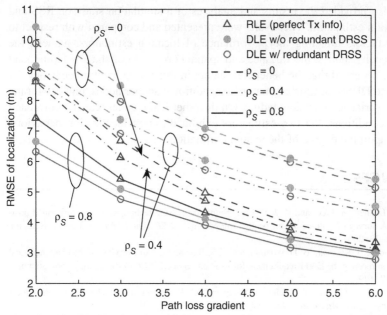

Figure 11.10   RMSE of RLE and DLE versus path loss gradient n_p ($\sigma_S = 5\,$dB and $m = 6$).

One may expect that larger values of n_p (i.e., higher rates of signal power loss over distance) would degrade the performance of location estimators. However, a larger PL rate tends to improve the RSS-based estimator performance because given some RSS uncertainty, the size of the corresponding uncertainty domain of a distance d becomes smaller with larger PL rates. This can be readily observed on the scatter plot of RSS (or PL) as a function of the distance d [30]. This conjecture is confirmed by Figure 11.10, where the performances of both RLE and DLE improve with higher values of n_p. From the figure, the effects of the correlation and DRSS redundancy over a range of n_p values can be found to be very similar to those observed previously. Since the number of anchors is small ($m = 6$), RLE generally outperforms DLE except for high correlation values.

11.7   CONCLUSION

In this chapter, we have presented the fundamental aspects of RSS-based position location (especially lateration techniques). Our emphasis was placed on two range-based approaches using RSS and DRSS measurements, and we also introduced other major RSS-based techniques; namely, RF fingerprinting and connectivity- or proximity-based localization. Our goal in this chapter was to present an overview of RSS position location, its possibilities and its limitations from both theoretical and practical perspectives. To accomplish this goal, we have covered the basics of RSS/DRSS-based positioning, location error sources, geometric interpretations, and theoretical and practical issues associated with location estimation.

We also provided the results and analyses of a simulation study on RSS and DRSS location estimation. The results were presented and compared with respect to four major factors affecting the performance of location estimators, namely, the number of anchor nodes, the variance and spatial correlation of shadow fading, and PL rate. We observed that the beneficial factors in improving location accuracy for both RSS and DRSS location estimators are additional anchor nodes, higher PL rate, and smaller variance of shadow fading. On the other hand, under realistic correlated shadowing conditions, the two estimators exhibit different localization behaviors, depending upon the degree of the spatial correlation.

REFERENCES

[1] G. Sun, J. Chen, W. Guo, and K. J. R. Liu, "Signal processing techniques in network-aided positioning: A survey of state-of-the-art positioning designs," *IEEE Signal Process. Mag.*, vol. 22, no. 4, pp. 12–23, 2005.

[2] J. H. Reed, K. J. Krizman, B. D. Woerner, and T. S. Rappaport, "An overview of the challenges and progress in meeting the E-911 requirement for location service," *IEEE Commun. Mag.*, vol. 36, no. 4, pp. 30–37, 1998.

[3] M. Vossiek, L. Wiebking, P. Gulden, J. Wieghardt, C. Hoffmann, and P. Heide, "Wireless local positioning," *IEEE Microwave Mag.*, vol. 4, no. 4, pp. 77–86, 2003.

[4] Y.-B. Ko and N. H. Vaidya, "Location-aided routing (LAR) in mobile ad hoc networks," in *Proc. of the 4th Annual ACM/IEEE Int. Conf. on Mobile Computing and Networking*, 1998, pp. 66–75.

[5] A. H. Sayed, A. Tarighat, and N. Khajehnouri, "Network-based wireless location: Challenges faced in developing techniques for accurate wireless location information," *IEEE Signal Process. Mag.*, vol. 22, no. 4, pp. 24–40, 2005.

[6] N. Patwari, J. N. Ash, S. Kyperountas, A. O. Hero III, R. Moses, and N. Correal, "Locating the nodes: Cooperative localization in wireless sensor networks," *IEEE Signal Process. Mag.*, vol. 22, no. 4, pp. 54–69, 2005.

[7] G. Mao, B. Fidan, and B. D. O. Anderson, "Wireless sensor network localization techniques," *ACM Comput. Net.*, vol. 51, no. 10, pp. 2529–2553, 2007.

[8] J. H. Lee and R. M. Buehrer, "Location estimation using differential RSS with spatially correlated shadowing," in *Proc. of the 28th IEEE Conf. on Global Telecommunications*, 2009, pp. 4613–4618.

[9] H. B. Lee, "A novel procedure for assessing the accuracy of hyperbolic multilateration systems," *IEEE Trans. Aerosp. Electron. Syst.*, vol. AES-11, no. 1, pp. 2–15, 1975.

[10] D. J. Torrieri, "Statistical theory of passive location systems," *IEEE Trans. Aerosp. Electron. Syst.*, vol. AES-20, no. 2, pp. 183–198, 1984.

[11] J. Smith and J. Abel, "Closed-form least-squares source location estimation from range-difference measurements," *IEEE Trans. Acoust.*, vol. 35, no. 12, pp. 1661–1669, 1987.

[12] P. Bahl and V. N. Padmanabhan, "RADAR: An in-building RF-based user location and tracking system," in *Proc. of the IEEE Conf. on Comp. Commun.*, vol. 2, 2000, pp. 775–784.

[13] N. Patwari and A. O. Hero III, "Using proximity and quantized RSS for sensor localization in wireless networks," in *Proc. of the 2nd ACM Int. Conf. on Wireless Sensor Networks and Applications*, 2003, pp. 20–29.

[14] J. Hightower, G. Boriello, and R. Want, "Spoton: An indoor 3D location sensing technology based on RF signal strength," University of Washington, Tech. Rep. CSE 2000-02-02, Feb. 2002.

[15] P. Castro, P. Chiu, T. Kremenek, and R. R. Muntz, "A probabilistic room location service for wireless networked environments," in *Proc. of the 3rd Int. Conf. on Ubiquitous Computing*, Sep. 2001, pp. 18–34.

[16] B. Parkinson and J. J. Spiker, Eds., *Global Positioning System: Theory and Applications*, vol. 1. Danvers, MA: American Institute of Aeronautics and Astronautics, 1996.

[17] G. M. Djuknic and R. E. Richton, "Geolocation and assisted GPS," *IEEE Comput.*, vol. 34, no. 2, pp. 123–125, 2001.

[18] R. J. Fontana and S. J. Gunderson, "Ultra-wideband precision asset location system," in *2002 IEEE Conf. on Ultra Wideband Systems and Technologies*, May 2002, pp. 147–150.

[19] J. J. Caffery and G. L. Stuber, "Subscriber location in CDMA cellular networks," *IEEE Trans. Veh. Technol.*, vol. 47, no. 2, pp. 406–416, 1998.

[20] J. Zhang, P. V. Orlik, Z. Sahinoglu, A. F. Molisch, and P. Kinney, "UWB systems for wireless sensor networks," *Proc. IEEE.*, vol. 97, no. 2, pp. 313–331, 2009.

[21] D. Dardari, A. Conti, U. Ferner, A. Giorgetti, and M. Z. Win, "Ranging with ultrawide bandwidth signals in multipath environments," *Proc. IEEE.*, vol. 97, no. 2, pp. 404–426, 2009.

[22] J. A. Pierce, "An introduction to Loran," *IEEE Aerosp. Electron. Syst. Mag.*, vol. 5, no. 10, pp. 16–33, 1990.

[23] A. Savvides, C.-C. Han, and M. B. Strivastava, "Dynamic fine-grained localization in ad-hoc networks of sensors," in *Proc. of the 7th Annual Int. Conf. on Mobile Computing and Networking*, 2001, pp. 166–179.

[24] N. B. Priyantha, A. Chakraborty, and H. Balakrishnan, "The Cricket location-support system," in *Proc. of the 6th Annual Int. Conf. on Mobile Computing and Networking*, 2000, pp. 32–43.

[25] Y. Zhao, "Standardization of mobile phone positioning for 3G systems," *IEEE Commun. Mag.*, vol. 40, no. 7, pp. 108–116, 2002.

[26] D. Niculescu and B. R. Badrinath, "Ad hoc positioning system (APS) using AOA," in *Proc. of the IEEE Conf. on Computer Communications*, Apr. 2003, pp. 1734–1743.

[27] P. Biswas, T.-C. Lian, T.-C. Wang, and Y. Ye, "Semidefinite programming based algorithms for sensor network localization," *ACM Trans. Sen. Netw.*, vol. 2, no. 2, pp. 188–220, 2006.

[28] S. Sakagami, S. Aoyama, K. Kuboi, S. Shirota, and A. Akeyama, "Vehicle position estimates by multibeam antennas in multipath environments," *IEEE Trans. Veh. Technol.*, vol. 41, no. 1, pp. 63–68, 1992.

[29] J. Kennedy and M. C. Sullivan, "Direction finding and 'smart antennas' using software radio architectures," *IEEE Commun. Mag.*, vol. 33, no. 5, pp. 62–68, 1995.

[30] T. S. Rappapport, *Wireless Communications: Principles and Practice*, 2nd ed. Upper Saddle River, NJ: Prentice-Hall, 2002.

[31] I. Guvenc and C.-C. Chong, "A survey on TOA based wireless localization and NLOS mitigation techniques," *IEEE Commun. Surveys Tuts.*, vol. 11, no. 3, pp. 107–124, 2009.

[32] P.-C. Chen, "A non-line-of-sight error mitigation algorithm in location estimation," in *Proc. of the IEEE Wireless Communications and Networking Conf.*, 1999, pp. 316–320.

[33] L. Cong and W. Zhuang, "Non-line-of-sight error mitigation in TDOA mobile location," in *Proc. of the IEEE Global Telecommunications Conf.*, vol. 1, 2001, pp. 680–684.

[34] L. Xiong, "A selective model to suppress NLOS signals in angle-of-arrival (AOA) location estimation," in *Proc. of the IEEE Symp. on Personal, Indoor and Mobile Radio Communications*, vol. 1, 1998, pp. 461–465.

[35] J. C. Liberti and T. S. Rappaport, "Statistics of shadowing in indoor radio channels at 900 and 1900mhz," in *Proc. of the IEEE Military Communications Conf.*, vol. 3, 1992, pp. 1066–1070.

[36] K. Zayana and B. Guisnet, "Measurements and modelisation of shadowing cross-correlations between two base-stations," in *Proc. of the IEEE Int. Conf. on Universal Personal Communications*, vol. 1, 1998, pp. 101–105.

[37] M. A. Spirito, "On the accuracy of cellular mobile station location estimation," *IEEE Trans. Veh. Technol.*, vol. 50, no. 3, pp. 674–685, 2001.

[38] D. B. Jourdan and N. Roy, "Optimal sensor placement for agent localization," *ACM Trans. Sen. Netw.*, vol. 4, no. 3, pp. 1–40, 2008.

[39] L. M. Kaplan, "Local node selection for localization in a distributed sensor network," *IEEE Trans. Aerosp. Electron. Syst.*, vol. 42, no. 1, pp. 136–146, 2006.

[40] L. M. Kaplan, "Global node selection for localization in a distributed sensor network," *IEEE Trans. Aerosp. Electron. Syst.*, vol. 42, no. 1, pp. 113–135, 2006.

[41] A. J. Coulson, A. G. Williamson, and R. G. Vaughan, "A statistical basis for lognormal shadowing effects in multipath fading channels," *IEEE Trans. Commun.*, vol. 46, no. 4, pp. 494–502, 1998.

[42] N. Patwari and P. Agrawal, "Effects of correlated shadowing: Connectivity, localization, and RF tomography," in *Proc. of the IEEE/ACM Int. Conf. on Information Processing in Sensor Networks*, 2008, pp. 82–93.

[43] M. Gudmundson, "Correlation model for shadow fading in mobile radio systems," *Electron. Lett.*, vol. 27, no. 23, pp. 2145–2146, 1991.

[44] Z. Wang, E. K. Tameh, and A. Nix, "Simulating correlated shadowing in mobile multihop relay/ ad-hoc networks," IEEE 802.16 Broadband Wireless Access Working Group, Tech. Rep., 2006.

[45] G. Golub and C. Van Loan, *Matrix Computations*, 3rd ed. Baltimore, MD: Johns Hopkins University Press, 1996.

[46] N. Patwari, A. O. Hero III, M. Perkins, N. S. Correal, and R. J. O'Dea, "Relative location estimation in wireless sensor networks," *IEEE Trans. Signal Process.*, vol. 51, no. 8, pp. 2137–2148, 2003.

[47] J. B. Tenenbaum, V. D. Silva, and J. C. Langford, "A global geometric framework for nonlinear dimensionality reduction," *Science*, vol. 290, no. 5500, pp. 2319–2323, 2000.

[48] S. T. Roweis and L. K. Saul, "Nonlinear dimensionality reduction by locally linear embedding," *Science*, vol. 290, no. 5500, pp. 2323–2326, 2000.

[49] I. Borg and P. Groenen, *Multidimensional Scaling, Theory and Applications*. New York: Springer, 1997.

[50] D. L. Donoho and C. Grimes, "Hessian eigenmaps: New locally linear embedding techniques for high-dimensional data," *Proc. Natl. Acad. Sci.*, vol. 100, no. 10, pp. 5591–5596, 2003.

[51] M. Belkin and P. Niyogi, "Laplacian eigenmaps for dimensionality reduction and data representation," *Neural Comput.*, vol. 15, no. 6, pp. 1373–1396, 2003.

[52] W. S. Torgeson, "Multidimensional scaling of similarity," *Psychometrika*, vol. 30, no. 1, pp. 379–393, 1965.

[53] Y. Shang, W. Ruml, Y. Zhang, and M. P. J. Fromherz, "Localization from mere connectivity," in *Proc. of the ACM Int. Symp. on Mobile Ad Hoc Networking and Computing*, 2003, pp. 201–212.

[54] N. Patwari and A. O. Hero III, "Manifold learning algorithms for localization in wireless sensor networks," in *Proc. of the IEEE Conf. on Acoustics, Speech, and Signal Processing*, vol. 3, May 2004, pp. 857–860.

[55] G. Peyre, "A toolbox for dimension reduction methods," 2006 . Available: http://www.mathworks .com/matlabcentral/fileexchange/

[56] S. Capkun, M. Hamdi, and J. P. Hubaux, "GPS-free positioning in mobile ad-hoc networks," in *Proc. of the Hawaii Int. Conf. on Systems Science*, 2001, pp. 3481–3490.

[57] H. Wu, C. Wang, and N.-F. Tzeng, "Novel self-configurable positioning technique for multihop wireless networks," *IEEE/ACM Trans. Networking*, vol. 13, no. 3, pp. 609–621, 2005.

[58] S. M. Kay, *Fundamentals of Statistical Signal Processing: Estimation Theory*. Upper Saddle River, NJ: Prentice-Hall, 1993.

[59] X. Li, "RSS-based location estimation with unknown pathloss model," *Wireless Commun. IEEE Trans.*, vol. 5, no. 12, pp. 3626–3633, 2006.

[60] A. O. Hero III, J. A. Fessler, and M. Usman, "Exploring estimator biasvariance tradeoffs using the uniform CR bound," *IEEE Trans. Signal Process.*, vol. 44, no. 8, pp. 2026–2041, 1996.

[61] S. Kay and Y. C. Eldar, "Rethinking biased estimation [lecture notes]," *IEEE Signal Process. Mag.*, vol. 25, no. 3, pp. 133–136, 2008.

[62] W. Wothke, "Nonpositive definite matrices in structural modeling," in *Testing Structural Equation Models*, K. Bollen and J. Long, Eds. Newbury Park, CA: Sage, 1993, pp. 256–293.

[63] R. H. Byrd, R. B. Schnabel, and G. A. Shultz, "Approximate solution of the trust region problem by minimization over two-dimensional subspaces," *Math. Programming*, vol. 40, no. 1, pp. 247–263, 1988.

[64] W. Hahn and S. Tretter, "Optimum processing for delay-vector estimation in passive signal arrays," *IEEE Trans. Inf. Theory*, vol. 19, no. 5, pp. 608–614, 1973.

ON THE PERFORMANCE OF WIRELESS INDOOR LOCALIZATION USING RECEIVED SIGNAL STRENGTH

Jie Yang,[1] Yingying Chen,[2] Richard P. Martin,[2]
Wade Trappe,[2] and Marco Gruteser[2]
[1]Florida State University, Tallahassee, FL
[2]Rutgers University, Piscataway, NJ

THIS CHAPTER focuses on received signal strength (RSS)-based indoor positioning. It first provides an overview of various indoor localization techniques employing RSS, including lateration methods, machine learning classification, probabilistic approaches, and statistical supervised learning techniques. It then compares their performance in terms of localization accuracy through measurement studies in real office building environments under representative Wi-Fi and ZigBee wireless networks. It further surveys techniques and methods that are developed to support robust wireless localization and to improve localization accuracy, including real-time RSS calibration via anchor verification, closely spaced multiple antennas, taking advantage of robust statistical methods to provide stability to contaminated measurements, utilizing linear regression to characterize the relationship between RSS and the distance to anchors, and exploiting RSS spatial correlation. Finally, it concludes with a discussion of popular location-based applications.

Handbook of Position Location: Theory, Practice, and Advances, Second Edition.
Edited by S. A. (Reza) Zekavat and R. Michael Buehrer.
© 2019 by the Institute of Electrical and Electronics Engineers, Inc.
Published 2019 by John Wiley & Sons, Inc.
Companion Website: www.wiley.com/go/zekavat/positionlocation2e

12.1 INTRODUCTION

The widespread deployment of indoor wireless technologies is resulting in a variety of location-based services. For instance, in the public arena, doctors want to use location information to track and monitor patients in medical facilities, and first responders need to track each other and locate victims during emergency rescues. On the other hand, in the enterprise domain, location-based access control is desirable for accessing proprietary corporate materials in restricted areas or rooms. For example, during corporate meetings, certain documents can only be received by laptops that reside within the involved conference rooms, which requires location-aware content delivery. In addition, asset tracking also relies on location information. To ensure the wide deployment of these pervasive applications, accurate positioning is important as the location is a critical input to many high-level tasks supporting these applications. In addition, in indoor environments such as shopping malls, hospitals, warehouses, and factories, where global positioning system (GPS) devices generally do not work, indoor localization systems promise the benefits of accurate location estimates for wireless devices such as handheld devices, electronic badges, and laptop computers.

Compared to various physical modalities for localization, such as time of arrival (TOA) [1], time difference of arrival (TDOA) [2], and angle of arrival (AOA), using RSS [3–5] is an attractive approach to perform localization since it can reuse the existing wireless infrastructures and thus presents tremendous cost savings over deploying localization-specific hardware. Also, all current standard commodity radio technologies, such as Wi-Fi, ZigBee, active radio frequency identification (RFID), and Bluetooth provide RSS measurements, and consequently the same algorithms can be applied across different platforms.

Performing RSS-based localization is a challenging task due to multipath effects in unpredictable indoor settings. These effects include shadowing (i.e., blocking a signal), reflection (i.e., waves bouncing off an object), diffraction (i.e., waves spreading in response to obstacles), and refraction (i.e., waves bending as they pass through different mediums). Thus, the RSS measurements will be attenuated in unpredictable ways due to these effects. To tackle these challenges, recent studies have resulted in a plethora of methods for localizing wireless devices using RSS. This chapter provides an overview of various indoor localization techniques employing RSS, including lateration methods, machine learning classification, probabilistic approaches, and statistical supervised learning techniques. We then study the localization performance by presenting representative evaluation metrics. We further compare the performance of various localization algorithms in terms of localization accuracy through real office building environments using prevailing Wi-Fi and ZigBee wireless networks.

Furthermore, we survey techniques and methods that are developed to support robust wireless localization and improve localization accuracy, including real-time RSS calibration via anchor verification, multiple closely spaced antennas, taking advantage of the robust statistical methods to provide stability to contaminated measurements, utilizing linear regression to characterize the relationship between RSS and the distance to anchors more accurately, and exploiting RSS spatial correlation.

Finally, we conclude the chapter by presenting current and emerging applications that leverage the location information.

12.2   RSS-BASED LOCALIZATION ALGORITHMS

An indoor positioning system consists of a set of anchor nodes (e.g., Wi-Fi access points or traffic sniffers), placed at known locations in the area of interest, and a wireless device carried by the person or attached to the object that needs to be localized. During the operation of localization, radio signals are transmitted between the wireless device and multiple anchors. Based on the received wireless signal at either the anchors or the wireless device, a localization algorithm estimates the position of the wireless device. We next describe a generalized localization model that the localization algorithms are based upon to map the observed RSS in signal space to the physical location in the physical space [5].

In a generalized localization model, let us suppose that we have a domain D in two dimensions, such as an office building, over which we wish to localize wireless devices. Within D, a set of n anchor points is available to assist in localization. A wireless device that transmits with a fixed power in an isotropic manner will cause a vector of n signal strength readings to be measured by the n anchor points. In practice, these n signal strength readings are averaged over a sufficiently large time window to remove statistical variability. Therefore, corresponding to each location in D, there is an n-dimensional vector of signal readings $\mathbf{s} = (s_1, s_2, \dots, s_n)$ that resides in range R.

This relationship between positions in D and signal strength vectors defines a fingerprint function $F: D \to R$ that takes our real-world position (x, y) and maps it to a signal strength reading, $\mathbf{s}$. F has some important properties. First, in practice, F is not completely specified, but rather a finite set of positions (x_j, y_j) is used for measuring a corresponding set of signal strength vectors, $\mathbf{s}_j$. Additionally, the inverse of F is a function G that is not well-defined: There are holes in the n-dimensional space in which R resides for which there is no well-defined inverse.

It is precisely the inverse function G, though, that allows us to perform localization. In general, we will have a signal strength reading $\mathbf{s}$ for which there is no explicit inverse (e.g., perhaps due to noise variability). Instead of using G, which has a domain restricted to R, we consider various pseudoinverses G_{alg} of F for which the domain of G_{alg} is the complete n-dimensional space. Here, the notation G_{alg} indicates that there are different *algorithmic* choices, that is, various localization algorithms, for the pseudoinverse.

We use signal strength to illustrate the generalized localization model because signal strength is a common wireless signal modality used by a widely diverse set of localization algorithms. For instance, radio frequency (RF) fingerprinting approaches utilize RSS [3, 6], and many lateration approaches [7] use it as well. Chapter 15 describes fingerprinting approaches in more detail and Chapter 11 describes lateration approaches in more detail. In spite of its several meter-level accuracy, using RSS is a natural choice because it can reuse the existing wireless infrastructure, and this feature presents tremendous cost savings over deploying localization-specific hardware. We next provide an overview of a representative set of algorithms using RSS to perform position estimation.

12.2.1 Approach Overview

There are several ways to classify localization algorithms that use signal strength: range-based schemes, which explicitly involve the calculation of distances to landmarks, and RF fingerprinting schemes, which originated from machine learning classification, whereby a radio map is constructed using prior measurements, and a device is localized by referencing this radio map through classification. We provide an overview of various indoor localization techniques employing RSS, including lateration methods, machine learning classification, probabilistic approaches, and statistical supervised learning techniques. They can be further categorized as range-based methods, such as lateration methods and Bayesian networks (BNs), and RF fingerprinting strategies, including fingerprint matching using machine learning classification and maximum likelihood estimation (MLE) using probabilistic approaches.

In lateration methods, the RSS measurements are used to perform ranging between the wireless device and anchor points based on fitting a signal propagation model, and then the location of the wireless device can be computed via trilateration.

For the machine learning classification via fingerprint matching, it first builds a priori radio signal strength maps of the localization region during the training phase based on the measured signal strengths (i.e., RF fingerprints) at different locations. By comparing the online measured RSS to the preconstructed signal map using some optimization criterion, the location can be deduced. For example, the minimum Euclidean distance in the signal strength vector space can be used as an optimization criterion as described in RADAR [3], which is a localization algorithm developed by Microsoft. In these approaches, location estimation is only considered through the value of the RSS measurements.

On the other hand, probabilistic approaches have been proposed where the RSS is treated as a random variable that may be modeled by a lognormal distribution [8] at a specific location. In these approaches, a signal strength probability distribution based on the RSS measurements is modeled. During the localization phase, the position of the targeted wireless device is estimated using probabilistic methods, such as MLE.

In addition, BNs and Kernel methods (which are discussed in Chapter 14) are methods that utilize the statistical supervised learning approach [9]. Statistical supervised learning infers a function from *supervised* training data. The training data consist of a set of training examples including input vectors, such as the RSS from a set of anchors and a desired output value (i.e., the location of the wireless device). The localization algorithms that perform supervised learning analyze the training data and produce an inferred function, which is called a classifier, which predicts the location of a wireless device for any valid input RSS vectors. We will now examine each class of algorithm in more detail.

12.2.2   Lateration Methods

Lateration is the most common method for deriving the location of a wireless device [10–12]. By estimating the distance from a wireless device to multiple anchors, lateration approaches derive the wireless device's location based on least squares methods.

In particular, RSS is employed to estimate the distance between a wireless device and an anchor. We next show an example on how to use an off-line training phase to estimate the propagation model parameters so as to derive ranging information from the measured RSS during the runtime localization phase. We note that there are some approaches that treat the propagation model parameters as nuisance parameters that are examined along with the position. Chapter 11 describes lateration approaches in more detail.

Example 12.1   Parameter Estimation in the Signal Propagation Model

In this example, we show that the radio propagation parameters can be estimated from the training data collected during the off-line training phase.

Solution

During the off-line training phase, RSS samples are collected at various known locations from multiple anchors, and distances are calculated from the known locations to anchors. The measured RSS readings and the corresponding distances are then used to fit the signal propagation model based on the signal–distance relationship [8, 13]:

$$P(d)[\text{dBm}] = P(d_0)[\text{dBm}] - 10\gamma \log_{10}\left(\frac{d}{d_0}\right) + X_\sigma, \qquad (12.1)$$

where $P(d_0)$ represents the received power of a wireless device at the reference distance d_0, d is the distance between the wireless device and the anchor, γ is the path loss exponent, and X_σ is the shadow fading, which follows a zero-mean Gaussian distribution with σ standard deviation. Given the RSS and distances, linear regression is usually used to fit the propagation model [14].

During the runtime localization phase, there are two steps: *ranging* and *lateration*. In the ranging step, according to the measured RSS from the targeting wireless device and the fitted signal–distance relationship (i.e., the propagation model parameters), the distances between the wireless device and multiple anchors can be calculated. In the lateration step, the location of the wireless device can be estimated according to distances between the wireless device and the anchors based on least squares methods. In the literature, there are two popular methods in the lateration step: *nonlinear least squares (NLS)* and *linear least squares (LLS)*.

NLS: Given the estimated distances d_i from the targeting device to anchors and known positions (x_i, y_i) of the anchors, the position (x, y) of the wireless device can be estimated by finding $(\hat{x}, \hat{y})$ satisfying

$$(\hat{x}, \hat{y}) = \arg\min_{x,y} \sum_{i=1}^{N} \left[\sqrt{(x_i - x)^2 + (y_i - y)^2} - d_i^2 \right], \tag{12.2}$$

where N is the number of anchors that are chosen to estimate the position of the wireless device. NLS can be viewed as an optimization problem where the objective is to minimize the sum of the square error. This is an NLS problem and usually involves some iterative searching technique, such as gradient descent or Newton method, to get the solution. Moreover, to avoid local minimum, it is necessary to rerun the algorithm using several initial starting points, and as a result, the computation is relatively expensive.

LLS: The LLS approach linearizes the NLS problem by introducing a constraint in the formulation and obtaining a closed-form solution of the location estimate. We next show an example of how the NLS problem can be linearized by introducing a constraint in the formulation.

Example 12.2   Transforming the NLS Problem into an LLS Problem

Solution

Start with the $N \geq 2$ equations:

$$(x_1 - x)^2 + (y_1 - y)^2 = d_1^2$$
$$(x_2 - x)^2 + (y_2 - y)^2 = d_2^2$$
$$\vdots \tag{12.3}$$
$$(x_N - x)^2 + (y_N - y)^2 = d_N^2.$$

Now, subtracting the constraint

$$\frac{1}{N} \sum_{i=1}^{N} \left[(x_i - x)^2 + (y_i - y)^2 \right] = \frac{1}{N} \sum_{i=1}^{N} d_i^2 \tag{12.4}$$

from both sides of each equation, the following set of linear equations can be obtained:

$$\left(x_1 - \frac{1}{N} \sum_{i=1}^{N} x_i \right) x + \left(y_1 - \frac{1}{N} \sum_{i=1}^{N} y_i \right) y$$
$$= \frac{1}{2} \left[\left(x_1^2 - \frac{1}{N} \sum_{i=1}^{N} x_i^2 \right) + \left(y_1^2 - \frac{1}{N} \sum_{i=1}^{N} y_i^2 \right) - \left(d_1^2 - \frac{1}{N} \sum_{i=1}^{N} d_i^2 \right) \right]$$
$$\vdots \tag{12.5}$$
$$\left(x_N - \frac{1}{N} \sum_{i=1}^{N} x_i \right) x + \left(y_N - \frac{1}{N} \sum_{i=1}^{N} y_i \right) y$$
$$= \frac{1}{2} \left[\left(x_N^2 - \frac{1}{N} \sum_{i=1}^{N} x_i^2 \right) + \left(y_N^2 - \frac{1}{N} \sum_{i=1}^{N} y_i^2 \right) - \left(d_N^2 - \frac{1}{N} \sum_{i=1}^{N} d_i^2 \right) \right].$$

Therefore, the previous equation can be easily rewritten using the form $\mathbf{Ax} = \mathbf{b}$ with

$$
\mathbf{A} = \begin{pmatrix}
x_1 - \dfrac{1}{N}\displaystyle\sum_{i=1}^{N} x_i & y_1 - \dfrac{1}{N}\displaystyle\sum_{i=1}^{N} y_i \\
\vdots & \vdots \\
x_N - \dfrac{1}{N}\displaystyle\sum_{i=1}^{N} x_i & y_N - \dfrac{1}{N}\displaystyle\sum_{i=1}^{N} y_i
\end{pmatrix}
\tag{12.6}
$$

and

$$
\mathbf{b} = \frac{1}{2} \begin{pmatrix}
\left(x_1^2 - \dfrac{1}{N}\displaystyle\sum_{i=1}^{N} x_i^2 \right) + \left(y_1^2 - \dfrac{1}{N}\displaystyle\sum_{i=1}^{N} y_i^2 \right) \\
- \left(d_1^2 - \dfrac{1}{N}\displaystyle\sum_{i=1}^{N} d_i^2 \right) \\
\vdots \\
\left(x_N^2 - \dfrac{1}{N}\displaystyle\sum_{i=1}^{N} x_i^2 \right) + \left(y_N^2 - \dfrac{1}{N}\displaystyle\sum_{i=1}^{N} y_i^2 \right) \\
- \left(d_N^2 - \dfrac{1}{N}\displaystyle\sum_{i=1}^{N} d_i^2 \right)
\end{pmatrix}.
\tag{12.7}
$$

Note that $\mathbf{A}$ is described by the coordinates of anchors only, while $\mathbf{b}$ is represented by the distances to the landmarks together with the coordinates of landmarks. Thus, the estimated location of a wireless device using LLSs is $\mathbf{x} = (\mathbf{A}^T\mathbf{A})^{-1}\mathbf{A}^T\mathbf{b}$. Due to the subtraction, the solution obtained from the linear (12.7) is not exactly the same as the solution of the original nonlinear (12.2). The calculation of the linear equation solution requires low computational power, and the obtained solution can serve as the starting point for the NLS problem. In general, nonlinear searching from the linear estimate produces more accurate results than the linear method at the price of a higher computational complexity.

12.2.3   Classification via Machine Learning

The fingerprint matching proposed in [3] can be viewed as a machine learning classification method. In [3], a k-nearest neighbor (KNN) method is used for positioning wireless devices based on the closest known locations (i.e., training points) in the preconstructed signal map. Thus, in the KNN method, the location estimation of the wireless device is assigned as the average location of its k-nearest locations in the signal map. If $k = 1$, then the position estimate of the wireless device is simply assigned as the location of its nearest known location in the signal map.

In particular, this kind of localization process through matching includes two phases: off-line training phase, which is used to collect the training data for constructing a radio map, and online localization phase, which is used to estimate the position of the wireless device based on the prebuilt radio map.

During the off-line phase, a mobile transmitter travels to known positions and broadcasts beacons periodically, and the RSS readings at each known position are measured at the set of anchors. The RSS readings together with the locations where the RSS readings are collected are called RF fingerprints; that is, $\{(x_j, y_j), ss_1, ss_2, \ldots, ss_N\}$, where (x_j, y_j), with $j = 1, 2, \ldots, m$ is the location of the mobile transmitter traveling, and $ss_i(x_j, y_j)$ with $i = 1, 2, \ldots, N$ is the RSS readings at the ith anchor. To mitigate the effect of noise, each ss_i is either the averaged value or the median value of multiple measurements collected over a time period. Collecting together the RSS readings from m different locations from a set of N anchors provides a radio map.

During the online localization phase, localization is performed by measuring the targeted wireless device's RSS at each anchor, and the vector of RSS values, a fingerprint, that is, $\{ss_1', ss_2', \ldots, ss_N'\}$ is compared to the prebuilt radio map. In the nearest neighbor method, the record in the radio map whose signal strength vector is the closest in the Euclidean sense to the observed RSS fingerprint is declared to correspond to the location of the transmitter:

$$(\hat{x}, \hat{y}) = \arg \min_{(x_j, y_j)} \sqrt{\sum_{i=1}^{N} (ss_i' - ss_i(x_j, y_j))^2}. \tag{12.8}$$

Other versions of this approach return the average position (e.g., centroid) of the top k-closest vectors, that is, KNN. For example, if $k = 2$, it takes the closest two candidates and returns the midpoint between them. Other techniques from the field of machine learning classification that have been used for indoor localization include neural networks [15, 16], decision trees [17, 18], and support vector machines [19, 20]. A disadvantage of this approach is that it requires a large number of training points to perform adequately and is labor intensive.

To reduce the training efforts, one approach is to use an interpolated map grid (IMG), which is used to build signal maps [4]. Since the quality of the signal map is sensitive to the number of known locations [21], the purpose of using an IMG is to improve the resolution of the signal map so as to obtain better localization accuracy. Directly measuring the RSS at a large number of known locations is expensive; the purpose of the interpolation approach is to improve the quality of the signal map based on the averaged RSS readings from a smaller number of known locations. We next show an example on how to build an IMG from collected fingerprints.

Example 12.3 Building an IMG Signal Map

Solution

The area of interest is divided into a regular grid of equal-sized tiles. The center of the tile is representative of its location. The tiles are a simple way to map the expected signal strength to locations. Building an IMG is thus similar to "surface fitting"; the goal is to derive an expected fingerprint for each tile from the collected fingerprints. There are several approaches in the literature for interpolating surfaces, such as splines and triangle-based linear interpolation. For instance, triangle-based linear interpolation, which divides the floor into triangular regions using a Delaunay triangulation [4], can be used in IMG. The expected signal strength at the center of each grid can be linearly interpolated.

When performing localization, given the observed RSS readings of a targeting wireless device, gridded-RADAR uses IMG to build the signal map and returns the (x, y) of the nearest neighbor in the IMG as the one to localize.

12.2.4   Probabilistic Approaches

We next examine the probabilistic approaches, where the RSS is treated as a random variable that can be modeled as a lognormal distribution at a physical location [8]; one category of probabilistic methods is MLE [22, 23]. Assuming the targeted wireless device is located at location $L_j = (x_j, y_j)$, given the online observed RSS values $\mathbf{s}$, that is, $\mathbf{s} = \{ ss'_1, ss'_2, \ldots, ss'_N \}$, the estimated location of the targeted wireless device based on the MLE is given by

$$(\hat{x}, \hat{y}) = \arg \max_{L_j} [p(L_j | \mathbf{s})], \tag{12.9}$$

where $p(L_j|\mathbf{s})$ denotes the probability that the wireless device is at location L_j.

Using Bayes' rule, the above equation is equivalent to finding the position L_j, which maximizes

$$P(L_j|\mathbf{s}) = \frac{P(\mathbf{s}|L_j) \times P(L_j)}{P(\mathbf{s})}. \tag{12.10}$$

Without a priori information about the position of the wireless device, we can assume that the probabilities that the wireless device is located at different places are equally likely. Therefore, Equation (12.10) can be rewritten as

$$P(L_j|\mathbf{s}) = c \times P(\mathbf{s}|L_j), \tag{12.11}$$

where $c = P(L_j)/P(\mathbf{s})$ is a constant.

Equation (12.11) can be further simplified by assuming conditional independence of the measurement from all anchors:

$$P(\mathbf{s}|L_j) = P(ss'_1|L_j) \cdot P(ss'_2|L_j) \cdots P(ss'_N|L_j). \tag{12.12}$$

Assuming the RSS measurements at each location follow a lognormal distribution, the expected RSS vector at each location, that is, $ss_i(x_j, y_j)$, $P(\mathbf{s}|L_j)$ for every location L_j can be computed. Finally, the MLE returns the location L_j, which maximizes $P(\mathbf{s}|L_j)$.

Rather than returning a single location, the position estimate can instead be given as an area of confidence. For example, the area-based probability (ABP) method [4] tries to return an area bounded by a predefined probability level that the wireless device is within the returned area. In this approach, the predefined probability is called the confidence level, which is an adjustable parameter, and is represented by the parameter α. The ABP method first divides the whole area of interest into a set of tiles. The RSS vector of each tile is represented by the expected RSS vector at the center of each. This can be done by using one of the aforementioned interpolation methods, such as linear interpolation. Given that the wireless device must be located within the area of interest, that is, $\sum_{j=1}^{m} P(L_j|s) = 1$, ABP returns the top probability tiles up to its confidence, α.

12.2.5 Statistical Supervised Learning Techniques

The statistical supervised learning method infers a function from supervised training data. A localization algorithm that utilizes statistical supervised learning analyzes the training data and produces an inferred function that predicts the location of a wireless device based on the input RSS vector. We next introduce BNs as an example of localization algorithms that employ statistical supervised learning.

BNs, also called *belief networks* or *probabilistic networks*, are graphical models for representing the interaction between variables visually. A BN is composed of nodes and arcs between the nodes. Each node corresponds to a random variable, v, and has a value corresponding to the probability of the random variable, $P(v)$. If there is a directed arc from node v to node w, this indicates that v has a *direct influence* on w. This influence is specified by the conditional probability $P(w|v)$. The network is a *directed acyclic graph* (DAG); namely, there are no cycles. The nodes and the arcs between the nodes define the structure of the networks, and the conditional probabilities are the *parameters* given the structure.

Adopting a BN for RSS-based localization [7], BN encodes the signal-to-distance propagation model into the Bayesian graphical model for location estimation. In DAG, the *parents* of a vertex v, $pa(v)$, are those vertices from which point into v. The *descendants* of a vertex v are the vertices that are reachable from v along a direct path. A vertex w is called a child of v if there is an edge from v to w. The parent(s) of v are taken to be the only nodes that have direct influence on v, so that v is independent of its nondescendant given its parents. In BN, the overall joint density of $v \in V$, where v is a random variable, only depends on the parents of v, denoted as $pa(v)$:

$$p(V) = \prod_{v \in V} p(v_i | \text{pa}(v_i)). \tag{12.13}$$

Once $p(V)$ is computed, the marginal distribution of any subset of the variables of the network can be obtained, as it is proportional to overall joint distribution.

Figure 12.1 presents two BN algorithms, M1 and M2. Each rectangle is a plate and shows a part of the network that is replicated; the nodes on each plate are repeated for each of the N anchors whose locations are known. The vertices X and Y represent location; the vertex s_i is the signal reading from the ith anchor; and the vertex D_i represents the Euclidean distance between the location specified by X and Y and the ith anchor. The value of s_i follows should a signal propagation model $s_i = b_{0i} + b_{1i} \log D_i$, where b_{0i} and b_{1i} are the parameters specific to the ith anchor. The distance $D_i = \sqrt{(X - x_i)^2 + (Y - y_i)^2}$ in turn depends on the location (X, Y) of the measured signal and the coordinates (x_i, y_i) of the ith anchor. The network models noise and

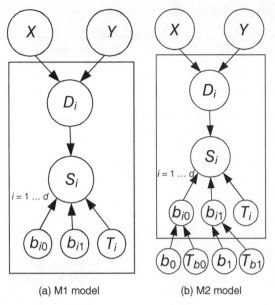

(a) M1 model   (b) M2 model

Figure 12.1   BN localization algorithm: Bayesian graphical models using plate notation.

outliers by modeling the s_i as a Gaussian distribution around the above propagation model, with variance τ_i:

$$s_i \sim N(b_{0i} + b_{1i} \log D_i, \tau_i). \tag{12.14}$$

The initial parameters (b_{0i}, b_{1i}, τ_i) of the model are unknown, and the training data are used to adjust the specific parameters of the model according to the relationships encoded in the network. Through Markov chain Monte Carlo (MCMC) simulation, the BN returns the sampled distribution of the possible location of X and Y as the localization result.

The M1 model utilizes a simple BN model, as depicted in Figure 12.1a, and requires location information in the training set in order to give good localization results. The M2 model extends this to a hierarchical model as shown in Figure 12.1b, which makes the coefficients of the signal propagation model have common parents. The BN M2 algorithm can localize multiple devices simultaneously with no training set, which can significantly reduce the labor-intensive efforts of training data collection during location estimation.

12.2.6   Summary of Localization Algorithms

We summarize RSS-based localization approaches and their main characteristics in Table 12.1. Compared to the other methods, lateration-based approaches are sensitive to RSS noise caused by environmental bias, for example, multipath effects. On the other hand, the machine learning classification and probabilistic approaches are robust to environmental bias and RSS noise. However, they either need to collect a

TABLE 12.1. Summary of RSS-Based Localization Approaches and Their Main Characteristics

Approach	Description	Algorithm Example
Lateration	Two steps are involved: ranging and lateration. During ranging, the distances between the wireless device and multiple anchors can be derived according to different measurement modalities. RSS is one of the approaches to be used for ranging. During lateration, the location of the wireless device can be estimated according to derived distances based on least squares methods. Pros: This method is simple to use and works well in free space. Cons: This method is sensitive to RSS variation, for example, multipath effects.	Nonlinear least squares (NLS) and linear least squares (LLS)
Machine learning classification	It builds a priori radio signal maps of the localization region during the training phase based on the measured signal strength (i.e., RSS fingerprints) at different locations. During localization, by comparing the online measured RSS to the preconstructed signal map, the location can be deduced. Pros: This approach is robust to RSS noise and environmental bias, for example, multipath effects. Cons: It is labor intensive to build signal maps.	RADAR, k-nearest neighbor (KNN), and support vector machine (SVM)
Probabilistic approaches	The RSS is treated as a random variable by modeling it with a lognormal distribution at each location. The position of the wireless device is returned as the most likely location with the highest probability. Pros: This approach is robust to RSS noise and environmental bias, given the prior knowledge of RSS distribution at a large number of locations. Cons: It needs to use the RSS distribution as a priori knowledge, for example, to obtain RSS distribution through training.	Maximum likelihood estimation
Statistical supervised learning	This approach derives the sample distribution of the estimation locations based on statistical learning. Pros: This approach has the capability of localizing multiple devices at the same time and returns reasonable localization accuracy even without training data. Cons: It can be computationally intensive.	Bayesian network

large number of training data to build RSS profiles and are thus labor intensive, or require prior knowledge of RSS distributions. Concerning localization using statistical supervised learning, it has the advantage of being able to localize multiple devices at the same time and can return reasonable localization accuracy even without training.

12.3   LOCALIZATION PERFORMANCE STUDY

The performance of each of the localization techniques needs to be evaluated through various aspects. In this section, we first outline the main evaluation metrics and then present a performance comparison across different localization algorithms using Wi-Fi and ZigBee networks.

12.3.1   Performance Metrics

The set of evaluation metrics that we overview includes *accuracy*, *precision*, *robustness*, *complexity*, and *stability*.

Accuracy. For a given localization attempt, accuracy is the Euclidean distance between the location estimate obtained from the localization system and the actual location of the targeted wireless device in the physical space. Accuracy is also referred to as localization error or distance error. Usually, the average or median distance error is adopted as the performance metric. The better the accuracy, the better the location technique. Thus, accuracy can be used to evaluate the overall performance of the localization techniques.

Precision. Precision is defined as the success probability of position estimations with respect to the predefined accuracy. Precision details the statistical characterization of the localization error, which varies over many localization trials. It measures how consistently the localization technique works and reveals the variation of localization errors. In some works, precision is defined as the standard deviation of the localization error or the geometric dilution of precision (GDOP) [24, 25]. However, generally, the cumulative distribution function (CDF) of the localization error is used for measuring the precision of an indoor localization system. When the accuracies of two location algorithms are the same, the algorithm that gives better precision is preferred. In practice, precision is described by either a CDF or the percentile format. For example, one typical indoor location system has a location precision of 97% within 30 ft (the localization error is less than 30 ft with a probability of 97%) and median error of 10 ft.

Robustness. A robust positioning should be able to function normally even when some signals are not available, or when some of the RSS patterns have not been seen before. For example, the signal at an anchor point may be blocked; thus, the RSS reading at that anchor point may be missing or incorrect. Sometimes, environmental changes (e.g., furniture relocation or the presence and mobility of human beings) will also affect the RSS patterns. In addition, some anchors could be out of function or damaged in a harsh environment. For example, under malicious attacks, the attackers can control anchor points and manipulate the RSS measurements. The positioning techniques have to use this incomplete information to localize the wireless devices. Therefore, it is important to measure the robustness of the localization algorithms. In all these cases, if a localization technique is robust, then it should be able to provide the wireless device's position information without reduced accuracy even if the quality of the measurement is reduced.

Complexity. The complexity of a localization system can be attributed to hardware, computing, and human intervention/efforts during deployment. The most commonly used measure of complexity is the computing complexity, which is the complexity of the localization algorithm. If the computation of the localization algorithm is performed in a distributed manner (i.e., on the wireless device side), then we would prefer localization algorithms with low complexity, which can extend the operation life of devices that have limited battery or power supply. Alternatively, we can use localization latency, the time it takes for a wireless device to localize. The first time and amortized latencies are different. For example, the first localization of a wireless device is long, but subsequent localization attempts in nearby areas are typically much faster.

Stability. Stability measures how much the location estimate changes in the physical space in response to small-scale movements of a wireless device. Stability is a desirable property in localization systems since a location estimate should not move too far in the physical space if there is a small-scale movement of the wireless device. For instance, when someone works at his office desk and moves his laptop 1 ft away, the localized position of the laptop should not change too much.

12.3.2   Performance Investigation Using Real Wireless Networks

We next compare the performance of different localization algorithms using Wi-Fi and ZigBee wireless networks by focusing on accuracy and precision. We first describe the experimental scenarios in real office building environments. We then examine the experimental results across localization algorithms under different experimental scenarios.

Experimental Scenarios: Figure 12.2 shows two experimental scenarios for localization performance comparison. The floor map in Figure 12.2a is the wireless network laboratory (WINLAB) at Rutgers University, which has a floor size of 219×169 ft. Both a Wi-Fi network and a ZigBee network are deployed in WINLAB. The ZigBee network is implemented using Tmote Sky motes. The second experimental setup shown in Figure 12.2b is the ORBIT testbed in WINLAB [26],

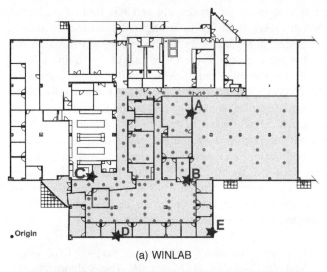

(a) WINLAB

(b) ORBIT room

Figure 12.2   Deployment of anchors and training locations on the experimental floors [27, 28].

a large-scale indoor wireless testbed, which consists of 400 wireless nodes in a 20×20 regular grid with an internode separation of 3 ft spanning a total area of $3600\,\text{ft}^2$. The Wi-Fi network is used in the ORBIT testbed.

The distinctive characteristics between these two experimental scenarios are the number of anchor points and the propagation environments. The WINLAB office setup is a typical indoor propagation environment with heavy multipath effects due to the walls, furniture, and people movement. There are five anchors shown as stars in Figure 12.2a and are denoted as A, B, C, D, and E. On the other hand, in the ORBIT testbed, there can be up to 400 anchor points and the propagation environment is free of major shadowing and has limited multipath effects. Each wireless node in Figure 12.2b can be configured as an anchor point.

Furthermore, the small dots in Figure 12.2a are the training points where the RSS readings are collected at each anchor. There are a total of 101 training locations in the WINLAB setup. On the other hand, in the ORBIT testbed, the RSS readings are measured at each grid point. Thus, there are 400 total training locations in the ORBIT testbed. At each location, about 350 packets of RSS are collected, and the averaged RSS for each location is used. To evaluate the different algorithms, the well-known leave-one-out approach is used to divide the data into training and testing sets. That means one location is chosen as the targeted location, whereas the rest of the locations are chosen as training data. For example, in the lateration-based algorithm, one location is randomly chosen to be localized, and the rest of the data (i.e., training data) are used to fit the propagation model (e.g., (12.1)) to get the propagation parameters, such as path loss parameters.

Performance Results: The algorithms under evaluation in the WINLAB office experimental scenario are *NLS*, *LLS*, *RADAR*, *ABP*, and *BN*. These algorithms are summarized in Table 12.2.

We note that when using ABP, because it returns the most likely area in which the targeted wireless device may reside, the localization error is defined as the median localization error from the returned area.

The localization results in both the Wi-Fi and ZigBee networks are depicted as the CDF of the localization error in Figure 12.3. We observed that the overall performance of localization algorithms under the Wi-Fi network is better than that of the ZigBee network due to the fact that Wi-Fi devices are more reliable than motes [27]. First, the median error from different algorithms ranges from 7 to 20 ft in the Wi-Fi network, whereas it is from 8 to 30 ft in the ZigBee network. For both networks, ABP performs the best because it uses a fine-grained signal map via the IMG technique, while the lateration-based method performs the worst because of the inaccurate ranging information derived from the RSS readings affected by the unpredictable indoor setups. In addition, RADAR and BN have comparable performance in both networks. The median error is about 10 ft in the Wi-Fi network and 20 ft in ZigBee network.

TABLE 12.2.   Performance Study: Localization Algorithms Used in a WINLAB Environment

Algorithm	Abbreviation	Description
Nonlinear least squares	NLS	Lateration-based algorithm; NLS is used
Linear least squares	LLS	Lateration-based algorithm; LLS is used
Fingerprint matching	RADAR	Classification-based method, returns the nearest neighbor: $k = 1$
Area-based probability	ABP	Probabilistic method using IMG; returns the most likely area: $\alpha = 0.75$; tile size in IMG: 10×5 in.
Bayesian network	BN	Statistical supervised learning Bayesian graphical model: M1

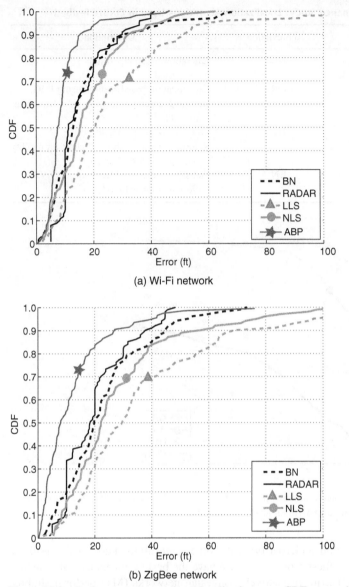

Figure 12.3   Performance comparison: localization error CDF across different algorithms under the WINALB experimental scenario [27].

We now turn to study the algorithms under evaluation in the ORBIT testbed. The algorithms under study are summarized in Table 12.3. Figure 12.4 shows the localization error CDF across different algorithms using the ORBIT testbed with 400 anchors [28]. With the high density of anchors and benign propagation, which is essentially line of sight (LOS), the performance of all the algorithms improves significantly. Again, the algorithms using IMG have the best performance, while the

TABLE 12.3. Performance Study: Localization Algorithms Used in ORBIT Testbed

Algorithm	Abbreviation	Description
Nonlinear least squares	NLS	Lateration-based algorithm NLS is used.
Fingerprint matching Gridded RADAR	GR	Classification-based method using IMG Returns the nearest neighbor: $k = 1$ tile size in IMG: 2×2 in.
Highest probability	H1	Probabilistic method using IMG Returns location with the highest probability tile size in IMG: 2×2 in.
Bayesian network	BN	Statistical supervised learning Bayesian graphical model: M1

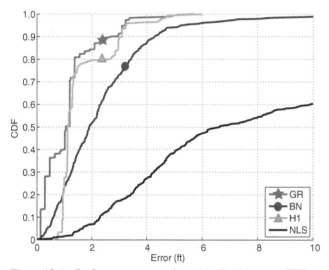

Figure 12.4 Performance comparison: localization error CDF across different algorithms under the ORBIT testbed experimental scenario [28].

lateration-based method underperforms the other algorithms. Particularly, gridded RADAR (GR) and highest probability (H1) have the best performance with median errors of 0.31 and 0.33 m, respectively. Bayesian networks (M1) perform slightly worse, with the median error at 0.58 m. NLS performs the worst with a median error of about 2 m. This is because the lateration algorithms generally are very sensitive to data from low-quality anchors among the large number of anchors that cannot be fitted in the propagation model [28].

From the results in these two experimental scenarios, the following important conclusions can be drawn:

1. The IMG technique can improve the localization accuracy of RADAR, ABP, and BN localization algorithms because it provides a fine-grained signal map that benefits these algorithms.

2. The lateration-based method is sensitive to the RSS variations caused by multipath effects, radio interference, or the off-the-shelf device diversity, and has the worst localization performance. This is because the theoretical propagation model either cannot describe complicated indoor environments (e.g., multipath effect) or needs to be calibrated due to device diversity (e.g., different transmitting power level), and thus the ranging information derived from RSS together with the propagation model is inaccurate.

3. Except for the lateration-based methods, all the other localization strategies, such as fingerprint matching and probabilistic and statistic learning, can achieve comparable localization accuracy and precision in a typical indoor environment.

4. Finally, a benign propagation environment or increasing the density of high-quality anchors can improve the localization performance significantly.

12.4   ENHANCING THE ROBUSTNESS OF LOCALIZATION

We next provide an overview of the methods that help to enhance the robustness of the localization results. These methods are either specific to one category of localization algorithms (e.g., robust statistical methods can be applied to lateration techniques and fingerprint matching methods, and revisiting linear regression and correlation methods are suitable for lateration methods) or generic to all the localization algorithms, for instance, employing multiple antennas to improve localization accuracy.

12.4.1   Real-Time Infrastructure Calibration

In an indoor environment, radio signal propagation is affected by several factors, such as multipath effects, environmental temperature, humidity variations, doors opening and closing, and the mobility of human beings. The RSS readings at a given location will be very different over time since the indoor environment will change from time to time. To overcome the degradation of the localization performance due to environmental changes, it is necessary to calibrate the parameters of the positioning system, such as the propagation parameters and signal maps. However, collecting RSS readings to calibrate the localization system from time to time manually is labor intensive. For example, collecting RSS readings at different locations to update the signal map periodically to maintain high accuracy is extremely cumbersome. Therefore, it is desirable that a localization system is able to adapt to the environmental changes and to autoconfigure itself in response to environmental dynamics.

The localization system architecture presented in [29] aims to achieve real-time infrastructure calibration [30]. Figure 12.5 depicts the proposed localization system architecture from [29], in which anchors are transceivers with known locations. These anchors record the signal of beacon broadcasts from each other. These signal links between anchors are made periodically to realize a fully automated and

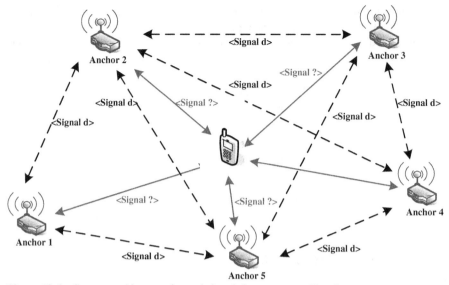

Figure 12.5   System architecture for real-time infrastructure calibration.

online calibration of the radio signal in the spatiotemporal domain. A mapping that characterizes the relationship of the signal measurements and the geographic distances to anchors is then created online with certain mapping rules.

To mitigate the effects of environmental dynamics, besides utilizing the measurements of signals between the wireless devices and anchors to localize the wireless devices, this localization system takes as an additional input the online measurements of signal strength between anchors. The online signal measurements between the anchors are used to capture the environmental changes and to create a mapping between the signal measurement and the actual geographic distance. To create a signal-distance mapping, the online measured RSS readings and the corresponding distances between anchors are used to fit the radio propagation model, which is presented in Section 12.1.

In addition, techniques such as the truncated singular value decomposition (SVD) pseudoinverse method [29] can be used to improve the robustness of signal-distance mapping. Depending on the mode in which localization will be performed, either a wireless device or the localization infrastructure measures the radio signal between the wireless device and its neighboring anchors. Finally, the localization process applies the signal-distance mapping, infers the wireless device's geographic distances to anchors, and estimates its location via lateration-based methods [14, 24].

12.4.2   Effects of Employing Multiple Antennas

As mentioned previously, using RSS to perform localization is attractive since it reuses the existing wireless infrastructure and is expected to provide tremendous cost savings over deploying localization-specific hardware. However, a significant problem with RSS is that small-scale multipath fading adds high-frequency components with

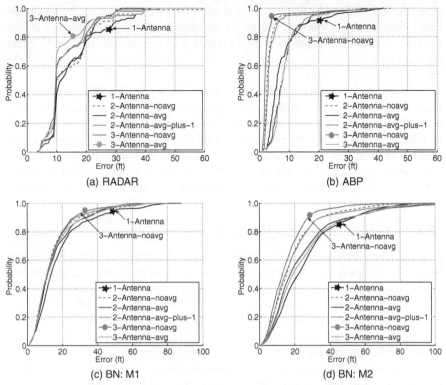

Figure 12.6   Localization error CDFs when using multiple antennas in the WINLAB experimental scenario under the Wi-Fi network [27].

large amplitudes to the signal at a given location. Thus, the RSS can vary by 5–10 dB with small (a few wavelengths) changes in location. Because the small-scale fading effects occur at the level of several wavelengths (about 12 cm at 2.4 GHz), and the granularity of the localization system is typically much larger (2–3 m), using multiple antennas spaced on the order of a few wavelengths presents the opportunity to smooth out these effects while maintaining the same number of anchors used by the localization system [27].

To demonstrate the potential benefits, we next compare the localization accuracy across different localization algorithms when using multiple antennas [31]. Figure 12.6 presents the localization error CDFs of different localization algorithms, including RADAR, ABP, and BNs when multiple antennas are applied. The experimental scenario is the WINLAB environment under the Wi-Fi network. The observation from Figure 12.6 is that in nearly all cases, the performance of localization algorithms improved when using multiple antennas. This is because for RADAR, the collected training points are directly used to build the signal map. The distances of the training points in the experiments range from 5 to 10 ft. The improvement for the RADAR algorithm comes from the reduced RSS variation when averaging RSS readings of three antennas. However, for ABP, a fine-grained interpolated signal map is used, where the tile size is around 10 in. per side. When using the fine-grained interpolated signal map, the physical distance between two RSS samples is much less (i.e., several inches) than the distance between two antennas (i.e., 1–2 ft in this experimental set) installed on one anchor. Thus, ABP treats each antenna as a separate anchor and achieves the best performance under the case of 3–*antenna–noavg*.

In addition, BNs present consistent behaviors when using different Bayesian graphical models (i.e., M1 and M2). The performance of 3–antenna–noavg is slightly better than that of others, and the 1–*antenna* performs the worst. This is due to the fact that BNs benefit from the smoothed out small-scale fading when averaging RSS readings of three antennas. In summary, we can conclude that the localization accuracy can be improved when employing multiple antennas in both cases of averaging and not averaging the RSS readings from multiple antennas at each anchor point.

12.4.3 Robust Statistical Methods

In an indoor environment, signal strength measurements may be significantly altered by opening doorways in a hallway or by the presence of passersby. These measurement errors caused by either unintentional human behaviors or intentional adversarial behaviors can be severe and consequently degrade the performance of a localization method. On the other hand, these measurement errors may only be present on a small subset of anchors (i.e., specific radio links between a wireless device and the affected anchors). The idea of the robust statistical method is to utilize the redundancy in the localization infrastructure; that is, only a portion of the RSS readings at certain anchors is affected and not all of the RSS readings are affected. The strategy is to enhance the robustness of the localization system so as to mitigate the effects of such measurement errors and adversarial behaviors by taking advantage of the redundancy in the deployment of the localization infrastructure to provide stability to contaminated measurements [32].

Statistical tools are developed to make localization techniques robust to measurement errors and adversarial data. In particular, the methods developed make use of the median as a resilient estimate of the average of aggregated data [33, 34]. For example, in lateration methods, rather than minimizing the summation of the residue squares, minimizing the median of the residue squares is used to reduce the effects of measurement outliers. The lateration algorithm is vulnerable to the "outliers."

Given the distance from the wireless devices to the anchors d_i together with the anchors' location (x_i, y_i), the device location estimate $(\hat{x}_0, \hat{y}_0)$ can be found by least squares (i.e., (12.2)). In order to achieve accurate localization estimation when there are measurement errors present, LMS can be used instead of least squares; that is, $(\hat{x}_0, \hat{y}_0)$ can be found, such that

$$(\hat{x}_0, \hat{y}_0) = \arg \min_{(x_0, y_0)} \text{med}_i \left[\sqrt{(x_i - x_0)^2 + (y_i - y_0)^2} - d_i \right]^2. \quad (12.15)$$

It is known that LMS tolerates up to 50% measurement errors among N total measurements. The exact solution for LMS is computationally prohibitive; however, an efficient and statistically robust alternative can be found in [35].

12.4.4 Revisiting Linear Regression

Radio propagation indoors is complicated by signal reflection, refraction, shadowing, and scattering due to the walls, furniture, and the movement of people. According to the connectivity between the wireless device and the anchor point, the signal propagation can be classified into LOS and non-line-of-sight (NLOS) scenarios [3, 36]. These two scenarios represent different signal propagation environments. Thus, under different scenarios, the propagation parameters are different, such as the path loss exponent and the shadow fading as described in (12.1). The training data collected in the area of interest usually includes both LOS and NLOS scenarios, and we cannot differentiate to which scenario the targeting wireless device belongs. Thus, in the off-line training phase, the fitted theoretical log-distance propagation model cannot characterize both LOS and NLOS scenarios simultaneously and will result in large errors in distance estimation during the ranging step of RSS-based lateration methods, and consequently, the localization accuracy is significantly affected. In Section 12.3, we observed that RSS-based lateration methods underperform other indoor localization algorithms, including machine learning classification techniques, probabilistic approaches, and BN localization.

To improve the applicability of RSS-based lateration methods in indoor environments and to further provide feasible mathematical analysis for indoor localization, a regression model may be used over the theoretical log-distance propagation model to better model the relationship between the RSS and distance, and to improve localization accuracy for lateration methods in real-world scenarios [14].

The polynomial regression model is adapted to model the RSS–distance relationship since the polynomials dominate the interpolation theory (Weierstrass's theorem) and they are easily evaluated [37].

Example 12.4 Using the Polynomial Regression Model to Model the Relationship between the RSS and Distance

Solution

Given the M training points (d_i, RSS_i) collected in the area of interest, where d_i is the distance between the wireless device and an anchor point, and RSS_i is the corresponding signal strength reading at the training point i, the nth-degree polynomial is fitted through the set of data and M sets of equations are obtained. The ideal nth-degree polynomial should satisfy

$$\hat{d}_i = a_0 + a_1 \times RSS_i + a_2 \times RSS_i^2 + \ldots + a_n \times RSS_i^n, \tag{12.16}$$

where a_j (with $j = 0, 1, 2, \ldots, n$) are the coefficients of the polynomial, RSS_i is the RSS, and $\hat{d}_i$ is the estimated distance. However, there are estimation errors e_i. Thus, we have

$$e_i = (d_i - \hat{d}_i) = (d_i - a_0 - a_1 * RSS_i - \ldots - a_n * RSS_i^n), \tag{12.17}$$

where $i = 1, 2, \ldots, M$.

We use least squares approximation, in which the coefficients can be obtained by minimizing the sum of the error squares, which is given by

$$E(a_0, a_1, \ldots, a_n) = \sum_{i=1}^{i=M} (e_i)^2. \tag{12.18}$$

Equation (12.18) is a function of variables a_0, a_1, ..., a_n, which minimizes the sum of error squares. We equate its partial derivatives to zero with respect to a_0, a_1, ..., a_n. Then, we get

$$\frac{\partial E}{\partial a_j} = \sum_{i=1}^{i=M} -2(RSS_i^j)(d_i - a_0 - a_1 * RSS_i - \ldots - a_n * RSS_i^n) = 0 \tag{12.19}$$

for each j, with $j = 0, 1, 2, \ldots, n$. By rearranging (12.19), we obtain the normal equations

$$\sum_{i=1}^{i=M} RSS_i^j d_i = a_0 \sum_{i=1}^{i=M} RSS_i^j + a_1 \sum_{i=1}^{i=M} RSS_i^{(j+1)} + \ldots + a_n \sum_{i=1}^{i=M} RSS_i^{(j+n)}, \tag{12.20}$$

for each j, with $j = 0, 1, 2, \ldots, n$. Now, the coefficients a_j, with $j = 0, 1, 2, \ldots, n$, can be solved by Gauss elimination of (12.20).

Figure 12.7 presents the localization results of the regression-based methods using a second degree of polynomial for both the Wi-Fi network and the ZigBee network under the WINLAB experimental scenario, respectively. We observed that the regression-based lateration methods result in a better performance than the original log-distance-based lateration methods. The overall improvement of median error is above 29% in both networks. In addition, the maximum localization error is significantly reduced. The overall improvement of maximum error is over 50% in the two networks. These results indicate that using regression-based lateration methods is effective in indoor localization.

12.4.5   Exploiting Spatial Correlation

Although radio propagation is complicated due to the placement of walls, obstacles, and movement of people indoors, the signal propagation from close-by locations in a local area to an anchor point is highly correlated, as nearby locations face a similar propagation environment. For instance, the locations in a regular-sized room are facing the same radio connectivity to the access points (i.e., LOS or NLOS) and similar distance to the anchor points. Thus, the signal propagation from a local area experiences similar signal attenuation (due to distance) and penetration losses through walls. Therefore, the signal propagation model may be better fitted if we just use the data collected in the local area. Further, experimental results have provided strong evidence that shadow fading is spatially correlated indoors [38] due to the local area facing similar obstacles. The correlation distance can range from several to many tens of meters [36, 38]. Therefore, the correlated RSS measurements collected in a local area can help to characterize the relationship between RSS and the distance, and consequently to improve the localization accuracy when applying lateration methods in unpredictable indoor environments [14].

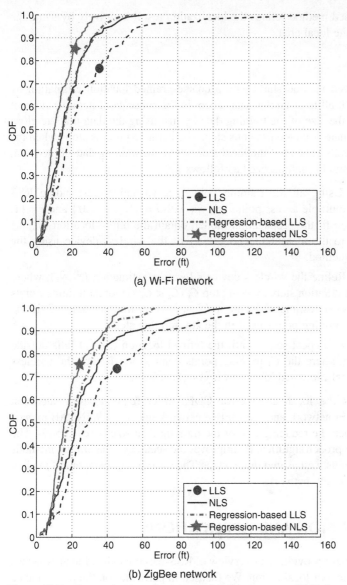

(a) Wi-Fi network

(b) ZigBee network

Figure 12.7   Regression-based approach: localization error CDFs in two networks [14].

Example 12.5   Using Spatial Correlation to Iteratively Refine the Localization Results

The correlation-based method is developed in [14] by utilizing the correlated RSS readings that are collected from a local area to fit the theoretical log-distance propagation model. The objective is to obtain more accurate distance estimations for the wireless device, whose location belongs to that local area, based on the fitted model.

The correlation-based method refines the localization result iteratively by using the correlated RSS in the local area.

Solution

The correlation-based method starts with a coarse-grained location estimation by using lateration with all the training data. Then, in the subsequent iterative steps, it adaptively reduces the size of the training data by just using the data that are close to the location estimate from the previous step. The iterative process stops when the estimated location falls outside of the local area where the training data come from. The correlation method is summarized as follows.

1. *Initialize.* Using all the training points (x_j, y_j) and corresponding RSS $\{RSS_{ij}\}$ at multiple access points (x_i, y_i), with $i = \{1, 2, \dots, M\}$ and $j = \{1, 2, \dots, N\}$, to fit the propagation model, obtain the initial location $(\hat{x}_0, \hat{y}_0)$ according to the measured RSS $\{RSS_i\}$ of the wireless device using the lateration method.

2. *Iteration.* Refine the wireless device's location estimation $(\hat{x}_k, \hat{y}_k)$, where k is the kth iteration step, by using top C_k $(C_k < C_{k-1})$ closest training points to the previous estimated location $(\hat{x}_{k-1}, \hat{y}_{k-1})$ as training data. C_k is the number of training points used in the kth iteration step.

3. *Termination.* Repeat step 2 until the refined location $(\hat{x}_k, \hat{y}_k)$ falls outside of the area where the C_k training points come from. Return the wireless device's final estimated location $(\hat{x}, \hat{y}) = (\hat{x}_{k-1}, \hat{y}_{k-1})$.

Figure 12.8 shows the localization error CDFs of correlation-based lateration methods in the Wi-Fi network and the ZigBee network under the WINLAB experimental scenario when the reducing rate of the training data size is set to 50%. The correlation-based approach significantly improves the accuracy of lateration methods in both median error as well as maximum error. The overall improvement of median error is over 33% in both networks.

12.5 CONCLUSION AND APPLICATIONS

In this chapter, we have provided an overview of RSS-based localization approaches and algorithms for indoor localization. We focused in particular on the performance of these localization techniques and advanced techniques to enhance the robustness of these algorithms. Generally, the average localization accuracy of the RSS-based localization techniques is around 10–20 ft. We compared the performance of these algorithms in terms of accuracy and precision, and showed by experimental evaluations that the localization accuracy can be improved via several approaches, such as the interpolated signal map and increased density of anchors. Because the radio propagation in indoors is complicated due to walls, furniture, and the movement of people, the robustness of localization algorithms is especially important in indoor environments. We thus described techniques in enhancing the robustness of localization, in which the enhanced localization techniques are more robust to the environmental dynamics.

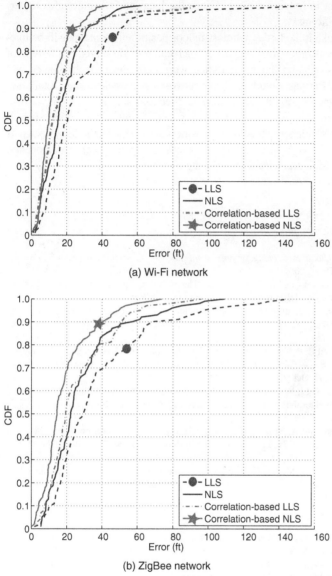

(a) Wi-Fi network

(b) ZigBee network

Figure 12.8   Correlation-based method: localization error CDFs in two networks [14].

Location information can directly support a variety of location-based services or serve as input to other high-level emerging pervasive applications. For example, in the healthcare domain, doctors can directly use location information to track and monitor patients in medical facilities, or activities can be inferred using the estimated position and higher-level decisions can be made accordingly. In the latter case, if a doctor and a nurse are both localized in the same room as a patient, then it will be concluded that this patient is getting treated. In the enterprise domain, location-based access control and asset tracking exploit location information directly. And the workflow management in industrial plants needs the location information as lower-level input for a specific task.

Future trends make it likely that wireless indoor positioning systems will be integrated into a unified localization infrastructure, which can provide spatial positioning of wireless devices across organizational boundaries, work with diverse technologies, and support numerous applications including social networking, manufacturing, retail, and security. Therefore, some important future research directions are the following: how to integrate the fragmented location systems belonging to different communities, such as indoor and outdoor positioning systems; how to work with diverse technologies, such as different wireless devices (i.e., Wi-Fi, ZigBee, Bluetooth, and mote sensors); and location techniques (i.e., data fusion and hybrid location algorithms).

REFERENCES

[1] P. Enge and P. Misra, *Global Positioning System: Signals, Measurements and Performance*. Lincoln, MA: Ganga-Jamuna Press, 2001.

[2] N. Priyantha, A. Chakraborty, and H. Balakrishnan, "The cricket location-support system," in *Proc. of the ACM Int. Conf. on Mobile Computing and Networking (MobiCom)*, Aug. 2000, pp. 32–43.

[3] P. Bahl and V. N. Padmanabhan, "RADAR: An in-building RF-based user location and tracking system," in *Proc. of the IEEE Int. Conf. on Computer Communications (INFOCOM)*, Mar. 2000, pp. 775–784.

[4] E. Elnahrawy, X. Li, and R. P. Martin, "The limits of localization using signal strength: A comparative study," in *Proc. of the First IEEE Int. Conf. on Sensor and Ad Hoc Communcations and Networks (SECON 2004)*, Oct. 2004, pp. 406–414.

[5] Y. Chen, K. Kleisouris, X. Li, W. Trappe, and R. P. Martin, "A security and robustness performance analysis of localization algorithms to signal strength attacks," *ACM Trans. Sens. Netw.*, vol. 5, no. 1, pp. 1–37, 2009.

[6] R. Battiti, M. Brunato, and A. Villani, "Statistical learning theory for location fingerprinting in wireless LANs," University of Trento, Informatica e Telecomunicazioni, Tech. Rep. DIT-02-086, Oct. 2002.

[7] D. Madigan, E. Elnahrawy, R. Martin, W. Ju, P. Krishnan, and A. S. Krishnakumar, "Bayesian indoor positioning systems," in *Proc. of the IEEE Int. Conf. on Computer Communications (INFOCOM)*, Mar. 2005, pp. 324–331.

[8] T. Rappaport, *Wireless Communications: Principles and Practice*. Upper Saddle River, NJ: Prentice Hall, 1996.

[9] T. Hastie, R. Tibshirani, and J. Friedman, *The Elements of Statistical Learning: Data Mining, Inference, and Prediction*. New York: Springer, 2001.

[10] K. Langendoen and N. Reijers, "Distributed localization in wireless sensor networks: A quantitative comparison," *Comput. Netw.*, vol. 43, no. 4, pp. 499–518, 2003.

[11] D. Niculescu and B. Nath, "Ad hoc positioning system (APS)," in *Proc. of the IEEE Global Telecommunications Conf. (GLOBECOM)*, 2001, pp. 2926–2931.

[12] J. Yang, Y. Chen, V. Lawrence, and V. Swaminathan, "Robust wireless localization to attacks on access points," presented at the IEEE Sarnoff Symposium 2009, Princeton, NJ, 2009.

[13] T. Sarkar, Z. Ji, K. Kim, A. Medouri, and M. Salazar-Palma, "A survey of various propagation models for mobile communication," *IEEE Antennas Propagation Mag.*, vol. 45, no. 3, p. 51, Jun. 2003.

[14] J. Yang and Y. Chen, "Indoor localization using improved RSS-based lateration methods," in *Proc. of the IEEE Global Telecommunications Conf. (GLOBECOM)*, 2009.

[15] R. Battiti, R. Battiti, A. Villani, T. L. Nhat, T. L. Nhat, R. Villani, and R. Villani, "Location-aware computing: A neural network model for determining location in wireless LANs," Tech. Rep., 2002.

[16] S.-H. Fang and T.-N. Lin, "Indoor location system based on discriminant-adaptive neural network in IEEE 802.11 environments," *IEEE Trans. Neural Netw.*, vol. 19, no. 11, pp. 1973–1978, 2008.

[17] Y. Chen, Q. Yang, J. Yin, and X. Chai, "Power-efficient access-point selection for indoor location estimation," *IEEE Trans. Knowl. Data Eng.*, vol. 18, no. 7, pp. 877–888, 2006.

[18] J. Yim, "Introducing a decision tree-based indoor positioning technique," *Expert Syst. Appl.*, vol. 34, no. 2, pp. 1296–1302, 2008.

[19] M. Brunato and R. Battiti, "Statistical learning theory for location fingerprinting in wireless LANs," *Comput. Netw.*, vol. 47, no. 6, pp. 825–845, 2005.

[20] C.-L. Wu, L.-C. Fu, and F.-L. Lian, "WLAN location determination in e-home via support vector classification," in *2004 IEEE Int. Conf. on Networking, Sensing and Control*, Taipei, Taiwan, 2004, pp. 1026–1031.

[21] J. Yang and Y. Chen, "A theoretical analysis of wireless localization using RF-based fingerprint matching," in *Proc. of the Fourth Int. Workshop on System Management Techniques, Processes, and Services (SMTPS)*, Apr. 2008.

[22] M. Youssef, A. Agrawal, and A. U. Shankar, "WLAN location determination via clustering and probability distributions," in *Proc. of the First IEEE Int. Conf. on Pervasive Computing and Communications (PerCom)*, Mar. 2003, pp. 143–150.

[23] M. Youssef and A. Agrawala, "Handling samples correlation in the horus system," in *IEEE Infocom*, 2004, pp. 1023–1031.

[24] H. Lee, "A novel procedure for assessing the accuracy of hyperbolic multilateration systems," *IEEE Trans. Aerosp. Electron. Syst.*, vol. AES-11, no. 1, pp. 2–15, 1975.

[25] D. Torrieri, "Statistical theory of passive location systems," *IEEE Trans. Aerosp. Electron. Syst.*, vol. AES-20, no. 2, pp. 183–198, 1984.

[26] WINLAB, "ORBIT wireless network testbed." Available: http://www.winlab.rutgers.edu/pub/docs/focus/ORBIT.html

[27] K. Kleisouris, Y. Chen, J. Yang, and R. P. Martin, "The impact of using multiple antennas on wireless localization," in *Proc. of the Fifth Annual IEEE Communications Society Conf. on Sensor, Mesh and Ad Hoc Communications and Networks (SECON)*, Jun. 2008.

[28] G. Chandrasekaran, M. Ergin, J. Yang, S. Liu, Y. Chen, M. Gruteser, and R. Martin, "Empirical evaluation of the limits on localization using signal strength," in *6th Annual IEEE Communications Society Conf. on Sensor, Mesh and Ad Hoc Communications and Networks, SECON '09*, Jun. 2009, pp. 1–9.

[29] H. Lim, L. Kung, J. Hou, and H. Luo, "Zero-configuration, robust indoor localization: Theory and experimentation," in *Proc. of the IEEE Int. Conf. on Computer Communications (INFOCOM)*, Mar. 2006.

[30] Y. Gwon and R. Jain, "Error characteristics and calibration-free techniques for wireless LAN-based location estimation," in *MobiWac '04: Proceedings of the Second International Workshop on Mobility Management & Wireless Access Protocols*, A. Boukerche and K. Sivalingam, Eds. New York: ACM, 2004, pp. 2–9.

[31] K. Kleisouris, Y. Chen, J. Yang, and R. Martin, "Empirical evaluation of wireless localization when using multiple antennas," *IEEE Trans. Parallel Distributed Syst.*, vol. 21, no. 11, pp. 1595–1610, 2010.

[32] Z. Li, W. Trappe, Y. Zhang, and B. Nath, "Robust statistical methods for securing wireless localization in sensor networks," in *4th Int. Conf. on Information Processing in Sensor Networks (IPSN)*, 2005, pp. 91–98.

[33] B. Przydatek, D. Song, and A. Perrig, "SIA: Secure information aggregation in sensor networks," in *ACM SenSys*, Nov. 2003.

[34] D. Wagner, "Resilient aggregation in sensor networks," in *Proc. of the Fourth ACM Workshop on Security of Ad Hoc and Sensor Networks (SASN)*, 2004, pp. 78–87.

[35] P. Rousseeuw and A. Leroy, *Robust Regression and Outlier Detection*. New York: John Wiley & Sons, 1987.

[36] M. Gudmundson, "Correlation model for shadow fading in mobile radio systems," *Electron. Lett.*, vol. 27, pp. 2145–2146, 1991.

[37] J. L. Buchanan and P. R. Turner, *Numerical Methods and Analysis*. New York: McGraw-Hill, 1992.

[38] N. Jalden, P. Zetterberg, B. Ottersten, A. Hong, and R. Thoma, "Correlation properties of large scale fading based on indoor measurements," in *IEEE Wireless Communications and Networking Conf., WCNC*, Mar. 2007.

CHAPTER 13

IMPACT OF ANCHOR PLACEMENT AND ANCHOR SELECTION ON LOCALIZATION ACCURACY

Yingying Chen,[1] Jie Yang,[2] Wade Trappe,[1] and Richard P. Martin[1]

[1]Rutgers University, Piscataway, NJ
[2]Florida State University, Tallahassee, FL

L**OCALIZATION OF** wireless devices in wireless networks is important because the location of wireless devices is a critical input for many higher-level networking tasks. Both the placement and the selection of anchors can affect the accuracy of location estimation significantly, and consequently impact the widespread deployment of location-based services (LBSs). Thus, in addition to developing methods to localize wireless devices, it is equally important to systematically investigate how the placement of the anchors impacts localization performance. This chapter investigates the problems of anchor placement and anchor selection. For anchor placement, how the geometric layout of the anchors affects the localization performance is studied. Different approaches that include heuristic search methods, acute triangular-based deployment, adaptive beacon placement, and optimal placement solutions via the maximum lambda and minimum error (maxL–minE) algorithm are discussed. Additionally, this chapter studies how smartly selected anchors can achieve comparable or better localization results as compared with using all the available anchors. A variety of effective strategies are investigated, including a joint clustering technique, entropy-based information gain, convex hull selection, and smart selection in the presence of high density of anchors.

13.1 INTRODUCTION

Localization of wireless devices in wireless networks is important for many applications because the location information of wireless devices is a critical input for many

Handbook of Position Location: Theory, Practice, and Advances, Second Edition.
Edited by S. A. (Reza) Zekavat and R. Michael Buehrer.
© 2019 by the Institute of Electrical and Electronics Engineers, Inc.
Published 2019 by John Wiley & Sons, Inc.
Companion Website: www.wiley.com/go/zekavat/positionlocation2e

higher-level networking tasks, such as healthcare monitoring, personnel and asset tracking, emergency rescue and recovery, and military netcentric warfare operations. In addition to recent efforts on developing a plethora of methods to localize wireless devices, a systematic study of how the placement of anchors with known locations (e.g., access points in Wi-Fi) impacts localization performance is an important task because it provides valuable insights about how to utilize the localization infrastructure effectively for wireless localization, and consequently to benefit the widespread deployment of LBS.

In this chapter, we first provide an overview of the problem of anchor deployment in terms of anchor placement and anchor selection. We then survey the existing approaches concerning anchor placement and anchor selection. We show that these approaches deviate from the traditional localization methods, which are focused on improving the localization algorithms themselves. Instead, these methods achieve effective localization through optimizing the deployment of anchors.

For anchor placement, we investigate how the geometric layout of the anchors affects the localization performance. We cover representative approaches, including heuristic search methods, acute triangular-based deployment, adaptive beacon placement, and optimal placement solutions via the maxL–minE algorithm. In particular, we show that in the optimal anchor placement solution approach, an upper bound on the maximum location error can be derived when the placement of anchors is determined by analyzing the popular lateration-based localization approaches. Based on this theoretical analysis, a scheme called maxL–minE has been developed for finding optimal patterns for anchor placement that minimizes the maximum localization error. To show the effectiveness of the optimal anchor placement patterns on localization performance, experimental results using both a Wi-Fi network as well as a ZigBee network in a real office building environment are presented.

For anchor selection, we discuss how smartly selected anchors can achieve comparable or better localization results as compared to using all the available anchors. We focus our attention on a variety of effective strategies, including a joint clustering technique, entropy-based information gain, convex hull selection, and smart selection in the presence of a high density of anchors. For anchor selection in high-density environments, comprehensive experimental results are presented using 400 anchor points in conjunction with a group of localization algorithms, such as fingerprint matching, a probabilistic strategy, and Bayesian networks (BNs).

The rest of the chapter is organized as follows: In Section 13.2, we provide an overview and survey existing approaches to anchor placement. We then present the impact of anchor selection on localization performance in Section 13.3. Finally, we discuss and conclude in Section 13.4.

13.2 ANCHOR PLACEMENT

13.2.1 Overview

A wide range of localization techniques are based on the availability of anchor points, whose positions are known as prior knowledge, such as multilateration methods [1–5] (e.g., global positioning system [GPS]), indoor localization approaches

[6–10], and anchor-based localization in wireless sensor networks (WSNs) [11–14]. In all these localization techniques, the geometric layout of anchors affects the localization performance. For example, the location accuracy obtained from GPS at a certain location may be different from time to time because the geometric layout of the visible satellites is different from time to time. Understanding the impact of anchor placement on localization performance is important, as it offers important insights into how to deploy anchors effectively to achieve better localization accuracy. In this section, we first present two initial studies on the impact of anchor placement to localization techniques employing received signal strength (RSS). We then study two approaches for a special case when three anchors are deployed in the area of interest. More specifically, the two methods are a heuristic search-based method and an acute triangular-based deployment. Both methods aim to find the optimal anchor placement with respect to the average localization error. We further overview a beacon placement approach that adapts to the noisy and unpredictable environmental conditions. Given an existing field of beacons, the adaptive beacon placement strategies address how to place additional beacons to improve the localization performance. Finally, we discuss the process of finding the optimal placement of anchors in well-defined regular regions. To do so, we present both a theoretical analysis of exploring optimal anchor patterns and an algorithm that finds the optimal pattern of anchor placement. The goal of the anchor placement is to minimize the maximum localization error. Table 13.1 summarizes the studies on anchor placement that are discussed in this section.

TABLE 13.1.   Summary of Studies on Anchor Placement

Study	Approach Description
Impact of anchor placement	Objective: to study the dependence of the location uncertainty based on various factors, such as the number of anchors and the distance between anchors [15, 16]
	Approach: A mathematical model is proposed to map the uncertainty in signal space to the uncertainty in physical space.
	Observation: If anchors are added to the same area, the uncertainty decreases. However, if the area of coverage is increased while increasing the number of anchors, the minimum uncertainty increases or remains stable for the values of n anchors and specific anchor placements under exploration.
Heuristic search	Objective: to find the optimal anchor placement with respect to the average localization error in the area of interest [17].
	Approach: explored various searching techniques, such as local search (LS) [20], reactive tabu search (RTS) [21], and simulated annealing (SA) [22], to find the optimal anchor placement that minimizes the overall errors
	Observation: The optimal placement found by the search techniques is not deterministic. The RTS and SA techniques outperform simple LS. And the localization error is rather low when close to the anchor points.

Continued

TABLE 13.1. Continued

Study	Approach Description
Acute triangular-based deployment	Objective: to find the optimum anchor deployment to minimize the overall localization estimation uncertainty in the area of interest when there are three anchors [23, 24]. Approach: The researchers designed an *uncertainty area* (UA) to measure the localization performance, given the RSS perturbation from measurements. Observation: Different types of triangles for anchor deployment do not make too much difference for the points inside the triangle. The equilateral deployment gives the best performance, and the localization performance for positions inside the triangle is much better than that outside the triangle. The optimal deployment for overall localization performance for the whole floor is found when the center of gravity of the equilateral triangle (formed by three anchors) coincides with that of the floor plane.
Adaptive beacon placement	Objective: Given an existing field of beacons, how should additional beacons be placed for best advantage [25]? Approach: Two strategies, *max* and *grid*, are proposed to improve the localization quality by adding addition anchors. Observation: The grid algorithm is clearly the best and superior to max and random deployment in the presence of noise. In addition, compared to the noise level, the anchor density rather has a higher impact on the performance of anchor placement algorithms in WSNs.
Optimal placement via maximum lambda and minimum error (maxL–minE)	Objective: to find the optimal placement of anchors in well-defined regular regions such as indoors to minimize the maximum localization error [6]. Approach: An upper bound for the localization error of the linear least squares (LLS) algorithm is derived. An algorithm called maxL–minE has been developed to find the pattern of anchor placement that targets to minimize the maximum localization error. Observation: For a small number of anchors, simple shapes such as equilateral triangles and squares result in placements with better localization performance. For a higher number of anchors, the simple shapes enclose one another (e.g., two enclosing equilateral triangles), resulting in placements with better localization performance.

13.2.2 Impact of Anchor Placement

Chen and Kobayashi [15] and Krishnakumar and Krishnan [16] conducted the first studies on the impact of anchor placement on the RSS-based localization techniques. Chen and Kobayashi [15] propose signal strength-based indoor geolocation. They show that indoor geolocation based on signal strength is feasible for locating wireless local area network (WLAN) users in an indoor environment through mathematical analysis and simulation experiments. In [15], it is depicted that a mobile user can be localized via the lateration approach based on the availability of the distance of a mobile user and multiple anchors. There are basically three steps in [15] to obtain the location of a wireless device:

1. Regression analysis on top of RSS readings together with corresponding distances is used to derive parameters in the signal propagation model.

2. The ranging information is derived using the propagation model and the RSS at the mobile device.

3. By applying the method of least squares estimation, the location of the mobile device can be estimated based on ranging information from multiple anchors.

The researchers presented the geometric relationship between the positions of the mobile user and the anchors and showed how the localization error is affected by the number and placement of anchors through simulation. However, they did not provide a quantitative relationship between the geometric layout of anchors and the localization accuracy.

Author in [16] addresses the problem of finding the inherent uncertainty of signal strength-based location estimation techniques. The researchers proposed a mathematical model for mapping the uncertainty in signal space to the uncertainty in physical space. The analysis is used to draw conclusions about the dependence of the minimum location uncertainty of various factors, such as the number of anchors, the distance between anchors, and the signal variance.

The researcher studied a region of interest, R, in the (x, y)-plane. Consider n anchors located at $(x_i, y_i) \in R$, $i = 1, 2, \ldots, n$. Let $(x, y) \in R$ be an arbitrary point. The RSS vector at (x, y) is $\vec{s} = \langle s_1, s_2, \ldots, s_n \rangle$. Assume $\vec{s}$ is denoted by $S(x, y)$ and the mean signal strength vector $\langle \bar{s}_1, \bar{s}_2, \ldots, \bar{s}_n \rangle$ by $\bar{S}(x, y)$. The RSS s_i from anchor i is a stochastic variable due to the effect of multipath fading. This is modeled in the log domain as a normal distribution around a *mean value* $\bar{s}_i$ (in decibel), with a *variance, σ_i^2*:

$$S_i = \bar{s}_i + \Delta s_i, \tag{13.1}$$

where $\Delta s_i \sim N(0, \sigma_i)$.

Let $M : R \to \bar{S}$ be the mapping from location to mean signal strength. For $0 \leq \alpha \leq 1$, the α-region is defined $A_\alpha \in R$ by the following relation: $\int_{A_\alpha} p(s|(x, y)) dx \, dy = \alpha$. The region A_α may be interpreted as a set of locations in R such that the total probability that the observed signal strength vector is due to an emitter located at some point in A_α is α. Therefore, given a desired level of confidence α, the region of uncertainty can be computed.

Given a location (x, y), the measured signal strength vector at this location is a multivariate normal variable in signal strength space centered at $\bar{S}(x, y)$. Since a variation in measured signal strength due to change in location is indistinguishable from a variation due to shadowing, a decision rule will map a set of locations in the neighborhood of (x, y) to (x, y). This is the uncertainty in the location estimate, caused by signal variance. Let $\Delta \vec{x} = [\Delta x \ \Delta y]^T$ be the vector of the change in location around (x, y). Here, v^T denotes the transpose of v. Similarly, let

$\Delta \vec{s} = [\Delta s_1 \quad \Delta s_2 \quad \ldots \quad \Delta s_n]^T$ be the vector of signal strength variations around $\vec{s}$, where $S(x, y) = \vec{s}$. Then,

$$\Delta \vec{s} = J \Delta \vec{x}, \tag{13.2}$$

where $J = t_{ij}$ is the Jacobian of mapping M from location to mean signal strength measurements and is an n matrix with $t_{i1} = \partial \vec{s} / \partial x$ and $t_{i2} = \partial \vec{s} / \partial y$. Note that J is a function of (x, y).

Theorem 1: Given a desired level of confidence α, the region of uncertainty in the (x, y)-plane is an ellipse with semiaxes given by

$$r_{\text{major,minor}} = \sqrt{\frac{-2d}{(a+b) \pm \sqrt{(a-b)^2 + c^2}}}, \tag{13.3}$$

where a, b, c, and d are defined as the following:

$$a = \sum_{i=1}^{n} \frac{t_{i1}^2}{\sigma_i^2}, b = \sum_{i=1}^{n} \frac{t_{i2}^2}{\sigma_i^2}, c = \sum_{i=1}^{n} \frac{2 t_{i1} t_{i2}}{\sigma_i^2}, d = -R_n^2. \tag{13.4}$$

In the above equation, R_n is a scaling factor related to the confidence level α. This relationship is given by

$$\alpha = \frac{\Gamma(n/2, R_n^2/2)}{\Gamma(n/2)}, \tag{13.5}$$

where $\Gamma(., .)$ is the *incomplete gamma function*. Given α and n, Equation (13.5) can be used to compute R_n.

Given (13.3), the *uncertainty* is further defined as the r_{major}, which is used to measure the location uncertainty. Generally, given a certain number of anchors, the minimum uncertainty (i.e., the best localization accuracy) that can be achieved decreases as the distance between anchors decreases. Moreover, it is depicted that the relationship between uncertainty (not just the minimum uncertainty) and distance between anchors scales linearly.

In addition, Krishnakumar and Krishnan [16] study the impact of the number of anchors on localization uncertainty under three different situations. It considers n anchors at the vertices of a set of a regular n-polygon inscribed in a circle as the region of interest. The variation of uncertainty over such a region for a circle with a fixed radius is considered in the following three cases:

1. *Fixed-Circle Case.* The anchors are at the vertices of a polygon inscribed in a circle of radius L. This models the case where more anchors are used for the same area.

2. *Fixed-Distance Case.* The adjacent anchors of the regular polygon are at a distance L from each other. The radius of the circle in which the polygon is inscribed grows with n.

3. *Fixed-Area-per-Anchor Case.* The n anchors are at the vertices of a polygon inscribed in a circle of radius L, such that $\pi L_n^2/n$ is a constant. In this case, the radius of the circle grows with n, however, proportional to $\sqrt{n}$.

The *fixed-distance case* and the *fixed-area-per-anchor case* provide two possible methods in which more anchors are used but also cover a larger area. The researchers further show that if anchors are added to the same area, the uncertainty decreases. However, if the area of coverage is increased while increasing the number of anchors, the minimum uncertainty increases or remains stable for the values of n and specific anchor placements under exploration.

From the above study, we obtain a general idea about how the anchor placement affects the localization accuracy. However, an important question remains unanswered; that is, given a certain number of anchors, how can the placement of those anchors help to achieve the best localization accuracy?

13.2.3 Heuristic Search

Battiti et al. [17] developed a mathematical model for the localization error based on the variability of signal strength measurements. The proposed mathematical model aims to find the optimal anchor placement with respect to the average localization error when there are three anchors deployed in the area of interest. The heuristic search methods are developed to find the optimal anchor placement. This model has been designed to be independent from the actual localization technique; therefore, it is only based on generic assumptions on the behavior of the localization algorithm employed.

Considering a planar environment with a local two-coordinate system, valid coordinates are inside a rectangular region A: $x \equiv (x, y) \in A = [x_{\min}, x_{\max}] [y_{\min}, y_{\max}] \subset \mathbb{R}^2$. Assuming the real location of a wireless device is $\hat{x}$, the estimated location of the wireless device x with the conditional probability distribution $P(x|\hat{x})$ can be represented as

$$\forall \hat{x} \in A \int_A P(x|\hat{x}) d\hat{x} = 1,$$
(13.6)

where $0 \leq P(x|\hat{x}) \leq 1$. This equation indicates the total probability that the wireless device is located within the area A is 1.

Defining the distance function between locations x and $\hat{x}$ as $d(x, \hat{x})$, the average error can be calculated as

$$E(\hat{x}) = \int_A d(x, \hat{x}) P(\hat{x}|x) dx$$
(13.7)

The overall expected error in the whole area A is obtained by

$$E = \frac{1}{\|A\|} \int_A E(\hat{x}) d\hat{x},$$
(13.8)

where $\|A\|$ is the area of A.

To minimize the objective function, which is the overall expected localization error over the whole area, the conditional probability $P(x|\hat{x})$ must be estimated. Suppose there are n anchor points deployed in the area of interest A. Let $d_{AP_i}(x)$ be the distance of location x from the ith anchor point, and let $\omega_{AP_i}(x)$ be the sum of the widths of all walls crossed by the segment joining the ith anchor point to x. According to the indoor signal propagation model [18], the average signal strength of the signal received at location x from the ith anchor point is

$$\mu_i(x) = \beta_0 + \beta_1 \log(d_{AP_i}(x)) + \beta_2 \omega_{AP_i}(x). \tag{13.9}$$

The propagation parameters β_0, β_1, and β_2 can be determined by a least squares fit to the experimental data. Chapter 4 discusses indoor and outdoor channel models in detail.

Considering that the signal strength in the log domain follows a Gaussian distribution with the expected value μ and the standard deviation σ at a given location, the probability density of the detected signal strength s corresponds to

$$S(s|\mu) = \frac{1}{\sigma\sqrt{2\pi}} e^{-\frac{1}{2}\left(\frac{s-\mu}{\sigma}\right)^2}. \tag{13.10}$$

Since the measured signal from anchor point i is $\mu_i(x)$, the density of the probability that the wireless device is located at $\hat{x}$ is $S[\mu_i(x)|\mu_i(\hat{x})]$. If we consider that the measurement errors from different anchor points are roughly independent,* the conditional probability $P(x|\hat{x})$ can be represented by [19]

$$P(x|\hat{x}) \approx \frac{\prod_{i=1}^{n} S[\mu_i(x)|\mu_i(\hat{x})]}{\int_A \prod_{i=1}^{n} S[\mu_i(\xi)|\mu_i(\hat{x})]d\xi}. \tag{13.11}$$

Given the position of the anchor points and the propagation parameters from (13.9), the estimation of the overall expected error within the area A can be calculated by substituting the above equation into (13.7) and (13.8). By placing the anchors at different locations, the overall localization error will be changed. The objective is to find a deployment of anchor points, which can minimize the objective function (13.8) (i.e., overall localization error). In [17], the researchers have explored various searching techniques, such as local search (LS) [20], reactive tabu search (RTS) [21], and simulated annealing (SA) [22], to find the optimal anchor placement that minimizes the overall errors.

Additionally, the researchers conducted experiments in an area of $750\,\text{m}^2$ to validate error minimization techniques via various search mechanisms. There are three anchors in total. In order to calculate the overall localization error as described in (13.8), positions of wireless devices and the estimated positions are discretized on a 10×10 point mesh. The experimental results show that the RTS and SA

---

*This assumption may not be true for other studies.

techniques outperform simple LS. The global minimum of the overall averaged localization error is found to be 5.8 m. Further, the localization error is rather low when close to the anchor points, while it may go up to 10 m in some locations away from the anchors.

13.2.4 Acute Triangular-Based Deployment

Yiming et al. [23, 24] propose acute triangular-based deployment by focusing on the fundamental deployment structure with only three anchors in indoor wireless localization using RSS. The detailed analysis and experimental results show that the best overall localization performance is obtained when the center of gravity of the equilateral triangle (formed by three anchors) coincides with that of the experimental floor plan. In addition, in order to provide optimal localization for all locations on a large floor, it is necessary to deploy more than three anchors in a semimesh style such that any position on the floor is always covered by three nearby anchors.

To analyze how the deployment of anchors affects the localization performance, in [24], the measured RSS for a given location at an anchor is assumed to be in

$$SS \in \left[SS_{\text{true}} \cdot (1 - \delta_s), SS_{\text{true}} \cdot (1 + \delta_s) \right], \tag{13.12}$$

where SS_{true} is the true signal strength value at that location and δ_s is the maximum perturbation from measurements. The RSS at a position uniquely maps to a range r_{true} from the wireless device to an anchor. δ_r represents the uncertainty of the radio range corresponding to the perturbation of the signal strength δ_s. The range r can be expressed as

$$r \in \left[r_{\text{true}} \cdot (1 - \delta_r), r_{\text{true}} \cdot (1 + \delta_r) \right]. \tag{13.13}$$

The researchers designated *uncertainty area* (UA) to measure the localization performance. UA is determined by the intersection area of a multiple annular region that is centered at these anchors as shown in Figure 13.1. The annular region of the anchor point is defined as the area between the range $[r_{\text{true}} \cdot (1 - \delta_r)]$ and $[r_{\text{true}} \cdot (1 + \delta_r)]$ of that anchor. In addition, the *average uncertainty distance* d_{AUD} is defined as the average distance of all points in the point set within the *UA*. Assuming an UA contains total m discrete points in the set, and d_{ij} is the distance between two points, i and j, in the area ($i, j \in [1, m]$, $i < j$ in the set), then $d_{\text{AUD}} = \text{average}\{d_{ij}\}$.

Example 13.1   Calculating the Average Uncertainty Distance

Solution

Assume the experimental floor size is $M \times N$ feet, and the uncertainty area is UA. The whole experimental floor can be divided into gridlike subareas. Setting the grid size as 1 in., the experimental floor can be divided into a $12M \times 12N$ grid. And the grid center represents each grid. For each grid center, we examine whether it falls into the UA or not and find out all the grid centers (i.e., points) fall into the UA. If there are total m points falling into the UA, we can calculate the distance between

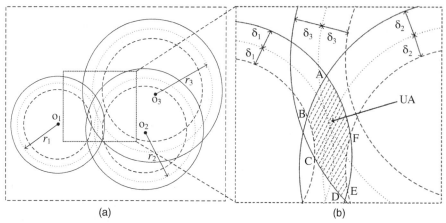

(a)   (b)

Figure 13.1   UA [24].

any two points that fall into the UA and obtain $m(m-1)/2$ distances d_{ij}, where i, $j \in [1, m]$, $i < j$. The average uncertainty distance d_{AUD} is then obtained at the average distance of all points in the point set within the UA: $d_{AUD} = \text{average}\{d_{ij}\}$.

Two problems are investigated under the above assumptions: (1) For an interested position, how does one deploy three anchors such that the estimation uncertainty for that position is minimized; and (2) for a given floor plan, what is the best deployment strategy for three anchors such that the average localization error for the whole floor is minimized?

Through simulations, when the anchors are placed as the vertices in an equilateral triangle, the minimal estimation uncertainty is found at the center of gravity of the equilateral triangle. To answer the second question, the average d_{AUD} at all points inside and outside of the triangles (i.e., the shape of the anchor deployment) is calculated in the simulation where three anchors are deployed in a square space of $40 \times 40\,\text{m}$. Simulation results reveal several key observations:

1. Different types of triangles for anchor deployment do not make too much difference for the location estimation of the points inside the triangle.

2. The equilateral deployment gives the best performance, and the localization performance for positions inside the triangle is much better than that outside the triangle.

3. The optimal deployment for the overall localization performance for the whole floor is found when the center of gravity of the equilateral triangle (formed by three anchors) coincides with that of the experimental floor plan.

In addition, the experimental results confirmed the following observations:

1. Equilateral triangle anchor deployment has the best performance.

2. Extra anchors could improve the localization performance, and the deployment of multiple equilateral triangles connected to each other is one of the best deployment methods.

3. Anchors should be maximally separated from each other if each individual anchor is able to provide effective signal coverage for the experimental floor plan.

13.2.5 Adaptive Beacon Placement

Bulusu et al. [25] point out the need for empirically adaptive beacon placement during localization and outline a general approach based on exploration and instrumentation of the terrain conditions by a mobile human or robot agent. Existing beacon placement approaches are basically fixed strategies that do not take into account environmental conditions that cannot be predicted a priori. It is virtually impossible to preconfigure to such terrain and propagation uncertainties and to compute an ideal (or even satisfying) beacon placement that uniformly achieves a desired quality of localization across the region. Thus, the beacon placement needs to adapt to the noisy and unpredictable environmental conditions. This work addresses the problem of adaptive beacon placement: Given an existing field of beacons, how should additional beacons be placed for the best advantage?

The approach to incremental improvement of localization through beacon placement is based on empirical adaptation. By adaptation, the researchers intend to improve the quality of localization by adjusting beacon placement or by adding a few beacons rather than by completely redeploying all beacons. By empirical, they mean the deployment of additional beacons is influenced by measurements of the operating localization system rather than by careful or complete off-line analysis of a complete system model.

In particular, two strategies are proposed to improve the localization quality by adding additional anchors. The first one is called *max*, in which additional anchors are placed at the positions where the localization errors are maximum. This method is predicated on the assumption that points with high localization error are spatially correlated. For example, with similar terrain conditions, the multipath effects in that area are similar; consequently, the localization error will be spatially correlated [26]. The advantage of this algorithm is that it can be computed in a very straightforward manner. However, it may be overly influenced by propagation effects or random noise that may cause very high localization error at one point, while localization error at points very close to it remains low. The second method is called *grid*, in which additional anchors are placed at points where the cumulative localization error is high. To calculate the cumulative localization error at each point, the whole area is partitioned into overlapping grids. The cumulative localization error of the points is the summation of the localization error of all points within the grid. This method is based on the observation that the newly added anchor affects its nearby area, not just the point where it is deployed.

The proposed methods are tested using the simple centroid-based localization method, in which the localized result of the wireless device is the averaged coordinates of the anchors who can be heard by the targeting wireless device [27]. The simulation is conducted in a $100 \times 100 \, \text{m}^2$ terrain. Each wireless node has a nominal radio range of 15 m. The number of anchors varies from 20 to 240 in the increments of 10 anchor nodes. The noise is introduced to affect the connectivity between the anchors and wireless nodes. The simulation results show that the grid algorithm is clearly the best and superior to max and random deployment in the presence of noise. In addition, the researchers confirm that compared with the noise level, the anchor density rather has a higher impact on the performance of anchor placement algorithms.

13.2.6 Optimal Placement via maxL–minE

Using both analysis and experiments, Chen et al. [6] aim to find optimal placement of anchors in well-defined regular regions such as those found indoors. From the analytic analysis, an upper bound for the localization error of the linear least squares (LLS) algorithm is derived. This bound reflects the placement of anchors as well as measurement errors at the anchors. The theoretical analysis centers on the least squares (LS) algorithm for two reasons. First, LS is a widely used multilateration algorithm, as is evidenced by its application as a step in many recent localization research works. Second, mathematical analysis of LLS is tractable, resulting in equations with closed-form solutions. For a myriad of other algorithms, closed-form solutions that describe the localization error as a function of anchor placement are not tractable and, as a result, heuristic search strategies must be used to find an optimal placement, as was done in [17].

Further, for a given number of anchors, an algorithm called *maxL–minE* has been developed to find the pattern of anchor placement that attempts to minimize the maximum localization error. The interesting result shown in [6] is that for a small number of anchors, simple shapes such as equilateral triangles and squares result in placements with better localization performance. Interestingly, for a higher number of anchors, the researchers showed that extensions of shapes with equal sides, for example, a hexagon, are nonoptimal. Rather, the simple shapes enclose one another, for example, two concentric equilateral triangles. We next review the theoretical analysis and present the experimental results using RSS and time of arrival (TOA) as ranging modalities, respectively, in a Wi-Fi network and a ZigBee network presented in [6].

Theoretical Analysis: Given the estimated distances d_i between the wireless device and multiple anchors and known positions (x_i, y_i) of the anchors, the position (x, y) of the wireless device can be estimated by finding $(\hat{x}, \hat{y})$ satisfying

$$(\hat{x}, \hat{y}) = \arg\min_{x,y} \sum_{i=1}^{N} \left[\sqrt{(x_i - x)^2 + (y_i - y)^2} - d_i \right]^2, \tag{13.14}$$

where N is the number of anchors. The above equation is a nonlinear least squares (NLS) problem, which requires significant complexity and is difficult to analyze. It can be linearized by introducing a geometric constraint in the formulation. Starting with the $N \geq 2$ equations

$$(x_1 - x)^2 + (y_1 - y)^2 = d_1^2$$
$$(x_2 - x)^2 + (y_2 - y)^2 = d_2^2$$
$$\vdots \tag{13.15}$$
$$(x_N - x)^2 + (y_N - y)^2 = d_N^2$$

and subtracting the constraint

$$\frac{1}{N}\sum_{i=1}^{N}\left[(x_i - x)^2 + (y_i - y)^2\right] = \frac{1}{N}\sum_{i=1}^{N}d_i^2 \tag{13.16}$$

from both sides, the following set of linear equations can be obtained:

$$\left(x_1 - \frac{1}{N}\sum_{i=1}^{N}x_i\right)x + \left(y_1 - \frac{1}{N}\sum_{i=1}^{N}y_i\right)y$$

$$= \frac{1}{2}\left[\left(x_1^2 - \frac{1}{N}\sum_{i=1}^{N}x_i^2\right) + \left(y_1^2 - \frac{1}{N}\sum_{i=1}^{N}y_i^2\right) - \left(d_1^2 - \frac{1}{N}\sum_{i=1}^{N}d_i^2\right)\right]$$

$$\vdots \tag{13.17}$$

$$\left(x_N - \frac{1}{N}\sum_{i=1}^{N}x_i\right)x + \left(y_N - \frac{1}{N}\sum_{i=1}^{N}y_i\right)y$$

$$= \frac{1}{2}\left[\left(x_N^2 - \frac{1}{N}\sum_{i=1}^{N}x_i^2\right) + \left(y_N^2 - \frac{1}{N}\sum_{i=1}^{N}y_i^2\right) - \left(d_N^2 - \frac{1}{N}\sum_{i=1}^{N}d_i^2\right)\right].$$

Therefore, the above can be easily solved linearly using the form $\mathbf{Ax} = \mathbf{b}$ with

$$\mathbf{A} = \begin{pmatrix} x_1 - \dfrac{1}{N}\displaystyle\sum_{i=1}^{N}x_i & y_1 - \dfrac{1}{N}\displaystyle\sum_{i=1}^{N}y_i \\ \vdots & \vdots \\ x_N - \dfrac{1}{N}\displaystyle\sum_{i=1}^{N}x_i & y_N - \dfrac{1}{N}\displaystyle\sum_{i=1}^{N}y_i \end{pmatrix} \tag{13.18}$$

and

$$\mathbf{b} = \frac{1}{2}\begin{pmatrix} \left(x_1^2 - \dfrac{1}{N}\displaystyle\sum_{i=1}^{N}x_i^2\right) + \left(y_1^2 - \dfrac{1}{N}\displaystyle\sum_{i=1}^{N}y_i^2\right) \\ -\left(d_1^2 - \dfrac{1}{N}\displaystyle\sum_{i=1}^{N}d_i^2\right) \\ \vdots \\ \left(x_N^2 - \dfrac{1}{N}\displaystyle\sum_{i=1}^{N}x_i^2\right) + \left(y_N^2 - \dfrac{1}{N}\displaystyle\sum_{i=1}^{N}y_i^2\right) \\ -\left(d_N^2 - \dfrac{1}{N}\displaystyle\sum_{i=1}^{N}d_i^2\right) \end{pmatrix}. \tag{13.19}$$

Note that $\mathbf{A}$ is only described by the coordinates of anchor points; $\mathbf{b}$ is represented by the distances to the anchor points together with the coordinates of anchor points; and $\mathbf{x}$ is the estimated location of the wireless device.

The objective is to minimize the location estimation error introduced by LLS. In an ideal situation $\mathbf{x} = [x, y]^T$ corresponds to

$$\mathbf{x} = \left(\mathbf{A}^T\mathbf{A}\right)^{-1}\mathbf{A}^T\mathbf{b}. \tag{13.20}$$

However, the estimated distances are impacted by noise, bias, and measurement error. By expressing the resulting distance estimation error $\mathbf{e}$ in terms of $\tilde{\mathbf{b}}$ with estimated distances and $\mathbf{b}$ with true distances as $\tilde{\mathbf{b}} = \mathbf{b} + \mathbf{e}$, the localization result is

$$\tilde{\mathbf{x}} = \left(\mathbf{A}^T\mathbf{A}\right)^{-1}\mathbf{A}^T\tilde{\mathbf{b}}. \tag{13.21}$$

We note that the error $\mathbf{e}$ is the difference between $\mathbf{b}$ obtained when d_i is perfect and when d_i is noisy.

The location estimation error is thus bounded by

$$\|\mathbf{x} - \tilde{\mathbf{x}}\| \le \|\mathbf{A}^+\|\|\mathbf{e}\|, \tag{13.22}$$

where the matrix $\mathbf{A}^+$ is the Moore–Penrose pseudoinverse of $\mathbf{A}$. It can be shown that, under the 2-norm,

$$\|\mathbf{A}^+\| = \frac{1}{\gamma_2},$$

where $\gamma_1 \ge \gamma_2$ are the singular values of $\mathbf{A}$ [28]. This means that for a certain size of error $\mathbf{e}$, the LS estimation error is stretched by

$$\frac{1}{\gamma_2}.$$

It can be proved that the eigenvalues of $\mathbf{A}^T\mathbf{A}$ are the squares of the singular values of $\mathbf{A}$ [28]. Therefore, the concern on the location estimation error is limited to the eigenvalues of $\mathbf{A}^T\mathbf{A}$, where $\mathbf{A}^T\mathbf{A}$ is a matrix of the form

$$\mathbf{A}^T\mathbf{A} = \begin{pmatrix} a & b \\ b & c \end{pmatrix}$$

with

$$a = \sum_{i=1}^{N}\left(x_i - \frac{1}{N}\sum_{i=1}^{N}x_i\right)^2, \tag{13.23}$$

$$b = \sum_{i=1}^{N}\left[\left(x_i - \frac{1}{N}\sum_{i=1}^{N}x_i\right)\left(y_i - \frac{1}{N}\sum_{i=1}^{N}y_i\right)\right], \tag{13.24}$$

and

$$c = \sum_{i=1}^{N}\left(y_i - \frac{1}{N}\sum_{i=1}^{N}y_i\right)^2. \tag{13.25}$$

Note that a, b, and c are only related to the coordinates of the anchors (x_i, y_i). The eigenvalues of $\mathbf{A}^T\mathbf{A}$ can be found to be the roots of

$$\lambda^2 - (a+c)\lambda + (ac - b^2) = 0.$$

Thus, λ is given by

$$\lambda = \frac{(a+c) \pm \sqrt{(a-c)^2 + 4b^2}}{2}, \qquad (13.26)$$

where the discriminant, $(a - c)^2 + 4b^2$, is nonnegative.

The goal is to derive the deployment patterns of anchors that minimize the total errors. Recall that there are two terms on the right side of (13.22). The approach is to choose x_i and y_i so as to make λ_2 (the smaller eigenvalue) as close to λ_1 as possible, because this will minimize the first term, $\|\mathbf{A}^+\|$. Having minimized the second term given the first term is minimized is clearly a local minima. Such a local minima is called an *optimal deployment*, because no movement of a single anchor can improve the error bound.

Returning to minimizing the first term $\|\mathbf{A}^+\|$, to minimize

$$\frac{1}{\sqrt{\lambda_2}},$$

a general strategy would be to make $(a - c)$ small or to make b small, or both. Interestingly, this is determined only by the coordinates of the anchors.

Finding the anchor positions that satisfy $\lambda_1 \cong \lambda_2$ could yield an optimal deployment. The researchers found that such an optimal anchor deployment setup follows some simple and symmetrical patterns [6]. This makes it not only possible to achieve but also easy to deploy practically. Figure 13.2 shows the patterns for an optimal anchor deployment setup when utilizing three to eight anchors. These patterns consist of squares, equilateral triangles, or the concentric placement of multiple such shapes. For a higher number of anchors, the extensions of shapes with equal sides, for example, a hexagon, do not satisfy $\lambda_1 \cong \lambda_2$, and thus are not optimal. Instead, simple shapes that enclose one another present optimal solutions.

Example 13.2   Deploying the Optimal Anchor Patterns in an Experimental Floor Plan

The previous discussion dealt with deploying anchors without considering the physical constraints of the wireless environment, and, as such, only provides a general guideline as to the "shape" of the deployment. Placing the anchors within a particular environment requires stretching/shrinking the deployment shape so that it fits within the confines of the environment. The stretching/shrinking should be done so as to minimize localization errors.

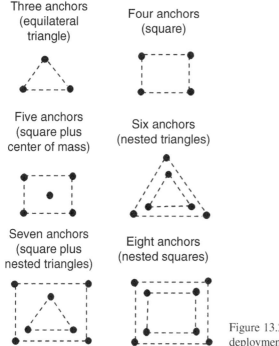

Three anchors
(equilateral
triangle)

Four anchors
(square)

Five anchors
(square plus
center of mass)

Six anchors
(nested triangles)

Seven anchors
(square plus
nested triangles)

Eight anchors
(nested squares)

Figure 13.2   Patterns for optimal anchor deployments [6].

Solution

Based on (13.22), the location estimation error is also impacted by $\|\mathbf{e}\|$, where $\tilde{\mathbf{b}} = \mathbf{b} + \mathbf{e}$. The term $\|\mathbf{e}\|$ is a result of distance estimation errors introduced by ranging. By considering this important effect, an algorithm, called maxL–minE, has been developed. maxL–minE helps to find the best anchor coordinates given the dimensions of the environment, the number of anchors, and the optimal anchor deployment pattern. Figure 13.3 shows the pseudocode that implements maxL–minE. The algorithm first minimizes $\|\mathbf{A}^{+}\|$ using geometry and then performs an iterative search. The search begins with a maximal-sized optimal pattern (e.g., a square) and simply keeps reducing the size of the pattern until such movements stop reducing the distance estimation error $\mathbf{e}$. The main steps are summarized as follows:

1. *Initialization.* Given the number of anchors, find the optimal anchor deployment pattern based on the results as shown in Figure 13.2 and fit the optimal pattern into the maximum floor size to obtain the initial coordinates of anchors.

2. *Iteration.* Generate sample localizing targeting nodes and random noise; calculate the localization errors under current settings; and reduce the size of the pattern by moving all the anchors toward the center of the floor plan one step.

3. *Termination.* Repeat step 2 until the reduced size of the pattern stops reducing the distance estimation error $\mathbf{e}$; return current anchor coordinates as the solution.

input $floorSize, numOfAnchor$
output *optimized anchor coordinates*

[**initialize**] get optimal pattern based on geometry;
fit optimal pattern into maximum floorsize;
generate initial anchor coordinates;
calculate λ_1 and λ_2.

$minError = maxNum$

loop until $thisError > minError$
  generate random localizing nodes
  for each localizing node **begin**
    apply random noise or bias
    $B = \|b - \tilde{b}\|$
  end for
  $thisError = \frac{avg(B)}{\sqrt{\lambda_2}}$
  if $thisError < minError$, $minError = thisError$
  [**anchor adjustment**] move towards the center of the floor one step
end loop
return *optimized anchor coordinates*

Figure 13.3   The maxL–minE algorithm.

Algorithm Overview and Experimental Evaluation: Although the previous analysis is based on LLS, the anchor deployment based on the maxL–minE algorithm can improve localization accuracy in widely diverse scenarios. To demonstrate the general applicability of the anchor deployment algorithm, in addition to NLS and LLS, the researchers investigated three different localization algorithms that employ signal strength measurements: RADAR [7], area-based probability (ABP) [29], and BN [9]. RADAR is a point-based, scene-matching algorithm. The user first builds a training set of RSS values from anchors matched to known locations. To localize, the object creates a vector of RSS values from the anchors and the algorithm returns the training point closest to the vector using Euclidean distance as the discriminating function [7]. ABP returns an area bounded by a probability that the wireless device is within the returned area [29]. The probability is called the confidence, α, and is adjustable by the user. ABP assumes the measured RSS for each anchor follows a lognormal distribution with a mean equal to the expected value of the RSS reading vector. The Gaussian error from each anchor is assumed to be independent. ABP then computes the probability of the wireless device being at each nonoverlapping small area on the floor using Bayes'rule. Finally, APB returns the most likely areas up to its confidence α as the location estimation of the wireless device. Taking the BN approach, the BN algorithm uses a Bayesian graphical model based on lateration to find the estimated location [9].

    A series of experiments were conducted in both Wi-Fi and ZigBee networks by using four anchors set up in an indoor office scenario. Two anchor placements

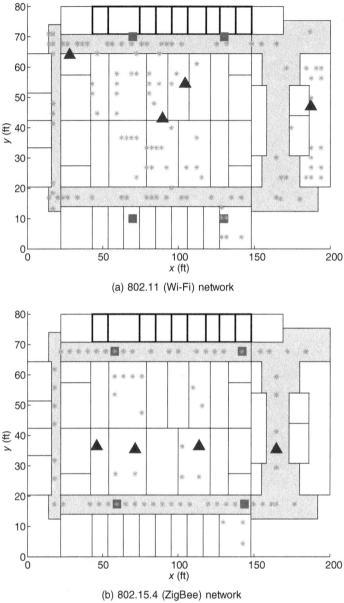

(a) 802.11 (Wi-Fi) network

(b) 802.15.4 (ZigBee) network

Figure 13.4   Deployment of anchors and training locations on the experimental floor [6].

were under study: *colinear* and *square*. Figure 13.4 shows the colinear anchor deployment setup* in triangles and the optimized square anchor deployment as squares for the Wi-Fi and the ZigBee networks, respectively.

Figure 13.5a, b presents the Wi-Fi accuracy cumulative distribution function (CDF) under colinear and square anchor deployments, respectively. Figure 13.5a

*The colinear anchor deployment is set up by the building IT department for maximum signal coverage.

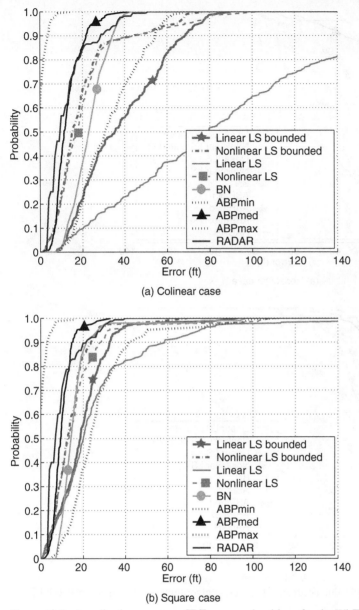

(a) Colinear case

(b) Square case

Figure 13.5   Localization accuracy CDFs across algorithms for the Wi-Fi network [6].

shows that under the horizontal-like deployment, LLS always performs very poorly, while NLS, RADAR, ABP, and BN are qualitatively similar. All the algorithms have long tails. Figure 13.6a showed a similar result when using the ZigBee, although in here, the perfect colinear deployment, the horizontal case, reduces the performance of the lateration approaches (BN, NLS, and LLS) compared to Wi-Fi.

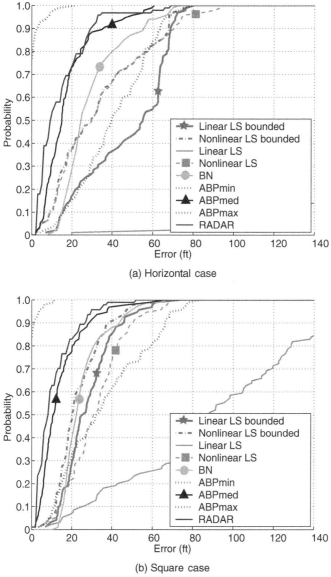

(a) Horizontal case

(b) Square case

Figure 13.6 Localization accuracy CDFs across algorithms for the ZigBee network [6].

Figures 13.5b and 13.6b show the key impact of anchor deployments. All of the CDFs are shifted up and to the left compared to those in Figures 13.5a and 13.6a. Thus, a significant fraction of the results is more accurate using the optimized deployments generated by the maxL–minE algorithm. In addition, for ABP, the gap between the min and max CDFs is much narrower, implying the returned areas are, on average, smaller than those in the horizontal deployments.

13.3   ANCHOR SELECTION

13.3.1   Overview

More anchor points participating in the localization process may yield better localization accuracy by assuming that the noise level at each anchor point is the same. However, in practice, the noise levels at different anchor points are different due to multipath effects caused by diffraction, reflection and shadowing, different reliability of devices used by anchor points (such as differences in connectors and thermal effect), and the position accuracy of anchors themselves [30]. For example, in indoor environments, the noise is not uniformly distributed within the area of interest. The noise level at anchor points close to the door is more likely affected by the door closing and opening; consequently, the signal measurement from such an anchor will be affected [31]. Moreover, when using the off-the-shelf devices in WLAN location systems, different kinds of devices may have different reliability, and consequently, the quality of the measured signal from different anchors may vary [32]. Further, in the case of localization in WSNs, estimation of the intermediate nodes (i.e., anchor points) has introduced errors already; thus, the accuracy of the anchors' positions is different [33]. Therefore, intelligently selecting anchors, which can provide lower noise levels, high device reliability, and high position accuracy, for wireless localization is important.

In this section, we review current studies on smart anchor selection methods, which can achieve better localization accuracy as opposed to using all the available anchors. We first study a joint clustering technique that intuitively chooses anchors with the strongest RSS readings by grouping locations that observe the same set of strongest anchors into the same cluster. We then look into a more sophisticated strategy called information gain-based anchor selection, which calculates the entropy of each anchor and selects the anchors with the strongest discriminative power. Further, we describe two methods that formulate convex hulls to select anchors for better localization in WSNs. To provide an empirical quantification of the accuracy limits of RSS localization using commodity wireless hardware, in the last part of the survey, we show an investigation of anchor selection based on data quality in a very high-density environment. Table 13.2 provides a summary of anchor selection methods that are discussed in this section.

13.3.2   Joint Clustering Technique

Youssef et al. [34] presented a joint clustering technique to select anchors for effective WLAN location determination. The joint clustering technique uses (1) signal strength probability distributions to address the noisy wireless channel and (2) clustering of locations to reduce the computational cost of searching the radio map. In particular, the researchers proposed to choose k anchors whose observed signal strength are the top k strongest among a set of available anchors to perform localization. The rationale behind selecting the top k strongest RSS is based on two observations of RSS: (1) At a fixed location, the signal strength from an anchor point varies with time; and (2) the number of anchor points covering a location varies with

TABLE 13.2. Summary of Anchor Selection Methods

Method	Description and Approach
Joint clustering technique	Approach: It first groups the anchors in the area of interest into several clusters based on the coverage of each subset of anchors. Then it chooses k anchors whose observed signal strengths are top k strongest among a set of available anchors to perform localization [34].
	Observation: The localization accuracy is better than that when selecting all the anchors to perform localization.
Entropy-based information gain	Approach: Top k anchors with the highest information gain values are selected to perform localization [35].
	Observation: It outperforms the traditional methods that only consider the strength of signals from the anchors and select the strongest ones.
Convex hull selection	Approach: Select the convex hull to localize the nodes in WSNs by taking the consideration of position accuracy of the anchor nodes [33].
	Observation: The simple convex hull selection method without considering the accuracy of the anchor nodes is better than advanced hull selection with the consideration of anchor nodes' accuracy in low-density networks, whereas the advanced hull selection method improves the simple selection slightly in high-density networks.
Selection from high density of anchors	Approach: Choose the anchors with high RSS quality that can fit the signal propagation model better [37].
	Observation: For lateration-based approaches, which use a signal-to-distance function, the quality of the RSS measurements is more important compared to the quantity of measurements. And a subset of high-quality anchors provide better localization performance.

time. Intuitively, the anchor points that appear most of the time in the area of interest should be chosen to perform localization. Because the coverage of the anchor varies with time, the top k anchors with strongest RSS readings are preferred since they will most likely cover the same area over time.

In the proposed localization system described in [34], the whole area of interest is first grouped into several clusters based on the coverage of a subset of anchors. To group the locations into clusters, the system defines the locations belonging to the same cluster as those locations that observe the same set of q strongest measurements from anchors. The localization is then performed within the cluster using top k strongest signal measurements from anchor points.

The researchers conducted experiments on a floor with an area of 224×85 ft. On average, there are four anchors covering each location on the floor. The experimental results showed that the values of the parameters $q = 3$ and $k = 4$ lead to the best localization performance; that is, the whole floor is grouped into about 15 clusters (where clustering is based on the strongest three anchors observed) and the four strongest anchors are selected to perform localization within each cluster. The localization error of the proposed system is within 7 ft over 90% of the time, which is better than the localization accuracy when using all the anchors (i.e., within 7 ft only at 38% of the time) to perform localization without using the joint clustering technique.

13.3.3   Entropy-Based Information Gain

By considering the discriminative ability of different anchor points toward locations and different signal values from the same set of anchors, Chen et al. [35] developed an information gain-based anchor selection method (*InfoGain* for short) based on the calculated entropy of each anchor. The objective of this study to perform anchor selection is to maintain as high a level of accuracy as possible by using a subset of anchors. By doing this, the overall energy consumption of the indoor localization process can be saved while maintaining high localization accuracy.

Suppose the area of interest is partitioned into n grids and there are m anchor points in total. Each anchor point (AP_i, $1 \leq i \leq m$) is viewed as a feature and each grid (G_j, $1 \leq j \leq n$) is described by these m features (i.e., anchors). The signal strength values at each grid from all the anchors are measured off-line. The criteria of InfoGain for anchor selection is used to evaluate the discriminative power of each anchor and selects the highest ones. Specifically, the discriminative power for AP_i is calculated as the reduction in entropy as follows:

$$\text{InfoGain}(AP_i) = H(G) - H(G \mid AP_i), \tag{13.27}$$

where $H(G) = -\sum_{j=1}^{n} Pr(G_j) \log Pr(G_j)$ is the entropy of the grids when AP_j's value is not known. $Pr(G_j)$ is the prior probability of grid G_j, which is modeled as a uniform distribution if there is no prior knowledge on where the wireless device is located. $H(G \mid AP_i) = -\sum_{v} \sum_{j=1}^{n} Pr(G_j, AP_i = v) \log Pr(G_j \mid AP_i = v)$ is the conditional entropy of grids, given AP_i's value v. After the information gain of each anchor is calculated, the top k anchors with the highest values are selected to perform localization.

To validate the proposed InfoGain anchor selection method, experiments were conducted in a real office building environment with 99 grids, each representing a 1.5-m^2 grid cell, and a total of 25 anchors. The experimental results show that the InfoGain selection method outperforms the traditional approaches that only consider the strength of signals from the anchors and select the strongest ones to perform localization.

13.3.4   Convex Hull Selection

Ermel et al. [33] developed two methods based on the simple convex hull and the advanced hull to select anchor nodes to improve the accuracy of location estimation in WSN. In WSN, there are two types of nodes: self-locating nodes and simple nodes. The self-locating nodes are defined as nodes with accurate position, for example, nodes equipped with GPS. Simple nodes are defined as nodes that have to estimate their positions using the location information from the self-locating nodes or other simple nodes whose positions have been estimated already. Let S be a simple node whose true and estimated positions are S_{real} and S_{est}, respectively. The accuracy of the node's estimated location is defined as a function of the localization error and can be represented as

$$C_{\text{acc}} = 1 - \frac{\left|\overline{S_{\text{real}}S_{\text{est}}}\right|}{R_{\max}}, \tag{13.28}$$

where $R_{\max}$ is the maximum theoretical transmission range of a node and $\left|\overline{AB}\right|$ is the distance between A and B.

Based on the definitions of a simple node and a self-locating node, the self-locating node has a position accuracy of 1, whereas the simple node has a position accuracy $0 \leq C_{\text{acc}} < 1$. Since the real position of the simple node is unknown, the simple node has to estimate its position accuracy using some existing methods [36]. Assuming a simple node can obtain its position accuracy C_{acc}, the R_{error} is defined as the maximum radius of the radio coverage of a node that takes into account the accuracy of its estimated position:

$$R_{\text{error}} = R_{\max} + R_{\max}(1 - C_{\text{acc}}). \tag{13.29}$$

Therefore, for a self-locating node, $R_{\text{error}} = R_{\max}$, while for a simple node, $R_{\max} < R_{\text{error}} < 2R_{\max}$.

To localize a simple node, the centroid method can be used by considering all the nodes that are one-hop neighbors of that simple node. These one-hop neighbors are defined as the anchor nodes that are either the self-locating nodes or the simple nodes whose location have been estimated. To improve the localization accuracy, the convex hull anchor selection method is developed by considering the geometry of the anchors. The basic idea of using a convex hull as the selection criterion is to choose only anchors that have the greatest distance from each other. Because, intuitively, the further the anchors are from each other, the better the position estimation accuracy will be [24].

The simple convex hull selection method only considers each anchor's physical position, but does not take into account the anchor's position accuracy C_{acc}. On the other hand, the advanced hull selection method considers both the physical position as well as the position accuracy C_{acc} of an anchor. The virtual accuracy parameter V_{acc} is designed to combine both the anchor's physical position as well as its position accuracy:

$$V_{\text{acc}}^B = \frac{\left|\overline{AB}\right|}{R_{\text{error}}^A} C_{\text{acc}}^B, \tag{13.30}$$

where $\left|\overline{AB}\right|$ is the distance between the simple node A and an anchor B. The advanced hull method selects the anchors close to the convex hull (i.e., simple convex hull) with the highest virtual accuracy. Nodes that do not belong to the simple hull compare their virtual position accuracy to their nearest hull nodes, and the highest virtual accuracy node is selected as a member of the advanced hull. If the virtual accuracy of the nodes is the same, the node with the highest accuracy C_{acc} is selected.

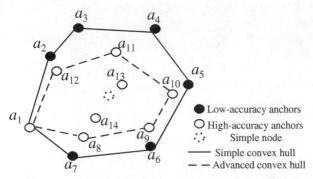

Figure 13.7   Convex hull: simple convex hull versus advanced convex hull.

Example 13.3   Determining the Simple Convex Hull and the Advanced Convex Hull

Solution

Given 14 available anchors as shown in Figure 13.7, suppose the anchors a_i with $i = 2, 3, 4, 5, 6, 7$ are low-accuracy anchors, whereas the anchors a_j with $j = 1, 8, 9, 10, 11, 12, 13, 14$ are high-accuracy anchors. Based on the simple convex hull selection method that considers only the distance metric to select hull nodes, the simple convex hull should consist of $a_1 - a_2 - a_3 - a_4 - a_5 - a_6 - a_7$. By taking into consideration both the distance and position accuracy of the anchors, the advanced convex hull is identified as $a_1 - a_8 - a_9 - a_{10} - a_{11} - a_{12}$.

Simulation results show that the localization accuracy of the simple convex hull selection method outperforms by 20% when using all one-hop anchors. Researchers further show that the accuracy of an estimated position only takes into account the distance between hull nodes, independent of their position accuracy. Thus, the simple convex hull selection method is better than the advanced hull selection in low-density networks, whereas the advanced hull selection method improves the simple selection slightly in high-density networks.

13.3.5   Selection from High Density of Anchors

Chandrasekaran et al. [37] proposed how to select anchors from a high-density deployment based on the data quality of the anchors. Specifically, in [37], they empirically quantify the accuracy limits of RSS localization using a Wi-Fi device. The researchers conducted experiments to capture RSS data in a controlled (and extremely dense) laboratory environment with a single transmitter and up to 369 anchors, which represents an ideal scenario for localization algorithms. Traces were collected using the ORBIT testbed [38], which is a 400-node indoor wireless experimental apparatus that has 800 antennas attached to 800

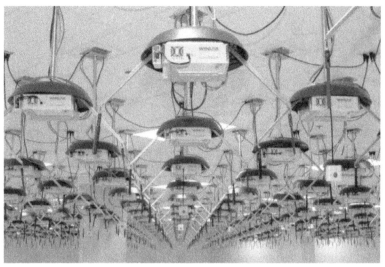

Figure 13.8   Experimental setup: ORBIT room [37].

IEEE 802.11 radios in a $3600\,\text{ft}^2$ area. Using the ORBIT platform as illustrated in Figure 13.8 enabled the researchers to capture long, high-quality packet traces in a dense environment free of major shadowing and with limited multipath effects.

The data quality of an anchor is defined as how well the signal measurements for the anchor match a signal propagation model. Since the measured signal strength from the anchors is used to perform localization, the RSS-to-distance propagation model is used in the paper [18, 26]:

$$P(d)[\text{dBm}] = P(d_0)[\text{dBm}] - 10\gamma \log_{10}\left(\frac{d}{d_0}\right) + X_\sigma, \qquad (13.31)$$

where $P(d_0)$ represents the transmitting power at the reference distance d_0; d is the distance between the wireless device and the anchor; γ is the path loss exponent; and X_σ is the shadow fading, which follows a zero-mean Gaussian distribution with σ standard deviation.

During the off-line training phase, RSS samples were collected at various known locations from each anchor in the given high-density environments, and distances were calculated from known locations to the anchors. For a certain anchor, the measured RSS readings and the corresponding distances were then used to fit the signal propagation model-based linear regression method [39]. The goodness of fit of the data is used to measure the data quality of the anchor. The goodness of fit is observable as the coefficient of determination or R^2 [40]. R^2 can take values from 0 to 1, with a value of 1 indicating a perfect fit to the model, and a value close to 0 indicating a poor fit. The R^2 value of each anchor is calculated by using all the RSS measurements from the anchor based on the linear regression method. Given the R^2

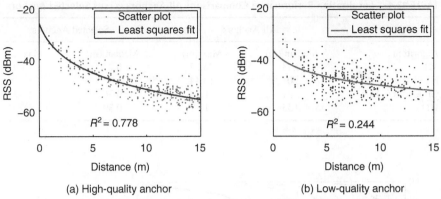

(a) High-quality anchor (b) Low-quality anchor

Figure 13.9 Plots representing the fit for the distance-to-RSS model for two different anchors: (a) anchor with good fit ($R^2 = 0.778$) and (b) anchor with poor fit ($R^2 = 0.2443$) [37].

TABLE 13.3. **Algorithms under Study to Perform Anchor Selection**

Algorithm	Abbreviation	Description
Nonlinear least squares [41]	NLS	Lateration-based algorithm NLS is used
Fingerprint matching [7] gridded RADAR	GR	Classification-based method using IMG Returns the nearest neighbor: $K = 1$ tile size in IMG: 2×2 in.
Highest probability [37]	H1	Probabilistic method using IMG Returns location with the highest probability tile size in IMG: 2×2 in.
Bayesian network [9]	BN	Statistical supervised learning Bayesian graphical model: M1

IMG, interpolated map grid.

values of all the anchors, the anchors with R^2 higher than a predefined threshold are chosen to perform localization.

Figure 13.9 plots the distance–RSS relationship together with the fitted propagation model for two different anchors using the same set of training points. It is observed that the quality of the fit in terms of R^2 differs significantly. Based on the anchor selection method, the anchors with a coefficient of determination of $R^2 \geq 0.5$ are selected. This includes 179 selected anchors in the ORBIT testbed environment with total 369 anchors.

Table 13.3 lists the group of localization algorithms studied and Table 13.4 compares the median and maximum errors with and without anchor selection. In order to have a closer look at how the anchor selection based on data quality affects the localization accuracy, Figure 13.10 plots the error CDF for BNs when varying

TABLE 13.4. Localization Performance Comparison: All Anchors versus Selected Anchors

	All Anchors		Selected Anchors	
Algorithm	Median (m)	Max (m)	Median (m)	Max (m)
M1	0.58	26.87	0.24	1.60
NLS	2.01	13.44	1.62	5.37
Gridded RADAR	0.31	1.74	0.36	1.97
H1	0.33	1.82	0.39	1.70

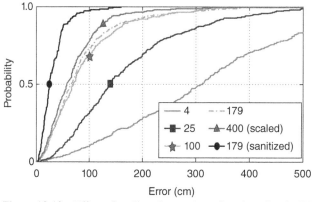

Figure 13.10 Effect of scaling the number of anchors for the BN algorithm (M1) [37].

the number of anchors from 4, 25, 100, 179 to 400. The key observations are the following:

1. RSS-based localization can achieve median errors as low as 24 cm, with a maximum error of 1.5 m. NLS has the worst performance for real RSS observations, with a mean error of 1.6 m and a maximum error of 5.4 m.

2. For lateration-based approaches, which use a signal-to-distance function, the quality of the RSS measurements is more important compared with the quantity of measurements. A subset of 179 landmarks whose data yield a good signal-to-distance fit provide the best localization performance. Simply increasing the number of receivers over this actually increased the median error from 24 to 58 cm.

3. Classification algorithms are qualitatively less sensitive to variances and noise in the input set than lateration-style algorithms. Given RSS measurements that deviate substantially from standard models, these algorithms maintained good average and worst-case performance. Given high-quality inputs, the results also agree with prior work [29], which confirms all algorithms have a similar performance, with the notable exception of NLS.

13.4 DISCUSSION AND CONCLUSION

In this chapter, we provided an overview of the studies that address the important problems of anchor placement and anchor selection for wireless localization. These approaches deviate from the traditional studies, which are focused on improving the localization algorithms themselves given a set of anchors. Instead, these methods achieve effective localization through optimizing the deployment of anchors or performing intelligent selection of the anchors. For anchor placement, we covered a heuristic search method, acute triangular-based deployment, adaptive beacon placement, and optimal placement solutions via the maxL–minE algorithm. In general, simple shapes such as equilateral triangles and squares resulted in placements with better localization performance, and the center of the placement shape achieved minimum localization error. For anchor selection, different methods were examined by considering different factors. In particular, strongest RSS-based selection considered the signal coverage; entropy-based information gain considered the discriminative power of anchors toward locations; and the convex hull-based selection considered the geometry of anchors. Finally, the data quality of the anchors in terms of fitting the propagation model was considered in a high-density environment during anchor selection.

REFERENCES

[1] P. Enge and P. Misra, *Global Positioning System: Signals, Measurements and Performance*. Lincoln, MA: Ganga-Jamuna Press, 2001.

[2] Z. Li, W. Trappe, Y. Zhang, and B. Nath, "Robust statistical methods for securing wireless localization in sensor networks," in *The Fourth Int. Conf. on Information Processing in Sensor Networks (IPSN)*, 2005, pp. 91–98.

[3] K. Langendoen and N. Reijers, "Distributed localization in wireless sensor networks: A quantitative comparison," *Comput. Netw.*, vol. 43, no. 4, pp. 499–518, 2003.

[4] D. Niculescu and B. Nath, "Ad hoc positioning system (APS)," in *Proc. of the IEEE Global Telecommunications Conf. (GLOBECOM)*, 2001, pp. 2926–2931.

[5] J. Yang, Y. Chen, V. Lawrence, and V. Swaminathan, "Robust wireless localization to attacks on access points," in *IEEE Sarnoff Symp. 2009*, Princeton, NJ, 2009.

[6] Y. Chen, J. Francisco, W. Trappe, and R. P. Martin, "A practical approach to landmark deployment for indoor localization," in *Proc. of the Third Annual IEEE Communications Society Conf. on Sensor, Mesh and Ad Hoc Communications and Networks (SECON)*, Sep. 2006.

[7] P. Bahl and V. N. Padmanabhan, "RADAR: An in-building RF-based user location and tracking system," in *Proc. of the IEEE International Conf. on Computer Communications (INFOCOM)*, Mar. 2000, pp. 775–784.

[8] N. Priyantha, A. Liu, H. Balakrishnan, and S. Teller, "The cricket compass for context-aware mobile applications," in *Proc. of the Seventh Annual ACM/IEEE International Conf. on Mobile Computing and Networking (Mobicom 2001)*, 2001, pp. 1–14.

[9] D. Madigan, E. Elnahrawy, R. P. Martin, W. Ju, P. Krishnan, and A. S. Krishnakumar, "Bayesian indoor positioning systems," in *Proc. of the IEEE International Conf. on Computer Communications (INFOCOM)*, Mar. 2005, pp. 324–331.

[10] A. Haeberlen, E. Flannery, A. M. Ladd, A. Rudys, D. S. Wallach, and L. E. Kavraki, "Practical robust localization over large-scale 802.11 wireless networks," in *MobiCom '04: Proc. of the 10th Annual International Conf. on Mobile Computing and Networking*, New York, 2004, pp. 70–84.

[11] D. Nicelescu and B. Nath, "DV based positioning in ad hoc networks," *Telecomm. Syst.*, vol. 22, no. 1–4, pp. 267–280, 2003.

[12] N. Patwari, J. N. Ash, S. Kyperountas, A. O. Hero III, R. L. Moses, and N. S. Correal, "Locating the nodes: Cooperative localization in wireless sensor networks," *IEEE Signal Process. Mag.* vol. 22, no. 4, pp. 54–69, 2005.

[13] T. He, C. Huang, B. Blum, J. A. Stankovic, and T. Abdelzaher, "Range-free localization schemes in large scale sensor networks," in *Proc. of the Ninth Annual ACM International Conf. on Mobile Computing and Networking (MobiCom'03)*, 2003, pp. 81–95.

[14] L. Doherty, K. S. J. Pister, and L. ElGhaoui, "Convex position estimation in wireless sensor networks," in *Proc. of the IEEE International Conf. on Computer Communications (INFOCOM)*, Apr. 2001, pp. 1655–1663.

[15] Y. Chen and H. Kobayashi, "Signal strength based indoor geolocation," in *Proc. of the IEEE International Conf. on Communications (ICC)*, Apr. 2002, pp. 436–439.

[16] A. S. Krishnakumar and P. Krishnan, "On the accuracy of signal strength-based location estimation techniques," in *Proc. of the IEEE Int. Conf. on Computer Communications (INFOCOM)*, Mar. 2005.

[17] R. Battiti, M. Brunato, and A. Delai, "Optimal wireless access point placement for location-dependent services," Tech. Rep. DIT-03-052, Department of Information and Communication Technology, University of Trento, Italy, Oct. 2003.

[18] T. K. Sarkar, Z. Ji, K. Kim, A. Medouri, and M. Salazar-Palma, "A survey of various propagation models for mobile communication," *IEEE Antennas Propagation Mag.* vol. 45, no. 3, pp. 51–82, 2003.

[19] A. Goldsmith, *Wireless Communications*. New York: Cambridge University Press, 2005.

[20] V. J. Rayward-Smith, I. H. Osman, C. R. Reeves, and G. D. Smith, *Modern Heuristic Search Methods*. John Wiley & Sons, 1996.

[21] R. Battiti, G. Tecchiolli, I. Nazionale, and F. Nucleare, "The reactive tabu search," 1993.

[22] S. Kirkpatrick, C. D. Gelatt Jr., and M. P. Vecchi, "Optimization by simulated annealing," *Science*, vol. 220, pp. 671–680, 1983.

[23] S. Biaz, Y. Ji, and P. Agrawal, "Impact of sniffer deployment on indoor localization," presented at the *2005 Int. Conf. on Collaborative Computing: Networking, Applications and Worksharing*, San Jose, CA, 2005.

[24] Y. Ji, S. Biaz, S. Wu, and B. Qi, "Optimal sniffers deployment on wireless indoor localization," in *Proc. of the 16th International Conf. on Computer Communications and Networks, ICCCN 2007*, pp. 13–16.

[25] N. Bulusu, J. Heidemann, and D. Estrin, "Adaptive beacon placement," in *21st Int. Conf. on Distributed Computing Systems*, Apr. 2001, pp. 489–498.

[26] T. S. Rappaport, *Wireless Communications: Principles and Practice*. Prentice Hall, 1996.

[27] N. Bulusu, J. Heidemann, and D. Estrin, "GPS-less low-cost outdoor localization for very small devices," *IEEE Pers. Commun. Mag.*, vol. 7, pp. 28–34, 2000.

[28] B. Noble and J. W. Daniel, *Applied Linear Algebra*. Englewood Cliffs, NJ: Prentice-Hall, 1988.

[29] E. Elnahrawy, X. Li, and R. P. Martin, "The limits of localization using signal strength: A comparative study," in *Proc. of the First IEEE Int. Conf. on Sensor and Ad Hoc Communcations and Networks (SECON 2004)*, Oct. 2004, pp. 406–414.

[30] S. Ghassemzadeh, L. J. Greenstein, A. Kavcic, T. Sveinsson, and V. Tarokh, "An empirical indoor path loss model for ultra-wideband channels," *J. Commun. Netw.*, vol. 5, pp. 303–308, 2003.

[31] H. Lim, L. Kung, J. Hou, and H. Luo, "Zero-configuration, robust indoor localization: Theory and experimentation," in *Proc. of the IEEE Int. Conf. on Computer Communications (INFOCOM)*, Mar. 2006.

[32] K. Kleisouris, Y. Chen, J. Yang, and R. Martin, "Empirical evaluation of wireless localization when using multiple antennas," *IEEE Trans. Parallel Distributed Syst.*, vol. 21, no. 11, pp. 1595–1610, 2010.

[33] E. Ermel, A. Fladenmuller, G. Pujolle, and A. Cotton, "On selecting nodes to improve estimated positions," in *Mobile and Wireless Communication Networks*, 2005, pp. 449–460.

[34] M. Youssef, A. Agrawal, and A. U. Shankar, "WLAN location determination via clustering and probability distributions," in *Proc. of the First IEEE International Conf. on Pervasive Computing and Communications (PerCom)*, Mar. 2003, pp. 143–150.

[35] Y. Chen, Q. Yang, J. Yin, and X. Chai, "Power-efficient access-point selection for indoor location estimation," *IEEE Trans. Knowl. Data Eng.*, vol. 18, no. 7, pp. 877–888, 2006.

[36] E. Ermel, A. Fladenmuller, G. Pujolle, and A. Cotton, "Estimation de positions dans des reseaux sans-fil hybrides," in *CFIP 2003*, 2003.

[37] G. Chandrasekaran, M. A. Ergin, J. Yang, S. Liu, Y. Chen, M. Gruteser, and R. P. Martin, "Empirical evaluation of the limits on localization using signal strength," in *6th Annual IEEE Communications Society Conf. on Sensor, Mesh and Ad Hoc Communications and Networks, SECON '09*, Jun. 2009, pp. 1–9.

[38] D. Raychaudhuri, I. Seskar, M. Ott, S. Ganu, K. Ramachandran, H. Kremo, R. Siracusa, H. Liu, and M. Singh, "Overview of the ORBIT radio grid testbed for evaluation of next-generation wireless network protocols," in *Proc. IEEE WCNC*, vol. 3, Mar. 2005, pp. 1664–1669.

[39] J. L. Buchanan and P. R. Turner, *Numerical Methods and Analysis*. New York: McGraw-Hill, 1992.

[40] N. J. D. Nagelkerke, "A note on a general definition of the coefficient of determination," *Biometrika*, vol. 78, no. 3, pp. 691–692, 1991.

[41] J. Yang and Y. Chen, "Indoor localization using improved RSS-based lateration methods," in *Proc. of the IEEE Global Telecommunications Conf. (GLOBECOM)*, 2009.

CHAPTER 14

KERNEL METHODS FOR RSS-BASED INDOOR LOCALIZATION

Piyush Agrawal and Neal Patwari
University of Utah, Salt Lake City, UT

THIS CHAPTER explores the features and advantages of kernel-based localization. Kernel methods simplify received signal strength (RSS)-based localization by providing a means to learn the complicated relationship between RSS measurement vector and position. We discuss their use in self-calibrating indoor localization systems. In this chapter, we review four kernel-based localization algorithms and present a common framework for their comparison. We show results from two simulations and from an extensive measurement data set, which provide a quantitative comparison and intuition into their differences. Results show that kernel methods can achieve a root mean square error (RMSE) up to 55% lower than a maximum likelihood estimator.

14.1 INTRODUCTION

Knowledge of a user's position is becoming increasingly important in applications that include medicine and health care [1], personalized information delivery [2, 3], and security. Indoor localization algorithms have been proposed using various methods, such as angle of arrival, time of flight, and RSS, of which RSS-based algorithms are the most common.

In general, existing RSS-based indoor localization algorithms can be classified into three main categories: (1) model-based algorithms, (2) kernel-based algorithms, and (3) RSS fingerprinting algorithms. Kernel-based algorithms, the subject of this chapter, are a "middle ground" between model-based and RSS fingerprinting algorithms.

Model-based algorithms [4–7] use standard statistical channel models to provide a functional relationship between distance and RSS. Using this functional

Handbook of Position Location: Theory, Practice, and Advances, Second Edition.
Edited by S. A. (Reza) Zekavat and R. Michael Buehrer.
© 2019 by the Institute of Electrical and Electronics Engineers, Inc.
Published 2019 by John Wiley & Sons, Inc.
Companion Website: www.wiley.com/go/zekavat/positionlocation2e

relationship, the location of a tag (unknown location device) is estimated from the RSS measured by in-range access points (APs) or anchors (known location devices) by first estimating the distances to the in-range APs using models and then using methods of lateration to determine the coordinates. Some research [8–10] also proposes using statistical models to create an entire radio map as a function of position, in which the location of the tag is estimated directly from the RSS measured by the in-range APs.

RSS fingerprinting methods [11, 12], on the other hand, work in two phases—an *off-line* training phase and an *online* estimation phase. In the off-line training phase, *radio frequency (RF) signatures* are collected at some known locations in the deployment region, which are then stored in a database. An RF signature is a vector of RSS values measured by some predetermined APs. In the online estimation phase, a location is searched from the constructed database whose RF signature matches closely with the RF signature of the tag.

Statistical channel models, in most cases, are unable to capture the complicated relationship between RSS and location in indoor environments. They also typically assume that shadow fading on links is mutually independent, even though environmental obstructions cause similar shadowing effects to many links that pass through them, an effect called correlated shadowing [13, 14].

RSS fingerprinting methods, on the other hand, do not assume any prior relationship between RSS and position, but the training phase consumes a significant amount of time and effort [11, 15]. To some extent, the training effort can be reduced via spatial smoothing [15–17], but this is possible only to distances at which the RSS is correlated. Some research has also suggested supplementing some of the measurements using predicted RSS using channel models [11]. Changes in the environment over time reduce the accuracy of the database, requiring recalibration.

In summary, model-based algorithms require the least training effort, but they rely heavily on the prior knowledge of the relationship between RSS and position. RSS fingerprinting algorithms, on the other hand, are not based on any prior knowledge of the relationship between RSS and position but require considerable training effort and time.

This chapter is an exploration of kernel-based algorithms, which provide the ability to mix the features of both model-based and RSS fingerprinting algorithms. Kernel-based algorithms encapsulate the complicated relationship between RSS and position, along with correlation in the RSS at proximate locations, in a *kernel*, which can be assumed as a parametrized "black box" that takes the measured RSS as inputs and gives a *measure* of position as output. In this chapter, we describe four different kernel-based RSS localization algorithms using a common mathematical framework and compare and contrast their performance (to each other and to a baseline model-based algorithm, the maximum likelihood estimator) using a simulation example and using an extensive experimental study. These algorithms include LANDMARC [18], Gaussian kernel localization [19], radial basis function localization [15], and linear signal-distance map localization [20].

The experimental study described in this chapter demonstrates that all four of the kernel-based localization algorithms outperform the maximum likelihood coordinate estimation (MLE) in a real-world environment. In fact, the improvement in average RMSE is as high as 55% compared to the MLE. In this chapter, we explain this improved performance of the kernel-based algorithms by using several numerical and simulation examples, in which kernel methods are shown to enable the tag's coordinate estimates to be robust to both shadowing and independent and identically distributed (i.i.d.) fading. The experimental evaluation also suggests that the complexities of the fading environment and the complicated nature of the large-scale deployment require more parameters than are available to typical model-based algorithms. In particular, in this chapter, we attempt to explain why the kernel-based algorithms perform better than model-based localization algorithms.

Standard kernel-based algorithms still require a training phase for calibration of *kernel parameters*. In this chapter, we discuss methods to minimize the calibration requirements of kernel-based algorithms by performing training simultaneously while the system is online, using pairwise measurements between APs. Specifically, several APs are deployed at some known locations throughout the building. Each AP is a transceiver and can measure the RSS of packets from other APs (although we note that we do not limit ourselves to Wi-Fi APs; we may use any standard, which allows peer-to-peer communication). These pairwise measurements constitute the training data for calibration purposes.

14.1.1 Outline of the Chapter

Prior to discussing the four kernel methods for RSS-based localization algorithms, we present a common mathematical framework for kernel-based localization algorithms in Section 14.2.2. The remainder of Section 14.2 discusses four kernel-based algorithms. To provide more intuitive understanding of the advantages of kernel methods, we present a simple numerical example in Section 14.3. In Section 14.4, we evaluate the algorithms using kernel methods on a real-world measurement data set collected in a hospital environment. Finally, we conclude this chapter in Section 14.5.

14.2 KERNEL METHODS

Kernel methods are a class of statistical learning algorithms in which the complicated relationship between the input (e.g., signal strength) and the output (e.g., physical coordinates) is encapsulated using kernel functions. A kernel function is a potentially nonlinear and parameterized function of input variables. The parameters control the functional dependencies between input and output, in our case, between signal strength and physical coordinates. A key feature of statistical learning is that it estimates the parameters based on some known input/output pairs, also called *learning* from known data. Models using kernel methods are typically *linear* with respect to the parameters, which gives them simple analytical properties, yet are nonlinear with respect to the input variables, for example, RSS.

In this section, we present an overview of coordinate estimation using statistical learning with kernel methods. We begin our discussion in this section by defining our problem statement and then proceed to present a general mathematical framework for coordinate estimation using kernel methods.

14.2.1   Problem Statement

In this chapter, we consider signal strength-based tag localization. Specifically, we wish to find a two-dimensional tag coordinate, $\hat{x}_t$, given the known two-dimensional reference coordinates of N APs, x_i, $\forall_i \in \{1, ..., N\}$, their pairwise RSS measurements, $s_{i,j}$, $\forall_i \neq j$, $i, j \in \{1, ..., N\}$, and the RSS measured by N reference APs for a signal transmitted by a tag, $s_{i,t}$, $\forall_i \in \{1, ..., N\}$. Also, let notation s_j indicate the RSS vector for AP j, where $s_j = [s_{1,j}, ..., s_{N,j}]^T$. Similarly, let notation s_t indicate the RSS vector for a tag, t, where $s_t = [s_{1,t}, ..., s_{N,t}]^T$.

Note that even though we consider a two-dimensional coordinate estimation here, the same methodology can readily be extended to a three-dimensional case. Before we proceed further, we clarify our notation for the signal strength $s_{i,j}$. A measurement, $s_{i,j}$, represents the decibel signal strength measured by AP i for the signal transmitted by AP j. Similarly, subscript t indicates that the measurement is for a tag (with an a priori unknown location).

The measurement $s_{i,i}$, which corresponds to the RSS measured by colocated APs, is unavailable. In practice, even if two APs are located at the same position, the RSS measured between them is nonzero, that is, $s_{i,i} \neq 0$, and depends on the transmit power of the APs [20]. Some localization algorithms require the value of $s_{i,i}$ to be known; thus, in this chapter, we assume when necessary that $s_{i,i} = -33\,\text{dBm}$ [20].

We do not assume full connectivity between links. Consequently, we define set $H(j)$ to be the set of APs that are in direct communication range of AP j. Set $H(j)$ does not include the AP j and $H(j) \subset \{1, ..., N\}$. Similarly, $H(t)$ is the set of APs that are in direct communication range of tag t.

An AP k that is not in the set $H(j)$ is *not* in the direct communication range of AP j and would not measure any RSS from AP j. It does not necessarily mean that AP k does not receive any signal from AP j. Rather, it simply means that the signal power from AP j was so low that AP k could not demodulate its signal. This "nonmeasurement" of RSS by AP k is known as the "censored data" problem in statistics. We know this RSS value is low, but we do not know the value of $s_{k,j}$. How should an algorithm represent $s_{k,j}$ for $k \notin H(j)$ in its RSS vector s_j? Most kernel-based approaches have not addressed this censored data issue and have simply assumed full connectivity between APs. One algorithm estimates the nonmeasured RSS values using expectation maximization [21]. In this article, we will provide for each kernel method a means to address nonmeasured RSS.

14.2.2   General Mathematical Formulation

In the framework of kernel methods, a function of the coordinate estimate of a tag, $f(\hat{x}_t)$, can be expressed as

$$f(\hat{x}_t) = \sum_{i \in \bar{H}(t)} \alpha_i \phi_i(s_t) + \alpha_0, \tag{14.1}$$

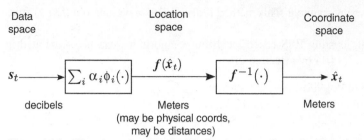

Figure 14.1   Flowchart showing the localization operation using kernel methods.

where $\alpha_i, \forall i \in \tilde{H}(t)$ is the *coordinate weight* of AP i, $\tilde{H}(t)$ denotes the set of APs that contribute to the kernel, and $\phi_i(\cdot)$ is known as the *kernel function* corresponding to AP i. The parameter α_0 is known as the *bias* parameter, which compensates for any fixed offset in the data [22]. In this section, we will show how the parameters $\{\alpha_i\}_{i \in \tilde{H}(t)}$ and α_0 are optimized and estimated, and how the kernel functions $\{\phi_i(\cdot)\}_{i \in \tilde{H}(t)}$ are chosen for different algorithms and techniques in the literature.

The parameters $\{\alpha_i\}_{i \in \tilde{H}(t)}$ are sometimes called "weights," in particular, when predetermined functions are used as kernel functions $\phi_i(\cdot)$, for example, Gaussian functions. However, these parameters represent the coordinates of the APs in "location space." Typically, these parameters are functions of the AP coordinates and are optimized to match the information given by the AP pairwise RSS measurements and their coordinates. The coordinate estimate of the tag $\hat{x}_t$ is determined by taking the inverse f^{-1} of $f(\hat{x}_t)$. Figure 14.1 shows the operation of coordinate estimation using kernel methods.

Algorithms in the class of kernel-based localization differ in the methods of optimization of $f(\hat{x}_t)$. Some algorithms set the kernel functions with predetermined functions and optimize the parameters $\{\alpha_i\}_{i \in \tilde{H}(t)}$ based on pairwise RSS measurements [15, 20, 23]. In contrast, other algorithms set the parameters $\{\alpha_i\}_{i \in \tilde{H}(t)}$ with some functions of the physical coordinates of a set of APs and optimize the kernel functions using their pairwise RSS measurements [18, 19].

Determination of Kernel Parameters: Typically, the kernel functions $\phi_i(\cdot)$ belong to a class of parametric nonlinear functions. The determination of kernel parameters is not a trivial task and has been extensively studied in the statistical learning literature [24]. A common technique used for their estimation is cross validation [25]. For the purposes of cross validation of localization algorithms, we use the data set collected between APs. In this case, the AP measurement data set is divided into two groups, one group containing $(N - 1)$ APs and the other group containing one "left-out" AP, where N is the total number of APs. Thus, there are N ways of dividing the data set. In cross validation, we estimate the location of the left-out AP as if its coordinate was unknown. The location error can be determined after coordinate estimation because every AP coordinate is, in fact, known. The average location error is computed by averaging overall left-out APs. The location error is a function of the kernel parameters. By repeating this procedure across a range of candidate values of the parameters, we can optimize the kernel parameters for the particular environment. This method is also called leave-one-out (LOO) cross validation.

In general, existing RSS-based localization algorithms can be formulated in the framework of (14.1) by selection of

- the function of coordinate estimates, $f(\cdot)$;
- a set of APs that contribute to the kernel, $\tilde{H}(t)$;
- coordinate weights, $\{\alpha_i\}_{i \in \tilde{H}_t}$;
- kernel functions, $\phi_i(\cdot)$; and
- the bias parameter, α_0.

In the remainder of this section, we show how the mathematical framework of (14.1) can be applied to different positioning algorithms. In particular, we select four different algorithms from the RSS-based localization literature and show how the developed framework is applied for each algorithm.

Example Framework: Consider an example of a wireless network with four APs deployed at known locations x_i, $\forall i \in \{1, 2, 3, 4\}$ as shown in Figure 14.2. Also, consider a tag whose actual location (in meters) is $x_t = [3, 2]^T$. The coordinates (in meters) of the four APs are $x_1 = [0.5, 0.5]$, $x_2 = [0.5, 3.5]$, $x_3 = [3.5, 3.5]$, and $x_4 = [3.5, 0.5]$. Let us assume that all the APs are in direct communication range of the other APs and the tag; that is, $|H(t)| = |H(j)| = 4$, $\forall j \in \{1, 2, 3, 4\}$. Using the known locations of the APs, their pairwise RSS measurements are generated using a log-distance path loss model. A brief description of the log-distance path loss model is given in Section 14.3. An instance of these pairwise RSS measurements is tabulated in Table 14.1. Each row of Table 14.1 represents the RSS measured by the four APs for the signal transmitted by the corresponding device. For example, the

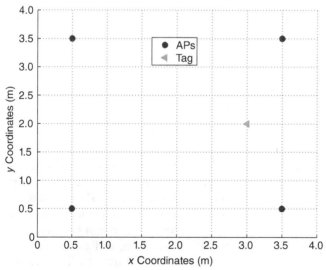

Figure 14.2   Position of APs and tags.

TABLE 14.1. Example RSS Values (in dBm) Measured by APs

	Receiver APs			
	AP 1	AP 2	AP 3	AP 4
Tx AP 1	n/a	−72	−62	−81
Tx AP 2	−70	n/a	−59	−85
Tx AP 3	−60	−65	n/a	−63
Tx AP 4	−59	−77	−72	n/a
Tag	−72	−69	−61	−60

n/a, not applicable.

TABLE 14.2. Lognormal Path Loss Parameter Description and Values Used in the Running Example Framework

Parameter	Description	Value
n_p	Path loss exponent	4.0
σ_{dB}	Fading standard deviation	6.0 dB
Π_0	Reference received power (at 1 m)	−50 dBm

TABLE 14.3. Table Summarizing the Similarities and Dissimilarities of LANDMARC (LM), Gaussian Kernel (GK), Radial Basis Function (RBF), and Linear Signal-Distance Map (SDM) Localization Algorithms

	LM	GK	RBF	SDM
$f(\hat{x}_t)$	$\hat{x}_t$	$\hat{x}_t$	$\hat{x}_t$	$\hat{\delta}_t$
α_i	Actual AP coordinates, x_i	Actual AP coordinates, x_i	Optimized coordinates in "location space"	Optimized coordinates in "data space"
$\phi_i(s_t)$	Unitless weight determined by (14.2)	Unitless weight determined by (14.5)	Unitless weight determined by (14.11)	RSS (dBm) from tag to AP i
α_0	**0**	**0**	Centroid of all deployed APs	**0**
$\tilde{H}(t)$ length of s_t	Top k APs N	All APs d using (14.8)	All APs length N	All in-range APs length N

RSS values in the first row represents the RSS measured by the deployed APs when AP1 was transmitting. Similarly, the APs' RSS measurements for the tag are generated, an instance of which is tabulated in the last row of Table 14.1.

For the purpose of this example, the values of various parameters are tabulated in Table 14.2.

The goal in this example is to estimate the location of the tag, $\hat{x}_t$, using (1) the known location of four APs, $\{x_1, x_2, x_3, x_4\}$; (2) their pairwise RSS measurements $s_{i,j}$, $\forall i \neq j$, $i, j \in \{1, 2, 3, 4\}$, tabulated in Tables 14.1 and 14.3; (3) and the RSS measured by the four APs for a signal transited by the tag, $s_{i,t}$, $\forall_i \in \{1, 2, 3, 4\}$, tabulated in the last row of Table 14.1.

We will revisit this example of a wireless network throughout this section when we consider each localization algorithm in detail.

14.2.3 LANDMARC Algorithm

LANDMARC [18] is an RSS-based localization algorithm in which a tag's coordinate estimate is given by a weighted average of the coordinates of k-closest APs that can hear the tag's transmission. In this section, we present how the LANDMARC algorithm can be expressed as a kernel method. Because LANDMARC is intuitive and can be explained with a single weighted average, it helps to demonstrate the concepts of kernel algorithms in an intuitive manner and shows how simple a kernel-based algorithm can be.

In LANDMARC, a tag's estimated coordinate is written using $f(\hat{x}_t) = \hat{x}_t$, $\alpha_i = x_i$, $\alpha_0 = \mathbf{0}$, and

$$\phi_i(s_t) = \frac{1/\|s_t - s_i\|^2}{\sum_{j \in \tilde{H}(t)} 1/\|s_t - s_j\|^2}. \tag{14.2}$$

Applying these relations in (14.1), we have

$$\hat{x}_t = \sum_{i \in \tilde{H}(t)} x_i \frac{1/\|s_t - s_i\|^2}{\sum_{j \in \tilde{H}(t)} 1/\|s_t - s_j\|^2}. \tag{14.3}$$

Here, $\tilde{H}(t)$ is the set of k APs that are closest to the tag. In LANDMARC, "closeness" is quantified by the Euclidean distance between the RSS vector of AP i, s_i, and the RSS vector of the tag, s_t; that is, $E_i = \|s_t - s_i\|$. We define the vector E as

$$E = [E_1, \ldots, E_N]^T. \tag{14.4}$$

The set of APs $\tilde{H}(t)$ in (14.2) and (14.3) is the set of k AP indices with the k-smallest E_i in the vector E. In LANDMARC, k is a variable parameter that determines the number of APs that contribute in the kernel. Any nonmeasured RSS in vectors s_t or s_i, as discussed in Section 14.2.1, is replaced by the minimum RSS observed over the duration of the experiment minus one.

Estimation of Parameters: The only parameter that needs to be estimated in LANDMARC is the set of APs that contribute in the kernel, $\tilde{H}(t)$, which in turn depends on the parameter k. If $k = 1$, then we choose the AP that is "closest" (smallest E_i) to the tag as the coordinate estimate of the tag. Similarly, if $k = 2$, then the two closest APs are considered and parameters are determined using (14.2). Unfortunately, there is no analytical solution for the optimal value of k, although it can be determined experimentally for a given environment or by using the cross-validation approach described in Section 14.2.2.

The above argument implicitly assumes that the optimal value of k is less than the number of neighbors of a tag, that is, $k < |H(t)|$, where $H(t)$ denotes the set of one-hop neighboring APs that hear the transmission from the tag. A question that arises is what would be the set $\tilde{H}(t)$ when k is greater than the number of one-hop APs, $H(t)$. A naive approach for this situation would be to place an upper threshold on the value of k. In other words, $k' = \min(k, |H(t)|)$, where k' is the actual number of neighbors used for a particular tag, t.

Example 14.1   (Revisit Example Framework)

Consider the wireless network of Figure 14.2. Estimate the tag's coordinate, $\hat{x}_t$, using the LANDMARC algorithm.

Solution

Let us assume $k = 3$. Using (14.4) and the definition of E_i, we can compute the Euclidean distance vector (in dBm) as

$$E = [44.36, 43.86, 30.71, 33.58]^T.$$

The set of APs $\tilde{H}(t)$ in (14.2) and (14.3) is the set of three AP indices with the smallest E_i in vector E. Thus, $\tilde{H}(t) = \{2, 3, 4\}$. The values of the kernel function, $\phi_i(\cdot)$, corresponding to each AP $i \in \tilde{H}(t)$, is computed using (14.2):

$$\phi_2(s_t) = 0.21, \quad \phi_3(s_t) = 0.43, \quad \phi_4(s_t) = 0.36.$$

Using these kernel values and the known AP coordinates $\{x_2, x_3, x_4\}$ in (14.3), the LANDMARC coordinate estimate of the tag $\hat{x}_t$ (in meters) is computed as

$$\hat{x}_t = [2.9, 2.4]^T.$$

Compared to the actual tag location, the LANDMARC coordinate estimate has an error of 0.44 m.

14.2.4   Gaussian Kernel Localization Algorithm

The Gaussian kernel localization algorithm is an RSS-based localization algorithm proposed by Kushki et al. [19]. Similar to LANDMARC, in the Gaussian kernel localization algorithm, the coordinate estimate of a tag is the weighted average of coordinates of the closest APs. However, the weights are determined by a Gaussian kernel, which gives a measure of distance between the RSS vector of the tag and the RSS vectors of the APs. Any stationary kernel function could be used in place of the Gaussian kernel to determine the weights. The authors chose to use the Gaussian kernel because it is widely used and studied in the literature. As presented by the authors, this algorithm uses the AP measurements over time, for time

$\tau = 1, \ldots, T$, denoted by $s_i^{(\tau)}$. In this section, we briefly describe the various aspects of this algorithm and how it fits into the framework developed in Section 14.2.2.

In the Gaussian kernel positioning algorithm, the coordinate estimate of the tag $\hat{x}_t$ is written as in (14.1) with $f(\hat{x}_t) = \hat{x}_t$, $\alpha_i = x_i$, $\alpha_0 = 0$, $\tilde{H}(t) = H(t)$, and

$$\phi_i(s_t) = \frac{1}{T(\sqrt{2\pi\sigma^2})^d} \sum_{\tau=1}^{T} \exp\left(\frac{-\|\tilde{s}_t - \tilde{s}_i^{(\tau)}\|^2}{2\sigma^2}\right), \tag{14.5}$$

where σ is a parameter called "width" of the kernel. The notation $\tilde{s}_t$ represents the "reduced" s_t, the RSS vector of the tag, and is given as $\tilde{s}_t = [s_{t,i_1}, \ldots, s_{t,i_d}]^T$, where $i_1, \ldots, i_d$ is a list of the elements in $H_d(t)$, a set of d predetermined APs. Similarly, $\tilde{s}_i^{(\tau)}$ represents the reduced RSS vector for AP i at a particular time instant, τ, $\forall \tau \in \{1, \ldots, T\}$. The estimation of the AP set $H_d(t)$ will be discussed next in the estimation of parameters section.

Simplifying, Equation (14.1) reduces to

$$\hat{x}_t = \sum_{i \in H(t)} x_i \phi_i(s_t). \tag{14.6}$$

The kernel functions $\{\phi_i (\ldots)\}$ of (14.5) may be normalized [19]. Normalization avoids the situation where there are "holes" in the RSS space, regions of $\tilde{s}_t$ where the sum in (14.5) has a low value. This would lead to fewer predictions at those regions of the RSS space [26]. In other words, normalization makes sure that the resulting kernel covers the whole range of RSS values measured by APs. The normalized kernel functions, $\phi_i(s_t)$, can be represented as

$$\phi_i(s_t) = \frac{1}{C}\left[\frac{1}{T(\sqrt{2\pi\sigma^2})^d} \sum_{\tau=1}^{T} \exp\left(\frac{-\|\tilde{s}_t - \tilde{s}_i^{(\tau)}\|^2}{2\sigma^2}\right)\right], \tag{14.7}$$

where

$$C = \sum_{i \in H(t)} \phi_i(s_t).$$

Estimation of Parameters: The Gaussian kernel localization algorithm requires estimation of the following parameters:

1. width of the kernel, σ, and
2. a predetermined set of d APs, $H_d(t)$.

Neighboring APs would report correlated RSS measurements. It is suggested that including all neighbors leads to redundancy as well as biased estimates [19]. One can minimize this effect by selecting a subset of d APs, $H_d(t)$, from the set of neighboring APs, $H(t)$, which have minimum redundancy. This set $H_d(t)$ is found to be the set that has minimum divergence. Let $V_{i,j}$ denote a vector time series of RSS measured by AP i for a signal transmited by AP j; that is, $V_{i,j} = [s_{i,j}^{(1)}, \ldots, s_{i,j}^{(T)}]$. The set of d APs, $H_d(t)$, is selected, such that

$$H_d(t) = \frac{\arg\min}{H_p(t) \subset H(t) : |H_p(t)| = d} \sum_{h_i, h_j \in H_p(t)} |h_i - h_j|, \tag{14.8}$$

where

$$|h_i - h_j| = \min_k \Delta(V_{i,k} - V_{j,k}),$$

where $\Delta(V_{i,k} - V_{j,k})$ is the *divergence* between two time series, $V_{i,k}$ and $V_{j,k}$, and is given by [19]:

$$\Delta(V_{i,k} - V_{j,k}) = \frac{1}{8}(\mu_{i,k} - \mu_{j,k})^2 \left[\frac{\sigma_{i,k} + \sigma_{j,k}}{2}\right]^{-1} + \frac{1}{2}\log\frac{0.5(\sigma_{i,k}^2 + \sigma_{j,k}^2)}{\sigma_{i,k}\sigma_{j,k}}, \tag{14.9}$$

where $\mu_{i,k}$ and $\mu_{j,k}$ are the means and $\sigma_{i,k}$ and $\sigma_{j,k}$ are the standard deviations for the time series $V_{i,k}$ and $V_{j,k}$, respectively. Although other divergence measures exist in the literature, the divergence represented in (14.9) is commonly used when the distribution of the time series is Gaussian.

Another parameter to be determined is σ, the kernel width parameter. Note that the kernel width parameter σ is a global parameter that depends on the pairwise measurements of the APs and is independent of the RSS measurement of the tag. In our evaluation in Section 14.4, σ is determined using the cross-validation approach described in Section 14.2.2.

Example 14.2   (Revisit Example Framework)

Let us return to the example wireless network of Figure 14.2. Estimate the tag's coodinate, $\hat{x}_t$, using the Gaussian kernel localization algorithm.

Solution

Note that the Gaussian kernel localization algorithm uses the AP measurements over time, which is used to determine the subset of APs that have minimum redundancy in their measured RSS. Our example framework cannot provide information about this redundancy because of the pairwise i.i.d. nature of the RSS simulation model used in this example. For the purposes of this example, let us assume the set $H_d(t) = \{1, 2, 4\}$.

The next parameter we need to determine is the width of the kernel, σ, which is estimated using the cross-validation approach described in Section 14.2.2. Based on the experimental evaluation in Section 14.4, the value of the kernel width is found to be $\sigma = 30\,dB$.

Using the RSS values from Table 14.1 along with $H_d = \{1, 2, 4\}$, $\sigma = 30\,dB$, and $T = 1$ in (14.7), the values of normalized kernel functions, $\phi_i(\ldots)$, corresponding to AP i, $\forall_i \in \{1, 2, 3, 4\}$, are determined to be

$$\phi_1(s_t) = 0.16, \quad \phi_2(s_t) = 0.16, \quad \phi_3(s_t) = 0.42, \quad \phi_4(s_t) = 0.27.$$

Using these values of the kernel functions and their corresponding coordinates in (14.6), the estimated tag coordinate, $\hat{x}_t$, (in meters) is

$$\hat{x}_t = [2.6, 2.2]^T.$$

Compared to the actual tag coordinate, the Gaussian kernel-based coordinate estimate has an error of 0.5 m.

14.2.5   Radial Basis Function-Based Localization Algorithm

In the statistical learning literature, radial basis functions have been widely used as the kernel functions $\phi_i(\cdot)$. Radial basis functions were introduced for the purpose of *exact* function interpolation. Given a set of training inputs and their corresponding outputs, the purpose of radial basis function interpolation is to create a smooth function that fits training data exactly [22].

Functions of this class have a property that the basis functions depend only on the *radial distance* (typically Euclidean) from a center, μ_i, such that

$$\phi_i(s_t) = h(\|s_t - \mu_i\|),$$

where $h(\cdot)$ is a radial basis function. Typically, the number of basis functions and the position of their centers, μ_i, are based on the input data set $\{s_i\}_{i=1,\ldots,N}$. A straightforward approach is to create a radial basis function centered at every $\{s_i\}_{i=1,\ldots,N}$.

In the context of tag localization, the use of radial basis functions was proposed by Krumm and Platt [15]. Primarily, they introduced interpolation using radial basis functions as a way to reduce the calibration effort of RADAR positioning [11] at the same time maintaining an acceptable location error. Specifically, the authors construct an interpolation function using radial basis functions that give the location of a tag t, $\hat{x}_t$, as a function of its RSS vector s_t.

Using the same notation as in Section 14.2.1, the coordinate estimate of a tag, $\hat{x}_t$, using radial basis functions, can be written in our common framework using $f(\hat{x}_t) = \hat{x}_t$, $\tilde{H}(t) = 1, \ldots N$, along with

$$\alpha_0 = \frac{1}{N} \sum_{j=1}^{N} x_j \tag{14.10}$$

and

$$\phi_i(s_t) = \exp\left(\frac{-\|s_t - s_i\|^2}{2\sigma_{\text{RBF}}^2}\right), \tag{14.11}$$

where N denotes the total number of deployed APs and σ_{RBF} is the width of the kernel. Applying these relations in (14.1), we have

$$\hat{x}_t = \sum_{i=1}^{N} \alpha_i \phi_i(s_t) + \alpha_0. \tag{14.12}$$

The parameters $\{\alpha_i\}_{i \in \{1,...,N\}}$ are estimated. Unlike LANDMARC and the Gaussian kernel positioning algorithm, in which $\alpha_i = x_i$ is the *actual* AP location, in the radial basis function localization algorithm, the parameters $\{\alpha_i\}_{i \in \{1,...,N\}}$ are "artificial" coordinates for each AP set such that $\hat{x}_t$ in (14.12) minimizes the location error for all training measurements.

Estimation of Parameters: There are two parameters that need to be estimated in the radial basis function-based localization algorithm:

- coordinate weights, $\{\alpha_i\}_{i \in \{1,...,N\}}$, and
- width of the radial basis function kernel, σ_{RBF}.

We begin with the estimation and optimization of coordinate weights $\{\alpha_i\}_{i \in \{1,...,N\}}$. Specifically, the coordinate weights $\{\alpha_i\}_{i \in \{1,...,N\}}$ are the coordinates in location space. There is no need to make them equal to the coordinates of APs. In radial basis function localization, these parameters are optimized such that the information given by the AP pairwise RSS measurements best matches the known AP locations. Substituting the AP pairwise RSS measurements and their corresponding coordinates in (14.12) and expressing them in matrix, we get

$$Z = \Phi A, \tag{14.13}$$

where $\mathbf{A}$ is the coordinate weight matrix $\mathbf{A} = [\alpha_1, ..., \alpha_N]^T$, and Z is a matrix whose *i*th row, z_i, is given as

$$z_i = [x_i - \alpha_0]^T,$$

and Φ is the kernel design matrix whose i, j element, $\Phi(i, j)$, is given as

$$\Phi(i, j) = \phi_i(s_j) = \exp\left(\frac{-\|s_j - s_i\|^2}{2\sigma_{RBF}^2} \right). \tag{14.14}$$

An optimal solution can be found by using the method of *least squares* [27]. Within the framework of least squares, the coordinate weight matrix $\mathbf{A} = [\alpha_1, ..., \alpha_N]^T$ can be estimated as

$$\mathbf{A} = \left(\Phi^T \Phi \right)^{-1} \Phi^T Z. \tag{14.15}$$

The term

$$\Phi^\dagger = \left(\Phi^T \Phi \right)^{-1} \Phi^T$$

in (14.15) is known as the *pseudoinverse* of the matrix Φ. The psuedoinverse is a generalized matrix inverse for nonsquare matrices [28].

The other parameter that needs to be estimated is the radial basis function kernel width, σ_{RBF}. Estimation of this parameter, similar to the estimation of kernel width for the Gaussian kernel position algorithm described in Section 14.2.4, is done via cross validation.

Example 14.3   (Revisit Example Framework)

Consider the wireless network of Figure 14.2. In this example, we will estimate the tag's coordinate, $\hat{x}_t$, using the radial basis function-based localization algorithm.

Solution

The first step is to estimate the value of parameter σ_{RBF}, which is estimated using cross validation as explained in Section 14.2.2. Based on the experimental evaluation in Section 14.4, the value (in decibel) of kernel width is found to be $\sigma_{\text{RBF}} = 30\,\text{dB}$.

Using the AP pairwise RSS from Table 14.1, the kernel design matrix Φ is determined using (14.14) as

$$
\Phi = \begin{bmatrix}
1 & 0.20 & 0.35 & 0.18 \\
0.20 & 1 & 0.27 & 0.06 \\
0.35 & 0.27 & 1 & 0.24 \\
0.18 & 0.06 & 0.24 & 1
\end{bmatrix}.
$$

The next step is to determine the coordinate weight matrix A using (14.15). In our example, $\alpha_0 = [2, 2]^T$, using (14.10). The coordinate weight matrix A is determined to be

$$
A = \begin{bmatrix}
-2.24 & -2.28 \\
-1.81 & 1.43 \\
2.43 & 2.32 \\
1.43 & -1.74
\end{bmatrix},
$$

in which row i corresponds to the coordinates of the APs i in location space, α_i^T. Using the estimated values of $\{\alpha_1, \alpha_2, \alpha_3, \alpha_4\}$, the coordinate estimate of the tag according to radial basis function-based localization algorithm is computed from (14.12) as

$$
\hat{x}_t = [2.8, 2.1]^T.
$$

Compared to the actual tag location, the radial basis function-based coordinate estimate has an error of 0.25 m.

14.2.6   Linear Signal-Distance Map Localization Algorithm

The linear signal-distance map localization algorithm differs from the kernel-based localization algorithms discussed so far, in which the physical coordinate of a device (tag/AP) is "directly" expressed as a weighted nonlinear function of the RSS vector s_t. In other words, the function of the coordinate estimate $f(\hat{x}_t)$ in (14.1) was the coordinate estimate itself; that is, $f(\hat{x}_t) = \hat{x}_t$. In the linear signal-distance map algorithm, $f(\hat{x}_t)$ is the log of the distance between $\hat{x}_t$ and each AP in $H(t)$; that is,

$$
f(\hat{x}_t) = [\log\|\hat{x}_t - x_{i_1}\|, \ \ldots, \ \log\|\hat{x}_t - x_{i_n}\|]^T,
$$

where $\{i_i, ..., i_n\} = H(t)$ [20]. In this algorithm, we first estimate $f(\hat{x}_t)$ and then use the estimated $f(\hat{x}_t)$ and the known coordinates of the APs to position the tag using methods like multilateration. Specifically, multilateration can be viewed as the inverse f^{-1} function. In this section, we present a mathematical formulation of this algorithm in the framework of kernel methods developed in Section 14.2.2.

A key aspect of this localization algorithm is determing the relationship between the pairwise RSS measurements between the APs and their geographical distances. Let the log-distance estimate between a tag, t, and AP, k, be represented by $\delta_{k,t}$, such that the estimated log-distance vector of the tag to the in-range APs, $H(t)$, is represented by $\delta_t = [\delta_{1,t}, ..., \delta|_{H(t)|,t}]$. Following the same notation of Section 14.2.1 and using (14.1), the estimated log-distance vector of the tag $\hat{\delta}_t$ is given by

$$f(\hat{x}_t) = \hat{\delta}_t = \sum_{i=1}^{N} \alpha_i \phi_i(s_t), \tag{14.16}$$

where $\alpha_I \in \mathbb{R}^{|H(t)|}$, N denotes the total number of deployed APs, $\tilde{H}(t) = \{1, ..., N\}$ and

$$\alpha_0 = \mathbf{0},$$
$$\phi_i(s_t) = e_i^T s_t, \tag{14.17}$$

where e_i is a column vector whose jth element, $e_i(j)$, is given as

$$e_i(j) = \begin{cases} 1, & \text{if } i = j \\ 0, & \text{otherwise} \end{cases}. \tag{14.18}$$

Note that α_i is a column vector of the same length as δ_t. Once the log distance to every known location AP in the set $H(t)$ is estimated, the coordinate of the tag can be determined using techniques like multilateration.

From (14.16), we can observe that the log-distance estimate between a tag and AP is a linear function of the raw RSS. The basis for this linearity comes from existing radio propagation models such as the log-distance path loss model [29]. A typical log-distance path loss model represents the received power P_r at a distance d from the transmitter as

$$P_r = \Pi_0 - 10n \log_{10} d.$$

So, one might represent the log distance as

$$\log_{10} d = \frac{\Pi_0 - P_r}{10n}, \tag{14.19}$$

which shows that the log distance is linear with respect to the RSS P_r. Technically, (14.19) is affine, while (14.16) is linear; however, Equation (14.19) shows some motivation for the formulation of log distance as linear with RSS.

Estimation of Parameters: In the linear signal-distance map algorithm, the only parameters that need to be estimated are the coordinate weights $\{\alpha_i\}$, $\forall_I \in \{1, \ldots, N\}$. A least squares approach is used. As in the radial basis function localization algorithm, let the coordinate weight matrix, $\mathbf{A}$, be represented as $\mathbf{A} = [\alpha_1, \ldots, \alpha_N]^T$. The least squares solution gives the estimate of coordinate weight matrix as

$$\mathbf{A} = (S^T S)^{-1} S^T \log(D), \qquad (14.20)$$

where S denotes the signal strength matrix, such that

$$S = [s_{i_1}, \ldots, s_{i_n}], \quad \text{where} \quad \{i_1, \ldots, i_n\} = H(t),$$

and D is the Euclidean AP distance matrix whose i, j element is the Euclidean distance between AP i and AP j, $\|x_i - x_j\|$, and $\log(D)$ is the element-wise logarithm on elements of matrix D.

Example 14.4 (Revisit Example Framework)

Consider the wireless network of Figure 14.2. The goal of this example is to estimate the coordinate of the tag $\hat{x}_t$ using the linear signal-distance map localization algorithm.

Solution

As in the previous examples, we start with the estimation of parameters. Using known coordinates of the APs, the distance matrix D is

$$D = \begin{bmatrix} 0 & 3.0 & 4.2 & 3.0 \\ 3.0 & 0 & 3.0 & 4.2 \\ 4.2 & 3.0 & 0 & 3.0 \\ 3.0 & 4.2 & 3.0 & 0 \end{bmatrix}.$$

Note: Typically, before taking the logarithm of matrix D, the diagonal is replaced by a small positive value ε in order to avoid taking the logarithm of zero. It has been recommended in [20] to take the value of $\varepsilon = d_{\min}/e$, where $d_{\min}$ is the minimum value of the off-diagonal elements of the matrix D. In our example, $d_{\min} = 3.0$ m and, hence, $\varepsilon = 1.1$ m.

The RSS matrix S in (14.20) is constructed from Table 14.1. Using the matrices S and D, the coordinate weight matrix A is determined using (14.20) as

$$A = \begin{bmatrix} -0.032 & 0.001 & 0.015 & -0.004 \\ -0.005 & -0.025 & -0.006 & 0.002 \\ 0.014 & 0.003 & -0.031 & 0.009 \\ -0.006 & 0.006 & -0.006 & -0.021 \end{bmatrix},$$

in which row i corresponds to the optimized coordinates of AP i in the "data space," α_i^T. Using the estimated values of $\{\alpha_1, \alpha_2, \alpha_3, \alpha_4\}$ in (14.16), the estimated log distance of the tag to the four APs, $\hat{\delta}_t$, is

$$\hat{\delta}_t = \left[1.48, 1.11, 0.75, 0.84\right]^T.$$

Using this estimated log distance to the APs, the coordinate estimate of the tag is determined using methods of multilateration, which are discussed in the other chapters of this book. Instead, in this example, we use a simple grid search method in which distances are computed between each grid point and the four APs. The grid point that gives the least squared error with the estimated distances to the four APs, $\hat{\delta}_t$, is the desired coordinate of the tag. Using this method, the coordinate estimate of the tag is computed as

$$\hat{x}_t = \left[3.5, 2.1\right]^T.$$

Compared to the actual tag coordinate, the linear signal-distance map-based coordinate estimate has an error of 0.51 m.

14.2.7 Summary

This section first presented a mathematical framework for localization using kernel methods. Next, we showed, through four RSS-based localization algorithms, how the common framework, represented in (14.1), can be applied to different positioning algorithms. The various parameters and their differences and similarities are presented in Table 14.3.

14.3 NUMERICAL EXAMPLES

For the purposes of obtaining an intuitive understanding of the advantages of kernel methods in localization algorithms, we show some numerical examples. Specifically, we compare the coordinate estimation of kernel-based localization algorithms, in an example setting, with that of an MLE using the lognormal shadowing model. Before we proceed with the example, we briefly describe the MLE algorithm.

14.3.1 MLE

A commonly used statistical model for radio propagation is the log-distance path loss model, in which the shadowing is modeled as lognormal (i.e., Gaussian if expressed in decibel). Within this model, the decibel RSS between devices i and j is represented as [29, 30]

$$s_{i,j} = \Pi_0 - 10n_p \log_{10}\|x_i - x_j\| + X_{i,j}, \tag{14.21}$$

where Π_0 represents RSS at a reference distance of $1\,\text{m}$, n_p represents the path loss exponent, and $X_{i,j}$ (in decibel) is the fading error, modeled as a Gaussian random variable with a standard deviation (in decibel) of σ_{dB}.

Estimating Coordinate from RSS: Given the path loss model parameters in (14.21), the coordinate of the tag can be estimated by maximizing the likelihood of s_t or by minimizing the negative log-likelihood. The negative log-likelihood of s_t is given by (14.22):

$$L(s_t|x_t, n_p, \Pi_0) = C + \frac{1}{2\sigma_{dB}^2} \sum_{j \in H(t)} [s_{t,j} - (\Pi_0 - 10n_p \log_{10}\|x_t - x_j\|)]^2, \quad (14.22)$$

where C is a constant. Note that the negative log-likelihood equation (14.22) assumes that the shadowing on links is mutually independent. This is a simplifying assumption, as it has been observed that geographically proximate links exhibit correlated shadowing [13].

The maximum likelihood coordinate estimate of the tag, $\hat{x}_t^{MLE}$, can be determined by minimizing the negative log-likelihood function of (14.22),

$$\hat{x}_t^{MLE} = \frac{\arg\min}{x_t} \sum_{j \in H(t)} [s_{t,j} - [\Pi_0 - 10n_p \log_{10}\|x_t - x_j\|]]^2. \quad (14.23)$$

In the likelihood equation of (14.22), the path loss parameters n_p and Π_0 were assumed to be known. These parameters can be estimated using the pairwise RSS measurements between APs and their known locations. Specifically, a linear regression is performed on the pairwise RSS measurements and log distances, computed using the known locations, to give the path loss parameters n_p and Π_0 [31].

Implementation Details: Due to the lack of an analytical solution to (14.23), the minimum is computed by using a *brute force* grid search over possible coordinates of x_t. The deployment area is divided into a grid of predetermined size, which determines the resolution of the coordinate estimate. At each grid point, the value of negative log-likelihood, $L(s_t|x_t, n_p, \Pi_0)$, is determined by using (14.22). The grid point that has the minimum negative log-likelihood is the MLE coordinate estimate of the tag $\hat{x}_t^{MLE}$.

The MLE of (14.23) suffers from computational disadvantage. Compared to the coordinate estimation of (14.1) using kernel methods, there is no closed-form solution for minimizing the the negative log-likelihood of (14.23). Typically, for real-time implementation, one must use numerical optimization methods [31].

14.3.2 Description of Comparison Example

In this section, we illustrate through example the advantages of kernel-based position estimation over model-based estimation. Consider a simple network of four APs placed at the corners of a square with a wall separating them, as shown in Figure 14.3. We consider two tag positions, one at the center of the network and the other at the edge.

Using this AP placement, we generate RSS values using (14.21) where $X_{i,j}$ includes the wall loss if the line between the transmitter and the receiver crosses through the wall. Specifically, the RSS, $s_{i,j}$, between devices i and j is computed using (14.21) with $n_p = 2$ and $X_{i,j}$ given as

$$X_{i,j} = \begin{cases} L_w + Y_{i,j}, & \text{if link } (i, j) \text{ passes through the wall} \\ Y_{i,j}, & \text{otherwise} \end{cases}, \qquad (14.24)$$

where $Y_{i,j}$ is the shadowing loss, modeled as a zero mean i.i.d. Gaussian in decibel random variable; that is,

$$Y_{i,j}[\text{dB}] \sim \mathcal{N}(0, \sigma_{\text{dB}}^2),$$

and L_w is the additional loss incurred when passing through a wall. The channel parameters used for generating the RSS vectors are tabulated in Table 14.4.

Note that the channel parameters given in Table 14.4 are assumed to be a priori unknown to the MLE algorithm. The path loss parameters, n_p and Π_0 in (14.23), are estimated using the pairwise RSS measurements between the APs and their known locations. Specifically, a linear regression is performed on the pairwise RSS measurements and log distances, determined using the known locations of the APs [31, 32]. In determining the path loss, the most recent RSS measurement between the AP pair is used.

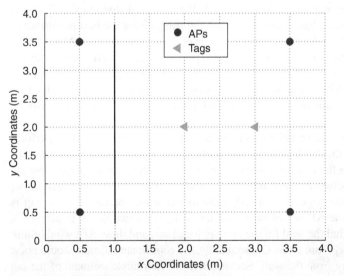

Figure 14.3   Position of APs and tags. There is a wall separating the network, represented by a black vertical line.

TABLE 14.4. Parameter Description and Values Used in the Simulation of a Network

Parameter	Description	Value
n_p	Path loss exponent	2
L_w	Loss across wall	5.0 dB
σ_{dB}	Fading standard deviation	0 and 6.0 dB
Π_0	Reference RX power	−40 dBm

The values are assumed to be a priori unknown to the localization algorithm.

Example 14.5

Consider the scenario when the shadowing standard deviation, $\sigma_{dB} = 0$. A noise variance of 0 dB is practically not possible, but, nevertheless, it helps in understanding the effects of shadowing on coordinate estimation, with no other fading losses. Links that pass through the wall suffer an additional loss of L_w dB due to transmission through the wall.

Solution

The coordinate estimates for different kernel algorithms discussed in the previous section are shown in Figure 14.4. In addition to kernel-based algorithms, we plot the coordinate estimate using MLE, described in Section 14.3.1. For a fair comparison between algorithms, we keep the set of the APs that contribute to the kernel, for different algorithms, the same and equal to four; that is, $|\tilde{H}(t)| = 4$ for all algorithms.

We also studied the performance of the coordinate estimation algorithms when different sets of APs contribute to the kernel. We observed that (1) the Gaussian kernel and (2) the radial basis function-based algorithms have the best performance when all the in-range APs contribute to the kernel. However, for the LANDMARC localization algorithm, the best AP set $\tilde{H}(t)$ depends on the relative location of the tag in the network. For tags located at the center of the network (e.g., Fig. 14.4a), taking all in-range APs as the set $\tilde{H}(t)$ performs best. On the other hand, for tags located on the edge of the network, taking the top three APs as the set $\tilde{H}(t)$ performs the best.

We observe that the kernel-based algorithms for coordinate estimation perform better compared to the MLE. One can also observe that the MLE coordinate estimates have errors in the direction away from the wall. This is intuitive because the presence of a wall between an AP and a tag would lower the RSS of the transmitting tag. Consequently, the lognormal propagation model would predict that the tag is further away from the AP, which is behind the wall. For example, in Figure 14.3, APs 1 and 2 are behind the wall for the two tag locations, and these APs would think the tag is further away from them. Consequently, the coordinate estimate would point in the direction away from the wall. Statistically, the coordinate estimate of the tag is said to have a bias, pointing away from the wall. More bias analysis is performed in the next example.

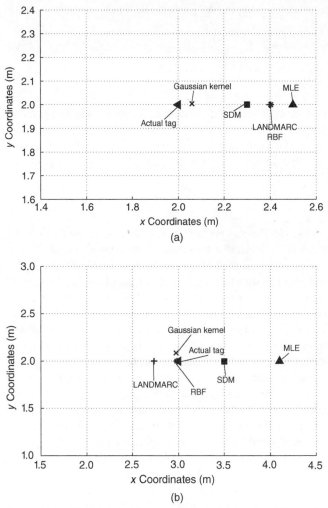

Figure 14.4   Plot showing the coordinate estimates for different localization algorithms along with the position of the tag. In (a), the tag is at the center of the network, and in (b), the tag is at the edge of the network.

This example clearly demonstrates the effect of shadowing and how kernel-based methods can overcome these effects. Specifically, shadowing due to walls or obstacles causes a reduction in RSS from the mean RSS. General propagation models, like the one in (14.21), would account this loss to the loss suffered because of distance, in order to minimize the error $X_{i,j}$. Consequently, even in the absence of noise variance, the estimates are biased in the direction away from the source of obstruction. Kernel-based algorithms, on the other hand, "learn" to adapt to this loss because they have more freedom in the parameters than a pure model-based approach and, thus, overcome the limitations of the model. Kernel methods interpolate the

RSS and the physical coordinates using the AP pairwise RSS measurements and AP known coordinates.

Example 14.6

In this example, in addition to the wall loss of Example 14.5, the effect of shadowing variance is added in the path loss equation (14.24); that is, $\sigma_{dB} > 0$. Independent Monte Carlo trials are run and the coordinate of the tag is estimated in each trial. A one-standard deviation covariance ellipse and bias performance of the location estimates is then determined. The one-standard deviation covariance ellipse is a useful representation of the magnitude and variation of the coordinate estimates [33].

Solution

The covariance ellipse, along with the bias performance, is shown in Figures 14.5 and 14.6. In the Figure 14.5, the tag is located at the center of the network (at $[2, 2]^T$ m), and in Figure 14.6, the tag is located at the edge of the network (at $[3, 2]^T$ m).

One of the most important lessons learned from this example is that the kernel-based localization algorithms perform better than the MLE in terms of average RMSE. This can be easily observed in Figures 14.5 and 14.6, where the MLE coordinate estimates, Figures 14.5e and 14.6e, suffer from both high bias and high variance. The performance improvement for kernel-based localization algorithms can be explained as follows. In the kernel-based methods, the estimated coordinate of a tag is a weighted average of some function of the coordinates of the APs that are in range of the tag. These weights are distinct for the distinct APs, and each AP maintains a table of its weights for all the other APs in the network. The determination of the weights is different for different algorithms. Consequently, the APs that are on a particular side of the wall would have lower weights for the APs that are on the other side. For example, in Figures 14.5 and 14.6, APs 1 and 2 have lower weights assigned to them by AP 3 as compared to the weight assigned to AP 4.

On the other hand, the MLE algorithm assumes a *common* statistical channel model for all the links in the network. Consequently, when minimizing the overall error, the path loss exponent, which signifies the slope of the decay in RSS with respect to log distance, is higher, similar to Example 14.5. Since all the links are weighted equally, this causes a high bias in the maximum likelihood coordinate estimates, pointing away from the wall.

In summary, the advantage of the kernel-based localization algorithms over MLE is that in the kernel-based algorithms, the APs, which naturally have significantly different RSS values compared to the tag, are weighted less compared to the other APs. On the other hand, in MLE, all the in-range APs have equal weights when computing the likelihood ratio and thus, the APs that have significantly different RSS values compared to the tag dominate the coordinate estimates pushing the tag further away from its actual location.

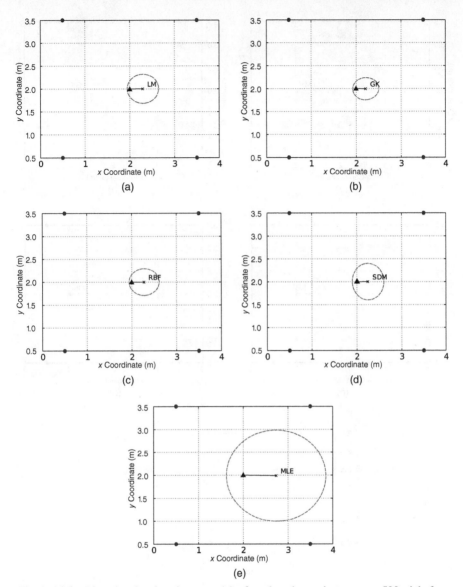

Figure 14.5   Bias plot showing the mean (×) of tag location estimates over 500 trials for LANDMARC (LM), Gaussian kernel (GK), radial basis function (RBF), and linear signal-distance map (SDM) localization algorithms. The actual tag location (▲) is connected to the mean location estimate (——). The plot also shows 1-σ covariance ellipse (– – –) for the coordinate estimates. The APs (•) are at the corners of the grid. The tag is located at the center of the network.

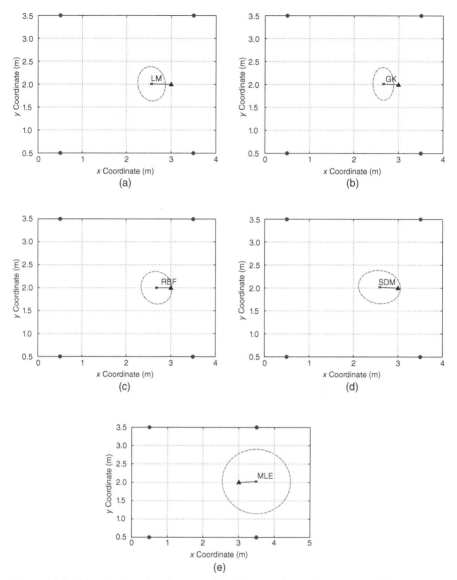

Figure 14.6   Bias plot showing the mean (×) of tag location estimates over 500 trials for LANDMARC (LM), Gaussian kernel (GK), radial basis function (RBF), and linear signal-distance map (SDM) localization algorithms. The actual tag location (▲) is connected to the mean location estimate (———). The plot also shows 1-σ covariance ellipse (– – –) for the coordinate estimates. The APs (•) are at the corners of the grid. The tag is located at the edge of the network.

14.4 EVALUATION USING MEASUREMENT DATA SET

In this section, we compare the performance of the different kernel-based localization algorithms introduced and formulated in the previous sections. Performance is quantified using two related measures:

- *Bias.* Bias is the difference between the average coordinate estimate (over many trials) and the actual coordinate. In this chapter, we show the bias using a bias plot, in which the actual coordinate and average coordinate estimate are plotted together for each tag. Bias is a consistent error in the coordinate estimate.

- *RMSE.* The RMSE is used to summarize both bias and variance effects. The "squared error" is the difference between the coordinate estimate and the actual location, squared, with units of square meter. The RMSE is then the square root of the average squared error (averaged over all tags in the deployment). Bias and error variance are two components of RMSE. The two together, quantified by RMSE, provide a good summary metric for quantifying localization performance.

In the rest of this section, we describe the environment along with the processing of the experimental data and the evaluation procedure for each data set.

14.4.1 Measurement Campaign Description

The measurement data consist of the pairwise RSS measured between 224 known location wireless APs deployed on a single floor of a hospital with an area of $16,700 \, \text{m}^2$. These APs are wireless transceivers that operate in the 2.4- to 2.48-GHz frequency band and transmit at a constant power. The APs have a limited range, and as such, the network formed by the deployed APs is not fully connected. Each AP has a limited set of neighboring APs which it can hear and on which it can make RSS measurements. The RSS values were collected for a period of 10 minutes, during which 40 RSS measurements were collected for each measurable link.

Since there are no "tags" in the measurement data, we simulate an unknown location tag using the LOO procedure. In the LOO procedure, whenever we need to "create" a known location tag, we "change" an AP into a tag for purposes of evaluation. We expect the RMSE for the LOO procedure to be higher than would be seen in deployed systems with tags. APs are deployed purposefully to be spatially separated from one another, for purposes of achieving coverage with a small number of APs. So, when one AP is converted to a tag, its nearest neighboring APs are relatively far from it, compared to the nearest neighbors of an actual tag that would be used in the system when no APs were "left out."

14.4.2 Evaluation Procedure

It was mentioned in Section 14.4.1 that the measurement data consist of pairwise RSS between APs only. In order to simulate a tag measurement, we employ the LOO approach. As mentioned before, an AP is assumed as a tag and its position is

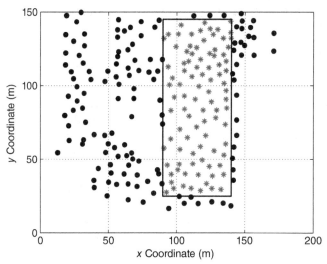

Figure 14.7 Coordinates of APs in the measurement analysis. The APs represented by the symbol * are said to be in sweet spot of the deployment.

estimated based on the remaining APs in the deployment. When we refer to a "tag" in this section, we mean the left-out AP, which is used as a known-location tag.

The RMSE for each particular tag is computed based on the RSS collected over a period of 10 minutes. This procedure is repeated for all the APs in the deployment. When reporting an average RMSE, we provide two numbers. First, we average over *all* the tag (left-out AP) locations. Second, we average over just the APs in what we consider to be the "sweet spot," that is, APs in the middle of the largest section of the floor plan shown by the symbol * in Figure 14.7. These APs should have lower bias because they are not at the edge of the building and therefore the edge of the network. The RMSE in the sweet spot provides intuition about location estimation in "good" areas, while the RMSE for all APs provides the average error result.

14.4.3 Results

In this section, we present the results of applying the four kernel-based localization algorithms, discussed in Section 14.2, on the measurement data set. Specifically, we quantify the algorithms with the two related measures, namely, (1) bias and (2) RMSE.

Bias Results: Figure 14.8 shows the bias plot for the four kernel-based localization algorithms and MLE. As mentioned before, bias is a consistent error in the coordinate estimates. In addition, we compute the average bias for each localization algorithm, which is shown in Table 14.5. We observe that the lowest bias is observed for the signal-distance map localization algorithm. Specifically, the average bias is 3.72 m. The coordinate estimates are more biased for the MLE algorithm, with an

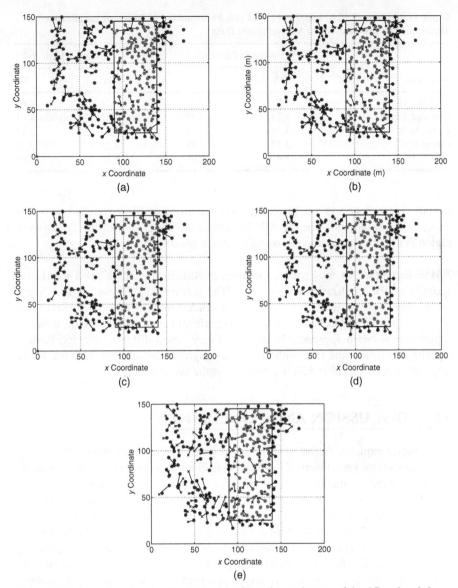

Figure 14.8   Bias plot showing the mean (×) of location estimates of the APs when it is emulated as a tag using (a) LANDMARC, (b) Gaussian kernel, (c) radial basis function, (d) linear signal-distance map, and (e) maximum likelihood localization algorithms. In all the plots, actual AP location (•) is connected to the mean location estimate (——). The APs in the sweet spot are marked with (•) and are inside the box.

TABLE 14.5. Overall (OA) and Sweet Spot (SS) Performance of Different Localization Algorithms for the Real-World Measurement Data

Algorithm	Average Bias (m)		Average RMSE (m)	
	OA	SS	OA	SS
LANDMARC	5.01	3.28	5.48	4.06
Gaussian kernel	5.25	3.30	5.87	4.04
Radial basis function	4.16	2.75	4.87	3.53
Linear SDM	3.72	2.49	4.31	3.18
MLE	5.41	5.01	6.87	7.04

overall bias of 5.41 and 5.01 m for the sweet spot region. Moreover, as one would expect, the average bias for the APs located in the sweet spot of the deployment region is lower compared to the overall average bias.

RMSE Results: In most cases, the average RMSE provides a good metric for quantifying the localization performance. The average RMSE results for different localization algorithms are tabulated in Table 14.5. From the table, we observe that all the kernel-based localization algorithms perform better than the MLE, which is a pure model-based approach. In fact, the linear signal-distance map localization algorithm performs the best with an overall improvement of 37% over the MLE, while the improvement is 55% for the APs in the sweet spot region.

14.5 DISCUSSION AND CONCLUSION

This chapter explores the advantages and features of a class of statistical learning algorithms, called kernel methods, as used in RSS-based localization. Kernel methods provide a simplified framework for localization without any a priori knowledge of the complicated relationship between the RSS and position. Instead, these relationships are encapsulated in parametrized nonlinear functions. Algorithms based on kernel methods inherently account for spatial correlation in the RSS, which most model-based approaches fail to capture. Kernel methods do not rely solely on a database of training measurements, like RSS fingerprinting algorithms, which must be measured very densely in space. In this chapter, a calibration scheme was presented, which attempts to minimize the calibration requirements of kernel-based algorithms. Specifically, in this scheme, training is performed simultaneously while the system is online, using the AP pairwise measurements.

A simulation example of a simple four-AP network was presented to provide better understanding of kernel methods. The results show that the kernel-based algorithms provide better location accuracy compared with the model-based algorithms in terms of average RMSE. This is because kernel methods provide an adaptive weighting scheme for the APs. Within this weighting scheme, the APs that have significantly different RSS values compared to the tag are weighted less compared with the other APs.

An extensive experimental evaluation was performed for all the kernel-based algorithms and the MLE using a data set collected from a large hospital facility. These real-world experimental results indicate that all four kernel-based algorithms perform better than the MLE. In fact, the linear signal-distance map localization algorithm has the best performance in terms of average RMSE. The linear signal-distance map localization algorithm has an overall RMSE reduction of 37% over the MLE, while the RMSE reduction is as high as 55% for the sweet spot areas of the deployment region. The complexities of the fading environment and the complicated nature of the large-scale real-world deployment require more parameters than are available to a single log-distance path loss model. In particular, even though the linear signal-distance map localization algorithm assumes linearity with respect to the log-distance relationship for RSS, the parameters of the linear relationship are learned and adapted locally to the RSS measured at each AP.

Another perspective of this analysis is that spatial correlation in the RSS can be particularly useful in wireless localization. Typically, geographically proximate links would encounter similar environmental obstructions, and the shadowing loss suffered on these links would be correlated. Better understanding of the area can be obtained when considering spatial correlations. Kernel methods are a strong candidate because the kernel in a kernel-based algorithm provides a spatial similarity measure. Additionally, kernel models are typically linear with respect to the parameters, allowing good analytical properties, yet are nonlinear with respect to the RSS measurements.

REFERENCES

[1] Awarepoint, "Real-time awareness solutions," 2010. Available: http://www.awarepoint.com/

[2] M. Hazas, J. Scott, and J. Krumm, "Location-aware computing comes of age," *Computer*, vol. 37, no. 2, pp. 95–97, 2004.

[3] G. Chen and D. Kotz, "A survey of context-aware mobile computing research," Tech. Rep., 2000.

[4] A. LaMarca, J. Hightower, I. Smith, and S. Consolvo, "Self-mapping in 802.11 location systems," *Lect. Notes Comput. Sci.*, vol. 3660, p. 87, 2005.

[5] A. Savvides, C. Han, and M. Strivastava, "Dynamic fine-grained localization in ad-hoc networks of sensors," in *Proceedings of the 7th Annual International Conference on Mobile Computing and Networking*. New York: ACM, 2001, pp. 166–179.

[6] D. Niculescu and B. Nath, "VOR base stations for indoor 802.11 positioning," in *Proceedings of the 10th Annual International Conference on Mobile Computing and Networking*. New York: ACM, 2004, pp. 58–69.

[7] Y. Ji, S. Biaz, S. Pandey, and P. Agrawal, "ARIADNE: A dynamic indoor signal map construction and localization system," in *Proceedings of the 4th International Conference on Mobile Systems, Applications and Services*. ACM, 2006, p. 164.

[8] T. Roos, P. Myllymäki, H. Tirri, P. Misikangas, and J. Sievänen, "A probabilistic approach to WLAN user location estimation," *Int. J. Wireless Inf. Netw.*, vol. 9, no. 3, pp. 155–164, 2002.

[9] M. Youssef and A. Agrawala, "The Horus location determination system," *Wireless Netw.*, vol. 14, no. 3, pp. 357–374, 2008.

[10] T. King, S. Kopf, T. Haenselmann, C. Lubberger, and W. Effelsberg, "COMPASS: A probabilistic indoor positioning system based on 802.11 and digital compasses," in *Proceedings of the 1st International Workshop on Wireless Network Testbeds, Experimental Evaluation and Characterization*. ACM, 2006, p. 40.

[11] P. Bahl and V. Padmanabhan, "RADAR: An in-building RF-based user location and tracking system," *IEEE INFOCOM*, vol. 2, pp. 775–784, 2000.

[12] M. Brunato and C. Kalló, "Transparent location fingerprinting for wireless services," presented at Med-Hoc-Net, Sardegna, Italy, 2002.

[13] P. Agrawal and N. Patwari, "Correlated link shadow fading in multi-hop wireless networks," *IEEE Trans. Wireless Commun.*, vol. 8, no. 8, pp. 4024–4036, 2009.

[14] N. Patwari and P. Agrawal, "Effects of correlated shadowing: Connectivity, localization, and RF tomography," in *Proceedings of the 7th International Conference on Information Processing in Sensor Networks*. Washington, DC: IEEE Computer Society, 2008, pp. 82–93.

[15] J. Krumm and J. C. Platt, "Minimizing calibration effort for an indoor 802.11 device location measurement system," Tech. Rep., Microsoft Coorporation, 2003.

[16] A. Howard, S. Siddiqi, and G. S. Sukhatme, "An experimental study of localization using wireless ethernet," in *Field and Service Robotics*, S. Yuta, H. Asama, E. Prassler, T. Tsubouchi, and S. Thrun, Eds. Berlin: Springer, 2006, pp. 145–153.

[17] J. Letchner, D. Fox, and A. LaMarca, "Large-scale localization from wireless signal strength," in *Proceedings of the National Conference on Artificial Intelligence*, vol. 20, Cambridge, MA: AAAI Press/MIT Press, 2005, p. 15.

[18] L. Ni, Y. Liu, Y. Lau, and A. Patil, "LANDMARC: Indoor location sensing using active RFID," *Wireless Netw.*, vol. 10, no. 6, pp. 701–710, 2004.

[19] A. Kushki, K. Plataniotis, and A. Venetsanopoulos, "Kernel-based positioning in wireless local area networks," *IEEE Trans. Mobile Comput.*, vol. 6, no. 6, pp. 689–705, 2007.

[20] H. Lim, L. Kung, J. Hou, and H. Luo, "Zero-configuration, robust indoor localization: Theory and experimentation," in *Proc. of IEEE Infocom*, 2006, pp. 123–125.

[21] T. Roos, P. Myllymäki, and H. Tirri, "A statistical modeling approach to location estimation," *IEEE Trans. Mobile Comput.*, vol. 1, no. 1, pp. 59–69, 2002.

[22] C. Bishop, *Pattern Recognition and Machine Learning*. New York: Springer, 2006.

[23] Y. Gwon and R. Jain, "Error characteristics and calibration-free techniques for wireless LAN-based location estimation," in *Proceedings of the Second International Workshop on Mobility Management & Wireless Access Protocols*. New York: ACM, 2004, pp. 2–9.

[24] C. Bishop, *Neural Networks for Pattern Recognition*. Oxford: Oxford University Press, 2005.

[25] M. Brunato and R. Battiti, "Statistical learning theory for location fingerprinting in wireless LANs," *Comput. Netw.*, vol. 47, no. 6, pp. 825–845, 2005.

[26] T. Hastie, R. Tibshirani, and J. H. Friedman, *The Elements of Statistical Learning: Data Mining, Inference, and Prediction*. New York: Springer, 2009.

[27] S. Kay, *Fundamentals of Statistical Signal Processing: Estimation Theory* (Prentice-Hall Signal Processing Series). Englewood Cliffs, NJ:Prentice-Hall, 1993, p. 595.

[28] G. Strang, Thompson Publications, Belmont, CA, USA, 4th Edition, 2006, *Linear Algebra and Its Applications*.

[29] T. Rappaport, *Wireless Communication: Principles and Practice*. Upper Saddle River, NJ: Prentice Hall, 1996.

[30] H. Hashemi, "The indoor radio propagation channel," *Proc. IEEE*, vol. 81, no. 7, pp. 943–968, 1993.

[31] N. Patwari, A. Hero, M. Perkins, N. Correal, and R. O'Dea, "Relative location estimation in wireless sensor networks," *IEEE Trans. Signal Process.*, vol. 51, no. 8, pp. 2137–2148, 2003.

[32] N. Patwari, Y. Wang, and R. O'Dea, "The importance of the multipoint-to-multipoint indoor radio channel in ad hoc networks," in *IEEE Wireless Communications and Networking Conf.* 17–21 Mar. 2002, Orlando, FL, USA, vol., pp. 608–612.

[33] L. Paradowski, M. Acad, and P. Warsaw, "Uncertainty ellipses and their application to interval estimation of emitter position," *IEEE Trans. Aeros. Electron. Syst.*, vol. 33, no. 1, pp. 126–133, 1997.

CHAPTER 15

FINGERPRINTING LOCATION TECHNIQUES

Rafael Saraiva Campos[1] and Lisandro Lovisolo[2]

[1]Centro Federal de Educação Tecnológica Celso
Suckow da Fonseca, Campus Petrópolis

[2]Universidade do Estado do Rio de Janeiro

I N THIS chapter, the principles of radio frequency (RF) fingerprinting methods, also known as database correlation methods (DCM), are presented. Although there is a wide variety of DCM implementations, they all share the same basic elements. These elements are identified and analyzed in this chapter. Different alternatives for building RF fingerprint correlation databases (CDBs) are presented, and the advantages and drawbacks of each of them are discussed, as well as their impact on the positioning accuracy of location services based on DCM. Different methods to numerically evaluate the similarity between each fingerprint within the CDB and the measured RF fingerprint are presented, including an artificial neural network based approach. Two alternatives to reduce the correlation or search space are analyzed, including an approach based on genetic algorithms. Finally, field tests are used to evaluate the positioning accuracy of fingerprinting techniques in both GSM (Global System for Mobile Communications) and Wi-Fi (wireless fidelity) 802.11b networks.

15.1 INTRODUCTION

RF fingerprinting location techniques are a class of mobile station (MS) positioning methods, which can be applied in any wireless network [9, 29]. In Part II (Chapter 6 to Chapter 10), one has explored different aspects of *direction of arrival* (DOA), *time of arrival* (TOA) or *time difference of arrival* (TDOA) and similar RF-based positioning techniques. From Chapter 11 to Chapter 14, one has seen how received

Handbook of Position Location: Theory, Practice, and Advances, Second Edition.
Edited by S. A. (Reza) Zekavat and R. Michael Buehrer.
© 2019 by the Institute of Electrical and Electronics Engineers, Inc.
Published 2019 by John Wiley & Sons, Inc.
Companion Website: www.wiley.com/go/zekavat/positionlocation2e

signal strength (RSS) can be used for location. Positioning methods as DOA, TOA, and TDOA in general rely on *line-of-sight* (LOS) geometric assumptions, as explained in Chapter 3, Chapter 27, and in [9] and [29].

Fingerprinting location techniques are based on comparing the characteristics of measured RF signals against typical templates that correspond to positions in a map or a set of coordinates. The more similar the patterns, the more likely the terminal is close to the position associated with the match. This provides fingerprinting methods with the capability to generalize the mapping between signal features (as, for example, RSS and round-trip time) and the transmitter-receiver geometry. Fingerprinting-based positioning algorithms do not require an a priori mapping between the RSS and the ray direction, path loss, or time of fly.

Even though there is a wide variety of fingerprinting location techniques (some fingerprinting applications are also described in Chapters 8, 10, 11, and 26), they all share the same basic elements:

- *RF Fingerprint*: This is a set of location-dependent signal parameters, available in the *radio access network* (RAN). Each RF fingerprint is associated with a specific position. Note that the more signals that are observed or the more parameters per signal that are observed, the more unique the fingerprint, and thus the better the location accuracy. If only RAN inherent parameters are included in the RF fingerprint, the fingerprinting location technique can be entirely network based, therefore not requiring any modification to existing MS [52]. The RF fingerprint is defined in Section 15.2.

- *CDB*: RF fingerprints are collected in field tests or generated using simulation models, and stored in a database called the CDB, which is directly accessible to the location server. Each RF fingerprint stored in the CDB is associated with a specific position. The CDB is defined in Section 15.3.

- *Location Server*: This is the network element responsible for receiving location requests, consulting the CDB, and estimating the MS location.

- *Reduction of the Search Space within the CDB*: The CDB might be quite large, and analyzing all RF fingerprints stored in it might be very time consuming. Therefore, all fingerprinting location techniques apply some method to reduce the search space within the CDB. As a consequence, the time required to produce a position fix is also reduced. Two alternatives of such techniques are presented in Section 15.4.

- *Pattern Matching*: In order to estimate the MS position, the location server must compare the RF fingerprint measured by the MS with a subset of the RF fingerprints stored in the CDB. This comparison or pattern matching might be done using different techniques, as discussed in Section 15.5.

Any fingerprinting location technique has two phases. The first is the training phase, when the CDB is built. The second is the test or operational phase, during which MS position estimates are produced [11].

A simplified diagram of a generic fingerprinting location solution is presented in Figure 15.1. This diagram corresponds to a MS-originated position request [16]. In step 1, the MS sends a position request to the location server, through the RAN.

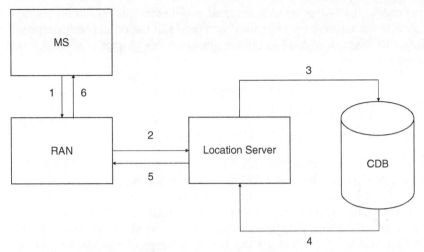

Figure 15.1   Schematic diagram of a fingerprinting location.

In step 2, the RAN communicates with the location server, usually through a *gateway*. The location server receives the position request, together with the RF fingerprint measured by the MS. In step 3, the location server then queries the CDB, obtaining in step 4 the RF fingerprints, which will be compared to the RF fingerprint measured by the MS. The location server then applies a location estimation or comparison function to obtain the MS position estimate, which is sent back to the MS through the RAN in steps 5 and 6.

15.2   RF FINGERPRINTS

A RF fingerprint is a set of location-dependent signal parameters, measured by the MS or the anchor cells [29]. Just like a human fingerprint, which carries the unique identification of a person, a RF fingerprint is expected to uniquely identify a geographic position. In order to do so, the number of signal parameters in the RF fingerprint must be high enough to allow for a unique correspondence with a given location. The selected signal parameters—or at least, their time averages—must have low variability in time at any given position. However, it is obvious that they will not be completely stable over time. Even though the use of mean values reduces small scale variations, changes in the RAN—like the addition of new cells, changes in transmission or reception antennas, changes in output power, etc.—might jeopardize the connection between a given RF fingerprint and a certain position. In such cases, it is necessary to obtain new RF fingerprints using the methods described in Section 15.3.2.

A RF fingerprint can be classified as either a *target* or *reference* fingerprint. A target RF fingerprint is the RF fingerprint associated with the MS that is to be localized, i.e., it contains signal parameters measured by the MS or by its anchor cells. The reference RF fingerprints are the RF fingerprints collected or generated during the training phase and stored in the CDB. Each reference RF fingerprint is associated with a unique set of geographic coordinates. Ideally, all the parameters used in the target fingerprint must be present in the reference fingerprint. The target RF fingerprint used in the remainder of this chapter is given by the $N_a \times 3$ matrix:

$$
\mathbf{F} = \begin{bmatrix} \text{ID}_1 & \text{RSS}_1 & \text{RTD}_1 \\ \vdots & \vdots & \vdots \\ \text{ID}_{N_a} & \text{RSS}_{N_a} & \text{RTD}_{N_a} \end{bmatrix},
\tag{15.1}
$$

where N_a is the number of anchor cells within range of the MS, ID_i and RSS_i are the *cell identity* (ID) and measured *RSS* from the i-th anchor, respectively. RTD_i is the *round trip delay* (RTD) between the MS and the i-th anchor cell. The rows are sorted in descending order of RSS, so $\text{RSS}_i \geq \text{RSS}_j$ if $i \leq j$. The RSS and RTD quantization steps and dynamic ranges, the maximum number of anchor cells, and the number of available RTD values are specific to each type of wireless network. In some networks, like GSM, the RTD value is available only for the best server [15], which usually is the anchor cell with the highest RSS. In other networks, like Wi-Fi, there is no RTD value available at all [26].

A wide variety of signal parameters can be selected to compose a RF fingerprint: RSS [6], RTD [7], power delay profile calculated using the channel impulse response [2], DOA [33], etc. These parameters are measured from (or to) a number of anchor cells. The more anchors that can be measured, the more unique the RF fingerprint. Ideally, the selected parameters should already be available in the network. The use of parameters ordinarily involved with call or session management prevents additional load at the RAN. Another benefit of this approach is that no modifications— hardware or software—are required in the MS, therefore making it possible to locate any MS within the network coverage area. That is why the most frequently used parameters are the RSS and RTD, which are the parameters selected for the target and reference RF fingerprints used in this chapter. The MS periodically measures the control channel RSS of cells to allow for cell selection and handover. Those values are reported to the core network through special messages called *network measurement reports* (NMRs). Only the RSS of channels transmitted with constant (or known) power should be inserted in a RF fingerprint [6], so the control channel in cellular networks, which is transmitted by all cells and whose output power is constant (i.e., power control is not applied), is the most obvious choice. The RTD is periodically measured by either the MS or the anchor cells in networks using *code division multiplexing* (CDM) or *time division multiplexing* (TDM) in the physical layer.

15.3 CDB

The CDB is a collection of reference RF fingerprints. As previously stated, each reference RF fingerprint is associated with a unique set of geographic coordinates. The CDB is built during the training phase of the RF fingerprinting location algorithm [19] using radio-propagation modeling, field measurements, or a combination of both alternatives.

15.3.1 CDB Structure

Each CDB element contains a reference RF fingerprint and a unique set of geographic coordinates. The planar—or spatial, for 3-dimensional location—distribution of these coordinates within the *service area* defines the CDB structure. The service area is the region where a location service is to be offered. In the remainder of this chapter, only 2-dimensional location is used, but the extension to the 3-dimensional case is straightforward.

Uniform Grid

If the CDB is organized as an *uniform grid*, all the reference coordinates are evenly spaced in the plane. One RF fingerprint is associated with each reference coordinate. The distance between adjacent reference coordinates defines the uniform grid spacing or planar resolution. The planar resolution should be comparable to the location method's expected precision [29]. The effect of different values of planar resolution on the location precision of fingerprinting methods is analyzed in Section 15.6. The uniform grid is the most suitable structure for CDBs built using propagation modeling and is presented in more detail in Section 15.3.2.

Indexed List

If the CDB is organized as an *indexed list*, the reference coordinates planar distribution do not follow any regular pattern. This structure is usually adopted for CDBs built using field measurements [11]. For example, if the CDB is built using drive test measurement routes, the irregular pattern of the street grid might prevent obtaining evenly spaced reference fingerprints.

Each element in the list contains a reference RF fingerprint and a set of reference geographic coordinates, obtained either by a GNSS (*global navigation satellite system* discussed in Part V) receiver or directly from a map or a manual entry—this is usually the case for CDBs built from indoor measurements.

15.3.2 Building the CDB

The CDB stores the reference RF fingerprints collected during field tests or generated with propagation models. The CDB is built during the training phase of the RF fingerprinting location method. This section discusses the main alternatives to build the CDB, pointing out the main advantages and drawbacks of each technique.

Field Measurements

The CDB can be built entirely from field measurements. This usually requires a MS, a software to collect and process radio interface measurements made by the MS, and a GNSS receiver, in the case of outdoor measurements. The software might be running on a laptop or palmtop connected to the MS. Periodically, location related parameters, like those listed in (15.1), are collected by the software and are stored for further processing. These parameters are either measured by the MS or the network. For each collected set of parameters, the ground truth position of the MS is registered by the GNSS receiver connected to the laptop or palmtop. For indoor measurements, it might be necessary to use a map of the building layout and manually register the MS reference position within the building. The MS reference coordinates and the measured location dependent parameters compose an entry in the CDB, internally structured as an indexed list, as described in Section 15.3.1.

The empirical CDB obtained by field measurements usually provides the highest location accuracy. However, it has a major drawback, especially when used in metropolitan area networks (MANs). In those networks, to keep empirical CDBs up-to-date, drive tests must be carried out after any change in RAN elements. These changes—deployment of new cells; changes in antenna models, azimuth, and inclination; increase or decrease of transmit power; etc.—occur constantly, especially in cellular networks, making this solution impractical for MANs. The detrimental effects due to the use of out-of-date network parameters in the pattern matching process in cellular networks have been evaluated in [11] and [21]. However, for indoor RF fingerprinting location systems, using CDBs built from measurements might be a feasible option if one considers the greater complexity of the indoor environment—which makes it difficult to accurately model RF propagation—and the smaller area to be covered by the measurement campaign.

Propagation Modeling

The main advantage of using a CDB built from propagation modeling is to allow for easy, fast, and inexpensive CDB updating. Whenever there are changes in the RAN elements, it is necessary only to rerun the propagation models with the new RAN parameters to obtain an updated CDB. However, the achieved location precision might be lower than the one obtained with field measurements. This degradation can be minimized by means of proper propagation model calibration [1].

There is a wide variety of mathematical models for radio propagation prediction, but they can be roughly grouped in two main classes: deterministic and empirical. Deterministic propagation models are based on ray tracing techniques. They describe the electromagnetic wave propagation using rays launched from the transmitting antenna. These rays are reflected and diffracted at walls and other obstacles. Ray tracing models require a very accurate knowledge of the environment and have a high computational load, resulting in a long computation time for the coverage predictions [49]. Empirical propagation models are based on extensive field measurements that, after statistical analysis, produce parametric path loss equations. Those parameters or coefficients can be adjusted, within some predetermined bounds, to better represent a particular propagation environment [50]. Empirical models are less computationally intensive and, even though they are usually less accurate than deterministic propagation models, they still provide an accuracy compatible with the average accuracy of most RF fingerprinting methods for outdoor positioning [7, 52]. Therefore, empirical models become the most suitable option to build a CDB with RF propagation modeling in outdoor environments.

The Okumura-Hata model [22] provides an empirical formula for propagation loss derived from extensive field measurements in urban areas. This model is applicable to system designs for UHF (*ultra high frequency*) and VHF (*very high frequency*), under the following conditions: frequency range 100–1500 MHz, distance 1–20 km, base station antenna height 30–200 m, and MS antenna height 1–10 m. The Okumura-Hata model is widely used for RF planning in cellular networks.

The basic Okumura-Hata propagation loss formula does not explicitly takes into account diffraction over terrain and buildings. In order to do so, the topography of the service area is represented by a matrix $\mathbf{H}$, also referred to as a digital elevation model (DEM) or digital topographical database [27]. Each matrix element $h_{i,j}$ stores the terrain height averaged over a $r_H \times r_H$ m^2 surface, and is referred to as a *pixel*. Parameter r_H is the $\mathbf{H}$ matrix planar resolution. The $\mathbf{H}$ matrix might also contain, added to the terrain height, the building heights. If the region covers a total surface of $l \times w$ m^2, then $\mathbf{H}$ has $\left\lceil \frac{l}{r_H} \right\rceil \times \left\lceil \frac{w}{r_H} \right\rceil$ elements (pixels).

To represent the service area surface as a plane, divided into evenly spaced pixels, it is necessary to apply a geographic coordinate system which uses a rectangular cartographic projection. The *Universal Transverse Mercator* (UTM) [42] is an example of such a system. Assume that the UTM system is being used and that $h_{1,1}$, the first element of $\mathbf{H}$, is placed at the northwest corner of the service area, as depicted in Figure 15.2. If the UTM coordinates $[x_1 \; y_1]^\mathrm{T}$ of $h_{1,1}$ are known, then the coordinates of $h_{i,j}$ are given by:

$$\begin{bmatrix} x_j \\ y_i \end{bmatrix} = \begin{bmatrix} x_1 + r_H(j-1) \\ y_1 - r_H(i-1) \end{bmatrix}, \tag{15.2}$$

where $i = 1, 2, \ldots, \left\lceil \frac{w}{r_H} \right\rceil$ and $j = 1, 2, \ldots, \left\lceil \frac{l}{r_H} \right\rceil$.

The terrain profile—including the building heights, if available—between the kth cell and pixel (i, j) is read from the DEM. Figure 15.3 shows a terrain profile and the boundary of the first Fresnel zone [50] in the radio link.

After obtaining the terrain profile, the diffraction losses must be calculated using a specific model, like Epstein-Peterson, Bullington, or Deygout [50]. The

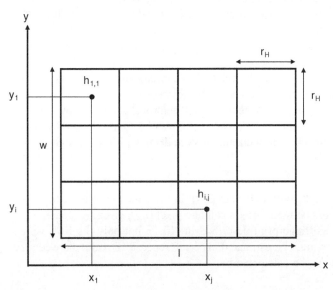

Figure 15.2   Service area surface represented as a matrix.

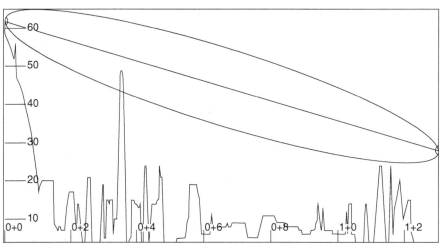

Figure 15.3   Terrain profile with building heights.

Okumura-Hata average propagation loss in dB between the kth cell in the service area and pixel (i, j), plus the additional diffraction loss $u_{i,j,k}$, is given by:

$$L_{i,j,k} = c_1 + c_2 \log_{10}(d_{i,j,k}) + c_3 \log_{10}(z_k) + c_4 u_{i,j,k} + c_5 \log_{10}(z_k) \log_{10}(d_{i,j,k}), \quad (15.3)$$

where z_k is the kth cell antenna effective height [35] in meters and $d_{i,j,k}$ is the distance in meters between the kth cell antenna and pixel (i, j). The model coefficients (c_1, c_2, c_3, c_4 and c_5) depend on the area morphology and transmission frequency. For a network using the 869–881 MHz band, like the one in Section 15.6.1, the model is applied at the central frequency of 875 MHz. The coefficient values are $c_1 = -12.1$, $c_2 = -44.9$, $c_3 = -5.83$, $c_4 = 0.5$, and $c_5 = 6.55$. All of these values are the standard Okumura-Hata values for urban environments, except c_4, which was empirically defined by the authors [6].

   The vertical ϕ and horizontal θ angles between the kth cell antenna and pixel (i, j) can be calculated using trigonometry, as the geographic coordinates in space—x, y, and z—of both the antenna and the pixel are known. The kth cell control channel transmission power is assumed to be known, as well as the connector and cable losses between the transmitter and the antenna. Therefore, with the antenna vertical and horizontal radiation patterns, it is possible to estimate the kth cell control channel *effective isotropic radiated power* (EIRP) in the direction of pixel (i, j). This direction is defined by angles ϕ and θ. The reference kth cell control channel RSS in dBm at pixel (i, j) is given by:

$$\text{RSS}_{i,j,k} = \text{EIRP}_{i,j,k} - L_{i,j,k}, \quad (15.4)$$

where $\text{EIRP}_{i,j,k}$ is the kth cell control channel EIRP in the direction of pixel (i, j), and $L_{i,j,k}$ is the propagation loss between the kth cell and pixel (i, j), given by (15.3).

   Considering the RF fingerprint described in (15.1), not only the RSS values but also the RTD values must be estimated. The reference RTD value between the kth cell and pixel (i, j) can be calculated by:

$$\text{RTD}_{i,j,k} = \left\lfloor \frac{2d_{i,j,k}}{c\text{T}_s} \right\rfloor, \quad (15.5)$$

where c is the speed of light in free space in meters per second, T_s is the symbol period in seconds, and $d_{i,j,k}$ is the distance in meters between the kth cell antenna and pixel (i, j). Equation (15.5) assumes LOS conditions between the transmitting antenna and the pixel, but this is hardly the case, especially in dense urban areas. To enhance the accuracy of the reference RTD value, the additional propagation delay due to non-line-of-sight (NLOS) conditions can be modeled as a random variable [7]; NLOS conditions and models are more profoundly discussed in Chapter 3 and in Part IV.

The reference RF fingerprint at (i, j) is completed after $RSS_{i,j,k}$ and $RTD_{i,j,k}$ have been calculated for $k = 1, 2, \ldots, N_{i,j}$, where $N_{i,j}$ is the number of cells whose predicted RSS values are above the minimum threshold at pixel (i, j). Note that $1 \le N_{i,j} \le N_c$, where N_c is the total number of cells in the service area. The reference RF fingerprint at pixel (i, j) is represented by:

$$\mathbf{S}_{i,j} = \begin{bmatrix} ID_{i,j,1} & RSS_{i,j,1} & RTD_{i,j,1} \\ \vdots & \vdots & \vdots \\ ID_{i,j,N_{i,j}} & RSS_{i,j,N_{i,j}} & RTD_{i,j,N_{i,j}} \end{bmatrix}, \tag{15.6}$$

where $ID_{i,j,k}$ is the kth cell ID at pixel (i, j). The rows are classified in descending order of RSS, i.e., $RSS_{i,j,k'} \ge RSS_{i,j,k''}$, if $k' \le k''$.

The CDB is complete after $\mathbf{S}_{i,j}$ has been calculated for all pixels in the service area—i.e., for $i = 1, 2, \ldots, \lceil \frac{w}{r_H} \rceil$ and $j = 1, 2, \ldots, \lceil \frac{l}{r_H} \rceil$. The structure of the CDB thus obtained is a uniform grid, as defined in Section 15.3.1.

The planar resolution of the CDB should be comparable to the location method expected precision [29]. So, if r_H is much smaller than the expected location precision, the CDB grid can be subsampled [7]. The resulting matrix will have $\lceil \frac{l}{r_S} \rceil \times \lceil \frac{w}{r_S} \rceil$ elements, where r_S is the new planar resolution of the CDB. Figure 15.4 shows an example where $r_S = 2r_H$. The new reference RF fingerprint $\mathbf{S}'_{1,1}$ of the first matrix element is obtained by averaging the original reference RF fingerprints $\mathbf{S}_{1,1}$,

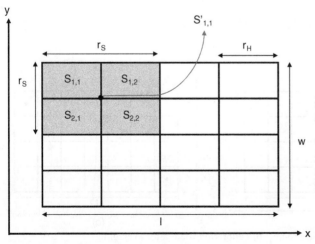

Figure 15.4   CDB organized as a uniform grid with subsampling.

$S_{1,2}$, $S_{2,1}$, and $S_{2,2}$. The process is repeated for all pixels. If the UTM coordinates $[x_1\ y_1]^\mathsf{T}$ of $S'_{1,1}$ are known, then the coordinates of $S'_{i,j}$ are given by:

$$\begin{bmatrix} x_j \\ y_i \end{bmatrix} = \begin{bmatrix} x_1 + r_S(j-1) \\ y_1 - r_S(i-1) \end{bmatrix}, \tag{15.7}$$

where $i = 1, 2, \ldots, \left\lceil \frac{w}{r_S} \right\rceil$ and $j = 1, 2, \ldots, \left\lceil \frac{l}{r_S} \right\rceil$.

If the CDB is built using propagation modeling, field tests might be used for fine tuning the empirical propagation models. This procedure is expected to enhance the location precision of a RF fingerprinting method using such a CDB. Consider that a calibration route is carried out across the service area. At each measurement point, the RSS of each detected cell is collected. The GNSS coordinates of the measurement points are also registered, allowing one to identify the CDB pixels where they are located, as shown in Figure 15.5a. An average of each cell RSS must be calculated for all measurement points located at the same pixel. After that, the test route is complete and each measurement point is identified by the 3-uple (i_n, j_n, M_n), as shown in Figure 15.5b. The pair (i_n, j_n) identifies the pixel where the nth measurement point is located. Note that $1 \le n \le N_m$, where N_m is the number of measurement points in the calibration route. Matrix M_n is the set of RSS measurements collected at the nth point and is given by:

$$M_n = \begin{bmatrix} \mathrm{ID}_{n,1} & \mathrm{RSS}_{n,1} \\ \vdots & \vdots \\ \mathrm{ID}_{n,N_n} & \mathrm{RSS}_{n,N_n} \end{bmatrix}, \tag{15.8}$$

where $\mathrm{ID}_{n,k}$ and $\mathrm{RSS}_{n,k}$ are the kth cell ID and RSS, respectively, at the the nth measurement point. Note that $1 \le k \le N_n$, where N_n is the number of cells detected at the n-th point and $1 \le N_n \le N_c$. The rows are classified in descending order of RSS, so $\mathrm{RSS}_{n,k'} \ge \mathrm{RSS}_{n,k''}$ if $k' \le k''$.

At the nth point in the calibration route, the difference between the predicted and measured RSS of the kth cell is given by:

$$b_{n,k} = M_n(k', 2) - S'_{i,j}(k'', 2) \tag{15.9}$$

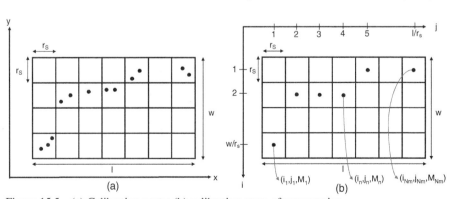

Figure 15.5 (a) Calibration route; (b) calibration route after averaging.

for

$$\mathbf{M}_n(k', 1) = \mathbf{S}'_{i,j}(k'', 1) = \mathrm{ID}_k \tag{15.10}$$

and

$$(i_n, j_n) = (i, j), \tag{15.11}$$

where ID_k is the kth cell identity. Observe that k' is the index of the line in $\mathbf{M}_n$ whose cell ID (value in the first column) is equal to the kth cell identity, i.e., ID_k. An analogous observation can be made regarding k'' and $\mathbf{S}'_{i,j}$. Note that $1 \le k' \le N_n$, $1 \le k'' \le N_{i,j}$, $1 \le k \le N_c$, and $1 \le n \le N_m$.

Calculating $b_{n,k}$ for all points on the calibration route where the kth cell has been detected, one obtains a $N_{m,k} \times 1$ matrix $\mathbf{B}_k$. The parameter $N_{m,k}$ is the number of calibrating points where the kth cell was detected. Note that $1 \le N_{m,k} \le N_m$.

The propagation model calibration is done on a per cell basis, defining a calibration factor C_k that shall be added to the path loss, given by (15.3), between the kth cell and any pixel (i, j) in the service area. The C_k factor minimizes the sum of the squared differences between the measured and predicted kth RSS values along the calibration route points where the kth cell was detected. The C_k factor in dB can be estimated using least-squares (LS) [30] as follows:

$$C_k = \left(\mathbf{V}^\mathrm{T}\mathbf{V}\right)^{-1}\mathbf{V}^\mathrm{T}\mathbf{B}_k, \tag{15.12}$$

where $\mathbf{V}$ is a $N_{m,k} \times 1$ matrix of ones [51].

Mixing Predicted and Measured Values

It is also possible to simultaneously use both predicted and measured reference fingerprints in the CDB. First, a CDB structured as a uniform grid is built using propagation models. Then, measurements are carried out to collect reference fingerprints. These measurements points might be isolated or located along drive test routes. At pixels where measurement points are available, the measured reference fingerprints replace the predicted fingerprints. To smooth discontinuities between measured and predicted RSS values, some form of interpolation might be used at pixels around the measurement points [51]. For pixels far from measurement points, the purely predicted reference RF fingerprints are used.

The insertion of measured reference RF fingerprints in the CDB is expected to increase the MS location accuracy, especially at and around pixels where those measured fingerprints are available. However, as in CDBs built entirely with measurements, the problem of CDB updating is also crucial in mixed CDBs. After any change in RAN elements, updated measurements must be obtained to prevent location accuracy degradation [11]. If the measurement points are obtained from drive test routes, this means that new routes must be carried out to update the CDB. This might be very time-consuming and expensive.

An alternative solution for updating a mixed CDB is the use of *passive listeners* [21], which are MS placed at fixed known locations. These MS perform measurements, which are sent to serving base stations through NMRs, allowing for an automatic update of the mixed CDB. It is possible to increase the location accuracy at a given zone by deploying a sufficiently high number of passive listeners at that zone. An algorithm to indicate the optimum distribution of passive listeners in a given area is proposed in [21].

15.4 TECHNIQUES TO REDUCE THE SEARCH SPACE

Each CDB element contains a reference RF fingerprint and a set of geographic coordinates. The *search space* is the set of CDB elements whose reference RF fingerprints are compared to the target RF fingerprint. The geographic coordinates of the search space elements are *location candidates* for the MS location problem.

Initially, the search space comprises all CDB elements. However, it is not feasible to compare the target fingerprint to all reference fingerprints stored in the CDB, as this would result in a very large computational load and in a correspondingly long time to produce a position fix. Therefore, some technique must be applied to reduce the search space without significantly impairing the location accuracy.

Two search space-reducing techniques are presented in this section: CDB filtering [7] and optimized search with genetic algorithms (GA) [20, 8]. For better understanding, the search space-reducing techniques are presented assuming that the CDB is organized as a uniform grid. However, the extension to a CDB structured as an indexed list is straightforward.

The *original search space* is represented by set $\mathcal{A}$ and comprises the whole CDB. If the CDB is organized as a uniform grid and the service area covers a total surface of $l \times w \, \text{m}^2$, then the cardinality—i.e., the number of elements—of set $\mathcal{A}$ is given by $\#\mathcal{A} = \left\lceil \frac{l}{r_S} \right\rceil \times \left\lceil \frac{w}{r_S} \right\rceil$, where r_S is the planar resolution of the CDB. Set $\mathcal{A}$ is defined as:

$$\mathcal{A} = \left\{ (x_j, y_i, \mathbf{S}'_{i,j}) \mid i = 1, 2, \ldots, \left\lceil \frac{w}{r_S} \right\rceil \text{ and } j = 1, 2, \ldots, \left\lceil \frac{l}{r_S} \right\rceil \right\}, \quad (15.13)$$

where $\mathbf{S}'_{i,j}$ is the reference RF fingerprint at pixel (i, j). The geographic coordinates (x_j, y_i) of pixel (i, j) are given by (15.7).

The *reduced search space* $\mathcal{D}$ is a subset of $\mathcal{A}$. The *search space reduction factor* can be defined as:

$$\gamma = 1 - \frac{\#\mathcal{D}}{\#\mathcal{A}}, \quad (15.14)$$

where $\#\mathcal{A}$ and $\#\mathcal{D}$ are the number of elements in the original ($\mathcal{A}$) and reduced ($\mathcal{D}$) search spaces, respectively. Note that $\mathcal{D} \subset \mathcal{A}$. For a service area of $10 \times 10 \, \text{km}^2$ and $r_S = 5 \, \text{meters}$, $\#\mathcal{A} = 4{,}000{,}000$ elements. Without a technique to reduce the search space, for every position fix, 4,000,000 reference fingerprints would have to be compared to the target RF fingerprint. For a search space-reducing technique with $\gamma = 99\%$, this number would drop to 40,000 reference fingerprints per position fix.

15.4.1 CDB Filtering

This technique progressively reduces the search space, applying three successive filtering steps to the CDB elements [7]. The whole service area is depicted in Figure 15.6a, where the best server area of each cell is shown, as well as the vectors representing the streets. The service area map includes all CDB elements and corresponds to the original search space $\mathcal{A}$.

First Filtering Step

In the first filtering step, the search space $\mathcal{A}$ is restricted to the CDB elements within the best server area of the sector with the highest RSS in the target RF fingerprint, obtaining:

$$\mathcal{B} = \left\{ (x_j, y_i, \mathbf{S}'_{i,j}) \,\middle|\, \mathbf{S}'_{i,j} \in \mathcal{A} \text{ and } \mathbf{S}'_{i,j}(1,1) = \mathbf{F}(1,1) \right\}, \qquad (15.15)$$

where $\mathbf{S}'_{i,j}(1,1)$ and $\mathbf{F}(1,1)$ are the cell IDs of the strongest received signal at the reference and target RF fingerprints, respectively. The reference RF fingerprint is defined by (15.6). The target RF fingerprint is defined by (15.1).

The best server area of an arbitrary cell, whose ID is indicated by $\mathbf{F}(1,1)$, is shown in Figure 15.6b. This area contains the elements of $\mathcal{B}$.

Second Filtering Step

In the second filtering step, the search space $\mathcal{B}$ is restricted to the elements whose best server RTD values are equal to the best server RTD value in the target RF fingerprint. The resulting set is represented by:

$$\mathcal{C} = \left\{ (x_j, y_i, \mathbf{S}'_{i,j}) \,\middle|\, \mathbf{S}'_{i,j} \in \mathcal{B} \text{ and } \mathbf{S}'_{i,j}(1,3) = \mathbf{F}(1,3) \right\}, \qquad (15.16)$$

where $\mathbf{S}'_{i,j}(1,3)$ and $\mathbf{F}(1,3)$ are the reference and target best server RTD values, respectively. This filtering step is omitted in networks where RTD values are not available, like in Wi-Fi networks. Figure 15.6c shows the pixels that belong to $\mathcal{C}$.

Third Filtering Step

In the third filtering step, the search space $\mathcal{C}$ is restricted to the elements whose reference RF fingerprints contain the first N cells listed in the target RF fingerprint $\mathbf{F}$. As the rows of $\mathbf{F}$ are classified in descending order of RSS, these N cells are the ones with the highest RSS values in $\mathbf{F}$. The set of the N cell IDs with the highest RSS values in the target RF fingerprint is given by:

$$\mathcal{I}_{T_N} = \left\{ \mathbf{F}(1:N,1) \,\middle|\, N \in [1, N_a] \right\}, \qquad (15.17)$$

where N_a is the number of anchor cells in $\mathbf{F}$.

The set of cell IDs in the reference RF fingerprint at pixel (i, j) is given by:

$$\mathcal{I}_{R,i,j} = \left\{ \mathbf{S}'_{i,j}(1:N_{i,j}, 1) \,\middle|\, \mathbf{S}'_{i,j} \in \mathcal{C} \right\}, \qquad (15.18)$$

where $N_{i,j}$ is the number of cells in $\mathbf{S}'_{i,j}$.

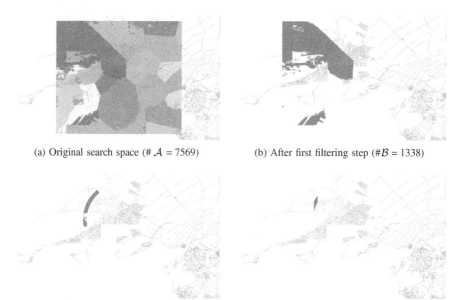

(a) Original search space (# $\mathcal{A}$ = 7569)   (b) After first filtering step (#$\mathcal{B}$ = 1338)

(c) After second filtering step (# $\mathcal{C}$ = 78)   (d) After third filtering step (#$\mathcal{D}$ = 8)

Figure 15.6   Reducing the search space with CDB filtering.

The cardinality of the set $(\mathcal{I}_{R,i,j} \cap \mathcal{I}_{T_N})$ informs how many of the cell IDs in the reference RF fingerprint of pixel (i, j) are among the N cell IDs with the highest RSS values at the target RF fingerprint. In the third filtering step, the search space $\mathcal{C}$ is restricted to the elements where $\#(\mathcal{I}_{R,i,j} \cap \mathcal{I}_{T_N}) \geq N$. The reduced search space thus obtained is represented by:

$$\mathcal{D} = \{(x_j, y_i, S'_{i,j}) \,|\, S'_{i,j} \in \mathcal{C} \text{ and } \#(\mathcal{I}_{R,i,j} \cap \mathcal{I}_{T_N}) \geq N \text{ and } N \in [1, N_a]\}. \quad (15.19)$$

Figure 15.6d shows the pixels that belong to $\mathcal{D}$. Note that $\mathcal{D} \subset \mathcal{C} \subset \mathcal{B} \subset \mathcal{A}$ and that $\#\mathcal{D} \ll \#\mathcal{A}$, which means that the CDB filtering technique achieves a high search space-reduction factor.

Example 15.1

Given a target RF fingerprint **F** defined by (15.20) and a sample 3×3 uniform grid CDB defined by (15.21), use CDB filtering and identify the elements in $\mathcal{D}$ for $N = 4$. Assume that only the best server RTD value is known, so all other RTD values are given a negative value to indicate that they are not available. Also assume that the RSS is quantized using 64 possible values in 1 dB steps from 0 (-110 dBm or below) to 63 (-48 dBm or above). Both assumptions are based on the GSM radio interface specifications [15].

$$F = \begin{bmatrix} 100 & 110 & 5 & 2 & 99 \\ 62 & 60 & 54 & 43 & 40 \\ 0 & -1 & -1 & -1 & -1 \end{bmatrix}^T, \quad (15.20)$$

$$S'_{1,1} = [100\ 55\ 1; 5\ 50 - 1; 110\ 49 - 1; 111\ 45 - 1; 10\ 34 - 1; 200\ 30 - 1; 201\ 29 - 1],$$

$$S'_{1,2} = [100\ 60\ 0; 110\ 50 - 1; 2\ 45 - 1; 5\ 40 - 1; 10\ 35 - 1],$$

$$S'_{1,3} = [100\ 59\ 1; 110\ 49 - 1; 2\ 50 - 1; 5\ 39 - 1; 10\ 36 - 1],$$

$$S'_{2,1} = [100\ 54\ 0; 5\ 50 - 1; 110\ 49 - 1; 111\ 45 - 1; 10\ 34 - 1; 200\ 30 - 1; 201\ 29 - 1],$$

$$S'_{2,2} = [100\ 61\ 0; 110\ 50 - 1; 2\ 45 - 1; 5\ 40 - 1; 10\ 35 - 1], \tag{15.21}$$

$$S'_{2,3} = [110\ 60\ 0; 2\ 52 - 1; 100\ 50 - 1; 5\ 39 - 1],$$

$$S'_{3,1} = 110\ 63\ 0; 2\ 52 - 1; 100\ 50 - 1; 5\ 38 - 1,$$

$$S'_{3,2} = [110\ 60\ 0; 100\ 52 - 1; 2\ 50 - 1],$$

$$S'_{3,3} = [110\ 59\ 1; 100\ 52 - 1; 2\ 50 - 1].$$

Solution

After the first filtering step is carried out, as defined in (15.15), the set $\mathcal{B}$ is obtained, being given by $\mathcal{B} = \{S'_{1,1}, S'_{1,2}, S'_{1,3}, S'_{2,1}, S'_{2,2}\}$. As the CDB is organized as a uniform grid, the $\mathcal{B}$ set (as well as the other sets obtained during the CDB filtering) might contain, instead of the reference fingerprints, the relative coordinates of the pixels within the CDB (i.e., their lines and columns). So, alternatively, this set would be expressed by $\mathcal{B} = \{(1, 1), (1, 2), (1, 3), (2, 1), (2, 2)\}$. Note that the cell IDs (i.e., the value in the first column) in the first line of all elements in $\mathcal{B}$ are equal to the best server ID in **F** (cell ID = 100). Then, the second filtering step is applied, as defined in (15.16), selecting the elements of $\mathcal{B}$ whose RTD value is equal to the one informed in **F** (RTD = 0), yielding the set $\mathcal{C} = \{(1, 2), (2, 1), (2, 2)\}$. Finally, the third filtering step, as defined in (15.19), is applied, yielding the reduced search space $\mathcal{D} = \{(1, 2), (2, 2)\}$. Note that $S'_{1,2}$ and $S'_{2,2}$ are the only reference fingerprints in the CDB that contain the 4 strongest cell IDs in **F** (cell IDs 100, 110, 5, and 2). Please refer to Appendix A for a MATLAB implementation of this solution.

15.4.2 Optimized Search Using GA

GA is an optimized and highly parallel search technique based on the principle of natural selection and genetic reproduction, which is expected to converge to a suboptimal solution after evaluating just a small subset of the entire search space [37]. Therefore, GA is a suitable option to reduce the search space in RF fingerprinting algorithms [32, 8]. Each candidate solution is an individual, represented by a numeric sequence called a chromosome. When using binary representation, each bit in a chromosome is referred to as a gene. The set of individuals at each cycle or generation is called the population. The individuals of a population are modified and combined by means of genetic operators—crossover, mutation, and elitism—producing a new population for the following generation. Crossover mixes segments of chromosomes of two individuals (parents), producing two new individuals (crossover children) for the next generation. Mutation is a random modification of one or more genes of a chromosome. Elitism is the technique of cloning the best individual of a generation into the next cycle [20]. The aptitude or fitness of an individual is assessed by means of an evaluation function. Better-fitted individuals have a higher probability of being selected for reproduction (crossover). The best individual in a population is the one who achieves the highest value at the evaluation function. This cycle continues until a stop criterion—maximum number of generations, fitness of the best individual, processing time, etc.—has been reached. The best individual of the last generation provides a suboptimal solution to the problem [37].

In the proposed GA application, each individual is a pixel. Each pixel has a reference RF fingerprint, which is used to evaluate the individual's fitness. The GA steps are:

1. initialize the first generation population, randomly selecting individuals within set $\mathcal{B}$, defined by (15.15);

2. evaluate the fitness of each individual of the current population, using a correlation function;

3. create chromosomes, converting the individual coordinates to a binary format;

4. apply genetic operators—crossover, mutation, and elitism—to create a new generation;

5. convert chromosomes to an integer format;

6. if stop criterion has been met, provide MS location, given by the coordinates of the fittest individual; otherwise, return to step 2.

The first step is an improvement of the formulation presented in [32], where the initial population is randomly selected among the pixels within $\mathcal{A}$. The proposed improvement is based on the assumption that the probability of an MS being located within the predicted best server area of its serving sector is higher than in any other pixels in the service area. Therefore, when initializing the first generation population, instead of randomly selecting individuals throughout the whole service area, the individuals should be randomly selected among the pixels within $\mathcal{B}$ [8]. As a result, the average fitness of the first generation population is higher—i.e., on average, the first generation individuals are closer to the real MS location—which means that, in comparison to [32], less generations are required to reach the suboptimal solution.

Each individual has a reference RF fingerprint. The higher the correlation between the reference RF fingerprint and the target RF fingerprint, the higher the fitness of this individual. The fittest individual in any given generation is the one who achieves the highest correlation. The correlation is calculated using one of the techniques described in Sections 15.5.1 and 15.5.3.

If the CDB is a uniform grid, the length of each chromosome created in the third step is the number of bits required to identify the position of a pixel—i.e., its row and column within the CDB—and is given by $\left\lceil \left(\log_2 \left\lceil \frac{l}{r_s} \right\rceil + \log_2 \left\lceil \frac{w}{r_s} \right\rceil \right) \right\rceil$, where $l \times w \, \text{m}^2$ is the service area surface and r_s is the CDB planar resolution.

The GA stops when one of the two conditions occurs: i) The maximum number of generation g_{max} is reached; and ii) the fitness of the best individual during α consecutive generations does not improve by a value higher than ε. The second condition is a modification of the common stop criterion based only on the maximum number of generations: If the aptitude of the best individual reaches a steady state, it might mean that the algorithm has reached a local maximum and therefore there is no need to evaluate new generations [25].

The reduced search space $\mathcal{D}$ contains the coordinates and the reference RF fingerprints of all individuals evaluated along all generations. The cardinality of set $\mathcal{D}$ is $\#\mathcal{D} = g \times \tau$, where g is the number of generations and τ is the number of individuals per generation. Note that $g \leq g_{max}$ and that $\mathcal{D} \subset \mathcal{B} \subset \mathcal{A}$.

15.5   PATTERN MATCHING OF RF FINGERPRINTS

After defining the RF fingerprint, the CDB, the search space, and the techniques to reduce it, it is necessary to specify how the reference RF fingerprints in the reduced search space are compared to the target RF fingerprint. The key idea is to find the reference RF fingerprint in the search space having the highest similarity or correlation with the target RF fingerprint.

If the correlation between the reference and target RF fingerprints is evaluated using *absolute RSS values*, this evaluation can be done in two ways: i) calculating the distance in the N-dimensional RSS space between the reference and target RF fingerprints [7] or ii) using a pattern matching algorithm based on artificial neural networks (ANNs) [9, 41]. If the correlation is evaluated using *relative RSS values*, this evaluation can be done, for example, using the *Spearman rank correlation coefficient* [45].

The MS is assumed to be located at the pixel whose reference RF fingerprint has the highest correlation with the target RF fingerprint. Alternatively, instead of selecting just the best match, it is possible to select the K-best matches, in which case the MS location is given by the arithmetic mean of the K-best matches coordinates. This method is called KNN (*K-nearest neighbors*) [39], and its effect in the location accuracy is analyzed in Section 15.6.

15.5.1   Distance in N-Dimensional RSS Space

The problem of defining the similarity between the reference and target RF fingerprints can be seen as one of determining the distance between those fingerprints in an N-dimensional RSS space [3]. Each dimension corresponds to a cell. The distance in the kth dimension is proportional to the difference between the reference and target RSS values of the kth cell. Figure 15.7 shows an example for $N = 3$. The Euclidean distance between the target RF fingerprint (black dot) and each reference RF fingerprint (white dots) in the 3-dimensional RSS space is indicated by the line segments.

The similarity or correlation between the reference and target RF fingerprints is inversely proportional to the N-dimensional distance between those fingerprints in the RSS space. This distance can be calculated using different metrics: Euclidean distance, *sum of absolute differences* (SAD), etc. The use of different metrics might yield significantly different location estimates. This effect is analyzed in Section 15.6.

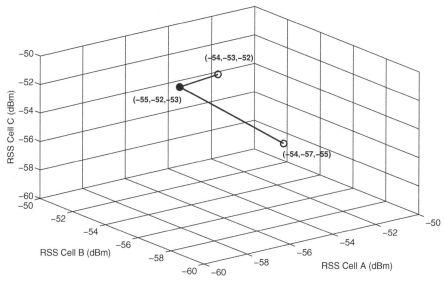

Figure 15.7 Euclidean distances between target (black dot) and reference (white dots) RF fingerprints in three-dimensional RSS space.

Two situations are considered when calculating the N-dimensional distance in RSS space: a *particular case*, to be applied specifically when CDB filtering is used [7], and the *generic case*, where a penalty term is introduced [29].

Particular Case

For simplicity, it is assumed here that the CDB is organized as a uniform grid and that the reduced search space $\mathcal{D}$ is obtained using CDB filtering, as described in Section 15.4.1. It is also assumed that $(i, j, \mathbf{S}'_{i,j}) \in \mathcal{D}$. The Euclidean distance between the target RF fingerprint $\mathbf{F}$ and the reference RF fingerprint $\mathbf{S}'_{i,j}$ in the N-dimensional RSS space is given by:

$$d_{i,j} = \sqrt{\sum_{k=1}^{N} \left(\left\lfloor \frac{\mathbf{S}'_{i,j}(n_k, 2) - \mathbf{F}(k, 2)}{\delta} \right\rfloor \right)^2}, \qquad (15.22)$$

where n_k is the index of the line in $\mathbf{S}'_{i,j}$, whose cell ID is equal to the cell ID in the kth line in $\mathbf{F}$, i.e., $\mathbf{S}'_{i,j}(n_k, 1) = \mathbf{F}(k, 1)$, with $n_k \in [1, N_{i,j}]$. The parameter $N_{i,j}$ is the number of rows in $\mathbf{S}'_{i,j}$. The parameter δ represents the MS inherent RSS measurement inaccuracy in dB [6]. In (15.22), any difference between target and reference RSS values smaller than δ is considered to be zero.

Generic Case with Penalty Term

It is assumed here that the CDB is organized as a uniform grid and that $(i, j, \mathbf{S}'_{i,j}) \in \mathcal{D}$. In the generic case with a penalty term, the Euclidean distance between the target

RF fingerprint $\mathbf{F}$ and the reference RF fingerprint $\mathbf{S}'_{i,j}$ in the N-dimensional RSS space is given by:

$$d_{i,j} = \sqrt{\sum_{k=1}^{N}\left(\left\lfloor\frac{S'_{i,j}(n_k,2)-F(m_k,2)}{\delta}\right\rfloor\right)^2 + 2\beta(N_a-N)}, \qquad (15.23)$$

where n_k and m_k are the indexes of the lines in $\mathbf{S}'_{i,j}$ and $\mathbf{F}$, respectively, whose cell IDs are equal to the cell ID of the kth element in $\mathcal{I} = \mathbf{S}'_{i,j}(:,1) \cap \mathbf{F}(:,1)$. Note that $n_k \in [1, N_{i,j}]$ and $m_k \in [1, N_a]$. The parameter N_a is the number of anchor cells in $\mathbf{F}$. The parameter β is the RSS dynamic range in dB units. In GSM networks, for example, $\beta = 63\,\text{dB}$ [15].

In (15.23), the parameter N informs how many of the cell IDs that are listed in $\mathbf{F}$ are also listed in $\mathbf{S}'_{i,j}$, i.e., $N = \#\mathcal{I}$. Unlike (15.22), these cell IDs are not necessarily the ones with the N highest RSS values in $\mathbf{F}$. For each cell ID listed in $\mathbf{F}$ and absent from $\mathbf{S}'_{i,j}$, a penalty term 2β is added. This value is twice the maximum difference between any target and reference RSS values. This ensures that, given two reference RF fingerprints, the one with the higher N will be closer to the target RF fingerprint, regardless of the value yielded by the first term of (15.23). Note that, for any given reference RF fingerprint, if $N = N_a$, Equations (15.22) and (15.23) yield the same result.

If instead of the Euclidean distance, the sum of absolute differences is used, then (15.23) becomes:

$$d_{i,j} = \sum_{k=1}^{N}\left\lfloor\frac{|S'_{i,j}(n_k,2)-F(m_k,2)|}{\delta}\right\rfloor + 2\beta(N_a-N). \qquad (15.24)$$

Example 15.2

Given a target RF fingerprint $\mathbf{F}$ and two reference RF fingerprints—$\mathbf{S}'_{10,20}$ and $\mathbf{S}'_{10,25}$—associated with the pixels (10, 20) and (10, 25), respectively, find the pixel whose reference RF fingerprint is closest to the target RF fingerprint using (15.22). Repeat using (15.24). The target and reference RF fingerprints are given by (15.25). Assume that only the best server RTD value is known, so all other RTD values are given a negative value to indicate that they are not available. Also assume that the RSS is quantized using 64 possible values in 1dB steps from 0 ($-110\,\text{dBm}$ or below) to 63 ($-48\,\text{dBm}$ or above). Both assumptions are based on the GSM radio-interface specifications [15].

$$F = \begin{bmatrix} 100 & 62 & 0 \\ 110 & 60 & -1 \\ 005 & 54 & -1 \\ 002 & 43 & -1 \end{bmatrix} \quad S'_{10,20} = \begin{bmatrix} 100 & 55 & 0 \\ 005 & 50 & -1 \\ 110 & 49 & -1 \\ 111 & 45 & -1 \\ 010 & 34 & -1 \\ 200 & 30 & -1 \end{bmatrix} \quad S'_{10,25} = \begin{bmatrix} 100 & 60 & 0 \\ 110 & 50 & -1 \\ 002 & 45 & -1 \\ 005 & 40 & -1 \\ 010 & 35 & -1 \end{bmatrix}. \qquad (15.25)$$

Solution

If the reduced search space $\mathcal{D}$ is obtained using CDB filtering, the sets $\mathcal{I}_{T_N}$ and $\mathcal{I}_{R,i,j}$ must be identified. They are defined by (15.17) and (15.18), respectively. From the RF fingerprints in (15.25) and assuming $N = 3$, one obtains:

$$
\begin{aligned}
\mathcal{I}_{R,10,20} &= \{005, 010, 100, 110, 111, 200\} \\
\mathcal{I}_{R,10,25} &= \{002, 005, 010, 100, 110\} \\
\mathcal{I}_{T_N} &= \{005, 100, 110\} \\
(\mathcal{I}_{R,10,20} \cap \mathcal{I}_{T_N}) &= (\mathcal{I}_{R,10,25} \cap \mathcal{I}_{T_N}) = \{005, 100, 110\}.
\end{aligned}
\tag{15.26}
$$

Therefore, $(x_{20}, y_{10}, S'_{10,20}) \in \mathcal{D}$ and $(x_{25}, y_{10}, S'_{10,25}) \in \mathcal{D}$, because:

$$
\begin{aligned}
S'_{10,20}(1,1) &= F(1,1), S'_{10,20}(1,3) = F(1,3), \#(\mathcal{I}_{R,10,20} \cap \mathcal{I}_{T_N}) = 3 \\
S'_{10,25}(1,1) &= F(1,1), S'_{10,25}(1,3) = F(1,3), \#(\mathcal{I}_{R,10,25} \cap \mathcal{I}_{T_N}) = 3.
\end{aligned}
\tag{15.27}
$$

Using (15.22) to calculate the N-dimensional distances in RSS space between each reference RF fingerprint and the target fingerprint, one obtains:

$$
d_{10,20} =
$$

$$
\sqrt{\left(\left\lfloor \frac{S'_{10,20\,(1,2)} - F_{(1,2)}}{\delta} \right\rfloor\right)^2 + \left(\left\lfloor \frac{S'_{10,20\,(3,2)} - F_{(2,2)}}{\delta} \right\rfloor\right)^2 + \left(\left\lfloor \frac{S'_{10,20\,(2,2)} - F_{(3,2)}}{\delta} \right\rfloor\right)^2}
$$

$$
d_{10,25} =
\tag{15.28}
$$

$$
\sqrt{\left(\left\lfloor \frac{S'_{10,25\,(1,2)} - F_{(1,2)}}{\delta} \right\rfloor\right)^2 + \left(\left\lfloor \frac{S'_{10,25\,(2,2)} - F_{(2,2)}}{\delta} \right\rfloor\right)^2 + \left(\left\lfloor \frac{S'_{10,25\,(4,2)} - F_{(3,2)}}{\delta} \right\rfloor\right)^2}.
$$

Assuming $\delta = 6\,\mathrm{dB}$ [15]:

$$
\begin{aligned}
d_{10,20} &= \sqrt{\left(\left\lfloor \frac{55-62}{6} \right\rfloor\right)^2 + \left(\left\lfloor \frac{49-60}{6} \right\rfloor\right)^2 + \left(\left\lfloor \frac{50-54}{6} \right\rfloor\right)^2} = 1.41 \\
d_{10,25} &= \sqrt{\left(\left\lfloor \frac{60-62}{6} \right\rfloor\right)^2 + \left(\left\lfloor \frac{50-60}{6} \right\rfloor\right)^2 + \left(\left\lfloor \frac{40-54}{6} \right\rfloor\right)^2} = 2.23.
\end{aligned}
\tag{15.29}
$$

If there are no other location candidates, then $\hat{x} = (x_{20}, y_{10})$, as $d_{10,20} < d_{10,25}$.

Using (15.24) to calculate the N-dimensional distances in RSS space between each reference RF fingerprint and the target fingerprint, one obtains:

$$
\begin{aligned}
d_{10,20} &= \left\lfloor \frac{|S'_{10,20\,(1,2)} - F_{(1,2)}|}{\delta} \right\rfloor + \left\lfloor \frac{|S'_{10,20\,(3,2)} - F_{(2,2)}|}{\delta} \right\rfloor \\
&\quad + \left\lfloor \frac{|S'_{10,20\,(2,2)} - F_{(3,2)}|}{\delta} \right\rfloor + 2\beta(4-3) \\
d_{10,25} &= \left\lfloor \frac{|S'_{10,25\,(1,2)} - F_{(1,2)}|}{\delta} \right\rfloor + \left\lfloor \frac{|S'_{10,25\,(2,2)} - F_{(2,2)}|}{\delta} \right\rfloor \\
&\quad + \left\lfloor \frac{|S'_{10,25\,(3,2)} - F_{(4,2)}|}{\delta} \right\rfloor + \left\lfloor \frac{|S'_{10,25\,(5,2)} - F_{(3,2)}|}{\delta} \right\rfloor + 2\beta(4-4).
\end{aligned}
\tag{15.30}
$$

Assuming $\delta = 6$ dB and $\beta = 63$ dB [15]:

$$d_{10,20} = \left\lfloor \frac{|55-62|}{6} \right\rfloor + \left\lfloor \frac{|49-60|}{6} \right\rfloor + \left\lfloor \frac{|50-54|}{6} \right\rfloor + 126 = 128$$

$$d_{10,25} = \left\lfloor \frac{|60-62|}{6} \right\rfloor + \left\lfloor \frac{|50-60|}{6} \right\rfloor + \left\lfloor \frac{|40-54|}{6} \right\rfloor + \left\lfloor \frac{|45-43|}{6} \right\rfloor = 3. \tag{15.31}$$

If there are no other location candidates, then $\hat{x} = (x_{25}, y_{10})$, as $d_{10,25} < d_{10,20}$. Note that by changing the distance evaluation function from (15.22) to (15.24), the MS location estimate $\hat{x}$ changes from (x_{20}, y_{10}) to (x_{25}, y_{10}).

Please refer to Appendix B for a MATLAB implementation of this solution.

15.5.2 Pattern Matching Using Artificial Neural Networks

ANNs are distributed parallel systems composed by interconnected processing units called *neurons*, which perform mathematical functions [23]. ANNs are usually applied to solve function approximation and classification problems.

An ANN can be used to approximate a function that represents the mapping between a RF fingerprint and a position (x, y) with the least squared error [44]. The proposed ANN topology has $3N_c$ inputs, M neurons in the hidden layer, and 2 neurons in the output layer [41], as shown in Figure 15.8.

In the input layer, there are three inputs per cell in the service area: the first is a boolean variable indicating if the cell is present or not in the RF fingerprint; the second is the normalized RSS value of that cell; and the third is the normalized RTD

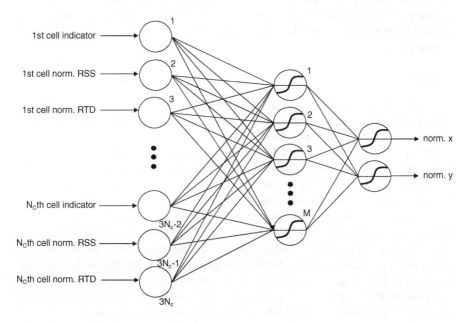

Figure 15.8   Schematic representation of ANN topology.

value of that cell, if available. The RSS and RTD values are normalized within a range that depends on the transfer function used on the neurons in the hidden layer.

Only one hidden layer is used with *M* neurons. This is in keeping with the *universal approximation theorem*, which states that a single hidden layer, if correctly dimensioned, is enough to approximate most nonlinear functions [41].

There are 2 neurons at the output layer, as only two-dimensional positioning is being considered. The MS estimated coordinates at the ANN output are normalized within a range that depends on the transfer function used in the neurons in the output layer.

The connections between the neurons are called *synapses*, each one having a numeric weight. Each neuron has a *bias*, which is a numeric value added to the output of its transfer function. The weights and biases are adjusted during the training phase to enable the multilayer feed-forward ANN to approximate a nonlinear function. The initial weights and biases are usually randomly selected within predetermined boundaries.

During the *supervised learning phase* [23], all reference RF fingerprints stored in the CDB are presented as inputs to the ANN. For each input $\mathbf{S}'_{i,j}$, the ANN yields an output $(\hat{x}_i, \hat{y}_i)$, which is then compared to the target output (x_j, y_i). The weights and biases of the network are iteratively adjusted to minimize the *mean squared error* (MSE) between the network outputs and the target outputs. The training ends when a target MSE has been reached or after a maximum number of epochs.

After being successfully trained, the ANN is ready to receive target RF fingerprints as inputs, yielding as outputs the MS location estimates.

Example 15.3

Given a test and training sets, with their respective target outputs (i.e., the MS truth ground positions associated with each fingerprint in the test and training sets), train and test a ANN in MATLAB. During the test phase, evaluate the resulting MS positioning error.

Solution

The MATLAB code in Appendix C shows a supervised training session of a backpropagation ANN to be used for MS location. First, it is necessary to load the input training patterns. Each input pattern follows the structure shown in Figure 15.8. There will be an input pattern for each reference RF fingerprint stored in the CDB. After normalization using the mapminmax command, the coordinates associated with each CDB element become the ANN output targets. The backpropagation ANN is created using the newff command. After the stop criteria has been met—maximum number of epochs, MSE value, or maximum number of validation failures—the trained ANN is stored for further use, as well as the training session record. The MATLAB code in Appendix D shows how the trained ANN is tested using the sim command. First, the test set must be loaded. It contains the target RF fingerprints—already in the format described by Figure 15.8. The test set is given to the ANN, and the achieved outputs are compared to the target MS coordinates. The location error is calculated by the 2-dimensional Euclidean distance between the ANN output and the MS target coordinate.

15.5.3 Spearman Rank Correlation Coefficient

Manufacturing variations among MS devices might affect the way they measure RSS values. As a result, at a same location and time, different MS devices might yield different RSS values for the same cell. If the CDB is built from field measurements using a certain MS device, and another MS device is to be localized, the location accuracy of a fingerprinting method using this CDB will deteriorate. This is even more likely to occur if the MS used to build the CDB and the MS used to test the location accuracy are from different manufacturers. To mitigate this *cross-device* effect [11], instead of using absolute RSS values in the correlation function, one might use the relative RSS values, i.e., their ranking. The ranking of RSS values is more robust to cross-device operation, which means that, while the absolute RSS values of a set of cells might be quite different when measured by different MS devices, their ranking is more likely to remain the same or at least, more similar. This is based on the assumption that the relationship between the input signal strength and the RSS is a monotonically increasing function [28].

For a better understanding, consider two MS devices, MS_1 and MS_2, both placed at the same location, and two base stations, BTS_A and BTS_B. The signals from these base stations reach both MS devices with input signal strengths s_A and s_B. The RSS values informed by MS_1 are RSS_{1A} and RSS_{1B}. The RSS values informed by MS_2 are RSS_{2A} and RSS_{2B}. If the relationship between the input signal strength and the RSS is a monotonically increasing function, then, if $s_A > s_B$, $RSS_{1A} > RSS_{1B}$ and $RSS_{2A} > RSS_{2B}$. While (RSS_{1A}, RSS_{1B}) might be different from (RSS_{2A}, RSS_{2B}), the ranking of BTS_A and BTS_B RSS values is the same on both MS devices.

The similarity between different rankings of the same set of cells might be evaluated by the Spearman rank correlation coefficient [45]. This coefficient might be used to calculate the correlation between the target RF fingerprint $\mathbf{F}$ and a reference RF fingerprint $\mathbf{S}'_{i,j}$. However, these fingerprints do not necessarily have the same number of cells nor the same cells. Therefore, before applying the Spearman correlation factor, some modification is required. Two $N_c \times 2$ matrices, $\mathbf{V}_F$ and $\mathbf{V}_S$ are created, with initial values defined by:

$$\mathbf{V}_F(n, 1) = \mathbf{V}_S(n, 1) = \mathrm{ID}_n$$
$$\mathbf{V}_F(n, 2) = \mathbf{V}_S(n, 2) = N_c, \tag{15.32}$$

where $n = 1, 2, \ldots, N_c$. The parameter N_c is the number of cells in the service area and ID_n is the cell ID of the nth cell in the service area.

The position of each cell in the RSS ranking in $\mathbf{F}$ must be inserted in the second column of the correspondent row in $\mathbf{V}_F$. As the rows in $\mathbf{F}$ are organized in descending order of RSS, the position of each cell in the RSS ranking is the row index, as defined by:

$$\mathbf{V}_F(n_k, 2) = k, \text{where } \mathbf{V}_F(n_k, 1) = \mathbf{F}(k, 1), n_k \in [1, N_c] \text{ and } k = 1, 2, \ldots, N_a. \tag{15.33}$$

The same procedure must be followed for $\mathbf{S}'_{i,j}$ and $\mathbf{V}_S$, as defined by:

$$\mathbf{V}_S\left(n_k, 2\right) = k, \text{where } \mathbf{V}_S\left(n_k, 1\right) = \mathbf{S}_{i,j}\left(k, 1\right), n_k \in \left[1, N_c\right] \text{ and } k = 1, 2, \ldots, N_{i,j}. \quad (15.34)$$

The Spearman rank correlation coefficient between the target RF fingerprint and the reference RF fingerprint at pixel (i, j) is given by:

$$\rho_{i,j} = \frac{\sum_{n=1}^{N_c}\left\{\left(\mathbf{V}_F\left(n, 2\right) - \bar{R}_F\right)\left(\mathbf{V}_S\left(n, 2\right) - \bar{R}_S\right)\right\}}{\sqrt{\sum_{n=1}^{N_c}\left\{\left(\mathbf{V}_F\left(n, 2\right) - \bar{R}_F\right)^2\right\}\sum_{n=1}^{N_c}\left\{\left(\mathbf{V}_S\left(n, 2\right) - \bar{R}_S\right)^2\right\}}}, \quad (15.35)$$

where $\bar{R}_F = \frac{1}{N_c}\sum_{n=1}^{N_c}\left\{\mathbf{V}_F\left(n, 2\right)\right\}$ and $\bar{R}_S = \frac{1}{N_c}\sum_{n=1}^{N_c}\left\{\mathbf{V}_S\left(n, 2\right)\right\}$. The Spearman distance can be defined as [45]:

$$d_{i,j} = 1 - \rho_{i,j}. \quad (15.36)$$

Note that, as $\rho_{i,j}$ ranges from -1 to 1, the Spearman distance, as defined in (15.36), ranges from 0 to 2.

Example 15.4

Calculate the Spearman distance between the target RF fingerprint $\mathbf{F}$ defined in (15.20) and the reference RF fingerprint $\mathbf{S}'_{2,2}$ defined in (15.21), assuming $N_c = 10$.

Solution

First, the second column of $\mathbf{V}_F$ and $\mathbf{V}_S$ must be filled with the ranking of the cell IDs in $\mathbf{F}$ and $\mathbf{S}'_{2,2}$, respectively. The matrices thus obtained are given by (15.37) and (15.38), with $\bar{R}_F = \bar{R}_S = 6.5$. Applying (15.37), (15.38), $\bar{R}_F$, and $\bar{R}_S$ to (15.36), one obtains the Spearman distance $d = 0.1379$. Please refer to Appendix E for a MATLAB implementation of this solution.

$$\mathbf{V}_F = \begin{bmatrix} 1 & 2 & 5 & 10 & 99 & 100 & 110 & 120 & 130 & 200 \\ 10 & 4 & 3 & 10 & 5 & 1 & 2 & 10 & 10 & 10 \end{bmatrix}^T \quad (15.37)$$

$$\mathbf{V}_S = \begin{bmatrix} 1 & 2 & 5 & 10 & 99 & 100 & 110 & 120 & 130 & 200 \\ 10 & 3 & 4 & 5 & 10 & 1 & 2 & 10 & 10 & 10 \end{bmatrix}^T. \quad (15.38)$$

15.6 EXPERIMENTAL PERFORMANCE

This section presents some experimental results obtained applying the RF fingerprinting techniques described in this chapter in two different RANs and environments: in a GSM network in an outdoor dense urban area and in Wi-Fi 802.11b/g networks in an indoor environment.

15.6.1   Outdoor 850-MHz GSM Network

Field tests were performed in a 850-MHz GSM network in the downtown region of Rio de Janeiro. The region is a $2.2 \times 2.2\,\text{km}^2$ dense urban area with 24 cells/km^2 [7]. The DEM used to represent the test area has a planar resolution $r_H = 5$ meters and includes building heights, which increases the accuracy of the propagation modeling used to build the CDB. The test set was composed of a GSM phone and a GNSS receiver, both connected to a laptop placed inside a moving vehicle. The MS was in active mode and for each transmitted NMR the current location was calculated by the GNSS receiver. The MS sends two NMRs per second containing the cell ID and RSS of the best server and up to the six strongest neighbor cells. The RTD values, known as *timing advance* (TA) in GSM systems, were registered every time a NMR was sent. A total of 4500 NMRs, TA values, and GNSS measurements were recorded for further processing. The GNSS-computed location was assumed to be the reference position, so, for each NMR and each location method, the positioning error is the Euclidean distance in meters between the GNSS position and the location provided by the respective method.

The location precision of five MS positioning methods is evaluated using the *cumulative distribution function* (CDF) of the location error. For each method, a *moving average filter* with length 20 is used to eliminate abrupt variations in location estimates between adjacent position fixes along the test route [40], so the current MS location estimate is given by the arithmetic mean of the previous 20 estimated positions.

Methods I to IV are fingerprinting location methods using a CDB built from propagation modeling, and Method V is a cell identity (CID) location method [19], where the MS is assumed to be located at the best serving cell coordinates. The main characteristics of Methods I to V are summarized in Table 15.1. The CID positioning method usually has poor accuracy and is included for comparison only.

The GA specific parameters in Method III are: 60% crossover ratio, roulette selection [25], 1% mutation ratio, elitism, 16-bit chromosomes, $g_{max} = 20$ generations, $\alpha = 5$ generations, $\varepsilon = 0.00001$, and $\gamma = 3\%$. The number of individuals per generation is given by $(\#\mathcal{B} \times \gamma)/g_{max}$. Set $\mathcal{B}$ is defined by (15.15), and γ is the search-space reduction factor, defined by (15.14) [8].

TABLE 15.1.   Positioning Methods in GSM Test

Method	Type	Search-Space Reduction Technique	Correlation Function	N	δ
I	Fingerprinting	CDB filtering	Euclidean distance (15.22)	5	6 dB
II	Fingerprinting	CDB filtering	Spearman distance (15.36)	n/a	n/a
III	Fingerprinting	GA	SAD with penalty term (15.24)	5	6 dB
IV	Fingerprinting	n/a	ANN	n/a	n/a
V	Cell identity	n/a	n/a	n/a	n/a

Method IV uses a ANN to estimate the MS position. The ANN topology is the one described in Section 15.5.2, with 15 neurons in the hidden layer. The neurons in the hidden and output layers have hyperbolic tangent sigmoid transfer functions [47]. The input patterns—i.e., the reference fingerprints stored in the CDB—are randomly divided into two groups: 95% are selected for the *training set* and 5% are selected for the *validation set*. The validation set is used to monitor the ANN learning and prevent *overtraining* [23]. The input patterns in the training set are presented three times to the ANN to reinforce its learning. The learning algorithm is the *Levenberg-Marquardt backpropagation* [5].

Methods I to IV are fingerprinting location methods using a CDB built from propagation modeling. This CDB was built using the Okumura-Hata path loss equation upon a DEM with 5 meters planar resolution. The CDB is available at three different planar resolutions, 5, 10, and 25 meters. Figure 15.9a shows Method I location error CDF obtained with the three different values of r_S. The location precision is slightly lower for $r_S = 25$ meters. For $r_S = 5$ meters and $r_S = 10$ meters, the location precision is approximately the same. Therefore, 10 meters is the better choice for the CDB planar resolution, as it provides the same location precision of $r_S = 5$ meters, while resulting in a CDB four times smaller, which reduces computational complexity and the time to produce each position fix.

Figure 15.9b shows the location error CDF of Methods I to IV, using a CDB with $r_S = 10$ meters and of Method V, which is a CID location method and is inserted for comparison. Method I achieves the best overall results, with location errors of 98, 128, and 205 meters for the 50th, 67th and 90th percentiles, respectively. Similar fingerprinting methods using CDBs built from field measurements in GSM networks in dense urban areas achieved 94 and 291 meters for the 50th and 90th percentiles in [11] and an average positioning error of 100 meters in [21]. These results show that a fingerprinting location method using a CDB built from properly tuned propagation models can achieve a location precision comparable to that achieved with a CDB built from field measurements.

Methods I to III greatly outperform the basic CID method, while Method IV has the worst performance. The low accuracy achieved by Method IV suggests that the ANN with the described topology is not able to approximate the nonlinear function mapping the RF fingerprints into geographic coordinates when reference RF fingerprints generated with propagation modeling are used in the training phase.

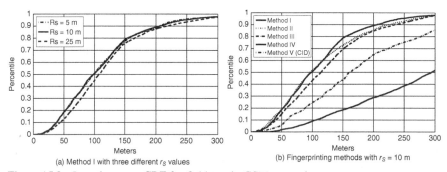

(a) Method I with three different r_S values

(b) Fingerprinting methods with $r_S = 10$ m

Figure 15.9 Location error CDF for field test in GSM network.

Nonetheless, good results have been achieved in [41] and [44] using ANN trained with measured reference RF fingerprints, which therefore held greater similarity with the target RF fingerprints used in the test phase.

15.6.2   Indoor Wi-Fi Networks

A total of 275 measurement positions has been selected on the fourth floor of the Universidade do Estado do Rio de Janeiro main building. Two Wi-Fi 802.11b/g adapters—an *Atheros* AR5005GS and a *TP-Link*—were used to collect the ID and RSS of *access points* (APs), at a rate of 1 measurement per second. Each measurement contains the ID and RSS of up to 20 APs. Each adapter collected between 90 and 120 measurements per point. The RSS value of each detected AP, averaged during the 90–120 second period, was inserted in a reference RF fingerprint and stored in the CDB. Therefore, two CDBs built from field measurements were obtained, one for each adapter, both with 275 elements. To test the fingerprinting location methods, a single measurement (1 second) per measurement point was randomly selected to compose a target RF fingerprint for each adapter.

Due to the small number of elements in the CDB, no search space-reduction technique was used. Three correlation functions were tested: N-dimensional Euclidean distance, SAD with penalty factor, and Spearman distance. The first two functions used a parameter $\delta = 5\,dB$ [26].

Figure 15.10a shows the location error CDF obtained when the test set and CDB are built using the *Atheros* adapter. All functions yield a median accuracy approximately equal to the measurement point grid spacing, i.e., 3 meters. However, the Spearman distance achieves the higher precision, with location errors of 2.9, 3.4, and 16 meters for the 50th, 67th and 90th percentiles, respectively. In the GSM test, Method II—which uses Spearman distance—does not yield the best results. Ranking correlation achieves the highest precision among other methods in Wi-Fi networks, but not in GSM networks, probably due to the higher dimensionality—i.e., number of anchor cells—of the Wi-Fi RF fingerprints: each RF fingerprint collected in the Wi-Fi test might have up to 20 APs, while in the GSM test the maximum number of cells in a fingerprint is 7. Ranking correlation—as the Spearman distance—compares sequences of cell IDs, ordered by RSS. A longer sequence is more likely to be unique, i.e., not to repeat itself at different geographic locations. In such

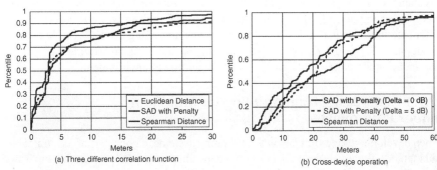

(a) Three different correlation function   (b) Cross-device operation

Figure 15.10   Indoor location error CDF.

conditions, the ranking correlation has a higher probability of correctly identifying the MS location, in comparison to correlation functions comparing absolute RSS values.

Figure 15.10b shows the location error CDF obtained in *cross-device operation*, i.e., when the test set is built using one adapter (*TP-Link*), and the CDB is built another adapter (*Atheros*). For the cross-device test, two laptops (one with a *TP-Link* Wi-Fi card, another with *Atheros* Wi-Fi card) running *NetStumbler* were placed upon a table with wheels. Each measurement for each card was taken at the same position, at the same time. The precision is clearly worse than in Figure 15.10a. In such conditions, the Spearman distance and SAD with penalty factor ($\delta = 5\,$dB) obtain higher precision, in comparison to the SAD with penalty factor ($\delta = 0\,$dB). These results indicate that, as discussed in Section 15.5.3, the ranking correlation mitigates the accuracy degradation due to cross-device operation.

15.7   SOME FINAL REMARKS ON FINGERPRINTING LOCATION TECHNIQUES

So far we have discussed planar (two-dimensional) positioning using fingerprinting techniques. We have described the main issues regarding fingerprinting and how to put them to work. Although, we have to point out to some developments. A collection of recent advances for Wi-Fi Fingerprinting localization is avaliable at [24]. Let us briefly illustrate some aspects. For example, in [14] the fingerprint is constructed based on Shannon's entropy. The different RSS levels received from an AP are defined as the events and their relative frequencies of falling within a RSS range are used to compute the entropy of an event—a set of RSS levels.

Some works propose using CIR (channel impulse response) or CSI (channel state information—the channel frequency response) to build large dimension fingerprints. These may be processed using the pattern matching techniques already described or using deep learning [48] to estimate the position of the mobile node. The use of such machine (deep/extreme) learning algorithms is increasing for fingerprint-based positioning [13] (maybe due to the availability of such systems even remotely on the cloud).

Multipath may produce fast RSS fading and thus RSS to vary rapidly what in turn makes it difficult to obtain an accurate position estimate from RSS (as explained in Chapters 11 and 12). As an alternative, other types of fingerprints may also be used. For example, in [18] it is proposed to estimate the receiver position from the difference in the arrival times between the waves impinging on the receiver by different paths. While this demands the capability to discern signal arriving paths, only one anchor node can be required in a heavily multipath environment.

15.7.1   RSS-Based Fingerprinting for Multifloor Indoor Positioning

When developing three-dimensional positioning systems, a great part of the strategies and methods already discussed can be used depending on the RF propagation environment. To overcome possible inaccuracies using the fingerprinting methodology, a set of consecutive RSS measurements, an outlier detection strategy [9], or a finer CDB grid may be used to counterbalance the fast-fading effects. That is, three-dimensional positioning could also employ some of the clustering techniques discussed to group RF fingerprints using the underlying assumption that they will as similar as their correspondent locations are closer.

However, for multifloor indoor localization the complexity of radio-wave propagation environment may severely hamper the above strategies. Consider an indoor localization system using Wi-Fi signals. The received RSS levels may arrive from APs that are placed in neighbor rooms or even in a different floor. That is, obstacles may exist in the RF link, whose attenuation depend on the material traversed by the RF wave. Consequently, there is no guarantee that the RSS levels from an AP measured at different locations inside a building will somehow reflect any proximity among locations, nor an architectural correspondence (floor, aisle, rooms, etc).

Consequently, the implicit assumption that similarities among RSS data points map to proximity may lead to unacceptable positioning estimates. In emergency calls, response teams (firemen, paramedics, policemen) need accurate position fixes of the target, particularly in what refers to floor identification, whose correctness should closely match 100% [46, 34]. Buildings are designed for people to move more frequently within a floor than between floors. The same applies for indoor positioning applications designed to help people with special needs, like children and elderly tracking [34], or to help the blind finding the floor of destination in multistory buildings [4].

To overcome floor identification requirements, some methods rely on using other sensor data to detect the displacement between floors [43], and fingerprinting is employed to estimate the position within the floor. The strategy of augmenting location information using other embedded sensors together with fingerprinting seems to be a choice for the localization of future IoT (Internet of Things) devices within buildings [31]. Other works approach the problem of providing a good floor estimate from a fingerprint using a mix of techniques. For example, in [10] unsupervised clustering (K-medians and Kohonen layer) is used together with majority voting committees of ANNs. The unsupervised clustering is employed for letting fingerprints group freely in the data space. This makes no assumption of architectural constraints and any natural arrangement of the collected fingerprints. Majority voting committee of backpropagation ANNs is used within each cluster for identifying the floor in which the MS is located from the RSS levels data.

15.8 CONCLUSION

In this chapter, the main elements of network-based RF fingerprinting methods have been presented: i) the target and reference RF fingerprints; ii) the CDB and the alternatives to build it, either using field measurements or propagation modeling, including an overview of empirical propagation model calibration; iii) the concept of search space and techniques to reduce it; iv) target and reference RF fingerprint correlation, using either ANNs, absolute RSS correlation or RSS ranking correlation.

The MS positioning precision of the presented location techniques was analyzed in field tests in a GSM network in an outdoor dense urban environment and in Wi-Fi networks in an indoor environment. Table 15.2 summarizes the main

TABLE 15.2. Location Methods Comparison

Test Environment	Search Space Reduction (and CDB Type)	Pattern Matching	Filtering	67th and 95th Location Errors (meters)	Performance Remarks	References
Outdoor GSM	CDB filtering (Predicted)	RSS correlation (Euclidean dist.)	Moving avg. and KNN	134 and 261	Highest precision in outdoor networks	[6, 7, 11]
Outdoor GSM	CDB filtering (Predicted)	Ranking correlation (Spearman dist.)	Moving avg. and KNN	133 and 267	Best suited for cross-device operation	[12, 28]
Outdoor GSM	GA (Predicted)	RSS correlation (SAD with penalty term)	Moving avg.	145 and 276	High computational complexity results in high delay	[8, 32]
Outdoor GSM	— (Predicted)	ANN	Moving avg.	352 and 493	Suitable only for use with CDBs built from field measurements	[38, 41, 44]
Indoor Wi-Fi	— (Measured)	RSS correlation (Euclidean dist.)	Moving avg.	5.3 and 37	Better precision than SAD and lower delay than Spearman	[3]
Indoor Wi-Fi	— (Measured)	RSS correlation (SAD with penalty term)	Moving avg.	5.9 and 33.9	Low computational complexity	[36]
Indoor Wi-Fi	— (Measured)	Ranking correlation (Spearman dist.)	Moving avg.	3.4 and 23.3	Best suited for cross-device operation	[12]

characteristics and experimental results of the analyzed fingerprinting location techniques.

The results obtained in the Wi-Fi test suggest that ranking correlation is capable of improving the location accuracy in cross-device operation conditions, especially for RF fingerprints with higher dimensionality, i.e., with a greater number of RSS values (at least 20). A similar result is obtained when the MS-inherent RSS measurement inaccuracy is considered in the evaluation function.

The results obtained in the GSM network and in the literature suggest that a fingerprinting location method using a CDB built from propagation modeling can achieve a MS location precision comparable to that achieved when a CDB built from field measurements is used. Model I (CDB filtering, N-dimensional Euclidean distance and KNN with $K = 5$, moving average filtering with $L = 20$) achieved the higher overall precision, with location errors of 98, 128, and 244 meters for the 50th, 67th, and 95th percentiles, respectively. The Federal Communications Commission (FCC) enhanced 911 (E911) location accuracy requirements for network-based methods are 100 and 300 meters for the 67th and 95th percentiles, respectively [17]. The results obtained by Method I meet the FCC E911 accuracy requirement for the 95th percentile but not for 67th percentile. Additional research and tests are necessary to improve Method's I location precision, possibly through the use of a mixed CDB, like the one described in Section 15.3.2, using propagation modeling and field measurements obtained by passive listeners [21]. Calibration of the propagation models might also improve Method's I positioning accuracy.

REFERENCES

[1] E. Aarnæs and S. Holm, "Tuning of empirical radio propagation models effect of location accuracy," *Wireless Pers. Commun.*, vol. 4, no. 2–4, pp. 267–281, Dec. 2004.

[2] S. Ahonen and H. Laitinen, "Database correlation method for UMTS location," in *Proc. of IEEE 57th Vehicular Technology Conf.*, Jeju, Korea, Apr. 2003, pp. 2696–2700.

[3] P. Bahl and V. N. Padmanabhan, "RADAR: An in-building RF-based user location and tracking system," in *Proc. of 19th Annual Joint Conf. of the IEEE Computer and Communications Societies*, Tel Aviv, Israel, Mar. 2000, pp. 775–784.

[4] Y. Bai, W. Jia, H. Zhang, Z.-H. Mao, and M. Sun, "Helping the blind to find the floor of destination in multistory buildings using a barometer," in *2013 35th Annual Int. Conf. of the IEEE Engineering in Medicine and Biology Society (EMBC)*, Osaka, Japan, Jul. 2013, pp. 4738–4741.

[5] D. M. Bates and D. G. Watts, *Nonlinear Regression and Its Applications.* John Wiley & Sons, 1988.

[6] R. S. Campos and L. Lovisolo, "Location methods for legacy GSM handsets using coverage prediction," in *Proc. of IEEE 9th Workshop on Signal Processing Advances in Wireless Communications*, Recife, Brazil, Jul. 2008, pp. 21–25.

[7] R. S. Campos and L. Lovisolo, "A fast database correlation algorithm for localization of wireless network mobile nodes using coverage prediction and round trip delay," in *Proc. of IEEE 69th Vehicular Technology Conf.*, Barcelona, Spain, Apr. 2009, pp. 1–5.

[8] R. S. Campos and L. Lovisolo, "Mobile station location using genetic algorithm optimized radio frequency fingerprinting," in *Proc. of ITS 2010—Int. Telecommunications Symp.*, Manaus, Brazil, Sep. 2010, pp. 1–5.

[9] R. S. Campos and L. Lovisolo, *RF Positioning: Fundamentals, Applications, and Tools.* Artech House, 2015.

[10] R. S. Campos, L. Lovisolo, and M. de Campos, "Wi-fi multi-floor indoor positioning considering architectural aspects and controlled computational complexity," *Expert Syst. Appl.*, vol. 41, no. 14, pp. 6211–6223, 2014.

[11] M. Chen, T. Sohn, D. Chmelev, D. Haehnel, J. Hightower, J. Hughes, A. LaMarca, F. Potter, I. Smith, and A. Varshavsky, "Practical metropolitan-scale positioning for GSM phones," in *Proc. of 8th Int. Conf. on Ubiquitous Computing,* Newport Beach, CA, Sep. 2006, pp. 225–242.

[12] Y.-C. Cheng, Y. Chawathe, and J. Krumm, "Accuracy characterization for metropolitan-scale Wi-Fi localization," in *Proc. of the 3rd Int. Conf. on Mobile Systems, Applications, and Services,* Seattle, WA, Jun. 2005, pp. 233–245.

[13] H. Dai, W.-H. Ying, and J. Xu, "Multi-layer neural network for received signal strength-based indoor localisation," *IET Commun.,* vol. 10, no. 6, pp. 717–723, 2016.

[14] M. Dashti and H. Claussen, "Extracting location information from RF fingerprints," in *2016 IEEE Globecom Workshops (GC Wkshps),* Dec. 2016, pp. 1–6.

[15] European Telecommunications Standard Institute, 1998.

[16] European Telecommunications Standard Institute, 2004.

[17] Federal Communications Commission, "Guidelines for testing and verifying the accuracy of wireless E911 location systems," Apr. 2000. Available: https://www.hsdl.org/?view&did=13705

[18] Y. Gao, Y. Chang, B. Su, J. Xue, and A. Li, "Analysis of localization using multipath characteristics as location fingerprint," in *IEEE Wireless Communications and Networking Conf. (WCNC),* 2017, pp. 1–5.

[19] S. Gezici, "A survey on wireless position estimation," *Wireless Pers. Commun.,* vol. 44, no. 3, pp. 263–282, Feb. 2008.

[20] D. Goldberg, *Genetic Algorithms in Search, Optimization and Machine Learning.* Addison-Wesley, 1989.

[21] M. A. Hallak, M. Safadi, and R. Kouatly, "Mobile positioning technique using signal strength measurement method with the aid of passive mobile listener grid," in *Proc. of 2nd Int. Conf. on Information and Communication Technologies: From Theory to Applications,* Damascus, Syria, Apr. 2006, pp. 105–110.

[22] M. Hata, "Empirical formula for propagation loss in land mobile radio services," *IEEE Trans. Veh. Technol.,* vol. 29, no. 3, pp. 317–325, Aug. 1980.

[23] S. Haykin, *Neural Networks: A Comprehensive Foundation.* Prentice Hall, 1994.

[24] S. He and S.-H. G. Chan, "Wi-fi fingerprint-based indoor positioning: Recent advances and comparisons," *IEEE Commun. Surveys Tuts.,* vol. 18, no. 1, pp. 466–490, 2016.

[25] C. R. Houck, J. Joines, and M. G. Kay, "A genetic algorithm for function optimization: A MATLAB implementation," *NCSU-IE TR,* vol. 95, no. 9, pp. 1–10, 1995.

[26] IEEE Computer Society, 2007.

[27] International Telecommunications Union, "Digital topographic database for propagation studies," May 1997.

[28] J. Krumm, G. Cermak, and E. Horvitz, "RightSPOT: A Novel sense of location for a smart personal object," in *Proc. of Ubicomp 2003,* Seattle, WA, Oct. 2003, pp. 36–43.

[29] H. Laitinen, J. Lahteenmaki, and T. Nordstrom, "Database correlation method for GSM location," in *Proc. of IEEE 53rd Vehicular Technology Conf.,* Rhodes, Greece, May 2001, pp. 2504–2508.

[30] C. L. Lawson and R. J. Hanson, *Solving Least Squares Problems.* Prentice Hall, 1974.

[31] K. Lin, M. Chen, J. Deng, M. M. Hassan, and G. Fortino, "Enhanced fingerprinting and trajectory prediction for IoT localization in smart buildings," *IEEE Trans. Autom. Sci. Eng.,* vol. 13, no. 3, pp. 1294–1307, 2016.

[32] M. J. Magro and C. J. Debono, "A genetic algorithm approach to user location estimation in UMTS networks," in EUROCON—*The Int. Conf. on Computer as a Tool,* Warsaw, Poland, Sep. 2007, pp. 1136–1139.

[33] G. Manara, M. Porreta, P. Nepa, and F. Giannetti, "Location, location, location: Use of deterministic propagation models for testing wireless networks location techniques," *IEEE Veh. Technol. Mag.,* vol. 3, no. 2, pp. 20–29, Jun. 2008.

[34] N. Moayeri, J. Mapar, S. Tompkins, and K. Pahlavan, "Emerging opportunities for localization and tracking," *IEEE Wireless Commun.,* vol. 2, pp. 8–9, 2011.

[35] MSI, "PLANET Technical Reference Guide," Version DMS 2.0, 1999.

[36] V. Otsason, A. Varshavsky, A. LaMarca, and E. de Lara, "Accurate GSM indoor localization," in *Proc. of the 7th Int. Conf. on Ubiquitous Computing,* Tokyo, Japan, Sep. 2005, pp. 141–158.

[37] P. M. Pardalos, L. Pitsoulis, T. Mavridou, and M. Resende, "Parallel search for combinatorial optimization: genetic algorithms, simulated annealing, tabu search and GRASP," in *Proc. of the 2nd Int.*

Workshop on Parallel Algorithms for Irregularly Structured Problems, Lyon, France, Sep. 1995, pp. 317–332.

[38] Z. Salcic, "GSM mobile station location using reference stations and artificial neural networks," *Wireless Pers. Commun.,* vol. 19, pp. 205–226, 2001.

[39] T. Seidl and H.-P. Kriegel, "Optimal multi-step K-nearest neighbor search," in *Proc. of the 1998 ACM SIGMOD Int. Conf. on Management of Data,* Seattle, WA, Jun. 1998, pp. 154–165.

[40] X. Shen, J. W. Mark, and J. Ye, "Mobile location estimation in cellular networks using fuzzy logic," in *Proc. of IEEE 52nd Vehicular Technology Conf.,* Boston, MA, Sep. 2000, pp. 2108–2114.

[41] M. A. Spirito, M. Caceres, and F. Sottile, "WLAN-based real time vehicle locating system," in *Proc. of IEEE 69th Vehicular Technology Conf.,* Barcelona, Spain, Apr. 2009, pp. 1–5.

[42] S. Dutch, "The universal transverse Mercator system," Jan. 2000. Available: http://www.uwgb.edu/dutchs/fieldmethods/utmsystem.htm

[43] L. Sun, Z. Zheng, T. He, and F. Li, "Multifloor wi-fi localization system with floor identification," *Int. J. Distrib. Sens. Netw.,* vol. 11, no. 7, pp. 1–8, 2015.

[44] C. Takenga, C. Xi, and K. Kyamakya, "A hybrid neural network-data base correlation positioning in GSM network," in *Proc. of IEEE 10th Int. Conf. on Communication Systems,* Singapore, Oct. 2006, pp. 1–5.

[45] The MathWorks, Inc., "Pairwise distance between pairs of objects," May 2010. Available: http://www.mathworks.de/access/helpdesk/help/toolbox/stats/pdist.html

[46] A. Varshavsky, A. LaMarca, J. Hightower, and E. de Lara, "The skyloc floor localization system," in *Fifth Annual IEEE Int. Conf. on Pervasive Computing and Communications,* Mar. 2007, pp. 125–134.

[47] T. P. Vogl, J. K. Mangis, A. K. Rigler, W. T. Zink, and D. Alkon, "Accelerating the convergence of the back-propagation method," *Biol. Cybern.,* vol. 59, pp. 257–263, 1988.

[48] X. Wang, L. Gao, S. Mao, and S. Pandey, "CSI-based fingerprinting for indoor localization: A deep learning approach," *IEEE Trans. Veh. Technol.,* vol. 66, no. 1, pp. 763–776, 2017.

[49] G. Wölfle, R. Hoppe, D. Zimmermann, and F. Landstorfer, "Enhanced localization technique within urban and indoor environments based on accurate and fast propagation models," in *European Wireless,* Florence, Italy, Feb. 2002, pp. 25–28.

[50] M. D. Yacoub, *Foundations of Mobile Radio Engineering.* CRC Press, 1993.

[51] J. Zhu and G. D. Durgin, "Indoor/outdoor location of cellular handsets based on received signal strength," *Electron. Lett.,* vol. 41, no. 1, pp. 24–26, Jan. 2005.

[52] D. Zimmermann, J. Baumann, A. Layh, F. Landstorfer, R. Hoppe, and G. Wölfle, "Database correlation for positioning of mobile terminals in cellular networks using wave propagation models," in *Proc. of IEEE 60th Vehicular Technology Conf.,* Los Angeles, CA, Sep. 2004, pp. 4682–4686.

LOS/NLOS LOCALIZATION – IDENTIFICATION – MITIGATION

NON-LINE-OF-SIGHT (NLOS) propagation is an important issue in localization systems since many techniques are based on lateration which requires range measurements and range measurements inherently assume LOS signal propagation. The fourth part of the handbook is composed of four chapters, Chapters 16–19, which examine NLOS identification, mitigation and localization methods. NLOS identification can avoid major positioning errors. Researchers have recently started working in this area and many novel techniques have been proposed. All three chapters come with critical MATLAB examples, and solutions.

Chapter 16 reviews the many NLOS identification and localization techniques and compares them in terms of complexity and performance. NLOS identification techniques can be categorized into cooperative and non-cooperative. Cooperative localization techniques use multiple nodes to identify NLOS measurements. Non-cooperative NLOS identification are based on single-node channel measurements. These techniques rely on: (1) the range (TOA) statistics; (2) channel characteristics, such as received signal power, Rician K-factor and features extracted from the power delay profile; and (3) the consistency between the TOA measurement and path loss for LOS/NLOS, and the consistency between the direction of departure (DOD) and DOA. In the second group, suitable channel characteristics used in narrow/wide band systems, ultra-wide-band (UWB) systems and systems equipped with an antenna array are discussed. NLOS localization techniques can also be cooperative and non-cooperative. These techniques are categorized into: (1) NLOS localization using TOA, DOA measurements and an environment map; (2) localization using the

Handbook of Position Location: Theory, Practice, and Advances, Second Edition.
Edited by S. A. (Reza) Zekavat and R. Michael Buehrer.
© 2019 by the Institute of Electrical and Electronics Engineers, Inc.
Published 2019 by John Wiley & Sons, Inc.
Companion Website: www.wiley.com/go/zekavat/positionlocation2e

measurements from reflectors. This chapter discusses the advantages and disadvantages of these techniques and the complexity and performance of each technique.

Chapter 17 introduces NLOS mitigation methods. Generally, the NLOS mitigation methods for geolocation can be grouped into four categories. Mitigation methods based on maximum likelihood, least squares and constrained optimization as well as robust statistics are discussed. These methods are then compared in terms of different performance measures. In addition, this chapter discusses a novel geolocation example using a single moving sensor. Numerical examples are then presented to demonstrate how position estimation can be achieved for the case of a single moving sensor with NLOS errors.

Chapter 18 investigates TOA estimation techniques for hybrid RSS-TOA localization in mixed LOS/NLOS environments. The chapter explains how to choose a sufficient number of TOA measurements so that the computational burden of the position estimation can be reduced without performance degradation. In addition, this chapter investigates how to determine a mobile's position using a number of TOAs while maintaining reasonable error performance. The results can help system designers to manage the tradeoff between accuracy and computational complexity.

Chapter 19 investigates the problem of mobile tracking in mixed line-of-sight (LOS)/non-line-of-sight (NLOS) conditions. The state-of-the-art methods in this field are first reviewed. Two types of Bayesian filters are studied: the Gaussian mixture filter (GMF) and the particle filter (PF). The modified extended Kalman filter (EKF) bank method, as one specific GMF, is described. The chapter also discusses the computation of a posterior Cramer-Rao lower bound (CRLB) for the mobile tracking problem.

CHAPTER *16*

NLOS IDENTIFICATION
AND LOCALIZATION

Wenjie Xu, Zhonghai Wang, and S. A. (Reza) Zekavat
Michigan Technological University, Houghton, MI

WHEN LINE of sight (LOS) is not available, that is, in non-line-of-sight (NLOS) conditions, direction-of-arrival (DOA) and time-of-arrival (TOA) techniques would involve considerable errors. To address this problem, many techniques have been proposed to identify LOS conditions. If the NLOS situations are identified, the corresponding measurements should be excluded from the localization process to eliminate the corresponding NLOS error. In addition, a number of NLOS localization methods have been proposed: multiple NLOS localization measurements are integrated to enable NLOS localization.

This chapter reviews many NLOS identification and localization techniques. NLOS identification techniques can be categorized into cooperative and noncooperative. Cooperative localization techniques use multiple nodes to identify NLOS measurements. Noncooperative NLOS identification is based on single-node channel measurement. These techniques are based on (1) the range (TOA) statistics; (2) channel characteristics, such as received signal power, Rician K-factor, and features extracted from the power delay profile; and (3) the consistency between the TOA measurement and path loss for LOS/NLOS, and the consistency between the direction of departure (DOD) and DOA. In the second group, suitable channel characteristics used in narrow/wideband systems, ultra-wideband (UWB) systems, and systems equipped with antenna arrays are discussed.

The NLOS localization techniques can be cooperative and noncooperative. These techniques are categorized into (1) NLOS localization using TOA, DOA measurements, and the environment map; and (2) localization using the measurements from reflectors.

Handbook of Position Location: Theory, Practice, and Advances, Second Edition.
Edited by S. A. (Reza) Zekavat and R. Michael Buehrer.
© 2019 by the Institute of Electrical and Electronics Engineers, Inc.
Published 2019 by John Wiley & Sons, Inc.
Companion Website: www.wiley.com/go/zekavat/positionlocation2e

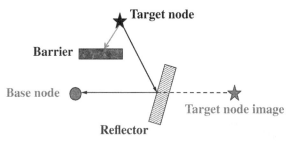

Figure 16.1   Localization error in NLOS propagation environment.

This chapter discusses the advantages and disadvantages of these techniques and the complexity and performance of each technique.

16.1   INTRODUCTION

Target localization has a wide range of military and civilian applications in wireless mobile networks. Examples include battlefield command and control [1], fire fighter tracking [2], Enhanced 911 (E-911) [3], road traffic alert [4], resource allocation in mobile ad hoc networks [5], and routing in sensor networks [6, 7].

Localization techniques are either based on TOA, DOA [8], or received signal strength (RSS) [9]. The techniques that are proposed based on TOA and DOA are very sensitive to the availability of the LOS, which is the direct path between the transmitter and receiver. If LOS is not available, that is, in NLOS propagation conditions, the received signal will travel longer distances compared with the LOS path. This results in a larger TOA and a wrong DOA estimation, as shown in Figure 16.1. Therefore, a large localization error is experienced [10].

This chapter introduces NLOS identification and localization techniques. In order to reduce the NLOS localization error, NLOS measurements should be identified and excluded from the localization process. Besides localization application, some NLOS identification methods can provide the LOS link quality information. Based on this information, more complex TOA estimators can be selected for low-quality links, and less complex TOA estimators shall be selected for high-quality links [11]. In addition, identifying LOS conditions allows optimal adjustment of the transmission mode of communication systems by switching to a higher order of modulation for LOS links to achieve higher data rates.

In essence, NLOS identification is a statistical detection problem, where NLOS and LOS conditions are considered as two hypotheses. The main task is to find out metrics that differentiate NLOS and LOS hypotheses and enable a binary hypothesis test to identify NLOS conditions.

NLOS identification techniques can be cooperative or noncooperative. Cooperative localization techniques use multiple nodes that are geographically distributed in an environment to identify a NLOS measurement. Noncooperative NLOS identification is based on single-node channel measurement. The corresponding techniques can be divided into three groups:

1. based on the range (TOA) statistics. The range is the product of TOA and the speed of light. If LOS is available, the estimated range is affected by the TOA estimation error and, therefore, is Gaussian distributed. But for NLOS situations, the estimated range is positively biased and has non-Gaussian distribution. In addition, NLOS range measurements tend to have a larger variance compared with the LOS one;

2. based on channel characteristics, such as received signal power, Rician K-factor, and features extracted from the power delay profile; and

3. hybrid approaches that explore the consistency between the TOA measurement and path loss for LOS/NLOS, and the consistency between the DOD and DOA.

NLOS identification techniques could improve the localization accuracy by removing the NLOS base nodes (BNs) and performing localization incorporating only the LOS BNs. However, a rich multipath environment has abundant scatterers in the proximity of both the BNs and the target. Therefore, almost all BNs are in the NLOS of the target except those located close to the target. As a result, there are no sufficient LOS BNs for localization. Note that, as discussed in Chapters 1 and 2 and chapters in Part III of this book, at least two LOS BNs are required for DOA-based localization methods, and three BNs are required for TOA-based localization methods. Thus, the localization performance might be unsatisfactory in rich scattering environments such as urban and indoor areas. For those cases, NLOS localization can overcome the problem by performing NLOS localization using NLOS measurements and geometrical information. The proposed NLOS localization techniques can be divided into two categories:

1. NLOS localization using TOA, DOA measurements, and the environment map; and

2. localization using the measurements from reflectors.

In the latter method, NLOS measurements are first identified, and then the geometrical relationship of the BNs, target node, and reflectors are used to localize the target node.

16.2 NLOS IDENTIFICATION

NLOS identification is a hypothesis testing problem and can be depicted using Figure 16.2. The source block in Figure 16.2 generates one of the possible outputs, that is, an NLOS or LOS hypothesis. Then, random observations are generated based on the conditional probability density function (PDF) $f(\cdot|H_0)$ or $f(\cdot|H_1)$. In the decision rule block, the likelihood ratio $\Lambda(r)$ is compared with a threshold, and then a decision is made on whether LOS or NLOS is available.

Example 16.1   NLOS Identification Based on Rician K-factor

Rician K-factor is the power ratio of LOS and NLOS components. It can be used to form a hypothesis testing to identify NLOS situations. Given the conditional PDF

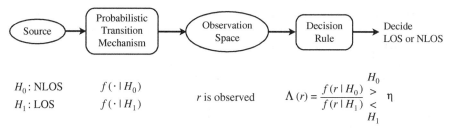

H_0: NLOS $f(\cdot \mid H_0)$

H_1: LOS $f(\cdot \mid H_1)$

r is observed

$$\Lambda(r) = \frac{f(r \mid H_0)}{f(r \mid H_1)} \underset{H_1}{\overset{H_0}{\underset{<}{>}}} \eta$$

Figure 16.2   A big picture of NLOS identification.

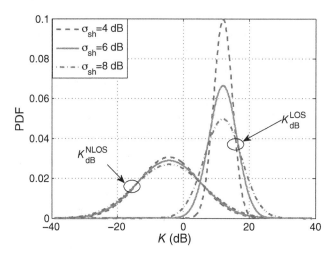

Figure 16.3   PDFs of K_{dB}^{NLOS} and K_{dB}^{LOS} for $\sigma_{sh} = 4$, 6, and 8 dB.

of K under LOS and NLOS situations derived in [12], determine the threshold K_{th}, and compute the probability of detection and probability of false alarm.

Solution

According to [12], K in the decibel scale is defined as $K_{dB} = 10 \log K$, and K_{dB} under LOS and NLOS are Gaussian. The mean and standard deviation (STD) of $f(K_{dB}|LOS)$ are μ_1 and σ_1. The mean and STD of $f(K_{dB}|NLOS)$ are μ_0 and σ_0; $\mu_1 = 12$ dB and $\mu_0 = -4.2$ dB; σ_1 is equal to STD of shadowing, that is, $\sigma_1 = \sigma_{sh}$; and $\sigma_0 = \sqrt{\sigma_{sh}^2 + 12.4^2}$.

For $\sigma_{sh} = 4$, 6, and 8 dB, the PDFs $f(K_{dB}|NLOS)$ and $f(K_{dB}|LOS)$ are shown in Figure 16.3. The decision rule is

$$\text{Decide NLOS} \quad \text{if } K < K_{th},$$
$$\text{Decide LOS} \quad \text{if } K > K_{th}. \tag{16.1}$$

The threshold K_{th} is the intersection of $f(K_{dB}|NLOS)$ and $f(K_{dB}|LOS)$ for each color and it is expressed as

TABLE 16.1.   Comparison of K_{th}, P_F, and P_D for Various σ_{sh}

	K_{th}	P_F	P_D
$\sigma_{sh} = 4\,dB$	6.54 dB	0.0863	0.7952
$\sigma_{sh} = 6\,dB$	5.17 dB	0.1276	0.7519
$\sigma_{sh} = 8\,dB$	4.24 dB	0.1662	0.7164

$$K_{th} = \frac{\left(\sigma_0^2\mu_1 - \sigma_1^2\mu_0\right) - \sqrt{\left(\sigma_0^2\mu_1 - \sigma_1^2\mu_0\right)^2 - \left(\sigma_0^2 - \sigma_1^2\right)\left(\sigma_0^2\mu_1^2 - \sigma_1^2\mu_0^2 - \sigma_1^2\sigma_0^2 \ln\frac{\sigma_0 P(H_1)}{\sigma_1 P(H_0)}\right)}}{\left(\sigma_0^2 - \sigma_1^2\right)}.$$

(16.2)

The detection probability of NLOS condition is the probability of deciding NLOS when NLOS hypothesis is true and it is computed by

$$P_D = \int_{-\infty}^{K_{th}} f\left(K_{dB}^{NLOS} = k\right) dk = 1 - Q\left((K_{th} - \mu_0)/\sigma_0\right).$$

(16.3)

The false alarm probability of NLOS condition is the probability of deciding NLOS when LOS hypothesis is true and it is computed by

$$P_F = \int_{-\infty}^{K_{th}} f\left(K_{dB}^{LOS} = k\right) dk = 1 - Q\left((K_{th} - \mu_1)/\sigma_1\right).$$

(16.4)

The threshold K_{th}; the detection probability of NLOS, P_D; and the false alarm probability of NLOS, P_F; are summarized in Table 16.1.

Refer to the file chapter_16_Example_1.m for MATLAB code used to generate the above results. MATLAB codes can be found online at ftp://ftp.wiley.com/public/sci_tech_med/matlab_codes.

In the following, we present a review on cooperative (multiple BNs) and noncooperative (single BN) NLOS identification techniques for different wireless systems.

16.2.1   Cooperative Methods

When multiple BNs are involved in determining the position of a target, BNs located in the LOS of a target produce consistent localization results, but BNs in the NLOS condition produce inconsistent localization results. Because inconsistent NLOS measurements tend to have large residuals, residual testing is an approach that can identify NLOS measurements. Many residual testing techniques have been proposed in the literature. Here we present a summary of those techniques.

DOA Residual Testing: This approach has been proposed in [13]. In this method, it is assumed that there are N BNs at known locations $(x_i, y_i) = i = 1 \ldots N$ as illustrated in Figure 16.4. All BNs can measure DOAs of the signal transmitted by the

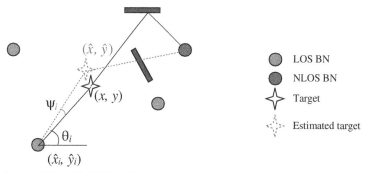

Figure 16.4   NLOS identification based on DOA residual testing.

target. Let θ_i be the DOA measured by BN i. The NLOS identification procedure consists of the following steps:

1. Find the maximum likelihood position $(\hat{x}, \hat{y})$ of the target using all DOA estimates.

2. Calculate the DOA residual ψ_i (see Fig. 16.4), which is the absolute difference of θ_i and the DOA obtained via the estimated target position $(\hat{x}, \hat{y})$.

3. Select the NLOS BNs whose $\psi_i > 1.5\text{RMS}(\psi)$. $\text{RMS}(\psi)$ denotes the root mean square of ψ_i.

After excluding the NLOS BNs, the maximum likelihood position of the target is estimated again, and the positioning accuracy is improved. The simulation result shows that when there are four LOS BNs and one NLOS BN, applying NLOS identification can cut the root mean square positioning error greatly, from 1.2 km to 125 m, which meets E-911 requirements.

Time-Difference-of-Arrival (TDOA) Residual: This approach, defined in [14], takes into account the fact that NLOS error is always positive. Assuming the measurement noise under the LOS condition is Gaussian distributed, the residual e_i is expressed as

$$e_i = 0.5 + 0.5 \, \text{erf}\left(\frac{m_i - f_i(\hat{\theta})}{\sqrt{2}\sigma_i} \right). \tag{16.5}$$

In (16.5), $\text{erf}(\cdot)$ is the error function defined as $\text{erf}(x) = \left(2/\sqrt{\pi}\right) \int_0^x e^{-t^2} \, dt$, m_i represents the TDOA measurement of BN i and the reference BN, $\hat{\theta}$ is the estimated target position using all measurements, $f_i(\hat{\theta})$ is the BN i's TDOA given the target position $\hat{\theta}$, and σ_i is the square root of the measurement noise variance. The higher this residual, the more likely the measurement m_i is biased by NLOS error(s). Then the residual is compared with the threshold λ and those BNs with residuals larger than λ are labeled as NLOS BNs. When there is only one NLOS BN out of a total of six BNs, the identification probability can attain 0.79. However, this probability decreases when fewer BNs are available or more NLOS BNs are present.

Residual Distribution Testing: This approach finds the set of LOS BNs [15]. Let there be N BNs, which use TOA estimates to locate target nodes. The minimum number of BNs required for obtaining an estimate of the target position is three. Therefore, there are a number of position estimates corresponding to different combinations of BNs and the total number is $S = \sum_{i=3}^{N} {}_N C_i$, where ${}_N C_i$ represents the number of i-element combination out of N-elements; that is, ${}_N C_i = N!/i!(N-i)!$. Then the normalized residuals are defined as

$$\chi_x^2(k) = \frac{[\hat{x}(k) - \hat{x}]^2}{B_x(k)},$$

$$\chi_y^2(k) = \frac{[\hat{y}(k) - \hat{y}]^2}{B_y(k)}, \quad k = 1, \ldots, S-1,$$

(16.6)

where $(\hat{x}(k), \hat{y}(k))$ represents the target position estimate using the kth BN combination, $(\hat{x}, \hat{y})$ is the estimation involving all BNs, and $B_x(k)$, $B_y(k)$ are the approximations of the Cramer–Rao lower bound (CRLB) on the estimation error of the respective x and y target coordinates. If all N BNs are in the LOS condition, $\hat{x}(k) - \hat{x}/\sqrt{B_x(k)}$ and $\hat{y}(k) - \hat{y}/\sqrt{B_y(k)}$ are approximately Gaussian distributed with zero mean and unit variance. Therefore, the normalized residuals in (16.6) have an approximate central χ^2 PDF. If one or more BNs are in NLOS condition, the means of $(\hat{x}(k) - \hat{x})$ and $(\hat{y}(k) - \hat{y})$ are biased by the NLOS measurement. Therefore, the normalized residuals have a noncentral χ^2 PDF. The noncentral χ^2 distribution can be detected using the fact that the probability of the central χ^2 distributed random variable being greater than 2.71 is 0.02. Thus, the appearance of a value higher than the threshold $TH = 2.71$ indicates that there is one or more NLOS measurements with high likelihood. The identification steps are as follows:

1. The normalized residuals defined in (16.6) are calculated for a total number of N BNs.
2. The residuals are compared with $TH = 2.71$. Then, the cases $\chi_x^2 > TH$ and $\chi_y^2 > TH$ are counted.
3. If less than 10% of the residuals are above TH, then the number of LOS BN is $D = N$. Otherwise, the test moves to $(N-1)$ BNs.
4. This process stops when it has determined a D, or when $D = 3$.
5. The excluded BNs are identified as NLOS.

Simulations show that the location mean square error (MSE) is close to the CRLB for more than four LOS BNs. But for three LOS BNs, identifying the NLOS BNs is difficult and the location MSE is higher.

The pros and cons of the cooperative NLOS identification approaches are summarized as follows:

Pros

- NLOS conditions are selected in a way that the position estimation error can be reduced.

Cons

- In order to correctly detect NLOS measurements in residual testing, there should be at least four LOS BNs, while in real environments rarely enough LOS BNs are available.
- The locations of all BNs are required.
- The computation complexity is high, and it increases with the number of BNs.

In Section 16.2.2, single-node identification methods are studied. Single-node methods do not require LOS BNs or the locations of any BN, and their complexity is not high in general.

16.2.2 Single-Node Methods Based on the Range Statistics

Range refers to the distance between the BN and the target. Range is computed by multiplying TOA and the speed of light. NLOS can be identified based on the features derived from the time series of the range estimates [16–19] or the features from the range estimates across different frequency bands [20].

In this section, first, the methods that are based on range measurements over time are investigated, and next, the methods that are based on range measurement over different frequency bands are studied.

Techniques Based on Range Measurements over Time: For LOS and NLOS situations, the ith range measurements can be modeled as

$$
\begin{aligned}
&\text{LOS:}\quad r_i = d_i + n_i,\\
&\text{NLOS:}\ r_i = d_i + n_i + e_i, \quad i = 1,\dots,N,
\end{aligned}
\tag{16.7}
$$

where d_i is the true LOS range, n_i is the measurement noise, and e_i is the NLOS error. In (16.7), n_i is modeled as Gaussian distributed with zero mean and variance σ^2, and e_i is modeled by a random variable, which is exponentially [21] or Gaussian distributed [17] with positive mean μ_e and variance σ_e^2. Normally, n_i and e_i are independent. Empirical measurement in [22] shows that the NLOS error is more irregular than the measurement noise, which results in $\sigma_e^2 > \sigma^2$. Therefore, a hypothesis test can be formed based on the PDF or the variance of the estimated range.

The testing based on range variance is used when there is no a priori information about the NLOS error (such as PDF, mean, and variance) [17]. The measurement noise variance σ^2 is usually known, as it is determined by the range estimation method. Let $\hat{\sigma}^2$ denote the estimated range variance, and the hypothesis testing is given by

$$
\begin{aligned}
&H_0:\quad \hat{\sigma}^2 = \sigma^2\,(\text{LOS condition}),\\
&H_1:\quad \hat{\sigma}^2 > \sigma^2\,(\text{NLOS conditon}).
\end{aligned}
\tag{16.8}
$$

The decision rule is

$$\hat{\sigma}^2 \overset{H_1}{\underset{H_0}{\gtrless}} \eta. \tag{16.9}$$

The threshold η may vary with the availability of a priori information: If only the noise variance is known, η is σ^2; if the NLOS error variance σ_e^2 is known, $\eta = \sigma_e^2/2$. The threshold η can also depend on the typically known maximal velocity of the object [23]. In reality, the true range d_i varies with time i, and a polynomial fitting is used to compute the range variance $\hat{\sigma}^2$ [16]. The true range is reconstructed via the polynomial fitting, and the reconstructed range is denoted by s_i. Then, the range variance can be calculated:

$$\hat{\sigma}^2 = \sqrt{\frac{1}{N} \sum_{i=1}^{N} (s_i - r_i)^2}. \tag{16.10}$$

The testing based on the range distribution can be divided into two groups: parametric and nonparametric. In parametric methods, part or complete a priori information is known, such as the likelihood of LOS and NLOS error PDF. Assuming Gaussian NLOS error, [17] discusses parametric NLOS identification and forms a couple of likelihood ratio tests for different levels of a priori information. For example, if the likelihood of LOS and the PDF of NLOS error are known, and the true LOS range is not known, a generalized likelihood ratio test (GLRT) can be used, which corresponds to

$$\Lambda_g(\mathbf{r}) = \frac{\max_{d+\mu_e} f_{\text{NLOS}}(r)}{\max_d f_{\text{LOS}}(r)} \overset{H_1}{\underset{H_0}{\gtrless}} \frac{P(\text{LOS})}{P(\text{NLOS})}, \tag{16.11}$$

where $\mathbf{r}$ is a vector of range measurements, $\mathbf{r} = [r_1, \ldots, r_N]$; d is the true LOS range; f_{NLOS} and f_{LOS} represent the PDFs of LOS and NLOS range measurement; and $P(\text{LOS})$ and $P(\text{NLOS})$ are the likelihoods of LOS and NLOS hypotheses, respectively.

Because the NLOS distribution is site specific, its characterization is very difficult. In these cases, the nonparametric technique, which does not assume the knowledge of NLOS error statistics, can be used in NLOS identification. In [18] and [19], it is assumed that it is only known that the NLOS error is not Gaussian. Tests for the normality of the range measurements are developed in [18]. In [19], a metric measuring the distance between two PDFs is introduced, and the distances between the candidate LOS range PDFs and the range measurements PDF are computed. Then, LOS is decided when the minimum PDF distance is smaller than the threshold, and NLOS is decided otherwise.

Techniques Based on the Range Measurements over Different Frequency Bands: Based on channel measurements in a typical indoor environment, the authors of [20] showed that under LOS condition, estimated ranges are similar across

sub-bands, but under NLOS condition, they are drastically different across sub-bands. The difference of ranges across sub-bands is due to the different propagation characteristics across sub-bands: higher operational frequency means lower penetration capabilities. In other words, signals at higher frequency bands may not penetrate blockage and experience an NLOS propagation, while signals at lower frequency bands may penetrate blockage and still experience a LOS propagation. Thus, the LOS ranging measurements over sub-bands have a small variance, and the NLOS ranging measurements over sub-bands have a large variance. Let σ denote the STD of the ranging measurements, and let f_{NLOS} and f_{LOS} be the PDFs of σ for respective hypothesis. The decision rule is

$$\frac{f_{\text{NLOS}}(\sigma)}{f_{\text{LOS}}(\sigma)} \underset{\text{LOS}}{\overset{\text{NLOS}}{\gtrless}} \sigma_{\text{sh}}, \tag{16.12}$$

where f_{LOS}, f_{NLOS} and the threshold σ_{sh} need to be determined experimentally. This method can be implemented on multiband orthogonal frequency division multiplexing (OFDM) systems. The multiband approach requires a frequency hopping capable radio-frequency (RF) front end, and therefore the cost and the complexity are higher. It would be a cheaper solution to combine radio ranging signal and low-frequency sound (such as in [24]). How their ranging is different under different channel conditions needs to be investigated.

A summary of this part is given in Table 16.2. A general disadvantage of range statistics-based methods is the latency (about 5 seconds) due to using time series of the range estimates. Therefore, in Section 16.2.3, faster NLOS detection methods are studied. When the BN and the target are both stationary, which means that the signal traveling path does not change, this method would fail because the range statistics will not differ considerably for LOS and NLOS situations.

16.2.3 Single-Node Methods Based on Channel Characteristics

This section investigates NLOS identification approaches based on channel characteristics. Almost all channel characteristics mentioned here are extracted from the power delay profile of the received signal.

TABLE 16.2. A Summary on Range Statistics-Based Methods

	Estimated Range PDF	Range Variance	Range Variance across Sub-Bands
LOS	Gaussian	Small	Small
NLOS	Non-Gaussian	Large	Large
Cons	Not for stationary BN and target	Not for stationary BN and target	Only for multiband UWB
	Latency		

Since the power delay profile exhibits differently for systems with different bandwidth, this discussion includes methods for narrow and wideband systems and UWB systems. The power delay profile is normally obtained via single antenna. If multiple antennas are used, the power delay profile has an extra spatial dimension. The potential of multiple antennas for high-performance NLOS identification is investigated in the last part of this section.

Narrow and Wideband Systems: In this case, the power envelope distribution of the received signal can be used to identify NLOS [25] because the power distribution of the first arriving path is usually modeled as Rayleigh fading for NLOS condition and Rician fading for LOS condition [26]. Here is the identification process:

1. Estimate the PDF of the first arriving path power. To correctly estimate this PDF, a set of independent fading coefficients are needed. The fading coefficients would be considered independent if they are separated by at least a coherence time.

2. Compare the estimated PDF with some reference PDFs, such as Rayleigh or Rician, via Pearson's test statistic [27] or Kolmogorov–Smirnov test [25].

3. Form a hypothesis test on the comparison result and make a decision.

4. This method has two disadvantages: (1) the observation interval should be long enough to compute the accurate PDF of the first path power. As reported in [27], this time interval is in the order of 1 second; and (2) when the LOS component is much smaller than the NLOS component in the first path, it is difficult to distinguish the power distribution under LOS condition from the distribution under NLOS condition, that is, Rayleigh.

In order to further reduce the observation time, an approach based on the Rician K-factor of the first arriving path is proposed in [17] and [28]. The Rician K-factor is defined as the ratio of LOS and NLOS component powers. When there is no LOS component (NLOS condition), $K = 0$ by definition. When LOS component exists, $K > 0$. In [27], the Rician K-factor is estimated, denoted by $\hat{K}$, and the LOS state is weighted according to a predefined scale:

$$\text{if } \hat{K} > K_{\max}, \qquad \text{decide LOS;}$$

$$\text{if } K_{\min} < \hat{K} < K_{\max}, \quad \text{the probability of LOS is } \left(\hat{K} - K_{\min}\right)\big/\left(K_{\max} - K_{\min}\right); \quad (16.13)$$

$$\text{if } \hat{K} < K_{\min}, \qquad \text{decide NLOS.}$$

In [28], a simplified hypothesis testing is used:

$$\hat{K} > 1, \quad \text{decide LOS;}$$
$$\hat{K} < 1, \quad \text{decide NLOS.} \qquad (16.14)$$

The time required to estimate K is around 10 ms, reported in [27].

Another approach to identify NLOS depends on the autocorrelation properties of each multipath component [29]. The autocorrelation of multipath components indicates how the corresponding fading coefficient varies with time. If fading coefficients vary fast, the autocorrelation is low. Otherwise, the autocorrelation is high. The NLOS multipath component coefficient usually varies fast, since it consists of numerous time-varying irresolvable paths. The presence of an LOS component in the first path provides higher autocorrelation as compared with the other paths that do not include an LOS component, because the LOS component coefficient has a deterministic structure and varies slowly. If there is no LOS component in the first path, the autocorrelation of the first path would be comparable to the following paths. This observation can be used to identify the existence of the LOS component.

UWB Systems: UWB enables precise ranging and localization via incorporating extremely short duration pulses. In this case, the multipath components of the received signal can be well resolved. Therefore, it is a very promising technique for indoor localization. Moreover, the UWB channel models have been intensively characterized for LOS and NLOS channel conditions [30], based on which some metrics distinguishing LOS and NLOS are studied.

In [23], a confidence metric is given as a function of the amplitude α_1 and the arrival time τ_1 of the first path, and the strongest path amplitude α_{max} and the respective arrival time τ_{max}. Based on the observation that compared with the first path, the subsequent multipath components should have lower power in the LOS case and vice versa in the NLOS case: the confidence metric would be high for the LOS case and low for the NLOS case. Another similar approach is proposed in [31]. Here, the first path power $|\alpha_1|^2$ and the delay of the strongest path, that is, $\tau_{max} - \tau_1$, are used to form a joint likelihood ratio test as follows:

$$J\left(|\alpha_1|^2, \tau_{max} - \tau_1\right) = \frac{f_{LOS}\left(|\alpha_1|^2\right)}{f_{NLOS}\left(|\alpha_1|^2\right)} \times \frac{f_{LOS}\left(\tau_{max} - \tau_1\right)}{f_{NLOS}\left(\tau_{max} - \tau_1\right)} \underset{NLOS}{\overset{LOS}{\gtrless}} 1. \qquad (16.15)$$

A disadvantage of these two methods is that they may mistakenly detect nondominant direct path (NDDP) channel condition as NLOS, because in NDDP cases the direct path (LOS) is not the strongest but still detectable by an appropriate receiver architecture.

A method based on the change of signal power is proposed in [23]. The principle is that a sudden decrease of the maximum signal power $|\alpha_{max}|^2$ could indicate the movement from a LOS into an NLOS condition, and vice versa. The LOS and NLOS states are detected when the transition between LOS and NLOS occurs; therefore, it is not suitable for the case when the channel stays at one state for a long time.

NLOS identification methods based on RSS test are proposed in [32] and [33]. RSS is defined as the total received power of the received signal. The received signal $h(t)$ is

$$h(t) = \sum_{l=1}^{L} \alpha_l \delta(t - \tau_l), \qquad (16.16)$$

where L is the total number of multipaths, α_l is the amplitude of the lth multipath, and τ_l is the delay of the lth multipath. Then, RSS is represented by

$$\text{RSS} = \sum_{l=1}^{L} |\alpha_l|^2. \tag{16.17}$$

RSS can be easily measured by most wireless devices. The estimated RSS has been modeled as a lognormal random variable with different variances in LOS and NLOS scenarios [32]. Then a likelihood ratio test similar to (16.12) can be applied to determine LOS or NLOS. In [33], RSS is modeled by Weibull distribution based on measurements.

Other metrics can be extracted from the received multipath signal $h(t)$, and similar hypothesis testings can be formed to identify NLOS. Those metrics include mean excess delay, delay spread, kurtosis, and skewness [32, 34–36], and they are defined as follows:

Mean excess delay

$$\tau_m = \frac{\displaystyle\int_{-\infty}^{\infty} t |h(t)|^2 \, dt}{\displaystyle\int_{-\infty}^{\infty} |h(t)|^2 \, dt}, \tag{16.18}$$

where $h(t)$ is defined in (16.16).

Delay spread

$$\tau_{\text{rms}} = \frac{\displaystyle\int_{-\infty}^{\infty} (t - \tau_m)^2 |h(t)|^2 \, dt}{\displaystyle\int_{-\infty}^{\infty} |h(t)|^2 \, dt}. \tag{16.19}$$

Here, τ_m is defined in (16.18).

Kurtosis

$$\kappa = \frac{E\left[\left(|h(t)| - \mu_{|h|} \right)^4 \right]}{\sigma_{|h|}^4}, \tag{16.20}$$

where $E(\cdot)$ denotes expectation over delay, and $\mu_{|h|}$ and $\sigma_{|h|}$ are the mean and STD of $|h(t)|$, respectively.

Skewness

$$s = \frac{E\left[\left(|h(t)| - \mu_{|h|} \right)^3 \right]}{\sigma_{|h|}^3}. \tag{16.21}$$

Some of the above metrics can be combined as shown in (16.15) to achieve higher identification performance [32–34, 36].

All metrics introduced for NLOS identification in UWB systems can be obtained from a snapshot of the received multipath signal. In other words, no statistics information over time (variance, mean, PDF, etc.) needs to be collected. Therefore, the NLOS identification process speed is very fast.

The PDFs of these metrics are required for likelihood ratio tests. In some cases, these PDFs are unavailable. In these scenarios, self-learning techniques used in classification problems can be applied. Examples are support vector machine [37] and neural network [33]. In those methods, a training set is needed, which is a group of channel characteristics data with known LOS/NLOS conditions. Then, the pattern of LOS data and NLOS data can be learned from the training set, and the recognized pattern is used to identify NLOS scenario.

Systems Using Antenna Array: The antenna array technique is supported by many wireless communication standards, such as wireless LAN 802.11n [38], WiMAX 802.16e [39], and third-generation mobile phone [40], since this technique can suppress interference via beamforming techniques, and support high-throughput and high-performance transmissions via diversity methods. Here, we introduce NLOS identification techniques for systems equipped with antenna array.

Spatial Correlation: Spatial correlation of channel coefficients across antenna elements is proposed as a metric for NLOS identification in [41]. The NLOS and LOS channel coefficient of receive antenna 1 is

$$h_1^{\text{NLOS}} = \sum_p a_p, \quad h_1^{\text{LOS}} = a_0 + h_1^{\text{NLOS}}, \tag{16.22}$$

where a_p represents the contribution of the pth scatterer, and a_0 represents the LOS component. The NLOS and LOS channel coefficient of receive antenna 2 is

$$h_2^{\text{NLOS}} = \sum_p b_p, \quad h_2^{\text{LOS}} = b_0 + h_2^{\text{NLOS}}, \tag{16.23}$$

where b_p and b_0 are similarly defined. Assuming that the scattered signal from different scatterers are uncorrelated, then the spatial correlation for NLOS and LOS is expressed as

$$E\left(h_1^{\text{NLOS}} h_2^{\text{NLOS*}}\right) = \sum_p E\left(a_p b_p^*\right),$$

$$E\left(h_1^{\text{LOS}} h_2^{\text{LOS*}}\right) = E\left(h_1^{\text{NLOS}} h_2^{\text{NLOS*}}\right) + a_0 b_0^* + a_0 E\left(h_2^{\text{NLOS*}}\right) + b_0^* E\left(h_1^{\text{NLOS}}\right). \tag{16.24}$$

Now the correlation defined in (16.24) is sketched for a scattering environment illustrated in Figure 16.5. The scattering environment encountered by users located within a valley is formed by two rough scattering planes represented by the top and bottom lines in Figure 16.5. The scattering surface is segmented equally, and the center of each part is denoted by the small circle on the upper and lower planes. The

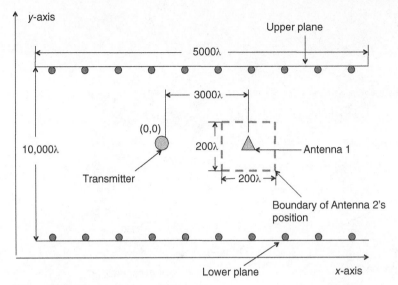

Figure 16.5   The scattering environment.

position of Antenna 2 varies in a square region bounded by the broken line illustrated in Figure 16.5.

The simulation parameters are listed as the wavelength of the transmitted signal $\lambda = 0.1$ m; the length of each surface scatterer is 500λ; the correlation distance of the rough surface $T = 25\lambda$; the roughness of the surface $\sigma = 0.5\lambda$; the power ratio of LOS and NLOS components, that is, Rician K-factor $K = 1$.

The plots of $E\left(h_1^{\mathrm{NLOS}} h_2^{\mathrm{NLOS*}}\right)$ and $E\left(h_1^{\mathrm{LOS}} h_2^{\mathrm{LOS*}}\right)$ are shown in Figure 16.6, and it is observed that the spatial correlation in LOS scenarios is greatly boosted compared with that of NLOS scenarios, since the LOS component is deterministic.

Variance of Phase Difference: An NLOS identification technique based on the statistics of the phase difference across two antenna elements is proposed in [12]. The phase difference $\Delta\phi$ is illustrated in Figure 16.7. In essence, this technique is based on the Rician K-factor, as the variance of $\Delta\phi$ is used to estimate the parameter K. This study investigates the PDFs of K under LOS and NLOS scenarios and derives a threshold of K depending on propagation parameters. A flowchart of this identification process is shown in Figure 16.8.

Example 16.2   *K*-Estimator Based on $\Delta\phi$

Use MATLAB to simulate and plot the root mean square error (RMSE) and bias of the K-estimator based on $\Delta\phi$, that is, the phase difference across two antenna elements.

Solution

The Rician K-factor is a function of variance of $\Delta\phi$. The K-factor estimation can be carried out by following the steps as shown in Figure 16.8. First, the phase samples are generated using the pseudorandom number generator. Then, the phase samples

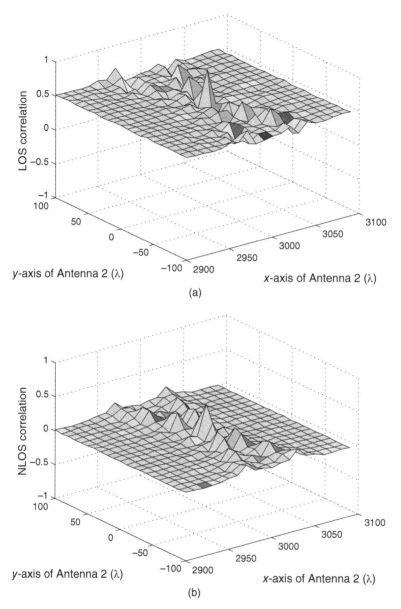

Figure 16.6   (a) LOS correlation $E\left(h_1^{\text{LOS}} h_2^{\text{LOS}*}\right)$. (b) NLOS correlation $E\left(h_1^{\text{NLOS}} h_2^{\text{NLOS}*}\right)$.

are wrapped to a proper range so as to yield a correct variance of $\Delta\phi$, that is, $\sigma_{\Delta\phi}^2$. Finally, $\sigma_{\Delta\phi}^2$ is mapped to the estimated value of K.

Now, we will have a look at how phase samples are generated. The phase sample of the first antenna is denoted by ϕ_1 and that of the second antenna is denoted by ϕ_2. The ϕ_1 and ϕ_2 are the phases of r_1 and r_2, the received signals of the two antennas. As shown in Figure 16.9, $r_1 = r_{\text{LOS1}} + r_{\text{DIF1}}$ and $r_2 = r_{\text{LOS2}} + r_{\text{DIF2}}$, where the

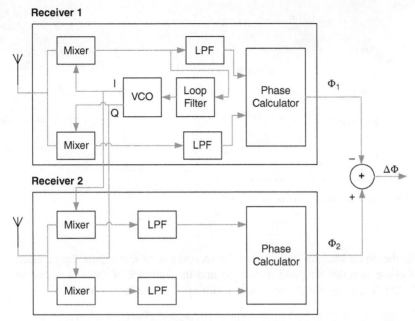

Figure 16.7   Phase difference. LPF, low pass filter; VCO, voltage control oscillator.

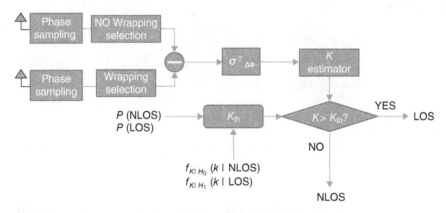

Figure 16.8   Flowchart of phase difference-based NLOS identification.

subscripts LOS and DIF denote the LOS and diffusive components, respectively. r_{LOS1} and r_{LOS2} are unknown and fixed during the sample collection period. r_{DIF1} and r_{DIF2} are independent zero-mean Gaussian. The power ratio of the LOS and NLOS components is $K: K = |r_{LOS1}|^2/E(|r_{DIF1}|^2) = |r_{LOS2}|^2/E(|r_{DIF2}|^2)$.

It is noted that the collected phase samples can be wrapped either in the range $[-\pi, \pi]$ or $[0, 2\pi]$. However, only one of the wrappings will result in correct phase difference variance. A wrapping selection algorithm is developed as follows:

1. For the phase samples collected at the first antenna, compute the variance of phase samples wrapped to $[-\pi, \pi]$ and the variance of those wrapped to $[0, 2\pi]$. Then, choose the wrapping resulting in a smaller variance.

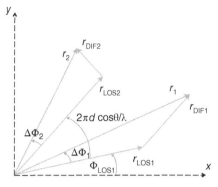

Figure 16.9   Received signals in vector space.

2. For the phase samples collected at the second antenna, compute the variance of phase samples wrapped to $[-\pi, \pi]$ and the variance of those wrapped to $[0, 2\pi]$. Then, choose the wrapping resulting in a smaller variance.

Then, $\sigma^2_{\Delta\phi}$ can be computed given the properly wrapped phase samples.

According to [12], the relationships of K and phase difference variance $\sigma^2_{\Delta\phi}$ are as follows:

$$\text{when } \sigma^2_{\Delta\phi} < 0.232, \quad \hat{K} = \frac{1/\sigma^2_{\Delta\phi} - \gamma}{\beta};$$

$$\text{when } \sigma^2_{\Delta\phi} > 0.232, \quad \hat{K} = \frac{-b + \sqrt{b^2 - 4a\left(c - 1/\sigma^2_{\Delta\phi}\right)}}{2a}. \tag{16.25}$$

So the RMSE of the K-estimator is defined as

$$\text{RMSE} = E\left(K - \hat{K}\right)^2, \tag{16.26}$$

and the bias is defined as

$$\text{Bias} = E\left(\hat{K} - K\right), \tag{16.27}$$

where K represents the true value and $\hat{K}$ represents the estimated value.

The RMSE and bias of the K-estimator are depicted in Figure 16.10.

Here are some MATLAB routines used in the simulation: randn generated Gaussian-distributed random number; wrapto2pi wraps the phase to the range $[0, 2\pi]$; wraptopi wraps the phase to the range $[-\pi, \pi]$; mean returns the mean of a sequence; and var returns the variance of a sequence. Refer to the file chapter_16_Example_2.m for the MATLAB codes. MATLAB codes can be found online at ftp://ftp.wiley.com/public/sci_tech_med/matlab_codes.

A summary of NLOS identification based on channel characteristics and their performance are tabulated in Table 16.3. The identification probability can be one of the following: the correct decision probability under LOS defined as $P(\text{LOS}|\text{LOS})$,

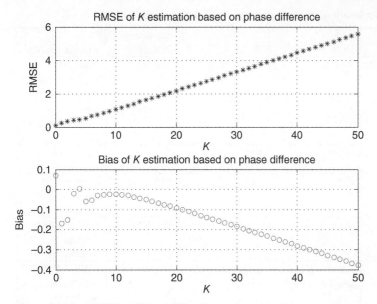

Figure 16.10   RMSE and bias of K-estimator based on $\Delta\phi$.

and the correct decision probability under NLOS defined as P(NLOS|NLOS), the overall correct decision probability defined as P(LOS)P(LOS|LOS) + P(NLOS) P(NLOS|NLOS), where P(LOS) and P(NLOS) are a priori probabilities of respective LOS and NLOS scenarios. In general, the performance order is self-learning techniques > combined metrics > single metric.

16.2.4 Single-Node Hybrid Approach

In this section, we introduce hybrid NLOS identification techniques. For either LOS or NLOS condition, a relationship can be maintained across different channel metrics. For example, in LOS condition, as the TOA increases, it is expected that the RSS decreases following LOS path loss model. Here, the consistency between different channel metrics is explored to perform NLOS identification.

TOA and RSS matching techniques are discussed in [25], [43], and [44]. The intuition behind this method is that if the measured TOA is for a LOS/NLOS BN, then the received power should obey the LOS/NLOS propagation channel model. In [25], the received power is computed from the LOS and NLOS Walfisch–Ikegami path loss model where the distance is substituted with the measured range. Then, the computed power is compared with the measured power (RSS) to see whether it is closer to the LOS model or to the NLOS model.

In [43], the likelihood ratio is given as

$$\frac{f\left(\hat{L}_p\big|\hat{d},H_n\right)}{f\left(\hat{L}_p\big|\hat{d},H_l\right)} \overset{H_n}{\underset{H_l}{\gtrless}} \kappa, \tag{16.28}$$

TABLE 16.3. A Summary of Channel Characteristics-Based Methods

Channel Characteristics	LOS	NLOS	Identification Probability	Application/Note
First arriving path power PDF	Rician	Rayleigh	N/A	Narrow and wideband [25, 27]
Rician K-factor	High	Low	71.6–80.0%	Narrow and wideband [12, 27, 28]
Multipath autocorrelation	High	Low	N/A	Narrow and wideband [29]
Spatial correlation	High	Low	N/A	Narrow band, require multiple antennas [42]
Confidence metric	High	Low	93–100%	UWB, NDDP detected as NLOS [23]
First path power $\|\alpha_1\|^2$ and delay of strongest path $\tau_{max} - \tau_1$	Large $\|\alpha_1\|^2$ and small $\tau_{max} - \tau_1$	Small $\|\alpha_1\|^2$ and large $\tau_{max} - \tau_1$	87.3–93.6%	UWB, NDDP detected as NLOS [31]
Change of power	Increases from NLOS to LOS	Decreases from LOS to NLOS	60.3–100%	UWB, only detect transition [23]
RSS	High	Low	78.30%	UWB [32, 33]
Delay spread	Small	Large	61.7–100%	UWB [23, 32, 34]
Mean excess delay	Small	Large	74.3–100%	UWB [34]
Kurtosis	High	low	66.3–98.4%	UWB [34, 35]
Skewness	High	Low	N/A	UWB [36]
Combined metrics			81.8–99.9%	UWB [32–34, 36]
Self-learning techniques			91–92%	UWB [33, 37]

N/A, not available.

where $\hat{L}_p$ is the estimated path loss, $\hat{d}$ is the estimated range, H_n and H_l are the respective hypotheses of NLOS and LOS, and κ is the threshold that depends on the preassigned false alarm probability.

Three kinds of channel link conditions can be identified in [44]. The likelihood ratio is performed on the conditional PDFs $f\left(C_i \mid \hat{d}, \hat{L}_p\right)$, $i = 0, 1, 2$, where C_0, C_1, and C_2 refer to the channel link conditions LOS, NDDP, and NLOS, respectively.

When both sides of the communication channel are equipped with an antenna array, DOA and DOD matching method can be used to identify NLOS [43]. This method is based on an observation of LOS component as illustrated in Figure 16.11: $\theta_0 = \phi_0$. This relationship does not hold for NLOS components.

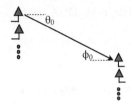

Figure 16.11   The relationship of DOD and DOA of LOS component.

TABLE 16.4.   Comparison of NLOS Identification Methods

	Hardware Complexity	Software Complexity	SNR Requirement	Processing Time	Performance
Cooperative methods	Low–medium	High	High	Medium–high	Good for enough BNs
Range statistics based	Low	Low	High	High	Fair
Channel characteristics based	Low	Low–medium	Low–medium	Low–medium	Varying
Using antenna array	Medium	Low	Low–medium	Low–medium	Fair
Hybrid methods	Low (high for DOD and DOA matching)	Low	High	Low	Fair–good

16.2.5   Comparison of NLOS Identification Methods

Table 16.4 compares different NLOS identification methods in terms of hardware complexity, software complexity, signal-to-noise ratio (SNR) requirement, processing time, and performance. When the antenna array is required, the hardware complexity is considered medium to high. When the DOA or TOA needs to be estimated, the SNR should be high. When the statistics information needs to be collected or the algorithm is complex, it needs a longer processing time. It is observed that channel characteristics-based and antenna array-based methods maintain a good trade-off between requirements and performance: fair-to-good performance may be achieved with low-to-medium requirements.

16.3   NLOS LOCALIZATION

In NLOS localization, the collected measurements are used to localize the target node. Available NLOS localization methods include received signal strength indicator (RSSI) [45], bidirectional TOA–DOA fusion [46], single BN TOA–DOA fusion with the assistance of the environment map [47], and multinode TOA–DOA fusion [48]. Here, we investigate those proposed techniques.

16.3.1 RSSI

The localization methods based on RSSI include connectivity fusion [49] and signature mapping [45]. The localization methods that are based on connectivity fusion require the availability of the LOS channel between reference nodes and target node. When the LOS channel is not available, the accuracy of connectivity fusion localization degrades. In the connectivity fusion, the RSS is transformed to the distance between the BN and the target node through the path loss model. Then, multiple distances between the target and BNs are used to calculate the target node position, for example, via minimum mean square error (MMSE) method.

The signature mapping technique is independent of the availability of a LOS channel between the target node and reference nodes. In this method, reference BNs with fixed positions broadcast a beacon signal and its ID information. The target node can measure the RSS from those BNs as well as identify them. Then, an RSS signature map of the environment is needed to localize the target. Let M denote the number of reference points and N denote the number of BNs. The RSS map includes the positions of reference points, P_i, $1 \le i \le M$, and the RSS signature, consisting of all received beacon signals at reference points, RSS_i, $1 \le i \le M$. Table 16.5 shows the structure of the RSS signature map.

When a target node receives beacon signals from the BNs, the received RSS signature, $\text{RSS}_r = (\hat{a}_{r,1}, \hat{a}_{r,2}, \hat{a}_{r,N})$, is collected. The distance between a signature i and the received signature is defined as

$$d_{r,i} = d(\text{RSS}_r, \text{RSS}_i). \tag{16.29}$$

The Manhattan or Euclidean distance can be used to calculate the distance between the two signatures. With Manhattan distance, we have the distance in the form of

$$d_{r,i} = \sum_{k=1}^{N} |a_{r,k} - a_{i,k}|. \tag{16.30}$$

When the Euclidean distance is applied, the distance of signatures is calculated using

TABLE 16.5. RSS Signature Map

Index of Reference Point	Position	RSS
1	$P_1 = (x_1, y_1)$	$RSS_1 = (a_{1,1}, a_{1,2}, \ldots, a_{1,N})$
2	$P_2 = (x_2, y_2)$	$RSS_2 = (a_{2,1}, a_{2,2}, \ldots, a_{2,N})$
$\vdots$	$\vdots$	$\vdots$
M	$P_M = (x_M, y_M)$	$RSS_M = (a_{M,1}, a_{M,2}, \ldots, a_{M,N})$

$$d_{r,i} = \sqrt{\sum_{k=1}^{N}(a_{r,k} - a_{i,k})^2}.$$ (16.31)

There are three methods to localize the target node with a received signature. In the first method, the target node is the reference point whose RSS signature is closest to the received signature. In other words, the target node is localized at P_k, where

$$k = \arg\min_{1 \leq i \leq M} d(\text{RSS}_r, \text{RSS}_i).$$ (16.32)

This method has low complexity, but the localization accuracy is low.

In the second scheme, the target node is localized at the centroid of α (α is predetermined) reference points whose signatures are close to the received signature, and the target coordinates are computed as

$$x = \frac{1}{\alpha}\sum_{m=1}^{\alpha}x_m, \quad y = \frac{1}{\alpha}\sum_{m=1}^{\alpha}y_m,$$ (16.33)

where (x_m, y_m) is the position corresponding to the signature of reference point m. With a large number of uniformly distributed reference points, a reasonable localization accuracy can be obtained for a fixed number α. However, if the reference points are sparsely distributed in the system coverage area, some reference points are close to the target and the others are far away from the target. If the far-away reference points are selected and used in (16.33), the estimated target node position would be biased toward the far-away reference points, and the localization accuracy is low.

The third method is similar to the second one. Here, the number of selected reference points is not fixed, and it depends on a threshold, c, which can be empirically determined [50]. In this technique, the signature that is the closest to the received signature is calculated as

$$\text{RSS}^* = \arg\min_{1 \leq i \leq M} d(\text{RSS}_r, \text{RSS}_i).$$ (16.34)

Then, the reference signature m is selected if it satisfies

$$\frac{d(\text{RSS}_r, \text{RSS}_m)}{d(\text{RSS}_r, \text{RSS}^*)} \leq c.$$ (16.35)

This results in the signatures closer to the received signature being selected. The scheme only selects reference points that are close to the target node and excludes the far-away reference points. Thus, it eliminates the bias generated by the second method and enhances the localization performance.

The advantage of RSS-based localization methods is that they do not need extra hardware (e.g., antenna array for DOA estimation) or complex software (e.g., cross-correlation calculation for TOA estimation). The RSS is measured in a receiver with simple calculation. In addition, RSS measurement does not require wideband signal and can be easily implemented on the available protocols such as Bluetooth [51] and 802.11 [52]. The shortcoming of RSS-based methods is that the localization accuracy depends on the number of reference points and the distribution of reference BNs (reference BN density). If higher accuracy is required, a large number of reference points with reasonable density in the desired area is necessary.

16.3.2 Bidirectional TOA–DOA Fusion

In bidirectional TOA–DOA fusion NLOS localization [46], it is assumed that BNs and the target node are capable of estimating each other's TOA and DOA. In NLOS scenarios shown in Figure 16.12, the signal transmitted by the target node T can be received by the BN B via reflector A (single-bounce specular reflection), and the TOA and DOA are measured, denoted by $\left(R_1^{(BT)}, \theta_1^{(BT)}\right)$. Conversely, the BN's signal can be received by the target node via the same reflector, and the TOA and DOA are measured as well, denoted by $\left(R_1^{(TB)}, \theta_1^{(TB)}\right)$. In this case, we say that the BN and the target node share the same reflector, and the wireless channel between them is reversible. For the shared reflector A, a line that the target node is located on can be constructed (in Fig. 16.12, the line is T″TB″). When there are more shared reflectors, more lines where the target node is can be constructed, and the target position is determined by the crossing point of those constructed lines.

Next, we explain how the line T″TB″ is constructed. It is determined by the TOA–DOA measurement pair $\left(R_1^{(BT)}, \theta_1^{(BT)}\right)$ and $\left(R_1^{(TB)}, \theta_1^{(TB)}\right)$. In Figure 16.12, point T′ is the image of T regarding to reflector A. The position of the point T′ in the BN's local rectangular coordinate is

$$x_{T'} = R_1^{(BT)} \cdot \cos\theta_1^{(BT)}, \quad y_{T'} = R_1^{(BT)} \cdot \sin\theta_1^{(BT)}. \tag{16.36}$$

Point B′ is the image of B regarding to reflector A, and the position of the point B′ in the target node's local rectangular coordinate is

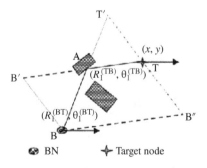

Figure 16.12 Bidirectional TOA–DOA fusion.

$$x_{B'} = R_1^{(TB)} \cdot \cos\theta_1^{(TB)}, \quad y_{B'} = R_1^{(TB)} \cdot \sin\theta_1^{(TB)}. \tag{16.37}$$

The quadrangle BB′TB″ in Figure 16.12 is a parallelogram, and T′TB″ forms a line because B′B is parallel with T′T. Then, the position of B″ in the BN's local rectangular coordinate is

$$x_{B''} = -x_{B'}, \quad y_{B''} = -y_{B'}. \tag{16.38}$$

Given the positions of points T′ and B″, the coordinates of the target node x_T and y_T in the BN's coordinate are subject to

$$\frac{x_T - x_{T'}}{-x_{B'} - x_{T'}} = \frac{y_T - y_{T'}}{-y_{B'} - y_{T'}}, \tag{16.39}$$

which is also the function of the line T′TB″. Substituting (16.36–16.38) into (16.39), we get

$$\overline{a}_1^{T} \cdot X = b_1, \tag{16.40}$$

where $\overline{a}_1 = [a_{11}, a_{12}]$, $X = [x_T, y_T]^T$, and

$$a_{11} = R_1^{(TB)} \sin\theta_1^{(TB)} + R_1^{(BT)} \sin\theta_1^{(BT)}, \tag{16.41}$$

$$a_{12} = -\left(R_1^{(TB)} \cos\theta_1^{(TB)} + R_1^{(BT)} \cos\theta_1^{(BT)}\right), \tag{16.42}$$

$$b_1 = R_1^{(TB)} R_1^{(BT)} \sin\theta_1^{(TB)} \cos\theta_1^{(BT)} - R_1^{(TB)} R_1^{(BT)} \cos\theta_1^{(TB)} \sin\theta_1^{(BT)}. \tag{16.43}$$

If there is a LOS path between B and T, the line T′TB″ will regenerate to one point and (16.40) still holds. When there are N shared reflectors, we have

$$AX = B, \tag{16.44}$$

where $A = [\overline{a}_1, \ldots, \overline{a}_N]^T$, $B = [b_1, \ldots, b_N]^T$. Linear estimation methods can be applied to (16.44) to obtain the target node position. The advantages of this method include (1) no LOS signal is necessary, and (2) when two or more single-bounce reflection NLOS channels are available, the target node can be localized. Its shortcoming is that each target node needs to be equipped with a smart antenna, and that is expensive if the system includes a large number of target nodes. The method can be applied to the system that does not have many target nodes.

16.3.3   Single BN TOA–DOA Fusion with the Assistant Environment Map

In this localization method, the BN can estimate TOA–DOA of the BN, and the electronic environment map is used to localize the omnidirectional target node [47]. It is assumed that this target node is active and responds to the BN inquiry.

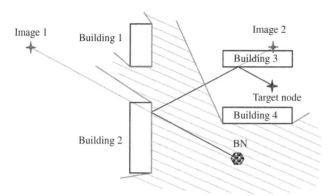

Figure 16.13 Single BN NLOS TOA–DOA fusion with environment map.

The system structure is shown in Figure 16.13. In the electronic environment map, each building's position, size, and reflection surface are known; thus, LOS area (the shadowed area in Fig. 16.13) in this region is known. If a target node is in the LOS area, TOA and DOA measurements can be directly used to perform localization. If a target node is in the NLOS area, then back ray tracing (BRT) analysis is required.

An example of BRT analysis is shown in Figure 16.13. This is a two-bounce reflection between the target node and the BN. Using the TOA and DOA of the reflected signal from Building 2, the target node would be localized at the position of Image 1. From the environment map, it is known that the point of Image 1 is in the NLOS area. Thus, Image 1 cannot be the position of the target node. Then, we trace back to the ray impinging on Building 2 with the knowledge of DOA at the BN and the reflection surface position of Building 2. Thus, the target is localized at the point of Image 2. The position of Image 2 and the reflection surface of Building 2 are at different sides of Building 3. Since the ray cannot penetrate buildings, Image 2 is not the target position either. Then, we trace back to the ray impinging on Building 3 again and obtain a target node position estimate. There are no buildings between the estimated target node position and the reflection surface of Building 3. Therefore, this estimate should be close to the true target node position, and the BRT process is stopped.

In a real localization process, the BN collects multiple TOAs and DOAs of the target node due to reflections from multiple buildings. Let the total number of reflections be K. Using BRT, multiple estimated target node positions, (x_i, y_i), $i = 1, 2, \ldots, K$, are obtained, and the target node is localized at the centroid of all estimated positions, that is,

$$\hat{x} = \frac{1}{K} \sum_{i=1}^{K} x_i, \quad \hat{y} = \frac{1}{K} \sum_{i=1}^{K} y_i. \tag{16.45}$$

16.3.4 Multinode TOA–DOA Fusion

This method [48] utilizes the scatterer's information to localize the target node when a LOS channel between the target node and the BN is not available, and single-bounce

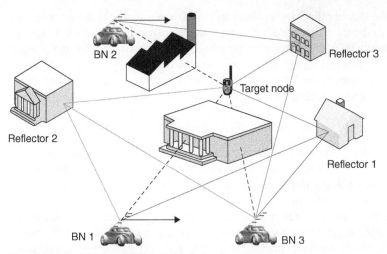

Figure 16.14   Multinode TOA–DOA fusion localization.

reflection channel is assumed. BNs' positions are known, and they are equipped with antenna arrays that are capable of estimating TOA and DOA of the received signal. The target node is equipped with omnidirectional antennas.

The structure of the localization system is shown in Figure 16.14. In this figure, the LOS channel between the target node and BNs are blocked by buildings, and the single-bounce reflection NLOS channel between the target node and multiple BNs are available. Here, it is assumed that the NLOS channels have been identified. Our goal is to identify the shared reflectors via which multiple BNs could receive signals (e.g., Reflector 1 is shared by BNs 1 and 3) and find out the positions of these shared reflectors, so as to use their positions to localize the NLOS target node.

BN i (i is the BN index) obtains TOAs and DOAs of the signal sent by the target node from different directions due to multiple reflectors. The TOA estimated by BN i via reflector k is $\tau_{i,k}$, and the corresponding DOA is $\theta_{i,k}$. Here, wideband system and high-precision TOA estimation is assumed. The signal-traveling distance can be computed as $R_{i,k} = \tau_{i,k}\cdot c$, where c is the wave propagation speed.

In Figure 16.14, Reflector 1 is shared by the target node and BNs 1 and 3. The position of the reflector can be obtained by fusing the two DOAs collected at the two BNs. Then, the distances between Reflector 1 and BNs 1 and 3 can be calculated, and the distances are denoted by $R_{1,1}^{(B,R)}$ and $R_{3,1}^{(B,R)}$, respectively. Then the distances between the target node and Reflector 1, denoted by $R_{i,k}^{(RT)}$, $i = 1, 3$, $k = 1$, are computed as $R_{i,k}^{(R,T)} = R_{i,k} - R_{i,k}^{(B,R)}$.

If reflector k is the shared reflector of BN i and j, $R_{i,k}^{(R,T)}$ and $R_{j,k}^{(R,T)}$ have to be subject to

$$\left| R_{i,k}^{(R,T)} - R_{j,k}^{(R,T)} \right| \le R_{\text{th}}, \tag{16.46}$$

where R_{th} is the threshold that determines whether the reflector is shared. R_{th} depends on the TOA estimation error and the reflector localization error due to the DOA fusion. For details of R_{th}, please refer to [47].

With the shared reflectors' position and the distance from the shared reflectors to the target node, the target node would be localized using the TOA fusion method. One point should be noted: A reflector may be shared by multiple BNs and a target node with multiple reflecting points; in other words, the multiple BNs do not receive the target node signal from the same point on the reflector. In this case, extra localization error would be generated.

Example 16.3 NLOS Localization via Multinode TOA–DOA Fusion

Use MATLAB simulation to evaluate the NLOS localization performance of multinode TOA–DOA fusion method in terms of localization error cumulative distribution function (CDF) when the target node is localized using two NLOS BNs and three effective shared reflectors (the reflectors can be localized via DOA fusion with reasonable accuracy). The BN TOA and DOA estimation errors are zero-mean Gaussian random variables with STD of $\sigma_R = 5$ m and $\sigma_\theta = 1°$, respectively. All the BNs, target node, and shared reflectors are randomly distributed in a $200\sigma_R \times 200\sigma_R$ square area. To generate the error CDF, 1000 geometrical distributions are needed.

Solution

The NLOS localization error CDF is the objective of the simulation, the localization error for each geometrical distribution should be calculated, and then the error CDF can be generated. In each geometrical distribution, the effective shared reflectors are first localized via DOA fusion; and then the distance between the shared reflector and the target node is calculated; finally, the target node is localized via TOA fusion. In the effective shared reflector selection process, the inner angle threshold (θ_{th}) is selected to be $\pi/10$. Refer to [48] for the detail of effective shared reflector selection method. The simulation is as follows:

1. Generate 1000 sets of node positions in the $200\sigma_R \times 200\sigma_R$ square area.
2. Find the effective shared reflectors.
3. Localize the shared reflectors via DOA fusion.
4. Calculate the distances between the target node and the shared reflectors.
5. Localize the target node via TOA fusion.
6. Calculate the localization error.
7. Generate the localization error CDF.

The NLOS localization error CDF is shown in Figure 16.15. It is seen that the localization error is smaller than 40 m with a probability of 74%, and it is less than 80 m with a probability of 88%. The file chapter_16_Example_3.m details the MATLAB codes needed for the generation of the CDF of Figure 16.15. MATLAB codes can be found online at ftp://ftp.wiley.com/public/sci_tech_med/matlab_codes.

16.3.5 Comparison

Table 16.6 compares the above four NLOS localization methods. RSSI-based methods use simple hardware and software. In addition, RSSI does not need high

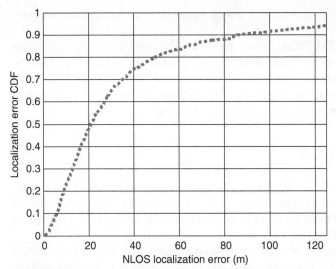

Figure 16.15   NLOS localization error CDF with two BNs and three shared reflectors.

TABLE 16.6.   NLOS Localization Methods Comparison

	Hardware Requirement	Software Requirement	Installation Requirement	SNR Requirement	Localization Accuracy
RSSI	Lowest	Low	High	Low	Varying with RSSI map
Bidirectional TOA–DOA	High	Low	Low	High	High
Single-node TOA–DOA with map	Low	High	High	High	Low
Multinode TOA–DOA	Medium	High	Low	High	Medium

SNR. However, RSSI-based methods require a predetermined RSSI signature map. The localization accuracy is high for a fine grid map and the accuracy is low for a coarse grid map.

The other three methods require DOA information to localize target node. High SNR is needed to ensure high-precision DOA estimation. In addition, high-precision TOA estimation requires high SNR and large bandwidth.

In bidirectional TOA–DOA fusion, each target node is equipped with an antenna array. Thus, it requires high complexity hardware. Its software and installation are simple. High localization accuracy can be achieved when the wireless channel is reversible.

For the single BN TOA–DOA fusion technique, only the single BN requires an antenna array. Thus, its hardware complexity is low. The software complexity is high, because one or more steps of back ray tracing analysis are needed. Due to the environment map's inaccuracy, its localization accuracy is low.

In multinode TOA–DOA fusion, multiple BNs need to be equipped with antenna arrays; thus, its hardware requirement is considered medium. Because the shared reflectors should be first localized, its software complexity is high. When multiple reflection points are shared by BNs, extra localization error is expected; thus, its localization accuracy is considered as medium.

16.4 CONCLUSION

This chapter discusses NLOS identification and NLOS localization techniques. NLOS identification could improve the localization accuracy and indicate the channel link conditions. When there are no LOS measurements, NLOS localization methods can be applied to find the target position.

There are a variety of NLOS identification methods with different complexity and performance levels. The cooperative NLOS identification techniques perform well only when there are enough LOS measurements, and the software complexity is high. The range statistics-based methods require a high SNR for TOA estimation and also require a long processing time to acquire statistics. Moreover, they may fail to correctly identify NLOS when BNs, the target node, and the scatterers are all stationary. On the contrary, the channel characteristics-based, antenna array-based, and hybrid methods provide a good trade-off between requirements and performance.

NLOS localization is possible when additional information is available, such as the RSSI signature map, the environment map, and the fact that all reflections are specular. In RSSI-based methods, RSS information is very easy to obtain in wireless systems, but the RSSI signature map has to be predetermined and the localization accuracy depends on the resolution of the RSSI map. The single-node TOA–DOA method needs the assistance of the environment map, and the localization accuracy is low due to the inaccuracy of the map. Bidirectional TOA–DOA and multinode TOA–DOA methods do not require any prior information. The bidirectional TOA–DOA method can be used when each node is equipped with an antenna array and the localization accuracy is high. In the multinode TOA–DOA method, the shared reflectors are identified and located, and then those reflectors are used as reference nodes to localize the target. Thus, this method can be used for NLOS localization when there exist at least three shared reflectors and those shared reflectors are in LOS with the target. An example of TOA–DOA localization systems is the wireless local positioning system (WLPS) as has been discussed in Chapter 1 briefly and in Chapter 34 in detail.

REFERENCES

[1] I. Amundson and X. Koutsoukos, "A survey on localization for mobile wireless sensor networks," presented at the 2nd Int. Workshop on Mobile Entity Localization and Tracking in GPS-Less Environments (MELT), Orlando, FL, Sep. 2009.

[2] S. Ingram, D. Harmer, and M. Quinlan, "Ultrawideband indoor positioning systems and their use in emergencies," in *Position Location and Navigation Symposium, PLANS 2004*, Apr. 2004, pp. 706–715.

[3] Federal Communications Commission, "Enhanced 911." Available: http://www.fcc.gov/pshs/services/911-services/enhanced911

[4] W. Jones, "Keeping cars from crashing," *IEEE Spectr.*, vol. 38, no. 9, pp. 40–45, 2001.

[5] K. Amouris, "Position-based broadcast TDMA scheduling for mobile ad-hoc networks (MANETs) with advantaged nodes," in *IEEE Military Communications Conf., MILCOM 2005*, vol. 1, Oct. 2005, pp. 252–257.

[6] V. Sumathy, P. Narayanasmy, K. Baskaran, and T. Purusothaman, "GLS with secure routing in ad-hoc networks," in *Conf. on Convergent Technologies for Asia-Pacific Region TENCON 2003*, vol. 3, Oct. 2003, pp. 1072–1076.

[7] M. Rahman, M. Mambo, A. Inomata, and E. Okamoto, "An anonymous on-demand position-based routing in mobile ad hoc networks," in *Int. Symp. on Applications and the Internet, SAINT 2006*, Jan. 2006, pp. 7–306.

[8] J. Caffery, *Wireless Location in CDMA Cellular Radio System*. Norwell, MA: Kluwer Academic Publisher, 1999.

[9] M. Bshara, U. Orguner, F. Gustafsson, and L. Van Biesen, "Fingerprinting localization in wireless networks based on received-signal-strength measurements: A case study on WiMAX networks," *IEEE Trans. Veh. Technol.*, vol. 59, no. 1, pp. 283–294, 2010.

[10] M. Silventoinen and T. Rantalainen, "Mobile station locating in GSM," in *IEEE Wireless Communication System Symp., 1995*, Nov. 1995, pp. 53–59.

[11] C. Falsi, D. Dardari, L. Mucchi, and M. Z. Win, "Time of arrival estimation for UWB localizer in realistic environment," *EURASIP J. Appl. Signal Process.*, vol. 2006, pp. 1–13, 2006.

[12] W. Xu, "Non-line-of-sight identification in wireless localization via phase difference statistics across two antenna elements," *IET Commun.*, vol. 3, pp. 1814–1822, Sep. 2011.

[13] X. Li, "A selective model to suppress NLOS signals in angle-of-arrival (AOA) location estimation," in *9th IEEE Int. Symp. on Personal, Indoor and Mobile Radio Communications*, Sep. 1998, pp. 461–465.

[14] L. Cong and W. Zhuang, "Nonline-of-sight error mitigation in mobile location," *IEEE Trans. Wireless Commun.*, vol. 4, no. 2, pp. 560–573, 2005.

[15] Y.-T. Chan, W.-Y. Tsui, H.-C. So, and P. C. Ching, "Time-of-arrival based localization under NLOS conditions," *IEEE Trans. Veh. Technol.*, vol. 55, no. 1, pp. 17–24, 2006.

[16] M. Wylie and J. Holtzman, "The non-line of sight problem in mobile location estimation," in *5th IEEE Int. Conf. on Universal Personal Communications*, Sep. 1996, pp. 827–831.

[17] J. Borras, P. Hatrack, and N. Mandayam, "Decision theoretic frame-work for NLOS identification," in *48th IEEE Vehicular Technology Conf., VTC 98*, vol. 2, May 1998, pp. 1583–1587.

[18] S. Venkatraman, J. J. Caffery, and H.-R. You, "Location using LOS range estimation in NLOS environments," in *55th IEEE VTC*, 2002, pp. 856–860.

[19] S. Gezici, H. Kobayashi, and H. V. Poor, "Non-parametric non-line-of-sight identification," in *Proc. of the IEEE 58th VTC*, vol. 4, Orlando, FL, Oct. 2003, pp. 2544–2548.

[20] N. Alsindi, M. Heidari, and K. Pahlavan, "Blockage identification in indoor UWB ranging using multi band OFDM signals," in *IEEE Wireless Communications and Networking Conf., WCNC 2008*, Mar. 31–Apr. 3, 2008, pp. 3231–3236.

[21] P.-C. Chen, "A non-line-of-sight error mitigation algorithm in location estimation," in *1999 IEEE Wireless Communications and Networking Conf., WCNC 1999*, vol. 1, 1999, pp. 316–320.

[22] S.-S. Woo, H.-R. You, and J.-S. Koh, "The NLOS mitigation technique for position location using IS-95 CDMA networks," in *52nd VTC 2000*, vol. 6, 2000, pp. 2556–2560.

[23] J. Schroeder, S. Galler, K. Kyamakya, and K. Jobmann, "NLOS detection algorithms for ultra-wideband localization," in *4th Workshop on Positioning, Navigation and Communication*, Mar. 2007, pp. 159–166.

[24] L. Mak and T. Furukawa, "A time-of-arrival-based positioning technique with non-line-of-sight mitigation using low-frequency sound," *Adv. Robot.*, vol. 22, no. 5, pp. 507–526, 2008.

[25] J. S. Al-Jazzar and J. Caffery, "New algorithms for NLOS identification," presented at the IST Mobile and Wireless Communications Summit, Dresden, Germany, 2005.

[26] H. Bolcskei, "Fundamentals of wireless communications," Lecture Handouts, Mar. 2005.

[27] A. Lakhzouri, R. Hamila, E. S. Lohan, and M. Renfors, "Extended Kalman filter channel estimation for line-of-sight detection in WCDMA mobile positioning," *EURASIP*, vol. 2003, no. 13, pp. 1268–1278, 2003.

[28] F. Benedetto, G. Giunta, A. Toscano, and L. Vegni, "Dynamic LOS/NLOS statistical discrimination of wireless mobile channels," in *65th IEEE VTC 2007*, Apr. 2007, pp. 3071–3075.

[29] S. Yarkan and H. Arslan, "Identification of LOS and NLOS for wireless transmission," in *1st Int. Conf. Cognitive Radio Oriented Wireless Networks and Communications*, Jun. 8–10, 2006, pp. 1–5.

[30] A. Molisch, D. Cassioli, C.-C. Chong, S. Emami, A. Fort, B. Kannan, J. Karedal, J. Kunisch, H. Schantz, K. Siwiak, and M. Win, "A comprehensive standardized model for ultrawideband propagation channels," *IEEE Trans. Antennas Propag.*, vol. 54, no. 11, pp. 3151–3166, 2006.

[31] A. Maali, H. Mimoun, G. Baudoin, and A. Ouldali, "A new low complexity NLOS identification approach based on UWB energy detection," *IEEE Radio and Wireless Symp., RWS '09*, Jan. 18–22, 2009, pp. 675–678.

[32] S. Venkatesh and R. Buehrer, "Non-line-of-sight identification in ultra-wideband systems based on received signal statistics," *Microwaves Antennas Propag. IET*, vol. 1, no. 6, pp. 1120–1130, 2007.

[33] M. Heidari, N. Alsindi, and K. Pahlavan, "UDP identification and error mitigation in TOA-based indoor localization systems using neural network architecture," *IEEE Trans. Wireless Commun.*, vol. 8, no. 7, pp. 3597–3607, 2009.

[34] I. Guvenc, F. W. C. Chong, and H. Inamura, "NLOS identification and weighted least-squares localization for UWB systems using multipath channel statistics," *EURASIP J. Adv. Signal Process.*, vol. 2008, no. 36, 2008.

[35] L. Mucchi and P. Marcocci, "A new parameter for UWB indoor channel profile identification," *IEEE Trans. Wireless Commun.*, vol. 8, no. 4, pp. 1597–1602, Apr. 2009.

[36] A. Abbasi and M. Kahaei, "Improving source localization in LOS and NLOS multipath environments for UWB signals," in *14th Int. CSI Computer Conf., CSICC 2009*, Oct 20–21, 2009, pp. 310–316.

[37] S. Marano, W. Gifford, H. Wymeersch, and M. Win, "Nonparametric obstruction detection for UWB localization," in *IEEE Global Telecommunications Conf., GLOBECOM 2009*, Nov. 2009, pp. 1–6.

[38] E. Jacobsen et al., High Throughput Task Group, "IEEE P802.11 wireless LANs," Tech. Rep. IEEE 802.11-03/940r4, May 2004.

[39] IEEE, "IEEE standard for local and metropolitan area networks, Part 16: Air interface for broadband wireless access systems," IEEE Std 802.16-2009 (Revision of IEEE Std 802.16-2004), May 2009.

[40] J. Salo, G. D. Galdo, J. Salmi, P. Kyosti, M. Milojevic, D. Laselva, and C. Schneider, "Spatial channel model for multiple input multiple output (MIMO) simulations," Sep. 2003. Available: http://ftp.3gpp.org/specs/html-info/25996.htm

[41] W. Xu, S. Zekavat, and H. Tong, "A novel spatially correlated multiuser MIMO channel modeling: Impact of surface roughness," *IEEE Trans. Antennas Propag.*, vol. 57, no. 8, pp. 2429–2438, 2009.

[42] W. Xu and S. Zekavat, "Spatially correlated multi-user channels: LOS versus NLOS," in *IEEE 13th Digital Signal Processing Workshop and 5th IEEE Signal Processing Education Workshop, DSP/SPE 2009*, Jan. 4–7, 2009, pp. 308–313.

[43] K. Yu and Y. Guo, "Statistical NLOS identification based on AOA, TOA, and signal strength," *IEEE Trans. Veh. Technol.*, vol. 58, no. 1, pp. 274–286, 2009.

[44] N. Alsindi, C. Duan, J. Zhang, and T. Tsuboi, "NLOS channel identification and mitigation in ultra wideband TOA-based wireless sensor networks," in *6th Workshop on Positioning, Navigation and Communication, WPNC 2009*, Mar. 19, 2009, pp. 59–66.

[45] L. Konrad and W. Matt, "Motetrack: A robust, decentralized approach to RF-based location tracking," in *Int. Workshop on Location and Context-Awareness*, vol. 3479, 2005, pp. 63–82.

[46] C. Seow and S. Tan, "Non-line-of-sight localization in multipath environments," *IEEE Trans. Mobile Comput.*, vol. 7, no. 5, pp. 647–660, 2008.

[47] H. Song, H. Wang, K. Hong, and L. Wang, "A novel source localization scheme based on unitary ESPRIT and city electronic maps in urban environments," *Prog. Electromagnetics Res. PIER*, vol. 94, no. 94, pp. 243–262, 2009.

[48] Z. Wang and S. A. Zekavat, "Omnidirectional mobile NLOS identification and localization via multiple cooperative nodes," *IEEE Trans. Mobile Comput.*, vol. 11, pp. 2047–2059, Dec. 2012.

[49] N. Bulusu, J. Heidemann, and D. Estrin, "GPS-less low-cost outdoor localization for very small devices," *IEEE Pers. Commun.*, vol. 7, no. 5, pp. 28–34, 2000.

[50] L. Konrad and W. Matt, "Motetrack: A robust, decentralized approach to RF-based location tracking," *Pers. Ubiquitous Comput.*, vol. 11, no. 6, pp. 489–503, 2006.

[51] M. Rodriguez, J. Pece, and C. Escudero, "In-building location using bluetooth," presented at Int. Workshop on Wireless Ad Hoc Networks, London, UK, 2005.

[52] A. Harder, L. Song, and Y. Wang, "Towards an indoor location system using RF signal strength in IEEE 802.11 networks," in *Int. Conf. on Information Technology: Coding and Computing, ITCC 2005*, vol. 2, Apr. 4–6, 2005, pp. 228–233.

CHAPTER 17

NLOS MITIGATION METHODS FOR GEOLOCATION

Joni Polili Lie,[1] Chin-Heng Lim,[1] and Chong-Meng Samson See[1,2]

[1]Nanyang Technological University, Singapore
[2]DSO National Laboratories, Singapore

THE PROBLEM of locating mobile sensors has received considerable attention, particularly in the field of wireless communications. It is well known that the presence of non-line-of-sight (NLOS) errors in the geolocation problem leads to severe degradation in the localization performance. Assuming that redundant location metrics measurements (e.g., received signal strength, time of arrival [TOA], time difference of arrival [TDOA], or angle of arrival) are available and the number of LOS measurements available satisfies the minimum requirement to compute the location estimates, the NLOS errors can then be identified or its effect can be mitigated. In Chapter 16, several NLOS identification techniques and the application to localization have been discussed extensively. In this chapter, NLOS mitigation methods for geolocation are introduced. Generally, the NLOS mitigation methods for geolocation can be grouped into four categories. They are the methods based on maximum likelihood (ML), least squares (LS), and constrained optimization, as well as robust statistics. These methods are then compared in terms of different performance measures. In addition, this chapter also discusses a novel geolocation example using a single moving sensor. Numerical examples are then presented to demonstrate how position estimation can be achieved for the case of a single moving sensor with NLOS errors.

Handbook of Position Location: Theory, Practice, and Advances, Second Edition.
Edited by S. A. (Reza) Zekavat and R. Michael Buehrer.
© 2019 by the Institute of Electrical and Electronics Engineers, Inc.
Published 2019 by John Wiley & Sons, Inc.
Companion Website: www.wiley.com/go/zekavat/positionlocation2e

17.1 INTRODUCTION

Localization technologies have received considerable attention in recent times [1], and they are applicable to many fields, such as emergency systems [2], health industries, and intelligent transport systems. The global positioning system (GPS) covers most of the earth's surface and GPS chipsets are continually decreasing in cost, making it feasible for them to be integrated into many mobile devices. However, GPS is unreliable when it comes to urban environments, like the "canyons" formed by high-rise buildings.

As discussed in earlier chapters, several geolocation techniques have been proposed in the literature [3–5]. The location parameters include received signal strength indicator (RSSI) measurements, direction of arrival (DOA), TOA, and TDOA, or some combinations thereof. The localization capability highly depends on the acquisition of accurate information that can be used for positioning.

The estimation accuracy of the aforementioned location parameters degrades significantly in the absence of direct or LOS propagation path. These scenarios are said to occur when there is no direct path between the transmitter (Tx) and receiver (Rx) due to the presence of materials that obstruct the direct propagation path, which can occur in both indoor and outdoor propagation.

In a common geolocation scenario, the number of Tx–Rx pairs that can be established is much more than the minimum required to compute the location estimates. For instance, in order to compute a two-dimensional (2-D) position, at least three range measurements (e.g., RSSI, TOA, or TDOA) or two direction-based measurements (e.g., DOA) are required. It is further assumed that in a NLOS geolocation scenario, the number of LOS measurements that is present among the NLOS ones satisfies this minimum requirement. In consequence of this condition, the NLOS errors can be identified or its effect can be mitigated. In Chapter 16, several NLOS identification techniques and the application to localization have been discussed extensively. This chapter focuses on the NLOS mitigation methods.

The effect of NLOS propagation on the performance of the localization techniques can be investigated from its physical characteristics. Generally, the NLOS measurements can be characterized by the excess propagation path length due to reflection and diffraction. As a result, it will introduce a positive bias to the range measurements, which ultimately results in an erroneous location estimate.

Due to these characteristics, it is possible to mitigate the NLOS measurements from the mixed LOS and NLOS measurements. The mitigation techniques may exploit the a priori information of these NLOS measurements [6], such as the conditional probabilities of the position estimation error due to NLOS positive bias, and derive an ML estimate of the position given these statistics. Other NLOS mitigation techniques are based on transforming the location metrics into overdetermined systems of equation, and then deriving its LS solution. These LS-based techniques can be further grouped based on whether the NLOS measurements identified are discarded or not [7]. Apart from the ML- and LS-based techniques, some NLOS mitigation techniques are based on constrained optimization and other robust estimators.

In the last section of this chapter, a case of a single moving sensor geolocation and the application of the NLOS mitigation techniques are discussed. In a single moving sensor geolocation, it is assumed that the geolocation of the mobile terminal (MT) is achieved from multiple location metrics measurements from the single sensor while it is moving. This is possible under the condition that the position of the sensor is known during the location metrics measurement. Besides reducing the wireless infrastructure complexity due to the use of a single sensor, the solution to such a geolocation problem could exploit the fact that the location metrics from one measurement to the next one does not change abruptly. Finally, numerical examples are presented to demonstrate the geolocation performance.

17.2 GEOLOCATION SYSTEM MODEL

Consider a wireless network made of N fixed terminals (FTs) and an MT whose location is to be determined [8]. Prior to estimating the position of the MT, it is assumed that location metric measurements between N FT–MT pairs have been acquired and contain errors due to measurement noise and NLOS propagation. Without loss of generality, it is assumed that the location metrics considered are the TOA and its range measurement between N FT–MT pairs, which can be expressed as

$$r_n = d_n + w_n + i_n, \tag{17.1}$$

for $n = \{1, 2, \ldots, N\}$, where r_n refers to the range measurement, which is obtained by multiplying the TOA measurement t_n with the speed of light; d_n is the true distance of the nth FT–MT pair; w_n is the measurement noise (modeled as a zero-mean Gaussian random variable with variance γ_n^2); and i_n is the NLOS error (also viewed as impulsive noise). In the absence of NLOS errors, $i_n = 0$.

Assume that (x, y) is the position of MT and (x_n, y_n) is the position of the nth FT; d_n can be written as a nonlinear function of (x, y):

$$d_n = \sqrt{(x_n - x)^2 + (y_n - y)^2}. \tag{17.2}$$

A way for solving the nonlinear equations directly is proposed in [5], but this is computationally intensive.

Alternatively, it is possible to linearize the nonlinear equations [9] as follows:

$$\begin{aligned} d_n^2 &= (x_n - x)^2 + (y_n - y)^2 \\ &= K_n - 2x_n x - 2y_n y + x^2 + y^2. \end{aligned} \tag{17.3}$$

By defining $K_n = x_n^2 + y_n^2$ and $R = x^2 + y^2$, Equation (17.3) can be rewritten through a set of linear expressions:

$$d_n^2 - K_n = -2x_n x - 2y_n y + R. \tag{17.4}$$

Let $\theta = [x, y, R]^T$ and express (17.4) in matrix form:

$$\mathbf{b} = \mathbf{H}\theta, \tag{17.5}$$

where

$$\mathbf{b} = \begin{bmatrix} b_1 \\ b_2 \\ \vdots \\ b_N \end{bmatrix} = \begin{bmatrix} d_1^2 - K_1 \\ d_2^2 - K_2 \\ \vdots \\ d_N^2 - K_N \end{bmatrix},$$

$$\mathbf{H} = \begin{bmatrix} \mathbf{h}_1^T \\ \mathbf{h}_2^T \\ \vdots \\ \mathbf{h}_N^T \end{bmatrix} = \begin{bmatrix} -2x_1 & -2y_1 & 1 \\ -2x_2 & -2y_2 & 1 \\ & \vdots & \\ -2x_N & -2y_N & 1 \end{bmatrix}.$$

For distance-based measurement, the actual DOA of the nth FT–MT pair can be expressed as

$$d_n = \tan^{-1} \frac{y - y_n}{x - x_n}, \qquad (17.6)$$

and its linearization can be derived as follows:

$$y_n - x_n \tan(d_n) = y - x \tan(d_n) \qquad (17.7)$$

and expressed in matrix form

$$\mathbf{b} = \mathbf{H}\theta, \qquad (17.8)$$

where $\theta = [x, y]^T$

$$\mathbf{b} = \begin{bmatrix} b_1 \\ b_2 \\ \vdots \\ b_N \end{bmatrix} = \begin{bmatrix} y_1 - \tan(d_1)x_1 \\ y_2 - \tan(d_2)x_2 \\ \vdots \\ y_N - \tan(d_N)x_N \end{bmatrix},$$

$$\mathbf{H} = \begin{bmatrix} \mathbf{h}_1^T \\ \mathbf{h}_2^T \\ \vdots \\ \mathbf{h}_N^T \end{bmatrix} = \begin{bmatrix} 1 - \tan(d_1) \\ 1 - \tan(d_2) \\ \vdots \\ 1 - \tan(d_N) \end{bmatrix}.$$

Note that the above solutions assume the noiseless case where d_n is available. In practice, only r_n is available, and it usually contains the NLOS errors i_n and Gaussian noise w_n. For the LOS case, one could derive the LS solution based on the above linear model by replacing d_n with r_n. For the NLOS case, the following NLOS mitigation techniques are required.

17.3 A REVIEW OF NLOS MITIGATION TECHNIQUES

17.3.1 ML-Based Techniques

Recall that the range measurement model in (17.1) contains the NLOS errors denoted as i_n. Assume that the NLOS errors due to the nth FT–MT pair are modeled as a constant. It is also further assumed that, instead of having only one measurement, M range measurements are collected for each FT–MT pair:

$$r_{m,n} = d_n(x,y) + w_{m,n} + b_k. \tag{17.9}$$

Under this assumption, the conditional density function (probability density function [PDF]) of the Mth range measurements for the nth FT–MT pair can be expressed as

$$p_n^{\text{NLOS}}\left(\mathbf{r}_n|(x,y),b_n\right) = \prod_{m=1}^{M} \frac{1}{\sqrt{2\pi}\gamma_n} e^{-\frac{1}{2}\left(\frac{r_{m,n}-d_n(x,y)-b_n}{\gamma_n}\right)^2}, \tag{17.10}$$

$$p_n^{\text{LOS}}\left(\mathbf{r}_n|(x,y)\right) = \prod_{m=1}^{M} \frac{1}{\sqrt{2\pi}\gamma_n} e^{-\frac{1}{2}\left(\frac{r_{m,n}-d_n(x,y)}{\gamma_n}\right)^2}, \tag{17.11}$$

where $\mathbf{r}_n = [r_{1,n}, \ldots, r_{M,n}]^T$. Notice that the earlier expression corresponds to the PDF when the nth pair is NLOS, while the later is LOS. Both expressions assume that the variance γ_n^2 is assumed to be known.

Assume that the algorithm considers the number of possible NLOS pairs are L among N pairs. This means that, of all N pairs, the number of NLOS pairs could be 0 or up to L pairs. Thus, the number of hypotheses to be tested, N_h, can be expressed as the sum of combinatorics for each possibility:

$$N_h = \sum_{l=0}^{L}\binom{N}{l} = \sum_{l=0}^{L}\frac{N!}{l!(N-l)!}, \tag{17.12}$$

where $\binom{N}{l}$ denotes the combinatorics expression for choosing l-element subsets of the N-element set.

Let q denote the index of the hypothesis under consideration and L_q denote its likelihood function. By grouping the set of NLOS pairs S_q^{NLOS} and the set of LOS pairs S_q^{LOS}, L_q can be expressed as

$$L_q\left(\mathbf{r}|(x,y),\mathbf{b}_q^{\text{NLOS}}\right) = \prod_{n\in S_q^{\text{NLOS}}} p_n^{\text{NLOS}}\left(\mathbf{r}_n|(x,y),b_n\right) \prod_{n\in S_q^{\text{LOS}}} p_n^{\text{LOS}}\left(\mathbf{r}_n|(x,y)\right), \tag{17.13}$$

where $\mathbf{r}$ is the $(M \times N)$-length vector that contains all the range measurements, $\mathbf{b}_q^{\text{NLOS}} = \{b_n\}_{n\in S_q^{\text{LOS}}}$ is a vector of NLOS biases whose dimension depends on the number of pairs in S_q^{NLOS}. It is worth mentioning that $L_q\left(\mathbf{r}|(x,y),\mathbf{b}_q^{\text{NLOS}}\right)$ depends on a different subset of unknown parameters, for example, (x,y) and $\mathbf{b}_q^{\text{NLOS}}$. For the case when γ_n^2 is unknown, the likelihood function also depends on γ_n^2.

Given these hypotheses, the essence of the ML-based NLOS mitigation techniques can be summarized in two stages. First, the algorithm finds the ML estimation of the unknown (x,y) and b_n for each hypotheses. The estimated parameters from each hypothesis are then used to formulate the conditional PDF, and finally, it chooses the hypothesis that results in maximum conditional PDF q_l. From the chosen hypothesis, the position estimation of (x,y) can be obtained from the earlier ML estimates of the chosen hypothesis.

Finding N_h ML Estimates of Unknown Parameters: Let $\phi_q = \left[x, y, \left(b_q^{\text{NLOS}}\right)^T\right]^T$
denote the unknown parameters in the qth hypothesis. These parameters include the
actual position of the MT and the NLOS biases from the set of NLOS pairs under
the qth hypothesis. Notice that the dimension of ϕ_q is not fixed, as the number of
NLOS pairs is different for different hypotheses. The ML estimates of ϕ_q can then
be expressed as

$$\hat{\phi}_q = \arg\max_{\phi_q} L_q\left(\mathbf{r}|(x, y), \mathbf{b}_q^{\text{NLOS}}\right). \tag{17.14}$$

Note that the implementation of the above optimization requires $(|\mathbf{b}_q^{\text{NLOS}}| + 2)$-dimen-
sional search, where $|\cdot|$ denotes the cardinality of a set or the number of elements
in a set. As there are N_h hypotheses, the optimization search has to be performed N_h
times. Therefore, the complexity of the algorithm will grow with the number of
considered possible NLOS pairs L.

Finding the Most Possible Hypothesis: Given $\hat{\phi} = \left[\hat{x}_q, \hat{y}_q, \left(\hat{b}_q^{\text{NLOS}}\right)^T\right]^T$
estimated from the previous computation, the likelihood function of the qth hypoth-
esis can now be computed by substituting $\hat{x}_q$, $\hat{y}_q$ and $\hat{\mathbf{b}}_q^{\text{NLOS}}$ to the likelihood expression in (17.13). Next, the algorithm finds the most possible hypothesis among the
N_h hypotheses as

$$q_{ml} = \arg\max_{q \in \{1, \cdots, N_h\}} L_q\left(\mathbf{r}|(\hat{x}_q, \hat{y}_q), \hat{\mathbf{b}}_q^{\text{NLOS}}\right). \tag{17.15}$$

And finally, the position estimation is chosen from the q_{ml}th estimate: $\hat{x} = \hat{x}_{q_{ml}}$ and
$\hat{y} = \hat{y}_{q_{ml}}$.

Other variants of ML-based geolocation in the presence of NLOS errors
include the works reported in [10] and [11], which extract the parameters of NLOS
channel impulse response information and subject them to a likelihood ratio test in
order to identify if the TOA measurement is due to LOS or NLOS propagation.
These parameters are the kurtosis, mean excess delay, and root mean square delay
spread test. Other statistical-based approaches are also described in [12]. When the
statistics of the NLOS errors are unknown, the authors in [13] attempted to solve
the identification problem using nonparametric density estimations. For a more
elaborate explanation on NLOS identification techniques, the readers are advised to
refer to Chapter 16.

Pros and Cons: In general, the main advantage of the ML-based technique is
that it is the asymptotically optimal solution under the condition that all the assumptions stated in deriving the model are true. Furthermore, it requires that not only the
PDF of both the NLOS and LOS measurements follow strictly the assumed model,
but also the parameters that shape these density functions are known a priori.

17.3.2 LS-Based Techniques

From the matrix representation of the linear systems in (17.8), the classical LS solution can be expressed as

$$\hat{\theta} = \arg\min_{\theta}\|\mathbf{b} - \mathbf{H}\theta\|_2^2$$

$$= \left(\mathbf{H}^\mathrm{T}\mathbf{H}\right)^{-1}\mathbf{H}^\mathrm{T}\mathbf{b}. \tag{17.16}$$

When NLOS errors are present in the range measurements r_n that forms $\mathbf{b}$, the resulting LS estimate $\hat{\theta}$ contains a large bias. In other words, the LS estimate is expected to be more accurate when r_n contains only LOS measurements. However, as the minimum number of range measurements required in order to avoid an underdetermined system is three, the number of possible combinations of taking only a subset of LOS measurements from the N measurement set can be computed as

$$N_{ls} = \sum_{l=3}^{N}\binom{N}{l} = \sum_{l=3}^{N}\frac{N!}{l!(N-l)!}. \tag{17.17}$$

It is worth noting that when both the range and directional measurements are available for each node considered in the geolocation system, the location estimation can be established with only a single-node utilizing joint DOA–TOA or DOA–TDOA measurement [14–19].

Let q denote the combination index and S_q^{LS} denote the corresponding subset that contains only LOS measurements; the LS estimate can be written as

$$\hat{\theta}_q = \left(\mathbf{H}_q^T\mathbf{H}_q\right)^{-1}\mathbf{H}_q^T\mathbf{b}_q, \tag{17.18}$$

where $\mathbf{b}_q = \left[b_{q,1},\ldots,b_{q,|S_q^{\mathrm{LS}}|}\right]$ is $|S_q^{\mathrm{LS}}|$-dimension vector with its element $b_{q,n} = d_n^2 - K_n$ and $\mathbf{H}q$ is $|S_q^{\mathrm{LS}}|\times 3$ matrix with its row vector $\mathbf{h}_{q,1} = \left[-2x_n, -2y_n, 1\right]^T$, defined for $n \in S_q^{\mathrm{LS}}$.

Given these LS estimates, the essence of the LS-based mitigation techniques is to find the estimate that results in minimum LS error. The LS error can be calculated as a function of the LS estimate and expressed as

$$\varepsilon\left(\hat{\theta}_q\right) = \left\|\mathbf{b}_q - \mathbf{H}_q\hat{\theta}_q\right\|_2^2. \tag{17.19}$$

And the solution can be obtained by finding the LS estimates that result in minimum normalized LS error, defined as

$$\bar{\varepsilon}\left(\hat{\theta}_q\right) = \varepsilon\left(\hat{\theta}_q\right)\big/\left|S_q^{\mathrm{LS}}\right|. \tag{17.20}$$

The reason for introducing the normalization term is to remove the dependency of the number of data from the residual LS error. The LS estimator that uses this normalized residual LS error is known as the residual minimum (RMIN) algorithm [20, 21]. The position estimate according to the RMIN algorithm is the one with the minimum normalized residual LS error, which can be written as

$$\hat{\theta}_{\mathrm{RMIN}} = \arg\min_{\theta_q}\bar{\varepsilon}\left(\hat{\theta}_q\right). \tag{17.21}$$

A slightly different algorithm considers averaging all the LS estimates weighted by its normalized residual LS error. This algorithm is known as residual weighting (RWGH) algorithm [20, 21]:

$$\hat{\theta}_{RWGH} = \left(\sum_{q=1}^{N_{ls}} \frac{\hat{\theta}_q}{\overline{\varepsilon}(\hat{\theta}_q)} \right) \bigg/ \left(\sum_{q=1}^{N_{ls}} \frac{1}{\overline{\varepsilon}(\hat{\theta}_q)} \right). \tag{17.22}$$

Note that both the RMIN and RWGH algorithms only use the LOS measurements while discarding the NLOS ones. In order to utilize the range information from both the LOS and NLOS measurements, the weighted least squares (WLS) algorithm is derived based on the idea that the LOS measurements are given more emphasis, while the NLOS ones are given less [22]. The weights are designed by exploiting the assumption that, by modeling the NLOS error i_n as a random variable, the variances of the range measurements are larger for NLOS pairs than the LOS pairs. Let $\hat{\gamma}_n^2$ denote the estimated variance of the nth pair range measurement; the WLS algorithm can be expressed as

$$\hat{\theta}_{WLS} = (\mathbf{H}^T \mathbf{W} \mathbf{H})^{-1} \mathbf{H}^T \mathbf{W} \mathbf{b}, \tag{17.23}$$

where $\mathbf{W}$ is an $N \times N$ diagonal matrix with its diagonal element $W_{nn} = 1/\hat{\gamma}_n^2$. It is worth noting that the variance estimate can only be calculated by taking multiple range measurements for each FT–MT pair.

Pros and Cons: The main advantage of LS-based NLOS mitigation techniques is their low complexity. Except for the WLS mitigation techniques, these methods generally do not need to know the a priori information related to the statistics of the location metrics. It is also worth mentioning that the RMIN algorithm requires at least three measurements in order to avoid solving for underdetermined system. However, this requirement does not apply to the recursive weighted least squares (RWLS) algorithm, as each possible combination of LOS–NLOS measurements is weighted by its normalized residual LS.

17.3.3 Constrained Optimization Techniques

The key idea of constrained optimization-based NLOS mitigation techniques is to minimize the residual errors subject to constraints that are derived by incorporating the NLOS measurement characteristics, such as the fact that the NLOS biases are always positive.

In [23], the authors formulate a simple convex optimization based on weighted LS derivation and use a quadratic programming technique for solving it. The optimization formulation is given by

$$\hat{\theta}_{qp} = \arg\min_{\theta} \quad (\mathbf{b} - \mathbf{H}\theta)^H \mathbf{W} (\mathbf{b} - \mathbf{H}\theta) \qquad \text{subject to} \qquad \mathbf{H}\theta \leq \mathbf{b}. \tag{17.24}$$

Because of the relaxation applied on the constraints, the solution to this optimization is expected to be more accurate than that obtained using the WLS algorithm. This can be explained by the fact that there exists a better solution that gives smaller residuals within the relaxed search region.

A different optimization formulation based on linear programming is also reported in [24] and [25]. It assumes perfect a priori identification of LOS and NLOS pairs. Unlike the LS-based approach, this approach utilizes the NLOS measurements to construct a linear feasible region of the MT using the linear programming approach. It shares the similar idea as the previous approach. The optimization objective function is derived using the LOS pairs, while the constraints are formulated by relaxing the NLOS measurements. Geometrically, this relaxation forms a linear feasible region in which the MT is present.

Other approaches also report a slight variation in the constraints formulation. For example, the authors in [26] redefined the constraints according to the geometrical constraint.

Pros and Cons: Among the NLOS mitigation techniques covered in this chapter, the constrained optimization techniques require the most computational complexity. The level of complexity also increases with the number of constraints considered in solving the optimization. Due to the availability of numerical tools dedicated to solve similar optimization formulations, one can easily include additional or change existing constraints (e.g., geometry [road] constraints or tracking-model constraints) and find the solutions to the resulting constrained optimization.

17.3.4   Robust Estimator Techniques

Robust estimator-based techniques belong to the class of NLOS mitigation that uses different residual formulation to achieve robustness against the NLOS biases. The key assumption is that the number of range measurements due to LOS propagation is more than that due to NLOS propagation. Some of the well-known robust estimators are Huber M-estimator and median-based estimator.

For a general robust M-estimator, the location estimation is obtained by solving the following optimization:

$$\hat{\theta}_M = \arg\min_{\theta} \quad \rho\big(\mathbf{b} - (\mathbf{H}\theta)^H \, \mathbf{W}(\mathbf{b} - \mathbf{H}\theta)\big), \tag{17.25}$$

where $\rho(\cdot)$ is a continuous function (also known as score function) defined according to

$$\rho(v) = \begin{cases} v^2/2 & |v| \le \xi \\ \xi|v| - \xi^2/2 & |v| > \xi \end{cases}. \tag{17.26}$$

Unlike the M-estimator, the median-based estimator belongs to the noncontinuous function and its application to the geolocation problem is described in the following optimization formulation:

$$\hat{\theta}_M = \arg\min_{\theta} \quad \{\text{med}(\mathbf{b} - \mathbf{H}\theta)\}^2, \tag{17.27}$$

where $\text{med}(\mathbf{v})$ is the median of the values in elements of vector $\mathbf{v}$. Other works that uses different noncontinuous functions known as Gaussian-tailed zero-memory nonlinearity (GZMNL) are reported in [27].

TABLE 17.1. Comparative Overview of Geolocation NLOS Mitigation Techniques

Method	Pros	Cons
Maximum likelihood (ML)	Provides asymptotically optimal solution	The performance degradation may be observed due to the deviation between the observed and presumed likelihood model.
Least squares (LS)	Provides lower complexity solution than the ML techniques. Generally, it does not require the a priori information related to the statistics of the location metrics.	Not able to solve the underdetermined system. Also, it does not exploit the location information from the NLOS measurements (except for RWLS algorithm).
Constrained optimization	Greater flexibility to include additional geolocation scenario-based constraints into the optimization formulation in order to improve on the geolocation performance	Requires the highest computational complexity. The level of complexity also increases as the number of constraints considered in solving the optimization.
Robust estimator	Provides the least computational complexity	Performance depends on the score function $\rho(\cdot)$; thus, it may vary depending on the geolocation scenario. Provides robustness up to 50% contamination of NLOS measurements.

Pros and Cons: Compared with the LS-based techniques, the robust estimator techniques are generally much simpler in terms of the computational complexity. Nevertheless, the score function $\rho(\cdot)$ needs to be configured to achieve better performance. In addition, the robust estimator techniques are deduced based on the assumption that the LOS measurements are more dominant than the NLOS ones.

In summary, Table 17.1 lists the pros and cons of the NLOS mitigation techniques covered in this chapter.

17.4 APPLICATION OF THE SINGLE MOVING SENSOR GEOLOCATION

The problem of the single moving sensor geolocation in a 2-D plane is best illustrated in Figure 17.1. Here, all the range measurements are obtained from the single sensor at different time instant while it is moving. Suppose that N measurements are collected and the position of the sensor when collecting the nth measurement (x_n, y_n) is assumed to be known. In a dense urban environment, there may be no direct path between the sensor and MT at some time instances. Thus, some measurements contain NLOS errors. Hence, it can be concluded that the single moving sensor geolocation shares the same signal model described in (17.1).

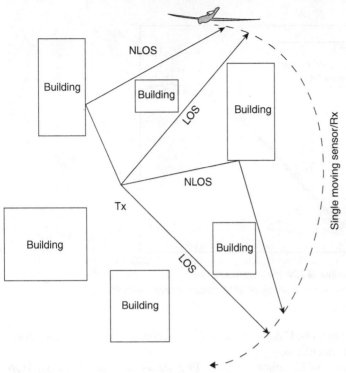

Figure 17.1   Single moving sensor setup.

The main advantages of considering the single moving sensor geolocation are twofold. First, the wireless infrastructure complexity is greatly reduced due to the use of a single sensor. One apparent reduction in the infrastructure is the exclusion of a data fusion center that establishes multiple communication links with the sensors and processes the measurements to perform geolocation [28]. Second, the solution to such a geolocation problem could exploit the fact that the measurements from one measurement to the next one do not change abruptly in LOS cases in order to detect NLOS measurements, discard them, and even reconstruct the LOS measurements.

On the other hand, the single moving sensor geolocation poses strict requirements on the geolocation algorithm. As the location estimation calculation needs to be performed by the moving sensor, the algorithm is required to be low in complexity. Therefore, algorithms based on ML, LS, and constrained optimization (CO) discussed in the previous sections are not considered for such geolocation.

17.4.1   Range Measurements Profile-Based Trimming

Generally, the range measurements exhibit a certain characteristic depending on the movement of the sensor. This characteristic can be modeled as Pth order polynomial function. To obtain such a polynomial function, a polynomial curve fit is performed

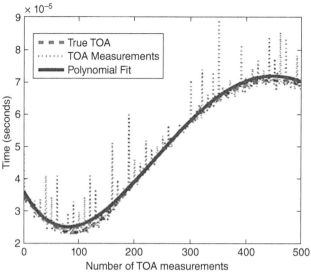

Figure 17.2   Polynomial curve fitting of TOA measurements, with $P = 3$.

on the range measurements. Using this curve fitting technique, the polynomial coefficients can then be calculated.

For the purpose of illustration, Figure 17.2 shows an example of the TOA measurements profile of the single moving sensor geolocation. The actual TOA measurements are simulated under the following setup. The initial position of the sensor is at $(10\,\text{km}, 10\,\text{km})$ while the MT is located at $(5\,\text{km}, 5\,\text{km})$. The sensor moves around the MT, forming a circle, and as it moves, it collects the TOA measurements at different instants. For this particular trajectory, the TOA profile shows a sinusoid, and a third-order polynomial function is sufficient to model the TOA profile. This can be seen in Figure 17.2 from the comparison between the polynomial fit (solid line) and the actual TOA profile (dashed line). It is worth mentioning that a similar polynomial fit was also proposed in [29].

Example 17.1

How to generate the TOA measurements for the single moving sensor geolocation scenario described in Section 17.4.1?

Solution

At the initial position, the sensor is $45°$ with respect to the MT. Assume that the sensor is moving around the MT and taking TOA measurement every $5°$ step size. The nth position when the sensor is taking measurement can be obtained by simple geometry analysis. Finally, the nth actual TOA is calculated as the distance divided by the propagation speed.

The noise components comprise the measurement noise (modeled as zero-mean Gaussian random variable) and NLOS errors (modeled as ε-contaminated

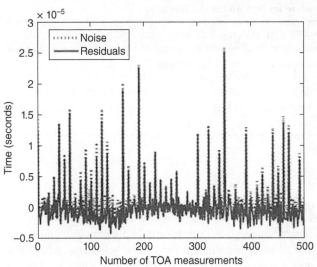

Figure 17.3   TOA residuals and noise components ($w_n + i_n$).

Gaussian mixture model [ε-cGMM]). The positive NLOS errors are generated by taking the magnitude of the random variables from the ε-cGMM with the second summand being the Gaussian density of mean 0 and variance $\kappa\gamma^2$ [30]:

$$f_\varepsilon(v) = (1-\varepsilon)f_G\left(v;\gamma^2\right) + \varepsilon f_G\left(v;\kappa\gamma^2\right), \qquad (17.28)$$

where ε represents the amount of contamination, f_G $(v;\gamma^2)$ is a Gaussian distribution with variance γ^2 representing the background noise and κ represents the relative strength of the impulsive component. Finally, the TOA measurements are generated from the addition of the actual TOA and the noise components.

For example, assume that the target is located at (5,5) km and the moving sensor is initially at (10,10) km. Also assume that the following parameters are used to generate the noisy TOA measurements: Ktheta = 50; noise_var = 1; epsilon = 0.5; kappa = 100;. Then, the TOA measurements can be obtained by calling the following function:

```
[Rx_pos,true_toa,meas_toa]=Chapter_17_Function_1
(Ktheta,noise_var,epsilon,kappa);
```

Let $\tilde{r}_n$ denote the TOA measurements generated from the Pth order polynomial function fitted earlier. The residual between $\tilde{r}_n$ and the TOA measurements r_n is given by $u_n = r_n - \tilde{r}_n$. These residuals are the difference between the smoothed curve and the TOA measurements (e.g., see the solid line in Fig. 17.3). Notice also that these residuals are a good estimate of the noise components ($w_n + i_n$) in the TOA measurements, as can be seen in Figure 17.3.

Thereafter, the method tries to trim off the outliers from these residuals as they are due to the NLOS errors. To do so, an iterative method is developed based on sorting the residuals and Gaussianity test. After sorting, the TOA residuals are given by

TABLE 17.2.   Iterative Procedure for Performing the Trimming

1. Set $h = 0$. Sort the entries within the TOA residuals vector $\mathbf{u}$ according to their magnitude.
2. Trim $g = k(h + 1)$ measurements at the higher-magnitude end of the sorted vector $\mathbf{u}_s$ to obtain $\mathbf{u}_s^{Z}$ and k is the initial number of measurements to be trimmed.
3. Calculate the test statistic according to (17.30) and its P-value (probability of observing the given result by chance given that the Gaussianity hypothesis is true) [38]. If $\alpha \geq P$-value, reject the hypothesis that the remaining TOA measurements are Gaussian distributed at significance level α. Increase h by 1 and go back to Step 2.

$$\mathbf{u}_s = \begin{bmatrix} u_{1s} & u_{2s} & \dots & u_{Ns} \end{bmatrix}^T, \tag{17.29}$$

where $u_{1s} \leq u_{2s} \leq \dots \leq u_{Ns}$. The measurements from the end with higher magnitude are trimmed, since the NLOS errors are positive (due to the extra distance traveled by the signal), and they are most likely to be contaminated. The remaining measurements are tested for Gaussianity using the Saphiro-Wilk (SW) goodness-of-fit W test statistic [31, 32] at a predefined level of significance α. If the Gaussianity hypothesis on the remaining measurements is not accepted, more measurements will be trimmed until the hypothesis is accepted.

It should be noted that the Gaussianity test works as a criterion for stopping the trimming. Due to the superposition of impulsive and Gaussian noise, our proposed method deals with the heavy-tailed components by trimming and does not attempt to transform the noise distribution. Other tests may be used (see, e.g., [33]), but we use the SW test because it is simple and powerful.

The W test statistic is given by [31, 34]:

$$W = \frac{\left(\Sigma_i c_i \mathbf{u}_s^{Z}(i) \right)^2}{\Sigma_i \left(\mathbf{u}_s^{Z}(i) - \overline{\mathbf{u}}_s^{Z} \right)^2}, \tag{17.30}$$

where $\mathbf{u}_s^{Z}$ denotes the trimmed $\mathbf{u}_s$; c_i's are tabulated coefficients derived from the means, variances, and covariances of the order statistics of a sample of size N from a Gaussian distribution [31, 34]; and $\overline{\mathbf{u}}_s^{Z}$ is the sample mean of $\mathbf{u}_s^{Z}$.

The procedure of the proposed trimming method is summarized in Table 17.2. The trimming process can be optimized [35], but this is tedious and is not covered here.

Example 17.2

How to apply the range profile-based trimming algorithm in Table 17.2 for geolocation in the presence of NLOS errors?

Solution

Assume that the TOA measurements considered are generated by the MATLAB codes in Chapter_17_Example_1.m. MATLAB codes can be found online at ftp:// ftp.wiley.com/public/sci_tech_med/matlab_codes. Next, the indices of the TOA measurements to be trimmed off can be calculated as follows:

```
[trim_indices]=Chapter_17_Function_2(meas_toa,Porder,alpha,
tail);
```

with the following parameters: Porder = 3; alpha = 0.05; tail = 0;. Lastly, the location estimation can be computed as the LS solution using the following function:

[xHat,yHat] = Chapter_17_Function_3(trimmed_TOA*3e8,Rx_pos);

where the following parameters are used: trimmed_TOA = meas_toa;

```
trimmed_TOA(trim_indices) = 0; Rx_pos(trim_
indices,:) = 0;
```

17.4.2 Reconstruction of Trimmed TOA Profile

As will be shown in Section 17.4.4, the proposed trimming method in Section 17.4.1 performs better than conventional methods. However, it is a "take-it-or-leave-it" approach, which may hamper the location estimation if the contaminated measurements (although corrupted by NLOS errors) can still be used or when the level of contamination is low. This trimming approach can be further enhanced by reconstructing the trimmed TOA profile. By tapping all the exploitable information from the measurements instead of simply discarding them, an improvement can be obtained by extracting the "good" data, especially when the "wrong" measurements are trimmed.

The reconstruction can be implemented by using another polynomial curve fit, after the trimming process. This will allow an estimate of the "zeroed" measurements to be obtained and result in a smoothed curve. Another example is shown in Figure 17.4, and the reconstructed TOA profile is indicated by the line labeled "TOA polyfit."

An alternative approach to reconstruct the TOA profile is by making use of geometry (between Tx and Rx) to improve the geolocation estimate. Instead of performing a polynomial curve fit to the TOA residuals after trimming, the "zeroed" TOA measurements are reconstructed by using the governing range equation (geometry between the estimated Tx position and the Rx position at that particular instant):

$$r_n^Z = \sqrt{(\hat{x} - x_n)^2 + (\hat{y} - y_n)^2},$$ (17.31)

where r_n^Z refers to the trimmed TOA measurements; $(\hat{x}, \hat{y})$ denotes the estimated Tx position and can be obtained via the LS method from the trimmed TOA measurements in Section 17.4.1.

These reconstructed TOA measurements, together with the untrimmed ones, can undergo the trimming process in Section 17.4.1 again followed by another reconstruction. This iterative procedure continues until a stopping condition is met and/or a fixed number of iterations have occurred. An example of this reconstruction, after the fifth iteration, is indicated by the line labeled "fifth iteration" in Figure 17.4. For brevity, this method will be referred to as "ReconGeo."

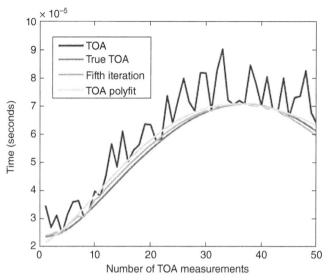

Figure 17.4   Reconstruction of trimmed TOA profile.

Example 17.3

How to apply ReconGeo to estimate the Rx position?

Solution

With the same parameters setting as Example 17.2, the ReconGeo algorithm computes the reconstructed TOA measurements using the function below:

```
[recon_TOA] = Chapter_17_Function_4(meas_toa, Rx_
pos, Porder, alpha, tail);
```

Then, the location estimation can be computed in a similar way as the previous example:

```
[xHat, yHat] = Chapter_17_Function_3(recon_TOA*3e8, Rx_pos);
```

17.4.3   Robust Trimming with Nonparametric Noise Density Estimator

The geolocation performance of the robust trimming method in Section 17.4.1 and the ReconGeo method in Section 17.4.2 can both be further improved by using the adaptive nonparametric (NP) noise density estimator [36] on the remaining TOA measurements. In this approach, an adaptive NP kernel density estimator is utilized to estimate the noise distribution directly from the data.

Robust estimation of signal parameters in the additive noise model is an important problem. Its relevance can be attributed to the realization that impulsive noise is present in many applications. Many estimators assume a parametric form for the noise distribution, chosen from one of many impulsive noise models, and then design an optimal or suboptimal detector. To decrease the sensitivity of the estimator to the

TABLE 17.3. Iterative Procedure for Estimating the Score Function

---

1. Initialization: Set $i = 0$. Obtain an initial estimate of Θ from the LS estimator, $\hat{\Theta}^0 = \hat{\Theta}^0_{LS}$.
2. Determine the residuals: $\hat{\mathbf{e}} = \mathbf{b} - \mathbf{H}\hat{\Theta}^i$.
3. Estimate the nonparametric score function: From $\hat{\mathbf{e}}$, estimate the density $\hat{f}_V(\mathbf{e})$ and evaluate the score function estimate as

$$\hat{\psi}(\mathbf{e}) = -\frac{\hat{f}_V'(\mathbf{e})}{\hat{f}_V(\mathbf{e})}, \qquad (17.32)$$

where $\hat{f}_V'(\mathbf{e})$ refers to the derivative of $\hat{f}_V(\mathbf{e})$.

4. Update the parameter estimates: Evaluate $\mathbf{z} = \Psi(\mathbf{e})$ and update the parameters as

$$\hat{\Theta}^{i+1} = \hat{\Theta}^i + \mu(\mathbf{H}^T\mathbf{H})^{-1}\mathbf{H}^T\mathbf{z}, \qquad (17.33)$$

where μ is the step size.

5. Check for convergence: If $\left\|\hat{\Theta}^{i+1} - \hat{\Theta}^i\right\| < \delta$, stop; else set $i = i + 1$, go to step 2. Herein, δ is a small constant.

---

underlying distribution, the theory of M-estimators can be used to implement a sub-optimal nonlinear detector that is robust to changes in the noise distribution.

This robust NP M-estimation algorithm makes minimal a priori assumptions on the noise model, requiring only a symmetric density. It does not require training data, as it relies on an iterative scheme to estimate the density and parameters of the estimator together from the observations. More details can be found in [36], and the iterative procedure for estimating $\hat{\Theta}$ is repeated in Table 17.3.

Example 17.4

How to incorporate the NP density estimator to further improve the location estimation of the ReconGeo algorithm?

Solution

At the end of Example 17.3, the ReconGeo algorithm gives the position estimate of the MT. To further improve on the accuracy, the NP density estimator can be applied as follows:

```
KVec = Rx_pos(:,1).^2 + Rx_pos(:,2).^2;
range_meas=recon_TOA*3e8;
hVec = (range_meas.^2-KVec);
GaMat = [-2*Rx_pos(:,1)  -2*Rx_pos(:,2)
ones(length(recon_TOA),1)];
theta0=[xHat yHat Rhat].';
[ZaVec_NP,ZaVec_NP_mat,fv_NP]=Chapter_17_Function_
5(hVec,GaMat,theta0,[],5,1e-4);
```

Then, the location estimation can be found in the final output ZaVec_NP.

17.4.4   Performance Analysis

To evaluate the performance of the NLOS mitigation techniques applied to the single moving sensor geolocation, several sets of numerical experiments are conducted based on $M_c = 10{,}000$-run Monte Carlo simulation. The performance metrics considered are based on the realization of the estimates $\hat{x}_m$ and $\hat{y}_m$, namely the mean Euclidean distance (MED)

$$\text{MED} = \sqrt{\frac{1}{M_c} \sum_{m=1}^{M_c} \left[(x - \hat{x}_m)^2 + (y - \hat{y}_m)^2 \right]} \qquad (17.34)$$

and the empirical cumulative distribution function (CDF) of the MED errors. The CDF plot is useful for examining the distribution of the errors and indicates the estimation error for a certain confidence level.

The NLOS mitigation techniques considered are the adaptive robust trimming method in Section 17.4.1 and its extension to the NP estimator in Section 17.4.3 (labeled respectively as "Trim" and "Trim-NP"), as well as the ReconGeo method and its NP extension (labeled as "ReconGeo-NP"). The ideal trimming case is also used as a comparison (labeled as "Ideal-Trim"), where perfect knowledge of the NLOS errors is available.

Figure 17.5 shows the accuracy of the location estimates when the number of NLOS measurements, out of a total of 50 TOA measurements, is increased. It can be seen that the trimming methods perform better than simply using an NP estimator (considered by the authors in [36]) and the case where there is no trimming. The benefit of using an adaptive NP estimator for the trimming method can also be noted

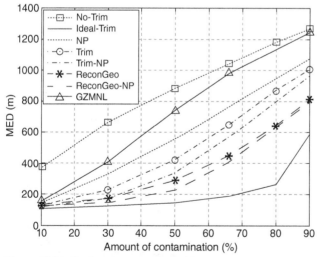

Figure 17.5   Geolocation accuracy: MED as a function of the number of NLOS errors out of 50 TOA measurements, for $\alpha = 5\%$, $\varepsilon = 1$, and $\kappa = 100$.

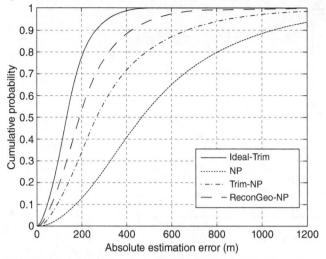

Figure 17.6   Cumulative error distribution for 25 NLOS errors out of 50 TOA measurements.

by comparing between Trim-NP and Trim or between ReconGeo and ReconGeo-NP, with the NP extensions giving better performance. This highlights the benefits of using an adaptive NP kernel to estimate the noise density after performing the trimming.

The cumulative error distribution is presented in Figure 17.6, which shows the MED distribution of the 10,000 Monte Carlo trials, for 25 NLOS errors out of 50 TOA measurements. This measure is more accurate, since an "extreme" value can cause the MED to be biased. In this case, it is observed that the ReconGeo-NP method clearly performs better than the rest of the techniques throughout. It has the closest geolocation performance to the Ideal-Trim curve. It indicates the improvement in geolocation performance when reconstruction of the trimmed TOA profile is performed. The availability of more TOA measurements leads to a better location estimate.

The effects of changing the parameters of ε and κ are shown in Figures 17.7–17.10. Unless stated, the simulation parameters are unchanged. Once again, it can be observed that the trimming methods compare favorably with the other methods and the advantages are especially evident in cases of high contamination. Near these regions, the ReconGeo-NP method performs very similar to the Ideal-Trim method. In addition, one can observe that the No-Trim case suffers severe deterioration in performance. This shows that mitigation for NLOS errors is an important issue for TOA geolocation.

It is interesting to note, from Figures 17.9 and 17.10, that the ReconGeo-NP method can actually outperform the Ideal-Trim method. With 66% of the available TOA measurements being contaminated, the location estimate from the remaining uncontaminated measurements may not be as good as the case when the trimmed measurements are reconstructed. This indicates that simply implementing

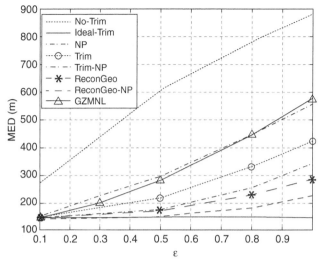

Figure 17.7 Geolocation accuracy: MED as a function of ε for 25 NLOS errors out of 50 TOA measurements, $\alpha = 5\%$ and $\kappa = 100$.

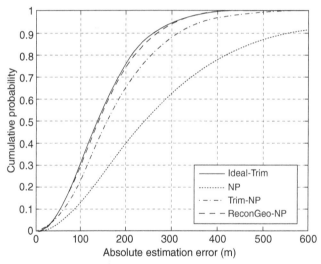

Figure 17.8 Cumulative error distribution for $\varepsilon = 0.5$, $\kappa = 100$, $\alpha = 5\%$, and 25 NLOS errors out of 50 TOA measurements.

the trimming approach (a "take-it-or-leave-it" approach) may not always be the ideal solution. This may hamper the localization performance, especially if the trimmed measurements can still be used when the individual contamination level is not very high. Instead of discarding the contaminated measurements, an improvement can be obtained by tapping all the exploitable information from these measurements.

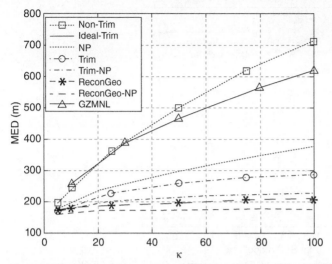

Figure 17.9   Geolocation accuracy: MED as a function of κ for 33 NLOS errors out of 50 TOA measurements, $\alpha = 5\%$ and $\varepsilon = 0.5$.

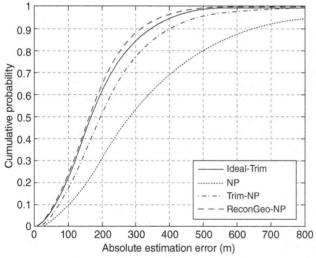

Figure 17.10   Cumulative error distribution for $\kappa = 75$, $\varepsilon = 0.5$, $\alpha = 5\%$, and 33 NLOS errors out of 50 TOA measurements.

To demonstrate the robustness of the NLOS mitigation techniques, different noise models are considered for simulating the NLOS errors. The positive NLOS errors are generated by taking the magnitude of the random variables from the symmetric alpha-stable (SAS) model, which is a good approximation to the Middleton Class B noise model [37]. The characteristic function of the SAS model is given as

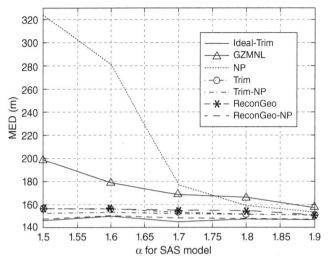

Figure 17.11 Geolocation accuracy: MED as a function of α_{SAS}, for 25 NLOS errors out of 50 TOA measurements, $\delta = 0\%$ and $\gamma = 1$.

$$\phi(\omega) = e^{j\delta\omega - \gamma|\omega|^{\alpha_{SAS}}}, \tag{17.35}$$

where α_{SAS} is the characteristic exponent indicative of the thickness of the tail of impulsiveness of the noise, δ is the localization parameter (analogous to mean), and γ is the dispersion parameter (analogous to variance). When $\alpha_{SAS} = 2$, this represents Gaussian noise and for $\alpha_{SAS} = 1$, the noise has a Cauchy distribution.

Figure 17.11 shows the geolocation performance accuracy when the α_{SAS} parameter of the SAS model is varied. It can be seen that as the severity of the NLOS errors increases, the geolocation accuracy remains relatively the same for all the NLOS mitigation techniques, except for the GZMNL method. This highlights the benefits of not only identifying the erroneous TOA measurements but also removing them for geolocation. The ReconGeo-NP method can also outperform the Ideal-Trim method in this NLOS-error model, which has been observed in an earlier simulation analysis.

Figure 17.12 depicts the geolocation performance accuracy versus α of the SW test. As α increases, the performance of both robust trimming methods improve and tend to the Ideal-Trimming bound. By choosing a higher α (type-I error), a lower β (type-II error) is obtained. The latter refers to accepting that the TOA measurements are Gaussian distributed when they are not. Therefore, by reducing the type-II error (higher α), more NLOS-contaminated measurements will be trimmed, and this leads to a better geolocation performance estimation. The trade-off in choosing a higher α is the higher probability of wrongly rejecting that the TOA measurements are Gaussian distributed. This may lead to noncontaminated measurements being trimmed. Although this is not as critical as having a low β, it will nevertheless result in loss of useful information.

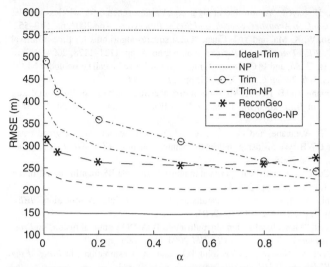

Figure 17.12   Geolocation accuracy: MED as a function of α for 25 NLOS errors out of 50 TOA measurements, $\varepsilon = 1$ and $\kappa = 100$. RMSE, root mean square error.

17.5   CONCLUSIONS

This chapter discussed the problem of locating MT using a network of wireless sensors under the ideal condition when LOS propagation is not guaranteed. Under NLOS propagation, the resulting range measurements are characterized by large positive bias. This fact is utilized in most of the NLOS mitigation techniques, as reviewed in Section 17.3. Following which, a unique single moving sensor geolocation scenario was introduced. Motivated by its simplicity, robust statistics-based trimming methods for NLOS mitigation were explained in detail. Several examples illustrating the implementation of these methods to solve for the single moving sensor geolocation were also covered across Section 17.4. The chapter is then concluded with discussion on the performance analysis from numerical simulation.

REFERENCES

[1] A. Dogandzic, J. Riba, G. Seco, and A. L. Swindlehurst, "Positioning and navigation with applications to communications," *IEEE Signal Process. Mag.*, vol. 2, no. 4, pp. 10–11, 2005.

[2] Federal Communications Commission, "Revision of the commissions rules to ensure compatibility with enhanced 911 emergency calling systems, RM-8143," CC Docket no. 94–102, Jul. 1996.

[3] T. Rappaport, *Wireless Communications: Principles and Practice*. Upper Saddle River, NJ: Prentice-Hall, 1996.

[4] M. McGuire and K. N. Plataniotis, "A comparison of radiolocation for mobile terminals by distance measurements," in *Proc. of Int. Conf. Wireless Communications*, 2000, pp. 1356–1359.

[5] J. Caffery and G. L. Stuber, "Subscriber location in CDMA cellular networks," *IEEE Trans. Veh. Technol.*, vol. 47, pp. 406–416, 1998.

[6] J. Borras, P. Hatrack, and N. Mandayam, "Decision theoretic framework for NLOS identification," *IEEE Veh. Technol. Conf.*, vol. 2, pp. 1583–1587, 1998.

[7] N. Khajehnouri and A. H. Sayed, "A non-line-of-sight equalization scheme for wireless cellular location," in *IEEE Int. Conf. on Acoustics, Speech, and Signal Processing*, Apr. 2003, pp. 549–552.

[8] K. W. Cheung, H. C. So, W. K. Ma, and Y. T. Chan, "Least squares algorithms for time-of-arrival based mobile location," *IEEE Trans. Signal Process.*, vol. 52, no. 4, pp. 1121–1128, 2004.

[9] Y. T. Chan, W. Y. Tsui, H. C. So, and P. C. Ching, "Time-of-arrival based localization under NLOS conditions," *IEEE Trans. Veh. Technol.*, vol. 55, no. 1, pp. 17–24, 2006.

[10] A. Rabbachin, I. Oppermann, and B. Denis, "ML time-of-arrival estimation based on low complexity UWB energy detection," in *IEEE 2006 Int. Conf. on Ultra-Wideband*, vol. 1, Sep. 24–27, 2006, pp. 599–604.

[11] I. Gven, C.-C. Chong, F. Watanabe, and H. Inamura, "NLOS identification and weighted least-squares localization for UWB systems using multipath channel statistics," *EURASIP J. Adv. Signal Process.*, vol. 2008, pp. 1–14, 2008.

[12] S. Venkatraman and J. Caffery, "A statistical approach to non-line-of-sight BS identification," in *Int. Symp. on WPMC*, vol. 1, Oct. 2002, pp. 296–300.

[13] S. Gezici, H. Kobayashi, and H. V. Poor, "Non-parametric non-line-of-sight identification," *IEEE Veh. Technol. Conf.*, vol. 4, pp. 2544–2548, Oct. 2003.

[14] Z. Wang and S. A. Zekavat, "Manet localization via multi-node TOA-DOA optimal fusion," in *Proc. of the IEEE Military Communications Conf., MILCOM 2006*, Oct. 2006, pp. 1–7.

[15] E. Lagunas, M. Najar, and M. Navarro, "UWB joint TOA and DOA estimation," in *Proc. of the IEEE Int. Conf. Ultra-Wideband, ICUWB 2009*, Sept. 2009, pp. 839–843.

[16] R. Ding, Z. Qian, and X. Wang, "Joint TOA and DOA estimation of IR-UWB system based on matrix pencil," in *Proc. of the Int. Forum on Information Technology and Applications, IFITA '09*, vol. 1, May 2009, pp. 544–547.

[17] H. El Arja, B. Huyart, and X. Begaud, "Joint TOA/DOA measurements for UWB indoor propagation channel using MUSIC algorithm," in *Proc. of the European Wireless Technology Conf., EuWIT 2009*, Sep. 2009, pp. 124–127.

[18] W. Li, W. Yao, and P. J. Duffett-Smith, "Comparative study of joint TOA/DOA estimation techniques for mobile positioning applications," in *Proc. of the 6th IEEE Consumer Communications and Networking Conf., CCNC 2009*, Jan. 2009, pp. 1–5.

[19] E. Lagunas, M. Najar, and M. Navarro, "Joint TOA and DOA estimation compliant with IEEE 802.15.4a Standard," in *Proc. of the 5th IEEE Int. Wireless Pervasive Computing (ISWPC) Symp.*, May 2010, pp. 157–162.

[20] Y. T. Chan and K. C. Ho, "A simple and efficient estimator for hyperbolic location," *IEEE Trans. Signal Process.*, vol. 42, no. 8, pp. 1905–1915, 1994.

[21] X. Li, "An iterative NLOS mitigation algorithm for location estimation in sensor networks," presented at the IST Mobile and Wireless Communications Summit, Myconos, Greece, Jun. 2006.

[22] J. J. Caffery and G. L. Stuber, "Overview of radiolocation in CDMA cellular systems," *IEEE Commun. Mag.*, vol. 36, no. 4, p. 3845, 1998.

[23] X. Wang, Z. Wang, and B. O'Dea, "A TOA-based location algorithm reducing the errors due to non-line-of-sight (NLOS) propagation," *IEEE Trans. Veh. Technol.*, vol. 52, no. 1, pp. 112–116, 2003.

[24] S. Venkatesh and R. M. Buehrer, "A linear programming approach to NLOS error mitigation in sensor networks," in *2006 5th Int. Conf. on Information Processing in Sensor Networks, IPSN 2006*, Nashville, TN, vol. 2006, no. 1, pp. 301–308, April 2006.

[25] S. Venkatesh and R. M. Buehrer, "NLOS mitigation using linear programming in ultrawideband location-aware networks," *IEEE Trans. Veh. Technol.*, vol. 56, no. 5, pp. 3182–3198, 2007.

[26] C. L. Chen and K. T. Feng, "An efficient geometry-constrained location estimation algorithm for NLOS environments," in *Proc. of the IEEE Int. Conf. on Wireless Networks, Communications, and Mobile Computing*, Wuhan, China, Jun. 2005, pp. 244–249.

[27] A. Swami and B. M. Sadler, "On some detection and estimation problems in heavy-tailed noise," *Signal Process.*, vol. 82, pp. 1829–1846, 2002.

[28] A. H. Sayed, A. Tarighat, and N. Khajehnouri, "Network-based wireless location: challenges faced in developing techniques for accurate wireless location information," *IEEE Signal Process. Mag.*, vol. 22, no. 4, pp. 24–40, 2005.

[29] M. P. Wylie and J. Holtzman, "The non-line of sight problem in mobile location estimation," in *IEEE Int. Conf. on Universal Personal Communications*, vol. 2, Sep. 1996, pp. 827–831.

[30] P. Huber, *Robust Statistics*. New York: John Wiley & Sons, 1981.

[31] W. Shapiro and M. Wilk, "An analysis of variance test for normality (complete samples)," *Biometrika*, vol. 52, pp. 591–611, 1965.

[32] R. B. D'Agostino and M. A. Stephens, *Goodness-of-Fit Techniques*. New York: Marcel Dekker, 1986.

[33] A. M. Zoubir and M. J. Arnold, "Testing Gaussianity with the characteristic function: The i.i.d. case," *Signal Process.*, vol. 53, no. 2, pp. 245–255, 1996.

[34] J. P. Royston, "An extension of Shapiro and Wilk's W test for normality to large samples," *Appl. Stat.*, vol. 31, pp. 115–124, 1982.

[35] A. M. Zoubir and D. R. Iskander, *Bootstrap Techniques for Signal Processing*. Cambridge: Cambridge University Press, 2004.

[36] C. H. Lim, A. M. Zoubir, C. M. S. See, and B. P. Ng, "A robust statistical approach to non-line-of-sight mitigation," in *IEEE Statistical Signal Processing Workshop*, Aug. 2007, pp. 428–432.

[37] D. Middleton, "Non-Gaussian noise models in signal processing for telecommunications," *IEEE Trans. Inf. Theory*, vol. 45, no. 4, pp. 1129–1149, 1999.

[38] J. P. Royston, "The W test for normality," *Appl. Stat.*, vol. 31, pp. 176–180, 1982.

CHAPTER 18

MOBILE POSITION ESTIMATION USING RECEIVED SIGNAL STRENGTH AND TIME OF ARRIVAL IN MIXED LOS/NLOS ENVIRONMENTS

Bamrung Tau Sieskul,[1] Feng Zheng,[2] and Thomas Kaiser[3]

[1]*University of Vigo, Vigo, Spain*
[2]*Leibniz University of Hannover, Hannover, Germany*
[3]*University of Duisburg Essen, Duisburg, Germany*

As DISCUSSED in Chapters 16 and 17, time-of-arrival (TOA) estimation and the inherent accuracy of the mobile position estimation have been considered for wireless non-line-of-sight (NLOS) geolocation. In order to increase the TOA estimation accuracy, it is often useful to exploit additional information, such as path attenuation or path loss, which can in principle be observed from the received signal strength (RSS). Previous works have indicated that the hybrid RSS–TOA approach improves performance compared to TOA method. In this chapter, we investigate the TOA estimation techniques for the hybrid RSS–TOA localization in mixed line-of-sight (LOS)/NLOS environments. The chapter presents how the number of TOA measurements should be selected to reduce computation complexity while improving the localization performance. More importantly, it will be shown how to determine a mobile position from such sufficient TOAs while the error performance attains certain performance bounds. The results shared in this chapter help a system designer select the number of nodes that should be used in the process of localization and NLOS mitigation. The selection is based on a trade-off between accuracy and computational complexity.

Handbook of Position Location: Theory, Practice, and Advances, Second Edition.
Edited by S. A. (Reza) Zekavat and R. Michael Buehrer.
© 2019 by the Institute of Electrical and Electronics Engineers, Inc.
Published 2019 by John Wiley & Sons, Inc.
Companion Website: www.wiley.com/go/zekavat/positionlocation2e

18.1 INTRODUCTION

Wireless geolocation commercially emerges from the incentives of locating vehicles, people, and parcels (see, e.g., [1–8]). For wireless carriers and vendors, the knowledge of a user position brings about more precision in service charges and plans. Applications of subscriber location continue to expand in wireless services, for example, global positioning system, intelligent transportation system, and mobile yellow pages (see, e.g., [9] and [10] for outdoor and [5] and [11] for indoor scenarios). In addition to data communications, user geolocation in wireless networks is deemed viable due to the emergency requirement by the Federal Communications Commission and the European Commission [7].

18.1.1 Background

The wireless geolocation is an operation that measures the radio signals traveling between a mobile station (MS) and a set of fixed stations. Such a schematic setting can also be adopted to a smaller volume, for example, ad hoc wireless sensor networks [12, 13]. On a two-dimensional plane, the LOS distances between a mobile and at least three participating base stations (BSs) can be used to locate the mobile terminal. In general, two methods can conceptually be applied to determine the user position: network-based and handset-based approaches.

In cellular systems, the determination of the MS position can be performed using the measurements of the mobile signal at multiple BSs [14]. The radiolocation information of the RSS, direction of arrival (DOA), TOA, time difference of arrival (TDOA), or their combinations (see, e.g., [15–17] and references therein) can be gathered. Such information is deployed to form a location estimate. Toward this end, position location may be calculated in a unilateral or a multilateral manner [1]. In a unilateral system, the signals propagate from the BS transmitters to the MS receiver. When the locations of the BSs are known, the MS estimates its own position by processing those signals. For a multilateral system, the MS position is estimated by using the signal that is transmitted from the MS to multiple BSs.

The major task in the wireless geolocation is to estimate the mobile position. Hence, the performance metrics can be the same as those for the evaluation of an estimation algorithm, including statistical error performance, for example, bias and root mean square error (RMSE), and algorithmic computational complexity. One of the most important problems in the wireless geolocation is the manipulation of the sources of errors [18], such as multipath propagation, NLOS propagation, and multiple access interference. For the localization in mixed LOS/NLOS environments, one may also have to consider the performance metrics in the view of the error behavior of the detection between the LOS and the NLOS components.

18.1.2 Literature Review

The initial step toward estimating the mobile position in the wireless geolocation is the characterization of LOS or NLOS paths. It is in fact sufficient to consider either the NLOS detection or the LOS detection. After identifying the LOS or the

NLOS components, a further step may be the NLOS mitigation (see, e.g., [19–27] and references therein). Afterward, the mobile position can be determined from those appropriate portions after the classification.

LOS/NLOS Detection: Based on statistical theory, the discrimination between the LOS and the NLOS components can be cast into a hypothesis-testing problem. This problem is an inquiry to determine to which of the two classes, LOS or NLOS, a given observation belongs. Various ideas have been proposed in [28–40] to decide whether the incoming path is LOS or NLOS. These works are based on range measurements. In what follows, we shall briefly discuss some of these works.

In [28], the range is modeled as a linear combination of history times. If the NLOS is present, the measured range will be larger than the modeled range. Under certain conditions, the rejection of LOS/NLOS hypothesis can be confirmed by analyzing a residual rank test.

In [29], four hypothesis-testing criteria are derived for the Gaussian NLOS probability model, where the prior probability density function (PDF) can be either known or unknown. Both the unknown mean and the unknown standard deviation of the NLOS are either deterministic or random.

In [30], the NLOS range error is modeled as a non-Gaussian random variable. If the probability of NLOS corruption is small and the NLOS involves a large traveling TOA, the range measurements are averaged for the test of outliers. If there is no NLOS component, the test variable is likely to be a normal-distributed random variable. The numerical results showed that the test of normality performs accurately.

In [31], the PDF of the range measurement is approximated using some window functions around the samples. The Kullback–Leibler distance is presented to fit the estimated PDF to a known measurement PDF in order to determine whether a given BS signal belongs to the LOS or the NLOS of the MS.

In [32], Pearson's test statistic is used to compare the estimated PDF of the first arriving path with certain referenced PDFs, such as Rayleigh, Rician, normal, and lognormal distributions.

In [33], four algorithms are proposed to identify the NLOS. The first method considers the cumulative distribution of the received signal envelopes, computes the test ratio, and then compares the computed test ratio with a bound. The second and the third approaches observe the level crossing rate and the average fade duration, respectively. The fourth algorithm takes into account the power of the received signal and then sees whether the computed signal power is close to an NLOS/LOS or not.

In [34], the amplitude of signal components is used to observe a sudden decrease of the signal-to-noise ratio (SNR) from the LOS to the NLOS condition.

In general, the fourth property of a Euclidean matrix is referred to as the triangular inequality, that is, $\sqrt{d_{ij}} \leq \sqrt{d_{ik}} + \sqrt{d_{kj}}$ for $i \neq j \neq k$, where d_{ij} is the Euclidean distance in $\mathbb{R}^{N_1 \times 1}$, which is defined as $d_{ij} = \left\| \mathbf{x}_i - \mathbf{x}_j \right\|_{\mathrm{E}}^2$, with $\|\cdot\|_{\mathrm{E}}$ denoting the Euclidean norm, and $\mathbf{x}_i$ being the ith column vector of a matrix $\mathbf{X} \in \mathbb{R}^{N_1 \times N_2}$. More details on theory and several properties of the Euclidean distance matrix can be found in [41] (chapter 5). In [36], the fourth property of the Euclidean distance matrix is invoked to formulate a test statistic that is the average of the relative distance difference.

In [37], the excess NLOS path length is modeled as an exponential random variable, while the TOA is drawn from a Gaussian random variable whose variance is proportional to the distance powered by the path loss exponent (PLE). The hypothesis test and the distance estimation are jointly performed.

In [38], the quantity of kurtosis is computed to shape the channel amplitude distribution, which is well fitted to a lognormal distribution under the Kolmogorov–Simirov goodness-of-fit test, whereas the mean excess delay and the root mean square (RMS) delay spread are computed from the measured distance between a fixed terminal and a mobile terminal. To identify the NLOS or the LOS, the likelihood ratio tests are formulated using the kurtosis, the mean excess delay, and the RMS delay spread.

In [39], the autocorrelation coefficient of the channel impulse response is used to determine the LOS or the NLOS.

In [40], the Neyman–Pearson tests are considered for the NLOS/LOS detection using the DOA and TOA measurements as well as the joint TOA and path loss measurements.

Wireless Geolocation: Recently, a two-step positioning approach is suggested to be comprised of

- the estimation of the fundamental parameters, such as the TOA and the DOA, and
- the succeeding position estimation from those parameters [42].

It is also reported therein that the TOA is well suited to localization, especially with the use of ultra-wideband signals.

In [43], the Cramer–Rao bound (CRB) has been analyzed for several geolocation schemes in the presence of the NLOS. It is shown that the Fisher information matrix (FIM) of a hybrid scheme using the RSS and the TDOA can be acquired by the superposition of the FIMs from both schemes. The hybrid scheme outperforms the schemes using only one feature in terms of estimation accuracy [44] and reliability [45]. The frameworks in [43] and [44] are composed of two separate techniques, which require two different measurements on baseband-received signal and mean signal strength, respectively. Unfortunately, this kind of combination inevitably makes the parameter estimation cumbersome.

In [46], the path loss is incorporated into the path gain. The inherent accuracy of the MS position estimation is studied in [47]. For a handset-based mutilateral geolocation system using the TOA, the CRB is derived. Although the theoretical performance reveals that the hybrid RSS–TOA method outperforms the usual TOA method, it is necessary to analyze its efficiency in practice.

18.1.3 Merits

This chapter introduces the mobile position estimators for the RSS–TOA method with the aim of achieving the expected theoretical error variance. The problem is formulated as determining the position of a mobile user on a general two-dimensional plane as well as on a specific cellular setting with a number of participating BS receivers. Each participating BS receiver is assumed to receive either a LOS or an NLOS component.

Several techniques, which transform a number of the TOAs into mobile position coordinates, are reviewed. The advantages and the disadvantages of these techniques are pointed out. The first algorithm uses the linearized least squares (LLS) based on first-order Taylor series. The feature of the first algorithm lies in that (1) the method is iterative and requires an initial guess, (2) no convergence with respect to the iteration is assured, and (3) it is computationally expensive. However, this approach does not sacrifice any degree of freedom; that is, the system does not spend any free TOA from a BS to support a new created parameter. The second method is based on LLS with additional parameterization. There exists a closed-form solution, through which the degree of freedom for the BS is reduced. In the third estimation algorithm, approximate maximum likelihood (AML), there is no loss of degree of freedom, but an iteration procedure is needed and an initial value is required since no closed-form solution is found.

We will also illustrate mobile position computation incorporating the TOAs of both the LOS and NLOS BSs. It is shown that the pure LOS TOAs contribute the same mobile position error variance as that of using both LOS and NLOS components. Similar to other problems in parameter estimation, classical approaches, such as least squares (LS), weighted least squares (WLS), and maximum likelihood (ML) estimators, are employed to estimate the mobile position from a sufficient number of the TOA estimates.

The error performance of the LS and WLS estimators is addressed. In general, the WLS estimator provides less error variance than the LS estimator. When the TOA error variances are identical, the WLS estimator performance is the same as that of the LS estimator. Compared with the ML estimator, the WLS estimator provides the same error performance, which attains the CRB. Numerical examples illustrate the statistical performance of these three estimators. It is concluded that the LS estimator, in general, cannot achieve the CRB, while the WLS and the ML estimators are statistically efficient.

18.1.4 Organization

This chapter is organized as follows. In Section 18.2, the wireless geolocation in mixed LOS/NLOS environments is introduced. In Section 18.3, we consider several aspects of the mobile position estimation based on the TOA estimates and then present the LS, the WLS, and the ML estimators. Section 18.4 provides a compact form of the CRBs for the estimation of the TOA and the mobile position, respectively. Numerical examples are conducted in Section 18.5 to illustrate the performance of these three estimators. In Section 18.6, the main conclusion found in this chapter is summarized.

18.2 SYSTEM MODEL

Let us consider an MS transmitting a radio signal through a wireless channel to a number of BSs. Let B be the number of all BSs, whose locations,

$$\mathbf{p}_b = \begin{bmatrix} x_b & y_b \end{bmatrix}^T, b \in \{1, 2, \ldots, B\}, \tag{18.1}$$

are known. Let $M < B - 1$ be the number of the BSs that receive a set $N = \{1, 2, \ldots, M\}$ of NLOS signals. Let τ_b be the TOA of the received signal at the bth BS, expressed as

$$\tau_b(x, y, l_b) = \frac{1}{c}\left(\sqrt{\tilde{x}_b^2 + \tilde{y}_b^2} + l_b\right), \tag{18.2}$$

where $\tilde{x}_b = x - x_b$ and $\tilde{y}_b = y - y_b$ are the relative distances with $\mathbf{p} = [x \ y]^T$ representing the mobile position coordinate and x_b and y_b introduced in (18.1), and $l_m \in \mathbb{R}_+$ is the additional propagation distance for the NLOS $m \in \{1, 2, \ldots, M\}$. Note that $l_b = 0$ implies the LOS component for $b \in \{M + 1, M + 2, \ldots, B\}$, meaning no additional propagation distance for the LOS transmission. The above model covers a general setting for wireless localization on the two-dimensional plane. In the numerical examples, the plane will be restricted to the cellular setting.

18.2.1 Existing Techniques for Mobile Position Estimation

Many estimation algorithms based on the TDOA are developed by means of spherical interpolation [48, 49], divide and conquer [50], Newton-type iteration [51], steepest descent method [52], two-step WLS [53], quadratic LS [54], LS calibration [55], modified multidimensional scaling [56], and other techniques in [57–60].

Although the solutions in [48, 49, 57], and [58] are developed as closed forms, their error performance is suboptimal. The divide-and-conquer method in [50] can achieve optimal performance, but it requires sufficiently large Fisher information. The approaches in [51] and [52] can also attain the optimal performance, but both are computationally intensive and require sufficiently precise initial estimates. In the two-step WLS method developed in [53], the relation between an extra variable and the MS position is exploited. In [54], a technique of Lagrange multipliers is employed. It is shown in [56] that for three BSs without NLOS error, the modified multidimensional scaling technique provides lower mean square position error than the two-step WLS estimator in [53], the quadratic LS estimator in [54], and an LLS estimator in [54]. However, when the number of BSs increases, the two-step WLS estimator in [53] outperforms the quadratic LS and the LLS estimators in [54] and the modified multidimensional scaling technique in [56].

In [59], a closed-form position estimate is found by solving the equation that is related to the hyperbolas associated with the TDOA measurements. The advantage

TABLE 18.1.   Mobile Position Estimation Techniques and Their Pros and Cons

Method	Advantages	Disadvantages	References
LLS based on the Taylor series	No loss of degree of freedom	Iteration, initial value requirement, no guarantee of convergence, computationally expensive	[61, 83, 84]
LLS with additional parameterization	Closed-form solution, no requirement of optimization	Loss of degree of freedom	[63]
AML	No loss of degree of freedom	Iteration, initial value requirement	[64]

of this algorithm is that it is optimal in the sense of being equivalent to the ML estimator. However, it works only when the minimum number of the TDOAs is satisfied; that is, at least two TDOA measurements are required for a two-dimensional localization problem.

In what follows, we consider several existing works that take into account the estimation of the mobile position and invoke the TOA. A comparison of these algorithms is reported in Table 18.1.

LLS Based on First-Order Taylor Series: A method in classical estimation of the position on two-dimensional plane is the LLS based on the first-order Taylor approximation [61]. The mobile position is estimated from any navigational measurements, such as range, and bearing angle. Since in this chapter we shall treat the TOA as the measurement, the model in [61] can be written as

$$d_b[i] = d_{b,i}(x, y, x_b, y_b) + e_b[i]; i \in \{1, 2, \dots, N\}, \tag{18.3}$$

where $d_b[i]$ is the ith range measurement to the bth station, $d_{b,i}(x, y, x_b, y_b)$ is the true range, which depends on the true position (x, y) and the bth BS position (x_b, y_b), $e_{b,i}$ is the error at the ith measurement, and N is the number of measurements.

Let $\check{\mathbf{p}} = [\check{x} \quad \check{y}]^T$ be a guessed value whose elements are represented by $\check{x} = x - \delta_x$ and $\check{y} = y - \delta_y$, where (δ_x, δ_y) is the corresponding error pair of the guessed position. By keeping only the terms up to the first order of the Taylor series, the linear range regression in (18.3) can be approximated as

$$d_b[i] \simeq d_{b,i}(\check{x}, \check{y}) + \dot{d}_{b,i}(\check{x})\delta_x + \dot{d}_{b,i}(\check{y})\delta_y + e_b[i], \tag{18.4}$$

where $\dot{d}_{b,i}(\check{x})$ and $\dot{d}_{b,i}(\check{y})$ are given by

$$\dot{d}_{b,i}(\check{x}) = \frac{\partial}{\partial x} d_{b,i}(x, y, x_b, y_b) \Big|_{x=\check{x}, y=\check{y}}, \tag{18.5a}$$

$$\dot{d}_{b,i}(\check{y}) = \frac{\partial}{\partial x} d_{b,i}(x, y, x_b, y_b) \Big|_{x=\check{x}, y=\check{y}}, \tag{18.5b}$$

For all N measurements, we have

$$\mathbf{d}_b \simeq \Delta \delta + \check{\mathbf{d}}_b + \mathbf{e}_b, \tag{18.6}$$

where $\Delta \in \mathbb{R}^{N \times 2}$, $\delta \in \mathbb{R}^{2 \times 1}$, $\mathbf{d}_b \in \mathbb{R}^{2 \times 1}$, $\check{\mathbf{d}}_b \in \mathbb{R}^{2 \times 1}$, and $\mathbf{e}_b \in \mathbb{R}^{2 \times 1}$ are defined as

$$\Delta = \begin{bmatrix} \dot{d}_{b,1}(\check{x}) & \dot{d}_{b,1}(\check{y}) \\ \dot{d}_{b,2}(\check{x}) & \dot{d}_{b,2}(\check{y}) \\ \vdots & \vdots \\ \dot{d}_{b,N}(\check{x}) & \dot{d}_{b,N}(\check{y}) \end{bmatrix}, \tag{18.7a}$$

$$\delta = \begin{bmatrix} \delta_x & \delta_y \end{bmatrix}^{\mathrm{T}}, \tag{18.7b}$$

$$\mathbf{d}_b = [d_b[1] \quad d_b[2] \cdots d_b[N]]^{\mathrm{T}}, \tag{18.7c}$$

$$\check{\mathbf{d}}_b = [d_{b,1}(\check{x}, \check{y}) \quad d_{b,2}(\check{x}, \check{y}) \cdots d_{b,N}(\check{x}, \check{y})]^{\mathrm{T}}, \tag{18.7d}$$

$$\mathbf{e}_b = [e_b[1] \quad e_b[2] \cdots e_b[N]]^{\mathrm{T}}, \tag{18.7e}$$

with $\dot{d}_{b,i}(\check{x})$ and $\dot{d}_{b,i}(\check{y})$ introduced in (18.5); $d_b[i]$, $d_{b,i}(\check{x}, \check{y})$, and $e_b[i]$ introduced in (18.4); and $(\cdot)^{\mathrm{T}}$ being the transpose of a matrix or a vector $\cdot$. If the measurement errors are independent of each other and have an identical variance, the solution of the above linear model in the LS sense can be expressed as

$$\begin{aligned} \hat{\delta}_{\mathrm{LLS}} &= \arg\min \| \Delta \delta - (\mathbf{d}_b - \check{\mathbf{d}}_b) \|_{\mathrm{E}}^2 \\ &= (\Delta^{\mathrm{T}} \Delta)^{-1} \Delta^{\mathrm{T}} (\mathbf{d}_b - \check{\mathbf{d}}_b), \end{aligned} \tag{18.8}$$

where $\|\cdot\|_{\mathrm{E}}$ is the Euclidean norm, $(\cdot)^{-1}$ is the inverse operator, Δ is defined in (18.7a), $\mathbf{d}_b$ is defined in (18.7c), and $\check{\mathbf{d}}_b$ is defined in (18.7d). Substituting the estimate $\hat{\delta}_{\mathrm{LLS}}$ into (18.7b), we obtain the position estimate $\check{\mathbf{p}} + \delta_{\mathrm{LLS}}$.

The solution can be refined by assigning $\check{\mathbf{p}} + \delta_{\mathrm{LLS}}$ as $\check{\mathbf{p}}$ for the next update. Substituting $\hat{\delta}_{\mathrm{LLS}}$ into (18.7a) and (18.7d), we obtain a recursive computation of (18.8), which should make $\left\| \hat{\delta}_{\mathrm{LLS}} \right\|_{\mathrm{E}}$ as small as possible so that $\check{\mathbf{p}} + \delta_{\mathrm{LLS}}$ reaches $\mathbf{p}$ as closely as possible. However, it is mentioned in [61] that (1) the method is iterative and requires an initial guess, (2) no convergence is assured, and (3) the iteration is computationally expensive.

The LLS estimator with another approximation is considered in [62] for three-dimensional (3-D) localization. The method approximates the 3-D equation in such a way that there is no additional parameterization and the solution can be represented in a closed form.

LLS with Additional Parameterization: The second method based on the LLS is discussed as follows. From the transceiver model, the range equation for the LOS can be expanded into

$$\begin{aligned} d_b^2 &= x_b^2 - 2x_b x + x^2 + y_b^2 - 2y_b y + y^2, \text{ i.e.,} \\ &- 2x_b x - 2y_b y + (x^2 + y^2) = d_b^2 - (x_b^2 + y_b^2). \end{aligned} \tag{18.9}$$

Let M be the number of NLOS BSs. A linear system can be constructed from the above expression as [63]:

$$\Upsilon q = \zeta, \tag{18.10}$$

where $\Upsilon \in \mathbb{R}^{(B-M) \times 3}$, $q \in \mathbb{R}^{3 \times 1}$, and $\zeta \in \mathbb{R}^{(B-M) \times 1}$ are defined as

$$\Upsilon = \begin{bmatrix} -2x_{M+1} & -2y_{M+1} & 1 \\ -2x_{M+2} & -2y_{M+2} & 1 \\ \vdots & \vdots & \vdots \\ -2x_B & -2y_B & 1 \end{bmatrix}, \tag{18.11a}$$

$$q = \begin{bmatrix} x & y & x^2 + y^2 \end{bmatrix}^T, \tag{18.11b}$$

$$\zeta = \begin{bmatrix} d_{M+1}^2 - \left(x_{M+1}^2 + y_{M+1}^2 \right) \\ d_{M+2}^2 - \left(x_{M+1}^2 + y_{M+2}^2 \right) \\ \vdots \\ d_B^2 - \left(x_B^2 + y_B^2 \right) \end{bmatrix}. \tag{18.11c}$$

The LLS estimate is given by

$$\hat{q}_{LLS} = \arg\min_q \left\| \Upsilon_q - \hat{\zeta} \right\|_E^2, \tag{18.12}$$

where Υ is defined in (18.11a), q is defined in (18.11b), and the bth element of $\hat{\zeta} \in \mathbb{R}^{(B-M) \times 1}$ is given by

$$\hat{\zeta}_b = \left(c\hat{\tau}_b \right)^2 - \left(x_b^2 + y_b^2 \right); \, b \in \{M+1, M+2, \ldots, B\}. \tag{18.13}$$

Differentiating $\left\| \Upsilon q - \zeta \right\|_E^2$ with respect to q, we obtain

$$\hat{q}_{LLS} = \left(\Upsilon^T \Upsilon \right)^{-1} \Upsilon^T \hat{\zeta}, \tag{18.14}$$

where the bth element of $\hat{\zeta}$ is introduced in (18.13).

The second LLS estimator provides a closed-form solution and requires no optimization. However, the LLS equation in (18.10) treats $x^2 + y^2$ as another unknown parameter, which is not mathematically justified. The LLS estimate in (18.14) can thus be computed, when Υ is of full-rank, that is, $B - M \geq 3$, which consumes a degree of freedom.

AML: In [64] (section III), the exact ML is derived regardless of the NLOS for three linear sensors. For any number of receivers and identical time delay error variance, the ML reduces to the LS from

$$\{\hat{x}, \hat{y}\}_{ML} = \arg\min_{x,y} \sum_{b=1}^{B} \left(\hat{d}_b - d_b(x, y) \right)^2, \tag{18.15}$$

where $\hat{d}_b$ is an estimate of the distance between the MS and the bth BS, and $d_b(x, y)$ is the theoretical distance given by $d_b(x, y) = \sqrt{(x_b - x)^2 + (y_b - y)^2}$ with the MS coordinates (x, y) and the bth BS coordinates (x_b, y_b).

Differentiating the objective function $\sum_{b=1}^{B}\left(\hat{d}_b - d_b(x, y)\right)^2$ with respect to the unknown x-coordinate x, we have

$$\frac{\partial}{\partial x}\sum_{b=1}^{B}\left(\hat{d}_b - d_b(x, y)\right)^2 = 2\sum_{b=1}^{B}\frac{1}{d_b}(x_b - x)\left(\hat{d}_b - d_b\right)$$

$$= 2\sum_{b=1}^{B}\frac{1}{d_b\left(\hat{d}_b + d_b\right)}(x_b - x) \qquad (18.16)$$

$$\left(\hat{d}_b^2 - \left(x_b^2 + y_b^2 + x^2 + y^2 - 2x_b x - 2y_b y\right)\right)$$

and

$$\frac{\partial}{\partial y}\sum_{b=1}^{B}\left(\hat{d}_b - d_b(x, y)\right)^2 = 2\sum_{b=1}^{B}\frac{1}{d_b\left(\hat{d}_b + d_b\right)}(y_b - y)$$

$$\left(\hat{d}_b^2 - \left(x_b^2 + y_b^2 + x^2 + y^2 - 2x_b x - 2y_b y\right)\right). \qquad (18.17)$$

At the critical points for x and y, that is, $\partial/\partial x\sum_{b=1}^{B}\left(\hat{d}_b - d_b(x, y)\right)^2 = 0$ and $\partial/\partial y\sum_{b=1}^{B}\left(\hat{d}_b - d_b(x, y)\right)^2 = 0$, it follows from [64] (26) that

$$2\sum_{b=1}^{B}\frac{1}{\left(\hat{d}_b + d_b\right)d_b}\begin{bmatrix}(x_b - x)x_b & (x_b - x)y_b \\ (y_b - y)x_b & (y_b - y)y_b\end{bmatrix}\begin{bmatrix}x \\ y\end{bmatrix}$$

$$= \sum_{b=1}^{B}\frac{1}{\left(\hat{d}_b + d_b\right)d_b}\begin{bmatrix}(x_b - x)\left(\hat{d}_b^2 - \left(x_b^2 + y_b^2 + x^2 + y^2\right)\right) \\ (y_b - y)\left(\hat{d}_b^2 - \left(x_b^2 + y_b^2 + x^2 + y^2\right)\right)\end{bmatrix}. \qquad (18.18)$$

The solution of the above equations for x and y is quite complicated, since both are highly nonlinear. If the initial values of x and y are given by $\hat{d}_b^2 = (x_b - x)^2 + (y_b - y)^2$, it yields

$$2\mathbf{P}\mathbf{p} = \xi, \qquad (18.19)$$

where $\mathbf{P} \in \mathbb{R}^{B\times 2}$, $\mathbf{P} \in \mathbb{R}^{2\times 1}$, and $\xi \in \mathbb{R}^{B\times 1}$ are given by

$$\mathbf{P} = \begin{bmatrix}x_1 & y_1 \\ x_2 & y_2 \\ \vdots & \vdots \\ x_B & y_B\end{bmatrix}, \qquad (18.20a)$$

$$\mathbf{p} = \begin{bmatrix}x \\ y\end{bmatrix}, \qquad (18.20b)$$

$$\xi = \begin{bmatrix}x_1^2 + y_1^2 + x^2 + y^2 - \hat{d}_1^2 \\ x_2^2 + y_2^2 + x^2 + y^2 - \hat{d}_2^2 \\ \vdots \\ x_B^2 + y_B^2 + x^2 + y^2 - \hat{d}_B^2\end{bmatrix}. \qquad (18.20c)$$

We can see that the expression of the AML system in (18.19) is equivalent to the expression of (18.10), whereas (18.19) treats only x and y as unknown parameters. However, the solution of (18.19) is not easy to find, since ξ depends on unknown **p**. It is suggested in [64] to solve (18.19) in an iterative manner.

More sophisticated methods are proposed to estimate the mobile position based on the range measurement, such as range-based LS, squared-range-based LS, unconstrained squared-range-based LS, squared-range-difference-based LS, and unconstrained squared-range-difference LS [65]. The performances of the LS and the squared-range-LS are addressed in [66].

18.2.2 Path Loss Model

We assume that there is no additional loss of energy, except the path loss attenuation, for the transmitted signal when radio waves propagate in a medium. The received energy at the bth BS can be expressed as (see, e.g., [67], p. 46 and [68], p. 38):

$$E_b = \frac{d_0^{\gamma_b}}{d_b^{\gamma_b}} \kappa E_s,$$

(18.21)

where d_0 is the close-in reference distance in the far-field region, d_b is the distance between the MS and the bth BS, γ_b is the PLE at the bth BS, $E_s = \int_{-\infty}^{\infty} |s(t)|^2 \, dt$ is the energy of a transmitted signal $s(t)$, and κ is the unitless constant that depends on antenna characteristics and propagation environment. For the free space propagation and omnidirectional antenna, κ is given by [69]:

$$\kappa = \frac{c^2}{16\pi^2 f_0^2 d_0^2},$$

(18.22)

with f_0 being the central frequency of the transmitted signal, and c being the speed of light.

The channel is assumed herein to be static such that (1) the large-scale fading is considered as a spatial average over the small-scale fluctuations of signals ([70], p. 847), and (2) the excess propagation distances $\{l_m\}_{m=1}^M$ are constant over the observation period.* Because we have $d_b = c\tau_b$, the energy based on (18.21) can be rewritten as

$$E_b = \frac{1}{\left(\sqrt{\left(\frac{\tilde{x}_b}{d_0}\right)^2 + \left(\frac{\tilde{y}_b}{d_0}\right)^2} + \frac{l_b}{d_0} \right)^{\gamma_b}} \kappa E_s.$$

(18.23)

Since (18.21) and (18.23) are valid only in the far field, it is assumed that d_0 is less than $\sqrt{\tilde{x}_b^2 + \tilde{y}_b^2}$. This means that within a circle of the radius d_0, there is no BS. It

---

*For random NLOS excess propagation paths, if the knowledge of the distribution of l_m is available, the parameter estimation can be cast into a *maximum a posteriori* approach [43].

should be noted that for the free space path loss, the PLE is $\gamma_b = 2$ (see, e.g., [71], p. 304 and [72], p. 88), while for a wireless environment, the PLE can be less than 2 ([67], p. 47). The range of the PLE is wide due to the attenuation caused by several scattering objects.

The received baseband signal can be written as [43]:

$$r_b(t) = a_b s(t - \tau_b) + n_b(t), \tag{18.24}$$

where $s(t)$ is a known waveform, a_b and τ_b are the amplitude and the TOA for the propagation to the bth BS, and $n_b(t)$ is the additive noise at the bth BS and assumed to be a circularly symmetric complex-valued zero-mean white Gaussian process with double-sided spectral density σ_n^2. Assume that the transmitted signal is nonzero only in the interval $(0, T_s)$, where T_s is the signal period.

The relation among the received energy, the transmitted energy, and the path gain can be postulated as $E_b = a_b^2 E_s$. The amplitude then can be written as

$$a_b = \frac{1}{\left(\sqrt{\left(\frac{\tilde{x}_b}{d_0} \right)^2 + \left(\frac{\tilde{y}_b}{d_0} \right)^2} + \frac{l_b}{d_0} \right)^{\frac{1}{2}\gamma_b}} \sqrt{\kappa}. \tag{18.25}$$

18.3 MOBILE POSITION ESTIMATION

The TOA estimation is the first step in the two-step technique for the mobile position estimation. After the TOAs are estimated, the obtained TOAs will be passed onto an estimator to determine the mobile position.

18.3.1 TOA Estimation

We assume in this chapter that the TOA and the mobile position are constant. The ML estimate of the TOA τ_b is given by [46]:

$$\hat{\tau}_b = \arg\min_{\tau} a_b^2(\tau) E_s - 2a_b(\tau) \int_0^{T_o} \Re(r_b^*(t) s(t - \tau)) \, dt, \tag{18.26}$$

where $a_b(\tau) = \sqrt{\kappa}(d_0/c\tau)^{\frac{1}{2}\gamma_b}$ is the path gain as a function of the TOA τ, $\Re(\cdot)$ is the real part of $\cdot$, and T_o is the observation period in which the TOA and the mobile position are constant. The above estimator is unbiased and its estimation error variance can be expressed as follows [46]:

$$E_{n_b(t)}\{\hat{\tau}_b - \tau_b\} = 0, \tag{18.27a}$$

$$E_{n_b(t)}\left\{(\hat{\tau}_b - \tau_b)^2\right\} = \frac{1}{\frac{E_s}{\sigma_n^2} 8\pi^2 \bar{\beta}^2 a_b^2 \left(1 + \frac{1}{16\pi^2 \bar{\beta}^2 \tau_b^2} \gamma_b^2 \right)}, \tag{18.27b}$$

where $E_{n_b(t)}\{\cdot\}$ is the expectation with respect to $n_b(t)$ and $\bar{\beta}$ is the effective (RMS) bandwidth defined as

$$\bar{\beta} = \sqrt{\dfrac{\displaystyle\int_{-\infty}^{\infty} f^2 \left|S(f)\right|^2 df}{\displaystyle\int_{-\infty}^{\infty} \left|S(f)\right|^2 df}},$$

(18.28)

with $S(f)$ being the Fourier transform of the transmitted $s(t)$. Let us introduce $\bar{\Phi} = \frac{\partial}{\partial \mathbf{p}} \bar{\mathbf{d}}^{\mathrm{T}}(\mathbf{p}) \in \mathbb{R}^{2 \times (B-M)}$, where $\bar{\mathbf{d}} = c\bar{\tau}$ is the LOS distance with the LOS TOA vector:

$$\bar{\tau} = \begin{bmatrix} \tau_{M+1} & \tau_{M+2} \cdots \tau_B \end{bmatrix}^{\mathrm{T}}.$$

(18.29)

It can be shown that

$$\bar{\Phi} = \begin{bmatrix} \cos(\phi_{M+1}) & \cos(\phi_{M+2}) & \cdots & \cos(\phi_B) \\ \sin(\phi_{M+1}) & \sin(\phi_{M+1}) & \cdots & \sin(\phi_B) \end{bmatrix},$$

(18.30)

where ϕ_b is defined as

$$\phi_b = \arctan\left(\frac{y_b - y}{x_b - x}\right),$$

(18.31)

with (x, y) being the mobile position and (x_b, y_b) being the position of the bth BS. Let us introduce the PLE vector and the LOS TOA error variance vector as

$$\gamma = \begin{bmatrix} \gamma_1 & \gamma_2 \cdots \gamma_B \end{bmatrix}^{\mathrm{T}},$$

(18.32)

$$\bar{\sigma}^2 = \begin{bmatrix} \sigma_{M+1}^2 & \sigma_{M+2}^2 \cdots \sigma_B^2 \end{bmatrix}^{\mathrm{T}},$$

(18.33)

where $\sigma_b^2 = E_{n_b(t)}\left\{(\hat{\tau}_b - \tau_b)^2\right\}$ is given by (18.27b).

Recall the notion "o" of $u\left(\|\hat{\mathbf{p}} - \mathbf{p}\|_{\mathrm{E}}\right) = o\left(\upsilon\left(\|\hat{\mathbf{p}} - \mathbf{p}\|_{\mathrm{E}}\right)\right)$, which stands for $\lim_{\hat{\mathbf{p}} \to \mathbf{p}} \frac{u\left(\|\hat{\mathbf{p}} - \mathbf{p}\|_{\mathrm{E}}\right)}{\upsilon\left(\|\hat{\mathbf{p}} - \mathbf{p}\|_{\mathrm{E}}\right)} = 0$ ([73], p. 1).

LOS Sufficiency: Using either $\hat{\bar{\tau}} = \begin{bmatrix} \hat{\tau}_{M+1} & \hat{\tau}_{M+2} \cdots \hat{\tau}_B \end{bmatrix}^{\mathrm{T}}$ or $\hat{\tau} = \begin{bmatrix} \hat{\tau}_1 & \hat{\tau}_2 \cdots \hat{\tau}_B \end{bmatrix}^{\mathrm{T}}$, the position estimates using the LOS TOAs, $\hat{\mathbf{p}}_{\bar{\tau}}$, and using both the NLOS and the LOS TOAs, $\hat{\mathbf{p}}_{\tau}$, yield zero mean and the same variance; that is,

$$
\begin{aligned}
E_{\mathbf{n}(t)}\left\{(\hat{\mathbf{p}}_{\tau} - \mathbf{p})(\hat{\mathbf{p}}_{\tau} - \mathbf{p})^{\mathrm{T}}\right\} &= E_{\bar{\mathbf{n}}(t)}\left\{(\hat{\mathbf{p}}_{\bar{\tau}} - \mathbf{p})(\hat{\mathbf{p}}_{\bar{\tau}} - \mathbf{p})^{\mathrm{T}}\right\} \\
&\approx c^2\left(\bar{\Phi}\bar{\Phi}^{\mathrm{T}}\right)^{-1}\bar{\Phi}\mathbf{D}\left(\bar{\sigma}^2\right)\bar{\Phi}^{\mathrm{T}}\left(\bar{\Phi}\bar{\Phi}^{\mathrm{T}}\right)^{-1},
\end{aligned}
$$

(18.34)

where $E_{\mathbf{n}(t)}\{\cdot\}$ is the expectation with respect to $\mathbf{n}(t)$, and $\mathbf{D}(\cdot)$ is the diagonal matrix whose diagonal vector is $\cdot$, $\approx$ is the approximation by neglecting $o\left(\|\hat{\mathbf{p}} - \mathbf{p}\|_{\mathrm{E}}\right)$, and $\bar{\Phi}$ is introduced in (18.30).

The proof of (18.34) is provided in the Appendix section "Detailed Derivation of (18.34)." Note that the error variance in (18.34) depends only on the LOS component. Since the TOA from both the LOS and the NLOS components, that is, τ, possesses a sufficient relation to the mobile position $\mathbf{p}$ and the NLOS excess path length $\mathbf{1} = [l_1 \; l_2 \ldots l_M]^{\mathrm{T}}$, the TOA estimate $\hat{\tau}$ will be transformed into the mobile position $\mathbf{p}$ by an estimator.

18.3.2 LS

The LS estimate of the mobile position can be calculated from

$$
\begin{aligned}
\hat{\mathbf{p}}_{\mathrm{LS}} &= \arg\min_{x,y} \left\| \begin{bmatrix} \hat{\tau}_{M+1} \\ \vdots \\ \hat{\tau}_B \end{bmatrix} - \frac{1}{c} \begin{bmatrix} \sqrt{(x-x_{M+1})^2 + (y-y_{M+1})^2} \\ \vdots \\ \sqrt{(x-x_B)^2 + (y-y_B)^2} \end{bmatrix} \right\|_{\mathrm{E}}^2 \\
&= \arg\min_{x,y} \sum_{b=M+1}^{B} \left(\hat{\tau}_b - \frac{1}{c}\sqrt{(x_b - x)^2 + (y_b - y)^2} \right)^2,
\end{aligned}
\tag{18.35}
$$

where $\hat{\tau}_b$ is the ML estimate given by (18.26). The detailed derivation of (18.35) is provided in the Appendix section "Detailed Derivation of (18.35)." Note that the concentrated LS estimate of the mobile position depends only on the LOS time delays. The LS estimate in (18.35) requires joint two-dimensional minimization.

18.3.3 WLS

The WLS estimate of **p** is given by

$$
\begin{aligned}
\hat{\mathbf{p}}_{\mathrm{WLS}} &= \arg\min_{\mathbf{p}} \left\| \hat{\boldsymbol{\tau}} - \boldsymbol{\tau} \right\|_{\bar{\mathbf{W}}}^2 \\
&= \arg\min_{x,y} \sum_{b=M+1}^{B} \frac{1}{\sigma_b^2(x,y)} \left(\hat{\tau}_b - \frac{1}{c}\sqrt{(x_b - x)^2 + (y_b - y)^2} \right)^2,
\end{aligned}
\tag{18.36}
$$

where $\|\upsilon\|_{\mathbf{W}}^2 = \upsilon^{\mathrm{H}}\mathbf{W}\upsilon$ is the weighted Euclidean norm with an unknown positive-definite Hermitian weighting matrix $\mathbf{W} \in \mathbb{C}^{B \times B}$ and $(\cdot)^{\mathrm{H}}$ denotes the Hermitian transpose, M is the number of the BSs that receive the NLOS component, B is the number of all BSs, $\sigma_b^2(x,y)$ is the error variance introduced in (18.27b), and $\hat{\tau}_b$ is defined in (18.26). The weighting matrix for the LOS component $\bar{\mathbf{W}} = \mathrm{E}_{\hat{\tau}}\left\{ \left(\hat{\boldsymbol{\tau}} - \boldsymbol{\tau}\right)\left(\hat{\boldsymbol{\tau}} - \boldsymbol{\tau}\right)^{\mathrm{T}} \right\}$ should be optimal such that the fitting error variance $\left\| \hat{\boldsymbol{\tau}} - \boldsymbol{\tau} \right\|_{\mathrm{E}}^2$ is minimized. The detailed derivation of (18.36) is provided in the Appendix section "Detailed Derivation of (18.36)."

18.3.4 ML

The ML estimate of **p** is calculated by

$$
\hat{\mathbf{p}}_{\mathrm{ML}} = \arg\min_{x,y} \sum_{b=M+1}^{B} \ln\left(\sigma_b^2(x,y)\right) + \frac{1}{\sigma_b^2(x,y)}\left(\hat{\tau}_b - \tau_b(x,y)\right)^2.
\tag{18.37}
$$

The detailed derivation of (18.37) is straightforward from the PDF of the estimated TOA, which is assumed to be Gaussian.

LS Error Variance: For the estimation performance, both the LS and the WLS criteria in (18.35) and (18.36) yield zero mean and the variances are given by

$$E_{n(t)}\left\{(\hat{\mathbf{p}}_{LS}-\mathbf{p})(\hat{\mathbf{p}}_{LS}-\mathbf{p})^T\right\}=c^2\left(\overline{\Phi}\overline{\Phi}^T\right)^{-1}\overline{\Phi}\mathbf{D}\left(\overline{\sigma}^2\right)\overline{\Phi}^T\left(\overline{\Phi}\overline{\Phi}^T\right)^{-1}, \quad (18.38)$$

$$E_{n(t)}\left\{(\hat{\mathbf{p}}_{WLS}-\mathbf{p})(\hat{\mathbf{p}}_{WLS}-\mathbf{p})^T\right\}=c^2\left(\overline{\Phi}\mathbf{D}^{-1}\left(\overline{\sigma}^2\right)\overline{\Phi}^T\right)^{-1}, \quad (18.39)$$

where c is the speed of the light, $\overline{\Phi}$ is defined in (18.30), and $\overline{\sigma}^2$ is defined in (18.33).

The proof of (18.38) and (18.39) is provided in the Appendix section "Detailed Derivation of (18.38) and (18.39)." From (18.38) and (18.39), we can show by using the Schur complements that (see, e.g., [74], Lemma 2c)

$$E_{n(t)}\left\{(\hat{\mathbf{p}}_{LS}-\mathbf{p})(\hat{\mathbf{p}}_{LS}-\mathbf{p})^T\right\}\succeq E_{n(t)}\left\{(\hat{\mathbf{p}}_{WLS}-\mathbf{p})(\hat{\mathbf{p}}_{WLS}-\mathbf{p})^T\right\}, \quad (18.40)$$

where $\hat{\mathbf{p}}_{LS}$ is given by (18.38), $\hat{\mathbf{p}}_{WLS}$ is given by (18.39), $\mathbf{A}_1 \succeq \mathbf{A}_2$ means that $\mathbf{x}^H(\mathbf{A}_1 - \mathbf{A}_2)\,\mathbf{x} \geq 0$ for any nonzero vector $\mathbf{x}$, and the equality holds when the variances of the TOA estimates are identical. It can be seen that

- the WLS and ML estimators achieve the same variance of the estimation error, which is lower than that given by the LS estimator,
- the use of only the LOS signal is sufficient to achieve the same performance as that using both the LOS and the NLOS TOAs, and
- the variance of the estimation error depends only on the LOS portion.

18.4 CRB FOR MOBILE POSITION ESTIMATION

Let $\hat{\boldsymbol{\theta}}$ be any unbiased estimate of $\boldsymbol{\theta}_0$. Then, the accuracy of $\hat{\boldsymbol{\theta}}$ is lower bounded by the Cramer–Rao inequality (see, e.g., [75–77]; [78], chapter 32; and [79], section 5a):

$$E_{n(t)}\left\{(\hat{\boldsymbol{\theta}}-\hat{\boldsymbol{\theta}}_0)(\hat{\boldsymbol{\theta}}-\hat{\boldsymbol{\theta}}_0)^T\right\}\succeq \mathbf{H}_{\theta\theta}^{-1}, \quad (18.41)$$

where the FIM $\mathbf{H}_{\theta\theta}$ is given by

$$\mathbf{H}_{\theta\theta}=-E_{n(t)}\left\{\frac{\partial^2}{\partial\boldsymbol{\theta}\,\partial\boldsymbol{\theta}^T}\ln(p(\mathbf{r}(t); t\in(0, T_o] \,|\, \boldsymbol{\theta}))\bigg|_{\theta=\theta_0}\right\}. \quad (18.42)$$

18.4.1 FIM of TOA Estimation

When the PLE $\boldsymbol{\gamma}$ is known, the FIM is given by

$$\begin{aligned}\mathbf{H}_{\tau\tau}&=-E_{n(t)}\left\{\frac{\partial^2}{\partial\boldsymbol{\tau}\,\partial\boldsymbol{\tau}^T}\ln(p(\mathbf{r}(t); t\in(0, T_o] \,|\, \boldsymbol{\tau}))\bigg|_{\tau=\tau_0}\right\}\\&=\frac{E_s}{\sigma_n^2}\mathbf{D}^2(\mathbf{a})\left(8\pi^2\overline{\beta}^2\mathbf{I}+\frac{1}{2}\mathbf{D}^2(\boldsymbol{\gamma})\mathbf{D}^{-2}(\boldsymbol{\tau})\right),\end{aligned} \quad (18.43)$$

where $\mathbf{I}$ is the identity matrix, $\mathbf{D}(\cdot)$ is the diagonal matrix given by the vector $\cdot$, $\boldsymbol{\gamma}$ is introduced in (18.32), and $\mathbf{a} \in \mathbb{R}^{B\times 1}$ and $\boldsymbol{\tau} \in \mathbb{R}_+^{B\times 1}$ are defined as

$$\mathbf{a}=\begin{bmatrix}a_1 & a_2 \cdots a_B\end{bmatrix}^T, \quad (18.44a)$$

$$\tau = \begin{bmatrix} \tau_1 & \tau_2 \cdots \tau_B \end{bmatrix}^{\mathrm{T}}. \tag{18.44b}$$

The proof of (18.43) is provided in the Appendix section "Detailed Derivation of (18.43)." We can see that each diagonal element of (18.43) corresponds to (18.27b).

18.4.2 CRB for TOA Estimation

Taking the inverse of (18.43), the unbiased CRB of the TOA is given by

$$\mathbf{B}_{\tau\tau} = \frac{1}{\dfrac{E_s}{\sigma_n^2} 8\pi^2 \overline{\beta}^2} \mathbf{D}^{-2}(\mathbf{a}) \left(\mathbf{I} + \frac{1}{16\pi^2 \overline{\beta}^2} \mathbf{D}^2(\gamma) \mathbf{D}^{-2}(\tau) \right)^{-1}. \tag{18.45}$$

18.4.3 CRB for Mobile Position Estimation

For the case of known γ, the unbiased CRB of the mobile position is given by

$$\mathbf{B}_{\mathbf{pp}} = \frac{1}{\dfrac{E_s}{\sigma_n^2} 8\pi^2 \overline{\beta}^2} c^2 \left(\overline{\Phi} \mathbf{D}^2(\overline{\mathbf{a}}) \left(\mathbf{I} + \frac{1}{16\pi^2 \overline{\beta}^2} \mathbf{D}^2(\overline{\gamma}) \mathbf{D}^{-2}(\overline{\tau}) \right) \overline{\Phi}^{\mathrm{T}} \right)^{-1}, \tag{18.46}$$

where $\overline{\mathbf{a}} \in \mathbb{R}^{(B-M)\times 1}$ and $\overline{\gamma} \in \mathbb{R}^{(B-M)\times 1}$ are defined as

$$\overline{\mathbf{a}} = \begin{bmatrix} a_{M+1} & a_{M+2} \cdots a_B \end{bmatrix}^{\mathrm{T}}, \tag{18.47a}$$

$$\overline{\gamma} = \begin{bmatrix} \gamma_{M+1} & \gamma_{M+2} \cdots \gamma_B \end{bmatrix}^{\mathrm{T}}. \tag{18.47b}$$

The explicit expression of $\mathrm{tr}(\mathbf{B}_{\mathbf{pp}})$ in (18.46) is available in [47]. In general, the ML estimator asymptotically achieves the CRB; that is, $\mathrm{E}_{\mathbf{n}(t)}\left\{(\hat{\mathbf{p}}_{\mathrm{ML}} - \mathbf{p})(\hat{\mathbf{p}}_{\mathrm{ML}} - \mathbf{p})^{\mathrm{T}}\right\} = \mathbf{B}_{\mathbf{PP}}$. Since we have

$$\mathrm{E}_{\mathbf{n}(t)}\left\{(\hat{\mathbf{p}}_{\mathrm{ML}} - \mathbf{p})(\hat{\mathbf{p}}_{\mathrm{ML}} - \mathbf{p})^{\mathrm{T}}\right\} = \mathrm{E}_{\mathbf{n}(t)}\left\{(\hat{\mathbf{p}}_{\mathrm{WLS}} - \mathbf{p})(\hat{\mathbf{p}}_{\mathrm{WLS}} - \mathbf{p})^{\mathrm{T}}\right\}, \tag{18.48}$$

the error variance ratio of the position estimate can be shown as

$$\sqrt{\frac{\mathrm{tr}\left(\mathrm{E}_{\mathbf{n}(t)}\left\{(\hat{\mathbf{p}}_{\mathrm{WLS}} - \mathbf{p})(\hat{\mathbf{p}}_{\mathrm{WLS}} - \mathbf{p})^{\mathrm{T}}\right\}\right)}{\mathrm{tr}\left(\mathbf{B}_{\mathbf{pp}}\right)}} = 1, \tag{18.49}$$

which can be inferred that the WLS estimator is statistically efficient.

18.5 NUMERICAL EXAMPLES

The following schematic setting and parameter assignment will be used for all the examples in the sequel. Let us consider a certain configuration of a cellular system operating at the center frequency $f_0 = 1.9\,\mathrm{GHz}$. In seven hexagonal cells, let the origin of the Cartesian coordinates lie at the center of the central cell as shown in Figure 18.1. The BSs are located at the center of each cell with

$$\mathbf{P}_{\mathrm{cell}} = r \begin{bmatrix} 0 & \dfrac{3}{2} & 0 & -\dfrac{3}{2} & -\dfrac{3}{2} & 0 & \dfrac{3}{2} \\ 0 & \dfrac{\sqrt{3}}{2} & \sqrt{3} & \dfrac{\sqrt{3}}{2} & -\dfrac{\sqrt{3}}{2} & -\sqrt{3} & \dfrac{-\sqrt{3}}{2} \end{bmatrix}^{\mathrm{T}}, \tag{18.50}$$

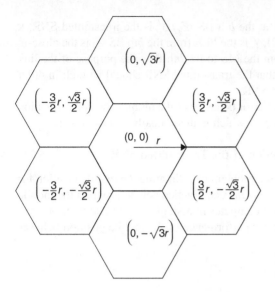

Figure 18.1   Cellular system with cell radius r.

where r is the cell radius. As introduced in (18.20b), the MS is assumed to locate at

$$\mathbf{p} = \frac{1}{2} r \cos\left(\frac{1}{6}\pi\right)\left[\cos\left(\frac{1}{6}\pi\right) \quad \sin\left(\frac{1}{6}\pi\right)\right]^{\mathrm{T}}. \tag{18.51}$$

The mobile is therefore $\frac{\sqrt{3}}{4} r$ m apart from the center of the central cell. With respect to the MS position, the associated angles of the BSs can be written as

$$\boldsymbol{\phi} = \begin{bmatrix} -150° & 30° & 103.9° & 160.9° & -150° & -100.9° & -43.9° \end{bmatrix}^{\mathrm{T}}. \tag{18.52}$$

The TOA estimate $\hat{\tau}_b$ is generated from a Gaussian random variable according to the ML error performance in (18.27). We assume that the first M BSs receive the NLOS signals. From the objective functions in (18.35–18.37), the mobile position is solved by the MATLAB built-in function "fminsearch," which is essentially realized by the simplex-search method whose initial value is given by the true value perturbed by a zero-mean unit-variance Gaussian random variable. The RMSE is calculated by the square root of the trace of the error variances from the LS, the WLS, and the ML estimators; that is,

$$\bar{\varepsilon} = \sqrt{\frac{1}{N_{\mathrm{R}}} \sum_{n_{\mathrm{R}}=1}^{N_{\mathrm{R}}} \left\| \hat{\mathbf{p}}[n_{\mathrm{R}}] - \mathbf{p} \right\|_{\mathrm{E}}^2}$$

$$= \sqrt{\frac{1}{N_{\mathrm{R}}} \sum_{n_{\mathrm{R}}=1}^{N_{\mathrm{R}}} \left(\hat{x}[n_{\mathrm{R}}] - x\right)^2 + \left(\hat{y}[n_{\mathrm{R}}] - y\right)^2}, \tag{18.53}$$

where N_{R} is the number of experimental realizations. From (18.21), the link budget can be computed from

$$10\log_{10}\left(\frac{E_b}{\sigma_{\mathrm{n}}^2}\right) = 10\log_{10}\left(\frac{E_{\mathrm{s}}}{\sigma_{\mathrm{n}}^2}\right) + 10\log_{10}(\kappa) + 10_{\gamma_b} \log_{10}\left(\frac{d_0}{d_b}\right), \tag{18.54}$$

where E_b/σ_n^2 is the received SNR at the bth BS, E_s/σ_n^2 is the transmitted SNR, κ is the constant introduced in (18.22), γ_b is the PLE from the bth BS, d_0 is the close-in distance, and d_b is the distance from the MS to the bth BS. The purpose of the link budget calculation is to point out that the transmitted SNR should be high in order to maintain an acceptable received SNR.

Examples 18.1–18.4 study the error behavior of the mobile position estimation to investigate the estimation efficiency, which is theoretically shown earlier.

Example 18.1 RMSE as a Function of the Transmitted SNR

In this example, we shall consider the problem of estimating the mobile position for several values of the transmitted SNR. One can practice how to program the mobile position estimation using a personal computer in a single file. In addition, one can learn from this example how the error performance of the mobile position behaves when the SNR varies.

Solution

In the Appendix section "The Main m-File" of this chapter, the MATLAB codes are set up for the numerical simulation for different transmitted SNRs. The main m-file consists of three subfunctions for three objective functions and a built-in function in MATLAB for the optimization of each objective function:

1. The result is shown in Figure 18.2. Without slight changes, the m-file can be used to simulate the results in other cases, such as the change in the effective bandwidth in Figure 18.3, the change in the number of the NLOS BSs in

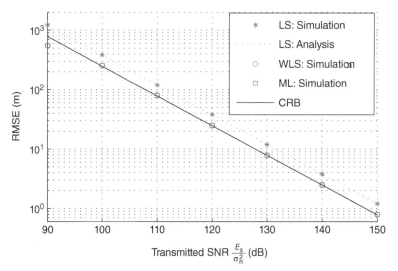

Figure 18.2 RMSE of the position estimate as a function of the transmitted SNR for $M = 0$ NLOS BSs, $\overline{\beta} = \left(1/\sqrt{3}\right)10\pi$ MHz, $r = 2000$ m, $\gamma_b = 4.5425$, and $N_R = 10{,}000$ independent simulation runs.

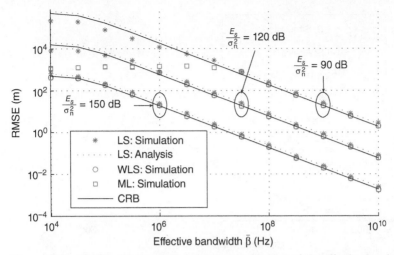

Figure 18.3   RMSE of the position estimate as a function of the effective bandwidth of the transmitted signal for $E_s/\sigma_n^2 = 90, 120, 150$ dB, $M = 1$ NLOS BSs, $r = 2000$ m, $\gamma_b = 4.5425$, and $N_R = 100$ independent simulation runs.

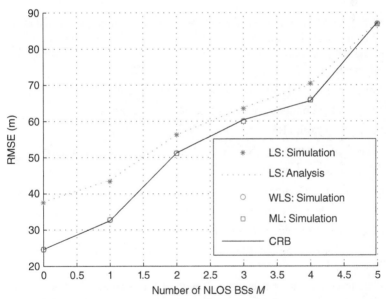

Figure 18.4   RMSE of the position estimate as a function of the number of the NLOS BSs for $E_s/\sigma_n^2 = 120$ dB, $\bar{\beta} = \left(1/\sqrt{3}\right)10\pi$ MHz, $r = 2000$ m, $\gamma_b = 4.5425$, and $N_R = 10{,}000$ independent simulation runs.

Figure 18.4, and the change in the cell radius in Figure 18.5. The programming structure of the main m-file is as follows:

- Several constants, such as the speed of light, the effective bandwidth, the cell radius, the mobile position, the number of all BSs, the number of the BSs that receive the NLOS components, the parameters for computing

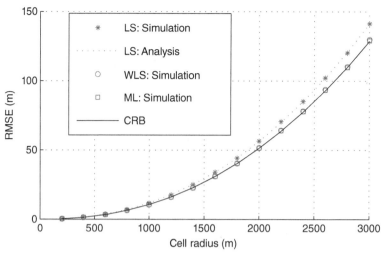

Figure 18.5   RMSE of the position estimate as a function of the cell radius for $E_s/\sigma_n^2 = 120$ dB, $\bar{\beta} = \left(1/\sqrt{3}\right)10\pi$ MHz, $M = 2$ NLOS BSs, $\gamma_b = 4.5425$, and $N_R = 10,000$ independent simulation runs.

the PLE γ_b according to [69], the central frequency, the close-in distance, the constant caused by antenna characteristics and average channel attenuation, the BS positions, and the angles of each BSs with respect to the MS are assigned.

- The for-loop of the SNR is set up and includes the following:
  * The PLEs are computed according to various scenarios.
  * The for-loop of the BS is set up and involves the following:
    a. The TOAs from either NLOS or LOS component are computed.
    b. The variance of the TOA perturbation is computed.
  * The for-loop of the noise realization is set up and involves the following:
    a. The for-loop of the BS is set up and involves the TOA perturbation.
    b. The LOS TOA vector is assigned.
    c. The initial value of the mobile position is assigned for the optimization.
    d. The mobile position is estimated by the LS criterion.
    e. The mobile position is estimated by the WLS criterion.
    f. The mobile position is estimated by the ML criterion.
    g. The square of the error or the difference between the TOA estimate by the LS criterion and the true value of the mobile position is computed.
    h. The square of the error or the difference between the TOA estimate by the WLS criterion and the true value of the mobile position is computed.

i. The square of the error or the difference between the TOA estimate by the ML criterion and the true value of the mobile position is computed.

* The simulated RMSE of the error square by the LS estimator is computed.
* The simulated RMSE of the error square by the WLS estimator is computed.
* The theoretical variance of the error for the LS estimator is computed.
* The theoretical RMSE of the LS estimator is computed.
* The experimental RMSE of the ML estimator is computed.
* The CRB or the theoretical error variance by the WLS and the ML estimators is computed.
* The square root of the CRB is computed.
* Simulated and theoretical RMSE as a function of the SNR is plotted.

2. Three subfunctions are set up for realizing the details of the three objective functions.

3. The MATLAB built-in function `fminsearch` is used to search for an optimized input at which the relevant objective function is minimized.

In Figure 18.2, the RMSE is shown as a function of the transmitted SNR from $E_s/\sigma_n^2 = 90$ dB to $E_s/\sigma_n^2 = 150$ dB. In this case, we have $E_1/\sigma_n^2 = E_s/\sigma_n^2 - 120.5873$ dB. It means that the central cell actually receives the SNR from $E_1/\sigma_n^2 \approx -30$ dB to $E_1/\sigma_n^2 \approx 30$ dB. From $E_s/\sigma_n^2 = 100$ dB to $E_s/\sigma_n^2 = 150$ dB, the LS estimate in Figure 18.2 well coincides with its expected error analysis but cannot attain the CRB, whereas the WLS estimate and the ML estimate can do so. For $E_s/\sigma_n^2 < 100$ dB, the WLS estimate and the ML estimate of the mobile position deviate from the CRB. This is because the ML error performance in (18.27) holds true only for a high received SNR. At a low received SNR, the time delay error variance σ_b^2 is inaccurate and then leads to an erroneous value of $\sigma_b^2(x, y)$ for the WLS estimator in (18.36) and for the ML estimator in (18.37).

Example 18.2   RMSE as a Function of the Effective Bandwidth

In this example, we shall consider the problem of estimating the mobile position for several values of the effective bandwidth. One can learn from this example how the error performance of the mobile position behaves when the effective bandwidth of the transmitted signal is changed.

Solution

For the numerical simulation in the aspect of the effective bandwidth, the main m-file in the Appendix section "The Main m-File" of this chapter will be used. It needs to be modified by (1) replacing the for-loop of the transmitted SNR by the for-loop of the effective bandwidth, and (2) replacing the assignment of the NLOS BSs by that of the number of the transmitted SNR. In more detail, we use

```
%beta_bar = 2 * pi * (1/sqrt(3)) * 5 * 10^6;
instead of
beta_bar = 2 * pi * (1/sqrt(3)) * 5 * 10^6;
```

and the following line of codes

```
Phi_tilde = Phi(:,1:M);
Phi_bar = Phi(:,M+1:B);
N_R = 10000;
SNR_range = [50:10:230]; %dB
n_SNR = 0;
for SNR_dB = SNR_range;
 n_SNR = n_SNR + 1;
 SNR = 10^(SNR_dB/10);
 n_scenario = 0;
 for scenario = 1 %scenario_range
 n_scenario = n_scenario + 1;
```

are replaced by

```
Phi_tilde = Phi(:,1:M);
Phi_bar = Phi(:,M+1:B);
N_R = 10000;
SNR_dB = 120; % or 90 or 150 dB
SNR = 10^(SNR_dB/10);
beta_bar_range = 10.^[0:.5:11]; %Hz
n_beta_bar = 0;
for beta_bar = beta_bar_range;
 n_beta_bar = n_beta_bar + 1;
 n_scenario = 0;
 for scenario = 1 %scenario_range
 n_scenario = n_scenario + 1;
```

In Figure 18.3, the error variance of the mobile position estimate decreases with the increase in the effective bandwidth. From a condition of the Taylor series expansion, the performance analysis of the time delay estimate $\hat{\tau}_b$ is accurate when the estimate $\hat{\tau}_b$ is close to the true value τ_b, that is, for small σ_b^2 in (18.27), which means a high SNR, and/or a large effective bandwidth. Therefore, for a low effective bandwidth, all the estimators based on the time delay cannot well match to their theoretical analysis, that is, theoretical LS performance and the CRB. However, for a high SNR, for example, $E_s/\sigma_n^2 = 150$ dB, the prediction by the analyzed LS performance and the CRB is accurate.

Example 18.3   RMSE as a Function of the Number of the LOS BSs

In this example, we shall consider the problem of estimating the mobile position for several values of the number of the LOS BSs, which may happen in various environments. One can learn from this example how the error performance of the mobile position behaves when the number of the LOS BSs varies.

Solution

For the numerical simulation in the aspect of the number of the NLOS BSs, the main m-file in the Appendix section "The Main m-File" of this chapter will be used. It needs to be modified by (1) replacing the for-loop of the transmitted SNR by the for-loop of the number of the NLOS BSs, and (2) replacing the assignment of the NLOS BSs by that of the number of the transmitted SNR. In more detail, we use

```
%M = 0; % base stations receiving NLOS
instead of
M = 0; % base stations receiving NLOS
```

and the following lines of codes

```
phi = atan2(P_diff(:,2),P_diff(:,1));
Phi = [cos(phi).';sin(phi).'];
Phi_tilde = Phi(:,1:M);
Phi_bar = Phi(:,M+1:B);
N_R = 10000;
SNR_range = [50:10:230]; %dB
n_SNR = 0;
for SNR_dB = SNR_range;
 n_SNR = n_SNR + 1;
 SNR = 10^(SNR_dB/10);
 n_scenario = 0;
 for scenario = 1 %scenario_range
 n_scenario = n_scenario + 1;
```

are replaced by

```
phi = atan2(P_diff(:,2),P_diff(:,1));
Phi = [cos(phi).';sin(phi).'];
N_R = 10000;
SNR_dB = 120;
SNR = 10^(SNR_dB/10);
M_range = [0:5]; %dB
n_M = 0;
for M = M_range;
n_M = n_M + 1;
Phi_tilde = Phi(:,1:M);
Phi_bar = Phi(:,M+1:B);
n_scenario = 0;
for scenario = 1 %scenario_range
 n_scenario = n_scenario + 1;
```

In Figure 18.4, the number of all BSs is kept as a constant, that is, $B = 7$, while the number of the LOS BSs varies from 0 to 5 according to the condition $B \geq M + 2$. We can see that the statistically efficient estimators, the WLS and the ML estimators, outperform the LS estimator, especially for a low M. The reason is that the WLS and the ML estimators adopt more information in their criteria.

Example 18.4   RMSE as a Function of the Cell Radius

In this example, we shall consider the problem of estimating the mobile position for several values of the cell radius, which can vary in various systems.

One can learn from this example how the error performance of the mobile position behaves when the cell radius is varied.

Solution

For the numerical simulation in the aspect of the cell radius, the main m-file in the Appendix section "The Main m-File" of this chapter will be used. It needs to be modified by (1) replacing the for-loop of the transmitted SNR by the for-loop of the cell radius, and (2) replacing the assignment of the cell radius by that of the transmitted SNR. In more detail, we use

```
%r = 2000; %4000/sqrt(3);
%p = (1/2) * r * cos(pi/6)* [cos(pi/6);sin(pi/6)];
B = 7;
.
. (here without the comment sign % to each line)
.
kappa = c^2 / (16 * pi^2 * f^2 * d_0^2);
%P = r * [0,3/2,0,-3/2,-3/2,0,3/2;0,sqrt(3)/2,sqrt(3),...
% sqrt(3)/2,-sqrt(3)/2,-sqrt(3),-sqrt(3)/2].';
%P_diff = P - repmat(p',B,1);
%phi = atan2(P_diff(:,2),P_diff(:,1));
%Phi = [cos(phi).';sin(phi).'];
%Phi_tilde = Phi(:,1:M);
%Phi_bar = Phi(:,M+1:B);
```

instead of

```
r = 2000; %4000/sqrt(3);
p = (1/2) * r * cos(pi/6)* [cos(pi/6);sin(pi/6)];
B = 7;
.
.
.
kappa = c^2 / (16 * pi^2 * f^2 * d_0^2);
P = r * [0,3/2,0,-3/2,-3/2,0,3/2;0,sqrt(3)/2,sqrt(3),...
 sqrt(3)/2,-sqrt(3)/2,-sqrt(3),-sqrt(3)/2].';
P_diff = P - repmat(p',B,1);
phi = atan2(P_diff(:,2),P_diff(:,1));
```

```
Phi = [cos(phi).';sin(phi).'];
Phi_tilde = Phi(:,1:M);
Phi_bar = Phi(:,M+1:B);
```

and the following lines of codes

```
SNR_range = [50:10:230];  %dB
n_SNR = 0;
for SNR_dB = SNR_range;
 n_SNR = n_SNR + 1;
 SNR = 10^(SNR_dB/10);
 n_scenario = 0;
 for scenario = 1 %scenario_range
 n_scenario = n_scenario + 1;
```

are replaced by

```
SNR_dB = 120;
SNR = 10^(SNR_dB/10);
r_range = [200:200:3000];
n_r = 0;
for r = r_range;
 n_r = n_r + 1;
 p = (1/2) * r * cos(pi/6)* [cos(pi/6);sin(pi/6)];
 p_true = p;
 P = r * [0,3/2,0,-3/2,-3/2,0,3/2;0,sqrt(3)/2,sqrt(3),...
 sqrt(3)/2,-sqrt(3)/2,-sqrt(3),-sqrt(3)/2].';
 P_diff = P - repmat(p',B,1);
 phi = atan2(P_diff(:,2),P_diff(:,1));
 Phi = [cos(phi).';sin(phi).'];
 Phi_tilde = Phi(:,1:M);
 Phi_bar = Phi(:,M+1:B);
 n_scenario = 0;
 for scenario = 1 %scenario_range
 n_scenario = n_scenario + 1;
```

In Figure 18.5, the RMSE is shown as a function of the cell radius, which depends on the distance between the MS and the BS. It can be seen that the use of more sophisticated estimators can achieve higher accuracy at the expense of more computation.

18.6 CONCLUSIONS

The estimation of the mobile position from the TOA has been considered for the NLOS geolocation exploiting the path attenuation. The use of only the LOS TOAs offers the same performance as that given by using both the LOS and the NLOS

TOAs. The LS, WLS, and ML estimators are adopted to estimate the mobile position. There is a noticeable gap between the error performance of the LS estimator and the CRB, whereas the WLS estimator and the ML estimator are statistically efficient when the SNR is high and the effective bandwidth is large. Even though the TOA is estimated by the ML estimator at first, the direct LS fit for determining the mobile position, in general, is suboptimal. Rather, the WLS estimator and the sophisticated ML estimator can achieve the optimal performance.

APPENDIX

Detailed Derivation of (18.34)

Pure LOS Component: Taking the first-order Taylor series around the true value $\overline{\tau}$ (see, e.g., [80], p. 66), we have

$$
\begin{aligned}
\hat{\overline{\tau}} &= \overline{\tau}(\mathbf{p}) + \left(\frac{\partial}{\partial \mathbf{p}} \overline{\tau}^{\mathrm{T}}(\mathbf{p}) \right)^{\mathrm{T}} (\hat{\mathbf{p}} - \mathbf{p}) + o\left(\|\hat{\mathbf{p}} - \mathbf{p}\|_{\mathrm{E}} \right) \\
&\simeq \overline{\tau}(\mathbf{p}) + \frac{1}{c} \overline{\Phi}^{\mathrm{T}} (\hat{\mathbf{p}} - \mathbf{p}).
\end{aligned}
\tag{18.55}
$$

From (18.55), the pseudoinverse results in

$$
\hat{\mathbf{p}} - \mathbf{p} \simeq c \left(\overline{\Phi} \overline{\Phi}^{\mathrm{T}} \right)^{-1} \overline{\Phi} \left(\hat{\overline{\tau}} - \overline{\tau}(\mathbf{p}) \right).
\tag{18.56}
$$

Taking the expectation of the error square matrix $(\hat{\mathbf{p}} - \mathbf{p})(\hat{\mathbf{p}} - \mathbf{p})^{\mathrm{T}}$ from (18.56) with respect to $\hat{\overline{\tau}}$, we have

$$
\begin{aligned}
&\mathrm{E}_{\hat{\overline{\tau}}} \left\{ (\hat{\mathbf{p}} - \mathbf{p})(\hat{\mathbf{p}} - \mathbf{p})^{\mathrm{T}} \right\} \\
&= c^2 \left(\overline{\Phi} \overline{\Phi}^{\mathrm{T}} \right)^{-1} \overline{\Phi} \mathrm{E}_{\hat{\overline{\tau}}} \left\{ \left(\hat{\overline{\tau}} - \overline{\tau}(\mathbf{p}) \right) \left(\hat{\overline{\tau}} - \overline{\tau}(\mathbf{p}) \right)^{\mathrm{T}} \right\} \overline{\Phi}^{\mathrm{T}} \left(\overline{\Phi} \overline{\Phi}^{\mathrm{T}} \right)^{-1} \\
&= c^2 \left(\overline{\Phi} \overline{\Phi}^{\mathrm{T}} \right)^{-1} \overline{\Phi} \mathbf{D}(\overline{\sigma}^2) \overline{\Phi}^{\mathrm{T}} \left(\overline{\Phi} \overline{\Phi}^{\mathrm{T}} \right)^{-1}.
\end{aligned}
\tag{18.57}
$$

Both LOS and NLOS Components: Let us introduce $\boldsymbol{\eta} \in \mathbb{R}^{(M+2) \times 1}$ as

$$
\boldsymbol{\eta} = \begin{bmatrix} \mathbf{p}^{\mathrm{T}} & \mathbf{l}^{\mathrm{T}} \end{bmatrix}^{\mathrm{T}}.
\tag{18.58}
$$

Next, we consider the analysis of $\tau(\boldsymbol{\eta})$ in such a way that

$$
\begin{aligned}
\hat{\tau} &= \tau(\boldsymbol{\eta}) + \left(\frac{\partial}{\partial \boldsymbol{\eta}} \tau^{\mathrm{T}}(\boldsymbol{\eta}) \right)^{\mathrm{T}} (\hat{\boldsymbol{\eta}} - \boldsymbol{\eta}) + o\left(\|\hat{\boldsymbol{\eta}} - \boldsymbol{\eta}\|_{\mathrm{E}} \right) \\
&\simeq \tau(\boldsymbol{\eta}) + \nabla_{\boldsymbol{\eta}\tau}^{\mathrm{T}} (\hat{\boldsymbol{\eta}} - \boldsymbol{\eta}),
\end{aligned}
\tag{18.59}
$$

where the Jacobian matrix $\nabla_{\boldsymbol{\eta}\tau} \in \mathbb{R}^{(M+2) \times B}$ is given by

$$
\nabla_{\boldsymbol{\eta}\tau} = \frac{1}{c} \begin{bmatrix} \tilde{\Phi} & \overline{\Phi} \\ \mathbf{I} & \mathbf{O} \end{bmatrix},
\tag{18.60}
$$

with $\tilde{\Phi} \in \mathbb{R}^{2 \times M}$ given by

$$\tilde{\Phi} = \begin{bmatrix} \cos(\phi_1) & \cos(\phi_2) & \cdots & \cos(\phi_M) \\ \sin(\phi_1) & \sin(\phi_2) & \cdots & \sin(\phi_M) \end{bmatrix}. \qquad (18.61)$$

From the first-order Taylor series expansion in (18.59), we have

$$
\begin{aligned}
\hat{\eta} - \eta &\approx \left(\nabla_{\eta\tau} \nabla_{\eta\tau}^{\mathrm{T}} \right)^{-1} \nabla_{\eta\tau} (\hat{\tau} - \tau) \\
&= \left(\frac{1}{c^2} \begin{bmatrix} \tilde{\Phi}\tilde{\Phi}^{\mathrm{T}} + \bar{\Phi}\bar{\Phi}^{\mathrm{T}} & \tilde{\Phi} \\ \tilde{\Phi}^{\mathrm{T}} & \mathbf{I} \end{bmatrix} \right)^{-1} \frac{1}{c} \begin{bmatrix} \tilde{\Phi} & \bar{\Phi} \\ \mathbf{I} & \mathbf{O} \end{bmatrix} (\hat{\tau} - \tau) \\
&= c \begin{bmatrix} \left(\bar{\Phi}\bar{\Phi}^{\mathrm{T}}\right)^{-1} & -\left(\bar{\Phi}\bar{\Phi}^{\mathrm{T}}\right)^{-1} \tilde{\Phi} \\ -\tilde{\Phi}^{\mathrm{T}}\left(\bar{\Phi}\bar{\Phi}^{\mathrm{T}}\right)^{-1} & \mathbf{I} + \tilde{\Phi}^{\mathrm{T}}\left(\bar{\Phi}\bar{\Phi}^{\mathrm{T}}\right)^{-1} \tilde{\Phi} \end{bmatrix} \begin{bmatrix} \tilde{\Phi} & \bar{\Phi} \\ \mathbf{I} & \mathbf{O} \end{bmatrix} (\hat{\tau} - \tau) \quad (18.62) \\
&= c \begin{bmatrix} \mathbf{O} & \left(\bar{\Phi}\bar{\Phi}^{\mathrm{T}}\right)^{-1} \bar{\Phi} \\ \mathbf{I} & -\tilde{\Phi}^{\mathrm{T}}\left(\bar{\Phi}\bar{\Phi}^{\mathrm{T}}\right)^{-1} \bar{\Phi} \end{bmatrix} (\hat{\tau} - \tau),
\end{aligned}
$$

which provides the error variance of $\hat{\eta}$ as

$$
\begin{aligned}
& \mathrm{E}_{\hat{\tau}} \left\{ (\hat{\eta} - \eta)(\hat{\eta} - \eta)^{\mathrm{T}} \right\} \\
&\approx c^2 \begin{bmatrix} \mathbf{O} & \left(\bar{\Phi}\bar{\Phi}^{\mathrm{T}}\right)^{-1}\bar{\Phi} \\ \mathbf{I} & -\tilde{\Phi}\left(\bar{\Phi}\bar{\Phi}^{\mathrm{T}}\right)^{-1}\bar{\Phi} \end{bmatrix} \begin{bmatrix} \mathbf{D}(\tilde{\sigma}^2) & \mathbf{O} \\ \mathbf{O} & \mathbf{D}(\bar{\sigma}^2) \end{bmatrix} \begin{bmatrix} \mathbf{O} & \mathbf{I} \\ \bar{\Phi}^{\mathrm{T}}\left(\bar{\Phi}\bar{\Phi}^{\mathrm{T}}\right)^{-1} & -\bar{\Phi}^{\mathrm{T}}\left(\bar{\Phi}\bar{\Phi}^{\mathrm{T}}\right)^{-1}\tilde{\Phi}^{\mathrm{T}} \end{bmatrix} \\
&= c^2 \begin{bmatrix} \mathbf{O} & \left(\bar{\Phi}\bar{\Phi}^{\mathrm{T}}\right)^{-1}\bar{\Phi} \\ \mathbf{I} & -\tilde{\Phi}\left(\bar{\Phi}\bar{\Phi}^{\mathrm{T}}\right)^{-1}\bar{\Phi} \end{bmatrix} \begin{bmatrix} \mathbf{O} & \mathbf{D}(\tilde{\sigma}^2) \\ \mathbf{D}(\bar{\sigma}^2)\bar{\Phi}^{\mathrm{T}}\left(\bar{\Phi}\bar{\Phi}^{\mathrm{T}}\right)^{-1} & -\mathbf{D}(\bar{\sigma}^2)\bar{\Phi}^{\mathrm{T}}\left(\bar{\Phi}\bar{\Phi}^{\mathrm{T}}\right)^{-1}\tilde{\Phi}^{\mathrm{T}} \end{bmatrix} \\
&= c^2 \begin{bmatrix} \left(\bar{\Phi}\bar{\Phi}^{\mathrm{T}}\right)^{-1}\bar{\Phi}\mathbf{D}(\bar{\sigma}^2)\bar{\Phi}^{\mathrm{T}}\left(\bar{\Phi}\bar{\Phi}^{\mathrm{T}}\right)^{-1} & -\left(\bar{\Phi}\bar{\Phi}^{\mathrm{T}}\right)^{-1}\bar{\Phi}\mathbf{D}(\bar{\sigma}^2)\bar{\Phi}^{\mathrm{T}}\left(\bar{\Phi}\bar{\Phi}^{\mathrm{T}}\right)^{-1}\tilde{\Phi}^{\mathrm{T}} \\ -\tilde{\Phi}\left(\bar{\Phi}\bar{\Phi}^{\mathrm{T}}\right)^{-1}\bar{\Phi}\mathbf{D}(\bar{\sigma}^2)\bar{\Phi}^{\mathrm{T}}\left(\bar{\Phi}\bar{\Phi}^{\mathrm{T}}\right)^{-1} & \mathbf{D}(\tilde{\sigma}^2) + \tilde{\Phi}\left(\bar{\Phi}\bar{\Phi}^{\mathrm{T}}\right)^{-1}\bar{\Phi}\mathbf{D}(\bar{\sigma}^2)\bar{\Phi}^{\mathrm{T}}\left(\bar{\Phi}\bar{\Phi}^{\mathrm{T}}\right)^{-1}\tilde{\Phi}^{\mathrm{T}} \end{bmatrix}.
\end{aligned}
$$

$$(18.63)$$

Therefore, we can see that

$$\mathrm{E}_{\hat{\tau}} \left\{ (\hat{\mathbf{p}}_\tau - \mathbf{p})(\hat{\mathbf{p}}_\tau - \mathbf{p})^{\mathrm{T}} \right\} \approx c^2 \left(\bar{\Phi}\bar{\Phi}^{\mathrm{T}}\right)^{-1} \bar{\Phi}\mathbf{D}(\bar{\sigma}^2)\bar{\Phi}^{\mathrm{T}} \left(\bar{\Phi}\bar{\Phi}^{\mathrm{T}}\right)^{-1}. \qquad (18.64)$$

For the mobile position estimation based on the time delay parameterization, the use of the LOS TOAs is sufficient to achieve the same performance as that obtained by the use of both the LOS and the NLOS TOAs.

Detailed Derivation of (18.35)

The LS between the received data and the theoretical model is given by $f_{LS}(\theta) = \left\|\hat{\tau} - \tau(\theta)\right\|_E^2$. From the Euclidean norm, we can write

$$
\begin{aligned}
\hat{\theta}_{LS} &= \arg\min_\theta f_{LS}(\theta) \\
&= \arg\min_\theta \sum_{b=1}^{B} \left(\hat{\tau}_b - \tau_b\right)^2 .
\end{aligned}
\tag{18.65}
$$

The LS estimator for $\mathbf{p}$ and $\mathbf{l}$ needs to employ both the NLOS and the LOS portions, because the additional NLOS path length $\mathbf{l}$ is inherent in the NLOS portion. Next, we shall try to get rid of the undesired parameter $\mathbf{l}$. To concentrate on $\mathbf{p}$, the objective function $f_{LS}(\theta)$ is differentiated with respect to $\mathbf{l}$ as

$$
\begin{aligned}
\frac{\partial}{\partial \mathbf{l}}\left\|\hat{\tau} - \frac{1}{c}\mathbf{D}(\theta)\right\|_E^2 &= 2\left(\frac{\partial}{\partial \mathbf{l}}\left(\hat{\tau} - \frac{1}{c}\mathbf{D}(\theta)\right)^{\mathrm{T}}\right)\left(\hat{\tau} - \frac{1}{c}\mathbf{D}(\theta)\right) \\
&= -\frac{1}{c}2\left(\frac{\partial}{\partial \mathbf{l}}\tilde{\mathbf{d}}^{\mathrm{T}}(\theta)\right)\left(\tilde{\hat{\tau}} - \frac{1}{c}\tilde{\mathbf{d}}(\theta)\right) \\
&= -\frac{1}{c}2\left(\frac{1}{c}\tilde{\mathbf{d}}(\theta) - \tilde{\hat{\tau}}\right) \\
&= \frac{1}{c}2\left(\frac{1}{c}\left(\tilde{\mathbf{d}}_0(\mathbf{p}) + 1\right) - \tilde{\hat{\tau}}\right),
\end{aligned}
\tag{18.66}
$$

where $\tilde{\mathbf{d}}_0(\mathbf{p})$ is the vector whose mth element ($m \in \{1, 2, \ldots, M\}$) is given by $\sqrt{\tilde{x}_m^2 + \tilde{y}_m^2}$, and $\tilde{\hat{\tau}}$ is the vector whose elements are drawn from the first M-elements of $\hat{\tau}$. Forcing $\partial/\partial \mathbf{l}\left\|\hat{\tau} - (1/c)\mathbf{D}(\theta)\right\|_E^2\Big|_{l=\hat{l}_{LS}} = \mathbf{0}$, we obtain

$$
\hat{\mathbf{l}}_{LS} = c\tilde{\hat{\tau}} - \tilde{\mathbf{d}}_0(\mathbf{p}).
\tag{18.67}
$$

Substituting (18.67) into (18.65), we obtain (18.35).

Detailed Derivation of (18.36)

Let us consider the loss function based on the WLS:

$$
\hat{\theta}_{WLS} = \arg\min_\theta \left\|\hat{\tau} - \tau(\theta)\right\|_W^2,
\tag{18.68}
$$

where $\|\mathbf{v}\|_W^2 = \mathbf{v}^H W \mathbf{v}$ is the weighted Euclidean norm with an unknown positive-definite Hermitian weighting matrix $\mathbf{W} \in \mathbb{C}^{B \times B}$. Let us partition the weighting matrix into

$$
\mathbf{W} = \begin{bmatrix} \tilde{\mathbf{W}} & \breve{\mathbf{W}} \\ \breve{\mathbf{W}}^{\mathrm{H}} & \bar{\mathbf{W}} \end{bmatrix},
\tag{18.69}
$$

where $\tilde{\mathbf{W}} \in \mathbb{C}^{M \times M}$, $\check{\mathbf{W}} \in \mathbb{C}^{M \times (B-M)}$, and $\bar{\mathbf{W}} \in \mathbb{C}^{(B-M) \times (B-M)}$ are sub-block matrices. In general, the weighting matrix, which provides the minimum variance of the estimation error, is the inverse of the mean dispersion error given by [81, 82]:

$$
\begin{aligned}
\mathbf{W} &= \left(\mathrm{E}_{\hat{\mathbf{\tau}}} \left\{ (\hat{\mathbf{\tau}} - \mathbf{\tau})(\hat{\mathbf{\tau}} - \mathbf{\tau})^{\mathrm{T}} \right\} \right)^{-1} \\
&= \mathbf{D}^{-1}(\sigma^2),
\end{aligned}
\tag{18.70}
$$

where $\sigma^2 \in \mathbb{R}^{B \times 1}$ is the vector of error variance given by

$$
\sigma^2 = \begin{bmatrix} \sigma_1^2 & \sigma_2^2 & \cdots & \sigma_B^2 \end{bmatrix}^{\mathrm{T}}.
\tag{18.71}
$$

Since the weighting matrix depends on the unknown parameters $\mathbf{l}$ and $\mathbf{a}$, it is difficult to perform the optimal WLS. Without loss of asymptotic performance, we neglect the dependence on $\mathbf{l}$ in $\mathbf{W}$. By differentiating the WLS function with respect to $\mathbf{l}$, the gradient yields

$$
\begin{aligned}
\frac{\partial}{\partial \mathbf{l}} \| \hat{\mathbf{\tau}} - \mathbf{\tau}(\mathbf{\theta}) \|_{\mathbf{W}}^2 &= 2 \left(\frac{\partial}{\partial \mathbf{l}} \left(\hat{\mathbf{\tau}} - \tilde{\mathbf{\tau}}(\mathbf{\theta}) \right)^{\mathrm{T}} \right) \tilde{\mathbf{W}} \left(\hat{\mathbf{\tau}} - \tilde{\mathbf{\tau}}(\mathbf{\theta}) \right) \\
&\quad + 2 \left(\frac{\partial}{\partial \mathbf{l}} \left(\hat{\mathbf{\tau}} - \tilde{\mathbf{\tau}}(\mathbf{\theta}) \right)^{\mathrm{T}} \right) \check{\mathbf{W}} \left(\hat{\mathbf{\tau}} - \bar{\mathbf{\tau}}(\mathbf{p}) \right) \\
&= -2 \tilde{\mathbf{W}} \left(\hat{\mathbf{\tau}} - \tilde{\mathbf{\tau}}(\mathbf{\theta}) \right) - 2 \check{\mathbf{W}} \left(\hat{\mathbf{\tau}} - \bar{\mathbf{\tau}}(\mathbf{p}) \right).
\end{aligned}
\tag{18.72}
$$

If we assign $\partial / \partial \mathbf{l} \| \hat{\mathbf{\tau}} - \mathbf{\tau}(\mathbf{\theta}) \|_{\mathbf{W}}^2 |_{l = \hat{l}_{LS}} = \mathbf{0}$, an estimate of the mobile position $\mathbf{p}$ should make $\hat{\mathbf{\tau}} - \bar{\mathbf{\tau}}(\mathbf{p})$ close to zero and provides

$$
\hat{\mathbf{l}}_{\mathrm{WLS}} = c \hat{\mathbf{\tau}} - \tilde{\mathbf{d}}_0(\mathbf{p}).
\tag{18.73}
$$

Substituting (18.73) into $\| \hat{\mathbf{\tau}} - \mathbf{\tau}(\mathbf{\theta}) \|_{\mathbf{W}}^2$, there exists such an estimate from (18.36).

Detailed Derivation of (18.38) and (18.39)

Let a generalized LS (GLS) be

$$
f_{\mathrm{GLS}}(\mathbf{p}) = \| \hat{\bar{\mathbf{\tau}}} - \bar{\mathbf{\tau}} \|_{\bar{\mathbf{W}}}^2.
\tag{18.74}
$$

The first derivative of the GLS is given by

$$
\frac{\partial}{\partial \mathbf{p}} f_{\mathrm{GLS}}(\mathbf{p}) = -\frac{1}{c} 2 \bar{\mathbf{\Phi}} \bar{\mathbf{W}} \left(\hat{\bar{\mathbf{\tau}}} - \bar{\mathbf{\tau}} \right).
\tag{18.75}
$$

Let the asymptotic Hessian matrix of $f_{\mathrm{GLS}}(\mathbf{p})$ be

$$
\begin{aligned}
\bar{\mathbf{H}}_{\mathbf{pp}} &= \mathrm{E}_{n(t)} \left\{ \frac{\partial^2}{\partial \mathbf{p} \partial \mathbf{p}^{\mathrm{T}}} f_{\mathrm{GLS}}(\mathbf{p}) \right\} \\
&= \frac{1}{c^2} 2 \bar{\mathbf{\Phi}} \bar{\mathbf{W}} \bar{\mathbf{\Phi}}^{\mathrm{T}}.
\end{aligned}
\tag{18.76}
$$

Let $\breve{\mathbf{p}}$ lie between the estimated value $\hat{\mathbf{p}}_{GLS}$ and the true value $\mathbf{p}$. The Taylor series can be expanded into

$$\frac{\partial}{\partial \mathbf{p}} f_{GLS}(\mathbf{p})\Big|_{\mathbf{p}=\hat{\mathbf{p}}_{GLS}} = \frac{\partial}{\partial \mathbf{p}} f_{GLS}(\mathbf{p})\Big|_{\mathbf{p}=\mathbf{p}_0} + \frac{\partial^2}{\partial \mathbf{p}\, \partial \mathbf{p}^T} f_{GLS}(\mathbf{p})\Big|_{\mathbf{p}=\breve{\mathbf{p}}} \left(\hat{\mathbf{p}}_{GLS} - \mathbf{p}\right). \quad (18.77)$$

Due to the critical point of the optimization, the gradient at the estimated value $\hat{\mathbf{p}}_{GLS}$ remains zero; that is, $(\partial/\partial\mathbf{p}) f_{GLS}(\mathbf{p})|_{\mathbf{p}=\hat{\mathbf{p}}_{GLS}} = \mathbf{0}$. Since $(\partial^2/\partial\mathbf{p}\,\partial\mathbf{p}^T) f_{GLS}(\mathbf{p})|_{\mathbf{p}=\breve{\mathbf{p}}}$ converges to $\bar{\mathbf{H}}_{\mathbf{pp}}$ with probability 1, we have

$$\hat{\mathbf{p}}_{GLS} - \mathbf{p} \simeq -\bar{\mathbf{H}}_{\mathbf{pp}}^{-1}\left(\frac{\partial}{\partial \mathbf{p}} f_{GLS}(\mathbf{p})\Big|_{\mathbf{p}=\hat{\mathbf{p}}_0}\right)$$

$$= c\left(\bar{\boldsymbol{\Phi}}\bar{\mathbf{W}}\bar{\boldsymbol{\Phi}}^T\right)^{-1}\bar{\boldsymbol{\Phi}}\bar{\mathbf{W}}\left(\hat{\bar{\boldsymbol{\tau}}} - \bar{\boldsymbol{\tau}}\right). \quad (18.78)$$

Substituting $\bar{\mathbf{W}} = \mathbf{I}$ for the LS and $\bar{\mathbf{W}} = \mathbf{D}^{-1}(\bar{\boldsymbol{\sigma}}^2)$ for the WLS into (18.78), we obtain (18.38) and (18.39), respectively.

Detailed Derivation of (18.43)

Let the desired signal vector $\mathbf{s}_a(t; \tau)$ be

$$\mathbf{s}_a(t; \tau) = \mathbf{a} \odot \mathbf{s}(t; \tau), \quad (18.79)$$

where $\odot$ is the Schur–Hadamard or element-wise product. The FIM can be written as

$$\mathbf{H}_{\tau\tau} = \frac{1}{\sigma_n^2} 2\int_0^{T_0} \Re\left(\left(\frac{\partial}{\partial \tau}\mathbf{s}_a^T(t; \tau)\right)\left(\frac{\partial}{\partial \tau}\mathbf{s}_a^T(t; \tau)\right)^H\right) dt, \quad (18.80)$$

where $(\cdot)^H$ is the Hermitian transpose. Consider the derivative

$$\frac{\partial}{\partial \tau}\mathbf{s}_a^T(t; \tau) = \nabla_{\tau s}\mathbf{D}(\mathbf{a}) + \left(\frac{\partial}{\partial \tau}\mathbf{a}^T\right)\mathbf{D}\left(\mathbf{s}(t; \tau)\right)$$

$$= \nabla_{\tau s}\mathbf{D}(\mathbf{a}) - \frac{1}{2}\mathbf{D}\left(\mathbf{a}\odot\boldsymbol{\gamma}\odot\mathbf{s}(t; \tau)\right)\mathbf{D}^{-1}(\tau), \quad (18.81)$$

where $\nabla_{\tau s} \in \mathbb{C}^{B\times B}$ is the Jacobian matrix defined as

$$\nabla_{\tau s} = \frac{\partial}{\partial \tau}\mathbf{s}^T(t; \tau). \quad (18.82)$$

By using $\int_0^{T_0} \nabla_{\tau s}\mathbf{D}\left(\mathbf{s}*(t; \tau)\right) dt = \mathbf{O}$, the FIM yields

$$\mathbf{H}_{\tau\tau} = \frac{1}{\sigma_n^2} 2\int_0^{T_0} \Re\Bigg(\nabla_{\tau s}\mathbf{D}^2(\mathbf{a})\nabla_{\tau s}^H - \frac{1}{2}\nabla_{\tau s}\mathbf{D}\left(\mathbf{s}*(t; \tau)\right)\mathbf{D}^2(\mathbf{a})\mathbf{D}^{-1}(\tau)\mathbf{D}(\boldsymbol{\gamma})$$

$$-\frac{1}{2}\mathbf{D}(\boldsymbol{\gamma})\mathbf{D}^2(\mathbf{a})\mathbf{D}^{-1}(\tau)\mathbf{D}\left(\mathbf{s}(t; \tau)\right)\nabla_{\tau s}^H$$

$$+\frac{1}{4}\mathbf{D}^2(\mathbf{a}\odot\boldsymbol{\gamma})\mathbf{D}^{-2}(\tau)\mathbf{D}\left(\mathbf{s}(t; \tau)\odot\mathbf{s}*(t; \tau)\right)\Bigg) dt \quad (18.83)$$

$$= \frac{1}{\sigma_n^2} 2\int_0^{T_0} \Re\left(\nabla_{\tau s}\mathbf{D}^2(\mathbf{a})\nabla_{\tau s}^H\right) dt + \frac{1}{2}\frac{E_s}{\sigma_n^2}\mathbf{D}^2(\mathbf{a}\odot\boldsymbol{\gamma})\mathbf{D}^{-2}(\tau).$$

Let us consider

$$\int_0^{T_o} \nabla_{\tau s} \mathbf{D}^2(\mathbf{a}) \nabla_{\tau s}^H dt = \sum_{b=1}^{B} a_b^2 \int_0^{T_o} \left(\frac{\partial}{\partial \tau} s(t - \tau_b) \right) \left(\frac{\partial}{\partial \tau} s(t - \tau_b) \right)^H dt. \quad (18.84)$$

The (b_1, b_2)-th element of $\int_0^{T_o} \left((\partial/\partial\tau) s(t - \tau_b) \right) \left((\partial/\partial\tau) s(t - \tau_b) \right)^H dt$ is given by

$$\left[\int_0^{T_o} \left(\frac{\partial}{\partial \tau} s(t - \tau_b) \right) \left(\frac{\partial}{\partial \tau} s(t - \tau_b) \right)^H dt \right]_{[b_1,b_2]}$$

$$= \int_0^{T_o} \left| \frac{\partial}{\partial \tau_b} s(t - \tau_b) \right|^2 dt \delta_{b,b_1} \delta_{b_1,b_2}$$

$$= \int_{-\tau_b}^{T_o - \tau_b} \left| \left(\frac{\partial}{\partial t} t' \right) \frac{\partial}{\partial t'} s(t') \right|^2 dt' \delta_{b,b_1} \delta_{b_1,b_2}; t' = t - \tau_b \quad (18.85)$$

$$= \int_{-\infty}^{\infty} |(j2\pi f) S(f)|^2 df \delta_{b_1,b_2}$$

$$= 4\pi^2 \bar{\beta}^2 E_2 \delta_{b_1,b_2},$$

where $j = \sqrt{-1}$ is the unit imaginary number, and $\delta_{.,.}$ is the Kronecker delta function. Therefore, it provides

$$\int_0^{T_o} \nabla_{\tau s} \mathbf{D}^2(\mathbf{a}) \nabla_{\tau s}^H dt = 4\pi^2 \bar{\beta}^2 E_s \mathbf{D}^2(\mathbf{a}). \quad (18.86)$$

Substituting (18.86) into (18.83), we obtain (18.43).

REFERENCES

[1] T. S. Rappaport, J. H. Reed, and B. D. Woerner, "Position location using wireless communications on highways of the future," *IEEE Commun. Mag.*, vol. 34, no. 10, pp. 33–41, 1996.
[2] J. Werb and C. Lanzl, "Designing a positioning system for finding things and people indoors," *IEEE Spectr.*, vol. 35, no. 9, pp. 71–78, 1998.
[3] J. J. Caffery and G. L. Stuber, "Overview of radiolocation in CDMA cellular systems," *IEEE Commun. Mag.*, vol. 36, no. 4, pp. 38–45, 1998.
[4] J. H. Reed, K. J. Krizman, B. D. Woerner, and T. S. Rappaport, "An overview of the challenges and progress in meeting the E-911 requirement for location service," *IEEE Commun. Mag.*, vol. 36, no. 4, pp. 30–37, 1998.
[5] K. Pahlavan, P. Krishnamurthy, and J. Beneat, "Wideband radio propagation modeling for indoor geolocation applications," *IEEE Commun. Mag.*, vol. 36, no. 4, pp. 60–65, 1998.
[6] K. Pahlavan, X. Li, and J.-P. Makela, "Indoor geolocation science and technology," *IEEE Commun. Mag.*, vol. 40, no. 2, pp. 112–118, 2002.
[7] S. S. Soliman and C. E. Wheatley, "Geolocation technologies and applications for third generation wireless," *Wireless Commun. Mobile Comput.*, vol. 2, no. 3, pp. 229–251, 2002.
[8] R. Subrata and A. Y. Zormaya, "Evolving cellular automata for location management in mobile computing networks," *IEEE Trans. Parallel Distrib. Syst.*, vol. 14, no. 1, pp. 13–26, 2003.
[9] B. Hofmann-Wellenhof, H. Lichtenegger, and J. Collins, *Global Positioning System: Theory and Practice*, 4th ed. New York: Springer, 1997.

[10] J. Caffery, *Wireless Location in CDMA Cellular Radio System.* Norwell, MA: Kluwer Academic Publisher, 1999.

[11] K. Pahlavan, X. Li, and J.-P. Mäkelä, "Indoor geolocation science and technology," *IEEE Commun. Mag.*, vol. 40, no. 2, pp. 112–118, 2002.

[12] C. Savarese, J. M. Rabaey, and J. Beutel, "Location in distributed ad-hoc wireless sensor networks," in *Proc. of the IEEE Int. Conf. on Acoustics, Speech and Signal Processing 2001 (ICASSP 2001)*, vol. 4, Salt Lake City, UT, May 2001, pp. 2037–2040.

[13] A. Savvides, C.-C. Han, and M. B. Srivastava, "Dynamic fine-grained localization in ad hoc networks of sensors," in *Proc. of the 7th Int. Conf. on Mobile Computing and Networking (ACM MOBICOM 2001)*, Rome, Italy, Jul. 2001, pp. 166–179.

[14] F. Gustafsson and F. Gunnarsson, "Mobile positioning using wireless networks," *IEEE Signal Process. Mag.*, vol. 22, no. 4, pp. 41–53, 2005.

[15] G. L. Stuber and J. Caffery, Jr., "Radiolocation techniques," in *The Mobile Communications Handbook*, 2nd ed., J. Gibson, Ed. Boca Raton, FL: CRC Press, 2000, ch. 24.

[16] R. Roberts, "Ranging subcommittee final report," Harris Corporation, Tech. Rep., Nov. 2004. Available: http://grouper.ieee.org/groups/802/15/pub/04/15-04-0581-07-004a-ranging-subcommittee-final-report.doc

[17] S. Gezici, Z. Tian, G. B. Giannakis, H. Kobayashi, A. F. Molisch, H. V. Poor, and Z. Sahinoglu, "Localization via ultra-wideband radios: A look at positioning aspects for future sensor networks," *IEEE Signal Process. Mag.*, vol. 22, no. 4, pp. 70–84, 2005.

[18] J. J. Caffery, Jr. and G. L. Stüber, "Overview of radiolocation in CDMA cellular systems," *IEEE Commun. Mag.*, vol. 36, no. 4, pp. 38–45, 1998.

[19] M. P. Wylie-Green and S. S. Wang, "Robust range estimation in the presence of the nonline-of-sight error," in *Proc. of the 54th IEEE Vehicular Technology Conf. 2001 (VTC2001 Fall)*, vol. 1, Atlantic City, NJ, Sept. 2001, pp. 101–105.

[20] S. Al-Jazzar, J. J. Caffery, Jr., and H.-R. You, "A scattering model based approach to NLOS mitigation in TOA location systems," in *Proc. of the 55th IEEE Vehicular Technology Conf. 2002 (VTC2002 Spring)*, vol. 2, Birmingham, AL, May 2002, pp. 861–865.

[21] J. Riba and A. Urruela, "A robust multipath mitigation technique for time-of-arrival estimation," in *Proc. of the 56th IEEE Vehicular Technology Conf. 2002 (VTC2002 Fall)*, vol. 4, Vancouver, Canada, Sep. 2002, pp. 2663–2267.

[22] N. Patwari, A. O. Hero, III, M. Perkins, N. S. Correal, and R. J. O'Dea, "Relative location estimation in wireless sensor networks," *IEEE Trans. Signal Process.*, vol. 51, no. 8, pp. 2137–2148, 2003.

[23] J. Riba and A. Urruela, "A non-line-of-sight mitigation technique based on ML-detection," in *Proc. of the IEEE Int. Conf. on Acoustics, Speech and Signal Processing 2004 (ICASSP 2004)*, vol. 2, Montreal, QC, May 2004, pp. 153–156.

[24] L. Cong and W. Zhuang, "Non-line-of-sight error mitigation in mobile location," *IEEE Trans. Wireless Commun.*, vol. 4, no. 2, pp. 560–573, 2005.

[25] J.-F. Liao and B.-S. Chen, "Robust mobile location estimator with NLOS mitigation using interacting multiple model algorithm," *IEEE Trans. Wireless Commun.*, vol. 5, no. 11, pp. 3002–3006, 2006.

[26] C. Ma, R. Klukas, and G. Lachapelle, "A nonline-of-sight error-mitigation method for TOA measurements," *IEEE Trans. Veh. Technol.*, vol. 56, no. 2, pp. 641–651, 2007.

[27] S. Venkatesh and R. M. Buehrer, "NLOS mitigation using linear programming in ultrawideband location-aware networks," *IEEE Trans. Veh. Technol.*, vol. 56, no. 5, pp. 3182–3198, 2007.

[28] M. P. Wylie and J. Holtzman, "The non-line of sight problem in mobile location estimation," in *Proc. of the IEEE Int. Conf. on Universal Personal Communications 1996 (ICUPC'96)*, vol. 2, Cambridge, MA, Sep. 1996, pp. 827–831.

[29] J. Borràs, P. Hatrack, and N. B. Manclayam, "Decision theoretic frame-work for NLOS identification," in *Proc. of the 48th IEEE Vehicular Technology Conf. 1998 (VTC'98)*, vol. 2, Ottawa, Canada, May 1998, pp. 1583–1587.

[30] S. Venkatraman and J. Caffery, Jr., "A statistical approach to non-line-of-sight BS identification," in *Proc. of the Int. Symp. on Wireless Personal Multimedia Communications 2002 (WPMC 2002)*, vol. 1, Honolulu, HI, Oct. 2002, pp. 296–300.

[31] S. Gezici, H. Kobayashi, and H. V. Poor, "Non-parametric non-line-of-sight identification," in *Proc. of the 58th IEEE Vehicular Technology Conf. 2003 (VTC2003 Fall)*, vol. 4, Orlando, FL, Oct. 2003, pp. 2544–2548.

[32] A. Lakhzouri, E. S. Lohan, R. Hamila, and M. Renfors, "Extended Kalman filter channel estimation for line-of-sight detection in WCDMA mobile positioning," *EURASIP J. Appl. Signal Process.*, vol. 13, pp. 1268–1278, 2003.

[33] S. Al-Jazzar and J. Caffery, Jr., "New algorithms for NLOS identification," in *Proc. of the IST Mobile & Wireless Communications Summit 2005 (IST Mobile Summit 2005)*, Dresden, Germany, Jun. 2005.

[34] J. Schroeder, S. Galler, K. Kyamakya, and K. Jobmann, "NLOS detection algorithms for ultra-wideband localization," in *Proc. of the Workshop on Positioning, Navigation and Communication 2007 (WPNC'07)*, Hanover, Germany, Mar. 2007, pp. 159–166.

[35] W.-K. Chao and K.-T. Lay, "NLOS measurement identification for mobile positioning in wireless cellular systems," in *Proc. of the 66th IEEE Vehicular Technology Conf. 2007 (VTC2007 Fall)*, vol. 4, Baltimore, MD, Sep./Oct. 2007, pp. 1965–1969.

[36] G. Destino, D. Macagnano, and G. T. F. de Abreu, "Hypothesis testing and iterative WLS minimization for WSN localization under LOS/NLOS conditions," in *Proc. of the Asilomar Conf. on Signals, Systems and Computers 2007 (ACSSC 2007)*, Pacific Grove, CA, Nov. 2007, pp. 2150–2155.

[37] S. Venkatesh and R. Buehrer, "Non-line-of-sight identification in ultrawideband systems based on received signal statistics," *IET Microwaves Antennas Propag.*, vol. 1, no. 6, pp. 1120–1130, 2007.

[38] İ. Güvenç, C.-C. Chong, F. Watanabe, and H. Inamura, "NLOS identification and weighted least-squares localization for UWB systems using multipath channel statistics," *EURASIP J. Adv. Signal Process.*, vol. 2008, pp. 1–14, 2008.

[39] S. Yarkan and H. Arslan, "Identification of LOS in time-varying, frequency selective radio channels," *EURASIP J. Wireless Commun. Netw.*, vol. 2008, pp. 1–14, 2008.

[40] K. Yu and Y. J. Guo, "Statistical NLOS identification based on AOA, TOA and signal strength," *IEEE Trans. Veh. Technol.*, vol. 58, no. 1, pp. 274–286, 2009.

[41] J. Dattorro, *Convex Optimization & Euclidean Distance Geometry*, version 2010.10.26. Palo Alto, CA: Meboo, 2005. Available: http://meboo.convexoptimization.com/access.html

[42] S. Gezici and V. Poor, "Position estimation via ultra-wide-band signals," *Proc. IEEE*, vol. 97, no. 2, pp. 386–403, 2009.

[43] Y. Qi, H. Kobayashi, and H. Suda, "Analysis of wireless geolocation in a non-line-of-sight environment," *IEEE Trans. Wireless Commun.*, vol. 5, no. 3, pp. 672–681, 2006.

[44] A. Catovic and Z. Sahinoglu, "The Cramér-Rao bounds of hybrid TOA/RSS and TDOA/RSS location estimation schemes," *IEEE Commun. Lett.*, vol. 8, no. 10, pp. 626–628, 2004.

[45] B.-C. Liu and K.-H. Lin, "Cellular geolocation employing hybrid of relative signal strength and propagation delay," in *Proc. of the IEEE Wireless Communications & Networking Conf. 2006 (WCNC 2006)*, vol. 43, Las Vegas, NV, Apr. 2006, pp. 280–283.

[46] B. T. Sieskul, F. Zheng, and T. Kaiser, "Time-of-arrival estimation in path attenuation," in *Proc. of the IEEE Int. Workshop on Signal Processing Advances for Wireless Communications 2009 (SPAWC 2009)*, Perugia, Italy, Jun. 2009, pp. 573–577.

[47] B. T. Sieskul, T. Kaiser, and F. Zheng, "A hybrid SS-TOA wireless NLOS geolocation based on path attenuation: Cramér-Rao bound," in *Proc. of the IEEE 69th Vehicular Technology Conf. 2009 (VTC2009 Spring)*, Barcelona, Spain, Apr. 2009, pp. 1–5.

[48] B. Friedlander, "A passive localization algorithm and its accuracy analysis," *IEEE J. Ocean. Eng.*, vol. 12, no. 1, pp. 234–245, 1987.

[49] J. O. Smith and J. S. Abel, "Closed-form least-squares source location estimation from range-difference measurements," *IEEE Trans. Acoust. Speech Signal Process.*, vol. 35, no. 12, pp. 1661–1669, 1987.

[50] J. S. Abel, "A divide and conquer approach to least-squares estimation," *IEEE Trans. Aerosp. Electron. Syst.*, vol. 26, no. 2, pp. 423–427, 1990.

[51] M. Hellebrandt, R. Mathar, and M. Scheibenbogen, "Estimating position and velocity of mobiles in a cellular radio network," *IEEE Trans. Veh. Technol.*, vol. 46, no. 1, pp. 65–71, 1997.

[52] J. J. Caffery, Jr. and G. L. Stuber, "Subscriber location in CDMA cellular networks," *IEEE Trans. Veh. Technol.*, vol. 47, no. 2, pp. 406–416, 1998.

[53] Y. T. Chan and K. C. Ho, "A simple and efficient estimator for hyperbolic location," *IEEE Trans. Signal Process.*, vol. 42, no. 8, pp. 1905–1916, 1994.

[54] A. J. Fenwick, "Algorithms for position fixing using pulse arrival times," *Proc. IEEE Radar Sonar Navig.*, vol. 146, no. 4, pp. 208–212, 1999.

[55] J. C. Chen, R. E. Hudson, and K. Yao, "Maximum-likelihood source localization and unknown sensor location estimation for wideband signals in the near field," *IEEE Trans. Signal Process.*, vol. 50, no. 8, pp. 1843–1854, 2002.

[56] K. W. Cheung and H. C. So, "A multidimensional scaling framework for mobile location using time-of-arrival measurements," *IEEE Trans. Signal Process.*, vol. 53, no. 2, pp. 460–470, 2005.

[57] J. S. Abel and J. O. Smith, "The spherical interpolation method for closed-form passive source localization using range difference measurements," in *Proc. of the IEEE Int. Conf. on Acoustics, Speech and Signal Processing 1987 (ICASSP'87)*, Dallas, TX, Apr. 1987, pp. 471–474.

[58] H. Schau and A. Robinson, "Passive source localization employing intersecting spherical surfaces from time-of-arrival differences," *IEEE Trans. Acoust. Speech Signal Process.*, vol. 35, no. 8, pp. 1223–1225, 1987.

[59] B. T. Fang, "Simple solutions for hyperbolic and related position fixes," *IEEE Trans. Aerosp. Electron. Syst.*, vol. 26, no. 5, pp. 748–753, 1990.

[60] J. Riba and A. Urruela, "Novel closed-form ML position estimator for hyperbolic location," in *Proc. of the IEEE Int. Conf. on Acoustics, Speech and Signal Processing 2004 (ICASSP 2004)*, vol. 2, Montreal, QC, May 2004, pp. 149–152.

[61] W. H. Foy, "Position-location solutions by Taylor-series estimation," *IEEE Trans. Aerosp. Electron. Syst.*, vol. 12, no. 2, pp. 187–194, 1976.

[62] K. Yu, "3-D localization error analysis in wireless networks," *IEEE Trans. Wireless Commun.*, vol. 6, no. 10, pp. 3473–3481, 2007.

[63] C.-H. Lim, A. M. Zoubir, C.-M. S. See, and B.-P. Ng, "A robust statistical approach to non-line-of-sight mitigation," in *Proc. of the IEEE Workshop on Statistical Signal Processing 2007 (SSP'07)*, Madison, WI, Aug. 2007, pp. 428–432.

[64] Y.-T. Chan, H. Y. C. Hang, and P. C. Ching, "Exact and approximate maximum likelihood localization algorithms," *IEEE Trans. Veh. Technol.*, vol. 55, no. 1, pp. 10–16, 2006.

[65] A. Beck, P. Stoica, and J. Li, "Exact and approximate solutions of source localization problems," *IEEE Trans. Signal Process.*, vol. 56, no. 5, pp. 170–178, 2008.

[66] E. G. Larsson and D. Danev, "Accuracy comparison of LS and squared-range LS for source localization," *IEEE Trans. Signal Process.*, vol. 58, no. 2, pp. 916–923, 2010.

[67] A. Goldsmith, *Wireless Communications*. New York: Cambridge University Press, 2005.

[68] T. S. Rappaport, *Wireless Communications: Principle and Practice*, 2nd ed. Englewood Cliffs, NJ: Prentice-Hall, 2002.

[69] V. Erceg, L. J. Greenstein, S. Y. Tjandra, S. R. Parkoff, A. Gupta, B. Kulic, A. A. Julius, and R. Bianchi, "An empirically based path loss model for wireless channels in suburban environments," *IEEE J. Sel. Areas Commun.*, vol. 17, no. 7, pp. 1205–1211, 1999.

[70] B. Sklar, *Digital Communications: Fundamentals and Applications*, 2nd ed. Upper Saddle River, NJ: Prentice-Hall, 2001.

[71] R. E. Collins, *Antennas and Radiowave Propagation*. Singapore: McGraw-Hill, 1985.

[72] C. A. Balanis, *Antenna Theory*. New York: John Wiley & Sons, 1997.

[73] R. J. Serfling, *Approximation Theorems of Mathematical Statistics*. New York: John Wiley & Sons, 1980.

[74] C. R. Rao, "Least squares theory using an estimated dispersion matrix and its application to measurement of signals," in *Proc. of the Fifth Berkeley Symposium on Mathematical Statistics and Probability*, vol. 1, 1967, pp. 355–372.

[75] H. Cramér, "A contribution to the theory of statistical estimation," *Skand. Akt. Tidskr.*, vol. 29, pp. 85–94, 1945.

[76] C. R. Rao, "Information and accuracy attainable in the estimation of statistical parameters," *Bull. Calcutta Math. Soc.*, vol. 37, pp. 81–91, 1945.

[77] C. R. Rao, "Minimum variance and the estimation of several parameters," *Proc. Cambridge Phil. Soc.*, vol. 43, pp. 280–283, 1946.

[78] H. Cramér, *Mathematical Methods of Statistics*, 5th ed. Bombay, India: Asia Publishing House, 1962.

[79] C. R. Rao, *Linear Statistical Inference and Its Applications*, 2nd ed. New York: John Wiley & Sons, 1972.

[80] K. Königsberger, *Analysis 2*, 5th ed. Heidelberg, Germany: Springer, 2004.

[81] G. S. Watson, "Linear least squares regression," *Ann. Math. Stat.*, vol. 38, no. 6, pp. 1679–1699, 1967.

[82] C. R. Rao and H. Toutenburg, *Linear Models: Least Squares and Alternatives*. New York: Springer, 1999.

[83] D. J. Torrieri, "Statistical theory of passive location systems," *IEEE Trans. Aerosp. Electron. Syst.*, vol. 20, no. 2, pp. 183–197, 1984.

[84] M. A. Spirito, "On the accuracy of cellular mobile station location estimation," *IEEE Trans. Veh. Technol.*, vol. 50, no. 3, pp. 674–685, 2001.

CHAPTER 19

MOBILE TRACKING IN MIXED LINE-OF-SIGHT/NON-LINE-OF-SIGHT CONDITIONS: ALGORITHMS AND THEORETICAL LOWER BOUND

Liang Chen,[1] Simo Ali-Löytty,[2] Robert Piché,[2] and Lenan Wu[3]

[1]Wuhan University, Wuhan, China
[2]Tampere University of Technology, Tampere, Finland
[3]Southeast University, Nanjing, China

THIS CHAPTER investigates the problem of mobile tracking in mixed line-of-sight (LOS)/non-line-of-sight (NLOS) conditions. The state-of-the-art methods in this field are first reviewed. Then, we consider the problem in the Bayesian estimation framework and focus on two types of Bayesian filters: the Gaussian mixture filter (GMF) and the particle filter (PF). In the GMF section, the approximation property and the convergence results are summarized. Then, the modified extended Kalman filter (EKF) banks method, as one specific GMF, is described. In the PF section, generic PF is first introduced, and a more effective PF, approximated Rao–Blackwellized particle filtering (ARBPF), is further discussed in detail. The chapter closes with a discussion of the computation of a posterior Cramer–Rao lower bound (CRLB) for this kind of mobile tracking problem.

19.1 INTRODUCTION

Precise positioning in NLOS conditions is a challenge. In NLOS conditions, the direct path between the transmitter and receiver has been blocked by buildings

Handbook of Position Location: Theory, Practice, and Advances, Second Edition.
Edited by S. A. (Reza) Zekavat and R. Michael Buehrer.
© 2019 by the Institute of Electrical and Electronics Engineers, Inc.
Published 2019 by John Wiley & Sons, Inc.
Companion Website: www.wiley.com/go/zekavat/positionlocation2e

and other obstacles. The propagating wave may travel excess path lengths due to reflection, refraction, and scattering. In terms of range-based measurements such as time of arrival (TOA), time difference of arrival (TDOA), and received signal strength (RSS), this extra propagation distance imposes positive biases on the true path, which causes large errors on the location estimations. In dense urban regions, the NLOS condition is very common. For example, a field test in a global system for mobile communications (GSM) network shows that the mean and standard deviation of NLOS range errors are on the order of 513 m and 436 m, respectively [1].

Generally, methods to deal with the NLOS problem can be divided into two categories: for static positioning systems and for mobility tracking systems. The methods for static positioning systems can be further divided into three ways:

- detect and identify the LOS signals for localization [2–4];
- mitigate the NLOS effect and minimize the estimate error, which includes weights or scaling factors [5, 6] and equalization method [7];
- model the NLOS propagation paths: scattering models [8, 9], multipath information in the time domain [10], or space-time domain [11].

In mobility tracking systems, the general idea is to exploit the redundant measurements in time series to mitigate the NLOS errors. Polynomial fit [12] and two-step Kalman filtering techniques [13] are applied to smooth range measurements and mitigate NLOS errors by assuming that the standard deviation of the range measurement in the case of NLOS is significantly larger than that of LOS. A Markov process is introduced to describe the LOS and NLOS condition as two interactive modes in [14]. A Kalman-based interacting multiple model (IMM) smoother is further proposed to estimate the range between the corresponding base station (BS) and mobile station (MS). It can track the true range distance more accurately than the rough LOS/NLOS smoother in [13], especially in the transitional intervals. The method in [15] uses a first-order homogeneous hidden Markov chain to simultaneously model the transition of LOS/NLOS condition and receiver position. Grid-based Bayesian estimation [15] and particle filtering [16] are used for ultra-wideband (UWB) indoor positioning.

In this chapter, we investigate the problem of mobile tracking in mixed LOS/NLOS conditions within the Bayesian estimation framework. Two types of Bayesian filters are considered: the GMF and the PF. We discuss the modified EKF banks and ARBPF in detail. A numerical computing method is presented to calculate the posterior CRLB in this kind of problem. Simulation results are further provided to compare the performance of the algorithms and the posterior CRLB. The filtering algorithms and the computing methodology for the theoretical lower bound are completely general for other platforms, such as UWB, satellite-based position, and so on.

19.2 SYSTEM DESCRIPTION

19.2.1 General Problem Formulation

The dynamic model of the mobility tracking in the mixed LOS/NLOS conditions can be represented as follows:

$$\mathbf{x}_k = \mathbf{\Phi}_{k-1}\mathbf{x}_{k-1} + \mathbf{w}_{k-1}, \tag{19.1a}$$

$$\mathbf{z}_k = \mathbf{h}_k(\mathbf{x}_k) + \mathbf{v}_k(\mathbf{s}_k), \tag{19.1b}$$

$$s_{i,k} \sim \mathrm{MC}(\pi_i, \mathbf{A}_i), \tag{19.1c}$$

where $\mathbf{x}_k$ is the state at time instant t_k, matrix $\mathbf{\Phi}_{k-1}$ is the known state transition matrix, and $\mathbf{w}_{k-1}$ is the state model noise with known statistics. Vector $\mathbf{z}_k$ represents the measurement, function $\mathbf{h}_k$ is a known measurement function, and $\mathbf{v}_k$ is the measurement noise, which depends on sight conditions $\mathbf{s}_k$. Sight condition $s_{i,k}$ is the ith component of vector $\mathbf{s}_k$, $i \in \{1, 2, \ldots, M\}$, $k \in \mathbb{N}$, and it is a Boolean variable $s_{i,k} \in \{0, 1\}$ to represent LOS/NLOS condition between the MS and BS_i at time instant t_k, with $s_{i,k} = 0$ for LOS and $s_{i,k} = 1$ for NLOS.

In mobile tracking, the sight conditions undergo dynamical transitions, which can be further modeled as a time-homogeneous first-order Markov chain $s_{i,k} \sim \mathrm{MC}(\pi_i, \mathbf{A}_i)$ with initial probability vector π_i and the transition probability matrix

$$\mathbf{A}_i = \begin{bmatrix} p_0 & 1 - p_0 \\ 1 - p_1 & p_1 \end{bmatrix}, \tag{19.2}$$

where $p_0 = \mathrm{P}(s_{i,k} = 0 | s_{i,k-1} = 0)$ and $p_1 = \mathrm{P}(s_{i,k} = 1 | s_{i,k-1} = 1)$. Note that the sight condition of each BS is governed by its own independent Markov chain.

Denote the total observation sequence up to time t_k as $\mathbf{z}_{1:k}$, where $\mathbf{z}_k \triangleq [z_{1,k}, z_{2,k}, \ldots z_{M,k}]^T$. The problem of mobile tracking in mixed LOS/NLOS conditions is to infer the current mobile state $\mathbf{x}_k$ from the observation $\mathbf{z}_{1:k}$, that is, to compute the marginal posterior $f(\mathbf{x}_k | \mathbf{z}_{1:k})$. The marginal posterior is the mixture

$$f(\mathbf{x}_k | \mathbf{z}_{1:k}) = \sum_{\mathbf{s}_k} f(\mathbf{x}_k, \mathbf{s}_k | \mathbf{z}_{1:k}), \tag{19.3}$$

where $\mathbf{s}_k$ go through all 2^M possibilities. Components of mixture (which are also mixtures) $f(\mathbf{x}_k, \mathbf{s}_k | \mathbf{z}_{1:k})$ can be determined recursively according to the following relations [17].

Prediction:

$$f(\mathbf{x}_k, \mathbf{s}_k | \mathbf{z}_{1:k-1}) = \sum_{\mathbf{s}_{k-1}} \mathrm{P}(\mathbf{s}_k | \mathbf{s}_{k-1}) \int f(\mathbf{x}_k | \mathbf{x}_{k-1}) f(\mathbf{x}_{k-1}, \mathbf{s}_{k-1} | \mathbf{z}_{1:k-1}) d\mathbf{x}_{k-1}, \tag{19.4}$$

Update:

$$f(\mathbf{x}_k, \mathbf{s}_k | \mathbf{z}_{1:k}) = \frac{f(\mathbf{z}_k | \mathbf{x}_k, \mathbf{s}_k) f(\mathbf{x}_k, \mathbf{s}_k | \mathbf{z}_{1:k-1})}{\sum_{\mathbf{s}_k} \int f(\mathbf{z}_k | \mathbf{x}_k, \mathbf{s}_k) f(\mathbf{x}_k, \mathbf{s}_k | \mathbf{z}_{1:k-1}) d\mathbf{x}_k}, \tag{19.5}$$

where the transition probability

$$\mathrm{P}(\mathbf{s}_k | \mathbf{s}_{k-1}) = \prod_{i=1}^{M} \mathrm{P}(s_{i,k} | s_{i,k-1}), \tag{19.6}$$

the transitional density $f(\mathbf{x}_k|\mathbf{x}_{k-1}) = f_{\mathbf{w}_{k-1}}(\mathbf{x}_k - \Phi_{k-1}\mathbf{x}_{k-1})$, and the likelihood $f(\mathbf{z}_k|\mathbf{x}_k, \mathbf{s}_k) = f_{\mathbf{v}_k(\mathbf{s}_k)}(\mathbf{z}_k - \mathbf{h}_k(\mathbf{x}_k))$. The solution to (19.5) cannot be derived analytically because measurement function $\mathbf{h}$ is nonlinear. Moreover, the number of mixture components grows exponentially with time. For this reason, we resort to two kinds of suboptimal solutions: one is a modified EKF banks method, a specific GMF; the other is a sample-based numerical approximate method.

19.2.2 Example of the State Model

One commonly used state model is the so-called constant velocity model, which is also used in simulation part in Section 19.6:

$$\mathbf{x}_{k+1} = \Phi_k \mathbf{x}_k + \mathbf{w}_k, \tag{19.7}$$

where $\mathbf{x}_k = [x_k, y_k, \dot{x}_k, \dot{y}_k]^T$, where $[x_k, y_k]^T$ corresponds to the east and north coordinates of the mobile position, and $[\dot{x}_k, \dot{y}_k]^T$ are the corresponding velocities. The state transition matrix $\Phi_k = \begin{bmatrix} I_2 & \Delta t_k I_2 \\ 0 & I_2 \end{bmatrix}$, with identity matrix $I_2 \in \mathbb{R}^{2\times 2}$ and $\Delta t_k = t_{k+1} - t_k$. The random process $\mathbf{w}_k$ is a white zero mean Gaussian noise, with covariance matrix

$$\mathbf{Q}_k = \begin{bmatrix} \dfrac{\Delta t_k^4}{4}\mathbf{Q} & \dfrac{\Delta t_k^3}{2}\mathbf{Q} \\ \dfrac{\Delta t_k^3}{2}\mathbf{Q} & \Delta t_k^2\mathbf{Q} \end{bmatrix}, \text{ where } \mathbf{Q} = \begin{bmatrix} \sigma_x^2 & 0 \\ 0 & \sigma_y^2 \end{bmatrix}. \tag{19.8}$$

19.2.3 Example of the Measurement Model

Range measurement is commonly used. Here in simulations we consider only two-dimensional cases. Suppose $d_{i,k}$ represents the true distance between the mobile position $[x_k, y_k]^T$ and the location of the ith BS $[x_{\text{bs}_i}, y_{\text{bs}_i}]^T$:

$$d_{i,k} \triangleq h_{i,k}(\mathbf{x}_k) = \sqrt{(x_k - x_{\text{bs}_i})^2 + (y_k - y_{\text{bs}_i})^2}, \tag{19.9}$$

where $i \in \{1, 2, \ldots, M\}$ and M is the number of BSs. In a LOS environment, the range measurement between MS and BS_i is only corrupted by the system measurement noise $n_{i,k}$, which can be modeled as an independent and identically distributed (i.i.d.) zero mean white Gaussian noise $N(0, \sigma_n^2)$. In NLOS conditions, the range measurement is corrupted by two sources of errors: the measurement noise $n_{i,k}$ and the NLOS error $e_{i,k}$. According to the field tests in [1], $e_{i,k}$ can be modeled as a positively biased distribution. In this chapter we assume that NLOS errors have Gaussian distribution $N(\mu_{\text{NLOS}}, \sigma_{\text{NLOS}}^2)$. We also assume that errors $n_{i,k}$ and $e_{i,k}$ are independent. Then the range measurement equations are

$$\text{LOS:} \quad z_{i,k} = d_{i,k} + n_{i,k}. \tag{19.10a}$$

$$\text{NLOS:} \quad z_{i,k} = d_{i,k} + n_{i,k} + e_{i,k}. \tag{19.10b}$$

Equation (19.10a, b) can then be written as (see (19.1b))

$$z_{i,k} = d_{i,k} + v(s_{i,k}),$$ (19.11)

where $v(s_{i,k}) \sim N(m[s_{i,k}], \mathbf{R}[s_{i,k}])$
and

$$m(s_{i,k}) = s_{i,k}\mu_{\text{NLOS}},$$ (19.12a)

$$\mathbf{R}(s_{i,k}) = \sigma_n^2 + s_{i,k}\sigma_{\text{NLOS}}.$$ (19.12b)

19.3 TRACKING ALGORITHM BASED ON GMF

19.3.1 The Development of GMF

In this subsection, we consider one conventional GMF, which solves the system (19.13). The system (19.13) is the same as the system (19.1) if we know or separately estimate the sight conditions. This GMF is able to handle the nonlinearity of the measurement model (19.13b) but is not optimized to real-time computation of LOS/NLOS problem. Another closely related GMF is presented in Section 19.3.2. In this section, we consider the system

$$\mathbf{x}_k = \mathbf{\Phi}_{k-1}\mathbf{x}_{k-1} + \mathbf{w}_{k-1},$$ (19.13a)

$$\mathbf{z}_k = \mathbf{h}_k(\mathbf{x}_k) + \mathbf{v}_k,$$ (19.13b)

where errors $\mathbf{w}_{k-1}$ and $\mathbf{v}_k$ are white, mutually independent, and independent of the initial state $\mathbf{x}_0$. Lower index $k \in \mathbb{N} \backslash \{0\}$ represents time t_k. We assume that errors $\mathbf{w}_{k-1}$, $\mathbf{v}_k$, and initial state $\mathbf{x}_0$ are Gaussian mixtures (GMs), with probability density functions (PDFs)

$$f_{\mathbf{w}_k}(\xi) = \sum_{i=1}^{n_{\mathbf{w}_k}} \alpha_{i,\mathbf{w}_k} N_{\hat{P}_{i,\mathbf{w}_k}}^{\hat{x}_{i,\mathbf{w}_k}}(\xi),$$ (19.14a)

$$f_{\mathbf{v}_k}(\xi) = \sum_{i=1}^{n_{\mathbf{v}_k}} \alpha_{i,\mathbf{v}_k} N_{\hat{P}_{i,\mathbf{v}_k}}^{\hat{x}_{i,\mathbf{v}_k}}(\xi),$$ (19.14b)

$$f_{\mathbf{x}_0}(\xi) = \sum_{i=1}^{n_0} \alpha_{i,0} N_{P_{i,0}}^{x_{i,0}}(\xi),$$ (19.14c)

respectively. Here $N_{\hat{\mathbf{P}}}^{\hat{x}}(\xi)$ is PDF of Gaussian distribution $N_n(\hat{\mathbf{x}}, \hat{\mathbf{P}})$:

$$N_{\hat{\mathbf{P}}}^{\hat{x}}(\xi) = \frac{1}{(2\pi)^{\frac{n}{2}}\sqrt{\det(\hat{\mathbf{P}})}} \exp\left(-\frac{1}{2}(\xi - \hat{\mathbf{x}})^T \hat{\mathbf{P}}^{-1}(\xi - \hat{\mathbf{x}})\right).$$ (19.15)

We use the abbreviations

$$\mathbf{w}_k \sim \mathrm{M}\left(\alpha_{i,\mathrm{w}_k}, \hat{\mathbf{x}}_{i,\mathrm{w}_k}, \hat{\mathbf{P}}_{i,\mathrm{w}_k}\right)_{(i,n_{\mathbf{w}_k})}, \tag{19.16a}$$

$$\mathbf{v}_k \sim \mathrm{M}\left(\alpha_{i,\mathrm{v}_k}, \mathbf{x}_{i,\mathrm{v}_k}, \mathbf{P}_{i,\mathrm{v}_k}\right)_{(i,n_{\mathbf{v}_k})}, \tag{19.16b}$$

$$\mathbf{x}_0 \sim \mathrm{M}\left(\alpha_{i,0}, \mathbf{x}_{i,0}, \mathbf{P}_{i,0}\right)_{(i,n_0)}. \tag{19.16c}$$

This is not a restrictive assumption because any density function may be approximated as density function of GM as closely as we wish [18, 19]. The GMF algorithm for system (19.13) is summarized as Algorithm 19.1, where the last time step is $t_{n_{\mathrm{meas}}}$.

Algorithm 19.1   GMF

Initial state at time t_0: $\mathbf{x}_0 \sim \mathrm{M}\left(\alpha_{i,0}, \mathbf{x}_{i,0}, \mathbf{P}_{i,0}\right)_{(i,n_0)}$
  for $k = 1$ to n_{meas} *do*

1. Prediction (see Section 19.3.1):

$$\mathbf{x}_k^- \sim \mathrm{M}\left(\alpha_{i*j,k}^-, \mathbf{x}_{i*j,k}^-, \mathbf{P}_{i*j,k}^-\right)_{(i*j,n_k^-)}.$$

2. Approximate $\mathbf{x}_k^-$ as a new GM $\overline{\mathbf{x}}_k^-$ if necessary (see Section 19.3.1):

$$\overline{\mathbf{x}}_k^- \sim \mathrm{M}\left(\overline{\alpha}_{i,k}^-, \mathbf{x}_{i,k}^-, \mathbf{P}_{i,k}^-\right)_{(i,\overline{n}_k^-)}.$$

3. Update (see Section 19.3.1 and Example 19.1):

$$\overline{\mathbf{x}}_k \sim \mathrm{M}\left(\overline{\alpha}_{i*j,k}, \overline{\hat{\mathbf{x}}}_{i*j,k}, \overline{\hat{\mathbf{P}}}_{i*j,k}\right)_{(i*j,\overline{n}_k)}.$$

4. Reduce the number of components (see Section 19.3.1):

$$\mathbf{x}_k \sim \mathrm{M}\left(\alpha_{i,k}, \hat{\mathbf{x}}_{i,k}, \hat{\mathbf{P}}_{i,k}\right)_{(i,n_k)}$$

end for

  Prediction, Step 1
  Prediction is based on (19.13a):

$$\mathbf{x}_k^- \sim \mathrm{M}\left(\alpha_{i*j,k}^-, \mathbf{x}_{i*j,k}^-, \mathbf{P}_{i*j,k}^-\right)_{(i*j,n_k^-)}, \tag{19.17}$$

where

$$n_k^- = n_{k-1} n_{\mathrm{w}_{k-1}}, \tag{19.18a}$$

$$\alpha_{i*j,k}^- = \alpha_{i,k-1} \alpha_{j,\mathrm{w}_{k-1}}, \tag{19.18b}$$

$$\hat{\mathbf{x}}_{i*j,k}^- = \Phi_{k-1}\hat{\mathbf{x}}_{i,k-1} + \hat{\mathbf{x}}_{j,\mathrm{w}_{k-1}}, \tag{19.18c}$$

$$\hat{\mathbf{P}}_{i*j,k}^- = \Phi_{k-1}\hat{\mathbf{P}}_{i,k-1}\Phi_{k-1}^T + \hat{\mathbf{P}}_{j,\mathrm{w}_{k-1}}. \tag{19.18d}$$

Approximate GM as a New GM, Step 2

Conventional EKF does not work properly if we have "highly nonlinear" measurements/components. We say that mixture component $N_{\hat{P}}^{\hat{x}}(\xi)$ is "highly nonlinear" if

$$\frac{\sqrt{\text{tr}\left(\mathbf{h}_i''(\hat{\mathbf{x}})\hat{\mathbf{P}}\mathbf{h}_i''(\hat{\mathbf{x}})\hat{\mathbf{P}}\right)}}{\sqrt{\mathbf{R}_{i,i}}} \gg 1, \text{ for some } i, \tag{19.19}$$

where $\mathbf{h}_i''(\hat{\mathbf{x}})$ is the Hessian matrix of the ith element of the measurement function $\mathbf{h}(\mathbf{x})$, and $\mathbf{R} = V(\mathbf{v})$ is the covariance matrix of the measurement error [20]. One possibility to overcome nonlinearity is to approximate the prior density function as a GM so that covariance matrices of components are smaller than the prior covariance matrices [18, 21, 22]. There are different methods to compute Step 2. These methods are described in [18].

Update, Step 3

The update is computed approximately using a bank of EKFs. So

$$\overline{\mathbf{x}}_k \sim M\left(\overline{\alpha}_{i*j,k}, \overline{\hat{\mathbf{x}}}_{i*j,k}, \overline{\hat{\mathbf{P}}}_{i*j,k}\right)_{(i*j,\overline{n}_k)}, \tag{19.20}$$

where

$$\overline{n}_k = n_{\mathbf{v}_k}\overline{n}_k^-, \tag{19.21a}$$

$$\overline{\alpha}_{i*j,k} = \frac{\overline{\alpha}_{i,k}^-\alpha_{j,\mathbf{v}_k} N_{\mathbf{H}_{i,k}\overline{\hat{\mathbf{P}}}_{i,k}^-\mathbf{H}_{i,k}^T+\hat{\mathbf{P}}_{j,\mathbf{v}_k}}^{\mathbf{h}_k\left(\overline{\hat{\mathbf{x}}}_{i,k}^-\right)+\hat{\mathbf{x}}_{j,\mathbf{x}_k}}(\mathbf{z}_k)}{\displaystyle\sum_{j=1}^{n_{\mathbf{v}_k}}\sum_{i=1}^{\overline{n}_k}\overline{\alpha}_{i,k}^-\alpha_{j,\mathbf{v}_k} N_{\mathbf{H}_{i,k}\overline{\hat{\mathbf{P}}}_{i,k}^-\mathbf{H}_{i,k}^T+\hat{\mathbf{P}}_{j,\mathbf{v}_k}}^{\mathbf{h}_k\left(\overline{\hat{\mathbf{x}}}_{i,k}^-\right)+\hat{\mathbf{x}}_{j,\mathbf{v}_k}}(\mathbf{z}_k)}, \tag{19.21b}$$

$$\mathbf{H}_{i,k} = \left.\frac{\partial\mathbf{h}_k(\xi)}{\partial\xi}\right|_{\xi=\overline{\hat{\mathbf{x}}}_{i,k}^-}, \tag{19.21c}$$

$$\overline{\hat{\mathbf{x}}}_{i*j,k} = \overline{\hat{\mathbf{x}}}_{i,k}^- + \mathbf{K}_{i*j,k}\left(\mathbf{z}_k - \mathbf{h}_k\left(\overline{\hat{\mathbf{x}}}_{i,k}^-\right) - \hat{\mathbf{x}}_{j,\mathbf{v}_k}\right), \tag{19.21d}$$

$$\overline{\hat{\mathbf{P}}}_{i*j,k} = \left(\mathbf{I} - \mathbf{K}_{i*j,k}\mathbf{H}_{i,k}\right)\overline{\hat{\mathbf{P}}}_{i,k}^-, \tag{19.21e}$$

$$\mathbf{K}_{i*j,k} = \overline{\hat{\mathbf{P}}}_{i,k}^-\mathbf{H}_{i,k}^T\left(\mathbf{H}_{i,k}\overline{\hat{\mathbf{P}}}_{i,k}^-\mathbf{H}_{i,k}^T + \hat{\mathbf{P}}_{j,\mathbf{v}_k}\right)^{-1}. \tag{19.21f}$$

Example 19.1   How to Compute Update Step 3 Using MATLAB

How to compute Update Step 3 using MATLAB? Note this is the crucial part of the algorithm.

Solution

The example MATLAB code for that part is given in separate file: Chapter_19_Example_19_1.m. MATLAB codes can be found online at ftp://ftp.wiley.com/public/sci_tech_med/matlab_codes.

Reduce the Number of Components, Step 4

One major challenge when using GMF efficiently is keeping the number of components as small as possible without losing significant information. There are many ways to do so. We use two different types of mixture reduction algorithms: forgetting and merging [21, 23–25].

Forgetting Components: We re-index the posterior approximation $\bar{\mathbf{x}}_k$ (19.20) such that

$$\bar{\mathbf{x}}_k \sim \mathrm{M}\left(\bar{\alpha}_{i,k}, \hat{\bar{\mathbf{x}}}_{i,k}, \hat{\bar{\mathbf{P}}}_{i,k}\right)_{(i,\bar{n}_k)}, \tag{19.22}$$

where $\bar{\alpha}_{i,k} \geq \bar{\alpha}_{i+1,k}$. Let $0 \leq \varepsilon_f \leq 1$ be an arbitrary threshold value. Let $\bar{n}_{k,f}$ be the first index such that $\sum_{i=1}^{\bar{n}_{k,f}} \bar{\alpha}_{i,k} \geq 1-\varepsilon_f$. We forget all mixture components whose index $i > \bar{n}_{k,f}$, and after normalization we get $\bar{\mathbf{x}}_{k,f}$. Now

$$\bar{\mathbf{x}}_{k,f} \sim \mathrm{M}\left(\bar{\alpha}_{i,k,f}, \hat{\bar{\mathbf{x}}}_{i,k,f}, \hat{\bar{\mathbf{P}}}_{i,k,f}\right)_{(i,\bar{n}_{k,f})}, \tag{19.23}$$

where

$$\bar{\alpha}_{i,k,f} = \frac{\bar{\alpha}_{i,k}}{\sum\limits_{j=1}^{\bar{n}_{k,f}} \bar{\alpha}_{j,k}}, \tag{19.24a}$$

$$\hat{\bar{\mathbf{x}}}_{i,k,f} = \hat{\bar{\mathbf{x}}}_{i,k}, \tag{19.24b}$$

$$\hat{\bar{\mathbf{P}}}_{i,k,f} = \hat{\bar{\mathbf{P}}}_{i,k}. \tag{19.24c}$$

Merging Components: Our merging procedure is iterative. We merge two components, say the i_1th component and the i_2th component, into one component using the moment matching method if they are sufficiently similar, that is, if (for simplicity we suppress indices k and f) both inequalities

$$\left\|\hat{\bar{\mathbf{x}}}_{i_1} - \hat{\bar{\mathbf{x}}}_{i_2}\right\| \leq \varepsilon_{m_1} \text{ and } \left\|\hat{\bar{\mathbf{P}}}_{i_1}^+ - \hat{\bar{\mathbf{P}}}_{i_2}^+\right\| \leq \varepsilon_{m_2} \tag{19.25}$$

hold. The threshold values $0 \leq \varepsilon_{m_1}$ and $0 \leq \varepsilon_{m_2}$ are arbitrary. The new component, which replaces components i_1 and i_2, is a component whose weight, mean, and covariance matrix are

$$\bar{\alpha}_{i_1,m} = \bar{\alpha}_{i_1} + \bar{\alpha}_{i_2}, \tag{19.26a}$$

$$\hat{\bar{\mathbf{x}}}_{i_1,m} = \frac{\bar{\alpha}_{i_1}}{\bar{\alpha}_{i_1,m}} \hat{\bar{\mathbf{x}}}_{i_1} + \frac{\bar{\alpha}_{i_2}}{\bar{\alpha}_{i_1,m}} \hat{\bar{\mathbf{x}}}_{i_2}, \tag{19.26b}$$

$$\hat{\bar{\mathbf{P}}}_{i_1,m} = \frac{\bar{\alpha}_{i_1}}{\bar{\alpha}_{i_1,m}} \left(\hat{\bar{\mathbf{P}}}_{i_1} + \left(\hat{\bar{\mathbf{x}}}_{i_1} - \hat{\bar{\mathbf{x}}}_{i_1,m}\right)\left(\hat{\bar{\mathbf{x}}}_{i_1} - \hat{\bar{\mathbf{x}}}_{i_1,m}\right)^T\right)$$
$$+ \frac{\bar{\alpha}_{i_2}}{\bar{\alpha}_{i_1,m}} \left(\hat{\bar{\mathbf{P}}}_{i_2} + \left(\hat{\bar{\mathbf{x}}}_{i_2} - \hat{\bar{\mathbf{x}}}_{i_1,m}\right)\left(\hat{\bar{\mathbf{x}}}_{i_2} - \hat{\bar{\mathbf{x}}}_{i_1,m}\right)^T\right), \tag{19.26c}$$

respectively. After re-indexing (forgetting component i_2) we merge iteratively more components until there are no sufficiently similar components, that is, components that statisfy inequalities (19.25). Then, after re-indexing, we get

$$\mathbf{x}_k \sim \mathrm{M}\left(\alpha_{i,k}, \hat{\mathbf{x}}_{i,k}, \hat{\mathbf{P}}_{i,k}\right)_{(i,n_k)}. \tag{19.27}$$

Convergence Result of GMF: It has been shown that when the number of the Gaussian mixture components is increased (Algorithm 19.1, Step 2), then the GM approximation converges weakly to the original prior distribution. Moreover, posterior approximations of GMF converge weakly to the correct posteriors at every time step $t_k \le t_{n_\mathrm{meas}}$ [18, 21, 22].

19.3.2 The Modified EKF Banks

Applying the idea of GMF, a modified EKF bank could first compute the intermediate state $\{\overline{\mathbf{x}}_{i,k}, \overline{\mathbf{P}}_{i,k}\}$ based on the range measurement $\mathbf{z}_{i,1:k}$ from the ith BS. Then, fusing all $\{\overline{\mathbf{x}}_{i,k}, \overline{\mathbf{P}}_{i,k}\}_{i=1}^M$, the state $\{\hat{\mathbf{x}}_k, \hat{\mathbf{P}}_k\}$ is estimated. The intermediate state $\{\overline{\mathbf{x}}_{i,k}, \overline{\mathbf{P}}_{i,k}\}$ is also the merge result of two Gaussian components, each of which is obtained by EKF in LOS or NLOS conditions.

Algorithm Description: Based on the measurements $\mathbf{z}_{i,1:k}$, the posterior of $\mathbf{x}_{i,k}$ can be expressed as

$$f\left(\mathbf{x}_{i,k} \mid \mathbf{z}_{i,1:k}\right) = \sum_{s_{i,k}=0}^{1} f\left(\mathbf{x}_{i,k} \mid \mathbf{z}_{i,1:k}, s_{i,k}\right) f\left(s_{i,k} \mid \mathbf{z}_{i,1:k}\right). \tag{19.28}$$

The desired estimation of $\mathbf{x}_{i,k}$ is the conditional mean given by

$$
\begin{aligned}
\overline{\mathbf{x}}_{i,k} = \mathrm{E}\left(\mathbf{x}_{i,k} \mid \mathbf{z}_{i,1:k}\right) &= \int \mathbf{x}_{i,k} f\left(\mathbf{x}_{i,k} \mid \mathbf{z}_{i,1:k}\right) \mathrm{d}\mathbf{x}_{i,k} \\
&= \sum_{s_{i,k}=0}^{1} \underbrace{\int \mathbf{x}_{i,k} f\left(\mathbf{x}_{i,k} \mid \mathbf{z}_{i,1:k}, s_{i,k}\right) \mathrm{d}\mathbf{x}_{i,k}}_{\tilde{\mathbf{x}}_{i,k}(s_{i,k})} f\left(\mathbf{x}_{i,k} \mid \mathbf{z}_{i,1:k}\right),
\end{aligned} \tag{19.29}
$$

where the mean $\tilde{\mathbf{x}}_{i,k}(s_{i,k})$ and the covariance matrix $\tilde{\mathbf{P}}_{i,k}(s_{i,k})$ can be computed by LOS EKF and NLOS EKF, respectively. $f(s_{i,k} \mid \mathbf{z}_{i,1:k})$ can be further updated via Bayes' rule:

$$f\left(s_{i,k} \mid \mathbf{z}_{i,k}\right) = \frac{f\left(z_{i,k} \mid s_{i,k}, \mathbf{z}_{i,1:k-1}\right) f\left(s_{i,k} \mid \mathbf{z}_{i,1:k-1}\right)}{f\left(z_{i,k} \mid \mathbf{z}_{i,1:k-1}\right)}, \tag{19.30}$$

where $f(z_{i,k} \mid s_{i,k}, \mathbf{z}_{i,1:k-1})$ involves high-dimensional integrals. We simplified it as $f\left(z_{i,k} \mid s_{i,k}, \hat{\mathbf{x}}_{k-1}\right)$, where $\hat{\mathbf{x}}_{k-1}$ is the state estimation at time $k-1$. Then, $f(z_{i,k} \mid s_{i,k}, \mathbf{z}_{i,1:k-1})$ can be approximated by Gaussian density:

$$f\left(z_{i,k} \mid s_{i,k}, \mathbf{z}_{i,1:k-1}\right) \approx \mathrm{N}\left(\mathbf{H}_{i,k}\hat{\mathbf{x}}_{k|k-1} + m\left(s_{i,k}\right), \mathbf{H}_{i,k}\hat{\mathbf{P}}_{k|k-1}\mathbf{H}_{i,k}^T + R\left(s_{i,k}\right)\right), \tag{19.31}$$

where $\mathbf{H}_{i,k} = \partial h_i / \partial \mathbf{x} |_{\mathbf{x} = \hat{\mathbf{x}}_{k|k-1}}$ and $\{\hat{\mathbf{x}}_{k|k-1}, \hat{\mathbf{P}}_{k|k-1}\}$ are the one-step prediction of the $\{\hat{\mathbf{x}}_{k-1}, \hat{\mathbf{P}}_{k-1}\}$.

The term $f(s_{i,k}|\mathbf{z}_{i,1:k-1})$ in (19.30) can be evaluated via the Chapman–Kolmogorov equation:

$$f\left(s_{i,k}|\mathbf{z}_{i,1:k-1}\right) = \sum_{s_{i,k-1}=0}^{1} f\left(s_{i,k}|s_{i,k-1}, z_{i,1:k-1}\right) f\left(s_{i,k-1}|\mathbf{z}_{i,1:k-1}\right)$$

$$= \sum_{s_{i,k-1}=0}^{1} f\left(s_{i,k}|s_{i,k-1}\right) f\left(s_{i,k-1}|\mathbf{z}_{i,1:k-1}\right). \tag{19.32}$$

Thus, utilizing $\tilde{\mathbf{x}}_{i,k}(s_{i,k})$ and $f(s_{i,k}|\mathbf{z}_{i,1:k-1})$, $\overline{\mathbf{x}}_{i,k}$ can be computed by (19.29). The corresponding estimation covariance matrix is given by

$$\overline{\mathbf{P}}_{i,k} = \sum_{s_{i,k}=0}^{1} \left\{ \tilde{\mathbf{P}}_{i,k}\left(s_{i,k}\right) + \left[\tilde{\mathbf{x}}_{i,k}\left(s_{i,k}\right) - \overline{\mathbf{x}}_{i,k}\right]\left[\tilde{\mathbf{x}}_{i,k}\left(s_{i,k}\right) - \overline{\mathbf{x}}_{i,k}\right]^{T} \right\} f\left(s_{i,k}|\mathbf{z}_{i,1:k}\right). \tag{19.33}$$

Fusing all the intermediate states $\{\overline{\mathbf{x}}_{i,k}, \overline{\mathbf{P}}_{i,k}\}$ estimated from M BS measurements, the state $\{\hat{\mathbf{x}}_k, \hat{\mathbf{P}}_k\}$ can be estimated as [26]:

$$\hat{\mathbf{P}}_k^{-1} = \hat{\mathbf{P}}_{k|k-1}^{-1} + \sum_{i=1}^{M} \left(\overline{\mathbf{P}}_{i,k}^{-1} - \hat{\mathbf{P}}_{k/k-1}^{-1}\right). \tag{19.34a}$$

$$\hat{\mathbf{x}}_k = \hat{\mathbf{P}}_k \left[\hat{\mathbf{P}}_{k|k-1}^{-1} \hat{\mathbf{x}}_{k|k-1} + \sum_{i=1}^{M} \left\{ \overline{\mathbf{P}}_{i,k}^{-1} \overline{\mathbf{x}}_{i,k} - \hat{\mathbf{P}}_{k|k-1}^{-1} \hat{\mathbf{x}}_{k|k-1} \right\} \right]. \tag{19.34b}$$

The algorithm is summarized in Algorithm 19.2 and is illustrated in Figure 19.1.

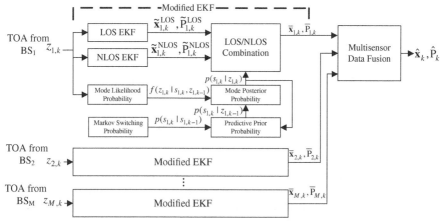

Figure 19.1   Structure of modified EKF banks.

Algorithm 19.2   Modified EKF banks and data fusion algorithm

Initial state at time $k = 0$:

Set $\left\{ \hat{\mathbf{x}}_0, \hat{\mathbf{P}}_0 \right\}$, transition probability $P(s_{i,k}|z_{i,k-1})$, and initial estimation $P(s_{i,k})$.

for k = 1 to T do

1. Predict the mean $\hat{\mathbf{x}}_{k|k-1}$ and covariance $\hat{\mathbf{P}}_{k|k-1}$.

2. **for** $i = 1, \dots, M$ and $s_{i,k} = 0, 1$

   - Predict the measurements in both LOS and NLOS conditions: $\hat{z}_{i,k|k-1}(s_{i,k}) = h(\hat{\mathbf{x}}_{k|k-1}) + m(s_{i,k})$.

   - EKF update:

$$\tilde{\mathbf{P}}_{i,k}^{-1}(s_{i,k}) = \hat{\mathbf{P}}_{k|k-1}^{-1} + \mathbf{H}_{i,k}^T R(s_{i,k})^{-1} \mathbf{H}_{i,k}$$

$$\mathbf{K}_{i,k}(s_{i,k}) = \tilde{\mathbf{P}}_{i,k}(s_{i,k}) \mathbf{H}_{i,k}^T R(s_{i,k})$$

$$\tilde{\mathbf{x}}_{i,k}(s_{i,k}) = \hat{\mathbf{x}}_{k|k-1} + \mathbf{K}_{i,k}(s_{i,k})\left[z_{i,k} - \hat{z}_{i,k|k-1}(s_{i,k}) \right].$$

   - Calculate the posterior $f(s_{i,k}|\mathbf{z}_{i,1:k})$: Equations (19.30–19.32).

   - Estimate $\left\{ \bar{\mathbf{x}}_{i,k}, \bar{\mathbf{P}}_{i,k} \right\}$: Equations (19.29) and (19.33).

   end for

3. Update the $\left\{ \hat{\mathbf{x}}_k, \hat{\mathbf{P}}_k \right\}$ according to (19.34).

end for

19.4   TRACKING METHOD BASED ON ARBPF

19.4.1   Generic PF

The basic idea behind PFs is as follows. Denote $\mathbf{y}_k \triangleq \begin{bmatrix} \mathbf{x}_k \\ \mathbf{s}_k \end{bmatrix}$, and suppose a set of N weighted samples $\left\{ \mathbf{y}_{k-1}^{(j)}, w_{k-1}^{(j)} \right\}_{j=1}^N$ is used to approximate the posterior $f(\mathbf{y}_{k-1}|\mathbf{z}_{1:k-1})$ at time t_{k-1} with the following point distribution:

$$f(\mathbf{y}_{k-1}|\mathbf{z}_{1:k-1}) \approx \sum_{j=1}^N w_{k-1}^{(j)} \delta\left(\mathbf{y}_{k-1} - \mathbf{y}_{k-1}^{(j)} \right), \tag{19.35}$$

where $\delta(\cdot)$ denotes the Dirac delta function.

Then, new samples are generated from a suitably designed proposal distribution, which may depend on the old state and the new measurement: $\mathbf{y}_k^{(j)} \sim \pi\left(\mathbf{y}_k | \mathbf{y}_{k-1}^{(j)}, \mathbf{z}_k \right)$. The new importance weights are set to

$$w_k^{(j)} \propto w_{k-1}^{(j)} \frac{f\left(\mathbf{z}_k | \mathbf{y}_k^{(j)} \right) f\left(\mathbf{y}_k^{(j)} | \mathbf{y}_{k-1}^{(j)} \right)}{\pi\left(\mathbf{y}_k^{(j)} | \mathbf{y}_k^{(j)}, \mathbf{z}_k \right)}. \tag{19.36}$$

Thus, a new set of samples $\left\{ \mathbf{y}_k^{(j)}, w_k^{(j)} \right\}_{j=1}^N$ is approximately distributed according to $f(\mathbf{y}_k|\mathbf{z}_{1:k})$ at time t_k by the above sequential Monte Carlo (SMC) procedure.

Since, in most cases, it is difficult or computationally too expensive to directly sample from the posterior, some trial sample densities can be used to draw particles. In standard particle filtering, transition priors are utilized as the proposal distribution

$$\pi\left(\mathbf{y}_k|\mathbf{y}_{k-1}^{(j)}, \mathbf{z}_k\right) = f\left(\mathbf{y}_k|\mathbf{y}_{k-1}^{(j)}\right) = f\left(\mathbf{x}_k|\mathbf{x}_{k-1}^{(j)}\right) f\left(\mathbf{s}_k|\mathbf{s}_{k-1}^{(j)}\right). \qquad (19.37)$$

Thus,

$$w_k^{(j)} \propto w_{k-1}^{(j)} f\left(\mathbf{z}_k|\mathbf{y}_k^{(j)}\right). \qquad (19.38)$$

However, in this problem, the mobile state $\mathbf{x}_k$ and the sight condition state $\mathbf{s}_k$ constitute a high dimensional state space (vector $\mathbf{s}_k$ has M independent components, so there are 2^M states of sight conditions). Thus, to obtain an accurate estimation, a large number of particles should be used, which prohibitively increases computational complexity.

The Rao-Blackwellized particle filter (RBPF) decomposes the state space into two parts, with one part being estimated by particle filtering and the other part being analytically calculated. As a result, the variance of the estimates can be reduced compared with the standard particle filtering [27]. RBPF has been previously investigated in mobile tracking in [28–30]. In Section 19.4.2, we use the RBPF method in our problem.

19.4.2 ARBPF

Factorize the posterior $f(\mathbf{x}_k, \mathbf{s}_k|\mathbf{z}_{1:k})$ according to Bayes' rule:

$$f\left(\mathbf{x}_k, \mathbf{s}_k|\mathbf{z}_{1:k}\right) = f\left(\mathbf{x}_k|\mathbf{s}_k, \mathbf{z}_{1:k}\right) f\left(\mathbf{s}_k|\mathbf{z}_{1:k}\right). \qquad (19.39)$$

If the posterior density of $f(\mathbf{s}_k|\mathbf{z}_{1:k})$ could be represented by a set of weighted samples $\left\{\mathbf{s}_k^{(j)}, w_k^{(j)}\right\}_{j=1}^{N}$, that is,

$$f\left(\mathbf{s}_k|\mathbf{z}_{1:k}\right) \approx \sum_{j=1}^{N} w_k^{(j)} \delta\left(\mathbf{s}_k - \mathbf{s}_k^{(j)}\right), \qquad (19.40)$$

then the marginal density $f(\mathbf{s}_k|\mathbf{z}_{1:k})$ can be approximately expressed by a mixture of densities:

$$f\left(\mathbf{x}_k|\mathbf{z}_{1:k}\right) \approx \sum_{j=1}^{N} w_k^{(j)} f\left(\mathbf{x}_k|\mathbf{s}_k, \mathbf{z}_{1:k}\right) \delta\left(\mathbf{s}_k - \mathbf{s}_k^{(j)}\right) = \sum_{j=1}^{N} w_k^{(j)} f\left(\mathbf{x}_k|\mathbf{s}_k^{(j)}, \mathbf{z}_{1:k}\right), \qquad (19.41)$$

where the mixture component $f\left(\mathbf{x}_k|\mathbf{s}_k^{(j)}, \mathbf{z}_{1:k}\right)$ approximately conforms to Gaussian distribution $N\left(\hat{\mathbf{x}}_k^{(j)}, \hat{\mathbf{P}}_k^{(j)}\right)$, which can be calculated by decentralized EKF, an extension to decentralized KF [26]:

$$\hat{\mathbf{x}}_k^{(j)} = \hat{\mathbf{x}}_{k|k-1}^{(j)} + \sum_{j=1}^{M} \mathbf{K}_{i,k}^{(j)} \left(z_{i,k} - \hat{z}_{i,k|k-1}^{(j)} \right), \tag{19.42a}$$

$$\hat{\mathbf{P}}_k^{(j)} = \left[\left(\hat{\mathbf{P}}_{k|k-1}^{(j)} \right)^{-1} + \sum_{i=1}^{M} \left(\mathbf{H}_{i,k}^{(j)} \right)^T \mathbf{R} \left(s_{i,k}^{(j)} \right)^{-1} \mathbf{H}_{i,k}^{(j)} \right]^{-1} \tag{19.42b}$$

where

$$\hat{\mathbf{x}}_{k|k-1}^{(j)} = \mathbf{\Phi} \hat{\mathbf{x}}_{k-1}^{(j)}, \tag{19.43a}$$

$$\hat{\mathbf{P}}_{k|k-1}^{(j)} = \mathbf{\Phi}_{k-1} \hat{\mathbf{P}}_{k-1}^{(j)} \mathbf{\Phi}_{k-1}^T + \mathbf{Q}, \tag{19.43b}$$

$$\hat{z}_{i,k|k-1}^{(j)} = h_i \left(\hat{\mathbf{x}}_{k|k-1}^{(j)} \right) + m \left(s_{i,k}^{(j)} \right), \tag{19.43c}$$

$$\mathbf{K}_{i,k}^{(j)} = \hat{\mathbf{P}}_{i,k}^{(j)} \left(\mathbf{H}_{i,k}^{(j)} \right)^T \mathbf{R} \left(s_{i,k}^{(j)} \right)^{-1} \tag{19.43d}$$

$$\mathbf{H}_{i,k}^{(j)} = \left. \frac{\partial h_i}{\partial \mathbf{x}} \right|_{\mathbf{x} = \hat{\mathbf{x}}_{k|k-1}^{(j)}}. \tag{19.43e}$$

To sample $\mathbf{s}_k^{(j)}$ from $f(\mathbf{s}_k | \mathbf{z}_{1:k})$, we choose the optimal trial distribution, which minimizes the variance of the importance weights, conditioned upon $\mathbf{s}_{k-1}^{(j)}$, $\mathbf{x}_{k-1}^{(j)}$ and $\mathbf{z}_k$ [27]:

$$\begin{aligned} \pi \left(\mathbf{s}_k | \mathbf{s}_{k-1}^{(j)}, \mathbf{x}_{k-1}^{(j)}, \mathbf{z}_k \right)_{\text{opt}} &= P \left(\mathbf{s}_k | \mathbf{s}_{k-1}^{(j)}, \mathbf{x}_{k-1}^{(j)}, \mathbf{z}_k \right) \\ &= \frac{f \left(\mathbf{z}_k | \mathbf{s}_k, \mathbf{s}_{k-1}^{(j)}, \mathbf{x}_{k-1}^{(j)} \right) P \left(\mathbf{s}_k | \mathbf{s}_{k-1}^{j} \right)}{f \left(\mathbf{z}_k | \mathbf{s}_{k-1}^{(j)}, \mathbf{x}_{k-1}^{(j)} \right)}, \end{aligned} \tag{19.44}$$

where $f \left(\mathbf{z}_k | \mathbf{s}_k, \mathbf{s}_{k-1}^{(j)}, \mathbf{x}_{k-1}^{(j)} \right)$ can be further approximated as

$$\begin{aligned} f \left(\mathbf{z}_k | \mathbf{s}_k, \mathbf{s}_{k-1}^{(j)}, \mathbf{x}_{k-1}^{(j)} \right) &= \int f \left(\mathbf{z}_k | \mathbf{s}_k, \mathbf{x}_k \right) f \left(\mathbf{x}_k | \mathbf{x}_{k-1}^{(j)} \right) d\mathbf{x}_k \\ &\approx \int f \left(\mathbf{z}_k | \mathbf{s}_k, \mathbf{x}_k \right) \delta \left(\mathbf{x}_k - \hat{\mathbf{x}}_{k|k-1}^{(j)} \right) d\mathbf{x}_k \\ &= f \left(\mathbf{z}_k | \mathbf{s}_k, \hat{\mathbf{x}}_{k|k-1}^{(j)} \right). \end{aligned} \tag{19.45}$$

Based on the independent transition of the M sight conditions (19.6), the trial distribution (19.44) can be further expressed as

$$\pi \left(\mathbf{s}_k | \mathbf{s}_{k-1}^{(j)}, \mathbf{x}_{k-1}^{(j)}, \mathbf{z}_k \right)_{\text{opt}} = \frac{\prod_{i=1}^{M} f \left(z_{i,k} | \mathbf{x}_{k|k-1}^{(j)}, s_{i,k} \right) P \left(s_{i,k} | s_{i,k-1}^{(j)} \right)}{f \left(\mathbf{z}_k | s_{k-1}^{(j)}, \mathbf{x}_{k-1}^{(j)} \right)}. \tag{19.46}$$

The likelihood $f \left(z_{i,k} | \hat{\mathbf{x}}_{k|k-1}^{(j)}, s_{i,k}^{(j)} \right)$ conforms approximately to a Gaussian distribution with mean $\hat{z}_{i,k|k-1}^{(j)}$ (19.43c) and covariance:

$$\hat{\mathbf{\Sigma}}_{i,k|k-1}^{(j)} = \mathbf{H}_{i,k}^{(j)} \hat{\mathbf{P}}_{k|k-1}^{(j)} \left(\mathbf{H}_{i,k}^{(j)} \right)^T + \mathbf{R} \left(s_{i,k}^{(j)} \right). \tag{19.47}$$

The importance weight corresponding to the optimal trial distribution can be calculated as

$$
\begin{aligned}
w_k^{(j)} &\propto w_{k-1}^{(j)} f\left(\mathbf{z}_k \mid \mathbf{s}_{k-1}^{(j)}, \mathbf{x}_{k-1}^{(j)}\right) = w_{k-1}^{(j)} \sum_{s_k}\left[f\left(\mathbf{z}_k \mid \mathbf{s}_k, \mathbf{x}_{k-1}^{(j)}\right) \mathrm{P}\left(\mathbf{s}_k \mid \mathbf{s}_{k-1}^{(j)}\right)\right] \\
&= w_{k-1}^{(j)} \sum_{s_k}\left[\prod_{i=1}^{M} f\left(z_{i,k} \mid \mathbf{x}_{k|k-1}^{(j)}, s_{i,k}\right) \mathrm{P}\left(s_{i,k} \mid s_{i,k-1}^{(j)}\right)\right].
\end{aligned}
\tag{19.48}
$$

From (19.48), the importance weight $w_k^{(j)}$ only depends on the current measurement $\mathbf{z}_k$ and the particles of t_{k-1}, that is,

$$
\left\{\hat{\mathbf{x}}_{k-1}^{(j)}, \hat{\mathbf{P}}_{k-1}^{(j)}, \mathbf{s}_{k-1}^{(j)}\right\}_{j=1}^{N},
$$

while $\mathbf{s}_k$ is marginalized out. Thus, to improve the sample effectiveness, the particles of t_{k-1} could be selected (resampled) based on current measurement $\mathbf{z}_k$, and the fittest particles could be allowed to propagate. Then, new particles could be sampled from $\mathbf{s}_k$ based on $\mathbf{s}_{k-1}^{(j)}$ and $\mathbf{z}_k$ according to (19.46). Algorithm 19.3 summarizes the whole scheme.

The advantage of the method is that by factorizing the posterior $f(\mathbf{x}_k, \mathbf{s}_k \mid \mathbf{z}_{1:k})$, we could first use PF to estimate the marginal posterior $f(\mathbf{s}_k \mid \mathbf{z}_{1:k})$, then use decentralized EKF to analytically compute the mean and covariance of the mobile state. As a result of the marginalization, the estimated covariance can be reduced compared with the standard PF.

In particle filtering, we choose the optimal trial distribution to achieve the minimum weight conditional variance of importance weights. The resampling step is implemented before sampling step, and the fittest particles are chosen to propagate. Thus, the particle effectiveness is improved. The method applies the EKF to calculate the mobile state $\mathbf{x}_t$, which introduces the approximation. Compared with the basic RBPF method, we call the method the ARBPF.

Algorithm 19.3   ARBPF method

for k = 1, 2… do

    for j = 1, 2, …, N do

    Compute predicted mean $\hat{\mathbf{x}}_{k|k-1}^{(j)}$ and covariance $\hat{\mathbf{P}}_{k|k-1}^{(j)}$ using (19.43a, b) and new weight $w_k^{(j)}$ using (19.48).

    end for

    Resample particles $\left\{w_k^{(j)}, \mathbf{s}_{k-1}^{(j)}, \hat{\mathbf{x}}_{k|k-1}^{(j)}, \hat{\mathbf{P}}_{k|k-1}^{(j)}\right\}_{j=1}^{N}$ using new weights $w_k^{(j)}$ to obtain $\left\{w_k^{(l)}, \mathbf{s}_{k-1}^{(l)}, \hat{\mathbf{x}}_{k|k-1}^{(l)}, \hat{\mathbf{P}}_{k|k-1}^{(l)}\right\}_{l=1}^{N}$, where $w_k^{(l)} = 1/N$.

    for l = 1, 2, …, N do

    1. Sample $\mathbf{s}_k^{(l)} \sim \mathrm{P}\left(\mathbf{s}_k \mid \mathbf{s}_{k-1}^{(l)}, \mathbf{z}_k\right)$ according to (19.46).

    2. Update using prior particles $\left\{\mathbf{s}_{k-1}^{(l)}, \hat{\mathbf{x}}_{k|k-1}^{(l)}, \hat{\mathbf{P}}_{k|k-1}^{(l)}\right\}_{l=1}^{N}$ and decentralized EKF according to (19.42a–19.43d) to obtain $\left\{\mathbf{s}_k^{(l)}, \hat{\mathbf{x}}_k^{(l)}, \hat{\mathbf{P}}_k^{(l)}\right\}_{l=1}^{N}$.

    end for

end for

19.5   LOWER BOUND OF PERFORMANCE

An error lower bound gives an indication of performance limitations. In time-invariant statistical models, a commonly used lower bound is the Cramer–Rao bound (CRB). In the time-varying systems context we deal with here, a CRB for random parameters is referred to as posterior CRLB [31]. In this section, the posterior CRLB of the described problem is derived under the assumption that LOS and NLOS transition history is known, which avoids the false detection of sight condition.

Let $\hat{\mathbf{x}}_k$ be an estimator of the vector $\mathbf{x}_k$ based on the measurements $\mathbf{z}_{1:k}$. Then, the estimate covariance $\mathbf{P}_k$ is bounded by the posterior CRLB $\mathbf{J}_k^{-1}$:

$$\mathbf{P}_k = \mathrm{E}\left\{[\hat{\mathbf{x}}_k - \mathbf{x}_k][\hat{\mathbf{x}}_k - \mathbf{x}_k]^T\right\} \geq \mathbf{J}_k^{-1}, \tag{19.49}$$

where $\mathbf{J}_k$ is the posterior Fisher information matrix (FIM):

$$\mathbf{J}_k = \mathrm{E}\left\{-\nabla_{\mathbf{x}_k}\nabla_{\mathbf{x}_k}^T \log f(\mathbf{x}_k, \mathbf{z}_k)\right\}. \tag{19.50}$$

$\nabla_{\mathbf{x}_k}$ is the first-order partial derivative operator with respect to $\mathbf{x}_k$, and $\geq$ means that the difference $\mathbf{P}_k - \mathbf{J}_k^{-1}$ is a positive semidefinite matrix.

Tichavsky et al. [32] show that FIM $\mathbf{J}_k$ can be recursively calculated as

$$\mathbf{J}_{k+1} = \mathbf{D}_k^{22} - \mathbf{D}_k^{21}\left(\mathbf{J}_k + \mathbf{D}_k^{11}\right)\mathbf{D}_k^{12}, \tag{19.51}$$

where

$$\mathbf{D}_k^{11} = \mathrm{E}\left\{-\nabla_{\mathbf{x}_k}\nabla_{\mathbf{x}_k}^T \log f(\mathbf{x}_{k+1}|\mathbf{x}_k)\right\}, \tag{19.52a}$$

$$\mathbf{D}_k^{12} = \left[\mathbf{D}_k^{21}\right]^T = \mathrm{E}\left\{-\nabla_{\mathbf{x}_k}\nabla_{\mathbf{x}_{k+1}}^T \log f(\mathbf{x}_{k+1}|\mathbf{x}_k)\right\}, \tag{19.52b}$$

$$\mathbf{D}_k^{22} = \mathrm{E}\left\{-\nabla_{\mathbf{x}_{k+1}}\nabla_{\mathbf{x}_{k+1}}^T \log f(\mathbf{x}_{k+1}|\mathbf{x}_k)\right\} + \mathrm{E}\left\{-\nabla_{\mathbf{x}_{k+1}}\nabla_{\mathbf{x}_{k+1}}^T \log f(\mathbf{z}_{k+1}|\mathbf{x}_{k+1})\right\}$$
$$= \mathbf{D}_k^{22,a} + \mathbf{D}_k^{22,b}, \tag{19.52c}$$

and the recursion (19.51) is initialized with

$$\mathbf{J}_0 = \mathrm{E}\left\{-\nabla_{\mathbf{x}_0}\nabla_{\mathbf{x}_0}^T \log f(\mathbf{x}_0)\right\}. \tag{19.53}$$

For the case of the linear dynamic Gaussian white noise acceleration state model in (19.1), (19.52) can be simplified as

$$\mathbf{D}_k^{11} = \Phi_k^T \mathbf{Q}_k^{-1} \Phi_k, \tag{19.54a}$$

$$\mathbf{D}_k^{12} = \left(\mathbf{D}_k^{21}\right)^T = -\Phi_k^T \mathbf{Q}_k^{-1}, \tag{19.54b}$$

$$\mathbf{D}_k^{22} = \mathbf{Q}_k^{-1} + \mathbf{D}_k^{22,b}. \tag{19.54c}$$

Substituting (19.54) into (19.51) and applying the inverse matrix lemma, we get

$$\mathbf{J}_{k+1} = \left(\mathbf{Q}_k + \Phi_k \mathbf{J}_k^{-1}\Phi_k^T\right)^{-1} + \mathbf{D}_k^{22,b}. \tag{19.55}$$

The expectation $\mathbf{D}_k^{22,b}$ relates to nonlinear measurement equation and thus has no analytically closed-form results. Monte Carlo random sampling approach could be used to circumvent the difficulty by converting the above integrals to summations [33]. Considering that the mobile state can be approximately estimated using decentralized EKF, we here propose a new method using sigma point set and unscented transformation, a deterministic sampling method with relatively low computation complexity.

Using linearization approximation, $\mathbf{D}_k^{22,b}$ can be further computed as

$$\mathbf{D}_k^{22,b} = \frac{1}{2}\mathrm{E}\left\{\nabla_{\mathbf{x}_{k+1}}\nabla_{\mathbf{x}_{k+1}}^T[\mathbf{z}_{k+1} - \mathbf{h}(\mathbf{x}_{k+1}) - m(\mathbf{s}_{k+1})]^T\Sigma_{k+1}^{-1}[\mathbf{z}_{k+1} - \mathbf{h}(\mathbf{x}_{k+1}) - m(\mathbf{s}_{k+1})]\right\}$$
$$\approx \mathrm{E}_{f(\mathbf{x}_{k+1}|\mathbf{z}_{k+1})}\left\{\mathbf{H}(\mathbf{x}_{k+1})^T\Sigma_{k+1}^{-1}\mathbf{H}(\mathbf{x}_{k+1})\right\},$$
$$(19.56)$$

where $\mathbf{H}(\mathbf{x}_{k+1}) = [\mathbf{H}_1(\mathbf{x}_{k+1}), \ldots, \mathbf{H}_M(\mathbf{x}_{k+1})]^T$ and $\mathbf{H}_i(\mathbf{x}_{k+1}) = \dfrac{\partial h_i(\mathbf{x})}{\partial \mathbf{x}}\Big|_{\mathbf{x}=\mathbf{x}_{k+1}}$. The matrix Σ_{k+1} is the measurement covariance matrix, which is time varying because of the changing LOS/NLOS.

Let

$$\Lambda_k^{22,b} \triangleq \mathbf{H}(\mathbf{x}_{k+1})^T\Sigma_{k+1}^{-1}\mathbf{H}(\mathbf{x}_{k+1}). \qquad (19.57)$$

Then, Equations (19.56) can be simplified as

$$\mathbf{D}_k^{22,b} = E_{f(\mathbf{x}_{k+1}|\mathbf{z}_{k+1})}\left\{\Lambda_k^{22,b}\right\}. \qquad (19.58)$$

Under the assumption that the LOS and NLOS condition between MS and each BS is known during the whole MS trajectory, the density $f(\mathbf{x}_k|\mathbf{z}_k)$ is Gaussian. Since the time-varying LOS and NLOS conditions have different mean and variance, we again apply the decentralized EKF method to compute the $f(\mathbf{x}_k|\mathbf{z}_k)$ approximately. The mean vector $\hat{\mathbf{x}}_k$ is:

$$\hat{\mathbf{x}}_k = \hat{\mathbf{x}}_{k|k-1} + \sum_{i=1}^M \mathbf{K}_{i,k}\left(z_{i,k} - \hat{z}_{i,k|k-1}\right), \qquad (19.59)$$

where

$$\hat{z}_{i,k|k-1} = h_i\left(\hat{\mathbf{x}}_{k|k-1}\right) + m\left(s_{i,k}\right), \qquad (19.60a)$$

$$\mathbf{K}_{i,k} = \hat{\mathbf{P}}_{i,k}\mathbf{H}_i\left(\mathbf{x}_{k|k-1}\right)^T \mathbf{R}\left(s_{i,k}\right)^{-1}, \qquad (19.60b)$$

$$\hat{\mathbf{P}}_{i,k} = \left[\hat{\mathbf{P}}_{k|k-1}^{-1} + \mathbf{H}_i\left(\mathbf{x}_{k|k-1}\right)^T \mathbf{R}\left(s_{i,k}\right)^{-1}\mathbf{H}_i\left(\mathbf{x}_{k|k-1}\right)\right]^{-1}, \qquad (19.60c)$$

$$\hat{\mathbf{P}}_k = \left[\hat{\mathbf{P}}_{k|k-1}^{-1} + \sum_{i=1}^M \mathbf{H}_i\left(\mathbf{x}_{k|k-1}\right)^T \mathbf{R}\left(s_{i,k}\right)^{-1}\mathbf{H}_i\left(\mathbf{x}_{k|k-1}\right)\right]^{-1}. \qquad (19.60d)$$

Assume a n_x dimension motion state variable $\mathbf{x}_k$ is estimated with mean $\hat{\mathbf{x}}_k$ and covariance $\hat{\mathbf{P}}_k$. To calculate the statistics of $\mathbf{D}_k^{22,b}$, we use unscented transformation method as follows: First, a set of $2n_x + 1$ sigma points $SS_k^{(j)} = \{W_k^{(j)}, \mathbf{x}_k^{(j)}\}$ can be deterministically sampled from the multivariate Gaussian distribution $\mathbf{x}_k^{(j)} \sim N(\hat{\mathbf{x}}_k, \hat{\mathbf{P}}_k)$. A symmetric set of sigma points can be generated according to Example 19.2. Parameter $\kappa > 0$ is a scaling parameter.

Example 19.2 Symmetric Sigma Points Set of $N[\hat{\mathbf{x}}_k, \hat{\mathbf{P}}_k]$. Write MATLAB code to generate the symmetric set of Sigma points described in the following table.

Index (j)	Weight ($W_k^{(j)}$)	Sigma point ($\mathbf{x}_k^{(j)}$)
0	$\dfrac{\kappa}{n_x + \kappa}$	$\hat{\mathbf{x}}_k$
$1, \ldots, n_x$	$\dfrac{1}{2(n_x + \kappa)}$	$\hat{\mathbf{x}}_k + \left(\sqrt{(n_x + \kappa)\hat{\mathbf{P}}_k}\right)_j$
$n_x + 1, \ldots, 2n_x$	$\dfrac{1}{2(n_x + \kappa)}$	$\hat{\mathbf{x}}_k - \left(\sqrt{(n_x + \kappa)\hat{\mathbf{P}}_k}\right)_j$

Solution

The example MATLAB code for generating Sigma point is given in separate file: Chapter_19_Example_19_2.m. MATLAB codes can be found online at ftp://ftp.wiley.com/public/sci_tech_med/matlab_codes.

Substituting $\mathbf{x}_k^{(j)}$ into (19.57), $\Lambda_{k-1}^{22,b(j)}$ can be calculated. Then, the expectation $\mathbf{D}_{k-1}^{22,b}$ can be computed as

$$\mathbf{D}_{k-1}^{22,b} = \sum_{j=0}^{2n_x} \Lambda_{k-1}^{22,b(j)} W_k^{(j)}. \tag{19.61}$$

A detailed scheme of posterior CRLB computation is given in Algorithm 19.4.

Algorithm 19.4 Posterior CRLB computation

for $k = 1, 2,\ldots$ do

1. Predict the mean and covariance of mobile state

$$\hat{\mathbf{x}}_{k|k-1} = \Phi_{k-1}\hat{\mathbf{x}}_{k-1}.$$

$$\hat{\mathbf{P}}_{k|k-1} = \Phi_{k-1}\hat{\mathbf{P}}_{k-1}\Phi_{k-1}^T + \mathbf{Q}_k.$$

2. Update the $\{\hat{\mathbf{x}}_k, \hat{\mathbf{P}}_k\}$ using decentralized EKF method according to (19.59–19.60d).
3. Deterministically choose a set of $2n_x + 1$ sigma points $SS_k^{(j)} = \{W_k^{(j)}, \mathbf{x}_k^{(j)}\}$ according to Example 19.2.
4. Compute the $\Lambda_{k-1}^{22,b(j)} = \left(\mathbf{H}\left(\mathbf{x}_{k+1}^{(j)}\right)\right)^T \Sigma_{k+1}^{-1} \mathbf{H}\left(\mathbf{x}_{k+1}^{(j)}\right)$ according to (19.57) and estimate the expectation according to (19.61).
5. Update $\mathbf{J}_k$ according to (19.55).

6. The position MSE bound is $\sqrt{\mathbf{J}_k^{-1}(1,\,1)+\mathbf{J}_k^{-1}(2,\,2)}$, where, $\mathbf{J}_k^{-1}(1,1)$ and $\mathbf{J}_k^{-1}(2,\,2)$ are the bounds on the MSE corresponding to x_k and y_k, respectively.

end for

Symmetric sigma set and unscented transformation method compute mean and covariance to second-order accuracy [34]. Thus, based on the analytical estimate of the mobile state, and the deterministic sampling method using sigma set and unscented transformation, the posterior CRLB can be effectively calculated.

19.6 NUMERICAL RESULTS

In the simulation, it is assumed that the MS can receive the signals from three BS all the time. The coordinates of BS are [−3.0 km, −1.0 km], [−3.0 km, 5.0 km], and [5.0 km, −1.0 km]. The mobile trajectories are generated according to the mobility model described in Section 19.2, in which the initial position of the MS is set to [−1.5 km, 1.5 km], and the initial velocity is set as [20 m/s, 0 m/s]. The random acceleration variances σ_x^2, σ_y^2 are both chosen to be 0.5 $(m/s^2)^2$. The simulated trajectory has $L = 1600$ time samples, and the sample interval $\Delta t = 0.2$ seconds. The simulated measurement data are generated by adding the measurement noise and the NLOS noise to the true distance from MS to each BS. The measurement noise is assumed to be a white random variable with zero mean and standard deviation $\sigma_m = 150$ m, whereas the NLOS measurement noise is also assumed to be a white random variable with positive mean $\mu_{NLOS} = 513$ m, and standard deviation $\sigma_{NLOS} = 409$ m [12]. The sight condition between the MS and each BS is changed every 200 samples [13, 35, 36].

The performance of the IMM [35], the modified EKF banks, and the ARBPF method are compared. Different number of particles are used in the ARBPF method, denoted as ARBPF(N) for brevity. The initial position is calculated by Chan's algorithm [37] using the first three range measurements. For vague prior of the mobile state, the initial velocity is set as [0 m/s, 0 m/s] and the covariance matrix

$$\mathbf{C}_{t0} = \begin{bmatrix} 150^2 \cdot \mathbf{I}_2 & \mathbf{0} \\ \mathbf{0} & 20^2 \cdot \mathbf{I}_2 \end{bmatrix},$$

corresponding to a standard deviation of 150 m for the position and 20 m/s (72 km/h) for the velocity of each coordinate. The initial estimation of sight condition is set to $P(s_{i,0} = 0) = P(s_{i,0} = 1) = 0.5$, where $i = 1, 2$, and 3.

The indexes used in the evaluation are

1. Root Square Error: $\text{RSE} \triangleq \sqrt{(\hat{x}_k - x_k)^2 + (\hat{y}_k - y_k)^2}$,

2. Root Mean Square Error (RMSE) at time k:

$$\text{RMSE}_k \triangleq \sqrt{\frac{1}{n_{mc}} \sum_{m=1}^{n_{mc}} \left[(\hat{x}_{k,m} - x_k)^2 + (\hat{y}_{k,m} - y_k)^2 \right]},$$

3. Average $\text{RMSE} \triangleq \frac{1}{L} \sum_{k=1}^{L} \text{RMSE}_k$.

TABLE 19.1.   Average RMSE versus Transition Probability of Sight Condition

	0.55	0.65	0.75	0.85	0.95
IMM	201.5	158.5	113.8	83.2	57.6
Modi-EKF	115.2	108.0	81.9	68.6	53.7
ARBPF(10)	109.6	92.4	73.3	50.8	48.6
ARBPF(100)	109.6	92.3	69.4	45.2	42.5

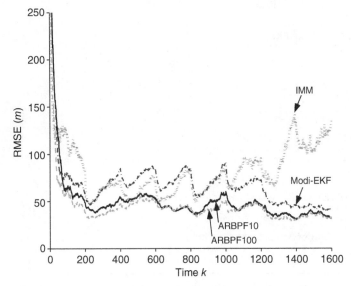

Figure 19.2   Position RMSE versus time k.

The mobile location error is calculated after 100 samples so as to ignore the error caused by the initial settings. Simulation results are obtained based on $n_{mc} = 50$ Monte Carlo realizations for all the algorithms.

19.6.1   Performance Comparison with Different Algorithms

Figure 19.2 shows the position RMSE versus time, and Figure 19.3 shows the cumulative distribution function (CDF) of RSE. The prior transition probability of LOS/NLOS is set as $p_0 = p_1 = 0.85$. From Figures 19.2 and 19.3, the ARBPF (100) has slightly better performance than ARBPF(10), while both ARBPF(100) and ARBPF(10) can track more accurately than the IMM and modified EKF method.

Table 19.1 shows the impact of different transition probabilities on the algorithms. Since the transition matrix is symmetric, the values are studied from 0.55 to 0.95. From Table 19.1, the larger setting errors in the transition probability (the prior errors) lead to larger tracking errors in all the algorithms. However, ARBPF can track more effectively than the others, which suggests that ARBPF can estimate the posterior distribution more correctly despite prior setting errors. The improvement may come from its optimal sampling distribution and effective particle selecting mechanisms.

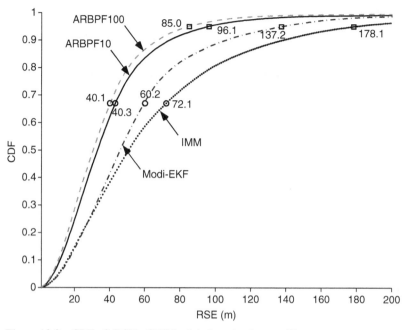

Figure 19.3   CDF of (RSE). GNSS, global navigation satellite system.

TABLE 19.2.   One Realization of Actual Sight Condition

	T_1	T_2	T_3	T_4	T_5	T_6	T_7	T_8
$s_{1,k}$	0	1	0	0	0	0	0	0
$s_{2,k}$	1	0	0	1	0	0	0	0
$s_{3,k}$	0	1	0	1	0	0	0	0

$T_p = \{200(p-1) + k | k = 1,2,\ldots,200\}$; $s_{i,k} = 0$ for LOS and $s_{i,k} = 1$ for NLOS.

19.6.2   Comparison with Posterior CRLB

To compare the algorithms' performance with posterior CRLB, the mobile trajectory and the sight conditions between MS and each BS are generated first and fixed in all the 50 Monte Carlo simulations. The sight condition is assumed to be known when computing the posterior CRLB. Table 19.2 shows one realization of the actual sight condition used in the simulation.

The position RMSE is presented in Figure 19.4. The error standard derivations of all the algorithms are higher than the posterior bound, especially in the time periods T_1, T_2, and T_4, which suggests that the NLOS condition between the MS and a certain BS interferes in the mobile estimation.

Since the first 100 estimations are ignored to decrease the impact of initial values, the CDF of RSE in posterior CRLB is almost a vertical line within [29.7 m, 30.7 m] in Figure 19.5. Figure 19.5 also suggests that, by comparing with the posterior CRLB, there is still room for improving the performance.

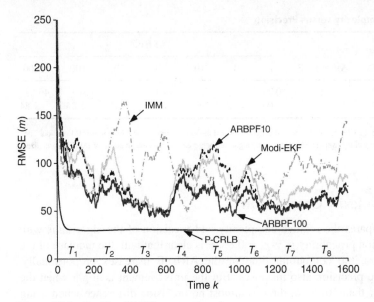

Figure 19.4   Position RMSE versus time k.

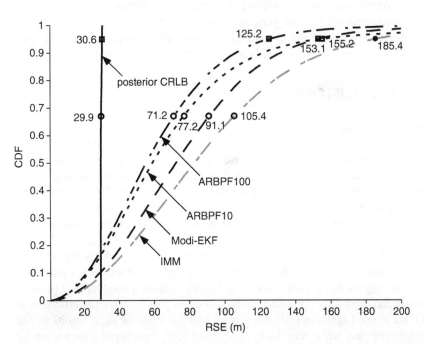

Figure 19.5   CDF of (RSE). IMU, inertial measurement unit; INS, inertial navigation system.

TABLE 19.3. Complexity versus Precision

	IMM	Modi-EKF	ARBPF						
			1	4	10	30	50	100	1000
Complexity	1	1.6	0.7	2	4.9	14.1	23.3	46.3	462.3
Precision	1	1.21	1.06	1.28	1.64	1.71	1.78	1.84	1.84

Complexity is based on the CPU running time of the algorithms and the value is proportional to that of IMM method. The precision is the reciprocal of the average RMSE that each algorithm achieves, and also the precision of IMM is normalized.

19.6.3 Complexity Comparison

Table 19.3 compares the relative complexity and precision of the algorithms with the prior transition probability $p_0 = p_1 = 0.85$. It is clear that with the increase of the particle numbers, the computing time of the ARBPF increases proportionally. Accordingly, the precision also increases. But the improvement is slight when the number is larger than 10. Also, there is almost no precision difference when using 100 and 1000 particles. Thus, it can be concluded that ARBPF(10) achieves a good trade-off between complexity and precision.

In this section, we compare the performance of three methods and also with posterior CRLB in LOS/NLOS transition conditions. Simulation results show that ARBPF(10) and ARBPF(100) can achieve more accurate estimation than modified EKF method and IMM. Also ARBPFs are more robust to prior setting errors. When the sight conditions can be correctly detected, the error performances of ARBPF(10) and ARBPF(100) are in good agreement with the posterior CRLB. Complexity analysis suggests that the ARBPF(10) achieves a good trade-off between complexity and precision.

19.7 CONCLUSIONS

The chapter investigated the problem of mobile tracking in mixed LOS/NLOS conditions. The problem was considered under the Bayesian estimation framework, which is based on the prior information of the mobile state model and the dynamic transition model of LOS/NLOS sight conditions. Two different algorithms were presented, that is, the modified EKF banks and the ARBPF method. The posterior CRLB for mobile tracking in this kind of condition was further studied. Simulation tests suggest that, among all the algorithms compared, the ARBPF method can be less influenced by setting errors of prior probability and achieves the most estimation accuracy. When the LOS and NLOS transition probability is estimated correctly, the modified EKF and ARBPF have close performance and both have smaller estimation errors than IMM methods. Complexity comparison suggests that the ten particles of ARBPF achieve a good tradeoff between complexity and precision. Simulation results also show that the error performances of ARBPF are in good agreement with the posterior CRLB when the sight conditions can be correctly detected, while considering the CDF of RSE, there is still a large room for improving the location performance, which deserves deep investigation to develop more efficient tracking methods in the further research.

REFERENCES

[1] M. I. Silventoinen and T. Rantalainen, "Mobile station emergency locating in GSM," in *Proc. of IEEE Int. Conf. on Personal Wireless Communications*, Dallas, TX, Jun. 1996, pp. 232–238.

[2] J. Borras, P. Hatrack, and N. B. Mandayam, "Decision theoretic framework for NLOS identification," in *Proc. of IEEE Vehicular Technology Conf.*, Ottawa, Canada, May 1998, pp. 731–734.

[3] J. Riba and A. Urruela, "A non-line-of-sight mitigation technique based on ML-detection," in *Proc. of IEEE Int. Conf. on Acoustics, Speech, and Signal Processing*, vol. 2, Montreal, QC, Canada, May 2004, pp. 153–156.

[4] Y. T. Chan, W. Y. Tsui, H. C. So, and P. C. Ching, "Time-of-arrival based localization under NLOS conditions," *IEEE Trans. Veh. Technol.*, vol. 55, no. 1, pp. 17–24, 2006.

[5] P. C. Chen, "A non-line-of-sight error mitigation algorithm in location estimation," in *Proc. of IEEE Wireless Communications Networking Conf.*, Sep. 1999, pp. 316–320.

[6] L. Cong and W. Zhang, "Non-line-of-sight error mitigation in TDOA mobile location," in *Proc. of IEEE GLOBECOM*, vol. 1, San Antonio, TX, Dec. 2001, pp. 680–684.

[7] N. Khajehnouri and A. H. Sayed, "A non-line-of-sight equalization scheme for wireless cellular location," in *Proc. of IEEE Conf. on Acoustics, Speech, and Signal Processing*, vol. 6, Apr. 2003, pp. 549–552.

[8] S. Al-Jazzar, J. J. Caffery, and H. R. You, "A scattering model based approach to NLOS mitigation in TOA location systems," in *Proc. of IEEE Vehicular Technology Conf.*, vol. 2, Birmingham, UK, 2002, pp. 861–865.

[9] S. Al-Jazzar and J. J. Caffery, "ML and Bayesian TOA location estimators for NLOS environments," in *Proc. of IEEE Vehicular Technology Conf.*, vol. 2, Birmingham, UK, May 2002, pp. 1178–1181.

[10] Y. Qi, H. Kobayashi, and H. Suda, "Analysis of wireless geolocation in a non-line-of-sight environment," *IEEE Trans. Wireless Commun.*, vol. 5, no. 3, 672–681, 2006.

[11] H. Miao, K. Yu, and M. J. Juntti, "Positioning for NLOS propagation: Algorithm derivation and Cramer-Rao bounds," *IEEE Trans. Veh. Technol.*, vol. 56, no. 5, pp. 2568–2580, 2007.

[12] M. P. Wylie and J. Holtzman, "The non-line of sight problem in mobile location estimation," in *Proc. of IEEE Int. Conf. on Universal Personal Communications*, vol. 2, 1996, pp. 827–831.

[13] B. L. Le, K. Ahmed, and H. Tsuji, "Mobile location estimatior with NLOS mitigation using Kalman filtering," in *IEEE Wireless Communications and Networking Conf.*, vol. 3, New Orleans, LA, 2003, pp. 1969–1973.

[14] J.-F. Liao and B.-S. Chen, "Robust mobile location estimator with NLOS mitigation using interacting multiple model algorithm," *IEEE Trans. Wireless Commun.*, vol. 5, no. 11, pp. 3002–3006, 2006.

[15] C. Morelli, M. Nicoli, V. Rampa, and U. Spagnolini, "Hidden Markov models for radio localization in mixed LOS/NLOS conditions," *IEEE Trans. Signal Process.*, vol. 55, no. 4, pp. 1525–1542, 2007.

[16] M. Nicoli, C. Morelli, and V. Rampa, "A Jump Markov particle filter for localization of moving terminals in multipath indoor scenarios," *IEEE Trans. Signal Process.*, vol. 56, no. 8, pp. 3801–3809, 2008.

[17] B. Ristic, S. Arulampalam, and N. Gordon, *Beyond the Kalman Filter, Particle Filters for Tracking Applications*. Boston, MA: Artech House, 2004.

[18] S. Ali-Löytty, "Gaussian mixture filter in hybrid positioning," Ph.D. dissertation, Tampere University of Technology, Aug. 2009. Available: http://URN.fi/URN:NBN:fi:tty-200905191055

[19] W. Cheney and W. Light, *A Course in Approximation Theory*. Pacific Grove, CA: Brooks/Cole Publishing Company, 2000.

[20] S. Ali-Löytty and N. Sirola, "Gaussian mixture filter and hybrid positioning," in *Proc. of ION GNSS 2007*, Fort Worth, TX, Sep. 2007, pp. 562–570.

[21] S. Ali-Löytty, "On the convergence of the Gaussian mixture filter," in *Gaussian mixture filter in hybrid positioning*, Ph.D. dissertation, Tampere University of Technology, Aug. 2009. Available: http://URN.fi/URN:NBN:fi:tty-200905191055

[22] B. D. O. Anderson and J. B. Moore, *Optimal Filtering*. Englewood Cliffs, NJ: Prentice-Hall, 1979.

[23] S. Ali-Löytty and N. Sirola, "Gaussian mixture filter in hybrid navigation," in *Proc. of the European Navigation Conf. GNSS 2007*, May 2007, pp. 831–837.

[24] H. W. Sorenson and D. L. Alspach, "Recursive Bayesian estimation using Gaussian sums," *Automatica*, vol. 7, no. 4, pp. 465–479, 1971.

[25] D. J. Salmond, "Mixture reduction algorithms for target tracking," in *IEEE Colloquium on State Estimation in Aerospace and Tracking Applications,* 1989, pp. 7/1–7/4.

[26] H. F. Durrant-Whyte, B. Y. S. Rao, and H. Hu, "Toward a fully decentralized architecture for multi-sensor data fusion," in *Proc. of IEEE Int. Conf. on Robotics and Automation*, Cincinnati, OH, May 1990, pp. 1331–1336.

[27] M. S. Arulampalam, S. Maskell, N. Gordon, and T. Clapp, "A tutorial on particle filters for online nonlinear/non-Gaussian Bayesian tracking," *IEEE Trans. Signal Process.*, vol. 50, no. 2, pp. 174–188, 2002.

[28] R. Chen and J. Liu, "Mixture Kalman filters," *J. R. Stat. Soc B*, vol. 62, pp. 493–508, 2000.

[29] F. Gustafsson and F. Gunnarsson, "Mobile positioning using wireless networks: possibilities and fundamental limitations based on available wireless network measurements," *IEEE Signal Process. Mag.*, vol. 22, no. 4, pp. 41–53, Jul. 2005.

[30] L. Mihaylova, D. Angelove, S. Honary, D. R. Bull, C. N. Canagarajah, and B. Ristic, "Mobility tracking in cellular networks using particle filtering," *IEEE Trans. Wireless Commun.*, vol. 6, no. 10, pp. 3589–3599, 2007.

[31] H. L. van Trees, *Detection, Estimation and Modulation Theory*, vol. I. New York: Wiley, 1968.

[32] P. Tichavsky, C. H. Muravchik, and A. Nehorai, "Posterior Cramer-Rao bounds for discrete-time nonlinear filtering," *IEEE Trans. Signal Process.*, vol. 46, no. 5, pp. 1386–1396, 1998.

[33] R. Taylor, B. R. Flanagan, and J. A. Uber, "Computing the recursive posterior Cramer-Rao bound for a nonlinear nonstationary system," in *Proc. of Int. Conf. on Acoustics, Speech and Signal Processing (ICASSP)*, Hong Kong, Apr. 2003, pp. 673–676.

[34] S. J. Julier and J. K. Uhlmann, "Unscented filtering and nonlinear estimation," *Proc. IEEE*, vol. 92, no. 3, pp. 401–422, 2004.

[35] B.-S. Chen, C.-Y. Yang, F.-K. Liao, and J.-F. Liao, "Mobile location estimator in a rough wireless environment using extended kalman-based IMM and data fusion," *IEEE Trans. Veh. Technol.*, vol. 58, no. 3, pp. 1157–1169, 2009.

[36] L. Chen and L. Wu, "Mobile positioning in mixed LOS/NLOS conditions using modified EKF banks and data fusion method," *IEICE Trans. Commun.*, vol. EB92, no. 4, pp. 1318–1325, 2009.

[37] Y. T. Chan and K. C. Ho, "A simple and efficient estimator for hyperbolic location," *IEEE Trans. Signal Process.*, vol. 42, no. 8, pp. 1905–1915, 1994.

PART V

GLOBAL POSITIONING

WITH THE rapid emergence of global positioning in autonomous vehicles, this part of the handbook includes four chapters, Chapters 20–23, that discuss the details of positioning based on Global Navigation Satellite Systems (GNSS).

Chapter 20 discusses the fundamentals of satellite navigation systems, i.e., GNSS. The chapter introduces the fundamentals of global position estimation using Time-of-Arrival estimation incorporating signals received from a cluster of satellites. The impact of the satellite constellation geometry on the positioning accuracy is assessed. Further, an overview of current and future GNSS, signal formats and modulations is presented. This chapter also offers MATLAB examples as well as the corresponding code.

Chapter 21 introduces the general architecture of a GNSS receiver. The signal processing performed in the various blocks of the receiver is described in detail. The different functions of the receiver are explained including signal detection and acquisition for the satellites in view. The chapter also discusses low complexity signal detection methods such as the Fast Fourier Transform (FFT), and Bayesian detection theory as implemented in typical GNSS receivers. This chapter also includes MATLAB examples and their corresponding code.

Chapter 22 discusses the implementation of Kalman filtering for satellite navigation. In particular, the structure of the KF is exploited in a GNSS receiver to compute the position of the user and to integrate the standard GNSS receiver with an Inertial Navigation System (INS). It discusses the characterization of the inertial devices in terms of both deterministic and stochastic noise, calibration and alignment, as fundamental pre-requisites for any fruitful integration process. The chapter presents three typical GNSS-INS integration approaches: (a) loosely integrated, (2) tightly integrated and (3) ultra-tightly integrated architectures. Finally, the results of a live vehicular test campaign to compare two such implementations are shown and discussed.

Handbook of Position Location: Theory, Practice, and Advances, Second Edition.
Edited by S. A. (Reza) Zekavat and R. Michael Buehrer.
© 2019 by the Institute of Electrical and Electronics Engineers, Inc.
Published 2019 by John Wiley & Sons, Inc.
Companion Website: www.wiley.com/go/zekavat/positionlocation2e

Chapter 23 presents strategies and techniques that increase the sensitivity of GNSS receivers, to make them usable in difficult environments, such as urban canyons, indoor scenarios, deep forests, or even space. The chapter discusses assisted GNSS receiver. It also introduces approaches to increase the sensitivity at the acquisition stage, including delay and Doppler shift estimation/compensation, and the intrinsic limitations to coherent and non-coherent integration time extension.

CHAPTER *20*

OVERVIEW OF GLOBAL POSITIONING SYSTEMS

Fabio Dovis,[1] Davide Margaria,[2] Paolo Mulassano,[2]
and Fabrizio Dominici[2]
[1]Politecnico di Torino, Italy
[2]Istituto Superiore Mario Boella, Italy

I**N THIS** chapter, fundamentals of satellite navigation are discussed. The chapter introduces the fundamentals of the global position estimation using time of arrival estimation of signals received from a cluster of satellites. The impact of the satellite constellation geometry on the global positioning accuracy is assessed. An overview of current and future global navigation satellite systems, signal formats, and modulations is presented. Examples of augmentation systems capable of improving the positioning performance are briefly presented. Examples of terrestrial and satellite systems complementary to the baseline satellite constellation and able to improve the positioning performance are briefly presented.

20.1 INTRODUCTION

Several countries worldwide are independently investing in the development and modernization of global navigation satellite systems (GNSS). Even though the most recent satellite constellations (such as the European Galileo and the Chinese BeiDou) introduce innovation at the system and signal levels, all of them share common theoretical and functional principles. Global positioning system (GPS) is the most popular among today's GNSS and is considered by system integrators as a mature technology. Most of our cars and mobile phones are equipped with low-cost mass-market receivers providing a sufficiently accurate user position. Accuracy, however, is not the only gure of merit for a good position estimation. In future GNSS, there is still a large potential for improving other performance metrics, such as reliability and service continuity. For this reason, GPS itself is still evolving and a modernization

Handbook of Position Location: Theory, Practice, and Advances, Second Edition.
Edited by S. A. (Reza) Zekavat and R. Michael Buehrer.
© 2019 by the Institute of Electrical and Electronics Engineers, Inc.
Published 2019 by John Wiley & Sons, Inc.
Companion Website: www.wiley.com/go/zekavat/positionlocation2e

plan is foreseen by the United States government. The GLobal NAvigation Satellite System (GLONASS), managed by the Russian federation, is also being modernized and upgraded. The reason for such worldwide interest in GNSS, apart from strategic decisions related to national security, is due to the growing commercial relevance of location-based services (LBS). After the revolution of the "wireless communication era" a "navigation era" recently started, and new LBS are being explored and deployed, boosting the growth of new applications based on user position. GNSS innovation is expected to grow in the near future in order to match new requirements stemming from the challenging safety-of-life services domain, in which reliability and system trust are key considerations.

In this evolving context, a clear view on the basic assumptions related to the general problem of knowing the position with respect to some reference frame or map needs to be understood. Keeping in mind that this problem can be approached in several ways, this chapter focuses on GNSS as the primary source of navigation signals, highlighting the innovative features of GNSS (with respect to the today's GPS) that will enable new services in the near future.

20.1.1 What Is Radio Navigation?

Most radio-navigation Plans (i.e., strategic documents written by most countries) define "navigation" as "driving in a safe and secure mode a mobile from a starting point to its nal destination." This statement underlines that navigation is by definition a "real-time" process, in the sense that it is a process that takes place on a time scale comparable to the speed of a mobile user. Navigation is then the determination of the successive positions of a mobile that must be obtained and maintained on a temporal scale consistent with the mobile speed. Considering radio navigation, such positions (and possibly real-time speed values) are obtained on the basis of observations of an electromagnetic signal, by estimating or measuring some parameters of the signal itself.

In the case of GNSS systems, these parameters are the propagation times from the transmitter (i.e., from each satellite in view) to the receiver. Such parameters are used in order to estimate the distances between the receiver and a reference point (i.e., the transmitter), the position of which is known in a predefined reference frame. In most satellite-based navigation systems, these distances are not directly estimated due to a lack of perfect synchronization between transmitters and receivers. Thus, *pseudoranges* (or range differences) are estimated instead. The user position estimate is then obtained by the intersection of geometrical loci solving a multidimensional equation system. Depending on the measurements performed, the location solution is provided by the intersection of a certain number of geometrical loci such as circles and hyperbolas (in a two-dimensional scenario) or spheres, hyperboloids, and cones (in a three-dimensional scenario).

20.1.2 Spherical Systems

In these kind of systems, the receiver estimates a parameter of the incoming signal from the sources whose value is proportional to the absolute distance from a set of transmitters placed at known locations.

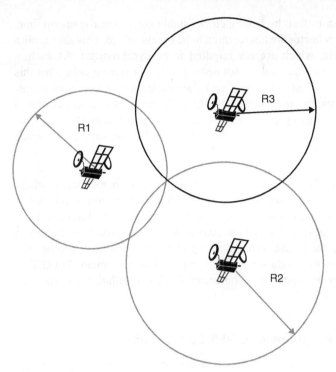

Figure 20.1   A spherical positioning system.

The measurement of an absolute distance from the sources allows for a user position estimate as the intersection among spheres. The centers of the spheres are the transmitters and the radii are equal to the distance estimates. These kind of systems are usually based on the received signal strength or a measurement of a time of arrival (TOA) as in Figure 20.1. In TOA systems, the time estimation can be estimated in different ways, implying different levels of complexity and different constraints on the timing systems.

Two-Way Measurements: The estimation of the propagation time is performed measuring the round-trip time, i.e., the signal transmitted is rebroadcast by the counterpart, and the transmitter evaluates the difference between the transmission time and the received time. Such a difference includes twice the propagation time between transmitter and user plus a possible latency time due to the retransmission (in case the counterpart is not acting like a pure reflector for the signal). The concept can be implemented on the user side, transmitting the signal to a set of reflectors and then obtaining the distance from each of them, or a set of reference transmitters can broadcast the signal to the user, which acts like a reflector. In the latter case, to intersect the geometrical loci, the reference transmitters must be interlinked in order to exchange the pieces of information collected by each of them. The main advantage of this type of time measurement is that transmission and reception time are measured on the same time scale. In fact, by having clocks stable during the round-trip time, the transmitter and the reflector are not required to be synchronized. This also applies to different transmitters, which are not required to be synchronized. As such, a common precise reference time scale is not needed. It is worth mentioning that this concept embeds a "privacy" issue that may not be desirable for sensitive applications. The interaction between the transmitter and the reflector is also evidence that a positioning procedure is taking place. In GNSS, due to constraints related to the distance between satellites and receivers (more than 20000 km), such an approach is not used.

One-Way Measurements: The one-way propagation time is estimated, where users only receive the signal broadcast by the set of reference transmitters (i.e., satellites in the GNSS case). Transmitters must be synchronous with high precision (within dozens of ns), having a common time-scale. This approach is chosen in GNSS systems, having the GNSS terminal acting only as a receiver of the satellite signals. Further sections will highlight how this approach is implemented in GNSS, presenting the problem of synchronization between the constellation and the user terminal.

20.1.3 Evolution Programs of GNSS Constellations

In addition to the legacy signals currently transmitted in L1 (1575.42 MHz) and L2 (1227.60 MHz) frequency bands by GPS and GLONASS systems (both in a modernization phase), a large number of signals are being transmitted by new GNSS, such as Galileo, the chinese BeiDou Navigation Satellite System (BDS), and other regional systems like the japanese Quasi-Zenith Satellite System (QZSS) and the Indian Regional Navigational Satellite System (IRNSS, also known as NavIC), sharing the same carrier frequencies and also using new bands (e.g., E5, E6, and S band). Table 20.1 gives an overview of the current GNSS scenario as of August 2017 (see [32] and references therein), taking into account the main characteristics of each system.

In this modern multisystem GNSS environment, a user able to process the signals from several constellations is able to obtain better accuracy, reduced geometrical constraints (i.e., a larger number of visible satellites in the sky portion) and increased reliability. The major technical innovations introduced in the last decade and related to GNSS can be summarized as follows:

- New services and interoperable signals are being implemented and deployed, enabling a new market for multifunction and multiband receivers in several application domains (e.g., new road toll collection schemes based on GNSS);
- new modulation and multiplexing schemes have been studied and implemented, allowing a more robust and reliable signal reception (e.g., reducing multipath errors);
- larger bandwidths (equal or larger than 10 times the bandwidth of the legacy GPS L1 C/A signal) are available, leading to better performance in the position estimation;

TABLE 20.1. GNSS Constellations and Features (as of August 2017 [32])

System	Number of Satellites	Main Features
Galileo	18 (15 operational + 2 test + 1 unusable); planned constellation with 30 satellites (24 operational + 6 spare)	• Global coverage (all MEO satellites) • First test satellite (GIOVE-A) launched in Dec. 2005 • Galileo initial Open Service declared operational in Dec. 2016 • Foreseen full operational capability from 2020 • Signals over 3 carrier frequencies (E1, E5, E6) • Ongoing studies for Galileo Second Generation (G2G)
GPS	35 (31 operational + 1 test + 3 reserve)	• Global coverage (all MEO) and full operational capability • First Block I test satellite (Navstar 1) launched in Feb. 1978 • Constellation in 24+3 (or "Expandable 24") configuration • Modernized L2C and L5 signals in preoperational phase • Foreseen new L1C signal (with TMBOC, in Block III) • First Block III satellite launch no earlier than 2018
GLONASS	27 (23 operational + 3 test + 1 spare)	• Global coverage (all MEO) and full operational capability • Frst test satellite (Kosmos-1413) launched in Oct. 1982 • FDMA-based legacy signals on L1 and L2 • New CDMA signals over 3 frequencies (L1, L2, L3) • 2 GLONASS-K1 satellites already in orbit • New GLONASS-K2 satellites in development
BDS (BeiDou)	23 (14 operational + 9 unusable); planned constellation with 37 satellites (5 GEO + 27 MEO + 5 IGSO)	• Initial regional system (BeiDou-1) • First test GEO satellite (BeiDou-1A) launched in Oct. 2000 • Second phase (BeiDou-2) with global coverage • Declared operational in China and surroundings in Dec. 2011 • Open service and authorized service • Launch of new satellites (BeiDou-3) started in 2015 • Forseen global coverage by 2020
QZSS	3 (1 operational + 2 test satellites); planned 4 satellites constellation (3 IGSO + 1 GEO) from 2018	• Regional coverage (Japan and surroundings) • First satellite (QZS-1, IGSO) launched in Sept. 2010 • Second satellite (QZS-2, IGSO) launched in June 2017 • Third satellite (QZS-3, GEO) launched on August 2017 • Interoperable with GPS signals (L1-C/A, L1C, L2C, L5) • New Centimeter Level Augmentation Service on L6 signal • Foreseen future expansion up to 7 satellites
IRNSS (NavIC)	7 (3 GEO + 4 IGSO satellites)	• Regional coverage (India and surroundings) • Also known as Navigation with Indian Constellation (NavIC) • First IGSO satellite (IRNSS-1A) launched in July 2013 • Signals at 1176.45 MHz (L5-band) and 2492.028 MHz (S-band) • System expected to be operational by early 2018 • Foreseen future expansion from 7 to 11 satellites

- pilot (dataless) channels are now transmitted, suitable to reduce the time to first fix and to enhance the receiver sensitivity;
- cryptographically protected signals (by means of authentication and/or encryption strategies) are becoming available also for civil applications, reducing the risk of attacks based on counterfeit navigation signals (e.g., spoofing [5]).

20.2 PRINCIPLES OF SATELLITE NAVIGATION

GNSS allows the users to estimate in real-time their position, velocity, and time (PVT) with respect to a reference frame in the space and time domains. Satellite navigation systems operate based on TOA estimation in order to determine the user position. The receiver of these systems measures the propagation time of signals broadcast by a set of satellites at known locations that are the reference points, as in Figure 20.1.

The propagation time has to be multiplied by the speed of light to obtain the distance between the receiver and the satellite. The obtained distances have to be combined to estimate the receiver position.

Assuming that the receiver clock is perfectly synchronized with the satellite transmitter's clock, the distance between each satellite and user can be calculated by measuring the transit time of the signal. If the jth satellite transmits a pulse at t_0, and it is received at time $t_0 + \tau$, the distance (R_j) between the transmitter (jth satellite) and the receiver can be estimated as:

$$R_j = c \cdot \tau, \tag{20.1}$$

where c is the speed of light.

In three-dimensional space, every distance R_j defines a spherical surface whose center is the position of the jth satellite. Through the intersection of at least three of these spheres (see Fig. 20.1), it is possible to compute one point that represents a precise user position. Strictly speaking, three spheres intersect in two points, but one intersection point can easily be rejected due to fact that it is located in an implausible location (e.g., in deep space).

However, in a real situation, the receiver clock is not synchronized with the transmitter. While all the satellite payloads host synchronous clocks, it is not possible to have user clocks aligned with the satellite time scale at low cost and complexity. Furthermore, GNSS are conceived to be *one-way* systems (i.e., not requiring the user to interact with the satellite constellation), thus not requiring two-way methods for keeping the satellites and the receivers aligned to the same time scale.

For this reason, the measure of the distance suffers a bias, as shown in Figure 20.2 by the ε term, that is common to each satellite. This bias represents the shift of the receiver time scale with respect to the GNSS time scale. The measurement performed by the receiver is then called *pseudorange* ρ, and it is defined as the sum of the true distance R_j and a term due to the time scale misalignment. Analytically the pseudorange for the jth satellite (ρ_j) can be written as

$$\rho_j = R_j + \varepsilon = R_j + c \cdot \delta t_u, \tag{20.2}$$

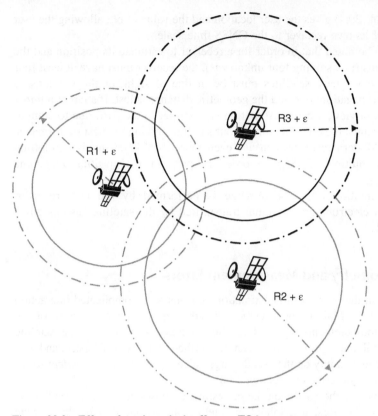

Figure 20.2   Effect of receiver clock offset on TOA measurements.

where c is the speed of the light and δt_u is the user clock bias. This term cannot be estimated using the data contained in the satellite signal; therefore, the intersection of another sphere generated on a further satellite is necessary. The generic jth pseudorange can be written as

$$\rho_j = \sqrt{(x_{sj} - x_u)^2 + (y_{sj} - y_u)^2 + (z_{sj} - z_u)^2} + b_{ut}, \tag{20.3}$$

where x_u, y_u, z_u are the user coordinates, x_{sj}, y_{sj}, and z_{sj} are the coordinates of the jth satellite, and $b_{ut} = c \cdot \delta t_u$ is the clock bias term. The intersection of four spheres from four satellites is then given by the following system of equations:

$$\begin{cases} \rho_1 = \sqrt{(x_{s1} - x_u)^2 + (y_{s1} - y_u)^2 + (z_{s1} - z_u)^2} + b_{ut} \\ \rho_2 = \sqrt{(x_{s2} - x_u)^2 + (y_{s2} - y_u)^2 + (z_{s2} - z_u)^2} + b_{ut} \\ \rho_3 = \sqrt{(x_{s3} - x_u)^2 + (y_{s3} - y_u)^2 + (z_{s3} - z_u)^2} + b_{ut} \\ \rho_4 = \sqrt{(x_{s4} - x_u)^2 + (y_{s4} - y_u)^2 + (z_{s4} - z_u)^2} + b_{ut}. \end{cases} \tag{20.4}$$

The solution of (20.4) gives the user location and the value of δt_u, allowing the user to synchronize its own receiver to the GNSS time scale.

It has to be noted that in order for a receiver to estimate its position and the clock bias terms (i.e., solving four unknowns), the receiver must have at least four satellites in view. These satellites must be in line of sight, or the relationship between the propagation time and the geometric distance is lost. If a larger number of satellites is in view, a better estimation is possible. In the past, due to computational constraints, the combination of four satellites giving the best performance was chosen. Modern receivers can use even more than 12 channels (exploiting signals from multiple GNSS) in order to perform the position and time estimation.

The estimation of a pseudorange is performed by the user receiver, processing an electromagnetic signal transmitted by the satellites as described in Chapter 21.

20.2.1   Geometry and Measurement Errors

The accuracy in the user position estimation depends on a complicated interaction of various factors. Mainly, the accuracy directly depends on the quality of the pseudorange measurements, as well as the accurate estimation of the satellite position. Several nonnegligible errors can be ascribed to the control, space, and user segments. The uncertainty on the pseudorange value is called *user equivalent range Error* (UERE), as described in Section 20.8.

The impact of the UERE on the nal estimated position is also dependent on some "geometric" factors, describing how satellites are displaced in the sky over the user. In Section 20.2.2 the relation between the UERE and the geometry of satellites will be derived, under proper assumptions.

20.2.2   Impact of Measurement Errors on User Position

Once a receiver has successfully locked four (or more) satellites, it can use the n measured pseudoranges for solving a set of equations in four unknowns to obtain the position coordinates x_u, y_u, and z_u, and the clock bias δt_u. Knowing an approximation of the true position* and bias $\left(\hat{x}_u, \hat{y}_u, \hat{z}_u, \hat{b}_{ut} \right)$, and expanding (20.3) in a Taylor series centered in $\left(\hat{x}_u, \hat{y}_u, \hat{z}_u, \hat{b}_{ut} \right)$ it is possible to obtain the position offset $(\Delta x_u, \Delta y_u, \Delta z_u, \Delta b_{ut})$ as a linear function of the known coordinates and of the pseudorange measurements.

---

*The rough knowledge of the position can always be assumed unless the receiver is operating in the so-called cold start mode (e.g., the first time it is turned on). The approximate point of linearization can be the last calculated solution or, in some cases, it could be provided by an external source of information. However, receivers usually apply recursive procedures for getting the solution, with faster convergence to a reliable result depending on how close the linearization point is to the true position.

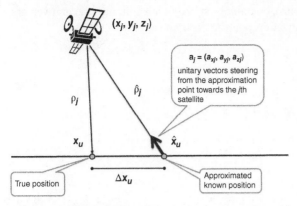

Figure 20.3   Example of a linearization in a one-dimensional scenario.

Figure 20.3 depicts the different variables in a simplified one-dimensional scenario. The linearized equation becomes:

$$
\begin{aligned}
\rho_j &= f\left(x_u, y_u, z_u, b_{ut}\right) \\
&= f\left(\hat{x}_u + \Delta x_u, \hat{y}_u + \Delta y_u, \hat{z}_u + \Delta z_u, \hat{b}_{ut} + \Delta b_{ut}\right) \\
&= f\left(\hat{x}_u, \hat{y}_u, \hat{z}_u, \hat{b}_{ut}\right) + \frac{\partial f\left(\hat{x}_u, \hat{y}_u, \hat{z}_u, \hat{b}_{ut}\right)}{\partial x_u}\Delta x_u + \frac{\partial f\left(\hat{x}_u, \hat{y}_u, \hat{z}_u, \hat{b}_{ut}\right)}{\partial y_u}\Delta y_u \\
&\quad + \frac{\partial f\left(\hat{x}_u, \hat{y}_u, \hat{z}_u, \hat{b}_{ut}\right)}{\partial z_u}\Delta z_u + \frac{\partial f\left(\hat{x}_u, \hat{y}_u, \hat{z}_u, \hat{b}_{ut}\right)}{\partial b_{ut}}\Delta b_{ut} + \ldots \\
&= \hat{\rho}_j - \frac{x_{sj} - \hat{x}_u}{\hat{r}_j}\Delta x_u - \frac{y_{sj} - \hat{y}_u}{\hat{r}_j}\Delta y_u - \frac{z_{sj} - \hat{z}_u}{\hat{r}_j}\Delta z_u + \Delta b_{ut} \\
&= \hat{\rho}_j - a_{x_j}\Delta x_u - a_{y_j}\Delta y_u - a_{z_j}\Delta z_u + \Delta b_{ut},
\end{aligned}
\tag{20.5}
$$

where $\hat{r}_j = \sqrt{(x_{sj} - \hat{x}_u)^2 + (y_{sj} - \hat{y}_u)^2 + (z_{sj} - \hat{z}_u)^2}$, and $\hat{\rho}_j = f\left(\hat{x}_u, \hat{y}_u, \hat{z}_u, \hat{b}_{ut}\right)$.

The expansion has been truncated at the first order to eliminate nonlinear terms. The delta-pseudorange $\Delta\rho_j = \hat{\rho}_j - \rho_j$ can be written as

$$
\Delta\rho_j = a_{xj}\Delta x_u + a_{yj}\Delta y_u + a_{zj}\Delta z_u - \Delta b_{ut},
\tag{20.6}
$$

where $1 \leq j \leq n$, n being the number of available satellites. These equations can be put in a matrix form defining:

1. $\boldsymbol{\Delta\rho} = (\Delta\rho_1 \ \Delta\rho_2 \ \ldots \ \Delta\rho_n)^T$ as the vector offset of the error-free pseudorange values corresponding to the user's actual position and the pseudorange values corresponding to the linearization point;

2. $\boldsymbol{\Delta x} = (\Delta x_u \ \Delta y_u \ \Delta z_u \ -\Delta b_{ut})^T$ as the vector offset from the position linearization point;

3. H as the geometrical matrix containing, in the first 3 columns, the unit vectors pointing from the linearization point to each jth satellite:

$$\mathbf{H} = \begin{pmatrix} a_{x1} & a_{y1} & a_{z1} & -1 \\ a_{x2} & a_{y2} & a_{z2} & -1 \\ \cdots & \cdots & \cdots & \cdots \\ a_{xn} & a_{yn} & a_{zn} & -1 \end{pmatrix}. \tag{20.7}$$

Hence,

$$\Delta\rho = \mathbf{H} \cdot \Delta\mathbf{x}, \tag{20.8}$$

which has the solution

$$\Delta\mathbf{x} = \mathbf{H}^{-1}\Delta\rho. \tag{20.9}$$

Equation (20.9) relates the displacement $\Delta\mathbf{x}$ in the user position and time bias, with respect to the linearization point to the offset in the error-free pseudoranges values $\Delta\rho$. When more than four pseudorange measurements are available, the method of least squares can be used to calculate the displacement $\Delta\mathbf{x}$.

The least squares method allows to compute the value of $\Delta\mathbf{x}$ so that $\mathbf{H}\Delta\mathbf{x}$ is the closest to $\Delta\rho$. For any particular value of $\Delta\mathbf{x}$, the vector quantity $\mathbf{r} = \mathbf{H}\Delta\mathbf{x} - \Delta\rho$ is called the residual.

The ordinary least squares solution is defined as the value of $\Delta\mathbf{x}$ that minimizes the square of the residual, which is

$$\mathbf{R}_{SE}(\Delta\mathbf{x}) = (\mathbf{H} \cdot \Delta\mathbf{x} - \Delta\rho)^2. \tag{20.10}$$

The solution can be obtained differentiating (20.10) with respect to $\Delta\mathbf{x}$ to obtain the gradient of $\mathbf{R}_{SE}$. Hence the gradient is set to zero and solved for $\Delta\mathbf{x}$ to seek a value minimizing $\mathbf{R}_{SE}$:

$$\nabla\mathbf{R}_{SE} = 2(\Delta\mathbf{x})^T \mathbf{H}^T \mathbf{H} - 2(\Delta\rho)^T \mathbf{H}. \tag{20.11}$$

Taking the transposed of (20.11) and setting it to zero, it is then possible to solve for $\Delta\mathbf{x}$, obtaining the formulation for overdimensioned problems. Provided that $\mathbf{H}^T\mathbf{H}$ is nonsingular, Equation (20.11) has solution

$$\Delta\mathbf{x} = (\mathbf{H}^T\mathbf{H})^{-1} \mathbf{H}^T \Delta\rho. \tag{20.12}$$

The condition that $\mathbf{H}^T\mathbf{H}$ is nonsingular is equivalent to the condition that the tips of the unit vectors from the linearization point to the satellites are not linearly dependent.

Usually any GNSS receiver searches for the minimum of (20.12) by using recursive methods or a Kalman filter (see Chapter 22). Once the unknowns are computed, the user's coordinates x_u, y_u, z_u, and the clock offset b_{ut} are then obtained.

20.3 THE IMPACT OF GEOMETRY

A formal derivation of the impact of the geometry relations needs to consider that the pseudorange measurements are not error free. In fact, the true user-to-satellite measurements are corrupted by independent errors, and their impact on the final solution has to be taken into account.

The set of equations to solve has then to be written as

$$\Delta\rho + \delta\rho = \mathbf{H}(\Delta\mathbf{x} + \delta\mathbf{x}). \tag{20.13}$$

The error in the pseudorange $\delta\rho$ can be related to the position error $\delta\mathbf{x}$ by observing the linearity of (20.13); thus

$$\delta\mathbf{x} = \left((\mathbf{H}^T\mathbf{H})^{-1}\mathbf{H}^T\right)\delta\rho, \tag{20.14}$$

where $\delta\rho$ represents the net error in the pseudorange values. The pseudorange errors are considered to be random variables and (20.14) expresses $\delta\mathbf{x}$ as a random variable functionally related to $\delta\rho$. The covariance of $\delta\mathbf{x}$ can be obtained by forming the product $\delta\mathbf{x}\delta\mathbf{x}^T$ and computing an expected value:

$$\text{cov}(\delta\mathbf{x}) = E\left\{(\mathbf{H}^T\mathbf{H})^{-1}\mathbf{H}^T\delta\rho\delta\rho^T\mathbf{H}(\mathbf{H}^T\mathbf{H})^{-1}\right\}, \tag{20.15}$$

$$\text{cov}(\delta\mathbf{x}) = (\mathbf{H}^T\mathbf{H})^{-1}\mathbf{H}^T\text{cov}(\delta\rho)\mathbf{H}(\mathbf{H}^T\mathbf{H})^{-1}. \tag{20.16}$$

Under the hypothesis of error contributions that can be modeled as Gaussian random variables with zero mean, identically distributed, independent, and with variance σ_{UERE}^2,

$$\text{cov}(\delta\rho) = \mathbf{I}_{n\times n}\sigma_{UERE}^2, \tag{20.17}$$

where $\mathbf{I}_{n\times n}$ is the identity matrix of $n \times n$ elements. σ_{UERE} is the standard deviation of the pseudorange error, a quantity obtained by combining all the error sources as described in Section 20.8, and assuming that errors on different pseudoranges are independent. The covariance matrix is then

$$\text{cov}(\delta\mathbf{x}) = (\mathbf{H}^T\mathbf{H})^{-1}\sigma_{UERE}^2 = \begin{pmatrix} \sigma_{x_u}^2 & \sigma_{x_u,y_u}^2 & \sigma_{x_u,z_u}^2 & \sigma_{x_u,b_{ut}}^2 \\ \sigma_{x_u,y_u}^2 & \sigma_{y_u}^2 & \sigma_{y_u,z_u}^2 & \sigma_{y_u,b_{ut}}^2 \\ \sigma_{x_u,z_u}^2 & \sigma_{x_u,z_u}^2 & \sigma_{z_u}^2 & \sigma_{z_u,b_{ut}}^2 \\ \sigma_{x_u,b_{ut}}^2 & \sigma_{y_u,b_{ut}}^2 & \sigma_{z_u,b_{ut}}^2 & \sigma_{b_{ut}}^2 \end{pmatrix}, \tag{20.18}$$

in which the off-diagonal elements indicate the level of cross-correlation between the variables.

The most general parameter used to assess the impact of the geometry on the final accuracy is the *geometrical dilution of precision* (GDOP) defined as

$$\text{GDOP} = \frac{\sqrt{\sigma_{x_u}^2 + \sigma_{y_u}^2 + \sigma_{z_u}^2 + \sigma_{b_{ut}}^2}}{\sigma_{UERE}}. \tag{20.19}$$

The square root term gives an overall characterization of the error in the solution. GDOP is the geometry factor, and it represents a scaling factor of the measurements error standard deviation onto the solution. Because this scaling factor is typically greater than one, it amplifies the pseudorange error, or dilutes the precision, of the position determination. By introducing the matrix $\mathbf{G} = (\mathbf{H}^T\mathbf{H})^{-1}$, the elements of which are g_{ij}, the GDOP term can be expressed as

$$\text{GDOP} = \sqrt{g_{11} + g_{22} + g_{33} + g_{44}} = \sqrt{tr\left\{(\mathbf{H}^T\mathbf{H})^{-1}\right\}}. \qquad (20.20)$$

The GDOP is a measure of how much the position error that results from measurement errors depends on the user/satellite relative geometry. This geometric effect can be understood analyzing Figure 20.4, where two different scenarios are shown. For simplicity, a two-dimensional case is considered. In both cases, the error-free rings are intersected at the user's location (dashed lines). Additional rings (continuous lines) represent the 1σ bound due to a Gaussian error on the range. The uncertainty region is the locus of possible navigation solutions depending on the instances of the error random process. The standard deviation of the ranging error is the same for both cases. However, the area of the uncertainty region is larger in the second case, where a worse case in terms of geometry is considered (higher GDOP), thus leading to higher position uncertainty.

The g_{ij} elements of matrix $\mathbf{G}$ depend only on the receiver-satellite geometry. In addition to the GDOP, several other partial DOP parameters are of common use and they are useful to estimate the accuracy of various components of the solution:

1. *Position dilution of precision (PDOP)*

$$\text{PDOP} \times \sigma_{UERE} = \sqrt{\sigma_{x_u}^2 + \sigma_{y_u}^2 + \sigma_{z_u}^2} \Rightarrow \text{PDOP} = \sqrt{g_{11} + g_{22} + g_{33}} \qquad (20.21)$$

2. *Horizontal dilution of precision (HDOP)*

$$\text{HDOP} \times \sigma_{UERE} = \sqrt{\sigma_{x_u}^2 + \sigma_{y_u}^2} \Rightarrow \text{HDOP} = \sqrt{g_{11} + g_{22}} \qquad (20.22)$$

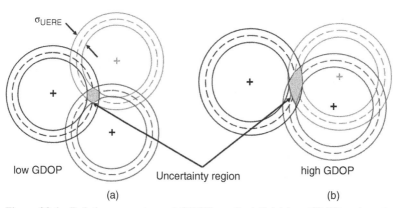

(a)   (b)

Figure 20.4   Relative geometry and GDOP: on the left (a) low GDOP, and on the right (b) high GDOP.

3. *Vertical dilution of precision (VDOP)*

$$\text{VDOP} \times \sigma_{UERE} = \sigma_{z_u} \Rightarrow \text{VDOP} = \sqrt{g_{33}} \tag{20.23}$$

4. *Time dilution of precision (TDOP)*

$$\text{TDOP} \times \sigma_{UERE} = \sigma_{b_{ut}} \Rightarrow \text{TDOP} = \sqrt{g_{44}} \tag{20.24}$$

The best case for the GDOP is when the tips of the four receiver-satellite unit vectors form a tetrahedron [19]. The larger the tetrahedron's volume, the smaller the DOPs. Of course, a GNSS receiver on the earth's surface cannot see below-horizon satellites, thus never reaching such an ideal case. If a number n of satellites, with $n > 4$, is used in the solution, than the tips of the n receiver-satellite unit vectors form a wider solid, whose volume is larger. Consequently, DOP values are lower, and hence the solution error is smaller (for the same ranging error).

20.3.1 GDOP as a Function of Position and Time

The DOP parameters are expressed as a combination of components of the matrix $\mathbf{G}$; hence they vary with the location of the user, of the satellites, and the time of the day. In particular, it can be demonstrated that the minimum GDOP value decreases directly as the inverse square root of the number of satellites used in the position computation. VDOP values are larger than the HDOP values, indicating that vertical position errors are larger than horizontal errors. The reason for this effect is that all of the satellites from which the user obtains signals are above the receiver. The horizontal coordinates do not suffer from a similar effect, as the user usually receives signals from all sides.

In Figure 20.5, the GDOP and the number of satellites are simulated using real orbital parameters for the GPS constellation on July 7, 2010, for a location at Lat = 45.2 deg and Long = 7.4 deg (in the north of Italy). It is possible to notice the presence of peaks in the outlines due to short periods of time in which the receiver experiences a bad geometry, and variations occur when satellites move in and out of view. In the GPS case, since the satellites of the constellation are in 12-hour orbits, the user-satellite geometry approximately repeats every 24 hours, and a 24-hour plot characterizes the DOP at a particular location.

The GDOP can be considered a performance parameter due to the relation between position dispersions and errors on pseudoranges, as described by (20.19). High values of GDOP denote unreliable solutions provided by the receiver. For this reason, a GDOP mask is usually set on receivers in order to exclude solutions obtained with GDOP values over a certain threshold, or at least to provide warnings to the users.

Because the various DOPs depend only on the receiver and satellite coordinates, they may be predicted ahead of time for any given set of satellites in view from a specified location using a satellite almanac. The DOP values strongly depend on the visibility of the constellation and, as a consequence, on the latitude at which the user is located. GNSS constellations are designed in order to assure a certain visibility on specific regions of the Earth. As an example, GPS constellations do not provide good visibility at higher latitudes, while Galileo, being designed also to serve

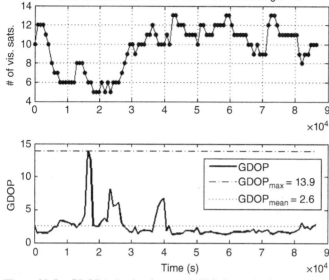

Figure 20.5   GDOP behavior for a real GPS Constellation in the north of Italy, at Lat = 45.2 deg and Long = 7.4 deg.

regions of northern Europe, has a different configuration of the constellation in order to provide a good coverage of such areas.

Furthermore, satellites' visibility can be limited in cities and canyons to high elevation satellites, with strong degradation of the DOP values. In Figure 20.6 the visibility of a GPS constellation in the north of Italy along 24 hours is simulated both for open sky (choosing a mask angle of 5 δεγρεεσ) and considering the visibility in a street 20 m wide and with buildings 15 m high, heading north-south. In the urban canyon case, the visible satellites are limited to a portion of the sky; not only the number of visible satellites is reduced but they are also close to each other, thus providing a worse GDOP condition (see Fig. 20.7). It can be noted that the number of visible satellites is often less than four, and no solution can be obtained (in these cases the GDOP would be "infinity," but for sake of presentation in the plot the value is set to 0). Due to the limited visibility, also when a sufficient number of satellites for getting a position is present, higher, not acceptable GDOP values are obtained.

Example 20.1

A terrestrial positioning system is based on the use of "GNSS-like" signal transmitters (i.e., beacons broadcasting signals with the same format of the GPS C/A code), placed as in Figure 20.8. Users are equipped with receivers that are kept synchronized to the transmitters.

1. Evaluate the geometric dilution of precision of the system for a user in P using the measurements from TX_1, TX_2, and TX_4; verify the results using MATLAB.
2. Repeat the evaluation assuming the user receiver is not synchronized with the transmitters.

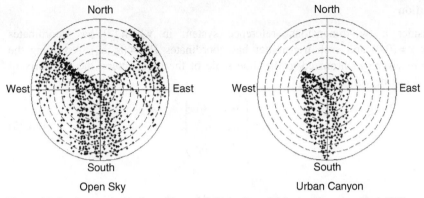

Figure 20.6   Comparison of satellites visibility, along 24 hours, for a nominal GPS constellation in the north of Italy, in case of open sky and urban canyon scenarios.

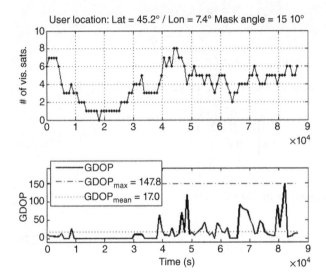

Figure 20.7   Number of visible satellites and GDOP values, along 24 hours, for a nominal GPS constellation in the north of Italy, for a urban canyon scenario.

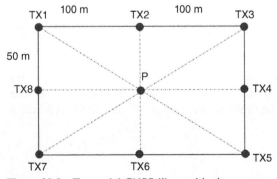

Figure 20.8   Terrestrial GNSS-like positioning system.

Solution

Consider a two-dimensional reference system in which P has coordinates $(x_P, y_P) = (0,0)$ and each transmitter has coordinates $TX_i = (a_{xi}, a_{yi})$. Assuming the user receiver synchronized to the time scale of the transmitters, the **H** matrix of (20.7) can be written as

$$\mathbf{H} = \begin{pmatrix} a_{x1} & a_{y1} \\ a_{x2} & a_{y2} \\ a_{x4} & a_{y4} \end{pmatrix}, \tag{20.25}$$

where

$$a_{xi} = \frac{x_i - x_P}{r_i}$$

$$a_{yi} = \frac{y_i - y_P}{r_i}$$

$$r_i = \sqrt{(x_i - x_P)^2 + (y_i - y_P)^2}.$$

Substituting the distance values of Figure 20.8, the geometrical matrix is

$$\mathbf{H} = \begin{pmatrix} -\frac{2}{\sqrt{5}} & \frac{1}{\sqrt{5}} \\ 0 & 1 \\ 1 & 0 \end{pmatrix}. \tag{20.26}$$

The GDOP values are obtained as in (20.19). Being

$$(\mathbf{H}^T\mathbf{H})^{-1} = \begin{pmatrix} 0.6 & 0.2 \\ 0.2 & 0.9 \end{pmatrix}, \tag{20.27}$$

a value of GDOP = 1.22 can be found.

In the second case, the time bias of the user receiver must be considered, thus having the geometrical matrix

$$\mathbf{H} = \begin{pmatrix} a_{x1} & a_{y1} & -1 \\ a_{x2} & a_{y2} & -1 \\ a_{x4} & a_{y4} & -1 \end{pmatrix}, \tag{20.28}$$

with the third column containing the coefficients of the time unknown in the pseudorange (20.6). Substituting the numerical values GDOP = 2.06 can be found. These results can be verified by means of the MATLAB script `Chapter_20_Example_1.m`.

Example 20.2

Let Consider the same system of Example 20.1 depicted in Figure 20.8. Under the hypothesis that the measurements are affected by errors that can be modeled as random variables with zero mean and standard deviation σ_{UERE}, evaluate the position estimated by a user close to the point P in case the user's receiver measures the pseudoranges:

- $\rho_1 = 112.5\,\text{m}$ from TX_1;
- $\rho_2 = 52\,\text{m}$ from TX_2;
- $\rho_3 = 99\,\text{m}$ from TX_4.

Consider both the case of user synchronized and not synchronized to the time scale of the positioning system.

Solution

Since the user is close to point P, an arbitrary position $(0; -10)$ can be chosen as linearization point. The geometrical matrix H has been evaluated in Example 20.1 for both the synchronized and not synchronized case. In order to obtain the displacement with respect to the linearization point, the least squares solution must be evaluated as

$$\Delta x = \left(\mathbf{H}^T \mathbf{H}\right)^{-1} \mathbf{H}^T \Delta \rho. \tag{20.29}$$

It has to be noticed that the measured values are different from the geometrical distances due to the errors and, if the user is not synchronized, also to the user clock bias. Being

$$\Delta \rho = \begin{pmatrix} r_1 - \rho_1 \\ r_2 - \rho_2 \\ r_3 - \rho_3 \end{pmatrix} = \begin{pmatrix} 4.12 \\ 8 \\ 1.5 \end{pmatrix} \tag{20.30}$$

and $\mathbf{H}$ can be recomputed as in Example 20.1, thus, the displacement is $\Delta x = (0.44, 8.17)$ in the synchronized case and $\Delta x = (0.27, 7.52, -0.48)$ in the nonsynchronous case. It must be remarked that in the synchronous case this is an overdimensioned problem, since more measurements are available with respect to the two unknowns, whereas in the second case, three measurements are needed in order to be able to obtain a position/time solution. The numerical values can be verified using the MATLAB script `Chapter_20.Example_2.m`.

20.4 OVERVIEW ON REFERENCE SYSTEMS

To formulate the mathematics of the satellite navigation problem, it is necessary to choose a reference coordinate system in which the states of both the satellite and the receiver can be represented. In this formulation, it is typical to describe satellite and receiver states in terms of position and velocity vectors measured in a Cartesian coordinate system. In order to compute the distance between a user and a satellite, it is necessary to have the position of both in a common reference system.

There are a number of commonly used Cartesian coordinate systems, including inertial and rotating systems. When choosing the "best" reference system, two requirements have to be taken into account:

- the position of the user is conventionally expressed in a coordinate system fixed to the earth and moves with it, so that a stationary object remain fixed;

- satellite motion is regulated by equations of motion, usually expressed in an inertial system (i.e., fixed in space or in uniform motion).

In this section, an overview of the coordinate systems used for GPS is provided. Other GNSS may choose different reference frames; however, the same basic principles have to hold.

20.4.1 Conventional Inertial Reference System

For the purposes of measuring and determining the orbits of the GNSS satellites, it is convenient to use an Earth-centered-inertial (ECI) coordinate system, in which the origin is the earth's center of mass. An ECI system is inertial in the sense that the equations of motion of an earth-orbiting satellite can be modeled as if the ECI system were unaccelerated. In other words, a GNSS satellite obeys Newton's laws of motion and gravitation in an ECI coordinate system. In typical ECI coordinate systems, the xy-plane is taken to coincide with Earth's equatorial plane, the +x-axis is permanently fixed toward the vernal (spring) equinox (i.e., direction of intersection of the earth's equatorial plane with the plane of earth's orbits around the sun), the +z-axis is taken normal to the xy-plane in the direction of the north pole along the rotation axis, and the +y-axes is chosen so as to form a right-handed coordinate system. Determination and propagation of the GNSS satellite orbits are carried out in an ECI coordinate system.

One issue in the definition of an ECI coordinate system arises due to irregularities in the earth's motion. The earth's shape is oblate, and largely due to the gravitational pull of the sun and the moon on the earth's equatorial bulge, the equatorial plane moves with respect to celestial sphere. Because the x-axis is defined relative to the celestial sphere and the z-axis is defined relative to the equatorial plane, the irregularities in the Earth's motion would cause the ECI frame as defined above not to be truly inertial. In fact, the center of mass of the earth moves around the sun (Kepler's second law). In addition, the axis of rotation of the earth is not fixed in space in relation to distant stars composite periodic components: precession and nutation due to the gravitational attraction of sun and moon that would be not present if the earth were spherical and homogeneous. However, precessions and nutation of the Earth's rotation axis are slow phenomena, having periods of 26000 years and 18.6 years, respectively, and are taken into account in the definition of the reference frames.

In the definition of the ECI system for GPS, the solution of this problem has been to define the orientation of the axes at a particular instant time, or epoch. The GPS ECI coordinate system uses the orientation of the equatorial plane at 12:00 hours UTC on January 1, 2000, as its basis. The x-axis is taken to point from the center of mass of the Earth to the direction of vernal equinox, and the y and z-axes are defined as described above, all at the aforementioned epoch. Since the orientation of the axes remains fixed, the ECI coordinate system defined in this way can be considered inertial for GPS purposes.

20.4.2 Conventional Terrestrial Reference System

For the purpose of computing the position of a GNSS receiver, it is more convenient to use a coordinate system that rotates with the earth, known as the earth-centered–earth-fixed (ECEF) system. In such a coordinate system, it is easier to compute the latitude, longitude, and height parameters that the receiver displays. As with the ECI coordinate system, the ECEF coordinate system that is used for GPS has its origin in the center of mass of the Earth and its xy-plane coincident with the Earth's equatorial plane. However, in the ECEF system, the x-axis heads to the direction of $0°$ longitude (passing through the intersection of the Greenwich meridian with the equatorial plane), and the $+y$-axes points in the direction of $90°$ East longitude. The x and y axes therefore rotate with the Earth and no longer describe fixed directions in the inertial space. In this ECEF system, the z-axis is chosen to be normal to the equatorial plane in the direction of the geographical north pole, thereby completing the right-handed coordinate system. Since the rotation axis is not fixed in relation to the solid Earth (polar motion), a conventional terrestrial pole (CTP) has been defined as the mean pole in years 1900–1905. In the ECEF system, then the origin in the center of mass of the earth, the z-axis passes through the CTP, the x-axis passes through the intersection of a reference meridian (Mean Greenwich meridian), with the CTP equatorial plane and the y-axis is chosen in the equatorial plane.

20.4.3 Ellipsoidal Coordinates

The surface of the earth is irregular and changeable, and the model of the Earth as a sphere is not reliable. Since 1735, Newton demonstrated it is flattened at the poles and that it can be approximated as an ellipsoid of revolution generated by revolving an ellipsoid about its minor axis (oblate ellipsoid). The ellipsoid is defined by means of the semimajor axis (a) and the semiminor axis (b), or in an equivalent way, by means of the eccentricity

$$e^2 = \frac{a^2 - b^2}{a^2}$$

(20.31)

and the flattening

$$f = \frac{a - b}{a}.$$

(20.32)

Referring to Figure 20.9 it is possible to define:

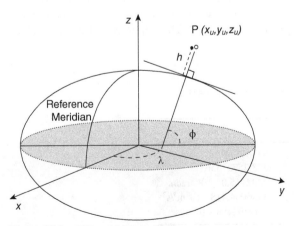

Figure 20.9   Ellipsoidal coordinates.

- geodetic latitude ϕ angle in the meridian plane through the point P between the equatorial (xy) plane of the ellipsoid and the line perpendicular to the surface of the ellipsoid passing in P (positive north from the equator);
- geodetic longitude λ angle in the equatorial plane between the reference meridian and the meridian plane through P (positive east);
- geodetic height h: measured along the normal to the ellipsoid through P.

20.4.4 The Geoid

The Earth is not spherical and not uniform in density. An absolute height is defined with respect to an idealized mean sea level. It is then necessary to define a surface that serves as the global zero reference for measurement of height. The geoid is defined as the locus of all points with the same gravity potential (equipotential surface) best fitting the average sea level globally. Such a surface is not regular but has a simple mathematical description and takes into account geological formation and topographic relief. It is usually represented numerically as a grid of points, and it is mapped relative to the reference ellipsoid. At each point a geoidal height is defined along the line perpendicular to the ellipsoid (from the ellipsoid to the geoid).

20.4.5 The Global Datum

The standard physical model of the Earth used for GPS application is the World Geodetic System 1984 (WGS84), and it represents a standard de facto. One part of the WGS84 is a detailed model of the Earth's gravitational irregularities. Such information is necessary to derive the accurate satellite ephemeris information; however, a typical GNSS receiver just aims to estimate its latitude, longitude, and height.

For this purpose, WGS84 provides an ellipsoidal model of the Earth's shape, as summarized in Table 20.2. In this model, cross-sections of the Earth parallel to the equatorial plane are circular. The equatorial cross-section of the Earth has radius 6378.137 km, which is the mean equatorial radius of the Earth. In the WGS84 Earth model, cross-sections of the Earth normal to the equatorial plane are ellipsoidal. In an ellipsoidal cross-section containing the z-axis, the major axis coincides with the equatorial diameter of the Earth. Therefore, the semimajor axis, a, has the same value as the mean equatorial radius given in Table 20.2. The minor axis of the ellipsoidal

TABLE 20.2. World Geodetic System 1984 Fundamental Parameters (1997 Revision)

Parameter	Value
Ellipsoid semimajor axis (a)	6378137.0 m
Ellipsoid reciprocal flattening	298.257223563
Earth's angular velocity	$7292115.0 \cdot 10^{-11}$ rad/s
Earth's gravitational constant	$3986004.418 \cdot 10^8$ m^3/s^2
Speed of light in vacuum	$2.99792458 \cdot 10^8$ m/s

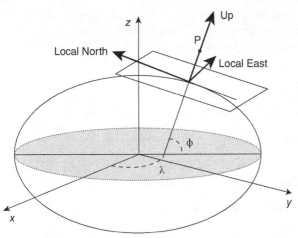

Figure 20.10   A local geodetic East-North-up (ENU) reference frame.

cross-section corresponds to the polar diameter of the Earth, and the semiminor axis, b, in WGS84 is taken to be 6356.7523142 km.

20.4.6   East-North-Up Reference Frame

The East-North-up (ENU) coordinate system is similar to the geocentric ECEF coordinate system. Both are Cartesian, which means that they have three mutually perpendicular axes and the coordinates are distances from the origin. Indeed, any particular ENU coordinate system is simply the *xyz* position represented on a reference system made of a plane tangent to the Earth surface in a local origin, with respect to which a local North, a local East, and the Up directions are defined (see Fig. 20.10). The ENU coordinate system can be very helpful when examining GNSS performance from the user perspective. As noted in Section 20.3, due to the impossible visibility of satellites below the horizon, GNSS measurements are less precise in the direction of the local vertical than horizontal (typically one and a half to twice as large). This effect is readily apparent when the positioning solutions are presented in an ENU coordinate system. The ENU representation of the performance provides the intuitive information that usually a regular user looks for when assessing "how good" a GNSS position is.

20.5   STRUCTURE OF THE SIGNAL-IN-SPACE

Nowadays, the GNSS environment is characterized by a constantly growing number of different signals. The GNSS frequency bands have been selected in the allocated spectrum for the radio navigation satellite services (RNSS). In addition to that, E5a/L5, E5b, and E1 bands are included in the allocated spectrum for aeronautical radio navigation services (ARNS), employed by civil aviation users and allowing dedicated safety-critical applications. Over these bandwidths, new signals-in-space (SIS) are going to be broadcast by modernized and new GNSS. In the following sections the main new features of the systems will be described.

20.5.1 GNSS Frequency Plan

A summary of the different frequency bands used (or foreseen) by Galileo, GPS, GLONASS, BDS (BeiDou), QZSS, and IRNSS (NavIC) are summarized in Table 20.3 [12, 14–16, 29, 3, 1, 2, 18, 32].

20.5.2 The Binary Offset Carrier Modulation

In this section the basic principles of the binary offset carrier (BOC) modulation scheme will be introduced.

TABLE 20.3. GNSS Frequency Bands (Current and Future).

System	Band	Carrier Frequency (MHz)	Bandwidth (MHz)
Galileo	E1	1575.42	24.552
	E6	1278.75	40.920
	E5 (E5a+E5b)	1191.795	51.150
GPS	L1	1575.42	20.460
	L2	1227.60	20.460
	L5	1176.45	24.000
GLONASS	L1 (FDMA)	1598.0625–1605.375	—
	L2 (FDMA)	1242.9375–1248.625	—
	L1 (CDMA)	1600.995	—
	L2 (CDMA)	1248.060	—
	L3 (CDMA)	1202.025	—
BDS (BeiDou)	B1 (Phase II)	1561.098	4.092
	B2 (Phase II)	1207.14	20.46
	B3 (Phase II)	1268.52	24
	B1 (Phase III)	1575.42	32.736
	B2 (Phase III)	1191.795	51.15
	B3 (Phase III)	1268.52	35.805
QZSS	L1	1575.42	30.69
	L2	1227.60	30.69
	L5	1176.45	24.90
	L6	1278.75	42.00
IRNSS (NavIC)	L5	1176.45	24
	S	2492.028	16.5

A BOC modulated signal is usually denoted as BOC(m, n), where m and n are two integer numbers representing:

- m: subcarrier frequency in multiples of 1.023 MHz;
- n: spreading code chip rate in multiples of 1.023 Mchip/s.

The generic BOC(m,n) signal can be written as:

$$s(t) = \sum_{i=-\infty}^{+\infty} \left\{ c_i d_i r \left(t - i \frac{T_R}{n} \right) \text{sign} \left[\sin \left(2\pi \frac{m}{T_R} t \right) \right] \right\}, \tag{20.33}$$

where $T_R = \frac{1}{f_R}$ and f_R = 1.023 MHz, c_i is the sequence of the code chips values, d_i are the data symbols, and $r(t)$ is a rectangular pulse shape of unitary amplitude and of duration $\frac{T_R}{n}$.

The signal

$$s_{\sin}(t) = \text{sign} \left[\sin \left(2\pi \frac{m}{T_R} t \right) \right] \tag{20.34}$$

represents the subcarrier. This kind of BOC modulation is also denoted as BOC$_{\sin}$(m,n) to point out that a sinusoidal square wave subcarrier is used. It must be noticed that also a different BOC modulation has been introduced in the literature, denoted as BOC$_{\cos}$(m,n) modulated signal and based on the use of the following subcarrier:

$$s_{\cos}(t) = \text{sign} \left[\cos \left(2\pi \frac{m}{T_R} t \right) \right]. \tag{20.35}$$

Typically, when not specified, a BOC$_{\sin}$(m,n) is considered.

As an example, Figure 20.11 depicts the construction of a BOC(10,5) signal starting from a ranging code with chip rate of 5.115 Mchip/s, and a squared subcarrier with a fundamental frequency of 10.23 MHz. Another example is depicted in Figure 20.12, where the BOC(1,1) signal is plotted.

Each BOC(m,n) signal has then a code chip duration $T_c = T_R/n$ and a period of the squared subcarrier $T_{sc} = T_R/m$. The duration of each subperiod of the subcarrier (i.e., $\frac{T_R}{2m}$ in the case of a sinusoidal subcarrier) is often denoted as a *slot*.

It has to be remarked that the effect of the multiplication with the subcarrier can be seen as the shaping of the pulse used for the transmission of the ranging code chips. This process is well known in literature in the communication field as *line coding*.

BOC Power Spectral Density: Analyzing the spectrum of the BOC signal, the maximum of the power spectrum is shifted with respect to the center frequency, and the energy is split into sidelobes shifted in frequency with respect to the carrier.

It is possible to theoretically evaluate the spectrum of a BOC(m,n) modulated signal with sinusoidal subcarrier as

$$G_{BOC(m,n)}(f) = \frac{1}{T_{sc}} \left(\frac{\sin \left(\frac{\pi f T_{sc}}{2} \right) \sin \left(\pi f T_c \right)}{\pi f \cos \left(\frac{\pi f T_{sc}}{2} \right)} \right)^2. \tag{20.36}$$

A comparison of the power spectra for different BOC modulations is provided in Figure 20.13.

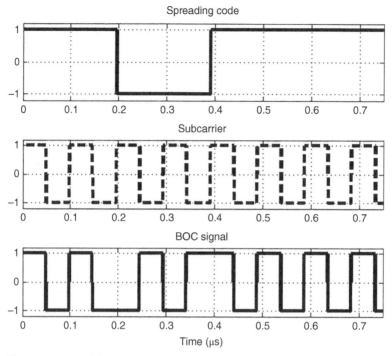

Figure 20.11   BOC(10,5) signal.

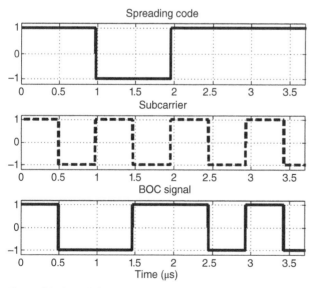

Figure 20.12   BOC(1,1) signal.

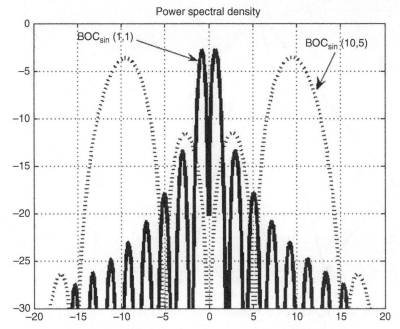

Figure 20.13 Power spectral densities of BOC(1,1) and BOC(10,5) signals.

Correlation Properties: The performance of GNSS receivers basically depends on the ability of the signal processing stage to align a locally generated code to the received psuedorandom noise (PRN) code. Such operation is obtained by means of correlation procedures, and the ability of the receiver to estimate the SIS propagation time depends on the shape of the correlation function of the signal. Due to the properties of shift-orthogonality of the spreading codes, the autocorrelation of a signal assumes in the interval $(-T_c, +T_c)$ the same shape of the autocorrelation of the chip shaping pulse.

Using a rectangular pulse for the chip shaping (i.e., binary phase-shift keying [BPSK]-like), as in the baseline GPS L1 C/A signal, the autocorrelation function is triangular, while for BOC signals it presents over $(-T_c, +T_c)$ a main peak and a number of side-peaks depending on the m and n values in (20.33). The BOC-modulated signals have a narrower correlation function that possibly allows a more accurate identification of the propagation time, as shown in Figure 20.14.

Figure 20.15 shows two other examples of autocorrelation functions for a BOC(4,1) and a BOC(10,1). It can be noted how, increasing the m/n ratio in the BOC(m, n), the number of side peaks increases as well, and the separation between the main peak and the side peaks is largely reduced. This effect introduces a potential ambiguity in the acquisition and tracking of the main correlation peak, especially in the case of degraded/filtered signals.

In conclusion, the narrower main peaks of BOC signals can potentially give better accuracy but, due to the presence of the side peaks, the improvement is traded off with the complexity of the digital receiver (needed to prevent/solve possible ambiguous tracking conditions).

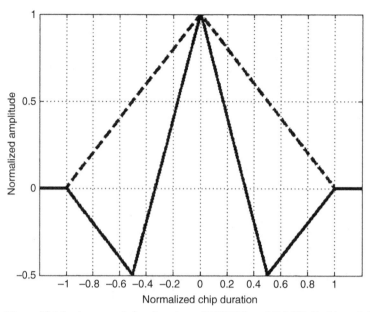

Figure 20.14   Autocorrelation function of BPSK(1) and BOC(1,1) chips of duration $T_c = 1$.

BOC vs. BOC$_{cos}$:   The use of a cosine-phased subcarrier, as in (20.35), results in a reduction of the secondary lobes in the frequency spectrum of the modulated signal. This fact can be useful to improve isolation with signals in the same band, reducing the spectral overlapping with other signals sharing the same carrier. As an example, Figure 20.16 compares the spectra for a BOC(15,2.5) and BOC$_{cos}$(15,2.5), which shows lower sidelobes.

20.5.3   The GNSS Transmitted Signal

The payload of a GNSS satellite generates all the components that are needed to create the SIS broadcast to the users. The basic requirement for the different modules of the payload is that they share a common reference time and frequency generator so that the synchronization of the different components is assured. Spreading codes, subcarriers and data are generated and combined to create *a channel* of the SIS. Different channels may be further combined through a multiplexing scheme, a examples of which will be described in Section 20.7.2 for Galileo. A multiplexed signal is then moved to a carrier frequency f_{RF} according to Table 20.3. Each transmitted channel can generically be represented as:

$$s_{RF}(t) = \sqrt{2P_T}\, c_b(t) d(t) \cos(2\pi f_{RF} t), \qquad (20.37)$$

where P_T is the power associated to the channel, $d(t)$ is the data signal (which could not be present in some channels, conventionally denoted as *pilot* channels), and

$$c_b(t) = c(t) \cdot s_b(t) \qquad (20.38)$$

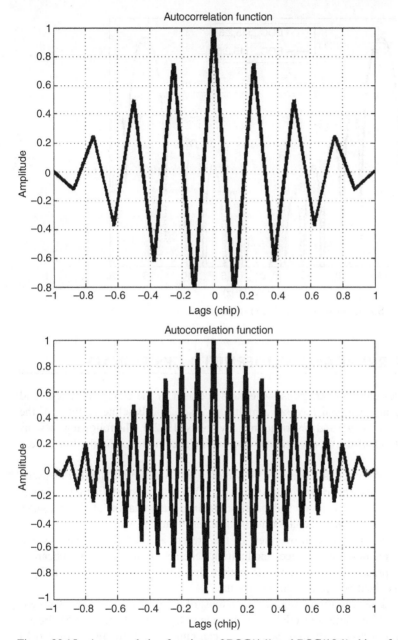

Figure 20.15   Autocorrelation functions of BOC(4,1) and BOC(10,1) chips of duration $T_c = 1$.

is the product of the code $c(t)$ and a proper subcarrier $s_b(t)$. Equation (20.27) is the most general form for a GNSS signal, including the signals in which no subcarrier is present (i.e., $s_b(t) = 1 \forall t$).

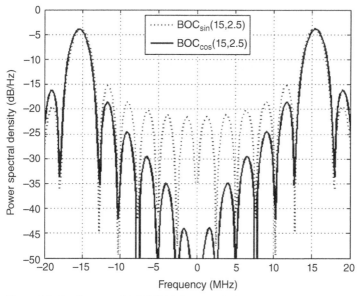

Figure 20.16 Spectra of BOC(15,2.5) and BOC$_{cos}$(15,2.5).

20.6 CURRENT AND MODERNIZED GPS SIGNALS

GPS provides global coverage, is owned and operated by the U.S. government, and is managed at a national level as a multi-use asset. The baseline features of the system are well known and described in many books [19, 22]. The following sections discuss features of the modernized system, with a specific focus on the free-access signals.

The GPS service is a one-way broadcast, with an unlimited number of users, using the frequency bands introduced in Table 20.3. Access to civilian GPS signals is free of direct user charges and public domain documentation of these signals [14–16] is available on an equal basis to users and industry.

Table 20.4 provides an overview of the present and future GPS signals and their characteristics (i.e., service, modulation scheme, occupied band, subcarrier frequency, spreading codes rate, navigation data rate, multiplexing technique, minimum received power, and current status).

At the time of writing (August 2017), most of these signals are already broadcast by the GPS satellite constellation, with some of them (i.e., L2C and L5) already transmitted in a *pre-operational phase* (i.e., can employed at the users' own risk) [16], while the others (i.e., L1C) will be broadcast in few years. In fact, given the recent expansion of the satellite navigation market, the United States is undertaking a GPS modernization phase, leading to a fully renovated system named GPS III.

With respect to the legacy signals, the GPS modernization has introduced a number of new signals, including a new carrier: the L5 carrier, (at 1176.45 MHz) in addition to the existing carriers at L1 (1575.42 MHz) and L2 (1227.6 MHz). The codes modulated on these carriers are modernized on both L1 and L2 frequencies,

TABLE 20.4.   Current and Modernized GPS Signals [14–16]

Band	Service	Type (*)	Modulation Scheme	Spect. Occup.(**) MHz		Code Rate chip/s	Navig. Data Rate bit/s	Min Rx Power dBW	Status (***)
L1	C/A	C	BPSK(1)	2.046		1.023e6	50	−158.5	T
L1/L2	P	M	BPSK(10)	20.46		10.23e6	50	−161.5	T
L1	L1C	C	TMBOC (6,1,4/33) BOC(1,1)	4.092 4.092	CP CD	1.023e6 1.023e6	no data 50	−158.25 −163	F
L2	L2C	C	BPSK(1)	2.046	CM CL	511.5e3 511.5e3	25 no data	−158.5	P
L1/L2	M	M	BOC(10,5)	30.69		511.5e6	N/A	N/A	T
L5	L5	C	QPSK(10)	20.46	I5 Q5	10.23e6 10.23e6	50 no data	−157 −157	P

REMARKS:

(*)   C = civil signal;

(**)   M = military signal.

(***)   null-to-null bandwidth

For BOC modulations, only the two main spectral lobes are considered

T = transmitted (full operation capability);

P = pre-operational broadcast;

F = foreseen signal.

introducing a novel modulation scheme on the L1C signal (described in Section 20.6.2) for civil user and a new military signal (the M-code, both on L1 and L2). New code sequences (C/A-like) will be also modulated on L2 frequency (i.e., the CM and CL codes for the L2C signal), while the new L5 will carry a P-Code-like signal for civil purpose, as outlined in Table 20.4.

The history and the design choices related to the modernized L1C signal are briefly reported in the next sections.

20.6.1   MBOC Signal Baseline

In June 2004 the European Commission (EC) and the USA reached an agreement on the compatibility and interoperability between Galileo and GPS (of present and future generations) with the aim of assuring the coexistence of the two systems [4]. The focus of this agreement was to design signals compatible and interoperable between each other. These two concepts are defined in [8] as:

- *Compatibility*: refers to the ability of space-based positioning and timing services to be used separately or together without interfering with each individual service or signal, and without adversely affecting navigation warfare;

- *Interoperability*: refers to the ability of civil space-based positioning and timing services to be used together to provide better capabilities at the user level than would be achieved by relying solely on one service or signal.

The USA/EC agreement clearly stated as a central point that the common signal baseline structure for the open access GPS and Galileo signals at the L1/E1 band was the BOC(1,1) modulation. In addition, the agreement also underlined that such common signals could be optimized by both parties in order to achieve better performance [17]. Possible innovative modulation strategies were then analyzed in view of the Galileo optimization and for the next generation of GPS signals, resulting in a further agreement that was signed by both parties in March 2006. The agreement foresaw a common signal baseline structure for the open access GPS and Galileo signals in the L1/E1 band for civilians and was announced on July 26, 2007 [8].

The agreement introduced the new concept of the *multiplexed BOC* (MBOC) as a combination of two different BOC modulations. MBOC is defined by means of its power spectral density (PSD) that is a mixture of BOC(1,1) spectrum and BOC(6,1) spectrum [17]. It can be denoted as MBOC(6,1,α/β). The term (6,1) refers to the BOC(6,1) and the ratio α/β represents the power repartition of the BOC(1,1) and BOC(6,1) spectrum components as given by:

$$G_{MBOC}(f) = \frac{\beta - \alpha}{\beta} G_{BOC(1,1)}(f) + \frac{\alpha}{\beta} G_{BOC(6,1)}(f), \qquad (20.39)$$

where $G_{BOC(m,n)}(f)$ is the unit PSD of a sine-phased BOC(m; n) modulation and $G_{MBOC}(f)$ is the resulting MBOC PSD. The spreading modulation design include the BOC(6,1) component in order to place a small amount of additional power at higher frequencies.

Figure 20.17 shows the comparison among the PSDs of the BOC(1,1) and the MBOC(6,1,1/11) implementation selected for the GPS L1C signal as well as for

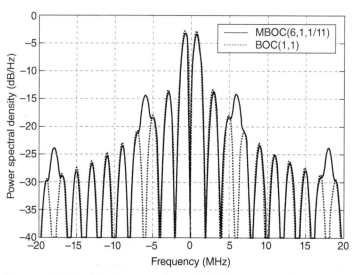

Figure 20.17   Unit PSDs comparison of BOC(1,1) and MBOC(6,1,1/11). Equivalent baseband representation of the L1/E1 carrier signals.

the Galileo E1 signal. From the figure it is quite evident the increase in higher frequency of the MBOC(6,1,1/11) power compared to the baseline BOC(1,1) solution. This feature provides a sharper code correlation peak for the receiver, allowing it to achieve improved signal tracking performance.

It is important to point out that MBOC modulations are defined only from the power spectrum point of view, and different implementations are possible trying to best fit the MBOC spectral mask. In fact, different time waveforms can be combined to produce a MBOC-like spectrum.

The *time-multiplexed BOC* (TMBOC) was chosen for the GPS L1C signal as the technique to "mix" BOC(1,1) and BOC(6,1), multiplexing signals in time domain, while in the Galileo framework the *composite BOC* (CBOC) was chosen for the OS E1 SIS combining the amplitudes of the signals [25]. The GPS L1C TMBOC implementation is then briefly outlined in the next section, whereas the CBOC will be described in Section 20.7.1.

20.6.2   TMBOC Modulation

As outlined in previous Section, the TMBOC is an implementation of the MBOC scheme. It has been optimized to provide improved performance and ensure compatibility with the Galileo signals also sharing the L1/E1 band. The TMBOC implementation implies that a simple BOC(1,1) modulates the entire data component and 29 out of 33 code chips of the pilot channel. However, 4 out of 33 pilot channel chips are modulated using a BOC(6,1) waveform, as depicted in Figure 20.18.

The lower part of the gure shows 33 pilot spreading code chips. Four of these (shaded slots) are the ones modulated using BOC(6,1) modulation. In the center of the figure a magnified view of one BOC(1,1) chip and one BOC(6,1) chip are depicted. The pilot component has 29/33 of its power in BOC(1,1) and 4/33 of its power in BOC(6,1). Since the pilot channels contains 75% of the total L1C signal power, then the percentage of the total BOC(6,1) power is then $\frac{1}{11}(=\frac{4}{33}\cdot 0.75)$, consistent with the desired MBOC(6,1,1/11) spectrum [17].

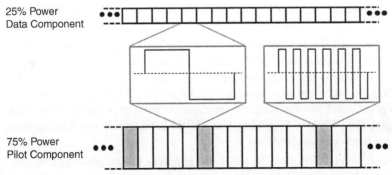

Figure 20.18   TMBOC(6,1,4/33) time series, 75%/25% power split [25].

20.7 GALILEO SYSTEM AND SIS

Galileo is a European project for a global satellite navigation system, approved and funded by the EC and the *European Space Agency* (ESA). The European venture in the satellite navigation field started with the European Geostationary Navigation Overlay System (EGNOS) program, developed to provide satellite-based augmentation signals to users in order to improve GPS performance.

Galileo is an open, global system, fully compatible with GPS but independent from it. Interoperability features with GPS, GLONASS, and other systems was a primary requirement to enable the provision of services based on the combined use of Galileo, GPS, and GLONASS signals, at least for mass-market applications. This means that users now able to locate themselves with a state-of-the-art receiver from any combination of GNSS satellites in view.

It is important to highlight that Galileo constituted the first satellite positioning and navigation system provided specifically for civilian purposes. Its baseline design is composed of a fully deployed constellation of 30 satellites (24 operational and 6 spares), positioned in three circular Medium Earth Orbit (MEO) planes [12], in order to ensure the entire coverage of the Earth's surface. At the time of writing (August 2017), 15 Galileo satellites (out of a total of 18 satellites) are already operational and usable by Galileo-enabled receivers, as previously highlighted in Table 20.1.

The adoption of Galileo leads to several improvements with respect to the legacy GPS performances, overcoming some of its limitations.

Galileo offers different categories of services oriented to several typologies of final users, with particular attention in terms of accuracy, availability, continuity, and reliability.

During the system design phase, four navigation services and one service to support search and rescue operations have been identified to cover the widest range of users needs, including professional users, scientists, massmarket users, safety of life, and public regulated domains. The following Galileo satellite-only services were intended to be provided worldwide and independently from other systems:

- The Open Service (OS), resulting from a combination of signals free of user charge, provides position and timing performances competitive with other GNSS (e.g. similar to the GPS Standard Positioning Service). The OS signals features unencrypted ranging codes and unencrypted freely accessible navigation message (F/NAV) and will be transmitted on the E5a, E5b and E1 carriers.

- The Safety-of-Life (SoL) Service was foreseen to provide improved accuracy, multipath robustness, and timely warnings to the user when it fails to meet certain margins of accuracy (integrity). SoL signals were planned to use the OS ranging codes and navigation data messages on E5b and E1 carriers, providing extended system integrity information with an integrity navigation message (i.e., I/NAV). Following a recent decision, the former Galileo SoL Service has been descoped;

- The Commercial Service (CS) will provide access to two additional signals, allowing a higher data rate throughput and enabling users to improve positioning accuracy. It is envisaged that a service guarantee will be provided for this service. CS signals will use encrypted ranging code sequences to transmit added-value commercial navigation messages on the E6 carrier (i.e., C/NAV). At the time of writing details of the CS are still under definition/validation.

- The Public Regulated Service (PRS) provides position and timing to specific users requiring a high continuity of service, with controlled access. PRS signals will use the encrypted PRS ranging code and navigation data messages on the E6 and E1 carriers. The PRS navigation signals will be available only to authorized users (military, police, etc.), and information about these signals is classified.

- The Search-and-Rescue (SAR) Service broadcast globally the alert messages received from distress emitting beacons. It will contribute to enhance the performances of the international COSPAS-SARSAT system based on the detection of emergency beacons.

These services were initially planned to be available to users of the Galileo system through 10 different navigation signals transmitted in E1, E6 and E5 frequency bands (see Table 20.3). Table 20.5 summarizes the Galileo signals and their characteristics (i.e., service, modulation scheme, occupied band, sub-carrier frequency, codes rate, data rate, multiplexing technique, minimum received power, and current status).

It must be noticed that, after the official *declaration of initial services* of Galileo in December 2016 [11], the OS, SAR, and PRS services can be considered as operational. In addition, with the recent adoption of the *Galileo Commercial Service implementing decision* in February 2017 [10], the European Commission confirmed the intention to deploy also the CS service. In detail, the Galileo CS service will provide both *high-precision* data for enhanced position accuracy and *authentication* data through access to encrypted codes. The provision of an authentication mechanism through the OS signals (denoted as OS NMA) is also foreseen and under implementation. Testing and implementation of E6 CS signals are ongoing and expected to be completed in 2018, followed by an *initial commercial operating phase* and then a *full commercial operating phase* from 2020 on [10], when the full Galileo constellation of 30 satellites is also expected to be completed [11].

More details about the CBOC modulation will be provided in next section, followed by a description of the AltBOC scheme, a novel multiplexing strategy adopted for the first time in the Galileo SIS.

20.7.1 E1 CBOC Modulation

As previously outlined in Section 20.6.1, the CBOC modulation was selected to implement the MBOC signal baseline in the Galileo E1 OS signals (i.e., the E1-B data channel and the E1-C pilot channel).

The main idea of the CBOC is to modulate every PRN code by a weighted combination of BOC(1,1) and BOC(6,1) spreading symbols as defined in the following:

$$s_{BOC(1,1)}(t) = \begin{cases} sign\left[\sin(2\pi t/T_C)\right] & 0 \le t \le T_C \\ 0 & elsewhere \end{cases} \tag{20.40}$$

TABLE 20.5. Current Galileo Signals [10–13]

Band	Service	Channel	Modulation Scheme	Spectral Occup.(*) MHz	Subcarrier Freq. (MHz)	Code Rate Mchip/s	Sec. Code Rate Chip/s	Nav. Data Symb/s	Multiplex. Tech.	Min. Rx Power dBW	Status(**)
E1	PRS	A	BOCcos(15,2.5)	35.805	15.345	2.5575	—	N/A		N/A	T
	OS/SAR	B	CBOC(6,1,1/11)	14.322	1.023/6.138	1.023	—	250	Const. env.	−157 (B+C)	T
	Pilot	C	CBOC(6,1,1/11)	14.322	1.023/6.138	1.023	250	no data			T
E6	PRS	A	BOCcos(10,5)	30.69	10.23	5.115	—	N/A		N/A	T
	CS	B	BPSK(5)	10.23	—	5.115	—	1000	Const. env.	−155 (B+C)	V
	Pilot	C	BPSK(5)	10.23	—	5.115	—	no data			V
E5a	OS	I	BPSK(10)	20.46	15.345	10.23	1000	50			T
	Pilot	Q	BPSK(10)	20.46	15.345	10.23	1000	no data	AltBOC(15,10)	−155 (I+Q)	T
E5b	OS	I	BPSK(10)	20.46	15.345	10.23	1000	250			T
	Pilot	Q	BPSK(10)	20.46	15.345	10.23	1000	no data		−155 (I+Q)	T

REMARKS:

*Null-to-null bandwidth;

For BOC modulations, only the two main spectral lobes are considered

**T = transmitted (initial services);

V = under test and validation.

50% power
Data Component ●●●

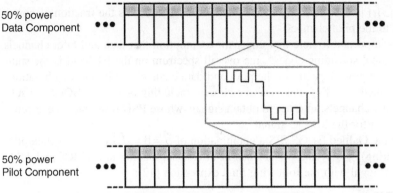

50% power
Pilot Component ●●●

Figure 20.19   Example of CBOC(6,1,1/11) time series 50%/50% power split [25].

and

$$s_{BOC(6,1)}(t) = \begin{cases} sign\left[\sin\left(12\pi t/T_C\right)\right] & 0 \leq t \leq T_C \\ 0 & elsewhere \end{cases}, \tag{20.41}$$

where $T_C = 1/(1.023 \cdot 10^6)$ [s] is the chip duration. An example of CBOC modulated signal in shown in Figure 20.19.

The notation usually reported for the CBOC signal is

$$CBOC(6, 1, \gamma/\rho), \tag{20.42}$$

where the parameters γ and ρ are related to the power splitting between the BOC(1,1) modulated signal and the BOC(6,1) contribution. However, such a notation does not highlight that the actual overall signal is obtained by combining data and pilot channels, then introducing a further degree of freedom while designing a signal matching the MBOC power spectrum. Furthermore, it is not mandatory that the BOC(6,1) contribution be present on both data and pilot channels, opening additional options for the implementation.

In the final Galileo E1 OS signal [12], a power sharing of 50% has been chosen for the data and pilot channels, implying a BOC(6,1) contribution in both channels. In addition, a different phase relation between the BOC(1,1) and BOC(6,1) components has been adopted for the data and the pilot channels (denoted as *in-phase* and *anti-phase* [12]).

Therefore, the time-domain signal on the E1-B data channel can be expressed as:

$$s_{E1-B}^{tx}(t) = e_{E1-B}(t) \cdot \alpha\left[\sqrt{\frac{\rho - \gamma}{\rho}} s_{BOC(1,1)}(t) + \sqrt{\frac{\gamma}{\rho}} s_{BOC(6,1)}(t)\right], \tag{20.43}$$

where $x_{E1-B}(t)$ is the product of the navigation message and the spreading code and $\alpha = \sqrt{\frac{1}{2}}$ is the fraction of power allocated to the data channel. In the same way, the E1-C pilot channel can be expressed as:

$$s_{E1-C}^{tx}(t) = e_{E1-C}(t) \cdot \beta\left[\sqrt{\frac{\rho - \gamma}{\rho}} s_{BOC(1,1)}(t) - \sqrt{\frac{\gamma}{\rho}} s_{BOC(6,1)}(t)\right], \tag{20.44}$$

where $x_{E1-C}(t)$ is the spreading code sequence and $\beta = \sqrt{\tfrac{1}{2}}$ is the fraction of power allocated to the pilot channel.

It is important to note that under the assumption that data and pilot channels use orthogonal spreading codes*, the overall spectrum on the E1 band is the summation of the power spectra of the pilot and data channels. Different combinations of the parameters α, β, γ and ρ can be chosen, including or not the BOC(6,1) subcarrier in the channels, in order to obtain signals whose PSD resembles the spectral mask defined for the MBOC signal.

For the Galileo E1 OS SIS a combination of $\alpha = \beta = \sqrt{\tfrac{1}{2}}$ (i.e., 50% data/pilot power sharing), $\gamma = 1$, and $\rho = 11$ has been chosen, leading to a CBOC(6,1,1/11) modulated signal with the following final expression [12]:

$$s(t) = \frac{1}{\sqrt{2}}\left\{e_{E1-B}(t)\cdot\left[\sqrt{\frac{10}{11}}s_{BOC(1,1)}(t) + \sqrt{\frac{1}{11}}s_{BOC(6,1)}(t)\right]\right.$$
$$\left. - e_{E1-C}(t)\cdot\left[\sqrt{\frac{10}{11}}s_{BOC(1,1)}(t) - \sqrt{\frac{1}{11}}s_{BOC(6,1)}(t)\right]\right\}. \qquad (20.45)$$

An example of a CBOC modulated signal has been already shown in Figure 20.19, whereas the modulation scheme of the complete E1 OS signal is shown by Figure 20.20. The advantages of CBOC with respect to the baseline BOC(1,1) are mainly given by the sharper autocorrelation peak, which allows better performance in terms of tracking and multipath robustness but requires a receiver with a wider front-end filter and more processing capability of the digital baseband

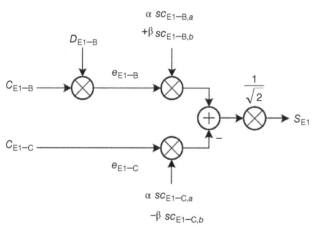

Figure 20.20   Modulation scheme for Galileo E1 CBOC signal [12].

---

*The residual cross-correlation between the spreading sequences chosen for Galileo can be considered negligible.

part. If received with a narrow band filter, cutting out the BOC(6,1) contribution, the signal can be processed as if it was a BOC(1,1) signal with limited performance losses [7, 13].

The E1 band is also used for the transmission of the Galileo public regulated service (PRS, i.e., E1-A component). As highlighted in Table 20.5, a BOCcos(15, 2.5) modulation (with a 15.345-MHz subcarrier frequency and a code rate of 2.5575 Mchip/s) is used and multiplexed together with the E1-B and E1-C CBOC modulated signals, taking advantage of a multiplexing scheme capable of ensuring a constant envelope of the resulting multiplexed signal (i.e., E1-A + E1-B + E1-C).

20.7.2 AltBOC Modulation and Multiplexing Scheme

The transmitted signals in the Galileo E5 band (1164–1215 MHz) are especially interesting due to the innovative features of the *Alternative BOC* (AltBOC), a novel transmission scheme that has been proposed and used for the first time by Galileo satellites.

The AltBOC is both a modulation scheme and a multiplexing technique, since it allows the transmission of four different channels in two adjacent sidebands, conventionally named E5a (1164–1191.795) and E5b (1191.795–1215 MHz), as depicted in Figure 20.21. The composite E5 signal is then made of the E5a and E5b signals, and it can be processed as a single large bandwidth signal (approx. 51 MHz) or as independent signals (approx. 20 MHz each one) with appropriate user receiver implementations (e.g., see [21] and [30]).

The ABOC Concept: The starting point in order to understand the AltBOC concept is the idea that the spectrum of two baseband real or complex signals $s_1(t)$ and $s_2(t)$ can be up and down converted around the zero frequency by writing

$$\tilde{s}(t) = s_1(t)e^{j2\pi f_b t} + s_2(t)e^{-j2\pi f_b t}. \tag{20.46}$$

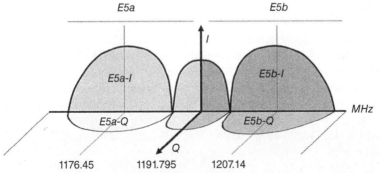

Figure 20.21 Power spectrum of Galileo E5 band (I and Q representation, with different shadings for the E5a and E5b sidebands) [31].

It is evident that $\tilde{s}(t)$ is generally a complex signal. A similar result is obtained if square waves are used instead of sinusoids in (20.46), that is if the exponential function $e^{\pm j2\pi f_b t}$ is substituted with a complex signal

$$r(t) = \text{sign}[\cos(2\pi f_b t)] \pm j \cdot \text{sign}[\sin(2\pi f_b t)], \tag{20.47}$$

whose line spectrum exhibits similar sideband characteristics as the complex exponential. The main difference is that, with $r(t)$, higher frequency contributions are present in the spectrum of $x(t)$; in fact, apart the contribution centered at $\pm f_b$, there are also secondary frequency peaks centered around $\pm 3 \cdot f_b$ due to the square waveform of the subcarrier. But these higher-order harmonics will be neglected in the following since they are potentially filtered out by the transmission filter of the satellite.

In cases of subcarriers with square waves the signal $\tilde{s}(t)$ can written as

$$\tilde{s}(t) = s_1(t)[\text{sign}[\cos(2\pi f_b t)] + j\text{sign}[\sin(2\pi f_b t)]]$$
$$+ s_2(t)[\text{sign}[\cos(2\pi f_b t)] - j\text{sign}[\sin(2\pi f_b t)]]. \tag{20.48}$$

If $s_1(t)$ and $s_2(t)$ are complex, one can write:

$$s_1(t) = s_{1I}(t) + js_{1Q}(t) \tag{20.49}$$
$$s_2(t) = s_{2I}(t) + js_{2Q}(t) \tag{20.50}$$
$$r_c(t) = \text{sign}[\cos(2\pi f_b t)] \tag{20.51}$$
$$r_s(t) = \text{sign}[\sin(2\pi f_b t)]. \tag{20.52}$$

$\tilde{s}(t)$ can be written as

$$\tilde{s}(t) = [s_{1I}(t) + s_{2I}(t) + j(s_{1Q}(t) + s_{2Q}(t))]r_c(t)$$
$$+ [s_{2Q}(t) - s_{1Q}(t) + j(s_{1I}(t) - s_{2I}(t))]r_s(t). \tag{20.53}$$

In this way, a complex AltBOC-like modulated signal expression has been obtained, featuring four different signal channels modulated in two adjacent sidebands.

It is possible to demonstrate [9] that the $\tilde{s}(t)$ expression in (20.53) features a nonconstant envelope constellation. The I and Q components of this signal can even be zero at the same time and some portions of the signal may be at null power, leading to an inefficient use of the satellite amplifier.

The Galileo E5 AltBOC Modulation: The final modulation scheme that has been chosen for the Galileo E5 signals is usually denoted as **E5 AltBOC(15,10)** in order to differentiate from the general AltBOC expression (without constant envelope). Since the AltBOC derives from the family of BOC modulations, the two numbers in its notation have the same meaning as in a BOC(m, n) modulation (see Section 20.5.2).

The AltBOC(15,10) allows to transmit four different channels in the E5 band (see Fig. 20.21):

- **E5a-I** data channel: $e_{E5a-I}(t) = d_{E5a-I}(t) \cdot c_{E5a-I}(t)$;

- **E5a-Q** pilot channel: $e_{E5a-Q}(t) = c_{E5a-Q}(t)$;

- **E5b-I** data channel: $e_{E5b-I}(t) = d_{E5b-I}(t) \cdot c_{E5b-I}(t)$;

- **E5b-Q** pilot channel: $e_{E5b-Q}(t) = c_{E5b-Q}(t)$.

E5a-I and *E5b-I* are *data channels*, since they carry navigation data, whereas *E5a-Q* and *E5b-Q* are *pilot channels* and are not data modulated. The data channels and the pilot channels are respectively transmitted as in-phase components (I) and quadrature phase components (*Q*) of the modulated signal (see also Table 20.5).

As far as the navigation messages are concerned, the *E5a-I* channel is used for the *Open Service (OS)* (F/NAV) and broadcasts navigation data at 50 symbols/s. On the other hand, the *E5b-I* channel carries I/NAV navigation data, transmitted at 250 symbols/s and initially intended to be dedicated to the *Safety-of-Life (SoL) Service*.

The wideband signal $s_{E5}(t)$, obtained with the E5 AltBOC(15,10) modulation, is reported in (20.54) using the notations present in [12] (normalized baseband complex envelope representation):

$$
\begin{aligned}
s_{E5}(t) = & \frac{1}{2\sqrt{2}}[e_{E5a-I}(t) + je_{E5a-Q}(t)]\left[sc_{E5-S}(t) - jsc_{E5-S}\left(t - \frac{T_{S,E5}}{4}\right)\right] \\
& + \frac{1}{2\sqrt{2}}[e_{E5b-I}(t) + je_{E5b-Q}(t)]\left[sc_{E5-S}(t) + jsc_{E5-S}\left(t - \frac{T_{S,E5}}{4}\right)\right] \\
& + \frac{1}{2\sqrt{2}}[\overline{e}_{E5a-I}(t) + j\overline{e}_{E5a-Q}(t)]\left[sc_{E5-P}(t) - jsc_{E5-P}\left(t - \frac{T_{S,E5}}{4}\right)\right] \\
& + \frac{1}{2\sqrt{2}}[\overline{e}_{E5b-I}(t) + j\overline{e}_{E5b-Q}(t)]\left[sc_{E5-P}(t) + jsc_{E5-P}\left(t - \frac{T_{S,E5}}{4}\right)\right]. \quad (20.54)
\end{aligned}
$$

In (20.54), the terms $e_{E5a-I}(t)$, $e_{E5a-Q}(t)$, $e_{E5b-I}(t)$, and $e_{E5b-Q}(t)$ corresponds to the signals transmitted in the four E5 channels, including PRN codes and navigation data. The four dashed terms $\overline{e}_{E5a-I}(t)$, $\overline{e}_{E5a-Q}(t)$, $\overline{e}_{E5b-I}(t)$, and $\overline{e}_{E5b-Q}(t)$ are the *product signals* (intermodulation products), defined by

$$\overline{e}_{E5a-I}(t) = e_{E5a-Q}(t) \cdot e_{E5b-I}(t) \cdot e_{E5b-Q}(t) \quad (20.55)$$

$$\overline{e}_{E5a-Q}(t) = e_{E5a-I}(t) \cdot e_{E5b-I}(t) \cdot e_{E5b-Q}(t) \quad (20.56)$$

$$\overline{e}_{E5b-I}(t) = e_{E5a-I}(t) \cdot e_{E5a-Q}(t) \cdot e_{E5b-Q}(t) \quad (20.57)$$

$$\overline{e}_{E5b-Q}(t) = e_{E5a-I}(t) \cdot e_{E5a-Q}(t) \cdot e_{E5b-I}(t). \quad (20.58)$$

Finally, $sc_{E5-S}(t)$ and $sc_{E5-P}(t)$ are four-level square-wave subcarriers depicted in Figure 20.22 used to perform frequency shifts, similarly to the effect obtainable with complex exponential multiplications, as in (20.53). It must be noted that the product signals are needed to obtain a constant envelope modulated signal [31]. In fact, the AltBOC complex baseband signal $s_{E5}(t)$ can be described as an *equivalent* 8-PSK signal, as depicted in Figure 20.23.

Despite the cumbersome implementation, the AltBOC(15,10) modulation allows one to use the E5 band as the two separate sub-bands E5a and E5b. A single data channel (equivalent to a BPSK signal) and a pilot channel (another BPSK

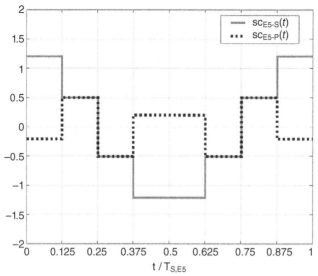

Figure 20.22 One period of the two AltBOC subcarriers.

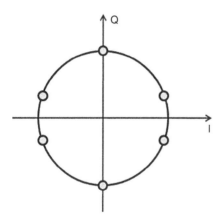

Figure 20.23 Equivalent 8-PSK phase-state diagram of E5 AltBOC signal.

signal) are in fact transmitted in each sideband as in-phase and quadrature components (see Fig. 20.21). This modulation scheme then corresponds to two separate QPSK modulations, placed respectively around the E5a and the E5b carriers. Thus, state-of-the-art Galileo receivers are able to use the signals transmitted from each satellite choosing to receive only one or both the sidebands (E5a and E5b) and then taking advantage of the correlation properties of up to four codes.

As a final remark, it is necessary to point out that the reception and the processing of AltBOC signals requires ad-hoc receiver architectures, with several differences with respect to conventional GPS receivers. Innovative solutions tailored to the AltBOC signals have been proposed for the acquisition [21], tracking [30], and data demodulation operations [20] as a trade-off between performance and cost. The

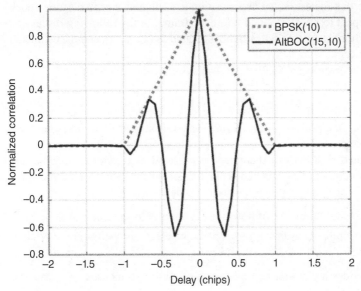

Figure 20.24   Comparison of the correlation functions used in BPSK-like or full AltBOC receiver architectures, simulated with infinite bandwidth [21].

possible receiver architectures for the E5 signals can be essentially classified into two categories: the *BPSK-like* or the *full AltBOC* architectures. In the BPSK-like schemes the signals in only one or both sidebands are separately received, in a non-coherent mode, and processed like simple BPSK signals. In such cases, the receiver takes advantage of a triangular correlation peak as in Figure 20.24 (dashed line), using similar signal processing stages as in standard GPS receivers.

On the other hand, if the whole E5 band is coherently received and processed (i.e., with a *coherent dual-band receiver*), it is possible to fully exploit the correlation properties of the AltBOC modulation (see the solid line in Fig. 20.24), achieving excellent performance but at the cost of an increased computational burden.

20.8   ERROR SOURCES FOR THE POSITION EVALUATION

It is well known that in a standalone GNSS system or in a terrestrial or hybrid positioning system, performance is affected by two main factors: the geometry, i.e., the relative position of the receiver and the reference point (e.g., GNSS satellites) and the error in the range or pseudorange measurements. The trilateration method provides the user position, working on some measured ranges. The errors performed in the measurement of the ranges represent a primary source of errors in the position, velocity and time (PVT) solution.

The position error is generally a random variable, which can be modeled in terms of bias, standard deviation, and probability density function. It is a metric evaluated at the position layer, that is, at the output of the PVT module.

We assume that the estimated user position is evaluated at each discrete epoch $t_k = kT_p$, with respect to a well-defined reference frame. In the following discussion the true user position, at each epoch, will be indicated as (x_k, y_k, z_k), and the estimated position as $(\hat{x}_k, \hat{y}_k, \hat{z}_k)$.

The position error in an ECEF reference system can be defined as:

$$\varepsilon_k^p = \sqrt{(\hat{x}_k - x_k)^2 + (\hat{y}_k - y_k)^2 + (\hat{z}_k - z_k)^2}. \tag{20.59}$$

For GNSS systems, the position error is often evaluated separately as horizontal and vertical components.

The estimated position can therefore be modeled as a random variable, the statistical distribution of which depends on the stochastic behavior of the error sources. Generally speaking, the metrics used are defined as:

- **accuracy**, being a measure of how close a point is to the true position;
- **Precision**, being a measure of how closely the estimated points are in relation to each other.

Figure 20.25 provides a pictorial representation of these two metrics.

20.8.1 GNSS Positioning

In the specific case of GNSS positioning, the position error in (20.59) is usually assumed to be a normal random variable with zero mean and a variance σ_{UERE}^2, which depends on the error sources.

The bias of the mean value with respect to the true value is a measure of the accuracy of the system, while the standard deviation can be taken as a precision index. Such a normal model for the error is usually considered for GNSS-standalone positioning (in three dimensions). However, the validity of the model is subject to some assumptions and some remarks are needed.

Impact of Ranging Errors on Position Metrics: As remarked in Section 20.2, the accuracy and precision in the user position estimation depends on both the ranging error and the geometry. A "rule of thumb" in GNSS standalone positioning is that the user precision (evaluated as a geometric average of the uncertainty over three dimensions) is

$$\sigma_{\varepsilon^p} = \text{PDOP} \times \sigma_{UERE}, \tag{20.60}$$

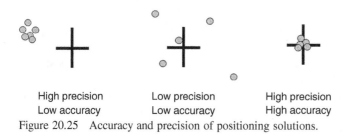

| High precision | Low precision | High precision |
| Low accuracy | Low accuracy | High accuracy |

Figure 20.25 Accuracy and precision of positioning solutions.

where PDOP is an amplifying factor related to geometry of the constellation and σ_{UERE} is the standard deviation of the pseudorange errors. However, such a relation depends on the fundamental assumption that the error contributions on the different pseudoranges are statistically independent and all with the same standard deviation σ_{UERE}.

The σ_{UERE} is usually obtained evaluating the error budget on the satellite-user link. The definition of an error budget for GNSS can be found in many books (see, e.g., [19]). Each error source has been duly investigated over the years, and many papers and publications address physical aspects, modeling, and mitigation of the effects. The error sources can be roughly classified into the following classes:

1. *control system*: ephemeris, clocks, codes, measurement errors;

2. *ionosphere*: the propagation delay depends on the frequency and on the density of electrons along the path. Large errors can occur in the case of abnormal ionospheric condition (i.e., scintillation);

3. *troposphere*: the propagation delay depends on the pressure, temperature, and humidity of the air;

4. *multipath*: error due to the possible presence of multiple reflected replicas in addition to the line-of-sight signal;

5. *receiver noise*: due to the thermal noise in the hardware components of the receiver;

6. *uncompensated relativistic effects*;

7. *Selective Availability* (SA): in the past the civil GPS signal has been intentionally disturbed to limit the accuracy for civil users. SA "ended" on May 1, 2000, and this error is likely not to be introduced anymore in the GPS signals.

As far as accuracy is concerned, nonzero-mean error contributions induce biases in the PVT solution with respect to the true position, and they are a dangerous error that is hard to detect (accuracy loss).

In order to improve accuracy of the solution, for known error sources, each GNSS receiver compensates as far as possible the effects of the "average" contribution of the error source. This result is obtained using models of the error sources or, in more complex receivers, specific processing algorithms. Only the error due to the ionosphere can be almost totally canceled if a double frequency receiver is used; however, an amplification of the other error contribution has to be taken into account [22]. Some correction parameters for the errors are directly provided by the GNSS system in the data channel of each system (e.g., bias of the satellite clock).

Under the hypothesis that proper models have been used, each *j*th *residual error* can be considered as a random variable, which is normal with zero mean and standard deviation σ_j, and it is independent from the other error sources. The standard deviation of the total error is then

$$\sigma_{UERE} = \sqrt{\sum_j \sigma_j^2}.$$

(20.61)

Furthermore, it is assumed that each pseudorange is independently affected by an error that can be modeled as a normal random variable with zero mean and a standard deviation σ_{UERE}.

It has to be remarked that such an error has then to be intended as a residual error, which still affects the measured pseudorange after the receiver has applied a correction to mitigate the effect of each error source. The application of these basic corrections is performed in any GNSS receiver, according to models embedded in the receiver and parameters provided by the system itself in the data message.

However, even assuming that the residual error has zero mean (i.e., the correction is effective for most of the time) accuracy losses are still due to unmodeled errors, such as multipath or radio-frequency interference [5]. Such error can also induce biases in the pseudorange evaluation.

An example of the behavior of the error is depicted in Figure 20.26. The position data evaluated over a period of time by a receiver are depicted with respect to the position of a georeferenced antenna (the "true" position). It can be noted how there is a direct proportionality of the error versus the GDOP and how both increase as the number of satellites decreases. When the number of satellites is not sufficient, the receiver is not able to get a position, and even when enough satellites are in view again, only after some time is it able to provide a solution (believed to be reliable) to the user.

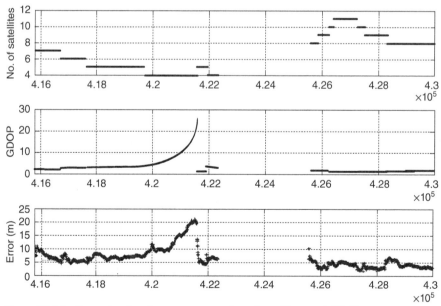

Figure 20.26   Example of real error w.r.t., number of satellites, and GDOP.

20.9 AUGMENTATIONS

When the performance of standalone GNSS positioning is not sufficient, external sources of information not belonging to the core GNSS architecture, have to be used. Generally speaking, these sources provide either better or new corrections that allow the GNSS receiver to improve its performance. The improvement can be primarily on accuracy and precision, but there are also other performance metrics that can benefit, such as the time to first fix (TTFF; see Chapter 23). This section provides an overview of differential correction and augmentation systems presenting both local and wide-area differential systems, as well as other augmentation strategies that are common in the current application scenario (i.e., assisted GNSS (AGNSS) via communication networks).

The augmentation systems aim to improve stand-alone GNSS performance, providing additional information to GNSS users. These systems are based on techniques generally referred as differential GNSS (DGNSS), that is, a method of improving the positioning performance of GNSS by using one or more reference stations at known locations, each equipped with at least one GNSS receiver. In this way, accuracy enhancement, integrity warning notifications or other data can be provided to user receivers via a data link. It must be noted that several different DGNSS techniques exist: they vary in sophistication and complexity from a single reference station, that calculates the errors at its position for use with nearby GNSS receivers, to worldwide networks that provide data for estimating errors from detailed error models at any position near Earth's surface. DGNSS systems can be classified in two basic types: ground based and satellite based. A ground-based augmentation system (GBAS) often uses radios to transmit the differential corrections and is a local area system, since the broadcast corrections are valid only near the reference station. On the other hand, a satellite-based augmentation System (SBAS) transmits corrections via satellites and aims at covering a wider area. Alongside the classic differential concepts, the use of other augmentation systems is growing, driven by the application needs. In particular, at the industrial level there is an increasing demand of reducing the TTFF and navigation outages. These goals are achieved by means of strategies, not only focused on precision augmentations and integrity, such as AGNSS.

20.9.1 Local Area Differential Corrections

As discussed in Section 20.8, GNSS systems have different error contributions that can be mitigated in order to increase the accuracy of the user's PVT computation. To improve the performance of a navigation system, it is possible to set up a local area differential GNSS (LADGNSS) system. This type of system exploits a fixed reference station (RS) that knows its own georeferenced position, and it can determine the difference between the measured distance RS satellite and the real one. A range error is computed to each satellite in view by the RS and this error is broadcast to users in the coverage area, transmitting the differential corrections over a communication channel using different protocols such as the RTCM2.3 [26] and the new RTCM3 [27]. The user equipment can reduce the errors, applying the received correction to each measured pseudorange. In principle, the RS could provide a global correction to the computed solution, but its effectiveness would be granted only if the same set of satellites is used for the PVT computation both by the RS and the user. For this reason, the RS transmits the correction of each pseudorange so that the receiver can remove the common errors from the pseudorange related to satellites that are in view to the user and to the RS.

It is known that many of the GNSS error sources are highly correlated over space and time. A differential system will ideally remove most of the error contributions (i.e., satellite clock errors, ephemeris errors, and troposphere and ionosphere errors) that are common to the RS and the user. Multipath and receiver noise errors are local to the receiver and are uncorrelated between receivers separated by even very short distances and then cannot be removed by the differential correction.

It must be noted that the differential corrections are effective in a specific region around the reference station. The effectiveness of the corrections and thus the final position accuracy will be degraded as the distance between the receiver and the base station (often referred as the *baseline*) grows. This is due to the decorrelation of the errors experienced by the user and the station.

In conclusion, the main limitation of a LADGNSS system is that it can be easily applied only on limited areas. Typical baselines for carrier phase positioning goes from 1 to 10 km, while baselines of 10 to 100 km can be used for code positioning only. The evolution trend for LADGNSS systems is to create networks of stations able to cover larger areas.

20.9.2 Wide Area Differential Corrections

Wide area differential GNSS (WADGNSS) is an augmentation strategy aiming to improve the GNSS performances providing differential corrections to users in a larger service area than LADGNSS systems. The architecture of a WADGNSS system, defined in [28], includes a network of reference stations for monitoring the GNSS signals, a central processing site, and a data link to provide corrections to users, typically taking advantage of one or more geostationary Earth orbiting (GEO) satellites. For this reason these systems are also known as satellite-based augmentation systems (SBAS) with a ground segment made of monitoring and processing sites and a space segment made of a set of GEO satellites. Currently there are three important SBAS in operation:

- *Wide Area Augmentation System* (WAAS), covering the United States;
- *European Geostationary Navigation Overlay System* (EGNOS), covering Europe;
- *Multifunctional Satellite-based Augmentation System* (MSAS), covering Japan.

It must be noted that India, China, and Russia have also started to develop their own SBASs, called *GPS-Aided Geo-Augmented Navigation* (GAGAN), *Satellite Navigation Augmentation System* (SNAS), and *System of Differential Correction and Monitoring* (SDCM), respectively.

The basic concept of WADGNSS is the categorization of error sources in the GNSS observables, breaking out the total pseudorange error experienced by the receiver into different components and estimating the variation of each component over a wide area where multiple RSs are located. Taking advantage of a model for each error source, the system is then able to create a so-called vector correction. Differently from the scalar correction transmitted by the LADGNSS, the vector corrections contain separated satellite clock errors, ephemeris errors, and ionosphere errors that have to be recombined to build the correction for each satellite. In detail, ionosphere corrections, which have the major impact on the augmentation of the position precision, are evaluated and broadcast for selected ionospheric grid points (IGPs), which are points of a virtual grid of lines with constant latitude and longitude at the height of the ionosphere. The receiver interpolates among these points to develop a vertical delay correction, able to accurately mitigate the ionosphere errors, for each visible satellite using their estimated elevation. The benefit of the vector correction is its improved ability to capture the spatial decorrelation of the error sources: in this way the achievable accuracy does not depend on the proximity of the user to a single RS, as in LADGNSS systems, but is nearly constant over the entire coverage area of the WADGNSS or SBAS system.

In summary, the main objectives of a SBAS are increasing the performance of the basic navigation system to ensure accuracy, integrity, continuity, and availability by providing the following information to the users:

- *differential corrections*, to improve accuracy of the system, typically from 10 or more meters (e.g., standalone GPS) to a couple of meters;
- *integrity monitoring*, to ensure that errors are within confidence bounds (tolerable limits) with a very high probability and thus ensure safety.

It has to be noted that SBAS systems have been mainly designed and developed with the goal of fulfilling the requirements of demanding aviation applications. Nowadays, the challenge is to redefine the target application field, moving from the pure aviation to professional road and maritime sectors. As an example, Europe recently set up a solution named EGNOS Data Access Service (EDAS) capable of broadcasting EGNOS corrections and additional data over Internet Protocol (IP). This will extend the use of EGNOS in critical environments (e.g., typical situations for road applications) in which the visibility of EGNOS geostationary satellite is limited. Since the integrity monitoring concept is the driving force behind SBAS, it will be discussed more in detail in the next section.

The Integrity Concept: Integrity represents the level of confidence that a user can have in the computed position and in the estimated error that may degrade it (protection level). This level of confidence is expressed as an *integrity risk*, which is the probability that the computed position falls outside the required protection levels at any computed point within the service area. The confidence bounds comes in two values: the vertical protection level (VPL) and horizontal protection level (HPL). A WADGNSS system designed according to the standard [28] assures integrity with respect to GNSS by providing the user with adequate warning of failure or exceeding of the protection level. Practically speaking, integrity can be seen from the user standpoint as:

- probability that GNSS information (i.e., pseudoranges and navigation messages) is correct (3 sigma);
- the delivery time of data warning should be received by users within 6 seconds.

The integrity is the most important factor in SoL applications and can be see as a thrust measure of GNSS signals. Note that in some cases, integrity is provided by receiver autonomous integrity monitoring (RAIM) algorithm, a technique capable of isolating and excluding from the PVT computation satellites with problems.

20.9.3 AGNSS and Cooperative Navigation

Assisted GNSS techniques have been developed to improve sensitivity of the GNSS SIS reception and to reduce TTFF in both mobile applications (i.e., supporting LBS) and embedded solutions (e.g., in-car black box for emergency call) as described in [6] and [23]. Regardless of the specific mobile LBS, the critical issues that have to be taken into account in the application design are the accuracy of the user position (not very stringent in most cases), the availability of the user position (usually very important,) and the elapse time needed for service provision (i.e., the already mentioned TTFF). This aiding strategy was specifically designed to increase the sensitivity of the receiver, and it is fully described in Chapter 23.

Note that a standalone GPS chip requires at least 30 seconds for a cold start, while with assistances such time reduces to few seconds. In particular, the AGNSS core idea is to provide aid to the terminal via a wireless network. Such aid includes (but are not limited to): precise ephemeris, reference position, reference time, ionospheric corrections, and acquisition parameters (estimate of the Doppler shift). A positioning server at the network level is in charge of generating the assistances, but also it can compute the user position on the basis of the observables (sent by the user to the server), improving the accuracy thanks to local differential corrections and or EGNOS (optional).

The communication between the terminal and the positioning server can be set up using two approaches:

- control plane, in which the assistances are sent via predefined cellular network signal structures like Global System for Mobile communications (GSM) and Wideband Code Division Multiple Access (WCDMA);
- user plane, in which the assistances are sent via a general TCP/IP data connection and so not requiring any wireless standard specific messages.

Cellular system agnostic solutions for the user-plane AGNSS approaches have been developed and standardized by the Open Mobile Alliance [24]. Note that the user-plane approach allows, in principle, one to create a local assistance infrastructure using, for example, a wireless ad hoc/sensor network having a positioning server available at the network management level. In this way, the same AGNSS paradigm used till now in the framework of mobile communication providers can be extended

TABLE 20.6. Example of Main Assistance Parameters.

Assistance	Description
Reference time	Reference time to time-stamp the assistance messages
Reference position	A rough estimate of the terminal position usually computed by the cellular network (e.g., cell-ID approach)
GPS navigation model	Mainly ephemeris to speed up the satellite positions computation
GPS almanac	Almanac of GPS constellation
GPS acquisition assistance	Mainly Doppler and code-phase estimation
GPS ionospheric model	Parameters for the estimate of the ionospheric delay

to cooperative navigation solutions (e.g., peer-topeer navigation) in which a node can disclose to other nodes assistance messages and/or relative positioning information. Even if some differences can be found in the GSM versus universal mobile telecommunications system (UMTS) specifications related to AGNSS, from a general standpoint the assistance parameters that can be potentially employed by GNSS receivers are reported in Table 20.6.

20.9.4 Trends in GNSS-Related Augmentation Solutions and Technologies

For the evolution of GNSS-related technologies as well as end-user applications, companies operating in the mass-market area are giving strong consideration to requirements and specifications derived from market expectations. In particular the main drivers are cooperative solutions, always-connected solutions, reduced startup time (i.e. reduced TTFF), seamless navigation (navigating everywhere and anytime) and reduced power consumptions. The key technological area that will complement any core GNSS (so integrating in various manners satellite constellations) are:

- augmentation broadcast through any available IP link (e.g., EDAS approach);
- integration between pure satellite navigation with digital video broadcasting solutions (i.e., DVB-T);
- merging the concepts of Internet of Things (IoT) and satellite navigation, since the knowledge of the position is a key element for IoT;
- cooperative navigation in which distributed sensors are connected and cooperating to reach the most suitable position information (of each network element);
- introduction of dead-reckoning and tight integration with inertial sensors for vehicular mass-market solutions;
- use of GNSS in complex embedded solutions.

20.10 CONCLUSIONS

In this chapter some of the basic concepts of satellite-based radio-navigation have been introduced. This general survey discussed the various factors affecting the performance of GNSS receivers and focused on the newer aspects of modernized and future signal and systems. Far from being exhaustive, the most interesting issues have been discussed in order to introduce the reader to this fascinating world. However, the systems and signals would be useless if good receivers able to process the signal coming from the space with high accuracy were not be available. The great potential of satellite navigation is embedded in the new signals format and in the features of modernized systems, and it represents an exciting technological and commercial challenge.

ACKNOWLEDGEMENTS

The authors would like to thank Dr. Christian Wullems for his valuable revision of the chapter.

REFERENCES

[1] Cabinet Office, Government of Japan, *Quasi-Zenith Satellite System Interface Specification, Centimeter Level Augmentation Service*, IS-QZSS-L6-001, Draft Edition, Mar. 28, 2017. Available: http://qzss.go.jp/en/technical/ps-is-qzss/ps-is-qzss.html

[2] Cabinet Office, Government of Japan, *Quasi-Zenith Satellite System Interface Specification, Satellite Positioning, Navigation and Timing Service*, IS-QZSS-PNT-001, Mar. 28, 2017. Available: http://qzss.go.jp/en/technical/ps-is-qzss/ps-is-qzss.html

[3] China Satellite Navigation Office, *BeiDou Navigation Satellite System, Signal in Space Interface Control Document, Open Service Signal*, Version 2.1, BDS-SIS-ICD-2.1, Nov. 2016. Available: http://www.beidou.gov.cn/attach/2016/11/07/21212.pdf

[4] USA/European Community, *Agreement on the Promotion, Provision and Use of Galileo and GPS Satellite-Based Navigation Systems and Related Applications*, Jun. 21, 2004.

[5] F. Dovis, *GNSS Interference Threats and Countermeasures*, Norwood, MA: Artech House, 2015.

[6] F. Dovis, R. Lesca, G. Boiero, and G. Ghinamo, "A test-bed implementation of an acquisition system for indoor positioning," *GPS Solut.*, vol. 14, no. 3, pp. 241–253, Jun. 2010.

[7] F. Dovis, L. Lo Presti, M. Fantino, P. Mulassano, and J. Godet, "Comparison between Galileo CBOC candidates and BOC(1,1) in terms of detection performance," *Int. J. Navig. Obs.*, pp. 1–9, 2008.

[8] J. A. Avila-Rodriguez, G. W. Hein, S. Wallner, J.-L. Issler, "The MBOC modulation," *Inside GNSS*, pp. 43–58, 2007.

[9] L. Ries, L. Lestarquit, E. Armengou-Miret, F. Legrand, W. Vigneau, C. Bourga, P. Erhard, and J. L. Issler, "A software simulation tool for GNSS2 BOC signals analysis," in *Proc. of ION GPS 2002*, Portland, OR, Sep. 24–27, 2002, pp. 2225–2239.

[10] European Commission, *Implementing Decision (EU) 2017/224 of 8 February 2017 Setting Out the Technical and Operational Specifications Allowing the Commercial Service Offered by the System Established Under the Galileo Programme to Fulfill the Function Referred to in Article 2(4)(c) of Regulation (EU) No 1285/2013 of the European Parliament and of the Council*, Feb. 8, 2017. Available: http://data.europa.eu/eli/dec_impl/2017/224/oj

[11] European Commission website, *Galileo Goes Live!*, Press release, Brussels, Dec. 14, 2016. Available: http://europa.eu/rapid/press-release_IP-16-4366_en.htm

[12] European Union, *European GNSS (Galileo) Open Service Signal-In-Space Interface Control Document*, OS SIS ICD Issue 1.3, Dec. 2016. Available: https://www.gsc-europa.eu/system/files/galileo_documents/Galileo-OS-SIS-ICD.pdf

[13] M. Fantino, P. Mulassano, F. Dovis, and L. Lo Presti, "Performance of the proposed Galileo CBOC modulation in heavy multipath environment," *Wireless Pers. Commun.,* vol 44, pp. 323–339, 2008.

[14] Global Positioning System Directorate, System Engineering and Integration, *Interface Specification, NAVSTAR GPS Space Segment/User Segment L1C Interfaces,* IS-GPS-800 Rev. D, Mar. 21, 2014. Available: http://www.gps.gov/technical/icwg/IS-GPS-800D.pdf

[15] Global Positioning System Directorate, System Engineering and Integration, *Interface Specification, NAVSTAR GPS Space Segment/User Segment L5 Interfaces,* IS-GPS-705 Rev. D, Mar. 21, 2014. Available: http://www.gps.gov/technical/icwg/IS-GPS-705D.pdf

[16] Global Positioning System Directorate, System Engineering and Integration, *Interface Specification, NAVSTAR GPS Space Segment/Navigation User Segment Interfaces,* IS-GPS-200 Rev. H, IRN003, Jul. 28, 2016. Available: http://www.gps.gov/technical/icwg/IRN-IS-200H-001+002+003_rollup .pdf

[17] G. W. Hein, J. A. Rodriguez, S. Walner, J. W. Betz, C. J. Hegarty, J. Rushanan, A. L. Kraay, A. R. Pratt, S. Lt. Lenahan, J. Owen, J. L. Issler, and T. A. Stansel, "The new optimized spreading modulation recommended for Galileo L1 OS and GPS L1C," *Inside GNSS,* vol. 1, no. 4, pp. 57–65, 2006.

[18] Indian Space Research Organization, *Indian Regional Navigation Satellite System, Signal in Space ICD for Standard Positioning Service,* Version 1.1, ISRO-IRNSS-ICD-SPS-1.1, Aug. 2017. Available: http://www.isro.gov.in/sites/default/files/irnss_sps_icd_version1.1-2017.pdf

[19] E. D. Kaplan and C. Hegarty, *Understanding GPS: Principles and Applications.* Artech House, 2006.

[20] D. Margaría, F. Dovis, and P. Mulassano, "An innovative data demodulation technique for Galileo AltBOC receivers," *J. Glob. Position. Syst.,* vol. 6, no. 1, pp. 89–96, 2007.

[21] D. Margaría, F. Dovis, and P. Mulassano, "Galileo AltBOC signal multiresolution acquisition strategy," *IEEE Aerosp. Electron. Sys. Mag.,* vol. 23, no. 11, pp. 4–10, 2008.

[22] P. Misra and P. Enge, *Global Positioning System: Signals, Measurements and Performance,* 2nd ed. Ganga-Jamuna Press, 2001.

[23] P. Mulassano, F. Dominici, and A. Defina, "Assisted-GNSS performances on embedded systems solutions," presented at the Proc. of IGNSS Symp. 2009, Surfers Paradise, Australia, Dec. 1–3, 2009.

[24] Open Mobile Alliance, *User Plane Location Protocol Candidate,* Version 1.0, OMA-TS-ULP-V1-0-20070122-C, Jan. 27, 2007.

[25] T. Pratt, *"MBOC: The new optimized spreading modulation for Galileo L1 OS and GPS L1,"* in *Signal Task Force Presentation at CNES Galileo Workshop,* Toulouse, France, Oct. 2006.

[26] Radio Technical Commission for Maritime Services, *RTCM Recommended Standards for Differential GNSS Service v2.3,* Aug. 20, 2001.

[27] Radio Technical Commission for Maritime Services, *RTCM Recommended Standards for Differential GNSS Service v3.0,* Oct. 27, 2006.

[28] RTCA, *Minimum Operational Performance Standards for Global Positioning System/Wide Area Augmentation System Airborne Equipment,* DO229D, Dec. 13, 2006.

[29] Russian Institute of Space Device Engineering, *GLONASS Interface Control Document, Navigational Radiosignal in Bands L1, L2,* ICD, Edition 5.1, 2008. Available: http://kb.unavco.org/kb/assets/727/ikd51en.pdf

[30] J. M. Sleewaegen, W. De Wilde, and M. Hollreiser, "Galileo altBOC receiver," in *Proc. of ENC GNSS 2004,* Rotterdam, Holland, May 16–19, 2004, pp. 1–9.

[31] M. Soellner and P. Erhard, "Comparison of AWGN code tracking accuracy for Alternative-BOC, Complex-LOC and Complex-BOC modulation options in Galileo E5-band," presented at the Proc. of GNSS 2003—ENC, Graz, Austria, Apr. 22–25, 2003.

[32] GPS World staff, "The almanac: Orbit data and resources on active GNSS satellites," *GPS World,* vol. 28, no. 8, pp. 44–48, 2017. Available: http://gpsworld.com/the-almanac/

CHAPTER 21

DIGITAL SIGNAL PROCESSING FOR GNSS RECEIVERS

Letizia Lo Presti,[1] Maurizio Fantino,[2] and Marco Pini[2]

[1]*Politecnico di Torino*
[2]*Istituto Superiore Mario Boella*

I**N THE** previous chapter we have introduced navigation satellite systems, whose task is to provide the position of a user device. This is obtained by means of a *global navigation satellite system* (GNSS) receiver, whose architecture consists in a number of blocks, or subsystems, each one devoted to a specific operation. This architecture is composed of two layers: the first one, called the *physical layer*, contains the subsystems that process the received signals in order to extract the parameters necessary to fix the user position, while the second layer, called the *range layer*, is devoted to the position computation. The latter generally employs standard techniques, which are not specific of GNSS, such as the Kalman filtering methods described in Chapter 22. Differently, the physical layer must take into account the specific structure of the GNSS *signal in space* (SIS), and requires the implementation of some sophisticated *signal processing* operations. Therefore, this chapter mainly focuses on the signal processing algorithms used in the physical layer of a GNSS receiver, while the range-layer operations are omitted, as they can be found in Chapter 20.

In order to illustrate the specific subject of the chapter more clearly, a general introduction of all the subsystems of a GNSS receiver is provided in the first two sections. The remainder of the chapter is organized into different sections, each one devoted to a specific aspect of the signal processing algorithms used in GNSS. In particular, the first operation that a receiver performs when it is switched on is described: the acquisition of the satellites in view and a first

Handbook of Position Location: Theory, Practice, and Advances, Second Edition.
Edited by S. A. (Reza) Zekavat and R. Michael Buehrer.
© 2019 by the Institute of Electrical and Electronics Engineers, Inc.
Published 2019 by John Wiley & Sons, Inc.
Companion Website: www.wiley.com/go/zekavat/positionlocation2e

estimate of certain SIS parameters. These estimates derive from the approximation of some results from *maximum likelihood* (ML) theory. Such estimates are solid, but coarse, and need to be refined by other blocks, known as *tracking systems* or *null seekers*, based on different signal processing techniques. The theoretical background of the tracking algorithms is the gradient method; therefore, the final sections of the chapter will be devoted to a description of the gradient theory and its application in the null seekers.

21.1   RECEIVED SIGNAL

In the study of GNSS receivers, the first key point to understand relates to the SIS, and, in particular, as it appears at the receiver antenna. It is interesting to observe that a GNSS receiver has an inherent similarity to *code division multiple access* (CDMA) communication receivers. However, the approach to the design and the performance evaluation is slightly different, because GNSS receivers prioritize time-of-arrival measurements over data estimation. This explains the increase in specific GNSS literature on this topic, [22, 28].

The *radio frequency* (RF) signal $s_{RF}(t)$ transmitted by each satellite is received by the antenna at the user terminal with a propagation delay τ_p. Therefore the received signal can be written as

$$r_{RF}(t) = \sqrt{2P_R}\, c_b(t - \tau_p) d(t - \tau_p) \cos\left(2\pi f_{RF}(t - \tau_p) + \varphi_{RF}\right), \qquad (21.1)$$

where $c_b(t) = c(t)\, s_b(t)$ is the signal which includes both the *psuedorandom noise* (PRN) code $c(t)$ and the subcarrier $s_b(t)$, P_R is the received power, f_{RF} is the RF carrier frequency, and φ_{RF} is a phase term. Notice that P_R represents the power of the signal received from a single satellite and associated with a specific PRN code. In practice the overall received power includes the contribution of the various signals transmitted in a given bandwidth from all the satellites in view.

In ideal conditions, the propagation delay τ_p depends on the distance D (called *range*) between the antennas of the satellite and the user terminal, which is

$$\tau_p = \frac{D}{c}, \qquad (21.2)$$

where c is the speed of light. In the expression of $r_{RF}(t)$ in (21.1), the term $f_{RF}\tau_p$ becomes

$$f_{RF}\tau_p = \frac{f_{RF}D}{c} = \frac{D}{\lambda_{RF}}, \qquad (21.3)$$

where $\lambda_{RF} = c/f_{RF}$ is the carrier wavelength. It is then possible to write the received signal in terms of range, that is,

$$r_{RF}(t) = \sqrt{2P_R}\, c_b(t - \tau_p) d(t - \tau_p) \cos\left(2\pi f_{RF}t - 2\pi\frac{D}{\lambda_{RF}} + \varphi_{RF}\right). \qquad (21.4)$$

21.1.1 The Doppler Effect in the Carrier

Because of the changing nature of the relative motion between the satellite and the user terminal, the distance and the propagation delay vary with time. It is then possible to write

$$\tau_p(t) = \frac{D(t)}{c}, \tag{21.5}$$

and the received signal becomes

$$r_{RF}(t) = \sqrt{2P_R}\, c_b(t - \tau_p) d(t - \tau_p) \cos\left(2\pi f_{RF} t - 2\pi \frac{D(t)}{\lambda_{RF}} + \varphi_{RF}\right). \tag{21.6}$$

The carrier component contained in the received SIS is a sinusoidal waveform with an *instantaneous phase*:

$$\Phi_{RF}(t) = 2\pi f_{RF} t - 2\pi \frac{D(t)}{\lambda_{RF}} + \varphi_{RF}. \tag{21.7}$$

From this equation, it is possible to introduce an *instantaneous frequency*:

$$f_{RF}(t) = \frac{1}{2\pi} \frac{d}{dt} \Phi_{RF}(t) = f_{RF} - \frac{1}{\lambda_{RF}} \frac{d}{dt} D(t), \tag{21.8}$$

which contains a time-varying contribution

$$f_d(t) = -\frac{1}{\lambda_{RF}} \frac{d}{dt} D(t), \tag{21.9}$$

known as a *Doppler frequency* component. It is evident that the Doppler contribution is only determined by the relative motion of the user terminal and the satellite, and it continuously modifies the carrier frequency value.

In most applications the Doppler frequency varies very slowly; therefore, during the operations performed by the receiver it is possible to write the received carrier frequency in the form of

$$f_{RX,d} = f_{RF} + f_d, \tag{21.10}$$

where f_d is a constant Doppler frequency offset. The received signal is written as

$$r_{RF}(t) = \sqrt{2P_R}\, c_b(t - \tau_p) d(t - \tau_p) \cos\left(2\pi(f_{RF} + f_d)t + \varphi_{RF}\right). \tag{21.11}$$

The range of possible values of the Doppler frequency depends on the relative speed of satellites and user. In terrestrial applications, the Doppler range seen by a GNSS receiver is in the range ± 5 kHz. On the other hand, in high-dynamic applications, where the GNSS receiver is mounted on rockets or *low Earth orbit* (LEO) satellites the Doppler range can increase up to 50 kHz. In [50] a precise description of the Doppler phenomenon in a *global positioning system* (GPS) system is given, together with the method for evaluating the Doppler range.

21.1.2 The Doppler Effect at Baseband

The variable delay should be also introduced in the baseband component of the RF signal $r_{RF}(t)$ to take into account the Doppler effect on the data. This means that in the baseband component, defined as

$$r_{BB}(t) = \sqrt{2P_R}\, c_b(t-\tau_p)d(t-\tau_p),\qquad (21.12)$$

the delay should be written as a function of t. However, this effect is much more significant on the carrier than on the data; therefore, for the time being, only the contribution of the Doppler on the carrier will be considered. Nevertheless, if the contribution at baseband has also to be taken into account, a simplified method to do this would consist in modifying the chip rate of the PRN code as

$$T_c' = \frac{T_c}{1+f_d/f_{RF}},\qquad (21.13)$$

where T_c is the nominal chip duration. A derivation of this result can be found in the Appendix.

21.2 THE GENERAL RECEIVER STRUCTURE

A radio receiver generally consists of two main parts connected by an *analog-to-digital converter* (ADC) as shown in Figure 21.1. These two parts have very different characteristics and requirements. The *analog* part (antenna and radio front-end) is placed before the *digital* platform that implements all the processing needed to extract the data carrying the *position, velocity, and timing* (PVT) information from the received signal. Generally, multibit front-ends use an automatic gain control prior to the ADC.

The main task of the front-end is to down-convert the RF signal received by the antenna to a *lower intermediate frequency* (IF), characterized by a carrier frequency $f_{IF,L}$ and a bandwidth B_{IF}. The IF signal can be written as

$$r_{IF}(t) = \sqrt{2P_R}\, c_b(t-\tau_p)d(t-\tau_p)\cos\left(2\pi\left(f_{IF,L}+f_d\right)t+\varphi_{IF}\right),\qquad (21.14)$$

assuming that the front-end filtering effect is negligible. However, if this effect is nonetheless necessary to take into account, a simple substitution of the term $c_b(t)d(t)$

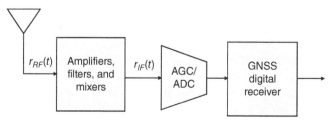

Figure 21.1 Schematic block diagram of the receiver chain.

with its filtered version would be sufficient. In addition, the power of the IF signal should be indicated as $P_{R,IF}$ to take into account the front-end gain. Section 21.2.2 will show that the system performance depends on the ratio between the signal and noise power, which implies that in (21.14) the receiver gain can be normalized to unity.

In most applications, the IF frequency is on the order of 1 to 10 MHz and the filter bandwidth is chosen so as to avoid frequency overlapping in down-conversion operations. Other configurations are possible, like those presented in [51], depending on the type of receiver. From now on, unless otherwise specified, only single down-conversion heterodyne schemes will be considered. Notice that the IF frequency can be written as

$$f_{IF,L} = f_{IF} + f_c, \tag{21.15}$$

where f_{IF} is a nominal IF frequency, and f_c is a term which takes into account the frequency shift of the local oscillator with respect to its nominal value. For low-cost GPS receivers, such a frequency shift can be on the order of thousands of hertz. Since it is an unknown and time-variant parameter (depending on the stability of the local oscillator), it has to be estimated whenever necessary.

When $r_{IF}(t)$ is applied to the ADC input, the sampler generates a sequence of samples with a sampling interval $T_s = 1/f_s$, where f_s is the sampling frequency. The output sequence $r_{IF}[n] = r_{IF}(nT_s)$ can be written as

$$r_{IF}[n] = \sqrt{2P_R}\, c_b(nT_s - \tau_p) d(nT_s - \tau_p) \cos\left(2\pi(f_{IF,L} + f_d)nT_s + \varphi_{IF}\right). \tag{21.16}$$

The notation $r_{IF}[n]$ is the standard notation used in the digital signal processing community to indicate discrete-time signals (see, e.g., [23]). The discrete-time signal passes through a quantizer, whose output can be written in the form

$$y_{IF,q}[n] = Q_k\{r_{IF}[n] + \xi_{IF}[n]\}, \tag{21.17}$$

where $\xi_{IF}[n]$ represents everything collected by the receiver antenna and passed through the bands of the front-end filters (the other satellites SISs, noise, interference, etc.). Because of the quantizer nonlinearity, the function $Q_k\{\cdot\}$ cannot be applied separately to the two components $r_{IF}[n]$ and $\xi_{IF}[n]$, and its effect cannot be studied by applying the superposition principle.

21.2.1   Sampling Frequency

According to the sampling theorem, the sampling frequency f_s should be determined on the basis of the maximum frequency present in the spectrum of the IF signal $r_{IF}(t)$. The trend of modern receivers is to generate an IF signal with a bandwidth B_{IF} on the order of $2f_{IF}$. In this case, the IF spectrum will cover the frequency range $(0, 2f_{IF})$, and consequently, the sampling frequency should satisfy the Nyquist criterion

$$f_s \geq 4f_{IF}.$$

The Nyquist rate $f_s = 4f_{IF}$ is enough for perfect signal reconstruction, and no oversampling is necessary to improve the reconstruction. The only advantage of oversampling is to facilitate the reconstruction process. However, signal reconstruction in a GNSS receiver is not as important as is an accurate measurement of the propagation delay τ_p that is inherent in the received signal. In theory, there is no limit to the accuracy requirements. Each improvement in accuracy is a new frontier in the field of GNSS applications.

It is worth noting that the accuracy in the PVT measurements is also related to the sampling frequency. In fact, to reach a fine estimation of the code delay, the number of samples per chip *must not result as an integer*. In other words, the sampling process should not be synchronized with the chip rate. This depends on the mechanism of "local code synchronization" performed inside the receiver, as is clearly shown in [35]. Indeed, with a sampling frequency multiple of the chip rate, each chip is represented with an integer number of samples. In this case, we have the ambiguity shown in Figure 21.2. Two versions of the sampled local codes match the incoming sequence of samples. This mainly leads to an error in the estimation of the code delay.

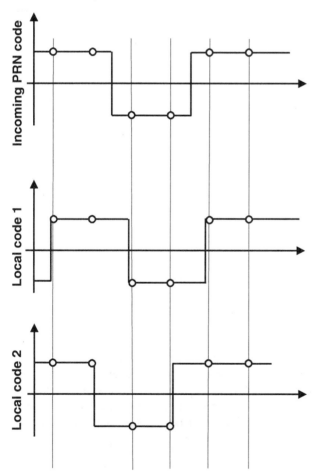

Figure 21.2   Ambiguity due to an integer number of samples per chip.

For the time being we can assume that the sampling frequency is chosen according to the sampling theorem, with an additional constraint that the number of samples per chip cannot be an integer, from which

$$f_s \cong 2B_{IF} \cong 4f_{IF},$$

and the exact value will be fixed to avoid synchronization with the chip rate.

It is also possible to undersample the filtered IF SIS. In this case some amount of aliasing will be generated. Aliasing is not necessarily an impairing factor in the GNSS application. In fact, the SIS bandwidth is very large, and the front-end filters introduce some truncation errors whose effects are visible on the chip transitions. Therefore, a trade-off between aliasing and truncation could be a valid alternative to the sampling at the Nyquist rate. An example of IF translation close to baseband is presented in [51], where a GPS receiver working with a baseband carrier frequency equal to 17.248 kHz is examined. In modern GNSS software-defined-radio receivers [29, 45], f_{IF} is on the order of a few megahertz, with no spectrum overlap around the origin and this shall be the situation considered in this chapter.

21.2.2   The Digital IF Signal

The signal at the ADC input is the result of the contribution of all the satellites in view. However, thanks to the CDMA characteristics of the SISs, each single satellite signal can be separately received and processed. For this reason, from now on, only a single SIS will be considered at the ADC output, unless otherwise specified. This is the signal $r_{IF}[n]$ given in (21.16), which is highly impaired by the noise contribution. To take into account this pervasive disturbing contribution, the received signal has to be written as

$$y_{IF}[n] = r_{IF}[n] + w_{IF}[n], \tag{21.18}$$

where the term $w_{IF}[n]$ represents a single realization of a random process $W_{IF}[n]$, which is the digital version of the continuous-time filtered noise at the input of the ADC block.

Carrier-to-Noise and Signal-to-Noise Ratios (SNR):   In the GNSS community, two parameters are generally introduced to characterize the noise contribution: the carrier-to-noise ratio and the *signal-to-noise ratio* (SNR). The carrier to noise ratio is defined as

$$\frac{C}{N_0} = \frac{P_R}{N_0},$$

where P_R is the signal power evaluated on the whole signal bandwidth, while N_0 can be considered as the noise power evaluated in a bandwidth of 1 Hz, or, in other terms, the noise power density. It follows that the carrier to noise ratio can be conveniently expressed in dBHz. In some sense this parameter compares two different quantities (power and density), and thus could be considered improper or incorrect from an engineering point of view. However, it is now generally adopted across the GNSS and communication communities because it can be seen as an "RF parameter" independent from the bandwidths of front-end filters.

In order to compare two homogeneous quantities, the SNR is introduced at the ADC input as

$$\text{SNR} = \rho_{IF} = \frac{P_R}{N_0 B_{IF}}, \tag{21.19}$$

where B_{IF} is the bandwidth, around f_{IF}, of the IF signal at the ADC input, due to the front-end filters, and

$$\sigma_{IF}^2 = N_0 B_{IF} \tag{21.20}$$

is the noise variance, that is, the noise power affecting the digital IF SIS in the whole signal bandwidth. Notice that both P_R and σ_{IF}^2 should be multiplied by the front-end gain, but this is irrelevant in the SNR evaluation. This is the reason why the front-end gain is generally normalized to unity. SNR can be written in terms of C/N_0 as

$$\text{SNR} = \rho_{IF} = \frac{P_R}{N_0 B_{IF}} = \frac{C}{N_0} \frac{1}{B_{IF}}. \tag{21.21}$$

Notice that C/N_0 is a parameter defined at the antenna output, where an *additive white gaussian noise* (AWGN) process $N(t)$, with a constant power spectrum equal to $N_0/2$, is conventionally defined. As it is well-known, $N(t)$ represents the contributions of the thermal noise generated in the various blocks of the front-end. In other terms $N(t)$ is an *equivalent* noise, which takes into account all the noise contributions generated inside the front-end and at the antenna input. The problem arises on how to *measure* the C/N_0. The main issue is that noise and signal are always mixed together, creating the necessity to find points on the receiver chain where one component dominates over another, in order to be able to separately measure the two contributions P_R and N_0, or at least some quantities related to P_R and N_0 with a known relationship. Methods to estimate the carrier-to-noise ratio can be found in [13–15], and in the references cited therein.

Since both P_R and N_0 are multiplied by the same gain factor throughout the front-end blocks, it is also possible to assign the definition of C/N_0 to the section of the front-end just before the ADC by neglecting the effect of the IF filters on the SIS power evaluation.

Example 21.1

One of the core operations performed by a GNSS receiver is to estimate the code-phase shift between the local and incoming spreading codes. The estimate is based on the evaluation of the code's cross correlation function, which depends on the characteristics of the spreading code and is fundamental in any CDMA system. This example shows how to estimate the code shift of a known PRN code. Consider the 20 chip-length code:

$$c_{loc}(t) = \sum_{i=0}^{19} \alpha_i r(t - iT_c),$$

where:

- $\alpha_i = \{1-1\,1-1-1-1-1\,1-1\,1-1-1-1\,1-1-1\,1\,1\,1\,1-1\}$;
- $r(t)$ is the unitary amplitude rectangular function;
- T_c is the chip width, which is the inverse of the code rate R_c equal to 0.5 MHz.

This represents the local code used by the signal acquisition of a GNSS receiver. Assuming that the periodic version of $c_{loc}(t)$, referred to as $c_{in}(t)$, represents the incoming signal, write a MATLAB script able to:

- generate the incoming signal, composed by 3 periods of $c_{loc}(t)$;
- sample $c_{in}(t)$ and $c_{loc}(t)$ at $f_s = 8$ MHz, obtaining the sequences of samples $c_{in}[n]$ $n = 1,\ldots, N$, and $c_{loc}[m]$ $m = 1,\ldots, 3N$, with $N = 20(f_s/R_c)$;
- plot their correlation function;
- plot their correlation function in case the initial code-phase of $c_{in}[n]$ is equal to 4 chips.

Solution

One of the properties of the autocorrelation function (ACF) of PRN sequences is to have a maximum when the two codes are perfectly aligned. In this case, a maximum is expected when $c_{loc}(t)$ is perfectly aligned with one period of $c_{in}(t)$. After sampling the PRN at the required sampling rate, it is easy to evaluate the correlation function by implementing the formula:

$$R[l] = \frac{1}{N}\sum_{n=1}^{N} c_{loc}[n]c_{in}[l+n] \quad l = 0, \ldots, N-1. \tag{21.22}$$

Referring to the MATLAB script in the Appendix, the correlation function results as shown in Figure 21.3, presenting a maximums at 0-chips delay.

If the initial code-phase shift of the incoming signal is equal to 4 chips, the cross correlation is as shown in Figure 21.4, with maximums at 4-chip delays.

Example 21.2

This example is based on the previous results and introduces the effect of AWGN on the estimate of the code-phase shift that maximizes the correlation function. In this case, the received signal can be written as

$$s[n] = c_{in}[n] + w[n] \quad n = 1, \ldots, N, \tag{21.23}$$

where:

- $c_{in}[n]$ represents the samples of the incoming signals;
- $w[n]$ is the AWGN
- $N = 60(f_s/R_c)$, assuming that f_s and R_c have the same values of the previous example.

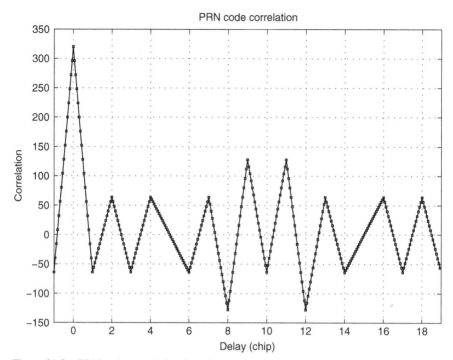

Figure 21.3 PRN code correlation function.

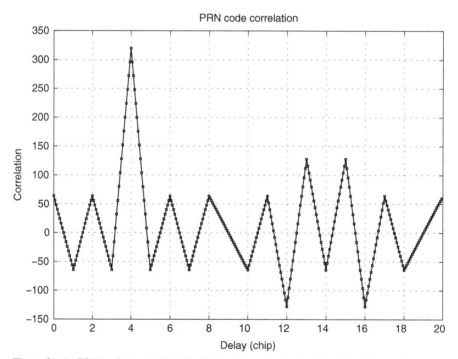

Figure 21.4 PRN code correlation function, with code-phase shift of 4 chips on the incoming signal.

Repeat the previous example, generating the AWGN noise with the *randn* MATLAB command. First simulate a noise standard deviation of 2.5, and then repeat with the standard deviation equal to 12. In Example 21.4, it is shown how to simulate signal and noise with a given C/N_0 value.

Solution

The presence of the AWGN on the incoming signal results in a noisy correlation function. As shown in Figure 21.5, with the noise standard deviation equal to 2.5, the maximum can be still detected but with an error on the estimated codephase shift equal to 1 sampling interval. Indeed, in this case, the first maximum is estimated at 3.975 chips while the real code-phase shift was 4 chips.

If the noise standard deviation is equal to 12, the maximum of the cross-correlation between the two spreading codes can be buried in the noise, as shown in Figure 21.6. In this case, the signal acquisition may result in an erroneous estimate of the code-phase shift.

As we will show in the next section, according to the maximum likelihood estimation theory [23], the delay that maximizes the correlation function can be estimated as described in (21.28).

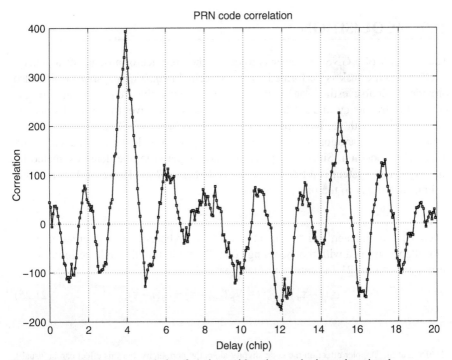

Figure 21.5   PRN code correlation function, with noise on the incoming signal.

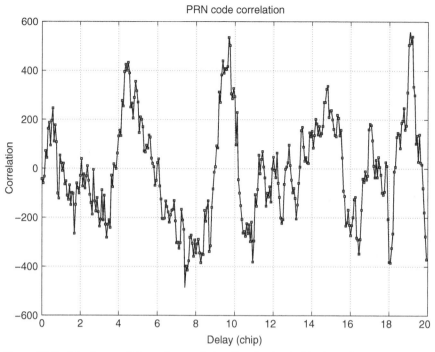

Figure 21.6 Maximum of the code correlation function buried in the noise.

21.3 ACQUISITION

The first task of a GNSS receiver is to detect the presence or absence of a generic satellite. Such a task is performed by the so-called *acquisition system*, which also provides a coarse estimation of two SIS parameters: the frequency shift of the nominal IF frequency, and a delay term that allows the receiver to create a local code that is aligned with the incoming code. In Section 21.8 we will see that this alignment is the key operation for precisely measuring the pseudoranges necessary to fix the receiver position. Since the spreading code is periodic, the delay estimated by the acquisition system is a fraction of the code period T_{code}. However, it is possible to link such delay to the propagation delay τ_p as

$$\tau_p = N_p T_{code} + \tau, \tag{21.24}$$

where N_p is an integer, and τ is the residual delay to be estimated in order to create a local code aligned with the incoming code. By exploiting the code periodicity, the term $c_b(t - \tau_p)$ in (21.14) can be written* as:

$$c_b\left(t - \tau_p\right) = c_b\left(t - N_p T_{code} - \tau\right) = c_b\left(t - \tau\right), \tag{21.25}$$

---

*Notice that the Doppler effect on the code slightly alters the periodicity, but this effect can be usually ignored in the time interval involved in the acquisition process.

from which

$$r_{IF}(t) = \sqrt{2P_R}\, c_b(t-\tau)d(t-\tau_p)\cos\left(2\pi(f_{IF,L}+f_d)t+\varphi_{IF}\right) \quad (21.26)$$

and

$$r_{IF}[n] = \sqrt{2P_R}\, c_b(nT_s-\tau)d(nT_s-\tau_p)\cos\left(2\pi(f_{IF,L}+f_d)nT_s+\varphi_{IF}\right). \quad (21.27)$$

The task of the acquisition system is to provide an estimate $\hat{\tau}^{(A)}$ of τ and $\hat{f}_d^{(A)}$ of f_d. These estimated parameters are then passed to a tracking system that represents the second stage of the signal processing unit of the receiver. This performs a local search for refining the estimates of τ and f_d. At this stage, the estimation of the carrier phase may be included. Once the signals of the detected satellites are tracked, the navigation message can be demodulated, the pseudoranges can be measured, and the PVT can be evaluated, if these operations are completed for at least four satellites. This is possible because the information on the propagation delay τ_p still remains within the navigation message $d(t)$.

The acquisition system is made of a number of functional blocks that conceptually operate independently. In real systems, these operations can be implemented simultaneously or "all in one." Therefore, it may not be an easy task to identify the specific functions by just looking at the circuit scheme of a common acquisition system. We believe that a functional analysis is far more useful in view of the new systems and signals that the GNSS worldwide scenario is laying out for us.

The signal at the input of the acquisition block is the IF sequence $y_{IF}[n]$ given in (21.18), where the signal component is represented by (21.27) and the noise term is a zero-mean white Gaussian sequence, with variance $\sigma_{IF}^2 = N_0 B_{IF}$, as in (21.20).

21.3.1   Detection and Estimation Main Strategy

There are two mathematical disciplines that govern the operation performed by an acquisition system: *signal detection theory* and *estimation theory*.

These two extensive theories are described in various literatures; applications and in-depth analysis can be found in so many papers that including a complete list here is not possible. However, we will mention a few books that tackle this topic with a language and style close to our approach. Milestones in this field are [20, 23, 24, 49]. A full text of examples can be found in [6], and finally a short and effective illustration of detection theory is contained in [19] and [42].

Parameter Estimation: By exploiting the concepts and the methodology of estimation theory it is possible to show that the ML estimate of the vector $\mathbf{p} = (\tau, f_d)$, whose elements are two unknowns of $y_{IF}[n]$, can be obtained by maximizing the function

$$\hat{\mathbf{p}}_{ML} = \arg\max_{\bar{\mathbf{p}}} \left| \frac{1}{L} \sum_{n=0}^{L-1} y_{IF}[n]\bar{r}_{IF}[n] \right|^2, \quad (21.28)$$

where L is the number of samples used to process the incoming signal $y_{IF}[n]$, $\bar{r}_{IF}[n]$ is a test signal, locally generated, of the type

$$\bar{r}_{IF}[n] = c_b(nT_s-\bar{\tau})\,e^{j2\pi(f_{IF}+\bar{f}_d)nT_s}. \quad (21.29)$$

$\overline{\mathbf{p}} = (\overline{\tau}, \overline{f_d})$ is a vector of the test variables $\overline{\tau}$, and $\overline{f_d}$ contained in $\overline{r}_{IF}[n]$, where $\overline{f_d} = f_c + \overline{f}_{d,v}$ is the estimate of the true Doppler $\overline{f}_{d,v}$ plus the frequency shift f_c,The test variables are defined on a proper support D_p, which contains all the possible values which can be assumed by the elements of $\mathbf{p} = (\tau, f_d)$,

The normalizing term $1/L$ is not strictly necessary. It is introduced only to simplify the comparison between different schemes and systems. The term

$$R_{y,r}(\overline{\mathbf{p}}) = R_{y,r}(\overline{\tau}, \overline{f_d}) = \sum_{n=0}^{L-1} y_{IF}[n]\overline{r}_{IF}[n] = \sum_{n=0}^{L-1} y_{IF}[n]c_b(nT_s - \overline{\tau})e^{j2\pi(f_{IF} + \overline{f_d})nT_s} \quad (21.30)$$

is an inner product and the result of (21.28) holds only if the energy of the test signal $\overline{r}[n]$ does not depend on $\overline{\mathbf{p}}$. A proof of this result can be found in [37]. It is possible to show [37] that (21.28) is insensitive to the phase term φ_{IF}, from which the name *noncoherent acquisition scheme* is used to identify the acquisition engine based on (21.28).

From a strictly theoretical standpoint, also, the data bit values are unknown and should be included among the parameters to be estimated. This choice would lead to an excessive increase in acquisition complexity and thus is not adopted in real receivers. The presence of a data bit transition impairs this estimation process. However, in real receivers this is taken into account, and specific algorithms have been proposed to perform the acquisition of a sequence of samples with no data-bit transition.

Detection: The same function introduced to estimate the SIS parameters can be used to decide if a specific satellite is in view or not. In fact, if the generalized *likelihood ratio test* (LRT) described in [24] is adopted for the satellite detection, the variable

$$S_{max} = \max \left| \frac{1}{L} \sum_{n=0}^{L-1} y_{IF}[n]c_b(nT_s - \overline{\tau})e^{j2\pi(f_{IF} + \overline{f_d})nT_s} \right|^2, \quad (21.31)$$

already introduced in (21.28), is used as test statistic, or decision variable, and compared against a threshold θ_t to test two possible hypotheses:

H_1: The satellite is in view;

H_0: The satellite is not in view.

If $S_{max} > \theta_t$, the satellite is considered present, otherwise it is absent. The event $S_{max} > \theta_t|H_1$ identifies the event "correct detection," while the event $S_{max} > \theta_t|H_0$ identifies the event "false alarm." The performance of the detector is evaluated in terms of *detection probability*; that is,

$$P_d = P(S_{max} > \theta_t | H_1), \quad (21.32)$$

and *false alarm probability*, that is,

$$P_{fa} = P(S_{max} > \theta_t | H_0). \quad (21.33)$$

The threshold θ_t is generally chosen according to the *Neyman–Pearson* (NP) theorem [24, 22]. An introduction to the NP-based detection methods is also provided in Section 21.5.

Notice that if different strategies are used to perform the detection, the two probabilities P_d and P_{fa} have to be defined differently [7]. Moreover, different measurements can be made before taking a decision, and different decision variables can be used to test the possible hypotheses. A possible approach based on multiple measurements is described in Section 21.5.

21.3.2 Cross Ambiguity Function

The term in (21.30) is based on the structure of a *cross ambiguity function* (CAF) [40] in the discrete time domain.

We can understand that a CAF is a sort of two-dimensional cross-correlation; for this reason it is often referred to as *correlation*. For a given value of $\bar{f}_d$ it is actually the cross-correlation between the two signals:

$$x_1[n] = y_{IF}[n]e^{j2\pi(f_{IF}+\bar{f}_d)nT_s};$$

$$x_2[n] = c_b(nT_s - \overline{\tau}).$$

It is important to highlight that the signal $c_b(nT_s - \tau) = c(nT_s - \tau)s_b(nT_s - \tau)$ contained in $y_{IF}[n]$ is a periodic sequence with a period equal to the code period* T_{code}; therefore, the delay τ can be estimated only in the range $(0, T_{code})$. Therefore, the test signal

$$\overline{r}_{IF}[n] = c_b(nT_s - \overline{\tau})e^{j2\pi(f_{IF}+\bar{f}_d)nT_s}$$

can be seen as an infinite sequence containing periodic repetitions of the satellite codes. In practice, only a portion of this infinite sequence enters into the sum in (21.30), the samples of the test signal for $n = 0, \ldots, L - 1$. This means that for a given value of $\bar{f}_d$, the correlation assumes the form of a *circular* correlation when the interval $(0, L - 1)$ contains an integer number of code periods. This remark is quite important in understanding why the methods based on the *fast Fourier transform* (FFT) can be conveniently used to evaluate the CAF. Indeed the FFT can be used to implement fast circular correlations. To be more precise, in $\overline{r}_{IF}[n]$ only the code and the subcarrier circulate in the correlation; however, this is sufficient to authorize the use of FFT in acquisition schemes.

The method so far described is sometimes called *active correlation* [1, 2] to indicate that the receiver has to activate a circular shift to the local test signal in order to evaluate the CAF.

---

*In the discrete-time domain the period of $c[n] = c_b(nT_s)$ should be expressed as $N_p = T_{code}/T_s$. To simplify the description of this paragraph we consider N_p an integer equal to the integration time L.

If only discrete values of the test variable $\bar{\tau} = iT_s$ are used, (21.30) can be modified by applying the delay to the input signal rather than to the local test signal. In fact, starting from

$$R_{y,r}\left(iT_s, \bar{f}_d\right) = \frac{1}{L} \sum_{n=0}^{L-1} y_{IF}[n] c_b((n-i)T_s) e^{j2\pi(f_{IF}+\bar{f}_d)nT_s},$$ (21.34)

we can then change $n - i = m$ to obtain

$$R_{y,r}\left(iT_s, \bar{f}_d\right) = \frac{1}{L} \sum_{m=-i}^{L-1-i} y_{IF}[m+i] c_b[m] e^{j2\pi(f_{IF}+\bar{f}_d)(m+i)T_s}.$$ (21.35)

For each value of i the term $e^{j2\pi(f_{IF}+\bar{f}_d)iT_s}$ is a phase term that does not affect the absolute value in (21.28) and (21.31). At this point the delay iT_s is applied only to the input signal. This suggests the introduction of a new decision function in the form of

$$R'_{y,r}\left(iT_s, \bar{f}_d\right) = \frac{1}{L} \sum_{n=0}^{L-1} y_{IF}[n+i] c_b[n] e^{j2\pi(f_{IF}+\bar{f}_d)nT_s},$$ (21.36)

which corresponds to the application of a linear variable delay $\bar{\tau} = iT_s$ to the signal $y_{IF}[n]$ rather than to the local code. Note that in the digital section of the receiver, no fractional delay (i.e., i noninteger) can be used to shift the input samples, as only samples at the ADC output are available. On the contrary, in the active method, any fractional delay can be in principle used for the local test signal.

The test signal, which is fixed and locally generated,

$$\bar{r}'_{IF}[n] = c_b[n] e^{j2\pi(f_{IF}+\bar{f}_d)nT_s},$$

is always the same for each value of tested delay. Therefore, the method is sometimes called *passive correlation* [1, 2].

The linear shift can be obtained by applying a moving window to $y_{IF}[n]$. With the moving window the input vector is updated sample by sample, adding a new input value and discarding the former one. The total number of input samples necessary to use the passive method becomes $2L - 1$. In fact, for $i = 0$ we need the input samples in the interval $n \in (0, L - 1)$; for $i = L - 1$, the interval becomes $n \in (L - 1, 2L - 2)$.

In GNSS literature the focus is often given to the evaluation and analysis applied to the CAF in order to detect satellites and estimate the SIS parameters, sometimes ignoring the common presence of the CAF in all the actual schemes where the squared envelope of the normalized CAF

$$S_{y,r}\left(\bar{\tau}, \bar{f}_d\right) = \left| \frac{1}{L} \sum_{n=0}^{L-1} y_{IF}[n] c_b(nT_s - \bar{\tau}) e^{j2\pi(f_{IF}+\bar{f}_d)nT_s} \right|^2$$ (21.37)

is the function evaluated in the so-called *search space* (SS).

The SS: In common acquisition schemes, the CAF envelope is evaluated over a grid of points $\bar{\tau}_i = i\Delta\tau$ and $\bar{f}_{d,l} = l\Delta f$, spanning the support D_p containing all the possible values that can be assumed by the unknown parameters $\bar{\tau}$ and $\bar{f}_d$. In the following pages, to keep our notation simple, we will omit the subscripts i and l. You should simply remember that $\bar{\tau}$ and $\bar{f}_d$ are two variables that are updated in the SS with discrete and noncontinuous steps. The subscripts will be utilized only when strictly necessary.

The support D_p is generally defined by two ranges: $\bar{\tau} \in (0, T_{code})$ and $\bar{f}_d \in (-f_{d,M}, +f_{d,M})$, where f_{dM} is the maximum Doppler frequency that can occur in a specific application.

Figure 21.7 shows a discretized SS, containing $N_\tau \times N_f$ cells, N_τ *delay bins*, and N_f *Doppler bins* (or *frequency bins*). The interval $\Delta\tau$ is the delay bin size and Δf is the Doppler frequency bin size. The number of delay bins is $N_\tau = T_{code}/\Delta\tau$ and the number of frequency bins is $N_f = 2f_{d,M}/\Delta f$.

It is quite evident that the system complexity increases with the SS dimensions. On the other hand, it is clear that performance will be enhanced if the CAF is represented in the SS with greater resolution. Therefore, the choice of the SS parameters will always be a compromise between performance and complexity. Another important issue in the setting of the bin sizes is in the capability of the tracking systems to refine the coarse estimate provided by the acquisition. Although the pull-in-range can be wide, they need to be properly initialized by the acquisition outputs. Therefore, the final choice of the bin sizes is the result of a joint design of acquisition and tracking systems.

Consideration on the Value of the Frequency Bin Size: The parameter Δf can be selected considering the shape of the CAF envelope in the right delay bin, that is, when $\bar{\tau} = \tau$. In this case, by neglecting the IF filtering effects, the possible sign transition, and the noise term, the squared CAF envelope becomes

$$ S_{y,r}(\tau, \bar{f}_d) = \left| \frac{\sqrt{2P_r}}{L} \sum_{n=0}^{L-1} \cos(2\pi(f_{IF} + f_d)nT_s + \varphi) e^{j2\pi(f_{IF} + \bar{f}_d)nT_s} \right|^2. \quad (21.38) $$

We recognize in this equation the *discrete time Fourier transform* (DTFT) of rectangular pulse in the discrete-time interval $(0, L-1)$ modulated by a cosine. Therefore, as can be verified, $S_{y,r}(\tau, \bar{f}_d)$ has a maximum when $\bar{f}_d = f_d$ and takes on smaller

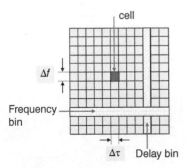

Figure 21.7   The SS.

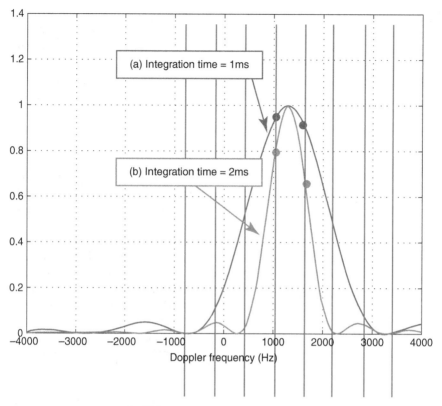

Figure 21.8 Plot of $S_{y,r}\left(\tau, \overline{f}_d\right)$ normalized with respect to its maximum.

values when $\overline{f}_d$ deviates from f_d. Details on this aspect can be found in [8, 9, 37]. An approximate expression of (21.38), valid in the delay and frequency bins crossing the right cell (τ, f_d), is given in [25].

Figure 21.8 shows the function $S_{y,r}\left(\tau, \overline{f}_d\right)$ (normalized with respect to its maximum) as a function of $\overline{f}_d$, considered as a continuous variable. Two curves are plotted, both related to a GPS C/A code affected by a Doppler frequency $f_d = 1250\,\text{Hz}$. Curve (a) has been obtained with an integration time of 1 ms, while 2 ms have been used for curve (b). The vertical lines indicate a possible choice for the frequency points $\overline{f}_d = l\Delta f$ where the CAF is evaluated. It is evident that the CAF samples in curve (a) are adequate for the detection task, as the value of the greatest sample is very close to the true CAF peak. This situation is not guaranteed in curve (b). The reason is that the width of the main lobe of the function in (21.38) (the CAF envelope in the bin $\overline{\tau} = \tau$) decreases as the integration time increases. In order to limit this loss, an empirical rule [16] is to choose

$$\Delta f = \frac{1}{LT_s}.$$ (21.39)

This value is a trade-off among various factors, such as the computational complexity, noise effects, and tracking pull-in ranges. Other authors [21], in the case of

GPS L1, suggest using a narrower bin size, $\Delta f = 2/(3LT_s)$. This choice mainly depends on the pull-in range of the frequency tracking loops. Finally, the navigation message can limit the value of L, as a bit transition may impair the CAF modifying the main lobe characteristics. For GPS, if the integration time is up to a maximum of 10 ms, it is guaranteed that in (at least) one data set there will be no bit transition. This consideration implies a limitation on the maximum value of the parameter L. The problem is different with GNSS signals with the same code and bit rates, such as the Galileo Open Service signal in the E1 band. In this case, a conservative choice is an integration time equal to the PRN code period; however, this choice leads to a correct result only if the CAF evaluation is performed by passive methods. This point will be more evident in Section 21.4, where the methods of CAF evaluation are addressed. Other details on the bit transition problem can be found in [9] and 36.

Consideration on the Value of the Delay Bin Size: The choice of $\Delta \tau$ depends on the shape of the ambiguity function along the correct frequency bin (where $\overline{f_d} = f_d$), which in turn depends on the ACF of a single chip with its subcarrier. The value of $\Delta \tau$ is generally equal to $M_\tau T_s$, where M_τ is an integer, so that $\overline{\tau} = i M_\tau T_s$. In this case the CAF at the right frequency bin becomes a classical cross-correlation function of type

$$S_{y,r}\left(\overline{\tau}, f_d\right)\Big|_{\overline{\tau}=iM_\tau T_s} \cong \left|\frac{1}{L}\sum_{n=0}^{L-1}c_b\left(nT_s - \tau\right)c\left(nT_s - iM_\tau T_s\right)s_b\left(nT_s - iM_\tau T_s\right)\right|^2, \quad (21.40)$$

where the double frequency is considered to be completely eliminated by the correlator, and there is no data signal. We know that the correlation peak is reached when $iM_\tau T_s = \tau$, which implies that τ/T_s should be an integer. This is obviously not guaranteed, because τ can assume any real value in the range $(0, T_{code})$

If we neglect the filtering effects on the chip, the ACF of a single chip with its subcarrier represents a very close approximation of (21.40). In Figure 21.9 the continuous curves represent the normalized absolute value of the ACF in GPS and Galileo

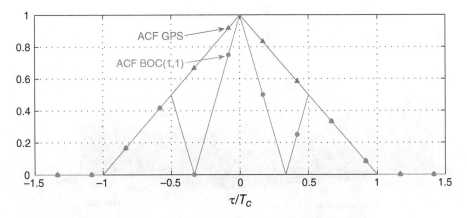

Figure 21.9 Absolute value of the ACF of GPS and Galileo BOC(1,1) chips.

BOC(1,1) chips, while the markers represent a possible situation in the SS if $T_s = T_c/4$ and $M_\tau = 1$. The true peak is lost in both cases and the maximum of the ACF exhibits a loss with respect to the true peak, a loss found to be greater in the BOC(1,1). We can thus see that the sharper the peak, the smaller the delay bin size has to be.

As already mentioned, another important constraint on the delay bin size is given by the capability of the tracking loop to improve the estimation provided by the acquisition block. Therefore, the choice of $\Delta\tau$ will be the result of a trade-off among many factors: acquisition performance, acquisition complexity, and tracking requirements on the initial value of the estimated delay.

SNR at the CAF Peak: It is possible to show that the SNR at the CAF peak is given by

$$\text{SNR}_p = \frac{P_R L}{N_0 B_{IF}} = \frac{C}{N_0} \frac{L}{2 B_{IF}} = \rho_{IF} \frac{L}{2}, \tag{21.41}$$

corresponding to an increase of ρ_{IF} (the SNR at the acquisition input, defined in (21.21)). The term $L/2$ represents the so-called *despreading gain*.

In Figures 21.10 and 21.11, the CAF is shown for a GPS SIS with $C/N_0 = 46\,\text{dBHz}$, $f_s = 1/T_s = 8.34\,\text{MHz}$, $f_d = 500\,\text{Hz}$, and $\tau = 0.521\,\text{ms}$. The parameters used to evaluate the SS are: 1 ms integration time, $\Delta f = 300\,\text{Hz}$, and $\Delta\tau = T_s$. The $\Delta f = 300\,\text{Hz}$ choice is quite conservative, and is used here to better show the CAF along the right delay bin. The CAF is impaired by a noise contribution scarcely affects the peak behavior due to despreading gain, which leads the SNR from a negative value (-19 dB, with a bandwidth of 4 MHz) to a positive value of about 17 dB.

Coherent and Noncoherent Integration: To reduce the high level of noise in the evaluated CAF, some noise reduction techniques can be applied; they mainly consist in averaging the CAF values.

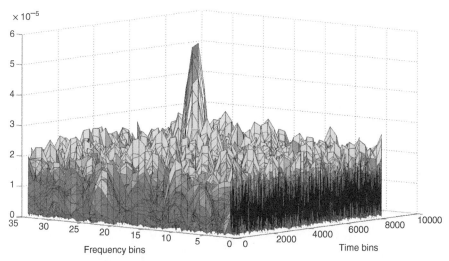

Figure 21.10   GPS CAF envelope in the SS for $C/N_0 = 46\,\text{dBHz}$, and 1 ms integration time.

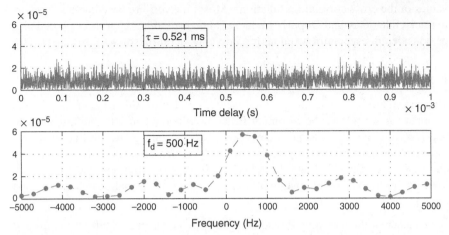

Figure 21.11   CAF envelope in the right delay and frequency bins.

If the average is made after the squared envelope of (21.40), a so-called *non-coherent integration* is performed. In this case, nonnegative random variables due to noise are averaged together, thus a residual noise term still remains. A more effective way of reducing the noise contribution is to adopt a so-called *coherent integration* (or pre-envelope integration). This consists de facto of a longer coherent integration (in other terms, a correlation with a local code containing more code periods), which can be implemented by using partial summations. The price to be paid for this improvement of the system performance is that the CAF becomes narrower, and the size of the frequency bins defined in (21.39) is reduced. Moreover, the duration of the coherent integration is limited by the presence of navigation data (or chips of the secondary codes), which could seriously impair the orthogonality characteristics of the primary codes. Reference [8] offers a deeper analysis on the impact of coherent and non-coherent integrations use.

21.3.3   Refinement of the Estimation of the SIS Parameters

After the acquisition system has coarsely estimated the Doppler shift ($\hat{f}_d^{(A)}$) and the code delay ($\hat{\tau}^{(A)}$), tracking systems are activated to refine the estimation of the two parameters f_d and τ. The fine estimation of τ is the key element of any GNSS receiver, as this parameter is strictly related to the measurement of the propagation delay, and thus to the pseudorange measurement. This measurement can also be refined by exploiting the information coming from the Doppler frequency by adopting techniques known as *carrier smoothing* [28, 34]. Therefore, a fine estimation of the f_d parameter needs to be performed. This estimation is also necessary to properly demodulate the SIS and to extract the navigation message.

Both the delay τ and the Doppler frequency f_d are generally extracted from the received signal by means of *closed loop* systems, based on the *null seeker* structures that are described in Section 21.7. The null seeking strategy is generally preferred to other synchronization methods because it is insensitive to the absolute values of the cross correlation functions. Most receivers are based on null seekers to track code delay and Doppler frequency; however, other approaches could also be considered, especially when taking into consideration software receivers [50].

Example 21.3

This example helps students get familiar with real acquisition systems. Previous examples assumed that the incoming PRN was already converted at baseband. In real receivers the signal acquisition processes the stream of samples at an IF and estimates the carrier frequency of the incoming carrier. This results in an unknown due to the Doppler effect and the frequency shift of the front-end local oscillator. Write a MATLAB script that implements a simplified serial search acquisition through the following steps:

- Generate the incoming spreading code with a code phase shift of 4 chips and modulated on a sinusoidal carrier at 2 MHz;
- Generate two local carriers, 90° phase shifted, to implement the complex exponential in (21.29);
- Correlate the incoming and local signals: for all possible code-phase shifts of the spreading codes and for all frequencies in a range of [-50 50] kHz, centered at 2 MHz, with frequency steps of 10 kHz;
- Plot the search space using the *surf* MATLAB command.

Solution

The cross-correlation is now a function of two variables (i.e., the code-phase shift and the carrier frequency). In satellite navigation, such a function is known as CAF. The CAF maximum is found when the local code is aligned with the incoming one and the local and incoming carriers have the same frequency. For each pair of values a new correlation is computed. The acquisition search is implemented in MATLAB with a double *for* structure:

```
for (all values of carrier frequency),
 (generate local carriers)
 for (all values of code phase shift),
  (correlate the incoming signal with local code and
  carriers)
 end
end
```

In our case, the cross-correlation maximum corresponds to a code phase shift of 4 chips, and a local carrier frequency of 2 MHz. Figure 21.12 shows the CAF computed with the MATLAB script reported in the Appendix.

Note that, as in real receivers, the algorithm proposed in this example is implemented through a noncoherent system, which employs in-phase and quadrature branches (i.e., they use two identical local carriers, 90° phase shifted) to recover the unknown phase of the incoming carrier.

Cross ambiguity function

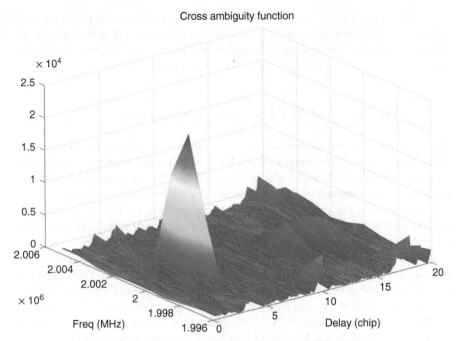

Figure 21.12   CAF resulting from the serial search acquisition algorithm.

21.3.4   Acquisition Performance

We have seen that an acquisition system is both a detector and an estimator. The performance of a generic detector may be completely expressed in a plot of P_d versus P_{fa}. The curves that depict P_d versus P_{fa} for various values of SNR (properly defined) are known as *receiver operating characteristics* (ROC). An example of derivation of ROC curves is in Section 21.5.1. Other examples of ROC curves in GNSS can be found in [47].

In general, the ROC curves are used to compare the performance of different detectors operating on the same input data. The curve closer to the upper left corner of the diagram generally identifies the best detector among the compared ones.

Another important metric to assess the acquisition performance is the *mean acquisition time* (MAT), which takes into account the time spent to detect a single satellite. The more time-consuming part of the detector activity is generally the CAF evaluation. Notice that more than a single CAF could be necessary to detect the presence of a given satellite. In fact, when a first decision about the satellite presence and a first estimation of the code delay and of the Doppler frequency are available, the system can refine these results. Thus multitrial techniques, based on the use of different CAFs, evaluated over subsequent portions of the input signal, can be employed. Two examples of these techniques are the M of N [22] and the Tong [48] methods.

Multitrial acquisition techniques generally do not require the computation of more than one complete CAF. In fact, every time the multitrial system decides to make a new analysis of the SS, the acquisition system starts all over again evaluating a new CAF, but this new CAF is not necessarily calculated with the parameters of the previous trial. Both the explored SS area and the bin sizes can be redefined with each trial.

The goal of the multitrial approach is to improve the acquisition performance in terms of detection probability, but this is obtained at the expense of an increased MAT, whose evaluation is generally quite complex and depends on the specific multitrial mechanism. An example of MAT evaluation can be found in [47].

21.4   THE ROLE OF FFT IN A GNSS RECEIVER

The squared CAF envelope defined in (21.37) can be implemented by using a serial scheme that evaluates the samples of the SS cell by cell. The detection operations can be either applied cell by cell, or after all points of the SS have been evaluated. In the former case, the sequence of the two operations (CAF evaluation and detection) can be interrupted the first time a value crosses the threshold θ_l. This is a suboptimal method, as it does not guarantee that the ML solution, based on the CAF maximum, is obtained. This approach is, however, much faster than the optimal one.

Other methods exist based on block processing techniques, which allow the evaluation of an entire row or column of the SS in a single shot. These methods exploit some properties of the CAF. In fact, if we properly group the signals involved in the CAF definition, we can easily recognize the presence of either a correlation or a Fourier transform inside the CAF structure. This will be explained in detail in Sections 21.4.1 and 21.4.2, but, for the time being, we can say that the FFT, the most famous signal-processing algorithm, can play a key role in speeding up the CAF evaluation. In fact, both the discrete-time Fourier transform and the correlation can be implemented with very efficient and fast algorithms based on FFT.

In the literature, different acquisition schemes are proposed [50, 33], with different characteristics. A brief and simple classification of these methods is reported in [9]. One of these methods is known as FFT in the time domain, which was originally proposed in [30], and is mentioned in [41], as a new trend in GNSS receivers for near-instantaneous signal detection. The method has attracted the interest of many researchers, and a number of variants have been proposed to further reduce its computational complexity. In [44] a technique to transform a long FFT into smaller FFTs is proposed. The same method has been applied in a real-time receiver implemented in Field Programmable Gate Array (FPGA) in [3]. In [4] FFT is used to optimize the signal acquisition operations, by applying one- or two-dimensional FFTs for both the code phase and Doppler frequency, and by reusing intermediate computations to reduce the number of redundant arithmetic operations. In [46], the FFT method is used as the basic algorithm for the proposed multi-PRN code acquisition technique.

21.4.1   FFT in the Time Domain

In this scheme, a vector $\mathbf{y} = [y_{IF}[0], y_{IF}[1], ..., y_{IF}[L-1]]$ of L samples is extracted by the incoming SIS $y_{IF}[n]$, and multiplied by $e^{j2\pi(f_{IF}+\bar{f}_d)nT_s}$, obtaining the sequence

$$q_l[n] = y_{IF}[n]e^{j2\pi(f_{IF}+\bar{f}_d)nT_s} \tag{21.42}$$

for each $\overline{f}_d$ value (that is for each frequency bin of the search space). At this point the term

$$R_{y,r}\left(\overline{\tau}, \overline{f}_d\right) = \sum_{n=0}^{L-1} q_l[n]c_b\left(nT_s - \overline{\tau}\right) \tag{21.43}$$

assumes the form of a circular *cross-correlation function* (CCF), which can be evaluated in the form

$$\tilde{R}_{y,r}\left(\overline{\tau}, \overline{f}_d\right) = \text{IDFT}\left\{\text{DFT}[q_l[n]]\,\text{DFT}[c_b(nT_s)]^*\right\}, \tag{21.44}$$

where DFT and IDFT stand for the well-known discrete Fourier transform and inverse discrete Fourier transform, respectively. Both DFT and IDFT can be evaluated by using FFT. A CCF evaluated with a moving window and the circular CCF coincide only in the presence of periodic sequences. This is the case when $\overline{f}_d = f_d$, except for the noise contribution and a negligible residual term due to a double frequency component contained in the term $q_l[n]$. In the other frequency bins, or in the right bin but with $\overline{f}_d$ not exactly equal to f_d ($\overline{f}_d \cong f_d$), the presence of a sinusoidal component could alter the periodicity of the sequence. These aspects are generally negligible, while it is evident that the presence of a sign transition in the vector **y** completely destroys the code periodicity, leading to serious peak impairments in the search space. However, some countermeasures can be taken as shown in [9] and [36].

21.4.2 FFT in the Doppler Domain

In this scheme, a moving vector can be extracted by the incoming SIS instant by instant and multiplied by $c_{Loc}[n] = c_b(nT_s)$, thus obtaining the sequence

$$q_i[m] = y_{IF}\left(\overline{\tau} + mT_s\right)c_{Loc}[m] \tag{21.45}$$

for each delay bin. A similar result can be obtained by extracting an input vector **y** every L sample and multiplying it by a delayed version of the local code $c_{Loc}[n]$. As mentioned before, this delay has to be obtained by applying a circular shift to the samples of $c_{Loc}[n]$. At this point the term

$$R_{y,r}\left(\overline{\tau}, \overline{f}_d\right) = \sum_{m=0}^{L-1} q_i[m]e^{j2\pi(f_{IF}+\overline{f}_d)mT_s} \tag{21.46}$$

assumes the form of an inverse DTFT. It is well-known that a DTFT can be evaluated by using an FFT if the normalized frequency $\left(f_{IF} + \overline{f}_d\right)T_s$ is discretized with a frequency interval

$$\Delta f = \frac{1}{L}$$

in the normalized frequency range (0, 1), which corresponds to the analog frequency range (0, f_s). The evaluated frequency points become $(f_{IF} + f_{d,l})T_s = l\Delta f$, for $l = 0$, 1,..., $L - 1$, from which

$$\overline{f}_{d,l} = \frac{l}{LT_s} - f_{IF}$$

and the CAF can be written as

$$R_{y,r}\left(\overline{\tau}, \overline{f}_{d,l}\right) = \sum_{m=0}^{L-1} q_i[m]e^{j\frac{2\pi}{L}lm}. \tag{21.47}$$

This is the expression of an inverse DFT, which can be evaluated by using a classic FFT. In this case the number of frequency points provided by the FFT is L, and the support of the SS along the frequency axis and the frequency bin size depend on the sampling frequency f_s and on the integration time L. For example, in the Galileo E1 case, with $T_s = 0.25\,\mu s$ (about four samples per chip), $L = T_{code}/T_s = 16000$ would give rise to a huge number of frequency points, the most of which would be in a Doppler range with no interest for our applications.

If the same support and bin size used in the previous method have to be used, the integration time has to be changed, and some demodulation and decimation (with prefilter) have to be adopted before applying the FFT. This modifies the input signal to be processed, and the comparison among the methods will be affected by the signal modifications. Notice that the maximum peak loss in the Doppler frequency domain is no longer a free parameter with this method, as it is governed by the FFT constraints, as shown in [8] and [27]. If it is necessary to mitigate this effect some zero-padding techniques can be used at the expense of interpolation loss.

Schemes with both passive and active correlations can be devised based on (21.47). They generally include an integrate-and-dump block followed by a decimation unit to reduce the number of samples on which the FFT is evaluated. This operation reduces the computational load but introduces a loss in the CAF quality [27, 8]. Moreover, we know that a decimation is equivalent to a sampling procedure; therefore, before decimation, the input signal has to be down-converted in a lower band compatible with the decimation factor. This can be realized by means of a digital down converter.

Finally, it has to be clear that these methods can be modified and hybrid schemes can be foreseen, especially if multitrial techniques are adopted.

Example 21.4

This example helps students get familiar with the CAF evaluation based on FFT. The signal acquisition processes the stream of samples at IF and estimates the Doppler frequency and the code delay of the incoming signal. Write a MATLAB script that implements the CAF evaluation through the following steps:

- Generate an incoming IF signal (GPS L1) with a code delay of 723 chips, an IF carrier at 4MHz, and a Doppler frequency of −2750 Hz;
- Add noise, with $C/N_0 = 48$dBHz;

- Implement the CAF evaluation by using FFT in the time domain through the following steps:
  - Create the local code (1 ms long) and its FFT;
  - Implement (21.42) in a frequency range [−5 +5] kHz, with frequency steps of 300 Hz.
  - Implement (21.44);
- Plot the search space using the *mesh* MATLAB command;
- Implement the CAF evaluation by using FFT in the Doppler domain through the following steps:
  - Define the filter parameters for decimation;
  - Multiply the signal by a complex sinusoid at the intermediate frequency (first step of down conversion);
  - Consider all possible code delays and multiply the signal obtained at the previous step by the delayed code (obtained with circular shift);
  - Filter (second step of down conversion) and decimate;
  - Consider applying zero padding;
  - Apply FFT;
- Plot the SS using the *mesh* MATLAB command.

Note: for the method of FFT in the frequency domain, the parameters have to be properly chosen.

Solution

First, the PRN code and the noisy signal are generated (SIS_seg).

The CAF evaluation based on FFT in the time domain is implemented in MATLAB with the following structure:

```
(evaluate the conjugate FFT of the PRN code, and obtain
DFTc) for (all values of Doppler frequency),
      (generate local carriers)
      (multiply the local carriers by SIS_seg)
      (evaluate FFT of the signal obtained at the previous
      step) (call it DFTq)
      (evaluate ifft(DFTc.*DFTq))
end
```

Figure 21.13 shows the CAF computed with the MATLAB script reported online.

The CAF evaluation based on FFT in the frequency domain is implemented in MATLAB with the following structure structure:

```
(Define the filter parameters)
(Create a complex sinusoid at the intermediate
frequency)
```

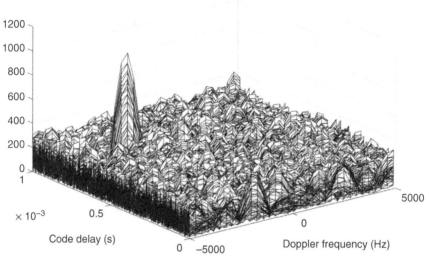

Figure 21.13   CAF evaluated by using FFT in the frequency domain.

```
(Multiply this complex sinusoid by SIS_seg)
(Call the obtained signal SIS_with_mixer)
(Process SIS_with_mixer with the for structure:
for (all values of code delay),
    (generate delayed local codes)
    (multiply the delayed code by SIS_with_mixer)
    (filter and decimate)
    (add zeros (zero padding))
    (evaluate FFT of the signal with zero padding)
end
```

Figure 21.14 shows the CAF computed with the MATLAB script reported on line.

21.5   METHODS FOR SIS DETECTION

The acquisition of any GNSS signal is one of the most important receiver operations. A receiver must be able to identify which satellites are in view as fast as possible in order to reduce the time to first fix. Moreover, the acquisition must be performed even in a weak signal power situation, which increases the challenge of this specific receiver function.

As already described in the previous sections, the acquisition procedure can be split into two main specific functions: SS evaluation and signal detection. Different ways of evaluating the SS have been described so far. In this section, the focus will be the problem of signal detection. In particular, the classical NP method and the Bayesian propagation method will be described.

CAF in the frequency domain

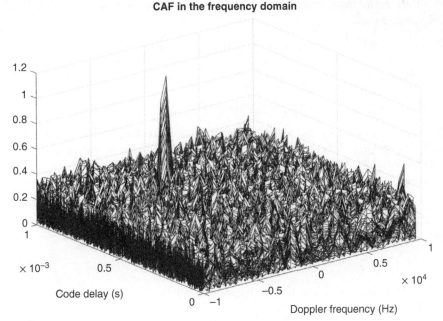

Figure 21.14   CAF evaluated by using the FFT in the frequency domain.

21.5.1   NP Approach

The *detection theory*, also known as *decision theory or hypothesis testing* [24], is the discipline at the base of the correct functioning of any CDMA acquisition engine, including GNSS. In fact, the aim of the detection theory is to take a decision among several hypotheses on the basis of collected measurements, and the detection performance depends on the capability of discriminating among these different hypotheses. In the simplest case of two hypotheses, the problem becomes one of choosing between H_0, which is termed the noise-only or null hypothesis, and H_1, which is the signal-present or alternative hypothesis.

The detection performance is measured in terms of *detection probability*, defined as the probability to make the decision for the event H_1 when the hypothesis H_1 is correct, thus

$$P_d = P(H_1|H_1), \tag{21.48}$$

and in terms of *false alarm probability*, defined as the probability to make the decision for the event H_1 when the correct hypothesis is H_0, thus

$$P_{fa} = P(H_1|H_0). \tag{21.49}$$

In Section 21.3.1, an example of detector is given where the decision is taken on the basis of a single variable, that is, the maximum of the CAF envelope.

Even more, in general, the decision can be taken on the basis of a collection of N measurements,

$$x_m = \{x_m[0], x_m[1], \cdots, x_m[N-1]\}, \tag{21.50}$$

which can be seen as a specific instance of a vector of stochastic variables

$$\mathbf{X} = \{X[0], X[1], \cdots, X[N-1]\} \tag{21.51}$$

statistically described by the two *conditional probability density functions* (PDF) $p(\mathbf{x}|H_0)$ and $p(\mathbf{x}|H_1)$, where $\mathbf{x} = \{x[0], x[1], \cdots, x[N-1]\}$ is the vector of instances of $\mathbf{X}$ defined in R^N. By splitting R^N into two subspaces R_1 and R_0, where

$$R_1 = \{\mathbf{x} : H_1 \text{ is decided}\} \tag{21.52}$$

and

$$R_0 = \{\mathbf{x} : H_0 \text{ is decided}\}, \tag{21.53}$$

both the false alarm and detection probabilities can be derived as:

$$P_{\text{fa}} = \int_{R_1} p(\mathbf{x}|H_0) d\mathbf{x} \tag{21.54}$$

and

$$P_{\text{d}} = \int_{R_1} p(\mathbf{x}|H_1) d\mathbf{x}. \tag{21.55}$$

In detection theory, a statistical test known as the LRT is introduced*, based on the function

$$L(\mathbf{x}_m) = \frac{p(\mathbf{x}_m|H_1)}{p(\mathbf{x}_m|H_0)} \tag{21.56}$$

called *likelihood ratio* (LR). An example of LRT is in Section 21.3.1. The LR is used in the NP detection method, based on a theorem that states that, for a given $P_{\text{fa}} = \alpha$, in order to maximize P_{d} it must be decided for H_1 when

$$L(\mathbf{x}_m) > \gamma, \tag{21.57}$$

where the threshold γ is derived by solving the integral equation

$$P_{\text{fa}} = \int_{\{\mathbf{x}:L(\mathbf{x})>\gamma\}} p(\mathbf{x}|H_0) d\mathbf{x} = \alpha. \tag{21.58}$$

NP Detection in GNSS: The traditional decision methods employed in the GNSS signal acquisition are generally based on the NP theorem: the SS cells are evaluated and a comparison against a threshold is performed, chosen on the basis of a given false alarm probability. Each cell measurement provides a variable y_j, which

---

*An example of LRT is in Section 21.3.1

is an instance of a stochastic variable Y_j, representing the samples of the CAF squared envelope defined in (21.37). When only white Gaussian noise disturbs the SIS (no multipath and interference are present), it is possible to show that $Y_j = X_{I,j}^2 + X_{Q,j}^2$, where $X_{I,j}$ and $X_{Q,j}$ are Gaussian random variables representing the real and imaginary parts of the CAF. They are zero-mean variables except in the *active* cell, that is, in the cell where the CAF peak is present*. The expected values of the Gaussian variables assume the form $E\{X_{I,j}\} = \mu_j \cos \varphi_j$ and $E\{X_{Q,j}\} = \mu_j \sin \varphi_j$, where $\mu_j = 0$ everywhere except in the active cell; the variance is the same for both variables and is denoted by σ^2. The values of μ_j and φ_j in the active cell can be easily found by evaluating (21.37) for $\bar{\tau} = \tau$ and $\bar{f}_d = f_a$. The value of μ_j in the active cell is denoted by C.

At this point, for each location j of the SS two simple hypotheses can be postulated:

$$H_{1,j} : \mu_j = C > 0; \tag{21.59}$$

$$H_{0,j} : \mu_j = 0. \tag{21.60}$$

Depending on the presence or absence of a GNSS signal, it can be proved that the random variable Y_j assumes the noncentral chi-squared distribution

$$g(y_j) = p(y_j | H_{1,j}) = \frac{1}{2\sigma^2} I_0 \left(\frac{\mu_j \sqrt{y_j}}{\sigma^2} \right) e^{-\frac{y_j + \mu_j^2}{2\sigma^2}} u(y_j) \tag{21.61}$$

for $\mu_j > 0$, and the chi-squared distribution

$$f(y_j) = p(y_j | H_{0,j}) = \frac{1}{2\sigma^2} e^{-\frac{y_j}{2\sigma^2}} u(y_j) \tag{21.62}$$

for $\mu_j = 0$, where $u(y_j)$ is the unit step function, that is, the discontinuous function whose value is 0 for negative argument and 1 for positive argument. The LR can be evaluated in each cell, as

$$L(y_j) = \frac{g(y_j)}{f(y_j)} \tag{21.63}$$

and the LRT could be applied cell by cell until the first threshold crossing is reached.

An example of chi-squared and noncentral chi-squared distributions is shown in Figure 21.15. In the curves the variance has been normalized to 1 (i.e., $\sigma^2 = 1$), while the CAF peak has been obtained by considering a received signal with $C/N_0 = 38$ dBHz, a bandwidth $B_{IF} = 4$ MHz, and an integration time of 1 ms. The same parameters have been used to obtained the ROC curves shown in Figure 21.16, but with three different values of C/N_0. From this figure we can easily observe the improvement, in terms of detector performance, obtained with increasing values of C/N_0.

---

*The zero-mean hypothesis is an approximation that is valid if the size of the bins of the SS are big enough and the code correlation values outside the peak can be neglected.

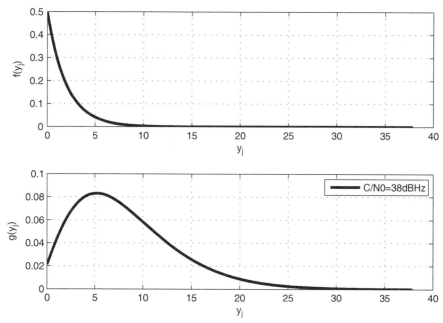

Figure 21.15   Top: chi-squared distribution. Bottom: noncentral chi-squared distribution for $C/N_0 = 38$ dBHz.

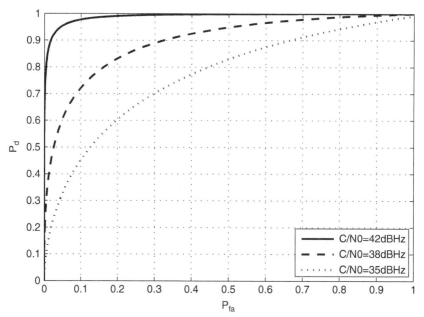

Figure 21.16   ROC curves.

21.5.2 Detection Based on the A Posteriori Probabilities

The method described in the previous section is based on the prior distributions $p(y_j|H_{1,j})$ and $p(y_j|H_{0,j})$, unlike other detection methods based on a posteriori probabilities, defined as

$$P(H_{0,j}|y_j) = P(H_{0,j}|Y_j = y_j) = P(\mu_j = 0|Y_j = y_j) \qquad (21.64)$$

$$P(H_{1,j}|y_j) = P(H_{1,j}|Y_j = y_j) = P(\mu_j = C|Y_j = y_j). \qquad (21.65)$$

A posteriori probabilities are generally more difficult to evaluate than prior probabilities. However, when dealing with measurement vectors, the use of *a* posteriori probabilities might become more advantageous.

In our GNSS application, using (21.61) and (21.62), the a posteriori probabilities can be rewritten by using Bayes, theorem, resulting in

$$P(H_{1,j}|Y_j = y_j) = \frac{g(y_j)P(H_{1,j})}{g(y_j)P(H_{1,j}) + f(y_j)P(H_{0,j})} \qquad (21.66)$$

$$P(H_{0,j}|Y_j = y_j) = \frac{f(y_j)P(H_{0,j})}{g(y_j)P(H_{1,j}) + f(y_j)P(H_{0,j})}. \qquad (21.67)$$

21.5.3 Bayesian Sequential Detection

The detection performance, especially when dealing with weak signals, can be improved if K measurements of the CAF are available. The detection could take place over the whole SS, but in order to reduce the detection complexity, only a subspace S_N made of N candidate cells can be considered, over which a *sequential detection* method can be applied. The selection of the candidate cells can be done in various ways. For example, the CAF samples can be sorted in descending order, from which only the first N cells are selected. At each step $k = 1, \dots, K$, a vector of N measurements is available

$$\mathbf{y}(k) = [y_1(k), \dots, y_j(k), \dots, y_N(k)], \qquad (21.68)$$

to which is associated a vector of stochastic variables

$$\mathbf{Y}(k) = [Y_1(k), \dots, Y_j(k), \dots, Y_N(k)]. \qquad (21.69)$$

Generally speaking, the way we use this information redundancy in the GNSS signal acquisition is by extending the *coherent or noncoherent* integration period. Another technique based on the recursive propagation of the a posteriori probabilities can also be used, as described in this section.

The assumption here is that the position of the active cell does not change during the K measurements. Moreover, the measurements are considered statistically independent. This assumption can be considered true as measurements derive from statistically independent error sources.

Sorting the vectors $\mathbf{Y}(k)$ in a matrix form

$$\mathbf{Y} = \begin{pmatrix} Y_1(1) & \cdots & Y_j(1) & \cdots & Y_N(1) \\ \cdots & \cdots & \cdots & \cdots & \cdots \\ Y_1(k) & \cdots & Y_j(k) & \cdots & Y_N(k) \\ \cdots & \cdots & \cdots & \cdots & \cdots \\ Y_1(K) & \cdots & Y_j(K) & \cdots & Y_N(K) \end{pmatrix}, \tag{21.70}$$

K column vectors

$$\mathbf{Y}_j(k) = \begin{pmatrix} Y_j(1) \\ \cdots \\ Y_j(k) \end{pmatrix} \quad k = 1, \ldots, K \tag{21.71}$$

can be extracted.

The a posteriori probabilities can be evaluated at each step k as

$$P[H_{1,j} | (\mathbf{Y}_j(k) = \mathbf{y}_j(k))] = P[(\mu_j = C) | (\mathbf{Y}_j(k) = \mathbf{y}_j(k))] \tag{21.72}$$

$$P[H_{0,j} | (\mathbf{Y}_j(k) = \mathbf{y}_j(k))] = P[(\mu_j = 0) | (\mathbf{Y}_j(k) = \mathbf{y}_j(k))]. \tag{21.73}$$

By observing that

$$\begin{aligned} \pi_{1,j}(k) &= P[H_{1,j} | (\mathbf{Y}_j(k) = \mathbf{y}_j(k))] \\ &= P\{[H_{1,j} | (\mathbf{Y}_j(k) = \mathbf{y}_j(k-1))] | y_j(k)\} \end{aligned} \tag{21.74}$$

$$\begin{aligned} \pi_{0,j}(k) &= P[H_{0,j} | (\mathbf{Y}_j(k) = \mathbf{y}_j(k))] \\ &= P\{[H_{0,j} | (\mathbf{Y}_j(k) = \mathbf{y}_j(k-1))] | y_j(k)\}, \end{aligned} \tag{21.75}$$

introducing the events

$$A_{1,j}(k-1) = \{H_{1,j} | (\mathbf{Y}_j(k) = \mathbf{y}_j(k-1))\} \tag{21.76}$$

and

$$A_{0,j}(k-1) = \{H_{0,j} | (\mathbf{Y}_j(k) = \mathbf{y}_j(k-1))\}, \tag{21.77}$$

it is possible to rewrite $\pi_{1,j}(k)$ by applying Bayes; theorem as

$$\pi_{1,j}(k) = \frac{P[y_j(k) | A_{1,j}(k-1)] P(A_{1,j}(k-1)}{P(y_j(k))}, \tag{21.78}$$

where

$$\begin{aligned} P(y_j(k)) = \; &P[y_j(k) | A_{1,j}(k-1)] P(A_{1,j}(k-1)) \\ &+ P[y_j(k) | A_{0,j}(k-1)] P(A_{0,j}(k-1)). \end{aligned} \tag{21.79}$$

It is then clear that $P(A_{1,j}(k-1)) = \pi_{1,j}(k-1)$. Considering also that the stochastic variable $Y_j(k)$ assumes real values, the three conditional probabilities $P(y_j(k)|A_{1,j}(k-1))$, $P(y_j(k)|A_{0,j}(k-1))$, and $P(y_j(k))$ must be substituted in the Bayes, formula by their associated distributions. Finally, assuming that the measurements are statistically independent, the equalities $P(y_j(k)|A_{1,j}(k-1)) = P(y_j(k)|H_{1,j})$ and $P(y_j(k)|A_{0,j}(k-1)) = P(y_j(k)|H_{0,j})$ are valid. It is now possible to put the pieces of the puzzle together, obtaining:

$$\pi_{1,j}(k) = \frac{g(y_j(k))\pi_{1,j}(k-1)}{g(y_j(k))\pi_{1,j}(k-1) + f(y_j(k))[1-\pi_{1,j}(k-1)]} \tag{21.80}$$

and

$$\pi_{0,j}(k) = \frac{f(y_j(k))\pi_{0,j}(k-1)}{f(y_j(k))\pi_{0,j}(k-1) + g(y_j(k))[1-\pi_{0,j}(k-1)]}, \tag{21.81}$$

which allows the evaluation of the a posteriori conditional probabilities for each step k, by using a recursive formula. Notice that similar results applied in a different context can be found in [11].

The probabilities in (21.80) and (21.81) can be rewritten in terms of the punctual LR, defined, according to (21.63), as

$$G(y_j(k)) = \frac{g(y_j(k))}{f(y_j(k))}, \tag{21.82}$$

obtaining

$$\pi_{1,j}(k) = \frac{\pi_{1,j}(k-1)}{\pi_{1,j}(k-1) + [1-\pi_{1,j}(k-1)]\dfrac{1}{G(y_j(k))}} \tag{21.83}$$

and

$$\pi_{0,j}(k) = \frac{\pi_{0,j}(k-1)}{\pi_{0,j}(k-1) + G(y_j(k))[1-\pi_{0,j}(k-1)]} \tag{21.84}$$

Sequential Detection in GNSS: Several ways can be imagined for using the a posteriori probabilities in detecting GNSS signals. A possible methodology that can be used in the signal acquisition is to restrict the propagation of the a posteriori probabilities over a subset of candidate cells. After that, the propagation of the probabilities $\pi_{1,j}(k)$ takes place only for a fixed number of measurements K, during which new values of the CAF are measured. At the end of these iterations, the detector applies a properly defined threshold to the a posteriori probabilities and decides for $H_{1,m}$ (i.e., the satellite is present and the CAF peak is at the cell $j = m$) if $\pi_{1,j}(K)$ crosses the threshold and if

$$P[H_{1,m}|(\mathbf{Y}_m(K) = \mathbf{y}_m(K))] > P[H_{1,i}|(\mathbf{Y}_i(K) = \mathbf{y}_i(K))] \quad \text{for } i \neq m. \tag{21.85}$$

This is clearly just a possible method of signal detection that can take advantage of the sequential propagation of a posteriori probabilities, and other methods can be envisaged. The complete analysis of these methods is beyond the purpose of this chapter and will not be considered. Examples can be found in [38] and [43].

21.6 GRADIENT METHOD FOR SIS PARAMETER ESTIMATION

As described above, the first step in GNSS processing is the signal acquisition: the satellites in view are detected and a first rough estimation of the Doppler shift and code phase is performed. The signal tracking follows the signal acquisition. Over each channel of the receiver, a *delay lock loop* (DLL) is used to synchronize the received spreading code and a local replica, while a *phase lock loop* (PLL) is generally employed to track the instantaneous phase of the incoming carrier.

The theory behind the tracking loops will be given in the following sections. For the moment, it is important to understand that the signal tracking has to perform a precise estimation of frequency and code delay. The frequency estimation of the incoming carrier is usually performed by a two-step process, which includes a rough estimation, followed by a fine search [50, 39, 31]. Conventional receiver architectures generally include a *frequency lock loop* (FLL) to refine the rough estimate performed by the signal acquisition. The FLL eases the PLL, reducing the transient time between the signal acquisition and the steady-state carrier/code tracking.

In this section, we focus on the transient between the signal acquisition and the signal tracking, since it requires a careful design in the development of real receivers. We introduce an instructive method, based on the gradient theory [26], for the refining of the initial Doppler shift and code delay estimate $\mathbf{p}^{(A)} = \left[\hat{\tau}^{(A)}, \hat{f}_d^{(A)} \right]$ performed in acquisition.

21.6.1 Transient between Signal Acquisition and Tracking

As explained in Section 21.2.2, the signal acquisition corresponds to the search of the correlation peak in the time and frequency domains. All acquisition systems for GNSS receivers are based on the ML estimation theory, and in practical implementations, they evaluate the code delay and Doppler shift, ignoring the unknown values of both navigation data bits and carrier phase. Considering classical acquisition schemes, the ML estimate can be written as in (21.28), which involves the evaluation of the CAF, as expressed in (21.30). Generally, the decision variable S_{max} in (21.31) is compared with predefined thresholds to determine the presence or absence of the signal.

Figure 21.10 shows the normalized CAF envelope obtained by processing a real GPS signal. In this case, the integration time was set equal to 1 ms, the frequency search scanned a range of 10.5 kHz centered on the nominal IF with a step of 300 Hz, while the delay bin size was equal to the sampling interval. It is evident that a first rough alignment between the incoming and the local signals can be found as the correlation peak rises above the noise floor. A decision logic determines if the correlation peak is detected and gives the first rough estimate of the code phase $\hat{\tau}^{(A)}$ and the Doppler shift $\hat{f}_d^{(A)}$ as output.

The correlation peak can be reasonably approximated to a convex paraboloid. One of the tasks of the tracking phase is to maintain the synchronization between the local and the incoming signals, which corresponds to continuously tracking the maximum of the correlation peak over time. As will be explained in Section 21.7, tracking loops can be seen as null-seeker systems. In fact, they use odd discrimination functions with a zero crossing point corresponding to the CAF maximum to maintain synchronization.

Ideally, the signal acquisition outputs $\hat{\tau}^{(A)}$ and $\hat{f}_d^{(A)}$ are close enough to the real values (i.e., $\left(\hat{\tau}^{(A)} - \tau\right)$ is on the order of few samples and $\left(\hat{f}_d^{(A)} - f_d\right)$ is within few Hz) in a reduced time interval (i.e., few milliseconds). With a good estimate of τ and f_d, the tracking loops can lock the signal with a short transient time. This is not the case in real receivers: the signal acquisition needs a nonnegligible processing time and a pull-in phase is often required to smooth the acquisition estimates and help the tracking subsystems lock the incoming signals. In other words, the objective of the pull-in phase consists in the refinement of the signal acquisition estimates.

From a mathematical standpoint, the problem is equivalent to find the maximum of a convex paraboloid (i.e., the correlation peak of the CAF), starting from an initial raw estimate. In Section 21.6.2 we introduce an algorithm based on the gradient method. Although it is not state of the art, it helps to understand the basic concepts behind the null-seeker operations.

21.6.2   Fundamentals of Gradient Theory

The gradient method (often referred to the method of steepest descent) [26, 17] is one of the oldest optimization algorithms. It is based on a comprehensive theory and is the core of other developments, like the conjugate gradient and Newton's methods [5]. In the past, gradient-based algorithms have attracted intense interest among researchers, both in theoretical optimization and in scientific applications [32], for maximizing or minimizing a function of several variables $F(x_1, x_2, \ldots, x_m)$.

Let us introduce the gradient method in a general manner. The optimization problem can be written as follows:

$$\mathbf{x}_{max} = \max_{\mathbf{x}} \left\{ F(\mathbf{x}) : \mathbf{x} \in \Omega \right\}, \tag{21.86}$$

where $\mathbf{x}$ is a n-dimensional vector, and $\mathbf{x}_{max}$ is a global maximum if Ω is the domain of $F(\mathbf{x})$ or a local maximum if Ω is a subset of the domain. The gradient method is generally applied to quadratic problems with good performance in terms of convergence. Consider the expression [26]:

$$F(\mathbf{x}) = \frac{1}{2}\mathbf{x}^T \mathbf{A}\mathbf{x} - \mathbf{b}^T \mathbf{x} + c, \tag{21.87}$$

where $\mathbf{A}$ is a matrix, $\mathbf{b}$ is a vector, and c is a constant. It can be seen that if $\mathbf{A}$ is symmetric and negative-definite, $F(\mathbf{x})$ is a convex paraboloid, which is maximized by the solution $\mathbf{Ax} = \mathbf{b}$. In fact, applying the gradient to (21.87), it is possible to obtain

$$\nabla F(\mathbf{x}) = \mathbf{Ax} - \mathbf{b}, \tag{21.88}$$

where

$$\nabla F(\mathbf{x}) = \left[\frac{\partial}{\partial x_1} F(\mathbf{x}), \frac{\partial}{\partial x_2} F(\mathbf{x}), \dots, \frac{\partial}{\partial x_n} F(\mathbf{x}) \right]. \tag{21.89}$$

Setting the gradient to zero $\nabla F(\mathbf{x}) = 0$, the resulting linear system to solve is $\mathbf{Ax} = \mathbf{b}$. The solution of such a system represents the maximum of $F(\mathbf{x})$.

The gradient method is often expressed and implemented in an iterative fashion:

$$\mathbf{x}[k+1] = \mathbf{x}[k] + a[k]\mathbf{g}[k], \tag{21.90}$$

where $a[k]$ is called step length and $\mathbf{g}[k]$ is called search direction. In order to find $\mathbf{x}$, which maximizes $F(\mathbf{x})$, the method starts from an arbitrary point $\mathbf{x}[0]$ and converges to the maximum after several steps. At each step, the algorithm computes a new value of $\mathbf{g}[k]$ until the kth value of $\mathbf{x}$ is close enough to the value of $\mathbf{x}$, which maximizes $F(\mathbf{x})$ (i.e., when the difference $\mathbf{x}[k] - \mathbf{x}[k-1]$ is below a predefined threshold).

One of the main drawbacks of gradient-based methods is the slow speed of convergence. However, with proper modeling of the problem, it is possible to build faster gradient schemes for various classes of problems [32]. It is important to note that the primary difference among methods (gradient method, Newton's method, etc.) resides in the definition of the vector $\mathbf{g}[k]$. Once $\mathbf{g}[k]$ is set, all the methods seek the maximum of $F(\mathbf{x})$, following their own convergence trend. In the gradient method $\mathbf{g}[k]$ is

$$\mathbf{g}[k] = \left[\nabla F(\mathbf{x}) \right]_{\mathbf{x}=\mathbf{x}[k]}. \tag{21.91}$$

In order to improve the converging speed, the step length $a[k]$ is generally chosen to maximize the equation $F(\mathbf{x}[k] - a[k]\mathbf{g}[k])$.

A two-dimensional function $F(\mathbf{x})$ and the iterative process in (21.90) are qualitatively illustrated in Figure 21.17, which shows contours of constant values of $F(\mathbf{x})$ and a typical sequence developed by the process. As mentioned, during the iterative process the subsequent direction vectors $\mathbf{g}[k]$ are selected, as are the subsequent gradients. Therefore, the directions are not specified beforehand; rather, they are determined sequentially at each step of the iteration. At step k the current gradient vector is evaluated. The result is multiplied by $a[k]$ and added to $\mathbf{x}[k]$ to obtain a new estimate $\mathbf{x}[k+1]$.

21.6.3 Application to GNSS Signals

The gradient method is a valid alternative to refine $\hat{\tau}^{(A)}$ and $\hat{f}_d^{(A)}$ after the signal acquisition. Referring to the notation introduced in Section 21.6.2, note that:

- $S_{y,r}\left(\overline{\tau}, \overline{f}_d\right)$ corresponds to $F(\mathbf{x})$, with $\mathbf{x} = \left(\overline{\tau}, \overline{f}_d\right)$;
- $\mathbf{x}[0]$ corresponds to the pair of initial estimates $\hat{\tau}^{(A)}$ and $\hat{f}_d^{(A)}$;

Function of two variables with a global maximum

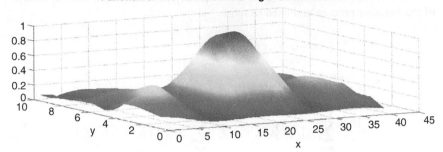

Contours and path obtained applying the gradient method for the estimate of the global maximum

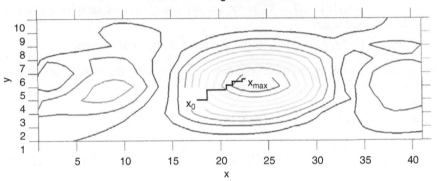

Figure 21.17   Top: A two-dimensional function $F(\mathbf{x})$. Bottom: Contours of constant values of $F(\mathbf{x})$ and the path obtained by gradient method [26].

- $\mathbf{x}_j$ represents the last estimated values $\hat{\tau}^{(R)}$ and $\hat{f}_d^{(R)}$, which will be passed to the tracking stage.

The fine estimation of the code phase and Doppler shift is equivalent to finding the CAF maximum, which requires the computation of the derivative of $S_{y,r}(\overline{\tau}, \overline{f}_d)$ in both time and frequency domains. In vector calculus, the gradient of a multidimensional function is a vector pointing at the direction of the greatest rate of increment, and whose magnitude is the greatest rate of change. The gradient of the CAF envelope $S_{y,r}(\overline{\tau}, \overline{f}_d)$ can be expressed as:

$$\nabla S_{y,r}(\overline{\tau}, \overline{f}_d) = \left[\frac{\partial}{\partial \overline{f}_d} S_{y,r}(\overline{\tau}, \overline{f}_d), \frac{\partial}{\partial \overline{\tau}} S_{y,r}(\overline{\tau}, \overline{f}_d) \right]. \tag{21.92}$$

Setting (21.92) to zero, the linear system can be solved as shown in (21.90). The two partial derivatives of (21.92) can be computed separately.

First, the partial derivative on the frequency domain is computed at each step of the iterative process and used to refine $\hat{f}_d^{(A)}$:

$$
\begin{aligned}
\Gamma_{f_d}\left(\overline{\tau}, \overline{f}_d\right) &\equiv \frac{\partial}{\partial \overline{f}_d} S_{y,r}\left(\overline{\tau}, \overline{f}_d\right) \\
&= -\frac{4\pi T_S}{L^2} \sum_{n=0}^{L-1} y_c\left[n, \overline{\tau}\right]\cos\left(\Psi[n, \overline{f}_d]\right) \sum_{n=0}^{L-1} ny_c\left[n, \overline{\tau}\right]\sin\left(\Psi[n, \overline{f}_d]\right) \\
&\quad + \frac{4\pi T_S}{L^2} \sum_{n=0}^{L-1} y_c\left[n, \overline{\tau}\right]\sin\left(\Psi[n, \overline{f}_d]\right) \sum_{n=0}^{L-1} ny_c\left[n, \overline{\tau}\right]\cos\left(\Psi[n, \overline{f}_d]\right),
\end{aligned}
\tag{21.93}
$$

where:

- $y_c[n, \overline{\tau}] \equiv y_{IF}[n]c_b(nT_s - \overline{\tau})$ is the incoming sequence of samples multiplied by the local code;
- $\Psi[n, \overline{f}_d] \equiv 2\pi(f_{IF} + \overline{f}_d)nT_s$ is the argument of the local carriers.

The process also refines the initial estimate $\hat{\tau}^{(A)}$, computing the partial derivative of the CAF envelope in the time domain:

$$
\begin{aligned}
\Gamma_{\tau}\left(\overline{\tau}, \overline{f}_d\right) &\equiv \frac{\partial}{\partial \overline{\tau}} S_{y,r}\left(\overline{\tau}, \overline{f}_d\right) \\
&= \frac{2}{L^2} \sum_{n=0}^{L-1} y_c\left[n, \overline{\tau}\right]\cos\left(\Psi[n, \overline{f}_d]\right) \sum_{n=0}^{L-1} y_c'\left[n, \overline{\tau}\right]\cos\left(\Psi[n, \overline{f}_d]\right) \\
&\quad + \frac{2}{L^2} \sum_{n=0}^{L-1} y_c\left[n, \overline{\tau}\right]\sin\left(\Psi[n, \overline{f}_d]\right) \sum_{n=0}^{L-1} y_c'\left[n, \overline{\tau}\right]\sin\left(\Psi[n, \overline{f}_d]\right),
\end{aligned}
\tag{21.94}
$$

where $y_c'[n, \overline{\tau}] \equiv y_{IF}[n]c_b'(nT_s - \overline{\tau})$ stands for the local code derivative in the time domain.

Note that the derivatives are computed in the discrete domain because the algorithm is applied to the stream of raw samples at the ADC output. While for $S_{y,r}\left(\overline{\tau}, \overline{f}_d\right)$ it is straightforward implementing (21.93) (i.e., both the sine and the cosine terms already exist in any GNSS receivers, since they are the local carriers of the carrier tracking loop), Equation (21.94) has to be carefully handled, but efficient implementations are possible. Equations (21.93) and (21.94) are key elements of the iterative method, both of which can be computed in software with moderate computational burdens. Furthermore, it must be noted that the term

$$
\sum_{n=0}^{L-1} y_{IF}[n]c_b\left(nT_s - \overline{\tau}\right)
\tag{21.95}
$$

is the correlation between the incoming signals and the local code. This term has to be evaluated both in signal acquisition and in signal tracking, because all GNSS signal processing is based on correlation. Equations (21.93) and (21.94) have been obtained in closed form and give an effective way of refining both the Doppler shift and the code phase estimates. When the difference between the last CAF envelope derivatives

$\left\{ \Gamma_{f_d}^k\left(\overline{\tau}, \overline{f}_d\right); \Gamma_{\tau}^k\left(\overline{\tau}, \overline{f}_d\right) \right\}$ and the previous values $\left\{ \Gamma_{f_d}^{k-1}\left(\overline{\tau}, \overline{f}_d\right); \Gamma_{\tau}^{k-1}\left(\overline{\tau}, \overline{f}_d\right) \right\}$ are below predefined thresholds, the algorithm stops refinement and provides the last frequency estimate to the carrier tracking loops and the last code phase estimate to the DLL.

The method can be assessed with real GPS signals, using PC-based software receivers. A possible implementation of the algorithm might foresee the computation of the gradient on the same segment of received data. Nonetheless, it must be observed that the real CAF is noisy. Computing the gradient on the same data set might induce the method to converge to a erroneous local maximum, resulting in a wrong estimate of the Doppler shift. A more appropriate implementation considers a new segment of incoming signal at each kth iteration. This way the algorithm processes a new realization of the noise every time it has to compute a new derivative, leading, on average, to the reduction of the noise contribution. This strategy has another advantage that is particularly appreciated in software implementations—with a new chunk of data at each iteration there is no need for any additional buffer for data storage, although new correlations between the incoming signal $y_{IF}[n]$ and local codes $c\left(nT_s - \overline{\tau}\right)$ are required.

As mentioned at the beginning of the section, the gradient method is not state of the art in real GNSS receivers. It has been introduced here as an educational example of refinement of the code delay and frequency estimates based on the evaluation of the CAF maximum. In practical implementations, this refinement is implemented by using the tracking systems described in the next sections.

21.7   NULL SEEKER AND TRACKING LOOPS

As seen in the previous section, the peak of the CAF can be accurately estimated with gradient-based methods, which seek for CAF derivative nulls. This type of algorithm is intrinsically iterative and stops when its output reaches a steady-state configuration. These basic concepts can be also found in the theory of classical tracking loops, which are commonly used in GNSS receivers to finely estimate the frequency shift and the code delay of the received signal.

There is a lot of literature devoted to tracking loops. The most famous is on the PLL, which is designed to estimate the instantaneous phase of a sinusoidal waveform. In the GNSS community the DLL is also important, which is designed for estimating the code delay. An analysis of these two systems is beyond the scope of this section; therefore, we will limit our attention to certain aspects, which are common to the two schemes and can be the starting points in tackling the problem of analyzing and designing PLLs and DLLs.

First of all, both PLLs and DLLs are *null seeker* systems and in this sense they implement the gradient method—that is, they seek a null to finely estimate the frequency shift and code delay. The input of a PLL has to be a sinusoidal waveform, while a DLL needs a PRN code. Therefore the IF signal $y_{IF}[n]$ is wiped off before entering the tracking loops. This is performed by a concatenated scheme, as shown in Figure 21.18, where two wiping systems are coupled together and linked to the estimators of the code delay and frequency shift. At each iteration, the estimators provide new estimates of frequency shift and code delay to the

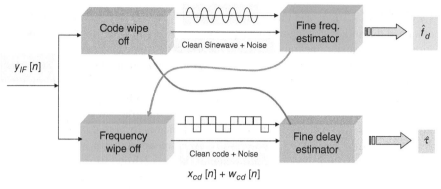

Figure 21.18   Concatenation of wipe-off and tracking systems.

wiping systems, which progressively improve the quality of the signals at their outputs.

It should be noted that this system is only a conceptual scheme, and the four blocks can be combined in the actual implementation of the receiver in different ways. Real receivers contain code and frequency wipe-off systems and code and frequency tracking loops, even if linked differently. The operations governing the concatenated systems depend on the specific implementation of the wipe-off/tracking scheme. An example of a real concatenated scheme is shown in Figure 21.19, where one can recognize the two loop control feedbacks and the code and carrier wipe-off that are performed by multiplication blocks and the integration and dump, as indicated in Figure 21.18. In the next sections, two simple schemes of DLL and PLL will be described. They share the same mathematical model, which will be described in Section 21.7.3.

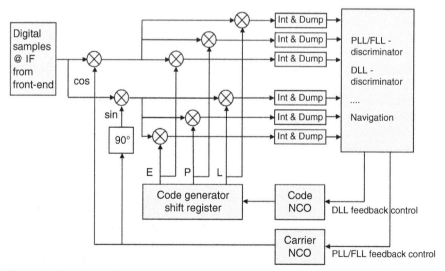

Figure 21.19   Block diagram of the digital tracking architecture typically implemented in GNSS receivers.

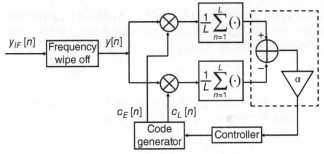

Figure 21.20   Basic scheme of a coherent DLL. The dashed box represents the so-called discriminator.

21.7.1   DLL

The main task of a DLL is to align a local code (called *prompt code*) to the code received from each satellite in view. Therefore a DLL can be considered a synchronizer system, which is typically implemented by means of a null-seeker structure. The coherent DLL described in this section is based on the hypothesis that the carrier phase and the frequency shift can be properly estimated and removed from the IF signal by means of a very precise carrier wipe-off system.

The coherent DLL is at the basis of any other DLL system. Its scheme, in its simplest form, is shown in Figure 21.20. The ranging performance enabled by CDMA signals largely depends on the working principle of this scheme. Its input is a baseband signal $y[n]$ obtained by frequency wipe-off of the IF signal $y_{IF}[n]$ given in (21.18). The carrier wipe-off can be obtained by multiplying $y_{IF}[n]$ by a local carrier of the type $\cos\left(2\pi\left(f_{IF}+\hat{f}_d\right)nT_s+\varphi_{IF}\right)$, where $\hat{f}_d$ is the best available estimate of the frequency shift. This gives rise to a baseband component and a double frequency component, which can be ignored, being filtered by the two DLL integrators, shown in Figure 21.20. Therefore, by adding the hypothesis $\hat{f}_d \cong f_d + f_c$, the baseband signal at the input of the DLL can be written as

$$y[n] = r_b[n] + W_m[n] \cong \sqrt{\frac{P_R}{2}}c_b\left(nT_s-\tau\right)d\left(nT_s-\tau\right)+w_m[n], \qquad (21.96)$$

where $w_m[n]$ is a noise component. The code generator in Figure 21.20 generates also two delayed versions of the code,

$$c_E[n] = c_b\left(nT_s-\left(\hat{\tau}^{(A)}+\overline{d}_\tau\right)+d_sT_c/2\right) \qquad (21.97)$$

and

$$c_L[n] = c_b\left(nT_s-\left(\hat{\tau}^{(A)}+\overline{d}_\tau\right)-d_sT_c/2\right), \qquad (21.98)$$

called, respectively, *early* and *late* codes, where:

- $\hat{\tau}^{(A)}$ is the delay code estimated by the acquisition system;
- $\overline{d}_\tau$ is the test variable of the null seeker. In particular it is a variable delay running within the loop. The value of $\overline{d}_\tau$ in steady-state working condition

represents the residual delay $\hat{d}_\tau$ estimated by the DLL. When the DLL reaches this condition, we say that the DLL is locked, and the test variable becomes the fine estimate of the delay, that is,

$$\overline{d}_\tau = \hat{d}_\tau,$$

- d_s is a *spacing* factor related to the delay between the early and late codes. In a software implementation the term $d_s T_c/2$ will be of the type

$$d_s \frac{T_c}{2} = d_s \frac{N_c T_s}{2},$$

where $d_s' = d_s N_c/2$ represents the spacing in terms of samples. Typically, $d_s' < 1$, as the uncertainty on $\hat{\tau}^{(A)}$ is generally of the order of $\pm T_s$, unless the delay bin size of the search space is greater than T_s.

Early and late local codes are used to compute the two correlations shown in Figure 21.20. To simplify the notations, and without loss of generality, we can set $\hat{\tau}^{(A)} = 0$, from which $\tau = d_\tau$. In this case, the two correlations can be expressed as

$$S_E = \sqrt{\frac{P_R}{2}} \frac{1}{L} \sum_{n=1}^{L} c_b(nT_s - d_\tau) c_b\left(nT_s - \overline{d}_\tau + d_s T_c/2\right) d(nT_s - d_\tau) \quad (21.99)$$

and

$$S_L = \sqrt{\frac{P_R}{2}} \frac{1}{L} \sum_{n=1}^{L} c_b(nT_s - d_\tau) c_b\left(nT_s - \overline{d}_\tau - d_s T_c/2\right) d(nT_s - d_\tau). \quad (21.100)$$

The DLL operations depend on the properties of (21.99) and (21.100) when no bit transition occurs during the integration. If the DLL works before the operations of bit synchronization, the value of L, referred to as *predetection integration time*, is typically less than the data bit period. After bit synchronization, we can use a *postdetection integration* time, which can cover a single data bit period. If the data bit sequence (or the secondary code of a pilot channel) is known, the bit sequence can be removed, and the integration time can be extended beyond the data bit interval.

Notice that while generating the early and late codes, the DLL generates also the *prompt* code

$$c_P[n] = c_b\left(nT_s - \left(\hat{\tau}^{(A)} + \overline{d}_\tau\right)\right) \quad (21.101)$$

to be used in the wipe-off system of Figure 21.19.

Discrimination Function: The *discriminator* is the block identified by a dashed box in Figure 21.20. It receives at its input the two correlation values S_E and S_L given in (21.99) and (21.100). The *discrimination function*, also called *S-curve*, is defined as a function of

$$\varepsilon_\tau = d_\tau - \overline{d}_\tau \quad (21.102)$$

The so-called *early-minus-late* (E-L) is the simplest discriminator, indicated in Figure 21.20, as

$$S(\varepsilon_\tau) = \sqrt{\frac{2}{P_R}} \{S_E - S_L\}. \tag{21.103}$$

This is not the sole possible definition of the discriminator function. A complete description of other possibilities is beyond the scope of this section, but can be found in [22], together with the introduction of the so-called noncoherent DLL scheme.

The definition in (21.103) implies that no bit transition is present, and a normalization factor $\alpha = \sqrt{2/P_R}$ is applied in the scheme of Figure 21.20. Examples of S-curves for different values of d_s are given in Figure 21.21 for the GPS C/A code signal, where it is possible to observe that there is a linear region around the origin where the S-curve can be written as

$$S(\varepsilon_\tau) = -\varepsilon_\tau \gamma_s. \tag{21.104}$$

Typically DLL works in this region, where the value of S is used as an error signal given to a controller, whose purpose is to modify the value of $\bar{d}_\tau$ in order to bring the value of S to zero; hence we can write

$$\hat{d}_\tau = \lim_{S \to 0} \bar{d}_\tau,$$

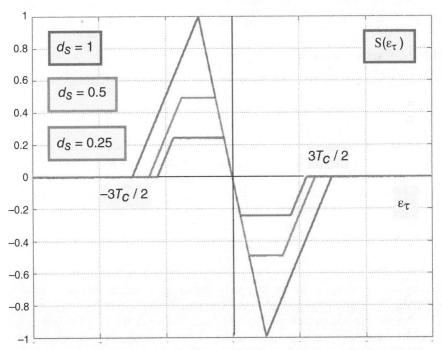

Figure 21.21   Comparisons of S-curves for GPS C/A, varying the spacing factor d_s.

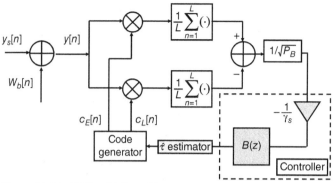

Figure 21.22   DLL with loop filter.

and the DLL can be considered as a null-seeker system, working step by step every time a new value of S is available. If the DLL falls in the nonlinear region, the system can lose lock, and a reacquisition becomes necessary.

In a real DLL, the output of the discriminator is affected by noise, whose effect can be reduced by applying some form of conditioning before producing an estimated delay. The general scheme becomes the one shown in Figure 21.22, where the multiplication by $(-1/\gamma_s)$ compensates for the S-curve slope, and $B(z)$ is a loop filter for the signal conditioning. The loop filter reduces the noise at the discriminator output and therefore should have a narrow band to remove the noise contribution as far as possible. However, the bandwidth should not be as small as to filter out signal contributions, resulting in a trade-off value to be assessed.

21.7.2   Carrier Tracking

The carrier tracking loop is a feedback loop able to finely estimate the frequency $(f_{IF} + f_d + f_c)$ of a noisy sinewave. It also tracks the frequency changes overtime. Most receivers also track the phase term φ_{IF} present in (21.27).

The most common scheme for frequency estimation is the PLL, which consists of a system capable of adjusting the frequency of a local oscillator to match the frequency of an input signal, which is generally a continuous wave or a frequency modulated signal. The local oscillator is called *voltage controlled oscillator* (VCO); however, this term is mainly related to analog PLL structures. In software implementation, the term VCO must be used in a wide sense to denote any local oscillator with a variable frequency controlled by an input parameter. To be more precise, in a digital tracking system a VCO is implemented by means of numerical controlled oscillator.

In GNSS applications, once the code is stripped from $y_{IF}[n]$, the PLL receives a continuous wave signal still modulated by the navigation data. This is the reason why PLLs insensitive to phase transitions are generally adopted [22].

Another type of carrier tracking loop, well known in the GNSS community, is the FLL capable of tracking the frequency of the carrier ignoring the phase term φ_{IF}. The FLL is often used as an aiding system of PLL. Once the FLL has locked a frequency $\hat{f}_{d,\text{FLL}}$, the PLL can refine the estimated frequency working in a narrow band around $\hat{f}_{d,\text{FLL}}$. The theory of PLL, FLL, and their variants (as such the Costas loop) is covered in many books, and [22, 28, 33], and [50] are excellent references for GNSS applications.

The basic principle of a PLL is similar to that of a DLL. Referring to Figure 21.20, we have a code wipe-off system for the PLL, which provides a sinusoidal waveform affected by noise at the input of the loop. The local generator provides two sinusoidal signals (i.e., sine and cosine tones) in the two branches of the loop, called *in-phase* (I) and *quadrature* (Q) components. The PLL aligns the instantaneous phase of the I component with the phase of the incoming signal. The integrators mitigate the noise effects and cut the term at 2 f_{IF}. At this point, a discriminator (different from the one of DLL) extracts the phase difference between the incoming and the local signals. A loop filter can be included to further reduce the effect of noise. Once the phase difference is approximately zero, the PLL reaches a steady-state condition and the local waveform is aligned with the incoming carrier. Details on PLL can be found in [22, 28, 33], and [50].

Although DLLs and PLLs have different schemes, especially at the discriminator level, they share the same loop structure. It is interesting to observe that they both admit a digital model, very useful for the design of the loop parameters. How to introduce this model is the subject of the next section.

21.7.3   Models of the Tracking Loops

The blocks for the fine estimation of carrier frequency and code delay generally consist of tracking loops, whose conceptual scheme is drawn in Figure 21.23. The signal $Ax(nT_s, \vartheta)$ is the noise-free signal component at the output of the wipe-off systems (thus a clean sinusoid at the output of the code wipe-off, and a clean PRN code at the output of the carrier wipe-off). This signal contains an unknown parameter to be estimated. Therefore, ϑ represents the unknown delay if $Ax(nT_s, \vartheta)$ is a clean code, and an unknown phase if $Ax(nT_s, \vartheta)$ is a clean carrier; in this last case ϑ could represent a residual frequency, a phase, or both of them (in this case ϑ is a 2-dimensional vector). The amplitude factor A is generally unknown, and can also be estimated if required. However the main purpose of the loop is the estimation of the variable ϑ. The signal at the loop input $y[n]$ is the observable (noisy) version of $Ax(nT_s, \vartheta)$.

The error evaluation block is a key element of the loop. It is properly designed with the aim of evaluating a quantity $g(k; \vartheta, \overline{\vartheta})$ related to the difference between the unknown ϑ and the parameter $\overline{\vartheta}$ of a local signal $x(nT_s, \overline{\vartheta})$, which exhibits the same structure of the incoming signal. The quantity $g(k, \vartheta, \overline{\vartheta})$ is evaluated at the epochs $t_k = kT_p$, where $T_p = LT_s$, and $L \geq 1$ is the discrete-time interval between two updates of the parameters of the local generator. Therefore, $g(k, \vartheta, \overline{\vartheta})$ can be seen as a discrete-time sequence working in the domain of the discrete-time k; Then it is filtered, and in turn is used to update the value of $\overline{\vartheta}$. If the loop is properly designed, $\overline{\vartheta} \to \hat{\vartheta}$, where $\hat{\vartheta}$ is the fine estimate of the unknown ϑ.

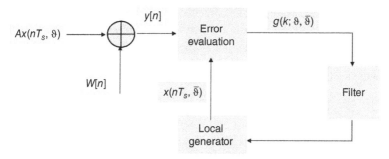

Figure 21.23   Physical model of a tracking loop.

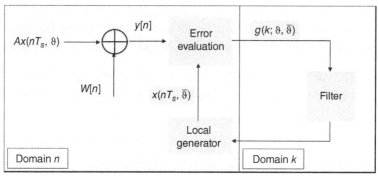

Figure 21.24   The two time scales in the tracking loop.

The exact behavior of the carrier and code loops cannot be explained on the basis of the scheme in Figure 21.23. It is necessary to consider the particular structure of both loops and to approach them separately. However, they have some important common aspects. The most relevant is that two time scales n and k work together in the loop, as highlighted in Figure 21.24. This allows for the introduction of a model working in the domain of the discrete-time k, as shown in Figure 21.25. The block $D(z)$ indicates the update of the parameter $\overline{\vartheta}[k]$ in the form

$$\overline{\vartheta}[k] = g_F[k-1] + \overline{\vartheta}[k-1],$$

and $v[k]$ is a noise term. This model is very important, as it allows a very precise analysis of the loop.

21.7.4   Tracking Loop Performance

We have seen that both DLLs and PLLs are null seekers, which reach the lock condition when the discriminator output is zero. Because of the presence of noise at the input of both schemes, the zero value is never obtained, and this introduces a noise component that is added to the two estimated quantities: the code delay and the frequency shift (we recall that these estimates are obtained when the tracking systems are locked). Such a noise component is a random process, generally called jitter,

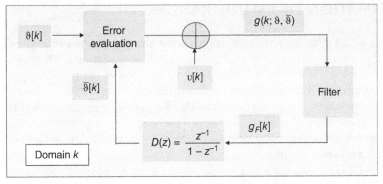

Figure 21.25   Mathematical model of a tracking loop in the k domain.

whose standard deviation is the parameter used to evaluate the performance of the tracking loops. We give now more details on the DLL jitter characterization. Similar considerations can be done for PLL.

The DLL performance is generally evaluated in terms of jitter standard deviation, or pseudorange error (when expressed in meters) σ_j. In the literature some theoretical expressions of σ_j are available and valid in a well-defined context. In [28] the result

$$\sigma_{j,\text{GPS}} = cT_s \sqrt{\frac{d_s}{C/N_0} \frac{B_{eq}}{2}} \qquad (21.105)$$

is given, valid for a coherent DLL, tracking a GPS C/A signal in the L1 band. The term B_{eq} is the noise equivalent bandwidth introduced by the loop filter, and c is the speed of light.

The evaluation of the jitter standard deviation is much more complex in the case of noncoherent DLL. Reference [22] gives an expression based on frequency domain quantities. Another expression, based on correlation domain quantities, can be found in [10].

The expressions of σ_j in both coherent and noncoherent schemes generally depend on the carrier-to-noise ratio C/N_0, which is not known, and needs to be estimated. The C/N_0 estimate is generally obtained from postcorrelation samples, where the signal-to-noise power ratio estimation methods are demonstrated to be more robust with respect to interference [18]. The classic approach to estimate C/N_0 is presented in [12], also known as the narrowband–wideband power ratio method. Nonetheless, several different methods can be applied, borrowed, and properly adapted from the digital communications world, where a wide literature is available for the problem of estimating the signal-to-noise of *M-ary phase shift keying* (M-PSK) modulations in additive white Gaussian noise. During the tracking phase, the navigation signal after the correlator driven by the prompt code of the DLL can be modeled as a sequence of binary phase-shift keying symbols in additive noise, with a symbol rate equal to the inverse of the correlators integration time. Thus, once the SNR of the signal after the correlator is estimated, the C/N_0 can be directly derived using (21.21). More details on this topic can be found in [13].

21.8 ESTIMATION OF PSEUDORANGES

The main task of a GNSS receiver is to compute the antenna position in a given reference frame. This is obtained through the measure of the psuedoranges between the receiver and the satellites. We will see that the estimation of the pseudoranges can be obtained after the tracking outputs are converted into navigation data, that is, when each satellite channel is in navigation mode. We say that a single channel is in navigation mode when it continuously tracks the spreading sequence and the carrier phase of the acquired satellite, and also correctly demodulates the navigation message broadcast by the satellite.

Once the DLL has created a clean code aligned with the SIS, the PRN code can be wiped-off from the SIS,* and the navigation message can be demodulated by applying a classical phase shift keying demodulation scheme on the wiped-off signal. These conceptual operations can be implemented in many different ways, and the actual scheme of the extraction of the navigation message depends on the particular structure of the DLL and PLL used in the receiver.

The pseudoranges are obtained by measuring propagation delays. This procedure can be divided in two different phases:

- estimating the initial biased set of propagation delays and converting them into distances;
- monitoring of the propagation delay evolution after the first position computation.

Referring to the GPS L1 C/A code, the key point for the pseudorange evaluation is the identification of the telemetry word (called TLM) in the navigation message of each received channel. The whole message contains 25 pages (or frames) of 30 seconds each. Each frame is subdivided into 5 subframes of 6 seconds each, and each subframe always starts with the TLM word. For the evaluation of the first set of initial propagation delays, the receiver must be able to perfectly locate the beginning of each subframe, verifying that after the eight-bit-long preamble the TLM word is present. Once each preamble is identified for all the available channels, it is possible to compare the times of arrival of each TLM. Since each subframe is broadcast by each satellite at the same time, and the distance between each satellite and the user is not generally constant, each TLM will be received with a different propagation delay, as shown in the example of Figure 21.26.

In order to determine the first, and only the first, set of pseudoranges, the channel with the earliest arriving subframe is assumed as a reference, and, for this first set, the minimum signal traveling time is assigned to the reference channel. Its value, for GPS, is between 65 and 85 ms. An arbitrary value in this range is then assigned to the reference channel (70 ms in Fig. 21.26). Then, the *difference of time of arrival* (DTOA) of the other TLM words with respect to the reference one is computed, as shown in Figure 21.26. This is performed by counting the number of codes, chips, and the fraction of last chip. In GPS L1 C/A each code corresponds to

---

*This operation is called *despreading.*

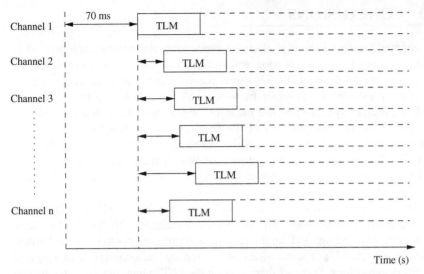

Figure 21.26   Channels misalignment related to the dierent propagation delay.

a 1 ms delay, each chip contributes with a 0.9775 μs delay, and the delay due to the fraction of chip can be evaluated from the DLL output. At this point, by multiplying the DToSs by the speed of light, we obtain the reference pseudorange (arbitrary), and all the other pseudoranges are derived by adding to the reference pseudorange the computed DTOAs of the other TLM words, multiplied by the speed of light.

When the procedure for evaluating the pseudorange is available for at least four satellites and the navigation messages of subframes from 1 to 3 are decoded for each of them, the receiver has all the necessary information to fix its first position. Indeed, subframes from 1 to 3 are necessary to retrieve the ephemeris of the satellites, from which it is possible to correctly locate each of them in space, while three satellites are necessary to provide enough equations in the least square PVT computation algorithms to get rid of the three x, y, and z unknowns and the fourth satellite for getting rid of the common clock bias error (difference between local and constellation timing reference), which creates a bias on all the pseudoranges estimates.

When the first position is correctly evaluated, this can be used to convert pseudoranges in ranges by evaluating the geometrical distances among satellites and receiver positions. Then, the arbitrary TOA reference can be substituted with the corresponding range estimate. In this way, the fourth unknown in the second PVT evaluation will just account (theoretically) for the clock bias error, and this information can be used for correcting the local clock and to provide also a very precise timing information to the users.

From this point on a specific set of pseudoranges can be estimated at any time by considering the displacement between the navigation message among all the different channels without waiting for the TLM word by simply periodically updating a specific counter for each channel.

21.9 CONCLUSIONS

This chapter presented the digital signal processing algorithms implemented in the physical layer (i.e., acquisition and tracking) of GNSS receivers. First, we started with the analysis of the RF signal received at the antenna and we highlighted the presence of a Doppler frequency shift, which has to be finally estimated in order to synchronize the incoming signal with local replicas. This is fundamental for the computation of the user's position. The frequency downconversion of the received signal and the sampling stage were also briefly illustrated. The general structure of the digital section of the receiver was presented, and details were given on the subsystems in charge of acquiring and tracking the signals in space.

We provided a survey of detection and estimation theory, which is fundamental to the acquisition methods. The operations associated with signal acquisition can be time consuming, and might require a significant computational burden. Therefore, we introduced some strategies to reduce the complexity of classical acquisition algorithms. Some of them rely on the FFT, which plays a fundamental role in the definition of these simpler strategies and is widely adopted in software-defined radio receivers, although other techniques exist. The chapter described both conventional FFT-based methods, Bayesian sequential detection, and gradient methods.

The transient between signal acquisition and tracking was addressed since it is a crucial point in the design of GNSS receivers, together with code and carrier tracking systems, described as particular cases of null seekers.

Finally, some performance metrics were described for both acquisition and tracking systems.

APPENDIX A: THE DOPPLER EFFECT AT BASEBAND

Let us consider the received baseband component, defined as

$$r_{BB}(t) = \sqrt{2P_R}\, c_b(t - \tau_p)\, d(t - \tau_p). \tag{A.1}$$

This expression can be rewritten by recalling that any signal with finite energy can be represented in a finite time interval T_F as a Fourier series of the type

$$r_{BB}(t) = \sum_{i=-\infty}^{+\infty} \mu_i e^{j2\pi \frac{i}{T_F}t} \quad t \in \left(-\frac{T_F}{2}, +\frac{T_F}{2}\right). \tag{A.2}$$

If T_F is large enough and the baseband signal is *de facto* bandlimited to a maximum frequency of the order of N_F/T_F, where N_F is an integer of the Fourier series summation, we can affirm that

$$r_{BB}(t) \cong \sum_{i=-N_F}^{+N_F} \mu_i e^{j2\pi \frac{i}{T_F}t} \tag{A.3}$$

is an expression valid for any value of t. From (21.9) we know that a generic frequency component $f_i = i/T_F = c/\lambda_i$ of the Fourier series will be affected by a Doppler effect of the type

$$f_{d,i}(t) = -\frac{1}{\lambda_i}\frac{d}{dt}D(t) = f_d(t)\frac{\lambda_{RF}}{\lambda_i} = f_d(t)\frac{f_i}{f_{RF}}. \tag{A.4}$$

In our application, any λ_i is much greater than λ_{RF}; therefore, any $f_{d,i}$ will be negligible with respect to the Doppler effect at the carrier. Notice that the same result could be obtained more formally by writing $r_{BB}(t)$ in terms of Fourier transform rather than Fourier series. However, the approach with the Fourier series has been preferred, since it is more intuitive.

By using (A.3) we can write the baseband component affected by Doppler as

$$r_{B,d}(t) \cong \sum_{i=-N_F}^{+N_F} \mu_i e^{j2\pi\left(f_i + f_d(t)\frac{f_i}{f_{RF}}\right)t} = \sum_{i=-N_F}^{+N_F} \mu_i e^{j2\pi\frac{i}{T_F}\left(1+\frac{f_d(t)}{f_{RF}}\right)t}. \tag{A.5}$$

In the time intervals, where $f_d(t)$ can be considered constant, we can write

$$r_{B,d}(t) \cong \sum_{i=-N_F}^{+N_F} \mu_i e^{j2\pi\frac{i}{T_F}\left(1+\frac{f_d(t)}{f_{RF}}\right)t} = \sum_{i=-N_F}^{+N_F} \mu_i e^{j2\pi\frac{i}{T'_F}t}, \tag{A.6}$$

where

$$T'_F = \frac{T_F}{1 + f_d/f_{RF}}. \tag{A.7}$$

This means that $r_{B,d}(t)$ can be seen as a new sequence of chips with the same structure of $r_{BB}(t)$. The Doppler shift will expand or compress all the periods of the sinusoids that form the baseband signal. Therefore intuitively, the net effect of Doppler on data can be seen as an expansion or compression of the chips. In [50] the amount of $f_{d,i}$ for the baseband frequency of 1.023 MHz (GPS chip rate) is evaluated as 3.2 Hz for a low-mobility vehicle and 6.4 Hz if the receiver moves at high speed.

A simplified method to take into account the Doppler effect at baseband is to modify the chip rate as

$$T'_c = \frac{1}{1/T_c + f_{d,T_c}} = \frac{T_c}{1 + f_d/f_{RF}}, \tag{A.8}$$

applying to T_c the same compression/expansion factor applied to T_F in (A.7). In (A.8) the term f_{d,T_c} is the Doppler shift that affects the frequency $1/T_c$.

REFERENCES

[1] C. L. Weber and A. Polydoros, "A unified approach to serial search spread spread-spectrum code acquisition—Part I: General theory," *IEEE Trans. Commun.*, vol. COM-32, no. 5, pp. 542–549, May 1984.

[2] C. L. Weber and A. Polydoros, "A unified approach to serial search spread spread-spectrum code acquisition—Part II: A matched filter receiver," *IEEE Trans. Commun.*, vol. COM-32, no. 5, pp. 550–560, May 1984.

[3] A. Alaqeeli, J. A. Starzyk, and F. van Graas, "Real-time acquisition and tracking for GPS receivers," in *Proc. of the 2003 Int. Symp. on Circuits and Systems, 2003, ISCAS '03*, vol. 4, May 2003, pp. 25–28.

[4] D. Alkopian, "Fast FFT based GPS satellite acquisition methods," *IEE Proc. Radar Sonar Nav.*, vol. 152, no. 4, pp. 277–286, Aug. 2005.

[5] K. A. Atkinson, *An Introduction to Numerical Analysis*. John Wiley & Sons, 1988.

[6] M. Barkat, *Signal Detection and Estimation*. Boston, MA: Artech House, 1991.

[7] D. Borio, L. Camoriano, and L. Lo Presti, "Impact of GPS acquisition strategy on decision probabilities," *IEEE Trans. Aerosp. Electron. Syst.*, vol. 44, pp. 996–1011, 2008.

[8] D. Borio, M. Fantino, and L. Lo Presti, "Acquisition analysis for Galileo BOC modulated signals: Theory and simulation," in *European Navigation Conf.*, Manchester, UK, May 2006, pp. 7–10.

[9] D. Borio, M. Fantino, and L. Lo Presti, "The impact of the Galileo signal in space in the acquisition system," presented at the Tyrrhenian International Workshop on Digital Communications (TIWDC'06)—Satellite Navigation and Communications Systems, Island of Ponza, Italy, Sep. 2006.

[10] D. Borio, M. Fantino, L. Lo Presti, and M. Pini, "Robust DLL discrimination functions normalization in GNSS receivers," in *Proc. of IEEE/ION PLANS 2008*, Monterey, CA, May 6–8, 2008, pp. 173–180.

[11] D. Castagnon, "Optimal search strategy in dynamic hypothesis testing," *IEEE Trans. Syst. Man Cybern.*, vol. 25, no. 7, pp. 1130–1138, Jul. 1995.

[12] A. J. Van Dierendonck, "GPS receivers," in *Global Positioning System: Theory and Applications*, vol. 1. Reston, VA: AIAA, 1996.

[13] E. Falletti, M. Pini, and L. Lo Presti, "Low complexity carrier to noise ratio estimators for GNSS digital receivers," *IEEE Trans. Aerosp. Electron. Syst.*, vol. 47, no. 1, pp. 420–437, Jan. 2001.

[14] E. Falletti, M. Pini, L. Lo Presti, and D. Margaria, "Assessment on low complexity C/No estimators based on M-PSK signal model for GNSS receivers," in *Proc. of IEEE/ION PLANS 2008*, Monterey, CA, May 6–8, 2008, pp. 167–172.

[15] E. Falletti, M. Pini, and L. Lo Presti, "Are carrier-to-noise algorithms equivalent in all situations?" *Inside GNSS*, vol. 5, no. 1, pp. 20–27, Jan.–Feb. 2010.

[16] M. Fantino, "Study of architectures and algorithms for software Galileo receivers," Ph.D. thesis, Electronics Department, Politecnico di Torino, May 2006.

[17] R. Fletcher and M. J. D. Powell, "A rapidly convergent descent method for minimization," *Comput. J.*, vol. 6, pp. 163–168, 1963.

[18] P. D. Groves, "GPS signal-to-noise measurements in weak signal and high-interference environments," *J. Inst. Navig.*, vol. 5, no. 2, pp. 83–94, 2005.

[19] J. C. Hancock and P. A. Wintz, *Signal Detection Theory*. New York: McGraw-Hill, 1966.

[20] C. W. Helstron, *Statistical Theory of Signal Detection*, 2nd ed. Oxford: Pergamon Press, 1968.

[21] E. D. Kaplan, *Understanding GPS: Principles and Applications*. Norwood, MA: Artech House, 1996.

[22] E. D. Kaplan and C. Hegarty, *Understanding GPS: Principles and Applications*, Artech House Mobile Communications Series. Boston, MA: Artech House, 2006.

[23] S. M. Kay, *Fundamentals of Statistical Signal Processing, Volume I: Estimation Theory*. Upper Saddle River, NJ: Prentice Hall, 1993.

[24] S. M. Kay, *Fundamentals of Statistical Signal Processing, Volume II: Detection Theory*. Upper Saddle River, NJ: Prentice Hall, 1998.

[25] L. Lo Presti and B. Motella, "The math of ambiguity: What is the acquisition ambiguity function and how is it expressed mathematically?" *Inside GNSS*, pp. 20–28, Jun. 2010.

[26] D. G. Luenberger, *Linear and Nonlinear Programming*, 2nd ed. Reading, MA: Addison-Wesley, 1984.

[27] H. Mathis, P. Flammant, and A. Thiel, "An analytic way to optimize the detector of a post-correlation FFT acquisition algorithm," in *Proc. of the 16th Int. Technical Meeting of the Satellite Division of the Institute of Navigation (ION GPS/GNSS 2003)*, Portland, OR, Sep. 9–12, 2003, pp. 689–699.

[28] P. Misra and P. Enge, *Global Positioning System: Signals, Measurements and Performance*, 2nd ed. Ganga-Jamuna Press, 2001.

[29] A. Molino, M. Nicola, M. Pini, and M. Fantino, "N-Gene GNSS software receiver for acquisition and tracking algorithms validation," presented at the 17th European Signal Processing Conf.— EUSIPCO, Glasgow, UK, Aug. 24–28, 2009.

[30] D. J. R. Van Nee and A. J. R. M. Coenen, "New fast GPS code-acquisition technique using FFT," *Electron. Lett.*, vol. 27, no. 2, pp. 158–160, 1991.

[31] B. Friedlander P. Stoica, R. L. Moses, and T. Soderstrom, "Maximum likelihood estimation of the parameters of multiple sinusoids from noisy measurements," *IEEE Trans. Signal Process.*, vol. 37, no. 3, pp. 378–392, Mar. 1989.

[32] D. P. Palomar and Y. C. Eldar, *Convex Optimization in Signal Processing and Communication.* Cambridge, UK: Cambridge University Press, 2010.

[33] B. W. Parkinson and J. J. Spilker Jr., *Global Positioning System: Theory and Applications*, Artech House Mobile Communications Series. Washington, DC: Artech House, 2006.

[34] M. Petovello, L. Lo Presti, and M. Visintin, "Can you list all the properties of the carrier-smoothing filter?" *Inside GNSS*, vol. 10, no. 4, pp. 32–37, 2015.

[35] M. Pini and D. M. Akos, "Effect of sampling frequency on GNSS receiver performance," *J. Inst. Navig.*, vol. 53, no. 2, pp. 85–95, 2006.

[36] L. Lo Presti, M. Fantino, P. Mulassano, and X. Zhu, "Acquisition systems for GNSS signals with the same code and bit rates," in *Proc. of IEEE/ION PLANS 2008,* Monterey CA, May 6–8, 2008, pp. 187–195.

[37] L. Lo Presti, X. Zhu, M. Fantino, and P. Mulassano, "GNSS signal acquisition in the presence of sign transition," *IEEE J. Sel. Topics Signal Process.*, vol. 3, no. 4, pp. 557–570, Aug. 2009.

[38] L. Lo Presti, M. Fantino, and M. Nicola, "Enhanced Bayesian detection for weak GPS and Galileo signal acquisition," in *Proc. of the 2010 Int. Technical Meeting (ITM) of the Institute of Navigation*, Jan. 25–27, 2010, pp. 773–783.

[39] B. G. Quinn, "Estimating frequency by interpolation using Fourier coefficients," *IEEE Trans. Signal Process.*, vol. 42, no. 5, pp. 1294–1268, May 1994.

[40] L. R. Rabiner and B. Gold, *Theory and Application of Digital Signal Processing.* Englewood Cliffs, NJ: Prentice-Hall, 1975.

[41] M. Sahmoudi, M. G. Amin, and R. Jr. Landry, "Acquisition of weak GNSS signals using a new block averaging pre-processing," in *Proc. of IEEE/ION PLANS 2008*, May 6–8, 2008, pp. 1362–1372.

[42] I. Selin, *Detection Theory.* Princeton, NJ: Princeton University Press, 1965.

[43] T. Seppo, "Acquisition of satellite navigation signals using dynamically chosen measurements," *IET Radar Sonar Nav.,* vol. 4, no. 1, pp. 49–61, Feb. 2010.

[44] J. A. Starzyck and Z. Zhu, "Averaging correlation for C/A code acquisition and tracking in frequency domain," in *Proc. of the 44th IEEE 2001 Midwest Symp. on Circuits and Systems, MWSCAS 2001*, vol. 2, Aug. 14–17, 2001, pp. 905–908.

[45] C. Stöber, M. Anghileri, A. S. Ayaz, D. Dötterböck, I. Krämer, V. Kropp, J.-H. Won, B. Eissfeller, D. S. Güixens, and T. Pany, "IpexSR: A real-time multi-frequency software GNSS receiver," in *Proc. Elmar—International Symposium Electronics in Marine*, Art. ID 5606128, pp. 407–416, 2010.

[46] C-C. Sun and S-S. Jan, "GNSS signal acquisition and tracking using a parallel approach," in *Proc. of IEEE/ION PLANS 2008*, May 6–8, 2008, pp. 1332–1340.

[47] T. Hai Ta, F. Dovis, D. Margaria, and L. Lo Presti, "Joint data/pilot strategies for high sensitivity Galileo E1 open service signal acquisition," *IET Radar Sonar Nav.*, vol. 4, no. 6, pp. 764–779, Dec. 2010.

[48] P. Tong, "A suboptimum synchronization procedure for pseudo-noise communication systems," in *Proc. of the National Telecommunications Conference*, 1973, pp. 26D-1–26D-5.

[49] H. L. Van Trees, *Detection, Estimation, and Modulation Theory: Part I.* New York: John Wiley & Sons, 1968.

[50] J. B. Tsui, *Fundamentals of Global Positioning System Receivers: A Software Approach*, 2nd ed. New York: John Wiley & Sons, 2005.

[51] W. Zhuang and J. Tranquilla, "Digital baseband processor for the GPS receiver modeling and simulation," *IEEE Trans. Aerosp. Electron. Syst.*, vol. 29, no. 4, pp. 1343–1349, Oct. 1993.

CHAPTER 22

KALMAN FILTER-BASED APPROACHES FOR POSITIONING: INTEGRATING GLOBAL POSITIONING WITH INERTIAL SENSORS

Emanuela Falletti and Gianluca Falco

Istituto Superiore Mario Boella, Torino, Italy

THE KALMAN filter (KF) theory is a fundamental milestone in signal processing and automatic control. This chapter aims at discussing the main techniques related to Kalman filtering for satellite navigation. In particular, we explain the structure of the KF exploited in a global navigation satellite system (GNSS) receiver to compute the position of the user and to integrate the standard GNSS receiver with an inertial navigation system (INS). The interest in the latter application is witnessed by the number of publications that have covered these topics over the last two decades. Furthermore, the advent of low-cost micro electromechanical systems (MEMSs) able to realize low-performance inertial measurement units (IMUs) has boosted the interest in developing smart and low-cost integrated systems for mass-market applications.

The first two sections of this chapter are devoted to define the KF and in particular its extended version, in order to show how it is used in a standalone GNSS receiver to compute the classic position-velocity-time (PVT) solution (Sections 22.2 and 22.3). Then, before introducing the KF as a means to realize the GNSS-INS integrated architectures, some fundamentals of inertial navigation are provided (Section 22.4). Sections 22.5 and 22.6 are devoted to discussing the characterization of the inertial devices in terms of deterministic and stocastic noises, calibration, and alignment as fundamental prerequisites for any fruitful

Handbook of Position Location: Theory, Practice, and Advances, Second Edition.
Edited by S. A. (Reza) Zekavat and R. Michael Buehrer.
© 2019 by the Institute of Electrical and Electronics Engineers, Inc.
Published 2019 by John Wiley & Sons, Inc.
Companion Website: www.wiley.com/go/zekavat/positionlocation2e

integration process. In Sections 22.7, 22.8, and 22.10 the three classic GNSS-INS integration principles are presented, namely the so-called *loosely integrated,* *tightly integrated,* and *ultra-tightly integrated,* architectures. Finally, the results of a live vehicular test campaign to compare two such implementations are shown and discussed in Section 22.9.

22.1 INTRODUCTION

GNSSs such as the global positioning system (GPS) provide the user with accurate estimates of his or her own position and velocity with a rate as high as 20 Hz. Unfortunately, high buildings, tunnels, foliage, interference, and many other obstacles and disturbances may alter the GNSS signal coming from satellites, making the estimated position highly noisy (and therefore erroneous) or even unavailable.

To limit these detrimental effects, it is possible to rely upon other navigation systems that can be used together with satellite navigation receivers. INSs represent perfect candidates to implement an integrated system together with the GNSS, thanks to their complementary characteristics [1, 2], which are listed in Table 22.1.

Three conceptual approaches can be identified for integrating INS-based positioning and GNSS-based positioning (*system hybridization*):

1. *loose integration;*
2. *tight integration;*
3. *ultra-tight integration.*

These solutions differ in the degree of integration of the two systems, i.e., the nature of the information extracted from the two systems and used in the hybridization process, as well as the architecture of the interactions between the two systems.

We believe that it is worth considering the terms integration and coupling as being distinct in order to better indicate different kinds of interactions among the involved technologies. In the following, when we use the term *integration* we indicate a system where the INS measurements are fed to an "hybridization engine" (usually, a KF) and corrected by means of different kinds of GNSS measurements. On the other hand, with the term *coupling* we point out a peculiar feature of the

TABLE 22.1. Pros and Cons of INSs and GNSSs

System	Pros	Cons
INS	Self-contained	Unlimited error growth
	Insensitive to external sources of errors (intentional	(drifts)
	or nonintentional, interference, environment)	
	High rate	
	Small errors in a short time	
GNSS	Bounded error in any instant	Not self-contained
		Sensitive to jamming
		Environment dependent

system, specifically the employment of the corrected PVT solution to drive the numerically controlled oscillators (NCOs) that are present in the phase lock loop (PLL) and, optionally, in the delay lock loop (DLL) present in the GNSS receiver.

In the following sections the algorithmic scheme that realizes every integrated system will be analyzed in detail. The mathematical tool that enables such an integration is the KF, thanks to its ability to blend different sources of noisy measurements in a single state-space description (*model*) of the system evolution. Although other mathematical tools for the system hybridization can be envisaged [35], the KF is historically and conceptually the principal approach, and for this reason it is addressed in this chapter.

22.2 REVIEW OF KALMAN FILTERING AND EXTENDED KALMAN FILTERING FOR NAVIGATION

The KF formulation is represented as a set of recursive equations that efficiently estimate the state of a system in order to minimize the mean squared error (MSE) between the real value and the estimated one (in this sense, it can be seen as a modification of the Wiener filter) [6, 7]. In GNSS applications the system is generally in motion over the Earth's surface, and its state may include position, velocity, acceleration, attitude, and also nuisance variables, due, for example, to noise sources. The equations involved can be defined in continuous-time or by means of discrete-time formulations. In general, since real applications of the recursive solution are usually implemented by discrete-time processors, the latter definition is preferred.

Here, the discrete KF is used to estimate the states of a discrete-time process represented by linear equations. The variables of interest are quantities, called states, that the KF aims to estimate by the recursive procedure. The states are described by means of a theoretical model that defines how their values change with time and by direct or indirect measurements. The information generated by the model and by measurements is combined by the KF to estimate the states. Note that the discrete-time formulation can be seen as a derivation of the continuous-time one.

22.2.1 State-Space Models

A general system of differential equations governing the *state-space model* on which the KF works can be written as*:

$$\dot{\mathbf{x}}(t) = \mathbf{f}(\mathbf{x}(t), \mathbf{u}(t), t) + \mathbf{w}(t)$$
$$\mathbf{z}(t) = \mathbf{h}(\mathbf{x}(t), t) + \mathbf{v}(t),$$
(22.1)

where $\mathbf{x}(t)$ is the *state vector* as a function of the time t, $\mathbf{u}(t)$ is the *deterministic forcing function*, $\mathbf{w}(t)$ is the *stochastic forcing function* (or *model noise* or *driving noise*), $\mathbf{z}(t)$ is the *measurement vector* (or *observation vector*), $\mathbf{v}(t)$ is the *measurement noise*, and $\mathbf{f}$ and $\mathbf{h}$ are two known linear or nonlinear functions. These two

---

*Note that in the following analysis, every vector will be considered as a column unless differently specified.

equations are referred to as the *state equation* and the *measurement equation* (or *observation equation*), respectively. Assuming a linear model, Equation (22.1) has the following form:

$$\dot{\mathbf{x}}(t) = \mathbf{F}(t)\mathbf{x}(t) + \mathbf{B}(t)\mathbf{u}(t) + \mathbf{w}(t)$$
$$\mathbf{z}(t) = \mathbf{H}(t)\mathbf{x}(t) + \mathbf{v}(t),$$
(22.2)

where $\mathbf{F}(t)$ is the *continuous-time state transition matrix*, $\mathbf{B}(t)$ is the matrix that relates the deterministic input to the states, and $\mathbf{H}(t)$ is the linear relationship between the states and the observations (*measurement matrix* or *observation matrix*). When the functions $\mathbf{f}$ and/or $\mathbf{h}$ in (22.1) are not linear, it is necessary to resort to the extended Kalman filter (EKF) model. In fact, the problem is nonlinear in all the applications of GNSS positioning and it is linearized using the EKF, thus reusing the model (22.2). This aspect will be addressed in Section 22.2.4.

Here, we focus on the discrete-time version of the system in (22.2). Considering the state variables at some discrete-time epochs $t_n = nT_c$, where T_c is the chosen *sampling interval*, the discrete-time state can be described by a *state vector* of the type

$$\mathbf{x}[n] = [x_1[n] \quad x_2[n] \quad \cdots \quad x_N[n]]^T,$$
(22.3)

where N is the number of state variables characterizing the system and ^{T} indicates the "transpose" operator.

A fundamental element of KF theory is the *state propagation model*, which takes into account the dynamics of the physical processes governing the system state evolution. The *discrete-time state-space equation* in the absence of deterministic forcing functions corresponds to:

$$\mathbf{x}[n+1] = \mathbf{\Phi}[n]\mathbf{x}[n] + \mathbf{w}[n],$$
(22.4)

where $\mathbf{x}[n+1]$ is the state vector at time $n+1$, $\mathbf{\Phi}[n]$ is the *discrete-time state transition matrix*, which relates the states at the time n to the states at the time $n+1$, and $\mathbf{w}[n]$ is a zero-mean white Gaussian noise vector of length n (*discrete-time model noise*), with known covariance

$$\mathbb{E}\{\mathbf{w}[n]\mathbf{w}[i]^T\} = \begin{cases} \mathbf{Q}[n] & n = i \\ \mathbf{0} & n \neq i \end{cases},$$
(22.5)

where $\mathbf{Q}[n]$ is the covariance matrix of the process noise and $\mathbb{E}\{.\}$ indicates statistical expectation. Notice that the vector $\mathbf{w}[n]$ can be written also as

$$\mathbf{w}[n] = \mathbf{\Gamma}[n]\mathbf{\eta}[n],$$
(22.6)

where $\mathbf{\eta}[n]$ is an N-length vector of sampled zero-mean white sequence and $\mathbf{\Gamma}[n]$ is the $N \times N$ *input matrix*. Because of the linearity of the operation, $\mathbf{w}[n]$ is a zero-mean white process too, as requested by KF theory.

At each epoch n a vector $\mathbf{z}[n]$ of M measured data can be defined as

$$\mathbf{z}[n] = [z_1[n] \quad z_2[n] \quad \cdots \quad z_M[n]]^T.$$
(22.7)

referring to the output of sensors that are used in estimating the system state. The measurement vector can be written as a linear combination of the state vector $\mathbf{x}[n]$, plus additive noise, that is

$$\mathbf{z}[n] = \mathbf{H}[n]\mathbf{x}[n] + \mathbf{v}[n], \tag{22.8}$$

where $\mathbf{H}[n]$ is the *discrete-time observation matrix*, which relates the state $\mathbf{x}[n]$ with the measurement $\mathbf{z}[n]$. The term $\mathbf{v}[n]$ is an additive noise component (of length M) with known statistical properties. Usually $\mathbf{v}[n]$ is supposed to be a zero-mean white Gaussian process, with known covariance $\mathbf{R}[n]$, defined by

$$\mathbb{E}\{\mathbf{v}[n]\mathbf{v}[i]^T\} = \begin{cases} \mathbf{R}[n] & n = i \\ \mathbf{0} & n \neq i \end{cases}. \tag{22.9}$$

The case with $N = M$ and $\mathbf{H}[n] = \mathbf{I}$, where $\mathbf{I}$ is the identity matrix, corresponds to the simplest case when all the state variables to be estimated are available for measurement. More generically, $\mathbf{H}[n]$ is an $M \times N$ matrix representing a linear transformation that the measuring system (or a transmission channel) operates on the states to be estimated. Note that the model noise and the observation noise are independent vectors; then

$$\mathbb{E}\{\mathbf{w}[n]\mathbf{v}[i]^T\} = \mathbf{0}. \tag{22.10}$$

It is evident that, given a system and its state, the first step is to verify whether its *dynamic model*, i.e., its evolution in time, can be written as in (22.4). If the state evolution is not governed by (22.4), the standard KF cannot be applied as is. The typical workaround is to adopt the EKF, as discussed in Section 22.2.4.

In the presence of a deterministic forcing function (i.e., when the state evolution is driven *also* by a deterministic signal, whose temporal evolution is known*), the system evolution and the measurements relation can be described by the more general forms:

$$\mathbf{x}[n+1] = \mathbf{\Phi}[n]\mathbf{x}[n] + \mathbf{B}[n]\mathbf{u}[n] + \mathbf{\Gamma}[n]\mathbf{\eta}[n] \tag{22.11}$$

$$\mathbf{z}[n] = \mathbf{H}[n]\mathbf{x}[n] + \mathbf{D}[n]\mathbf{u}[n] + \mathbf{v}[n], \tag{22.12}$$

where $\mathbf{\eta}[n]$ is the discrete-time *stochastic input noise*, usually considered white and Gaussian; $\mathbf{u}[n]$ is the discrete-time deterministic input; $\mathbf{B}[n]$ is the matrix that relates the deterministic input to the states; $\mathbf{\Gamma}[n]$ is the matrix that relates the stochastic input noise to the states; and $\mathbf{D}[n]$ is a matrix describing the connection between the measurement and the deterministic input.

Equation (22.11) can be obtained by the superimposition principle and decomposed in the sum of (1) a linear deterministic system

$$\mathbf{x}_u[n+1] = \mathbf{\Phi}[n]\mathbf{x}_u[n] + \mathbf{B}[n]\mathbf{u}[n]; \tag{22.13}$$

---

*An example of such a situation is a position-velocity state model, with an acceleration component assumed known in average at each instant. In this case, the evolution of state velocity is driven by a nonzero-mean, time-varying noise process, whose mean value is deterministic.

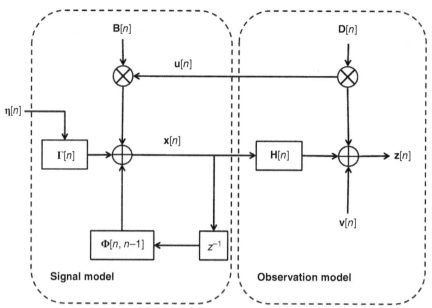

Figure 22.1 State model assumed by KF theory.

and (2) a linear stochastic system given by

$$\mathbf{x}_\eta[n+1] = \boldsymbol{\Phi}[n]\mathbf{x}_\eta[n] + \boldsymbol{\Gamma}[n]\boldsymbol{\eta}[n], \tag{22.14}$$

where the subscripts u and η are used to account for the contribution to the state $\mathbf{x}[n+1]$ from the deterministic input and stochastic input, respectively: $\mathbf{x}[n+1] = \mathbf{x}_u[n+1] + \mathbf{x}_\eta[n+1]$.

Also the measurement equation (22.12) can be decomposed into the sum of a deterministic input and a noise contribution (subscript v) as

$$\mathbf{z}[n] = \mathbf{z}_u[n] + \mathbf{z}_v[n], \tag{22.15}$$

where

$$\mathbf{z}_u[n] = \mathbf{H}[n]\mathbf{x}_u[n] + \mathbf{D}[n]u[n], \tag{22.16}$$

and

$$\mathbf{z}_v[n] = \mathbf{H}[n]\mathbf{x}_\eta[n] + \mathbf{v}[n] \tag{22.17}$$

The block diagram of Figure 22.1 is a graphical representation of the system described by (22.11) and (22.12).

22.2.2 Continuous-Time to Discrete-Time Transformation

Motion equations typically involved in navigation problems are usually written and handled in the continuous-time domain. However, the automatic computation of the navigation solution needs their translation into a discrete-time form (22.4) and

(22.8), which—for a differential problem like (22.2)—might be not straightforward for a KF newcomer. For this reason we devote this section to going through the transformation process of the differential problem (22.2), in order to give the reader the opportunity of thoroughly mastering the system statement.

Equations (22.4) and (22.8) can be derived from (22.2) in several ways. The most common approach is described in this section [6, 7].

Starting from the continuous-time state equation (22.2) in the absence of forcing functions and invoking the superposition principle so as to separate for a moment the noise contribution, Equation (22.2) can be rewritten as

$$\dot{\mathbf{x}}(t) = \mathbf{F}(t)\mathbf{x}(t), \tag{22.18}$$

whose solution can be determined as follows:

$$\mathbf{x}(t) = \mathbf{x}(0) \cdot e^{\mathbf{F}(t)t} = \mathbf{x}(0) \cdot [\mathbf{I} + \mathbf{F}(t)\,t + \cdots], \tag{22.19}$$

where the Taylor series expansion is valid if $\mathbf{F}(t)$ can be considered constant during the integration time $(0, t)$. Equation (22.19) can be evaluated in the time instant t_n in order to compute the value of the state at time $t_{n+1} = t_n + T_c$, obtaining

$$\mathbf{x}(t_{n+1}) = \mathbf{x}(t_n) \cdot e^{\mathbf{F}_{n+1,n}T_c} = \mathbf{x}(t_n) \cdot [\mathbf{I} + \mathbf{F}_{n+1,n}T_c + \cdots], \tag{22.20}$$

where $\mathbf{F}_{n+1,n}$ is the value assumed by $\mathbf{F}(t)$ in the interval (t_n, t_{n+1}); as a consequence, it is possible to identify the *discrete-time state transition matrix*

$$\mathbf{\Phi}[n] = \mathbf{I} + \mathbf{F}_{n+1,n}T_c + \frac{(\mathbf{F}_{n+1,n}T_c)^2}{2!} + \cdots. \tag{22.21}$$

We stress that such an approximation is valid if the the continuous time transition matrix is constant during a sampling interval T_c. Thereby from (22.20) and (22.21), the discrete-time formulation of the model equation in the absence of forcing functions is

$$\mathbf{x}[n+1] = \mathbf{\Phi}[n]\mathbf{x}[n]. \tag{22.22}$$

Equation (22.21) shows that different approximations are possible for the discrete-time state transition matrix. For instance, an approximation to the first order from time n to time $n + 1$, $\forall n \geq 0$, is

$$\mathbf{\Phi}[n] = \mathbf{I} + \mathbf{F}_{n+1,n}T_c. \tag{22.23}$$

In order to get a better approximation, a second-order approximation might be considered, obtaining

$$\mathbf{\Phi}[n] = \mathbf{I} + \mathbf{F}_{n+1,n}T_c + \frac{(\mathbf{F}_{n+1,n}T_c)^2}{2!}. \tag{22.24}$$

It is worth noticing that (22.23) and (22.24) are two different but possible formulations for the discrete-time state transition matrix, whose choice depends on the accuracy needed to represent the state model. The procedure explained in this paragraph to define the discrete-time matrix $\mathbf{\Phi}[n]$ is the most common one, but other methods can be applied, obtaining similar results.

While the discretization of the transition matrix is an easy task, the definition of the covariance matrix of the sampled noise term $\mathbf{w}[n] = \mathbf{\Gamma}[n]\eta[n]$ in (22.4) may be non trivial. The (approximate) analytical solution of (22.2) in the presence of the model noise $\mathbf{w}(t)$ only can be written as per state-space methods [6] as

$$\mathbf{x}(t_{n+1}) = \mathbf{\Phi}[n]\mathbf{x}(t_n) + \int_{t_n}^{t_{n+1}} \mathbf{F}\left(t_{n+1}, \tau\right)\mathbf{w}(\tau)\,d\tau. \tag{22.25}$$

It is then possible to write the relation between $\mathbf{w}[n]$ and $\mathbf{w}[t]$ as follows:

$$\mathbf{w}[n] = \int_{t_n}^{t_{n+1}} \mathbf{F}\left(t_{n+1}, \tau\right)\mathbf{w}(\tau)\,d\tau. \tag{22.26}$$

Note that the discrete-time noise vector $\mathbf{w}(n)$ is the result of the accumulation of the continuous-time noise $\mathbf{w}(t)$ over the sampling interval $T_c = t_{n+1} - t_n$. Therefore, it has the same kind of statistical distribution of $\mathbf{w}(t)$ but different variance.

As a consequence, the model noise covariance matrix that characterizes $\mathbf{w}[n]$ can be obtained as:

$$\begin{aligned}
\mathbf{Q}[n] &= \mathbb{E}\left\{\mathbf{w}[n]\mathbf{w}^H[n]\right\} \\
&= \mathbb{E}\left\{\left[\int_{t_n}^{t_{n+1}} \mathbf{F}\left(t_{n+1}, \tau\right)\mathbf{w}(\xi)\,d\xi\right]\left[\int_{t_n}^{t_{n+1}} \mathbf{F}\left(t_{n+1}, \tau\right)\mathbf{w}(\tau)\,d\tau\right]^H\right\} \\
&= \int_{t_n}^{t_{n+1}}\int_{t_n}^{t_{n+1}} \mathbf{F}\left(t_{n+1}, \tau\right)\mathbb{E}\left\{\mathbf{w}(\xi)\mathbf{w}^H(\tau)\right\}\mathbf{F}^H\left(t_{n+1}, \tau\right)[n]d\xi d\tau.
\end{aligned} \tag{22.27}$$

Equation (22.27) cannot be evaluated easily. However, if the transition matrix $\mathbf{F}(t_{n+1},\tau)$ can be considered constant during the sampling interval, which is our case, then (22.27) can be drastically simplified, obtaining:

$$\mathbf{Q}[n] = \mathbf{\Phi}[n]\mathbf{Q}_w\mathbf{\Phi}^H[n]T_c, \tag{22.28}$$

where the matrix $\mathbf{Q}_w = \mathbb{E}\{\mathbf{w}(\xi)\mathbf{w}^H(\tau)\}$ is a diagonal matrix due to the nature of the model noise, assumed uncorrelated in time.

Finally, the discrete-time version of the measurement equation is simply defined by sampling the continuous time expression at the instants $t_n = nT_c$:

$$\mathbf{z}[n] = \mathbf{H}[n]\mathbf{x}[n] + \mathbf{v}[n]. \tag{22.29}$$

22.2.3 Recursive Estimation and Initial Conditions

In the KF approach, the state variables are estimated by means of a recursive itera-tive process, based on two steps. The state estimate $\hat{\mathbf{x}}[n]$ at each discrete time n is a mixture of two main contributions:

1. *prediction*, a predicted estimate $\hat{\mathbf{x}}^-[n+1]$, obtained by applying the state transi-tion matrix, in the first step of the iterative procedure;
2. *update*, a contribution determined from the measurement vector, called cor-rection, or update.

This procedure is implemented in MATLAB in Appendix A.1.

The basic idea is to correct the prediction by using a function of the fresh measurement at each epoch n. The prediction of the state at the $(n + 1)$th instant is obtained from the state equation in the absence of model noise:

$$\hat{\mathbf{x}}^-[n] = \mathbf{\Phi}[n-1]\hat{\mathbf{x}}[n-1] \tag{22.30}$$

The symbol $\hat{\mathbf{x}}^-[n]$ specifies that such a vector (*a priori state*) does not represent the state estimate at time n but only its prediction, carried out using both the old estimation and the matrix $\mathbf{\Phi}[n - 1]$. The error between the current state $\mathbf{x}[n]$ and its predicted estimate $\hat{\mathbf{x}}^-[n]$ is defined as

$$\mathbf{e}^-[n] = \mathbf{x}[n] - \hat{\mathbf{x}}^-[n], \tag{22.31}$$

with covariance (*a priori error covariance matrix*)

$$\mathbf{P}^-[n] = \mathbb{E}\left\{\mathbf{e}^-[n]\mathbf{e}^-[n]^T\right\}. \tag{22.32}$$

Similarly, $\mathbf{e}[n]$ is defined as the error between the state $\mathbf{x}[n]$ and the estimate $\hat{\mathbf{x}}[n]$ (*a posteriori state estimation error*):

$$\mathbf{e}[n] = \mathbf{x}[n] - \hat{\mathbf{x}}[n], \tag{22.33}$$

with covariance (*a posteriori state error covariance matrix*)

$$\mathbf{P}[n] = \mathbb{E}\left\{\mathbf{e}[n]\mathbf{e}[n]^T\right\}. \tag{22.34}$$

Without going into derivation details, which can be found in many classic books of signal processing and control, the *a posteriori* state estimate $\hat{\mathbf{x}}[n]$ is computed as

$$\hat{\mathbf{x}}[n] = \hat{\mathbf{x}}^-[n] + \mathbf{K}[n]\big(\mathbf{z}[n] - \mathbf{H}[n]\hat{\mathbf{x}}^-[n]\big), \tag{22.35}$$

where the term $\alpha[n] = \big(\mathbf{z}[n] - \mathbf{H}(n)\hat{\mathbf{x}}^-[n]\big)$ is called *innovation* or *residual* and $\mathbf{K}[n]$ is the *Kalman gain*. $\mathbf{K}[n]$ is the weighting factor that implements the optimal mixing between the measured data $\mathbf{z}[n]$ and the predicted observation values based on the system model. The definition of $\mathbf{K}[n]$ is based on the minimization of the mean squared state error $\mathbb{E}\{\mathbf{e}[n]^H\mathbf{e}[n]\}$; the minimization procedure leads to the following expression:

$$\mathbf{K}[n] = \mathbf{P}^-[n]\mathbf{H}[n]^T\big(\mathbf{H}[n]\mathbf{P}^-[n]\mathbf{H}[n]^T + \mathbf{R}[n]\big)^{-1} \tag{22.36}$$

and to the associated formula for updating the error covariance matrix $\mathbf{P}[n]$:

$$\mathbf{P}[n] = [\mathbf{I} - \mathbf{K}[n]\mathbf{H}[n]]\mathbf{P}^-[n]. \tag{22.37}$$

It must be recalled that the linear structure of (22.35) offers the optimal solution when dealing with Gaussian noise processes [7].

Finally, in order to start the iterative procedure, the KF must be initialized with some initial conditions that may be arbitrary. Considering $n = 0$ as the initial epoch, both $\hat{\mathbf{x}}[-1]$ and $\mathbf{P}(-1)$ quantities must be known. The initialization of $\mathbf{P}[-1]$ must take into account the uncertainty on the state knowledge before the measuring process starts. A possible choice can be:

$$\hat{\mathbf{x}}[-1] = \mathbb{E}\{\mathbf{x}[-1]\} \tag{22.38}$$

$$\mathbf{P}[-1] = Var\{\mathbf{x}[-1]\}. \tag{22.39}$$

22.2.4   The EKF

The EKF allows one to use a KF to solve nonlinear problems. This implies the approximation of the nonlinear equation (22.1) by a first-order Taylor series computed about a "point" that belongs to the so-called *nominal trajectory*.*

The nominal trajectory has to be intended as a known trajectory that represents an approximation of the real trajectory. It can be a known approximated trajectory, or it can be provided by a theoretical motion model, such as the velocity random walk or the acceleration Gauss–Markov process or by a navigation system, like an INS.

The continuous-time differential equations governing the state-space model are written in (22.1) and reported here for major clarity:

$$\dot{\mathbf{x}}(t) = \mathbf{f}(\mathbf{x}(t), \mathbf{u}(t), t) + \mathbf{w}(t)$$

$$\mathbf{z}(t) = \mathbf{h}(\mathbf{x}(t), t) + \mathbf{v}(t).$$

As introduced in Section 22.2.1, it is necessary to resort to the EKF every time either $\mathbf{f}$ and $\mathbf{h}$ is nonlinear.

Assuming that a nominal trajectory $\breve{\mathbf{x}}(t)$ is available, it is possible to write the actual trajectory as follows:

$$\mathbf{x}(t) = \breve{\mathbf{x}}(t) + \Delta\mathbf{x}(t), \tag{22.40}$$

where $\Delta\mathbf{x}(t)$ is the *increment* between the nominal and the actual trajectory.

Substituting (22.40) in the system (22.1), it results in:

$$\dot{\breve{\mathbf{x}}}(t) + \Delta\dot{\mathbf{x}}(t) = \mathbf{f}\big(\breve{\mathbf{x}}(t) + \Delta\mathbf{x}(t), \mathbf{u}(t), t\big) + \mathbf{w}(t)$$

$$\mathbf{z}(t) = \mathbf{h}\big(\breve{\mathbf{x}}(t) + \Delta\mathbf{x}(t), t\big) + \mathbf{v}(t), \tag{22.41}$$

while for the nominal trajectory the state-space equations are:

$$\dot{\breve{\mathbf{x}}}(t) = \mathbf{f}\big(\breve{\mathbf{x}}(t), \mathbf{u}(t), t\big)$$

$$\breve{\mathbf{z}}(t) = \mathbf{h}\big(\breve{\mathbf{x}}(t), t\big). \tag{22.42}$$

---

*Trajectory has to be intended in this moment as the generic evolution in time of the state vector. In practice, for a positioning problem like that addressed later on in this chapter, it is the time series of the quantities that describe the body motion in a certain reference frame. The trajectory may contain position, velocity, sometimes acceleration, time (the latter is considered in GNSS receivers, where the bias and drift of the local clock with respect to the satellites clock are parameters of interest), and body attitude.

If the nominal trajectory is a good approximation of the actual trajectory, then the increment is small. In this case the two nonlinear equations $\mathbf{f}$ and $\mathbf{h}$ can be approximated by means of a first-order Taylor series:

$$\breve{\mathbf{x}}(t) + \Delta\dot{\mathbf{x}}(t) \approx \mathbf{f}\big(\breve{\mathbf{x}}(t), \mathbf{u}(t), t\big) + \left[\frac{\partial \mathbf{f}(t)}{\partial \mathbf{x}}\right]_{\mathbf{x}=\breve{\mathbf{x}}} \Delta\mathbf{x}(t) + o(\Delta\mathbf{x}(t)) + \mathbf{w}(t)$$

$$\mathbf{z}(t) \approx \mathbf{h}\big(\breve{\mathbf{x}}(t), t\big) + \left[\frac{\partial \mathbf{h}(t)}{\partial \mathbf{x}}\right]_{\mathbf{x}=\breve{\mathbf{x}}} \Delta\mathbf{x}(t) + o(\Delta\mathbf{x}(t)) + \mathbf{v}(t), \tag{22.43}$$

where the Jacobians of the nonlinear functions $\mathbf{f}$ and $\mathbf{h}$, namely $\left[\frac{\partial \mathbf{f}(t)}{\partial \mathbf{x}}\right]$ and $\left[\frac{\partial \mathbf{h}(t)}{\partial \mathbf{x}}\right]$, are employed. Then, using (22.42), the system (22.43) becomes

$$\Delta\dot{\mathbf{x}}(t) = \left[\frac{\partial \mathbf{f}(t)}{\partial \mathbf{x}}\right]_{\mathbf{x}=\breve{\mathbf{x}}} \Delta\mathbf{x}(t) + \mathbf{w}(t) = \mathbf{F}(t)\Delta\mathbf{x}(t) + \mathbf{w}(t)$$

$$\Delta\mathbf{z}(t) \triangleq \mathbf{z}(t) - \breve{\mathbf{z}}(t) = \left[\frac{\partial \mathbf{h}(t)}{\partial \mathbf{x}}\right]_{\mathbf{x}=\breve{\mathbf{x}}} \Delta\mathbf{x}(t) + \mathbf{v}(t) = \mathbf{H}(t)\Delta\mathbf{x}(t) + \mathbf{v}(t), \tag{22.44}$$

which is *linear* in the *incremental states* $\Delta\mathbf{x}(t)$, provided that the following definitions hold:

$$\mathbf{F}(t) = \left[\frac{\partial \mathbf{f}(t)}{\partial \mathbf{x}}\right]_{\mathbf{x}=\breve{\mathbf{x}}} ; \tag{22.45}$$

$$\mathbf{H}(t) = \left[\frac{\partial \mathbf{h}(t)}{\partial \mathbf{x}}\right]_{\mathbf{x}=\breve{\mathbf{x}}} ; \tag{22.46}$$

$$\breve{\mathbf{z}}(t) = \mathbf{h}\big(\breve{\mathbf{x}}(t)\big). \tag{22.47}$$

The analogy of (22.44) with (22.2) is evident and allows us to apply all the results discussed in the previous sections to this problem, provided that the incremental states $\Delta\mathbf{x}(t)$ are considered, as well as the incremental observations $\Delta\mathbf{z}[n]$. As a consequence, the KF applied to the discrete-time version of problem (22.44) corresponds to:

1. *Prediction:* the *a priori state* is estimated on the basis of the state-space model (for simplicity, we drop hereafter the $\hat{}$ notation used in Section 22.2.3):

$$\Delta\mathbf{x}^-[n] = \boldsymbol{\Phi}[n]\Delta\mathbf{x}[n-1], \tag{22.48}$$

and the *a priori error covariance matrix* is computed:

$$\mathbf{P}^-[n] = \boldsymbol{\Phi}[n]\mathbf{P}[n-1]\boldsymbol{\Phi}[n]^T + \mathbf{Q}[n-1], \tag{22.49}$$

where $\mathbf{P}[n]$ is the error covariance matrix computed on the incremental states $\Delta\mathbf{x}[n]$.

2. Update: the innovation due to a new measurement is computed and—through the Kalman gain—is applied to the predicted state to obtain its *a posteriori* estimate:

$$\alpha[n] = \mathbf{z}[n] - \mathbf{h}(\mathbf{\bar{x}}[n]) - \mathbf{H}[n]\Delta\mathbf{x}^-[n] \quad (innovation); \tag{22.50}$$

$$\mathbf{K}[n] = \mathbf{P}^-[n]\mathbf{H}^T[n]\big(\mathbf{H}[n]\mathbf{P}^-[n]\mathbf{H}^T[n] + \mathbf{R}[n]\big)^{-1} \quad (Kalman\ gain); \tag{22.51}$$

$$\Delta\mathbf{x}[n] = \Delta\mathbf{x}^-[n] + \mathbf{K}[n]\alpha[n] \quad (a\ posteriori\ state); \tag{22.52}$$

$$\mathbf{P}[n] = (\mathbf{I} - \mathbf{K}[n]\mathbf{H}[n])\mathbf{P}^-[n] \quad (a\ post.\ error\ covariance\ matrix). \tag{22.53}$$

The matrix $\mathbf{R}[n]$ in (22.51) is the *observation noise covariance matrix* defined as $\mathbf{R}[n] = \mathbb{E}\{v[n]v^H[n]\}$.

It is worth noticing that the term $\mathbf{H}[n]\Delta\mathbf{x}^-[n]$ in (22.50) is an a priori estimate $\Delta\mathbf{z}^-[n]$ of the increment $\Delta\mathbf{z}[n]$, computed without any information about the current measurement. At time n it is possible to consider the innovation as the residual error due to the information stored in the new observation, which is in general unpredictable using the measurement from time 0 to time $n - 1$.

Indicating with $\mathbf{x}[n]$ the *estimated* trajectory vector at time n and recalling that $\mathbf{\bar{x}}[n]$ is the known nominal trajectory at the same instant, it is possible to correct the latter by adding the a posteriori state:

$$\mathbf{x}[n] = \mathbf{\bar{x}}[n] + \Delta\mathbf{x}[n]. \tag{22.54}$$

Thus, Equation (22.54) concludes the nth iteration of the algorithm, reporting the incremental states to the estimated trajectory.

The system (22.44) describes a so-called *complementary* KF. This kind of filter is used to estimate the difference between the reference trajectory and an external measurement of the actual trajectory. The error estimate $\Delta\mathbf{x}(t)$ is then used to correct the reference trajectory, as in (22.54). Complementary KFs represents a valid solution for the design of multisensor navigation architectures [6].

Linearized and Extended Architectures: As expressed before, a complementary KF, characterized by an incremental error state, is a linear filter that can also be applied to nonlinear problems. In this case, the designer has to linearize the problem around a point on a known reference trajectory. Thereby the linearization represents an approximation and is affected by errors. A *corrected trajectory* is then computed by means of (22.54).

Whenever the corrected trajectory is kept away from the recursion, the system runs a *linearized KF*. Such a filter would perform well if the reference trajectory is reliable, because the linearization is taken about the reference trajectory.

If linearization takes place about the filter's estimated trajectory, the corrected trajectory is periodically fed back into the recursion (22.54), so that an *EKF* is realized. The corrected trajectory is a function of the estimated state, that is, a function of the measurements. This filter gets unstable in scenarios where initial uncertainties and measurements error are large, but it is a good choice whenever the corrected trajectory is reasonably better than the reference trajectory.

Example 22.1

Despite the fact that different kinds of KFs might be implemented, each one is characterized by a recursive step where the Kalman gain is computed. The reader is requested to program a simple MATLAB routine to realize the recursive step. It must be noticed that KFs are natively implemented in MATLAB, but in order to better understand these filters, it is important to design them from the beginning.

Solution

The purpose of the recursive step of a KF is to update the current state and covariance matrix using the available measurement matrix and vector. In this process, also the matrices that are characteristic of a KF, i.e., the state transition matrix, the state noise covariance matrix, and the measurement noise covariance matrix, are involved. In the following, we will refer to the MATLAB code provided in the Appendix A.1.

1. The current state is propagated to the next time using the state transition matrix, obtaining the a priori estimate of the state;
2. Also the state error covariance matrix is propagated using the state transition matrix. The $\mathbf{Q}$ matrix is used to accumulate the uncertainty due to the noise that affects the model;
3. The Kalman gain can be computed using (22.51);
4. The Kalman gain is then used to correct the a priori estimate of the state and of the state error covariance matrix;

22.2.5 The Unscented KF

The unscented KF (UKF) is an alternative to the EKF that provides superior performance at an equivalent computational complexity. In fact, the UKF addresses the approximation issues of the EKF, in which the KF states are propagated through the "first-order" linearization of the nonlinear system. This simple approximation can generate large error on the state estimates when the system is highly nonlinear. Conversely, the UKF fixes the nonlinear problem by using a minimal set of carefully chosen sample points. The states of the UKF are still assumed to be affected by additive white Gaussian noise (AWGN) as in the case of EKF. These sample points are chosen in such a way that they capture the true mean and covariance of the AWGN noise of the system. It can be proved that the UKF is able to approximate any nonlinearity to the second order of a Taylor series. The UKF leverages the unscented transformation (UT) [8].

The Unscented Transformation: The UT is a method proposed in [8] to calculate the statistics of a random variable that undergoes a nonlinear transformation. For instance, consider a random variable $\mathbf{x}$ of N dimensions that is passed through the nonlinear function $\mathbf{y} = f(\mathbf{x})$. The variable $\mathbf{x}$ is assumed to have a mean $\bar{\mathbf{x}}$ and covariance $\mathbf{P}_x$. To get the statistics of the output variable $\mathbf{y}$ we have to form a matrix X of $2N + 1$ *sigma vectors* X_i, such that

$$X_0 = \bar{\mathbf{x}};$$
$$X_i = \bar{\mathbf{x}} + \left(\sqrt{(N+\lambda)\mathbf{P}_{\mathbf{x}}}\right)_i, \quad \forall i = 1,\dots,N; \tag{22.55}$$
$$X_i = \bar{\mathbf{x}} - \left(\sqrt{(N+\lambda)\mathbf{P}_{\mathbf{x}}}\right)_{i-N}, \quad \forall i = N,\dots,2N,$$

where $\lambda = \alpha^2(N + k) - N$ is a scaling parameter. The term α determines the spread of the *sigma points* around $\bar{\mathbf{x}}$ and is usually set to a small positive value (e.g., $1e-3$), while k is a secondary scaling parameter and is usually set equal to 0. Moreover, $\left(\sqrt{(N+\lambda)\mathbf{P}_{\mathbf{x}}}\right)_i$ is the ith column of the matrix square root of $\mathbf{P}_{\mathbf{x}}$ (e.g., lower triangular Cholesky factorization [9, 10]).

For each sigma point, we define a predicted state point and a predicted measurement point by substituting the sigma point into the system model and the measurement model as stated in (22.2). The sigma points are propagated through the function $f(\ )$ as

$$Y_i = f(X_i) \quad i = 0,\dots,2N, \tag{22.56}$$

and the mean and covariance for $\mathbf{y}$ are approximated using a weighted sample mean and covariance of the posterior sigma points, i.e.,

$$\bar{\mathbf{y}} \approx \sum_{i=0}^{2N} W_i^{(m)} Y_i$$
$$\mathbf{P}_{\mathbf{y}} \approx \sum_{i=0}^{2N} W_i^{(c)} [Y_i - \bar{\mathbf{y}}] \cdot [Y_i - \bar{\mathbf{y}}]^T, \tag{22.57}$$

where

$$W_0^{(m)} = \lambda/(N+\lambda)$$
$$W_0^{(c)} = \lambda/(N+\lambda) + (1-\alpha^2 + \beta)$$
$$W_i^{(m)} = W_i^{(c)} = 1/[2(N+\lambda)] \quad i = 0,\dots,2N.$$

The coefficient β includes prior knowledge of the distribution of the states vector $\mathbf{x}$ (e.g., in case of Gaussian distribution $\beta = 2$).

Note that this method differs substantially from the general sampling methods (e.g., Monte Carlo methods such as particle filters [11]), which require high orders of magnitude of sample points in an attempt to propagate an accurate (possibly non-Gaussian) distribution of the state. As said before, the UT results in approximations that are accurate to the third order for Gaussian inputs for all nonlinearities. For non-Gaussian inputs, approximations are accurate to at least the second order, with the accuracy of third and higher order moments determined by the choice of α and k parameters.

From the Unscented Transformation to the UKF: The UKF is a straightforward extension of the UT where the state vector at the nth discrete-time instant is indicated as $\mathbf{x}^a[n]$. The UT sigma points generation is applied to $\mathbf{x}^a[n]$ by following the scheme stated in (22.55) and a sigma matrix $X^a[n]$ is obtained. The UKF equations do not need any Jacobians or Hessians to be computed. At the discrete-time $n = 0$, the mean of the UKF state vector $x^a[0]$ and its covariance matrix are initialized as:

$$\bar{\mathbf{x}}^a[0] = \mathbb{E}[\bar{\mathbf{x}}^a] = \mathbb{E}[\mathbf{x}[0]]^T \tag{22.58}$$

$$\mathbf{P}^a[0] = \mathbb{E}[(\mathbf{x}^a[0] - \bar{\mathbf{x}}^a[0]) \cdot (\mathbf{x}^a[0] - \bar{\mathbf{x}}^a[0])^T].$$

Then, for any discrete-time $n-1$, the sigma points matrix $X^a[n-1]$ is computed according to (22.55) as:

$$X^a[n-1] = [\bar{\mathbf{x}}^a[n-1], \bar{\mathbf{x}}^a[n-1] \pm \sqrt{(N+\lambda)\mathbf{P}^a[n-1]}]. \tag{22.59}$$

Then, the *prediction* and *measurements update* steps of the KF are computed as:

1. *Prediction.* In the prediction stage the sigma points are fed to the system process:

$$X^{a-}[n] = f(X^a[n-1], \mathbf{w}[n-1])$$

$$\bar{\mathbf{x}}^{a-}[n] = \sum_{i=0}^{2N} W_i^{(m)} \cdot X_i^{a-}[n] \tag{22.60}$$

$$\mathbf{P}^-[n] = \sum_{i=0}^{2N} W_i^{(c)} (X_i^{a-}[n] - \bar{\mathbf{x}}^{a-}[n]) \cdot (X_i^{a-}[n] - \bar{\mathbf{x}}^{a-}[n])^T + \mathbf{Q}[n],$$

where $\mathbf{Q}[n]$ is the covariance matrix of the noise associated to the discrete-time system model as reported in (22.5).

2. *Update.* If the measurements are available at time n, the relationship between predicted sigma points and propagated observations is computed as

$$Y[n] = h(X^{a-}[n])$$

$$\bar{\mathbf{y}}[n] = \sum_{i=0}^{2N} W_i^{(m)} Y_i[n]. \tag{22.61}$$

The covariance matrix of the measurements is

$$\mathbf{P}_{\bar{\mathbf{y}}[n],\bar{\mathbf{y}}[n]} = \sum_{i=0}^{2N} W_i^{(c)} (Y_i[n] - \bar{\mathbf{y}}[n]) \cdot (Y_i[n] - \bar{\mathbf{y}}[n])^T + \mathbf{R}[n], \tag{22.62}$$

where $\mathbf{R}[n]$ is the measurements noise covariance. The cross-covariance matrix between the measurements and the sigma points is computed as:

$$\mathbf{P}_{\mathbf{x}[n],\bar{\mathbf{y}}[n]} = \sum_{i=0}^{2N} W_i^{(c)} (Y_i[n] - \bar{\mathbf{y}}[n]) \cdot (X_i^a[n] - \bar{\mathbf{x}}^{a-}[n])^T. \tag{22.63}$$

The Kalman gain $\mathbf{K}$ at the discrete-time n is:

$$\mathbf{K}[n] = \mathbf{P}_{\mathbf{x}[n],\bar{\mathbf{y}}[n]} \cdot \mathbf{P}_{\bar{\mathbf{y}}[n],\bar{\mathbf{y}}[n]}^{-1}. \tag{22.64}$$

The update of the states and its covariance matrix is calculated as:

$$
\begin{aligned}
\overline{\mathbf{x}}^a[n] &= \overline{\mathbf{x}}^{a-}[n] + \mathbf{K}[n] \cdot (\mathbf{y}[n] - \overline{\mathbf{y}}[n]) \\
\mathbf{P}[n] &= \mathbf{P}^-[n] - \mathbf{K}[n]\mathbf{P}_{\overline{\mathbf{y}}[n],\overline{\mathbf{y}}[n]}\mathbf{K}^T[n].
\end{aligned}
\tag{22.65}
$$

22.3 EKF-BASED PVT COMPUTATION IN A STAND-ALONE GNSS RECEIVER

As previously introduced, the EKF can be adopted in a stand-alone GPS receiver to compute the PVT solution. This section presents the necessary measurements, the state variables, their initialization and their role in the PVT computation. We introduce hereafter the EKF-based PVT computation for a stand-alone GNSS receiver because it is fundamental to thoroughly understanding the hybrid architectures discussed in the rest of the chapter.

22.3.1 State-Space Model

The considered KF model is characterized by the following incremental states (corrections to be applied to the nominal quantities):

$$
\Delta\mathbf{x} = \left[\Delta x, \Delta y, \Delta z, \Delta\tau, \Delta v_x, \Delta v_y, \Delta v_z, \Delta v_\tau\right]^T,
\tag{22.66}
$$

where Δx, Δy, and Δz are the corrections to be applied to the nominal Earth-centered-Earth-fixed (ECEF) coordinates defining the receiver position on the Earth surface, $\Delta\tau$ is the correction to be applied to the nominal GNSS receiver clock bias, Δv_x, Δv_y, and Δv_z are the corrections to be applied to the nominal ECEF velocities, and Δv_τ is the correction to be applied to the nominal GNSS receiver clock drift.

The state vector is then an eight-dimensional vector, where the first four components represent position corrections in meters and the last four components represent velocity corrections in meters per second. This choice is fair because the considered reference system has three spatial dimensions and one temporal dimension, which can be anyway expressed in meters once it has been multiplied by the speed of light. A block scheme of the system is shown in Figure 22.2.

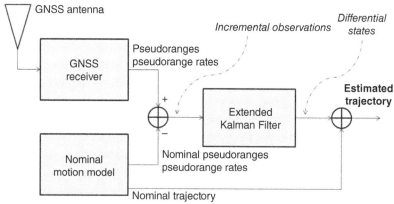

Figure 22.2   Block scheme for a stand alone GNSS receiver based on an EKF.

The state-space model usually considered in such a PVT estimation problem is a simple linear system written as

$$\begin{cases} \dot{x}(t) = v_x(t) \\ \dot{v}_x(t) = w_{v_x}(t) \end{cases}$$
(22.67)

for each ECEF coordinate x, y, z and

$$\begin{cases} \dot{\tau}(t) = v_\tau(t) + w_\tau(t) \\ \dot{v}_\tau(t) = w_{v_\tau}(t) \end{cases}$$
(22.68)

for the clock components. The model noises are assumed zero-mean random processes with variances σ_p^2 for the three acceleration components $w_{v_x}(t)$, $w_{v_y}(t)$, $w_{v_x}(t)$, variance σ_τ^2 for the component $w_\tau(t)$, and variance σ_d^2 for the component $w_{v_\tau}(t)$. In the discrete-time domain, the (time-invariant) state transition matrix is then:

$$\Phi[n] = \begin{bmatrix} 1 & 0 & 0 & 0 & T_c & 0 & 0 & 0 \\ 0 & 1 & 0 & 0 & 0 & T_c & 0 & 0 \\ 0 & 0 & 1 & 0 & 0 & 0 & T_c & 0 \\ 0 & 0 & 0 & 1 & 0 & 0 & 0 & T_c \\ 0 & 0 & 0 & 0 & 1 & 0 & 0 & 0 \\ 0 & 0 & 0 & 0 & 0 & 1 & 0 & 0 \\ 0 & 0 & 0 & 0 & 0 & 0 & 1 & 0 \\ 0 & 0 & 0 & 0 & 0 & 0 & 0 & 1 \end{bmatrix},$$
(22.69)

where T_c is the sampling interval, while the model noise vector is

$$\mathbf{w}[n] = [0, 0, 0, w_4[n], w_5[n], w_6[n], w_7[n], w_8[n]]^T,$$
(22.70)

with a (time-invariant) model noise covariance matrix structured as

$$\mathbf{Q}[n] = \mathbb{E}\{\mathbf{w}[n]\mathbf{w}^H[n]\} =$$

$$\begin{bmatrix} \sigma_p^2 \frac{T_c^3}{3} & 0 & 0 & 0 & \sigma_p^2 \frac{T_c^2}{2} & 0 & 0 & 0 \\ 0 & \sigma_p^2 \frac{T_c^3}{3} & 0 & 0 & 0 & \sigma_p^2 \frac{T_c^2}{2} & 0 & 0 \\ 0 & 0 & \sigma_p^2 \frac{T_c^3}{3} & 0 & 0 & 0 & \sigma_p^2 \frac{T_c^2}{2} & 0 \\ 0 & 0 & 0 & \sigma_\tau^2 T_c + \sigma_d^2 \frac{T_c^3}{3} & 0 & 0 & 0 & \sigma_d^2 \frac{T_c^2}{2} \\ \sigma_p^2 \frac{T_c^2}{2} & 0 & 0 & 0 & \sigma_p^2 T_c & 0 & 0 & 0 \\ 0 & \sigma_p^2 \frac{T_c^2}{2} & 0 & 0 & 0 & \sigma_p^2 T_c & 0 & 0 \\ 0 & 0 & \sigma_p^2 \frac{T_c^2}{2} & 0 & 0 & 0 & \sigma_p^2 T_c & 0 \\ 0 & 0 & 0 & \sigma_d^2 \frac{T_c^2}{2} & 0 & 0 & 0 & \sigma_d^2 T_c \end{bmatrix},$$
(22.71)

obtained as described in Section 22.2.2.

22.3.2 Linearization of the Measurement Equation

In the stand-alone PVT estimation problem based on the EKF, the measurement is defined as the column vector that stores the *pseudorange measurements* ρ in meters and the *pseudorange rates measurements* r in meters per second:

$$\mathbf{z}[n] = [\rho_1[n], \ldots, \rho_{N_{sat}}[n], r_1[n], \ldots, r_{N_{sat}}[n]]^T, \tag{22.72}$$

where N_{sat} is the number of visible satellites.

First Measurement Component: Pseudoranges: Pseudoranges are related to the Euclidean distance between the satellites and the user and are defined as follows:

$$\rho_i = \mathbf{h}_i(\mathbf{x}) = \sqrt{(x_i - x)^2 + (y_i - y)^2 + (z_i - z)^2} - \tau, \tag{22.73}$$

where the triplet (x_i, y_i, z_i) is the ith satellite position, while the triplet (x, y, z) indicates the user position. These are not *actual* ranges but *pseudo* ranges because they are distances that also account for the receiver clock misalignment τ, expressed in meters. When the pseudorange is computed with respect to the nominal trajectory, then $\breve{\rho}_i[n] = \mathbf{h}_i(\breve{\mathbf{x}}[n])$.

Unfortunately, the *geometric pseudorange* defined in (22.73) differs from the measured pseudorange. This happens because the geometric pseudorange does not account for effects such as satellites clock corrections, and ionospheric and tropospheric errors that affect the received GNSS signal. In fact, the measured *raw* pseudorange can be expressed as [12]:

$$\rho_i^{raw} = \rho_i + c\tau_i^{satclock} + e_i^{tropo} + e_i^{iono} + e^{loc}, \tag{22.74}$$

where c is the speed of light, $\tau_i^{satclock}$ is the ith satellite clock misalignment with respect to the GPS time in seconds, e_i^{tropo} is the delay due to the troposphere propagation in meters, e_i^{iono} is the delay due to the ionosphere propagation in meters, and e^{loc} is a nuisance due to other local sources of error (e.g., multipath), plus noise [12]. These errors should be predicted and possibly corrected to obtain ρ_i; basically, $t_i^{satclock}$, e_i^{tropo}, and e_i^{iono} are estimated from the data broadcast in the navigation message and applying tropospheric/ionospheric models. Of course, it is not possible to obtain a perfect prediction of the measured pseudorange values because of random uncorrelated error components and nonprecise error models [12, 13].

According to EKF theory, the nonlinear pseudorange equation (22.73) requires to be linearized using a first-order Taylor series. The Jacobian of the nonlinear relationship $\mathbf{h}[n]$ (22.73) between the user position and clock $\mathbf{x}^{(p)} \triangleq [x, y, z, \tau]^T$ and the N_{sat} pseudoranges $\rho_1, \ldots \rho_{N_{sat}}$ results in:

$$\frac{\partial \mathbf{h}[n]}{\partial \mathbf{x}^{(p)}} = \begin{bmatrix} \dfrac{\partial h_1}{\partial x} & \dfrac{\partial h_1}{\partial y} & \dfrac{\partial h_1}{\partial z} & \dfrac{\partial h_1}{\partial \tau} \\[2mm] \dfrac{\partial h_2}{\partial x} & \dfrac{\partial h_2}{\partial y} & \dfrac{\partial h_1}{\partial z} & \dfrac{\partial h_1}{\partial \tau} \\[2mm] \vdots & \vdots & \vdots & \vdots \end{bmatrix} = \begin{bmatrix} \dfrac{x - x_1}{d_1} & \dfrac{y - y_1}{d_1} & \dfrac{z - z_1}{d_1} & -1 \\[2mm] \dfrac{x - x_2}{d_2} & \dfrac{y - y_2}{d_2} & \dfrac{z - z_2}{d_2} & -1 \\[2mm] \vdots & \vdots & \vdots & \vdots \end{bmatrix}, \tag{22.75}$$

where d_j is the norm of the vector $[x - x_j, \ y - y_j, \ z - z_j]^T$ (distance). Then, it is possible to define the *measurement matrix*:

$$
\mathbf{H}_1[n] \triangleq \left[\frac{\partial \mathbf{h}[n]}{\partial \mathbf{x}^{(p)}} \right]_{\mathbf{x}=\breve{\mathbf{x}}} = \begin{bmatrix} \dfrac{\breve{x} - x_1}{d_1} & \dfrac{\breve{y} - y_1}{d_1} & \dfrac{\breve{z} - z_1}{d_1} & -1 \\ \dfrac{\breve{x} - x_2}{d_2} & \dfrac{\breve{y} - y_2}{d_2} & \dfrac{\breve{z} - z_2}{d_2} & -1 \\ \vdots & \vdots & \vdots & \vdots \end{bmatrix}, \tag{22.76}
$$

which can be also visualized as

$$
\mathbf{H}_1[n] = \begin{bmatrix} \breve{\mathbf{a}}_1[n] & \breve{\mathbf{a}}_2[n] & \cdots & \breve{\mathbf{a}}_{N_{sat}[n]} \\ -1 & -1 & \cdots & -1 \end{bmatrix}^T_{\mathbf{x}=\breve{\mathbf{x}}}, \tag{22.77}
$$

where $\breve{\mathbf{a}}_j[n] \triangleq \frac{1}{d_j[n]}[\breve{x}[n] - x_j[n], \breve{y}[n] - y_j[n], \breve{z}[n] - z_j[n]]^T$ is the unitary vector applied to the user position and pointing towards the jth satellite in the time instant n. As a consequence, the measurement equation (22.44) referring to pseudoranges at the time instant n can be linearized as

$$
\Delta \mathbf{z}^{(p)}[n] = \begin{bmatrix} \rho_1[n] - \breve{\rho}_1[n] \\ \vdots \\ \rho_{N_{sat}}[n] - \breve{\rho}_{N_{sat}}[n] \end{bmatrix} = \mathbf{H}_1[n] \Delta \mathbf{x}^{(p)}[n] + \mathbf{v}^{(p)}[n], \tag{22.78}
$$

where $\breve{\rho}_j[n]$, $\forall j = 1, \ldots N_{sat}$ are the nominal pseudoranges corresponding to the nominal trajectory where the linearization takes place (22.73), $\rho_j[n]$ are the measured pseudoranges after the correction (22.74), and $\mathbf{v}^{(p)}[n]$ is the component of the observation noise referred to positions.

Second Measurement Component: Pseudorange Rates: Pseudorange rates model the variations in time of the pseudoranges, and they are strictly related to the relative motion between satellites and user and to the Doppler shift in the carrier frequency that this motion implies. The formula that connects the Doppler shift to the radial velocity between the user and the jth satellite is

$$
\frac{c(f_j - f_{T_j})}{f_{T_j}} + \mathbf{v}_j[n] \cdot \mathbf{a}_j[n] = \mathbf{v}[n] \cdot \mathbf{a}_j[n] - \frac{f_j v_\tau[n]}{f_{T_j}}, \tag{22.79}
$$

where $\mathbf{v}_j[n]$ is the three-dimensional velocity vector relative to the jth satellite, $\mathbf{v}[n]$ is the three-dimensional user velocity vector, $v_\tau[n]$ is the clock drift in meters per second, f_{T_j} is the frequency transmitted by the jth satellite, and f_j is the Doppler-affected frequency of the jth satellite that is observed by the user.

Observing that the ratio $f_j / f_{T_j} \approx 1$, (22.79) can be simplified into

$$
\delta_j[n] = \mathbf{v}[n] \cdot \breve{\mathbf{a}}_j[n] - v_\tau[n], \tag{22.80}
$$

where $\delta_j[n] = c(f_j - f_{T_j})/f_{T_j} + \mathbf{v}_j[n] \cdot \breve{\mathbf{a}}_j[n]$ is known and the nominal user position is assumed in $\breve{\mathbf{a}}_j[n]$.

If the user velocity is known, it is possible to estimate the pseudorange rate relative to every single satellite, which is equal to the radial component of the difference between the jth satellite velocity and the user velocity

$$r_j[n] = \breve{\mathbf{a}}_j[n]^T \cdot \left(\mathbf{v}[n] - \mathbf{v}_j[n] \right) - v_{\tau}[n]. \qquad (22.81)$$

Considering the nominal pseudorange rate corresponding to the nominal trajectory as

$$\breve{r}_j[n] = \breve{\mathbf{a}}_j[n]^T \cdot \left(\breve{\mathbf{v}}[n] - \mathbf{v}_j[n] \right) - \breve{v}_{\tau}[n], \qquad (22.82)$$

then it is possible to write the incremental measurement on the pseudorange rate relative to the jth satellite as

$$r_j[n] - \breve{r}_j[n] = \breve{\mathbf{a}}_j[n]^T \cdot \left(\mathbf{v}[n] - \breve{\mathbf{v}}[n] \right) - \left(v_{\tau}[n] - \breve{v}_{\tau}[n] \right). \qquad (22.83)$$

Then, the measurement equation (22.44), referring to pseudorange rates at the time instant n can be linearized as

$$\Delta \mathbf{z}^{(v)}[n] = \begin{bmatrix} r_1[n] - \breve{r}_1[n] \\ \vdots \\ r_{N_{sat}}[n] - \breve{r}_{N_{sat}}[n] \end{bmatrix} = \mathbf{H}_1[n] \Delta \mathbf{x}^{(v)}[n] + \mathbf{v}^{(v)}[n], \qquad (22.84)$$

where $\Delta \mathbf{x}^{(v)}[n] = [\Delta v_x[n], \Delta v_y[n], \Delta v_z[n], \Delta v_{\tau}[n]]^T$, and $v^{(v)}[n]$ is the component of the measurement noise referred to velocities.

As a consequence of the above considerations, the complete *linearized observation equation* can be written by composing together (22.78) and (22.84) as:

$$\Delta \mathbf{z}[n] = \begin{bmatrix} \mathbf{H}_1[n] & \mathbf{0} \\ \mathbf{0} & \mathbf{H}_1[n] \end{bmatrix} \Delta \mathbf{x}[n] + \mathbf{v}[n], \qquad (22.85)$$

where it is possible to define the complete *observation matrix* for this problem:

$$\mathbf{H}[n] \triangleq \begin{bmatrix} \mathbf{H}_1[n] & \mathbf{0} \\ \mathbf{0} & \mathbf{H}_1[n] \end{bmatrix}, \qquad (22.86)$$

where $\mathbf{H}_1[n]$ is defined in (22.76).

22.3.3 Error Covariance Matrices

The model used in the EKF structure requires the initialization of the PVT solution and of the covariance matrices of the model noise, $\mathbf{Q}[n]$, observation noise, $\mathbf{R}[n]$, and of the state error, $\mathbf{P}[n]$.

The *model noise covariance matrix* $\mathbf{Q}[n]$ is time invariant and has been written in (22.71).

The *observation noise covariance matrix* $\mathbf{R}[n] = \mathbb{E}\{v[n]v^H[n]\}$ can be assumed as diagonal and it is time varying. The diagonal structure indicates that every pseudorange and pseudorange rate measurement error is considered independent from the other errors. The temporal variation accounts for the possibility of modifying during time the "confidence" associated with the observation error of each satellite, depending on the current satellite conditions. For instance, it is a common choice to weight each satellite using the inverse of the sine of its elevation, but other weights might be defined. A good choice might consider a balanced contribution of elevation and C/N_0.

Finally, the *state estimation error covariance matrix* $\mathbf{P}[n] = \mathbb{E}\{\mathbf{e}[n]\mathbf{e}^H[n]\}$, where $\mathbf{e}[n]$ is the estimation error on $\mathbf{\Delta x}[n]$, as determined by the KF, is a timevarying matrix that represents an index of correctness of the state. It must be initialized together with the state and its entries depend upon the level of knowledge of the initial state. It is a reasonable and conservative choice to initialize $\mathbf{P}[-1]$ slightly greater than its expected value, as the KF algorithm can tune it while running.

Example 22.2

This example is focused on the implementation of a script that is capable of computing the PVT solution by means of a KF. The reader should refer to the theory addressed in Section 22.3. Our purpose is to understand which measurements are required to compute the PVT solution and how they can be predicted. Moreover, a few expedients, like the relativistic correction, must be taken into account.

The reader is requested to list the measurements required to compute the PVT solution and how they must be processed, before being provided with an EKF.

Solution

The MATLAB code for this example is provided in Appendix A.2. The Kalman routine described in Appendix A.1 is also used, but it can be easily replaced by any user defined Kalman script.

The measurements that are needed to compute the PVT solution are the pseudorange and Doppler ones. Moreover, the estimated position and the velocity of the user and of the satellites must be available. While in the case of the satellites this information can be extracted from the navigation message, in the case of the user, they can be obtained from the adopted state model.

The first operation to be performed is the definition of the measurement matrix, which is named U in the script, and the prediction of the range and pseudorange rates, which can be obtained exploiting the estimated position and velocity of the satellites and of the user.

In the script, dX, dY, and dZ represent the components of the range between the *noSat*th satellite and the user in the ECEF frame, while dXv, dYv, and dZv represents the velocity vector between the user and the *noSat*th satellite along the connecting radius. Finally, the vector $\mathbf{AAA}$ is used to store the relativistic corrections, due to the Earth rotation during the propagation of the signal from the satellite to the user.

All these quantities are used to build up the main block of the observation matrix. This has been shown to be equal to the matrix of the unitary vectors pointing from the user toward the satellites and adding a fourth column equal to the 1 vector to take the clock receiver misalignment into account.

Once the predicted measurements are available, it is possible to compute the innovation. This operation is simply performed computing the dierence between the measurements and the predictions, and the result is stored in the *KFmeasurement* vector. Finally, the last lines before the Kalman routine are devoted to the construction of the block matrix that represents the measurement matrix to be fed to the KF.

22.4 INERTIAL NAVIGATION FUNDAMENTALS

The background behind IMUs' operation, related to basic physical principles, is briefly recalled in this section. Moreover, the structure of a generic *strapdown IMU* is discussed and the so-called *mechanization equations* are introduced.

22.4.1 Structure of an IMU

The main idea behind the inertial navigation is represented by *Newton's First Law*, which is stated as follows:

A body will continue in its state of rest, or of uniform motion in a straight line, unless an external force is applied to it.

Inertial navigation is based upon the measurement of this *external force* applied to the body. Every force applied to the body generates an acceleration, which may be integrated in time in order to obtain the body velocity and, after one more integration, the body position, both with respect to an initial condition. An *accelerometer* measures the acceleration it is subject to, along a predefined direction (in fact, three-axis acceleration is measured by a triplet of orthogonal accelerometers, also indicated as three-axis accelerometer).

A still body is subject to at least the gravity force. Accelerometers do not sense directly the gravity acceleration $\mathbf{g}$, but they do sense the reaction to this acceleration, applied by the surface where the body lies.* In fact, a three-axis accelerometer measures the so-called *specific force* $\mathbf{f}$, which is the acceleration of the body comprising the (reaction to the) gravity component:

$$\mathbf{f}(t) = \mathbf{a}(t) - \mathbf{g}(t), \qquad (22.87)$$

where $\mathbf{a}(t)$ is the total acceleration that determines the body motion in the inertial frame (i.e., $\mathbf{a}(t) = \mathbf{0}$ for a still body in the inertial frame, while the accelerometer measures $\mathbf{f}(t) = -\mathbf{g}(t)$ in the body reference frame). In order to extract the motion of the body from the measured acceleration, the gravity component sensed by the accelerometer must be removed from the total acceleration acting on the body. The

---

*This phenomenon can be easily understood if one considers a spring-mass system whose elongation axis is parallel to the gravity direction. The mass is attracted toward the center of the Earth, but the spring measures a force equal to the weight of the mass but in the *opposite* direction. The same criterion applies to the accelerometers, on whatever technology they rely.

gravity component is also present when the body is in motion and, in order to remove its contribution, the inertial device needs other sensors to measure the body attitude.

Gyroscopes measure angular speed (turn rate) the sensor is subject to along a predefined axis. The angular speed, once integrated, provides angular displacement (orientation). Thus, the angular orientation of the body can be calculated by integrating the angular rate measurements provided that an initial orientation of the sensor axis with respect to a reference is given.

As long as the accelerometers are rigidly attached to the body in movement, their reference frame is the body frame (i.e., a frame integral with the body and centered in the center of the accelerometer system), which is usually insufficient for providing the body movement with respect to an external reference frame. In fact, the rotation (angular displacement) of the body with respect to the external reference frame, i.e., the body attitude, must be computed from the instantaneous angular orientation of the body, obtained from three gyroscopes rigidly mounted on the body along the orthogonal accelerometers axes. Thereby, the attitude information is used to resolve the accelerometer measurements into the reference frame.

An IMU has to be intended as a combination of three accelerometers and three gyroscopes displaced along three orthogonal axes. The output of an IMU is generally compensates for errors like scale factors and biases, but KFs are instruments that allow the estimate of residual errors.

A system where the inertial sensors are mounted directly on the vehicle and move integrally with it is known as *strapdown* system. Otherwise the inertial sensors can be placed on a platform stabilized in space and then isolated from the body rotation (*gimballed* system). Strapdown systems are far more common than gimballed ones because of the simplicity of their mechanical realization.

A set of inertial sensors (accelerometers and gyroscopes, plus sometimes a magnetometer and/or an altimeter) forms an *IMU*. An IMU coupled with a computational unit to resolve the body position and velocity in a certain reference frame *(system mechanization)* is called an *INS inertial navigation system (INS)*.

22.4.2 The Coriolis Theorem

The Coriolis theorem relates the velocity of the body with respect to the Earth measured in an inertial frame $\mathbf{v}^i(t)$ to the velocity expressed in the rotating frame $\mathbf{v}^e(t)$.*

$$\mathbf{v}^e(t) = \dot{\mathbf{p}}^e(t) = \mathbf{v}^i(t) - \boldsymbol{\omega}_{ie} \times \mathbf{p}(t), \tag{22.88}$$

where $\mathbf{v}^e(t)$ is the ground speed, $\mathbf{v}^i(t)$ is the speed with respect to the inertial frame, $\boldsymbol{\omega}_{ie}$ is the turning rate of the ECEF frame with respect to the inertial frame, × denotes

---

*Hereafter we use the superscript ^{b}, ^{i}, and ^{e} to indicate quantities expressed in the *body*, *inertial*, and *Earth* reference frame, respectively.

vector cross product, and $\mathbf{p}(t)$ is the position of the vehicle on the Earth [14]. The derivative of (22.88) gives ([14], page 27):

$$
\begin{aligned}
\left.\frac{d\mathbf{v}^e(t)}{dt}\right|_i &= \dot{\mathbf{v}}^i(t) = \ddot{\mathbf{p}}^i(t) - \boldsymbol{\omega}_{ie}^e \times \dot{\mathbf{p}}^i(t) = \\
&= \ddot{\mathbf{p}}^i(t) - \boldsymbol{\omega}_{ie}^e \times \mathbf{v}^i(t) = \\
&= \ddot{\mathbf{p}}^i(t) - \boldsymbol{\omega}_{ie}^e \times \left[\mathbf{v}^e(t) + \boldsymbol{\omega}_{ie}^e \times \mathbf{p}^i(t)\right] = \\
&= \ddot{\mathbf{p}}^i(t) - \boldsymbol{\omega}_{ie}^e \times \mathbf{v}^e(t) - \boldsymbol{\omega}_{ie}^e \times \left[\boldsymbol{\omega}_{ie}^e \times \mathbf{p}^i(t)\right].
\end{aligned}
\tag{22.89}
$$

where the Earth turn rate has been assumed constant, i.e., $\dot{\boldsymbol{\omega}}_{ie}^e = \mathbf{0}$. In (22.89), the term $\boldsymbol{\omega}_{ie}^e \times \mathbf{v}^e(t)$ is known as *Coriolis acceleration* and represents the acceleration caused by the body velocity over the surface of a rotating Earth, while the term $\boldsymbol{\omega}_{ie}^e \times [\boldsymbol{\omega}_{ie}^e \times \mathbf{p}^i(t)]$ defines the *centripetal acceleration* experienced by the body owing to the rotation of the Earth.

22.4.3 Mechanization Equations

The system of equations used to compute the body trajectory (i.e., the instantaneous position, velocity, and attitude) in the selected reference frame from the inertial sensor measurements is called *mechanization*.

These equations are presented hereafter in the order they are used in an INS:

1. computation of the current body attitude in the proper reference frame;
2. computation of the current body velocity in the proper reference frame;
3. derivation of the current body position in the proper reference frame.

The considered reference frame is the ECEF one, whose axes have their origin in the center of the Earth. The x-axis is oriented toward the intersection between the equator and the Greenwich meridian, the z-axis along the Earth rotation axis, and the y-axis is oriented in order to obtain a right-handed triplet of orthogonal axes.

Computation and Tracking of the Body Attitude: The Direction Cosine Matrix (DCM): The gyroscopes output is a vector of three angular rates:

$$
\boldsymbol{\omega}^b = [\omega_{bx}, \omega_{by}, \omega_{bz}]^T,
\tag{22.90}
$$

from which the body attitude can be formally obtained via time integration. In order to express the angular rate vector measured by the gyros in the e frame, the following transformation must be computed ([14], p. 29):

$$
\boldsymbol{\omega}_{eb}^b(t) = \boldsymbol{\omega}^b(t) - \mathbf{C}_b^{eT}(t)\boldsymbol{\omega}_{ie}^e,
\tag{22.91}
$$

where $\boldsymbol{\omega}_{eb}^b(t)$ represents the turn rate of the body with respect to the e frame (subscript $_{eb}$) as measured by the gyros (superscript b), $\boldsymbol{\omega}_{ie}^e$ is the Earth rotation rate expressed in body axes, and $\mathbf{C}_b^e(t)$ is the rotation matrix (DCM) from the body frame to the Earth frame. In particular, the DCM may be calculated from $\boldsymbol{\omega}_{eb}^b(t)$ using the relationship ([14], page 29):

$$
\dot{\mathbf{C}}_b^e(t) = \mathbf{C}_b^e(t)\boldsymbol{\Omega}_{eb}^b(t),
\tag{22.92}
$$

where $\dot{\mathbf{C}}_b^e(t)$ is the time derivative of $\mathbf{C}_b^e(t)$, and $\Omega_{eb}^b(t)$ is the skew-symmetric matrix derived from $\boldsymbol{\omega}_{eb}^b(t)$.*

Discrete-time propagation of the DCM can be obtained by means of the Taylor approximation, as it will be discussed in detail in the following sections:

$$\mathbf{C}_b^e[n+1] = \left(\mathbf{I} + T_c \Omega_{eb}^b[n]\right)\mathbf{C}_b^e[n]. \tag{22.98}$$

Computation and Tracking of the Velocity: The body velocity in the e frame, $\mathbf{v}^e(t)$, is obtained from integration of the corresponding acceleration $\mathbf{a}^e(t)$, which, in turn, is a function of the specific force. However, to obtain the total body acceleration $\mathbf{a}^e(t) = \dot{\mathbf{v}}^e(t)$ it is necessary to rotate the accelerometer measurement $\mathbf{f}^b(t)$ into the proper frame and subtract the components due to gravity, Coriolis effect, and centripetal force.

From the Coriolis theorem expressed in Section 22.4.2, it is possible to relate the inertial frame and the ECEF frame velocity variations by means of the following equation, obtained by differentiating (22.88) with respect to the time [14]:

$$\dot{\mathbf{v}}^e(t) = \dot{\mathbf{v}}^i(t) - \boldsymbol{\omega}_{ie}^e \times \mathbf{v}^e(t). \tag{22.99}$$

The velocity variation $\dot{\mathbf{v}}^i(t)$ in the inertial frame can be written as in (22.87):

$$\ddot{\mathbf{p}}^i(t) = \mathbf{f}(t) + \mathbf{g}(t), \tag{22.100}$$

that, substituted into (22.89), gives

$$\dot{\mathbf{v}}^i(t) = \mathbf{f}(t) - \boldsymbol{\omega}_{ie}^e \times \mathbf{v}^e(t) + \mathbf{g}_{\ell}(t), \tag{22.101}$$

where $\mathbf{f}(t)$ represents the applied specific force, $\boldsymbol{\omega}_{ie}^e \times \mathbf{v}^e(t)$ is the Coriolis acceleration, and

$$\mathbf{g}_{\ell}(t) \triangleq \mathbf{g}(t) - \boldsymbol{\omega}_{ie}^e \times \left[\boldsymbol{\omega}_{ie}^e \times \mathbf{p}^e(t)\right] = \mathbf{g}(t) - \Omega_{ie}^e \Omega_{ie}^e \mathbf{p}^e(t) \tag{22.102}$$

---

*A *skew-symmetric matrix* is a square matrix whose transpose is also its negative; that is, it satisfies the equation $\mathbf{A}^T = -\mathbf{A}$. Given a vector

$$\mathbf{a} = [a_1 \ a_2 \ a_3]^T, \tag{22.93}$$

it is possible to build the skew-symmetric matrix associated to the vector

$$\mathbf{A} = \begin{bmatrix} 0 & -a_3 & a_2 \\ a_3 & 0 & -a_1 \\ -a_2 & a_1 & 0 \end{bmatrix}. \tag{22.94}$$

The following relationships hold between vector cross-product and vector-matrix product:

$$\mathbf{a} \times \mathbf{b} = \mathbf{A}\mathbf{b}, \tag{22.95}$$

$$\mathbf{a} \times \mathbf{b} = -\mathbf{b} \times \mathbf{a} = -\mathbf{B}\mathbf{a}, \tag{22.96}$$

$$\mathbf{a} \times [\mathbf{a} \times \mathbf{b}] = \mathbf{A}\mathbf{A}\mathbf{b}. \tag{22.97}$$

is the so-called *local gravity vector*, representing the sum of the accelerations caused by the mass attraction force ($\mathbf{g}(t)$, the gravitational acceleration vector) and the centripetal acceleration due to Earth's rotation.

Equation (22.101) can be substituted in (22.99), obtaining the differential equation for velocity variations in the Earth frame:

$$
\begin{aligned}
\dot{\mathbf{v}}^e(t) &= \left[\mathbf{f}(t) - \boldsymbol{\omega}_{ie}^e \times \mathbf{v}^e(t) + \mathbf{g}_\ell(t) \right] - \boldsymbol{\omega}_{ie}^e \times \mathbf{v}^e(t) = \\
&= \mathbf{C}_b^e \mathbf{f}^b(t) - 2\boldsymbol{\omega}_{ie}^e \times \mathbf{v}^e(t) + \mathbf{g}_\ell(t),
\end{aligned}
\tag{22.103}
$$

where $\mathbf{C}_b^e \mathbf{f}^b(t)$ is the frame rotation of the specific force from b to e.

Computation and Tracking of the Position: The position propagation is described by the well-known relationship between space and velocity:

$$
\dot{\mathbf{p}}^e(t) = \mathbf{v}^e(t),
\tag{22.104}
$$

where $\mathbf{p}^e(t)$ is the three-dimensional position of the body in the e frame.

Example 22.3

This example explains the main steps to implement the inertial mechanization equations of an IMU. This process is fundamental to achieve an integration between an INS module and a GNSS receiver. The reader is requested to implement a MATLAB routine to calculate the attitude, the velocity, and the position components starting from the acceleration and angular rates provided by the IMU. The INS navigation equations have to be computed at a medium rate (e.g., 10 Hz) bounded between the high rate (e.g., 100 Hz) at which the INS measurements are provided and the low rate (e.g., 1–5 Hz) of the GNSS measurements, respectively.

Solution

In Appendix A.3 the formulas considered in this example are implemented in a MATLAB script. The INS navigation equations are calculated with respect to the ECEF frame. The parameters used as input are the following:

- attitude DCM (3×3) from body to ECEF frame at previous (mediumrate) time $[n-1]$;
- velocity array (3×1) in ECEF frame at previous (medium-rate) time $[n-1]$;
- position array (3×1) in ECEF frame at previous (medium-rate) time $[n-1]$;
- coning and sculling calculated at high rate $[n]$;
- velocity and rotation angle computed in the *body* frame at high rate $[n]$;
- position array (3×1) in Latitude Longitude Height (LLH) frame rad at previous (medium-rate) time $[n-1]$.

The initial value of the DCM matrix will be explained in the section devoted to the initial alignment of the IMU. Then it has to be converted to the ECEF frame. A routine to do such conversion is included. The velocity and position arrays in ECEF can be set equal to zero at the first iteration.

A function for the calculation of the position array from the local frame to the ECEF frame has been included too, as well as a function for the computation of coning and sculling parameters. As far as coning and sculling factor are concerned, they need the IMU measurements clean from deterministic bias and additional effects (e.g., the vehicle's vibrations).

A proper lter is designed to reduce the vibrations experienced by the IMU when the vehicle is moving. In summary, three main steps are implemented within the INS navigation equations:

1. computation of the attitude in the body frame and conversion in the ECEF frame;

2. update of the velocity in the body frame and then integration in the ECEF frame;

3. position update in the ECEF frame.

22.5 NOISE CHARACTERIZATION OF AN INERTIAL UNIT

The major limitation of an INS is its *long-term stability*, compromised by the noise that characterizes all its sensors (i.e., gyroscopes and accelerometers). The errors accumulated with integration of the inertial data in the navigation algorithm will cause the navigation solution (as described in the section about mechanization equations) to drift from the true one. In the case of an accelerometer, a theoretically rigorous error model can be found in [15] that can be summarized according to [16] as

$$\delta \mathbf{f}^b = b_f + c_T \delta T + \sum_{i=1}^{2} c_i a_i + \sum_{i=1}^{3} m_i a_i + s_f + w_f, \qquad (22.105)$$

where b_f is a deterministic constant bias, $c_T \delta T$ is the temperature sensitivity effect, a_i is the acceleration along the ith axis of the sensor, $c_i a_i$ are anisoelasticity effects, m_i are misalignment errors, s is the scale factor term, and w_f is the stochastic noise that affects the accelerometer. Similarly, in the case of a gyroscope the error model becomes

$$\delta \omega_{ib}^b = b_\omega + c_T \delta T + \sum_{i=1}^{3} c_i B_i + \sum_{i=1}^{3} m_i \omega_i + s_\omega + w_\omega, \qquad (22.106)$$

where b_ω is a deterministic constant bias, $c_T \delta T$ is the temperature sensitivity effect, ω_i is the rotation rate along the ith axis of the sensor, $c_i B_i$ are magnetic sensitivity terms, m_i are misalignment errors, s is the scale factor term, and w_ω is the stochastic noise that affects the gyroscope.

From these two models it is possible to divide such errors into two main groups: *deterministic errors* and *stochastic errors*. The group of *deterministic errors* comprises: systematic bias, scale factor, thermal terms, and misalignment errors.

Conversely, the *stochastic errors* include random noises. Both deterministic and random noise will be discussed in details in the following section.

22.5.1   IMU Deterministic Errors

In this subsection the main deterministic sources that affect an IMU are described and their negative impact on the position, velocity, and attitude computation are shown as well.

Bias: In the ideal case, when no input is applied, the output signal should display 0. However, this does not happen in the real world due to the fact that an offset is always present. The units of an accelerometer sensor is g (the gravity acceleration) or m/s^2, while a gyroscope bias is typically expressed as degrees per hour or per second, $°/h$ or $°/s$. An example of a measure of gyroscope bias for a tri-axial MEMS IMU is depicted in Figure 22.3.

An uncompensated gyroscope bias generates an angle error $\delta\theta$ that increases linearly over time according to:

$$\delta\theta(t) = \int_0^t b_\omega d\tau = b_\omega t, \tag{22.107}$$

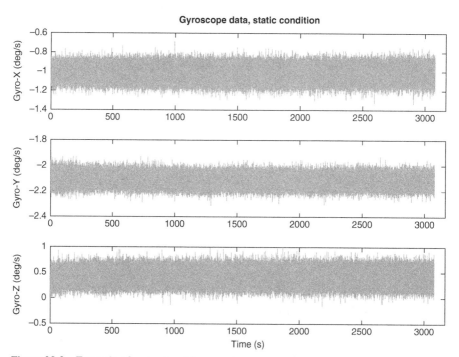

Figure 22.3   Example of gyroscope biases.

where b_ω is the gyro bias. This small error generates an effect on the acceleration equal to $\delta a(t) = g\sin(\delta\theta) \approx g \cdot \delta\theta = g \cdot b_\omega t$ [17]. Thus, the velocity and position drifts can be computed as:

$$\delta v(t) = \int_0^t \delta_a(\tau)d\tau = \int_0^t gb_\omega t d\tau = \frac{1}{2}b_\omega g t^2$$

$$\delta p(t) = \int_0^t v(\tau)d\tau = \int_0^t \int_0^t \frac{1}{2}b_\omega \tau^2 dt = \frac{1}{6}b_\omega g t^3.$$

(22.108)

Thus, a linear error on the angle causes a quadratic error in velocity and cubic error in position.

Similarly, an uncompensated acceleration bias of the accelerometer, b_f, generates an error in velocity proportional to time t and proportional to t^2 in the position. In fact:

$$\delta v(t) = \int_0^t b_f d\tau = b_f t$$

$$\delta p(t) = \int_0^t \delta v(\tau)d\tau = \int_0^t \int_0^t b_f \tau dt = \frac{1}{2}b_f t^2.$$

(22.109)

Scale Factor: By definition, the *scale factor* is the ratio of a change in output to a change in the intended input to be measured [18]. An ideal sensor has a scale factor equal to 1. Therefore, a scale factor is the difference between an ideal sensor and a realistic one and such an error is typically expressed in parts per million (ppm) or in percentage (%) in case of low-cost IMUs. For instance, the scale factor of a gyroscope is obtained by rotating such a sensor at some predefined rotation rates.

A scale factor on an accelerometer or on a gyroscope causes the same effect of the bias on the position and velocity.

Misalignment Errors: This type of error is related to an imperfection in mounting the sensors that are not perfectly orthogonal among each other [19]. This can be due to a manufacturing imperfection (as shown in Fig. 22.4) or to user error in placing the INS sensors with a certain misalignment angle with respect to the true reference axes (i.e., inertial ones). This kind of error is depicted in Figure 22.5.

Misalignment errors are typically expressed in ppm. Let us assume that the accelerometers have been mounted with a small roll tilt angle $\delta\theta$. Such an error makes each accelerometer measure a component of Earth gravity g: in the z-direction it measures $f_z = g\cos\delta\theta$, while in the y-direction it measures $f_y = g\sin\delta\theta$. For very small tilt angles we can approximate $f_z = g\cos\delta\theta \approx g$ and $f_y = g\sin\delta\theta \approx g\,\delta\theta$. The error in velocity and position results in:

$$\delta v(t) = \int_0^t g\delta\theta d\tau = g\delta\theta t$$

$$\delta p(t) = \int_0^t v(\tau)d\tau = \int_0^t \int_0^t g\delta\theta\tau dt = \frac{1}{2}g\delta\theta t^2.$$

(22.110)

IMU Temperature Error: The IMU internal temperature varies over time. If not properly compensated, it generates bias and scale factor variations [14]. An example of

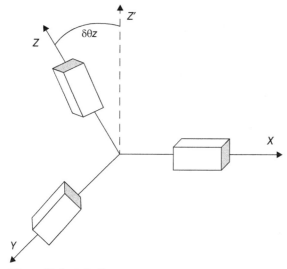

Figure 22.4   Misalignment due to manufacturing imperfections.

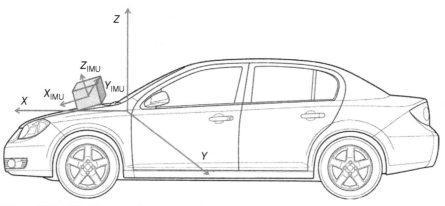

Figure 22.5   Misalignment due to mounting errors.

x-axis gyroscope bias variation due to the IMU temperature is plotted in Figure 22.6. This effect of temperature-dependent variation can be quite big in case of low-cost MEMS IMU, and mitigation techniques are required.

22.5.2   Modelling the IMU Stochastic Noise

The noise terms w_f and w_ω of the accelerometers and gyroscopes in (22.105) and (22.106) comprise all the nondeterministic effects that affect the IMU sensors. The detection and modeling of such error sources is a is a challenging task. Therefore, several works in the scientific literature have addressed the estimation of the stochastic model parameters [20–22]. In this section, we describe the most-used methods for noise identification and stochastic modeling of inertial sensors errors.

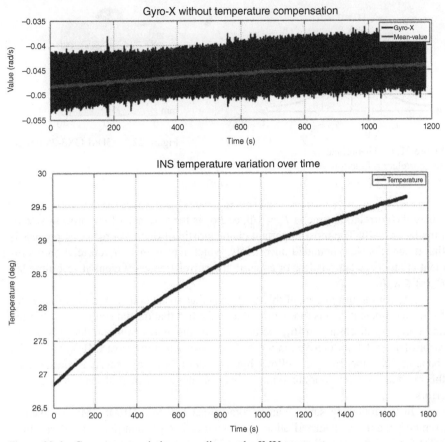

Figure 22.6   Gyroscopes variation according to the IMU temperature.

Autocorrelation Method: The autocorrelation function has been used in previous works to analyze the stochastic error of inertial sensors [22, 23] and also to obtain the parameters for modeling using first-order Gauss–Markov (GM) process. This process has been used widely for modeling random errors not only because it seems to fit a large number of physical processes with reasonable accuracy, but also because it has a relatively simple mathematical description [6].

For a random process x with zero mean, correlation time T_c, and driving noise w, the first-order GM process is described by the following continuous-time equation:

$$\dot{x} = -\frac{1}{T_c} x + w. \tag{22.111}$$

The parameters needed to implement this process can be extracted from its autocorrelation function (see Fig. 22.7), which is given by

$$R_{xx}(\tau) = \sigma^2 e^{\beta|\tau|}, \tag{22.112}$$

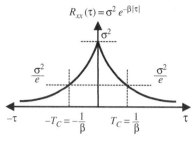

Figure 22.7 Theoretical autocorrelation function of the first-order GM process.

Figure 22.8 3DM-GX3-25 device.

where the correlation time is $T_c = 1/\beta$, and σ^2 is the variance of the process at zero time lag ($\tau = 0$). The most important characteristic of the first-order GM process is that it can represent bounded uncertainty, which means that any correlation coefficient at any time lag τ is less or equal to the correlation coefficient at zero time lag $R_{xx}(\tau) \leq R_{xx}(0)$ [7].

One of the limitations of this method is that an accurate autocorrelation curve from experimental data is rarely done due to the fact that the data collected is limited and finite. As discussed in [6], the accuracy of the autocorrelation depends on the record-length data. Moreover, in most of the cases when low-cost IMU are used, the shape of the autocorrelation follows higher-order GM processes. As a consequence this method is not appropriate to model in an accurate way inertial sensors stochastic errors.

In the following subsections a commercial low-cost MEMS IMU is taken to exemplify the noise characterization of a low-grade inertial unit [24]. The MicroStrain 3DM-GX3-25 device [25] shown in Figure 22.8 combines a triaxial accelerometer, tri-axial gyro, tri-axial magnetometer, and a temperature sensor. All quantities are temperature-compensated and are mathematically aligned to an orthogonal coordinate system. The characterization data provided by the manufacturer can be seen in Table 22.2.

Seven hours, static data at a sampling frequency of 100 Hz were collected in order to analyze the inertial sensor data with the autocorrelation method. According to [24], the analysis of the random errors of the aforementioned IMU by using the autocorrelation method has been performed after having removed the turn-on bias for each sensor, as well as the approximate measure of the gravity provided by the

TABLE 22.2. Sample IMU Characterization Data

IMU model	3DM-GX3-25
Acc bias stability	±0.5g for ± 5g
Acc nonlinearity	0.2%
Gyro bias stability	±0.2°/sfor ± 360°/s
Gyro repeatability	0.2°
Gyro nonlinearity	0.2%

output of the z-axis accelerometer. Then, the high-frequency terms were attenuated by applying a wavelet denoising technique [26]. Thus, the autocorrelation is calculated and the corresponding parameters are extracted from the curve; in the case of the first-order GM process the parameters are T_c and σ, respectively.

Figure 22.9a depicts the normalized autocorrelation function of the accelerometers before applying the denosing procedure, while Figure 22.9b corresponds to the autocorrelation curve when the wavelet technique is applied with 6 levels of decomposition and using Daubechies 4 as wavelet mother. This autocorrelation result shows that the noise component that affects the x-axis accelerometer is mainly white noise. For inertial sensors based on MEMS technology, the assumption that the

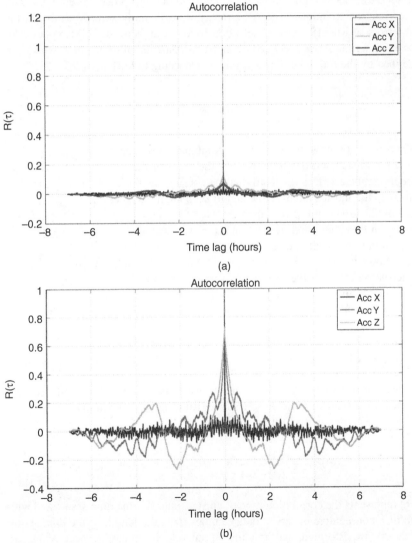

Figure 22.9   3DM-GX3-25: Autocorrelation processes for the accelerometers' measurements.

stochastic error follows a first-order GM process is not valid in most situations. This can be observed by comparing Figure 22.7 with Figure 22.9b, where it can be seen that they are different from the autocorrelation function of the first-order GM process. This is because these sensors are composed of more complex noise types; the first-order GM process is only part of this complex noise mixture, as discussed in the following subsections.

Autoregressive Processes: In order to overcome the problem of inaccurate modeling of INS errors, as in the case of autocorrelation method, other techniques can be applied. One of them is based on auto-regressive (AR) models that have been described in-depth in [22, 23, 27]. Through an AR method is possible to model the inertial sensor errors as higher order than the first-order GM process used in the case of autocorrelation technique. The AR strategy is particularly suitable to model the random errors of MEMS sensors that often follow a higher-order GM process. An AR process is a time series produced by linear combination of past values, which can be described by the following linear equation according to [24] and [28]:

$$x[n] = \sum_{k=1}^{p} \alpha_k x[n-k] + \beta_0 w[n], \tag{22.113}$$

where $x[n]$ is the process output that is a combination of past outputs, plus a white noise $w[n]$ with standard deviation β_0; p is the order of the AR process; and α_k are the model parameters. It is assumed that the coefficients β_0 and α_k are computed in order to keep the linear system stable and to keep the model stationary [7]. It should be noted in (22.113) that if $p = 1$, then the AR process approximates first-order GM processes. On the other hand, if $p = 1$ and $\alpha_1 = 1$, it becomes a random walk, and if $\alpha_1 = 0$ it would be a white noise.

As shown in Figure 22.9b, a high-order AR model could be used to model the x-axis accelerometer, but the use of this sort of AR model in the integration filter would drastically increase its computational cost [22]. Concerning the other accelerometers, a second-or third order AR process may be suitable to model their noises.

Allan Variance: The Allan variance (AV) is a time-domain analysis originally developed to study the frequency stability of oscillators. It has been successfully applied to the modeling of inertial sensors [29, 30] by determining the characteristics of the random processes that give rise to the measurement noise of the sensors. Indeed, AV is often used to identify the noise characteristics in an observed data set [31].

The AV is computed as follows:

$$\sigma^2(\Theta) = \frac{1}{2\Theta^2(N-2n)} \sum_{k=1}^{N-2n} (\psi_{k+2n} - 2\psi_{k+n} + \psi_k)^2, \tag{22.114}$$

where Θ represents the correlation time or cluster time, i.e., the time associated with a group of n consecutive observed data samples, N is the length of the data set on which the AV is performed, and ψ is the output velocity in case of accelerometers or output angle in case of gyroscopes.

The basic idea to estimate the AV is to take a long sequence of data samples (*N*), collected from a stationary IMU. Once the data is obtained the deterministic bias is removed and a denoising filter is applied. Then, the output of the inertial sensors is integrated in order to obtain ψ and, eventually, the AV is computed through (22.114). When an AV technique is used, the uncertainty in the data is assumed to be generated by noise sources of specific character [24], like, for example: rate random walk, angle random walk, bias instability, etc. In order to obtain the covariance of each noise source affecting the sensor output, it is necessary to analyze the AV result computed by (22.114). This is usually done by plotting a log-log AV curve as is depicted in Figure 22.10, from which the covariance values for each error are usually extracted.

The computation of AV needs a finite number of clusters that can be generated from the raw data measurements of the sensors. Depending on the size of these clusters, the AV can identify any noise term that affects the data sensor [24]. It is important to mention that the estimation accuracy of the AV for a given Θ depends on the number of independent clusters within the data set [31]. The bigger the number of independent clusters, the better the estimation accuracy is. According to [20], the percentage error of AV, $\sigma(\delta)$, for a certain $\sigma(\Theta)$ and with a data set of *N* points is given by:

$$\sigma(\delta) = \frac{1}{\sqrt{2\left(\dfrac{N}{n} - 1\right)}},$$ (22.115)

where *N* is the number of samples collected from the sensors and *n* is the number of points within a certain cluster and used to estimate $\sigma(\Theta)$. Equation (22.115) shows that the estimation errors in the region of short cluster length Θ are small as the number of independent cluster in these regions is large. On the other hand, the error estimation in the region of long cluster length Θ are large as the number of independent

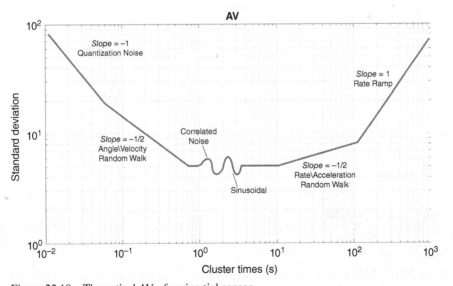

Figure 22.10   Theoretical AV of an inertial sensor.

clusters in these regions is small [20, 31]. For example, according to [24], if 360,000 samples are collected from a static IMU sensor, and if we want to obtain a bias instability with a correlation time equal to 10 minutes, clusters of 60,000 points are used. According to (22.115) the percentage error of the AV for this random process would be approximately 32%.

As a practical example of AV calculation, we have used the same data collection from the 3DM-GX3-25 Microstrain IMU also used for the autocorrelation method. The log-log plot of AV standard deviation versus cluster times (Θ) is obtained by applying (22.114). The results are plotted in Figure 22.11a for the accelerometer and Figure 22.12 for gyro data, respectively.

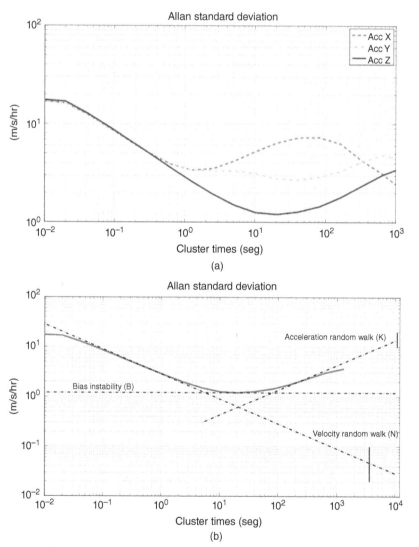

Figure 22.11   3DM-GX3-25: (a) AV for accelerometers measurements; (b) noise sources evaluation for the z-axis accelerometer.

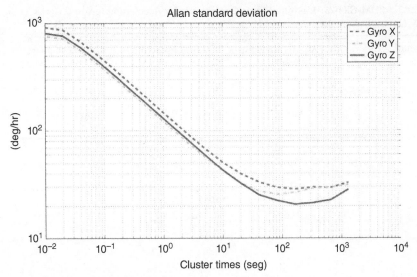

Figure 22.12   3DM-GX3-25: AV for the gyros' measurements.

Figure 22.11a shows the AV estimated on the 3DM-GX3-25 accelerometers. These accelerometers are affected by three types of error: velocity random walk (N), bias instability, (B) and acceleration random walk (K), which are identifiable through their corresponding slopes. Figure 22.11b depicts the AV for each noise component of the z-axis accelerometer: it has N, B, and K with slopes—1/2, 0, and 1/2, respectively. It can be seen that the dominant noise in short cluster times is the velocity random walk, while the dominant error in long cluster time is the acceleration random walk. From the straight line with slope—1/2 fitted to the beginning of the N noise, a value σ = 0.047 (m/s/hr) at a cluster times of 1 hr can be read. Since the velocity random walk N is present where the number of independent cluster is very large, the estimation accuracy of the AV is approximately 1.1%, which is quite reliable. Thus, the velocity random walk N for the z-axis accelerometer is determined as

$$N = 0.047 \pm 0.00050 \left(m/s/\sqrt{hr} \right). \tag{22.116}$$

The AV standard deviation versus cluster times Θ for gyro data is depicted in Figure 22.12, where two types of noises can be recognized: angle random walk N for short cluster times and bias instability B for long cluster times.

For the x-axis gyro (dark dashed curve), the bias instability is present in the time range between 321.92 s and 654.01 s. The value of this error can be measured with a flat line at 29.57 deg/hr. Dividing this standard deviation by the factor 0.664 as suggested in [31], the B coefficient can be achieved as:

$$B = 44.533 \pm 5.14 \, (deg/h). \tag{22.117}$$

More details on the procedure to analyze the AV curve can be found in [20] and [31]. Table 22.3 summarizes the error coefficients with their respective uncertainty

TABLE 22.3.   **3DM-GX3-25: Identified Error Coefficients for Accelerometers and Gyros With AV [24]**

Acceleration Axis	Velocity Random Walk (N) (m/s/$\sqrt{h}$)	Bias Instability (B) (m/s/h)	Acceleration Random Walk (K) (m/s/h^3/^{2})
x	0.045 ± 0.00023	5.1581 ± 0.0370	166.30 ± 4.6398
y	0.045 ± 0.00022	4.5507 ± 0.0506	24.95 ± 2.8368
z	0.047 ± 0.00050	1.8336 ± 0.0524	13.53 ± 1.8685
Rotation Axis	Angle Random Walk (N) (deg/$\sqrt{h}$)	Bias Instability (B) (deg/h)	
x	2.420 ± 0.0974	44.533 ± 5.14	
y	1.988 ± 0.0565	38.810 ± 2.51	
z	2.164 ± 0.0599	31.717 ± 2.29	

TABLE 22.4.   **3DM-GX3-25: Identified Θ_c and Standard Deviation [24]**

	x-Axis Acc.	y-Axis Acc.	z-Axis Acc.	x-Axis Gyro	y-Axis Gyro	z-Axis Gyro
Correlation Time						
Θ_c (s)	1.29	41.09	20.74	654.01	83.27	166.49
Standard Deviation	0.0068	0.0065	0.0063	0.0055	0.0045	0.0048

for accelerometers and gyro data. The correlation time (Θ_c) of the bias instability B and the standard deviation for each sensor of the IMU 3DM-GX3-25 is shown in Table 22.4.

Power Spectral Density: The power spectral density (PSD) is another important descriptor of a random process because it provides information not easy to extract from a time-domain technique (e.g., AV). The PSD is related to the autocorrelation function as

$$S_x(jw) = F[R_{xx}(\tau)] = \int_{-\infty}^{\infty} R_{xx}(\tau)e^{-jwt}d\tau, \qquad (22.118)$$

where $S_x(jw)$ is the PSD of the process x, as a function of the angular frequency $w = 2\pi/$, $F[\cdot]$ indicates Fourier transform; and $R_{xx}(\tau)$ is the autocorrelation of the process x. Basically, the PSD is used to identify the stochastic errors of the inertial sensors from the frequency components and the parameters obtained from the PSD are eventually used in the KF model when the INS is integrated with other sensors (e.g., GNSS). Figure 22.13 depicts a theoretical single-sided PSD for an inertial sensor.

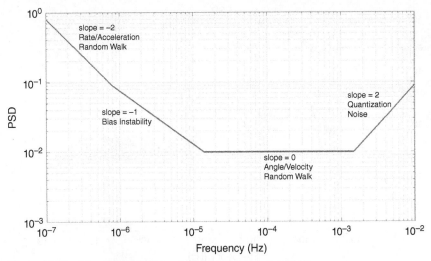

Figure 22.13   Theoretical PSD in single-sided form of an IMU.

According to the curve shown in Figure 22.13, the noise sources can be identified by considering the different slopes of the noise terms. Obviously, the number of random noises that might be present are strictly dependent on the type of sensors. The derivation of each noise terms are well detailed in [20] and [31]. A noise data analysis has been carried out by exploiting the PSD method, and it can be also useful to check the validity of the noise coefficients obtained with AV. The PSD has been implemented using Welch's method. Figure 22.14a depicts the one-side PSD for accelerometer data. This log-log plot shows very noisy high-frequency components, which makes it difficult to identify the correct noise terms and to obtain accurate parameters of the stochastic model. The variance of these high-frequency parts of the spectrum may be decreased by averaging adjacent frequencies of the estimated PSD. This task can be accomplished by using a technique that is called *frequency averaging*. Further details of this technique can be found in [19].

Figure 22.14b shows the PSD curve after applying frequency averaging: it can be noticed that the noise terms identification is more clear than in Figure 22.14a, and although the low-frequency part of the PSD plot has a high uncertainty, it still conveys some information [19]. According to the ideal trend of the PSD as shown in Figure 22.13, it is possible to detect three types of noise for the IMU under investigation: the acceleration random walk (K), the bias instability (B), and the velocity random walk (N). Figure 22.14b shows that the z-axis accelerometer has a bias instability smaller than the other two accelerometers, and the velocity random walk is almost the same for all the accelerometers (slope 0), which is coherent with the results obtained with the AV technique (slope $-1/2$).

The values for each noise parameter have been extracted similarly to the procedure adopted for the AV, but in this case the interceptions are different. For instance, the PSD curve for z-axis accelerometer is plotted in Figure 22.15 with straight lines for each noise N, B, and K, with their respective slopes $0, -1, -2$.

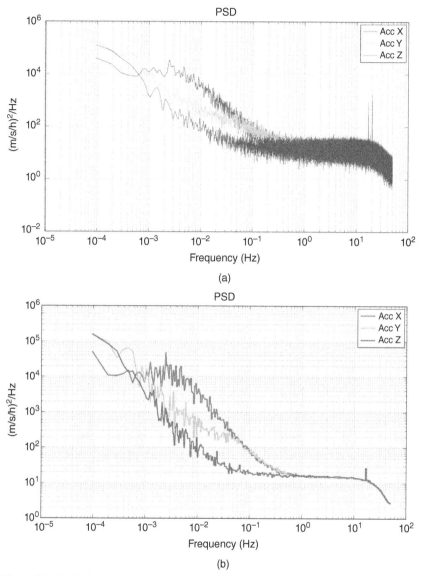

Figure 22.14   3DM-GX3-25: (a) PSD of the accelerometers' data; (b) PSD with frequency averaging.

The acceleration random walk K is present in the low-frequency components between $1 \cdot 10^{-4}$ Hz and $2.29 \cdot 10^{-3}$ Hz. It is obtained at the intersection point between a straight line with a slope of -2 starting from $1 \cdot 10^{-4}$ Hz and a vertical line centered at 1 Hz. For details about the procedure to determine the different noise parameters see [31]. The bias instability B is the dominant noise between $2.29 \cdot 10^{-3}$ Hz and $7.1 \cdot 10^{-2}$ Hz with a slope of -1, while the velocity random walk (N) is present

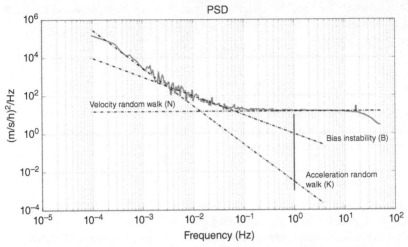

Figure 22.15   3DM-GX-25: Analysis of PSD of the *z*-axis accelerometer's data.

between 0.1248 Hz and 20 Hz. After 20 Hz there is an attenuation because of the digital moving average filter, which is used to minimize high-frequency spectral noise produced by the MEMS sensors. A similar procedure can be adopted for the gyroscopes as well.

Eventually, Table 22.5 summarizes the different errors (and their order of magnitude) that affect the inertial sensors of the IMU under investigation by using the PSD method.

The estimated values reported in Table 22.5 are used to check that the results obtained through the AV analysis in Table 22.4 are consistent: consistency is confirmed by the fact that most of the estimated values are within the confidence interval of the AV results (Table 22.3).

TABLE 22.5.   3DM-GX3-25: Identified Error Coefficients for Accelerometers and Gyros [24]

Acceleration Axis	Velocity Random Walk (N) (m/s/($\sqrt{h}$)	Bias Instability (B) (m/s/h)	Acceleration Random Walk (K) (m/s/h$^{3/2}$)
x	0.045	4.6447	168.60
y	0.044	4.6700	26.66
z	0.047	1.7733	14.60

Rotation Axis	Angle Random Walk (N) (deg/$\sqrt{h}$)	Bias Instability (B) (deg/h)	
x	2.297	43.438	
y	1.937	39.614	
z	2.058	30.705	

In conclusion, for the IMU under investigation in this example, it can be stated that the accelerometer stochastic error w_b can be modeled as

$$w_b = \text{WN}(N) + 1^{st}\,\text{GM}(B) + \text{RW}(K), \qquad (22.119)$$

where the noise term associated to N is modeled as a white noise (WN), the bias instability is modeled as first-order GM process (1st GM), and the noise term associated to K is considered a random walk (RW) process.

Furthermore, the model of the gyro stochastic error w_f associated with the considered IMU can be written as

$$w_f = \text{WN}(N) + 1^{st}\,\text{GM}(B), \qquad (22.120)$$

where the noise term associated to N is modeled as white noise (WN) and the bias instability B is modeled as a first-order GM process (1st GM).

This noise modelling plays a fundamental role in the design of the KF in charge of blending the measurements coming from the IMU and the GNSS module. The design of such system hybridization will be discussed with mathematical details in Sections 22.7 and 22.8.

22.6 CALIBRATION AND INITIAL ALIGNMENT OF THE INERTIAL NAVIGATION SYSTEM

Calibration is the process of comparing instrument outputs with known reference information and determining the coefficients that force the outputs to agree with the reference information over a range of output values [17]. In our context, the calibration process is focused on the evaluation of the deterministic biases and scale factors of the IMU sensors.

22.6.1 Calibrating a High-Grade IMU

One method that is suitable for high-grade IMUs has been proposed in [32], where the authors describe an error calibration model based on mounting the IMU on a leveled turntable and aligning the IMU axes alternately up and down with respect to different angle rotations. Then the biases b_f, b_ω, and scale factor errors s_a, s_ω in (22.105) and (22.106) are estimated by mounting the inertial system with its axes pointing alternatively up and down. In formula, it means:

$$
\begin{aligned}
b_f &= \frac{l_f^{up} + l_f^{down}}{2} \\
s_f &= \frac{l_f^{up} - l_f^{down} - 2k}{2k},
\end{aligned}
\qquad (22.121)
$$

where l_f^{up} is the sensor measurement when the sensitive axis is pointed upward, l_f^{down} is the sensor measurement when the sensitive axis is pointed downward, k is the known reference signal that, in case of accelerometers, is equal to the gravity acceleration.

Thus, for a triad of accelerometers a six-position method will be adopted according to (22.121), where each sensor will be placed in two different positions.

22.6.2 Calibrating a Low-Grade IMU

Unfortunately, the technique described in the previous section cannot be used for a consumer/automotive-grade MEMS IMU, since the bias instability and noise levels of its sensors are too high (e.g., in case of gyroscopes they completely mask the Earth rotation rate) and are also affected by changes in the environmental conditions, especially the temperature. Thus, additional calibration methods, targeted to low-cost IMUs, have been proposed in the scientific literature. One of the most common techniques is the so-called multiposition calibration method (MPCM), that was developed in [33] for this purpose. The main advantage of this method is it can be applied in the field and does not require any additional instrument to align the IMU to any certain directions. Therefore, it can be applied in the field directly and at any time during the operation in order to estimate the IMU systematic errors (e.g., the bias and scale factor).

Accelerometer Calibration: According to [33], calibration models for accelerometer errors are as follows:

$$f = a_x^2 + a_y^2 + a_z^2 - \|g\|^2 = 0, \tag{22.122}$$

where g is the gravity reference. IMU measurements taken from the x-axis accelerometer can be further expanded as:

$$l_{f,x} = b_{f,x} + (1 + s_x)a_x + \eta_{f,x}, \tag{22.123}$$

where l is the measurement data and η_f is the stochastic noise. After having substituted (22.123) into (22.122), the calibration model for accelerometers can be written as:

$$f = \left(\frac{l_{f,x} - b_{f,x}}{1 + s_x} \right)^2 + \left(\frac{l_{f,y} - b_{f,y}}{1 + s_y} \right)^2 + \left(\frac{l_{f,z} - b_{f,z}}{1 + s_z} \right)^2 - \|g\|^2 = 0. \tag{22.124}$$

The implicit model described in (22.124) can be solved through a combined case least squares method according to [34]:

$$\mathbf{A_0}\hat{\boldsymbol{\delta}} + \mathbf{A_1}\hat{\mathbf{r}} + \mathbf{f_1} = 0$$
$$\hat{\mathbf{x}} = \mathbf{x} + \hat{\boldsymbol{\delta}} \tag{22.125}$$
$$\hat{\mathbf{l}} = \mathbf{l} + \hat{\mathbf{r}},$$

where $\mathbf{l}$ is the vector of observations: $\mathbf{l} = [\cdots \quad l_{f,x} \quad l_{f,y} \quad l_{f,z} \quad \cdots]^T$, $\mathbf{f}_1$ is the misclosure vector: $\mathbf{f}_1 = [\cdots \quad f(\underline{x}, \underline{l}) \quad \cdots]^T$, and the design matrices $\mathbf{A}_0$ and $\mathbf{A}_1$ for the accelerometers can be expressed as:

$$\mathbf{A}_0 = \begin{bmatrix} \cdots & \cdots & \cdots & \cdots & \cdots & \cdots \\ \dfrac{\delta f}{\delta b_{f,x}} & \dfrac{\delta f}{\delta b_{f,y}} & \dfrac{\delta f}{\delta b_{f,z}} & \dfrac{\delta f}{\delta s_x} & \dfrac{\delta f}{\delta s_y} & \dfrac{\delta f}{\delta s_z} \\ \cdots & \cdots & \cdots & \cdots & \cdots & \cdots \end{bmatrix}$$

$$\mathbf{A}_1 = \begin{bmatrix} \cdots & \cdots & \cdots & 0 & 0 & 0 & 0 & 0 & 0 \\ 0 & 0 & 0 & \dfrac{\delta f}{l_{f,x}} & \dfrac{\delta f}{l_{f,y}} & \dfrac{\delta f}{l_{f,z}} & 0 & 0 & 0 \\ 0 & 0 & 0 & 0 & 0 & 0 & \cdots & \cdots & \cdots \end{bmatrix}$$

(22.126)

and $\hat{\delta}$ and $\hat{\mathbf{r}}$ are the correction vector and the vector of residuals, respectively:

$$\hat{\delta} = \begin{bmatrix} \delta b_{f,x} & \delta b_{f,y} & \delta b_{f,z} & \delta s_x & \delta s_y & \delta s_z \end{bmatrix}^T$$

$$\hat{\mathbf{r}} = \begin{bmatrix} \cdots & \cdots & \cdots & r_{a,x} & r_{a,y} & r_{a,z} & \cdots & \cdots & \cdots \end{bmatrix}^T$$

(22.127)

The mathematical expression of the partial derivatives of (22.126) can be found in [33]. At each step of the least squares the correction vector $\hat{\delta}$ is estimated as

$$\mathbf{C} = \mathbf{A}_1 \mathbf{P}^{-1} \mathbf{A}_1^T$$

$$\mathbf{N} = \mathbf{A}_0^T \mathbf{C}^{-1} \mathbf{A}_0^T$$

$$\hat{\delta} = -\mathbf{N}^{-1} \mathbf{A}^T \mathbf{C}^{-1} \mathbf{f}_1,$$

(22.128)

where $\mathbf{P}$ is the measurements noise covariance matrix. Since the observations can be assumed uncorrelated, $\mathbf{P}$ can be written as a diagonal matrix $\mathbf{P} = \sigma^2 \mathbf{I}$, where σ^2 can be assumed to be the velocity random walk noise term characterized in the previous section. Then, the estimated vector of residuals $\hat{\mathbf{r}}$ is computed as:

$$\hat{\mathbf{r}} = -\mathbf{P}^{-1} \mathbf{A}_1^T \mathbf{C}^{-1} (\mathbf{f}_1 + \mathbf{A}_0 \hat{\delta}).$$

(22.129)

The principle of the least square is to minimize $\hat{\mathbf{r}}^T \mathbf{P} \hat{\mathbf{r}}$ by following an iterative approach. Thus, Equations (22.128) and (22.129) will be repeatedly calculated, and after a certain amount of iterations the term $\hat{\delta}$ will go to zero, and the estimated vector of residuals $\hat{\mathbf{r}}$ will keep stable.

In order to avoid singularities in the computation of (22.128), the IMU has to be placed at different attitudes, and the data storage of the accelerometers has to be executed for each IMU orientation. In detail, at least six different IMU attitude measurements are required in order to be able to estimate the bias and the scale factor. An example of six possible IMU orientations is depicted in Figure 22.16.

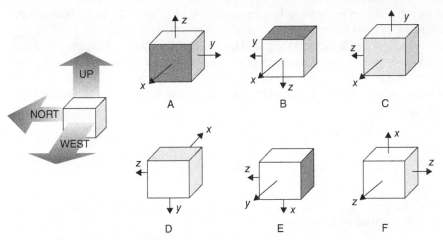

Figure 22.16   Sequence of different IMU orientations for accelerometer calibration.

Furthermore, the data corresponding to the accelerometer triad needs to be averaged within each attitude in order to reduce the effect of noise in the determination of the systematic parameters (e.g., bias and scale factors). Thus, it yields a 3×1 vector for each attitude that is then included in the observation vector $\mathbf{l}$, and the least squares algorithm can be executed.

Gyro Calibration: Gyro bias errors have an enormous impact on the overall performance of an INS system. In fact, a bias in the gyro output, when integrated, will cause an orientation error that grows linearly with time. Therefore, the gyro bias has to be compensated accurately.

In case of static IMU, a theoretical gyro model can be written as:

$$\omega = \omega_x^2 + \omega_y^2 + \omega_z^2 - \|\omega_{ie}\|^2 = 0. \tag{22.130}$$

However, in case of low-cost IMUs, the gyroscopes are not sensitive to the Earth rotation rate. Thus, the only possible way to estimate the bias is to assume that the gyro triad should be zero while the sensor is sitting stationary and not rotating. Therefore, Equation (22.130) can be approximated as:

$$\omega = \omega_x^2 + \omega_y^2 + \omega_z^2 \approx 0. \tag{22.131}$$

Thus, the gyro bias is simply obtained as the mean value of a static data collection of such sensor, i.e., $\hat{b}_{\omega,j} = \frac{1}{n}\sum_i \omega_{j,i}, \ j = \{x,y,z\}, i = 0, \cdots n-1$. It is important to remind that the gyro bias of an IMU is affected by a random walk component (i.e., *bias instability*) that cannot be calibrated. Such a component can be evaluated through noise characterization techniques, such as the one illustrated in Figure 22.11b and discussed in Section 22.5.2. In case of multisensor fusion, the random walk noise that afects both the gyros' and accelerometers' biases has to be considered as an extra state to be estimated by the centralized KF, as it will be described in the following of this chapter.

The gyro scale factor is measured by rotating the gyro around each sensor axis at known angular rates. A linear least squares, similar to the one used for accelerometer calibration, is then implemented to calculate the best fit linear scale factor to fit such data. In case of gyroscopes, the misclosure function can be written similarly to (22.124):

$$\omega = \left(\frac{l_{\omega,x} - \hat{b}_{\omega,x}}{1 + s_x} \right)^2 + \left(\frac{l_{\omega,y} - \hat{b}_{\omega,y}}{1 + s_y} \right)^2 + \left(\frac{l_{\omega,z} - \hat{b}_{\omega,z}}{1 + s_z} \right)^2 = 0, \qquad (22.132)$$

where the only three unknowns are the scale factor parameters, whilst $\hat{b}_{\omega,j}$ is the previously estimated gyro bias.

Temperature Effects on IMU Calibration: The errors of the low-cost sensors, especially the MEMS ones, change with time and are highly dependent on environmental conditions, and temperature in particular. In fact, the actual values of the deterministic errors can be different from those obtained in calibration due to the diference between the operational and calibration temperatures [17]. Therefore, the calibration process has to be repeated at different temperatures, for example by placing the IMU into a temperature-controlled chamber and repeating the calibration procedure at different temperatures. In this way one can obtain the trend of bias and scale factor for a certain temperature range. An example that shows how the accelerometers' bias and scale factor vary according to the internal temperature of the IMU is shown in Figure 22.17a and 22.17b, respectively: notice, for example, the variation of 0.2 g m/s^2 of the z-axis acceleration bias.

An interpolation polynomial function is a simple method to model the bias variation over the temperature by exploiting the discrete set of known data points, as shown in Figure 22.17a.

IMU Turn-On Bias: An additional effect that influences every IMU is the non-deterministic bias of accelerometers and gyros shift every time the device is switched on. An example of turn-on bias in accelerometers is depicted in Figure 22.18 in the case of a MEMS IMU. It can be only estimated at run time via software routines and possibly mitigated.

22.6.3   Initial Alignment

Before an INS system is ready to navigate, it must compute the displacement of the local frame with respect to the navigation frame. This operation consists of estimating the initial attitude angles on the basis of the acceleration measured by the accelerometers and of the measurements of other sensors (gyros, magnetometer, baro-altimeter, etc.).

One of the underlying assumptions used in the presentation of the mechanization equations as given in (22.98) and (2.103) was that the initial condition of the system (in terms of orientation) was already known. While position and velocity are

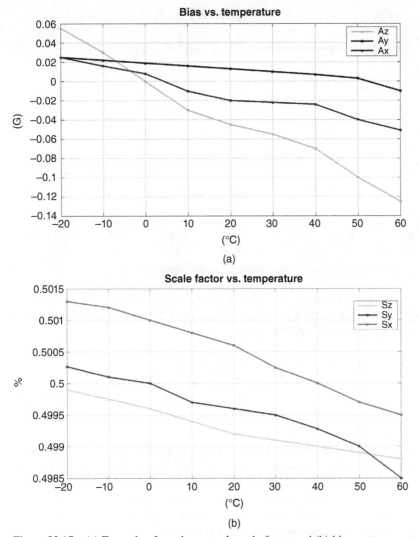

Figure 22.17 (a) Example of accelerometer's scale factor and (b) bias vs. temperature.

usually easily input by the user, the initial orientation of the system is not so readily available. For this reason, the INS usually has to execute an initial alignment procedure, typically performed in two steps: a *coarse alignment* followed by a *fine alignment*.

High-Grade IMU: High-grade IMUs include gyroscopes that are sensitive to the Earth rotation (i.e., their internal noise is significantly lower than the Earth rotation constant); in this case it is possible first to compute a *coarse estimation* of the three attitude angles (Euler angles), then to refine the initial alignment of such angles through a *fine alignment* process.

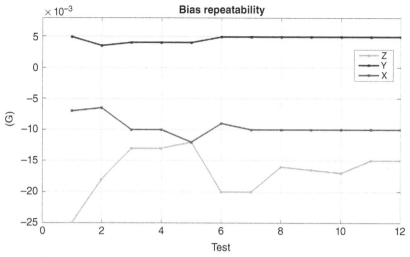

Figure 22.18 MEMS accelerometere. Example of bias repeatability.

Under static conditions, the Euler angles associated with the x- and y-axes of the body frame can be roughly estimated by exploiting the accelerometers; measurements as in the following equations [16]:

$$
\begin{aligned}
\hat{\phi} &= -\sin^{-1}\left(\frac{\overline{a_x^b} \cdot \Delta t}{\gamma T_{align}} \right) \\
\hat{\theta} &= \sin^{-1}\left(\frac{\overline{a_y^b} \cdot \Delta t}{\gamma T_{align}} \right)
\end{aligned}
\tag{22.133}
$$

where $\overline{a^b}$ is the time averaged acceleration along the specified axis, Δt is the time at which the accelerometer measurement is provided by the IMU (e.g., 100 Hz), $\gamma = \sqrt{a_x^2 + a_y^2 + a_z^2}$ is the squared root of the accelerometers that is equal to g according to (22.122), and T_{align} is the time interval devoted to averaging (e.g., 90 s). Then, the coarse estimation of the third angle about the z-axis can be obtained by rotating the average angular increments into a horizontal frame as follows [16]:

$$
\overline{\theta_{ib}^h} = \mathbf{C}_1(-\hat{\theta}) \cdot \mathbf{C}_2(-\hat{\phi}) \cdot \overline{\theta_{ib}^b},
\tag{22.134}
$$

where $\theta_{ib}^b = \omega_{ib}^b \cdot t$, $\mathbf{C}_1 = \begin{bmatrix} 1 & 0 & 0 \\ 0 & \cos\hat{\theta} & -\sin\hat{\theta} \\ 0 & \sin\hat{\theta} & \cos\hat{\theta} \end{bmatrix}$ and $\mathbf{C}_2 = \begin{bmatrix} \cos\hat{\phi} & 0 & \sin\hat{\phi} \\ 0 & 1 & 0 \\ -\sin\hat{\phi} & 0 & \cos\hat{\phi} \end{bmatrix}$.

Then a coarse estimation of the DCM (22.98) can be computed.

Such matrix can be further refined by implementing a fine alignment process. It is based on updates of an EKF that estimates the error states of a system; in Section 22.7 we will introduce a system model typically adopted in such case (22.162) and (22.163). The so-called zero velocity updates (ZUPT) are the most common observations used in the fine alignment process.

Low-Grade IMU: In case of MEMS IMUs, a self-alignment can just provide a coarse estimation of two angles, as the gyroscopes cannot be used to estimate the rotation about the z-axis, MEMS sensors being unable to sense the rotation of the Earth [32]. Thus, the first two Euler angles can be estimated under static conditions using (22.133). However, since a valid third angle cannot be self-estimated, it is necessary to resort to external information; this is typically represented by the North and East velocity components computed by a GNSS receiver when the vehicle starts moving. In such a case the third angle can be simply estimated as

$$\widehat{\psi} = \tan^{-1}\left(\frac{v_{East}}{v_{North}}\right) \tag{22.135}$$

It is important to stress that (22.135) becomes meaningless when the speed of the vehicle is close to zero.

A fine alignment of the coarse DCM obtained by the angles (22.133) and (22.135) is possible using the mentioned EKF with ZUPT measurements.

Example 22.4

This example explains the process of data collection of the accelerometers of an IMU. This process is required for the horizontal alignment of the platform. This means that we compute the roll and pitch angles with respect to the East-North-up (ENU) frame.

At this stage the vehicle is still, mainly for two reasons:

- a mean of the output of each accelerometer can be computed in a predetermined time gap. This operation allows to soften the effect of measurement noise;

- the accelerometers measure *only* the reaction to the gravity acceleration; It is then possible to determine the displacement of the local frame with respect to the navigation frame.

In this example **a** is the three-component vector obtained as an arithmetic mean of the output of the accelerometers. The entries of this vector are the means obtained for the x-, the y- and the z-axis accelerometers, respectively. We focus on the operation of *horizontal alignment* or *leveling*.

In real environments, due to the presence of noise and because gravity is not constant on the surface of the Earth, the modulus of **a** is not determined a priori. For ease of comprehension, in this example the modulus of the mean acceleration vector is equal to one.

The column vector $\mathbf{a} = [a_x; a_y; a_z]^T = [-1\sqrt{2}; 0.5; 0.5]^T$ is considered. This vector is used to determine the displacement of the IMU with respect to the horizontal plane.

The reader must determine the roll and pitch angles to level the IMU. Once the platform is horizontally aligned, the vector a has a nonnull entry only along the vertical axis (in this case the up axis), i.e., $\mathbf{a}_{ENU}[0; 0; 1]^T$.

Solutionw

In Appendix A.3 the formulas considered in this example are implemented in a MATLAB script. Since this script runs using real data, an initial averaging of the outputs of the accelerometers is applied to reduce the noise effects (code Block Nr.1). No particular toolbox or command have been used in the code. The reader is suggested to start from this simple example to get in touch with the notions required to handle real data.

At first, we rotate the reference system around the roll axis (x). To this purpose, we consider the components of the acceleration along the y- and z-axes.

Our objective is to rotate the local frame around the roll axis, so that after this rotation the component of the **a** vector along the yz plane is only along the z-axis. The reader can refer to Figure 22.19.

It is then necessary to estimate the roll angle ϕ (code Block Nr.2):

$$\hat{\phi} = -\sin^{-1}\left(\frac{a_y}{\sqrt{a_x^2 + a_y^2 + a_z^2}}\right), \tag{22.136}$$

tanking into account the ambiguity of the $\sin^{-1}$ function.

Considering the given vector **a**, the estimated roll angle will be equal to $-\pi/4$, so that the reference system must be rotated by $\pi/4$.

Now the pitch angle (around y) is estimated in a similar way. The acceleration is now distributed on the x and z accelerometers, since a first rotation has been performed. We can avoid computing a midway rotation, estimating the pitch angle θ starting from the initial data (code Block Nr.3):

$$\hat{\theta} = \sin^{-1}\left(\frac{a_x}{\sqrt{a_x^2 + a_y^2 + a_z^2}}\right). \tag{22.137}$$

Also in this case, the estimated angle will be equal to $-\pi/4$ and the reference system must be rotated by $\pi/4$.

Now that we found the roll and pitch value, we can use these values to compute a DCM matrix (already discussed in this chapter) to be applied to the vector **a**. This operation will provide the vector $\mathbf{a}_{ENU} = [0; 0; 1]$, proving that the local frame has been aligned with the ENU frame.

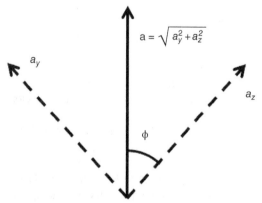

Figure 22.19   Vectors involved in the first rotation.

22.7 GENERAL ARCHITECTURE FOR THE LOOSE INTEGRATION

The three hybridization architectures between GNSS and INS cited in the introduction of this chapter share in principle the same basic state-space model of the involved quantities. Minor differences are introduced in the number of the involved states and in the forcing functions models; on the other hand, substantial differences determine the three observation models, which lead to substantially different integration strategies.

A loosely-integrated GNSS+INS system uses the *trajectory* measured from the GNSS receiver to compute the *corrections* to be applied to the trajectory estimated by the INS computer and to estimate, if necessary, the biases that affect the accelerometers and the gyroscopes. Loose integration is based on the definition of a state-space model of the hybrid system and the application of an EKF to compute the corrections necessary to refine the INS-based trajectory.

It is worth noticing that *in the loose integration case the GNSS information is used as a refinement of the INS information:* the GNSS information is used to counteract the intrinsic derivation of the INS solution, correcting the INS trajectory. Such corrections are computed via an EKF, whose state-space model is described hereafter.

The architecture of a loose integration can be represented as in Figure 22.20: the IMU measurements are used to predict the nominal position, velocity, and attitude by means of the navigation equations in the INS block. The INS-predicted nominal trajectory is then employed to feed the KF, which uses the differences in trajectory (GPS-measured minus nominal INS-predicted) as its observations.

In the following, the body trajectory will be expressed in the ECEF frame and every vector in this frame will be labeled with the superscript^e. Other references such

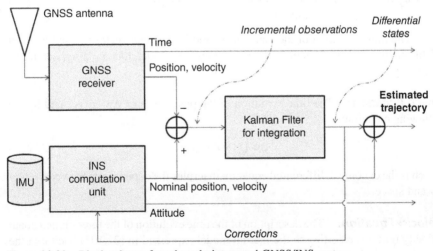

Figure 22.20 Block scheme for a loosely integrated GNSS/INS system.

ENU, North-East-down (NED) or wander azimuth can be adopted. The IMU outputs are given in the body frame, labeled by the superscript[b].

22.7.1 Loose Integration: State-Space Model

The definition of the state-space model needed to implement a loosely integrated system for GNSS-INS hybridization is shown hereafter. This model is basically common for *all* the integration architectures, therefore, it will be recalled with due minor changes in Section 22.8 and 22.10. The model equations are initially written in the continuous-time domain, then they are translated in the discrete-time domain using the methods discussed in Section 22.2.

In order to implement a KF-based integration architecture, the set of the system states *(incremental states)* has to be firstly identified. The structure adopted here is the following:

$$\Delta \mathbf{x}(t) = [\Delta \mathbf{p}^e(t)^T, \Delta \mathbf{v}^e(t)^T, \Delta \boldsymbol{\psi}^e(t)^T, \mathbf{b_a}^b(t)^T, \mathbf{b_g}^b(t)^T]^T \in \mathbb{R}^{15,1}, \quad (22.138)$$

where the superscripts ^{e,b} indicates Earth frame and body frame, respectively. The state vector store the following components:

$\Delta \mathbf{p}^e(t) \in \mathbb{R}^{3,1}$ is the *corrections vector* to be applied to the *nominal body position* at the time instant t, expressed in the Earth frame;

$\Delta \mathbf{v}^e(t) \in \mathbb{R}^{3,1}$ is the *corrections vector* to be applied to the *nominal body velocity* at the time instant t, expressed in the Earth frame;

$\Delta \boldsymbol{\psi}^e(t) \in \mathbb{R}^{3,1}$ is the vector of misalignment angles along each axis (*attitude corrections*) at the time instant t, expressed in the Earth frame;

$\mathbf{b_a}^b(t) \in \mathbb{R}^{3,1}$ is the vector of the biases of the accelerometers at the time instant t, expressed in the body frame;

$\mathbf{b_g}^b(t) \in \mathbb{R}^{3,1}$ is the vector of the biases of the gyroscopes at the time instant t, expressed in the body frame;

The time evolution of the elements stored in the state vector is ruled by a set of differential equations, namely the *state-transition model*, that are discussed in the following.

Space Equation: The time evolution of the position error $\Delta \mathbf{p}^e(t)$ is ruled by the following equation:

$$\Delta \dot{\mathbf{p}}^e(t) = \Delta \mathbf{v}^e(t), \quad (22.139)$$

which is the canonical differential equation that rules the dependency between velocity and space.

Velocity Equation: The description of the time evolution of the three-dimensional velocity requires resorting to the inertial model expressed in (22.103). Defining the velocity correction $\Delta \mathbf{v}^e(t)$ as the difference between the true velocity and the nominal velocity, obtained from the INS, i.e., $\Delta \mathbf{v}^e(t) = \breve{\mathbf{v}}^e(t) - \breve{\mathbf{v}}^e(t)$, Equation (22.103) allows expressing the time evolution of the velocity correction as

$$\Delta \dot{\mathbf{v}}^e(t) = \left[\mathbf{C}_b^e \mathbf{f}^b - \breve{\mathbf{C}}_b^e \breve{\mathbf{f}}^b \right] - \left[2\Omega_{ie}^e \mathbf{v}^e(t) - 2\breve{\Omega}_{ie}^e \breve{\mathbf{v}}^e(t) \right] + \left[\mathbf{g}_\ell - \breve{\mathbf{g}}_\ell \right],$$

where the symbol $\breve{}$ refers to quantities provided by the INS (and taken as nominal). The misalignment $\Delta \psi^e(t)$ that affects the measured DCM $\breve{\mathbf{C}}_b^e$ with respect to the actual DCM $\mathbf{C}_b^e$ relates the two as follows [14]:

$$\breve{\mathbf{C}}_b^e = \left[\mathbf{I}_{3\times 3} - \Delta \psi^e(t) \times \right] \mathbf{C}_b^e. \tag{22.140}$$

Substituting (22.140) in (22.103) one obtains:

$$\begin{aligned} \Delta \dot{\mathbf{v}}^e(t) &= \mathbf{C}_b^e \mathbf{f}^b - \mathbf{C}_b^e \breve{\mathbf{f}}^b + \Delta \psi^e(t) \times \mathbf{C}_b^e \breve{\mathbf{f}}^b - 2\Omega_{ie}^e \Delta \mathbf{v}^e(t) + \Delta \mathbf{g}_\ell(t) \\ &= \mathbf{C}_b^e \left(\mathbf{f}^b - \breve{\mathbf{f}}^b \right) + \Delta \psi^e(t) \times \breve{\mathbf{f}}^e - 2\Omega_{ie}^e \Delta \mathbf{v}^e(t) + \Delta \mathbf{g}_\ell(t) \\ &= \mathbf{C}_b^e \mathbf{b}_\mathbf{a}^b(t) - \mathbf{F}\Delta \psi^e(t) - 2\Omega_{ie}^e \Delta \mathbf{v}^e(t) + \Delta \mathbf{g}_\ell(t), \end{aligned} \tag{22.141}$$

where the bias on the accelerometers has been expressed as $\mathbf{b}_\mathbf{a}^b(t) = \mathbf{f}^b - \breve{\mathbf{f}}^b$, the difference $\Delta \Omega_{ie}^e = \Omega_{ie}^e - \breve{\Omega}_{ie}^e = 0$ due to the knowledge of the Earth rotation rate, and the mathematical equality (22.96) has been used to express $\Delta \psi^e(t) \times \breve{\mathbf{f}} = -\mathbf{F}\Delta \psi^e(t)$.

As far as the term $\Delta \mathbf{g}_\ell(t) = \mathbf{g}_\ell(t) - \breve{\mathbf{g}}_\ell(t)$ is concerned, it can be written explicitly by resorting to the definition of the local gravity component, Equation (22.102). First, the dependence of the gravity vector $\mathbf{g}(t)$ on the position of the body with respect to the Earth can be expressed through the *tensor of the gravitational field* Υ, namely:

$$\mathbf{g}(t) = \Upsilon \mathbf{p}^e(t), \tag{22.142}$$

where Υ is assumed constant in time. Then, from (22.102) it is possible to write

$$\Delta \mathbf{g}_\ell(t) = \Upsilon \Delta \mathbf{p}^e(t) - \Omega_{ie}^e \Omega_{ie}^e \Delta \mathbf{p}(t) = \mathbf{N}^e \Delta \mathbf{p}^e(t). \tag{22.143}$$

The term $\mathbf{N}^e \triangleq \Upsilon - \Omega_{ie}^e \Omega_{ie}^e$ is called *the tensor of gravity gradients* [16], which includes the gravitational and centripetal acceleration components, so that $\mathbf{N}^e \Delta \mathbf{p}^e(t)$ is the position-dependent *local gravity component perturbation* affecting the acceleration. An expression for $\mathbf{N}^e$ can be found in [16].

This way, the differential equation governing the evolution of the incremental velocity $\Delta \mathbf{v}^e(t)$ can be written as:

$$\Delta \dot{\mathbf{v}}^e(t) = \mathbf{N}^e \Delta \mathbf{p}^e(t) - 2\Omega_{ie}^e \Delta \mathbf{v}^e(t) - \mathbf{F}\Delta \psi^e(t) + \mathbf{C}_b^e \mathbf{b}_\mathbf{a}^b(t) + \mathbf{C}_b^e \boldsymbol{\eta}_a(t), \tag{22.144}$$

where $\boldsymbol{\eta}_a(t)$ is a driving noise term acting on the accelerometers in the body frame.

Attitude Misalignment Equation: Equation (22.140) is the starting point for the derivation of the attitude misalignment equation; it can be rearranged as follows:

$$\Delta \psi^e(t) \times = \mathbf{I}_{3\times 3} - \breve{\mathbf{C}}_b^e \mathbf{C}_b^{eT}, \tag{22.145}$$

where it has been exploited the property of every DCM matrix to be orthonormal. Differentiating (22.145) yields:

$$\Delta\dot{\psi}^e(t)\times = -\breve{\mathbf{C}}_b^e\dot{\mathbf{C}}_b^{eT} - \dot{\breve{\mathbf{C}}}_b^e\mathbf{C}_b^{eT}. \tag{22.146}$$

Recalling (22.92) and substituting the appropriate quantities in (22.146), we write

$$\begin{aligned}
\Delta\dot{\psi}^e(t)\times &= -\breve{\mathbf{C}}_b^e\left[\mathbf{C}_b^e\Omega_{eb}^b\right]^T - \left[\breve{\mathbf{C}}_b^e\breve{\Omega}_{eb}^b\right]\mathbf{C}_b^{eT} \\
&= \hat{\mathbf{C}}_b^e\left[\Omega_{eb}^b - \breve{\Omega}_{eb}^b\right]\mathbf{C}_b^{eT} \\
&= \hat{\mathbf{C}}_b^e\Delta\Omega_{eb}^b\mathbf{C}_b^{eT},
\end{aligned} \tag{22.147}$$

where the skew-symmetric matrix property $\mathbf{A}^T = -\mathbf{A}$ has been exploited. Inserting (22.140) into the last row of (22.147) yields:

$$\begin{aligned}
\Delta\dot{\psi}^e(t)\times &= \left[\mathbf{I}_{3\times3} - \Delta\psi^e(t)\times\right]\mathbf{C}_b^e\Delta\Omega_{eb}^b\mathbf{C}_b^{eT} \\
&= \mathbf{C}_b^e\Delta\Omega_{eb}^b\mathbf{C}_b^{eT} - \Delta\psi^e(t)\times\mathbf{C}_b^e\Delta\Omega_{eb}^b\mathbf{C}_b^{eT} \\
&\approx \mathbf{C}_b^e\Delta\Omega_{eb}^b\mathbf{C}_b^{eT}
\end{aligned} \tag{22.148}$$

since the products between incremental corrections result in small second-order terms, which can be ignored. It is possible to observe that (22.148) can be expressed in vector form as [15]:

$$\Delta\dot{\psi}^e(t) = \mathbf{C}_b^e\Delta\omega_{eb}^b. \tag{22.149}$$

The rotation between the Earth and the body frame can be regarded as the sum of two contributions, namely the rotation of the body with respect to the inertial frame and the rotation of the ECEF with respect to the inertial frame, leading to:

$$\omega_{eb}^b = \omega_{ib}^b - \mathbf{C}_e^b\omega_{ie}^e. \tag{22.150}$$

Using (22.150) for both the true and the nominal rotations and recalling (22.140), it is easy to write

$$\begin{aligned}
\Delta\omega_{eb}^b &= \Delta\omega_{ib}^b - \left(\mathbf{C}_e^b - \breve{\mathbf{C}}_e^b\right)\omega_{ie}^e \\
&= \Delta\omega_{ib}^b - \left(\Delta\psi^e\times\mathbf{C}_b^e\right)\omega_{ie}^e,
\end{aligned} \tag{22.151}$$

where the Earth rotation term ω_{ie}^e has been considered perfectly known. Equation (22.151) can now be substituted into (22.149). Exploiting the properties of the skew-symmetric matrices, it is possible to obtain:

$$\begin{aligned}
\Delta\dot{\psi}^e(t) &= \mathbf{C}_b^e\left[\Delta\omega_{ib}^b - (\Delta\psi^e\times\mathbf{C}_b^e)\omega_{ie}^e\right] \\
&= \mathbf{C}_b^e\left[\Delta\omega_{ib}^b + \mathbf{C}_e^b\Delta\psi^e\times\omega_{ie}^e\right] \\
&= \mathbf{C}_b^e\Delta\omega_{ib}^b + \Delta\psi^e\times\omega_{ie}^e \\
&= \mathbf{C}_b^e\Delta\omega_{ib}^b - \Omega_{ie}^e\Delta\psi^e.
\end{aligned} \tag{22.152}$$

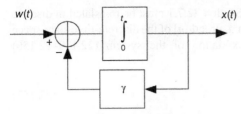

Figure 22.21   Block scheme for a GM continuous-time system.

Now, expressing the bias on the gyroscopes as $\mathbf{b_g}^b(t) = -\Delta\omega_{ib}^b$ and considering the presence of a driving noise $\mathbf{\eta}_g(t)$ on the gyros themselves, the differential equation governing the attitude misalignment is written as:

$$\Delta\dot{\psi}^e(t) = -\Omega_{ie}^e\Delta\psi^e(t) - \mathbf{C}_b^e\mathbf{b_g}^b(t) - \mathbf{C}_b^e\mathbf{\eta}_g(t). \tag{22.153}$$

Accelerometers Bias Equation: The accelerometers; bias vector $\mathbf{b_a}^b(t)$ is modeled as a continuous-time GM process:

$$\dot{\mathbf{b}}_a^b(t) = \mathbf{D}_a\mathbf{b}_a^b(t) + \mathbf{\eta}_{aa}(t), \tag{22.154}$$

where $\mathbf{D}_a$ is the time-constant diagonal matrix that defines a first-state GM model and $\mathbf{\eta}_{aa}(t)$ is the driving noise for the gyro biases, as represented in Figure 22.21 [16]. The matrix $\mathbf{D}_a$ stores in its diagonal the time constants γ_a of the GM model:

$$[\mathbf{D}_a]_{ij} = \begin{cases} e^{-\gamma_a} - 1 \approx -\gamma_a, & i = j \\ 0, & i \neq j. \end{cases} \tag{22.155}$$

The noise $\mathbf{\eta}_{aa}(t)$ is an AWGN process with zero mean and variance $\sigma_{aa}^2 = 2\sigma_{ba}^2\gamma_a$, where σ_{ba}^2 is the variance of the GM process.

Gyroscopes Bias Equation: Analogously to $\mathbf{b_a}^b(t)$, the gyroscopes bias vector $\mathbf{b_g}^b(t)$ is modeled as a continuous-time GM process:

$$\dot{\mathbf{b}}_g^b(t) = \mathbf{D}_g\mathbf{b}_g^b(t) + \mathbf{\eta}_{gg}(t), \tag{22.156}$$

where the matrix $\mathbf{D}_g$ stores in its diagonal the time constants, $-\gamma_g$, of the GM model. The noise $\mathbf{\eta}_{gg}(t)$ is an AWGN process with zero mean and variance $\sigma_{gg}^2 = 2\sigma_{bg}^2\gamma_g$, being σ_{bg}^2 the variance of the GM process.

22.7.2   Loose Integration: State Transition Matrix

Once the equations that describe the system are determined, it is necessary to provide the definition of the equivalent problem in the discrete-time domain, so as to implement in software the discrete-time version of the complementary KF discussed at the beginning of this chapter.

The differential equations discussed in Section 22.7.1 must be translated in discrete time in order to define the state transition matrix $\Phi[n]$ of the discrete-time state-space model. Thus, the discrete-time approximation of the system (22.139–22.156) becomes

$$\Delta \mathbf{p}^e[n+1] = \Delta \mathbf{p}^e[n] + T_c \Delta \mathbf{v}^e[n]; \tag{22.157}$$

$$\Delta \mathbf{v}^e[n+1] = \mathbf{N}^e \Delta \mathbf{p}^e[n] + \left(\mathbf{I}_3 - 2T_c \Omega_{ie}^e \right) \Delta \mathbf{v}^e[n] + \\ - T_c \mathbf{F}[n] \Delta \mathbf{\psi}^e[n] + T_c \mathbf{C}_b^e[n] \mathbf{b}_{\mathbf{a}}^b[n] \qquad + T_c \mathbf{C}_b^e[n] \mathbf{\eta}_a[n]; \tag{22.158}$$

$$\Delta \mathbf{\psi}^e[n+1] = \left(\mathbf{I}_3 - T_c \Omega_{ie}^e \right) \Delta \mathbf{\psi}^e[n] - T_c \mathbf{C}_b^e[n] \mathbf{b}_{\mathbf{g}}^b[n] \quad - T_c \mathbf{C}_b^e[n] \mathbf{\eta}_g[n]; \tag{22.159}$$

$$\mathbf{b}_{\mathbf{a}}^b[n+1] = \left(\mathbf{I}_3 + T_c \mathbf{D}_a \right) \mathbf{b}_{\mathbf{a}}^b[n] \qquad\qquad + T_c \mathbf{\eta}_{aa}[n]; \tag{22.160}$$

$$\mathbf{b}_{\mathbf{g}}^b[n+1] = \left(\mathbf{I}_3 + T_c \mathbf{D}_g \right) \mathbf{b}_{\mathbf{g}}^b[n] \qquad\qquad + T_c \mathbf{\eta}_{gg}[n], \tag{22.161}$$

so that the discrete-time state-space model is written as

$$\Delta \mathbf{x}[n+1] = \Phi[n] \Delta \mathbf{x}[n] + \Gamma[n] \mathbf{\eta}[n], \tag{22.162}$$

where:

$$\mathbf{\eta}[n] = \left[\mathbf{\eta}_a (nT_c)^T, \mathbf{\eta}_g (nT_c)^T, \mathbf{\eta}_{aa} (nT_c)^T, \mathbf{\eta}_{gg} (nT_c)^T \right]^T \in \mathbb{R}^{12,1}; \tag{22.163}$$

$$\Phi[n] = \begin{bmatrix} \mathbf{I}_3 & T_c \mathbf{I}_3 & \mathbf{0} & \mathbf{0} & \mathbf{0} \\ \mathbf{N}^e & \mathbf{I}_3 - 2T_c \Omega_{ie}^e & -T_c \mathbf{F}[n] & T_c \mathbf{C}_b^e[n] & \mathbf{0} \\ \mathbf{0} & \mathbf{0} & \mathbf{I}_3 - T_c \Omega_{ie}^e & \mathbf{0} & -T_c \mathbf{C}_b^e[n] \\ \mathbf{0} & \mathbf{0} & \mathbf{0} & \mathbf{I}_3 + T_c \mathbf{D}_a & \mathbf{0} \\ \mathbf{0} & \mathbf{0} & \mathbf{0} & \mathbf{0} & \mathbf{I}_3 + T_c \mathbf{D}_g \end{bmatrix} \in \mathbb{R}^{15,15}; \tag{22.164}$$

$$\Gamma[n] = \begin{bmatrix} \mathbf{0} & \mathbf{0} & \mathbf{0} & \mathbf{0} \\ T_c \mathbf{C}_b^e[n] & \mathbf{0} & \mathbf{0} & \mathbf{0} \\ \mathbf{0} & -T_c \mathbf{C}_b^e[n] & \mathbf{0} & \mathbf{0} \\ \mathbf{0} & \mathbf{0} & T_c \mathbf{I}_3 & \mathbf{0} \\ \mathbf{0} & \mathbf{0} & \mathbf{0} & T_c \mathbf{I}_3 \end{bmatrix} \in \mathbb{R}^{15,12}. \tag{22.165}$$

Note that, comparing (22.162) with (22.4), here the stochastic forcing function is defined as $\mathbf{w}[n] = \Gamma[n] \mathbf{\eta}[n]$, so that the model noise covariance matrix $\mathbf{Q}[n]$ in (22.27) and (22.28) must be modified accordingly:

$$\mathbf{Q}[n] = \Phi[n] \Gamma[n] \mathbf{Q}_\omega \Gamma^T[n] \Phi^T[n] T_c.$$

22.7.3   Loose Integration: Measurement Equation

Differently than the state transition model derived in the previous subsections, the observation model defined for the loosely integrated architecture is peculiar for this specific approach.

The incremental observation vector $\Delta \mathbf{z}[n] = \mathbf{z}[n] - \breve{\mathbf{z}}[n]$ in (22.44) for the complementary KF used to implement the loose integration is defined through the terms:

$\mathbf{z}[n] = \left[\mathbf{p}^e[n]^T \, \mathbf{v}^e[n]^T \right]^T \in \mathbb{R}^{6,1}$, vector of the GNSS-measured trajectory at the time instant n;

$\breve{\mathbf{z}}[n] = \left[\breve{\mathbf{p}}^e[n]^T \, \breve{\mathbf{v}}^e[n]^T \right]^T \in \mathbb{R}^{6,1}$, INS-estimated trajectory at time n.

The predicted incremental observation $\Delta \mathbf{z}^-[n]$ relative to the state-space model is obtained from the a priori incremental state $\Delta \mathbf{x}^-[n] = \mathbf{\Phi}[n] \Delta \mathbf{x}[n-1]$ as

$$\Delta \mathbf{z}^-[n] = \mathbf{H}_{lo}[n] \Delta \mathbf{x}^-[n], \tag{22.166}$$

where for this problem the observation matrix $\mathbf{H}_{lo}[n]$ is constant in time and evidently equal to:

$$\mathbf{H}_{lo} = \begin{bmatrix} \mathbf{I}_3 & \mathbf{0}_3 & \mathbf{0}_3 & \mathbf{0}_{3\times 6} \\ \mathbf{0}_3 & \mathbf{I}_3 & \mathbf{0}_3 & \mathbf{0}_{3\times 6} \end{bmatrix} \in \mathbb{R}^{6,15}. \tag{22.167}$$

Therefore, it is possible to compute the innovation $\boldsymbol{\alpha}[n]$ in (22.50) as:

$$\boldsymbol{\alpha}[n] = (\mathbf{z}[n] - \breve{\mathbf{z}}[n]) - \mathbf{H}_{lo} \Delta \mathbf{x}^-[n] \tag{22.168}$$

With the above definitions, the complementary KF iterations outlined in Section 22.2.4 can be used to implement the loose integration algorithm.

Example 22.5

This example explains the main steps to implement a loose integration between INS and GNSS modules. The reader is requested to implement a MATLAB routine to calculate the attitude, the velocity, and the position components by combining the measurement the comes from such two devices.

Solution

In Appendix A.5 the formulas considered in this example are implemented in a MATLAB script. The INS/GPS loosely coupled algorithm is structured as in the following;

- The implementation is divided into three parts that are executed at different rates. In the high-rate part (e.g., 100 Hz) the reader has to implement the coning and sculling computation, as well as to update the antivibration filter as explained in Appendix A.3. At the medium rate (e.g., 10–20 Hz) we have the implementation of the INS navigation equations as well as the update of the KF model. Eventually, at low rate, (1–2 Hz) we have the coupling of the INS measurements (velocity and position) with the ones coming from the GPS module.

- As far as the KF prediction is concerned, the reader has to implement the equations described within Section 22.7.2. An example of implementation is provided in MATLAB.

- At low rate, the information coming from GPS and INS are coupled together. Before the integration, the distance between the IMU and the GPS antenna is taken into account by the lever-arm parameter. The loosely coupled provided within the MATLAB script follow the equations detailed in Section 22.7.3.

22.8 GENERAL ARCHITECTURE FOR THE TIGHT INTEGRATION

A tightly integrated system uses the *pseudorange and pseudorange rate information* extracted from the GNSS receiver to compute the corrections to be applied to the trajectory estimated by the INS computer and to estimate, if necessary, the biases that affect the accelerometers and the gyroscopes. Tight integration is based again on the definition of a state-space model of the hybrid system and the application of an EKF to compute the corrections necessary to refine the INS-based nominal trajectory.

Similarly to the loose integration, in the tight integration case the GNSS information is used as a refinement of the INS information; also in this case the GNSS information is used to counteract the intrinsic derivation of the INS solution, correcting the INS trajectory. Such corrections are computed via an EKF, whose state-space model is described hereafter.

The architecture of a tight integration can be represented as in Figure 22.22: the IMU measurements are used to predict the nominal position, velocity, and

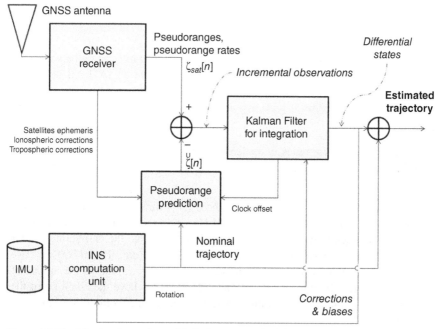

Figure 22.22 Block scheme for a tightly integrated GNSS/INS system.

attitude by means of the navigation equations in the INS block. The nominal trajectory is then employed to predict the pseudoranges of all the visible satellites. The KF uses the pseudoranges and pseudorange rates incremental observations (GNSS-measured minus INS-predicted) as its measurement inputs.

22.8.1 Tight Integration: State-Space Model

As done in the previous section, in order to implement a tightly integrated architecture, the set of the system states (*incremental states*) has to be firstly identified. The structure considered here is the following:

$$\Delta \mathbf{x}(t) = [\Delta \mathbf{p}^e(t)^T, \Delta\tau(t), \Delta \mathbf{v}^e(t)^T, \Delta v_\tau(t), \Delta \mathbf{\psi}^e(t)^T, \mathbf{b}_a^b(t)^T, \mathbf{b}_g^b(t)^T]^T \in \mathbb{R}^{17,1}. \quad (22.169)$$

It can be easily noticed that the state vector (22.169) for tightly integrated systems is identical to (22.138), previously set for the loosely integrated system, plus two additional components:

$\Delta\tau$ $(t) \in \mathbb{R}^{1,1}$ is the receiver clock bias expressed in meters;

Δv_τ $(t) \in \mathbb{R}^{1,1}$ is the receiver clock drift expressed in meters per second.

The time evolution of the new elements stored in the state vector is ruled by the differential equations presented hereafter.

Clock Misalignment Equation: The time evolution of the clock misalignment error $\Delta\tau(t)$ is ruled by:

$$\Delta\dot{\tau}(t) = \tau v_\tau(t), \quad (22.170)$$

whose differential structure is consistent with (22.139).

Clock Drift Equation: The clock drift error $\Delta v_\tau(t)$ evolution is described by the following equation:

$$\Delta\dot{v}(t) = v_\tau(t), \quad (22.171)$$

where $v_\tau(t)$ is a white noise. Equation (22.171) defines, therefore, a random walk upon clock drift.

22.8.2 Tight Integration: State Transition Matrix

The tightly integrated system is characterized by the same transition matrix of the loosely integrated system plus two more lines to characterize the clock bias and drift. The discrete-time domain equations are defined starting from (22.170) and (22.171):

$$\Delta\tau[n+1] = \Delta\tau[n] + \Delta v_\tau[n]T_c \quad (22.172)$$

$$\Delta v_\tau[n+1] = \Delta v_\tau[n] + v_\tau T_c. \quad (22.173)$$

It is now easy to determine the matrices we are interested in:

$$
\Phi[n]=
\begin{bmatrix}
\mathbf{I}_3 & \mathbf{0}_{3\times1} & T_c\mathbf{I}_3 & \mathbf{0}_{3\times1} & \mathbf{0}_3 & \mathbf{0}_3 & \mathbf{0}_3 \\
\mathbf{0}_{1\times3} & 1 & \mathbf{0}_{1\times3} & T_c & \mathbf{0}_{1\times3} & \mathbf{0}_{1\times3} & \mathbf{0}_{1\times3} \\
\mathbf{N}^e & \mathbf{0}_{3\times1} & \mathbf{I}_3-2T_c\Omega^e_{ie} & \mathbf{0}_{3\times1} & -T_c\mathbf{F}[n] & T_c\mathbf{C}^e_b[n] & \mathbf{0}_3 \\
\mathbf{0}_{1\times3} & 0 & \mathbf{0}_{1\times3} & 1 & \mathbf{0}_{1\times3} & \mathbf{0}_{1\times3} & \mathbf{0}_{1\times3} \\
\mathbf{0}_3 & \mathbf{0}_{3\times1} & \mathbf{0}_3 & \mathbf{0}_{3\times1} & \mathbf{I}_3-T_c\Omega^e_{ie} & 0 & -T_c\mathbf{C}^e_b[n] \\
\mathbf{0}_3 & \mathbf{0}_{3\times1} & \mathbf{0}_3 & \mathbf{0}_{3\times1} & \mathbf{0}_3 & \mathbf{I}_3+T_c\mathbf{D}_a & \mathbf{0}_3 \\
\mathbf{0}_3 & \mathbf{0}_{3\times1} & \mathbf{0}_3 & \mathbf{0}_{3\times1} & \mathbf{0}_3 & \mathbf{0}_3 & \mathbf{I}_3+T_c\mathbf{D}_g
\end{bmatrix},
\tag{22.174}
$$

where $\Phi[n]\in\mathbb{R}^{17,17}$, and

$$
\Gamma[n]=
\begin{bmatrix}
\mathbf{0}_{3\times1} & \mathbf{0}_3 & \mathbf{0}_3 & \mathbf{0}_3 & \mathbf{0}_3 \\
0 & \mathbf{0}_{1\times3} & \mathbf{0}_{1\times3} & \mathbf{0}_{1\times3} & \mathbf{0}_{1\times3} \\
\mathbf{0}_{3\times1} & T_c\mathbf{C}^e_b[n] & \mathbf{0}_3 & \mathbf{0}_3 & \mathbf{0}_3 \\
T_c & \mathbf{0}_{1\times3} & \mathbf{0}_{1\times3} & \mathbf{0}_{1\times3} & \mathbf{0}_{1\times3} \\
\mathbf{0}_{3\times1} & \mathbf{0}_3 & -T_c\mathbf{C}^e_b[n] & \mathbf{0}_3 & \mathbf{0}_3 \\
\mathbf{0}_{3\times1} & \mathbf{0}_3 & \mathbf{0}_3 & T_c\mathbf{I}_3 & \mathbf{0}_3 \\
\mathbf{0}_{3\times1} & \mathbf{0}_3 & \mathbf{0}_3 & \mathbf{0}_3 & T_c\mathbf{I}_3 \\
\mathbf{0}_{3\times1} & \mathbf{0}_3 & \mathbf{0}_3 & \mathbf{0}_3 & \mathbf{0}_3 \\
\mathbf{0}_{3\times1} & \mathbf{0}_3 & \mathbf{0}_3 & \mathbf{0}_3 & \mathbf{0}_3
\end{bmatrix}\in\mathbb{R}^{17,13},
\tag{22.175}
$$

with the definition of the model noise vector $\eta[n]=[v_\tau[n],\ \eta_a[n]^T,\ \eta_g[n]^T,\ \eta_{aa}[n]^T,\ \eta_{gg}[n]^T]^T$, $\eta[n]\in\mathbb{R}^{13,1}$.

22.8.3   Tight Integration: Measurement Equation

The incremental observation vector of the complementary KF, $\Delta\mathbf{z}[n]$, is defined as

$$
\Delta\mathbf{z}[n]=\zeta_{sat}[n]-\breve{\zeta}[n],
\tag{22.176}
$$

where

$\zeta_{sat}[n]=[\rho[n]^T\ \mathbf{r}[n]^T]^T\in\mathbb{R}^{2N_{sat},1}$ is the vector of the pre-corrected GNSS-measured pseudoranges $\rho[n]$ and pseudorange rates $\mathbf{r}[n]$ at the time instant n: and

$\breve{\zeta}[n]=[\breve{\rho}[n]^T\ \breve{\mathbf{r}}[n]^T]^T$ is the nominal pseudorange and pseudorange rate vector.

The predicted incremental observation dependent on the state-space model only is written from the a priori state $\Delta\mathbf{x}^-[n]=\Phi[n]\Delta\mathbf{x}[n-1]$ as follows:

$$
\Delta\mathbf{z}^-[n]=\mathbf{H}_{ti}[n]\Delta\mathbf{x}^-[n],
\tag{22.177}
$$

where the observation matrix $\mathbf{H}_{ti}[n]$ is time varying and depends on the GNSS-only observation matrix $\mathbf{H}[n]$ in (22.86), being equal to

$$
\mathbf{H}_{ti}[n]=
\begin{bmatrix}
\mathbf{H}[n] & \mathbf{0}_{N_{sat}\times3} & \mathbf{0}_{N_{sat}\times3} & \mathbf{0}_{N_{sat}\times8} \\
\mathbf{0}_{N_{sat}\times3} & \mathbf{H}[n] & \mathbf{0}_{N_{sat}\times3} & \mathbf{0}_{N_{sat}\times8}
\end{bmatrix}\in\mathbb{R}^{2N_{sat},17}.
\tag{22.178}
$$

Therefore, combining (22.176) and (22.177), it is possible to compute the innovation $\alpha[n]$:

$$\alpha[n] = \Delta \mathbf{z}[n] - \mathbf{H}_{ti}[n] \Delta \mathbf{x}^-[n]. \tag{22.179}$$

As in the case of loose integration, with the above definitions the complementary KF iterations outlined in Section 22.2.3 can be used to resolve the tight integration algorithm.

22.9 PERFORMANCE COMPARISON BETWEEN LOOSE AND TIGHT INTEGRATION

In order to assess the performance of the loose and tight GNSS/INS integration architecture discussed so far, we show in this section the results of a live vehicular test campaign conducted in downtown Turin, Italy [35].

The reference trajectory (ground truth) for the test campaign is shown in Figure 22.23. We identify two zones of analysis as the most meaningful for our purposes:

1. *Zone #1*: a *car parking* area that is characterized by good visibility of the sky and high number of satellites in view;

2. *Zone #2*: a *dense urban area* characterized by narrow streets and densely packed buildings that limit the number of satellites in visibility.

The vehicle used to carry out the test campaign was equipped with a commercial mass-market grade GNSS module, a commercial MEMS IMU loosely integrated with a mass-market GPS receiver (commercial off the shelf (COTS) device), and a processing board where the tight-integration algorithm runs in real-time. Furthermore, a tactical grade INS coupled with a highly precise GPS real-time kinematic (RTK) receiver was used as reference.

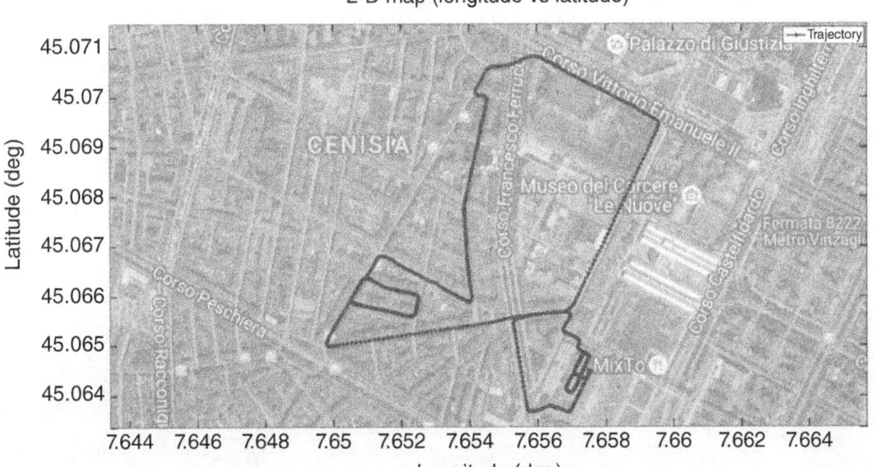

Figure 22.23   Reference trajectory of the live vehicular test campaign in Turin.

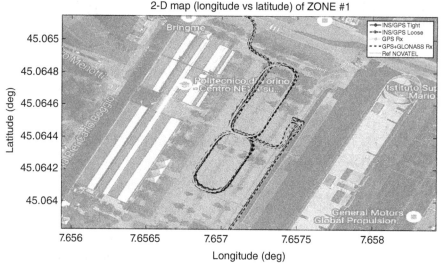

Figure 22.24   A view of the zone #1 path.

22.9.1   Zone #1: Open-Sky Car Parking

In this scenario the vehicle was driven across a car parking in full sky view, to form several eight-shaped figures. A zoomed view of this area is shown in Figure 22.24.

The horizontal position error for the configurations under test versus the reference trajectory is plotted in Figure 22.25. It is possible to appreciate how the tightly integrated solution provides the best performance in terms of position accuracy with

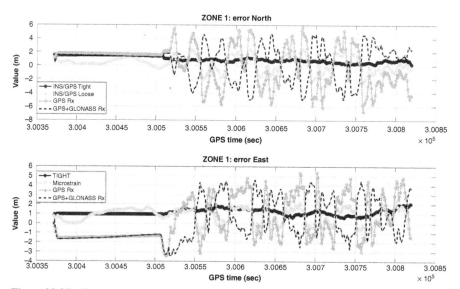

Figure 22.25   Zone #1: Horizontal position error for different sensors under investigation.

respect to the other navigation modules. The COTS device running the loose integration has an error that varies from −2 to 2 m both in the North and East axes, while the GPS module shows a position error with a range from −5 to 5 m along the North coordinate and from −3 to 4 along the East coordinate. The benefits of a multiconstellation system cannot be noted in this scenario due to the high number of GPS satellites in view, so that the performance of the GPS+GLONASS receiver is just slightly better than the one obtained by the GPS-only receiver.

The along-track and cross-track errors can be seen in Figure 22.26. The tightly integrated navigation solution provides an along-track error within 1 m, while the loosely integrated COTS device produces an along-track error that varies from −2 to 2 m. On the contrary, in case of cross-track error the tightly integrated solution is biased with 1 m error with respect to the true solution.

The loosely integrated module does not experience such bias, and the solution changes over time around zero but with a higher variance with respect to the tight integration algorithm.

22.9.2   Zone #2: Dense Urban Area

This section describes the results obtained in the most challenging scenario: a dense urban area characterized by narrow streets, packed buildings, and reduced sky visibility. A snapshot of the path is visible in Figure 22.27.

In this challenging scenario the number of satellites in view plays a fundamental role in obtaining accurate navigation performance. In fact, the tightly integrated algorithm can work, on average, with 5–6 satellites only, with frequent drops to 4 satellites in view. On the contrary, a dual constellation system can still guarantee a higher number of satellites in view, as in the case of a GPS+GLONASS receiver: during this data collection, the multi-GNSS receiver was able to track signals from 15 satellites for most of the time.

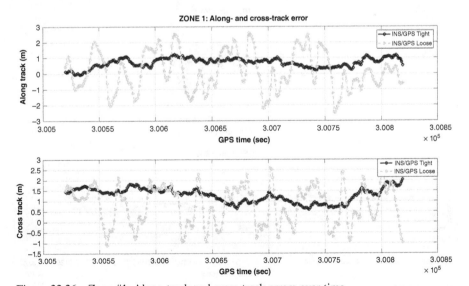

Figure 22.26   Zone #1: Along-track and cross-track errors over time.

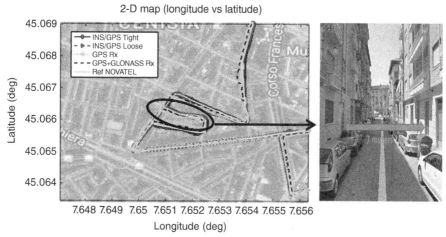

Figure 22.27   A view of the zone #2 path, characterized by narrow streets and packed buildings.

The position error is shown in Figure 22.28 for all the receivers under investigation. The advantage of using a tight integration with respect to loose one is quite clear. This improvement can be easily explained by the fact that the tight integration strategy is able to provide accurate performance also in poor visibility conditions. Furthermore, if we compare the performance of the single constellation GPS stand-alone receiver and multiconstellation one, we can clearly see the benefits of using more than one GNSS system since the higher number of available

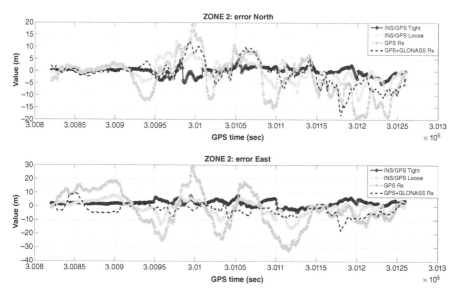

Figure 22.28   Zone #2: Horizontal position error for different sensors under investigation.

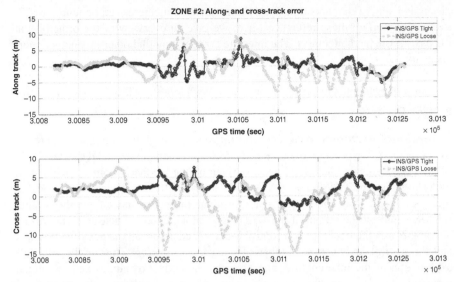

Figure 22.29   Zone #2: Along-track and cross-track error over time.

satellites reflects into a better position accuracy. The along-track and cross-track errors over time are shown in Figure 22.29: they are lower in the case of the tightly integrated algorithm than the loosely integrated one. The reason is due to the tight integration being intrinsically more robust in poor visibility conditions, since it compensates for the degraded signal tracking performance with the inertial information in the EKF.

22.10 GENERAL ARCHITECTURE FOR THE ULTRA-TIGHT INTEGRATION

Ultra-tight coupling is fundamentally different than the previous types of GNSS/INS integration in that the traditional core GNSS receiver architecture is modified to include information coming from the inertial unit: in this way the INS mechanization becomes an integral part of the GNSS processing [36]. Such a deep level of interaction enhances the robustness of the positioning system in high dynamics and degraded signal conditions, such as interference, jamming, or attenuated signal. This comes at the cost of a remarkable implementation complexity, which involves the necessity of modifying the receiver architecture itself; furthermore, the level of performance highly depends on the quality of the IMU integrated in the system. For these reasons only ad hoc developed devices can be used, which increase the cost of such solutions for a final user.

The term "ultra-tight integration," or *deep integration*, may point to several kinds of architecture, all characterized by the common element of a feedback control on the tracking loops driven by the information produced with the aid of the inertial system. Although several implementation flavors are possible, a specific and detailed discussion about reference ultra-tightly integrated architectures can be found in [37] and [38]; based on those references, here we give an overview of the principles at the basis of such architectures. The interested reader may find other instructive references in [39–43].

The principal classification of the ultra-tight coupling architectures is between (1) *centralized* and (2) *federated* approaches. We adopt this classification in the following discussion.

Centralized Architecture. The *centralized* ultra-tightly integrated approach uses a single KF to process together the GNSS signal and the inertial data; the NCOs of the GNSS signals tracking loops are driven by the feedback from the KF, which blend together the GNSS raw information from *all* the visible satellites and the inertial system. The code and carrier residual measurements of all the visible satellites, built at the output of the signal correlators, enter as measurements of the KF. In this way the tracking loop control is correlated across all the tracking channels and the control architecture is inherently a *vector tracking*. This implies the full exploitation of the inherent coupling between the body's dynamics and the signal dynamics observed in the tracking loops. However, although the centralized architecture is optimal in the statistical sense and relatively plain in design, it may be numerically intensive and experience numerical instabilities. For this reason it is considered a valid performance benchmark, but it is not the preferred solution in many practical situations.

Federated Architecture. On the other hand, the *federated* architecture includes a *master filter* and a number of *residual filters* (or *prefilters*), one per tracking channel. The prefilters may converge in a unified vector architecture, managed in a common residual filter. The master filter estimates the full state vector for the body position, velocity, attitude, clock, and IMU biases, and its estimates are used to compute the feedback control to the code and carrier local oscillators. The residual filter, in turn, processes the postcorrelation GNSS measurements and estimates the prediction errors produced by the master filter. The state estimates of the residual filter are used as input for the master filter, which, in this way, does not handle directly the the raw postcorrelation measurements. In the federated approach both scalar and vector tracking architectures may be used; however, the scalar tracking tends to lose part of the deep integration benefits.

22.10.1 Ultra-Tight Integration: Centralized Architecture

The relevance of the centralized version of an ultra-tight integration architecture is in the fact that it achieves the best integration performance from a statistical point of view and represents a benchmark for other federated architectures. A remarkable example of its implementation in a very challenging user scenario is reported in [44].

Centralized Architecture: State-Space Model and State Transition Matrix:
The vector of incremental states of the centralized KF is the same introduced for the

tight integration architecture in (22.169) and reported here for convenience. It is a subset of the states used in [44].

$$\Delta \mathbf{x}_C[n] = [\Delta \mathbf{p}^e[n]^T, \Delta \tau[n], \Delta \mathbf{v}^e[n]^T, \Delta v_\tau[n], \Delta \mathbf{\psi}^e[n]^T, \mathbf{b}_a{}^b[n]^T, \mathbf{b}_g{}^b[n]^T]^T \in \mathbb{R}^{17,1},$$
(22.180)

where the subscript C indicates the "centralized" architecture. As a consequence, the same state transition matrix $\mathbf{\Phi}[n]$ and input matrix $\mathbf{\Gamma}[n]$, respectively reported in (22.174) and (22.175), are applied in the state equation of the ultra-tightly coupled centralized KF.

Centralized Architecture: Measurement Equation: The incremental observation vector is represented by the pseudorange and pseudorange rate residuals produced in each tracking channel through an ad hoc *discrimination function*, for the code and carrier tracking error, respectively, i.e.,

$$\Delta \mathbf{z}_C[n] = \left[\tilde{\mathbf{\varepsilon}}_{code}[n]^T, \tilde{\mathbf{\varepsilon}}_{carr}[n]^T \right]^T \in \mathbb{R}^{2N_{sat},1},$$
(22.181)

where $\tilde{\mathbf{\varepsilon}}_\bullet[n]$ is the vector gathering all the outputs of either the code or carrier discrimination functions at the time instant n, properly scaled and corrected to directly express pseudorange and pseudorange rate errors. As a consequence, the measurement matrix used to compute the innovation vector can be identical to the one written in (22.178), i.e.,

$$\mathbf{H}_{uti,C}[n] = \mathbf{H}_{ti}[n] \in \mathbb{R}^{2N_{sat},17}.$$
(22.182)

What differs in this architecture from the tight integration discussed in Section 22.8 is the fact that the incremental observations (pseudorange and pseudorange rate residuals) are generated from the code and carrier discriminators, while the feedbacks to the NCOs are computed from the prediction of the code rate and Doppler frequency derived from the inertial system corrected through the bias estimates of the EKF. In this way the satellite signal processing channels are coupled together through the common feedback from the filter states. In order to keep a stable and reactive tracking process, the feedback must be generated at "correlation rate" (i.e., between 50 Hz and 1 kHz), which may be a challenging requisite for the centralized filter. A block diagram that exemplifies this architecture is shown in Figure 22.30.

22.10.2   Ultra-Tight Integration: Federated Architecture

Since vector tracking in a federated architecture is recognized as achieving better performance than scalar [37, 38], only this approach is mentioned in this section. The federated filtering approach relaxes the demand for high-rate update (i.e., correlation rate) posed to the KF in the centralized architecture: it apportions the integrated signal tracking task between the *master filter* and the *residual filter*, which is directly part of the tracking channels.*

---

*As pointed out in [37], the difference between a federated ultra-tight integration and some tight integrations with aiding on the tracking loops can be quite subtle.

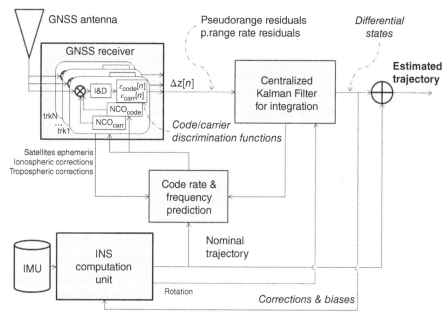

Figure 22.30   Block diagram of an ultra-tight integration architecture with a centralized KF.

The trajectory states of the master filter are used in the feedback loop to predict the received pseudoranges and pseudorange rates at "low rate" (e.g., 1 Hz); such predictions drive the NCOs of the tracking loops through low-rate estimates of the code delay and carrier frequency. This low-rate feedback loop is indicated in Figure 22.31 with the label "1 Hz." On the other hand, the residual filter estimates the *pseudorange and pseudorange rate residuals* in each tracking channel at "correlation rate" (loop labeled as "> 50 Hz" in Figure 22.31): the quantities $\Delta\rho[k]$, $\Delta r[k]$ are the estimated errors in the predictions produced by the master filter between two consecutive low-rate updates. In fact, the states of the residual filter are employed as *measurements* of the master filter, and then are *zeroed* at each low-rate cycle after being input to the master filter.

Federated Architecture: State-Space Model: Since the IMU mechanization is handled by the master filter only, the state-space model of the *residual filter* is exactly the one introduced at the beginning of this chapter for the EKF architecture in a stand-alone GNSS receiver (Section 22.3), i.e.,

$$\Delta\mathbf{x}_R[k] = \left[\Delta\mathbf{p}_R^e[k]^T, \Delta\tau_R[k], \Delta\mathbf{v}_R^e[k]^T, \Delta v_{\tau R}[k]\right]^T \in \mathbb{R}^{8,1}, \qquad (22.183)$$

which corresponds to (22.66); the subscript R denotes the "residual" filter, while the time index k proceeds at "correlation rate" (≥ 50 Hz). The purpose of the residual filter states is to track the errors in the master filter's estimates of the body trajectory between two consecutive cycles of the master filter. Hence, since it is assumed that

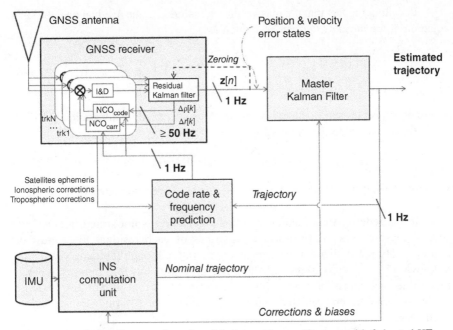

Figure 22.31   Block diagram of an ultra-tight integration architecture with federated KFs.

the master filter produces the best unbiased estimate of the body trajectory at every iteration, the residual filter states $\Delta\mathbf{x}_R[k]$ are *zeroed* at each iteration of the master filter.

On the other hand, the states of the *master filter* represent the complete navigation and attitude states of the receiver, plus the IMU biases and an additional incremental state vector that represent the *errors in the residual filters' state estimate*, i.e.,

$$\mathbf{x}_M[n] = \left[\mathbf{p}^e[n]^T, \tau[n], \mathbf{v}^e[n]^T, v_\tau[n], \mathbf{\psi}^e[n]^T, \ldots \right.$$
$$\left. \mathbf{b}_a^b[n]^T, \mathbf{b}_g^b[n]^T, \delta\mathbf{x}_R[n]^T \right]^T \in \mathbb{R}^{25,1}, \tag{22.184}$$

where $\delta\mathbf{x}_R[n] \in \mathbb{R}^{8,1}$ represents the estimation errors of the a posteriori states of the residual filter, i.e.,

$$\delta\mathbf{x}_R[n] = \Delta\mathbf{x}_R[n] - \Delta\mathbf{x}_R^+[n], \tag{22.185}$$

where $\Delta\mathbf{x}_R[n]$ are the true residuals. Thus, each state of the residual filter $\Delta\mathbf{x}_R[k]$ has a corresponding error state in the master filter $\mathbf{x}_M[n]$. Notice that the time index n in (22.184) is purposely different than the k used in the residual filter states formulation above to indicate different sample rate.

Federated Architecture: State Transition Matrix: The state transition matrix of the *residual filters*, $\mathbf{\Phi}_R$, is simply the EKF $\mathbf{\Phi}$ matrix reported in (22.69), where now $T_c = 1/$(correlation rate). Similarly, the structure of (22.71) applies for the model noise covariance matrix $\mathbf{Q}$ of the residual filter.

Recalling the fact that the master filter is cascaded to the residual one, the state equation of the *master filter* can be written as

$$
\begin{bmatrix} \mathbf{x}_M^{(1:17)}[n+1] \\ \delta\mathbf{x}_R[n+1] \end{bmatrix} = \begin{bmatrix} \boldsymbol{\Phi}_{ti}[n] & \mathbf{0}_{17\times8} \\ \mathbf{0}_{8\times17} & \boldsymbol{\Phi}_{\delta}[n] \end{bmatrix} \begin{bmatrix} \mathbf{x}_M^{(1:17)}[n] \\ \delta\mathbf{x}_R[n] \end{bmatrix} + \mathbf{w}_M[n,n+1], \quad (22.186)
$$

where $\boldsymbol{\Phi}_{ti}[n]$ is the same state transition matrix used in the tight integration (22.174), while $\boldsymbol{\Phi}_{\delta}[n]$ can be derived as in [38]:

$$
\boldsymbol{\Phi}_{\delta}[n] = \boldsymbol{\Phi}_R(\mathbf{I}_8 - \mathbf{K}_R[n]\mathbf{H}_R[n]), \quad (22.187)
$$

where $\mathbf{K}_R[n]$ and $\mathbf{H}_R[n]$ are the Kalman gain and measurement matrix of the residual filter.

In the federated architecture the noise modeling is not straightforward for several reasons: first, the covariance matrices of each residual filter must be appropriately reinitialized after each state zeroing: although the matrix structure is known (22.71), the variance of the noise processes has to be properly determined. Second, since the master filter is cascaded to the residual filter, its noise model $\mathbf{w}_M$ may be complicated and the representation of its covariance matrix needs care.

Federated Architecture: Measurement Equation: The innovation equation (22.50) of the *residual filter* is

$$
\boldsymbol{\alpha}_R[k] = \Delta\mathbf{z}_R[k] - \mathbf{H}_R[k]\Delta\mathbf{x}_R^-[k] \in \mathbb{R}^{2N_{sat},1}, \quad (22.188)
$$

where $\Delta\mathbf{z}_R[k]$ are the pseudorange and pseudorange rate residuals measured at the discriminators' output for all the satellites in tracking, as for the centralized filter in (22.181), while the measurement matrix $\mathbf{H}_R[k]$ has the format defined in (22.86): $\mathbf{H}_R[k] = \mathbf{H}[k], \mathbf{H}_R[k] \in \mathbb{R}^{2N_{sat},8}$. Notice that in (22.188) the innovation is computed at correlation rate.

The measurements of the *master filter* are the states of the residual filter, i.e.,

$$
\mathbf{z}_M[n] = \Delta\mathbf{x}_R[n] = \mathbf{H}_M[n]\mathbf{x}_M[n] + \mathbf{v}[n], \quad (22.189)
$$

where $\mathbf{H}_M[n]$ has to be determined. Since $\Delta\mathbf{x}_R[n] = \mathbf{x}_M^{(1:8)}[n] - \delta\mathbf{x}_R[n]$, then $\Delta\mathbf{x}_R[n] = [\mathbf{I}_8 \ \mathbf{0}_{8\times9} \ -\mathbf{I}_8]\, \mathbf{x}_M[n]$, so that the measurement matrix of the master filter results in:

$$
\mathbf{H}_M = [\mathbf{I}_8 \ \mathbf{0}_{8\times9} \ -\mathbf{I}_8] \in \mathbb{R}^{8,25}. \quad (22.190)
$$

For a detailed modelling and discussion of the noise performance of such architectures, the interested reader may refer to [38] and [45].

It is worth noticing that a simplified version of the master filter architecture can be obtained by avoiding the use of the states $\delta\mathbf{x}_R[n]$ in (22.186), i.e., the errors in the residual filter.

22.11   CONCLUSIONS

This chapter has discussed applications of the KF in the satellite navigation field. The reader has been lead through a path starting from the simplest Kalman-based navigation system and ending with complex integrated systems that include IMUs.

The first application presented was the use of the KF in a stand-alone GNSS receiver for the PVT computation. The underlying KF model was presented, defining all the required matrices and providing some tips for implementing the solution; this architecture is the foundation for the more complex schemes discussed later.

Then, the focus was shifted to the basic concepts of inertial navigation: a mathematical description of the basic mechanization equations was given, followed by an extensive discussion about IMU calibration and initialization processes.

Finally, the reader encoutered the three principal families of integrated GNSS/INS architectures. An interesting point is that the KF model adopted by the three architectures is almost the same, while the main difference is represented by the considered measurements and the related matrices.

The proposed examples allow the reader to practice with the topics addressed in the chapter by means of simple MATLAB scripts developed and fruitfully used for educational purposes by G. Falco.

APPENDIX A: MATLAB Code

This appendix is devoted to the discussion of the MATLAB code mentioned in this chapter.

A.1   A Simple KF Routine

Since the main tool described in this chapter is represented by the KF, in the following we show a script that may be used to implement the filter and to compute the correction to be applied to the estimate.

The input of this script is represented by:

- the current measurement;
- the last a posteriori estimate of the state (which is zero in the case of an EKF);
- the a posteriori error covariance matrix;
- the measurement matrix;
- the measurement error covariance matrix;
- the state transition matrix;
- the state noise covariance matrix.

The script computes the current estimate of the state and the current a posteriori error covariance matrix. Each line is commented and the variables are named to be self-explanatory.

[See file Chapter_22_Appendix_A_1.m]

A.2 EKF for a GNSS Receiver

The code shown in this appendix reflects what is described in Section 22.3.

The input of this script is represented by:

- a KF structure that stores Kalman data;
- a matrix whose rows are time instants and whose columns are the ID of the satellite, the pseudorange measurement, and the Doppler measurement, respectively;
- a vector storing the predicted position of the user;
- a vector storing the predicted velocity of the user;
- a matrix whose rows are the ECEF position components of the visible satellites and whose columns are the visible satellites;
- a matrix whose rows are the ECEF velocity components of the visible satellites and whose columns are the visible satellites;
- a scalar representing the predicted clock bias;
- a scalar representing the predicted clock drift;
- a vector storing the relativistic correction for each satellite.

The output is represented by the estimated correction and by the a posteriori error covariance matrix.

[See file Chapter_22_Appendix_A_2.m]

The code is structured in subsequent steps:

- definition of the constants and of the number of available satellites;
- computation of the matrix made up by the unitary vectors pointing from the estimated user position towards the satellites; these positions are corrected using the relativistic corrections, computed on the basis of the pseudorange measurements;
- pseudorange and pseudorange rate estimates are computed in the for loop;
- measurements are redefined;
- the innovation vector is defined;
- the measurement matrix is built;
- the KF is run.

The only obscure point in the script may be represented by the *kron* function. It allows for block products, so that each element of the first input matrix is multiplied by the whole matrix given as the second input.

A.3 INS Mechanization Equations

The code shown in this appendix describes how to calculate the inertial mechanization equations in Section 22.4.

The input of this script is represented by

- a (3 × 3) matrix representing the attitude DCM from BODY to ECEF frame computed at medium rate;

- a (3 × 1) vector representing the velocity in ECEF frame computed at medium rate;
- a (3 × 1) vector representing the position in ECEF frame computed at medium rate;
- a (3 × 1) vector representing the Velocity in BODY frame computed at high rate;
- a (3 × 1) vector representing the sculling effect computed at high rate;
- a scalar representing the integration time of the medium rate;
- a (3 × 1) vector representing the angular rate in BODY frame computed at high rate;
- a (3 × 1) vector representing the coning effect computed at high rate;
- a (3 × 1) vector representing the position in Latitude, Longitude, Height (LLH) frame computed at medium rate.

[See file Chapter_22_Appendix_A_3.m]
The output of this script is represented by:

- a (3 × 3) matrix representing the updated attitude DCM from BODY to ECEF frame computed at medium rate;
- a (3 × 1) vector representing the updated velocity in ECEF frame computed at medium rate;
- a (3 × 1) vector representing the updated position in ECEF frame computed at medium rate.

A.4 IMU Coarse Alignment: Horizontal Plane

As shown in this chapter, in order to realize an INS/GNSS integrated system, an IMU must be exploited, and this device needs to collect some data before being used for the navigation, since an alignment process is required.

The alignment process is divided into a *coarse alignment* and a *refinement step*. In the following, the MATLAB code for the coarse alignment along the horizontal plane is reported. Once this step is finished, a KF can be exploited to perform the refinement.

- a matrix that stores the logged inertial data;
- a scalar that sets the duration of the coarse alignment operation;
- a scalar that points out the sampling period;
- other inputs may be needed for further operations, like fine alignment.

The output is represented by the DCM that allows the horizontal alignment and by the estimated Euler angles.

[See file Chapter_22_Appendix_A_4.m]
The code is structured in three different steps, on the basis of simple trigonometric assumptions:

1. a mean total gravity is computed considering the whole data coarse alignment time interval;
2. a coarse roll estimate is obtained exploiting the mean measurement of the y- and z-axes accelerometers;

3. a coarse pitch estimate is obtained exploiting the mean measurement of the *x*- and *z*-axes accelerometers.

Further references can be found in [16].

A.5 INS/GPS Integration: Loosely Coupled Algorithm

Different kinds of integration between IMU and GPS are possible, as explained in Sections 22.7, 22.8, and 22.10. In this example we limit to the loose integration only.

    [See file Chapter_22_Appendix_A_5.m]

The main function of loose function has the following parameters as inputs:

- a (3 × 1) vector representing the INS position in ECEF frame and calculated at low rate;
- a (3 × 1) vector representing the GPS position in ECEF frame and calculated at low rate;
- a (3 × 1) vector representing the INS velocity in ECEF frame and calculated at low rate;
- a (3 × 1) vector representing the GPS velocity in ECEF frame and calculated at low rate;
- a scalar representing the number of states of the KF;
- a (15 × 15) matrix representing the noise covariance matrix associated to the states of the KF;
- a (15 × 15) matrix representing the noise covariance matrix associated to the model variables of the KF.

As output, the following parameters are provided:

- a *struct* variable containing the updates of the states of the KF and of the noise variance matrix;

Other subroutines are included. The main goals of them are:

1. to calculate the noise variance associated to the GPS measurements;
2. to calculate and predict the system error model of the INS mechanization equations;
3. the transformation of the system error model from continuous-time to discrete-time domain.

REFERENCES

[1] P. Groves, *Principles of GNSS, Inertial, and Multisensor Integrated Navigation Systems.* Artech House, 2008.

[2] N. Hjortsmarker, "Experimental system for validating GPS/INS integration algorithms," Ph.D. dissertation, Lulea Univeristy of Technology, 2005.

[3] A. Giremus, A. Doucet, V. Calmettes, and J. Tourneret, "A Rao-Blackwellized particle filter for INS/GPS integration," in *IEEE Int. Conf. on Acoustics, Speech, and Signal Processing*, vol. 3, 2004, pp. 964–967.

[4] R. van der Merwe and E. A. Wan, "Sigma-point Kalman filters for integrated navigation," in *60th Annual Meeting of the Institute of Navigation*, 2004, pp. 641–654.

[5] J. L. Crassidis, "Sigma-point Kalman filtering for integrated GPS and inertial navigation," in *AIAA Guidance, Navigation and Control Conf.*, 2005, pp. 1–24.

[6] R. G. Brown and P. Y. C. Hwang, *Introduction to Random Signal and Applied Kalman Filtering*, 3rd ed. John Wiley & Sons, 1997.

[7] A. Gelb, *Applied Optimal Estimation*. MIT Press, 1974.

[8] S. J. Julier and J. K. Uhlmann, "A new extension of the Kalman filter to nonlinear systems," in *AeroSense, Aerospace/Defence Sensing, Simulation and Controls Conf.*, vol. 3068, 1997, pp. 182–193.

[9] G. H. Golub and C. G. Van Loan, *Matrix Computations*, 2nd ed. John Hopkins University Press, 1996.

[10] R. A. Horn and C. R. Johnson, *Matrix Analysis*, 2nd ed. Cambridge University Press, 2012.

[11] J. de Freitas, M. Niranjan, A. Gee, and A. Doucet, "Sequential Monte-Carlo methods for optimisation of neural networks models," Tech. Rep. CUES/F-INFENG/TR-328, University of Cambridge, 1998.

[12] E. D. Kaplan and C. Hegarty, Eds., *Understanding GPS: Principles and Applications*, 2nd ed. Artech House Publishers, 2005.

[13] P. Misra and P. Enge, *Global Positioning System: Signals, Measurements and Performance*. Ganga Jumuna Press, 2006.

[14] D. Titterton and J. Weston, *Strapdown Inertial Navigation Technology*, 2nd ed. Institution of Electrical Engineers, 2004.

[15] C. Jekeli, *Inertial Navigation Systems with Geodetic Application*. De Gruyter, 2001.

[16] M. G. Petovello, "Real-time integration of a tactical-grade IMU and GPS for high-accuracy positioning and navigation," Ph.D. dissertation, University of Calgary, 2003.

[17] P. Aggarwal, Z. Syed, A. Noureldin, and N. El-Sheimy, *MEMS-Based Integrated Navigation*. Artech House, 2010.

[18] C. Hide, "Integration of GPS and low-cost INS measurements," Ph.D. dissertation, Institude of Engineering, Survey and Space Geodesy, University of Notthingham, 2005.

[19] *IEEE Standard Specification Format Guide and Test Procedure for Linear, Single-Axis, Non-gyroscopic Accelerometers*, 2011.

[20] N. El-Sheimy, H. Hou, and X. Niu, "Analysis and modeling of inertial sensors using Allan variance," *IEEE Trans. Instrum. Meas.*, vol. 57, pp. 140–142, Jan. 2008.

[21] R. Vaccaro and A. Zaki, "Statistical modeling of rate gyros," *IEEE Trans. Instrum. Meas.*, vol. 61, no. 3, pp. 673–684, Mar. 2012.

[22] J. Georgy, A. Noureldin, M. Korenberg, and M. Bayoumi, "Modeling the stochastic drift of a MEMS-based gyroscope in gyro/odometer/GPS integrated navigation," *IEEE Trans. Intell. Transp. Syst.*, vol. 11, no. 4, pp. 856–872, Dec. 2010.

[23] M. Park and Y. Gao, "Error and performance analysis of MEMS-based inertial sensors with a low-cost GPS receiver," *Sensors*, vol. 8, no. 4, pp. 2240–2261, Apr. 2008.

[24] A. Quinchia, C. Ferrer, G. Falco, E. Falletti, and F. Dovis, "A comparison between different error modeling of MEMS applied to GPS/INS integrated systems," *Sensors*, vol. 13, no. 8, pp. 9549–9588, Jul. 2013.

[25] *3DM-GX3-25 Technical Product Overview*.

[26] A. Noureldin, J. Armstrong, A. El-Shafie, T. Karamat, D. McGaughey, M. Korenberg, and A. Hussain, "Accuracy enhancement of inertial sensors utilizing high resolution spectral analysis," *Sensors*, vol. 58, no. 9, pp. 11638–11660, Jul. 2012.

[27] N. El-Sheimy, S. Nassar, K. P. Schwarz, and A. Noureldin, "Modeling inertial sensor errors using autoregressive (AR) models," *NAVIGATION*, vol. 51, no. 4, pp. 259–268, Winter 2005.

[28] A. Noureldin, T. Karamat, M. Eberts, and A. El-Shafie, "Performance enhancement of MEMS-based INS/GPS integration for low-cost navigation applications," *IEEE Trans. on Veh. Technol.*, vol. 58, no. 3, pp. 1077–1096, Mar. 2009.

[29] M. El-Diasty, "Calibration and stochastic modelling of inertial navigation sensor errors," *J. Global Positioning Systems*, vol. 7, no. 4, pp. 170–182, May 2008.

[30] J. Hidalgo, P. Poulakis, J. Khler, J. Del-Cerro, and A. Barrientos, "Improving planetary rover attitude estimation via MEMS sensor characterization," *Sensors*, vol. 12, no. 2, pp. 2219–2235, Feb. 2012.

[31] *IEEE Standard Specification Format Guide and Test Procedure for Single-Axis Interferometric Fiber Optic Gyro*, 2011.

[32] P. Aggarwal, Z. Syed, X. Niu, and N. El-Sheimy, "A standard testing and calibration for low cost MEMS inertial sensors and units," *J. Naviga.*, vol. 61, no. 2, pp. 323–336, Apr. 2008.

[33] E.-H. Shin and N. El-Sheimy, "A new calibration method for strapdown inertial navigation systems," *Z. Vermess*, vol. 127, pp. 1–10, Apr. 2002.

[34] E. J. Krakiwsky, *The Method of Least Squares: A Synthesis of Advances,* 1st ed. Calgary, AB: Department of Geomatics Engineering, The University of Calgary, 1990.

[35] G. Falco, M. Pini, and G. Marucco, "Loose and tight GNSS/INS integrations: Comparison of performance assessed in real urban scenarios," *Sensors*, vol. 17, no. 2, pp. 1–25, Jan. 2017.

[36] G. Gleason and D. Gebre-Egziabher, Eds., *GNSS Applications and Methods.* Artech House, 2009.

[37] M. Lashley, D. M. Bevly, and J. Y. Hung, "Analysis of deeply integrated and tightly coupled architectures," in *IEEE/ION Position, Location and Navigation Symp.*, May 2010, pp. 382–396.

[38] M. Lashley, "Modeling and performance analysis of GPS vector tracking algorithms," Ph.D. dissertation, Auburn University, 2009.

[39] M. G. Petovello and G. Lachapelle, "Comparison of vector-based software receiver implementations with application to ultra-tight GPS/INS integration," in *Proc. of the 19th Int. Technical Meeting of the Satellite Division of the Institute of Navigation (ION GNSS 2006)*, 2006, pp. 1790–1799.

[40] M. Petovello, C. O'Driscoll, and G. Lachapelle, "Ultra-tight integration of an IMU with GPS/ GLONASS," in *13th IAIN World Congress,* Stockholm, Oct. 2009, pp. 1–10.

[41] J.-H. Won, D. Dotterbock, and B. Eissfeller, "Performance comparison of different forms of Kalman filter approaches for a vector-based GNSS signal tracking loop," *NAVIGATION*, vol. 57, pp. 185–199, Fall 2010.

[42] T. Pany and B. Eissfeller, "Use of a vector delay lock loop receiver for GNSS signal power analysis in bad signal conditions," in *2006 IEEE/ION Position, Location, and Navigation Symp.*, Apr. 2006, pp. 893–903.

[43] X. Tang, G. Falco, E. Falletti, and L. Lo Presti, "Performance comparison of a KF-based and a KF-VDFLL vector tracking-loop in case of GNSS partial outage and low-dynamic conditions," in *2014 7th ESA Workshop on Satellite Navigation Technologies and European Workshop on GNSS Signals and Signal Processing (NAVITEC)*, Dec. 2014, pp. 1–8.

[44] T. M. Buck, J. Wilmot, and M. J. Cook, "A high G, MEMS based, deeply integrated, INS/GPS, guidance, navigation and control flight management unit," in *2006 IEEE/ION Position, Location, and Navigation Symp.*, Apr. 2006, pp. 772–794.

[45] F. Qin, X. Zhan, and G. Du, "Performance improvement of receivers based on ultra-tight integration in GNSS-challenged environments," *Sensors*, vol. 13, no. 12, pp. 16406–16423, 2013.

CHAPTER 23

AN OVERVIEW ON GLOBAL POSITIONING TECHNIQUES FOR HARSH ENVIRONMENTS

Nicola Linty and Fabio Dovis

Politecnico di Torino, Italy

T HE SCOPE of this chapter is to present strategies and techniques used to increase the sensitivity of global navigation satellite system (GNSS) receivers in order to make them usable in harsh environments, such as urban canyons, light indoor scenarios, deep forests, or space. In such scenarios, the received signal experiences a lower signal-to-noise ratio, coupled with distortions and presence of spurious contributions, which affect the quality of the GNSS position, and sometimes even deny the positioning service. The chapter discusses the assistance that can be provided to the GNSS receiver through communication channels to ease the acquisition and tracking processes. Assisted GNSS is a consolidated standard, but other kinds of assistance and signal processing techniques can improve the ability of the receiver to process the signal at a low signal-to-noise ratio. The chapter introduces the common approaches to increase the sensitivity at the acquisition stage, discussing the impact on the accuracy of the delay and Doppler shift estimation, and the intrinsic limitation to coherent and noncoherent integration time extension. As far as the tracking stage is concerned, techniques to increase robustness to low signal-to-noise ratio scenarios are presented, considering the structure of new and modernized GNSS signals.

23.1 INTRODUCTION

Chapter 20 provides a general outline of the basic principles of global navigation satellite systems. It is worth recalling that the task of a GNSS receiver is to provide an estimate of the position, velocity, and time (PVT) of a user. This is achieved by

Handbook of Position Location: Theory, Practice, and Advances, Second Edition.
Edited by S. A. (Reza) Zekavat and R. Michael Buehrer.
© 2019 by the Institute of Electrical and Electronics Engineers, Inc.
Published 2019 by John Wiley & Sons, Inc.
Companion Website: www.wiley.com/go/zekavat/positionlocation2e

processing radio frequency (RF) electro magnetic signals transmitted by a constellation of satellites by means of a specifically designed communication receiver. Briefly, their architecture is composed of a *physical layer* in charge of processing the received signal to detect and estimate some of its parameters, and by a *range layer* responsible for position computation [1–3]. Chapter 21 gives a detailed description of the digital signal processing techniques employed in a standard GNSS receiver physical layer. Such traditional acquisition and tracking schemes offer users satisfactory performance when the quality of the received signal is sufficiently high.

However, the rapid and widespread diffusion of GNSS for a large variety of location-based applications (LBAs) has pushed the performance requirements of GNSS receivers to extreme levels. Moreover, the present use of GNSS is rather different from its original conception. The global positioning system (GPS) was designed in the early 1970 by the Department of Defense (DoD) of the United States (US). Similarly, the GLObal NAvigation Satellite System (GLONASS) was developed by the Russian Federation. Both systems have been designed mainly for military applications: they were required to work outdoors, with a clear view of the sky and ideal line-of-sight conditions between the users and the satellites for most of the time. There were indeed no plans to use it in indoor environments. Furthermore, a start-up time of about one minute seemed to be adequate considering the maturity of the technology of the time. The size, cost and, in some cases, power consumption of such devices, were not primary drivers. Nowadays, thanks to the evolution of computational platforms, to the availability of more powerful processors, to the design of advanced signal processing techniques and, in parallel, to the spread of LBAs, GNSS is used by billions of people all around the world, and the number of civilian receivers outnumbers military devices by orders of magnitude. The purpose of GNSS has considerably changed: a GNSS receiver is nowadays required to work almost instantly, anywhere on the Earth, and to be small, cheap, and with low power consumption [4].

The need to use GNSS-based positioning also in environments where the signal-to-noise ratio is below the nominal level experienced in clear sky, and where the ideal propagation conditions are not fulfilled, led to the design of *high sensitivity* GNSS (HS-GNSS) receivers. High sensitivity receivers perform sophisticated signal processing operations in order to assure a PVT solution in those environments and situations in which a regular GNSS receiver would fail. Such scenarios include, but are not limited to, urban canyons, light indoor environments, deep forests, and space. The scope of this chapter is to present strategies and signal processing techniques used to increase the sensitivity of GNSS receivers in order to make them usable in such harsh environments.

At a high level, high sensitivity techniques involve two main strategies:

1. the increase of the signal processing gain by extending integration times, thus decreasing the minimum carrier-to-noise density power ratio (C/N_0) required for acquisition and tracking (i.e., increasing receiver sensitivity);

2. the use of GNSS assistance data (such as almanac, ephemeris, satellite health, GPS time and coarse user position) to assist and facilitate the receiver operations.

First, this introduction gives a general overview of high-sensitivity requirements. Then, in Section 23.2, the concepts of noise and correlation gain are described. These concepts are extended in Section 23.3, in which the two main strategies for increasing the sensitivity of a receiver are presented: the coherent and noncoherent integration time extension. Afterwards, Section 23.4 describes the consequences and drawbacks of the extension of the integration time and provides solutions; in particular, the problems of data bit transitions, Doppler shift, and Doppler rate are considered. In Section 23.5, the design and implementation of a high-sensitivity GNSS receiver is addressed, focusing on the acquisition and tracking stages. A few examples of acquisition and tracking results are also provided. Section 23.6 introduces the topic of GNSS assistance, which is of paramount importance when designing high-sensitivity receivers. Finally, Section 23.7 gives a brief overview of the advantages of the new and modernized GNSS signals as far as high sensitivity is concerned.

23.2 SIGNAL POWER, NOISE AND CORRELATION GAIN

One of the main features of any GNSS is the very low signal power received by the user (Chapter 20). GNSS signals are transmitted by a constellation of satellites, which are more than 20,000 km from the Earth. As an example, according to the GPS Interface Control Document (ICD) [5], the minimum received power level for the legacy GPS L1 C/A signal is equal to −158.1 dBW (assuming a unity gain right hand circularly polarized (RHCP) antenna [1]). Typical values for the noise are around −140 dBW, which means that the useful GNSS signal is completely buried in the noise floor. Multipath reflections, radio-frequency interference (RFI), obstacles, buildings, or additional propagation losses further affect the link budget. As an example, three layers of dry bricks or 20 cm of steel add an extra attenuation of about 30 dB. A more detailed analysis of the noise derivation is out of the scope of this chapter and can be found in [4].

Chapter 21 (Section 21.1) carefully describes the GNSS received signal. In particular, the signal at the input of the signal processing blocks of a receiver (acquisition and tracking), denoted $y_{IF}[n]$ and reported in (23.1), is the sum of two different contributions:

- a *useful* signal component, $r_{IF}[n]$, representative of the GNSS transmitted signal;
- a noise component, $w_{IF}[n]$;

$$y_{IF}[n] = r_{IF}[n] + w_{IF}[n]. \tag{23.1}$$

The subscript IF recalls the fact the the signal has been down-converted to an intermediate frequency (IF) by the receiver front-end. At this stage of the receiver, i.e., at the analog-to-digital converter (ADC) output, the signal is a sequence of samples taken at $t = nT_s$, where $T_s = 1/f_s$ and f_s is the sampling frequency. The notation $y_{IF}[n]$ is then used to indicate the discrete-time signal, where $y_{IF}(n) = y_{IF}(nT_s)$ [6]. It has to be noted that only a single signal-in-space (SIS) contribution is considered, as each single satellite signal can be separately received and processed thanks to the code division multiple access (CDMA) characteristics of GNSS signals. In other words, the intrasystem interference generated by other satellites of the constellation can be neglected.

The signal-to-noise ratio (SNR) is commonly used in communications to characterized the noise contribution. It is defined as the ratio of the useful signal power to the noise power. Conversely, in GNSS the C/N_0 is used, being a quantity independent from the front-end filter bandwidth and thus from the receiver design (see Chapter 21, Section 21.2.2). The C/N_0 is defined as

$$C/N_0 = \frac{P_R}{N_0},$$ (23.2)

where P_R is the signal power, evaluated on the whole signal bandwidth, and N_0 is the noise power density.

Considering legacy open GPS signals, in open-sky conditions, in absence of impairments and using a common antenna, the nominal value of the C/N_0 ranges between 45.5 and 55 dB-Hz [2]. Therefore, the typical sensitivity of a common GNSS receiver is around 45.5 dB-Hz. Typical values of C/N_0 for harsh environments are in the range 5 to 45.5 dB-Hz. The following classification is considered in the rest of this chapter:

- C/N_0 above the nominal value: $C/N_0 \geq 45.5$ dB-Hz. This is the case of "regular" open-sky signals, for which high sensitivity is not required.
- C/N_0 below the nominal value: 30 dB-Hz $\leq C/N_0 < 45.5$ dB-Hz. This is the case of weak signals, requiring high sensitivity receivers.
- Very low C/N_0: $C/N_0 < 30$ dB-Hz. This is the case of very weak signals, requiring advanced high-sensitivity techniques.

The SNR at the front-end output, denoted ρ_{IF}, can be derived by dividing the C/N_0 defined in (23.2) by the front-end filter bandwidth B_{IF}:

$$\text{SNR} = \rho_{IF} = \frac{C/N_0}{B_{IF}} = \frac{P_R}{N_0 B_{IF}}.$$ (23.3)

A typical value of the bandwidth of a mass-market single frequency GPS receiver is 4 MHz. Thus, the resulting sensitivity of a common receiver in terms of SNR is −20.5 dB, while a high-sensitivity receiver is required to acquire signals down to −36 dB. A high-sensitivity professional receiver, with a bandwidth of 20 MHz, is able to acquire signals at a SNR as low as −43 dB.

23.2.1   The Correlation Gain

The first task of any GNSS receiver is to extract the signal from the noise floor in order to correctly detect and estimate the signal parameters. In the case of high-sensitivity receivers, this task is even harder. This is achieved by exploiting the CDMA structure of GNSS signals. The spreading codes used in GNSS are periodic sequences, modulated on top of the signal at a rate R_c. In the case of GPS L1 C/A,

$R_c = 1.023$ Mchip/s, and the code is 1023 chips long*. The length of the sequence, T_{code}, is then equal to

$$T_{code} = \frac{1023}{R_c} = \frac{1023}{1.023 \cdot 10^6} = 1 \text{ ms.} \tag{23.4}$$

The despreading is performed through the *correlation* operation. A portion of the received SIS is cross-correlated with a replica of the signal generated locally. Thanks to the excellent correlation properties of the spreading sequences used in GNSS, a correlation peak emerges in correspondence of the right delay between the received and local signals. The more evident and sharp the peak, the better the receiver performance. The ratio between the correlation peak and the correlation floor is the postcorrelation SNR. The longer the accumulation time within the correlation function, the higher the correlation peak, and therefore the higher the despreading gain that can be obtained.

Chapter 21 reports the expression of the postcorrelation SNR (21.41):

$$\rho_a = \frac{P_R L}{N_0 B_{IF}} = \frac{C}{N_0} \frac{L}{2 B_{IF}} = \rho_{IF} \frac{L}{2}. \tag{23.5}$$

$L/2$ is the correlation despreading gain. It is then determined by the term L, which corresponds to the number of samples of the incoming signal used to compute the correlation. Consequently, the product LT_s corresponds to the total length in time of the correlation procedure. LT_s is usually denoted *accumulation time, integration time*, or *predetection integration time*. In most implementations, LT_s is fixed to the length of a code period T_{code}, e.g., 1 ms for GPS L1 C/A legacy signals. In fact, when considering one full code period of the incoming signal and one full code period of the local signal, a complete correlation is performed.

In order to improve the despreading gain and, in turn, the postcorrelation SNR, different options can be considered:

- Increasing the sampling frequency f_s so as to increase the number of samples L. However, the correlated nature of adjacent noise samples leads to an increase of the noise contribution at the correlator output. This effect is magnified at high sampling frequencies [4].
- Using longer codes so as to increase the code period T_{code}. This is the case of modernized and new GNSS signals. For example, the code length of the Galileo E1b signal is four times the code length of the legacy GPS L1 C/A. Example 23.1 below demontrates such a gain.
- Extending the observation period of the incoming signal to more than one code period. This procedure is known as coherent integration time extension and is the core of high-sensitivity GNSS receivers.

Example 23.1:   Extension of the Code Length

The goal of this example is to prove that a longer code benefits from a larger despreading gain, allowing the correct estimation of a peak even in the presence of a high noise standard deviation.

*Parameters for other signals can be found in Chapter 20 and in Section 23.7

In Examples 21.1 and 21.2, the cross-correlation of a code is evaluated, respectively, in the absence and in the presence of noise. A 20-chip long code is considered:

$$c_{loc} = \sum_{i=0}^{19} \alpha_i r(t - iT_c). \tag{23.6}$$

Similarly, a 31-chip-long code can be defined:

$$d_{loc} = \sum_{i=0}^{29} \beta_i r(t - iT_c), \tag{23.7}$$

where

- $\beta_i = [1\ 1\ -1\ 1\ -1\ 1\ 1\ -1\ -1\ 1\ 1\ 1\ 1\ 1\ -1\ -1\ \ldots$
  $-1\ 1\ 1\ -1\ 1\ 1\ 1\ -1\ 1\ -1\ 1\ -1\ -1\ -1\ -1\ 1]$;
- $r(t)$ is the unitary amplitude rectangular function; and
- T_c is the chip length, the inverse of the code rate $R_c = 0.5$ MHz.

The tasks to be performed are:

1. generate the incoming signals, c_{in} and d_{in}, composed by three periods of c_{loc} and d_{loc}.
2. sample the incoming and local codes at $f_s = 8$ MHz, obtaining the sequences of samples $c_{in}[n]$, $c_{loc}[m]$, $d_{in}[n]$, and $d_{loc}[m]$, with $n = 1, \ldots, N$ and $M = 1, \ldots, 3N$, and $N = 20(f_s/R_c)$;
3. generate three realizations of additive white Gaussian noise (AWGN) noise $w[n]$, with standard deviation equal to 1, 4, and 8 respectively, and sum them to the incoming signal:

$$s[n] = c_{in}[n] + w[n], \tag{23.8}$$
$$r[n] = d_{in}[n] + w[n], \tag{23.9}$$

with $n = 1, \ldots, N$;

4. plot the correlation function, where the initial code phase of $c_{in}[n]$ and $d_{in}[n]$ is equal to 4 chips, and for the different values of the noise standard deviation;
5. compare the results in terms of peak detection.

Solution

Figure 23.1 shows the results when considering a 20-chip-long input code. As in the example reported in Chapter 21, the the correlation peak is clearly visible when the noise variance is equal to 1. When increasing the noise variance to 4, the correlation peak is still visible, but the noise floor is larger and an error of a few fractions of a chip is introduced. On the contrary, for higher noise variances, such as eight, the peak is buried in the noise floor, and the peak estimation procedure fails.

As expected, a longer code gives a larger correlations gain. Figure 23.2 reports the same correlation results as above, obtained using a 31-chip-long code. The correlation peak is correctly estimated also for a noise variance of eight. In addition,

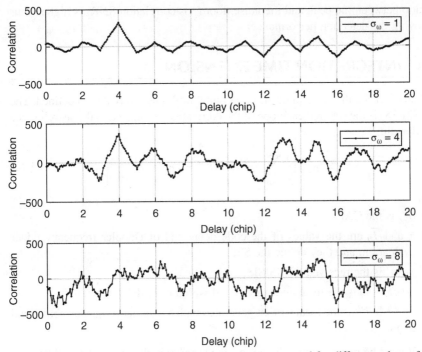

Figure 23.1   Correlation peak for a 20-chip-long sequence and for different values of the noise variance.

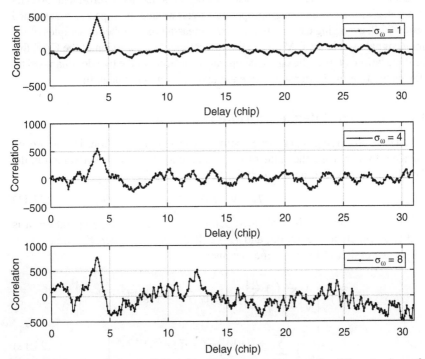

Figure 23.2   Correlation peak for a 31-chip-long sequence and for different values of the noise variance.

the ratio between the peak and the noise floor (postcorrelation SNR) is higher, thus improving the detection performance.

23.3 INTEGRATION TIME EXTENSION

In Chapter 21.3.2 the concept of the cross ambiguity function (CAF) is defined. The CAF (21.38) is the squared envelope of the correlations evaluated in the search space (SS) domain:

$$S_{y,r}\left(\bar{\tau}, \bar{f}_D\right) = \left| \frac{1}{L} \sum_{n=0}^{L-1} y_{\text{IF}}[n] c_b\left(nT_s - \bar{\tau}\right) e^{j2\pi\left(f_{\text{IF}} + \bar{f}_D\right)nT_s} \right|^2, \tag{23.10}$$

where:

- $S_{y,r}$ is the CAF;
- $\bar{\tau}$ and $\bar{f}_D$ are the values of the code delay and of Doppler frequency under evaluation, and they define the SS;
- $y_{\text{IF}}[n]$ is the received GNSS SIS;
- $c_b(nT_s - \bar{\tau})$ is the local code, generated with a code delay equal to $\bar{\tau}$.
- $e^{j2\pi(f_{\text{IF}} + \bar{f}_D)nT_s}$ is the local carrier, generated with a Doppler frequency shift equal to $\bar{f}_D$.

It must be remarked that the noise samples $w[n]$ in the received signal have an effect on each cell of the SS, altering the result in the CAF evaluation. Such noise contribution must be reduced. The noise can be modelled as a zero-mean AWGN random process; therefore averaging operations can reduce its impact. As explained in Section 23.2, averaging can be obtained by increasing the number of samples of the incoming signal used in the CAF evaluation. Recalling the expression of the CAF, this can be done before or after the envelope operation, leading to the two dominant strategies for integration time extension: *coherent* and *noncoherent* integration.

23.3.1 Coherent Integration Time Extension

In the first case, averaging is performed before the envelope operation. This approach is equivalent to increasing the value of L. N_c coherent sums can be performed in the CAF evaluation, thus leading to an integration time equal to:

$$T_c = N_c \cdot LT_s. \tag{23.11}$$

As the averaging operation is performed before the modulus square operation, it is referred to as coherent integration, and T_c is called *coherent integration time*.

The expression of the CAF the becomes:

$$S_{y,r,c}\left(\bar{\tau}, \bar{f}_D\right) = \left| \frac{1}{N_c} \sum_{m=0}^{N_c} \left(\frac{1}{L} \sum_{n=0}^{L-1} y_{\text{IF}}[n] c_b\left(nT_s - \bar{\tau}\right) e^{j2\pi\left(f_{\text{IF}} + \bar{f}_D\right)nT_s} \right) \right|^2$$

$$= \left| \frac{1}{N_c L} \sum_{n=0}^{N_c L - 1} y_{\text{IF}}[n] c_b\left(nT_s - \bar{\tau}\right) e^{j2\pi\left(f_{\text{IF}} + \bar{f}_D\right)nT_s} \right|^2. \tag{23.12}$$

The advantage of coherent summations is that the signal power is increased by a factor proportional to N_c, while the noise power is constant, as long as noise is zero mean, white, and Gaussian. Therefore, the postcorrelation SNR increases by a factor N_c. For instance, if coherent summation over two code periods is carried out ($N_c = 2$), the SNR increases by a factor of two, corresponding to about 3 dB. The derivation of the noise variance and of the theoretical distributions is provided in Section 23.5.1.

As a quality metric of the GNSS signal at the acquisition stage, the coherent output SNR is considered, and under ideal conditions (infinite front-end filter bandwidth and no quantization and frequency mismatching loss) it can be defined as

$$\rho_c = \frac{C}{N_0} \frac{LN_c}{2B_{IF}} = \rho_{IF} \frac{LN_c}{2} = \rho_\alpha N_c, \tag{23.13}$$

where ρ_{IF} is the SNR at the acquisition input, defined in (23.3). The term $LN_c/2$ accounts for both the despreading gain $L/2$ and the gain of the coherent integration (N_c) contributions. However, in real conditions, the front-end filtering effect leads to a nonnegligible loss in the achievable coherent gain, since the pseudorandom noise (PRN) code cannot be considered a pure square wave, thus impacting the shape of the correlation peak, which appears rounded. Additional losses to the coherent SNR, such as quantization effect, code alignment, and frequency mismatch are deeply addressed in [4].

The longer the coherent integration time is, the lower the noise floor will be, and the better the detection of the CAF peak will be. However, there are several limits in the extension of the predetection integration time, as clarified in Section 23.4. To overcome these issues, noncoherent accumulations can be performed.

Example 23.2: Extension of Integration Time

In Example 23.1, cross-correlation of a code in the presence of noise is evaluated, respectively, for a short and for a long sequence, proving that long sequences guarantee a larger despreading gain.

The goal of this example is to prove that a correlation gain can be obtained also by extending the integration time. The tasks to be performed are

1. generate the incoming signals, c_{in}, composed by 20 periods of c_{loc};
2. sample the incoming and local codes at $f_s = 8$ MHz, obtaining the sequences of samples $c_{in}[n]$ and $c_{loc}[m]$, with $n = 1, \ldots, N$ and $M = 1, \ldots, 20N$, and $N = 20(f_s/R_c)$;
3. generate three realizations of AWGN $w[n]$, with standard deviation equal to 1 and 12.5, respectively, and sum them to the incoming signal:

$$s[n] = c_{in}[n] + w[n], \tag{23.14}$$

with $n = 1, \ldots, N$;

4. generate the correlation for 10 subsequent portions of the incoming signal and sum them;

5. plot the correlation function, where the initial code phase of $c_{in}[n]$ and $d_{in}[n]$ is equal to 4 chips, and for the different values of the noise standard deviation;

6. compare the results in terms of peak detection.

Solution

Figure 23.3 reports the correlation results for the standard case ($N_c = 1$), as shown in Example 21.2, and for the case $N_c = 10$. As expected, in the second case it is possible to correctly detect the correlation peak at the right delay of 4 chips. Larger values of N_c allow detection of the correlation peak also for higher values of the noise variance.

23.3.2 Noncoherent Integration Time Extension

An averaging operation can be performed also after the envelope. N_n noncoherent sums are performed in the CAF evaluation, thus leading to an overall integration time equal to:

$$T_{nc} = N_n \cdot T_c = N_n \cdot N_c \cdot LT_s. \tag{23.15}$$

This kind of operation is denoted noncoherent integration, as the phase of the signal is lost in the square operation. T_{nc} is called *noncoherent integration time*.

The expression of the CAF then becomes:

$$S_{y,r,n}\left(\overline{\tau}, \overline{f}_D\right) = \frac{1}{N_n} \sum_{p=1}^{N_n} \left| \frac{1}{N_c} \sum_{m=1}^{N_c} \left(\frac{1}{L} \sum_{n=0}^{L-1} y_{IF}[n] c_b \left(nT_s - \overline{\tau}\right) e^{j2\pi\left(f_{IF} + \overline{f}_D\right)nT_s} \right) \right|^2. \tag{23.16}$$

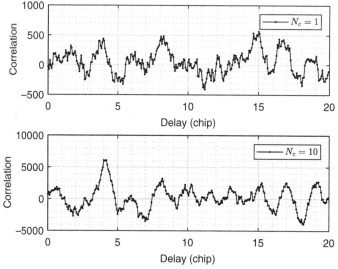

Figure 23.3 Correlation peak for different values of the integration time (N_c sums).

In this case, each block of $N_n N_c L$ samples is not treated as a single block but rather split into N_n sub-blocks, which are separately processed. Noncoherent accumulations are obtained summing noncoherently several instances of coherent integration. This strategy improves the capabilities of detecting weaker signals, at the expense of an increased number of operations.

It has to be noted that the Doppler frequency shift has to be taken into account when the coherent results are summed together. In the acquisition process, the Doppler effect on the spreading code is normally neglected. Therefore, the initial phase of the incoming code can change slightly with respect to the locally generated sequence, depending on the absolute value of the Doppler shift. In order to compensate for this, the coherent correlations need to be properly time-shifted during the noncoherent accumulation process.

When performing the summation after the envelope, both the signal and the noise power are increased. The theoretical distributions are provided in Section 23.5.1. The signal increases by a factor N_n, while the noise increases by a factor $\sqrt{N_n}$ [4]. For instance, if a noncoherent summation over two code periods is carried out ($N_n = 2$), the signal power increases by a factor of two and the noise by a factor of $\sqrt{2}$. This is due to the fact that the noise after the envelope is no longer zero mean. Therefore, the postcorrelation SNR increases by a factor of $2/\sqrt{2} = \sqrt{2}$, corresponding to about 1.5 dB. This is known as *squaring loss* [7].

An equivalent coherent SNR can be defined as the coherent SNR required for a noncoherent detector based on N_n accumulations in order to achieve the same performance of a pure coherent detection. The equivalent SNR can be defined as

$$\rho_n\big|_{dB} = \rho_c\big|_{dB} - L_s + 10\log_{10} N_n, \tag{23.17}$$

where L_s is the squaring loss. Analytical derivation for the squaring loss can be found in [4] and [8].

Given a target scenario, in terms of sensitivity, signal dynamic, and availability of assistance information, a trade-off between the coherent integration time and the number of noncoherent accumulations has to be analyzed. In absence of other issues, which will be considered in Section 23.4, coherent integration is more efficient than noncoherent integration in terms of SNR gain, and thus preferable. Being the extension of the coherent integration time limited by several factors, as detailed later on, noncoherent accumulations are exploited to further increase the SNR gain.

Figure 23.4 summarizes the concept of coherent and noncoherent integration.

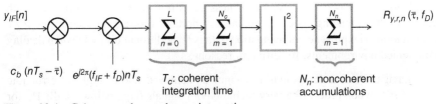

Figure 23.4 Coherent and noncoherent integrations.

23.3.3 Differential Combination

Although being much less common in consumer receivers, differential combination can be included into the high sensitivity techniques. It was first proposed by [9] and [10]. Differential correlation can be seen as an extension of noncoherent intregration, in which, before the square envelope operation, the correlation output is multiplied by the conjugate of the correlation output obtained at the previous integration interval and accumulated. Denoting the $R[m]$ the correlation output at epoch m,

$$R[m] = \frac{1}{L} \sum_{n=0}^{L-1} y_{\text{IF}}[n] c_b (nT_s - \overline{\tau}) e^{j2\pi(f_{\text{IF}} + \overline{f}_D)nT_s}, \qquad (23.18)$$

the differential combination integration assumes the form

$$S_{y,r,d} = \left| \sum_{m=2}^{M} R[m] R^*[m-1] \right|^2. \qquad (23.19)$$

A slight performance improvement can be obtained when compared to noncoherent integration, because the signal components remain highly correlated in two consecutive correlation intervals, whereas the noise components tends to be decorrelated. A second advantage is the computational burden required, due to the fact that the SS size does not change, as described in [11]. However, the despreading gain is lower than the gain achievable exploiting a pure coherent integration scheme. Furthermore, it suffers from the phase reversal due to bit transition. Advanced and generalized differential combination techniques have been studied, and a summary is reported in [12].

23.4 EFFECTS OF INCREASED INTEGRATION TIME

The integration time extension is the natural and straightforward solution to increase the receiver sensitivity. However, integration time extension introduces new issues, and high-sensitivity GNSS receivers are required to tackle several problems. The main issues are described in detail hereafter.

23.4.1 Data Transition

Most of the GNSS signals, such as GPS L1 C/A and Galileo E1b, are modulated by a navigation message. The navigation message is transmitted at a lower rate with respect to the spreading code and contains important information used in the PVT computation. The presence of navigation data is an additional obstacle when dealing with denied environments, for different reasons:

- navigation data demodulation cannot be achieved due to high bit error rate (BER) when processing very feeble GNSS signals (e.g., below 24 dB-Hz for GPS legacy signals) [3];

- at low C/N_0 also the data bit synchronizing could fail, thus preventing the generation of pseudoranges;
- the data bit length T_b limits the coherent integration time extension, since data bit transitions introduce shifts in the carrier phase within the integration window, leading to partial or total cancellation of the correlation power.

In Section 23.3 we described how extending the coherent integration time can improve the postcorrelation SNR. However, this is valid as long as no phase transitions due to the navigation message bits occur.

Figure 23.5 depicts a standard GNSS receiver acquisition stage, in which GPS L1 C/A signals are considered. The code period is $T_{code} = 1$ ms and the data bit duration is $T_b = 20$ ms. The navigation message bits, the SIS received code, and the local code are drawn, respectively, in green, blue, and red. The correlation between 1 ms of received signal and 1 ms of local code does not include a bit transition. As a result, the monodimensional CAF depicted in the figure exhibits a clear correlation peak.

When extending the integration time, for instance to $T_c = 10$ ms, two situations might occur. If we are lucky enough, no data bit transition will occur within the integration period. This happens if the 10 ms are entirely included in the same bit of the navigation message, of if they are across two bits with the same sign. Also in this case, no side effects occur. The first situation is depicted in Figure 23.6. $N_c = 10$ code periods of the incoming signal are coherently accumulated and correlated with

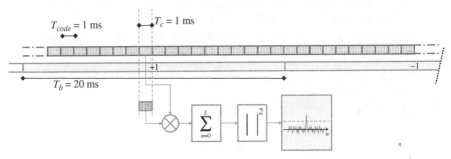

Figure 23.5   Pictorial representation of the correlation procedure in the case $T_c = 1$ ms.

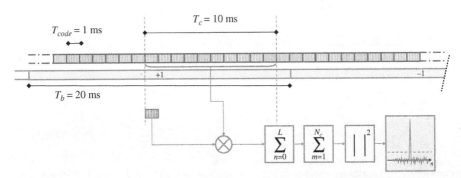

Figure 23.6   Pictorial representation of the correlation procedure in the case $T_c = 10$ ms.

the local code. The CAF peak is clear and higher than in the previous case because of the coherent integration gain. The CAF can be envisaged as the sum of 10 CAFs, each of them containing a positive correlation peak similar to the one in Figure 23.5.

On the contrary, if a data transition occurs within the coherent integration integration period, the signal detection might fail. The example reported in Figure 23.7 shows the worst case, i.e., the case in which a bit transition falls exactly in the middle of the coherent integration period. The correlation peak is absent, because the phase transition induced by the −1 bit annihilates the correlation. The resulting CAF can be envisaged as the sum of 10 correlations, 5 of them containing a positive peak and 5 of them containing a negative peak. Intermediate conditions can occur, leading to similar results.

This last example shows why the bit duration poses a severe limit to the integration time extension. In order to avoid correlation cancellation, it is required to have $T_c \leq T_b$, which means that the coherent integration time is limited to 20 ms for GPS L1 C/A signals.

It has to be reminded that navigation message bit synchronization has to also be achieved to avoid correlation losses. The time instant in which the navigation bit transition occurs has to be known by the receiver. This is normally the case in the tracking stage, but it never happens for acquisition, unless some aiding is provided. To minimize the probability of correlation annihilation due to bit transition, a common solution is to limit the coherent integration time to half the bit length, i.e., $T_c = 10$ ms. To further reduce this probability, it is also possible to perform two acquisition trials, using two consecutive 10-ms-long sequences. This assures that at least in one of the two CAFs no transitions occurred.

In order to overcome the data transition problem, several solutions are possible:

- the use of pilot dataless signals, such as Galileo E1c, as described in Section 23.7.2;
- the use noncoherent sums, as described in Section 23.3.2;
- the wipe-off of the navigation message, exploiting assistance information, as described in Section 23.4.1;
- the prediction of the navigation message data bits, as detailed in Section 23.4.1.

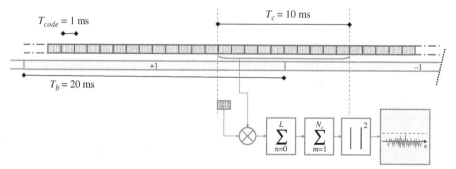

Figure 23.7   Pictorial representation of the correlation procedure in the case $T_c = 10$ ms, and a bit transition occurs at half the integration period.

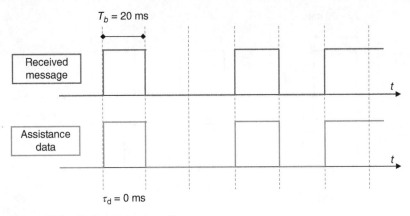

Figure 23.8   Perfect data wipe-off.

Navigation Message Wipe-Off   The bits of the navigation message can be wiped-off, assuming that:

1. the sequence of bits is known;
2. precise time information to synchronize the wipe-off operation is available.

The stream of data bits is normally included in the information of Assisted GNSS (A-GNSS). Reference stations are able to predict the content of the message and to send it to enabled receivers. In this way, on one side the receiver does not have to demodulate the message, as all the information needed for the PVT computation are provided by A-GNSS. On the other side, an ideally perfect navigation message wipe-off can be performed on the SIS, allowing arbitrarily long coherent integration times.

The navigation data bits can also be provided by a reference receiver, located close to the high-sensitivity receiver, as is normally done in geodesy. Also peer-to-peer and cooperative networks can be exploited, as described in [13].

When the accuracy time information is limited, an additional correlation loss is introduced. In [14], a set of semi-analytical simulations, emulating different levels of timing accuracy, has been performed to assess the impact of the assistance timing error. A simulated GPS L1 C/A signal, characterized by a data bit transition every 20 ms is considered. Coherent integration at $T_c = 20$ ms is performed on the input signal. An A-GNSS assistance system is emulated: the sequence of bits is assumed always correct, while a delay τ_d in the range 0 to 10 ms, is artificially introduced. As expected, in the ideal case of $\tau_d = 0$ ms the wipe-off is perfect. Figure 23.8 depicts this situation: the incoming data bits and the assistance message are perfectly synchronized. Consequently, all the signal power contributes to the correlation peak.

On the contrary, when an error τ is present, part of the power of the correlation operation is lost, as shown in Figure 23.9, where $\tau \approx 4$ ms. The worst case is when the bit transition happens exactly at half the bit length, which is for $\tau_d = 10$ ms; in this case, the phase change completely cancels the correlation result.

Figure 23.10 depicts the correlation SNR during a simulation for different values of the assistance timing error and for different values of signal C/N_0, confirming the theoretical assumptions. As long as the timing error increases, both the magnitude of

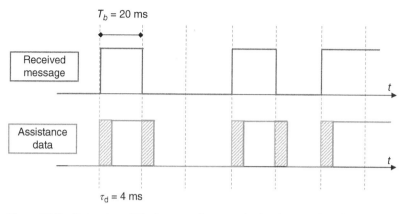

Figure 23.9   Data wipe-off in the case of an error in the time synchronization.

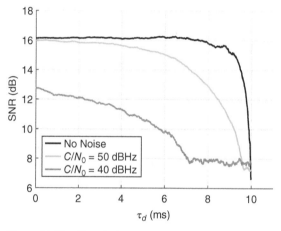

Figure 23.10   Correlation peak magnitude versus assistance timing error.

the peak and the noise floor level decease, with different trends. In the ideal case (no noise, black line), the SNR is constant until an error of about 8 ms; after this value the SNR starts falling to zero. On the contrary, for real signals, with C/N_0 equal to 50 and 40 dB-Hz, respectively, the SNR decreases much faster as long as the error increases. In the latter case, an error equal to 7 ms already completely cancels out the correlation. After this value, the correlation peak is completely buried in the noise floor.

Data Bits Prediction and Estimation   As an alternative, data bits can be predicted or estimated.

In the first case, it is possible to foresee the incoming navigation message bits by exploiting the regular structure of the navigation message. For instance, the preamble, once determined for the first time, is always identical. Similarly, in the case of GPS legacy signals, ephemeris and clock parameters are updated every 2 hours, while the almanac is updated at least every 6 days. The correspondence bits can thus can be propagated from a frame to the following one. Furthermore, some information common to all GNSS satellites can be demodulated from the signals with high C/N_0

TABLE 23.1   Length of the Data Bit for the Most Common OS Navigation Signal.

Signal	T_b
GPS L1 C/A	20 ms
GPS L1C	10 ms
GPS L2C	20 ms
GPS L5 I	10 ms
GPS L5 Q	dataless
Galileo E1b	4 ms
Galileo E1c	dataless
Galileo E5A-I	20 ms
Galileo E5A-Q	dataless
Galileo E5B-I	4 ms
Galileo E5B-Q	dataless

and exploited to increase the coherent integration time for low power signals. This makes the overall navigation message partially deterministic.

A second possibility is offered by estimation techniques. The most simple algorithm consists in trying all the possible combination of bits and then choosing the one offering the highest postcorrelation SNR. However, this is possible only if bit synchronization has already been achieved. The main drawback is the exponential increase of the computational complexity.

23.4.2   Secondary Code Synchronization

Modernized and new GNSS signals contain a secondary code in addition to the primary code. For example, in the case of Galileo E1c, the code is tiered to the primary code, 25 chips long, thus spanning a total time of 100 ms. The sequence is known and periodic; however, it must be wiped off from the signal in order to extend the integraion time. Synchronization must be achieved, leading to an increased dimension of the SS in the code delay domain and thus to a worsening of the targeted false alarm probability of the overall acquisition system, as shown in [15].

The secondary code can be removed either through self-techniques, such as maximum likelihood estimator (MLE) techniques, or exploiting assistance data. However, such techniques are not optimal at low C/N_0.

More details about high-sensitivity techniques for new and modernized GNSS signals are reported in Section 23.7.

23.4.3   Considerations on the Doppler Frequency

A complete description and derivation of the Doppler frequency affecting a GNSS signal has been provided in Chapter 21. The Doppler frequency at the receiver is caused by the combination of three different factors: the motion of the satellite, the motion of the user, and any uncompensated frequency offset in the receiver reference oscillator.

Good high sensitivity performance can be achieved only if the user dynamics are limited. The optimal situation is when the receiver is stationary or moving at relatively low constant velocity or very low acceleration, so that the only contribution of the Doppler shift is due to the satellites' motion.

During their pass, GPS satellites are moving toward and away from a static receiver on the Earth surface. They reach the maximum relative speed during rising and setting, when they are at zero elevation. For GPS satellites' orbit, $v_{max} = 930$ m/s. The maximum Doppler frequency is then equal to

$$f_{D,\max} = \frac{f_{L1} \cdot v_{\max}}{c} = \frac{1575.42 \times 10^6 \cdot 930}{c} = 4.9 \text{ kHz,} \tag{23.20}$$

where c is the speed of light. Similarly, the minimum Doppler frequency is equal to -4.9 kHz. Therefore, the Doppler range to be evaluated amounts to

$$\delta f = f_{D,\max} + |f_{D,\min}| = 9.8 \text{ kHz.} \tag{23.21}$$

For terrestrial applications, receiver motion is very small compared to the satellite speeds, so its effect is normally negligible. There is up to approximately 1.5 Hz of Doppler frequency for each 1 km/h of receiver speed [4]. Nevertheless, for high-sensitivity receivers featuring long integration times this should be considered.

For space applications, the relative speed between receiver and satellite can increase a lot, as described in [16]. Therefore, the maximum Doppler frequency can reach values as high as 20 kHz, and $\delta f = 40$ kHz. The main problem to face with such a large SS is that the number of frequency bins to be evaluated in the SS significantly increases, thus raising the acquisition time.

The total number of bins N_{bin} to be evaluated in the Doppler domain search depends indeed on the size of the SS and on width of the frequency bin:

$$N_{\text{bin}} = \left\lceil \frac{\delta f}{\Delta f} \right\rceil + 1. \tag{23.22}$$

The size of a Doppler bin Δf depends in turn on the coherent integration time, taking into account the *mistuning loss*. Chapter 21.3.2 (Fig. 21.7) describes the topic in detail and gives the following empirical rule to minimize the losses:

$$\Delta f = \frac{2}{3T_c}. \tag{23.23}$$

The correlation peak in frequency domain varies with frequency as a sinc function, as depicted in Figure 23.11. A larger T_c improves the sensitivity, but increases the width of the sinc function. If T_c is doubled, then sinc function width is halved and the number of frequency bins is doubled. It should be noted that, while a mistuning loss of a few dBs can be acceptable when working with signals at a nominal C/N_0, in the case of a high-sensitivity receiver the value of Δf might be further lowered to reduce the loss [3].

For large values of the integration times, Δf becomes extremely low. As an example, a coherent integration time equal to 3 s leads to a value of $\Delta f = 0.22$ Hz. Even though the Doppler resolution improves, the number of bins to be evaluated

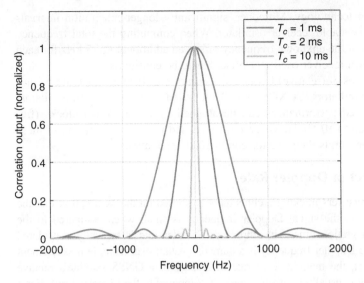

Figure 23.11   Correlation peak in the frequency domain for various values of T_c.

in the Doppler domain search excessively increases. Considering a Doppler range $\delta f = 9.8$ kHz, the value of N_{bin} is approximately 44,000, which is a nonreasonable value for terrestrial GNSS applications. In space environment, due to a wider Doppler range domain, the issue is magnified.

In a fast fourier transform (FFT)-based acquisition scheme parallel in time domain, the time required for the CAF computation depends on the number of frequency bins and on the overall observation time. It is called mean acquisition time (MAT), and it is expressed as:

$$T_a = T_c \cdot N_n \cdot N_{\text{bin}}. \tag{23.24}$$

This has clearly an impact in both the computational burden of the receiver and the time to first fix (TTFF). TTFF is defined as the time a receiver need from the moment it is switched on to the moment it provides the first PVT solution, and it depends on size of the SS, on the detection probability, on the probability of false alarm, and on the integration time.

Large banks of correlators or advanced digital signal processing operations are required to maintain an affordable MAT. Alternatively, it is necessary to reduce the frequency SS range by exploiting some sort of Doppler frequency aiding, such as assistance information, or relying on integration with other sensors. Even if a precise Doppler frequency is available the code alignment requires an accurate timing synchronization, better than 1 ms. An example is reported in Section 23.6.1.

23.4.4   Clock Errors and Stability

It should be mentioned that, even in presence of high accurate Doppler aiding, residual Doppler may be still present due to the finite accuracy of the local oscillator frequency, which usually differs from the nominal value by an amount related to the adopted oscillator technology. Moreover, significantly longer integration intervals also require a very stable receiver oscillator. When computing the total frequency range uncertainty of the unknown frequency offset, an additional 1.575 kHz for each 1 ppm of unknown receiver oscillator offset has to be considered.

Typical figures of (relative) frequency accuracy are 10^{-6} for temperature compensated crystal oscillators (TCXOs) and 10^{-8} for oven controlled crystal oscillators (OCXOs). Improved performances can be attained by using atomic clocks (frequency stability up to 10^{-26}). However, cost, size, and power consumption of these types of oscillators limits their use for commercial applications.

23.4.5. Impact of Doppler Rate

Standard GNSS receivers tracking architectures are based on the assumption that the maximum change in the signal Doppler frequency is small when compared to the inverse of the integration time. The impact of Doppler rate is typically considered negligible, and the Doppler frequency is assumed constant over a single measurement epoch [17]. In fact, the magnitude of rate of change of a GNSS satellite's relative Doppler frequency is usually relatively small, if compared to the Doppler itself. For a static user, considering GPS and Galileo legacy signal in L1, it can increase up to 0.8 Hz/s depending on the satellite elevation and on the user's latitude [4]. However, when either the receiver velocity or the integration time significantly increase, the satellite Doppler rate becomes potentially significant, and the sole knowledge of the instantaneous Doppler might be not sufficient. The capability of providing the GNSS receiver with an accurate Doppler rate aiding allows the local carrier generator to follow the Doppler variation over time, thus making the receiver more robust to the high dynamic.

Requirements of such an aiding, in terms of accuracy, have to be assessed while taking into account the coherent integration time adopted at the acquisition stage, which determines the CAF resolution Δf. As an example, supposing that the receiver is provided with an ideal estimate of the Doppler frequency and an accurate Doppler rate aiding, the required accuracy of the Doppler rate estimate has to be such that the final mismatching between the true average Doppler frequency and the estimated average Doppler frequency during the overall acquisition period lies within half of the CAF resolution. Figure 23.12 shows the theoretical frequency mismatching between the average estimated Doppler frequency and the true average Doppler frequency in the case of very weak GPS (left plot) and Galileo (right plot) signal acquisition at 10 dB-Hz for different values of the Doppler rate bias, which is the shift between the true Doppler rate affecting the GNSS signal and the Doppler rate estimate from the aiding source.

Unmodelled receiver velocity introduces a further component in the relative Doppler and Doppler rate. Any 19 cm movement (i.e., the wavelength at L1) in the direction of a satellite induces an unexpected Doppler rate and introduces a phase shift through 360°, leading to signal annihilation. Figure 23.13 shows the value of the coherent integration time for which a complete phase shift occurs, for any velocity from 0 to 2 m/s. Receiver velocities up to 2 m/s (typical pedestrian velocity) can already annihilate the signal for an integration time of 100 ms. When it is increased to 1 s, the unmodelled velocity should be lower than 0.2 m/s to keep tracking the SIS.

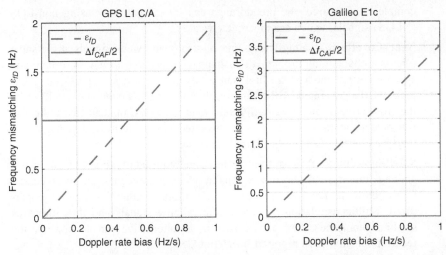

Figure 23.12   Theoretical Doppler rate error requirements for weak GNSS signal acquisition.

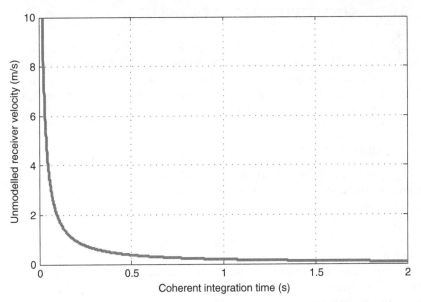

Figure 23.13   Maximum allowable unmodelled receiver velocity as a function of coherent integration time.

23.4.6   Other Factors Impacting Receiver Sensitivity

- **Interference.** RFI is caused by electronic equipment (such as television, mobile, and very high frequency [VHF] transmitters) radiating in the proximity of the GPS frequency band, or by intentional transmission of in-band RF signals (e.g., jamming and spoofing).
- **Multipath reflections.** Multipath causes additional power losses (fading) and measurement biases. In some cases, where the line-of-sight (LOS) signal is

completely obscured, the preudorange can be estimated by using non-LOS (NLOS) signals, which are inherently weaker.

- **Near–far effect.** Dealing with a mix of strong and weak signals at the same time is a challenging problem. This happens especially in indoor, urban, and space environments, where high-sensitivity techniques are employed. False locks can be generated by the autocorrelation of a PRN code with itself, or by the cross-correlation between different ranging codes. In such cases, the risk is to detect and track a peak from an unwanted and much stronger signal, rather than the correct peak of the desired weak signal.

- **Satellite visibility.** At least four satellites are required to produce a valid navigation solution. In all those situations in which some signals are received at a C/N_0 below the receiver sensitivity, the solution availability may be jeopardized. Multiconstellation provides more ranging signals and thus improves satellite geometries. This can be seen as an alternative way to improve the receiver sensitivity, as reported in Section 23.7.

23.5 HIGH-SENSITIVITY RECEIVER DESIGN

The design of GNSS receivers for high-sensitivity focuses mainly on the acquisition and tracking stages. The main challenge is the determination of the correlation peak and thus the estimation of the signal code phase. In particular, in the acquisition stage the risk is to fail in identifying the correct correlation peak. In the tracking stage, low C/N_0 leads to loss of tracking lock.

23.5.1 Acquisition Stage

Standard acquisition schemes are designed to work in open-sky conditions, where the antenna is expected to capture nominal received signal strength. In harsh environments, GNSS receivers have to deal with a reduced power level. Exploiting the direct-sequence spread spectrum (DSSS) nature of CDMA signals, a higher despreading gain can be achieved. This results in an SNR improvement in the CAF after the correlation process.

Therefore, the first strategy for increasing the acquisition sensitivity is the integration time extension. However, as briefly mentioned in Section 23.3.1, this is limited by several factors:

- the presence of unknown navigation data bit transitions, which imposes the condition $T_c \leq T_b$;
- the presence of a secondary code in pilot signals, which, although being a known sequence, still requires proper synchronization;
- the reduction of the acquisition Doppler bin size and thus the increase of the SS size and of the computational complexity.

Therefore, a trade-off between the sensitivity improvement and the complexity increase should be considered when changing the value of the integration time. Most of these limitations can be mitigated with the noncoherent combination of short coherent integrations, as described in Section 23.3.2. However, this approach leads to lower processing gains and, therefore, a second trade-off is usually made between the coherent integration time and the number of noncoherent combinations.

In the acquisition stage, the presence of a bit transition acts as a sort of subcarrier, which splits the main lobe of the spectrum in two side peaks, symmetric with respect to the correct bin, leading to an incorrect acquired Doppler frequency. The shape of the two lobes and the relative amplitudes of the peaks depend on the position of the bit transition. An example of GPS SIS acquisition, characterized by a phase transition in the middle of the observation windows, is reported in Figure 23.19 and will be described later on.

Statistical Characterization of the Acquisition

As remarked in Section 21.3.1, GNSS signal acquisition can be seen as a detection problem. Therefore, each value X of the search space, as defined in (21.37), i.e., the correlation output for each for each bin of the SS, can be modeled as a random variable. The maximum of the SS (S_{max}, Eq. (21.31)) is then used as a test statistic and compared against a threshold θ_t, to test two hypothesis:

- H_0, the null hypothesis, if only noise is present;
- H_1, the so-called alternative hypothesis, if the signal is present, along with noise, and is correctly aligned.

Recalling the expression of the CAF for coherent and noncoherent schemes, reported respectively in Sections 23.3.1 and 23.3.2, and assuming Gaussian noise, its theoretical distribution can be computed. Thus, the probability density function (PDF) of the random variable can be derived, both for coherent and noncoherent schemes. An exhaustive theoretical derivation can be found in [8].

The expression of the CAF in the case of coherent integration was reported in (23.12). $S_{y,r,c}(\overline{\tau}, \overline{f}_D)$ is the square absolute value of a complex Gaussian random variable, with independent real and imaginary parts $Y_I(\overline{\tau}, \overline{f}_D)$ and $Y_Q(\overline{\tau}, \overline{f}_D)$. Their variance can be computed as:

$$
\begin{aligned}
&\mathrm{Var}\left[Y_I(\overline{\tau}, \overline{f}_D)\right] \\
&= \mathrm{Var}\left[\Re\left\{\frac{1}{N_c L}\sum_{n=0}^{N_c L-1} y_{\mathrm{IF}}[n]c_b(nT_s - \overline{\tau})e^{j2\pi(f_{\mathrm{IF}}+\overline{f}_D)nT_s}\right\}\right] \\
&= \mathrm{Var}\left[\frac{1}{N_c L}\sum_{n=0}^{N_c L-1} y_{\mathrm{IF}}[n]c_b(nT_s - \overline{\tau})\cos(2\pi(f_{\mathrm{IF}}+\overline{f}_D)nT_s)\right] \\
&= \frac{1}{N_c^2 L^2}\sum_{n=0}^{N_c L-1}\mathrm{Var}\left[y_{\mathrm{IF}}[n]c_b(nT_s - \overline{\tau})\cos(2\pi(f_{\mathrm{IF}}+\overline{f}_D)nT_s)\right] \\
&= \frac{1}{N_c^2 L^2}\sum_{n=0}^{N_c L-1}\frac{\sigma_{IF}^2}{2} = \frac{\sigma_{IF}^2}{2N_c L},
\end{aligned}
\tag{23.25}
$$

where σ_{IF}^2 is the noise variance of the digital IF SIS, as defined in (21.20). Similarly, it can be proven that $\text{Var}[Y_Q(\overline{\tau}, \overline{f}_D)] = \dfrac{\sigma_{IF}^2}{2N_c L}$, and thus the coherent noise variance is:

$$\text{Var}\left[Y_I\left(\overline{\tau}, \overline{f}_D\right)\right] = \text{Var}\left[Y_Q\left(\overline{\tau}, \overline{f}_D\right)\right] = \frac{\sigma_{IF}^2}{2N_c L} = \sigma_c^2. \tag{23.26}$$

It is easy to observe that the variance of the noise is inversely proportional to N_c, confirming the benefits of coherent acquisition schemes.

Under the null hypotesis, $S_{y,r,c}(\overline{\tau}, \overline{f}_D)$ is zero mean. Thus, $S_{y,r,c}(\overline{\tau}, \overline{f}_D) \mid H_0$ follows a central χ^2 distribution (or Rayleigh distribution), with PDF:

$$f_c(s \mid H_0) = \frac{s}{\sigma_c^2} e^{-\frac{s^2}{2\sigma_c^2}}. \tag{23.27}$$

Under the alternative hypotesis, $S_{y,r,c}(\overline{\tau}, \overline{f}_D)$ is no longer zero mean. The sum of the square of two nonzero mean independent Gaussian random variables leads to a noncentral χ^2 distribution (or Rice distribution), with PDF

$$f_c(s \mid H_1) = \frac{s}{\sigma_c^2} e^{-\frac{s^2 + A^2}{2\sigma_c^2}} I_0\left(\frac{sA}{\sigma_c^2}\right) u(s), \tag{23.28}$$

A being the root mean square signal amplitude, $I_0(\cdot)$ the modified Bessel function of the first kind and zero order, and $u(s)$ the unit step function.

Considering N_n noncoherent integrations, the CAF $S_{y,r,n}(\overline{\tau}, \overline{f}_D)$, given in (23.12), is the sum of N_n independent χ^2 random variables with 2 degrees of freedom. Thus, under the null hypothesis, $S_{y,r,n}(\overline{\tau}, \overline{f}_D)$ is a central χ^2 random variable with $2N_n$ degrees of freedom, with PDF:

$$f_n(s \mid H_0) = \frac{1}{2\sigma_c^2} \frac{1}{\Gamma(N_n)} \left(\frac{s}{2\sigma_c^2}\right)^{N_n - 1} e^{-\frac{s}{2\sigma_c^2}}, \qquad s \geq 0. \tag{23.29}$$

Under the alternative hypotesis, $S_{y,r,n}(\overline{\tau}, \overline{f}_D)$ is a noncentral χ^2 random variable with $2N_n$ degrees of freedom, with PDF:

$$f_n(s \mid H_1) = \frac{1}{2\sigma_c^2} \left(\frac{s}{A}\right)^{\frac{N_n - 1}{2}} e^{-\frac{s + A}{2\sigma_c^2}} I_{N_n - 1}\left(\frac{\sqrt{sA}}{\sigma_c^2}\right), \qquad s \geq 0. \tag{23.30}$$

Detection and False Alarm Probabilities The acquisition scheme performance can then be evaluated by means of probabilities. For a specific value of θ_t, there are four possible outcomes [1]:

- **detection probability, P_d,** the probability of correctly detecting the signal:

$$P_d = P(S_{max} > \theta_t \mid H_1); \tag{23.31}$$

- **miss detection probability, P_{md},** the probability of missing the detection of the signal in the case of presence of the signal:

$$P_{md} = P(S_{max} \leq \theta_t \mid H_1) = 1 - P_d; \tag{23.32}$$

- **false alarm probability,** P_{fa}, the probability of detecting a wrong signal, in the case in which the true signal is not present

$$P_{fa} = P(S_{max} > \theta_t \mid H_0);$$ (23.33)

- **correct dismissal probability,** P_{cd}, the probability of correctly not detecting a signal in the case where the true signal is not present

$$P_{cd} = P(S_{max} \leq \theta_t \mid H_0) = 1 - P_{fa}.$$ (23.34)

In Figure 23.14, the PDFs for the H_0 and H_1 conditions are drawn respectively in pink and in green. The same figure reports the threshold θ_t and the different probabilities defined above, corresponding to the shaded area. If $f(s \mid H_1)$ and $f(s \mid H_0)$ are respectively the PDFs in the two test hypothesis, then the single cell detection and false alarm probabilities are defined as

$$P_d = \int_{\theta_t}^{+\infty} f(s \mid H_1) \, ds$$ (23.35)

and

$$P_{fa} = \int_{\theta_t}^{+\infty} f(s \mid H_0) \, ds.$$ (23.36)

In the case of coherent correlation, by substituting (23.27) and (23.28) into (23.36) and (23.35), respectively and by exploiting properties of central and non-central χ^2 random variables [8, 18], the following false alarm and detection probabilities are obtained:

$$P_{fa,c}(\theta_t) = e^{-\frac{\theta_t^2}{2\sigma_c^2}},$$ (23.37)

and

$$P_{d,c}(\theta_t) = Q_1\left(\sqrt{\frac{A}{\sigma_c^2}}, \sqrt{\frac{\theta_t}{\sigma_c^2}}\right),$$ (23.38)

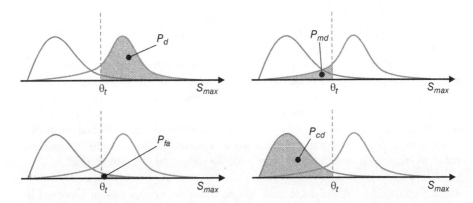

Figure 23.14   PDF in the case of H_0 and H_1 and related acquisition probability metrics.

where $Q_K(a,b)$ is the generalized Marcum Q-function, defined as:

$$Q_K(a,b) = \frac{1}{a^{K-1}} \int_b^{+\infty} x^K e^{-\frac{a^2+x^2}{2}} I_{K-1} ax \, dx. \tag{23.39}$$

Similarly, in the case of noncoherent correlation, the following false alarm and detection probabilities are obtained:

$$P_{fa,n}(\theta_t) = e^{-\frac{\theta_t}{2\sigma_c^2}} \sum_{i=0}^{N_n-1} \frac{1}{i!} \left(\frac{\theta_t}{2\sigma_c^2} \right)^i, \tag{23.40}$$

and

$$P_{d,n}(\theta_t) = Q_K \left(\sqrt{\frac{KA}{\sigma_c^2}}, \sqrt{\frac{\theta_t}{\sigma_c^2}} \right). \tag{23.41}$$

The acquisition threshold is determined according to a required target P_{fa}, by using (23.36), or to a maximum MAT. Therefore, for the case of coherent integration, the threshold corresponds to:

$$\theta_t = \sqrt{-2\sigma_c^2 \ln(P_{fa})}. \tag{23.42}$$

Once the acquisition threshold is set, the single cell detection probabilities can be evaluated, exploiting (23.35).

It is straightforward to notice that higher coherent integration times (i.e., higher values of N_c) lead to lower values of σ_c^2, and in turn to lower false alarm probablities P_{fa} and to higher detection probabilities P_d, for a given threshold θ_t. The same considerations hold for N_n. From a graphical point of view, it can be shown that for higher values of N_c and N_n the probability distributions depicted in Figure 23.14 move away one from each other.

ROC Curves   Once P_d and P_{fa} are obtained, they can be used to plot the receiver operating characteristic (ROC) curve, which depicts the trend of the false alarm probability versus the detection probability for different values of the threshold.

Figure 23.15 provides an example of a ROC curve for a GPS L1 C/A signal at 30 dB-Hz. The different points of the blue curve correspond to different values of θ_t. The curve highligths the fact that the lower the false alarm probability, the lower the detection probability.

The goal of a high-sensitivity receiver is indeed to move the working point in the area above the curve, so as to obtain low false alarm events and good detection probabilities at the same time. This can be achieved by exploiting coherent and noncoherent acquisition schemes. An example is provided in Figure 23.16, where it is shown that increasing the number N_c of coherent integrations, improves the performance, thanks to the reduction of the noise variance. When compared to noncoherent integration, coherent integration assures higher curves for the same total integration time.

Peak Separation   Acquisition peak-to-floor ratio metrics have been proposed in [12] to assess the performance of single acquisition trials. They are able to highlight

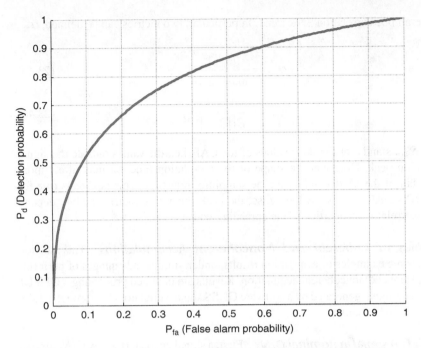

Figure 23.15   ROC curve for $C/N_0 = 30$ dB-Hz and for a single integration time period.

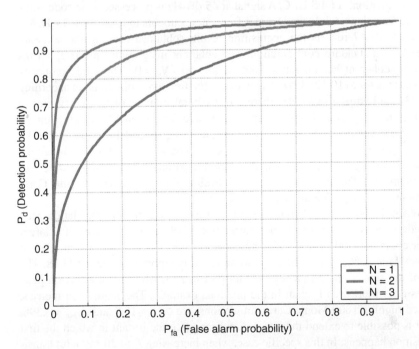

Figure 23.16   ROC curve for $C/N_0 = 30$ dB-Hz and different coherent integration values N_c.

the overall postcorrelation SNR trend. In particular, two metrics can be defined, α_{max} and α_{mean}:

$$\alpha_{max} = \frac{|S_{max}|^2}{\max|S_{floor}|^2},$$ (23.43)

$$\alpha_{mean} = \frac{|S_{max}|^2}{E\left[|S_{floor}|^2\right]},$$ (23.44)

where S_{floor} stands for the floor values of the CAF, i.e., the values outside the signal correlation peak. Note that the shape of the correlation function may affect more than 1 bin of the SS depending on the sampling frequency. This is always true for the BOC signals. In case of low noise, the peak separation metrics mainly depend on the correlation properties of the spreading codes.

Examples of Coherent and Noncoherent Accumulations This section reports some examples of acquisition results, underlying the advantages of performing coherent and noncoherent acquisition. Results are obtained processing a GPS L1 C/A legacy signal generated by a software GNSS signal simulator [19] through a fully software GNSS receiver.

GPS L1 C/A signal at nominal C/N₀ First, a signal characterized by C/N_0 around the nominal value is considered, to analyze the impact of larger integration times in the CAF evaluation. A GPS L1 C/A signal at 45 dB-Hz is processed. The code delay is equal to 512 chips, while the Doppler frequency amounts to 1500 Hz + 0.1 Hz/s.

Figure 23.17 reports the acquisition results, obtained running the software receiver with a standard configuration. The coherent integration time is $T_c = 1$ ms and no noncoherent accumulations are performed ($N_n = 1$). The total frequency search space is set to 10.125 kHz. The width of the frequency bin Δf is set according to (23.33), and equal to 667 Hz; therefore, the total number of frequency bins is $N_{bin} = 15$. The left part of the figure reports the three-dimensional CAF, while on the right the monodimensional code delay domain and frequency delay domain correlations are reported. As expected, the receiver can successfully acquire the signal; the peak clearly emerges from the noise floor and the parameters $\hat{f}_D$ and $\hat{\tau}$ can be correctly estimated. The acquisition metrics, evaluated running 1000 Monte Carlo simulations, are $\alpha_{max} = 5.3$ and $\alpha_{mean} = 912$.

When doubling the coherent integration time, $T_c = 2$ ms, the correlation peak amplitude increases, as does the postcorrelation SNR, at the expenses of a larger MAT or computational burden. Δf is now equal to 333 Hz, and the total number of frequency bins is $N_{bin} = 30$. The acquisition result is reported in Figure 23.18. The width of the sinc in the frequency domain plot halved, while the correlation peak emerges even more clearly than in the previous example. The acquisition metrics, evaluated running 1000 Monte Carlo simulations, are $\alpha_{max} = 16.5$ and $\alpha_{mean} = 2398$.

It is possible to extend the integration time up to the instant in which the first bit transition happens. In this specific case, when increasing T_c to 20 ms, a bit transition induces a failure in the acquisition process. As detailed in Section 23.5.1, the correlation peak is doubled because of the subcarrier effect of the phase reversal. In

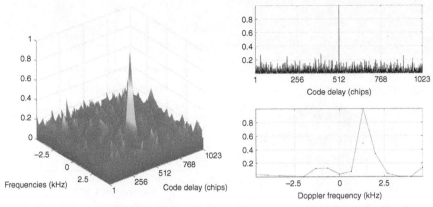

Figure 23.17   Acquisition results for a GPS L1 C/A signal at 45 dB-Hz, with $T_c = 1$ ms and $N_n = 1$.

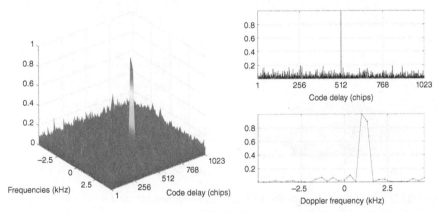

Figure 23.18   Acquisition results for a GPS L1 C/A signal at 45 dB-Hz, with $T_c = 2$ ms and $N_n = 1$.

this case it is not possible to correctly estimate code delay and Doppler frequency. Acquisition results for this case are reported in Figure 23.19 (the three-dimensional plot is enlarged to better identify the correlation peak).

If, rather than 20 coherent summations, 20 noncoherent summations are performed, the bit transition problem is avoided and the peak is correctly identified, as reported in Figure 23.20. Being $T_c = 1$ ms, the width of the sinc function is equal to the case reported in Figure 23.17.

GPS L1 C/A Signal at C/N_0 below the Nominal Value

When moving to signals characterized by a lower C/N_0 it is harder to acquire them using a standard scheme. In this example, a signal at $C/N_0 = 35$ dB-Hz is considered. A standard acquisition scheme with $T_c = 1$ ms and $N_n = 1$ fails in acquiring the signal, as reported in Figure 23.21.

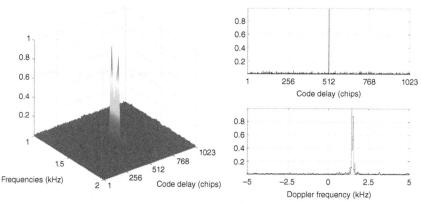

Figure 23.19 Acquisition results for a GPS L1 C/A signal at 45 dB-Hz, with $T_c = 20$ ms and $N_n = 1$.

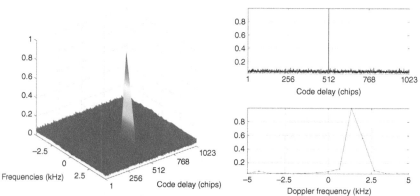

Figure 23.20 Acquisition results for a GPS L1 C/A signal at 45 dB-Hz, with $T_c = 1$ ms and $N_n = 20$.

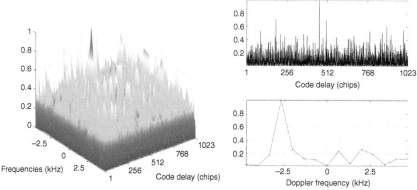

Figure 23.21 Acquisition results for a GPS L1 C/A signal at 35 dB-Hz, with $T_c = 1$ ms and $N_n = 1$.

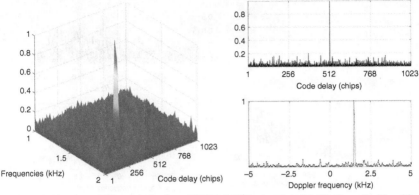

Figure 23.22   Acquisition results for a GPS L1 C/A signal at 35 dB-Hz, with $T_c = 20$ ms and $N_n = 1$.

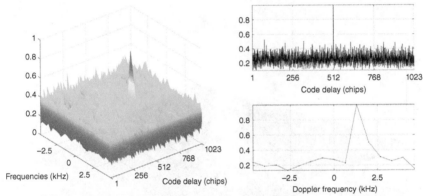

Figure 23.23   Acquisition results for a GPS L1 C/A signal at 35 dB-Hz, with $T_c = 1$ ms and $N_n = 20$.

When increasing the coherent integration time T_c to 20 ms, and assuming a correct bit synchronization, the signal is correctly acquired, as shown in Figure 23.22. In this case $\alpha_{max} = 13.5$ and $\alpha_{mean} = 2459$.

Alternatively, it is possible to acquire the signal by performing noncoherent integrations. For example, with settings $T_c = 1$ ms and $N_n = 20$, the total observation time is equal to 20 ms, as in the previous case. However, the noncoherent gain is lower, resulting in lower acquisition metrics, $\alpha_{max} = 3.2$ and $\alpha_{mean} = 15.6$. Also from visual inspection of the acquisition results, reported in Figure 23.23, it is clear that the postcorrelation SNR is lower, although it is being correctly detected.

A further step can be done, moving to signals characterized by a lower C/N_0, around 30 dBHz. In this case it is still possible to acquire the signal by fixing the coherent integration time to 20 ms and performing $N_n = 20$ noncoherent summations, thus spanning a total integration time of 400 ms. The acquisition results are reported in Figure 23.24. Also in this case the CAF plot is enlarged in the frequency range.

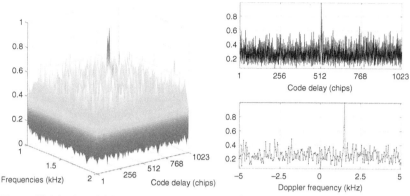

Figure 23.24 Acquisition results for a GPS L1 C/A signal at 30 dB-Hz, with $T_c = 20$ ms and $N_n = 10$.

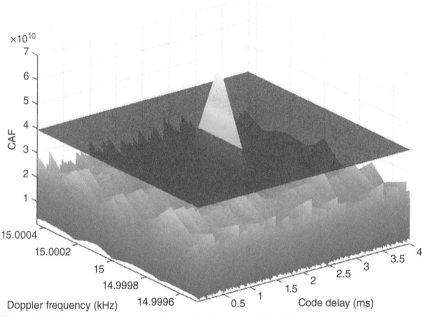

Figure 23.25 Acquisition results for a Galileo E1c at 5 dB-Hz, with $T_c = 3$ s and $N_n = 8$.

Galileo E1c Signal at Very Low C/N_0 When dealing with very weak signals, it is suitable to exploit dataless pilot components of new and modernized GNSS, such as Galileo E1c. When dealing with very low C/N_0, around 5 and 10 dB-Hz, it is necessary to considerably extend the coherent integration time and the number of noncoherent sums. As described in Section 23.4, some further compensations regarding Doppler, Doppler rate, and clock stability have to be taken into account.

Figure 23.25 reports the acquisition results of a Galileo E1c signal at 5 dB-Hz. The coherent integration time has been extended to $T_c = 3$ s, while $N_n = 8$ noncoherent

TABLE 23.2. Coherent Integration Time and Noncoherent Accumulations for Very Weak GNSS Signals Acquisition

Signal	5 dB-Hz	10 dB-Hz
GPS L1 C/A	$T_c = 2$ s; $N_n = 6$	$T_c = 0.5$ s; $N_n = 8$
Galileo E1c	$T_c = 3$ s; $N_n = 8$	$T_c = 0.7$ s; $N_n = 10$

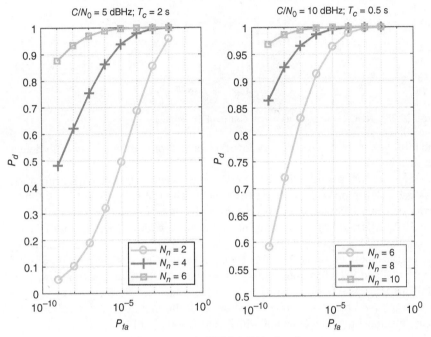

Figure 23.26   ROC curve for very weak GPS L1 C/A signals.

sums were considered, thus spanning a total integration time equal to 24 seconds. The Doppler frequency range has been considerably reduced to achieve a reasonable MAT. The correlation peak clearly emerges from the noise floor, both in the frequency and in the code delay domains.

Table 23.2 provides a summary of the minimum coherent integration time T_c and noncoherent accumulations N_n allowing acquisition of a very weak GNSS signal, with detection probability $P_d = 90\%$ for a given false alarm probability $P_{fa} = 10^{-8}$.

Figure 23.26 provides the ROC curve for GPS L1 C/A code acquisition at very low C/N_0 (5 dB-Hz in the left plot, 10 dB-Hz in the right plot). The curves prove that, in order to achieve correct detection of the main correlation peak with high probability, extension of the coherent integration time up to 2 s and 0.5 s is needed for acquisition at 5 and 10 dB-Hz, respectively. In addition, a proper number of noncoherent accumulations is needed in order to achieve a detection

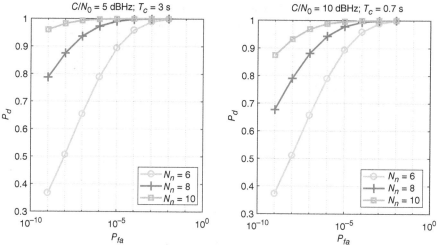

Figure 23.27   ROC curve for very weak Galileo E1c signals.

probability over 90% even at very low false alarm probability (very high detection threshold).

Figure 23.27 provides the ROC curve for the Galileo E1c channel acquisition at 5 dB-Hz (left plot) and 10 dB-Hz (right plot). The full Composite Binary Offset Carrier (CBOC) modulation has been considered, thus assuming $f_s = 50$ MHz and $B_{IF} = 20.46$ MHz. A larger coherent integration time with respect the case of GPS L1 acquisition has to employed, due to the increased front-end filter bandwidth which makes more noise power going through the correlator. A coherent integration time up to 3 and 0.7 s has to be employed for the full CBOC acquisition down to 5 and 10 dB-Hz, respectively. It has to be mentioned that such ROC curves do not take into account possible losses due to the front-end filtering, quantization, and/or frequency mismatching, the details of which can be found in [4].

Acquisition in the Presence of High Doppler Rate   The theoretical requirements on the Doppler rate aiding error reported in Figure 23.12 have been confirmed in [14] by a set of simulations exploiting a fully software GNSS receiver. A fully software GNSS signal generator [19] has been adopted for generating a GPS L1 C/A and a Galileo E1c signal at 10 dB-Hz, affected by Doppler frequency equal to 15 kHz and by a code delay equal to half of the primary code period (0.5 ms and 2 ms, respectively).

Figure 23.28 shows the GPS L1 acquisition performance at 10 dB-Hz for different values of the bias between the Doppler rate aiding and the true Doppler rate affecting the simulated signals. The two plots on top show the acquisition metrics α_{mean} and α_{max}. The two plots on bottom show the code delay and Doppler frequency estimated by the aided acquisition process. It can be observed that correct acquisition of the code delay is achieved for Doppler rate bias values up to 0.44 Hz/s. However, it can be observed also that, increasing the Doppler rate bias, a loss on the correlation amplitude is detected due to the increasing frequency mismatching error.

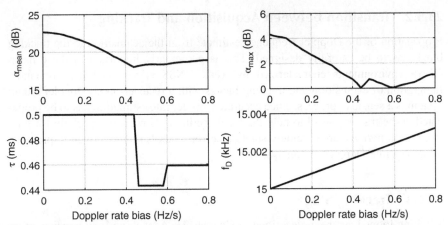

Figure 23.28 GPS L1 C/A code acquisition performance with respect the Doppler rate bias.

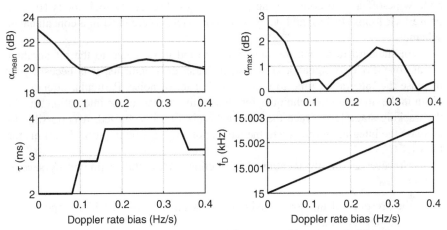

Figure 23.29 Galileo E1c code acquisition performance with respect the Doppler rate bias.

Figure 23.29 shows the Galileo E1c acquisition performance at 10 dB-Hz for different values of the Doppler rate bias. In this case, correct acquisition is achieved up to a Doppler rate error equal to 0.08 Hz/s, as confirmed by the correct estimation of the code delay.

As expected, the acquisition performance reported in Figures 23.28 and 23.29 is slightly worse with respect to the theoretical results reported in Figure 23.12 due to the presence of additional factors like noise, cross-correlation with other received PRNs, and quantization losses.

23.5.2 Transition between Acquisition and Tracking

Propagation of the Doppler frequency estimate from the acquisition to the tracking block has to be carefully designed. This is particularly important in those harsh GNSS environments characterized by weak GNSS signals and high dynamic. Integration of acquisition and tracking blocks with a receiver-based Doppler aiding system is a sensitive process, since accurate Doppler rate estimation has to be propagated at the tracking level to help the local oscillator. However, especially for long integration periods, an extremely stable receiver clock is required even in presence of an accurate Doppler aiding system.

23.5.3 Tracking Stage

Tracking loops have been described in Chapter 21. A tracking loop consists of an integrator, a discriminator, a loop filter, and a numerical controlled oscillator (NCO), and is able to provide, at each iteration, an estimate of the quantity to be tracked. In the case of GNSS, tracking loops are used to track the frequency, phase, and code delay of the incoming signal in order to perform carrier wipe-off and code wipe-off and to compute the pseudoranges. Frequency lock loops (FLLs), phase lock loops (PLLs), and delay lock loops (DLLs) are used in a concatenated scheme.

In standard receivers, the integration period is set equal to the bit duration (1 ms for GPS C/A signals). However, in harsh environments, it is necessary to increase the integration time to reduce the impact of noise. Particular attention is given in the literature to techniques for coherently increasing the integration time in the tracking stage [20]. Once the bit synchronization is achieved, it is possible to increase the integration time up to the navigation message bit duration T_b, 20 ms for GPS L1 C/A.

However, as briefly mentioned in Section 23.3.1, this is limited by several factors:

- the presence of unknown navigation data bit transitions, which imposes the condition $T_c \leq T_b$;
- the presence of a secondary code in pilot signals, which, although being a known sequence, still requires proper synchronization;
- the reduction of the acquisition Doppler bin size and thus the increase of the SS size and of the computational complexity;
- the instability of tracking loop filters that are usually designed for low integration times and relatively large loop bandwidths;
- the loss of coherency, introduced by the varying Doppler frequency, the clock instability, the varying code rate, and the receiver dynamics throughout the coherent integration period.

The sensitivity of a tracking loop can be evaluated measuring the *tracking jitter*. The tracking jitter is a measure of the amount of noise transferred from the input signal to the output of the loop on the final paramter estimate. It is the first measure of a loop performance and allows one to quantify the impact of the thermal noise. It is defined as:

$$\sigma_j = \frac{\sigma_r}{G_d}\sqrt{2B_n T_c} \, , \tag{23.45}$$

where

- σ_r is the standard deviation of the noise at the output of the discriminator;
- B_n is the equivalent bandwidth of the loop;
- G_d is the discriminator gain.

The tracking jitter is a normalized version of the noise standard deviation, taking into account the bandwidth of the complete loop. The theoretical expression of the tracking jitter in a standard PLL is:

$$\sigma_j = \sqrt{\frac{B_n}{C/N_0}\left(1 + \frac{1}{2C/N_0\, T_c}\right)}. \tag{23.46}$$

Figure 23.30 shows the tracking jitter for different C/N_0 values, summarizing the performance of a PLL/DLL tracking stage. The experimental tracking jitter is estimated according to (23.45), performing Monte Carlo simulations, exploiting a software-based tracking stage and using signals generated by a software GNSS signal generator [19]. GPS L1 C/A signals are considered, characterized by a Doppler rate equal to about 1 Hz/s. The coherent integration time is set equal to $T_c = 20$ ms, while the loop bandwidth of the PLL and of the DLL is equal to 10 Hz and 2 Hz, respectively. The vertical line that corresponds to $C/N_0 = 33$ dB-Hz marks the point in which the loop is no longer in lock conditions. This means that the tracking sensitivity is equal to 34 dB-Hz. To track signals at a lower C/N_0 it is necessary to further extend the integration time.

In order to go beyond the limit imposed by the bit transitions, a solution is given by bit estimation and recovery techniques. On one hand, it is possible to foresee the value of the next navigation bit. On the other hand, any possible

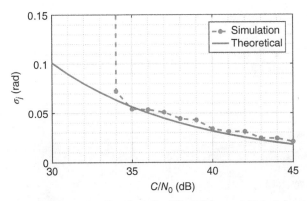

Figure 23.30 Tracking jitter for $T_c = 20$ ms and $N_n = 1$.

combination of the navigation message bits over the whole integration period can be tested and compared. The coherent integration time can indeed be extended, at the expense of a higher computational load, growing exponentially with the integration time. In addition, the bit estimation becomes unreliable at low C/N_0. This problem can be overtaken if assistance data are available. In this case, a perfect data wipe-off can be performed, as described in Section 23.4.1.

An alternative solution is to use discriminators insensitive to bit transitions, based on nonlinear operations, allowing, in principle, to extend arbitrarily the integration time. In [21], a noncoherent architecture for GNSS tracking loops is proposed and analyzed, deriving a noncoherent phase discriminator from the maximum likelihood (ML) principle. The ML phase estimator in the presence of sign transitions is derived, under the assumption of data bits randomly distributed. Under this hypothesis, the navigation message bits can be removed by squaring the signal.

Figure 23.31 reports the comparison between theoretical and experimental tracking jitter obtained using a tracking structure employing the discriminator described in [21] and [22]. $N_n = 4$ noncoherent sums are considered. T_c is equal to 20 ms as in the previous example, and therefore, the global noncoherent integration time is equal to 80 ms. The PLL bandwidth has been reduced to 2.5 Hz in order to keep the product $B_n N_n T_c$ constant and equal to 0.2, and to maintain the loop filters stable. The input GNSS signals are the same as above. From the figure it is possible to see that the noncoherent gain pushes the tracking sensitivity down to 27 dB-Hz.

The main drawbacks of larger integration times in the tracking stage are summarized in the following:

- the need to compensate for the Doppler change within the integration period: Indeed, when the observation time increases, the Doppler rate effect on the code is no longer negligible and has to be corrected. If assistance information is available, a more accurate frequency wipe-off can be performed. Noncoherent schemes are more robust to frequency residual errors than coherent architectures.

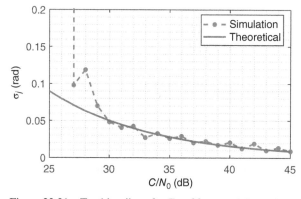

Figure 23.31   Tracking jitter for $T_c = 20$ ms and $N_n = 4$.

- the need to design proper discriminators, such as the ones described in [21] and [22];
- The loop filters stability: Loop filters are designed to be stable for short coherent integration times. Alternative loop filters formulations have to be adopted, such as the controlled root formulation proposed in [23] and [24].

23.6 HIGH SENSITIVITY AND A-GNSS

The topic of high sensitivity and harsh environment positioning is often linked to the argument of assisted GPS and GNSS. A-GNSS has been indeed designed not only to reduce the TTFF but also to increase the receiver sensitivity. Moreover, the two objectives are linked: the same aiding used to reduce the acquisition time can actually be exploited to improve the acquisition sensitivity. Most of the problems related to high-sensitivity acquisition and tracking presented above can be solved or eased exploiting proper external aiding. GNSS receivers, sensitivity can be enhanced if some a priori estimates of observables such as frequency offset, accurate time, code delay, and receiver position are known [4].

The core idea of A-GNSS is to assist the receiver providing all the information it would normally have obtained by processing the SIS through an alternative communication channel. The receiver keeps receiving and processing satellites signals, but thanks to assistance information it can do it faster and easier. Such aiding includes, but is not limited to:

- constellation almanac, to determine a priori the list of satellites in view and their approximate azimuth and elevation;
- precise ephemeris, to calculate the position of the satellites;
- receiver position estimate;
- approximate date and time (GPS time);
- ionospheric correction;
- accurate frequency reference to calibrate the local oscillator;
- satellite clock corrections;
- acquisition parameters (estimated Doppler shift, estimated delay).

For example, expected signal Doppler and Doppler rates can be used to reduce the frequency acquisition SS, and thus to make the acquisition process more robust to high dynamics, when large coherent integration times are employed. Similarly, some assistance servers are able to provide the content of the navigation message, which can be used to remove the data bits from the received signals, thus allowing extension of the coherent integration time beyond the data bit duration.

23.6.1 Example of Frequency Aiding in Acquisition

The availability of a rough estimates of the Doppler frequency is an added value for high-sensitivity receivers. In Section 23.4.3 it has been shown that for very large integration times, required to acquire very weak signals, the acquisition time becomes extremely high, often unaffordable. As T_c increases, the number of frqeuency bins to be evaluated increases. A-GNSS gives a solutions to this: by reducing the total frequency SS, the number of bins can be reduced, and in turn, the MAT can be dramatically reduced. If the receiver is provided with a Doppler aiding accurate up to ϵ_{f_D}, the value of N_{bin} in the Doppler domain search can be reduced as follows:

$$N_{bin} = \left\lceil \frac{2\epsilon_{f_D}}{\Delta f} \right\rceil + 1. \tag{23.47}$$

Recalling the example reported in Section 23.4.3, a coherent acquisition time of 3 s leads to a total number of frequency bins $N_{bin} \simeq 44{,}000$. If frequency assistance is provided with an accuracy $\epsilon_{f_D} = \pm 50$ Hz, the frequency range in the SS is reduced. The number of bins decreases to about 450, as does the MAT.

In parallel, it has to be said that a small frequency SS also improves the acquisition performance. In [15], the authors introduce a set of probability metrics at the SS level: the false alarm probability P_{FA} and the detection probability P_D. The P_{FA} at the SS level depends on the P_{fa} at cell level, defined in (23.33), according to the following equation:

$$P_{FA} = 1 - \left[1 - P_{fa}(\theta_t)\right]^N, \tag{23.48}$$

θ_t being the acquisition threshold and N the total number of cells in the SS. According to (23.48), fixing a required false alarm at cell level P_{fa}, a reduction of the SS size allows the achievement of better P_{FA} values.

As an example, Figure 23.32 shows the trend of the P_{FA} at the SS level with respect the P_{fa} at cell level for different values of the Doppler aiding accuracy ϵ_{f_D}. Fixing $P_{fa} = 10^{-8}$, a Doppler aiding accurate up to 1.5 Hz is needed to achieve a $P_{FA} = 10^{-2}$, in the case of Galileo E1c acquisition at 10 dB-Hz, exploiting $T_{int} = 0.7$ s combined with $N_n = 10$ noncoherent accumulations.

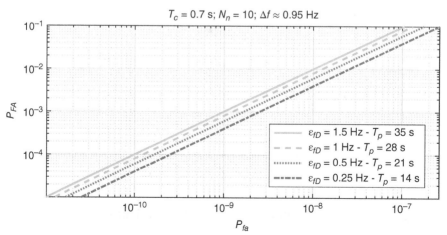

Figure 23.32   Galileo E1c noncoherent acquisition: acquisition metrics P_{FA} vs P_{fa} for different aiding scenarios.

23.7 NEW AND MODERNIZED GNSS

New and modernized GNSS signals offer users great opportunities for increasing the availability and accuracy of the positioning solutions. On one side, GPS and GLONASS are undergoing a modernization process. On the other side, new constellations, such as Galileo, Beidou, and the regional Indian and Japanese systems are by now operational. A complete description of the new signals is provided in Chapter 20. Throughout this chapter, only the GPS L1 C/A legacy signal has been considered. This section briefly outlines the main advantages of GPS and Galileo new civil signals from the perspective of high sensitivity.

23.7.1 Longer Code Periods

In Section 23.2.1 the impact of the code length in the correlation process has been evaluated. In particular, Example 23.1, showed how a longer spreading code can improve the correlation performance. This is the main reason why some of the new signals employ longer codes. For instance, Galileo E1b signals are modulated by 4 ms long memory codes. Other signals, such as GPS L2C, feature a more complex code structure, with a *moderate* and a *long* code multiplexed. While the moderate code is 20 ms long, the long code lasts for 1500 ms. GPS L5 signals, both in the in-phase and in the quadrature component, and Galileo E5a-I and E5a-Q signals, use 1-ms-long codes. However, they are transmitted at a higher code rate, 10.23 MHz.

A higher despreading gain can therefore be obtained, at the expense of a higher complexity and computational burden. It is also important to recall that signals characterized by a larger code rate require a larger front-end bandwidth, allowing more noise to enter the acquisition and tracking stages.

23.7.2 Pilot Channels

The main advantage of new and modernized signals for what concerns high-sensitivity capabilities is the presence of dataless signals, called *pilot* signals. Pilot signals are not modulated by the navigation message and can be used to assist the acquisition of and tracking of data signals. The absence of the navigation data bits allows the extension of the coherent integration time, avoiding any problem related to bit transitions. Dataless signals are broadcast by GPS (L1C, L5Q) and by Galileo (E1c, E5a-Q, E5b-Q, E6).

In addition, pilot signals are modulated by a secondary code. In order to extend the coherent integration time, synchronization with the secondary code has to be achieved, as detailed in Section 23.4.2.

23.7.3 Encoded Naviagtion Data

Modernized GNSS signals also feature encoded navigation data. As an example, GPS L2C navigation bits are encoded with a continuous FEC at a rate of 1/2. This allows to reach lower BERs at a lower C/N_0, improving the performance in harsh environments.

23.7.4 Different SS Definition

The definition of the number of bins of the SS to be evaluated during the acquisition phase changes for each signal. As described in (23.20), the maximum Doppler frequency depends on the maximum satellite velocity, which in turn depends on the constellation orbits. The radius of the satellite orbit is equal to 26,560 km for GPS and to 29,600 km for Galileo. Therefore, the maximum Doppler shift experienced by Galileo signals is slightly lower, around 4.2 kHz. However, as the code period of Galileo satellites is higher (e.g., 4 ms for Galileo E1b), the bin size is about four times smaller. Therefore, even though the total Doppler range to be evaluated is lower, more bins are required, thus increasing the MAT.

23.8 CONCLUSIONS

In this chapter some of the basic concepts of signal processing for high-sensitivity GNSS receivers were given. To counteract for the low C/N_0 of the GNSS signals in harsh environments, particular strategies and techniques shall be considered. First, a recall on the signal power, signal noise and correlation gain has been provided. Then, the concept of integration time extension has been outlined by describing coherent and noncoherent solutions and by providing examples and by proposing solutions to the main issues revealed. A focus on the acquisition and tracking stages of a high-sensitivity receiver was then provided, along with an overview on assistance techniques and on new and modernized GNSS signals. The examples and the case studies presented show how high-sensitivity receivers are able to acquire and to track signals even at very low C/N_0, allowing the computation of a PVT solution in environments in which standard receivers would fail.

REFERENCES

[1] E. D. Kaplan and C. J. Hegarty, *Understanding GPS: Principles and Applications*. Norwood, MA: Artech House, 2006.

[2] P. Misra and P. Enge, *Global Positioning System: Signals, Measurements and Performance*. Lincoln, MA: Ganga-Jamuna Press, 2006.

[3] J. B. Y. Tsui, *Fundamentals of Global Positioning System Receivers: A Software Approach*. Wiley-Interscience, 2005.

[4] F. S. T. Van Diggelen, *A-GPS: Assisted GPS, GNSS, and SBAS*. Boston, MA: Artech House, 2009.

[5] US Air Force, "Navstar GPS space segment/navigation user interfaces," 2004.

[6] S. M. Kay, *Fundamentals of Statistical Signal Processing*. Englewood Cliffs, NJ: Prentice Hall, 1993.

[7] C. Strässle, D. Megnet, H. Mathis, and C. Bürgi, "The squaring-loss paradox," in *GNSS 20th Int. Technical Meeting of the Satellite Division of the Institute of Navigation*, Fort Worth, TX, 2007, pp. 2715–2722.

[8] D. Borio, "A statistical theory for GNSS signal acquisition," Ph.D. thesis, Politecnico di Torino, 2008.

[9] H. Elders-Boll and U. Dettmar, "Efficient differentially coherent code/Doppler acquisition of weak GPS signals," in *2004 IEEE Eighth Int. Symp. on Spread Spectrum Techniques and Applications*, 2004, pp. 731–735.

[10] A. Schmid and A. Neubauer, "Performance evaluation of differential correlation for single shot measurement positioning," in *17th Int. Technical Meeting of the Satellite Division of the Institute of Navigation (ION GNSS 2004)*, 2001, pp. 1998–2009.

[11] W. Yu, B. Zheng, R. Watson, and G. Lachapelle, "Differential combining for acquiring weak GPS signals," *Signal Process.*, vol. 87, no. 5, pp. 824–840, 2007.

[12] F. Dovis and T. H. Ta, "High sensitivity techniques for GNSS signal acquisition," in *Global Navigation Satellite Systems: Signal, Theory and Applications*, S. Jin, Ed., Rijeka, Croatia: InTech, 2012.

[13] R. Garello, L. Lo Presti, G. E. Corazza, and J. Samson, "Peer-to-peer cooperative positioning, Part I: GNSS aided acquisition," *INSIDE GNSS*, vol. 7, no. 4, pp. 55–63, 2012.

[14] N. Linty, L. Musumeci, and F. Dovis, "Assistance requirements definition for GNSS receivers in hostile environments," in *2014 Int. Conf. on Localization and GNSS (ICL-GNSS 2014)*, Piscataway, NJ, Jun. 2014, pp. 1–6.

[15] D. Borio, L. Camoriano, and L. Lo Presti, "Impact of GPS acquisition strategy on decision probabilities," *IEEE Trans. Aerosp. Electron. Syst.*, vol. 44, pp. 996–1011, Jul. 2008.

[16] P. Silva, H. Lopes, T. Peres, J. Silva, J. Ospina, F. Cichocki, F. Dovis, L. Musumeci, D. Serant, T. Calmettes, I. Pessina, and J. Perelló, "Weak GNSS signal navigation to the Moon," in *26th Int. Technical Meeting of the Satellite Division of the Institute of Navigation (ION GNSS+ 2013)*, 2013, pp. 3357–3367.

[17] N. Sokolova, D. Borio, B. Forssell, and G. Lachapelle, "Doppler rate measurements in standard and high sensitivity GPS receivers: Theoretical analysis and comparison," in *2010 Int. Conf. on Indoor Positioning and Indoor Navigation (IPIN)*, Sep. 2010, pp. 1–9.

[18] D. A. Shnidman, "The calculation of the probability of detection and the generalized Marcum Q-function," *IEEE Trans. Inf. Theory*, vol. 35, no. 2, pp. 389–400, 1989.

[19] E. Falletti, D. Margaria, M. Nicola, G. Povero, and M. Gamba, "N-FUELS and SOPRANO: Educational tools for simulation, analysis and processing of satellite navigation signals," in *Frontiers in Education Conf., FIE*, 2013, pp. 303–308.

[20] P. L. Kazemi and C. O'Driscoll, "Comparison of assisted and stand-alone methods for increasing coherent integration time for weak GPS signal tracking," in *Proc. of the 21st Int. Technical Meeting of the Satellite Division of the Institute of Navigation (ION GNSS 2008)*, 2008, pp. 1730–1740.

[21] D. Borio and G. Lachapelle, "A non-coherent architecture for GNSS digital tracking loops," *Ann. Telecommun*, vol. 64, no. 9–10, pp. 601–614, 2009.

[22] D. Borio, N. Sokolova, and G. Lachapelle, "Memory discriminators for non-coherent integration in GNSS tracking loops," in *European Navigation Conf. (ENC 2009)*, vol. 9, 2009, pp. 1–12.

[23] S. Stephens and J. Thomas, "Controlled-root formulation for digital phase-locked loops," *IEEE Trans. Aerosp. Electron. Syst.*, vol. 31, no. 1, pp. 78–95, 1995.

[24] N. Kassabian, "Design of pilot channel tracking loop systems for high sensitivity Galileo receivers," Ph.D. thesis, Politecnico di Torino, 2014.

PART VI

NETWORK LOCALIZATION

CLASSICAL POSITION location is based on locating a target/mobile based on measurements to known anchors. In many applications what is needed are the positions of many nodes, some of which many not have connectivity to anchor nodes. In such cases, utilizing measurements between unlocalized nodes can be used to localize all of the nodes in the network. This is often termed *network localization, cooperative localization* or *collaborative localization* and is the subject of Part VI.

This part includes five chapters, Chapter 24–28, that detail several topics in the area of network-based or cooperative localization. The section also introduces techniques such as infrastructure-free local positioning system and wireless local positioning systems. Further, this part overviews the error characteristics of *ad hoc* positioning systems.

Chapter 24 describes the fundamentals of network (collaborative) position location, that is position location techniques where the nodes to be localized collaborate to determine their positions. This is sometimes also called cooperative position location or network position location to distinguish it from traditional position location techniques which localize a single node. After a brief introduction and definition of the problem, the chapter examines two main bounds for collaborative positioning: namely the Cramer-Rao Lower Bound (CRLB) and the weighted least squares solution. Although the nature of the bounds is somewhat different, they both provide guidance as to the achievable performance. Several sub-optimal approaches are described and compared in terms of their performance in various network configurations. Lastly, the chapter examines the impact of non-line-of-sight propagation and describes techniques to mitigate its effects.

In **Chapter 25** the authors review a series of results obtained in the field of localization that are based on polynomial optimization. First, they provide a review of a set of polynomial function optimization tools including the sum of squares (SOS) technique. Then they present several applications of these tools to various sensor network localization tasks. As a first application, they propose a method based

Handbook of Position Location: Theory, Practice, and Advances, Second Edition.
Edited by S. A. (Reza) Zekavat and R. Michael Buehrer.
© 2019 by the Institute of Electrical and Electronics Engineers, Inc.
Published 2019 by John Wiley & Sons, Inc.
Companion Website: www.wiley.com/go/zekavat/positionlocation2e

on SOS relaxation for node localization using noisy measurements and describe the solution through semi-definite programming (SDP). Later, they examine the network localization problem of relative reference frame determination based on range and bearing measurements.

Chapter 26 considers cooperative localization using probabilistic inference. These techniques are able to obtain not only location estimates, but also a measure of the uncertainty of those estimates. Since these methods are computationally very expensive, message-passing methods, which are also known as belief propagation (BP) methods have been proposed. To reduce algorithmic complexity, the use of particle-based approximation via nonparametric belief propagation (NBP) can make BP acceptable for localization in sensor networks. In this chapter, after a brief introduction to cooperative localization, the authors describe BP/NBP techniques and generalizations (GBP) for loopy networks. Due to the poor performance of BP/NBP methods in loopy networks, they introduce three methods: GBP based on Kikuchi approximation (GBP-K), nonparametric GBP based on junction tree (NGBPJT), and NBP based on spanning trees (NBP-ST).

Chapter 27 examines network localization techniques using multi-hop techniques that are based on the idea of distance vector routing to find the positions of all nodes in the network. Specifically, this chapter examines the performance of four basic multi-hop algorithms: range/angle-free algorithms, range-based algorithms, angle-based algorithms and multimodal algorithms. The error characteristics of all four classes are examined under various conditions using both theoretical analysis and simulations. The authors discuss the trade-offs involved among the capabilities used in each node, the density of the network, the ratio of anchors (landmarks) and the quality of the position estimates obtained.

Chapter 28 examines the problem of localizing multiple nodes using bearing measurements. The chapter begins by treating the problem of localizing three agents moving in a plane when the inter-agent distances are known, and in addition, the angle subtended at each agent by lines drawn from two landmarks at known positions is also known. It is shown that there are in general a finite number of possible sets of positions for the three agents. Afterwards, the generalization of the result for more than three agents is presented. Examples are given to show the applicability of the methods.

CHAPTER 24

COLLABORATIVE POSITION LOCATION

R. Michael Buehrer and Tao Jia
Virginia Tech, Blacksburg, VA

T HIS CHAPTER describes *collaborative* position location, that is, position location techniques where the nodes to be localized *collaborate* to determine their positions. This is sometimes also called *cooperative* position location or *network* position location to distinguish it from traditional position location techniques which localize a single node. After a brief introduction and definition of the problem, we will examine two main bounds for collaborative positioning, namely, the Cramer–Rao lower bound (CRLB) and the weighted least squares (LS) solution. Although the bounds are somewhat different, they both provide guidance as to the achievable performance. We will then describe several suboptimal approaches and compare their performance in various network configurations. Lastly, we will examine the impact of non-line-of-sight (NLOS) propagation and describe one particular technique to mitigate its effects.

24.1 INTRODUCTION

In a conventional infrastructure-based position location system (see, e.g., Part V of this text), such as global positioning systems (GPSs) [1] or cellular enhanced 911 (E-911) [2], wireless devices measure signals from GPS satellites or cellular base stations (i.e., nodes with known location) to obtain their location estimate. However, in some applications, such as sensor networks, equipping every node with a GPS receiver or deploying substantial preexisting infrastructure turns out to be an expensive and impractical solution. Therefore, in such applications, only a small portion of all network nodes can be

Handbook of Position Location: Theory, Practice, and Advances, Second Edition.
Edited by S. A. (Reza) Zekavat and R. Michael Buehrer.
© 2019 by the Institute of Electrical and Electronics Engineers, Inc.
Published 2019 by John Wiley & Sons, Inc.
Companion Website: www.wiley.com/go/zekavat/positionlocation2e

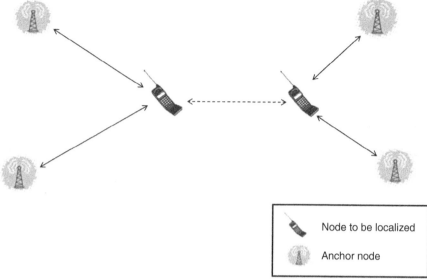

Figure 24.1   The two nodes have only two connections to anchors (solid lines) and thus cannot be localized without ambiguity. By collaborating (dashed line), their ambiguities are resolved and localization is possible for both nodes.

expected to have known locations. These nodes are commonly referred to as *anchors* or reference nodes, and those that initially do not know their locations are called unlocalized nodes. Additionally, it is usually assumed that wireless nodes have a limited communication range for the sake of energy efficiency and longer battery life. These two facts together lead to a unique challenge to position location, namely, that the majority of nodes do not have enough anchors within their communication range to make measurements. Consequently, these nodes cannot *directly* estimate their locations, and the position location scheme has to rely on other mechanisms to propagate the known location information throughout the network.

The solution to this location problem is based on the concept of *collaborative position location* (sometimes also called relative position location, cooperative position location, or network localization [3–6]), which is shown in Figure 24.1. In this illustration, two nodes wish to determine their respective positions using distance measurements to two distinct pairs of anchors. In this scenario, neither node has a sufficient number of connections to uniquely determine its location.* However, if the nodes also use their connection to each other, the two nodes can *collaboratively* determine both positions. Thus, we can see that in this case, the use of collaboration improves network position location *coverage*. That is, nodes that could not be localized using traditional position location can now be localized, increasing the number of localizable nodes. In addition to improving localization coverage, collaboration

---

*For two-dimensional localization, nodes must have connections to a minimum of three anchors when using distance-based measurements. Please see Chapter 6 for a more detailed explanation.

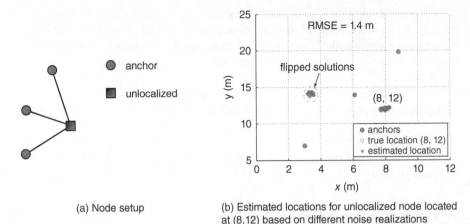

(a) Node setup

(b) Estimated locations for unlocalized node located at (8,12) based on different noise realizations

Figure 24.2   Illustration of the effect of node collaboration on the localization accuracy. Case 1: noncollaborative localization with range measurements to three anchors (*Note*: RMSE = root mean square error is 1.4 m) [15].

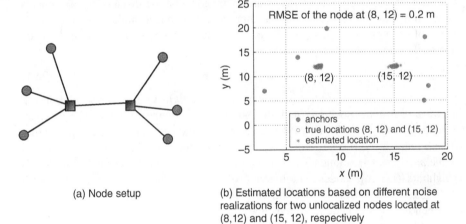

(a) Node setup

(b) Estimated locations based on different noise realizations for two unlocalized nodes located at (8,12) and (15, 12), respectively

Figure 24.3   Illustration of the effect of node collaboration on the localization accuracy. Case 2: collaborative localization using both anchor connections and connection between unlocalized node (*Note*: RMSE = root mean square error is 0.2 m) [15].

can improve *location accuracy*, particularly in poor geometric conditions. Consider the example shown in Figure 24.2. In this case, despite the fact that the node has a sufficient number of connections to determine its position, due to the fact that the geometry of the anchors is poor (nearly colinear as seen in Fig. 24.2a), the performance is poor. This is manifested in the fact that depending on the noise realization, the solution is near one of two possible locations (as shown in Fig. 24.2b). However, if two nodes can collaborate, as shown in Figure 24.3, the performance is improved substantially by eliminating the ambiguity. The root mean square error (RMSE) is reduced from 1.4 m to 0.2 m through collaboration.

Over the past several years, there has been an increasing interest in collaborative approaches of this kind. Summaries of existing algorithms can be found in [3, 5, 7]. Among them, a notable common notion is that the system should use not only the measurements between an unlocalized node and anchors but also those between unlocalized nodes to estimate their locations. This concept has been referred to as *collaborative* to emphasize the use of additional measurements among unlocalized nodes to improve both localization accuracy and coverage. In contrast, we refer to the conventional infrastructure-based position location approaches (e.g., GPS, E-911, or similar systems), where an unlocalized node estimates its location based on measurements to directly connected anchors, as *noncollaborative* position location. The measurements can be related to distance estimates, angle estimates, or even simple connectivity information as in [8, 9], although the latter leads to rather coarse localization. In this chapter we will focus on the use of distance measurements.

24.2 PROBLEM DEFINITION

Consider a set of n nodes whose two-dimensional positions* are defined by the $n \times 2$ matrix Θ. We wish to estimate Θ based on observations $\mathbf{r}$ and the positions of m known anchors $\mathbf{A}$. The observations are assumed to be the distance measurements between all nodes (including both anchors and unlocalized nodes) that are within range of each other. In general, the observations are a known function of the node and anchor positions plus observation error:

$$\mathbf{r} = f(\Theta, \mathbf{A}) + \mathbf{n}, \tag{24.1}$$

where $\mathbf{n}$ is the observation/measurement error, and $f(\Theta, \mathbf{A})$ is a known relationship between the node positions and the observations in the absence of any error. Specifically, $\Theta = [\theta_1; \theta_2; \ldots ; \theta_n]$ is an $n \times 2$ matrix with the ith row representing the ith unlocalized node's position $\theta_i = [x_i, y_i]$. Further, $\mathbf{A} = [\theta_{n+1}; \theta_{n+2}; \ldots ; \theta_{n+m}]$ is the $m \times 2$ matrix of anchor positions. Let $\mathcal{N}(i)$ represent the set of the ith node's neighbors (i.e., the nodes within range of node i). Then, $\mathbf{r}$ is vector of length $N_r = \sum_{i=1}^{n} \mathcal{N}(i)$. Finally, if the kth element of $\mathbf{r}$ is the observed range estimate between node i and node j, we have the functional relationship

$$f_k(\Theta, \mathbf{A}) = \sqrt{(x_i - x_j)^2 + (y_i - y_j)^2}. \tag{24.2}$$

Approaches to estimate Θ can be broken down into two basic categories: Bayesian and non-Bayesian. Non-Bayesian approaches are simpler and are more prevalent. The approach used generally depends on the available information and the complexity of the technique. For example, the approach that requires the least information (which is also one of the most prevalent) is to find the solution which minimizes the squared error between the observations and the known relationship between the data and node positions. This is also known as the LS solution and can be written as

$$\hat{\Theta} = \arg\min_{\Theta}(\mathbf{r} - f(\Theta, \mathbf{A}))^2. \tag{24.3}$$

*Three-dimensional localization is a straightforward extension.

However, it should be noted that due to the fact that $f(\Theta, \mathbf{A})$ is generally non-linear and nonconvex, even directly implementing this approach can be quite complex. Thus, approximations to this approach are often implemented.

When the observation error has a known distribution, we can improve our estimate. Specifically, we can maximize the likelihood function (the probability density function or PDF of $\mathbf{r}$ given Θ), which is known as the maximum likelihood estimate (MLE):

$$\hat{\Theta} = \arg\min_{\Theta}\left\{p(\mathbf{r}|\Theta, \mathbf{A})\right\}. \tag{24.4}$$

Note that when the error is random with independent Gaussian distributions for each observation, the MLE is a weighted LS solution:

$$\hat{\Theta} = \arg\min_{\Theta} \mathbf{W}\left(\mathbf{r} - f(\Theta, \mathbf{A})\right)^{2}, \tag{24.5}$$

where $\mathbf{W}$ is a diagonal matrix, with w_k being one over the variance of the kth observation r_k. Additionally, if the variance of each Gaussian error is equal to a common variance (often a reasonable assumption of the distances are nearly equal), the MLE collapses to the LS solution of (24.3). This provides some justification for the use of the simpler LS estimate.

If a priori information is available about the nodes' positions, Bayesian approaches are also possible, which treat the unknown positions as random variables with a known power density function (PDF) $p_\Theta(\Theta)$ instead of an unknown deterministic value. The maximum a posteriori or MAP approach maximizes the probability of Θ given the observations $\mathbf{r}$:

$$\hat{\Theta} = \arg\min_{\Theta}\left\{p(\Theta|\mathbf{r}, \mathbf{A})\right\} \tag{24.6a}$$

$$= \arg\max_{\Theta}\left\{\frac{p(\mathbf{r}|\Theta, \mathbf{A})\,p_\Theta(\Theta)}{p(\mathbf{r})}\right\} \tag{24.6b}$$

$$= \arg\max_{\Theta}\left\{p(\mathbf{r}|\Theta, \mathbf{A})\,p_\Theta(\Theta)\right\}, \tag{24.6c}$$

where $p(\mathbf{r})$ is treated as an unknown constant. The MAP approach maximizes (i.e., finds the mode of) the a posteriori distribution. On the other hand, the minimum mean square error (MMSE) approach sets the estimate to the *mean* of the distribution:

$$\hat{\Theta} = E\left\{p(\Theta|\mathbf{r}, \mathbf{A})\right\}. \tag{24.7}$$

As mentioned, due to the fact that a priori information is often not available or difficult to obtain, the most common approach is to use estimates based on an LS solution. However, since determining the minimum of a nonlinear, nonconvex objective function (such as that in (24.3)) can be quite complex, most practical approaches are *approximations* of the true LS solution. We will describe an approach to solve this nonlinear LS problem in Section 24.3.2.

To evaluate the performance of a position location scheme, we need to define a metric for comparison. Given a noise realization and the resulting location estimate

$\hat{\Theta}$, the *network-average* error Ω is defined as the square root of the average squared localization error:

$$\Omega \triangleq \sqrt{\frac{1}{n}\sum_{i=1}^{n}\left[\left(\hat{x}_i - x_i\right)^2 + \left(\hat{y}_i - y_i\right)^2\right]}, \tag{24.8}$$

where $\hat{\theta}_i = \left[\hat{x}_i, \hat{y}_i\right]$ denotes the ith node's estimated location and $\hat{\Theta} = \left[\hat{\theta}_1; \hat{\theta}_2; \dots \hat{\theta}_n\right]$. The *mean* localization error $\bar{\Omega}$, on the other hand, refers to the square root of the network-average mean squared error (MSE), that is,

$$\bar{\Omega} \triangleq \sqrt{\frac{1}{n}\sum_{i=1}^{n} E\left\{\left(\hat{x}_i - x_i\right)^2 + \left(\hat{y}_i - y_i\right)^2\right\}}, \tag{24.9}$$

where $E\{\cdot\}$ denotes the expectation over different noise realizations. The *residual error* is defined as how well the estimated locations fit the original observations. Although this measure is not a true measure of the accuracy of the scheme, it is measurable by the position location algorithm and is commonly used to guide the algorithm. The residual square error is defined as

$$e = \sum_{i=1}^{n}\sum_{\substack{j=i+1 \\ j \in \mathcal{N}(i)}}^{n+m}\left(r_{ij} - \|\theta_i - \theta_j\|\right)^2. \tag{24.10}$$

24.3   PERFORMANCE BOUNDS

The previous section outlines the collaborative position location problem by defining various optimal estimators. In general, optimal estimators are computationally complex, and thus practical estimates will be approximations of an optimal approach. However, optimal estimates are useful since they provide bounds that allow us to benchmark the performance of suboptimal approaches. To date, two such benchmarks have been developed for collaborative position location. Specifically, the CRLB, which is a lower bound (LB) on the variance of any unbiased estimator, has been developed for collaborative position location [3, 10, 11]. Additionally, the maximum likelihood solution of (24.4) has been formulated [3], and a technique to solve the MLE has been developed in [12]. We will review these two approaches for determining LBs on estimation error in this section.

24.3.1   CRLB

The CRLB is an LB on the variance of any *unbiased* estimator. Without any knowledge about NLOS range measurements, an unbiased location estimator will discard any NLOS range measurement to maintain the unbiasedness of the estimator

[13]. Therefore, to determine the LB, we assume all range measurements are line-of-sight (LOS). For LOS measurements, the observation error is assumed to be Gaussian, and the estimated range between nodes i and j r_{ij} is related to the true distance d_{ij} as*

$$r_{ij} = d_{ij} + n_{ij}, \tag{24.11}$$

where n_{ij} is independent Gaussian noise.

Based on these assumptions, if there exists a range measurement r_{ij} between the ith and jth nodes, its PDF, conditioned on the locations, θ_i and θ_j, can be written as

$$f\left(r_{ij}|\theta_i, \theta_j\right) = \frac{1}{\sqrt{2\pi}\sigma_{ij}} \exp\left(-\frac{\left(r_{ij} - d_{ij}\right)^2}{2\sigma_{ij}^2}\right), \tag{24.12}$$

where d_{ij} and σ_{ij}^2 are the true distance and the variance of measurement noise, respectively. That is,

$$d_{ij} = \|\theta_i - \theta_j\| = \sqrt{\left(x_i - x_j\right)^2 + \left(y_i - y_j\right)^2}, \tag{24.13}$$

and let the variance of the distance measurement be related to the distance as $\sigma_{ij}^2 = K_E d_{ij}^{\beta_{ij}}$, where K_E is scalar constant related to the receiver and β_{ij} is the path loss exponent. Substituting these values into (24.12), the log-likelihood function of (24.12) is given by

$$\log f\left(r_{ij}|\theta_i, \theta_j\right) = -\log\sqrt{2\pi K_E} - \frac{\beta_{ij}}{4}\log\left[\left(x_i - x_j\right)^2 + \left(y_i - y_j\right)^2\right] \tag{24.14a}$$

$$-\frac{1}{2K_E} \times \frac{\left[r_{ij} - \sqrt{\left(x_i - x_j\right)^2 + \left(y_i - y_j\right)^2}\right]^2}{\left[\left(x_i - x_j\right)^2 + \left(y_i - y_j\right)^2\right]^{\beta_{ij}/2}}. \tag{24.14b}$$

Assuming range measurements are corrupted by independent noise,[†] the joint PDF of the set of all range measurements is given by

$$f(\mathbf{r}|\Theta) = \prod_{i=1}^{n}\prod_{\substack{j>i \\ j\in\mathcal{N}(i)}}^{n+m} f\left(r_{ij}|\theta_i, \theta_j\right), \tag{24.15}$$

where $j > i$ in the second summation is to ensure each range measurement is only counted once, while $j \in \mathcal{N}(i)$ indicates that the jth node is a neighbor of the ith node. Therefore, the log-likelihood of the joint PDF of all range measurements is given by

---

*To be consistent with existing literature, we focus on the Gaussian distributed noise in this section.

[†]In this chapter, we limit our attention to independent range estimation noise. The effect of correlated noise is beyond the focus of this chapter.

$$\log f(\mathbf{r}|\Theta) = \sum_{\substack{i=1 \\ }}^{n} \sum_{\substack{j=i+1 \\ j \in \mathcal{N}(i)}}^{n+m} \log f(r_{ij}|\theta_i, \theta_j)$$

(24.16a)

$$= \sum_{\substack{i=1 \\ }}^{n} \sum_{\substack{j=i+1 \\ j \in \mathcal{N}(i)}}^{n+m} \left\{ -\log \sqrt{2\pi K_E} - \frac{\beta_{ij}}{4} \log\left[(x_i - x_j)^2 + (y_i - y_j)^2 \right] \right.$$

$$\left. - \frac{1}{2K_E} \times \frac{\left[r_{ij} - \sqrt{(x_i - x_j)^2 + (y_i - y_j)^2} \right]^2}{\left[(x_i - x_j)^2 + (y_i - y_j)^2 \right]^{\beta_{ij}/2}} \right\}.$$

(24.16b)

The CRLB depends on the Fisher information matrix (FIM) J_Θ, which is defined in [14]:

$$J_\Theta = -\mathbf{E}_\Theta \left\{ \nabla_\Theta [\nabla_\Theta \log f(\mathbf{r}|\Theta)]^T \right\}$$

(24.17a)

$$= \begin{pmatrix} J_{1,1} & J_{1,2} & \cdots & J_{1,2n-1} & J_{1,2n} \\ J_{2,1} & J_{2,2} & \cdots & J_{2,2n-1} & J_{2,2n} \\ \vdots & \vdots & \ddots & \vdots & \vdots \\ J_{2n-1,1} & J_{2n-1,2} & \cdots & J_{2n-1,2n-1} & J_{2n-1,2n} \\ J_{2n,1} & J_{2n,2} & \cdots & J_{2n,2n-1} & J_{2n,2n} \end{pmatrix}.$$

(24.17b)

Note that the FIM has a dimension of $2n \times 2n$, since there are totally $2n$ location parameters to be estimated.

Using (24.14b) and (24.16b), for $i = 1, 2, \ldots, n$, the main diagonal elements and the first diagonal below/above the main diagonal of the FIM are derived as (see [11, 15] for a detailed derivation):

$$J_{2i-1,2i-1} = \sum_{j \in \mathcal{N}(i)} \frac{w_{ij} \cos^2(\alpha_{ij})}{\sigma_{ij}^2},$$

(24.18)

$$J_{2i,2i} = \sum_{j \in \mathcal{N}(i)} \frac{w_{ij} \sin^2(\alpha_{ij})}{\sigma_{ij}^2},$$

(24.19)

$$J_{2i-1,2i} = J_{2i,2i-1} = \sum_{j \in \mathcal{N}(i)} \frac{w_{ij} \cos(\alpha_{ij}) \sin(\alpha_{ij})}{\sigma_{ij}^2},$$

(24.20)

where

$$w_{ij} = 1 + \frac{\beta_{ij}^2 K_E}{2} d_{ij}^{\beta_{ij}-2}$$

(24.21)

is a distance-dependent unitless scaling factor always greater than 1 and α_{ij} is the angle between nodes i and j. Note that due to its definition, K_E is unitless only when $\beta_{ij} = 2$.

On the other hand, for $i, j = 1, 2, \ldots , n$, $j \neq i$ and $j \in \mathcal{N}(i)$, the nondiagonal elements, except the first diagonal below/above the main diagonal, of the FIM are given by

$$J_{2i-1,2j-1} = J_{2j-1,2i-1} = -\frac{w_{ij}\cos^2(\alpha_{ij})}{\sigma_{ij}^2}, \tag{24.22}$$

$$J_{2i,2j} = J_{2j,2i} = -\frac{w_{ij}\sin^2(\alpha_{ij})}{\sigma_{ij}^2}, \tag{24.23}$$

$$J_{2i-1,2j} = J_{2j,2i-1} = J_{2i,2j-1} = J_{2j-1,2i} = -\frac{w_{ij}\cos(\alpha_{ij})\sin(\alpha_{ij})}{\sigma_{ij}^2}. \tag{24.24}$$

If $j \notin \mathcal{N}(i)$, the above three nondiagonal elements simply are zeros.

Finally, the CRLB for the localization error of the ith node can be described in [11]:

$$\mathbf{E}\left[(\hat{x}_i - x_i)^2 + (\hat{y}_i - y_i)^2\right] \geq J_{2i-1,2i-1}^{-1} + J_{2i,2i}^{-1}, \tag{24.25}$$

where $[\hat{x}_i\ \hat{y}_i]^T$ is an unbiased estimate of the true coordinate $[x_i\ y_i]^T$, and $J_{i,j}^{-1}$ denotes the (i, j)th element of the inverse of the FIM J_Θ. A potentially more useful bound is given by

$$\sqrt{\mathbf{E}\left[(\hat{x}_i - x_i)^2 + (\hat{y}_i - y_i)^2\right]} \geq \sqrt{J_{2i-1,2i-1}^{-1} + J_{2i,2i}^{-1}}. \tag{24.26}$$

We emphasize that the FIM derived here (and in [11]) differs from the result in [10] by having the additional distance-dependent scaling factor w_{ij} given in (24.21), which captures the impact of having the knowledge of range estimation noise variance model on the localization accuracy.

24.3.2   MLE/Weighted LS

The previous section described the lowest variance achievable by an unbiased estimator. While providing insight into the performance of collaborative position location, at times the CRLB can be misleading. Thus, we also would like to examine the weighted LS estimator, which for the case of Gaussian measurement error is also the MLE.

Consider a square network with a size of $(\mathcal{L} \times \mathcal{L})\text{m}^2$, consisting of m anchors at known locations and n unlocalized nodes whose locations are to be estimated. If the distance between the ith and jth nodes is less than certain communication range R, we say they can communicate with each other and thus obtain a (noisy) range estimate r_{ij} of the true distance d_{ij}. Further, we assume range estimate is symmetric, that is, $r_{ij} = r_{ji}$.

Again, assuming that range estimates are corrupted by independent Gaussian noise, the range estimate between the ith and jth nodes is modeled by (24.11), where $n_{ij} \sim \mathcal{N}(0, \sigma_{ij}^2)$ is a zero-mean Gaussian random variable with a variance of

σ_{ij}^2. Based on this model, the MLE of collaborative position location from (24.4) is formulated using (24.14b) as

$$\hat{\Theta} = \arg\min_{\Theta} \sum_{i=1}^{n} \sum_{\substack{j=i+1 \\ j \in \mathcal{N}(i)}}^{n+m} \frac{1}{\sigma_{ij}^2} \left(r_{ij} - \sqrt{\left(x_i - x_j\right)^2 + \left(y_i - y_j\right)^2} \right)^2 \qquad (24.27)$$

$$\text{s.t.:} \quad 0 \leq x_i \leq \mathcal{L}, 0 \leq y_i \leq \mathcal{L}, \text{ for } i = 1, 2, \ldots, n,$$

where $\mathcal{N}(i) = \{q | q = 1, 2, \ldots, n + m, d_{iq} \leq R\}$. Strictly speaking, it is possible the global optimal solution falls outside the network depending on noise realizations and the nodes' true locations. In our problem, we set $\mathcal{L}$ to be reasonably large and the nodes' locations/noise are generated such that the global optimum always satisfies the constraints in (24.27). This solution is also known as the weighted LS solution since we are minimizing the weighted squared residuals. Also, if $\sigma_{ij}^2 = \sigma^2$, the MLE does not include σ_{ij}^2, as shown in [4], and collapses to the simpler LS solution.

Example 24.1

Consider a small network with two unlocalized nodes θ_1, θ_2 and four anchors θ_3, θ_4, θ_5, θ_6. The anchors are located at $\theta_3 = [0, 0]$, $\theta_4 = [0, 10]$, $\theta_5 = [10, 10]$, and $\theta_6 = [0, 10]$. Node 1 is able to measure its range to anchors θ_3 and θ_4 as well as its range to Node 2. Additionally, Node 2 is able to measure its range to anchors θ_5 and θ_6 as well as its range to Node 1. Further, the nodes exchange messages to obtain the same range estimate to each other. If the measured ranges are $r_{12} = r_{21} = 4.25$, $r_{13} = 6.9$, $r_{14} = 6.8$, $r_{25} = 5.1$, and $r_{26} = 5.5$, determine the LS solution to the two localized nodes' positions.

Solution

If we define $\Theta = [\theta_1; \theta_2] = [x_1, y_1; x_2, y_2]$, the LS cost function is

$$f(\Theta) = \|\theta_1 - \theta_2\|^2 + \|\theta_1 - \theta_3\|^2 + \|\theta_1 - \theta_4\|^2 + \|\theta_2 - \theta_5\|^2 + \|\theta_2 - \theta_6\|^2.$$

The solution to this can be found using the MATLAB function lsqnonlin. MATLAB codes can be found online at ftp://ftp.wiley.com/public/sci_tech_med/matlab_codes. The result is $\theta_1 = [4.6051, 5.0661]$, $\theta_2 = [8.7776, 5.2076]$. The actual positions are $\theta_1 = [4, 5]$, $\theta_2 = [5, 8]$, and the average error is 0.7 m. The result is plotted in Figure 24.4. *Note*: The MATLAB function lsqnonlin is not guaranteed to find the LS solution, although it works pretty well if the cost function is well-behaved (as in this case). In the next section, we will examine an algorithm that is guaranteed to find the LS solution.

The Branch-and-Bound (BB)/Reformulation-Linearization Technique (RLT) Algorithm   In general, for a nonlinear and nonconvex programming problem as in (24.27), gradient search-based methods or metaheuristic approaches cannot offer any performance guarantee on the final solution. In other words, we are not guaranteed to find the global optimum. The BB/RLT is an effective technique to find a probably global optimal solution for nonconvex programming problems [16].

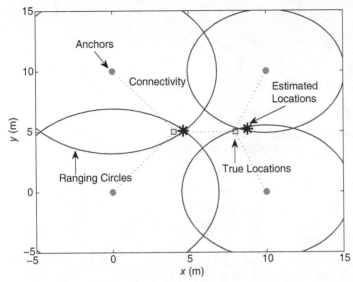

Figure 24.4 Anchor positions (filled circles), measured ranging circles (lines), true (squares), and estimated (asterisks) node positions from Example 24.1 (dashed lines represent node connectivity).

The approach was applied to collaborative position location in [12]. We repeat that description here. The principal idea of the BB/RLT is that, via the RLT, we can construct a linear programming (LP) relaxation to the original nonlinear minimization problem. The objective value of this LP problem serves as an LB for the original problem. Depending upon how the linear relaxation is carried out, the LP solution is either a feasible solution, or can be used as a starting point to perform a local search in order to find a feasible solution, to the original nonlinear programming problem. In our problem, the local search is not necessary since we induce the same constraints on x_i and y_i given in (24.27) into the LP, and the LP solution is thus always feasible. The objective value of the original problem using the LP solution is then an upper bound (UB) for the original minimization problem. With the ease of solving the LP problem and the guaranteed optimal solution to the LP, the BB strategy is brought in to partition the search space while tightening the UB and LB until LB $> (1 - \varepsilon)$UB, where ε is the target optimality parameter. In fact, it has been proved that as long as the partitioning intervals are compact, the BB/RLT converges to the global optimum [16]. A detailed description of the BB/RLT can be found in [12, 16].

Reformulation and Linearization of the MLE   In the RLT, we first linearize the objective function in (24.27) by making the following substitutions:

$$Z_{ij} = \sqrt{W_{ij}}. \tag{24.28}$$

$$W_{ij} = (x_i - x_j)^2 + (y_i - y_j)^2, \tag{24.29}$$

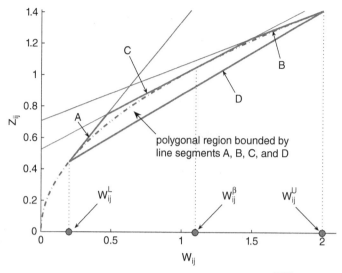

Figure 24.5 Polyhedral outer approximation to $Z_{ij} = \sqrt{W_{ij}}$ [12].

and the linearized objective function $\ell(Z,W)$ is then given by

$$\ell(Z, W) = \sum_{i=1}^{n} \sum_{\substack{j=i+1 \\ j \in \mathcal{N}(i)}}^{n+m} \frac{1}{\sigma_{ij}^2} \left(r_{ij}^2 - 2r_{ij}Z_{ij} + W_{ij} \right), \tag{24.30}$$

where $Z = \{Z_{ij}\}$ and $W = \{W_{ij}\}$. Second, we need to linearize the constraint given by (24.28). Specifically, for the square root function involved in (24.28), we relax it by defining a polyhedral bounding region shown in Figure 24.5 [12]. Given lower and upper bounds W_{ij}^L and W_{ij}^U for W_{ij}, two tangential lines of the square root function can be defined, containing the line segments A and C, respectively. If the middle point between W_{ij}^L and W_{ij}^U is denoted by W_{ij}^β, that is,

$$W_{ij}^\beta = \frac{1}{2} \left(W_{ij}^L W_{ij}^U \right), \tag{24.31}$$

another tangential line can be defined at W_{ij}^β. In addition, a chord (line segment D) can be defined by connecting the two end points on the square root function, corresponding to W_{ij}^L and W_{ij}^U, respectively. With all four line segments, the polyhedral bounded region is given as

$$-W_{ij} + 2\sqrt{W_{ij}^L} Z_{ij} \leq W_{ij}^L, \tag{24.32}$$

$$-W_{ij} + 2\sqrt{W_{ij}^U} Z_{ij} \leq W_{ij}^U, \tag{24.33}$$

$$-W_{ij} + 2\sqrt{W_{ij}^\beta} Z_{ij} \leq W_{ij}^\beta, \tag{24.34}$$

$$W_{ij} - \left(\sqrt{W_{ij}^U} + \sqrt{W_{ij}^L} \right) Z_{ij} \leq -\sqrt{W_{ij}^L W_{ij}^U}. \tag{24.35}$$

Now, we need to linearize the quadratic constraint given by (24.29), which is related to the estimated distance between the ith and jth node. For that, we define

$$U_i = x_i^2, \tag{24.36}$$

$$V_i = y_i^2, \tag{24.37}$$

for $i = 1, 2, \ldots, n$. In addition, if $j \leq n$, that is, node j is an unlocalized node, we define

$$S_{ij} = x_i x_j, \tag{24.38}$$

$$T_{ij} = y_i y_j. \tag{24.39}$$

Substituting (24.36–24.39) into (24.29), we obtain the following linearized constraints for $i = 1, 2, \ldots, n, j = i + 1, \ldots, n + m, j \in N\,(i)$:

1. If $j \leq n$, that is, node j is an unlocalized node:

$$W_{ij} - U_i + 2S_{ij} - U_j - V_i + 2T_{ij} - V_j = 0. \tag{24.40}$$

2. If $j \geq n + 1$, that is, node j is an anchor:

$$W_{ij} - U_i + 2x_i x_j - x_j^2 - V_i + 2y_i y_j - y_j^2 = 0. \tag{24.41}$$

With (24.40) and (24.41), we now need to linearize the quadratic constraints defined by (24.36–24.39). In particular, assuming the lower and upper bounds for x_i and y_i are given by x_i^L, x_i^U and y_i^L, y_i^U, respectively, we have, for x_i,

$$x_i - x_i^L \geq 0; \ x_i^U - x_i \geq 0. \tag{24.42}$$

Adopting the RLT [16], we can derive the following so-called bounding-factor constraints:

$$x_i^2 - 2x_i^L x_i \geq -\left(x_i^L\right)^2, \tag{24.43}$$

$$x_i^2 - 2x_i^U x_i \geq -\left(x_i^U\right)^2, \tag{24.44}$$

$$x_i^2 - \left(x_i^L + x_i^U\right)x_i \leq -x_i^L x_i^U. \tag{24.45}$$

Now, substituting (24.36) into (24.43–24.45), we obtain three linear constraints on U_i and x_i. Similar bounding-factor constraints can be derived for V_i via y_i. Furthermore, for S_{ij}, we have, for x_i and x_j,

$$x_i - x_i^L \geq 0; \ x_i^U - x_i \geq 0, \tag{24.46}$$

$$x_j - x_j^L \geq 0; \ x_j^U - x_j \geq 0. \tag{24.47}$$

Again using RLT, the bounding-factor constraints are given by

$$x_i x_j - x_i^L x_j - x_j^L x_i \geq -x_i^L x_j^L, \tag{24.48}$$

$$-x_i x_j + x_i^L x_j + x_j^U x_i \geq x_i^L x_j^U, \tag{24.49}$$

$$-x_i x_j + x_i^U x_j + x_j^L x_i \geq x_i^U x_j^L, \tag{24.50}$$

$$x_i x_j - x_i^U x_j - x_j^U x_i \geq -x_i^U x_j^U. \tag{24.51}$$

Substituting (24.38) into (24.48–24.51), we obtain four linear constraints on S_{ij}. Similar bounding-factor constraints can be derived for T_{ij} via y_i and y_j.

Finally, the combined relaxed LP problem is given as the following:

$$\underline{\phi}^* = \arg\min_{\phi} \sum_{i=1}^{n} \sum_{\substack{j=i+1 \\ j \in \mathcal{N}(i)}}^{n+m} \frac{1}{\sigma_{ij}^2} \left(r_{ij} - 2r_{ij}Z_{ij} + W_{ij} \right) \tag{24.52}$$

s.t.:   polygonal bounding constraints on (Z_{ij}, W_{ij}) from (24.32–24.35),

equality constraints in (24.40) and (24.41),

linear constraints on U_i derived from (24.43–24.45),

linear constraints on V_i, similar to U_i,

linear constraints on S_{ij} derived from (24.48–24.51),

linear constraints on T_{ij}, similar to S_{ij}.

In (24.52), the underline in the notation $\underline{\phi}^*$ emphasizes that the resulting LP objective value is an LB on the global optimum of the original MLE. The vector of optimization variables ϕ includes U_i, V_i, x_i, and y_i for $i = 1, 2, \ldots, n$. Furthermore, if r_{ij} exists, ϕ includes $\{\Theta, Z, W, U, V, S, T\}$.

Partitioning Variables, Relaxation Errors, and Partitioning Strategies   The partitioning variables in the BB process are those involved in the nonlinear terms for which new RLT variables have been defined. Specifically, the partitioning variables in our problem are x_i and y_i, while the RLT variables are U_i, V_i, S_{ij}, T_{ij}, W_{ij}, and Z_{ij}. During the BB/RLT algorithm, the bounding intervals for the partitioning variables are iteratively partitioned once their initial bounding intervals are given. In our problem, the initial bounding intervals for x_i and y_i can be simply set to be lower bounded by zero and upper bounded by the network edge length. Given those, the initial bounding intervals for W_{ij} can be easily computed using (24.29).

In each iteration of the BB/RLT algorithm, the partitioning variable resulting in the maximum relaxation error in the corresponding RLT variable is selected to be partitioned. The relaxation error of an RLT variable is defined to be the absolute difference between its value in the LP solution $\underline{\phi}^*$ and what is calculated using the underlying partitioning variable(s). For example, the relaxation error of RLT variable S_{ij} is given by

$$\Delta S_{ij} = \left| \underline{S}_{ij}^* - \underline{x}_i^* \underline{x}_j^* \right|,$$

where $\underline{S}_{ij}^*$, $\underline{x}_i^*$, and $\underline{x}_j^*$ are the values of U_i, x_i, and x_j in the LP solution $\underline{\phi}^*$, respectively. Relaxation errors for other RLT variables are defined in a similar fashion. If ΔS_{ij} happens to be the maximum relaxation error, we then examine the current lower and upper bounds for x_i and x_j. If $\left(x_i^U - x_i^L \right) > \left(x_j^U - x_j^L \right)$, x_i is selected, and its two new bounding intervals are $\left[x_i^L \ \underline{x}_i^* \right]$ and $\left[\underline{x}_i^* \ x_i^U \right]$; otherwise, x_j is selected, and its two new bounding intervals are $\left[x_j^L \ \underline{x}_j^* \right]$ and $\left[\underline{x}_j^* \ x_j^U \right]$. We finally emphasize that the choice of partitioning variable does not affect the final result. However, it does affect the algorithm convergence speed. Computational complexity is one drawback of this approach. Specifically, ensuring the global optimum comes at the high price of extremely high computational complexity (relative to linear LS approaches or other suboptimal approaches).

24.3.3   Numerical Results

In Figure 24.6, we compare the MLE with the linear least squares (LLS) location estimator, as well as the CRLB, for the case of noncollaborative position location, in order to demonstrate some properties of the MLE. The LLS is based on a Taylor series approximation to (24.3) and demonstrates the performance of a relatively simple estimator. This is described in detail in Chapter 6. In Figure 24.6a, we plot the mean localization error $\bar{\Omega}$ of the MLE and the LLS estimator for 1000 noise realizations with respect to σ^2. The unlocalized node's true location is (20, 20) and the four anchors' locations are at the four corners (0, 0), (40, 0), (0, 40), and (40, 40), respectively, which is a good node geometry. As can be seen, the MLE slightly outperforms the LLS, and is very close to the CRLB. This suggests that in the case of good node geometry, that is, the unlocalized node being well-surrounded by the anchors, the MLE performs as well as what the CRLB implies, and the LLS location estimator performs about as well as the much more sophisticated nonlinear MLE. On the other hand, in Figure 24.6b, the four anchors' locations are almost collinear, at (0, 0), (10, 11), (28, 29), and (40, 40), and the unlocalized node at (20, 20) lies almost on the same line roughly formed by connecting the four anchors, which is an extremely bad node geometry. Here, the MLE not only significantly outperforms the LLS but also is far more robust to range estimation noise than the LLS estimator. This is not too surprising since it is well-known that the LLS approach performs very poorly in bad anchor geometry. More surprisingly, however, the MLE performs even better than the CRLB. Despite being counterintuitive, this has partly been observed in [17, 18]. Specifically, in the case of bad node geometry, for example, almost but not exactly collinear, the CRLB does exist but becomes unreasonably large due to the fact that the FIM becomes nearly singular. In fact, it has been demonstrated in [17] that using a specially designed position location method, the localization error can be significantly reduced. As shown in [12], the phenomenon can be interpreted by the fact that the time-of-arrival (TOA)-based MLE in (24.27) is in general a *biased* estimator. Since the TOA-based MLE is indeed biased and the MSE of a biased estimator can be much smaller than the CRLB, as in [19, 20], this observation is easily justified.

A collaborative localization example is given in Figure 24.7. In this example we see that for networks where node geometry is bad, for example, some anchors or unlocalized nodes are close to collinear, the gradient search-based method may lead to flip ambiguities. Specifically, for a particular noise realization, a gradient search-based method (i.e., similar to LLS) can be trapped into local minimum with an objective value of 4.2 and thus has incorrect flipped solution with a localization error of 2.99 m. On the other hand, the BB/RLT is always able to find the global minimum of the MLE, which is 3.5 in this case and consequently gives a solution

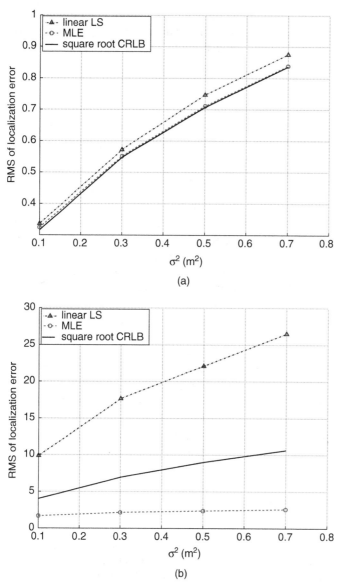

Figure 24.6   Mean localization errors of the LLS and the MLE versus the noise variance σ^2 with good geometry (a) and bad geometry (b). In (a), four anchors are placed at four corners $(0, 0)$, $(0, 40)$, $(40, 0)$, $(40, 40)$, and the unlocalized node is at $(20, 20)$. In (b), four almost-colinear anchors are placed at $(0, 0)$, $(10, 11)$, $(28, 29)$, $(40, 40)$, and the unlocalized node is at $(20, 20)$ [15].

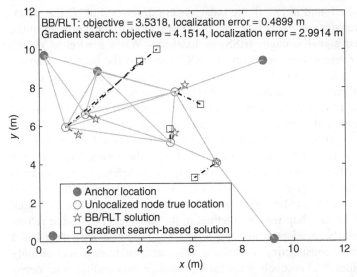

Figure 24.7 The MLE solution computed by the BB/RLT and the gradient search-based method for a particular noise realization, for a bad network geometry, leading to flip ambiguities when using the simpler gradient search approach [15].

that has a much smaller localization error of 0.49 m. Finally, we need to mention that for network collaborative position location, the convergence of the BB/RLT takes much longer than the gradient search approach. For the MLE in Figure 24.7, the number of iterations it takes to converge is over 20,000. Nevertheless, the BB/RLT solution to the MLE can still be used as a benchmark for evaluating any practical collaborative position location scheme.

24.4 AN OVERVIEW OF SUBOPTIMAL ALGORITHMS

The previous section examined performance bounds to collaborative position location. One such bound is the estimate that minimizes the square error between what was measured (observables) and the values implied by the position estimates. The performance of the approach is optimal in an LS sense and a ML sense if the error is Gaussian. However, the complexity of the approach is very high. Due to the difficulty in solving this optimization problem, most existing work focuses on low-complexity algorithms which can only approximate the optimal performance. In this section, we will briefly describe existing suboptimal techniques based on when they were introduced.

Beginning of the decade: Most of the early developments have not explicitly used the term collaborative position location and focused only on position location without the help of GPS [9, 21]. For example, in [8], the authors proposed a simple localization method where anchors broadcast beacons into the network, and each unlocalized node uses the centroid of the anchors from which it has received beacons as its location estimate. Similar work but with improved performance can be found in [9]. In the radar system developed in [22], a prede-ployment measurement campaign was conducted to establish a mapping from the vector of received signal strength (RSS) to locations. When a wireless device comes into the coverage area, it reports its RSS vector to the central database, which will return a location that best matches with the RSS vector to the device as its location estimate. The distance vector (DV)-hop approach proposed in [23] first calculates the average distance per hop by flooding anchor beacons within the network. Then, each unlocalized node uses this distance information, as well as the number of hops it is from any anchor that has been learned in the beacon-flooding stage, to estimate its location by performing trilateration. This can also be viewed as an effort to improve the work in [8]. Nonetheless, its performance still suffers from the average-hop distance estimation step. To improve this, in [24] the authors adopted a local trilateration-based refinement stage to refine the initial solution obtained by the "hop-terrain" method in the first stage. A similar refine-ment approach has been adopted in [25], where the proposed algorithm has a preprocessing step of constructing so-called collaborative subtrees to include only those nodes considered as uniquely localizable, although the conditions for deter-mining unique localizability only consider connectivity information and are thus questionable in practice. Another well-formulated approach is the work in [26], where the authors proposed a centralized convex programming approach to deter-mine node locations using the connectivity information of the entire network. In general, these earlier developments are mostly coarse-grained localization systems and focus more on demonstrating the feasibility of collaborative position location, and therefore have not carefully considered practical issues such as range estima-tion noise and NLOS propagation.

Around the middle of the decade: Later, an approach using only connectivity information was proposed in [47], where multidimensional scaling (MDS) technique was applied to the network connectivity matrix to construct a relative node location map. Furthermore, in the following work of Shang et al. [27] and in some parallel work [28, 29], distributed MDS using distance estimates was proposed to improve the localization accuracy as well as to address the algorithm scalability problem associated with centralized MDS localization. In [4], the authors described the MLE for sensor localization, based on the assumption that the TOA range estimation noise is Gaussian distributed. Despite the fact that the MLE is known to be optimal in terms of the MSE, a closed-form expression for the global optimal solution is unavailable due to the associated nonlinear and nonconvex optimization problem. In [30, 31], the authors resorted to a semidefinite programming (SDP) relaxation to the original MLE of collaborative position location and demonstrated its robust perfor-mance across different network topologies. In [32], a second-order cone program-ming (SOCP) relaxation is adopted, aiming at faster computation than the SDP approach at the cost of some localization accuracy degradation. However, both the SDP and the SOCP are no longer the original MLE and do not provide the true optimal performance of collaborative position location. In [33], the authors used a grid-based numerical solution search strategy to solve the MLE for collaborative position location based on the RSS measurements, but the solution is again not guaranteed to be optimal. In [34], the authors proposed the use of robust quadrilaterals to particularly handle flip ambiguity in node locations. However, the requirement of having many quadrilaterals in the network can be too stringent. The authors in [35] focused on a sequential localization method and developed a measurement selection procedure to improve localization accuracy and handle the problem of error propagation. One interesting research direction taken by the authors in [6, 36] is the use of graph rigidity theory to investigate node localizability. However, one obvious drawback of the approach is that it does not consider the presence of range estimation noise or NLOS bias. Another notable and distinguishing approach is the use of the nonparametric belief propagation (NBP) for collaborative position location [37] and some subsequent similar works [38, 39]. In the NBP approach, the network localization problem is formulated as a probabilistic inference problem, where the goal is to obtain an approximated probability distribution for each unlocalized node's location, instead of obtaining a snap-shot location estimate. Belief propagation (BP) refers to the process of nodes communicating and exchanging information with each other. The resulting approximated probability distribution naturally captures the uncertainty of the location estimate and can be used to determine node localizability in the presence of range estimation noise. However, the biggest drawback with NBP-based or other probabilistic localization methods is the prohibitively high computational complexity, which can be problematic for low-cost and low-power sensor deployment. Overall, these developments have greatly advanced the techniques of collaborative position location as well as drawn special attention to topics such as the optimal performance, algorithm efficiency and scalability, localization accuracy in NLOS conditions, and node localizability in practical situations.

   The past three years or so: More recently, there has been an increased amount of effort spent on the topics we just mentioned above. In particular, the authors in [40] derived the MLE in the presence of NLOS biases and developed a numerical method to solve it. However, the global optimality of the solution is still not guaranteed. In [12], the authors developed a BB/RLT framework to solve the MLE for distance-based collaborative position location. This is the first work that guarantees the global optimality of the solution to the MLE. Based on this, we identified cases where the MSE performance indicated by the CRLB can be surpassed by the MLE due to the fact that the MLE is biased and concluded that the optimal performance of the MLE is a more meaningful performance benchmark in such cases. To address the issue of algorithm efficiency and scalability, the authors in [41] developed a distributed SOCP approach to network localization. Similar efforts were described in [42], where a localization scheme based on sum of squares (SOS) relaxation is proposed. Also, the authors in [43] proposed a curvilinear component analysis (CCA)-based method to improve the MDS-MAP approach. However, these methods still have considerable computational complexity and do not scale well as the number of nodes and average node connectivity increase. Regarding NLOS conditions, in [44], we developed a collaborative quasilinear programming framework to consider both node collaboration and NLOS mitigation. However, the performance is still not satisfactory, especially when sequential localization has to be used. In terms of node localizability, the works in [45, 46] have the SDP and graph rigidity approaches, respectively. However, both have not considered the issue of range estimation noise and are thus not applicable in practical situations. To summarize, these more recent works have spent much effort to address several important practical issues we described above, but significant work still remains ahead.

24.4.1 A Taxonomy of Existing Algorithms

To better our understanding of existing localization algorithms, we now provide several different ways to classify them and briefly describe the pros and the cons associated with each class of methods. Within each class, we only strive to describe the key ideas of the most popular methods, which shall be only considered as being representative rather than being comprehensive.

Type of Measurement Data: Distance, Angle of Arrival (AOA), and RSS Fingerprinting Depending on the measurement data being used, localization algorithms can be classified as distance-based (via RSS) or TOA, connectivity-based, AOA-based, RSS fingerprinting, and hybrid methods. For example, algorithms proposed in [8, 23, 26, 27, 47, 48] are all connectivity-based localization, while algorithms such as those in [24, 25, 30, 35, 49, 50] are all distance-based localization. In general, connectivity-based localization has lower device and computational complexity but only generates coarse-grained location information. Distance-based localization, on the other hand, leads to more accurate location estimation but requires more sophisticated devices and has heavier computational load. Some algorithms, for example, [51], use connectivity-based location estimation as a starting point, and then apply distance-based location refinement.

Where the Computation Is Performed: Centralized or Distributed Another way to classify localization algorithms is based on where the computation is performed. In *centralized* approaches [4–9, 12, 15, 23–32, 40, 47, 49–53], all the anchor locations and measurement data are forwarded to a central processor to compute the unlocalized nodes' positions in a joint manner. In *distributed* approaches [23–25, 28, 29, 34, 35, 41, 44, 50, 54], the computation is spread over the entire network and thus, only local information exchange is required. Distributed approaches have the advantage of being scalable and more robust to node failures. Centralized approaches, on the other hand, utilize the information about the entire network and are supposed to yield more accurate location estimation. However, an efficient solver for large-scale nonlinear optimization problems associated with centralized approaches is needed to truly achieve the global optimum. Considering all the trade-off and complexity issues, distributed approaches have attracted much more attention and are more practical for the applications of interest.

How the Computation Is Performed: Sequential or Concurrent Depending on how the location information is *propagated* through the network, localization algorithms can also be classified as *sequential* localization or *concurrent* localization. In sequential approaches [35, 44, 50], each unlocalized node updates itself as a *virtual* anchor to assist other unlocalized nodes to localize themselves. In the sequential approaches, only measurements with neighboring nodes are used. An obvious drawback of the sequential approaches is the propagation of localization error [55, 56], which refers to the fact that virtual anchors themselves have localization error and will affect any unlocalized node that uses range estimates to them to perform location estimation. On the other hand, in the concurrent approaches, unlocalized nodes do not act as virtual anchors even after they have been localized. Examples include all the centralized approaches and the distributed algorithms proposed in [25, 29, 34]. Generally speaking, although sequential location estimation appears to be a simple way of disseminating location information throughout the network, the problem of localization error propagation has seriously limited its performance.

How the Problem Is Formulated: Probabilistic or Nonprobabilistic
Another way to classify existing algorithms is regarding whether the position location problem is formulated in a *probabilistic* manner. Specifically, depending on the nature of the framework as well as the returned location estimates, existing algorithms can be divided into *deterministic* and *probabilistic* approaches. Deterministic approaches solve position location problems in different ways, including pure optimization-based algorithms [26, 30, 32, 51, 57], estimation-based methods [4, 58], and other ad hoc algorithms, for example, in [8, 9, 23–25, 29, 34, 49, 50, 59]. A common feature of these algorithms is that they only compute a deterministic *one-shot* location estimate for nodes that have been localized. In other words, there is no additional information about the quality or the reliability of this solution. Metrics such as distance residual, geometric dilution of precision (GDOP), and error ellipses, which do provide some sense of reliability, may not be appropriate for collaborative position location, especially in the presence of NLOS propagation. On the other hand, in probabilistic localization approaches [37–39], the position location problem is formulated as a probabilistic inference problem, with the goal of computing (approximating) the probability distribution of each unlocalized node's location. Under such a framework, a position location algorithm returns a set of possible location estimations, each of which is preferably, although not necessarily, associated with a weight quantifying the uncertainty, or equivalently the reliability, about the corresponding location estimate. In this sense, probabilistic approaches return a more direct representation of the solution quality. However, existing algorithms under the probabilistic localization framework have much higher computational complexity than one-shot estimate-based algorithms, which can be a limiting factor for low-power sensor deployment.

24.5 SPECIFIC SUBOPTIMAL APPROACHES

The previous section briefly outlined various approaches to collaborative localization. In this section, we will provide additional details for three basic suboptimal approaches: (1) optimization-based approaches with relaxations to the nonlinear LS solution, (2) MDS, and (3) projection onto convex sets. It should be noted that we will describe the centralized approach for the first two for ease of explanation, but it should be kept in mind that these approaches have distributed implementations.

In the next section, we will then examine the performance of the distributed versions of these techniques. Note that Bayesian techniques, which are probabilistic approaches, are not described in this chapter. However, they are described in detail in Chapter 26.

24.5.1 Sequential LS

Perhaps the simplest approach to collaborative localization is what can be called *sequential* LS collaborative localization. The basic idea behind this approach is for each node, which has a minimum of three connections to anchors, to determine their location and then act as *virtual* anchors for other nodes. Specifically,

Algorithm 1: Sequential LS

1. All nodes estimate the range to all nodes within range.
2. All nodes with three or more range measurements to anchors determine their positions using standard LS solution for single-node localization.
3. All nodes that have determined their position become virtual anchors.
4. If all nodes are localized or no new nodes have been localized this round, stop.
5. Return to 2.

As an example, consider the network in Figure 24.8 (top-left). This network has five anchors and four unlocalized nodes. The first step in the algorithm is for all nodes

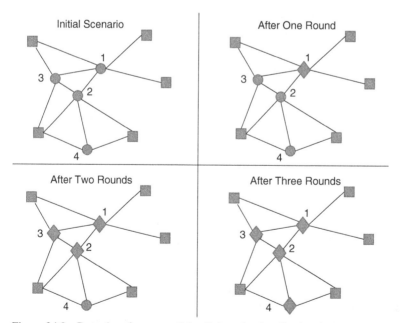

Figure 24.8 Procedure for sequential collaborative localization (squares represent anchors, circles represent unlocalized nodes, diamonds represent virtual anchors, and lines represent connections or range measurements).

to determine distance estimates to all nodes within range. These connections are shown by the lines in the figure. Once the distance estimates are completed, only Node 1 has estimates to three anchors and can thus estimate its position. Node 1 now becomes a virtual anchor as shown in Figure 24.8 (top-right). Nodes 2 and 3 are now able to estimate their locations due to fact that Node 1 is now a virtual anchor (bottom-left). Finally, Node 4 can determine its position as shown (bottom-right).

This algorithm is perhaps among the simplest collaborative algorithms. However, it has the disadvantage that coverage is limited. In other words, it requires that at least some of the nodes have connections with three or more anchors. For example, the scenario shown in Figure 24.1 would not allow a solution with this algorithm since neither of the nodes can localize themselves independently. Additionally, the sequential nature of the algorithm means that error can build quickly. For example, a large error in the estimate of Node 1 in Figure 24.8 would cause substantial error to the other nodes. This is termed *error propagation* [56]. We will investigate the coverage problem in the following example.

Example 24.2

Consider the four example networks shown in Figure 24.9. How many nodes could be localized in each network using conventional (noncollaborative) localization? How many could be localized in each network using a sequential localization algorithm? What about using a concurrent algorithm?

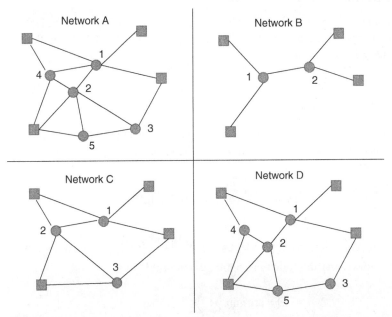

Figure 24.9   Four networks of Example 24.2 (squares represent anchors, circles represent unlocalized nodes, lines represent connections or range measurements).

Solution

Let us begin with Network A. In this network, only one unlocalized node (Node 1) has sufficient (i.e., three) connections with anchors to determine its position. However, using a sequential algorithm, Node 1 can act as an anchor to Node 2 and Node 4. This allows Node 4 to determine its position and act as an anchor to Node 2. Node 2 sequentially has two virtual anchors (Nodes 1 and 3) and one true anchor and can thus determine its position and act as an anchor to Nodes 3 and 5. However, neither Node 3 nor Node 5 have sufficient connections to anchors or virtual anchors at that point in time in order to localize themselves. Thus, using a sequential approach, three of five unlocalized nodes can be localized, compared to only one with conventional approaches. Although we have not shown this, using a parallel approach, all five nodes could be localized.

Examining Network B, we can see that using a conventional, noncollaborative approach, none of the nodes can be localized. Consequently, none of the nodes can be localized using a sequential approach either. This highlights the fact that there must be at least one node with three anchor connections for a sequential approach to be effective. However, as we saw previously, using a parallel approach, both nodes can be localized.

Similarly, examining Network C, we find that 1,3, and 3 nodes (out of three) could be localized using noncollaborative, sequentially collaborative, and parallel collaborative approaches, respectively. For Network D, the values are 1,1,3 out of five nodes.

24.5.2   Optimization-Based Approaches

A large number of centralized approaches are based on viewing the localization problem as an optimization problem. Specifically, let us recall the MLE that is formulated as

$$\hat{\Theta} = \arg\min_{\Theta} \sum_{i=1}^{n} \sum_{\substack{j=i+1 \\ j \in \mathcal{N}(i)}}^{n+m} \frac{1}{\sigma_{ij}^2} \left(r_{ij} - \|\theta_i - \theta_j\| \right)^2. \tag{24.53}$$

To understand the various optimization approaches, let us take the MLE and assume that the variances on each of the observations is the same, $\sigma_{ij} = \sigma$, $\forall i, j$. This results in an LS objective:

$$\hat{\Theta} = \arg\min_{\Theta} \sum_{i=1}^{n} \sum_{\substack{j=i+1 \\ j \in \mathcal{N}(i)}}^{n+m} \left(r_{ij} - \|\theta_i - \theta_j\| \right)^2. \tag{24.54}$$

Now, modifying the criterion slightly, let us use $\left| r_{ij}^2 - \|\theta_i - \theta_j\|^2 \right|$ as the error criterion. Thus, we can now write the following optimization problem:

$$\hat{\Theta} = \arg\min_{\Theta} \sum_{i=1}^{n} \sum_{\substack{j=i+1 \\ j \in \mathcal{N}(i)}}^{n+m} \left| r_{ij}^2 - y_{ij} \right| \tag{24.55}$$

$$s.t. \quad y_{ij} = \|\theta_i - \theta_j\|^2,$$

which is a nonconvex optimization problem due to the equality constraint. A very simple relaxation is to relax the equality constraint to obtain a convex constraint. Specifically, the problem

$$\hat{\Theta} = \arg\min_{\Theta} \sum_{i=1}^{n} \sum_{\substack{j=i+1 \\ j \in \mathcal{N}(i)}}^{n+m} |r_{ij}^2 - y_{ij}|$$

(24.56)

$$s.t. \quad y_{ij} = \|\theta_i - \theta_j\|^2$$

is a convex optimization problem. Specifically, it is an SOCP problem [32, 41].

A slightly tighter restriction is developed in [75]. Specifically, following [75], if we define $\Theta = [\theta_1, \theta_2, \dots, \theta_n]$ and $\mathbf{A} = [\theta_{n_u+1}, \dots, \theta_{n+m}]$, then

$$\|\theta_i - \theta_j\|^2 = (\theta_i - \theta_j)^T (\theta_i - \theta_j).$$

(24.57)

Further, if we define e_i as a vector of zeros with a one in the ith position and

$$b_{ij} = \begin{bmatrix} \mathbf{I}_n & 0 \\ 0 & \mathbf{A} \end{bmatrix} (e_i - e_j),$$

(24.58)

where $\mathbf{I}_n$ is an $n \times n$ identity matrix, then we can write

$$\|\theta_i - \theta_j\|^2 = (e_i - e_j)^T \begin{bmatrix} \Theta^T \\ \mathbf{A}^T \end{bmatrix} [\Theta \quad \mathbf{A}](e_i - e_j)$$

$$= b_{ij}^T \begin{bmatrix} \Theta^T \\ \mathbf{I}_2 \end{bmatrix} [\Theta \mathbf{I}_2] b_{ij}$$

(24.59)

$$= \left\langle b_{ij} b_{ij}^T, \begin{bmatrix} \Theta^T \Theta & \Theta^T \\ \Theta & I_2 \end{bmatrix} \right\rangle_F,$$

where $<\mathbf{A}, \mathbf{B}>_F = trace(\mathbf{AB})$. Now, it can be shown that if $\begin{bmatrix} \mathbf{Y} & \mathbf{X}^T \\ \mathbf{X} & \mathbf{I}_2 \end{bmatrix} \geq 0$ with rank 2, then $\mathbf{Y} = \mathbf{X}^T\mathbf{X}$. Thus, we can rewrite the original nonconvex optimization problem as

$$\min_{\mathbf{Z}} \sum_{i=1}^{n} \sum_{\substack{j=i+1 \\ j \in \mathcal{N}(i)}}^{n+m} |\langle b_{ij} b_{ij}^T, \mathbf{Z} \rangle_F - r_{ij}^2|$$

$$s.t. \quad [\mathbf{Z}_{ij}]_{i,j \geq n+m} = \mathbf{I}_2$$

$$\mathbf{Z} \geq 0$$

$$rank(\mathbf{Z}) = 2.$$

(24.60)

Now, if we relax the rank constraint and replace the absolute value with the standard slack variables, we have a convex optimization problem

$$\min_{Z} \sum_{i=1}^{n} \sum_{\substack{j=i+1 \\ j \in \mathcal{N}(i)}}^{n+m} \langle b_{ij} b_{ij}^{T}, \mathbf{Z} \rangle_{F} - r_{ij}^{2} + v_{ij} - u_{ij}$$

$$s.t. \quad [Z_{ij}]_{i,j \geq n_u + n_a} = \mathbf{I}_2$$

$$\mathbf{Z} \geq 0$$

$$u_{ij}, v_{ij} \geq 0, \tag{24.61}$$

which is an SDP problem. In general, this is more restrictive (less of a relaxation) than the SOCP but more computationally expensive. A similar approach can be taken using only *connectivity* information. Specifically, if we only use the fact that nodes are neighbors (i.e., can communicate with each other), we can adapt the above SOCP as

$$\hat{\Theta} = \arg\min_{\Theta} \sum_{i=1}^{n} \sum_{\substack{j=i+1 \\ j \in \mathcal{N}(i)}}^{n+m} |R^2 - y_{ij}| \tag{24.62}$$

$$s.t. \quad y_{ij} \geq \|\theta_i - \theta_j\|^2,$$

where R is the communication range.

24.5.3 MDS

A distinct centralized approach is based on MDS. MDS is a set of statistical methods that attempt to arrange objects in a space with a particular number of dimensions so as to represent a matrix of observed similarities between the objects. In particular, MDS can be used to create a set of two- or three-dimensional locations for nodes based on the observed distances between them. In classic MDS, we expect a complete similarity matrix (which corresponds to full connectivity in our scenario). Since full connectivity is unlikely in most network localization problems, we can approximate the distances between far-away nodes using the shortest path between nodes.

A nice feature of this approach is that the technique can use either distance measurements between nodes or simple connectivity information. Additionally, the technique can be applied to subnetworks separately, whose results can then be stitched together. This allows for the approach to be somewhat scalable. MDS will result in network locations that are relative. That is, they include an arbitrary scaling, rotation, translation, or flip from the true location. To handle this problem, the final solution can be transformed based on the knowledge of the anchor node locations.

The basic MDS approach can be summarized using the following steps:

1. Create a proximity matrix of size $(n + m) \times (n + m)$ for the network. The proximity matrix corresponds to the distances between all nodes in the network. If the network had full connectivity and the distances could be measured, all entries in the matrix would correspond to the true distances between all nodes in the network. In the typical case, however, only those nodes that are neighbors can calculate the distance between themselves. Thus, the measured

distances are used for neighboring nodes only. For nodes that are not neighbors, the proximity value is based on the shortest route between nodes summing the distances along the path. Classic shortest path algorithms can be used to accomplish this. In the case where only connectivity information is available, each distance between neighboring nodes can be represented by a single unit of distance, and the shortest path corresponds to the number of hops between nodes.

2. The next step is to perform MDS on the distance matrix. Specifically, we first create a matrix **P** of the *squared* proximity measures. The matrix is then double-centered by subtracting the mean of each row and each column from every value in the matrix. Then the overall mean (i.e., grand mean) is added back to each element. Performing a singular value decomposition on **P** (i.e., **P** = **UVU**^{T}), we can find the two-dimensional location of nodes by taking the two largest singular values and the corresponding two singular vectors and creating the vectors

$$\mathbf{X} = \mathbf{U}_2 \mathbf{V}_2, \tag{24.63}$$

where **U**$_2$ is an $n \times 2$ matrix of the two superior singular vectors, and **V**$_2$ is a 2×2 diagonal matrix of the maximum two singular values.

3. The first two steps result in *relative* locations of the nodes but not their absolute positions. That is, the result of step two is a set of positions that are arbitrarily scaled, rotated, translated, and flipped relative to the true locations. However, given the positions of a sufficient number of anchor nodes, the relative locations can be transformed into the absolute locations.

The approach can be modified in the following ways. First, the solution can be used as the initialization of an LS refinement. This can improve the solution substantially. Additionally, instead of creating a matrix for the entire network, a matrix for a local area can be created. The resulting positions can be combined with other local maps to create the entire global map.

The MDS approach has been shown to work well when the network is uniform but performs poorly when the network has an irregular shape. In this case, there is not a strong correspondence between the shortest path between two nodes and the Euclidean distance between them. Additionally, the approach is more sensitive to noise than other techniques. We will demonstrate both of these characteristics shortly. On the positive side, the approach can be used with pure connectivity and it is less sensitive to NLOS errors, as will be seen.

Example 24.3

Repeat Example 24.1 using MDS to solve.

Solution

The first step is to create a distance matrix between all nodes in the network (including the anchors). We have the measured distances and can calculate the exact distances between the anchors. The resulting distance matrix is:

$$\mathbf{R} = \begin{bmatrix} 0 & 4.25 & 6.90 & 6.80 & 0 & 0 \\ 4.25 & 0 & 0 & 0 & 5.11 & 5.51 \\ 6.90 & 0 & 0 & 10 & 14.14 & 10 \\ 6.80 & 0 & 10 & 0 & 10 & 14.14 \\ 0 & 5.11 & 14.14 & 10 & 0 & 10 \\ 0 & 5.51 & 10 & 14.14 & 10 & 0 \end{bmatrix}. \tag{24.64}$$

Note that the diagonal elements are all zero since they represent the distance of a node to itself. Additionally, there are several other zeros in the matrix due to the fact that there is not a direct connection between the nodes (as seen in Fig. 24.4), and the distance cannot be directly measured. We must determine the minimum multihop distance between the corresponding nodes. For example, considering node θ_1, we can estimate its distance to anchor θ_5 as the two-hop distance through Node 2:

$$R_{15} = R_{51} = r_{12} + r_{25} = 9.35. \tag{24.65}$$

Additionally, $R_{16} = R_{61} = r_{12} + r_{26} = 9.76$, $R_{23} = R_{32} = r_{21} + r_{13} = 11.15$, and $R_{24} = R_{42} = r_{21} + r_{14} = 11.05$. These values are added into the matrix $\mathbf{R}$. The next step is to create the matrix $\mathbf{P}$ by squaring each element in $\mathbf{R}$ and double-centering it. The result is

$$\mathbf{R} = \begin{bmatrix} -20.58 & -6.91 & -19.16 & -19.93 & 30.45 & 36.12 \\ -6.91 & -29.35 & 53.20 & 51.58 & -35.39 & -33.13 \\ -19.16 & 53.20 & -113.04 & -12.44 & 96.71 & -5.27 \\ -19.93 & 51.58 & -12.44 & -111.84 & -2.69 & 95.32 \\ 30.45 & -35.39 & 96.71 & -2.69 & -93.55 & 4.47 \\ 36.12 & -33.13 & -5.27 & 95.32 & 4.47 & -97.52 \end{bmatrix}. \tag{24.66}$$

Performing the singular value decomposition on $\mathbf{P}$ and obtaining the maximum two singular values, we have $\lambda_1 = 247.7$ and $\lambda_2 = 200.1$. Using the corresponding singular vectors, the estimated positions of all six nodes in the network are $\hat{\theta}_1 = [-48.32, 4.35]$, $\hat{\theta}_2 = [89.44, 2.03]$, $\hat{\theta}_3 = [-123.46, -100.83]$, $\hat{\theta}_4 = [-121.71, 99.14]$, $\hat{\theta}_5 = [102.66, 97.71]$, and $\hat{\theta}_6 = [101.40 - 102.41]$. Using the known locations of the anchors, we can determine a transformation $\theta_i = \mathbf{T}\hat{\theta}_i + \mathbf{b}$, where $\mathbf{T}$ is a linear transformation, and $\mathbf{b}$ is an offset to transform the estimated anchor locations to the true locations. Doing so results in

$$\mathbf{T} = \begin{bmatrix} 0.0469 & 0.0003 \\ -0.0003 & 0.0469 \end{bmatrix} \tag{24.67}$$

and $\mathbf{c} = [5.4819, 5.0782]^T$. Using this resulting transformation on the unknown node values results in $\hat{\theta}_1 = [3.2123, 5.2664]^T$ and $\hat{\theta}_2 = [9.6793, 5.2031]^T$ for an average estimation error of 1.3 m, which is about one-half a meter worse than nonlinear LS. The estimates are plotted in Figure 24.10.

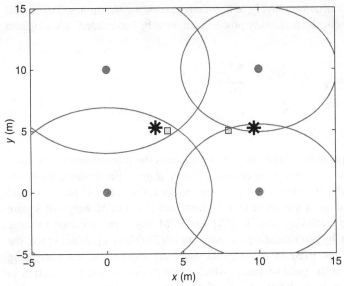

Figure 24.10   Anchor positions (filled circles), measured ranging circles (lines), true (squares), and estimated (asterisks) node positions from Example 24.3.

24.5.4   Set-Theoretic Approach: Iterative Parallel Projection Method (IPPM)

As a fourth example of collaborative position location, we describe the IPPM [60, 61]. Specifically, we first briefly describe the parallel projection method (PPM) and its application to noncollaborative position location. Then, we elaborate on how it can be developed into an iterative approach to solve the collaborative position problem.

The Modified Parallel Projection Method (MPPM)   The set-theoretic signal recovery problem is to find a signal that satisfies all feasibility sets defined by prior knowledge and the observed data [62]. The mathematical formulation of the convex set feasibility signal recovery problem can be written as

$$\text{Find} \quad a^* \in \bigcap_{k=1}^{K} S_k, \tag{24.68}$$

where S_k is the kth feasibility set and is assumed to be closed and convex. Methods such as projection onto convex sets (POCS), which sequentially projects an initial estimate onto each convex set until convergence, have been developed (see [63] for details) to solve (24.68).

In practice, due to inaccurate prior knowledge and noisy observations, it is possible that the feasibility sets are inconsistent, meaning that there exists one or more conflicting feasibility sets such that $\bigcap_{k=1}^{K} S_k = \varnothing$. In such cases, the convergence behavior of POCS methods is not known and the formulation in (24.68) is no longer reasonable. Alternatively, since there exists no solution satisfying all convex feasibility sets, the feasibility problem can now be formulated as a weighted LS problem [63]:

$$\min_{a} : \Phi(a) = \sum_{k=1}^{K} w_k d_k(a)^2,$$

$$s.t.: \sum_{k=1}^{K} w_k = 1, \quad \text{and} \quad \forall k, w_k > 0, \tag{24.69}$$

where $\Phi(a)$ is the proximity function, which measures the infeasibility of the solution a, and $\{w_k\}$ are a set of strictly convex weights. $d_k(a)$ is the distance from a to its projection onto the kth convex set S_k. As can be seen, the objective now is to find a best feasible solution in the sense that it minimizes the sum of weighted square distances to all the feasibility sets. In [63], the PPM has been proposed to solve (24.69). In addition, by reformulating the problem of (24.69) in a product space, the author proved that the PPM in the original space corresponds to the alternating projection method in the product space, which leads to the desired LS solution to (24.69). In particular, if we denote the set of solutions to (24.69) as G, it is proved in [63] that as long as one of the convex sets S_k is bounded, for any initial estimate a^0, the sequence of a^l given by

$$a^{l+1} = a^l + \lambda_l \left(\sum_{k=1}^{K} w_k P_k(a^l) - a^l \right), \quad \text{for} \quad l \geq 0 \tag{24.70}$$

will converge to a point in G, where $\lambda^l \in [\varepsilon, 2 - \varepsilon]$ for $0 < \varepsilon < 1$ is called the relaxation parameter, and $P_k(a^l)$ denotes the projection of a^l onto the kth convex set S_k, defined by

$$P_k(a^l) \in S_k, \forall b \in S_k, \|a^l - P_k(a^l)\| \leq \|a^l - b\|, \tag{24.71}$$

where $\|\cdot\|$ denotes the norm. Note that when $a^l \in S_k$, its projection $P_k(a^l) = a^l$. Also, the choice of the relaxation parameter λ_l can affect the convergence speed of the PPM, and there exist optimal values of λ^l [63]. For ease of discussion, we simply select $\lambda^l = 1$, which also saves the computation involved in obtaining an optimal value of λ^l in each iteration. In fact, choosing $\lambda^l = 1$ is called the unrelaxed form of (24.70), rewritten below as

$$a^{l+1} = \sum_{k=1}^{K} w_k P_k(a^l), \quad \text{for} \quad l \geq 0. \tag{24.72}$$

The above formulation is naturally suited for the noncollaborative position location problem. Consider an unlocalized node making range estimates to a few surrounding anchors with known locations. Each range estimate defines a ranging circle that the unlocalized node should lie on. However, if we relax the constraint that the unlocalized node should lie *on* the circle to a constraint that the unlocalized node should lie *within* the circle (which essentially transforms the original nonconvex constraint into a convex constraint), we can directly apply the PPM. However, it was shown that returning to the original constraint results in better localization performance in general [61]. Thus, the projection, distance measure, and proximity function are defined in the MPPM as

$$P_k^{ncl}(\hat{\theta}) = A_k + r_k \frac{\hat{\theta} - A_k}{\left\| \hat{\theta} - A_k \right\|}, \tag{24.73}$$

$$d_k^{ncl}(\hat{\theta}) = \left(r_k - \left\| \hat{\theta} - A_k \right\| \right)^2, \tag{24.74}$$

$$\Phi^{ncl}(\hat{\theta}) = \frac{1}{K} \sum_{k=1}^{K} d_k^{ncl}(\hat{\theta}), \tag{24.75}$$

where $\hat{\theta}$ is the unlocalized node's current estimated location, A_k represents its kth connecting anchor and r_k denotes the range estimate between the unlocalized node and A_k. K is the number of anchors and $w_k = 1/K$ for $k = 1, 2, \ldots , K$ represents the fact that all range estimates are treated equally. We can assign different weights to range estimates according to our confidence in their reliability.

We can easily see that if θ is outside of the ranging circle defined by r_k with A_k as the center, the MPPM is coincident with the original PPM. If θ is inside the ranging circle, MPPM differs from PPM in the sense that the projection will be *pushed* away from the center onto the ranging circle. Also note that in the original PPM, when the current solution is inside the kth ranging circle, the kth term in the proximity function becomes zero, while the kth term in (24.75) of the MPPM is not zero. Essentially, the objective of (24.75) is the sum of the weighted squared difference between range estimates and the distances calculated using the estimated location, which is in fact the same as the objective in the MLE. The major difference here is the approach to solve the problem. In the standard nonlinear LS solver, such as *lsqnonlin*, the trust-region method or the Levenberg–Marquardt method is used. The PPM approach is merely another numerical method to solve the nonlinear LS problem, for example, as in [64]. The main motivation for us to use it lies in the fact that the projection process can be easily modified to fit our application, especially for position location in NLOS conditions, as will be shown in Section 24.7. Additionally, the method is computationally very simple. The MPPM for noncollaborative position location is given as follows.

Algorithm 2: MPPM for Noncollaborative Position Location

1. Initialize the unlocalized node at $\hat{\theta}$, set $l = 0$, $\Phi_l = \Phi^{ncl}(\hat{\theta})$, and δ a small positive number;
2. update $\hat{\theta} \leftarrow (1/K) \sum_{k=1}^{K} P_k^{ncl}(\hat{\theta})$ and let $\Phi_{l+1} = \Phi^{ncl}(\hat{\theta})$;
3. if $|\Phi_{l+1} - \Phi_l| > \delta$, let $l = l + 1$ and go to 2);
4. terminate and return $\hat{\theta}$.

We emphasize that, due to the modified projection in (24.73), the final solution in many cases will depend on the initial guess. Figure 24.11 gives two examples of applying the MPPM to position location where an unlocalized node estimates its location using range estimates to three anchors, with four different initial solutions

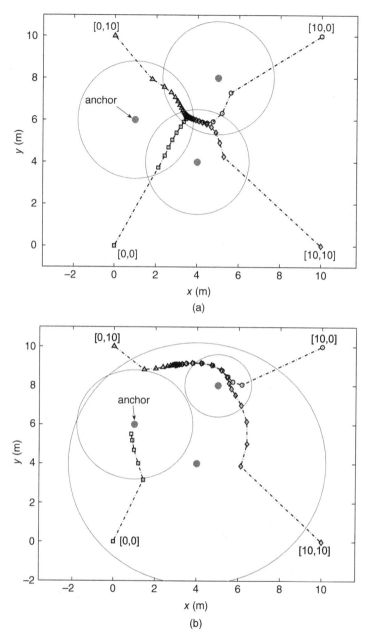

Figure 24.11 (a) Application of the modified PPM to noncollaborative position location for (top) less noisy range estimates where the intersection region of the range estimates contains the true location; (b) (bottom) one range estimate is extremely noisy. The blue circles denote ranging circles; dash dot lines represent trajectories along which the solution converges; initial guesses are denoted by [x, y] [15].

at [0, 0], [0, 10], [0, 10], and [10, 10], respectively. The three anchors are located at [1, 6], [4, 4], and [5, 8], respectively, while the unlocalized node is at [3.5, 6]. Figure 24.11a represents the case where the three range estimates are less noisy and have an intersection region containing the true location, while in Figure 24.11b, one of the three range estimates is extremely noisy and is much larger. In either case, the MPPM is able to converge to a final solution after a few iterations. As we can see, in the second case, the final solution depends on the initial guess. Note that we do not see this effect in the first case when the range estimates are less noisy.

Example 24.4

Consider a single node that has estimated its range to three anchors located at $\theta_2 = [0, 0]$, $\theta_3 = [0, 10]$, and $\theta_4 = [0, 10]$ as $r_{12} = 6.4$, $r_{13} = 7.8$, and $r_{14} = 6.4$. Using a starting location estimate of $[-5, -3]$, determine the first five iterations of the PPM technique. What is the final estimate? Repeat for an initial estimate of $[5, 10]$.

Solution

With an initial estimate of $\hat{\theta}_1^{(0)}$, the three projections onto the ranging circles can be calculated as:

$$P_2\left(\hat{\theta}_1^{(0)}\right) = [0, 0] + \frac{6.4}{\sqrt{5^2 + 3^2}}[-5, -3] \tag{24.76a}$$

$$= [-5.52, -3.31] \tag{24.76b}$$

$$P_1\left(\hat{\theta}_1^{(0)}\right) = [0, 10] + \frac{6.4}{\sqrt{5^2 + 13^2}}[-5, -13] \tag{24.77a}$$

$$= [-2.30, 4.03] \tag{24.77b}$$

$$P_3\left(\hat{\theta}_1^{(0)}\right) = [10, 0] + \frac{7.8}{\sqrt{15^2 + 3^2}}[-15, -3] \tag{24.78a}$$

$$= [2.35, -1.53]. \tag{24.78b}$$

The new position estimate is then

$$\hat{\theta}_1^{(1)} = \frac{1}{3}\sum_{i=1}^{3} P_i\left(\hat{\theta}_1^{(0)}\right) \tag{24.79a}$$

$$= [-1.82, -0.27]. \tag{24.79b}$$

Repeating this procedure, we obtain the sequence $\hat{\theta}_1^{(0)} = [-5, -3]$, $\hat{\theta}_1^{(1)} = [-1.82, -0.27]$, $\hat{\theta}_1^{(2)} = [-1.75, 0.86]$, $\hat{\theta}_1^{(3)} = [-1.58, 2.37]$, $\hat{\theta}_1^{(4)} = [-0.83, 3.54]$, $\hat{\theta}_1^{(5)} = [0.10, 4.10]$. Additional iterations are plotted in Figure 24.12. We can see that the final estimate is very close to the true location $\theta_1 = [4, 5]$. Iterations starting from [10, 5] are also plotted in Figure 24.12. Again, the iterations converge to the true location. It can be shown that this iteration will always converge to a local minimum. In this case, there is only a single local minimum, and it corresponds to the global minimum.

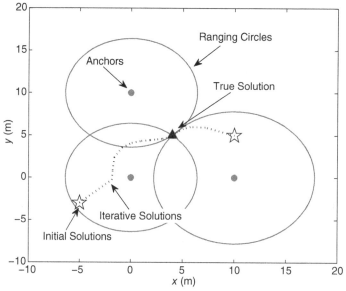

Figure 24.12   Illustration of the solution to Example 24.4.

IPPM for Collaborative Position Location   We now focus on the more general problem of collaborative position location. In particular, we use the MPPM as a basic element and extend it to an iterative and distributed numerical framework [60, 65]. The overall framework involves an initialization step and an iterative update step, where only local communications are necessary in both steps.

In the initialization step, each unlocalized node obtains an initial solution for its location. The approach we adopt is called closest-anchor initialization. Specifically, if an unlocalized node has connecting anchor(s), it will use the location of its closest anchor, determined from range estimates, as its initial solution. If an unlocalized node does not have a connection to any anchors, it will use the average location of its surrounding nodes' initial solution as its initial solution. For isolated nodes without any connections, we simply use the network center as the initial solution, and due to the absence of range estimates, their estimates will thus not be updated in the ensuing iterative update step. The initialization step continues until every unlocalized node obtains an initial solution. We emphasize that there exist other methods to obtain initial solutions [25], including using the solution from the MDS as an initial solution, but those generally involve substantially more computation. Instead, we use the simple initialization method described above.

In the iterative update step, each unlocalized node uses MPPM to update its location estimate, based on its range estimates to neighboring nodes, either anchors or other unlocalized nodes, and its current location estimate. In particular, if the ith and the jth nodes are neighbors, the projection of $\hat{\theta}_i$ onto the feasibility set given by the range estimate r_{ij} is

$$P_{ij}^{col}\left(\hat{\theta}_i\right)=\hat{\theta}_j+r_{ij}\frac{\hat{\theta}_i-\hat{\theta}_j}{\left\|\hat{\theta}_i-\hat{\theta}_j\right\|}. \tag{24.80}$$

Since only local information exchange is needed, the iterative update process can be distributed and the computational complexity scales linearly with network size. Each unlocalized node examines whether or not its residual changes over the previous iteration. The ith unlocalized node's residual based on its and its neighbors' current estimated locations is defined as

$$\Phi^{col}\left(\hat{\theta}_i\right)=\frac{1}{|\mathcal{N}(i)|}\sum_{j\in\mathcal{N}(i)}\left(r_{ij}-\left\|\hat{\theta}_i-\hat{\theta}_j\right\|\right)^2, \tag{24.81}$$

where $\mathcal{N}(i)$ is the set of the ith node's neighboring nodes, and $|\cdot|$ denotes cardinality. If its residual has not changed more than the precision parameter δ for κ consecutive iterations, the ith node will quit the iterative update step and mark itself as *localized*. The overall update process terminates after all of the unlocalized nodes have been marked as localized. IPPM for collaborative position location is described as the following.

Algorithm 3: IPPM for Collaborative Position Location

- **Initialization:**

Obtain initial guess $\Theta=[\hat{\theta}_1,\hat{\theta}_2,\ldots,\hat{\theta}_N]$ using the closest-anchor initialization;
Set $[\hat{\theta}_{N+1},\hat{\theta}_{N+2},\ldots,\hat{\theta}_{N+M}]=\mathbf{A}=[\theta_{N+1},\theta_{N+2},\ldots,\theta_{N+M}]$, which remain unchanged during the main loop.

Set $l=0$, δ as a small positive number, and κ as a positive integer;
Let $F_i=0$ and $W_i=0$, for $i=1,2,\ldots,N$;
$\Phi_{i,l}=\Phi^{col}(\hat{\theta}_i)$, for $i=1,2,\ldots,N$;

- **Main loop:**

While (any of F_i is equal to 0) {

   For $i=1,2,\ldots,N$; *if* $F_i=0$

$$\hat{\theta}_i\leftarrow(1/|\mathcal{N}(i)|)\sum_{j\in\mathcal{N}(i)}P_{ij}^{col}(\hat{\theta}_i),\text{ and }\Phi_{i,l+1}=\Phi^{col}(\hat{\theta}_i);$$

   If $|\Phi_{i,l}-\Phi_{i,l+1}|<\delta$

      Let $W_i=W_i+1$; *If* $W_i\geq\kappa$, *Let* $F_i=1$;

   Otherwise

      Let $W_i=0$;

   $l=l+1$;

}

In the above algorithm, F_i indicates whether the ith node has been localized, and W_i records the number of consecutive iterations that the ith node's residual has not decreased more than δ. Once $W_i\geq\kappa$, we set $F_i=1$ and consider the ith unlocalized node as localized. It is obvious that anchor locations, that is, θ_j, for $j=n+1$,

$n + 2, \ldots, n + m$, will remain unchanged during the whole process. Again, as in the case of MPPM for noncollaborative position location, the final solution of IPPM depends on the initial guess.

Compared with the case of noncollaborative position location, we can easily see that the major difference here is that we use W_i as a means of accumulating observations over multiple iterations regarding whether an unlocalized node is indeed localized. This is understandable since during the iterative update process, if an unlocalized node has updated its location, its neighboring unlocalized nodes may be affected in terms of their residuals and therefore need to be reexamined. Another advantage of IPPM is that the computational load involved is low, mainly because of the simple operation involved in updating node locations, which scales linearly with n.

Example 24.5

Repeat Example 24.1 using IPPM, assuming that the initial estimates are averages of the closest anchors.

Solution

The initial estimates are chosen to be the average of the two connected anchors for each node resulting in $\hat{\theta}_1^{(0)} = [0, 5]$ and $\hat{\theta}_2^{(0)} = [10, 5]$. Using the initial estimate of θ_2 as an anchor for θ_1, node θ_1 has three anchors for the first iteration of PPM. The same is true for θ_2. After one set of iterations of PPM, the estimates converge to

$$\hat{\theta}_1^{(1)} = [5.21, 5.06] \tag{24.82a}$$

$$\hat{\theta}_2^{(1)} = [9.40, 5.20]. \tag{24.82b}$$

Again, we use these new estimates as anchors for the other node. The second round of IPPM results in $\hat{\theta}_1^{(2)} = [4.93, 5.07]$ and $\hat{\theta}_2^{(2)} = [9.10, 5.20]$. After 10–20 iterations, we find that the solutions do not change. Thus, the final estimates are $\hat{\theta}_1^{(\infty)} = [4.61, 5.07]$ $\hat{\theta}_2^{(\infty)} = [8.78, 5.21]$. The mean location error using this approach is 0.7 m in this example, approximately what is obtained from nonlinear LS and somewhat better than MDS. The final estimates are plotted in Figure 24.13.

The number of anchors (i.e., the density within the network) has an impact on the network localization performance. Obviously, a larger number of anchors improves the performance. In Figure 24.14, we plot the localization results from applying the IPPM approach to three networks with different numbers and configurations of anchors. In the three networks, the number of unlocalized nodes is 50, while the numbers of anchors are 9, 4, and 3 for (a), (b), and (c), respectively. The communication range is $R = 8$ m for all three networks, and the resulting average node connectivity, that is, the average number of neighbors per node, is approximately 9 for all 3 networks. In Figure 24.14a, b, there are 4 anchors at the 4 corners of a 20×20 m^2 network, while in Figure 24.14a, there are 5 additional anchors randomly distributed within the network. In Figure 24.14c, there are only 3 anchors located within the network forming an equilateral triangle. Clearly, Figure 24.14a represents

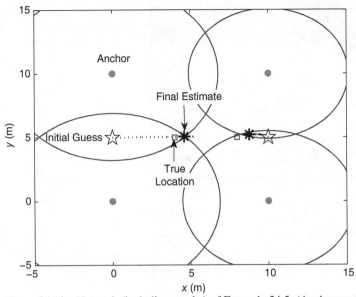

Figure 24.13   Network (including results) of Example 24.5. (Anchor positions are represented by filled circles, measured ranging circles are lines, the true and estimated node positions are represented by squares and asterisks, respectively.)

the most favorable anchor deployment, while Figure 24.14c is the least favorable deployment. We assume a pure LOS scenario with path loss exponent $\beta = 2$ and noise factor (see Section 24.3.1) $K_E = 0.01$. This is equivalent to range estimate noise, which is approximately 10% of the internode distance. We can observe that in Figure 24.14c, a few nodes get flipped solutions, and therefore result in a much larger localization error (RMSE) of 2.37 m, while Figure 24.14a has the smallest localization error of 0.34 m, due to the larger number of, as well as better distributed, anchors. The RMSE in Figure 24.14b is 0.63 m.

In Figure 24.15, we plot the mean localization error of the IPPM approach when localizing the three networks in Figure 24.14a–c, versus range estimation noise indicated by the value of K_E. As before, $\beta = 2$ and $R = 8$ m, while varying the value of K_E from 0.0025 to 0.16. As mentioned earlier, such a range of values for K_E corresponds to range estimation noise equivalent to 5% up to 40% of the internode distance. The larger K_E, the more range estimation noise. For each value of K_E, 500 noise realizations are simulated. As can be seen, the mean localization errors in all three cases increase as the amount of range estimation noise increases, among which the localization error is the smallest for Figure 24.14a due to more and better anchor deployment, while Figure 24.14c has the largest localization error due to fewer and poor anchor deployment. It is obvious that the effect of bad anchor deployment is much more prominent than the increase of range estimation noise. Again, we emphasize that due to the closest-anchor initialization step adopted in our approach, its performance depends upon how well the anchors are distributed over the entire network. For example, if all anchors are concentrated within a small region, the

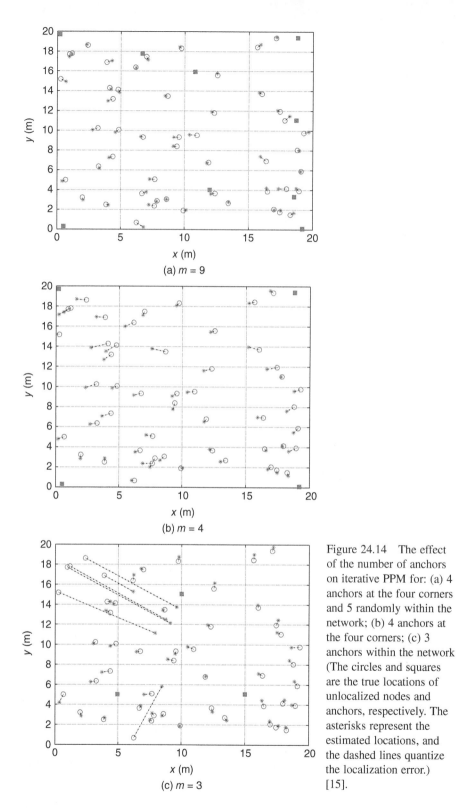

Figure 24.14 The effect of the number of anchors on iterative PPM for: (a) 4 anchors at the four corners and 5 randomly within the network; (b) 4 anchors at the four corners; (c) 3 anchors within the network (The circles and squares are the true locations of unlocalized nodes and anchors, respectively. The asterisks represent the estimated locations, and the dashed lines quantize the localization error.) [15].

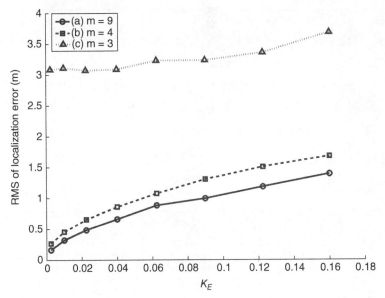

Figure 24.15   The corresponding mean localization errors versus noise variance $(K_E, \sigma_{ij}^2 = K_E d_{ij}^{\beta_{ij}})$ for the three cases in Figure 24.14 [15].

closest-anchor initialization is not a smart choice. Instead, using random initialization may be a better choice.

24.6   NUMERICAL COMPARISON OF APPROACHES

24.6.1   Localization Accuracy

In this section, we compare four approaches to collaborative position location with an emphasis on three. Specifically, we compare one sequential approach: sequential LS and three concurrent approaches (1) MDS-MAP (the distributed version of MDS), (2) SDP, and (3) IPPM. Specifically, we compute the mean localization error using the RMSE given by (24.9). For the SDP approach, we use the centralized full SDP solver provided by [53]. For the MDS-MAP, we used the patch-based MDS followed by the LS refinement, that is, MDS-MAP(P, R), with a radius of $2R$ for patching local maps and centralized LS refinement, which gives the best performance among all the four MDS approaches according to [27]. The experimental results show that the full SDP solver provides robust performance for different network shapes with reasonably distributed anchors. To the best of our knowledge, and also as mentioned in [41], the MDS-MAP(P, R) appears to be among the best distributed methods with a reasonably short running time. In addition to examining the impact of Gaussian estimation noise, we also examine the robustness of the three concurrent localization methods to NLOS propagation. For IPPM, we use $\delta = 10^{-4}$ and $\kappa = 10$, as using a smaller δ and a larger κ does not further improve the performance of IPPM.

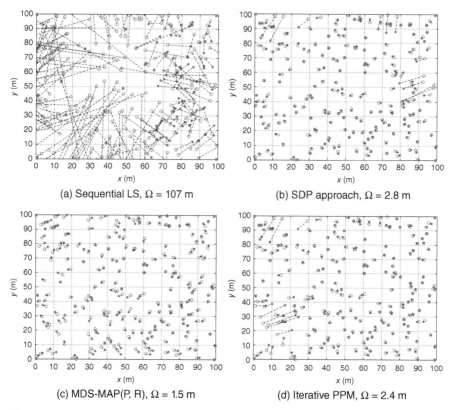

(a) Sequential LS, $\Omega = 107$ m

(b) SDP approach, $\Omega = 2.8$ m

(c) MDS-MAP(P, R), $\Omega = 1.5$ m

(d) Iterative PPM, $\Omega = 2.4$ m

Figure 24.16   Results of applying different localization schemes to a random network for one particular noise realization in the pure LOS scenario with $K_E = 0.01$ and $\beta = 2$ (the noise variance is $\sigma_{ij}^2 = K_E d_{ij}^{\beta_{ij}}$) [15].

We compare the four approaches (with an emphasis on the concurrent approaches) on a large network with 200 unlocalized nodes and 20 anchors, randomly distributed (with uniform distribution) within a $100 \times 100\,\mathrm{m}^2$ network. In Figure 24.16, we plot the estimated locations for one particular noise realization in the pure LOS scenario with $K_E = 0.01$ and $\beta = 2$. Note that the dashed lines quantize the amount of localization error. It is easily seen that in random large networks like this example, the paradigm of applying noncollaborative position location schemes in a sequential fashion suffers significant performance loss due to the combined effect of error propagation and potentially bad node geometry. We have found this to be generally true.

We now examine the localization error versus range estimation noise, for both pure LOS and NLOS scenarios. Since the sequential LS estimator performs very badly, we only present results for the three concurrent approaches. The communication range is assumed to be $R = 15\,\mathrm{m}$, the resulting average node connectivity is 12.9, and on average, each unlocalized node is connected to only 1.1 anchors. Compared with simulation results in [27, 41], where the average node connectivity is usually larger than 15 and sometimes increases to 30, these results represent a case where

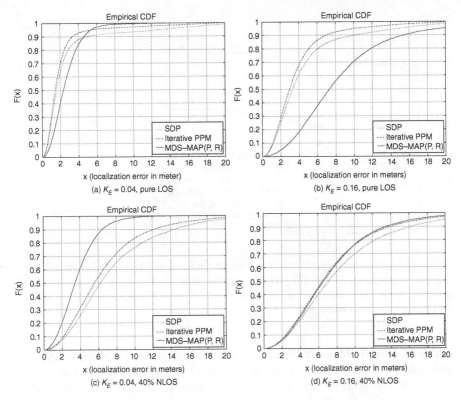

Figure 24.17   Cumulative density function of the localization errors of SDP, MDS-MAP(P, R), and iterative PPM for both pure LOS and 40% NLOS scenarios, with $K_E = 0.04$ and 0.16 (the noise variance is $\sigma_{ij}^2 = K_E d_{ij}^{\beta_{ij}}$, $\beta_{ij} = 2$ for LOS and $\beta_{ij} = 3$ for NLOS), respectively [15].

sensor nodes have shorter communication range, which is perhaps more practical due to power consumption concerns. Nevertheless, with larger node connectivity, the localization accuracy of all three methods will be improved. For each value of K_E, 50 noise realizations were averaged. We also compare the three methods in terms of their robustness to NLOS propagation. As for the NLOS scenario, we assume any range estimate has a 40% chance of being NLOS. If a range estimate is NLOS, a uniformly distributed positive bias is added onto that range estimate, with $b_{\min} = 4\,\text{m}$ and $b_{\max} = 8\,\text{m}$. The minimum NLOS bias of $b_{\min} = 4\,\text{m}$ is equivalent to 27% of the communication range and represents the fact that in practice, NLOS bias is usually much larger than the range estimation noise. In Figure 24.17, we plot the empirical cumulative density function (CDF) of the localization error of SDP, MDS-MAP(P, R), and IPPM, for pure LOS and 40% NLOS scenarios, with $K_E = 0.04$ and $K_E = 0.16$, respectively. As shown in Figure 24.17a, the MDS-MAP(P, R) is better than the other two in the sense that it reduces the large localization error. However, as the range estimation noise increases, as shown in Figure 24.17b, MDS-MAP(P, R) performs worse than the other two. On the other hand, in the presence of 40% NLOS range

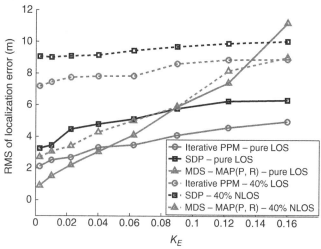

Figure 24.18 The mean localization error versus the amount of range estimation noise (the noise variance is $\sigma_{ij}^2 = K_E d_{ij}^{\beta_{ij}}$), for the network shown in Figure 24.16, with pure LOS and 40% NLOS range estimates. K_E varies from 0.0025 to 0.16, corresponding from 5% to 40% range estimation noise [61].

estimates, both IPPM and SDP suffer performance degradation. MDS-MAP(P, R), although it performs better in the case of small range estimation noise, that is, $K_E = 0.04$, we observe similar performance loss when range estimation noise increases, that is, $K_E = 0.16$.

In Figure 24.18, we plot the mean localization error, that is, the RMS values of the localization error, of the three methods with respect to different range estimation noise. For the case of pure LOS range estimates, IPPM outperforms SDP for all values of K_E. The MDS-MAP(P, R) performs better for a small amount of range estimation noise but performs worse than both the SDP and the IPPM approach as range estimation noise increases. It is obvious that both SDP and IPPM are more robust to increasing range estimation noise than MDS-MAP(P, R). This can be attributed to the fact that the MDS step in the MDS-MAP(P, R) is impacted by an increase of range estimation noise more than the other two methods. As for the 40%-NLOS scenario, it is observed that both the SDP and the IPPM approach are strongly affected by the NLOS bias, with IPPM still outperforming the SDP approach. For instance, for $K_E = 0.09$ (30% noise), the localization error of IPPM increases from about 4 m to more than 8 m, while that of the SDP increases from about 5.5 m to more than 9 m. MDS-MAP(P, R), on the other hand, shows more robustness to NLOS bias. As can be seen, its localization error in the case of 40% NLOS scenario is worse than the pure LOS scenario but becomes comparable to the pure LOS scenario as K_E increases. Another important observation is that as range estimation noise increases, the improvement of the MDS-MAP(P, R) over the SDP and the IPPM diminishes. This is again due to the degraded performance of the MDS-MAP(P, R) at higher levels of ranging noise. In general, for uniformly distributed networks with reasonably distributed anchors, we observe the following trends: (1)

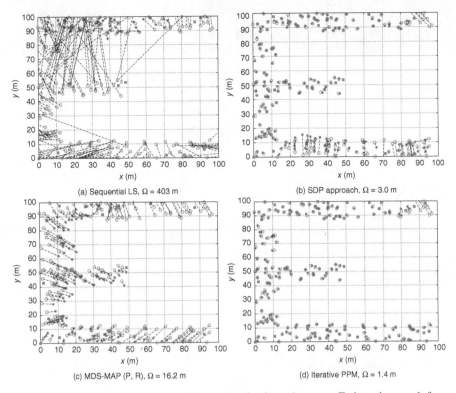

Figure 24.19   Results of applying different localization schemes to E-shaped network for one particular noise realization in the pure LOS scenario with $K_E = 0.01$ and $\beta = 2$ (the noise variance is $\sigma_{ij}^2 = K_E d_{ij}^{\beta_{ij}}$) [15].

MDS is more susceptible to increasing Gaussian range error than IPPM or SDP; (2) For relatively low levels of Gaussian error, MDS is more robust to NLOS bias; (3) Robustness is lost when Gaussian error increases beyond a certain level; and (4) SDP and IPPM are more susceptible to NLOS error than Gaussian range error.

In a practical sensor network deployment, it is likely that the network shape is irregular due to the presence of physical obstacles such as buildings. In Figure 24.19, we give an example of an E-shaped network, with 200 unlocalized nodes and 20 anchors. The communication range is assumed to be $R = 15\,\text{m}$, and the resulting average node connectivity is 19.9, and on average, each unlocalized node is connected to 1.9 anchors. We use the same parameters for the range estimation noise and NLOS bias and present localization results of all the methods in Figure 24.19 for one particular noise realization in the pure LOS scenario with $K_E = 0.01$ and $\beta = 2$. It is clear that the sequential LS estimator suffers a huge performance loss again due to the combined effect of potentially bad neighboring node geometry and error propagation. This effect is even more severe in the case of irregular network shape since it is more likely that node geometry becomes ill-conditioned. Another important observation is that the MDS-MAP(P,R) approach

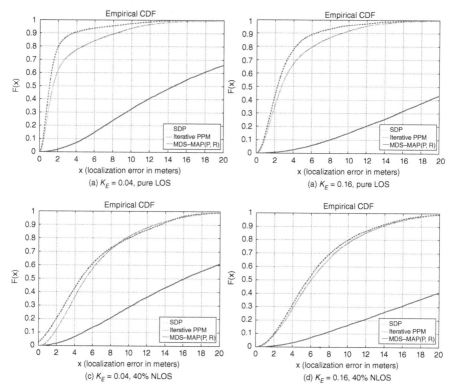

Figure 24.20   CDF of the localization errors of SDP, MDS-MAP(P, R), and iterative PPM for both pure LOS and 40% NLOS scenarios, with $K_E = 0.04$ and 0.16 (the noise variance is $\sigma_{ij}^2 = K_E d_{ij}^{\beta_{ij}}$, $\beta_{ij} = 2$ for LOS and $\beta_{ij} = 3$ for NLOS), respectively, applying to the E-shaped network [15].

also performs badly. This phenomenon has been reported in [27], because the step of building the complete distance matrix in the MDS algorithm becomes unreliable when the network shape is anisotropic. In Figure 24.20, we plot the CDF of the localization errors of the three methods, for pure LOS and 40% NLOS scenarios, with $K_E = 0.04$ and $K_E = 0.16$, respectively. As we can see, MDS-MAP(P, R) performs extremely poorly in this irregular network shape. SDP and IPPM, on the other hand, have similar performance due to their better robustness against anisotropic network shapes.

In Figure 24.21, we present the corresponding simulation results on the RMS localization error of SDP, MDS-MAP(P, R), and our proposed IPPM approach. It is observed that for both the pure LOS and 40% NLOS scenarios, the mean localization error for IPPM approach and SDP approach are comparable to that of the regular network shape in Figure 24.16. Both methods demonstrate similar robustness against range estimation noise. In particular, as K_E increases from 0.0025 to 0.16, the localization error will generally increase less than 2 m for both pure LOS and 40% NLOS scenarios. On the other hand, MDS-MAP(P, R) performs badly with an

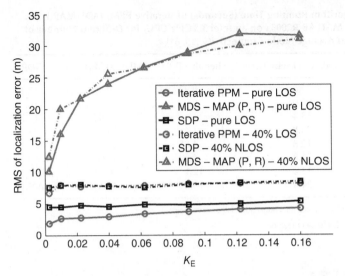

Figure 24.21   The mean localization error versus the amount of range estimation noise (the noise variance is $\sigma_{ij}^2 = K_E d_{ij}^{\beta_{ij}}$), for the network shown in Figure 24.19, with pure LOS and 40% NLOS range estimates. K_E varies from 0.0025 to 0.16, corresponding from 5% to 40% range estimation noise [61].

irregular network shape, whether it is a pure LOS or NLOS scenario. As shown in Figure 24.21, the localization error of MDS-MAP(P, R) is always larger than 10 m and increases significantly as range estimation noise increases. This is, as mentioned in [27], due to the inherent nature of the MDS step, which suffers from irregular network shapes despite the distributed approach used here, that is, MDS-MAP(P, R).

24.6.2   Computational Complexity

So far, we have focused on localization accuracy. Another important aspect of collaborative position location algorithms is the algorithm complexity. In Table 24.1, we give typical algorithm running time for the three methods on an Intel 3.3 GHz CPU in MATLAB R2008a [15, 61]. In general, the number of unlocalized nodes n and the average node connectivity, denoted by n_c, jointly determine the algorithm running time. Therefore, we compare the running times of the three methods for different combinations of n and n_c. The IPPM approach only involves simple computation and its complexity scales as $O(nn_cL)$, where L is the number of iterations for the IPPM approach to converge. The SDP approach, on the other hand, takes a much longer time to run since it has a complexity of $O(n^3)$. The MDS-MAP(P, R), although faster than the SDP, still needs considerably more time than our proposed approach and its complexity is $O(nn_c^3)$ for the MDS step and $O(n^3)$ for the global LS refinement. As observed from Table 24.1, the IPPM method has a significantly shorter running time and scales considerably better for large networks. Note that the running time presented here is the total algorithm running time.

TABLE 24.1.   Total Algorithm Running Time (Seconds) of Iterative PPM, MDS-MAP(P, R), and SDP, Obtained in MATLAB R2008a on an Intel 3.3 GHz CPU, for Different Number of Unlocalized Nodes and Average Node Connectivity [15, 61]

# of Nodes (N)	Ave. Node Connectivity (n_c)	Iterative PPM	MDS-MAP(P, R)	SDP
50	12.6	0.77	2.59	2.11
50	25.0	0.97	6.27	10.7
100	13.4	1.42	5.92	13.3
100	25.8	1.45	24.6	89.5
200	12.8	4.05	15.9	114
200	24.8	4.69	76.0	1280
500	13.4	12.1	N/A^a	3001
500	24.8	13.2	N/A^a	N/A^b

^{a}The MDS-MAP(P, R) solver failed since the computation is out of memory.
^{b}The SDP solver failed due to insufficient memory.

24.7   NLOS PROPAGATION

As we mentioned earlier, NLOS propagation is a prominent phenomenon for the envisioned position location applications. From the previous section, it was observed that the presence of NLOS propagation can significantly degrade the localization accuracy of any collaborative position location method. Therefore, it is a necessity to develop methods to mitigate NLOS propagation.

In fact, there exist many approaches and algorithms for mitigating the adverse effect of NLOS propagation but mainly in the context of infrastructure-based or noncollaborative position location (viz. cellular systems). For example, in [66], the authors proposed two NLOS mitigation algorithms for time difference of arrival (TDOA) and TDOA/AOA position location schemes. The authors in [57] developed an LP framework to incorporate NLOS range estimates into TOA-based position location without degrading the localization accuracy, based on the assumption that NLOS identification has been performed to a certain accuracy. In [67], the authors proposed a technique utilizing the statistics about mean excess delay, first detected path power to subtract the statistical value of NLOS error from the measurement range. In [68], the authors proposed a weighted LS scheme based on NLOS identification using multipath channel statistics. Also, techniques based on Kalman filters [69] have have been developed for NLOS mitigation. A brief survey and comparison of various NLOS identification and mitigation techniques can be found in [70].

However, the major difficulty of directly applying state-of-the-art NLOS mitigation techniques to the problem of interest is the high computational complexity or additional requirements. We emphasize that the desired NLOS mitigation method is to be deployed for large number of low-cost and low-power wireless sensors. Consequently, it is not practical to use complicated methods developed in the context of cellular systems, which either significantly increase the computational complexity [71] or put additional constraints on the minimum number of LOS range estimates

[57]. In that sense, the method developed in [44] is also less practical due to its possible use of the range scaling algorithm (RSA) method developed in [71] in order to handle the degenerate cases encountered in [57].

In view of these, in this section, we show how a simple, yet effective NLOS mitigation method, based on how much a priori knowledge of NLOS conditions we have, can be incorporated into the IPPM method examined in Section 24.5.4. The approach has scalable computational complexity depending on the utilization of the knowledge about NLOS conditions.

24.7.1 Knowledge about the NLOS Propagation

The range estimate model assumed is as follows:

$$r_{ij} = d_{ij} + n_{ij}, \qquad \text{if it is LOS}, \tag{24.83a}$$

$$r_{ij} = d_{ij} + n_{ij} + b_{ij}, \quad \text{if it is NLOS}, \tag{24.83b}$$

where the very last term b_{ij} represents an unknown NLOS bias, which is positive and much larger than the noise standard deviation σ_{ij}. The most common way to statistically describe NLOS bias is to model b_{ij} as being uniformly distributed. Note that exponentially distributed NLOS bias is also sometimes observed [72], and some work has also used Rayleigh distributed NLOS bias [73].

In practice, it is possible that some knowledge about the NLOS conditions are either known a priori or collected online by examining the received signal statistics. For example, it is demonstrated in [74] that using the received signal statistics such as root-mean-square delay spread (RDS), NLOS links can be successfully identified with a reasonably high probability (about 90%). Furthermore, it is conceivable that an extensive measurement campaign can provide some rough knowledge of the mean and the variance of the NLOS bias in certain propagation environments. Some may even assume that the complete statistics of the NLOS bias are known and derive the associated MLE [40, 66].

24.7.2 NLOS Mitigation Example

Despite the additional complexity associated with the corresponding signal processing, NLOS identification is a practically useful step to mitigate the effect of NLOS propagation. To demonstrate this, we will examine NLOS mitigation, which incorporates NLOS identification, and knowledge about the mean or minimum NLOS bias. We specifically focus on incorporating NLOS mitigation method into the IPPM technique developed in Section 24.5.4 without significantly increasing computational complexity [65].

It has been demonstrated that NLOS bias is usually positive and is much larger than range estimation noise [57]. In other words, if a range estimate is NLOS, it will appear to be much larger than the true distance, even in the presence of range estimation noise. This fact, coupled with NLOS identification, has been utilized in the method developed by the authors in [57] and has been demonstrated to provide good performance. A similar methodology was used to revise IPPM to incorporate NLOS knowledge [65]. In particular, the update step in the IPPM is modified based on whether a range estimate is NLOS or LOS. If it is NLOS, it further examines whether the current solution is within the ranging circle defined by the corresponding NLOS range estimate. If the current solution is outside the ranging circle, it proceeds as in the ordinary IPPM. However, if the current solution is already within the ranging circle, it will no longer contribute to the average of the projections. Mathematically, we define

$$\mathcal{N}(i) = \mathcal{N}_L(i) \cup \mathcal{N}_N(i), \tag{24.84}$$

where $\mathcal{N}_L(i)$ and $\mathcal{N}_L(i)$ are the sets of the ith node's neighbors with LOS and NLOS range estimates, respectively. We then define

$$\mathcal{N}_{\hat{\theta}_i}^A(i) = \mathcal{N}_L(i) \cup \left\{ j \mid j \in \mathcal{N}_N(i), \left\| \hat{\theta}_i - \hat{\theta}_j \right\| \geq r_{ij} \right\}. \tag{24.85}$$

The revised algorithm (which we will refer to as *Algorithm 3*) is identical to Algorithm 2 except that $\mathcal{N}(i)$ is replaced by $\mathcal{N}_{\hat{\theta}_i}^A(i)$, which is essentially the set of the ith node's neighbors whose range estimates will participate in the update of $\hat{\theta}_i$. More specifically, the set of the ith node's neighbors with LOS range estimates will always participate, while any of its neighbors with NLOS range estimate will participate only if $\left\| \hat{\theta}_i - \hat{\theta}_j \right\| \geq r_{ij}$ is satisfied, that is, the current solution of θ_i is still outside the corresponding ranging circle. The subscript of $\mathcal{N}_{\hat{\theta}_i}^A(i)$ is to emphasize that the fact that $\mathcal{N}_{\hat{\theta}_i}^A(i)$ depends on the current estimate $\hat{\theta}_i$.

Another form of knowledge (besides LOS/NLOS) that may be available is an estimate of the minimum NLOS bias. Denoting the estimated minimum NLOS bias as b_L, we revise the set of nodes contributing to the update as

$$\mathcal{N}_{\hat{\theta}_i}^B(i) = \mathcal{N}_L(i) \cup \left\{ j \mid j \in \mathcal{N}_N(i), \left\| \hat{\theta}_i - \hat{\theta}_j \right\| \geq r_{ij} - b_L \right\}. \tag{24.86}$$

The projection onto the ranging circle is also revised in order to utilize the knowledge about the minimum NLOS bias, as follows:

$$P_{ij}^{\text{col,NLOS-B}}(\hat{\theta}_i) = \begin{cases} \hat{\theta}_j + r_{ij} \dfrac{\hat{\theta}_i - \hat{\theta}_j}{\left\| \hat{\theta}_i - \hat{\theta}_j \right\|}, & j \in N_L(i) \\[2em] \hat{\theta}_j + (r_{ij} - b_L) \dfrac{\hat{\theta}_i - \hat{\theta}_j}{\left\| \hat{\theta}_i - \hat{\theta}_j \right\|}, & j \in \left\{ k \mid k \in N_N(i), \left\| \hat{\theta}_i - \hat{\theta}_k \right\| \geq r_{ik} - b_L \right\}. \end{cases} \tag{24.87}$$

As we can see, the above modified projection is basically trying to correct the NLOS range estimates by subtracting the estimated minimum NLOS bias from each NLOS range estimate. This has the effect of mitigating NLOS bias, depending on the accuracy of the estimated minimum NLOS bias.

The corresponding position location algorithm, which we will term *Algorithm 4* is again given by Algorithm 2 with the two modifications that the projection $P_{ij}^{\text{col}}(\hat{\theta}_i)$ is replaced by $P_{ij}^{\text{col,NLOS-B}}(\hat{\theta}_i)$ and the set of nodes participating in the update $\mathcal{N}(i)$ is replaced by $\mathcal{N}_{\hat{\theta}_i}^B(i)$.

The estimate of the minimum NLOS bias can be connection-specific, and the accuracy of the estimates affects the performance of the algorithm. Compared with Algorithm 3, the NLOS mitigation method given in Algorithm 4 is more aggressive in terms of using the knowledge about NLOS conditions. By subtracting the estimated minimum NLOS bias from each NLOS range estimate, it further provides correction to NLOS range estimates and thus is expected to perform better than Algorithm 3, as long as the estimated minimum NLOS bias is sufficiently accurate.

In practice, it is also possible to have a priori knowledge of the mean NLOS bias in certain environments, although it may be inaccurate. With such knowledge, we can revise our algorithm accordingly. In particular, denoting the mean NLOS bias as $\bar{b}$, we subtract the mean NLOS bias from the identified NLOS range estimates when making the decision as to whether to include a particular NLOS range estimate into the solution update procedure. In other words, we simply replace b_L with $\bar{b}$ in (24.86) and (24.87).

Considering that the actual NLOS bias may be larger or smaller than the mean NLOS bias, the approach may actually overestimate or underestimate the NLOS bias contained in each individual NLOS range estimate. Compared with the two previous algorithms, this algorithm differs in the sense that it corrects NLOS range estimates even more aggressively. However, its performance may depend on the actual statistics of the NLOS bias. For example, if the distribution of the actual NLOS bias concentrates around its mean value, this algorithm can be more effective than Algorithm 4.

24.7.3   Simulation Results

In this section, we present simulation results of NLOS mitigation techniques integrated into IPPM. To help understand the benefit of our NLOS mitigation method, we examine one of the networks examined in the previous section. The network has 200 unlocalized nodes and 20 anchors uniformly distributed over a $100 \times 100\,\mathrm{m}^2$ square area, forming a uniformly random network. As for the localization accuracy, we again compute the mean localization error using the RMSE given by (24.9). For NLOS scenarios, we will consider different percentages of NLOS range estimates, that is, 20%, 40%, 60%, and 80%, corresponding to various degrees of NLOS conditions. The range estimation noise and NLOS bias models are as described earlier, that is, zero-mean independent Gaussian distributed range estimation noise and uniformly distributed NLOS bias between $[B_{\min}, B_{\max}]$. To more accurately model range estimation noise, we set path loss exponents for LOS and NLOS range estimates differently; that is, $\beta_{\mathrm{LOS}} = 2$ and $\beta_{\mathrm{NLOS}} = 3$. For IPPM, we use the same parameters as used in the previous section: $\delta = 10^{-4}$ and $\kappa = 10$. First, we present simulation results of the proposed NLOS mitigation method with perfect NLOS identification; that is, we know exactly which range estimates are LOS or NLOS. For notational simplicity, we define the following terms:

- **IPPM:** the IPPM approach without using any NLOS mitigation, as described in Section 24.5.4;

- **IPPM-NM(ID):** the IPPM approach with NLOS mitigation based on NLOS identification only;
- **IPPM-NM(ID, min):** the IPPM approach with NLOS mitigation based on NLOS identification plus an estimate of the minimum NLOS bias;
- **IPPM-NM(ID, mean):** the IPPM approach with NLOS mitigation based on NLOS identification plus an estimate of the mean NLOS bias.

Also, following the notation defined previously, we use Ω defined in (24.8) to refer to the *network-average localization error* (averaged over all unlocalized nodes) for one particular noise and NLOS bias realization, and use $\bar{\Omega}$ defined in (24.9) to refer to the *mean localization error* average over all unlocalized nodes as well as different noise and NLOS bias realizations.

In Figure 24.22, we present the localization results of integrating NLOS mitigation methods into IPPM to localize the randomly generated network shown in Figure 24.16.* In particular, we present the localization results of using IPPM, IPPM-NM(ID), IPPM-NM(ID, min), and IPPM-NM(ID, mean) in the presence of 40% NLOS range estimates that are randomly selected from all the available range estimates, whether they are between unlocalized nodes and anchors, or among unlocalized nodes, for one particular range estimation noise and NLOS bias realization. The range estimation noise is generated with $K_E = 0.000625$. In the case of LOS range estimates, $\beta_{LOS} = 2$, which corresponds to noise equivalent to 2.5% of the inter-node distance. For NLOS range estimate, $\beta_{NLOS} = 3$, and the noise is a bit larger. The NLOS bias values are generated according to a uniform distribution with $B_{min} = 4$ and $B_{max} = 8$, corresponding to about 26% and 53% of the communication range R. In IPPM-NM(ID, min), we use $b_L = B_{min}$, and in the IPPM-NM(ID, min), we use $\bar{b} = (B_{min} + B_{max})/2$. For the sake of comparison, we also present the localization result of IPPM in a pure LOS scenario; that is, there are no NLOS range estimates. The anchor locations are shown by solid squares, and the true locations of the unlocalized nodes are shown by circles, while the estimated locations are denoted by the stars. The dashed lines connecting the true and the estimated locations quantize the localization errors. As shown in Figure 24.22a, in a pure LOS scenario, IPPM performs very well and results in a network-average localization error equal to 1.8 m. However, in the presence of 40% NLOS range estimates, the performance of IPPM without any NLOS mitigation degrades and results in a network-average localization error equal to 6.8 m, as shown in Figure 24.22b. On the other hand, if we apply IPPM-NM(ID), the network-average localization error is reduced almost by half to 3.5 m, as shown in Figure 24.22c. Furthermore, using additional knowledge of the minimum NLOS bias, that is, applying IPPM-NM(ID, min), the network-average localization error can be reduced even more to 2.8 m, as shown in Figure 24.22d. IPPM-NM(ID, mean) provides slightly worse performance than IPPM-NM(ID, min), with a network-average localization error equal to 2.9 m, shown in Figure 24.22e. These results clearly demonstrate that with some knowledge about NLOS conditions, IPPM is able to significantly improve the localization performance. More importantly,

---

*To review, in this network the communication range is $R = 15$ m, the average node connectivity is 6.9, and the average number of connected anchors is 1.1.

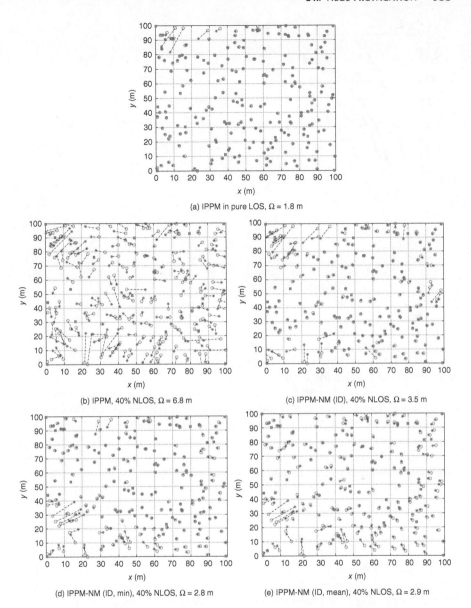

(a) IPPM in pure LOS, $\Omega = 1.8$ m

(b) IPPM, 40% NLOS, $\Omega = 6.8$ m

(c) IPPM-NM (ID), 40% NLOS, $\Omega = 3.5$ m

(d) IPPM-NM (ID, min), 40% NLOS, $\Omega = 2.8$ m

(e) IPPM-NM (ID, mean), 40% NLOS, $\Omega = 2.9$ m

Figure 24.22   Using our proposed NLOS mitigation methods to localize the network in Figure 24.16 for one noise and 40% NLOS bias realization. The dashed lines quantize the localization errors, and Ω is the network-average localization error [15].

having further knowledge such as minimum NLOS bias or mean NLOS bias beyond just simple NLOS identification definitely helps the localization. In Figure 24.23, we plot the CDF of the localization errors presented in Figure 24.22. It demonstrates that IPPM-NM(ID, min) has the best performance, and IPPM has the worst performance since it does not mitigate NLOS bias. Compared to IPPM-NM(ID),

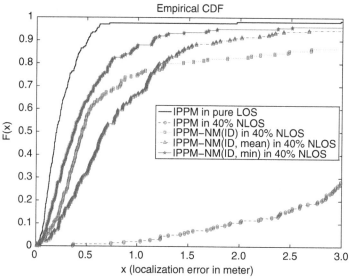

Figure 24.23   CDF of the localization errors of the methods in Figure 24.22 [15].

IPPM-NM(ID, mean) has the advantage of reducing large localization errors, leading a smaller mean localization error.

In Figure 24.24, we plot the mean localization error using the abovementioned NLOS mitigation methods versus the percentage of NLOS range estimates for the network in Figure 24.16, averaged over 100 range estimation noise and NLOS bias realizations. For each tested percentage of NLOS range estimates, the corresponding number of NLOS range estimates are randomly selected from all the available range estimates in the network and are kept unchanged over different noise and NLOS bias realizations. The range estimation noise and the NLOS bias parameters remain the same as before. From Figure 24.24, it is clearly demonstrated that without NLOS mitigation, IPPM suffers significant performance degradation as the percentage of NLOS range estimates increases. In particular, in the pure LOS scenario, that is, the percentage of NLOS range estimates is equal to zero, the IPPM has a mean localization error of 2.3 m, which increases to 12.8 m in the presence of 40% NLOS range estimates and 17.3 m in the presence of 80% NLOS range estimates. On the other hand, all three NLOS mitigation methods are capable of reducing the mean localization error. For example, with only the knowledge of NLOS identification, IPPM-NM(ID) is able to reduce the mean localization error to 3.6 m in the presence of 40% NLOS range estimates and 7.4 m in the presence of 80% NLOS range estimates, respectively. IPPM(ID, mean) is able to further provide 0.5 m and 2.3 m error reduction in the two cases, respectively, compared to IPPM-NM(ID). IPPM(ID, min), on the other hand, offers the best performance in the sense that the mean localization error is reduced to 2.6 m and 4.7 m in the presence of 40% and 80% NLOS range estimates, respectively. In other words, IPPM-NM(ID, min) offers the best protection against NLOS propagation such that even with 80% NLOS range

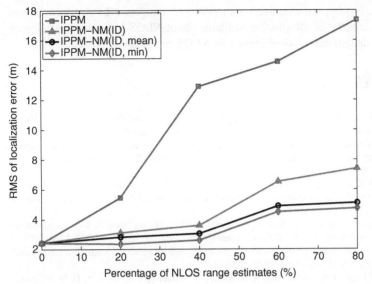

Figure 24.24   Mean localization error (averaged over 100 noise and NLOS bias realizations) versus the percentage of NLOS range estimates, when using our proposed NLOS mitigation methods to localize the network in Figure 24.16, with NLOS bias uniformly distributed between $B_{min} = 4\,m$ and $B_{max} = 8\,m$ [15].

estimates, the localization error is only increased by 2.4 m compared with the pure LOS scenario, which is only 16% of the communication range $R = 15\,m$. We believe this can be attributed to the fact that the minimum NLOS bias represents more accurate knowledge compared with the mean NLOS bias. Subtracting the mininum NLOS bias from NLOS range estimate is more trustworthy than subtracting the mean NLOS bias, since the actual NLOS bias may be smaller or larger than the mean NLOS bias. We emphasize that these simulation results are based on the perfect knowledge about NLOS identification, the minimum NLOS bias, and the mean NLOS bias. The impact of imperfect knowledge is given in [15].

24.8   SUMMARY

In this chapter we have presented collaborative position location techniques where nodes have limited connectivity to fixed infrastructure. Specifically, we have examined deterministic approaches to the problem that exploit distance measurements. Chapter 26 presents an in-depth discussion of *probabilistic* approaches, including BP and particle filtering approaches. We have focused on three main approaches that are examined in the literature, including SDP, MDS, and iterative parallel projection. We also examined two optimal bounds which provide benchmarks for the suboptimal approaches, but have impractical computational complexity. In particular, the factors impacting performance include the number, but more importantly the distribution, of anchor nodes, and the presence of bias (typically due to NLOS propagation) errors in the distance measurements. Finally, we examined approaches to mitigate NLOS errors. It was found that the amount of information available about NLOS links (identification or statistical knowledge) has a strong impact on NLOS mitigation approaches.

REFERENCES

[1] B. W. Parkinson, J. J. Spilker, Jr., P. Axelrad, and P. Enge, *The Global Positioning System: Theory and Applications*. Reston, VA: American Institute of Aeronautics and Astronautics, 1995.

[2] J. Reed, K. Krizman, B. Woerner, and T. Rappaport, "An overview of the challenges and progress in meeting the E-911 requirement for location service," *IEEE Pers. Commun. Mag.*, vol. 36, pp. 30–37, 1998.

[3] N. Patwari, J. Ash, S. Kyperountas, A. O. Hero III, R. Moses, and N. Correal, "Locating the nodes: Cooperative localization in wireless sensor networks," *IEEE Signal Process. Mag.*, vol. 22, no. 4, pp. 54–69, 2005.

[4] N. Patwari, A. O. Hero III, M. Perkins, N. S. Correal, and R. J. O'Dea, "Relative location estimation in wireless sensor networks," *IEEE Trans. Signal Process.*, vol. 51, no. 8, pp. 2137–2148, 2003.

[5] G. Mao, B. Fidan, and B. D. O. Anderson, "Wireless sensor network localization techniques," *Comput. Netw.*, vol. 51, no. 10, pp. 2529–2553, 2007.

[6] J. Aspnes, T. Eren, D. K. Goldenberg, A. S. Morse, W. Whiteley, Y. R. Yang, B. D. O. Anderson, and P. N. Belhumeur, "A theory of network localization," *IEEE Trans. Mobile Comput.*, vol. 5, no. 12, pp. 1663–1678, 2006.

[7] H. Wymeersch, J. Lien, and M. Z. Win, "Cooperative localization in wireless networks," in *Proc. IEEE*, vol. 97, 2009, pp. 427–450.

[8] N. Bulusu, J. Heidemann, and D. Estrin, "GPS-less low-cost outdoor localization for very small devices," *Special Issue on Smart Spaces and Environments, IEEE Pers. Commun.*, vol. 7, pp. 28–34, Oct. 2000.

[9] T. He, C. Huang, B. M. Blum, J. A. Stankovic, and T. Abdelzaher, "Rangefree localization schemes for large scale sensor networks," in *Proc. of ACM Int. Conf. on Mobile Computing and Networking (MobiCom)*, 2003, pp. 81–95.

[10] C. Chang and A. Sahai, "Estimation bounds for localization," in *Proc. of IEEE Conf. on Sensor, Mesh and Ad Hoc Communications and Networks (SECON)*, Oct. 2004, pp. 415–424.

[11] T. Jia and R. Buehrer, "A new Cramer-Rao lower bound for TOA-based localization," in *Proc. of the 2008 IEEE Military Communications Conf. (MILCOM'08)*, Nov. 16–19, 2008, pp. 1–5.

[12] T. Jia and R. Buehrer, "On the optimal performance of collaborative position location," *IEEE Trans. Wireless Commun.*, vol. 9, pp. 374–383, 2010.

[13] Y. Qi and H. Kobayashi, "Cramer-Rao lower bound for geolocation in nonline-of-sight environment," in *Proc. of the Int. Conf. on Acoustics, Speech and Signal Processing (ICASSP)*, vol. 3, 2002, pp. 2473–2476.

[14] H. L. V. Trees, *Detection, Estimation, and Modulation Theory, Part I*. New York: Wiley-Interscience, 2001.

[15] T. Jia, "Collaborative position location for wireless networks in harsh environments," Ph.D. thesis, Virginia Tech, 2010.

[16] H. D. Sherali and W. P. Adams, *A Reformulation-Linearization-Technique for Solving Discrete and Continuous Nonconvex Problems*. Boston, MA: Kluwer Academic Publishing, 1999.

[17] Y. Qi, T. Asai, H. Yoshino, and N. Nakajima, "On geolocation in ill-conditioned BS-MS layouts," in *IEEE Int. Conf. on Acoustics, Speech, and Signal Processing*, vol. 5, Mar. 2005, pp. 697–700.

[18] S. O. Dulman, A. Baggio, P. J. M. Havinga, and K. G. Langendoen, "A geometrical perspective on localization," in *Proc. of ACM Workshop on Mobile Entity Localization and Tracking in GPS-Less Environments*, 2008, pp. 85–90.

[19] A. O. Hero, J. A. Fessler, and M. Usman, "Exploring estimator bias-variance tradeoffs using the uniform CR bound," *IEEE Trans. Signal Process.*, vol. 44, no. 8, pp. 2026–2041, 1996.

[20] Y. C. Eldar, "Minimum variance in biased estimation: Bounds and asymptotically optimal estimators," *IEEE Trans. Signal Process.*, vol. 52, no. 7, pp. 1915–1930, 2004.

[21] N. Priyantha, A. Chakraborty, and H. Balakrishnan, "The Cricket location-support system," in *Proc. of ACM Int. Conf. on Mobile Computing and Networking (MobiCom)*, Aug. 2000, pp. 32–43.

[22] P. Bahl and V. N. Padmanabhan, "RADAR: An in-building RF-based user location and tracking system," in *Proc. of IEEE Conf. on Computer Communications (INFOCOM)*, vol. 2, Mar. 2000, pp. 775–784.

[23] D. Niculescu and B. Nath, "Ad hoc positioning system (APS)," in *Proc. of IEEE GLOBECOM*, 2001, pp. 2926–2931.

[24] C. Savarese and J. Rabaey, "Robust positioning algorithms for distributed ad-hoc wireless sensor networks," in *Proc. of USENIX Annual Technical Conf.*, 2002, pp. 317–327.

[25] A. Savvides, H. Park, and M. B. Srivastava, "The bits and flops of the n-hop multilateration primitive for node localization problems," in *Proc. of Int. Workshop on Sensor Networks Application*, 2002, pp. 112–121.

[26] L. Doherty, K. Pister, and L. E. Ghaoui, "Convex position estimation in wireless sensor networks," in *Proc. of IEEE Conf. on Computer Communications (INFOCOM)*, vol. 3, 2001, pp. 1655–1663.

[27] Y. Shang, W. Ruml, Y. Zhang, and M. Fromherz, "Localization from connectivity in sensor networks," *IEEE Trans. Parallel Distrib. Syst.*, vol. 15, pp. 961–974, 2004.

[28] X. Ji and H. Zha, "Sensor positioning in wireless ad-hoc sensor networks using multidimensional scaling," in *Proc. of IEEE Conf. on Computer Communications (INFOCOM)*, vol. 4, Mar. 2004, pp. 2652–2661.

[29] J. A. Costa, N. Patwari, and A. O. Hero III, "Distributed weighted multidimensional scaling for node localization in sensor networks," *ACM Trans.*, vol. 2, no. 1, p. 39, 2006.

[30] P. Biswas and Y. Ye, "Semidefinite programming for ad hoc wireless sensor network localization," in *Proc. of Int. Symp. on Information Processing in Sensor Networks (IPSN)*, 2004, pp. 46–54.

[31] P. Biswas, T.-C. Lian, T.-C. Wang, and Y. Ye, "Semidefinite programming based algorithms for sensor network localization," *ACM Trans. Sens. Netw.*, vol. 2, no. 2, pp. 188–220, 2006.

[32] P. Tseng, "Second-order cone programming relaxation of sensor network localization," *SIAM J. Optim.*, vol. 18, no. 1, pp. 156–185, 2006.

[33] D. Dardari and A. Conti, "A sub-optimal hierarchical maximum likelihood algorithm for collaborative localization in ad-hoc network," in *Proc. of IEEE Conf. on Sensor, Mesh and Ad Hoc Communications and Networks (SECON)*, Oct. 2004, pp. 425–429.

[34] D. Moore, J. Leonard, D. Rus, and S. Teller, "Robust distributed network localization with noisy range measurements," in *Proc. of ACM Conf. on Embedded Networked Sensor Systems*, 2004, pp. 50–61.

[35] J. Liu, Y. Zhang, and F. Zhao, "Robust distributed node localization with error management," in *Proc. of ACM Int. Symp. on Mobile Ad Hoc Networking and Computing (MobiHoc)*, 2006, pp. 250–261.

[36] T. Eren, D. K. Goldenberg, W. Whiteley, Y. R. Yang, A. S. Morse, B. D. O. Anderson, and P. N. Belhumeur, "Rigidity, computation, and randomization in network localization," in *Proc. of IEEE Conf. on Computer Communications (INFOCOM)*, vol. 4, Mar. 2004, pp. 2673–2684.

[37] A. T. Ihler, J. W. Fisher III, R. L. Moses, and A. S. Willsky, "Nonparametric belief propagation for self-localization of sensor networks," *IEEE J. Sel. Areas Commun.*, vol. 23, no. 4, pp. 809–819, 2005.

[38] D. Marinakis and G. Dudek, "Probabilistic self-localization for sensor networks," in *Proc. of the AAAI National Conf. on Artificial Intelligence*, Jul. 2006, pp. 780–785.

[39] R. Huang and G. V. Záruba, "Incorporating data from multiple sensors for localizing nodes in mobile ad hoc networks," *IEEE Trans. Mobile Comput.*, vol. 6, pp. 1090–1104, 2007.

[40] A. J. Weiss and J. S. Picard, "Network localization with biased range measurements," *IEEE Trans. Wireless Commun.*, vol. 7, pp. 298–304, 2008.

[41] S. Srirangarajan, A. Tewfik, and Z.-Q. Luo, "Distributed sensor network localization using SOCP relaxation," *IEEE Trans. Wireless Commun.*, vol. 7, pp. 4886–4895, 2008.

[42] J. Nie, "Sum of squares method for sensor network localization," *Comput. Optim. Appl.*, vol. 43, no. 2, pp. 151–179, 2009.

[43] L. Li and T. Kunz, "Cooperative node localization using nonlinear data projection," *ACM Trans. Sens. Netw.*, vol. 5, no. 1, pp. 1–26, 2009.

[44] T. Jia and R. M. Buehrer, "A collaborative quasi-linear programming framework for ad hoc sensor localization," in *Proc. of IEEE Wireless Communications and Networking Conf.*, Apr. 2008, pp. 2379–2384.

[45] S. Severi, G. Abreu, G. Destino, and D. Dardari, "Understanding and solving flip-ambiguity in network localization via semidefinite programming," in *Proc. of IEEE GlobeCom 2009*, Dec. 2009, pp. 3910–3915.

[46] Z. Yang, Y. Liu, and X.-Y. Li, "Beyond trilateration: On the localizability of wireless ad-hoc networks," in *Proc. of IEEE Conf. on Computer Communications (INFOCOM)*, Apr. 2009, pp. 2392–2400.

[47] Y. Shang, W. Ruml, Y. Zhang, and M. P. J. Fromherz, "Localization from mere connectivity," in *Proc. of Mobile Ad Hoc Networking and Computing (MobiHoc)*, 2003, pp. 201–212.

[48] S. Lederer, Y. Wang, and J. Gao, "Connectivity-based localization of large scale sensor networks with complex shape," in *Proc. of IEEE Conf. on Computer Communications (INFOCOM)*, Apr. 2008, pp. 789–797.

[49] S. Capkun, M. Hamdi, and J. Hubaux, "GPS-free positioning in mobile ad hoc networks," in *Proc. of Hawaii Int. Conf. on System Sciences*, 2001, pp. 3481–3490.

[50] J. Albowicz, A. Chen, and L. Zhang, "Recursive position estimation in sensor networks," in *Proc. of IEEE Int. Conf. on Network Protocols*, Nov. 2001, pp. 35–41.

[51] X. Li, "Collaborative localization with received-signal strength in wireless sensor networks," *IEEE Trans. Veh. Technol.*, vol. 56, pp. 3807–3817, 2007.

[52] R. L. Moses, D. Krishnamurthy, and R. M. Patterson, "A self-localization method for wireless sensor networks," *EURASIP J. Appl. Signal Process.*, vol. 4, pp. 348–358, 2003.

[53] Z. Wang, S. Zheng, S. Boyd, and Y. Ye, "Further relaxations of the SDP approach to sensor network localization," Tech. Rep., Stanford University, May 2007. Available: http://www.stanford.edu/~yyye/relaxationsdp9.pdf

[54] P. Zhang and M. Martonosi, "LOCALE: Collaborative localization estimation for sparse mobile sensor networks," in *Proc. of Int. Conf. on Information Processing in Sensor Networks (IPSN)*, Apr. 2008, pp. 195–206.

[55] S. Basagni, M. Battelli, M. Iachizzi, C. Petrioli, and S. Masoud, "Limiting the propagation of localization errors in multi-hop wireless network," in *IEEE Pervasive Computing and Communications Workshops*, Mar. 2006, pp. 6–11.

[56] R. M. Buehrer, S. Venkatesh, and T. Jia, "Mitigation of the propagation of localization error using multi-hop bounding," in *Proc. of IEEE Wireless Communications and Networking Conf. (WCNC)*, Apr. 2008, pp. 3009–3014.

[57] S. Venkatesh and R. M. Buehrer, "NLOS mitigation in UWB location-aware networks using linear programming," *IEEE Trans. Veh. Technol.*, vol. 56, pp. 3182–3198, 2007.

[58] J. J. Caffery, "A new approach to the geometry of TOA location," in *Proc. of IEEE Vehicular Technology Conf.*, vol. 4, 2000, pp. 1943–1949.

[59] D. K. Goldenberg, P. Bihler, M. Cao, J. Fang, B. D. O. Anderson, A. S. Morse, and Y. R. Yang, "Localization in sparse networks using sweeps," in *Proc. of ACM Int. Conf. on Mobile Computing and Networking (MobiCom)*, Sep. 2006, pp. 110–121.

[60] T. Jia and R. M. Buehrer, "Collaborative position location for wireless networks using iterative parallel projection method," presented at the IEEE Global Communications Conf., Miami, FL, Dec. 2010.

[61] T. Jia and R. M. Buehrer, "A set-theoretic approach to collaborative position location for wireless networks," *IEEE Trans. Mobile Comput.*, vol. 10, no. 9, pp. 1264–1275, 2010.

[62] P. L. Combettes, "Signal recovery by best feasible approximation," *IEEE Trans. Image Process.*, vol. 2, pp. 269–271, 1993.

[63] P. L. Combettes, "Inconsistent signal feasibility problems: Least-squares solutions in a product space," *IEEE Trans. Signal Process.*, vol. 42, pp. 2955–2966, 1994.

[64] M. A. Diniz-Ehrhardt and J. M. Martinez, "A parallel projection method for overdetermined non-linear system equations," *Numer. Algorithms*, vol. 4, pp. 241–262, 1993.

[65] T. Jia and R. M. Buehrer, "Collaborative position location with NLOS mitigation," in *Proc. of 2010 Workshop on Advances in Positioning and Location-Enabled Communications (APLEC)*, Sep. 2010, pp. 267–271.

[66] L. Cong and W. Zhuang, "Nonline-of-sight error mitigation in mobile location," *IEEE Trans. Wireless Commun.*, vol. 4, no. 2, pp. 560–573, 2005.

[67] M. Heidari, F. O. Akgul, and K. Pahlavan, "Identification of the absence of direct path in indoor localization systems," in *Proc. of IEEE Int. Symp. on Personal, Indoor and Mobile Radio Communications (PIMRC)*, Sep. 2007, pp. 1–6.

[68] I. Guvenc, C.-C. Chong, F. Watanabe, and H. Inamura, "NLOS identification and weighted least-squares localization for UWB systems using multipath channel statistics," *EURASIP J. Adv. Signal Process.*, vol. 2008, p. 36, 2008.

[69] C.-D. Wann and C.-S. Hsueh, "NLOS mitigation with biased Kalman filters for range estimation in UWB systems," in *Proc. of IEEE Region 10 Conf. TENCON*, 2007, pp. 1–4.

[70] J. Khodjaev, Y. Park, and A. S. Malik, "Survey of NLOS identification and error mitigation problems in UWB-based positioning algorithms for dense environments," *Ann. Telecomm.*, vol. 65, pp. 301–311, 2010.

[71] S. Venkatraman, J. J. Caffery, and Y. H. Ryeol, "A novel TOA location algorithm using LOS range estimation for NLOS environments," *IEEE Trans. Veh. Technol.*, vol. 53, no. 5, pp. 1515–1524, 2004.

[72] B. Alavi and K. Pahlavan, "Modeling of the distance error for indoor geolocation," in *Proc. of IEEE Wireless Communications and Networking Conf. (WCNC)*, vol. 1, Mar. 2003, pp. 668–672.

[73] K. Yu and Y. J. Guo, "NLOS error mitigation for mobile location estimation in wireless networks," in *Proc. of IEEE 65th Vehicular Technology Conf.*, Apr. 2007, pp. 1071–1075.

[74] S. Venkatesh and R. M. Buehrer, "NLOS identification in ultra-wideband systems based on received signal statistics," *Special Issue on Antenna Systems and Propagation for Future Wireless Communications, IEE Proc. Microw. Antennas Propag.*, vol. 1, pp. 1120–1130, Dec. 2007.

[75] O.-H. Kwon and H.-J. Song, "Localization through map stitching in wireless sensor networks," *IEEE Trans. Parallel Distrib. Syst.*, vol. 19, pp. 93–105, 2008.

CHAPTER 25

POLYNOMIAL-BASED METHODS FOR LOCALIZATION IN MULTIAGENT SYSTEMS

Iman Shames,[1] Barış Fidan,[2] Brian D. O. Anderson,[1] and Hatem Hmam[3]

[1]*The Australian National University and The University of Melbourne*
[2]*University of Waterloo, Waterloo, Canada*
[3]*Defence Science & Technology Organisation, Edinburgh, Australia*

I**N THIS** chapter we review a series of results obtained in the field of localization that are based on polynomial optimization. First, we provide a review of a set of polynomial function optimization tools, including sum of squares (SOS). Then we present several applications of these tools in various sensor network localization tasks. As the first application, we propose a method based on SOS relaxation for node localization using noisy measurements and describe the solution through semidefinite programming (SDP). Later, we apply this method to address the problems of target localization in the presence of noise and relative reference frame determination based on range and bearing measurements. Some simulation and experiment results are provided to show the applicability of the methods proposed here.

25.1 INTRODUCTION

The problem of target localization has gained much attention recently. This problem involves fusion of different measurements, for example, angle of arrival, distance, and time difference of arrival, obtained from different sensing nodes with known global positions, called anchors, in a network to estimate the position of the target. Solving the localization problem in the case where there is no noise is heavily studied

Handbook of Position Location: Theory, Practice, and Advances, Second Edition.
Edited by S. A. (Reza) Zekavat and R. Michael Buehrer.
© 2019 by the Institute of Electrical and Electronics Engineers, Inc.
Published 2019 by John Wiley & Sons, Inc.
Companion Website: www.wiley.com/go/zekavat/positionlocation2e

in the network localization literature; see for example, [1] and [2]. However, in almost all engineering applications, the assumption of having noiseless measurements is not realistic. And while the analysis of the problem of localization in the presence of measurement noise is still in its infancy, recently, in addition to trying to formally define and understand this problem (see, e.g., [2]), many algorithms have been proposed to tackle this problem. For example, in [3], a linear algorithm to address this problem is proposed. Later in [4], a nonlinear method using the *Cayley–Menger* determinant is proposed, and its performance is compared with that of [3], which is shown to have better results. Another paper that has considered the application of Cayley–Menger determinant to solve localization problems is [5], where an in-depth error analysis for the case of having three anchors is presented. In addition to the abovementioned methods, one can name other methods based on convex optimization [6–11], SOS relaxation [13], graph connectivity [14, 15], and multidimensional scaling [16].

To make the problem of noisy target localization clearer, consider the following illustration of a range-based localization scenario in the presence of noise. Consider a set of nodes scattered in an N-dimensional space, $N \in \{2, 3\}$. Consider $N_a > N$ anchor nodes, labeled $1, 2, \ldots, N_a$. Assume that each (anchor) node $I \in \{1, \ldots, N_a\}$ is at a generic position $\mathbf{a}_i \in \mathbb{R}^N$; $\mathbf{a}_i = [x_i, y_i]^\mathsf{T}$ for $N = 2$ and $\mathbf{a}_i = [x_i, y_i, z_i]^\mathsf{T}$ for $N = 3$. Here, by having generic positions, we mean that for $N = 2$, no triple of anchor nodes are colinear, and for $N = 3$, no quadruple of anchor nodes are coplanar. Consider a nonanchor sensor node, node 0, which is placed at an unknown position $\mathbf{x} = [x, y]^\mathsf{T}$, which can measure its distance from anchor nodes $1, 2, \ldots, N_a$, that is, it measures $d_i^* = \|\mathbf{a}_i - \mathbf{x}\|$, for each $i \in \{1, \ldots, N_a\}$. Note that in this paper we denote the actual distance $\|\mathbf{a}_i - \mathbf{x}\|$ by d_i^* and the corresponding measurement by d_i. The problem of interest is finding an accurate estimate $\hat{\mathbf{x}}$ for $\mathbf{x}$ using the measurements d_i.

If the measurements are precise ($d_i = d_i^*$) the solution is trivial: $\hat{\mathbf{x}} = \mathbf{x}$ is the unique point of intersection of the N_a circles, C_i:

$$C_i : \|a_i - x\|^2 - d_i^2 = 0, \quad i \in \{1, \cdots, N_a\}. \tag{25.1}$$

The setting (for the case where the measurements are noisy) is illustrated in Figure 25.1. However, unfortunately, it is never the case that the measurements agree with the actual distances. Furthermore, we know that when the measurements are noisy, the circles may not have a common point of intersection, so one needs another method to solve the problem. All the methods mentioned earlier in a sense try to solve this problem.

In this chapter we will report several results related to sensor networks where optimization tools based on SOS, SDP formulation and relaxation, and algebraic geometry have been applied. First, we revisit the problem of localization of nodes in sensor networks and propose a solution using the tools from algebraic geometry and specifically SOS relaxation: this relaxation requires solving an SDP problem. The SOS approach has been already implemented to solve the problem of localization in sensor networks where range measurements are available; see [12] and [17]. However, here we propose a method for localization of the nodes using not only range measurements but other types of measurements as well, and in particular we

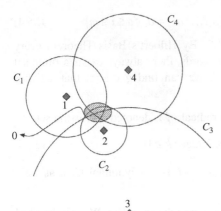

Figure 25.1   Distance-based localization problem.

formulate the problem of localization using noisy range-difference measurements. Additionally we intend to introduce methodologies about how to combine different measurement types obtained from heterogeneous measurement devices to achieve the task of localization.

Later, we propose a solution to the robot pose determination problem, that is, determining the relative pose, or relative position and orientation, of a pair of robots that move on a plane, using the same methodology in two-dimensional and three-dimensional space, when the agents have access to distance or bearing measurements from each other. This problem is very important in many practical applications, such as target tracking [18] or sensor fusion [19].

25.2 POLYNOMIAL FUNCTION OPTIMIZATION

Before proceeding we present some mathematical notations and definitions that will be used in this chapter; the reader may refer to [20] for further information. We use $\mathbb{R}[z]$, where $z = [z_1, \ldots, z_n]$, to denote the ring of all polynomials in n indeterminants (variables), $z_1, z_2, \ldots, z_n$, with real coefficients. The set $I \subset \mathbb{R}[z]$ is an *ideal* if $q.h \in I$ for any $q \in I$ and $h \in \mathbb{R}[z]$, where $q.h$ denotes the multiplication of polynomial q by polynomial h. Given polynomials $g_1, \ldots, g_r$ we write $(g_1, \ldots g_r)$ to represent the set of all polynomials that are polynomial linear combinations of $g_1, \ldots, g_r$.

Definition 1: Let $f_1, \ldots, f_m$ be polynomials in $\mathbb{R}[z_1, \ldots, z_n]$. Let the set V be

$$V(f_1, \ldots, f_m) = \{(a_1, \ldots, a_n) \in \mathbb{C} : f_i(a_1, \ldots, a_n) = 0, \forall 1 \le i \le m\}. \quad (25.2)$$

We call $V(f_1, \ldots, f_m)$ the *variety* defined by $f_1, \ldots, f_m$. Then, the set of polynomials that vanish in a given variety, that is,

$$I(V) = \{f \in \mathbb{R}[z_1, \ldots, z_n] : f(a_1, \ldots, a_n) = 0 \ \forall (a_1, \ldots, a_n) \in V\}, \quad (25.3)$$

is an ideal, called the *ideal of V*. Furthermore, the subset $V_{\mathbb{R}}(f_1, \ldots, f_m)$,

$$V_{\mathbb{R}}(f_1, \ldots, f_m) = \{(a_1, \ldots, a_n) \in \mathbb{R} : f_i(a_1, \ldots, a_n) = 0, \forall a \leq i \leq m\} \qquad (25.4)$$

is called the *real variety* defined by $f_1, \ldots, f_m$. By Hilbert's Basis Theorem every ideal $I \subset \mathbb{R}[z]$ is finitely generated. In other words, there always exists a finite set $\{f_1, \ldots, f_m\} \subset \mathbb{R}[z]$ such that for every $f \in I$, we can find $g_i \in \mathbb{R}[z]$ that satisfies $f = \sum_{i=1}^m g_i f_i$.

Definition 2: Let $I \subset \mathbb{R}[z]$ be an ideal. The radical of I, denoted $\sqrt{I}$, is the set

$$\{f | f^k \in I \quad \text{for some integer } k \geq 1\}. \qquad (25.5)$$

It is clear that $I \subset \sqrt{I}$, and it can be shown that $\sqrt{I}$ is a polynomial ideal as well. We call an ideal I radical if $I = \sqrt{I}$.

Definition 3: Consider a polynomial function f over the ring $\mathbb{R}[z]$. We call the ideal $I_\nabla(f) = \left\langle \dfrac{\partial f}{\partial z_1}, \cdots, \dfrac{\partial f}{\partial z_n} \right\rangle$ the gradient ideal of f.

Problem 1: Consider a polynomial function $F(z)$ of degree $2d$, over the polynomial ring $\mathbb{R}[z]$. Find F^* and $\mathbf{z}^* = [z_1^*, \ldots, z_n^*]$ such that $F^* = F(z^*) = \min F$.

An obvious method to find the solution to Problem 1 is to find all the critical points of F. The usual methods to solve such a system of nonlinear equations use Newton's method, and there is no guarantee that the solutions of the system will be found completely. However, there are other tools to find the the global minimum and minimizer of polynomial functions. We describe two of these methods briefly in what follows.

25.2.1 Polynomial Continuation (Homotopy) Methods

One exact method to find the global minimum of a polynomial optimization problem is to apply differentiation and then solve the resulting system of polynomial equations using the continuation (or homotopy) method. This method is suitable for finding isolated roots, and it basically defines a trivial polynomial system $Q(z)$ and a real variable, t, to form a family of curves by solving

$$0 = H(\mathbf{z}, t) = (1 - t)Q(\mathbf{z}) + tP(\mathbf{z}).$$

Several properties need to be satisfied by $H(z)$ [21] to ensure that all solutions of $P(z) = 0$ are reached at $t = 1$ by continually deforming the trivial solutions of $Q(z) = 0$ obtained at $t = 0$. An upper bound of the number of curves (or paths) that need to be continually traced from $t = 0$ to $t = 1$ is given by the Bézout number, which is the product of the system polynomial degrees. This upper bound is often too large for the actual number of solutions that need to be considered. Numerous research papers have been published that provide much lower bound estimates and construct smaller polynomial systems, $Q(z)$. If the system is sparse [22], then polyhedral homotopies which use the Newton polytopes of $P(z)$ to construct the continuation paths are more efficient. The mixed volume of the Newton polytopes is usually much smaller than the Bézout number. Homotopy methods are often regarded as slow (because they require tracing an unacceptably large number of paths), but proponents of these methods always argue that they are well suited to parallel computing.

25.2.2 SOS and SDP Approaches

Another method to find the global minimum of a polynomial function, whose application is the focus of this chapter, is to use SOS relation. A polynomial function $F(z)$ of degree $2d$ over the polynomial ring $\mathbb{R}[z]$ is an SOS if one can write

$$F(z) = \sum_{i=1}^{q} Q_i^2(z), \tag{25.6}$$

where $q \in \mathbb{Z}^+$ and $Q_i(z)$ are the polynomials over $\mathbb{R}[z]$. Denote the global minimizer and global minimum of a polynomial function $F(z)$, respectively, by z^* and $\gamma = F(z^*)$. z^* can be calculated solving the following optimization problem:

$$\begin{aligned} &\text{maximize} \quad \gamma \\ &\text{subject to} \quad F(z) - \gamma \ge 0. \end{aligned} \tag{25.7}$$

One can relax (25.7) and write it as

$$\begin{aligned} &\text{maximize} \quad \gamma \\ &\text{subject to} \quad F(z) - \gamma \ \text{ is SOS.} \end{aligned} \tag{25.8}$$

Remark 1: The relaxed problem is often computationally much easier to solve, and may yield the same solution. However, in general (25.7) and (25.8) are not identical, since there are positive polynomials that are not in the SOS form. For more information, see [22].

We know that any SOS polynomial $F(z)$ of degree $2d$, with z as n-tuple of variables, can be written as $F(z) = Z^{\mathrm{T}}QZ$, where Z is a vector of all monomials of degree up to d obtained from the variables in z with the first entry equal to one, and Q is a positive semidefinite matrix obtained by solving a set of linear matrix inequalities (LMIs) [23]. So one can reformulate (25.8) as

$$\begin{aligned} &\text{maximize} \quad \gamma \\ &\text{subject to} \quad Q - \hat{E}\gamma \ge 0, \end{aligned} \tag{25.9}$$

where $\hat{E}$ is a matrix with $E_{11} = 1$ and the rest of the entries are zero. The problem stated in (25.9) is an SDP problem and can be solved by SDP techniques [22]. By solving the dual problem of the SDP problem stated in (25.9), one can obtain the minimizer of F as well, using the procedure in [24].

Remark 2: For polynomials with two variables and degree of 4, (25.7) and (25.8) are equivalent [23].

In addition, we consider the following constrained optimization problem with polynomial F, g_i, and h_j:

$$\begin{aligned} \text{minimize} \quad & F(z) \\ \text{subject to} \quad & g_i(z) \geq 0 \quad i = 1, \ldots, M \\ & h_j(z) = 0 \quad j = 1, \ldots, N. \end{aligned} \qquad (25.10)$$

Assume there exists a set of SOS $\sigma_i(z)$, and a set of polynomials $\lambda_j(z)$, such that

$$F(z) - \gamma = \sigma_0(z) + \sum_j \lambda_j(z) h_j(z) + \sum_i \sigma_i(z) g_i(z)$$

$$+ \sum_i \sigma_{i_1, i_2}(z) g_{i_1}(z) g_{i_2}(z) + \sigma_M \prod_{i=1}^{M} g_i(z). \qquad (25.11)$$

Then, γ is a lower bound for (25.10) [23]. So by maximizing γ as before, one can get a lower bound that gets tighter as the degree of (25.11) increases. There are well-known SDP-based solutions to the aforementioned problem. For more information, one may refer to [25] and references therein. We use this method to optimize the cost functions introduced in this chapter. There are a few implementations of this method as software packages that are readily available. The main two of such packages are GloptiPoly 3 [26] and SOSTOOLS [27]. In what follows, we provide an example for polynomial optimization using GloptiPoly 3 from [26].

Example 25.1   Global Minimization of the Two-Dimensional Six-Hump Camel Back Function

Consider the following minimization problem:

$$\min_{z \in \mathbb{R}^2} F(z) \ 4z_1^2 + z_1 z_2 - 4z_2^2 - 2.1z_1^4 + 4z_2^4 + \frac{1}{3}z_1^6.$$

This problem has six local minima, two of them being global minima. The problem can be solved via the following procedure in MATLAB:

```
>> mpol z1 z2
>> F = 4*z1^2+z1*z2-4*z2^2-2.1*z1^4+4*z2^4+z1^6/3;
>> P = msdp(min(F));
>> [status,obj] = msol(P)
>> status
status =
1
>> obj
obj =
-1.0316
>> z = double([z1 z2]);
z(:,:, 1) =
0.0898 -0.7127
z(:,:, 2) =
-0.0898 0.7127,
```

where `status=1` means that the problem is solved and `obj=-1.0316` is the global minimum. The two solutions for z correspond to two global minimizers of the function.

25.3 NOISY TARGET LOCALIZATION

In this section, application of the method in Section 22.2 to target localization using (1) distance measurements and (2) range-difference measurements is presented. The formal problem definitions for these two application cases are given below, respectively, in Problem 2 and Problem 3.

Problem 2 (**Range-Based Localization**): Consider N_a anchor nodes in $\mathbb{R}^N$ ($N \in \{2, 3\}$) at the known positions $\mathbf{a}_i$, $i \in \{1, \dots, N_a\}$ and a node 0 at the unknown position $\mathbf{x}$. Let the noisy measurement d_i of the distance of node 0 to each node i *for* $i \in \{1, \dots, N_a\}$ be available to node 0. The task of node 0 is to produce the least-square estimate $\hat{\mathbf{x}}$ of $\mathbf{x}$ using the noisy distance measurements $d_1, \dots, d_n$.

Problems similar to Problem 2 have been already well-considered in the signal processing context, for example, in [3, 7] and [28]. A solution to this problem can be obtained by finding the point $\hat{\mathbf{x}}$ that solves the following minimization problem:

$$minimize \quad J_r(\mathbf{x}) = \sum_{i=1}^{n} \left(\|\mathbf{x} - \mathbf{a}_i\|^2 - d_i^2 \right)^2. \tag{25.12}$$

In general $J(\mathbf{x})$ is not convex (or concave), so ordinary convex optimization methods will not yield the desired result. In what follows some examples on target localization using distance measurements in two- and three-dimensional space are presented.

Example 25.2   Localization of One Node with Distance Measurements to Three Anchors (Calculating the Variety)

We use the same setting as in [4]: Consider three anchor nodes at $\mathbf{a}_1 = [0, 0]^T$, $\mathbf{a}_2 = [7, 43]^T$, and $\mathbf{a}_3 = [0, 47]^T$, and a sensor node zero using distance measurements from these anchors to localize itself. The actual distances that are not available are $d_1^* = 34.392$, $d_2^* = 44.1106$, and $d_3^* = 41.2608$, while the noisy distance measurements by sensor 0 are $d_1 = 35$, $d_2 = 42$, and $d_3 = 43$. Hence,

$$J(\mathbf{x}) = (x^2 + y^2 - 1225)^2 + ((x - 43)^2 + (y - 7)^2 - 1764)^2 + ((x - 47)^2 + y^2 - 1849)^2, \tag{25.13}$$

and solve:

$$\frac{\partial J}{\partial x} = 4(x^2 + y^2 - 1225)x + 2((x - 43)^2 + (y - 7)^2 - 1764)(2x - 86)$$
$$+ 2((x - 47)^2 + y^2 - 1849)(2x - 94)$$
$$= 0. \tag{25.14}$$

$$\frac{\partial J}{\partial y} = 4(x^2 + y^2 - 1225)y + 2((x-43)^2 + (y-7)^2 - 1764)(2y - 14)$$

$$+ 4((x-47)^2 + y^2 - 1849)y$$

$$= 0.$$

(25.15)

The real solutions to this system of equations construct the set

$$V_\mathbb{R} = \{[3.43, 0.85]^T, \ [13.90, 31.79]^T, \ [18.22, -29.24]^T\},$$

and by inspection $\mathbf{x}$ is found to be $[18.22, -29.24]^T$ (this point is the global minimizer of the function.). Comparing with the actual position of sensor node 0, $\mathbf{x} = [17.9719, -29.3227]^T$, it is seen that the estimate is considerably accurate. The result obtained here is the same as the one calculated in [4]. However, here, the steps for calculating the solution are less than those used in [4]. Furthermore, smaller number of floating point operations are used here, which increases the robustness of the method to numerical perturbations. See Chapter_25_Example_2.m for more details. MATLAB codes can be found online at ftp://ftp.wiley.com/public/sci_tech_med/matlab_codes.

Example 25.3 Localization of One Node with Distance Measurements to Four Anchors in $\mathbb{R}^3$

In this example the three-dimensional case is considered. Consider four anchors at positions $\mathbf{a}_1 = [10, 13, 14]^T$, $\mathbf{a}_2 = [5, 20, 40]^T$, $\mathbf{a}_3 = [12, 15, -10]^T$, and $\mathbf{a}_4 = [0, 5, 32]^T$. Furthermore, sensor node 0 senses its distance to these anchors to be $d_1 = 22.4388$, $d_2 = 46.0740$, $d_3 = 22.8327$, and $d_4 = 33.3214$. Here,

$$V = \{[19.73, 4.92, -5.27]^T, \ [29.95, 20.31, -1.32]^T [-1.29, -0.79, -0.38]^T\},$$

and $\hat{\mathbf{x}} = [-1.29, -0.79, -0.38]^T$ is the global minimum of J. Our estimate is very close to the actual position of 0, $[0, 0, 0]^T$. See Chapter_25_Example_3.m for more details. MATLAB codes can be found online at ftp://ftp.wiley.com/public/sci_tech_med/matlab_codes.

Example 25.4 Localization of One Node with Distance Measurements to Three, Four, and Five Anchors (SOS)

In this example we compare the methods from [3] and [4], and the one introduced here. Consider five anchors, $1, \ldots, 5$, are positioned at $\mathbf{a}_1 = [10, 31]^T$, $\mathbf{a}_2 = [12, 45]^T$, $\mathbf{a}_3 = [-9, 10]^T$, $\mathbf{a}_4 = [30, -3]^T$, and $\mathbf{a}_5 = [-7, 53]^T$. The measured distances by the sensor node 0, at the unknown position $[0, 0]^T$, to each of the anchors corrupted with a zero mean Gaussian noise with variance of $1\,\mathrm{m}^2$ are $d_1 = 32.1404$, $d_2 = 44.9069$, $d_3 = 13.5789$, $d_4 = 30.4373$, and $d_5 = 52.3138$. The actual position of node 0 is $\mathbf{x} = [0, 0]^T$. The estimated values for each of the methods using 3, 4, and 5 distance measurements are accessible from Tables 25.1 and 25.2. It can be noted that the linear method performs significantly worse than the nonlinear methods. In particular, for the case where three measurements are available, the improvement obtained by

TABLE 25.1.   The Estimated Position of Node 0

No. Anchors	Linear	C-M	Geometric
3	$\begin{bmatrix} -4.8493 \\ 5.1347 \end{bmatrix}$	$\begin{bmatrix} 0.8825 \\ 1.0590 \end{bmatrix}$	$\begin{bmatrix} 0.8825 \\ 1.0590 \end{bmatrix}$
4	$\begin{bmatrix} 1.8971 \\ -0.0505 \end{bmatrix}$	N/A	$\begin{bmatrix} -0.0150 \\ 1.2858 \end{bmatrix}$
5	$\begin{bmatrix} -0.6306 \\ -0.1361 \end{bmatrix}$	N/A	$\begin{bmatrix} 0.2211 \\ 0.0710 \end{bmatrix}$

The Method from [3] Is Labeled "Linear," the Method from [4] Is Labeled
"C-M," and the Method Introduced in This Paper Is Labeled "Geometric"

**TABLE 25.2.   The Error between the Estimated Position of
Node 0 and Its Actual Position**

No. Anchors	Linear	C-M	Geometric
3	7.0626	1.3785	1.3785
4	1.8977	N/A	1.2859
5	0.6452	N/A	0.2323

using the geometric method is very significant; this is due to the sensitivity of linear equations to perturbation in the coefficient. Moreover, for more than 4 measurements, the method based on Cayley–Menger determinant gets increasingly complicated and cannot be applied directly. See Chapter_25_Example_4.m for more details. MATLAB codes can be found online at ftp://ftp.wiley.com/public/sci_tech_med/matlab_codes.

Example 25.5   Localization of One Node with Distance Measurements to Five Anchors

In this example we compare the result obtained by applying our approach to the example introduced in [11]. Consider $N_a = 5$ anchors in $\mathbb{R}^2$ at $\mathbf{a}_1 = [4, \ 6]^T$, $\mathbf{a}_2 = [0, \ -10]^T$, $\mathbf{a}_3 = [5, \ -3]^T$, $\mathbf{a}_4 = [1, \ -4]^T$, and $\mathbf{a}_5 = [3, \ -3]^T$. The distance measurements to the sensor at the unknown position $[0, \ 0]^T$ corrupted by a zero-mean Gaussian noise with variance equal to 0.1 are $d_1 = 8.0051, d_2 = 13.0112, d_3 = 10.1138, d_4 = 7.7924$, and $d_5 = 8.0210$. The result obtained from applying range-based least square (R-LS) and squared-range-based least square (SR-LS) from [7] for $\hat{\mathbf{x}}$ *are* $[-1.9907, 3.0474]^T$ and $[-2.018, 2.9585]^T$. Applying the method from this study we obtain the same value as SR-LS. It can be seen that the optimization method proposed in [7] and the one based on SOS relaxation results in the same solution. See Chapter_25_Example_5.m for more details. MATLAB codes can be found online at ftp://ftp.wiley.com/public/sci_tech_med/matlab_codes.

Now, we introduce another problem whose formulation was presented in [7] in what immediately follows.

Problem 3 (**Range-Difference-Based Cooperative Localization**): Consider N_a anchor nodes in $\mathbb{R}^N$ ($N \in \{2, 3\}$) at the known positions $\mathbf{a}_i$, $i \in \{1, \dots, N_a\}$, another node $N_a + 1$ at the origin, and node 0 at the unknown position $\mathbf{x}$. Let δ_i, the noisy measurement of $\delta_i^* = d_i^* - \|\mathbf{x}\|$ for $i \in \{1, \dots, N_a\}$, be available (to node 0). The task (of node 0) is to produce the estimate $\hat{\mathbf{x}}$ of $\mathbf{x}$ using the noisy range-difference measurements $\delta_1, \dots, \delta_n$.

To solve the problem one is interested in, solve the following minimization problem:

$$\text{minimize } J_d(p) = \sum_{i=1}^{n} \left(\delta_i^2 - \|\mathbf{a}_i\|^2 + 2\delta_i \|\mathbf{x}\| + 2\mathbf{a}_i^T \mathbf{x} \right)^2. \tag{25.16}$$

Denoting $\|\mathbf{x}\|$ by D, and considering that $D^2 - \|\mathbf{x}\|^2 = 0$, Equation (25.16) can be rewritten as

$$\text{minimize } \sum_{i=1}^{n} \left(\delta_i^2 - \|\mathbf{a}_i\|^2 + 2\delta_i D + 2\mathbf{a}_i^T \mathbf{x} \right)^2, \tag{25.17}$$
$$\text{subject to } D^2 - \|\mathbf{x}\|^2 = 0, D \geq 0.$$

By setting, $z = \mathbf{x}$, $F(z) = J_r(\mathbf{x})$ and using (25.9), we can find the exact solution to (25.12). For (25.17), $F(z) = J_d(\mathbf{x})$, inequality constraints do not exist, and the only equality constraint is $h(\mathbf{x}) = \|\mathbf{x}\|^2 - D^2 = 0$, and we can find the solution to it by using the extended version of the methods introduced in the previous section. Now we present an example for this problem.

Example 25.6   Localization of One Node with Range-Difference Measurements to Five Anchors

In this example we consider another example presented in [7]. Consider five anchors and an extra anchor at the origin. The range-difference measurements corrupted by a zero-mean Gaussian noise with the variance of 0.2 are $\delta_1 = 11.8829$, $\delta_2 = 0.1803$, $\delta_3 = 4.6399$, $\delta_4 = 11.2402$, and $\delta_5 = 10.8183$. The result for $\mathbf{x}$ is $\hat{\mathbf{x}} = [-4.9798, 10.2786]^T$, which agrees with the one obtained in [7]. See Chapter_25_Example_6.m for more details. MATLAB codes can be found online at ftp://ftp.wiley.com/public/sci_tech_med/matlab_codes.

25.4   RELATIVE REFERENCE FRAME DETERMINATION

In this section we consider the problem of determining the rotation and translation associating the relative reference frames of a pair of robots. In the first subsection we consider the case where the robots can measure their distance from each other, and in the next one we consider the case where only relative angle measurements are available to one of the robots.

25.4.1   Relative Reference Frame Determination with Distance Measurements

In [29], the problem of determining the relative reference frames of a pair of robots that move on a plane while measuring distance to each other is studied. Later, in [30], the authors introduced the same problem in three-dimensional space. In what comes next, we state this problem.

Problem 4 (**Distance-Based Relative Reference [Pose] Determination**): Consider two agents (robots) A_1 and A_2 in $\mathbb{R}^N$, $N = 2, 3$ whose initial reference frames are indicated by Σ_1 and Σ_2, respectively. The two agents move through a sequence of n unknown different positions with a reference frame associated with each position, $\Sigma_1, \Sigma_3, \dots, \Sigma_{2n-1}$ for A_1 and $\Sigma_2, \Sigma_4, \dots, \Sigma_{2n}$ for A_2, where $n \in \mathbb{N}$. Their interagent distance, $d_{i,i+1}$, is measured at each of these positions, where $i \in \{1, 3, \dots, 2n - 1\}$. In addition, each agent is capable of estimating its current reference frame orientation and displacement with respect to its initial reference frame using dead-reckoning (odometry). In other words, A_1 and A_2 estimate the position vectors $\mathbf{x}_3^1, \dots, \mathbf{x}_{2n-1}^1$ and $\mathbf{x}_4^2, \dots, \mathbf{x}_{2n}^2$, respectively, where $\mathbf{x}_j^i$ is the position of the origin of Σ_j in Σ_i. Additionally, they know the rotation matrices that relate the orientations of their initial reference frame and all the others in their own sequence. Here we consider the rotation matrix $\mathbf{R}_j^i$ to be the rotation matrix relating Σ_i to Σ_j. The task is to find $\mathbf{x}_2^1$ and $\mathbf{R}_2^1$ using this information.

First, we consider the case $N = 2$ and assume that the origins of the reference frames of the agents at each time are the vertices of a graph, and if the distance between any pair of the origins is known, there is an edge connecting them together (see Fig. 25.2a). We call the resulting graph G_{a-a}. Without loss of generality, we select Σ_1 as our reference frame for solving the problem. As a result of this selection, the origins of the reference frames $\Sigma_1, \Sigma_3, \dots, \Sigma_{2n-1}$ can be calculated, and they can be considered as anchor points for the formation with G_{a-a} as its underlying graph

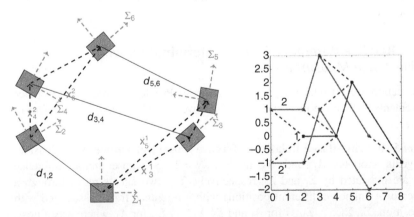

(a) The setting considered in Problem 4

(b) A nongeneric case where two solutions exist for the pose determination problem

Figure 25.2   Pose-determination problem and a nongeneric scenario.

(in addition, the rotation matrix relating each of them to Σ_1 can be calculated as well). The goal here is to find the positions of the origins of $\Sigma_2, \Sigma_4, \ldots, \Sigma_{2n}$ in Σ_1. We further can compute $\mathbf{R}_2^1$ using any equation of the following form:

$$\mathbf{x}_{2i}^1 = \mathbf{x}_2^1 + \mathbf{R}_2^1 \mathbf{x}_{2i}^2, \quad i \in \{1, \ldots, n\}. \tag{25.18}$$

It is obvious that the origins of the reference frames $\Sigma_1, \ldots, \Sigma_{2n-1}$ form a complete graph, as do the origins of $\Sigma_2, \ldots, \Sigma_{2n}$. For $n \geq 4$, these two complete graphs are connected to each other with $n \geq 4$ edges. (Note that these edges do not share any vertex with each other.) From [31] we know that the resulting graph is a globally rigid graph. Hence, the resulting formation when the origins of $\Sigma_1, \ldots, \Sigma_{2n-1}$ are considered anchors has a unique localization solution. As a result, due to the fact that there is a unique rotation relating Σ_1 to Σ_2, the agent-to-agent relative pose can be uniquely determined. We note that in [29] the authors have suggested a maximum of four solutions for Problem 4 when $N = 2$ and $n = 4$, and with probability one, only one of the solutions can be valid. Now, we formally present the following result.

Proposition 1: For $N = 2$ and $n \geq 4$, Problem 4 has a unique solution (generically).

One nongeneric scenario is when both agents move on a straight line [29]. Another nongeneric scenario is when the agents have the same displacement between any two consecutive measurements (see Fig. 25.2b). In these cases, there are two distinct solutions for Problem 4 and $N = 2$. Now we consider the three-dimensional case.

For the case $N = 3$, the origins of the reference frames $\Sigma_1, \ldots, \Sigma_{2n-1}$ form a complete graph, as do the origins of $\Sigma_2, \ldots, \Sigma_{2n}$. For $n \geq 4$, these two complete graphs are connected with n edges. (Note that these edges do not share any vertex with each other.) From [31] we know that the resulting graph is a globally rigid graph if (and only if) $n \geq 7$. Hence, we have the following proposition.

Proposition 2: For $N = 3$ and $n \geq 7$, Problem 4 has a unique solution (generically).

A nongeneric scenario is when the movement of each of agents is confined to a single plane.

25.4.2 Relative Reference Frame Determination with Relative Angle Measurements

In this section we introduce a problem similar to Problem 4, replacing distance measurements with angle measurements. First, we formally define the problem of interest.

Problem 5 **(Bearing-Based Relative Reference [Pose] Determination):** Consider two agents (robots) A_1 and A_2 in $\mathbb{R}^N$, $N = 2, 3$ whose initial reference frames are indicated by Σ_1 and Σ_2, respectively. The two agents move through a sequence of n unknown different positions with a reference frame associated with each position, $\Sigma_1, \Sigma_3, \ldots, \Sigma_{2n-1}$ for A_1 and $\Sigma_2, \Sigma_4, \ldots, \Sigma_{2n}$ for A_2, where n is a positive integer. Agent A_1 measures a unit vector corresponding to the bearing of agent A_2, θ_{2i}^{2i-1}, at each of these positions, where $i \in \{1, 2, \ldots, n\}$. Specifically,

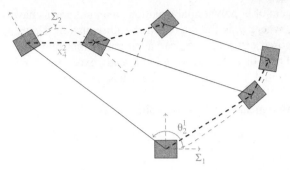

Figure 25.3   The setup of Problem 5 with $n = 3$ measurements.

$\theta_{2i}^{2i-1} = \left[\overline{\theta}_{2i-1,2i}, \overline{\theta}'_{2i-1,2i}, \overline{\theta}''_{2i-1,2i} \right]^{\mathrm{T}}$ is the unit vector associated with the line connecting $\mathbf{x}_{2i-1}^1$ and $\mathbf{x}_{2i}^1$ and in the direction toward $\mathbf{x}_{2i}^1$ in case $N = 3$, with obvious variation when $N = 2$. In addition, each agent is capable of estimating its current reference frame orientation and displacement with respect to its initial reference frame using dead-reckoning (odometry). In other words A_1 and A_2 estimate the position vectors $\mathbf{x}_3^1, \ldots, \mathbf{x}_{2n-1}^1$ and $\mathbf{x}_4^2, \ldots, \mathbf{x}_{2n}^2$, respectively, where $\mathbf{x}_j^i$ is the position of the origin of Σ_j in Σ_i. Additionally, they know the rotation matrices that relate the orientations of their initial reference frame and all the others in their own sequence. Here we consider the rotation matrix $\mathbf{R}_j^i$ to be the rotation matrix relating Σ_i to Σ_j. The task is to find $\mathbf{x}_2^1$ and $\mathbf{R}_2^1$ using this information.

We have the following result for three measurements.

Lemma 1: For $N = 2, 3$ and $n = 3$, Problem 5 has at most eight solutions (Fig. 25.3).

Proof: Each of the measurements θ_{2i}^{2i-1}, $i = 1, 2, 3$ corresponds to a line in Σ_1 such that $\exists t_{2i} \in \mathbb{R}$:

$$t_{2i}\theta_{2i}^{2i-1} + \mathbf{x}_{2i-1}^1 = \mathbf{x}_{2i}^1, \tag{25.19}$$

where t_{2i} are unknown. The problem of interest is to find t_{2i}. Furthermore, $\mathbf{x}_2^1$, $\mathbf{x}_4^1$, and $\mathbf{x}_6^1$ should satisfy the following set of equations:

$$\left\| \mathbf{x}_4^2 \right\|^2 = \left\| \mathbf{x}_2^1 - \mathbf{x}_4^1 \right\|^2, \tag{25.20}$$

$$\left\| \mathbf{x}_6^2 \right\|^2 = \left\| \mathbf{x}_2^1 - \mathbf{x}_6^1 \right\|^2, \tag{25.21}$$

$$\left\| \mathbf{x}_4^2 - \mathbf{x}_6^2 \right\|^2 = \left\| \mathbf{x}_4^1 - \mathbf{x}_6^1 \right\|^2. \tag{25.22}$$

Using (25.19), we obtain from this equation set the following three equations, each of degree 2 and with three unknowns, viz., t_2, t_4, t_6:

$$\left\| \mathbf{x}_4^2 \right\|^2 = \left\| t_2\theta_2^1 - t_4\theta_4^3 - \mathbf{x}_3^1 \right\|^2, \tag{25.23}$$

$$\left\| \mathbf{x}_6^2 \right\|^2 = \left\| t_2\theta_2^1 - t_6\theta_6^5 - \mathbf{x}_5^1 \right\|^2, \tag{25.24}$$

$$\left\| \mathbf{x}_4^2 - \mathbf{x}_6^2 \right\|^2 = \left\| t_4\theta_4^3 + \mathbf{x}_3^1 - t_6\theta_6^5 - \mathbf{x}_5^1 \right\|^2. \tag{25.25}$$

From [32] we know that a system of m polynomial equations in m unknowns have at most $\Pi_{j=1}^{m} q_j$ real solutions, where q_j is the degree of the jth equation. Hence, the system of equations composed of (25.23–25.25) has at most eight real solutions. Now, since we already know that one real solution already exists (the solution that corresponds to the actual scenario), the system has at least one real solution as well.

For more than three measurements, we have the following result.

Proposition 3: For $N = 2$, 3 and $n \geq 4$, Problem 5 generically has a unique solution.

Proof: It is enough to show that Problem 5 has a unique solution for $n = 4$. For $n = 4$ we have

$$\left\|\mathbf{x}_4^2\right\|^2 = \left\|t_2\boldsymbol{\theta}_2^1 - t_4\boldsymbol{\theta}_4^3 - \mathbf{x}_3^1\right\|^2, \tag{25.26}$$

$$\left\|\mathbf{x}_6^2\right\|^2 = \left\|t_2\boldsymbol{\theta}_2^1 - t_6\boldsymbol{\theta}_6^5 - \mathbf{x}_5^1\right\|^2, \tag{25.27}$$

$$\left\|\mathbf{x}_8^2\right\|^2 = \left\|t_2\boldsymbol{\theta}_2^1 - t_8\boldsymbol{\theta}_8^7 - \mathbf{x}_7^1\right\|^2, \tag{25.28}$$

$$\left\|\mathbf{x}_4^2 - \mathbf{x}_6^2\right\|^2 = \left\|t_4\boldsymbol{\theta}_4^3 + \mathbf{x}_3^1 - t_6\boldsymbol{\theta}_6^5 - \mathbf{x}_7^1\right\|^2, \tag{25.29}$$

$$\left\|\mathbf{x}_6^2 - \mathbf{x}_8^2\right\|^2 = \left\|t_6\boldsymbol{\theta}_6^5 + \mathbf{x}_5^1 - t_8\boldsymbol{\theta}_8^7 - \mathbf{x}_7^1\right\|^2, \tag{25.30}$$

$$\left\|\mathbf{x}_4^2 - \mathbf{x}_8^2\right\|^2 = \left\|t_4\boldsymbol{\theta}_4^3 + \mathbf{x}_3^1 - t_8\boldsymbol{\theta}_8^7 - \mathbf{x}_7^1\right\|^2. \tag{25.31}$$

Calculating t_4, t_6, t_8 from (25.26–25.28) in terms of t_2 and substituting them in (25.29–25.31) and squaring the equations twice to get rid of the square roots, we obtain 3 degree of 8 equations in t_2. Using a similar argument as in section IV of [29], it can be shown that such a system of polynomial equations has generically a unique solution. Following the same procedure for t_4, t_6, and t_8 we can find unique values for them as well, hence the problem has only one solution for $n = 4$. The result for $n > 4$ follows immediately.

25.4.3   Noisy Relative Reference Frame Determination

We already mentioned that Problem 4 for $N = 2$ has a unique solution, where there are four or more measurements. Here we consider the case where we have $n \geq 4$ measurements. The following set of equations governs the system for calculating $\mathbf{x}_2^1$ when we have range measurements available:

$$\begin{aligned}
\left\|\mathbf{x}_{2i-1}^1 - \mathbf{x}_{2i}^1\right\|^2 &= d_{2i-1,2i}^2, \quad i \in \{1, \dots, n\} \\
\left\|\mathbf{x}_{2j}^1 - \mathbf{x}_{2i}^1\right\|^2 &= \left\|x_{2j}^2\right\|^2, \quad i, j \in \{1, \dots, n\}.
\end{aligned} \tag{25.32}$$

The solution to (25.32), $\hat{\mathbf{x}} = [\hat{\mathbf{x}}_2^{1T}, \dots, \hat{\mathbf{x}}_{2n}^{1T}]$, is a root of the following polynomial as well:

$$P' = \sum_{i=1}^{n} \left(\|\mathbf{x}_{2i-1}^1 - \mathbf{x}_{2i}^1\|^2 - d_{2i-1,2i}^2 \right)^2 + \sum_{i,j \in \{1,...,n\}} \left(\|\mathbf{x}_{2j}^1 - \mathbf{x}_{2i}^1\|^2 - \|\mathbf{x}_{2j}^2 - \mathbf{x}_{2i}^2\|^2 \right)^2. \quad (25.33)$$

And similarly to before, it is easy to show that this solution is the global minimizer of (25.33) as well. So the solution $\hat{\mathbf{x}}$ is obtained by

$$\hat{\mathbf{x}} = \arg\min P'. \quad (25.34)$$

Again, in the presence of noise, (25.32) does not have a solution, and the best estimate can be obtained by solving (25.34).

To calculate $\mathbf{R}_2^1 = \begin{bmatrix} \cos\phi & \sin\phi \\ -\sin\phi & \cos\phi \end{bmatrix}$, we proceed as follows. Each vector equation of type (25.18) consists of two scalar equations linear in $\cos\phi$ and $\sin\phi$, of the form

$$\Phi_{i1}(\phi) = 0, \Phi_{i2}(\phi) = 0 \quad (25.35)$$

for $i \in \{4, \ldots, 2n\}$. Furthermore, consider the following cost function:

$$J_\phi(\phi) = \sum_{i \in \{4,...,2n\}} \left(\Phi_{i1}^2(\phi) + \Phi_{i2}^2(\phi) \right). \quad (25.36)$$

The best estimate for ϕ, $\hat{\phi}$ is obtained by

$$\hat{\phi} = \arg\min J_\phi. \quad (25.37)$$

Moreover, substituting $\sin\phi$ and $\cos\phi$ with x_s and x_c, we can rewrite (25.37) as

$$[\hat{x}_s, \hat{x}_c]^T = \arg\min J_\phi$$
$$\text{subject to } x_s^2 + x_c^2 = 1, \quad (25.38)$$

where $\sin\hat{\phi} = \hat{x}_s$ and $\cos\hat{\phi} = \hat{x}_c$.

Furthermore, in order to answer Problem 4 having $n \geq 7$ measurements, we construct the following cost function:

$$P_3' = \sum_{i=1}^{n} \left(\|\mathbf{x}_{2i-1}^1 - \mathbf{x}_{2i}^1\|^2 - d_{2i-1,2i}^2 \right)^2 + \sum_{i,j \in \{1,...,n\}} \left(\|\mathbf{x}_{2j}^1 - \mathbf{x}_{2i}^1\|^2 - \|\mathbf{x}_{2j}^2 - \mathbf{x}_{2i}^2\|^2 \right)^2 \quad (25.39)$$

The global minimizer of P_3', $\hat{\mathbf{x}} = [\hat{\mathbf{x}}_2^{1T}, \ldots, \hat{\mathbf{x}}_{2n}^{1T},]$ gives us the solution to the first part of Problem 4 for $N = 2$. To calculate the rotation matrix $\mathbf{R}_2^1$, first we note that this rotation matrix can be written as found in [33]:

$$\begin{bmatrix} s^2 + t^2 - u^2 - v^2 & 2(tu - sv) & 2(tv + su) \\ 2(tu + sv) & s^2 - t^2 + u^2 - v^2 & 2(uv - st) \\ 2(tv - su) & 2(uv + st) & s^2 - t^2 - u^2 + v^2 \end{bmatrix}, \quad (25.40)$$

where

$$s^2 + t^2 + u^2 + v^2 = 1, \quad (25.41)$$

and $s, t, u, v \in \mathbb{R}$. Then each of the equations of the form

$$\mathbf{x}_{2i}^1 = \mathbf{x}_2^1 + \mathbf{R}_2^1 \mathbf{x}_{2i}^2, \quad i \in \{1, \cdots, n\} \tag{25.42}$$

results in three scalar equations $\Phi_{ij} = 0$, $j = 1, 2, 3$ in indeterminants $s, t, u,$ and v. We construct the objective function,

$$J_\Phi(s, t, u, v) = \sum_{i=1}^{n \geq 7} \sum_{j=1}^{3} \Phi_{ij}^2. \tag{25.43}$$

Then, the global minimizer of J_Φ subject to the constraint $s^2 + t^2 + u^2 + v^2 = 1$ is the estimate for $\mathbf{R}_2^1$. One can use the method based on SOS that was introduced earlier to find the global minimizers of J_p and J_Φ subject to $s^2 + t^2 + u^2 + v^2 = 1$.

For pose determination with angle measurements, as before, we construct the following cost function for $n \geq 4$ angle measurements:

$$P'_{AOA} = \sum_{i,j \in \{1,\dots,n\}} \left(\left\| \mathbf{x}_{2j}^1 - \mathbf{x}_{2i}^1 \right\|^2 - \left\| \mathbf{x}_{2j}^2 \right\|^2 \right)^2. \tag{25.44}$$

Or equivalently,

$$P'_{AOA} = \sum_{i,j \in \{1,\dots,n\}} \left(\left\| t_{2i} \theta_{2i}^{2i-1} + \mathbf{x}_{2i-1}^1 - t_{2j} \theta_{2j}^{2j-1} - \mathbf{x}_{2j-1}^1 \right\|^2 - \left\| \mathbf{x}_{2j}^2 \right\|^2 \right)^2. \tag{25.45}$$

For $\hat{\mathbf{x}}_2^1$ we have

$$\hat{\mathbf{x}}_2^1 = \hat{t}_{2i} \theta_{2i}^{2i-1} + \mathbf{x}_1^1. \tag{25.46}$$

To calculate the rotation matrix, we construct and minimize the same cost functions as we used for the range measurement case.

We minimize the polynomial cost functions introduced here and find their global minimizers using SOS relaxation. We conclude this section by presenting some examples for reference-frame determination using noisy distance and bearing measurements.

Example 25.7   Pose Determination with Four Distance Measurements

In the first scenario that we consider, $\mathbf{x}_3^1 = [2.0407, -0.9862]^T$, $\mathbf{x}_5^1 = [3.0233, 1.9054]^T$, $\mathbf{x}_7^1 = [4.0186, 0.9239]^T$, $d_{12} = 7.1626$, $d_{34} = 5.9512$, $d_{56} = 6.6932$, $d_{78} = 5.0052$, $\mathbf{x}_4^2 = [1.0555, -1.9487]^T$, $\mathbf{x}_6^2 = [5.0235, 0.9997]^T$, and $\mathbf{x}_8^2 = [7.0963, -3.1266]^T$. The distance measurements are noisy, and the noise is considered to be a random Gaussian variable with zero mean and variance equal to $0.1 \, \text{m}^2$. Solving the optimization problem corresponding to it we obtain $\hat{\mathbf{x}}_2^1 = [1.2479, 7.0525]^T$, and $\hat{\phi} = 0.0312$ rad. Comparing with the real values, $\mathbf{x}_2^1 = [1, 7]^T$, and $\phi = 0$, we observe that estimates are very close to the real values. See Chapter_25_Example_7.m for more details. MATLAB codes can be found online at ftp://ftp.wiley.com/public/sci_tech_med/matlab_codes.

Example 25.8   The Effect of Different Noise Levels on Pose-Determination Accuracy

In the second scenario, we aim to study the effect of different levels of noise in distance measurements and odometry readings on the solution. First, we set the distance measurement variance equal to $0.1\,\mathrm{m}^2$ and set the noise variance on odometry readings $0.0001\,\mathrm{m}^2$. After repeating the procedure for 100 times, the average of the position estimate error magnitude, $\|\hat{\mathbf{x}}_2^1 - \mathbf{x}_2^1\|$, is $0.2665\,\mathrm{m}$, and the average of the angle estimate error, $|\hat{\phi} - \phi|$, is $0.0311\,\mathrm{rad}$. Then we set the distance measurement variance equal to $0.0001\,\mathrm{m}^2$ and set the noise on odometry readings $0.1\,\mathrm{m}^2$. After repeating the procedure for 100 times, the average of the position estimate error magnitude, $\|\hat{\mathbf{x}}_2^1 - \mathbf{x}_2^1\|$, is $0.4078\,\mathrm{m}$, and the average of the angle estimate error, $|\hat{\phi} - \phi|$, is $0.0385\,\mathrm{rad}$. While the average errors of the angle estimate in the two cases are close, the magnitude of the error of the position estimate is somewhat larger in the second scenario with larger odometry error, suggesting it to be more problematic. A reason for this phenomenon might be the larger number of odometry measurements used in the construction of the cost functions compared to the number of distance measurements. The code to test this scenario is very similar to the one presented in the MATLAB codes that can be found online at ftp://ftp.wiley.com/public/sci_tech_med/matlab_codes.

Example 25.9   Pose Determination with Four Bearing Measurements

In this scenario we test frame determination using four angle measurements. We consider $\mathbf{x}_3^1 = [2.3, 5]^T$, $\mathbf{x}_5^1 = [4, 2]^T$, $\mathbf{x}_7^1 = [9\ 0]^T$, $\boldsymbol{\theta}_2^1 = [0.9806, 0.1961]^T$, $\boldsymbol{\theta}_4^3 = [0.2873, 0.9578]^T$, $\boldsymbol{\theta}_6^5 = [0.1961, 0.9806]^T$, $\boldsymbol{\theta}_8^7 = [0.9487, -0.3162]^T$, $\mathbf{x}_4^2 = [-3, 3]^T$, $\mathbf{x}_6^2 = [0, 6]^T$, and $\mathbf{x}_8^2 = [-2, 1]^T$. The angle measurements are noisy, and the noise is considered to be a random Gaussian variable with zero mean and variance equal to $0.01\,\mathrm{rad}^2$. Solving the optimization problem corresponding to it we obtain\ $\hat{\mathbf{x}}_2^1 = [5.0585, 1.0747]^T$, and $\hat{\phi} = 0.0254\,\mathrm{rad}$. Comparing with the real values, $\mathbf{x}_2^1 = [5, 1]^T$, and $\phi = 0$, we observe that estimates are very close to the real values. See Chapter_25_Example_9.m for more details. MATLAB codes can be found online at ftp://ftp.wiley.com/public/sci_tech_med/matlab_codes.

25.4.4   Algorithmic Comparison with Some Existing Methods

The formulated localization problem, as in [4], with N_a anchors will now be described. Let ε_i be the error in the estimated squared distances between sensor 0 and anchor i. We want to minimize the sum of squared errors:

$$J_{CM} = \varepsilon_1^2 + \varepsilon_2^2 + \cdots + \varepsilon_n^2, \tag{25.47}$$

subject to $N_a - 2$ equality constraints

$$f_i(\varepsilon_1, \varepsilon_2, \varepsilon_i) = 0, \quad i = 3, 4, \ldots, n, \tag{25.48}$$

where f_i are obtained from writing different Cayley–Menger determinants [4] for different sets of anchor nodes. Using the Lagrange multiplier method, this minimization is equivalent to minimizing

$$H(\varepsilon_1, \ldots, \varepsilon_n, \lambda_1, \ldots, \lambda_{n-2}) = \sum_{i=1}^{n} \varepsilon_i^2 + \sum_{i=1}^{n-2} \lambda_i f_{i+2}(\varepsilon_1, \varepsilon_2, \varepsilon_{i+2}). \qquad (25.49)$$

One way of solving the above minimization problem is to differentiate (25.49) with respect to variables ε_i and λ_i, and to find the zeros of these differentials. In this process we have $2n - 2$ equations, each with degree 2, that need to be solved using a nonlinear root finding method (see Example 25.4). Hence, since in our proposed method adding anchors has no effect on the number of equations and variables, we always have a system of polynomial equations comprising two equations and two unknowns.

For three anchors, the results obtained using the method introduced here and the one in [4] are the same and are clearly better than the result obtained from linear algorithm of [3].

The result obtained by applying the SR-LS method from [7] in Example 25.5 and the method introduced here give the same numerical result. However, in solving SR-LS, positive definiteness of a certain matrix is assumed while here no assumption is required.

For more than three anchors, the method introduced in [4] does not result in a solution using the `fsolve` routine in MATLAB, which uses a quasi-Newton algorithm to solve a set of nonlinear equations. The method introduced in this paper, however, yields a result that is more accurate compared with the one obtained by the linear method in [3].

In the case of localization using range difference measurements, the biggest difference between the method introduced here and the one in [7] is the number of steps to reach the final result. In this paper we only need to solve one minimization problem, while to find the result using the technique from [7], one needs to go through a four-stage algorithm.

Comments on the Complexity of SOS Methods: With the addition of more unknowns (hence introduction of more variables and consequently higher dimension LMIs) the complexity of the SOS-based solution increases rapidly, but it still remains reasonable for up to 10 agents (20 variables). However, if one is interested in solving problems with a large number of variables, one should take advantage of the sparsity in the cost function to reduce the size of the underlying LMIs (see [12]). For more information on the size of LMIs to be solved, the reader may refer to [34].

To reduce the numerical sensitivity of the methods, one may want to decrease the condition number of the matrices involved in solving the optimization problems; one way to do this is to normalize the coefficients of the cost functions. The choice of the polynomial δ in (25.11) does not follow any rigid guidelines, but as a rule of thumb, the higher the order one chooses for δ, the better the bound one obtains on the optimal solution but at the expense of a considerable increase of the problem size (i.e., increase in matrix dimensions).

25.4.5 Colinear Anchors

In this section we discuss the situation (in $\mathbb{R}^2$) where the anchors are colinear. Consider N_a colinear anchors labeled $1, 2$ to N_a at positions $\mathbf{a}_i \in \mathbb{R}^2$, where $i \in \{1, \ldots, N_a\}$. Furthermore, sensor node 0 can measure its distance d_i from the anchor i. The geometric representation of a problem with five colinear anchors in $\mathbb{R}^2$ is depicted in Figure 25.4. In the case where the distances are exact, there are two points of intersection for the circles; however, in the presence of noise, the circles will not necessarily intersect at any common intersection point. Defining $J(p)$ as before and solving the related optimization problem, we can compute an estimate for the position of node 0, and since the anchors are colinear, the other possible position for node 0 is the mirror of this point where the line connecting the anchors is the mirror axis.

Using the method here, one can have estimates for the two possible positions of node 0.

Example 25.10   Localization with Colinear Anchors

Consider the problem of localization as depicted in Figure 25.4. The anchors are placed at $\mathbf{a}_1 = [10, 31]^T$, $\mathbf{a}_2 = [17, 45]^T$, $\mathbf{a}_3 = [-17, -23]^T$, $\mathbf{a}_4 = [-10, -9]^T$, and $\mathbf{a}_5 = [19, 49]^T$. Again, it is considered that distance measurements are being taken by sensor node 0 in the presence of a zero-mean Gaussian noise with variance of $1\,\text{m}^2$. The measured distances from the sensor to the anchors are $d_1 = 32.8674$, $d_2 = 46.7679$, $d_3 = 29.3150$, $d_4 = 15.0772$, and $d_5 = 51.8630$. The actual position of sensor node 0 is at $[0, 0]^T$. The estimates of the node 0 position (the ones closer to the actual position) using 3, 4, and 5 distance measurements are presented in Table 25.3. The code to generate this scenario is similar to that of Example 25.4.

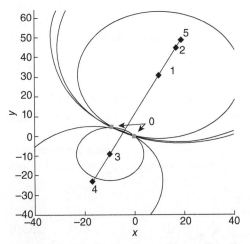

Figure 25.4   The localization problem with colinear anchors [13].

TABLE 25.3. The Error between the Estimated Position of Node 0 and Its Actual Position

No. Anchors	3	4	5
$\hat{\mathbf{x}}$	$\begin{bmatrix} -3.8399 \\ 2.8925 \end{bmatrix}$	$\begin{bmatrix} -4.0187 \\ 3.0142 \end{bmatrix}$	$\begin{bmatrix} -3.8906 \\ 2.8935 \end{bmatrix}$

25.5 AN EXTENSION OF THE SOS APPROACH

There are situations where the equivalence stated in Remark 2 does not necessarily hold, for example, in the three dimensional case, which means the solution obtained by SOS relaxation is not necessarily the global optimum. However, according to [25], one can construct a finite sequence of values that converges to the global minimum of a polynomial function if the gradient ideal of that polynomial function is radical. Next, we establish that this condition holds for almost all the polynomials of degree d in $\mathbb{R}[z]$.

Lemma 2: For almost all polynomials f of degree d in the ring $\mathbb{R}[z]$, the gradient ideal $I_\nabla(f)$ is radical and the gradient variety $V_\nabla(f)$ is a finite subset of $\mathbb{C}^n$.

Before proceeding to prove the lemma, we present the following theorem.

Theorem 1 [35]: Let I be a zero-dimensional ideal in $\mathbb{C}[z]$, and let $A = \mathbb{C}[z]/I$. Then $\dim_{\mathbb{C}}(A)$ is greater than or equal to the number of points in $V(I)$. Moreover, equality occurs if and only if I is a radical ideal.

This means that in the case where I is not radical, there are multiplicities at each point in $V(I)$ so that the sum of the multiplicities is equal to $\dim_{\mathbb{C}}(A)$.

Proof of Lemma 2: Consider the vector space S of all polynomials f in n variables of degree d. By Theorem 1, the condition for a polynomial f to have either infinitely many critical points or to have a critical point with multiplicity ≥ 2, or equivalently have a nonradical gradient ideal, is a closed condition. In order to prove the lemma, we simply need to show that the complementary open set in S is nonempty. For that purpose it suffices to show that there exists one polynomial that has a finite number of critical points, each with multiplicity 1. One such polynomial is

$$f(z) = z_1^d + z_2^d + \cdots + z_n^d - dz_1 - dz_2 - \cdots - dz_n.$$

It has a finite number of critical points, and its gradient ideal $I_\nabla(f) = \langle z_1^{d-1} - 1, z_2^{d-1} - 1, \dots, z_n^{d-1} - 1 \rangle$ is radical because it has $n(d-1)$ distinct complex roots. Thus, generically, the gradient ideal of any f is radical.

Having established Lemma 2, we present the following optimization problem formulation that [25] used to design an algorithm to construct a sequence of values converging to the global minimum:

$$\text{maximize} \quad \gamma$$

$$\text{subject to} \quad F(z) - \gamma - \sum_{j=1}^{n} \phi_j(z) \frac{\partial F}{\partial z_j} \text{ is SOS,} \tag{25.50}$$

where $\phi_j(z) \in \mathbb{R}[z]$ is a polynomial of maximum degree $2N - d + 1$. Call $F_{\nabla,N}^*$ the optimal value obtained by solving (25.50). $F_{\nabla,N}^*$ is a lower bound for F^*, which gets tighter as N increases. Furthermore [25] shows that if I_∇ is radical, there exists an integer N such that $F_{\nabla,N}^* = F^*$.

However, while for generic cost functions one can solve the optimization problem with the abovementioned method, the situations that we introduced throughout this chapter cannot be considered as generic, since the coefficients of the polynomial cost functions are related to each other through algebraic relations, which are the direct consequence of the measurements. To clarify, consider a polynomial cost function constructed from $\bar{m}$ measurements $\mu_1, \dots \mu_m$ with coefficients $c_1, \dots c_n$. These coefficients can be written as the functions of the measurements, that is, $c_i = f_i(\mu_1, \dots \mu_m)$ for some function $f_i(\cdot)$. This in turn results in the existence of an algebraic map g_i such that $c_i = g_i(\{c_j\}_{j=1, j \neq i}^{\bar{n}})$. Nonetheless, this does not eliminate the possibility of the ideals of these cost functions being radical. Hence, the problem of showing that the ideals of the cost functions introduced here are radical and the applicability of the method introduced in this section to minimize these cost functions remains as an open problem.

25.6 CONCLUSIONS

In this chapter we have briefly introduced the idea of localization and pose determination in the presence of noisy measurements using polynomial optimization. We have introduced two methods for optimizing polynomials and particularly implemented SOS relaxations to solve some common problems in localization and pose determination. Some examples were provided to show the applicability of the methods. In the end, we provided a condition for having exact global optimums for polynomial functions using SOS relaxation and established that, for generic polynomials, this condition is satisfied.

ACKNOWLEDGMENTS

This work is supported by NICTA, which is funded by the Australian government as represented by the Department of Broadband, Communications and the Digital Economy, and the Australian Research Council through the Information and Communication Technology Centre of Excellence program. Furthermore, the authors want to thank Prof. Bernd Sturmfels, Dr. James Borger, and Dr. Özgür Kişisel for their helpful comments on proving Lemma 2 and understanding the algebraic geometric aspects of the problems.

REFERENCES

[1] G. Mao, B. Fidan, and B. D. O. Anderson, "Wireless sensor network localization techniques," *Comput. Netw.*, vol. 51, no. 10, pp. 2529–2553, 2007.

[2] G. Mao, B. Fidan, and B. D. O. Anderson, "Localization," in *Sensor Networks and Configuration: Fundamentals, Techniques, Platforms and Experiments*, N. P. Mahalik, Ed. New York: Springer, 2006, ch. 13, pp. 281–315.

[3] A. H. Sayed and A. Tarighat, "Network-based wireless location," *IEEE Signal Process. Mag.*, vol. 22, no. 4, pp. 24–40, Jul. 2005.

[4] M. Cao, B. D. O. Anderson, and A. S. Morse, "Sensor network localization with imprecise distances," *Syst. Control Lett.*, vol. 55, pp. 887–893, 2006.

[5] F. Thomas and L. Ros, "Revisiting trilateration for robot localization," *IEEE Trans. Rob.*, vol. 21, no. 1, pp. 93–101, 2005.

[6] L. Doherty, K. S. J. Piste, and L. El Ghaoui, "Convex position estimation in wireless sensor networks," *IEEE INFOCOM*, vol. 3, pp. 1655–1663, 2001.

[7] A. Beck, P. Stoica, and J. Li, "Exact and approximate solutions of source localization problems," *IEEE Trans. Signal Process.*, vol. 56, pp. 1770–1778, 2008.

[8] M. W. Carter, H. H. Jin, M. A. Saunders, and Y. Ye, "An adaptive subproblem algorithm for scalable wireless sensor network localization," *SIAM J. Optim.*, vol. 17, pp. 1102–1128, 2006.

[9] P. Biswas, T.-C. Lian, T.-C. Wang, and Y. Ye, "Semidefinite programming based algorithms for sensor network localization," *ACM Trans. Sens. Netw.*, vol. 2, no. 2, pp. 188–220, 2006.

[10] P. Biswas and Y. Ye, "Semidefinite programming for ad hoc wireless sensor network localization," in *ACM/IEEE Conference on Information Processing in Sensor Networks, 2004*, Apr. 2004, pp. 46–54.

[11] Y. Ding, N. Krislock, J. Qian, and H. Wolkowicz, "Sensor network localization, Euclidean distance matrix completions, and graph realization," *Optim. Eng.*, vol. 11, pp. 45–66, 2008.

[12] J. Nie, "Sum of squares methods for sensor network localization," *Comput. Optim. Appl.*, vol. 43, pp. 151–179, 2009.

[13] I. Shames, P. T. Bibalan, B. Fidan, and B. D. O. Anderson, "Polynomial methods in noisy network localization," in *IEEE 17th Mediterranean Conf. on Control and Automation*, 2009, pp. 1307–1312.

[14] Y. Shang, W. Ruml, Y. Zhang, and M. P. J. Fromherz, "Localization from mere connectivity," in *Proc. of the 4th ACM Int. Symp. on Mobile Ad Hoc Networking and Computing*, 2003, pp. 201–212.

[15] S. Lederer, Y. Wang, and J. Gao, "Connectivity-based localization of largescale sensor networks with complex shape," *ACM Trans. Sens. Netw.*, vol. 5, no. 4, pp. 1–32, 2009.

[16] J. A. Costa, N. Patwari, and A. O. Hero III, "Distributed weighted-multidimensional scaling for node localization in sensor networks," *ACM Trans. Sens. Netw. (TOSN)*, vol. 2, no. 1, pp. 39–64, 2006.

[17] J. Nie and J. Demmel, "Sparse SOS relaxations for minimizing functions that are summations of small polynomials," *SIAM J. Optim.*, vol. 19, no. 4, pp. 1534–1558, 2006.

[18] R. R. Brooks, P. Ramanathan, and A. M. Sayeed, "Distributed target classification and tracking in sensor networks," *Proc. IEEE*, vol. 91, no. 8, pp. 1163–1171, 2003.

[19] A. N. Bishop and P. N. Pathirana, "Sensor fusion based localization of a mobile user in a wireless network," in *TENCON 2005*, pp. 1–6.

[20] D. A. Cox, J. B. Little, and D. O'Shea, *Ideals, Varieties, and Algorithms: An Introduction to Computational Algebraic Geometry and Commutative Algebra*. New York: Springer, 1997.

[21] T. Y. Li, "Numerical solution of multivariate polynomial systems by homotopy continuation methods," *Acta Numerica*, vol. 6, pp. 399–436, 2008.

[22] D. Han, "Global optimization with polynomials," in *High Performance Computation for Engineered Systems, HPCES 2004*.

[23] P. Parrilo, "Semidefinite programming relaxations for semi-algebraic problems," *Math. Program.*, vol. 96, no. 2, pp. 293–320, 2003.

[24] D. Henrion and J. Lasserre, "Detecting global optimality and extracting solutions in gloptipoly," in *Positive Polynomials in Control*, Lecture Notes on Control and Information Sciences, D. Henrion and A. Garulli, Eds. New York: Springer, 2005.

[25] J. Nie, J. Demmel, and B. Sturmfels, "Minimizing polynomials via sum of squares over the gradient ideal," *Math. Program. A*, vol. 106, pp. 587–606, 2005.

[26] D. Henriona, J. B. Lasserrea, and J. Löfbergd, "Gloptipoly 3: Moments, optimization and semidefinite programming," *Optim. Methods Softw.*, vol. 24, no. 4, pp. 761–779, 2009.

[27] S. Prajna, A. Papachristodoulou, and P. A. Parrilo, "Introducing SOSTOOLS: A general purpose sum of squares programming solver," in *IEEE Conf. on Decision and Control*, Las Vegas, NV, 2002, pp. 741–746.

[28] P. Stoica and J. Li, "Source localization from range-difference measurements," *IEEE Signal Process. Mag.*, vol. 23, pp. 63–65, 69, 2006.

[29] X. S. Zhou and S. I. Roumeliotis, "Robot-to-robot relative pose estimation from range measurements," *IEEE Trans. Robot.*, vol. 24, no. 6, pp. 1379–1393, 2008.

[30] N. Trawny, X. S. Zhou, K. Zhou, and S. I. Roumeliotis, "D relative pose estimation from distance-only measurements," in *IEEE/RSJ Int. Conf. on Intelligent Robots and Systems (IROS'07)*, 2007, pp. 1071–1078.

[31] C. Yu, B. Fidan, and B. D. O. Anderson, "Principles to control autonomous formation merging," in *American Control Conf.*, Minneapolis, MN, Jun. 2006, pp. 762–768.

[32] C. B. Garcia and T. Y. Li, "On the number of solutions to polynomial systems of equations," *SIAM J. Numer. Anal.*, vol. 17, no. 4, pp. 540–546, 1980.

[33] B. K. P. Horn, "Closed-form solution of absolute orientation using unit quaternions," *J. Opt. Soc. Am.*, vol. 4, pp. 629–642, 1987.

[34] P. A. Parrilo and B. Sturmfels, "Minimizing polynomial functions," *DIMACS Ser. Discrete Math. Theor. Comput. Sci.*, vol. 60, pp. 83–99, 2003.

[35] D. A. Cox, J. B. Little, and D. O'Shea, *Using Algebraic Geometry*. New York: Springer, 1998.

CHAPTER 26

BELIEF PROPAGATION TECHNIQUES FOR COOPERATIVE LOCALIZATION IN WIRELESS SENSOR NETWORKS

Vladimir Savic[1,2] and Santiago Zazo[1]

[1]*Signal Processing Applications Group, Technical University of Madrid, Spain*
[2]*Dept. of Computer Science and Engineering, Chalmers University of Technology, Sweden*

A NUMBER OF applications in wireless sensor networks (WSN) require sensor nodes to obtain their absolute or relative positions. Equipping every sensor with a global positioning system (GPS) receiver may be expensive, energy prohibitive, and limited to outdoor applications. Therefore, we consider *cooperative* localization, where each sensor node with unknown location obtains its location by cooperating with neighboring sensor nodes. In this chapter, we apply probabilistic inference to the problem of cooperative localization. These techniques are able to obtain not only location estimates but also a measure of the uncertainty of those estimates. Since these methods are computationally very expensive, we need to use message-passing methods, which are also known as *belief propagation* (BP) methods. BP is a way of organizing the global computation of marginal beliefs in terms of smaller local computations within the graph. It is one of the best-known probabilistic methods for distributed inference in statistical physics, artificial intelligence, computer vision, error-correcting codes, localization, etc. The whole computation takes a time proportional to the number of links in the graph, which is significantly less than the

Handbook of Position Location: Theory, Practice, and Advances, Second Edition.
Edited by S. A. (Reza) Zekavat and R. Michael Buehrer.
© 2019 by the Institute of Electrical and Electronics Engineers, Inc.
Published 2019 by John Wiley & Sons, Inc.
Companion Website: www.wiley.com/go/zekavat/positionlocation2e

exponentially large time that would be required to compute marginal probabilities naively. However, due to the presence of nonlinear relationships and highly non-Gaussian uncertainties, the standard BP algorithm is undesirable. Nevertheless, a particle-based approximation via *nonparametric belief propagation* (NBP) makes BP acceptable for localization in sensor networks. In this chapter, after an introduction to cooperative localization, we describe BP/NBP techniques and its generalizations (GBP) for the loopy networks. Due to the poor performance of BP/NBP methods in loopy networks, we describe three improved methods: GBP based on Kikuchi approximation (GBP-K), nonparametric GBP based on junction tree (NGBP-JT), NBP based on spanning trees (NBP-ST), and uniformly-reweighted NBP (URW-NBP).

26.1 INTRODUCTION TO COOPERATIVE LOCALIZATION IN WSN

Sensor localization, i.e., obtaining estimates of each sensor's position as well as accurately representing the uncertainty of that estimate, is an important step for the effective application of sensor networks to almost all tasks. Equipping every sensor with a GPS receiver (or equivalent technology) may be expensive, energy prohibitive, and limited to outdoor applications (for more details on GPSs, see Chapter 20). Therefore, we consider the problem in which some small number of sensors, called *anchor nodes*, obtain their coordinates via GPS or by installing them at points with known coordinates, and the rest, *unknown nodes*, must determine their own coordinates using the anchor nodes and measured intersensor distances. If reference sensors were capable of high-power transmission, they would be able to make measurements with all anchor nodes. This represents *single-hop* localization. However, we prefer to use energy-conserving devices without the energy necessary for long-range communication. In this case, each sensor only has available noisy measurements of its distance to several neighboring sensors (not necessarily anchor nodes). In other words, we still allow unknown nodes to make measurements with anchor nodes (if possible), but now we additionally allow unknown nodes to make measurements with other unknown nodes. It is still necessary that there is minimum of three (for two dimensional [2-D]) or four (for three dimensional [3-D]) anchor nodes in the network but not necessarily directly connected to all unknown nodes. This technique is known as *multihop* (or *cooperative*) localization (see Fig. 26.1), which is the main topic of this chapter (and also Chapter 24). In the following subsections, we classify these methods, describe typical measurement techniques, and present some motivating applications.

26.1.1 Classification of Cooperative Localization Methods

Range-Based versus Range-Free Methods: *Range-free* or *connectivity-based* localization methods [1–3, 6, 11] rely on connectivity between the nodes. The principle of these algorithms is to determine whether or not a sensor is in the transmission

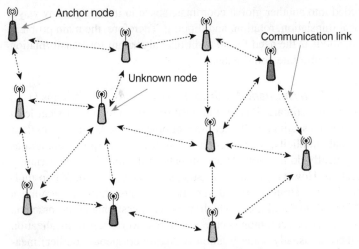

Figure 26.1 Illustration of cooperative localization in sensor networks.

range of another sensor. The most attractive feature of the range-free algorithms is their simplicity. However, they can only provide a coarse grained estimate of each node's location, which means that they are only suitable for applications requiring an approximate location estimate. *Range-based* or *distance-based* localization algorithms [1, 4, 5, 7, 10] use the intersensor distance measurements in a sensor network to locate the entire network. This type of algorithms is usually more accurate, but sensitive to measurement errors.

Centralized versus Distributed Methods: Based on the approach of processing the individual intersensor data, localization algorithms can be also considered in two main classes: *centralized* and *distributed* algorithms. Centralized algorithms [2, 4] utilize a single central processor (i.e., fusion center) to collect all the individual intersensor data and produce a map of the entire sensor network, while distributed algorithms [1, 3–7, 9–11] rely on self-localization of each node in the sensor network using the local information it collects from its neighbors. From the perspective of location estimation accuracy, centralized algorithms are likely to provide more accurate location estimates than distributed algorithms. However, centralized algorithms suffer from the scalability problem and generally are not feasible to be implemented for large scale WSN.

Anchor-Based versus Anchor-Free Methods: *Anchor-based* [1, 4, 6, 10, 11] methods assume that a certain minimum number of nodes know their position, e.g., by manual placement or using some other location mechanism such as GPS. This localization method has the limitation that it needs another method to bootstrap the anchor node positions, and cannot be easily applied to any context in which another location system is unavailable. In contrast, *anchor-free* [2, 3, 5, 7] algorithms use local distance information to attempt to determine node coordinates when no nodes have predefined positions. Of course, any such coordinate system will not be unique and can be embedded into another global coordinate space in infinitely many ways, depending on global translation, rotation, and flipping. Therefore, the main problem with anchor-free methods is the need for an additional algorithm for transformation from the relative to the absolute coordinates.

Probabilistic versus Deterministic Methods: *Deterministic* algorithms [1–6] use the measurements to estimate directly the positions by applying classical least squares, multidimensional scaling, multilateration, or other methods. In favor of their relative computational simplicity, they often lack a statistical interpretation, and as one consequence typically do not provide an estimate of the remaining uncertainty in each sensor location. However, iterative least-squares methods, like N-hop multilateration [4], do have a straightforward statistical interpretation but assume a Gaussian model for all uncertainties, which may be questionable in practice. Non-Gaussian uncertainty is a common occurrence in real-world sensor localization problems, where there is usually some fraction of highly erroneous (outlier) measurements. On the other hand, *probabilistic* (typically, Bayesian) methods [7–11] take into account uncertainty of the measurements, so given the likelihood of, e.g., measured distance and a prior probability density function (PDF) of the positions of all unknown nodes, they estimate the posterior PDF of the positions of all unknown nodes. However, the main drawback of the probabilistic methods is the high computational and communication cost which, in some applications, makes these methods unacceptable in low-power WSN. Nevertheless, the particle-based approximation via nonparametric representation, and an appropriate factorization of the PDFs, makes probabilistic methods acceptable for inference in sensor networks. In addition, nonparametric representation enables us to estimate any PDF that does not exist in analytical (parametric) form.

Model-Based versus Fingerprinting Methods: *Model-based* algorithms use a physical model (such as received signal strength (RSS) log-distance path-loss) to estimate the range from the measurements and then apply optimization to locate the target. These methods are usually very general but not very accurate due to the irregular signal propagation and the interference. On the other hand, *fingerprinting* methods use an off-line training to learn the wireless channel at predefined locations and store it in memory. Then, the target location is estimated in an online phase by matching the measured fingerprint with the fingerprints stored in the database. They can provide much higher accuracy than model-based methods since they can capture a complex wireless channel, including multipath and interference. However, they are less robust to environmental changes and the calibration cost may be too high. A recent trend is to use kernel-based machine learning methods [14, 15] for this problem since they are able to model arbitrary nonlinear relationships.

26.1.2   Measurement Techniques

Measurement techniques [14, 15] in WSN localization can be broadly classified into two categories: distance-related techniques (time of arrival [TOA], time difference of arrival [TDOA] and RSS), and angle of arrival (AOA) techniques.

RSS techniques (see also Chapters 11 and 12) are based on a standard feature found in most wireless devices, a received signal strength indicator (RSSI). They are attractive because they require no additional hardware and are unlikely to significantly impact local power consumption, sensor size, and thus cost. It is accepted on the basis of empirical evidence [16] that RSS ($P_r(d)$ [mW]) can be modeled as a random and log-normally distributed random variable with a distance dependent mean value. Thus, converted to dBm, it is given by:

$$P_r(d)[dBm] = P_0(d_0)[dBm] - 10n_p \log_{10}\left(\frac{d}{d_0}\right) + X_\sigma, \qquad (26.1)$$

where $P_0(d_0)$ is a known reference power value in dBm at a reference distance from the transmitter; n_p is the path loss exponent that measures the rate at which the RSS decreases with distance, typically between two and four depending on the specific propagation environment and X_σ is a zero-mean Gaussian distributed random variable with standard deviation σ, which accounts for the random effects of shadowing. In addition to the path loss, measured RSS is also a function of the calibration of both the transmitter and receiver. Depending on the quality of the manufacturing process, RSSI circuits and transmit powers will vary from device to device. Moreover, transmit powers can change as batteries deplete. Therefore, RSS measurements provide very coarse estimates.

Distances between neighboring sensors can be estimated from the TOA measurements between transmitter and receiver, using two types of these measurements: one way and round trip. *One-way* propagation time measurements measure the difference between the sending time of a signal at the transmitter and the receiving time of the signal at the receiver. This requires the local time at the transmitter and the local time at the receiver to be accurately synchronized. This requirement may add to the cost of sensors by demanding a highly accurate clock or increase the complexity of the sensor network by demanding a sophisticated synchronization mechanism. *Round-trip* propagation time measurements measure the difference between the time when a signal is sent by a sensor and the time when the signal returned by a second sensor is received at the original sensor. Since the same clock is used to compute the round-trip propagation time, there is no synchronization problem. The major error source in round-trip propagation time measurements is the delay required for handling the signal in the second sensor. This internal delay can be removed via a *priori* calibration. Taking time differences of TOA measurements (TDOA) eliminates the clock bias nuisance parameter. TDOA is done between one transmitter and a number of receivers. This measurement technique is able to achieve better accuracy than RSS and TOA. However, the accuracy is achieved at the expense of higher equipment cost. TOA-based localization techniques are described in detail in Chapter 6.

A recent trend in propagation time measurements is the use of ultra-wideband (UWB) signals for accurate distance estimation [17]. A UWB signal is either a signal whose fractional bandwidth (the ratio of its bandwidth to its center frequency) is larger than 0.2 or a signal with a total bandwidth of more than 500 MHz. UWB can achieve higher accuracy because its bandwidth is very large, and therefore its pulse has a very short duration. An overview of UWB localization and ranging techniques can be found in Chapter 8.

By providing information about the direction to neighboring sensors rather than the distance to neighboring sensors, AOA measurements provide localization information complementary to the TOA and RSS measurements discussed above. AOA can be divided into two subclasses: those making use of the receiver antenna's amplitude response and those making use of the receiver antenna's phase response. The accuracy of AOA measurements is limited by the directivity of the antenna, by shadowing, and by multipath reflections. AOA measurements rely on a direct line-of-sight (LOS) path from the transmitter to the receiver. However, a multipath component may appear as a signal arriving from an entirely different direction and can lead to very large errors. The solution for this problem requires a complex multidimensional search, and is thus computationally very intensive. A description of the fundamentals of these techniques is available in Chapter 9.

26.1.3 Motivating Applications

WSN localization has attracted significant research interest. We review here a few interesting applications. In environmental monitoring applications such as bush fire surveillance, precision agriculture, and water quality monitoring, the measurement data are meaningless without knowing the location from where the data are obtained. For the purposes of biological research, it is very useful to track animals over time. Using multihop routing of the data through the network enables low transmit powers from the animal tags. Furthermore, interanimal distances, which are of particular research interest, can be estimated using pairwise measurements and cooperative localization methods (without resorting to GPS). The main result of the longer battery lifetimes is less frequent recollaring of the animals. As another example, we can consider deploying a sensor network in a manufacturing floor. The monitoring and control of equipment has traditionally been wired, but making them wireless reduces the high cost of cabling and makes the manufacturing floor more dynamic. In addition, these sensors monitor storage conditions (temperature and humidity) and help control the heating, ventilation, and air conditioning system. Sensors on mobile equipment report their location when the equipment is lost or needs to be found, and can even contact security if the equipment is about to leave the building. Moreover, location estimation may enable many applications, such as inventory management, intrusion detection, road traffic monitoring, health monitoring, reconnaissance, surveillance, etc. A number of interesting applications of sensor networks can be found in [14] and [18].

26.2 PROBABILISTIC LOCALIZATION BASED ON BP

In this section, after a detailed introduction to probabilistic localization, we illustrate the well-known method for cooperative localization, belief propagation (BP). Then, we describe particle-based (nonparametric) approximation of BP, NBP. Finally, we discuss a new variant of NBP method, nonparametric boxed belief propagation (NBBP) [24]. Deterministic localization techniques are not part of this chapter since they are already described in detail in Chapter 24.

26.2.1 Introduction to Probabilistic Localization

In contrast to deterministic methods [1–6], *probabilistic* methods [7–11] take into account uncertainty of measurements, so given the likelihood of for example, measured distance, and prior PDF of the positions of all unknown nodes, they estimate the posterior PDFs of the positions of the all unknown nodes. In order to find location estimates, it is sufficient to find either the mean value, or maximum a posteriori (MAP) of this posterior PDF. These methods are also well-known as *Bayesian* methods. The main drawback of the probabilistic methods is the high complexity, which makes these methods unacceptable in low-power sensor networks. Nevertheless, the particle-based approximation [19, 23], and appropriate factorization using some message-passing method [22], makes probabilistic methods acceptable for localization in sensor networks.

Statistical Framework for Probabilistic Localization: Let us assume that we have N_s sensors (N_a anchors and N_u unknowns) scattered randomly in a planar region and denote the 2-D location of sensor t by x_t. The unknown node u obtains a noisy measurement d_{tu} of its distance from node t with some probability $P_d(x_t, x_u)$:

$$d_{tu} = \|x_t - x_u\| + v_{tu}, \quad v_{tu} \sim p_v(d_{tu} - \|x_t - x_u\| \mid \Theta_{tu}), \qquad (26.2)$$

where, for noise v_{tu}, we assume a Gaussian distribution p_v (with parameter $\Theta_{tu} = \{\mu_{tu}, \sigma_{tu}^2\}$). However, it is straightforward to change it to any desired distribution (e.g., obtained by running training experiments in the deployment area). The binary variable o_{tu} will indicate whether this observation is available or not:

$$o_{tu} = \begin{cases} 1, & d_{tu} \ observed, \\ 0, & otherwise. \end{cases} \qquad (26.3)$$

Finally, each sensor t has some prior distribution denoted $p_t(x_t)$. This prior could be an uninformative one (i.e., with uniform distribution in whole deployment area) for the unknowns and the Dirac delta function for the anchors. Then, the joint distribution is given by:

$$p(x_1, \dots, x_{N_u}, \{o_{tu}\}, \{d_{tu}\}) = \prod_{(t,u)} p(o_{tu} \mid x_t, x_u) \prod_{(t,u)} p(d_{tu} \mid x_t, x_u) \prod_t p_t(x_t). \qquad (26.4)$$

We also need to define probability of detection, from which we can draw the variable o_{tu}. For large-scale sensor networks, it is reasonable to assume that only a subset of pairwise distances will be available, primarily between sensors that are located within some radius R. The ideal model of probability of detection is given by:

$$P_d(x_t, x_u) = \begin{cases} 1, & for \ \|x_t - x_u\| \le R, \\ 0, & otherwise. \end{cases} \qquad (26.5)$$

Better approximations of $P_d(x_t, x_u)$ can be obtained using real experiments in the deployment area of interest, and this is especially advisable for indoor scenarios.

However, we can also use the exponential model [7], which represents a better approximation of real-world systems:

$$P_d(x_t, x_u) = \exp\left(-\frac{1}{2}\|x_t - x_u\|^2 / R^2\right). \tag{26.6}$$

Our goal is to compute the *posterior marginal distribution* $p(x_t, | \{o_{tu}\}, \{d_{tu}\})$ (for each unknown node t) by marginalizing the joint posterior PDF, which is not tractable for the localization problem. Practically, this marginalization would have exponential complexity (with respect to the number of unknown nodes). Therefore, we need to factorize the joint posterior PDF using some message-passing method.

26.2.2 BP

BP [22] is a way of organizing the "global" computation of marginal beliefs in terms of smaller local computations within the graph. It is one of the best-known probabilistic methods for distributed inference in statistical physics, artificial intelligence, computer vision, localization, etc.

In the standard BP algorithm, the belief at a node t (posterior marginal) is proportional to the product of the local evidence at that node $\psi_t(x_t)$, and all the messages coming into node t:

$$M_t(x_t) = k\psi_t(x_t)\prod_{u \in G_t} m_{ut}(x_t), \tag{26.7}$$

where x_t is a state of node t, k is a normalization constant, and G_t denotes the neighbors of node t. The messages are determined by the message update rule:

$$m_{ut}(x_t) = \int_{x_u} \psi_u(x_u)\psi_{tu}(x_t, x_u) \prod_{g \in G_u \setminus t} m_{gu}(x_u)dx_u, \tag{26.8}$$

where $\psi_{tu}(x_t, x_u)$ is the pairwise potential between nodes t and u. On the right-hand side, there is a product over all messages going into node u except for the one coming from node t. In other words, the message from node u to node t represents the "opinion" of node u about location of node t. The messages and beliefs are, of course, represented as PDF, but not necessarily normalized.

In practical computation, one starts with nodes at the edge of the graph and only computes a message when one has available all the messages required. It is easy to see [20] that each message needs to be computed only once for the graphs without loops. That means that the whole computation takes a time proportional to the number of links in the graph, which is significantly less than the exponentially large time that would be required to compute marginal posteriors naively.

Graphical Model: We use an undirected graph [21] $G = (V, E)$ consisting of a set of nodes or vertices V that are joined by a set of edges E. In order to define an undirected graphical model, we place at each node a random variable x_s taking values in some space. In the case of localization, this random variable represents the 2-D location, and each edge represents the measured distance. If we exclude anchor nodes, the graph is obviously undirected.

The relationship between the graph and joint distribution may be quantified in terms of potential functions ψ, which are defined over each of the graph's cliques. A clique (C) is a subset of nodes such that for every two nodes in C, there exists a link connecting the two. So the joint distribution is given by

$$p(x_1, \ldots, x_{N_u}) \propto \prod_{cliques\ C} \psi_C(\{x_i : i \in C\}). \tag{26.9}$$

We only need potential functions defined over variables associated with single nodes and pairs of nodes. Single-node potential (prior information about position) at each node t, and the pairwise potential (probabilistic information about distance) between nodes t and u, are respectively given by:

$$\psi_t(x_t) = p_t(x_t), \tag{26.10}$$

$$\psi_{tu}(x_t, x_u) = \begin{cases} P_d(x_t, x_u)p_v(d_{tu} - \|x_t - x_u\|), & \text{if } o_{tu} = 1, \\ 1 - P_d(x_t, x_u), & \text{otherwise.} \end{cases} \tag{26.11}$$

Example 26.1

Compute pairwise potential between an anchor and an unknown node. Assume that the measurement noise has zero-mean Gaussian distribution.

Solution

Pairwise potential around anchor node can be easily computed at grid of some deployment area. In Figure 26.2, we show pairwise potential $\psi_{tu}(x_t^*, x_u) = f(x_u)$, where x_t^* is anchor node. Hence, function $f(x_u)$, which has a 2-D argument x_u,

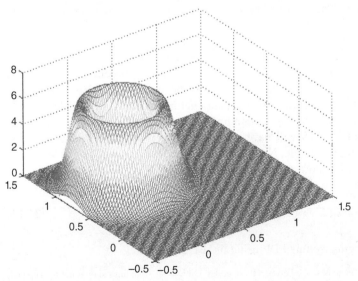

Figure 26.2 Pairwise potential $\psi_{tu}(x_t^*, x_u)$ around the anchor node x_t^*.

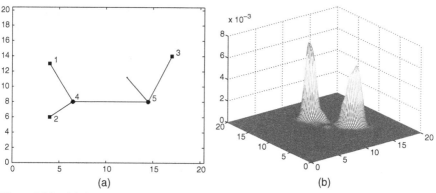

Figure 26.3 (a) An example of a 5-node network (Nodes 4 and 5 are unknown) and estimated location of node 5 (marked with dot); (b) belief of Node 5: the belief is bimodal because Node 5 has only two neighbors.

represents a circular Gaussian distribution around the anchor node x_t^*. Note also that, in order to avoid numerical problems, computation should start in log-domain. See MATLAB codes (available online at ftp://ftp.wiley.com/public/sci_tech_med/matlab_codes) for more details.

In addition, it is necessary to exchange information between the nodes that are not directly connected. Let us define a pair of nodes s and t to be one-hop neighbors of one another if they observe their pairwise distance d_{st}. Then, we define two-hop neighbors of node s to be all nodes t such that we do not observe the d_{st} but do observe d_{su} and d_{ut} for some node u. We can follow the same pattern for the three-hop neighbors, and so forth. These n-hop neighbors ($n > 1$) contain some information about the distance between them. Therefore, if two nodes do not observe the distance between them, they should be far away from each other.[*] In our case, we will include all one-hop and two-hop neighbors; others could be neglected without losing accuracy in the results. This additional information especially helps in the bimodal case when, due to low connectivity (<3), there are two possible solutions. We illustrate this in the 5-node network in Figures 26.3 and 26.4. We used 3 anchors and 2 unknown nodes. If we use only one-hop neighbors, the belief of Node 4 will be bimodal (i.e., with 2 local maximums), as shown in Figure 26.3b. However, adding two-hop neighbors will provide additional information, e.g., Node 1 "tells" Node 5 that it must be far away (assuming that there are no obstacles in between). Now the belief of Node 5 will have only one mode (Fig. 26.4b) so the position estimate will be more accurate. Note that we used the ideal probability of detection. See Example 26.3 for more details.

Finally, we can write the joint posterior PDF as function of potentials:

$$p(x_1, \ldots, x_{N_u} | \{o_{tu}, d_{tu}\}) \propto \prod_t \psi_t(x_t) \prod_{t,u} \psi_{tu}(x_t, x_u). \qquad (26.12)$$

Now we can marginalize this PDF using the BP algorithm.

---

[*]Assuming that there is no obstacle between these nodes. Otherwise, these links should not be included, or a more complex model should be used.

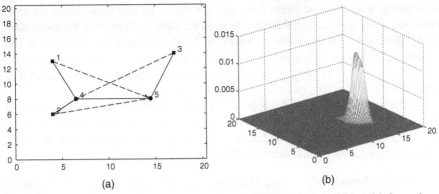

(a)   (b)

Figure 26.4   (a) The network from previous figure, which includes additional information from two-hop neighbors (marked with dashed lines); (b) belief of Node 5; the belief is now unimodal because Nodes 1 and 2 provided additional information to Node 5.

Description of the Algorithm: Having defined the graphical model, we can now estimate the sensor locations by applying the BP algorithm. We apply BP to estimate each sensor's posterior marginal and use the mean value of this marginal and its associated uncertainty to characterize the sensor positions. We will use a different form of BP algorithm in order to adapt it to the iterative localization scenario, where it is more practical to compute beliefs in each iteration. This form can be easily found by replacing (26.7) in (26.8).

Each node t computes its belief $M_t^i(x_t)$, the posterior marginal of its 2-D position x_t at iteration i, by taking a product of its local potential ψ_t with the messages from its set of neighbors G_t:

$$M_t^i(x_t) \propto \psi_t(x_t) \prod_{u \in G_t} m_{ut}^i(x_t). \tag{26.13}$$

The messages m_{ut}, from node u to node t, are computed by:

$$m_{ut}^i(x_t) \propto \int_{x_u} \psi_{ut}(x_t, x_u) \frac{M_u^{i-1}(x_u)}{m_{tu}^{i-1}(x_u)} dx_u. \tag{26.14}$$

In the first iteration of this algorithm it is necessary to initialize $m_{ut}^1 = 1$ and $M_t^1 = p_t$ for all u, t, and then repeat computation using (26.13) and (26.14) until the values sufficiently converge. For tree-like graphs, the number of iterations should be at most the length of the longest path in the graph. However, it is usually sufficient to run until all unknown nodes obtain information from minimum 3 noncollinear anchor nodes.

26.2.3   NBP

The presence of nonlinear relationships and potentially (highly) non-Gaussian uncertainties in sensor localization makes standard BP undesirable. Moreover, to obtain acceptable spatial resolution for the sensors, the discrete space (grid) in the deployment area must be made too large for BP to be computationally feasible. However, using particle-based representations via NBP [7, 8] enables the application of BP to localization in sensor networks. In NBP, the belief and message update equations, Equation (26.13) and (26.14), are performed using stochastic approximations, in two phases: (1) first, drawing particles from the belief $M_t^i(x_t)$; and (2) then using these particles to approximate each outgoing message m_{tu}^i.

The main advantage of this approach is the ability to provide information about location estimation uncertainties (in contrast to deterministic approaches), which are not necessarily Gaussian. Furthermore, it is a naturally distributed method, and it converges after a very small number of iterations.

Computing Messages: Given N weighted particles $\{W_t^{j,i}, X_t^{j,i}\}$ from the belief $M_t^i(x_t)$ obtained at iteration i, we can compute weighted particles of the outgoing BP message m_{tu}^i. We first consider the case of observed edges (1 hop) between unknown nodes. The distance measurement d_{tu} provides information about how far sensor u is from sensor t but no information about its relative direction (see Fig. 26.2). To draw a particle of the message $(x_{tu}^{j,i+1})$, given the particle $X_t^{j,i}$, that represents the position of sensor t, we simply select a direction $\theta^{j,i}$ at random ($i = 1$), chosen uniformly in the interval $[0, 2\pi)$. However, starting from the second iteration, we can also include angular information [8] from the previous iteration using already computed beliefs. We then shift $X_t^{j,i}$ in the direction of $\theta^{j,i}$ by an amount that represents the estimated distance between nodes u and t ($d_{tu} + v^j$):

$$x_{tu}^{j,i+1} = X_t^{j,i} + (d_{tu} + v^j)[\sin(\theta^{j,i}) \quad \cos(\theta^{j,i})]. \quad (26.15)$$

For example, if node t is an anchor, the unknown node u is located on noisy circle around node t. That means that the particles from node u are distributed according to Figure 26.2. Assuming that there is detection between sensor nodes t and u with probability $P_d(x_t, x_u)$, the particles are weighted by the reminder[†] of message update rule (26.14):

$$w_{tu}^{j,i+1} = P_d(X_t^{i,j}, x_u)\frac{W_t^{j,i}}{m_{ut}^i(X_t^{i,j})}. \quad (26.16)$$

The optimal value for bandwidth h_{tu}^{i+1}, the parameter that we use to find the *kernel density estimate* (KDE)[‡] of the message (necessary for denominator of (26.16)), can be obtained in a number of techniques. The simplest way is to apply the "rule of thumb" estimate [25]:

$$h_{tu}^{i+1} = N^{-\frac{1}{3}}Var(\{x_{tu}^{i+1}\}). \quad (26.17)$$

It is also necessary to define messages coming from anchor nodes, using (26.14) and the belief of the anchor node $x_t^*(M_t^i(x_t) = \delta(x_t - x_t^*))$, so:

$$m_{tu}^{i+1}(x_u) \propto \psi_{tu}(x_t^*, x_u). \quad (26.18)$$

[†]We just need to remove pairwise potential from which we already have drawn the particles.

[‡]Approximation of distribution $p(x)$: $\hat{p}(x) = \sum_j w^j K^h(x - x^j)$ for a kernel $K^h(x)$. The most common kernel function (K^h) is the spherically symmetric Gaussian kernel, $K^h(x) = N(x, 0, hI)$, where bandwidth h controls the variance [8, 25].

Messages along unobserved edges (two-hop, three-hop...) must be represented in parametric form since their potentials have the form $1 - P_d(x_t, x_u)$, which is typically not normalizable because it tends to 1 as the distance becomes large. Using the probability of detection P_d and particles from the belief M_t^i, an estimate of the outgoing message to node u is given by:

$$m_{tu}^{i+1}(x_u) = 1 - \sum_j \frac{W_t^{j,i}}{m_{ut}^i(X_t^{i,j})} P_d(X_t^{j,i}, x_u). \tag{26.19}$$

Finally, the messages along unobserved edges coming from anchor nodes $(W_t^{j,i} = 1/N)$ are given by:

$$m_{tu}^{i+1}(x_u) = 1 - P_d(x_t^*, x_u). \tag{26.20}$$

Computing Beliefs: To estimate the belief $M_u^{i+1}(x_u)$ using (26.13), we draw particles from the product of several KDEs of the messages and (eventually) analytic messages. In our case, it is very difficult to draw particles from this product, so we use the *proposal* distribution $q(x)$, the sum of the messages, and then reweight all particles. This procedure is well-known as *mixture importance sampling* [8].

Denote the set of neighbors of u, having observed edges to u and not including anchors, by G_u^0, and the set of of all neighbors by G_u. In order to draw N particles, we create a collection of kN weighted particles (where $k \geq 1$ is a parameter of the sampling algorithm) by drawing $kN/|G_u^0|$ particles from each message m_{tu} with $t \in G_u^0$ and assigning each particle a weight equal to the ratio:

$$W_u^{j,i+1} = \frac{\prod_{v \in G_u} m_{vu}^{i+1}}{q(x)}, \tag{26.21}$$

where the proposal distribution $q(x)$ is given by:

$$q(x) = \sum_{v \in G_u^0} m_{vu}^{i+1}. \tag{26.22}$$

Some of these calculated weights are much larger than the rest, especially after more iterations. This means that any particle-based estimate will be dominated by the influence of a few of the particles, and the estimate could be erroneous. To avoid this, we then draw N values independently from the collection $\{W_t^{j,i+1}, X_t^{j,i+1}\}$ with probability proportional to their weight, using *resampling with replacement* [8]. This means that we create N equal-weight particles drawn from the product of all incoming messages.

Convergence of NBP: A node is located when a convergence criteria is met. We use Kullback–Leibler (KL) divergence [8], a common measure of difference between two distributions. For the particle based beliefs in our algorithm, KL divergence between beliefs in two consecutive iterations, is given by:

$$KL_u^{i+1} = \sum_j W_u^{j,i+1} \log \frac{W_u^{j,i+1}}{M_u^i(X_u^{j,i+1})}. \tag{26.23}$$

Note that the set of particles is different in two consecutive iterations, so we had to use the parametric form of the belief from previous iteration computed at particles from the current iteration (see denominator in (26.23)). When KL_u^{i+1} drops below the predefined threshold, the node u is located and starts to behave as an anchor. In this way, we can locate all nodes incrementally. The execution is over when KL drops below the threshold for all nodes or when the maximum number of iteration is reached. In any case, the estimated positions of all unknowns and their uncertainties will be available.

26.2.4 NBBP

NBBP is a variant of the NBP algorithm. The main goal is to increase the performance of the algorithm by adding boxes that constrain the area from which the particles are drawn. These boxes, which are created almost without any additional communication between nodes, are also used to filter erroneous particles of the beliefs. We also use an incremental approach in order to decrease the computational and the communication cost.

Modifications: The following modifications are added to the already-described NBP algorithm:

- The particles are drawn from bounding box (e.g., a rectangle) that covers the region where the anchors' ranges overlap (Fig. 26.5);
- in each iteration, erroneous particles of the beliefs (all the particles that are outside of the appropriate box) are filtered out;
- nodes are located in an incremental way: As soon as the belief sufficiently converges, we characterize sensor positions with mean value and uncertainty, and from that point we consider this node as an anchor.

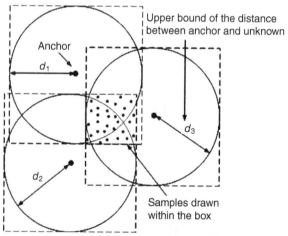

Figure 26.5   Drawing particles within the box that covers the region where anchors' ranges overlap.

Example 26.2

Given four anchor nodes, draw (within a bounded box) the initial set of particles of an unknown node. Given this set of particles, draw particles from the message for some neighboring node.

Solution

The bounding box can be constructed by finding four borders using min-max method as illustrated in MATLAB code. Then, all particles are drawn uniformly within the borders (Fig. 26.6a). In addition, we never draw particles out of deployment area. In order to draw particles from the message, we just need to shift each initial particle in a random direction by an amount that represents the estimated distance between these two nodes (Fig. 26.6b). These particles represent the partial information about the position of another (neighboring) node.

Performance Analysis: We placed 50 nodes randomly in 2×2 area, 40 of them are unknowns. The values of parameters are set as follows: standard deviation of the Gaussian noise ($\sigma = 0.1$), transmission radius ($R = 30\%$ of the diagonal length of the deployment area: $d_{max} = 2\sqrt{2}$ m), number of particles ($N = 50$ and $N = 100$), number of iterations ($N_{iter} = 3$), and KL threshold ($KL_{min} = 0.02$). The error is defined as a distance between true and estimated location. Finally, each point in the simulations represents the average over 20 Monte Carlo runs.

Using the defined scenario, we run the simulation for both algorithms (NBP and NBBP), and obtained the results shown in Figure 26.7. As expected, the location estimates for the NBBP are more accurate since all the estimates are placed within its box, so it limits the maximum error. Of course, since the nodes near the edges suffer from low connectivity, the error for those nodes is higher.

In Figure 26.8a and 26.8b, we have compared the average error and coverage with respect to the transmission radius. The coverage is defined as a percentage of located nodes with error less than predefined tolerance. In our case we set it to 5%,

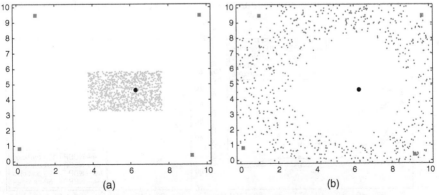

Figure 26.6   (a) Initial set of particles, and (b) particles from the message for the neighboring node. Anchor nodes are marked with red squares and true position of the unknown node with black circles.

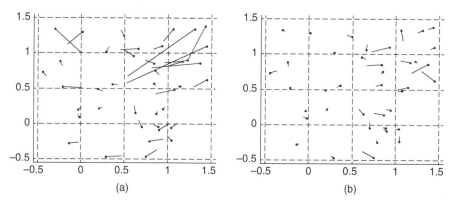

Figure 26.7   Comparison of the results for a 50-node network (a) NBP, (b) NBBP. The line between the true and the estimated positions represents the error.

but note that this is an application dependent parameter. As we can see in Figure 26.8, for high values of the transmission radius, the accuracy and the coverage are nearly constant since the nodes start to receive redundant information caused by high connectivity.

Finally, in Figure 26.9, we show a comparison of the computational and communication costs with respect to the number of particles for two different densities (we use 20 and 10 unknown nodes, respectively). To measure the communication cost, we count *elementary messages*, where one elementary message is defined as simple scalar data (e.g., one coordinate of the particle). The main conclusion is that the NBBP algorithm performs better than NBP in all terms. This result is expected because constructed boxes and filtering increase accuracy, and the incremental approach decreases the computational and communication cost.

Example 26.3

Implement the NBP localization algorithm for the arbitrary network. As a special case, consider two localization scenarios shown in Figure 26.3a and Figure 26.4a.

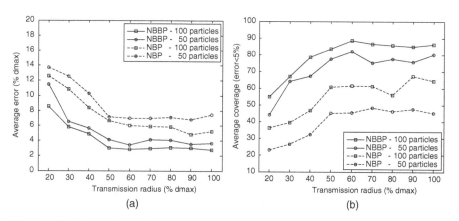

Figure 26.8   Comparison of (a) accuracy and (b) coverage.

Solution

The full code is included with the book. The code consists of the following parts:

- **NBP algorithm:** This is the main file that runs the NBP localization algorithm, computes errors, and prints results in the command window. After running this file, all location estimates, along with their beliefs (i.e., weighted particles) will be available. It also plots the network with estimates, and prints all location estimates, root mean square error (RMSE) and elapsed time.

- **Pairwise potential:** This function computes pairwise potential between two nodes, according to (26.11). It is a generalization of Example 26.1.

- **KDE:** This function computes KDE at given points (grid points or particles). The bandwidth is computed using "rule of thumb" according to (26.17). Using a similar idea as for pairwise potential, this function must be made numerically stable.

- **Plot figure:** This function plots network (anchors, unknowns, and edges).

- **Initialization:** This script defines all parameters and sets initial states of all variables, including all weighted particles of beliefs and messages. Moreover, it builds the bounded boxes (in contrast to Example 26.2, we use one-hop and two-hop anchors). There are two placement modes: random and deterministic. In the latter one, the user should carefully place all nodes manually, one by one. Default placement is according to Figure 26.4a (to adapt to scenario from Fig. 26.3a, just change the state of variable *two_hop*). We also encourage reader to simulate a number of random large-scale networks.

26.3   GBP METHODS

The BP algorithm, defined by (26.7) and (26.8) in the previous section, does not make a reference to the topology of the graph that it is running on. Thus, there is nothing to stop us from implementing it on a graph that has loops. One starts with some initial set of messages and simply iterates the message-update rules (26.8) until they eventually converge, and then one can read off the approximate beliefs from the belief equations (26.7). But if we ignore the existence of loops and permit the nodes to continue communicating with each other, messages may circulate indefinitely around these loops, and the process may not converge to a stable equilibrium. One can find examples of graphical models with loops, where, for certain parameter values, the BP algorithm fails to converge or predicts beliefs that are inaccurate. On the other hand, the BP algorithm can be successful in graphs with loops, e.g., error-correcting codes defined on Tanner graphs that have loops [26].

In this section, we describe three solutions for this problem: GBP-K, NGBP-JT, NBP-ST and URW-NBP.

26.3.1   Correctness of BP

Let us consider the example network in Figure 26.10. In this network, there are 3 unknown nodes (A, B and C) and three anchor nodes (E_A, E_B, and E_C), which

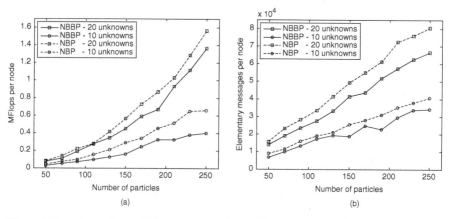

Figure 26.9   Comparison of (a) computational and (b) communication costs.

represent the local evidence. The message-passing algorithm (BP) can be thought of as a way of communicating local evidence between nodes such that all nodes calculate their beliefs given all the evidence.

In order for BP to be successful, it needs to avoid *double counting* [22, 32]— a situation in which the same evidence is passed around the network multiple times and mistaken for new evidence. Of course, this is not possible in a single-connected network because when a node receives some evidence, it will never receive that evidence again (see (26.7) and (26.8)). In a loopy network, double counting can not be avoided. For example, in Figure 26.10a, node B will send A's evidence to C, but in the next iteration, C will send that same information back to A. Thus, it seems that BP in such a network will give the wrong answer.

However, BP could still lead to correct inference if all evidence is "double counted" in equal amounts. This could be analyzed by an unwrapped network corresponding to the loopy network. The unwrapped network is a single-connected network constructed such that performing BP in the unwrapped network is equivalent to performing BP in the loopy network. The basic idea is to replicate the nodes as shown in Figure 26.10b. For example, the message received by node B after three iterations of BP in the loopy network are identical to the final messages received by node B in the unwrapped network. In this way, we can create an infinite network. The importance of the unwrapped network is that since it is single-connected, BP on it is guaranteed to give the correct beliefs. However, the usefulness of these beliefs depends on the similarity between the probability distribution induced by the unwrapped problem and the original loopy problem. And this similarity is satisfied in a single-loop network after a finite number of iterations. In the general case, BP will converge when the addition of these additional nodes at the boundary will not alter the posterior probability of the node in the center.

Finally, in Gaussian networks [33], the factor that determines the goodness of the approximation and the convergence rate is the amount of statistical independence between the root nodes and the leaf nodes in the unwrapped network. In Gaussian

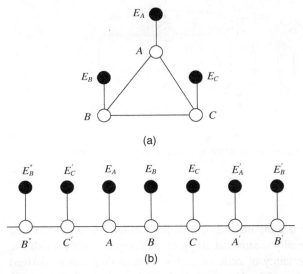

Figure 26.10   (a) A simple loopy network; (b) corresponding unwrapped network for the first three iterations.

networks with multiple loops the mean at each node is guaranteed to be correct, but the confidence around that mean may be incorrect (usually, overconfident). These results give a theoretical justification for applying BP in certain networks with multiple loops. This may enable fast, approximate probabilistic inference in a range of new applications, where measurement error could be similar to the Gaussian model.

For an extensive analysis of this topic, we refer the reader to [8] and [34], where many useful theorems are provided.

26.3.2   GBP-K

In standard BP, all messages are always going from a single node to another single node. It is natural to expect that messages from groups of nodes to other groups of nodes could be more informative, and thus lead to better inference.[*] That is the basic idea behind GBP-K [20]. Therefore, standard BP is just a special case with two-node "regions" as maximum (also known as *Bethe* approximation). In Figure 26.11, we show the basic clusters for both approximations.

The GBP-K algorithm nearly always improves, at least slightly, over the performance of standard BP, and it can significantly outperform standard BP if the graphical model under consideration has only short loops. However, the complexity of GBP-K grows exponentially with the size of the chosen basic clusters. If we include all loops as the basic clusters, the GBP-K algorithm is exact but computationally unacceptable. This is the reason why this method is not appropriate for the localization problem at hand. For the detailed description of the GBP-K method and its variations, the reader is referred to [20].

[*]The exception is a tree-like graph in which node-to-node messages are enough for optimal inference.

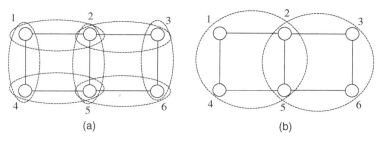

Figure 26.11 The basic clusters in the (a) Bethe approximation and (b) Kikuchi approximation.

26.3.3 NGBP-JT

Definition: The junction tree algorithm is a standard method for exact inference in graphical model. That literally means that all posterior marginals will provide the true information about uncertainty of node estimates. This method can be derived using an *elimination* procedure [27]. The graph is first *triangulated* (i.e., "virtual" edges are added so that every loop of length more than three nodes has a chord). Given a triangulated graph, with cliques C_i and potentials $\psi_{C_i}(x_{C_i})$, and given the corresponding junction tree that defines links between the cliques, we send the following message from clique C_i to clique C_j by the message update rule:

$$m_{ij}(x_{S_{ij}}) = \sum_{C_i \setminus S_{ij}} \psi_{C_i}(x_{C_i}) \prod_{k \in G_i \setminus j} m_{ki}(x_{S_{ki}}), \qquad (26.24)$$

where $S_{ij} = C_i \cap C_j$, and where G_i are the neighbors of clique C_i in the junction tree. The belief at clique C_i is proportional to the product of the local evidence at that clique and all the messages coming into clique i:

$$M_i(x_{C_i}) = k \psi_{C_i}(x_{C_i}) \prod_{j \in G_i} m_{ji}(x_{S_{ji}}). \qquad (26.25)$$

Beliefs for single nodes can be obtained via further marginalization:

$$M_i(x_i) = \sum_{C_i \setminus i} M_i(x_{C_i}) \text{ for } i \in C_i. \qquad (26.26)$$

Equations (26.24–26.26) represent the GBP-JT algorithm, which is valid for arbitrary graphs. The BP algorithm defined with (26.7) and (26.8) is a special case of GBP-JT, obtained by noting that the original tree is already triangulated, and has only pairs of nodes as cliques. In that case, sets S_{ij} are single nodes, and marginalization using (26.26) is unnecessary.

Example Network: Let us show how it works in our example in Figure 26.12. The network has 10 nodes, five anchors (Nodes 6–10), and 5 unknowns (Nodes 1–5). There is a loop 1-2-4-5-3, so we have to triangulate it by adding two more edges (2–3 and 3–4). Then we define eight cliques in the graph: $C_1 = \{x_1, x_2, x_3\}$, $C_2 = \{x_2, x_3, x_4\}$, $C_3 = \{x_3, x_4, x_5\}$, $C_4 = \{x_4, x_9\}$, $C_5 = \{x_5, x_{10}\}$, $C_6 = \{x_1, x_6\}$,

$C_7 = \{x_2, x_7\}$, $C_8 = \{x_3, x_8\}$. The appropriate potentials of the three-node cliques are given by:

$$\psi_{C_1}(x_1, x_2, x_3) = \psi_{12}(x_1, x_2)\psi_{13}(x_1, x_3) \tag{26.27}$$

$$\psi_{C_2}(x_2, x_3, x_4) = \psi_{24}(x_2, x_4) \tag{26.28}$$

$$\psi_{C_3}(x_3, x_4, x_5) = \psi_{35}(x_3, x_5)\psi_{45}(x_4, x_5). \tag{26.29}$$

Note that "virtual" edges do not appear in these equations since they are used only to define cliques. Other cliques, defined over pairs of nodes, are nothing other than the potential functions between two nodes already known from standard BP:

$$\psi_{C_4}(x_4, x_9) = \psi_{49}(x_4, x_9), \ \psi_{C_5}(x_5, x_{10}) = \psi_{510}(x_5, x_{10}),$$
$$\psi_{C_6}(x_1, x_6) = \psi_{16}(x_1, x_6), \ \psi_{C_7}(x_2, x_7) = \psi_{27}(x_2, x_7), \ \psi_{C_8}(x_3, x_8) = \psi_{38}(x_3, x_8). \tag{26.30}$$

The junction tree corresponding to the network in Figure 26.12 is shown in Figure 26.13. As we can see, "anchor cliques" (C_4–C_8) do not receive messages, so this graph does not contain loops. Actually, these "anchor cliques" also include one unknown node so we can send them messages, but this node could be also located by marginalizing the belief of some other clique. In the next step, we can compute all messages using (26.24). The complete set of messages is given by:

$$m_{61}(x_1) = \psi_{16}(x_1, x_6^*), \ m_{53}(x_5) = \psi_{510}(x_5, x_{10}^*), \ m_{71}(x_2) = m_{72}(x_2) = \psi_{27}(x_2, x_7^*),$$
$$m_{42}(x_4) = m_{43}(x_4) = \psi_{49}(x_4, x_9^*), \ m_{81}(x_3) = m_{82}(x_3) = m_{83}(x_3) = \psi_{38}(x_3, x_8^*), \tag{26.31}$$

$$m_{12}(x_2, x_3) = \psi_{27}(x_2, x_7^*)\psi_{38}(x_3, x_8^*)\sum_{x_1}\psi_{16}(x_1, x_6^*)\psi_{C_1}, \tag{26.32}$$

$$m_{32}(x_3, x_4) = \psi_{49}(x_4, x_9^*)\psi_{38}(x_3, x_8^*)\sum_{x_5}\psi_{510}(x_5, x_{10}^*)\psi_{C_3}, \tag{26.33}$$

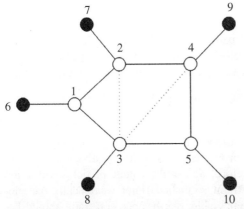

Figure 26.12 Example of 10-node network with loop with 5 anchors (Nodes 6–10), and five unknowns (Nodes 1–5). The network is already triangulated by adding two more edges (marked with dashed lines).

$$m_{21}(x_2, x_3) = \psi_{27}(x_2, x_7^*)\psi_{38}(x_3, x_8^*)\sum_{x_4} \psi_{49}(x_4, x_9^*)\psi_{C_2} m_{32}, \qquad (26.34)$$

$$m_{23}(x_3, x_4) = \psi_{49}(x_4, x_9^*)\psi_{38}(x_3, x_8^*)\sum_{x_2} \psi_{27}(x_2, x_7^*)\psi_{C_2} m_{12}, \qquad (26.35)$$

where an asterisk denotes the known location of the anchor node, and the messages from "anchor cliques" are directly replaced by the appropriate potential function. For clarity, we used simplified notation for messages and clique potentials on the right side of equations (e.g., $m_{12} = m_{12}(x_2, x_3)$, $\psi_{C_2} = \psi_{C_2}(x_2, x_3, x_4)$).

The beliefs of cliques are computed using (26.25):

$$M_1(x_1, x_2, x_3) = \psi_{C_1}\psi_{16}(x_1, x_6^*)\psi_{27}(x_2, x_7^*)\psi_{38}(x_3, x_8^*)m_{21}, \qquad (26.36)$$

$$M_2(x_2, x_3, x_4) = \psi_{C_2}\psi_{27}(x_2, x_7^*)\psi_{38}(x_3, x_8^*)\psi_{49}(x_4, x_9^*)m_{12}m_{32}, \qquad (26.37)$$

$$M_3(x_3, x_4, x_5) = \psi_{C_3}\psi_{38}(x_3, x_8^*)\psi_{49}(x_4, x_9^*)\psi_{510}(x_5, x_{10}^*)m_{23}. \qquad (26.38)$$

Now it is easy to compute beliefs of single nodes by marginalizing beliefs of cliques using (26.26). Obviously, it is sufficient to know beliefs of C_1 and C_3, since these cliques include all unknown nodes. Marginalization of C_2 provides a degree of freedom and could be used to check the estimated positions of some nodes (in our case, for Nodes 2, 3, and 4).

Nonparametric Approximation: Due to the same reasons as for BP (i.e., computational problems, presence of nonlinear relationships, and highly non-Gaussian uncertainties), we need to use nonparametric approximation of GBP-JT method (NGBP-JT). In order to use this method for localization, we use the same statistical model as for NBP: measured distance defined by (26.2), probability of detection defined by (26.5) or (26.6), and potential functions defined by (26.10) and (26.11). In this way, we can make this method tractable in small-scale networks. For detailed description of this method, see [28] and [29].

Example 26.4

Given four anchors, draw initial set of particles from three-node clique (i.e., three unknown nodes connected with each other) by: a) drawing independently within bounded boxes and b) adding a distance constraint in order to make particles more informative.

Solution

To draw particles independently, we can repeat Example 26.1 and obtain particles as illustrated in Figure 26.14a. However, we also have measured distance between each pair of the unknown nodes. Therefore, we can draw node particles within its boxes and accept the particle if the constraint is satisfied. If not, we reject the particle and try again. To make this procedure tractable, the distance constraint should be relaxed (e.g., distance between any pair of particles must be $\mu \pm 3\sigma$). Illustration of the improved set of initial particles is shown in Figure 26.14b.

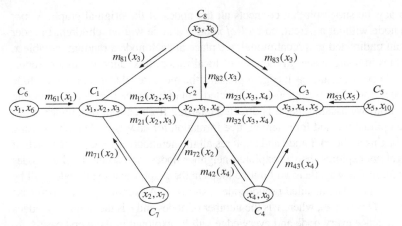

Figure 26.13   The junction tree corresponding to the network in Figure 26.12.

26.3.4   NBP-ST

The previous methods are still very complex for large-scale ad hoc/sensor networks. Moreover, the connectivity in these networks is very high, which introduces computational and communication burdens for low-power applications. Therefore, we will describe a technique to simplify the algorithm by breaking the loops using NBP-ST [30] created by a *breadth first search* (BFS) method [31].

Algorithm 26.1: BFS

 1: **Input**: list of nodes Q and root node *root*

 2: Set current root: $r \leftarrow root$

 3: **while** Q is not empty **do**

 4:    **for all** nodes $t \in G_r$ **do**

 5:       **if** $t \in Q$ **then**

 6:          Remove t from Q

 7:          Insert t in Q_r

 8:          Insert d_{rt} in S

 9:       **end if**

10:    **end for**

11:    Set current root: $r \leftarrow$ first unused node from Q_r

12: **end while**

13: **Output**: spanning tree $\{Q, S\}$

Spanning Tree Formation: We start by reviewing the basics of graphical models. An undirected graph $G = (V, E)$ consists of a set of nodes or vertices V that are joined by a set of edges E. A *loop* or *cycle* is a sequence of distinct edges forming a path from a node back to itself. A *tree* is a connected graph without any loops. A *spanning* *tree* is an acyclic subgraph that connects all the nodes of the original graph. A *root node* is a node without a parent, and a *leaf node* is a node without children. In order to define an undirected graphical model, we place at each node a random variable x_s taking values in some space. In the case of localization, this random variable represents the 2-D position, and each edge represents the measured distance. If we exclude anchor nodes, the graph is obviously undirected, but only for the first phase (spanning tree formation) we assume that it is directed (starting from chosen root node).

The optimal method for spanning tree formation for unweighted graphs is using a BFS. It begins at the root node and explores all the neighboring nodes. Then each of those neighbors explores their unexplored neighbor nodes, and so on, until all nodes are explored. In this way, there will not be loops in the graph because all nodes will be explored just once. The detailed pseudocode is shown in Algorithm 1. The worst case complexity is $O = (v + e)$, where v is the number of nodes and e is the number of edges in the graph, since every node and every edge will be explored in the worst case.

Example 26.5

Construct spanning tree using BFS method in the network with large number of nodes.

Solution

Spanning tree is created using already described BFS method (see Algorithm 26.1). In order to have it adapted for localization, the output is a set of distances in spanning tree. Note also that each running of the code generates a random network, and appropriate spanning tree. Some examples are shown in Figure 26.15.

In the case of NBP localization, we exclude all the anchors from the BFS algorithm since they do not form the loops in the graph (they just send, and never receive the messages). A graph generally has a large number of spanning trees, but since our graph is unweighted we choose a few (minimum of two) of them in a partly random way. The first root node we choose randomly from the set of all

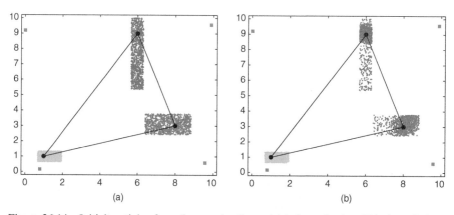

Figure 26.14 Initial particles from three-node clique: (a) independently within bounded boxes and (b) improved approach. Anchor nodes are marked with red squares, and true positions of the unknown nodes with black circles.

unknown nodes. In order to maximize the difference between two spanning trees, the second root node has to be as far as possible from the first root node. Thus, it should be one of the leaf nodes. If we want to form more spanning trees, the analog constraint will be used. Note that, using BFS, it is not possible to form two spanning trees with completely different edges and that usually some of the edges will be out of both spanning trees. If we want to include all edges, we have to add more spanning trees but it is usually not necessary since it will only provide us redundant information. This is especially the case in the networks with high connectivity.

The NBP method is naturally distributed through the graph, which means that there is no central unit that will handle all computations. Therefore, the proposed BFS method has to be done in a distributed way. This can be simply done if each unknown node initially broadcasts its ID to all neighbors, which will continue to broadcast to others, and so on, until each unknown node has a list of all unknown nodes in the graph. One node (e.g., with the lowest ID) has to be assigned to choose the root node from the list and give him permission (by multihop broadcasting) to start the BFS algorithm. Then, the chosen root node has all initial data to start the BFS algorithm, and, when it is necessary, has only to broadcast all data (i.e., variables from Algorithm 26.1) to all its neighbors. The last visited node will have all the data about the final spanning tree.

Finally, the NBP-ST algorithm represents two (or more) independent runnings of the NBP algorithm based on formed spanning trees. Each running will provide weighted particles of the node beliefs computed by (26.13). The simplest way to fuse these beliefs is to draw particles from the product of the beliefs from different spanning trees. This can be done using mixture importance sampling (see Section 26.2.3). The collection of weighted particles from all spanning trees represents our final output, from which we can easily extract any parameter that we need (e.g., mean value or variance of the location estimate). The pseudocode in Algorithm 26.2 illustrates the NBP-ST method. Given this algorithm, along with Examples 26.3 and 26.5, implementation of this method in MATLAB is straightforward.

Algorithm 26.2: NBP-ST Method for Localization

1: for all nodes **do**

2:   Take sensing actions

3:   Set all parameters to the initial values

4:   Broadcast own and all received IDs and listen for other sensor broadcasts (until receive all IDs)

5: end for

6: Set a list of nodes for BFS (excluding anchors): Q

7: Choose randomly root node from the list Q: *root*

8: for all *spanning trees* **do**

9:   Run BFS (Algorithm 26.1)

10:   Run NBP on defined spanning tree

11:   Choose root node as far as possible from the previous roots

12: end for

13: Fuse all beliefs into one and compute location estimates

Performance Analysis: To illustrate the performance of this method, we conducted several simulations. We placed 100 unknown and 10 anchor nodes randomly in a 20 m × 20 m area. Since the unknown nodes near the edges of the deployment area suffer from low connectivity, we include one realistic constraint: four anchors are randomly deployed within four areas of 4 m × 4 m near the edges, respectively. The standard deviation of the Gaussian noise on the distance estimate is set to *sigma* = 0.3 m and the number of iterations is set to N_{iter} = 3. All simulations are done for $N = 50$ and $N = 100$ particles with respect to the transmission radius ($R = 4$–10 m). The error is defined as a distance between true and estimated location. To measure the communication cost, we count *elementary messages*, where one elementary message is defined as simple scalar data. Finally, each point in the simulations represents the average over 20 Monte Carlo runs.

Using the defined scenario, we compared NBP and NBP-ST algorithms. For NBP-ST, we used 2 spanning trees. The original network and 2 spanning trees created by BFS are illustrated in Figure 26.15. Regarding accuracy and coverage (percentage of located nodes with error less than predefined tolerance) in Figure 26.16, NBP-ST performs better than NBP for $R > 7$ m, approximately. Obviously, for these values

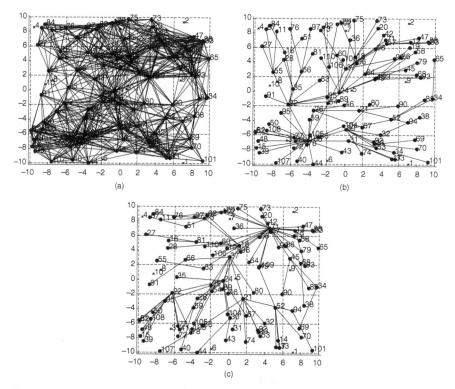

Figure 26.15   (a) Original network with 100 unknown nodes (marked with black circles) and 10 anchors (marked with asterisks) for $R = 6$ m, and (b), (c) two corresponding spanning trees. The roots are Node 103 and 71, respectively.

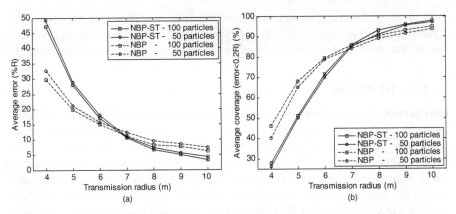

Figure 26.16 Comparison of (a) accuracy and (b) coverage.

of R there is a large number of loops in the network t decreases the performance of the NBP method. For lower values of R, we could expect that NBP-ST performs with higher (or same) accuracy, but we cannot forget that, by using only 2 spanning trees, we don't include all information (removed edges) that we have. Thus, the NBP over-performs NBP-ST in this case.

Regarding the computational/communication cost (Fig. 26.17), NBP-ST performs better than NBP for $R > 8$ m and $R > 9$ m, respectively. In order to explain this we have to remember two main things we have taken into account: (1) removing the edges in order to form the spanning trees and (2) running NBP two times in these spanning trees. The first operation decreases the computational/communication costs, but the second one increases it. Therefore, in low-connected networks the second operation predominates, but in high-connected networks the first one predominates. The additional contribution is that (for high transmission radius), computational and communication costs are nearly constant. This feature provides us

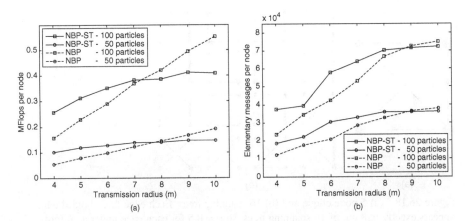

Figure 26.17 Comparison of (a) computational and (b) communication cost.

more precise information about battery life. The final conclusion is that the NBP-ST algorithm performs better than NBP in all terms, for $R > R_{min}$. In our case $R_{min} = 9\,\text{m}$, but this parameter depends on the density in the network (i.e., average node degree).

26.3.5 URW-NBP

For the methods from previous sections, we need to make some kind of graph transformations before applying message passing. This leads to reduced robustness of the whole algorithm, since the failure of just one node can have a significant effect on the localization error. In addition, it is necessary to synchronize the whole network, which is not always possible. Since standard NBP is robust to node failures, and able to run in asynchronous networks, we would like to find an improved method with the same features. Therefore, we consider a method based on tree-reweighted BP (TRW-BP) proposed in [35], and adapted to cooperative localization in [36].

In the standard TRW-BP algorithm, the belief at a node t is, similarly to BP, proportional to the product of the local potential $\psi_t(xt)$ and all *reweighted* messages coming into node t:

$$M_t(x_t) \propto \psi_t(x_t) \prod_{u \in G_t} m_{ut}(x_t)^{\rho_{ut}}, \qquad (26.39)$$

where $\rho_{tu} = \rho_{ut}$ is the appearance probability of the edge (t, u). The messages are determined by the following message update rule:

$$m_{ut}(x_t) \propto \int \psi_u(x_u) \psi_{tu}(x_t, x_u)^{1/\rho_{tu}} \prod_{k \in G_u \backslash t} \frac{m_{ku}(x_u)^{\rho_{ku}}}{m_{tu}(x_u)^{1-\rho_{tu}}} dx_u. \qquad (26.40)$$

On the right-hand side, there is a product over all reweighted messages going into node u except for the one coming from node t. The update-rule (26.40) is carried out over the network. Upon convergence, the beliefs are computed

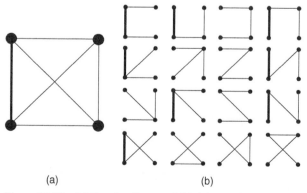

| (a) | (b) |

Figure 26.18 (a) 5-node clique and (b) 16 spanning trees. Each edge (e.g., bolded edge) appears exactly in 8 out of 16 spanning trees, so $\rho = 0.5$ for each edge under a uniform distribution over the spanning trees.

through (26.39). As for NBP, it is more convenient to compute the beliefs at every iteration i. This leads to an equivalent form of TRW-BP: by replacing (26.39) in (26.40), we find that the belief and message-update rule of TRW-BP are, respectively, given by:

$$M_t^i(x_t) \propto \psi_t(x_t) \prod_{u \in G_t} m_{ut}^i(x_t)^{\rho_{ut}} \tag{26.41}$$

$$M_{ut}^i(x_t) \propto \int \psi_{ut}(x_t, x_u)^{1/\rho_{tu}} \frac{M_{ut}^{i-1}(x_u)}{m_{tu}^{i-1}(x_u)} dx_u. \tag{26.42}$$

We can now apply TRW-BP to the localization problem. In the first iteration of this algorithm it is necessary to initialize $m_{ut}^1 = 1$ and $M_t^1 = p_t$ (i.e., information from anchors, if any) for all u, t, and then repeat computation using (26.41) and (26.42) until convergence or a preset number of iterations is attained. In a practical implementation, we have to use nonparametric version of TRW-BP (TRW-NBP). Therefore, the beliefs and message update equations, Equations (26.41) and (26.42), are performed using particle-based approximations. Since this approximation is done in the same way as for NBP, see Section 26.2.3 for more details.

We now describe how appropriate values for ρ_{tu} can be found. Given a graph G, let S be the set of all spanning trees T over G. Let $\vec{\rho}$ be a distribution over all spanning trees:

$$\vec{\rho} \triangleq \left\{ \rho(T),\ T \in S \middle| \rho(T) \geq 0,\ \sum_{T \in S} \rho(T) = 1 \right\}. \tag{26.43}$$

Note that there are many such distributions. For a given $\vec{\rho}$ and a given (undirected) edge (t, u), $\rho_{tu} = P_{\vec{\rho}}\{(t, u) \in T\}$; i.e., ρ_{tu} is the probability that the edge (t, u) appears in a spanning tree T chosen randomly under $\vec{\rho}$. Thus, ρ_{tu} represents *edge appearance probability* of the edge (t, u). A valid set of edge appearance probabilities must correspond to a valid distribution over spanning trees. For example, $\rho_{tu} = 1$ for all edges is a valid set if the graph G is a tree.

Finding a valid collection ρ_{tu} is infeasible since there is a large number of spanning trees even in small graphs. For example, in a four-node clique there are 16 different spanning trees (Fig. 26.18), and each edge appears exactly in 8 of them. If $\vec{\rho}$ is uniform over all spanning trees, then $\rho_{tu} = 0.5$ for every edge. Discovering all spanning trees, and then computing all ρ_{tu}, would require significant communication overhead, even for small networks. Moreover, determining a valid collection ρ_{tu} requires a procedure similar to routing, which we prefer to avoid in order to make this method robust to node failures. We note that the choice $\rho_{tu} = 1$ for all edges corresponds to standard BP. In TRW-BP on graphs with cycles, it is easy to see that $\rho_{tu} \leq 1$ for all edges. Hence, by removing the restriction of valid ρ_{tu} and making ρ_{tu} uniform, we intuit that we can combine the benefits of NBP (distributed and robust implementation) and of TRW-NBP (improved performance in loopy graphs). This leads to the novel method, URW-NBP. This method is applied for cooperative localization in [36] with the same model as for NBP and shown that it can outperform NBP if ρ is empirically modeled as a function of average node degree (that depends on the number of loops in the graph).

26.4 CONCLUSIONS

In this chapter, we have presented cooperative localization techniques based on BP and its generalizations. The main advantage of these methods is the capability to provide information about location estimation uncertainties (in contrast to deterministic approaches), which are not necessarily Gaussian. Furthermore, it is a naturally distributed method and converges after a very small number of iterations. The main drawback is high complexity, which sometimes makes these methods unacceptable in low-power sensor networks. Nevertheless, the particle-based approximation via nonparametric representation makes BP methods acceptable for localization in sensor networks. However, BP/NBP methods predict beliefs that are inaccurate in loopy networks, so we described a few solutions. The complexity of GBP-K grows exponentially with the size of the chosen basic clusters. The NGBP-JT method, which converges well in loopy networks, according to our current results, has acceptable complexity only in small-scale sensor networks. NBP-ST, which outperforms NBP in highly-connected networks, is computationally feasible in large-scale ad hoc/sensor networks. Finally, we discussed URW-NBP, which is much more robust to failures since it does not require any graph transformations.

ACKNOWLEDGEMENTS

This work is supported by the FPU fellowship from Spanish Ministry of Science and Innovation. We also thank partial support by: ICT project FP7-ICT-2009-4-248894-WHERE-2, program CONSOLIDER-INGENIO 2010 under grant CSD2008-00010 COMONSENS, Spanish National Project under Grant TEC2009-14219-C03-01, Controlando las redes de comunicaciones electromagneticas submarinas mediante despliegues autoconfigurables (Herakles, TEC2016-76038-C3-1-R), COMONSENS EXCELLENCE NETWORK (TEC2015-69648-REDC), and Area of Advance Transport, Chalmers University of Technology.

REFERENCES

[1] D. Niculescu and B. Nath, "Ad hoc positioning system (APS)," in *IEEE GLOBECOM '01*, vol. 5, Nov. 2001, pp. 2926–2931.

[2] Y. Shang, W. Ruml, Y. Zhang, and M. Fromherz, "Localization from connectivity in sensor networks," *IEEE Trans. Parallel Distrib. Syst.*, vol. 15, no. 11, pp. 961–974, Nov. 2004.

[3] Y. Shang and W. Ruml, "Improved MDS-based localization," in *Proc. of IEEE INFOCOM 2004*, vol. 4, Mar. 2004, pp. 2640–2651.

[4] A. Savvides, H. Park, and M. B. Srivastava, "The bits and flops of the n-hop multilateration primitive for node localization problems," *in Proc. of the 1st ACM Int. Workshop on Wireless Sensor Networks and Application*, Sep. 2002, pp. 112–121.

[5] N. B. Priyantha, H. Balakrishnan, E. Demaine, and S. Teller, "Anchor-free distributed localization in sensor networks," Tech. Rep., MIT Laboratory for Computer Science, Apr. 2003.

[6] V. Vivekanandan and V. W. S. Wong, "Concentric anchor beacon localization algorithm for wireless sensor networks," *IEEE Trans. Veh. Technol.*, vol. 56, no. 5, pp. 2733–2744, Sep. 2007.

[7] A. T. Ihler, J. W. Fisher III, R. L. Moses, and A. S. Willsky, "Nonparametric belief propagation for self-localization of sensor networks," *IEEE J. Sel. Areas Commun.*, vol. 23, no. 4, pp. 809–819, Apr. 2005.

[8] A. T. Ihler, "Inference in sensor networks: graphical models and particle methods," Ph.D. thesis, Department of Electrical Engineering and Computer Science, MIT, Jun. 2005.

[9] H. Wymeersch, J. Lien, and M. Z. Win, "Cooperative localization in wireless networks," in *Proc. of IEEE*, vol. 97, no. 2, Feb. 2009, pp. 427–450.

[10] R. Peng and M. L. Sichitiu, "Robust, probabilistic, constraint-based localization for wireless sensor networks," in *IEEE Proc. of 2nd Conf. on Sensor and Ad Hoc Communications and Networks—SECON*, Sep. 26–29, 2005, pp. 541–550.

[11] A. Baggio and K. Langendoen, "Monte Carlo localization for mobile wireless sensor networks," in *Ad-Hoc Networks*, vol. 6, no. 5, pp. 718–733, Jul. 2008.

[12] H. Wymeersch, S. Marano, W. M. Gifford, and M. Z. Win, "A machine learning approach to ranging error mitigation for UWB localization," *IEEE Trans. Commun.*, pp. 1719–1728, Jun. 2012.

[13] V. Savic and E. G. Larsson, and J. Ferrer-Coll, and P. Stenumgaard "Kernel methods for accurate UWB-based ranging with reduced complexity," *IEEE Trans. Wireless Commun.*, pp. 1783–1793, Mar. 2016.

[14] N. Patwari, J. N. Ash, S. Kyperountas, A. O. Hero III, R. L. Moses, and N. S. Correal, "Locating the nodes: Cooperative localization in wireless sensor networks," *IEEE Signal Process. Mag.*, vol. 22, no. 4, pp. 54–69, Jul. 2005.

[15] G. Mao, B. Fidan, and B.D.O. Anderson, "Wireless sensor network localization techniques," *Comput. Netw.*, vol. 51, no. 10, pp. 2529–2553, Jul. 2007.

[16] T. S. Rappaport, *Wireless Communications: Principles and Practice*, 2nd ed. Upper Saddle River, NJ: Prentice Hall, 2001.

[17] S. Gezici, Z. Tian, G. B. Giannakis, H. Kobayashi, A. F. Molisch, H. V. Poor, and Z. Sahinoglu, "Localization via ultra-wideband radios: A look at positioning aspects for future sensor networks," *IEEE Signal Process. Mag.*, vol. 22, no. 4, pp. 70–84, Jul. 2005.

[18] C.-Y. Chong and S. P. Kumar, "Sensor networks: evolution, opportunities, and challenges," *Proc. of the IEEE*, vol. 91, no. 8, pp. 1247–1256, Aug. 2003.

[19] M. S. Arulampalam, S. Maskell, N. Gordon, and T. Clapp, "A tutorial on particle filters for online nonlinear/non-Gaussian bayesian tracking," *IEEE Trans. Signal Process.*, vol. 50, no. 2, pp. 174–188, Feb. 2002.

[20] J. S. Yedidia, W. T. Freeman, and Y. Weiss, "Understanding belief propagation and its generalizations," in *Exploring Artificial Intelligence in the New Millennium*, vol. 8, pp. 239–269, Jan. 2003.

[21] M. J. Wainwright and M. I. Jordan, *Graphical Models, Exponential Families, and Variational Inference: Foundations and Trends in Machine Learning*, Hanover, MA: Now Publishers, 2008.

[22] J. Pearl, *Probabilistic Reasoning in Intelligent Systems: Networks of Plausible Inference*. San Francisco, CA: Morgan Kaufmann, 1988.

[23] P. M. Djuric, J. H. Kotecha, J. Zhang, Y. Huang, T. Ghirmai, M. F. Bugallo, and J. Miguez, "Particle filtering," *IEEE Signal Process. Mag.*, vol. 20, no. 5, pp. 19–38, Sep. 2003.

[24] V. Savic and S. Zazo, "Nonparametric boxed belief propagation for localization in wireless sensor networks," in *IEEE Proc. of SENSORCOMM*, Jun. 2009, pp. 520–525.

[25] B. W. Silverman, *Density Estimation for Statistics and Data Analysis*. New York: Chapman & Hall, 1986.

[26] B. J. Frey, "A revolution: belief propagation in graphs with cycles," in *Proc. of Advances in Neural Information Processing Systems*, vol. 10, 1997, pp. 479–185.

[27] M. I. Jordan and Y. Weiss, "Graphical model: Probabilistic inference," in *The Handbook of Brain Theory and Neural Networks*, 2nd ed. Cambridge, Micheal A. Arbib, Ed., MA: MIT Press, 2002.

[28] V. Savic and S. Zazo, "Sensor localization using nonparametric generalized belief propagation in network with loops," in *IEEE Proc. of Information Fusion*, Jul 2009, pp. 1966–1973.

[29] V. Savic and S. Zazo, "Pseudo-junction tree method for cooperative localization in wireless sensor networks," presented at the IEEE Proc. of Information Fusion, Edinburgh, UK, Jul. 2010.

[30] V. Savic and S. Zazo, "Nonparametric belief propagation based on spanning trees for cooperative localization in wireless sensor networks," *Proceedings of 2010 IEEE Vehicular Technology Conference—Fall*, Jul. 2010.

[31] D. A. Bader and K. Madduri, "Designing multithreaded algorithms for breadth-first search and st-connectivity on the Cray MTA-2," in *IEEE Proc. of Parallel Processing—ICPP*, Aug. 2006, pp. 523–530.

[32] Y. Weiss, "Correctness of local probability propagation in graphical models with loops," *Neural Comput.*, vol. 12, no. 1, pp. 1–41, Jan. 2000.

[33] Y. Weiss and W. T. Freeman, "Correctness of belief propagation in Gaussian graphical models of arbitrary topology," *Neural Comput.*, vol. 13, no. 10, pp. 2173–2200, Oct. 2001.

[34] J. M. Mooij and H. J. Kappen, "Sufficient conditions for convergence of the sum-product algorithm," *IEEE Trans. Inf. Theory*, vol. 53, no. 12, pp. 4422–4437, Dec. 2007.

[35] M. J. Wainwright, T. S. Jaakkola, and A. S. Willsky, "Tree-reweighted belief propagation algorithms and approximate ML estimation by pseudo-moment matching," presented at the Proc. of AISTATS, Key West, FL, 2003.

[36] V. Savic, H. Wymeersch, F. Penna, and S. Zazo, "Optimized edge appearance probability for cooperative localization based on tree-reweighted nonparametric belief propagation," in *Proc. of IEEE Int. Conf. on Acoustics, Speech and Signal Processing (ICASSP)*, May 2011, pp. 3028–3031.

CHAPTER 27

ERROR CHARACTERISTICS OF AD HOC POSITIONING SYSTEMS

Dragoş Niculescu

ETTI, University Politehnica of Bucharest, Bucharest, Romania

POSITION AND orientation information of individual nodes in ad hoc and sensor networks is useful for both service and application implementation. Services that can be enabled by the availability of position include routing, querying, and discovery. At the application level, position is required in order to label the reported data in a sensor network, whereas position and orientation enable tracking. Nodes in an ad hoc network may have local capabilities such as the possibility of measuring ranges to neighbors, direction of arrival, or global capabilities, such as GPS and digital compasses. Ad hoc positioning system (APS) is a family of multihop positioning schemes using the basic idea of distance vector (DV) routing to find positions in an ad hoc network using only a fraction of landmarks, or anchors. All nodes within a network are assumed to have the capability of measuring ranges, direction of arrival, orientation, or a combination of them. In this chapter, a positioning error is computed in a multihop network for a range/angle-free algorithm, and the error characteristics of four classes of multihop APS algorithms under various conditions are examined, using theoretical analysis and simulations. Analysis of range/angle-free, range-based, angle-based, and multimodal algorithms maintains a complex trade-off between network density, ratio of landmarks, and the estimated position performance.

27.1 INTRODUCTION

In many applications data collected by sensors should be associated with the position information. An example is the meteorological data collected by aircrafts or the image information collected and submitted by endoscopy capsules (as discussed in

Handbook of Position Location: Theory, Practice, and Advances, Second Edition.
Edited by S. A. (Reza) Zekavat and R. Michael Buehrer.
© 2019 by the Institute of Electrical and Electronics Engineers, Inc.
Published 2019 by John Wiley & Sons, Inc.
Companion Website: www.wiley.com/go/zekavat/positionlocation2e

Chapter 10). Position information along with images taken in endoscopy enables physicians and surgeons to better threat or operate on a patient. Meteorological data collected by an aircraft along with the location information help meteorologists to better monitor climate and maintain ties across climate and regional or global human activities and assess the impact of human activities or pollution on environment.

For some ad hoc networks, algorithms such as Cartesian routing [1], geocast [2] or GoAFR [3] enable routing with reduced or no routing tables. These algorithms are appropriate for devices such as the Rene mote [4], with only half a kilobyte of RAM. Location-aided routing [5] is an improvement that can be applied to some ad hoc routing schemes when position is available by limiting the search for a new route to a smaller destination zone. Positioning and orientation algorithms are appropriate for indoor location-aware applications, when the network's main feature is not the unpredictable, highly mobile topology but rather temporary and ad hoc deployment. These networks would not justify the cost of setting up an infrastructure to support positioning, as is proposed in [6–8].

When deploying an ad hoc network that uses positions at application or service levels, the important decisions to be made are with respect to the node density, node capabilities versus the desired quality of the positions. "One-hop" solutions, such as GPS, in which each node is within direct communication of the satellites, are desirable for their simplicity but may not be applicable in many setups because of these challenges: form factor, power, cost, or line-of-sight conditions. To deal with the challenges, we may apply positioning capability in a small fraction of the nodes, called landmarks or anchors. Anchor nodes distribute position computation through the network in a multihop, collaborative fashion. The collaborative positioning methods that are the object of this chapter, while being more appropriate with respect to the mentioned conditions, face problems of error propagation and provisioning. Error buildup is inherent in a multihop environment because angle and distance measurements are affected by error at each hop. Range measuring methods that can be used for these applications include received signal strength indicator (RSSI) and time of arrival (TOA) based on difference in propagation speed between ultrasound and radio frequency (RF) signals. Angle measurements could be provided by antenna arrays, or by time difference of arrival (TDOA) methods, as shown in the prototype of the Cricket compass [9].

Provisioning of large ad hoc networks includes not only the networking-related aspects but also the management of the error, when positioning is derived in a multihop fashion. As we show in this chapter, there is a trade-off between the capabilities of the nodes (angle and range measuring hardware), the density of the network, the placement of landmarks, and the quality of the positions estimates obtained. By analyzing four multihop algorithms that make use of different capabilities—range/angle free, angle, range, and angle + range—we examine the relationship between position error and the way network is provisioned.

First, we derive a Cramer–Rao lower bound (CRLB) for positioning error of a multihop range/angle-free positioning algorithm (APS/*DV-hop*)—see (27.11). Its importance is in bounding the position error achievable by this multihop algorithm for a given network setup. The simulated algorithm closely follows the trend of the theoretical bound. Second, we analyze a multimodal algorithm that uses both angle and range measurements. Although it has higher hardware requirements, its behavior is predictable through analysis, and its performance is bounded in the parameter space by three other algorithms: range/angle free, range based, and angle based. Its complex relationship with the other analyzed algorithms provides insights with respect to the quality of hardware needed to obtain high-quality positions in a multihop environment (Fig. 27.12). An important insight is that for the analyzed algorithms, if the statistical measurement error is beyond a certain limit, then it is no better than not measuring at all. When provisioning an ad hoc network in terms of density or measurement hardware used, this result helps in choosing a solution that gives the lowest positioning error for the given conditions.

The chapter is organized as follows: the next section has a short review of the APS family of algorithms, which are the subject of this analysis. Section 27.3 introduces error models for trilateration (27.3.2), *DV-hop* (27.3.4), and *DV-position* (27.3.5) algorithms. In Section 27.4, we comparatively analyze four algorithms with different hardware requirements. For better flow of the presentation, most mathematical derivations are detailed in the appendices at the end of the chapter.

27.2 APS ALGORITHMS

In most multihop solutions, it is assumed that positioning systems, measure location with respect to a node's neighbor. A node may measure *distance* or *range* relative to its neighbors, with a given accuracy. Alternatively, it may measure *bearing* with respect to the node's axis, or *angle* from which the signal is received, sometimes called direction-of-arrival (DOA). In extreme cases, a node has neither of the mentioned capabilities and has to rely only on simple connectivity. In the case where it has them both, it can make more precise estimations, and we refer to this as "multimodal position estimation."

The positioning problem that we intend to address is as follows: *given imprecise bearing/range measurements to neighbors in a connected ad hoc network where a small fraction of the nodes have self-positioning capability, estimate the positions for all the nodes in the network.* The difficulty of the problem stems from the fact that the capable nodes (landmarks) comprise only a small fraction of the network, and most regular nodes are not in direct contact with enough landmarks.

APS [10, 11] is a hybrid between two major concepts: DV routing, and beacon-based positioning (GPS). What makes APS similar to DV routing is the fact that information is forwarded in a hop by hop fashion independent of each landmark. What makes it similar to GPS is that eventually each node estimates its own position based on the landmark readings it gets. The APS concept has been shown to work using simple connectivity, range measurements [10], angle measurements [11], and a combination of them [12].

As in DV, each node independently estimates its distance to a landmark and then further propagates this estimate to other nodes. In this fashion, further-away nodes have the chance to estimate their own distance/bearing with respect to the landmark merely using measurements to neighbors. All propagations work very much like a mathematical induction proof. The fixed point nodes immediately adjacent to a landmark get their bearings/ranges directly from the landmark. The induction step: Assuming that a node has some neighbors with bearing/range for a landmark, it will be able to compute its own bearing/range with respect to that landmark, and forward it further into the network. What remains to be found is a method to compute this induction step, for any combination of local capabilities: connectivity only, ranging, angle measurements, and multimodal = angle + range.

If for some reason a node does not get enough ranges/bearings in order to triangulate/trilaterate its position, it could wait for the neighbors to successfully position themselves and use local measurements to get a position, or simply use a weighted average of the positions of those neighbors. Even if the position is available at a node, smoothing by incorporating position measurements made by neighbors has been reported beneficial in certain cases [13–15].

27.2.1 *DV-Hop* Propagation Method

This is the most basic scheme, which does not require any hardware to measure ranges or angles to neighbors. It only relies on the connectivity of the underlying graph and it comprises three nonoverlapping stages.

I. First, it employs a DV exchange so that all nodes in the network computer shortest paths, in hops, to the landmarks. Each node maintains a table $\{X_i, Y_i, h_i\}$ and exchanges updates only with its neighbors. X_i, Y_i are the coordinates of landmark i, and function here as a name for the landmark as well. h_i is the shortest distance in hops to that landmark. This phase is the classical Bellman–Ford distributed shortest path algorithm.

II. In the second stage, after it accumulates distances to other landmarks, a landmark estimates an average size for one hop, which is then deployed as a correction to the nodes in its neighborhood.

III. After receiving the correction, an arbitrary node may estimate distances to landmarks, in meters, which can be used to perform the trilateration. The correction a landmark (X_i, Y_i) computes is

$$c_i = \frac{\sum \rho_j}{\sum h_j}, i \neq j, \text{all landmarks } j \text{ heard by } i, \qquad (27.1)$$

where h_j is the shortest distance, in hops from node i to node j, and $\rho_j = \sqrt{(X_i - X_j)^2 + (Y_i - Y_j)^2}$ is the straight-line distance between landmarks i and j.

Example 27.1 Numerical Illustration for *DV-Hop*

In Figure 27.1, nodes L_1, L_2, and L_3 are landmarks, with the known distances indicated. The goal is for node A to find ranges to landmarks by following the stages of the algorithm.

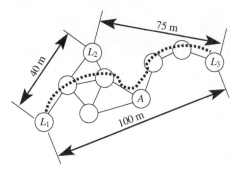

Figure 27.1 Example of propagation and correction computation for *DV-hop*.

Solution

Following stage I of *DV-hop*, L_1 has both the Euclidean distance to L_2 and L_3, and the path length of two hops and six hops, respectively. L_1 then computes the correction $(100 + 40)/(6 + 2) = 17.5$, which is in fact the estimated average size of one hop, in meters. L_1 has the choice of either computing a single correction to be broadcast into the network or preferentially sending different corrections along different directions. In our experiments, we are using the first option. In a similar manner, L_2 computes a correction of $(40 + 75)/(2 + 5) = 16.42$ and L_3 a correction of $(75 + 100)/(6 + 5) = 15.90$. During stage II, a regular such as the A node gets an update from one of the landmarks, and it is usually the closest one, depending on the deployment policy and the time the correction phase starts at each landmark. Corrections are distributed by controlled flooding, meaning that once a node gets and forwards a correction, it will drop all the subsequent ones. This policy ensures that most nodes will receive only one correction, from the closest landmark. When networks are large, a method to reduce signaling would be to set a time to live (TTL) field for propagation packets, which would limit the number of landmarks acquired by a node. Here, controlled flooding helps keep the corrections localized in the neighborhood of the landmarks they were generated from, thus accounting for anisotropies across the network. In the above example, assume A gets its correction from L_2—its estimate distances to the three landmarks would be as follows: to L_1, 3×16.42; to L_2, 2×16.42; and to L_3, 3×16.42. These values are then plugged into the triangulation procedure described in the previous section for A to get an estimate position.

The advantages of *DV-hop* are that it is completely distributed and localized, and it only relies on connectivity. The drawback is that it will only work for isotropic networks, that is, when the properties of the graph are the same in all directions, so that the corrections that are deployed reasonably estimate the distances between hops. Also, the shortest path to the landmark is assumed to be obtained by the DV procedure. This might not be the case in variable duty cycle sensor networks, mobile networks, or when processing is slow. When this happens, a landmark proceeds to stage II, or a node to stage III with a less than optimal path. These problems can be alleviated with careful provisioning of the time-outs between the stages in order to accommodate proper operating ranges for the density and the size of the network.

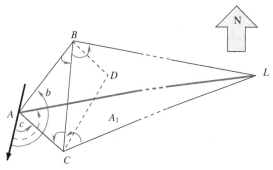

Figure 27.2 *DV-Euclidean, DV-bearing* propagation.

27.2.2 *DV-Euclidean* and *DV-Radial*

The range-based scheme *DV-Euclidean* works by propagating the estimated Euclidean distance to the landmark, so this method is the closest to the nature of GPS. *DV-Euclidean* makes use of range measurements to neighbors. The angle-based scheme *DV-bearing* propagates in a similar fashion, but each node computes and forwards a bearing—the angle under which it "sees" a landmark. It therefore needs angle measurement hardware, and can benefit from the presence of a compass in each node of the network.

I. An arbitrary node *A* needs to have at least two neighbors, *B* and *C*, that have estimates—angles or ranges—for the landmark *L* (Fig. 27.2). For *DV-Euclidean*, *A* also has measured estimates of distances for *AB*, *AC*, and *BC*, so there is the condition that either *B* and *C*, besides being neighbors of *A*, are neighbors of each other, or *A* knows distance *BC*, from being able to map all its neighbors in a local coordinate system. For the quadrilateral *ABCL*, all the sides are known and the diagonal *BC* is also known. This allows node *A* to compute the second diagonal *AL*, which in fact is the Euclidean distance from *A* to the landmark *L*. For *DV-bearing*, *A* knows all the angles in triangles $\triangle ABC$ and $\triangle BCL$, indicated with continuous arrows. This allows for the computation of the bearing of *A* with respect to *L*, indicated with a dashed arrow. This step is the induction step, which propagates the capability from nodes *B* and *C* to node *A* that is one hop further from landmark *L*.

II. Once a node has ranges or angles to three landmarks, it may, by itself, estimate its position by applying a trilateration procedure for the range-based, or a triangulation procedure for the angle-based method. The trilateration procedure is described in Section 27.3.1.

The advantage of the range/angle-based methods is that they provide better accuracy under certain conditions, and there is only one communication stage (no corrections). On the downside, when compared with *DV-hop*, they require additional hardware, which makes them susceptible to measurement error.

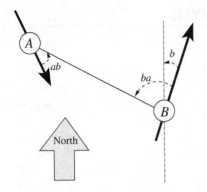

Figure 27.3   *DV-position* propagation method.

27.2.3   *DV-Position*

If ranging and angle measurements (DOA) are both available at all nodes, *DV-position* can be used as a method that makes simultaneous use of both capabilities for increased accuracy.

If compasses are available at landmarks, neighbors of landmarks can position themselves in one step using *DV-position*. In Figure 27.3, if landmark *B* has a compass and can measure angle $\hat{b}$, and also DOA capability to measure the angle *ba* at which *A* is seen, then it is possible to compute the equation line on which *A* is placed. If the range *AB* is known, node *A* can find its own position. The absolute orientation of *A* can also be found, as $2\pi - (\widehat{ba} + \pi - \widehat{ab}) + \hat{b}$. Using further propagation, node *A* can then behave as a less accurate landmark for further-away nodes.

DV-position is a much simpler scheme than the previous ones, but requires more hardware—both angle and range measurements. Nevertheless, its analysis is meaningful since its provides some insights into the behavior of the other algorithms and the quality of the hardware that is needed to produce good positions. Prototypes of small form factor nodes that feature the DOA capability (such as the Cricket compass [9]) and good ranging capability (such as the Medusa node [14]) make multimodal positioning a feasible solution.

The schemes examined can be compared from different points of view. One is hardware requirements—while *DV-hop* does not have any (beyond the required communication connectivity), *DV-position* requires a combination of distance and angle measurements. This can potentially be expensive both in terms of cost and in terms of battery consumption. Another aspect is the computation requirement at each node, and in this respect most schemes fare similarly, as they end with a trilateration or triangulation procedure that may involve iterations. *DV-position* propagates position so it could potentially require less computation. In terms of communication overhead, the DV part of the propagation amounts to one network-wide flooding per landmark, followed by the equivalent of a single flooding to propagate the corrections. *DV-position*, on the other hand, requires the equivalent of a single network-wide flooding, which is substantially less than the other ones.

27.3   POSITIONING ERROR ANALYSIS

27.3.1   Trilateration Review

In GPS [16], trilateration (in reality multilateration) uses range estimates to at least three known satellites to find the coordinates of the receiver, and four satellites to also find the clock bias of the receiver. For our ad hoc positioning purposes, we are using a simplified version of the trilateration, as we only deal with distances, and there is no need for clock synchronization. The trilateration procedure starts with a priori estimated position that is later corrected toward the true position. Let $\hat{\mathbf{x}}$ be the estimated position, $\mathbf{x} = [xy]^T$ the real position, $\mathbf{x}_i = [x_iy_i]^T$ the known coordinates of the satellites, $\rho = \|\mathbf{x}_i - \mathbf{x}\|$ the true range, and $\hat{\rho}_i = \|\mathbf{x}_i - \hat{\mathbf{x}}\|$ the measured range to satellite i. The problem is to find $\mathbf{x}$ given ρ_i and $\mathbf{x}_i$. We start with an initial guess $\hat{\mathbf{x}}$, which provides associated $\hat{\rho}_i$. The distance equation to each satellite is $\rho_i = \sqrt{(x_i - x)^2 + (y_i - y)^2}$. The correction of the range, $\Delta\rho$ is approximated linearly using Taylor expansion. If J_i is the unit vector of $\hat{\rho}_i$, $\hat{J} = (\mathbf{x}_i - \hat{\mathbf{x}})/\hat{\rho}_i$ and $\Delta\mathbf{x} = \hat{\mathbf{x}} - \mathbf{x}$, then the approximation of the correction in the range is: $\Delta\rho = \hat{\rho}_i - \rho_i \simeq \hat{J}_i \cdot \Delta\mathbf{x}$. Performing the above approximation for each satellite independently leads to a linear system in which the unknown is the position correction $\Delta\mathbf{x} = [\Delta x \Delta y]$:

$$\Delta\rho = J\Delta\mathbf{x}$$

$$
\begin{bmatrix} \Delta\rho_1 \\ \Delta\rho_2 \\ \dots \\ \Delta\rho_n \end{bmatrix} = \begin{bmatrix} \hat{J}_{1x} & \hat{J}_{1y} \\ \hat{J}_{2x} & \hat{J}_{2y} \\ \dots & \dots \\ \hat{J}_{nx} & \hat{J}_{ny} \end{bmatrix} \begin{bmatrix} \Delta x \\ \Delta y \end{bmatrix}.
\tag{27.2}
$$

When all measurements have the same uncertainty, the system is solved as with ordinary least squares (OLS): $\Delta\rho = J\Delta\mathbf{x} \Rightarrow \Delta\mathbf{x} = J^+\Delta\rho$, where $J^+ = (J^TJ)^{-1}J^T$ is the pseudo-inverse of J. If independent uncertainties σ_i are available for range estimates, the above system is solved using weighted least squares (WLS) with weights $W = \text{diag}\{1/\sigma_i^2\} = \text{Cov}[\Delta\rho]^{-1}$, J^+ becomes

$$J^+ = (J^TWJ)^{-1}J^TW \tag{27.3}$$

$$
\begin{aligned}
\text{Cov}[\Delta\mathbf{x}] &= E[\Delta\mathbf{x}\Delta\mathbf{x}^T] \\
&= E[J^+(\Delta\rho\Delta\rho^T)(J^+)^T] \\
&= J^+\text{Cov}[\Delta\rho](J^+)^T \\
&= J^+W^{-1}J^{+T}.
\end{aligned}
\tag{27.4}
$$

After each iteration, the corrections Δx and Δy are applied to the current position estimate. The iteration process stops when the correction in position is below a chosen threshold. The uncertainties W are obtained as the estimated errors of the ranges computed by *DV-hop* and *DV-Euclidean*. If W is known, it is actually possible to give a lower bound for the covariance of **x**, or position error, that can be achieved by trilateration.

27.3.2 CRLB for Trilateration

The CRLB is a method that sets a lower bound on the variance of *any* unbiased estimator. Its provides a benchmark to evaluate the performance of an estimator. In our case, the trilateration problem is cast as an estimation problem by considering the true position **x** as a parameter to be estimated. The distribution of errors of distances to landmarks $\hat{\rho}$ is assumed to be Gaussian.

Using notations introduced in the previous section, we now estimate the error in the obtained position given the range to landmark estimation error. This approach is applicable to all the algorithms that use trilateration as a final phase (*DV-hop* and *DV-Euclidean*). The hop-by-hop nature of multihop algorithms always produces normal errors for a sufficiently large number of hops—we confirmed this by simulation for all algorithms mentioned here.

In the Appendix section "Trilateration CRLB," we show that the CRLB of the variance in the estimated parameter x is:

$$\text{CRLB}(\text{Cov}[\mathbf{x}]) = (J_0^{\mathsf{T}} W J_0)^{-1}, \tag{27.5}$$

where $J_0 = (\mathbf{x}_i - \mathbf{x})/\rho_i$, using the true positions and ranges. This means that for a known setup, when landmark constellation is given as $\mathbf{x}_i$, true position as $\mathbf{x}$, and errors in ranges as W, the best estimation error will be limited by $\text{Var}[x] \geq \text{CRLB}(\text{Cov}[\mathbf{x}])_{11}$ and $\text{Var}[y] \geq \text{CRLB}(\text{Cov}[\mathbf{x}])_{22}$. While we do not have an analytical model for *DV-Euclidean*'s uncertainties W, they are propagated through the network and used in the triangulation phase (details can be found in [10]).

27.3.3 *DV-Hop* Range Error

In *DV-hop*, the main source of range error stems from the fact that a node translates the length of the shortest path to a landmark into a Euclidean distance by assuming that every hop produces progress c_i—(27.1). Greedy geographic forwarding has been shown in [17] to produce paths with low dilation, that is, good shortest path approximates. This means that if we apply a geographic forwarding policy such as the most forwarding within radius (MFR) [18] in a *uniform, dense* network, the progress made by each hop is a good approximation for c_i. If the network is not dense enough, areas with low connectivity result in either dropping the packet by the greedy forwarding scheme or requiring the use of an obstacle-avoiding algorithm. The analysis provided here only covers the cases when the network is dense enough so that simple greedy forwarding succeeds. High density, however, is a requirement for a network to be connected in the case where nodes are distributed randomly in a large network [19]. Kleinrock and Silvester [18] show that for greedy forwarding, the progress only depends on the node density. Their analysis is general and considers the cases when no advancement is possible along the direction from the landmark, and the case with isolated nodes:

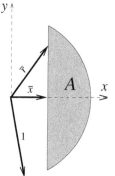

Figure 27.4   Progress per hop for the MFR policy. In a unit disk graph, each node can communicate with nodes placed at distances less than 1. Forwarding along horizontal direction produces advancement $\bar{x}$ by using the node at a distance $\bar{r}$, which is the rightmost inside the circle.

$$\hat{c}_i = 1 + e^{-\lambda\pi} - \int_{-1}^{1} e^{-\lambda\left(\arccos t - t\sqrt{1-t^2}\right)}\,dt. \tag{27.6}$$

We aim for a simplified version where we ignore corner cases, but show that the estimation of one hop is reasonably accurate. For *DV-hop*, this implies that in isotropic (uniform density) networks, nodes may locally estimate the correction based on their local perception of density, and thus eliminate the need for the second stage of the algorithm. Assuming a Poisson spatial distribution of nodes with rate λ, and a wireless range of radius 1, the expected number of neighbors of a node will be $\pi\lambda$. Throughout the chapter, we will use both the average number of neighbors, or the rate λ when referring to density. A node chooses a next hop that produces the most progress $\bar{x}$ toward a destination on the horizontal axis (Fig. 27.4). There are no nodes in area A, which allows for the following derivation of the distribution of $\bar{x}$:

$$A(\bar{x}) = 2\int_{\bar{x}}^{1} \sqrt{1-t^2}\,dt = \arccos(\bar{x}) - \bar{x}\sqrt{1-\bar{x}^2}$$

$$F_{\bar{x}}(\bar{x}) = Pr\{\bar{X} \le \bar{x}\} = e^{-\lambda A(\bar{x})}$$

$$f_{\bar{x}}(\bar{x}) = Pr\{\bar{X} = \bar{x}\} = -\lambda A'(\bar{x})e^{-\lambda A(\bar{x})} \tag{27.7}$$

$$= 2\sqrt{1-\bar{x}^2}\,e^{-\lambda\left(\arccos(\bar{x}) - \bar{x}\sqrt{1-\bar{x}^2}\right)}$$

$$E[\bar{x}] = \int_{-1}^{1} \bar{x}f(\bar{x})\,d\bar{x}$$

$$= 2\lambda\int_{-1}^{1} \bar{x}\sqrt{1-\bar{x}^2}\,e^{-\lambda\left(\arccos(\bar{x}) - \bar{x}\sqrt{1-\bar{x}^2}\right)}\,d\bar{x} \tag{27.8}$$

$$V[\bar{x}] = \int_{-1}^{1} \bar{x}^2 f(\bar{x})\,d\bar{x} - \left(\int_{-1}^{1} \bar{x}f(\bar{x})\,d\bar{x}\right)^2.$$

Unfortunately, there are no closed forms for the first and second moments $E[\bar{x}]$ and $V[\bar{x}]$ of this distribution, but it can be approximated as a beta distribution. In

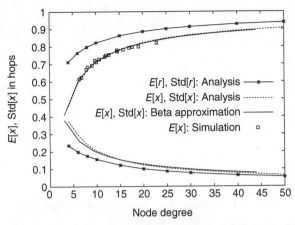

Figure 27.5   Mean and standard deviation of $\bar{x}, \bar{r}$. Node degree is the mean number of neighbors for each node (for a unit radius, $N = \pi\lambda$).

Figure 27.5, numerically evaluated variance $V[\bar{x}]$ and expectation $E[\bar{x}]$ as functions of density λ are shown together with the approximation to the beta distribution. Also shown in Figure 27.5 are values of c_i obtained from simulation* in a network with 1000 nodes with increasing densities, which match very closely the values obtained from the numerical evaluation for the respective densities. Since these values only depend on the density of the network, we precomputed a table with all the values necessary for the experiments in this chapter, namely for densities up to 20 neighbors.

In *DV-hop*, range estimates are obtained as $\rho_I = h_i c_i$, where h_i is the number of hops to landmark i, and c_i is approximated by $E[\bar{x}]$. If a geographic forwarding model is used to approximate the shortest path to the landmark, after jumping h_i times (assumed to be independent), the obtained range is $E[\rho_i] = h_i E[\bar{x}]$ and its variance is

$$V[\rho_i] = h_i V[\bar{x}]. \tag{27.9}$$

Referring to Figure 27.4, we now explore the behavior of $\bar{r}$, the actual distance traveled in one hop by the MFR policy. Its distribution is actually needed for the analysis of *DV-position*, but its characterization is based on the probability density function (PDF) of $\bar{x}$. The PDF of $\bar{r}$ is obtained as a sum of probabilities of a next hop, summed over the entire circle of radius $\bar{r}$.

$$f_{\bar{R}}(\bar{r}) = \int_0^{2\pi} \frac{f_{\bar{x}}(\bar{r}\cos t)}{2\sqrt{1 - \bar{r}^2 \cos^2 t}} dt \tag{27.10}$$

*We implemented our own null MAC, unit disk graph, event-based simulator with minimal support for positioning. The simulator uses an error model for angle and range measurements, and a broadcast primitive with perfectly circular coverage. The decision not to use an existing simulator such as the *ns-2* was based on (1) the need to use large topologies (10,000 nodes), (2) the chapter's focus on the quality of the positions rather than on the communication aspects of the positioning protocols, and (3) the reduced number of required primitives: node broadcast and DV shortest paths.

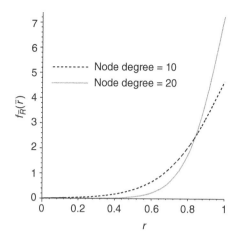

Figure 27.6   Probability distribution of $\bar{r}$ for two node densities.

$f_{\bar{R}}(\bar{r})$ is not available in closed form, but in Figure 27.6 we can see its behavior through numerical evaluation for two different densities. As expected, larger jumps are much more likely than short jumps, which means larger errors if $\bar{r}$ is to be used as a measure of distance. In Figure 27.5, the mean and standard deviation of $\bar{r}$ are shown as a function of density.

Example 27.2   Range Derivation—Density or Stage II of *DV-Hop*?

The question is to determine for a particular density how accurate is the range ρ if we derive it from density as opposed to stage II of the algorithm.

Solution

If we consider a network of 1000 nodes, for an average degree of 10.48, for this node degree $\mathrm{Std}[\bar{x}] \approx 0.2$, from Figure 27.5. Conforming to (27.9), and assuming each hop to have an independent error, the range error ρ should follow $\mathrm{Std}[\rho] = \sqrt{h}\mathrm{Std}[\bar{x}]$ if we estimate it solely by density, using $\bar{x}$. To verify this hypothesis, we generated 100 random networks with an average density of 10.48, and ran *DV-hop* to find ranges to landmarks using stages I and II of the algorithm. In Figure 27.7, we plot the range standard deviation as a function of distance in hops. With a continuous line, we plot the $\mathrm{Std}[\rho]$ based on density, namely $0.2\sqrt{h}$. With dashed lines, we plot the error obtained by *DV-hop* using stage-II corrections for two different densities of landmarks. The behavior is similar whether for corrections we use the second stage of *DV-hop* c_i, or the same $E[\bar{x}]$ for all nodes. The independence assumption between successive jumps in the geographic forwarding procedure fails to account for holes in the network, assuming each new jump starts on the horizontal axis. This is especially visible for longer paths, when the shortest path nature of *DV-hop* has the chance of optimizing routes around holes, in reality using a different $\bar{x}$ than the analysis. Higher fractions of landmarks have a beneficial effect since each landmark generates a specific correction for its area, which gets distributed around it.

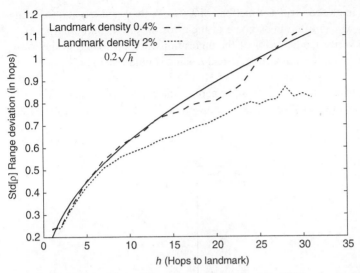

Figure 27.7   Dashed lines: Range deviation obtained by *DV-hop* in stage II, in an isotropic network. Solid line: Range deviation derived from network density. Graph was obtained by averaging 100 random networks with 1000 nodes and average density of 10.48.

In the remaining sections, the model for range error is assumed to be linear in the number of hops. This can be seen from (27.9), but also in a more intuitive way as a sum of independent and identically distributed (i.i.d.) variables $\bar{x}$. We also assume that the model extends to continuous distances; that is, the variance in range error is linear with the distance $V[\rho_i] = (\rho_i)/(E[\bar{x}])V[\bar{x}]$. This allows us to estimate the range uncertainties W that can be merged in the trilateration CRLB (27.5).

27.3.4   CRLB for *DV-Hop* Positioning

The last step of the analysis is to evaluate how range measurements are assembled into a position estimate, and what the covariance of this estimate is. We assume that *DV-hop* uses a TTL value of h for all the aspects of its protocol and that the fraction of nodes that are landmarks is f. We also assume that the landmarks location follow a spatial Poisson process with the rate $f\lambda$. A node is then able to contact on average $n = \pi(hE[\bar{x}])^2 f\lambda$ landmarks situated in a circle of radius $hE[\bar{x}]$. After applying the trilateration procedure (Section 27.3.1), the covariance of the position estimate is bounded using (27.5). The actual derivation is detailed in Appendix section "*DV-Hop* CRLB," and the obtained covariance is

$$\text{CRLB}(\text{Cov}[r_u]) = \frac{1}{f\pi h}\frac{V[\bar{x}]}{\lambda E^2[\bar{x}]}I_2. \qquad (27.11)$$

The CRLB of the positioning error covariance is not only inversely proportional to the fraction of landmarks deployed and to the TTL used by the method but also depends on the density of deployment λ (second fraction). Even if $V[\bar{x}]$ and $E[\bar{x}]$

cannot be shown in closed form as functions of λ (27.7), their approximations using the beta distribution can offer an estimate of the error for provisioning purposes.

In order to verify the behavior of the algorithm when compared to the lower bound, we traced the three parameters f, h, and λ using simulation. Figure 27.8 shows the values of the x coordinate of position standard deviation ($\text{Cov}[r_u]_{11}$ in hops) obtained by varying each parameter and keeping the other two constant, both in simulation and using (27.11). The values used were $\lambda = 10/\pi$, $h = 15$, and $f = 0.025$,

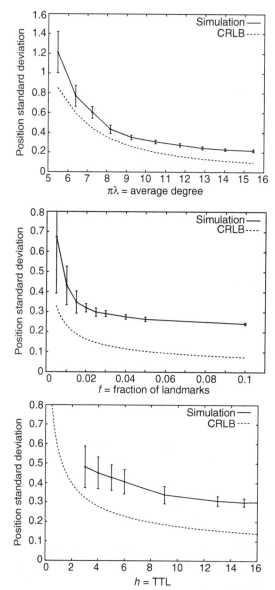

Figure 27.8   *DV-hop* positioning error as a function of density, landmark ratio, and TTL—simulation versus lower bound analysis: $\sqrt{\text{CRLB}(\lambda)}$, $\sqrt{\text{CRLB}(f)}$, and $\sqrt{\text{CRLB}(h)}$.

and the simulation was run in a network with 1000 nodes. Although the bound is not tight, it properly describes the trend of positioning performance with respect to the three parameters. The difference is a constant factor, some of which is due to our simplifying approximations. One such approximation is the independence assumption between paths to different landmarks, which may not always be disjoint. This, however, is only degrading the bound, not increasing it above the tight lower bound, which means *DV-hop* is actually closer to the bound than shown in our estimation.

A closed form of the lower bound can be determined if we do not use weighting in the multilateration process by setting $W = I$. In this case, Equation (27.22) becomes

$$J_0^T W J_0 = \begin{bmatrix} \sum \dfrac{x_i^2}{\rho_i^2} & \sum \dfrac{x_i y_i}{\rho_i^2} \\ \sum \dfrac{x_i y_i}{\rho_i^2} & \sum \dfrac{y_i^2}{\rho_i^2} \end{bmatrix} \tag{27.12}$$

$$\mathrm{CRLB}(\mathrm{Cov}[r_u]) = \frac{2}{n} I_2,$$

where n is the number of landmarks contacted by a node. It is of course a weaker bound (lower than (27.11)), but it gives an easier interpretation of the achievable accuracy by confirming the intuition that position estimates are improved when more landmarks are provided.

27.3.5 *DV-Position* Error

In order to propagate the uncertainty at each step, we assume that each node's position has an associated uncertainty covariance matrix. In order to assume normal spatial errors, DOA errors and range errors are approximated, assuming their independence. In Figure 27.9, assume the uncertainty at node B is U_B, and node A, using an DOA reading with uncertainty $\sigma_a^2 = V[a]$ and a range reading with uncertainty $\sigma_r^2 = V[r]$ infers its position with uncertainty U_A

$$U_A = U_B + \begin{bmatrix} \overline{r_x} \\ \overline{r_y} \end{bmatrix}^T \begin{bmatrix} \sigma_x^2 & 0 \\ 0 & \sigma_y^2 \end{bmatrix} \begin{bmatrix} \overline{r_x} \\ \overline{r_y} \end{bmatrix}. \tag{27.13}$$

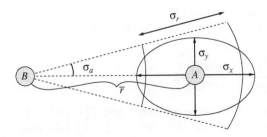

Figure 27.9 Normal approximation of uncertainty.

When angular errors are small, the eigenvalues of the covariance matrix are given by $\sigma_x^2 = \sigma_r^2$ and $\sigma_y^2 = r^2 \sin^2(\sigma_a)$, where r is the actual range measured. The unit eigenvectors $\bar{r}_x$ and $\bar{r}_y$ are given by the direction of the vector $\bar{r}$, or they can be seen as the columns of a rotation matrix. One advantage of expressing the uncertainty using a covariance matrix is that it can easily be accumulated as the position gets propagated. A second advantage is that positions obtained from different landmarks can be merged in order to get the best position estimate, as well as an estimate of its uncertainty. If $[x_i y_i]$ is the estimate from landmark i, and U_i is the uncertainty accumulated along the path from landmark i, the final uncertainty is

$$U = \left(\sum_{i=1}^{n} U_i^{-1} \right)^{-1},$$

(27.14)

and the position estimate is:

$$[xy] = \left(\sum_{i=1}^{n} [x_i \quad y_i] U_i^{-1} \right) U.$$

(27.15)

This is a simplified form of a Kalman filter that combines several "readings" with associated uncertainties in order to produce the most likely estimate and its uncertainty. Intuitively, the uncertainty grows with each error-affected measurement (27.13) but can be reduced if several landmarks are used (27.14). The actual position estimate is in fact a weighted estimate considering all the independent estimates and their uncertainties (27.15).

Since *DV-position* uses the shortest path to propagate position estimates, the range measured between two successive nodes may be approximated as a function of deployment density using the random variable for range advancement $\bar{r}$ in the MFR algorithm ((27.10) and Fig. 27.5).

In order to get a closed form for the final uncertainty U, we make the simplifying assumptions that angular errors and range errors produce a circular error ellipse. This avoids the need to apply rotation matrices at each step of the analysis when moving from a landmark to a node, and also to apply rotation matrices to combine estimates from different landmarks. Taking $\sigma_r = \sigma r$ assumes that σ is a relative deviation, meaning that uncertainty in measured range increases with the actual range, which is consistent with behavior of most radio hardware. If we model angle error as $\sin(\sigma_a) = \sigma$, for a standard angular deviation of $5.72° \approx 0.1$, a range error with a standard deviation of 10% of the measured range would be used. Again, this circular covariance is assumed just for the purposes of obtaining a closed form position error, as the actual algorithm operates with arbitrary covariances σ_a and σ_r. Using the same assumptions of Poisson distribution of nodes, and TTL limited operation of APS, the derivation of U is detailed in the Appendix section "Uncertainty for *DV-Position*" and shown to be

$$U = \frac{1}{2f\pi h} \sigma^2 \frac{E[\bar{r}^2]}{\lambda E^2[\bar{x}]} I_2.$$

(27.16)

Taking into account that σ is relative to the range measured, and $E[\bar{r}^2] = V[\bar{r}] + E^2[\bar{r}]$, relation (27.16) is similar to the CRLB derived for *DV-hop* (27.11), exhibiting the same type of dependence on the TTL h, the fraction of landmarks f, and density λ.

We simulated *DV-position* in a large network with 10,000 nodes in order to verify the validity of (27.16). In Figure 27.10, one parameter is varied, while

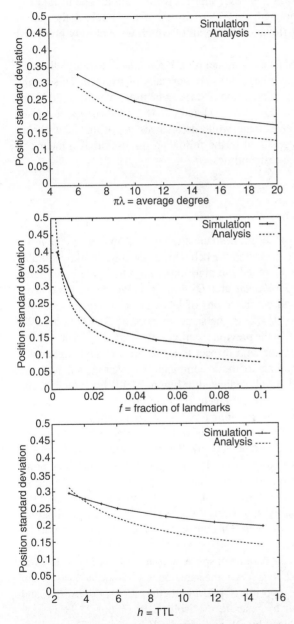

Figure 27.10 *DV-position* positioning error as a function of density, landmark ratio, and TTL for a circular error covariance ($\sigma_r = \sigma_a = 0.32$)—simulation versus analysis: $\sqrt{U_{11}(\lambda)}$, $\sqrt{U_{11}(f)}$, $\sqrt{U_{11}(h)}$.

keeping the other two constant at $\sigma = 0.32$, $h = 5$, $f = 0.01$, and $\lambda = 20/\pi$—the vertical axis shows the standard deviation of the x coordinate of the achieved position, in hops. The small difference between the analysis and the simulation is justified by the possible lack of independence assumed by (27.14) and (27.15). Even when a high degree is used, 20 in this case, shortest paths from a node to faraway landmarks may share one or more edges. We confirmed this by tracking the values of U_i (27.21), or by using a single landmark, both of which were closely predicted by simulation.

We have not yet developed an error model for *DV-Euclidean*, although through simulation we found that ranges to landmarks are normally distributed. This means that the same CRLB used for the trilateration (27.5) can be used, which would likely provide uncertainty estimates similar to (27.11) and (27.16). *DV-bearing*, however, uses triangulation as the final phase, and while its analysis is not included in this study, simulation-based results are used in the following discussion as a means to cover all options for the hardware capabilities.

27.4 DISCUSSION

A closed form for the general case *DV-position* uncertainty U (27.14) would be more laborious to obtain, therefore we evaluated the behavior of the algorithm by simulation. We also compared it with the other two algorithms for which we do not have closed form uncertainties (*DV-Euclidean* and *DV-bearing*). We tracked the error produced by the algorithm for all combinations of DOA error and range error for values of $\sigma_r = [0, 0.9]$ and $\sigma_a = [0, \pi/2]$ in the same network of 10,000 nodes. We now fix the three parameters used in the previous section at $f = 0.01$, $h = 15$, $\lambda = 10/\pi$, and look at the algorithm's behavior with respect to range error σ_r and angle error σ_a. The error surface corresponding to these combinations is projected into the parameter space with a few error levels indicated in Figure 27.11. For example, the curve labeled $U_{11} = 0.1$ indicates combinations of standard deviations for angle and range measurements that produce a positioning error standard deviation of 0.1 hops. An interesting isoline is the one for positioning error of 0.3 because that is the performance achieved by *DV-hop* in the same network. This curve almost fits a straight line that partitions the parameter space between the two algorithms. There is no point in using *DV-position* to get worse errors than what *DV-hop* gets, the latter doing so without any measurement hardware. For these particular conditions, precision of the hardware must satisfy the inequality $\sigma_r + 1.3\sigma_a < 0.75$ in order to use anything else than a range-free algorithm like *DV-hop*. Of course, this demarcation line will be different for different conditions.

In order to further partition the parameter space, we ran *DV-radial* (an improved version of *DV-bearing*) and *DV-Euclidean* in the same network in order to ascertain in which situations these two algorithms perform better than *DV-position* and *DV-hop*. The same surface used to generate the isolines in Figure 27.11 is now intersected with the curves produced by *DV-radial* and *DV-Euclidean*. The resulted intersection lines span different error levels, but their projection into parameter space is indicated in Figure 27.12. The dashed line marks points (σ_a, σ_r) in which

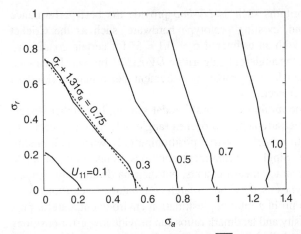

Figure 27.11 *DV-position* position error $\sqrt{U_{11}}$: The error surface is projected onto the plane of parameters. The 0.3 isoline indicates performance of *DV-hop* in the same network ($h = 15, f = 0.01, \lambda = 10/\pi$). σ_a is the error in angle measurement, and σ_r is the error in range measurement.

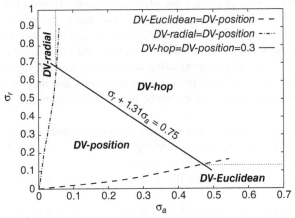

Figure 27.12 Partitioning of parameter space between four algorithms: *DV-hop* (range/ angle free), *DV-radial* (angle based), *DV-Euclidean* (range based), and *DV-position* (multimodal: range and angle based). σ_a is the error in angle measurement, and σ_r is the error in range measurement.

DV-position using angle errors of σ_a and range errors of σ_r performs as well as *DV-Euclidean*, which only makes use of range measurements with the same error σ_r. Due to the increased sensitivity to propagated errors, *DV-radial* and *DV-Euclidean* use parameters $f = 0.05, h = 8, \lambda = 10/\pi$.

The common line of *DV-hop* and *DV-position* is the same as in Figure 27.11, an isoline indicating positioning error of 0.3. This partitioning of the parameter space shows that in absence of high-performance measuring hardware for both angles and ranges, range-free algorithms such as *DV-hop* provide the best performance trade-off. On the other hand, existing prototype hardware, such as the Cricket compass [9], provide DOA with an estimated $\sigma_a = 0.1 \simeq 5°$ for certain conditions. This is not enough to choose the angle-only algorithm *DV-radial*, but may be enough for *DV-position*, given a satisfactory range measurement performance (Medusa nodes [14] achieve centimeter accuracy).

This study confirms the intuition that multimodal sensing has higher performance compared to single measurement approaches (angle only, or range only). In addition, it can compete with range/angle-free positioning only when high-quality measurements are available. Single measurement solutions (range only, or angle only) require even higher accuracy measurements, but their hardware requirements are available today in prototypes used by the research community. Even if measurement error inherently builds up in a multihop estimation environment, careful provisioning of the network density and landmark ratios can provide low-error positions for most combinations of capabilities.

27.5 RELATED WORK

Positioning in ad hoc networks has received a lot of attention, mostly with the advent of sensor networks. Doherty [20] uses convex optimization to determine positions based exclusively on connectivity. Also, using centralized methods, but accepting both mere connectivity and range measurements, multidimensional scaling—map (MDS-MAP) [21] finds positions using multidimensional scaling. The centralized solutions may not be desirable in certain situations, but they can be used as performance benchmarks since they make use of the knowledge of the entire topology.

One-hop solutions are those in which nodes can directly contact a landmark, Bulusu [6] proposed the use of a grid of strong landmarks, so that any node can contact some landmarks. This solution is distributed, but landmarks here play more the role of an infrastructure, having to completely cover the entire network.

In the class of multihop algorithms, AhLOS [14] is a method that divides the network into groups of nodes containing enough landmarks to solve a nonlinear system positioning all the nodes in the group. The method is distributed and localized, and is closest in spirit to APS.

Analytic characterization of performance of a positioning algorithm is achieved for a simplified model of a square neighborhood in [22]. An idea similar to APS/ *DV-hop* has been independently explored in the context of amorphous computing by Nagpal [13], who has also given an upper bound on the accuracy as $\pi r/4N_{avg}$, where r is the wireless radius and N_{avg} is the average number of neighbors of a node. For the AhLOS method, Reference [23] computes the CRLB of the covariance of achieved positions. Moses [24] investigates the use of a calibration phase before the actual positioning, and gives the CRLB for calibration uncertainty. While most approaches address the average case for uniform deployment of nodes, Bischoff et al. [25] show that hop-based algorithms are not competitive in the worst-case deployment. We shortly analyzed another case of nonuniform deployment in a previous study [10]. For a particular case of nonuniform density, *DV-hop* is shown to be less predictable in performance than *DV-Euclidean*.

Although Taylor series is the most common method for position estimation, including in GPS [16], closed-form estimators that perform close to the CRLB are available, one of which is presented in [26].

27.6 CONCLUSION

We analyzed the error characteristics of range/angle-free, range-based, angle-based, and multimodal positioning algorithms that use DV as their main propagation method. Using a simplified network model, we assume a uniform deployment of the nodes, without obstacles, and Gaussian error measurements, in order to provide error bounds for APS algorithms. Range-free and multimodal methods were analyzed and shown to produce errors that are inversely proportional with the fraction of landmarks and with the TTL associated with the methods. The dependence of the covariance error is characterized in terms of deployment density, even if no closed form is available. Partitioning of the parameter space between the four analyzed algorithms provides insights with respect to the hardware performance needed to achieve good positioning in a multihop environment.

APPENDICES

Expectations

Given a circle of radius R, with points that are Poisson distributed, we wish to find the expectations of $(\cos^2(\alpha))/\rho$, $(\sin^2(\alpha))/\rho$, and $(\sin^2(2\alpha))/2\rho$, where ρ and α are the polar coordinates. Referring to Figure 27.13, the joint distribution function for the polar coordinates is

$$F_{\rho\alpha} = P\{a < \alpha, r < \rho\} = \frac{\alpha\rho^2}{2}\frac{1}{\pi R^2}$$

$$f_{\rho\alpha} = \frac{\partial^2 F_{\rho\alpha}}{\partial\rho\,\partial\alpha} = \frac{\rho}{\pi R^2}$$

$$f_\rho = \frac{2\rho}{R^2}$$

$$f_\alpha = \frac{1}{2\pi},$$

meaning that the polar coordinates are independent with a Poisson deployment. Let $E = 1/\rho$, with $E \in [(1/R), \infty)$.

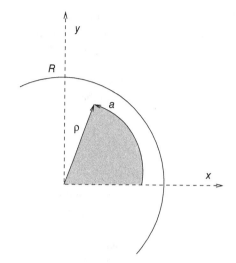

Figure 27.13   Joint distribution of ρ and α.

$$F_{\frac{1}{\rho}}(e) = P\{E \le e\} = P\left\{\frac{1}{\rho} \le e\right\}$$

$$= P\left\{\rho \ge \frac{1}{e}\right\} = 1 - P\left\{\rho \le \frac{1}{e}\right\}$$

$$= 1 - \frac{1}{R^2 e^2}. \tag{27.17}$$

$$f_{\frac{1}{\rho}}(e) = \frac{2}{R^2 e^3}$$

$$E\left[\frac{1}{\rho}\right] = \int_{\frac{1}{R}}^{\infty} \frac{2}{R^2 e^2}\, de = \frac{2}{R}.$$

Let $C = \cos(\alpha)$, with $C \in [-1,1]$.

$$F_C = P\{\cos(\alpha) < c\} = 1 - \frac{\arccos(c)}{\pi}$$

$$f_C = \frac{1}{\pi\sqrt{1 - c^2}}$$

$$f_{C^2}(x) = \frac{1}{\pi\sqrt{x(1-x)}} \tag{27.18}$$

$$E\left[\cos^2(\alpha)\right] = \frac{1}{2}.$$

In a similar fashion, distributions for $\sin(\alpha)$ and $\sin^2(\alpha)$ are derived and shown to be identical with $\cos(\alpha)$ and $\cos^2(\alpha)$. Let $S = \sin(2\alpha)$, with $S \in [-1,1]$.

$$P_s = \{\sin(2\alpha) < s\} = \frac{1}{2} + \frac{\arcsin(|s|)}{\pi}$$

$$f_s = \frac{1}{\pi\sqrt{1-s^2}} \tag{27.19}$$

$$E[\sin(2\alpha)] = 0.$$

Using (27.17–27.19), we conclude that

$$E\left[\frac{\cos^2(\alpha)}{\rho}\right] = E\left[\frac{\sin^2(\alpha)}{\rho}\right] = \frac{1}{R}$$

$$E\left[\frac{\sin(2\alpha)}{\rho}\right] = 0. \tag{27.20}$$

Uncertainty for *DV-Position*

Assuming there are n_i hops to landmark i, and a circular error covariance:

$$U_i = \sum_{j=1}^{n_i} \begin{bmatrix} r_j^2 \sigma^2 & 0 \\ 0 & r_j^2 \sin^2(\sigma_a) \end{bmatrix}$$

$$\approx \sigma^2 \sum_{j=1}^{n_i} r_j^2 I_2 \tag{27.21}$$

$$= \sigma^2 E[\bar{r}^2] n_i I_2$$

$$= \sigma^2 E[\bar{r}^2] \frac{\rho_i}{E[\bar{x}^2]} I_2.$$

Using derivations in the Appendix section "Expectations,"

$$\sum_{i=1}^{n} U_i^{-1} = \frac{E[\bar{x}]}{\sigma^2 E[\bar{r}^2]} \sum \frac{1}{\rho_i} I_2$$

$$= \frac{E[\bar{x}]}{\sigma^2 E[\bar{r}^2]} \frac{2n}{h E[\bar{x}]} I_2$$

$$\left(\sum_{i=1}^{n} U_i^{-1}\right)^{-1} = \frac{h \sigma^2 E[\bar{r}^2]}{2n} I_2$$

$$U = \frac{1}{2f\pi h} \sigma^2 \frac{E[\bar{r}^2]}{\lambda E^2[\bar{x}]} I_2.$$

Trilateration CRLB

Since the Fisher information matrix uses derivatives of the log-likelihood, we will need the following partial differentiations:

$$\frac{\partial \rho_i}{\partial x} = \frac{x - x_i}{\rho_i}$$

$$\frac{\partial^2 \rho_i}{\partial x \partial y} = -\frac{(x - x_i)(y - y_i)}{\rho_i^3}.$$

$\mathbf{x}$ is the position to be estimated, seen as a parameter here, whereas $\hat{\rho}$ is the normally distributed ranges to a landmark. Their expectations are the true ranges ρ, and their covariance is given by the diagonal matrix W.

$$p(\hat{\rho}; \mathbf{x}) = \sqrt{\frac{\det(W)}{(2\pi)^n}} \exp\left\{\frac{-1}{2}[\hat{\rho} - \rho]^T W [\hat{\rho} - \rho]\right\}$$

$$L(\mathbf{x}; \hat{\rho}) = -\frac{1}{2}\left\{\ln\left(\frac{(2\pi)^n}{\det(W)}\right) + [\hat{\rho} - \rho]^T W [\hat{\rho} - \rho]\right\}$$

$$= -\frac{1}{2}\left[\ln\left(\frac{(2\pi)^n}{\det(W)}\right) + \sum \frac{(\hat{\rho}_i - \rho_i)^2}{\sigma_i^2}\right]$$

$$\frac{\partial L}{\partial x} = \sum \frac{(\hat{\rho}_i - \rho_i)}{\sigma_i^2} \frac{\partial \rho_i}{\partial x}$$

$$\frac{\partial^2 L}{\partial x^2} = \sum \frac{1}{\sigma_i^2}\left[-\left(\frac{\partial \rho_i}{\partial x}\right)^2 + (\hat{\rho}_i - \rho_i)\frac{\partial^2 \rho_i}{\partial x^2}\right]$$

$$\frac{\partial^2 L}{\partial x \partial y} = \sum \frac{1}{\sigma_i^2}\left[-\frac{\partial \rho_i}{\partial x}\frac{\partial \rho_i}{\partial y} + (\hat{\rho}_i - \rho_i)\frac{\partial^2 \rho_i}{\partial x \partial y}\right].$$

Fisher information matrix $I(\mathbf{x})$ is defined as

$$I(\mathbf{x}) = -\int_{-\infty}^{\infty}\begin{bmatrix} \dfrac{\partial^2 L}{\partial x^2} & \dfrac{\partial^2 L}{\partial x \partial y} \\ \dfrac{\partial^2 L}{\partial x \partial y} & \dfrac{\partial^2 L}{\partial y^2} \end{bmatrix} p(\hat{\rho}; \mathbf{x}) d\hat{\rho}$$

$$-\int_{-\infty}^{\infty}\frac{\partial^2 L}{\partial x^2} p(\hat{\rho}; \mathbf{x}) d\hat{\rho}$$

$$= \int_{-\infty}^{\infty} p(\hat{\rho}; \mathbf{x}) \sum \frac{1}{\sigma_i^2}\frac{(x - x_i)^2}{\rho_i^2} d\hat{\rho}$$

$$+ \int_{-\infty}^{\infty} p(\hat{\rho}; \mathbf{x}) \sum \frac{(\hat{\rho}_i - \rho_i)}{\sigma_i^2}\frac{(x - x_i)(y - y_i)}{\rho_i^3} d\hat{\rho}$$

$$= \sum \frac{1}{\sigma_i^2}\frac{(x - x_i)^2}{\rho_i^2}$$

because $\int_{-\infty}^{\infty} p(\hat{\rho}; \mathbf{x}) d\hat{\rho} = 1$; $\int_{-\infty}^{\infty} p(\hat{\rho}; \mathbf{x})\hat{\rho} d\hat{\rho} = \rho$.

In the same manner it is shown that

$$-\int_{-\infty}^{\infty} \frac{\partial^2 L}{\partial x \partial y} p(\hat{\rho}; \mathbf{x}) d\hat{\rho} = \sum \frac{1}{\sigma_i^2} \frac{(x - x_i)(y - y_i)}{\rho_i^2},$$

which produces

$$I(x) = \begin{bmatrix} \sum \dfrac{1}{\sigma_i^2} \dfrac{(x - x_i)^2}{\rho_i^2} & \sum \dfrac{1}{\sigma_i^2} \dfrac{(x - x_i)(y - y_i)}{\rho_i^2} \\ \sum \dfrac{1}{\sigma_i^2} \dfrac{(x - x_i)(y - y_i)}{\rho_i^2} & \sum \dfrac{1}{\sigma_i^2} \dfrac{(y - y_i)^2}{\rho_i^2} \end{bmatrix}$$

$$= \begin{bmatrix} \dfrac{x - x_1}{\rho_1} & \dfrac{y - y_1}{\rho_1} \\ \dfrac{x - x_2}{\rho_2} & \dfrac{y - y_2}{\rho_2} \\ \cdots & \cdots \end{bmatrix}^{\mathrm{T}} W \begin{bmatrix} \dfrac{x - x_1}{\rho_1} & \dfrac{y - y_1}{\rho_1} \\ \dfrac{x - x_2}{\rho_2} & \dfrac{y - y_2}{\rho_2} \\ \cdots & \cdots \end{bmatrix}$$

$$= J_0^{\mathrm{T}} W J_0.$$

The matrix J_0 is in fact the Jacobian of the true ranges with respect to the true position.

$$\mathrm{CRLB}(\mathrm{Cov}[\mathbf{x}]) = (J_0^{\mathrm{T}} W J_0)^{-1}$$

DV-Hop **CRLB**

Without loss of generality, assume the true position to be in the origin $r_u = 0$. The variances in ranges are known to be inversely proportional with the distance in hops.

$$J_0 = \begin{bmatrix} \dfrac{x_1}{\rho_1} & \dfrac{y_1}{\rho_1} \\ \dfrac{x_2}{\rho_2} & \dfrac{y_2}{\rho_2} \\ \cdots & \cdots \end{bmatrix}$$

$$W = \begin{bmatrix} \dfrac{E[\bar{x}]}{\rho_i V[\bar{x}]} & 0 & 0 \\ 0 & \dfrac{E[\bar{x}]}{\rho_2 V[\bar{x}]} & 0 \\ 0 & 0 & \cdots \end{bmatrix}$$

$$
J_0^{\mathrm{T}} W J_0 = \frac{E[\bar{x}]}{V[\bar{x}]}
\begin{bmatrix}
\sum \dfrac{x_i^2}{\rho_i^3} & \sum \dfrac{x_i y_i}{\rho_i^3} \\[2ex]
\sum \dfrac{x_i y_i}{\rho_i^3} & \sum \dfrac{y_i^2}{\rho_i^3}
\end{bmatrix}
\tag{27.22}
$$

$$
= \frac{E[\bar{x}]}{V[\bar{x}]}
\begin{bmatrix}
\sum \dfrac{\cos^2(\alpha_i)}{\rho_i} & \sum \dfrac{\sin(2\alpha_i)}{2\rho_i} \\[2ex]
\sum \dfrac{\sin(2\alpha_i)}{2\rho_i} & \sum \dfrac{\sin^2(\alpha_i)}{\rho_i}
\end{bmatrix}
$$

$$
= \frac{E[\bar{x}]}{V[\bar{x}]} n
\begin{bmatrix}
E\left[\dfrac{\cos^2(\alpha)}{\rho}\right] & E\left[\dfrac{\sin(2\alpha)}{2\rho}\right] \\[2ex]
E\left[\dfrac{\sin(2\alpha)}{2\rho}\right] & E\left[\dfrac{\sin^2(\alpha)}{\rho}\right]
\end{bmatrix},
$$

where the sums in (27.22) are overall the landmarks inside the TTL circle of radius $hE[\bar{x}]$, and α_i is polar angle corresponding to (x_i, y_i). Using the results in the Appendix section "Expectations," it is shown (27.20) that $E[\cos^2(\alpha_i)/\rho_i] = E[\sin^2(\alpha_i)/\rho_i] = 1/(hE[\bar{x}])$ and $E[\sin(2\alpha_i)/2\rho_i] = 0$.

$$
\mathrm{CRLB}(\mathrm{Cov}[r_u]) = \frac{V[\bar{x}]}{E[\bar{x}]n}
\begin{bmatrix}
\dfrac{1}{hE[\bar{x}]} & 0 \\[2ex]
0 & \dfrac{1}{hE[\bar{x}]}
\end{bmatrix}^{-1}
$$

$$
= \frac{hV[\bar{x}]}{n} I_2
$$

$$
= \frac{1}{f\pi h} \frac{V[\bar{x}]}{\lambda E^2[\bar{x}]} I_2,
$$

where I_2 is the identity matrix of size 2.

REFERENCES

[1] G. Finn, "Routing and addressing problems in large metropolitan-scale internetworks," Tech. Rep., ISI Research Report ISI/RR-87-180, University of Southern California, Mar. 1987.

[2] J. C. Navas and T. Imielinski, "Geographic addressing and routing," presented at the ACM MobiCom, Budapest, Hungary, Sep. 26–30, 1997.

[3] F. Kuhn, R. Watternhofer, Y. Zhang, and A. Zollinger, "Geometric ad-hoc routing: Of theory and practice," presented at the 22nd ACM Symp. on the Principles of Distributed Computing (PODC), Boston, MA, Jul. 2003.

[4] J. Hill, R. Szewczyk, A. Woo, S. Hollar, D. Culler, and K. Pister, "System architecture directions for networked sensors," presented at the ASPLOS-IX, Cambridge, MA, Nov. 2000.

[5] Y.-B. Ko and N. H. Vaidya, "Location-aided routing (LAR) in mobile ad hoc networks," presented at the ACM MobiCom, Oct. 1998.

[6] N. Bulusu, J. Heidemann, and D. Estrin, "GPS-less low cost outdoor localization for very small devices," *Special Issue on Smart Spaces and Environments, IEEE Pers. Commun. Mag.*, vol. 7, pp. 28–34, 2000.

[7] P. Bahl and V. N. Padmanabhan, "RADAR: An in-building RF-based user location and tracking system," presented at IEEE INFOCOM, Tel Aviv, Israel, Mar. 2000.

[8] N. B. Priyantha, A. Chakraborty, and H. Balakrishnan, "The cricket location-support system," presented at the ACM MobiCom, Boston, MA, Aug. 2000.

[9] N. B. Priyantha, A. Miu, H. Balakrishnan, and S. Teller, "The cricket compass for context-aware mobile applications," presented at the ACM MobiCom, Rome, Italy, Jul. 2001.

[10] D. Niculescu and B. Nath, "DV based positioning in ad hoc networks," *Telecomm. Syst. Kluwer*, vol. 22, no. 1–4, pp. 267–280, 2003.

[11] D. Niculescu and B. Nath, "Ad hoc positioning system (APS) using AOA," presented at the IEEE INFOCOM, San Francisco, CA, Apr. 2003.

[12] D. Niculescu and B. Nath, "Position and orientation in ad hoc networks," *Elsever Ad Hoc Netw.*, vol. 2, no. 2, pp. 133–151, 2004.

[13] R. Nagpal, "Organizing a global coordinate system from local information on an amorphous computer," Tech. Rep. 1666, MIT AI Lab, 1999.

[14] A. Savvides, C.-C. Han, and M. Srivastava, "Dynamic fine-grained localization in ad-hoc networks of sensors," presented at the ACM MobiCom, Rome, Italy, 2001.

[15] K. Langendoen and N. Reijers, "Distributed localization in wireless sensor networks a quantitative comparison," Tech. Rep. PDS-2002-3, Delft University of Technology, The Netherlands, 2002.

[16] B. W. Parkinson and J. J. Spilker, *Global Positioning System: Theory and Application*. Washington, DC: American Institute of Astronautics and Aeronautics, 1996.

[17] P. Bose, P. Morin, I. Stojmenović, and J. Urrutia, "Routing with guaranteed delivery in ad hoc wireless networks," presented at the 3rd International Workshop on Discrete Algorithms and Methods for Mobile Computing and Communications, Seattle, WA, Aug. 1999.

[18] L. Kleinrock and J. Silvester, "Optimum transmission radii for packet radio networks or why six is a magic number," in *IEEE National Telecommunications Conf.*, Birmingham, AL, 1978 pp. 4.3.1–4.3.5.

[19] T. K. Philips, S. S. Panwar, and A. N. Tantawi, "Connectivity properties of a packet radio networks model," *IEEE Trans. Inf. Theory*, vol. 35, pp. 1044–1047, 1989.

[20] L. Doherty, L. E. Ghaoui, and K. S. J. Pister, "Convex position estimation in wireless sensor networks," presented at the IEEE INFOCOM, Anchorage, AK, Apr. 2001.

[21] Y. Shang, W. Ruml, Y. Zhang, and M. Fromherz, "Localization from mere connectivity," presented at the ACM MobiHoc, Annapolis, MD, Jun. 1–3, 2003.

[22] S. Simić and S. Sastry, "A distributed algorithm for localization in random wireless networks," Tech. Rep., UC Berkeley, EECS, 2002.

[23] A. Savvides, W. Garber, S. Adlakha, R. Moses, and M. B. Srivastava, "On the error characteristics of multihop node localization in ad-hoc sensor networks," presented at the IPSN03, International Workshop on Information Processing in Sensor Networks, PARC, Palo Alto, CA, Apr. 22–23, 2003.

[24] R. Moses, D. Krishnamurthy, and R. Patterson, "A self-localization method for wireless sensor networks," *Eurasip J. Appl. Signal Process.*, vol. 4, pp. 348–358, 2002.

[25] R. Bischoff and R. Wattenhofer, "Analyzing connectivity-based multi-hop ad-hoc positioning," presented at the IEEE Percom, Orlando, FL, Mar. 2004.

[26] Y. T. Chan and K. C. Ho, "A simple and efficient estimator for hyperbolic location," *IEEE Trans. Signal Process.*, vol. 42, no. 8, pp. 1905–1915, 1994.

CHAPTER 28

SELF-LOCALIZATION OF UAV FORMATIONS USING BEARING MEASUREMENTS

Iman Shames,[1] Barış Fidan,[2] Brian D. O. Anderson,[1]
and Hatem Hmam[3]
[1]The Australian National University and
The University of Melbourne
[2]University of Waterloo, Waterloo, Canada
[3]Defence Science and Technology Organisation, Edinburgh, Australia

THIS CHAPTER begins by treating a localization problem recently encountered in an operational context; that is, localizing three agents moving in a plane when the interagent distances are known, and in addition, the angle subtended at each agent by lines drawn from two landmarks at known positions is also known. It is shown as a key conclusion that there are in general a finite number greater than one of possible sets of positions for the three agents. In addition, the generalization of the result for more than three agents is presented. Examples are presented to show the applicability of the methods proposed here.

28.1 INTRODUCTION

In many multiagent applications, it is desired to know the positions of the agents. For example, in bushfire surveillance or a search-and-rescue operation, sensing the data without knowing the position of the sensing agents is not enough to accomplish the task at hand. A trivial solution to this problem is to install global positioning systems (GPSs) on each agent. However, when GPS signals are lost or corrupted [1], or when the agents operate indoors, use of GPSs for localization purposes may

Handbook of Position Location: Theory, Practice, and Advances, Second Edition.
Edited by S. A. (Reza) Zekavat and R. Michael Buehrer.
© 2019 by the Institute of Electrical and Electronics Engineers, Inc.
Published 2019 by John Wiley & Sons, Inc.
Companion Website: www.wiley.com/go/zekavat/positionlocation2e

be infeasible or limited [2]. Thus, it is imperative to design other methods for estimating the positions of the sensing agents when a GPS signal is not available. There are also many other scenarios when GPS signals are not available. Examples include outer space localization and localization on the surface of other planets. The vehicle itself can serve as a pair or more of landmarks, which the robots can use for absolute navigation and referencing, as long as the vehicle remains in the line of sight (LOS) of the robot.

In the literature, there are many methods that achieve the task of localizing agents via measuring distances [3, 4], time difference of arrival [5], and direction of arrival (DOA) [6].

In this chapter, we address the localization problem where self-localization is achieved cooperatively among a team of bearing and distance sensing agents (e.g., unmanned aerial vehicles [UAVs] or robots) and where the position information of the two landmarks is available, as are either interagent distances or the bearings measured to each agent of the agents in the formation.

The problem considered here is very similar to the problems considered in [6–10]. The cooperative localization task is then to put the pieces of information together, for example, interagent distances, subtended angles, and landmark positions, and localize the agents. Note here that the localization is to be achieved instantaneously; we are not envisaging collecting information from agents at a number of successive instants of time and using them to infer position at a single instant of time (the connection with Kalman filtering is explored further below). In the distance-based localization literature, the nodes with known position are called anchors (beacons) [11], and since each anchor is determined exactly, they can be considered as anchors as well. Due to the nature of the problem mentioned above, these anchors are collinear. However, the results obtained in this chapter are not limited to collinear anchors and can be applied to scenarios with arbitrary arrangement of the anchors. Note also that we later extend the results to systems with more than three agents capable of measuring the subtended angle by the landmarks.

It is shown in this chapter that the solution count for the above localization problem (involving three agents) is less than 12. If there is no unique solution, then what is the point of this analysis? There are in fact several ways in which it can be relevant. First, GPS data may be intermittently available to an agent. When it is available, localization is obviously unique. When it ceases to be available, the fact that the agents are moving continuously with their initial position known means that there will be a basis for making the correct selection out of a finite number of localization possibilities at subsequent times.* Second, given that the number of possible solutions to the localization problem is finite, it may be that additional very crude data indeed (e.g., agent T' is located in this general region) will be enough to disambiguate the multiple solutions. Third, if more agents are available, then the additional information will generically allow unique localization, and indeed, when measurements are noisy, will in general allow some amelioration of the distorting effects of the noisy measurements. Lastly, the method described in this chapter can

---

*At least until the agents move to positions corresponding to a double solution of the localization problem, after which two tracks would have to be followed assuming no disambiguating information.

be considered as batch processing of the agent locations serving to initialize and improve the updates of a Kalman-based filter that tracks the agent positions as the agents move in their environment. Kalman-based filters are also prone to errors when the agent's motion model deviates significantly from the actual agent motion. Our batch processing method can guard against such behavior and reinitialize the filter.

There is a vast body of literature using Bayesian methods that deals with localization problems. For example, in [12], a collective localization problem based on Kalman filtering is proposed, and [13] uses a Gaussian sum filter to solve the initialization problem in bearing-only simultaneous localization and mapping (SLAM). Differently from these approaches, the tools for obtaining our results are drawn from two nonconventional sources. The first is the theory of rigid graphs (see, e.g., [14–17]). In recent years, its relevance to localization of sensor networks has come into prominence [11, 18]. A good deal of the theory of rigid graphs deals with the question of when localization is possible [19, 20]. The second source we draw on is the mechanical engineering literature dealing with four-bar linkage mechanisms [21, 22]. As it turns out, this literature has developed ideas for describing the loci of points that are part of a planar mechanism made up from pin joints and rigid bars that provide distance constraints between the joints. Moreover, for the purposes of this chapter, we assume that we have access to noiseless interagent distance measurements. The significance of considering the noiseless scenario is that it equips us with a knowledge on the number of solutions to the problem and methods to calculate these solutions, where later methods to deal with noisy case are built upon.

The outline of this chapter is as follows. In the next section, we formally set up the problem. In Section 28.3, we first introduce the results related to network localization in the field of rigidity and then establish the connection of the problem described in Section 28.2 with the rigidity theory. In Section 28.4, we study four-bar linkage mechanisms, and in Section 28.5, we propose a solution to the problem of interest using the mathematical methods developed to analyze these mechanisms. In Section 28.6, we consider the localization problem for formations with a larger number of agents than initially conisdered. In Section 28.7, we study the situation where information about an extra landmark is available. In Section 28.8, we introduce another localization scenario where the agents can measure all the subtended angles (not only at the landmarks). Concluding remarks come at the end.

28.2 PROBLEM SETUP

The arrangement to be considered, and the one on which the later developments are based, is depicted in Figure 28.1. Specifically, we consider n mobile agents (in this case, $n = 3$), designated T_1, T_2, and T_3, that are to be localized. The agents detect two landmarks located at positions L_1 and L_2, which are known to the agents. The landmarks can be radars, radio frequency (RF) beacons, or some visible features if an imaging sensor is used. Each agent collects the bearing angle information to each of the landmarks. *However, with no GPS information (no knowledge about the absolute heading reference), there is no absolute heading reference for each agent, and the bearing angle information cannot be used directly for localization purposes.*

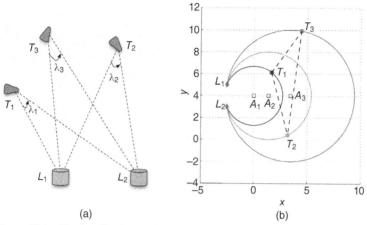

Figure 28.1   The localization problem setting and the agents loci. (a) Landmarks L_1, L_2, formation $T_1T_2T_3$, and angles λ_i in Problem 1. (b) Agent loci for a sample setting.

Nevertheless, using the angle difference (i.e., the difference between two bearings), the need for knowing the heading is removed. This angle difference is the angle subtended at each agent by the two landmarks (λ_i, $i = 1,2,3$) (see Fig. 28.1) and knowing its value. It is straightforward from the inscribed angle theorem to determine that each agent i is located on a circle of known center, A_i, and radius, R_i, that passes through the two landmarks. Note that both values of A_i and R_i can be determined by λ_i and the landmark positions. The centers, A_i ($i = 1,2,3$), lie on the perpendicular bisector of the line joining the two landmarks. A priori information is assumed to be available that positions all agents on the same known side of the line joining the two landmarks. This can be ensured by choosing landmarks on the boundary of the agents region of operation.

We adopt the naming convention from the distance-based localization literature and call the points with the known positions anchors (beacons) [11]. Since each A_i is determined exactly (from the knowledge of positions of the landmarks and the angles λ_i), they can be considered as (pseudo-)anchors as well. Due to the nature of the problem mentioned above, these anchors are collinear. However, the results obtained in this chapter are not actually limited to collinear anchors and can be applied to scenarios with arbitrary arrangement of the anchors. First, we formalize the definition of the *underlying graph* of a formation.

Definition 1 (Underlying Graph of a Formation): A formation of point agents in the plane, $\mathcal{F}$, can be represented by a graph $\mathcal{G}(\mathcal{V}, \mathcal{E})$, with vertex set $\mathcal{V}$ and edge set $\mathcal{E}$; the vertices in $\mathcal{V}$ correspond to the agents, and an edge in the graph exerts between two vertices $v_1, v_2 \in \mathcal{V}$ only when the distance between the corresponding agents of the formation is known. We call $\mathcal{G}$ the underlying graph of $\mathcal{F}$.

We can generalize the problem described above as the following problem.

Problem 1: Consider a formation $\mathcal{F}$ with the underlying graph $\mathcal{G}(\mathcal{V}, \mathcal{E})$, where $\mathcal{V} = \mathcal{T} \cup \mathcal{A}$ is the set of vertices with $\mathcal{T} = \{T_1, T_2, T_3\}$, $\mathcal{A} = \{A_1, A_2, A_3\}$. The agents in $\mathcal{A}$ are known as anchors, and those in $\mathcal{T}$ as targets. Furthermore, $\mathcal{E} = \{T_1T_2, T_1T_3, T_2T_3, A_1A_2, A_1A_3, A_2A_3, A_1T_1, A_2T_2, A_3T_3\}$ is the set of edges. Knowing the exact length of all edges in $\mathcal{E}$, and the exact positions of the anchors,

(i) Can one localize the targets, uniquely, or to one of a finite number of sets of positions?

(ii) If so, what are the possible localization solutions?

Remark 1: In this problem there is no assumption of collinearity of A_i, $i = 1, 2, 3$. The case where A_i are collinear can be treated as a special case of this problem.

In the sequel, we propose an answer to the first question posed in Problem 1. However, first we introduce some basic concepts from rigidity theory and its relationship to the localization problem.

28.3 A RIGID GRAPH THEORETICAL FRAMEWORK FOR FORMATION LOCALIZATION

In this section, we review some aspects of the problem of localizing, that is, determining the positions of agents in a formation where a number of interagent distances are known, and additionally, some absolute position data is available. We appeal to the literature on rigid graph theory and its application to sensor network localization [11, 18–20].

Definition 2 (Graph Realization Problem): Consider a formation $\mathcal{F}$ with underlying graph $\mathcal{G}(\mathcal{V}, \mathcal{E})$. The task of assigning coordinate values to each vertex of a graph, $\mathcal{G}(\mathcal{V}, \mathcal{E})$, so that the Euclidean distance between any two adjacent vertices is equal to the edge length associated with the edge joining these two vertices, is the graph realization problem.

Given one solution to the graph realization problem, it is trivial that any translation, rotation, or reflection of this solution is another solution. All such solutions are congruent. Hence, it is relevant to ask whether there can be two solutions that are not congruent, and whether, disallowing translations, rotations, or reflections, the number of distinct solutions is finite or infinite. When there can be only one family of congruent solutions, one can say that the graph realization problem has a unique solution. This problem is also known as the problem of localizability in the literature (see [23] for more information).

Hendrickson [19] presents necessary conditions for a graph to be uniquely realizable in $\mathbb{R}^2$, that is, with one family of congruent solutions, and the same conditions were proved later by Jackson and Jordan [20] to be necessary and sufficient.

These conditions involve two concepts, namely redundant rigidity of a graph and three-connectedness of a graph. The concept of redundant rigidity requires a prior concept of rigidity, that is as follows.

Definition 3 (Rigid Formations): A formation $\mathcal{F}$ is called rigid if by explicitly maintaining distances between all the pairs of agents whose representative vertices are connected by an edge in $\mathcal{E}$, the distances between all other pairs of agents in $\mathcal{F}$ are consequentially held fixed as well.

The reader may refer to [14–16] for detailed information on rigid formations and rigidity.

Definition 4 (Redundantly Rigid and Minimally Rigid Formations): A redundantly rigid formation is one that remains rigid when any single edge constraint is removed. By way of contrast, a minimally rigid formation is one which ceases to be rigid when any single edge constraint is removed.

For the sake of simplicity, the underlying graph of a formation is called rigid, redundantly rigid, and minimally rigid if the formation is, respectively, rigid, redundantly rigid, and minimally rigid.

The notion $\mathcal{G}$ of a three-connected graph is standard (see, e.g., [24]). Such a graph has the property that between any two vertices, three nonintersecting paths can be found.

Jackson and Jordan's result [20] is as follows.

Theorem 1: Consider a two-dimensional formation $\mathcal{F}$ with underlying graph $\mathcal{G}(\mathcal{V}, \mathcal{E})$. Then the graph realization problem is uniquely solvable for generic values of the formation edge lengths (interagent distances) if and only if $\mathcal{G}$ is redundantly rigid and three-connected.

From this result, we can have a definition for *globally rigid* graphs.

Definition 5 (Globally Rigid Graphs): A graph $\mathcal{G}$ with the two properties in Theorem 1, that is, redundant rigidity and three-connectedness, is termed globally rigid.

For a formation that is rigid but not globally rigid, one of at least of two ambiguities known as flip ambiguity or discontinuous flex ambiguity can occur [19]; these ambiguities are depicted in Figure 28.2. The reader may refer to [16] and [19] and

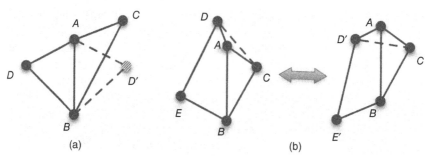

Figure 28.2 Illustration of (a) flip ambiguity: vertex D can be flipped over the edge (A, B) to a symmetric position D' and the distance constraints remain the same; (b) discontinuous flex ambiguity: temporarily removing the edge (C, D), the edge triple (D, A), (D, E), (E, B) can be flexed to obtain positions E' and D', such that the edge length (C, D) equals the edge length (C, D'), and then all the distance constraints are the same.

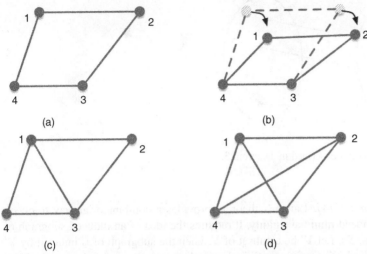

Figure 28.3 Rigid and nonrigid formations. The formation represented in panel (a) is not rigid. It can be deformed by a smooth motion without affecting the distance between the agents connected by edges, as shown in panel (b). The formations represented in panels (c) and (d) are rigid, as they cannot be deformed by any such move. In addition, the formation represented in panel (c) is minimally rigid because the removal of any edge would render it nonrigid. That of panel (d) is not minimally rigid; any edge may be removed without losing rigidity.

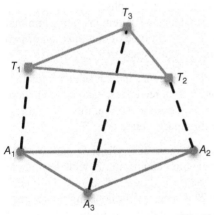

Figure 28.4 Graph representation of Problem 1.

references therein for further information on these ambiguities. Examples of non-rigid, rigid, and globally rigid graphs are presented in Figure 28.3.

Now we provide an answer the first question posed in Problem 1. An example of the formation $\mathcal{F}$ described in Problem 1 is depicted in Figure 28.4. A crucial concept pertinent to answering the first question posed in Problem 1 is a minimally rigid formation. Two ways are presented in the following paragraphs to see this fact.

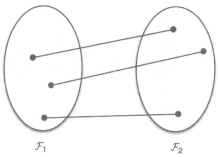

Figure 28.5 Setting described in Theorem 3.

Laman's Theorem [14]: Laman's theorem provides a combinatorial way to check rigidity, and indeed minimal rigidity. It requires the idea of an induced subgraph of a graph $\mathcal{G} = (\mathcal{V}, \mathcal{E})$. Let $\mathcal{V}'$ be a subset of $\mathcal{V}$. Then the subgraph of $\mathcal{G}$ induced by $\mathcal{V}'$ is the graph $\mathcal{G}' = (\mathcal{V}', \mathcal{E}')$, where $\mathcal{E}'$ includes all those edges of $\mathcal{E}$ that are incident on a vertex pair in $\mathcal{V}'$.

Theorem 2 (Laman's Theorem [14]): A graph $\mathcal{G} = (\mathcal{V}, \mathcal{E})$ *in* $\mathbb{R}^2$ of $\mathcal{V}$ vertices and $\mathcal{E}$ edges is rigid if and only if there exists a subgraph $\mathcal{G}' = (\mathcal{V}, \mathcal{E}')$ with $2\mathcal{V}$-3 edges such that for any subset $\mathcal{V}''$ of $\mathcal{V}$, the induced subgraph $\mathcal{G}'' = (\mathcal{V}'', \mathcal{E}'')$ of $\mathcal{G}'$ obeys $\mathcal{E}'' \leq 2\mathcal{V}'' - 3$. It is minimally rigid if $\mathcal{E} = 2\mathcal{V} - 3$.

It is easy to check for the graph of Figure 28.4 that $\mathcal{E} = 2|\mathcal{V}| - 3$; one takes $\mathcal{G}' = \mathcal{G}$ and can verify the counting condition for all induced subgraphs.

Combination of Rigid Formations [17]: Another way of demonstrating minimal rigidity of a formation is by showing that it is a certain type of combination of two minimally rigid formations. The key theorem is as follows.

Theorem 3 ([17]): Let $\mathcal{F}_1$ and $\mathcal{F}_2$ be two rigid formations, and consider a formation $\mathcal{F}$ formed by connecting these two formations with three edges, each edge incident on one vertex of $\mathcal{F}_1$ and one of $\mathcal{F}_2$, with no two edges incident on the same vertex. Then $\mathcal{F}$ is rigid. Moreover, if $\mathcal{F}_1$ and $\mathcal{F}_2$ are minimally rigid, so is $\mathcal{F}$.

The setting described in Theorem 3 is depicted in Figure 28.5. Observe that a triangle is obviously minimally rigid. The theorem then applies identifying $\mathcal{F}_1$ with the triangle formed by A_1, A_2, and A_3 and $\mathcal{F}_2$ with the triangle formed by T_1, T_2, and T_3. In light of the minimal rigidity of the formation of Figure 28.4, there will be noncongruent formations meeting the distance constraints. Then, even though the positions of A_1, A_2, and A_3 are fixed, the positions of T_1, T_2, and T_3 will not be uniquely determinable.

28.4 FOUR-BAR LINKAGE MECHANISMS

Four-bar linkage mechanisms perform a wide variety of motions with a few simple parts. Furthermore, due to their ease of design calculations, they have gained much popularity in mechanical machine design. An example of such a mechanism is presented in Figure 28.6.

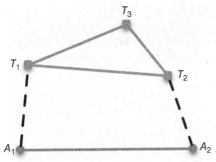

Figure 28.6   The four-bar linkage mechanism obtained after deletion of A_3T_3.

In a four-bar linkages mechanism, there is a fixed link that is called the *frame*. There are two *side links* that can revolve around the ends of the frame, and the remaining (fourth) link is called the *coupler*. A four-bar linkage mechanism is characterized by the length of each of its links, and the *configuration* of its coupler, that is, open or crossed configuration [25]. Furthermore, we have the following concepts in four-bar linkage mechanisms [26]:

- A side link that can fully revolve around (360°) the end point of the frame is called a *crank*.
- Any link that cannot fully revolve is called a *rocker*.

Although we do not use the following theorem extensively in this chapter, we present it as background information that will assist in interpreting the subsequent examples.

Theorem 4 (Grashof's Theorem [21]):   A four-bar mechanism has at least one crank link if and only if

$$s + l \le p + q, \tag{28.1}$$

and all three mobile links are rockers if

$$s + l > p + q. \tag{28.2}$$

Here, l and s are the lengths of the longest link and the shortest link, respectively, and p and q are the lengths of the other two links. For example, in Figure 28.6, s and l are A_2T_2 and A_1A_2, respectively. The inequality (Eq. 28.1) is known as *Grashof's criterion*.

In the mechanism depicted in Figure 28.6, T_3, which is a fixed point on a rigid body attached to the coupler link T_1T_2, is termed the *coupler point*. The curve that this coupler point moves on in both open and cross configurations is called the *coupler curve*. Generally, coupler curves are closed curves, but for some special mechanisms we will have open coupler curves. These correspond to situations where the area enclosed by the coupler curve shrinks to zero.

A coupler curve K_C may comprise either a single part or a bipartite curve (a bipartite curve is one with two branches, like a hyperbola). In the case where K_C is bipartite, we denote the branches as K_{C_1} and K_{C_2}, where K_{C_1} is the curve obtained by the coupler point in open configuration, and K_{C_2} is constructed by the curve in cross configuration.

For a given four-bar linkage mechanism, as in Figure 28.6, the equation of the coupler curve (bipartite or single part), K_C, where the center of Cartesian coordinates system is placed on A_1, and A_1 and A_2 are placed on the x-axis, is [21]:

$$
\begin{aligned}
&r_{23}^2((x-k)^2+y^2)(x^2+y^2+r_{13}^2-R_1^2)^2 \\
&-2r_{23}r_{13}((x^2+y^2-kx)\cos\eta_3+ky\sin\eta_3)(x^2+y^2+r_{13}^2-R_1^2) \\
&+((x-k)^2+y^2+r_{23}^2-R_2^2)+r_{13}^2(x^2+y^2)((x-k)^2+y^2+r_{23}^2-R_2^2)^2, \\
&-4r_{23}^2r_{13}^2((x^2+y^2-kx)\sin\eta_3-ky\cos\eta_3)^2=0
\end{aligned}
\tag{28.3}
$$

where k is the length of the frame link, $\overline{A_1A_2}$, $\eta_3 = \angle T_1T_3T_2$, r_{ij} is the distance between agents T_i and T_j, and $R_i = d/(2\sin\lambda_i)$.

In addition, another coupler curve, K_C', can be obtained from the reflection of K_C, when A_1A_2 is the image axis. In general, the equation describing K_C' can be obtained by substituting $-y$ for y in (28.3). In the case of a bipartite K_C, we denote the branches of K_C' as K_{C_1}' and K_{C_2}'. As a result, the locus of the coupler point is made up of two polynomial curves, each with degree of six. More specifically, we have the following result from [21] for bipartite coupler curves:

Proposition 1: Bipartite coupler curves occur when and only when Grashof's criterion holds; all other cases yield coupler curves that always consist of a single part.

28.5   A LOCALIZATION ALGORITHM BASED ON FOUR-BAR LINKAGE MECHANISMS

Let us revisit the problem described in Section 28.2. It has been assumed that the agents form a triangular formation, where T_i is the ith agent, with $\mathbf{x}_i = [x_{T_i}, y_{T_i}]^T \in \mathbb{R}^2$ as its coordinates. The distance between two agents T_i and T_j is known and equal to r_{ij} (or r_{ji}). For a given agent, T_i, and two landmarks with known positions, L_1 and L_2, the locus for the position of T_i when $\angle L_1T_iL_2 = \lambda_i$ is a part of a circle denoted by $C(\mathbf{a}_i, R_i)$, where $R_i = d/2\sin(\lambda_i)$ is the radius of the circle and A_i is the center. Furthermore, assuming that the origin of the global coordinates frame is the middle of L_1L_2, $d = \overline{L_1L_2}$ and the x-axis coincides with the perpendicular bisector of L_1L_2, the position of A_i is considered to be $\mathbf{a}_i = [x_i, y_i]^T = [d/2\tan\lambda_i, 0]^T$. Note that T_i, L_1, and L_2 lie on the same circle described above.

In Figure 28.1b, each mobile platform, T_i, $i = 1,2,3$, and the associated circles are depicted. In this case, the centers of the circles, A_i, serve as the virtual anchors since we know their exact positions in the plane. Hence, the agents, T_i $(i \in \{1,2,3\})$, in the formation and these virtual anchors, A_j $(j \in \{1,2,3\})$, form a graph that satisfies the conditions presented in Problem 1. Here, first, we make the relationship between the localization problem described in Problem 1 and the concept of four-bar linkage mechanism clear, and then using this concept we present an upper bound for the number of localization possibilities.

Note that the graph representation of Problem 1 in Figure 28.4 can be further iterated to obtain the virtual four-bar linkage mechanism representation in Figure 28.6 by representing the distance constraints on $|A_1A_2|$, $|A_1T_1|$, $|A_2T_2|$, $|T_1T_2|$ as bars, and decoupling the distance constraint on A_3T_3 from this representation. Because of this decoupled distance constraint, T_3 is on the circle $C(a_3, R_1)$ with A_3 as its center and R_1 as its radius. Hence, the possible solutions for the localization problem can be obtained from the calculation of intersections of the circle, $C(a_3, R_3)$, and the two coupler curves K_C and K'_C defined in Section 28.4.

In [22], the number of intersections of coupler curves and a circle is computed using concepts of circularity and order of the curves. The result, according to [22], in our context is as follows:

Lemma 1: The circle, $C(a_3, R_3)$, and the coupler curve described by (28.3) have at least two and at most six real points of intersection.

Using Lemma 1, we establish that the maximum number of localization solutions is 12 in the following theorem:

Theorem 5: The maximum number of (real) localization solutions for Problem 1 is 12. For generic values of distances and angles, the minimum number of localization solutions is four.

Proof: Equation (28.3) corresponds to the coupler curves for the four-bar linkages mechanism depicted in Figure 28.6 corresponding to Problem 1 [21]. Since there are two (single-part or bipartite) coupler curves (the second one is the image of the first one when the frame link is the image axis) and for each coupler curve based on Lemma 1 there are a maximum of six and a minimum of two possible solutions, we have at most 12 possible solutions, and at least four localization solutions.

Returning to the problem presented in Section 28.2, based on the procedure introduced earlier in this section for constructing a four-bar mechanism by deleting edge A_3T_3, we can have the linkage mechanism depicted by solid lines in Figure 28.1b. In addition, here for the coupler curve equation, we have $k = (d/[2 \tan \lambda_2]) - (d/[2 \tan \lambda_1])$, $R_1 = d/(2 \sin \lambda_1)$, and $R_2 = d/(2 \sin \lambda_2)$. From Theorem 5, we can have up to 12 localization solutions. We are now ready to provide a localization algorithm to address Problem 1(ii). Algorithm 28.1 can be used to find up to twelve localization solutions (up to six pairs of mirror solutions with respect to the line connecting A_1 to A_2).

Algorithm 28.1   Three Agent and Two Landmark Localization

Find $m \le 6$ real intersection points of (28.3) and $(x - x_3)^2 + (y - y_3)^2 - R_3^2$, $x_{3i} = [x_{iT_3}, y_{iT_3}]^T$, $i = 1, \dots, m$.
    for $i = 1$ to m **do**

Solve the system of equation for x_{T_1}, y_{T_1}, x_{T_2}, and y_{T_2}:

$$(x_{iT_3} - x_{T_1})^2 + (y_{iT_3} - y_{T_1})^2 - r_{13}^2 = 0$$

$$(x_{iT_3} - x_{T_2})^2 + (y_{iT_3} - y_{T_2})^2 - r_{23}^2 = 0$$

$$(x_{T_2} - x_{T_1})^2 + (y_{T_2} - y_{T_1})^2 - r_{12}^2 = 0$$

$$(x_1 - x_{T_1})^2 + (y_1 - y_{T_1})^2 - R_1^2 = 0$$

$$(x_2 - x_{T_2})^2 + (y_2 - y_{T_2})^2 - R_2^2 = 0$$

$$s_i \leftarrow [x_{T_1}, y_{T_1}, x_{T_2}, y_{T_2}, x_{iT_3}, y_{iT_3}] \quad s_i' \leftarrow [x_{T_1}, -y_{T_1}, x_{T_2}, -y_{T_2}, x_{iT_3}, -y_{iT_3}]$$

end for

Return S and S' as the sets of $2m$ localization solutions where s_i and s_i' are their rows, respectively.

Example 28.1   Simulations with Three Agents and Two Landmarks

The angle and distance values used in each simulation scenario are presented in Table 28.1. In addition, it is worth mentioning that after running several simulations, a case with 12 localization solutions was never encountered.

The important characteristics of each simulation result and the number of localization solutions in each scenario are presented in Table 28.2. The legends used in the simulation results are described in Table 28.3. In Figures 28.7 and 28.8, two scenarios where, respectively, four and eight localization solutions exist are presented. A nongeneric case is presented in Figure 28.9a, where there are repeated localization solutions. A bad geometry is identified in Figure 28.9b, where an infinite number of localization solutions exists. This infinite localization ambiguity occurs if the three sensors and the two landmarks lie on a common circle. Invalid localization solutions occur in the scenario that is depicted in Figure 28.10a. Another nongeneric case arises when the angles subtended at any two of the agents by the landmarks are equal, and this is depicted in Figure 28.10b. Note that for all the scenarios the landmarks L_1 and L_2 are placed at $[0,1]^T$ and $[0,-1]^T$, respectively.

TABLE 28.1.   The Angle and Distance Values in Each Scenario

Scenario	λ_1 (rad)	λ_2 (rad)	λ_3 (rad)	$\overline{T_1T_2}$ (m)	$\overline{T_1T_3}$ (m)	$\overline{T_2T_3}$ (m)
Figure 28.7	1.0472	0.6283	0.5236	1	1	1
Figure 28.8	0.3805	0.2526	0.1674	3.1623	5.099	2.8284
Figure 28.9a	1.0472	0.8976	0.7854	1.5	0.9765	1.25
Figure 28.9b	0.2487	0.2487	0.2487	2.0859	2.7552	4.6188
Figure 28.10a	0.7854	0.6283	0.5236	3	2	2
Figure 28.10b	0.7854	0.7854	0.5236	1	1	1

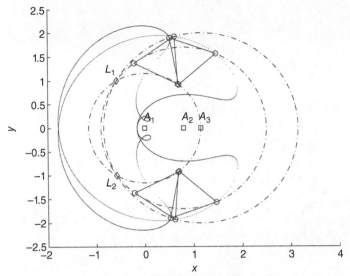

Figure 28.7   The possible localization solutions for scenario 1.

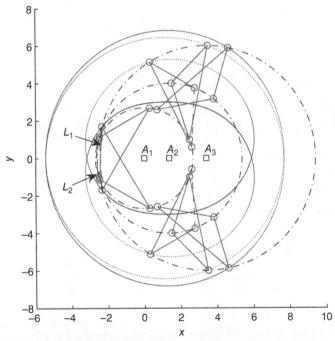

Figure 28.8   The possible localization solutions for scenario 2.

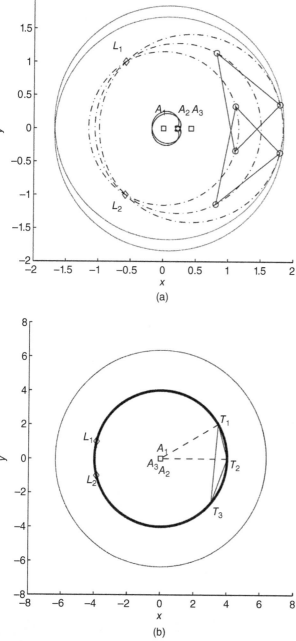

Figure 28.9   Simulation results. (a) Repeated localization solutions (scenario 3 as described in Table 28.2). (b) Infinite number of localization solutions for scenario 4 as described in Table 28.2. The locus of T_3 coincides with one of the branches of the coupler curve.

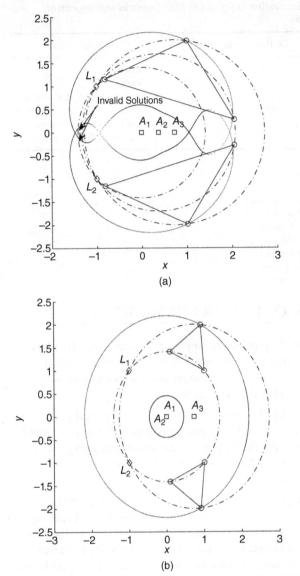

Figure 28.10   Simulation results. (a) Occurrence of invalid localization solutions (scenario 5 as described in Table 28.2). (b) Localization solutions shen two subtended angles λ_1 and λ_2 are equal (scenario 6 as described in Table 28.2)

TABLE 28.2. The Number of Localization Solutions in Each Scenario and Important Characteristics of Each Scenario

Scenario	No. of Distinct Solutions	Characteristic
1 (Fig. 28.7)	4	Generic
2 (Fig. 28.8)	8	Generic
3 (Fig. 28.9a)	2	Nongeneric/repeated solutions
4 (Fig. 28.9b)	Infinite	Nongeneric/infinite ambiguity
5 (Fig. 28.10a)	2	Generic/invalid solutions
6 (Fig. 28.10b)	2	Nongeneric/two equal angles

TABLE 28.3. The Legends Used in Simulation Results

Agent	Small solid circle
Landmark	Solid diamond
Formation	Solid triangles
Agent locus	Dashed circles
Coupler curve	Dotted curves

28.6 LOCALIZATION OF LARGER FORMATIONS

The methodology presented in the previous sections can be extended to localization of certain formations having more than three agents, as will be elaborated in what follows. The following theorem gives the maximum possible number of solutions for formations with globally rigid underlying graphs:

Theorem 6: Consider a formation $\mathcal{F}$ with the underlying globally rigid graph $\mathcal{G}(\mathcal{V}, \mathcal{E})$, and the three agents, T_1, T_2, and T_3, in the formation that form a triangle. Assuming that these three agents are the only agents capable of measuring the angle subtended at them by the two landmarks L_1 and L_2, with known positions, then there are at most 12 possible localization solutions for the formation.

Proof: Theorem 5 states that the upper bound for the number of localization solutions of a triangular formation using the value of the angles subtended at each agent by two landmarks is 12. On the other hand, in [20], it has been shown that the necessary and sufficient condition for unique localization of a formation is that the associated graph is globally rigid and there are three nodes with exactly known positions. As a result, for each possible localization of three agents, there is a localization solution for the whole formation, so there are up to 12 possible localization possibilities for the formation.

Now, assume that we have another agent, T_4, that can measure its distance from agents T_1, T_2, and T_3 and the angle subtended at itself by the two landmarks, λ_4. Knowing this angle, we can characterize another anchor node, A_4 with known position, similar to A_1, A_2, and A_3, for the time being assuming that A_i are not collinear. We can calculate the distance between T_4 and A_4. A_i, $i = 1, \ldots, 4$, form a complete (and therefore globally rigid) graph, and we already have implicitly assumed that T_1, T_2, T_3, and T_4 also form a complete graph. We know from [27] that by connecting a globally rigid graph to another one using four edges, the resulting graph is globally rigid as well, and as a result, for generic positions of the agents there is a unique realization. Thus, the formation described above has a unique localization solution for generic positions of agents and landmarks, and it is the addition of an agent capable of measuring the angle subtended at itself by the two landmarks that disambiguates the multiple localization solutions. However, the important point here is that the anchor nodes A_1, A_2, A_3, and A_4 are collinear, which violates genericity. This collinearity results in having mirror localization solutions at different sides of the line that the anchors are placed on.

We conclude this section with the following remark, which is an extension of Theorem 5:

Remark 2: Consider a formation F with the underlying globally rigid graph $\mathcal{G}(\mathcal{V},\mathcal{E})$, and the four agents, T_1, T_2, T_3, and T_4, in the formation that form a complete graph. Assuming that these four agents are the only agents capable of measuring the angle subtended at them by the two landmarks L_1 and L_2, with known positions, it can be shown that there are at most two possible localization solutions for the formation.

Example 28.2

Consider four agents T_i, $i = 1, \ldots, 4$, at unknown positions with $\lambda_1 = 1.4407$, $\lambda_2 = 1.3051$, $\lambda_3 = 0.9193$, $\lambda_4 = 0.8635$, $\overline{T_1T_2} = 0.4310$, $\overline{T_1T_3} = 0.6236$, $\overline{T_1T_4} = 0.9402$, $\overline{T_2T_3} = 0.9481$, $\overline{T_2T_4} = 0.7957$, $\overline{T_3T_4} = 0.9306$, $L_1 = (0,1)^T$, and $L_2 = [0,-1]^T$. The two possible positions for the agents are depicted in Figure 28.11. A simple way to calculate these solutions is to use Algorithm 28.1 to find a set of possible solutions for the positions of T_1, T_2, and T_3. Calculate a candidate solution for the position of T_4 using the interagent distances for each of these solutions sets, and discard all the solutions that are not consistent with the angle measurements associated with T_4.

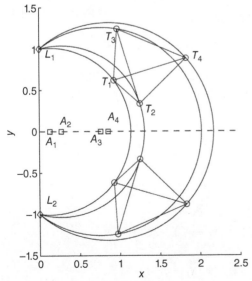

Figure 28.11   The possible localization solutions where four agents measure the subtended angle by the landmarks.

28.7 LOCALIZATION WITH EXTRA LANDMARKS

In this section, we consider the cases where bearing measurements from another landmark are available. Suppose another landmark, L_3, is positioned at a known position in the plane, and further imagine that agent T_1 can measure the subtended angle by landmarks L_1 and L_2, L_1 and L_3, and L_2 and L_3. These three angle measurements result in having three circles with the common point of intersection exactly at T_1 for generic positions for T_1. Hence, having another landmark generically disambiguates the multiple localization solutions.

There are certain geometries for which a unique solution for the agent cannot be obtained; for instance, if the agent is located on the circumcircle of the triangle $\triangle L_1 L_2 L_3$ one cannot localize it. In what follows, we address the problem of localization of a formation of three agents measuring the subtended angle at them by three landmarks in the presence of noise.

When there are three agents and two landmarks, there is no special technique for dealing with the noise. One simply proceeds using noisy measurements in lieu of noiseless measurements. However, when the noiseless equations are overdetermined, as for example with three landmarks, a modified method is required to handle the equations, temporarily assuming there is no measurement noise. As described before for each agent and any selection of two landmarks, we can define a circle. Call the center and the radius of the circle defined by T_i, L_1, and L_2 $a_i = [x_i, y_i]^T$ and R_i, respectively; term the center and the radius of the circle defined by T_i, L_1, and L_3 $a_i' = [x_i', y_i']^T$ and R_i', and term the center and the radius of the circle defined by T_i, L_2, and L_3, $a_i'' = [x_i'', y_i'']^T$ and R_i''. All the equations that govern the system can be written as

$$
\begin{aligned}
(x_{T_i} - x_i)^2 + (y_{T_i} - y_i)^2 - R_i^2 = 0, & \quad i \in \{1,2,3\} \\
(x_{T_i} - x_i')^2 + (y_{T_i} - y_i')^2 - R_i'^2 = 0, & \quad i \in \{1,2,3\} \\
(x_{T_i} - x_i'')^2 + (y_{T_i} - y_i'')^2 - R_i''^2 = 0, & \quad i \in \{1,2,3\} \\
(x_{T_i} - x_{T_j})^2 + (y_{T_i} - y_{T_j})^2 - R_{ij}^2 = 0, & \quad i,j \in \{1,2,3\}
\end{aligned}
\tag{28.4}
$$

where $x_{T_i} = [x_{T_i}, y_{T_i}]^T$ is the position of agent T_i. It is obvious that (28.4) has a unique set of solutions for the positions of the T_i. This solution is a root of the following polynomial.

$$
\begin{aligned}
P = & \sum_{i=1}^{3} \left((x_{T_i} - x_i)^2 + (y_{T_i} - y_i)^2 - R_i^2 \right)^2 \\
& + \sum_{i=1}^{3} \left((x_{T_i} - x_i')^2 + (y_{T_i} - y_i')^2 - R_i'^2 \right)^2 \\
& + \sum_{i=1}^{3} \left((x_{T_i} - x_i'')^2 + (y_{T_i} - y_i'')^2 - R_i''^2 \right)^2 \\
& + \sum_{i,j \in \{1,2,3\}} \left((x_{T_i} - x_{T_j})^2 + (y_{T_i} - y_{T_j})^2 - R_{ij}^2 \right)^2
\end{aligned}
\tag{28.5}
$$

Furthermore, it is easy to show that this solution is the global minimizer of (28.5) as well. So the solution $\hat{\mathbf{x}}T = \left[\hat{\mathbf{x}}_{T_1}^T, \hat{\mathbf{x}}_{T_2}^T, \hat{\mathbf{x}}_{T_3}^T\right]^T$ is obtained by

$$\hat{\mathbf{x}}_T = \operatorname{argmin} P. \tag{28.6}$$

Now assume that the measurement is noisy, hence (28.4) does not necessarily have a solution; however, in the light of the introduction of (28.5), one can solve the minimization equation to obtain an estimate for the solution. There are readily available software packages that enable us to minimize such cost functions, for example, Reference [28]. In the next section, we introduce some numerical simulations that illustrate the applicability of the methods introduced in this chapter to some possible scenarios.

Example 28.3   Simulations with Three Agents and Three Landmarks

Here we consider that three landmarks are placed at $[-1,0]^T$, $[1,0]^T$, and $[0,0]^T$. The exact position of agents 1, 2, and 3 are, respectively, $[3,6]^T$, $[5,8]^T$, and $[2,3]^T$ (the positions are in meters). We consider that interagent distance measurements are corrupted by a Gaussian noise with a variance equal to $0.25\,\mathrm{m}^2$, and the angle measurements are corrupted by a Gaussian noise with a variance equal to $0.0005\,\mathrm{rad}^2$. Running the simulation for 20 times and solving (28.6), we obtain $\bar{\mathbf{x}}_{T_1} = [2.9492, 5.9528]^T$, $\bar{\mathbf{x}}_{T_2} = [4.9306, 7.9594]^T$, and $\bar{\mathbf{x}}_{T_3} = [1.9290, 2.9609]^T$ as average values for the positions of T_1, T_2, and T_3, respectively. As is clear, the average estimates are very close to the actual values. Furthermore, the variances of all the solutions for the x coordinates of T_1, T_2, and T_3 are 0.0267, 0.0466, and 0.0443, and the variances of all the solutions for the y coordinates of T_1, T_2, and T_3 are 0.0155, 0.0095, and 0.0097, respectively. We have used Gloptipoly 3 [28] to solve the optimization problem generated from this scenario.

28.8   AVAILABILITY OF MORE ANGLE MEASUREMENTS FOR THREE AGENTS

So far, in this chapter, each agent is assumed to be able to detect the subtended angle by the landmarks at them only and not to be able to measure relative bearings to other agents or to one agent and one landmark. This scenario is more common if the agents are equipped with dedicated RF angle-of-arrival devices that measure bearings to transmitting beacons or detect other signals transmitted by other RF emitters of opportunity such as broadcast TV. Another scenario is that some or all agents have optical on-board sensors with a narrow field of view. In this case, the agents may not necessarily have other agents within their sensor field of view. However, use of, for example, a 360° camera sensor gives rise to a different scenario where the agents in the formation are capable of measuring not only the angle subtended by the landmarks at them but also the angle subtended at them by any pair of the agents or the landmarks.

    In this section, we briefly consider this latter scenario, in which (as before) there are three agents and two landmarks. For agent T_i, we denote the angle subtended by L_k and agent T_j, μ^i_{jk}, and the angle subtended by T_j and T_k by η_i. Measurements of these subtended angles as well as λ_i, as before, make it possible to localize up to two possible positions. We formally show this in the following theorem.

Theorem 7: Knowing all the angles λ_i, $i = 1,2,3$, η_i, $i = 1,2,3$, and $\mu^i_{k,j}$, $k = 1,2$, $i,j = 1,2,3$ and the positions of landmarks L_1 and L_2, one can find two solutions for the position of the agents T_i, $i = 1,2,3$, which are reflections through the line connecting the landmarks.

Proof: Consider a polygon with vertices defined by T_i, $i = 1,2,3$, and L_j, $j = 1,2$. By all the angles λ_i, $i = 1,2,3$, η_i, $i = 1,2,3$, and $\mu^i_{k,j}$, $k = 1,2$, $i,j = 1,2,3$, we can characterize all similar polygons with the same angles. Furthermore, knowing the distance between L_1 and L_2 fixes the scale of the polygon, knowing the position of either L_1 or L_2 fixes the two-dimensional translation, and knowing the position of the other landmark fixes the orientation of the polygon; and hence, the solution set is composed of up to two polygons that can be constructed via symmetry from each other, where L_1L_2 is the mirror axis.

Remark 3: Since we have already fixed on which side of L_1L_2 the formation is located, the only possibility is the polygon on the a priori known side of L_1L_2.

We conclude this section by the following example.

Example 28.4

Consider three agents T_i, $i = 1, \ldots, 3$, at unknown positions with $\lambda_1 = 0.2075$, $\lambda_2 = 0.1419$, $\lambda_3 = 0.0821$, $\eta_1 = 0.1419$, $\eta_2 = 2.8890$, and $\eta_3 = 0.1107$, $\mu^1_{1,2} = 2.8023$, $\mu^1_{1,3} = 2.9442$, $\mu^1_{2,1} = 2.5948$, $\mu^1_{2,3} = 2.7367$, $\mu^2_{1,1} = 0.1419$, $\mu^2_{1,3} = 3.0309$, $\mu^2_{2,1} = 0.2838$, $\mu^2_{2,3} = 3.1104$, $\mu^3_{1,1} = 0.0476$, $\mu^3_{1,2} = 0.0631$, $\mu^3_{2,1} = 0.1297$, $\mu^3_{2,1} = 0.0190$, $L_1 = (0,1)^T$, and $L_2 = [0,-1]^T$. Using elementary geometric arguments one can calculate all the angles of the pentagon formed by T_1, T_2, T_2, L_1, and L_4. Using these angles along with the exact positions of L_1 and L_2, one can calculate two mirrored solutions with L_1L_2 as the reflection axis for the target positions. The two possible sets of positions for the agents are $T_1 = [\pm 2,4]^T$, $T_2 = [\pm 6,7]^T$, and $T_3 = [\pm 10,12]^T$.

28.9 CONCLUSIONS

In this chapter, we have demonstrated that there are up to 12 possible localization solutions to the problem of localization of a formation composed of three agents, collecting bearing measurements to two landmarks at known positions and measuring their interagent distances. Furthermore, we extend this result to a more general case where a larger formation is to be localized using the same information as before. In addition, the effect of having extra landmarks on the number of the solutions to the problem was studied as well, and in this case a method to improve the accuracy of the localization solution was proposed. Some simulation results were presented to show the applicability of the methods. In the end, we briefly considered another closely related localization problem, where the agents can measure all the subtended angles at them, and show the uniqueness of the localization for this case.

ACKNOWLEDGMENTS

This work is supported by NICTA, which is funded by the Australian government as represented by the Department of Broadband, Communications, and the Digital Economy, and the Australian Research Council through the ICT Centre of Excellence program.

REFERENCES

[1] P. W. Ward, "GPS receiver RF interference monitoring, mitigation and analysis techniques," *Navigation*, vol. 41, pp. 367–391, 1994–1995.

[2] G. Dedes and A. G. Dempste, "Indoor GPS positioning: Challenges and opportunities," *IEEE Conf. on Vehicular Technology*, Dallas, TX, Sep. 2005, pp. 412–415.

[3] A. H. Sayed and A. Tarighat, "Network-based wireless location," *IEEE Signal Process. Mag.*, vol. 22, pp. 24–40, 2005.

[4] M. Cao, B. D. O. Anderson, and A. S. Morse, "Sensor network localization with imprecise distances," *Syst. Control Lett.*, vol. 55, pp. 887–893, 2006.

[5] P. Stoica and J. Li, "Source localization from range-difference measurements," *IEEE Signal Process. Mag.*, vol. 23, pp. 63–65, 69, 2006.

[6] I. Shimshoni, "On mobile robot localization from landmark bearings," *IEEE Trans. Rob. Autom.*, vol. 13, pp. 971–976, 2002.

[7] M. Betke and L. Gurvits, "Mobile robot localization using landmarks," *IEEE Trans. Rob. Autom.*, vol. 13, pp. 251–263, 1997.

[8] H. Hmam, "Mobile platform self-localisation," in *Proc. of Information Decision and Control*, Adelaide, Australia, Feb. 2007, pp. 242–247.

[9] J. S. Esteves, A. Carvalho, and C. Couto, "Generalized geometric triangulation algorithm for mobile robot absolute self-localization," in *IEEE Int. Symp. on Industrial Electronics*, Rio de Janeiro, Brasil, Jun. 2003, pp. 346–351.

[10] J. Ryde and H. Hu, "Fast circular landmark detection for cooperative localisation and mapping," in *Int. Conf. on Robotics and Automation*, Barcelona, Spain, Apr. 2005, pp. 2745–2750.

[11] T. Eren, D. Goldenberg, W. Whiteley, Y. Yang, A. S. Morseand, and B. D. O. Anderson, "Rigidity, computation and randomization in network localization," in *Proc. of Network Localization, Joint Conf. of IEEE Computer and Communication Societies*, Hong Kong, Mar. 2004, pp. 2673–2684.

[12] S. I. Roumeliotis and G. A. Bekey, "Collective localization: A distributed Kalman filter approach to localization of groups of mobile robots," in *Proc. of Int. Conf. on Robotics and Automation*, San Francisco, CA, Apr. 2000, pp. 2958–2965.

[13] N. M. Kwok, G. Dissanayake, and Q. P. Ha, "Bearing-only SLAM using a SPRT based Gaussian sum filter," in *Proc. of Int. Conf. on Robotics and Automation*, Barcelona, Spain, Apr. 2005, pp. 1109–1114.

[14] G. Laman, "On graphs and rigidity of plane skeletal structures," *J. Engrg. Math.*, vol. 11, pp. 331–340, 1970.

[15] T. Tay and W. Whiteley, "Generating isostatic frameworks," *Struct. Topology*, vol. 11, pp. 21–69, 1985.

[16] B. D. O. Anderson, C. Yu, B. Fidan, and J. M. Hendrickx, "Rigid graph control architectures for autonomous formations," *IEEE Control Syst. Mag.*, vol. 28, pp. 48–63, 2008.

[17] C. Yu, B. Fidan, J. M. Hendrickx, and B. D. O. Anderson, "Merging multiple formations: A meta-formation prospective," in *Proc. of CDC 2006*, San Diego, CA, Dec. 2006, pp. 4657–4663.

[18] G. Mao, B. Fidan, and B. D. O. Anderson, "Localization," in *Sensor Networks and Configuration: Fundamentals, Techniques, Platforms and Experiments*, New York: Springer, 2006, ch. 13.

[19] B. Hendrickson, "Conditions for graph unique realizations," *SIAM J. Comput.*, vol. 21, pp. 65–84, 1992.

[20] B. Jackson and T. Jordan, "Connected rigidity matroids and unique realizations of graphs," *J. Comb. Theory B*, vol. 94, pp. 1–29, 2005.

[21] R. Beyer, *Kinematic Synthesis of Mechanisms,* Translated from German by H. Kuenzel. London: Chapman & Hall, 1963.

[22] W. Chung, "The characteristics of a coupler curve," *Mech. Mach. Theory*, vol. 40, no. 10, pp. 1099–1106, 2005.

[23] D. K. Goldenberg, A. Krishnamurthy, W. C. Maness, Y. R. Yang, A. Young, A. S. Morse, A. Savvides, and B. D. O. Anderson, "Network localization in partially localizable networks," in *Proc. IEEE INFOCOM 2005. 24th Annual Joint Conf. of the IEEE Computer and Communications Societies*, 2005, pp. 313–326.

[24] W. T. Tutte, "A theory of 3-connected graphs," *Indagationes Math.*, vol. 23, pp. 441–455, 1961.

[25] R. L. Norton, *Design of Machinery*. New York: McGraw-Hill, 1992.

[26] Y. Zhang, S. Finger, and S. Behrens, "Rapid design through virtual and physical prototyping," Carnegie Mellon University. Available: http://www.cs.cmu.edu/~rapidproto/mechanisms/chpt5.html

[27] C. Yu, B. Fidan, and B. D. O. Anderson, "Principles to control autonomous formation merging," in *Proc. of American Control Conf.*, Minneapolis, MN, Jun. 2006, pp. 762–768.

[28] D. Henrion, J. B. Lasserre, and J. Loefberg, "Gloptipoly 3: Moments, optimization and semidefinite programming," *Optim. Methods Softw.*, vol. 24, p. 761, 2009.

PART **VII**

SPECIAL TOPICS AND APPLICATIONS

THE **FINAL** section of the book includes seven chapters, Chapters 29–35, that review several novel applications of position location systems. Many techniques and methods including GNSS, and RFID-based localization systems are discussed in detail. Moreover, an interesting application of localization to wild-life tracking is discussed. Next, an example of a remote positioning system, called Wireless Local Positioning System (WLPS) is introduced. Finally, the application of localization to autonomous driving is also discussed in two different chapters in this section.

Chapter 29 highlights localization technologies for the emerging field of autonomous driving. It reviews all localization technologies that are key to autonomous driving. It offers a detailed review of image-based localization methods. Localization using range sensors, vision sensors, and SLAM are highlighted. In addition, cooperative location estimation, filtering and sensor fusion are discussed. The chapter also presents localization techniques in use or in active research and briefly presents future directions of localization in autonomous driving.

Chapter 30 introduces the use of RFID for navigation of robots. Mobile robots are becoming intelligent, autonomous systems, and gradually able to accomplish assigned tasks such as guidance, transportation, and human-robot cooperation. For these tasks, a robust and stable navigation system is a key requirement. The chapter investigates how RFID enables high precision navigation of such robots.

Chapter 31 offers an overview of the technologies that have paved the way for visible light communication and positioning systems. The chapter also focuses on visible light positioning (VLP) systems. It also discusses how arrangements of different light detectors can improve VLP system accuracy and usability. Many techniques to use visible light for communications are also discussed.

Chapter 32 examines fourth generation cellular systems and describes the positioning standard for the Long Term Evolution (LTE) cellular system. The chapter reviews the LTE architecture and air-interface, as well as the positioning requirements, positioning architectures and signaling. The simplest LTE-based techniques,

Handbook of Position Location: Theory, Practice, and Advances, Second Edition.
Edited by S. A. (Reza) Zekavat and R. Michael Buehrer.
© 2019 by the Institute of Electrical and Electronics Engineers, Inc.
Published 2019 by John Wiley & Sons, Inc.
Companion Website: www.wiley.com/go/zekavat/positionlocation2e

the cell ID method, is discussed first. This method is often augmented with timing advance and angle of arrival information in what is called enhanced cell ID (E-CID) class of methods. Two approaches to fingerprinting-based positioning are described, including pattern matching and the self-learning adaptive enhanced cell ID (AECID) method. The LTE standard also supports a downlink observed time difference of arrival (OTDOA) approach which is discussed. Finally, the satellite navigation functionality provided by A-GNSS is described.

Chapter 33 describes the application of position location techniques to wildlife tracking. Specifically, the chapter begins with a discussion of currently available wildlife tracking devices (tags), and explains why tag mass is the primary design constraint. Current manual direction-finding methods are described, as are several automated implementations. The authors also discuss the need for generic asset (non-wildlife) tracking tags that are light and inexpensive, and review current asset tracking methods based on cellular and satellite platforms in this context. Two popular satellite-based location-tracking technologies, GPS and Argos, are described, as are several other tracking techniques. The shortcomings of existing systems motivate the need for a new approach that offers GPS-like accuracy, with vastly reduced energy consumption. Terrestrial TOA tracking methods are discussed as a lower energy cost solution, with specific sections dedicated to explaining the various concepts and their interplay in an integrated system. The chapter also introduces a few basic concepts needed for system analysis including spectrum utilization, autocorrelation, cross-correlation, signal to noise ratio, link budget and processing gain. Various positioning methods are compared using real and simulated data. The signal processing discussion details the computational requirements of real-time matched filtering, including the impact of Doppler shift. Several techniques used to implement real-time TOA receivers in embedded devices with limited computing resources, including the use of frequency domain processing via the FFT, and the intelligent reuse of data via time-shifting techniques are also described. The chapter concludes with a summary of the current performance of a TOA wildlife tracking system that the authors have implemented, its design limitations, and likely areas for improvement.

Chapter 34 introduces an active target remote positioning system called wireless local positioning system (WLPS) that has been recently developed and patented. The structure and block diagram of WLPS is discussed in the chapter. The chapter studies the implementation of the WLPS receiver and its performance. In addition, the implementation of DOA and TOA estimation techniques in WLPS are presented. One of the features of WLPS signals, called cyclo-stationary, is investigated and its impact on DOA estimation and the detection process is discussed. The chapter also presents a variety of civilian and military applications of WLPS and its impact on safety and security.

Chapter 35 is a complementary chapter to autonomous driving. It offers near-ground channel path-loss modeling that is key to both autonomous driving sensors (such as those installed on vehicle bumpers), as well as the general concept of near ground sensor localization. An accurate channel pathloss is key to RSSI-based localization methods and this chapter details.

CHAPTER 29

LOCALIZATION FOR AUTONOMOUS DRIVING

Ami Woo, Baris Fidan, and William W. Melek
University of Waterloo, ON, Canada

T HIS CHAPTER reviews state-of-the-art sensors, instrumentation and algorithms used for localization of autonomous vehicles. The current localization approaches for autonomous driving involve localizing by satellite navigation systems, vehicle motion sensors, range sensors, and vision sensors. The chapter presents current localization approaches, which are categorized as (1) global localization, (2) relative localization, and (3) simultaneous localization and mapping (SLAM). In relative localization, visual odometry (VO) is specifically highlighted with details. Two main approaches of VO, appearance-based and feature-based approaches, are described. Three main approaches of SLAM, Kalman filter, particle filter, and graph-based approaches, are presented. Afterwards, the chapter presents estimation, filtering, and sensor fusion techniques for cooperative localization. The chapter finally reviews some current localization techniques in use and discusses potential solutions to these gaps, as well as future directions for localization in autonomous driving.

29.1 INTRODUCTION

In the past three decades, there has been great interest and progress in the field of intelligent vehicles for both researchers and industries. Various intelligent vehicle systems have been proposed to reduce road accidents and traffic congestions. The continuous development of sensing and computation technologies has led to the development of various driver assistant systems. Some of these driver assistant systems, such as lane keeping, lane changing, adaptive cruise control, and highway driver assistant systems have already been installed on vehicles. There are six different levels of driver assistant systems ranging from fully human operated to fully

Handbook of Position Location: Theory, Practice, and Advances, Second Edition.
Edited by S. A. (Reza) Zekavat and R. Michael Buehrer.
© 2019 by the Institute of Electrical and Electronics Engineers, Inc.
Published 2019 by John Wiley & Sons, Inc.
Companion Website: www.wiley.com/go/zekavat/positionlocation2e

automated [1]. The current commercial driver assistant systems are limited to aiding the driver while the driver is responsible for overall control of the vehicle or some autonomous driving functions under certain driving conditions. There are still many challenges to be solved to allow autonomous vehicles to emerge into the market. One of the key challenges for the development of fully autonomous vehicles is localizing the vehicle in known, unknown, or uncertain environments.

For an autonomous vehicle to navigate without human support, the vehicle needs to locate itself with respect to a map. Without knowing its location within an area and the location of other objects at any instant, the vehicle cannot reach its destination nor be aware of where it is heading to. Depending on the objective of the vehicle, the degree of accuracy required for autonomous driving varies. An autonomous vehicle that aims to travel from one place to another needs to be aware of its location globally. The autonomous vehicle needs to perceive its immediate environment and localize itself relative to static and dynamic obstacles, including trees, buildings, other vehicles, pedestrians, or cyclists. In addition, the vehicle should also be aware of its location in the road, such as its relative location to lane marks, road marks, or curbs. This is required to ensure that the vehicle knows when to drive, stop, change lanes, or make turns. To avoid collision, the accuracy required for relative localization (remote localization) should be much higher compared to absolute localization (self-localization).

In the following sections, the sensors and the algorithms developed for autonomous vehicle are discussed. Localization using range sensors, vision sensors, and SLAM are specifically highlighted, as are cooperative location estimation, filtering, and sensor fusion. The concluding sections address localization techniques in use or in active research and briefly present future directions of localization in autonomous driving.

29.2 SENSORS AND INSTRUMENTATION

A variety of sensors and instrumentation are utilized in intelligent vehicles for localization. These sensors can be classified into two classes: proprioceptive and exteroceptive. To enhance the performance of localization algorithms and achieve reliability and accuracy, multiple sensors are used and fused together. This section discusses different types of sensors used to estimate vehicles' states, such as position and heading.

29.2.1 Exteroceptive Sensors

The exteroceptive sensors aim to measure information from the vehicle environment. Most of the exteroceptive sensors are used for localization relative to the immediate surrounding of the vehicle, although via feature extraction some of the relative localization sensors can also be used for absolute localization.

Global Navigation Satellites: The development of global navigation satellite systems was initially motivated by defense applications. Currently, these systems are widely used in commercial applications for positioning, navigation, and tracking. Absolute geo spatial positioning information obtained through the use of global navigation satellite systems, such as global positioning system (GPS) and globalnaya navigatsionnaya sputnikovaya Sistema (GLONASS) can augment other sensors to perform relative localization [2]. The thorough explanation and discussion on the topic is discussed in Chapters 20–23.

Range Sensors: There are a variety of range sensors commercially available to enable different driver assistant functionalities for smart and autonomous vehicles. This section presents laser and radar systems, which are the most commonly used range sensors. Further details on range sensors and range based localization can be found in Chapters 1 and 2.

The use of a laser scanner robustly provides the vehicle with information about what is ahead [3]. Laser scanner-based sensors emit light impulses of electromagnetic waves and use time of arrival (TOA) to determine the range and bearing of an object. A laser-based range sensor consists of a light source, which emits collimated light beams, and a receiver, which receives the light reflected from surrounding objects. Often this type of sensor is referred to as LiDAR (light detection and ranging). There are two-dimensional (2-D) and three-dimensional (3-D) LiDAR sensors that all work on the same principle of TOA, but the ways to measure the distance vary among the manufactures. Most of the LiDAR sensors collect the light beam using an oscillating mirror for 2-D and a rotating mirror for 3-D. Some methods use the elapsed time to calculate the distance. Some other methods observe the phase shift between the transmitted and received light. LiDAR can observe wide field-of-view and long ranges, and is precise. However, it is sensitive to atmospheric conditions. LiDAR sensors have been utilized by many research groups and industries involved with autonomous vehicles with promising results. Some believe LiDAR is essential for autonomous vehicles. However, it is very expensive compared to other sensors, and it is currently very bulky. There are debates whether LiDAR is a practical sensor for autonomous driving. As an alternative, radar (radio detection and ranging) sensors have also been considered for autonomous vehicle localization.

Radar sensors are LiDAR-like sensors that utilize radio waves instead of light. They are generally used for obstacle detection. The radio and light waves travel at the same speed, but the radio waves have much lower frequencies. To measure the distance to an object, radar sensors estimate signal TOA. Unlike LiDAR, the radar takes into account the Doppler shift of the echo; hence it can observe the velocity of the moving object without further numerical processing. It operates over longer distances compared to LiDAR and is less sensitive to weather conditions, performing well in extreme weather conditions. Most radar sensors are less expensive than LiDAR sensors, making them more practical. However, the wavelength of radio waves is significantly smaller compared to light waves; hence, the resolution of radar is not as high as LiDAR. The field of view of radar is usually wider compared to LiDAR, and the bearing measurement of radar is less accurate. Therefore, the radar outperforms LiDAR when the objective of the functionality is to detect or track objects, whereas the recognition of an object cannot be done as accurate as LiDAR.

Some industries believe radar, along with other sensors such as vision sensors, can accomplish the tasks accomplished by LiDAR, while some believe LiDAR is key to autonomous driving.

Vision Sensors: The prototypes and commercialized computer vision technology have been developed as the key components of autonomous driving and intelligent vehicles. The applications of computer vision include lane detection and tracking [4], road sign detection and understanding [5], traffic light detection [6], and many more. Originally, there were challenges in using vision for localization, since vision transforms the 3-D world into 2-D. Various techniques are proposed to recover 3-D features from images. However, this does not address the problem of high computational complexity with vision sensors.

GPS, radar, and LiDAR are dominant sensors for intelligent vehicles since they enable real-time navigation and localization. Even though range sensors are much more expensive compared to vision sensors, these sensors are more favourable due to their high accuracy and less computational power. The recent advances in computer technology and computer vision brought back the attention to vision sensors for navigation purposes. With the abundant information that they provide, vision sensors are considered excellent tools to observe details of immediate surrounding, to detect objects in front, and to recognize curbs, lane marks, traffic lights, and traffic signs.

One of the commonly used vision sensors is a stereo vision camera, a camera that uses two lenses. With a well-calibrated stereo camera, the 3-D structural information can be retrieved by comparing the same scene from two different perspectives. One commonly used technique to recover the 3-D structure through is the triangulation method [7, 8]. In the triangulation method, the estimation of location of a point from the scene can be determined using the intersection rays from both cameras, focal length of the cameras, and relative positions of the cameras. With the depth information from a single frame, the image scale can be instantaneously retrieved. Another approach is to use monocular camera that uses a single camera. Unlike stereo camera, which can retrieve the depth information for scenes with different perspectives, it is more difficult to observe depth information using a single camera. One technique to estimate depth is by realizing that the image properties change as a function of the motion. For practical implementation, the monocular camera is typically mounted to face the direction of the movement of the vehicle. As the vehicle advances, more than two images are available for stereo and monocular vision systems, and triangulation method can be used to estimate the location of objects observed by camera. The uncertainty in position estimation using monocular vision is usually higher than with stereo vision.

Compared to monocular vision systems with limited depth recover capability, stereo vision allows more efficient and accurate triangulation. However, it is more expensive compared to monocular vision and requires a more intensive calibration process. Nevertheless, both the monocular and stereo vision sensors are more affordable and offer smaller-scale solutions compared to range sensors. In addition, they provide large amounts of useful information, including the ability to see colors and textures of the surrounding objects and recognizing features more accurately than other sensors. However, at the current stage of technologies, vision-based localization is more computationally expensive and is very sensitive to weather conditions. New and interesting approaches to using camera for localization are continuously being developed, including exploiting deep learning (DL) for feature extraction from camera images. It is expected that these methods will be developed and implemented for autonomous driving.

Omnidirectional cameras with 360° field of view are also being used in autonomous driving. They outperform conventional cameras in certain situations [9]. There are different types of omnidirectional cameras, some have larger fields of view, and some smaller. Larger field of view for vehicle is an asset in autonomous driving, especially for localization, as the vehicle field of view remains for longer periods of time. However, modelling omnidirectional cameras is more complex compared to perspective cameras, and the resolution of a pixel is less. For simplicity, only the perspective camera model is presented in this chapter, but the localization techniques discussed in the remainder of the chapter are applicable to omnidirectional camera models.

3-D to 2-D Perspective Projection: The camera model proposed for the discussion is the pinhole camera model, which is a simple model to approximate the image processing. This model assumes that all points of an object possess sufficient amount of illumination, and radiate light rays to the small openings of a pinhole. The pinhole model only allows small amount of light to pass through, requiring long exposure times compared to larger holes, which allows more light but cause blurry images. The camera model uses a lens instead of a pinhole, which controls lens aperture opening and accordingly the amount of light that passes through. The same projection system as the pin-hole camera is applicable; the lights pass through the center of point (COP) of the camera and focuses on the image plane. The image projected on the image plane is upside down as the radiated ray passes through. For convenience, the virtual image plane is placed in front of COP; hence, the image projected is not inverted.

Mapping the 3-D world into the 2-D world can be accomplished by the following equation in homogeneous coordinates [8]:

$$
p = \begin{bmatrix} p_u \\ p_v \\ 1 \end{bmatrix} = \gamma \begin{bmatrix} R & t \end{bmatrix} P = \begin{bmatrix} f_u & \alpha & p_{u0} \\ 0 & f_v & p_{v0} \\ 0 & 0 & 1 \end{bmatrix} \begin{bmatrix} R & t \end{bmatrix} \begin{bmatrix} p_x \\ p_y \\ p_z \\ 1 \end{bmatrix}, \tag{29.1}
$$

where $P = [p_x, p_y, p_z, 1]^T$ is the homogeneous vector of the 3-D world point relative to the camera reference frame, and $p = [p_u, p_v, 1]^T$ is the homogeneous vector of the projection point on the image plane. The skew parameter α is for non rectangular pixels. Most of the current cameras have rectangular pixels, for which $\alpha = 0$. γ is the intrinsic camera parameter matrix, f_u and f_v are focal lengths, and p_{u_0} and p_{v_0} are the center points in x and y direction on the image plane, respectively. Normally the focal lengths f_u and f_v are assumed equal; hence, throughout the chapter they will be denoted as f. Rotation matrix $R \in \mathrm{SO}(3)$ and translation vector $t \in \mathbb{R}^{3 \times 1}$ map the 3-D world into the camera frame.

The knowledge of the camera parameters and transformation matrix allow the estimation of points from the real world in the image. This is only possible when the image is on focus. In real life, cameras have distortion, which is caused by imperfection of lenses, especially when the rays pass through the edges of the lens. In addition, images are exposed to noise. More comprehensive discussions and techniques on such problems and models can be found in [7] and [8].

Vehicular Network Sensors: Vehicular communication systems allow localization that improves safety and maximizes road space utilization. It improves such tasks by enhancing the technical applications such as platooning and collision warning. A vehicular ad hoc network (VANET) is a form of mobile ad hoc network (MANET) that allows sharing information within moving traffic and with the road side infrastructure. This includes wireless communication of vehicle to infrastructure (V2I), vehicle to vehicle (V2V), and vehicle to everything (V2X) [10]. In addition, the wireless positioing systems explained in Chapter 34 enables simultaneous localization and communication. Dedicated short-range communications (DSRC), a special form of radio frequency identification (RFID) dedicated for automotive use, has been commonly allocated the 5.8/5.9 GHZ band and is being widely used in smart toll collection. Additional information from other vehicles or infrastructure allows vehicles to perceive not only its immediate surroundings but also further distances that vehicle cannot directly sense from other sensors [11]. Further discussions on ad hoc positioning and RFID navigation can be seen in Chapter 27 and 30, correspondingly.

29.2.2 Proprioceptive Sensors

The proprioceptive sensors internally measure and collect data, such as velocity and steering angle of the vehicle. Most of the proprioceptive sensors used for localization and navigation system purposes are onboard motion sensors. As explained in Chapter 1, these sensors can only allow self-localization and cannot be used to find the location of other objects (remote localization).

Vehicle Motion Sensors: Most of the vehicle motion sensors that are being used in autonomous driving are low cost and high resolution. These sensors are self-localization systems, and they can only offer the location with respect to a bearing point. However, these sensors are subject to errors. To enable high-performance self-localization, they are integrated with other positioning systems such as GPS. As GPS does not perform well in downtown or indoor areas (e.g., tunnels), merger of GPS and motion sensors also enables high-performance localization in these areas.

Two types of encoders, velocity and steering, are commonly used to measure the internal states of the vehicle. Most popular encoders for both velocity and steering are optical encoders. A typical optical encoder contains a grid disk, which is attached to the axle, a light source, which illuminates on the disk, and an optical detector [12]. Many optical encoders that measure velocity are incremental. As the rotor rotates, the slit varies, resulting in varying amount of light observed by the optical detector. The output is usually a series of square waves in which high and low represents the presence and absence of light, respectively. The counts of falling or rising edge pulses are used to determine the rotational velocity of the axle. If a second illumination and detector pair is available, the direction can also be measured. This is possible if the sensors are placed a half slit-width apart, which will result in one sensor always being triggered first in one direction and the second sensor in opposite direction.

If each wheel in the vehicle is implemented with an encoder, difference in speed among the wheels can be observed. This measurements can be used to estimate the heading of the vehicle without external sensors. If an individual encoder is not available, steering encoder can be used to measure the heading. The steering encoder measures the angle of the front wheels with respect to the forward direction of the vehicle. The common steering encoder is an absolute encoder, which contains a coded disk that is imprinted with rows of broken arcs. Each arc is unique and it is arranged in a pattern to be encoded. Each row is placed with a light source and a sensor to observe the unique pattern properly.

An odometer measures the travelled distance of the vehicle. An odometer often uses an incremental encoder. Full or partial rotations of the wheel are used to compute the curvilinear displacement of the vehicle. The number of rotations of the wheel can be easily calculated by the number of output pulses during time slot multiplied by the full pulses per revolution. This ratio is then multiplied by the circumference of the wheel to estimate the curvilinear displacement of the wheel.

The resolution of the encoders depends on the number of slits or rows. The encoders usually have high accuracy, but under certain circumstances it can be very poor. Such scenarios and sources of errors include: wheel slips uneven road surfaces, skidding, changes in wheel diameter, different wheel diameters among wheels, etc. [13, 14]. Errors due to variation of wheel diameter could be reduced by calibrating and estimating the parameters in sensor integration. The other sources of errors could be predicted through more complex estimation methods. Some estimation methods include Rao–Blackwellized particle filter to estimate wheel slip and velocity [15], lateral tire-force measurement, and a recursive least square algorithm with a forgetting factor to estimate sideslip angle and roll angle [16].

Inertial Sensors: Traditionally, inertial sensors were developed for military applications such as aircraft, missiles, submarines and armoured vehicles. Due to its bulky and heavy structure, it was not mechanically nor economically feasible to implement on vehicles. The continuous development of sensing and computation technology, especially on microelectromechanical systems (MEMS), have opened up new applications for the automobiles, and they are being used in current vehicle systems. Inertial sensors are motion and rotation sensors that provide information about linear and rotational motion of the vehicle. An inertial measurement Unit (IMU) is commercially available inertial sensor that is generally used for inertial navigation. Typical IMUs include three orthogonal accelerometers, three orthogonal rate gyroscopes, and sometimes magnetometers. As highlighted in Chapter 1, the accuracy of inertial sensors degrades with time due to error propagation or accumulation, and they should be integrated with other localization systems, such as GPS. The operation and details of accelerometers and gyroscopes can be found in Chapters 1 and 22.

An IMU example is a magnetometer. A magnetometer is an old technology and has been traditionally used in a compass. It measures the strength and direction of magnetic field by Hall effect or magnetoresistive effect. Most magnetometers use Hall effect, which incorporates the properties of electrons in the presence of magnetic field. The electrons through a conductor do not run straight in the presence of a magnetic field. They experience a force called transverse force, which results in gathering on one side of the material. A magnetometer measures the way the motion of the electrons is disturbed by the Earth's magnetic field to provide inertial orientation. The heading of the vehicle relative to the Earth's magnetic north can be observed through the measurement [17].

The difference in angle between the true north of Earth and the magnetic north is called magnetic declination. To compensate for the difference, the orientation of the sensor must be known. The declination angle varies depending on the location of the vehicle and varies with time. Unfortunately, the magnetic field of the Earth is not very strong and metals and magnets can corrupt the magnetic field and measurement. The magnetic sensor can have good magnetic field sensing precision, but the orientation measurement accuracy can still be very poor due to such magnetic disturbances and declination angle effects. Hence, it is debatable whether or not the magnetometer can be used for vehicle navigation.

29.3 LOCALIZATION SYSTEM DESIGNS FOR AUTONOMOUS DRIVING

29.3.1 Global Localization

Having the knowledge of the global location is essential for a vehicle to be able to travel to other places. Even if the vehicle itself is aware of its location in immediate surroundings, it will never find its way to a destination unless it knows its own location. Absolute localization refers to the methods of localizing the vehicle in global sense. Most commonly used absolute localization techniques are free from accumulation of error with time or the distance travelled.

Satellite Navigation Systems: Most of the recently manufactured vehicles contain in-car localization systems using a satellite navigation with a digital map. Depending on the types of the satellite navigation system implemented, the localization accuracy and update rate can vary. Commonly, the absolute coordinate information is used in combination with a digital map. The digital map provides geometrical and spatial information of road scene to improve road interpretation [18]. Through a map matching process the coordinates from the satellite are often constrained by the existing streets in the map [19].

The satellite navigation system has limitations. Its most important drawback is requiring a clear view of the sky, which is infeasible to meet in urban driving, where tall buildings or tall trees appear, or under tunnels or inside buildings. Due to signal scattering or attenuation in such environments, the accuracy, reliability, and robustness of the localization reduces. Accordingly, vehicle navigation is mostly integrated with vehicle motion and/or inertial sensors, which enables localization in urban areas or within tunnels. More details and discussion on satellite navigation systems are provided in Chapter 20. Since the satellite navigation maps may not be up to date and due to the presence of other neighbouring vehicles, pedestrians, and animals in the dynamic environment, satellite navigation is unable to provide the environment details and cannot be a standalone solution.

Landmark-Based Navigation: Landmarks are objects that can be detected by sensors. If landmarks are known and fixed, the absolute location is measurable by relative positioning with respect to the landmark set. Landmark-based localization is one of the oldest absolute localization methods used by researchers in the field of robotics. It typically outperforms satellite navigation-based localization in areas with signal interference.

The most important challenge in the implementation of landmark-based localization for autonomous driving is the need to frequently and accurately detect the landmarks. LiDAR and vision sensors are two common methods to identify the landmark features. The information provided by LiDAR is accurate but only provides distance to an object. LiDAR can provide geometric features of the landmarks. However, it is difficult to use for differentiating landmarks with the same geometric shape. Using vision sensors to recognize the landmark is a potential approach to improve the drawbacks of LiDAR. Different vision techniques (discussed in Section 29.4) can be used to recognize or detect features of the immediate surroundings. Irrespective of the implementation approach, landmark-based localization requires a priori knowledge of the positions of uniquely identifiable landmarks in the driving environment. Hence, landmark-based localization may be a good solution for some small-scale well-known environments but cannot be a practical solution to drive around cities or highways.

29.3.2 Relative Localization

Dead Reckoning: Dead reckoning is one of the oldest methods to estimate relative position. It is the current solution for some manufactured vehicles. The localization is done by fusing onboard sensor information about current direction of motion, such as first and second derivative of attitude and heading, with its previous position information. The sensors used are relatively inexpensive compared to other localization sensors and are free from signal interference. Dead reckoning is a good compensation when there is signal intervention and a feasible solution to urban, tunnel, and indoor environments. However, this cannot perceive the dynamic environment and requires accurate initial position information.

Without the initial position, dead reckoning cannot be performed precisely. Initial position errors as well as structured and unstructured uncertainties can be reduced by precise calibration. They are caused by inaccuracies in kinematic vehicle and measurement models as well as sensor uncertainties. The unstructured uncertainties are random, more difficult to predict, and vary significantly depending on the driving conditions. Some are caused by the interaction of the vehicle with an unpredictable surrounding environment such as dry, wet, slippery, or icy roads [13, 14]. The associated errors grow and accumulate without bound as the vehicle travels unless it is compensated.

Inertial Navigation: Inertial navigation is a type of dead reckoning technique but is limited to inertial sensors, including accelerometers and gyroscopes. The acceleration measurement is double integrated with respect to time to determine the position of the vehicle.

There are two methods to attach these sensors to the vehicle: using stable platforms and strap-down systems [20]. Depending on the method, the measurements are processed differently. In stable platform systems, the sensor is mounted in a way that the sensor is isolated from external rotation motion. This allows measuring the roll, pitch, and yaw of the vehicle with respect to navigation frame. In strap-down systems, the sensors are rigidly attached to the navigation platform in a way that the rotations of the vehicle influence the system. The measurements are further processed to transform the coordinates of vehicle into navigation frame. The strap-down systems are less mechanically complicated and are lightweight but come with computational complexity. Inertial navigation systems are not exposed to unsystematic errors such as tire slip. Other benefits and limitations are similar to dead reckoning.

To maintain inertial navigation's practicality, different combinations of sensors are fused to estimate the pose of the vehicle. Commonly, inertial sensors are fused with a global satellite navigation system [21, 22]. Full details of the computation and implementation of this fusion is presented in Chapter 22.

Visual Odometry: The term visual odometry (VO) was first used by Nister [23]. VO is similar to dead reckoning, but it is a more accurate approach and has gained attention in research and industrial field with the advancement of computational power and computer vision [24, 25]. It is a particular case of a technique called structure from motion (SFM). The SFM reconstructs 3-D scenes and camera poses from an unordered set of images and usually is implemented off-line. VO, on the other hand, is often accomplished in real time but using an ordered image sequence. The first known real-time VO was implemented in 1980 using a sliding camera [26]. VO was incorporated in 2004 on a Mars mission by Mars rover autonomous vehicles Spirit and Opportunity [27]. Since then, VO gained more interest, and different research groups have been developing ways to apply it to varying applications.

Using the images captured at each instant, the goal of VO is to compute relative transformation from one image to the next and to find the full trajectory of the camera. Unlike dead reckoning, VO is free from any errors that are terrain or vehicle parameter-dependent, such as a vehicle wheel variations. VO is a much cheaper solution compared to LiDAR-based localization, which can perform similar tasks but with higher computational power. Unlike other sensing methodologies, VO works under an assumption that there is sufficient illumination in the surrounding of the cameras to observe enough texture in the environment. This chapter assumes that the images captured are sufficient enough to observe necessities to do motion estimation of the vehicle. To gain insight of VO involves knowledge of computer vision. Further discussion on VO is found in Section 29.4.

29.3.3 SLAM

SLAM incrementally constructs a map with information from sensors while simultaneously localizing itself to the map. Originally, SLAM was an active research field for robotic navigation. Achievements from the robotic field and immense development in sensing technology led SLAM to appear in autonomous driving. Global navigation sensors in combination with motion, range, or vision sensors offer good global and local localization. However, they suffer from drift of error. SLAM compensates for drift of error problems by building a global map and recognizing the revisited places. More discussion and details on SLAM can be found in Section 29.5.

29.4 VO

Vision-based odometry has been studied and used in many applications in both outdoor and indoor environments [25, 28, 29]. As images are full of useful information, there are many approaches to estimate the poses. The two main approaches, the appearance-based method and feature-based method, have been developed using various techniques. Both approaches estimate motion by estimating the movement of a point (a pixel or feature) from one image frame to the next. This requires the application of computer vision techniques. This section reviews general techniques. For more details on these computer vision methods, the readers are encouraged to consult [8, 30, 31].

29.4.1 Appearance-Based Method

The appearance-based (also referred to as direct or featureless) method have recently gained interest from the field of VO as opposed to the feature-based method. This method eliminates preprocessing steps such as feature extraction described in next section. Instead, it uses displacement of every pixel intensity from a sub- or full image to recover the pose by minimizing an error called photometric error. This process is more robust in low-textured environments but usually requires a very time-consuming optimization algorithm such as Gauss-Netwon.

Image Similarity: Finding similarities between images or aligning images can be accomplished by searching the full or parts of an image that correspond from one image to the next in a sequence of images. Image-to-image correspondence is often accomplished at the pixel level of images, using the luminous (light) intensity information on each pixel. In practice, a full search of the image is time consuming; instead, part of an image from one image, referred to as template, is used. Usually, the size of templates is much smaller compared to the original image or the image that is being compared. With the small template, a method referred to as template matching finds the matching point on one image, image 1, with image matrix g_1, in the next image, image 2, with image matrix g_2, by finding a motion vector $t_g = [u_g, v_g]^T$ between these two images. Here, the image matrix g_j is a matrix function mapping each pixel index to the value of a certain image quantity, that is, the intensity I_j. The motion vector t_g maps each point (p_u, p_v) on image 1 to a similar point $(p_{u'}, p_{v'})$ on image 2 that maximizes similarities of the corresponding image matrix value. Finding the similarity of points across images 1 and 2 is defined as

$$\begin{bmatrix} p_{u'} \\ p_{v'} \end{bmatrix} = \begin{bmatrix} u_g \\ v_g \end{bmatrix} + \begin{bmatrix} p_u \\ p_v \end{bmatrix}, \tag{29.2}$$

where $[p_u, p_v]^T$ and $[p_{u'}, p_{v'}]^T$ are points on images 1 and 2, respectively, assuming the pixel mapping from image 1 to image 2 is linear. To measure the similarity of images, a particular error matrix is minimized. Some common error matrices include sum of squared difference (SSD), defined as [7]:

$$SSD(t_g) = \sum_{p_u, p_v} (g_1(p_u, p_v) - g_2(p_u + u_g, p_v + v_g))^2, \tag{29.3}$$

and sum of absolute differences (SAD), defined as [7]:

$$SAD(t_g) = \sum_{p_u, p_v} |g_1(p_u, p_v) - g_2(p_u + u_g, p_v + v_g)|. \tag{29.4}$$

Another similarity measure, which is desired to be maximized, is the sum of cross-correlation coefficients over all possible locations, defined as [7]:

$$CC(t_g) = -\sum_{p_u, p_v} g_1(p_u, p_v) g_2(p_u + u_g, p_v + v_g). \tag{29.5}$$

Measuring the similarities using (29.3), (29.4), or (29.5) is limited by the fact that these measures are not invariant to the changes in the environment conditions, for example, illumination. This issue can be solved by normalization, for example, of the cross-correlation coefficients, as [7]:

$$NCC(t_g) = \frac{-\sum_{p_u, p_v} (g_1(p_u, p_v) - \bar{g}_1)(g_2(p_u + u_g, p_v + v_g) - \bar{g}_2)}{\sqrt{\sum_{p_u, p_v} (g_1(p_u, p_v) - \bar{g}_1)^2 \sum_{p_u, p_v} (g_2(p_u + u_g, p_v + v_g) - \bar{g}_2)^2}}, \tag{29.6}$$

where $\bar{g}_1$ and $\bar{g}_2$ are the mean intensities in each image.

The template matching can be implemented via exhaustive search of the entire image. A more efficient implementation is accomplished by constructing a pyramid of images [32]. Using an image, the pyramid is constructed by subsampling, often by a factors of two, as shown in Figure 29.1. A popular method of subsampling incorporates a Gaussian filter. An image in the lower level of the pyramid contains more information compared to its immediate upper-level image. The template match using the pyramid is firstly implemented on the highest level of the pyramid. This enables initialization for its immediate lower level and the procedure repeats to the original image. Whether it is an exhaustive search or image pyramid method, the template method is not invariant to scaling or rotation because both approaches are implemented at a pixel level. The template matching is a practical method for scenarios where there are two images with variations in translation and illumination, such as stereo vision cameras.

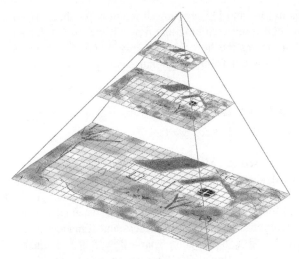

Figure 29.1   Pyramid of images.

Another approach to accelerate the process is alignment via analyzing the image in frequency domain using Fourier transform. Similar to exhaustive search and a pyramid of images, this method has been mostly implemented for translational shift of the point of interest, but rotations and scales using such a technique have also been proposed [33]. Another popular method for finding the motion vector involves optical flow and is discussed in the following section.

Motion Field and Optical Flow: Motion field describes the real-world motion and assigns a 3-D velocity vector to every point in an image. Optical flow describes the changes in the intensity of the pixels in a 2-D image. The goal of the optical flow is to approximate the motion field. The motion field and optical flow do not always correspond to each other, but the rest of the chapter treats them as similar.

In 1981, two famous works that pioneered optical flow estimation were proposed by Horn and Schunck [34] and Lucas and Kanade [35]. Both techniques assume small motion between the two images and brightness constancy between corresponding points in consecutive images, referred to as brightness constancy assumption:

$$I(p_u, p_v, t) \approx I(p_u + \dot{x}dt, p_v + \dot{y}dt, t + dt), \tag{29.7}$$

where $I(p_u, p_v, t)$ is the intensity value at points (p_u, p_v) at time instant t and the same point after $t + dt$ with displacement $[u_g, v_g]^T = [\dot{x}dt, \dot{y}dt]^T$. This assumption can be approximated by Taylor series and further simplified to a result that is often referred to as optical flow constraint and corresponds to [34, 35]:

$$I_u \dot{x} + I_v \dot{y} + I_t \approx 0, \tag{29.8}$$

where $I_u = \frac{\partial I}{\partial p_u}$, $I_v = \frac{\partial I}{\partial p_v}$, and $I_t = \frac{\partial I}{\partial t}$, and $\dot{x} = \frac{dp_u}{dt}$, $\dot{y} = \frac{dp_v}{dt}$ are optical flows. This constraint leads to an equation with two unknowns, $\dot{x}$ and $\dot{y}$. To solve for the unknowns, the two tehcniques in [34] and [35] employ different approahces. Horn and Schunck imposed a further constraint assuming that flow will be smooth for all pixels [34]. The assumption formulates the flow as an energy function that needs minimization. Using techniques adopted from variational calculus to the energy function leads to [34]:

$$(I_u\dot{x} + I_v\dot{y} + I_t)I_u + \Lambda\Delta^2\dot{x} = 0,$$

$$(I_u\dot{x} + I_v\dot{y} + I_t)I_v + \Lambda\Delta^2\dot{y} = 0,$$

(29.9)

where Λ is a Lagrange multiplier used to enforce the constraints, $\Delta^2\dot{x} = \frac{\partial^2\dot{x}}{\partial u^2} + \frac{\partial^2\dot{x}}{\partial v^2}$ and $\Delta^2\dot{y} = \frac{\partial^2\dot{y}}{\partial u^2} + \frac{\partial^2\dot{y}}{\partial v^2}$. This method iteratively finds $\dot{x}$ and $\dot{y}$ and results in dense optical flow, that is, intensity variations in almost all the pixels in the image. In reality, the assumption proposed for dense optical flow is not valid, especially when there are objects moving in different directions. Instead, Lucas and Kanade assume constant motion in a neighborhood [35] and producing sparse flow, that is, intensity variations in only certain pixels of the image. The flow is computed using least squares fit that corresponds to [35]:

$$\begin{bmatrix} \sum_{p_{u_i}, p_{v_i}} I_{u_i}^2 & \sum_{p_{u_i}, p_{v_i}} I_{u_i} I_{v_i} \\ \sum_{p_{u_i}, p_{v_i}} I_{u_i} I_{v_i} & \sum_{p_{u_i}, p_{v_i}} I_{v_i}^2 \end{bmatrix} \begin{bmatrix} \dot{x} \\ \dot{y} \end{bmatrix} = \begin{bmatrix} -\sum_{p_{u_i}, p_{v_i}} I_{u_i} I_{t_i} \\ -\sum_{p_{u_i}, p_{v_i}} I_{v_i} I_{t_i} \end{bmatrix}.$$

(29.10)

The sparse algorithm usually requires feature extraction steps to compute sparse points that need to be tracked. It is challenging to say which method performs better. In practice, the assumptions on both methods are violated; hence, the performance depends on the application.

29.4.2 Feature-Based Methods

Currently, the feature-based methods are the dominant practical solution for VO and SLAM methods. Instead of using all of the pixels, the motion of the camera is computed by analyzing the displacement of the features in feature-based methods. Features have properties that make them distinctive from their surroundings in terms of luminous intensity, color, texture, or any other characteristics that make them different. A feature can be as simple as a point, line, edge, or a corner, and can be more distinctive, such as a tree, building, or other vehicle. To use features for VO, the features must be invariant to translation, rotation, scaling, and luminous intensity. This section presents different types of feature detection, matching, and tracking techniques that have been used by computer vision community.

Epipolar Geometry: To estimate the motion of a camera using features, the essential step is to find the corresponding features from one image to another. The simple naive way to solve this problem is to fully search the entire image. Using epipolar geometry [7, 8] the search space reduces from 2-D (full image) to one-dimensional ([1-D]; line). Epipolar geometry describes geometric relations in

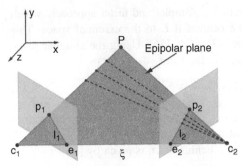

Figure 29.2 Epipolar geometry and epipolar constraint.

image pairs, as shown in Figure 29.2. A point, P, observed by two different image planes through projection center c_i is shown. For each correspondence in the images, there is an epipolar plane, which is a plane that contains P, c_i, and c_2. Within the epipolar plane, there is an epipolar axis, ξ, which is a line connecting two projection centers, c_1, and c_2. Epipole, e, is the projection center of one image that is projected on another image. For instance, e_1 is the projection of c_2 on image 1. The line that connects the projection center to the corresponding point in one image, projected onto another image, is the epipolar line; l_1 is the projected ray of Pc_2 onto the left image. The line contains every parts of a ray from a point, P, to the projection center c_i of one image and is projected on the second image.

Using epipolar geometry, the epipolar constraint is [7, 8]:

$$p_2^T E p_1 = 0, \tag{29.11}$$

where $p_1 = [p_u, p_v, 1]^T$ is a normalized projection point, P, in the first image, p_2 is its normalized correspondence in the second image, and E is the essential matrix. The essential matrix describes the rotation and unknown scale translation of images. There is an overall scale ambiguity; the true scale of the observed scene is unknown unless additional information is available. The essential matrix with multiplicative scalar denoted by the symbol $\simeq$ is:

$$E \simeq \hat{t} R, \tag{29.12}$$

where R is the rotation matrix and $\hat{t}$ is a skew-symmetric matrix of the form

$$\hat{t} = \begin{bmatrix} 0 & -t_z & t_y \\ t_z & 0 & -t_x \\ -t_y & t_x & 0 \end{bmatrix}, \tag{29.13}$$

where t_x, t_y, t_z denote translations in corresponding directions. Note that the essential matrix is used for calibrated cases only. The essential matrix has five degrees of freedom; two for translation and three for rotation. The fundamental matrix, which contains seven degrees of freedom, is used for non calibrated cases [7, 8].

The matrix E computed from a set of epipolar constraint, Equation (29.11), in general is not an essential matrix. Calculation of the actual essential matrix, $\hat{E}$, involves non linear constrained optimization. A simpler and faster approach, which is discussed in this chapter, projects the estimated E to the essential space. This approach minimizes the Frobenius norm $\|\hat{E} - E\|_f$ [36]. Using the singular value decomposition (SVD), E can be factorized in the form

$$E = U \, diag\{s_1, s_2, s_3\} V^T, \tag{29.14}$$

where $s_1 \geq s_2 \geq s_3 \geq 0$ are the singular values of E (square roots of the eigenvalues of $E^T E$), and U and V are unitary matrices of eigenvectors of EE^T and $E^T E$, respectively. Then $\hat{E}$ that minimizes the Frobenius norm is given by [36]:

$$\hat{E} = U \, diag\{\sigma, \sigma, 0\} V^T, \tag{29.15}$$

where $\sigma = \frac{s_1 + s_2}{2}$. Often σ in (29.15) is normalized to 1 to project onto the normalized essential space [7].

Triangulation: Obtaining depth and features from a 3-D point to 2-D from image is often required to use the camera data for vision-based localization. This is usually accomplished via triangulation of two images. Given correspondence pairs, the simplest method to implement triangulation is through the stereo normal case. The stereo normal case assumes that the images are fixed and share the same baseline, b, as shown in Figure 29.3.

The 3-D points can be determined by the intersection of the projected rays as shown in Figure 29.3. The epioplar plane cuts through the image plane and forms an epipolar line on the image plane. The same point on the left image plane and right image plane must pass through the epipolar line. This justifies the simplicity of the parallel alignment of the camera compared to other geometries; the epipolar line is horizontal, whereas if the cameras are not horizontally aligned, the epipolar line may not be horizontal.

Using the intercept theorem, the depth of a 3-D point, $P = [p_x, p_y, p_z, 1]^T$, is given by

$$p_z = \frac{f\xi}{\delta}, \tag{29.16}$$

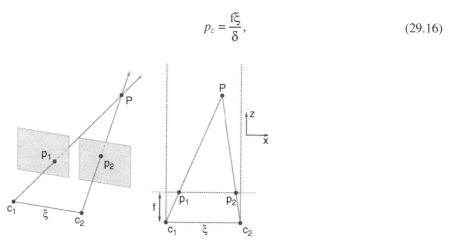

Figure 29.3 Triangulation of stereo vision.

where f is the focal length, ξ is the length of epipolar axis, δ denotes the disparity, the difference between a point in x direction from left to the right image ($\delta = p_{u_L} - p_{u_R}$). This shows that the depth is inversely proportional to the disparity. Then, triangle similarity allows one to map the 3-D real-world points into a 2-D image using the following equations:

$$p_x = \frac{p_z}{\mathrm{f}} u_L, \quad p_y = \frac{p_z}{\mathrm{f}} v_L, \quad (29.17)$$

where (p_{u_L}, p_{v_L}) is the corresponding 2-D point on the left image plane.

Feature Detector: To estimate the motion of a vehicle, frame by frame changes of the vehicle over a point of interest, referred to as features, should be determined. The vehicle must detect such features, usually via a low-level image processing operation. Feature detectors search for salient keypoints with respect to the local neighborhood [12]. Some desired properties of feature detectors include robustness, invariance under geometric transformation, precise localization of the features, and sometimes support of image interpretation. Many features, detectors in the VO field have been proposed, including corner detectors (Moravec [37], Harris [38], Shi-Tomasi [39] and FAST [40]) and blob detectors (SIFT [41] and SURF [42]). Depending on the environment conditions and the primary goal, one detector can outperform the other. This section reviews some of the feature detectors.

In general, corner detectors are computationally faster than blob detectors. The corner detector introduced by Moravec calculates the variation in the gradient in x and y directions by shifting the rectanguar windows [37, 43]. This can be expressed as [37, 43]:

$$S(t_g) = \sum_{p_u, p_v} w(p_u, p_v)[I(p_u + u_g, p_v + v_g) - I(p_u, p_v)]^2, \quad (29.18)$$

where w is the rectangular window function. The displacement in x and y directions is expressed by u_g and v_g, respectively. Intuitively, $S(t_g)$ in (29.18) would be close to 0 in constant areas, while the value would be large in distinctive areas such as corners. Moravec's corner detector works reasonably well under certain conditions, but it suffers from many problems such as noisy and anisotropic response. In 1988, Harris and Stephen improved upon the Moravecs's corner detector, producing one of the most well-known corner detectors in current days, and it is still being widely used [38]. Instead of shifting the rectangular window function, a Harris detector uses a Gaussian as window function. Discrete shift is avoided by considering the partial derivatives using the Taylor expansion. It can be shown that (29.18) can be expressed as [38]:

$$S(t_g) \approx \begin{bmatrix} u_g & v_g \end{bmatrix} V \begin{bmatrix} u_g \\ v_g \end{bmatrix}, \quad (29.19)$$

where

$$V = \sum_{p_u, p_v} w(p_u, p_v) \begin{bmatrix} I_u^2 & I_u I_v \\ I_u I_v & I_v^2 \end{bmatrix}, \quad (29.20)$$

and I_u and I_v represent the image derivatives. The eigenvalues λ_1, λ_2 of v distinguish the region as flat, edge, or corner. Small λ_1, λ_2 implies no interesting feature, one large λ_j is an edge, which implies one dominant direction of an image gradient, and large λ_1, λ_2 is a corner, which implies two dominant directions. The popular region of interest for a Harris detector are edges and corners. The computation of eigenvalues requires relatively high computation power. A Harris detector instead computes traces and determinants, which are less expensive. This allows fast computation and high repeatability. The measure of the features are calculated by [38]:

$$D(p_u, p_v) = \det v - \kappa \, \text{tr}^2(v), \qquad (29.21)$$

where κ is an empirically determined constant. $D(p_u, p_v)$ is measured for each (p_u, p_v) in the image coordinates and depends only on the eigenvalues of v. If the value of D is large, it implies that the selected area is a corner. If it is negative with large magnitude, it is an edge, and small $|D|$ represents a flat region. A Harris detector is rotation and partially intensity invariant. The major drawback of a Harris detector is that it is not invariant to image scale. For example, a feature classified as an edge can be classified as a corner when the image is zoomed out.

A very similar method to a Harris detector was proposed in 1994 by Shi-Tomasi [39]. Instead of using (29.21) to measure a corner, a Shi-Tomasi detector (also referred to as a Kanade-Tomasi detector) evaluates a corner by [39]:

$$D(p_u, p_v) = \min(\lambda_1, \lambda_2). \qquad (29.22)$$

If $D(p_u, p_v)$ is smaller than a threshold, it is considered a corner. Similar to a Harris detector, it is invariant to rotation but not invariant to image scale.

Blob detectors are slower compared to corner detectors but are more distinctive. In 2004, Lowe proposed a method that is scale and rotation invariant and partial invariant to illumination changes [41]. The proposed method known as scale-invariant feature transform (SIFT) not only detects the features but also describes them. This was the continuation of his previous work on invariant feature detection [44]. The detector and descriptor in SIFT have distinctive roles.

In feature detection stage, SIFT smooths the original image using different sizes of Gaussian filters, which yields different levels of blurred images. The blurred images are subtracted from one another for which the result is referred to as difference of Gaussian (DoG) images. The parts of the image that are plain will not be much different among different levels of blurred images. On the other hand, the regions that are more distinctive will have changes in the intensity values. This process is repeated on different scales of images, which are subsampled from the original image, often by a factor of 2. The potential interest points are searched by finding local maxima or minima at different scales. This task is implemented by comparing the nearest eight neighboring pixels on the selected scale image and nine neighbors on its immediate adjacently scaled images. These locally distinct features include edges, corners, and blobs. Among the features, edges can be problematic. Given an image, edge can be easily localized in one direction but can be difficult to find on other directions. Hence, the last step of SIFT performs extremum suppression on points that cannot be

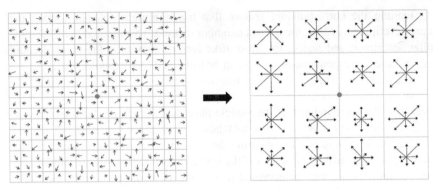

Figure 29.4   Image gradient and key descriptor of SIFT.

localized well. Because of this step, SIFT in general performs better in blob-rich regions rather than corner-rich regions.

The feature descriptor is used to uniquely describe a point of interest. Most of the feature descriptors, including SIFT, are based on building histograms around the point of interest. SIFT uses a 16×16 window of image gradient at the selected scale. This window is divided into 4×4 subblocks, as shown in Figure 29.4. Each subblock encodes gradient and orientation as a histogram with 8 bins, which describes 8 orientations. These values describe the feature and are saved as a vector. Hence, each key point contains $4 \times 4 \times 8 = 128$ dimensions describing the feature. The vector is further processed to make the descriptor rotation invariant and to reduce the effects of linear and non-linear illumination changes.

SIFT by far is the most robust such method and has become popular in different applications. However, SIFT is very slow. In 2006 a faster version that is comparable to SIFT named speed up robust features (SURF) was proposed by Bay et al. [42]. SURF still works on the same principle as SIFT. It builds different levels of images and has detector and descriptor pairs. However, instead of using DoG, it is approximated with box filters and works based on Haar wavelet responses. This makes SURF more computationally feasible but at the cost of robustness.

Feature Matching and Tracking: Correctly identifying corresponding features is the critical task in VO and this allows accurate motion estimation of the vehicle. This task can be categorized into two methods: feature matching and feature tracking. The feature-matching method finds the features from one image and matches them to the next image based on certain criteria, whereas the feature-tracking method tracks certain identified features.

In the absence of feature descriptors, the corresponding features can be found by comparing the similarity of pixel information, as discussed in Section 29.1. If the descriptors of the features are available, feature matching can be accomplished by comparing the similarity of these feature descriptors. The simple descriptor matching eventually uses the same similarity measures or their variants highlighted in Section 29.4.1. For SIFT or SURF descriptors this would minimize or maximize different metrics, such as Euclidean distance, depending on the method.

Finding the corresponding feature that has minimum distance among the feature descriptors is called the nearest neighbor method. This method finds the most similar descriptor, and hence, the most alike feature. However, the method could result in false correspondences, which can be mitigated by thresholding.

Thresholding would claim the features correspond only if the similarity measure is under certain threshold. This yields more accurate feature matching. However, if the surroundings of the vehicle have many repetitive features, such as a building with many windows, false matches are still unavoidable. The distinctive matching can be computed by comparing the ratio of the closest and second-closest features. This ratio will be close to 1 if the first and second features are similar. The corresponding feature will be accepted if the ratio is within predefined threshold. The exhaustive search with the descriptors can be computationally heavy, especially for scenes that are rich with features. Various methods for efficient matching such as hashing [45] and its variants [46] have been proposed.

Another method to match features is to keep track of the features from one frame to another. This approach involves an optical flow estimation with an assumption that sequential images are taken from nearby points. Often the tracking is implemented by Kanade-Lucas-Tomasi (KLT) tracker, which is highly efficient and can be processed in real time on CPU [35]. The basis of the optical flow is discussed in Section 29.4.1. The KLT tracker aims to find features on the first frame and then keeps the track of the features from frame to frame. If the error exceeds a certain threshold, the tracking of the feature is complete, and if there are not enough features to track, it reinitializes the new features.

Given a set of samples, it is essential to correctly identify correspondences and remove outliers. Determining corresponding features using a feature descriptor or feature tracking results in reasonable matches between the samples. However, the points could be exposed to noise, and the result could contain mismatched samples. In 1981, Fischler and Bolles proposed a technique to robustly reject outliers, referred to as random sample consensus (RANSAC) [47]. RANSAC is an iterative approach for finding the outliers. The RANSAC samples s number from the data points to estimate the model, such as essential matrix, E. Using the sampled points, model parameters such as R and t are computed assuming the chosen samples are correct points. With the estimated model parameters, the rest of the samples are further processed by classifying them as inliers or outliers by checking if other matches agreed with the chosen samples within certain a threshold. The process is recursively applied to find the best model with the highest consensus. To guarantee convergence to the correct solution, the number of iterations, $\mathcal{N}$, required can be calculated by [47]:

$$\mathcal{N} = \frac{log(1 - p_{good})}{log(1 - (1 - p_{fail})^s)}, \qquad (29.23)$$

where s is the number of corresponding samples needed, p_{fail} is the probability that a point is an outlier, and p_{good} is the desired probability to identify inliers correctly. According to the (29.23), the number of trials, $\mathcal{N}$, is greatly influenced by s. The algorithm that requires lower number of s is more feasible and computationally efficient.

29.4.3 Motion Estimation

The main goal of motion estimation is to find camera motion $T_{k,k-1} \in \mathbb{R}^{4 \times 4}$ between the two sequential images g_k, g_{k-1}. The transformation contains rotation matrix $R_{k,k-1} \in SO(3)$ and translation vector $t_{k,k-1} \in \mathbb{R}^{3 \times 1}$ and can be represented by [7, 8, 24]:

$$T_{k,k-1} = \begin{bmatrix} R_{k,k-1} & t_{k,k-1} \\ 0 & 1 \end{bmatrix}. \tag{29.24}$$

The transformation computes the transformation of the camera poses from one view, C_{k-1}, to the next, C_k, using $C_k = C_{k-1} T_{k,k-1}$. The final goal is to recover the trajectory of the vehicle using the transformation matrix $T_{1:n} = \{T_{1,0}, \ldots, T_{n,n-1}\}$.

For convenience of notation, the rest of the chapter denotes $T_{k,k-1}, R_{k,k-1}$, and $t_{k,k-1}$ as T_k, R_k, and t_k, respectively. There are three different methods to estimate the motion: 2-D to 2-D, 3-D to 2-D, and 3-D to 3-D. The 2-D to 2-D estimates the poses of the camera using the points $p = [p_u, p_v, 1]^T$ from the images, while 3-D to 2-D uses the points information in 3-D, $P = [p_x, p_y, p_z, 1]^T$, and 2-D points. The 3-D to 3-D uses only the points in 3-D to estimate the poses.

2-D to 2-D Method: This method estimates the motion using n features and its correspondences from two images g_1 and g_2. There are many famous solutions for this set up, such as an eight-point algorithm [48] and five-point algorithm [49]. For simplicity, an eight-point algorithm is discussed in this chapter.

2-D to 2-D method incorporates epipolar constraint discussed in Section 29.4.2. Expanding (29.11) yields [7, 8]:

$$\begin{aligned} & p_{u_1} p_{u_2} e_{11} + p_{v_1} p_{u_2} e_{12} + p_{u_2} e_{13} \\ & + p_{u_1} p_{v_2} e_{21} + p_{v_1} p_{v_2} e_{22} + p_{v_2} e_{23} \\ & + p_{u_1} e_{31} + p_{v_1} e_{32} + 1 e_{33} = 0, \end{aligned} \tag{29.25}$$

where (p_{u_i}, p_{v_i}) is a normalized point on each image plane, and e_{ii} is each element in essential matrix, E. Defining the two vectors, $a = [p_{u_1} p_{u_2}, p_{v_1} p_{u_2}, \ldots, 1]^T$, $E(:) = [e_{11}, e_{12}, \ldots, e_{33}]^T \in \mathbb{R}^9$, Equation (29.25) corresponds to:

$$a^T E(:) = 0 \tag{29.26}$$

This is one linear equation for one correspondence in image pairs. With n multiple correspondences for the same image pairs results in n linear equations which can be expressed as

$$\chi^T E(:) = 0, \tag{29.27}$$

where $\chi = [a_1, a_2, \ldots, a_n]$. The first correspondence is $a_1^T E = 0$, second is $a_2^T E = 0$, and so on. For an eight-point algorithm, Equation (29.27) is unique (up to a scaling factor) if there are at least $n \geq 8$ correspondences [48]. By solving the system (29.27) through different techniques, such as via SVD, E can be computed.

The second part of the algorithm involves retrieving the rotation and translation parts of the images from the estimated E. Unless additional information such as information about absolute distance of any part in the image is available, the recovered translation is only up to scale. The rotation and translation that can be decomposed from essential matrix is [7, 24]:

$$R = \pm U W^T V^*,$$

$$\hat{t} = \pm U W \, diag\{1, 1, 0\} U^T,$$

$$W^T \in \left\{ \begin{bmatrix} 0 & 1 & 0 \\ -1 & 0 & 0 \\ 0 & 0 & 1 \end{bmatrix}, \begin{bmatrix} 0 & -1 & 0 \\ 1 & 0 & 0 \\ 0 & 0 & 1 \end{bmatrix} \right\}, \tag{29.28}$$

where U, V^* are from SVD as discussed in 29.4.2. The sign of E cannot be recovered. In principal, each E yields two possible assignments of R and $\hat{t}$, as shown in (29.28). Therefore, there are four possible solutions for R and $\hat{t}$. The correct solution can be identified by selecting the two solutions with $\det R = 1$, triangulating the points, and selecting the solution with largest number of points in front of the camera.

3-D to 2-D Method: This method estimates the pose from features in 3-D, P_{k-1}, and its corresponding 2-D points, p_k. The goal of 3-D to 2-D motion estimation is to find transformation T_k that minimizes the image reprojection error, i.e., [24, 50]:

$$\underset{T_k}{\operatorname{argmin}} \sum_i \| p_k^i - \hat{p}_{k-1}^i \|^2, \tag{29.29}$$

where $\hat{p}_{k-1}^i$ is the reprojection point of P_{k-1} into image g_k with transformation T_k, and p_k is its corresponding 2-D point in image g_k. This problem is often known as perspective-n-point problem (PnP). There are many PnP solvers, such as direct least squares [51], nonlinear minimization using Levenberg-Marquardt optimization, and minimal case invovling three correspondences known as P3P [52]. P3P solution is robust and performs well with RANSAC outlier rejection [47]. This involves a large number of elimination steps. For simplicity, P6P is presented in this chapter (works with $n \geq 6$). Six-point correspondence results in a linear system of equation in $LT' = 0$ form of [53]:

$$\begin{bmatrix} 0 & -P_1^T & p_{v_1} P_1^T \\ P_1^T & 0 & -p_{u_1} P_1^T \\ \vdots & \vdots & \vdots \\ 0 & -P_6^T & p_{v_1} P_6^T \\ P_6^T & 0 & -p_{u_1} P_6^T \end{bmatrix} \begin{bmatrix} T_1'^T \\ T_2'^T \\ T_3'^T \end{bmatrix} = 0, \tag{29.30}$$

where $P_i = [p_{x_i}, p_{y_i}, p_{z_i}]^T$ and (p_{u_i}, p_{v_i}) are 3-D and 2-D points of feature i, respectively, and $T' = [R \mid t]$. $T_j'^T$ is the jth row of T' matrix. The elements of T' can be found by computing the nullvectors of L. The rotation matrix R and translation vector t can be retrieved from T'.

The 3-D to 2-D method works both on monocular and stereo vision. For monocular vision, a triangulation technique is required to observe 3-D points. This approach requires three views to compute motion estimation. The transformation of correspondences are first estimated by 2-D to 2-D method using two views, then by 3-D to 2-D method. For stereo case, two views are enough, as depth can be recovered with a single view of two cameras. However, 3-D to 2-D method works with an assumption that points comes from only one camera. There have been approaches that use both images at the same time, while some applications pick one camera to do estimation.

3-D to 3-D Method: As opposed to 3-D to 2-D method, which minimizes the reprojection error to estimate the motion, 3-D to 3-D method estimates the motion by minimizing the 3-D feature position errors, i.e., [24, 54]:

$$\underset{T_k}{\text{argmin}} \sum_i \| P_k^i - T_k P_{k-1}^i \|, \tag{29.31}$$

where P_k, P_{k-1} are the 3-D points of a feature and its correspondence, and T_k is the transformation matrix. The method is mostly used for stereo vision, as the depth can be easily observed by triangulation of features at each instant. The same method can also used for LiDAR data and its point cloud sets.

Similar to other two methods, (2-D to 2-D, 3-D to 2-D) many solutions have been investigated to estimate the poses using 3-D points. A famous approach considers when number of correspondences is $n \geq 3$ [24, 55]. One approach is to estimate the translation of points as [24, 55]:

$$t_k = \bar{P}_k - R\bar{P}_{k-1}, \tag{29.32}$$

where $\bar{P}_k$, and $\bar{P}_{k-1}$ are the centroid of the features in the feature set, and R is the rotation matrix. The rotation matrix can be computed by SVD as [24, 55]:

$$(P_{k-1} - \bar{P}_{k-1})(P_k - \bar{P}_k)^T = U\mathcal{S}V^*$$
$$R_k = VU^T. \tag{29.33}$$

29.5 SIMULTANEOUS LOCALIZATION AND MAPPING

One particular approach to update the maps for autonomous vehicle localization and embed the dynamic environment information in such maps is SLAM. The solution was primarily developed by Durrant-Whtye and Lenonard in 1991 [56] that was based on work by Smith et al. in 1990 [57]. SLAM is considered one of the key autonomous localization methodologies. When the vehicle is in an unknown environment, SLAM incrementally builds the map while localizing itself and navigating through the environment. SLAM research has been substantially active for past 30 years in robotics. Depending on the main task the vehicle is trying to accomplish, SLAM can be distinguished in many ways.

Full SLAM refers to posterior estimation of the entire path [58]. Online SLAM refers to the estimation of the current posterior only [58]. Depending on the types of maps that SLAM builds, it can be referred to as feature-based SLAM or volumetric SLAM [58]. The map is built based on points of interest, such as features, using feature-based SLAM, while the map is built at high resolution, which allows photo-realistic reconstruction, with volumetric SLAM [58]. Other methods of distinction are metric and topological mapping [58]. Metric SLAM has metric information in environments such as metric information from one place to another, while topological SLAM provides qualitative information about the places [58]. SLAM can also be distinguished by static and dynamic SLAM. Static SLAM assumes that the environment does not contain any dynamics of the surrounding, and if anything changes treats the changes as noise [58]. Dynamic SLAM takes into account dynamics and sometimes tracks the changes [58]. Depending on the objective of SLAM, different approaches and variations to tackle the problem at hand have been proposed.

29.5.1 Formulation of SLAM

The formulation of SLAM is depicted in Figure 29.5. At time step k, states of the vehicle, including position and orientation, are represented by x_k, and the control input applied at time step $k-1$ to move the vehicle state to x_k is denoted as u_k. The observable map points, and the measurements by sensors are denoted, respectively, by m_k and z_k. Note that the representation of the map can depend on the implemented sensors on the vehicle. In this chapter, the map is assumed as observable points, such as location of landmark. The state of the vehicle, map, control input, and measurement sequences are denoted as $x_{1:k} = \{x_1, x_2, ..., x_k\}$, $m_{1:M} = \{m_1, m_2, ..., m_M\}$, $u_{1:k} = \{u_1, u_2, ..., u_k\}$, and $z_{1:k} = \{z_1, z_2, ..., z_k\}$, respectively. The goal of SLAM is to find $x_{1:k}$ and $m_{1:M}$ from $u_{1:k}$ and $z_{1:k}$.

29.5.2 Paradigms of SLAM

The three main paradigms to solve the SLAM problems are Kalman filter, particle filter, and graph-based SLAM. Many effective variations of the three methods have

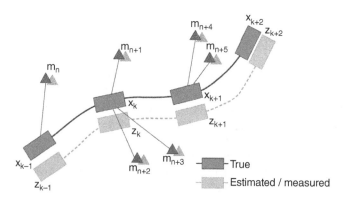

Figure 29.5 Formulation of SLAM.

been proposed and implemented in the literatures. The following dynamic state-space model notation is used to describe the vehicle motion and sensor model:

$$x_k = f(x_{k-1}, u_k, \epsilon_k)$$
$$z_k = h(x_k, m, +\iota_k), \tag{29.34}$$

where $f(\cdot), h(\cdot)$ model nonlinear vehicle kinematics and measurement, correspondingly, and x, m, u, z, ϵ, and ι are the state of the vehicle, map, control input, measurement, process noise and measurement noise, resepctively.

Bayes; filter is the basis of the two filter-based SLAMs. It uses all known inputs $u_{1:k}$, and measurements $z_{1:k}$, to estimate the joint posterior probability density function (PDF) of the states of the vehicle $x_{1:k}$ and map m. This probability distribution is denoted as:

$$p(x_k, m \mid z_{1:k}, u_{1:k}). \tag{29.35}$$

The belief about the current state prior to measurement and after the measurement, is denoted as:

$$\bar{\beta}(x_k, m) = p(x_k, m \mid z_{1:k-1}, u_{1:k})$$
$$\beta(x_k, m) = p(x_k, m \mid z_{1:k}, u_{1:k}). \tag{29.36}$$

Using the Bayes' rule, Markov property, and law of total probability, Bayes' filter algorithm recursively calculates the PDF in two-step, as shown in Table 29.1.

Table 29.1. Bayes' Filter Algorithm

Prediction:	$\bar{\beta}(x_k, m) = \int p(x_k \mid x_{k-1}, u_k)\beta(x_{k-1})dx_{k-1}$
Update:	$\beta(x_k, m) = \eta p(z_k \mid x_k)\bar{\beta}(x_k)$

In Table 29.1 $p(x_k \mid x_{k-1}, u_k)$ is the motion model, $p(z_k \mid x_k)$ is the sensor model, and η is the normalizing constant that does not depend on the state. There are many solutions to Bayesian filter for different systems.

Kalman or extended Kalman filter (EKF) can be the optimal estimation in terms of minimizing the mean square error sense. Kalman and EKF assumes the posterior density is Gaussian distributed with added zero-mean Gaussian process noise ϵ and measurement noise has ι and its covariance represented by $\mathcal{R}_k$ and $\mathcal{Q}_k$, respectively 60. EKF SLAM was been the earliest approach, introduced in [57], [60], and [61]. The EKF algorithm has been summarized in Table 29.2. The basics and further details on Kalman filters are discussed in Chapter 5; for more applications also see Chapters 19 and 22.

The state estimates from Table 29.2 include the poses of the vehicle and the features of the environment. As the vehicle moves, the state vector and its covariance matrix are updated. When a new feature is observed, it is added to the state vector. EKF SLAM and its variations have been implemented for many mobile robot navigation applications. In practice, the autonomous vehicles often exhibit nonlinear motion dynamics. Kalman filter and its variant filters do not perform

Table 29.2. EKF Algorithm

Prediction:	
Predicted State:	$\hat{\mu}_{k|k-1} = f(\mu_{k-1|k-1}, u_k)$
Predicted Covariance:	$\Sigma_{k|k-1} = \partial f_k \Sigma_{k-1|k-1} \partial f_k^T + \mathcal{R}_k$
Update:	
Kalman Gain:	$K_k = \Sigma_{k|k-1} \partial h_k^T (\partial h_k \Sigma_{k|k-1} \partial h_K^T + \mathcal{Q})^{-1}$
Estimated Covariance:	$\Sigma_{k|k} = (I - K_k \partial h_k) \Sigma_{k|k-1}$
Estimated States:	$\mu_{k|k} = \hat{\mu}_{k|k-1} + K(z_k - h(\hat{\mu}_{k|k-1}))$

where $\mu = (x, m)$, $\partial f_k = \dfrac{\partial f(\hat{\mu}_{k-1}, u_{k-1})}{\partial \mu}$, and $\partial h_k = \dfrac{\partial h(\hat{\mu}_k)}{\partial \mu}$.

well during nonlinear systems, and it has many limitations, resulting in two other methods becoming more popular.

Particle filter is another popular filter, which is uses N number of particles to model arbitrary distribution of true posterior [62, 63, 64]. The ith particle pose, x^i, and its normalized weight, w^i, are represented as $X = \{x^i, w^i\}_{i=1,...,N}$. These particles are sampled from a distribution such as posterior distribution. However, it is often difficult to sample from the true posterior density; instead the particles are sampled from a proposal distribution $\pi(x_k | X_{0:k-1}, z_{0:k})$, chosen by the designer. The idea behind the particle filter is to use the particles from a proposal distribution to approximate the target distribution (posterior density). To compensate for the difference between target distribution, importance weight, w^i, is computed. The samples that are less likely represent the state, x, are replaced by more likely samples through a resampling process. The algorithm of particle filter is presented in Table 29.3, and further details on particle filter can be found in Chapter 19.

Monte Carlo localization (MCL) is a specific form of particle filter-based localization that is being used widely in mobile robot localization [65]. In MCL, the proposal distribution is the motion model; the particles are propagated through a motion model. The weight of the particles are computed via the measurement model. This type of localization and its variants perform well in low-dimensional spaces, but as the size of the state space increases particle filter becomes impractical. This is the major drawback of particle filter-based localization: as the environment gets complicated, more particles are required for pose estimation and the computational complexity grows.

In 2002, Montemerlo et al. introduced FastSLAM to allows less computational power to solve SLAM problem using particle filter [66]. FastSLAM was able to reduce the high-dimensional state space by independently estimating the landmarks. By exploiting the dependences between variables, the trajectory of the vehicle is estimated by samples, while mapping is computed analytically using a 2-D Kalman filter.

There are two versions of FastSLAM. FastSLAM 1.0 uses the motion model for proposal distribution, and the particle of weights are computed according to the

Table 29.3.   Particle Filter Algorithm

Sampling:

for $i = 1, ..., N$

Draw $x_k^{(i)}$ from $\pi(x_k \mid X_{0:k-1}, z_{0:k})$

$$w_k^{(i)} = \frac{p(x^{(i)})}{\pi(x_k^{(i)})}$$

$$X_{0:k}^{(i)} = X_{0:k-1}^{(i)} + (x^{(i)}, w_k^{(i)})$$

Resampling:

for $i = 1, ..., N$

Draw $i \in 1, ..., N$ with probability $w_k^{(i)}$

Add $x_k^{(i)}$ to X_k

marginalized observation model [66]. FastSLAM 2.0 takes into account the current observation for the proposal distribution, and the particle weight is calculated according to the importance function [67].

Both filtering based localization techniques have been active research areas and important developments in the SLAM literature. For further reading readers are encouraged to review [68–69].

The current state of the art is solving SLAM problem by constructing a graph. The graph-based SLAM approach was proposed in 1997 by Lu and Milios [71]. The basic idea of graph-based SLAM is to represent nodes with the location of vehicles and the edges as spatial measurements between the nodes. The edge can be represented by odometry measurements taken from consecutive frames or by observation, which results in virtual measurements about the position of x_j from x_i, when the vehicle observes the same part of measurements. With the graph, the localization of the vehicle is implemented by finding the poses that satisfy the constraint. The architecture of graph-based SLAM is shown in Figure 29.6, having two main processes, front-end and back-end. The front-end process interprets sensor data and

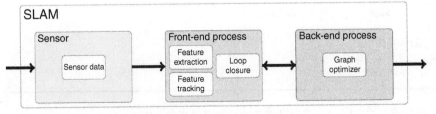

Figure 29.6   Graph-based SLAM architecture.

constructs the graph by defining nodes and edges, while the back-end performs inference on the graph by optimizing it.

The front-end heavily depends on the types of available sensors. The back-end formulates SLAM as a maximum a posterioir (MAP) estimation. In MAP, x is estimated as

$$x^* = \operatorname*{argmin}_x p(x \mid z) = \operatorname*{argmin}_x p(z \mid x) p(x). \qquad (29.37)$$

Assuming measurements are independent, and measurement noise is Gaussian with information matrix Ω, Equation (29.37) can be rewritten as

$$x^* = \operatorname*{argmin}_x -log\Big(p(x)\prod_{k=1}^{m} p(z_k \mid x_k)\Big) = \operatorname*{argmin}_x \sum_{k=1}^{m} \varepsilon^T \Omega_k \varepsilon, \qquad (29.38)$$

where $\varepsilon = \|h_k(x_k) - z_k\|$. Equation (29.38) minimizes the square of error, which is suitable for the least squares problem. Depending on the nature of available sensors, the meaning of error is different.

The basic graph-based SLAM is shown in Table 29.4. SLAM algorithms are mainly based on iterative nonlinear optimization, assuming that a good initial guess is available and local minima is in the neighborhood. The matrix H is symmetric, positive semidefinite, and sparse. With these properties, different solvers can be used to compute Δx [72–76].

Compared to other SLAM techniques, graph-based SLAM is capable of generating higher-dimensional maps. For more details about graph-based SLAM and

Table 29.4.   Graph-Based Algorithm

while (not converged)

  for all $(\varepsilon_{ij}, \Omega_{ij})$

  $A_{ij} = \dfrac{\partial \varepsilon_{ij}(x)}{\partial x_i}\Big|_{x=\hat{x}}$ $B_{ij} = \dfrac{\partial \varepsilon_{ij}(x)}{\partial x_j}\Big|_{x=\hat{x}}$

  $H_{ii} \mathrel{+}= A_{ij}^T \Omega_{ij} A_{ij}$ $H_{ij} \mathrel{+}= A_{ij}^T \Omega_{ij} B_{ij}$

  $H_{ji} \mathrel{+}= B_{ij}^T \Omega_{ij} A_{ij}$ $H_{jj} \mathrel{+}= B_{ij}^T \Omega_{ij} B_{ij}$

  $b_i \mathrel{+}= A_{ij}^T \Omega_{ij} \varepsilon_{ij}$ $b_j \mathrel{+}= B_{ij}^T \Omega_{ij} \varepsilon_{ij}$

  $\Delta x = -H^{-1} b$

  $\hat{x}^* \mathrel{+}= \Delta x$

where $\varepsilon_{ij} = z_{ij} - h_{ij}(x_i, x_j)$, the virtual measurement of node j from node i.

the state of the art for least-squares error minimization, readers are referred to [77] and [78].

29.5.3 Loop Closure

The front-end process of SLAM uses a technique, called "loop closure," to recognize previously visited places in order to make SLAM robust and long term. After some period of time running the SLAM algorithm, the map or the graphs are continually added and the error accumulates. Nevertheless, if the vehicle visits the places it visited before and recognizes the place, this could reduce the uncertainty and the error of SLAM. Therefore, it is clear that having a reliable and accurate technique is essential, as wrong loop closure could result in failure of SLAM.

The loop closure techniques are sensor dependent. The early approaches of loop closure were involved with laser scan [71, 79, 80]. The technique of overlapping the scan with respect to the reference scan is often referred to as scan matching. Various scan-matching algorithms were proposed in the last decade; among them one of the most widely used is iterative closest point (ICP) [81]. ICP, which iterates to align the scan, has been the basis for many variations on scan matching, and the improvements are promising [82–85].

A relatively recent approach to loop closure involves using cameras often referred to as place recognition or image/visual matching. Some approaches of image matching involved using the feature descriptors, while some proposed to use patches of image to recognize the places, also known as bag of words. For further readings and details on place recognition, readers are encouraged to read [86].

29.6 COOPERATIVE ESTIMATION, FILTERING, AND SENSOR FUSION

The objective of sensor fusion in localization is to enhance the position estimation accuracy by combining information obtained via similar or dissimilar sensors. Exteroceptive and proprioceptive sensors discussed in Section 29.2 are often used together in autonomous vehicle localization.

Sensor fusion approaches can be categorized as centralized and decentralized. In centralized fusion, a single filter is used to fuse the measurements from different sensors, as shown in Figure 29.7. Such an approach is often implemented when

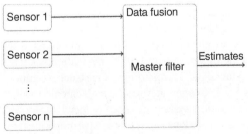

Figure 29.7   Centralized filter architecture.

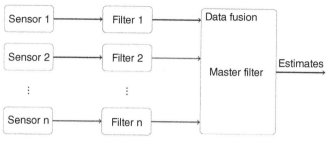

Figure 29.8   Decentralized filter architecture.

similar sensors are involved and can be heterogeneously implemented, for example, by combining GPS data with odometery [87] or inertial measurement reading [88]. Because all the information is synchronously combined at once, centralized approaches lose less information from the sensors as opposed to the decentralized ones, providining optimal estimation under ideal conditions [89, 90].

In decentralized fusion approaches, a collection of subfilters are used, each to provide a partial localization solution. The filtered partial results are then fused using a master filter to provide integrated navigation solution. Similar to centralized approaches, GPS data with inertial measurement readings can be used to estimate position via decentralized fusion [91]. As this approach involves filtering the measurements individually, it has lower computation complexity. However, decentralization may result in the reduction of accuracy compared to centralized filter [91].

The filters discussed in Section 29.5.2 and their variations are common filters that are implemented for sensor fusion. Some variations include adaptive filters for systems with high modelling and filter parameter uncertainty [92–94]. These filters require some information about the system, including measurement models and their parameters. The system model can be omitted by heuristic methods, such as neural network-based ones to fuse the information [95–97].

29.7   LOCALIZATION SYSTEMS IN USE FOR AUTONOMOUS DRIVING

In the last decade, many successful autonomous vehicle road tests have been made. Many vehicle companies are planning to release autonomous vehicles within the next decade. Satellite-based localization systems have constituted the most popular methods for localization in these tests and design plans. Most likely, it will be the essential component in fully autonomous vehicle localization tasks. The combinations with other sensing methodologies overcome the limitations of satellite navigation, such as signal interference and awareness to immediate surroundings. One such common combination is GPS with dead reckoning. This approach can be implemented using GPS in combination with an inertial navigation system [91] or odometer on top of GPS and INS [98]. Commercial Siemens car navigation systems have implemented map matching with the information from GPS, odometers, and gyroscopes [99].

While satellite navigation is the main global localization technique, different companies approach perception of the immediate surroundings in different ways. Waymo, originally and more publicly known as Google car, began developing autonomous vehicles in 2009. To produce more commercially affordable autonomous vehicles, Waymo, with collaboration of Fiat Chrysler Automobile, started producing self-driving Chrysler Pacifica Hybrid minivans [100]. This vehicle was equipped with custom-built sensors that are specifically designed for its software, which reduces the cost of sensors [100]. Among the newly integrated systems, LiDAR is still the heart of the sensing. Integration of medium-range and two custom-made (short, long-range) LiDARs give the vehicle uninterrupted view of front, behind, down, and side of the vehicles [100]. The vision system in the vehicle is used to detect small objects, traffic lights, stop signs, and parking lots [100]. The vehicle also comes with radar that tracks other vehicles and objects [100]. For absolute localization and tracking of the movement of the vehicle, the vehicle also has IMU and wheel encoder [101]. The data from the sensors are fused to build a 3-D map of the environment. To ensure that the vehicle can drive around, the accurate localization of the vehicle is implemented via the sensor data map and a high-resolution map of the area of interest [101].

Uber also uses LiDAR as its main sensor for localization [102]. Equipped with seven LiDARs, 20 cameras, and sonars, Uber tested its autonomous vehicle in Pittsburgh in 2016 with regular passengers [102]. Tesla aims to have autonomous driving technology by 2019. Tesla took a different approach for the main sensor to localize [103]. Instead of using expensive LiDARs, it relies on radar, sonar, IMU, and omnidirectional cameras to localize itself [104]. The commercial Tesla vehicles already have an "autopilot" function, which can steer, adjust speed, and detect objects in highway. Other car manufactures, such as Audi, BMW, Ford, GM, Honda, Hyundai, Mercedes-Benz, Nissan, and Toyota are also testing their autonomous localization and mapping technologies to bring the autonomous vehicles one step closer to the streets.

29.7.1 Autonomous Driving Competitions

The autonomous vehicle field was able to make an immense progress through driving competitions. One of the most well known is the Defense Advanced Research Project Agency (DARPA) Grand Challenge started in 2004. The unprecedented challenge to leverage the development of autonomous vehicle technologies involved building an autonomous vehicle that was capable of driving different trajectories. The competition in 2004 involved driving in desert region along the path of the US Interstate Highway 15. None of the vehicles could successfully complete the challenge, but this competition demonstrated the capabilities of the available autonomous vehicle technologies. The 2005 DARPA Grand Challenge involved passing through narrow tunnels and sharp turns. The challenge was completed by five vehicles. The third competition, known as the "Urban Challenge", was held in 2007. The challenge involved more realistic driving scenarios, including obeying the traffic rules and interacting with other vehicles. There were six vehicles that successfully finished the series of tests.

Boss, the car developed by Carnegie Mellon University, won first place. The vehicle was equipped with exteroceptive and proprioceptive sensors. The localization on the road was implemented by GPS measurements, with LiDARs to detect lane boundaries [105]. Numerous LiDARs on Boss were used to detect static and dynamic obstacles of the surrounding and to generate obstacle maps [105]. The road shape estimation function was also available but was not enabled during the competition. By using LiDAR and cameras, the vehicle was also able to drive on roads with unknown shapes [105].

Junior, the car developed by Stanford University, was the second car in the competition. The vehicle was equipped with LiDAR, GPS, INS, and radar for the competition [106]. The global localization was implemented by GPS, while a laser scanner detected lane markings and curb-like obstacles [106] to localize within the road. The static and dynamic obstacle detection and tracking was also implemented by laser sensors [106].

Odin, developed by Virginia Tech [107], and Talos, developed by MIT [108, 109], won third and fourth place in the challenge, respectively. Skynet [110], and Little Ben [111], developed by Cornell University and University of Pennsylvania, respectively, also finished the course but were over the 6-hour limit. All of the teams who successfully finished the course showcased the most advanced intelligent technology at the time. Since then many teams have worked collaboratively with other industries on developing more enhanced intelligent systems.

29.8 FUTURE OF LOCALIZATION IN AUTONOMOUS DRIVING

Cars have been continuously evolving ever since their invention. The design, which used to be completely mechanical, has advanced in many directions successively to the point that it currently supports drivers. Vehicles are inevitably becoming more and more intelligent everyday. An autonomous car is no longer science fiction, and the roadway systems are about to undergo a significant transformation. However, the path to reach an autonomous vehicle is not fully known and could be different depending on the application of the vehicle; a family car, taxi, bus, and truck could be equipped with different sensors and algorithms.

Apart from advancement of stronger computational technologies, the development of new sensors and sensor-equipped vehicles are becoming more popular. The introduction of 3-D LiDAR allowed more robust and practical SLAM for autonomous driving. Different types of vision sensors, such as depth, event-based, and omnidirectional cameras allow more efficient and robust algorithms to be developed. Sensors that share information with traffic and their potential to reduce accidents are also getting attention. Other well-established sensors are becoming more accurate, more compact, cheaper and more accessible.

Artificial intelligence (AI) and, especially, DL have gained significant interest in the research field of autonomous vehicles. Some believe that DL will truly allow fully autonomous vehicles, while others believe in the traditional approaches. AI mimics the neural network of human brain. The current research field uses DL for perception of the vehicle. Instead of using the computer vision techniques to recognize and observe the environment, DL enables machines to differentiate patterns, for example, those of pedestrians from other objects. Useful interpretation from the DL algorithms is outputted at promising rate, but the process occurring with the algorithms is not fully understood at this stage, one of the main reasons for debates between traditional and DL approach for fully autonomous vehicle. Whether it is smarter sensors or new algorithms, the race to commercialize affordable autonomous vehicles continues, and the first generation of autonomous vehicle is expected soon.

29.9 CONCLUSIONS

The spectrum of localization techniques and applications is significantly wide. This chapter has presented key characteristics and limitations of state-of-the-art localization sensors for autonomous driving. Various localization techniques have been presented in the chapter, specifically highlighting progress in vision-based localization. Since fully autonomous driving involves localization in uncertain environments, this chapter has also presented different SLAM techniques and sensor fusion algorithms.Various localization systems in use in research and industry fields have been presented, as have possible solutions for the future of autonomous driving.

REFERENCES

[1] S. O. Committee, *Taxonomy and Definitions for Terms Related to On-Road Motor Vehicle Automated Driving Systems, Levels of Driving Automation*, 2014.

[2] D. Titterton and J. L. Weston, *Strapdown Inertial Navigation Technology*. IET, 2004.

[3] D. Göhring, M. Wang, M. Schnürmacher, and T. Ganjineh, "Radar/LiDAR sensor fusion for car-following on highways," in *Proc. IEEE Conf. on Automation, Robotics and Applications*, 2011, pp. 407–412.

[4] E. D. Dickmanns, "The development of machine vision for road vehicles in the last decade," in *Proc. IEEE Intelligent Vehicles Symp.*, 2002, pp. 268–281.

[5] D. M. Gavrila and V. Philomin, "Real-time object detection for 'smart' vehicles," in *Proc. IEEE Int. Conf. on Computer Vision*, 1999, pp. 87–93.

[6] U. Franke and A. Joos, "Real-time stereo vision for urban traffic scene understanding," in *Proc. IEEE Intelligent Vehicles Symp.*, 2000, pp. 273–278.

[7] Y. Ma, S. Soatto, J. Kosecka, and S. S. Sastry, *An Invitation to 3-D Vision: From Images to Geometric Models*. Springer, 2012.

[8] R. Hartley and A. Zisserman, *Multiple View Geometry in Computer Vision*. Cambridge University Press, 2003.

[9] P. Chang and M. Hebert, "Omni-directional structure from motion," in *Proc. IEEE Workshop on Omnidirectional Vision*, 2000, pp. 127–133.

[10] S. Zeadally, R. Hunt, Y.-S. Chen, A. Irwin, and A. Hassan, "Vehicular ad hoc networks (VANETs): Status, results, and challenges," *Telecommun. Syst.*, vol. 50, no. 4, pp. 217–241, 2012.

[11] N. M. Drawil and O. Basir, "Intervehicle-communication-assisted localization," *IEEE Trans. Intell. Transp. Syst.*, vol. 11, no. 3, pp. 678–691, 2010.

[12] R. Siegwart, I. R. Nourbakhsh, and D. Scaramuzza, *Introduction to Autonomous Mobile Robots*. MIT Press, 2011.

[13] J. Borenstein and L. Feng, "Measurement and correction of systematic odometry errors in mobile robots," *IEEE Trans. Robot. Autom.*, vol. 12, no. 6, pp. 869–880, 1996.

[14] C. R. Carlson, J. C. Gerdes, and J. D. Powell, "Error sources when land vehicle dead reckoning with differential wheelspeeds," *Navigation*, vol. 51, no. 1, pp. 13–27, 2004.

[15] K. Berntorp, "Joint wheel-slip and vehicle-motion estimation based on inertial, GPS, and wheel-speed sensors," *IEEE Trans. Control Syst. Technol.*, vol. 24, no. 3, pp. 1020–1027, 2016.

[16] K. Nam, S. Oh, H. Fujimoto, and Y. Hori, "Estimation of sideslip and roll angles of electric vehicles using lateral tire force sensors through RLS and Kalman filter approaches," *IEEE Trans. Industr. Electron.*, vol. 60, no. 3, pp. 988–1000, 2013.

[17] E. Abbott and D. Powell, "Land-vehicle navigation using GPS," *Proc. IEEE*, vol. 87, no. 1, pp. 145–162, 1999.

[18] I. Skog and P. Handel, "In-car positioning and navigation technologies: A survey," *IEEE Trans. Intell. Transport. Syst.*, vol. 10, no. 1, pp. 4–21, 2009.

[19] M. A. Quddus, W. Y. Ochieng, and R. B. Noland, "Current map-matching algorithms for transport applications: State-of-the art and future research directions," *Transport. Res. Part C Emerg. Technol.*, vol. 15, no. 5, pp. 312–328, 2007.

[20] A. Lawrence, *Modern Inertial Technology: Navigation, Guidance, and Control*. Springer Science+Business Media, 2012.

[21] D. A. Grejner-Brzezinska, Y. Yi, and C. K. Toth, "Bridging GPS gaps in urban canyons: The benefits of ZUPTs," *Navigation*, vol. 48, no. 4, pp. 216–226, 2001.

[22] N. El-Sheimy, E.-H. Shin, and X. Niu, "Kalman filter face-off: Extended vs. unscented kalman filters for integrated GPS and MEMS inertial," *Inside GNSS*, vol. 1, no. 2, pp. 48–54, 2006.

[23] D. Nistér, O. Naroditsky, and J. Bergen, "Visual odometry," in *Proc. IEEE Computer Society Conf. on Computer Vision and Pattern Recognition*, vol. 1, 2004, pp. I-652–I-659.

[24] D. Scaramuzza and F. Fraundorfer, "Visual odometry [tutorial]," *IEEE Robot. Autom. Mag.*, vol. 18, no. 4, pp. 80–92, 2011.

[25] A. Howard, "Real-time stereo visual odometry for autonomous ground vehicles," in *Proc. IEEE/RSJ Conf. on Intelligent Robots and Systems*, 2008, pp. 3946–3952.

[26] H. P. Moravec, "Obstacle avoidance and navigation in the real world by a seeing robot rover," Stanford University, Department of Computer Science, Tech. Rep., 1980.

[27] Y. Cheng, M. W. Maimone, and L. Matthies, "Visual odometry on the mars exploration rovers—A tool to ensure accurate driving and science imaging," *IEEE Robot. Autom. Mag.*, vol. 13, no. 2, pp. 54–62, 2006.

[28] A. I. Comport, E. Malis, and P. Rives, "Real-time quadrifocal visual odometry," *Int. J. Robot. Res.*, vol. 29, no. 2–3, pp. 245–266, 2010.

[29] B. Williams and I. Reid, "On combining visual slam and visual odometry," in *Proc. IEEE Conf. on Robotics and Automation*, 2010, pp. 3494–3500.

[30] A. Gruen and T. S. Huang, *Calibration and Orientation of Cameras in Computer Vision*. Springer Science+Business Media, 2013.

[31] O. Faugeras, *Three-Dimensional Computer Vision: A Geometric Viewpoint*. MIT Press, 1993.

[32] J. R. Bergen, P. Anandan, K. J. Hanna, and R. Hingorani, "Hierarchical model-based motion estimation," in *European Conf. on Computer Vision*, 1992, pp. 237–252.

[33] E. De Castro and C. Morandi, "Registration of translated and rotated images using finite Fourier transforms," *IEEE Trans. Pattern Anal. Mach. Intell.*, vol. PAMI-9, no. 5, pp. 700–703, 1987.

[34] B. K. Horn and B. G. Schunck, "Determining optical flow," *Artif. Intell.*, vol. 17, no. 1–3, pp. 185–203, 1981.

[35] B. D. Lucas and T. Kanade, "An iterative image registration technique with an application to stereo vision," *Int. Joint Conf. on Artificial Intelligence*, vol. 2, 1981, pp. 674–679.

[36] T. S. Huang and O. D. Faugeras, "Some properties of the E matrix in two-view motion estimation," *IEEE Trans. Pattern Anal. Mach. Intell.*, vol. 11, no. 12, pp. 1310–1312, 1989.

[37] H. P. Moravec, "Towards automatic visual obstacle avoidance," in *Proc. 5th Int. Joint Conf. on Artificial Intelligence*, vol. 2, 1977, p. 584.

[38] C. Harris and M. Stephens, "A combined corner and edge detector," in *Proc. of the 4th Alvey Vision Conf.*, Manchester, UK, 1988, pp. 147–151.

[39] J. Shi, "Good features to track," in *Proc. IEEE Computer Society Conf. on Computer Vision and Pattern Recognition*, 1994, pp. 593–600.

[40] E. Rosten and T. Drummond, "Machine learning for high-speed corner detection," in *Proc. 9th European Conf. on Computer Vision*, 2006, pp. 430–443.

[41] D. G. Lowe, "Distinctive image features from scale-invariant keypoints," *Int. J. Comput. Vis.*, vol. 60, no. 2, pp. 91–110, 2004.

[42] H. Bay, T. Tuytelaars, and L. Van Gool, "Surf: Speeded up robust features," in *Proc. European Conf. on Computer Vision*, 2006, pp. 404–417.

[43] H. P. Moravec, "Visual mapping by a robot rover," in *Proc. 6th Int. Joint Conf. on Artificial Intelligence*, 1979, pp. 598–600.

[44] D. G. Lowe, "Object recognition from local scale-invariant features," in *Proc. IEEE Int. Conf. on Computer Vision*, vol. 2, 1999, pp. 1150–1157.

[45] A. Gionis, P. Indyk, and R. Motwani, "Similarity search in high dimensions via hashing," in *Proc. of the 25th Int. Conf. on Very Large Data Bases*, 1999, pp. 518–529.

[46] J. Wang, H. T. Shen, J. Song, and J. Ji, "Hashing for similarity search: A survey," 2014, *arXiv:1408.2927*.

[47] M. A. Fischler and R. C. Bolles, "Random sample consensus: A paradigm for model fitting with applications to image analysis and automated cartography," *Commun. ACM*, vol. 24, no. 6, pp. 381–395, 1981.

[48] H. C. Longuet-Higgins, "A computer algorithm for reconstructing a scene from two projections," *Nature*, vol. 293, no. 5828, pp. 133–135, 1981.

[49] D. Nistér, "An efficient solution to the five-point relative pose problem," *IEEE Trans. Pattern Anal. Mach. Intell.*, vol. 26, no. 6, pp. 756–770, 2004.

[50] W. Förstner, "A feature based correspondence algorithm for image matching," *Int. Arch. Photogram. Remote Sens.*, vol. 26, no. 3, pp. 150–166, 1986.

[51] J. A. Hesch and S. I. Roumeliotis, "A direct least-squares (DLS) method for PnP," in *Proc. IEEE Int. Conf. on Computer Vision*, 2011, pp. 383–390.

[52] L. Kneip, D. Scaramuzza, and R. Siegwart, "A novel parametrization of the perspective-three-point problem for a direct computation of absolute camera position and orientation," in *Proc. IEEE Conf. on Computer Vision and Pattern Recognition*, 2011, pp. 2969–2976.

[53] M. Calonder, V. Lepetit, C. Strecha, and P. Fua, "Brief: Binary robust independent elementary features," in *European Conf. on Computer Vision*, 2010, pp. 778–792.

[54] C. F. Olson, L. H. Matthies, H. Schoppers, and M. W. Maimone, "Robust stereo ego-motion for long distance navigation," in *Proc. IEEE Conf. on Computer Vision and Pattern Recognition*, 2000, pp. 453–458.

[55] K. S. Arun, T. S. Huang, and S. D. Blostein, "Least-squares fitting of two 3-D point sets," *IEEE Trans. Pattern Anal. Mach. Intell.*, vol. PAMI-9, no. 5, pp. 698–700, 1987.

[56] J. J. Leonard and H. F. Durrant-Whyte, "Mobile robot localization by tracking geometric beacons," *IEEE Trans. Robot. Autom.*, vol. 7, no. 3, pp. 376–382, 1991.

[57] R. Smith, M. Self, and P. Cheeseman, "Estimating uncertain spatial relationships in robotics," in *Autonomous Robot Vehicles*. Springer, 1990, pp. 167–193.

[58] S. Thrun and J. J. Leonard, "Simultaneous localization and mapping," in *Springer Handbook of Robotics*. Springer, 2008, pp. 871–889.

[59] S. Thrun, W. Burgard, and D. Fox, *Probabilistic Robotics*. MIT Press, 2005.

[60] R. C. Smith and P. Cheeseman, "On the representation and estimation of spatial uncertainty," *Int. J. Robot. Res.*, vol. 5, no. 4, pp. 56–68, 1986.

[61] P. Moutarlier and R. Chatila, "An experimental system for incremental environment modelling by an autonomous mobile robot," in *Experimental Robotics I*. Springer, 1990, pp. 327–346.

[62] G. Kitagawa, "Monte Carlo filter and smoother for non-Gaussian nonlinear state space models," *J. Comput. Graph. Stat.*, vol. 5, no. 1, pp. 1–25, 1996.

[63] A. Doucet, "On sequential simulation-based methods for Bayesian filtering," Tech. Rep., 1998.

[64] M. K. Pitt and N. Shephard, "Filtering via simulation: Auxiliary particle filters," *J. Am. Stat. Assoc.*, vol. 94, no. 446, pp. 590–599, 1999.

[65] D. Fox, W. Burgard, F. Dellaert, and S. Thrun, "Monte Carlo localization: Efficient position estimation for mobile robots," in *Proc. of the 16th Nat. Conf. on Artificial Intelligence (AAAI'99)*, Jul. 1999, no. 343–349.

[66] M. Montemerlo, S. Thrun, D. Koller, and B. Wegbreit, "FastSLAM: A factored solution to the simultaneous localization and mapping problem," in *Proc. 8th National Conf. on Artificial Intelligence*, 2002, pp. 593–598.

[67] M. Montemerlo, S. Thrun, D. Koller, and B. Wegbreit, "FastSLAM 2.0: An improved particle filtering algorithm for simultaneous localization and mapping that provably converges," in *Proc. of 18th Int. Joint Conf. on Artificial Intelligence*, Acapulco, Mexico, 2003, pp. 1151–1156.

[68] H. Durrant-Whyte and T. Bailey, "Simultaneous localization and mapping: Part I," *IEEE Robot. Autom. Mag.*, vol. 13, no. 2, pp. 99–110, 2006.

[69] T. Bailey and H. Durrant-Whyte, "Simultaneous localization and mapping (SLAM): Part II," *IEEE Robot. Autom. Mag.*, vol. 13, no. 3, pp. 108–117, 2006.

[70] S. Thrun, W. Burgard, and D. Fox, "A probabilistic approach to concurrent mapping and localization for mobile robots," *Auton. Robot.*, vol. 5, no. 3–4, pp. 253–271, 1998.

[71] F. Lu and E. Milios, "Globally consistent range scan alignment for environment mapping," *Auton. Robot.*, vol. 4, no. 4, pp. 333–349, 1997.

[72] F. Dellaert and M. Kaess, "Square root SAM: Simultaneous localization and mapping via square root information smoothing," *Int. J. Robot. Res.*, vol. 25, no. 12, pp. 1181–1203, 2006.

[73] G. Grisetti, C. Stachniss, and W. Burgard, "Nonlinear constraint network optimization for efficient map learning," *IEEE Trans. Intell. Transport. Syst.*, vol. 10, no. 3, pp. 428–439, 2009.

[74] M. I. Lourakis and A. A. Argyros, "SBA: A software package for generic sparse bundle adjustment," *ACM Trans. Math. Softw.*, vol. 36, no. 1, p. 2, 2009.

[75] R. Kümmerle, G. Grisetti, H. Strasdat, K. Konolige, and W. Burgard, "g^{2o}: A general framework for graph optimization," in *Proc. IEEE Int. Conf. on Robotics and Automation*, 2011, pp. 3607–3613.

[76] M. Kaess, H. Johannsson, R. Roberts, V. Ila, J. J. Leonard, and F. Dellaert, "ISAM2: Incremental smoothing and mapping using the Bayes' tree," *Int. J. Robot. Res.*, vol. 31, no. 2, pp. 216–235, 2012.

[77] G. Grisetti, R. Kummerle, C. Stachniss, and W. Burgard, "A tutorial on graph-based SLAM," *IEEE Intell. Transport. Syst. Mag.*, vol. 2, no. 4, pp. 31–43, 2010.

[78] C. Cadena, L. Carlone, H. Carrillo, Y. Latif, D. Scaramuzza, J. Neira, I. Reid, and J. J. Leonard, "Past, present, and future of simultaneous localization and mapping: Toward the robust-perception age," *IEEE Trans. Robot.*, vol. 32, no. 6, pp. 1309–1332, 2016.

[79] I. J. Cox, "Blanche—An experiment in guidance and navigation of an autonomous robot vehicle," *IEEE Trans. Robot. Autom.*, vol. 7, no. 2, pp. 193–204, 1991.

[80] J.-S. Gutmann and C. Schlegel, "Amos: Comparison of scan matching approaches for self-localization in indoor environments," in *Proc. of the First Euromicro Workshop on Advanced Mobile Robot*, 1996, pp. 61–67.

[81] P. J. Besl and N. D. McKay, "A method for registration of 3-D shapes," *IEEE Trans. Pattern Anal. Mach. Intell.*, vol. 14, no. 2, pp. 239–256, 1992.

[82] J. Nieto, T. Bailey, and E. Nebot, "Recursive scan-matching SLAM," *Robot. Auton. Syst.*, vol. 55, no. 1, pp. 39–49, 2007.

[83] D. Borrmann, J. Elseberg, K. Lingemann, A. Nüchter, and J. Hertzberg, "Globally consistent 3D mapping with scan matching," *Robot. Auton. Syst.*, vol. 56, no. 2, pp. 130–142, 2008.

[84] A. Censi, "An ICP variant using a point-to-line metric," in *Proc. IEEE Int. Conf. on Robotics and Automation*, 2008, pp. 19–25.

[85] M. Bosse and R. Zlot, "Continuous 3D scan-matching with a spinning 2D laser," in *Proc. IEEE Int. Conf. on Robotics and Automation*, 2009, pp. 4312–4319.

[86] S. Lowry, N. Sünderhauf, P. Newman, J. J. Leonard, D. Cox, P. Corke, and M. J. Milford, "Visual place recognition: A survey," *IEEE Trans. Robot.*, vol. 32, no. 1, pp. 1–19, 2016.

[87] C. Boucher and J.-C. Noyer, "A hybrid particle approach for GNSS applications with partial GPS outages," *IEEE Trans. Instrum. Meas.*, vol. 59, no. 3, pp. 498–505, 2010.

[88] J. Wendel and G. F. Trommer, "Tightly coupled GPS/INS integration for missile applications," *Aerosp. Sci. Technol.*, vol. 8, no. 7, pp. 627–634, 2004.

[89] Y. Gao, E. Krakiwsky, M. Abousalem, and J. McLellan, "Comparison and analysis of centralized, decentralized, and federated filters," *Navigation*, vol. 40, no. 1, pp. 69–86, 1993.

[90] S.-L. Sun and Z.-L. Deng, "Multi-sensor optimal information fusion Kalman filter," *Automatica*, vol. 40, no. 6, pp. 1017–1023, 2004.

[91] S. Sukkarieh, E. M. Nebot, and H. F. Durrant-Whyte, "A high integrity IMU/GPS navigation loop for autonomous land vehicle applications," *IEEE Trans. Robot. Autom.*, vol. 15, no. 3, pp. 572–578, 1999.

[92] A. Mohamed and K. Schwarz, "Adaptive Kalman filtering for INS/GPS," *J. Geodesy*, vol. 73, no. 4, pp. 193–203, 1999.

[93] K. Nummiaro, E. Koller-Meier, and L. Van Gool, "An adaptive color-based particle filter," *Image Vis. Comput.*, vol. 21, no. 1, pp. 99–110, 2003.

[94] Y. Geng and J. Wang, "Adaptive estimation of multiple fading factors in Kalman filter for navigation applications," *GPS Solutions*, vol. 12, no. 4, pp. 273–279, 2008.

[95] I. L. Davis and A. Stentz, "Sensor fusion for autonomous outdoor navigation using neural networks," in *Proc. IEEE/RSJ Int. Conf. on Intelligent Robots and Systems*, 1995, pp. 338–343.

[96] R. Sharaf, A. Noureldin, A. Osman, and N. El-Sheimy, "Online INS/GPS integration with a radial basis function neural network," *IEEE Aerosp. Electron. Syst. Mag.*, vol. 20, no. 3, pp. 8–14, 2005.

[97] J. Wang, Y. Zhang, and K. T. Chong, "GPS/DR navigation data fusion research using neural network," in *Proc. Int. Conf. on Natural Computation*, 2009, pp. 58–61.

[98] E. M. Hemerly and V. R. Schad, "Implementation of a GPS/INS/odometer navigation system," in *ABCM Symposium Series in Mechatronics*, vol. 3. ABCM, 2008, pp. 519–524.

[99] D. Obradovic, H. Lenz, and M. Schupfner, "Fusion of sensor data in Siemens car navigation system," *IEEE Trans. Veh. Technol.*, vol. 56, no. 1, pp. 43–50, 2007.

[100] Waymo Team, "Introducing Waymo's suite of custom-built, self-driving hardware." Available: https://mediumcom/waymointroducing-waymos-suite-of-custom-built-self-driving-hardware-c47d1714563

[101] E. Guizzo, "How Google's self-driving car works." Available: https://spectrum.ieee.org/automaton/robotics/artificial-intelligence/how-google-self-driving-car-works

[102] W. Knight, "What to know before you get in a self-driving car." Available: https://www.technologyreview.com/s/602492/what-to-know-before-you-get-in-a-self-driving-car/

[103] F. Lambert, "Elon Musk clarifies Tesla's plan for level 5 fully autonomous driving: 2 years away from sleeing in the car." Available: https://electrek.co/2017/04/29/elon-musk-tesla-plan-level-5-full-autonomous-driving/

[104] H. Reese, "Tesla's autopilot: The smart person's guide." Available: http://www.techrepublic.com/article/teslas-autopilot-the-smart-persons-guide/

[105] C. Urmson, J. Anhalt, D. Bagnell, C. Baker, R. Bittner, M. Clark, *et al.*, "Autonomous driving in urban environments: Boss and the urban challenge," *J. Field Robot.*, vol. 25, no. 8, pp. 425–466, 2008.

[106] M. Montemerlo, J. Becker, S. Bhat, H. Dahlkamp, D. Dolgov, S. Ettinger, *et al.*, "Junior: The Stanford entry in the urban challenge," *J. Field Robot.*, vol. 25, no. 9, pp. 569–597, 2008.

[107] A. Bacha, C. Bauman, R. Faruque, M. Fleming, C. Terwelp, C. Reinholtz, *et al.*, "Odin: Team Victortango's entry in the DARPA urban challenge," *J. Field Robot.*, vol. 25, no. 8, pp. 467–492, 2008.

[108] A. S. Huang, M. Antone, E. Olson, L. Fletcher, D. Moore, S. Teller, and J. Leonard, "A high-rate, heterogeneous data set from the DARPA urban challenge," *Int. J. Robot. Res.*, vol. 29, no. 13, pp. 1595–1601, 2010.

[109] J. Leonard, J. How, S. Teller, M. Berger, S. Campbell, G. Fiore, *et al.*, "A perception-driven autonomous urban vehicle," *J. Field Robot.*, vol. 25, no. 10, pp. 727–774, 2008.

[110] I. Miller, M. Campbell, D. Huttenlocher, F.-R. Kline, A. Nathan, S. Lupashin, *et al.*, "Team Cornell's skynet: Robust perception and planning in an urban environment," *J. Field Robot.*, vol. 25, no. 8, pp. 493–527, 2008.

[111] J. Bohren, T. Foote, J. Keller, A. Kushleyev, D. Lee, A. Stewart, P. Vernaza, J. Derenick, J. Spletzer, and B. Satterfield, "Little Ben: The Ben Franklin racing team's entry in the 2007 DARPA urban challenge," *J. Field Robot.*, vol. 25, no. 9, pp. 598–614, 2008.

CHAPTER 30

RFID-BASED AUTONOMOUS MOBILE ROBOT NAVIGATION

Sunhong Park,[1] Guillermo Enriquez,[2] and Shuji Hashimoto[2]

[1]*Smart Vehicle Technology Research Center, Korea Automotive Technology Institute, Cheonan, Korea*
[2]*Advanced Science and Engineering, Waseda University, Tokyo, Japan*

MANY EXISTING methods for robust and stable mobile robot navigation employ sensors such as vision, sonar, or inertial sensors; however, these often have problems with reliability. A common, but sometimes overlooked, solution is the use of radio frequency identification (RFID). Using RFID, one can dynamically calculate the robot's pose from the information of RFID tags and navigate autonomously with precise localization, even without all of the aforementioned sensors. Presented in this chapter are methods for estimating location and pose and how RFID can be used to navigate a mobile robot with not only regular patterns but also randomly distributed patterns of tags. In addition, a new concept, called "read time," is introduced, which can reduce localization error in RFID-based navigation. There is also explanation of how the system can be extended to perform obstacle avoidance, and possible applications for RFID-based navigation systems.

30.1 ROBUST RFID-BASED NAVIGATION SYSTEM

30.1.1 Basic Navigation Concepts

For mobile robots, accurate localization, pose estimation, and environment recognition are core navigation technologies essential in safely providing services for humans or performing assigned tasks. In relation to these abilities, mobile robot

Handbook of Position Location: Theory, Practice, and Advances, Second Edition.
Edited by S. A. (Reza) Zekavat and R. Michael Buehrer.
© 2019 by the Institute of Electrical and Electronics Engineers, Inc.
Published 2019 by John Wiley & Sons, Inc.
Companion Website: www.wiley.com/go/zekavat/positionlocation2e

navigation systems can be roughly broken down into three types, each with advantages and disadvantages.

1. *Inertial Systems*. Measuring the movement or distance traveled to estimate location.
2. *Absolute Systems*. Detecting landmarks in the environment to estimate position.
3. *Hybrid or Fusion Systems*. Some combination of inertial and absolute approaches.

Inertial systems are the most simple, in both hardware and software. For example, when the robot begins to move, encoders on the wheels can be used to count how many revolutions are made, and thereby the distance traveled can be estimated. While this approach would seem to be very robust, in reality, it suffers from error which is compounded over the distance traveled. This error can be caused by lateral drift, wheel slippage, or counting errors. Conversely, absolute systems continuously update the localization estimation by recognizing landmarks in the environment whose positions are known, often through the use of cameras or laser range finders. However, these systems often rely on recognition algorithms, which can be computationally intense and suffer from environmental problems such as inconsistent lighting or landmarks being obscured by obstacles.

In an attempt to solve these problems, many systems have been presented that utilize some combination of inertial and absolute approaches, a simple example of which would be an inertial system that also uses landmark recognition in order to reduce its localization estimation error. The drawback to these hybrid, or fusion, systems is that while they are often able to increase accuracy, they also increase the system cost, in terms of both hardware and computational complexity.

There are two schools of thought when it comes to solving the problem of mobile robot navigation. The first is based around the idea of upgrading the mechanical and software systems of the robot. However, it often concerns the increase of sensors mounted on the robot in order to reduce localization error. For example, if we are utilizing a scanning laser to detect wall features, the addition of more scanning lasers will allow us to compare the results from all the sensors and reduce noise and error. In general, the addition of more sensors will provide you with more accurate results; however, this comes at a higher cost, again, in both the hardware and software. The second school of thought is based on improving the environment in which the robot will operate. This can involve the installation of cameras or other sensors to detect the location of the robot in the navigable area. An added bonus of this is that these upgrades can be used by multiple robots/systems. RFID can be an excellent tool to improve the navigable area. By placing RFID tags throughout the area, robots can utilize these tags to determine their location and perform navigation.

RFID is an automatic identification technology using a radio frequency [1] to identify and track a product, animal, or person. The main function of an RFID system is to retrieve information, generally a serial number, from an RFID tag (also known as a transponder) and send it to the PC. A basic RFID system consists of three components:

1. *Reader*: Hardware to convert radio information to PC-readable data.

2. *Antenna*: To broadcast/receive radio waves.

3. *RFID Tags*: Transponders which provide information.

RFID systems can also include software such as drivers and middleware. The reader emits a radio frequency in ranges of anywhere from 1 in. to 10 m or more, depending on its power and the used radio frequency. When an RFID tag passes through the electromagnetic zone, it is detected by the reader. Then, the reader decodes the data in the RFID tag's integrated circuit and the data are passed to the PC for processing. The communication between an antenna and RFID tags uses inductive coupling and each has a coil. The energy is transferred from an antenna to RFID tags by means of mutual inductance between the two circuits. In general, there are four frequencies used for RFID system. Each frequency has unique features as follows. Therefore, when using an RFID system, we should select the frequency most suitable for the application.

1. *Low frequency* (LF; 30–500 KHz): LF is least affected by the presence of liquids, metal, or dirt. However, it has a very low data read rate when compared with the other operating frequencies.

2. *High frequency* (HF; 10–15 MHz): HF has higher transmission rates and ranges than LF; it also offers better performance near metals or liquids than UHF.

3. *Ultrahigh frequency* (UHF; 850–950 MHz): UHF has the longest range and the transmission rate is also very high, which allows it to read a single tag in a very short time. Therefore, UHF is most commonly used for item tracking and supply-chain management applications. A disadvantage is that UHF signals are susceptible to interference (which can reduce accuracy) when tags or antenna are in close proximity to liquids or metal surfaces.

4. *Microwave frequency* (2.4–2.5 GHz): Transmission rate of this frequency is very high, but it is weak to metal. There are several downsides to the microwave frequency. One is that comparatively, it consumes more energy than its lower-frequency counterparts. Microwave tags are typically more expensive than UHF tags as well. Another problem is that wireless 802.11 b/g(WiFi) networks may interfere with microwave RFID systems.

Generally, RFID tags for RFID systems can be classified into three types in relation to power or energy: passive, semi-passive (also known as semi-active), and active.

1. *Passive*: Passive RFID tags have no internal power supply. The minute electrical current induced in the antenna by the incoming radio frequency signal provides power for the Integrated Circuit (IC) in the tag to power up and transmit a response. Although the communication range of passive tags is short, they are compact and require no battery maintenance.

2. *Semi-passive*: Semi-passive RFID tags are similar to passive tags except for the addition of a small battery. This battery allows the RFID tag to be powered

when the tag is activated by the reader. Semi-passive RFID tags are faster in response and can produce a stronger signal when compared with passive tags.

3. *Active*: Unlike passive and semi-passive RFID tags, active RFID tags have their own internal power source which is used to power the IC and generate the outgoing signal. They are often called beacons because they broadcast their own signal. They may have longer range and larger memories than passive tags, as well as the ability to store additional information sent by the transceiver. However, its size is increased due to the need for a power supply and it requires regular maintenance for battery exchange.

The approaches discussed in this chapter involve navigation approaches for a mobile robot in indoor environments, using only an RFID system. Given these conditions, an HF, passive RFID system is used in all algorithms and experimental examples presented. An HF, passive system gives us better performance around metals and liquids, and passive tags are resilient to covers, wear, dirt, or vibration, which allows the tags to be placed under carpeting or flooring and still be able to be read by a reader mounted on the underside of the robot. Additionally, passive RFID tags require little to no maintainance and are less expensive than other types of tags.

Now we can begin to examine what capabilities this provides the robot. The first is the inherent ability to perform localization. Let us assume we have an area in which we have deployed RFID tags and the location of each tag is known; that is, we have a database that links an RFID tag serial number to its (x, y) coordinates in the navigable area. When the robot moves through the field, the reader detects tags that pass into the sensible area of the antenna, and so we know that the robot is at the coordinates of that tag.

There is, however, an error proportional to the sensible area of the antenna. This is caused by the nature of RFID readers and antennas, which can only detect whether an RFID tag is present or not. Therefore, when an antenna detects a tag, we cannot know exactly where the tag is in relation to the center of the antenna. There are some approaches that have attempted to reduce this error. For example, if plural, overlapping antennas are used, the probable area the detected tag can be in is reduced. If we increase RFID tag density, we can have multiple tags read at once, and use an average to reduce error. It is important to remember, however, that while these approaches can reduce the error, they also increase the overall cost of the system by increasing the extra computation to handle another antenna's signal or the calculations for multiple tags.

30.1.2   Estimating Robot's Pose

While localization estimation is relatively simple using RFID, the problem of estimating the robot's pose, or the direction it is facing, without the aid of other sensors is more complex. We begin by assuming that along with the location of the tags, the start and goal locations are known by the robot. Then, it starts to navigate toward the goal based on a calculated rotation angle between the start and goal. The initial pose of the robot is $0°$, representing the positive y-axis. Next, the robot estimates the orientation angle from the current and previous location information. The robot recalculates the orientation of the goal location and determines the rotation angle of the robot to reach the goal again. The antenna attempts to read new RFID tags at an interval of 0.2 second, and the robot rotates according to the calculated rotation angles periodically. Otherwise, the robot keeps moving forward. The robot speed is determined by the RFID system's sensing ability. If the robot moves too quickly, some tags will fail to be read. In the event the robot reads the same tag after a rotation, the robot proceeds in the same direction without a course adjustment. The rotation angle for navigation to the goal is calculated by the following method. First, the robot starts to move without considering the initial pose. The orientation of the goal, $\theta_{start \rightarrow goal}$, is derived from the initial and goal position by:

$$\theta_{star \rightarrow goal} = \arctan\left(\frac{y_{goal} - y_{init}}{x_{goal} - x_{init}}\right) \tag{30.1}$$

$$\theta_{rotation} = \theta_{start \rightarrow goal} - \theta_0. \tag{30.2}$$

In this equation, the term (x_{goal}, y_{goal}) represents the coordinates of the goal position. Likewise, (x_{init}, y_{init}) represents the coordinates of the start position. In the experiment, the initial pose of the robot θ_0 is $0°$. The robot obtains the coordinates of new RFID tags from the antenna when it moves. The pose of the robot, $\theta_{previous \rightarrow current}$, is obtained using:

$$\theta_{previous \rightarrow current} = \arctan\left(\frac{y_{current} - y_{previous}}{x_{current} - x_{previous}}\right), \tag{30.3}$$

where $(x_{previous}, y_{previous})$ and $(x_{current}, y_{current})$ denote the coordinates of the position scanned previously and presently, respectively. According to (30.4), the angle between the present position and the goal position, $\theta_{current \rightarrow goal}$ is obtained by:

$$\theta_{current \rightarrow goal} = \arctan\left(\frac{y_{goal} - y_{current}}{x_{goal} - x_{current}}\right). \tag{30.4}$$

From the above (30.3) and (30.4), the rotation angle, $\theta_{rotation}$, to turn toward the goal is provided as

$$\theta_{rotation} = \theta_{previous \rightarrow current} - \theta_{current \rightarrow goal}. \tag{30.5}$$

The current position and orientation are updated in real time. This procedure stops when the antenna detects the RFID tag representing the goal.

This approach can be used to further reduce localization error with equations based on polar/Cartesian conversion. When the robot detects a new RFID tag Q, as shown in Figure 30.1, we can calculate the robot's location $(x_{current}, y_{current})$ in RFID tag field using:

$$x_{current} = r \cos \theta' + x_1 \tag{30.6}$$

$$y_{current} = r \sin \theta' + y_1. \tag{30.7}$$

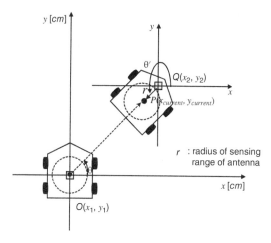

Figure 30.1   Proposed method of localization.

Radius r is 17 cm, the same as the radius of the sensing range of the antenna. Let angle(θ') denote an incident angle in relation to the newly detected RFID tag. The angle θ' is given by the accumulation of the rotation angle of the robot. Equations (30.6) and (30.7) are also used in the procedure for navigation. We are able to precisely estimate the orientation of a mobile robot from the relation between detected previous and current RFID tag, as shown in (30.8). When the robot detects a new RFID tag from $O(x, y)$ to $Q(x_1, y_1)$, the orientation can be obtained as follows:

$$\theta_{orientation} = \arctan\left(\frac{y_{current} - y}{x_{current} - x}\right) = \arctan\left(\frac{dy}{dx}\right). \tag{30.8}$$

30.1.3   Experimental Verification: Grid-Like Pattern

In our experiments, we used a robot system named UBIquitous RObot (UBIRO). UBIRO consists of three main parts: the PC for control, the RFID system for obtaining the location information of the robot, and the mobile base for navigation. UBIRO was developed based on an electric wheelchair (EMC-230) for elderly and disabled persons as a mobility device. The PC mounted on the electric wheelchair acts as a central controller that handles information of RFID tags from the RFID system and sends commands to the mobile base in order to reach the goal. The RFID system reads RFID tags on the floor, which allows the PC to roughly deduce the location and pose of the robot based on the proposed algorithms. Finally, the mobile base is controlled by the PC according to the results of calculations run on information provided by the RFID system. UBIRO has external sensors, such as distance sensors and touch sensors (bumper switches), for the detection of obstacles. The robot's front wheels are free rotating casters, which unfortunately causes some instability when moving. UBIRO is shown in Figure 30.2, and the specifications are given in the Table 30.1.

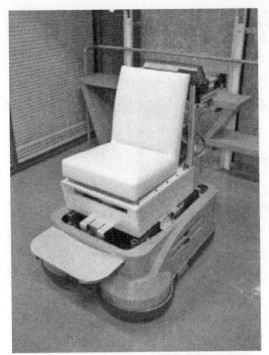

Figure 30.2   Mobile robot: UBIRO (Wheelchair type).

TABLE 30.1.   Specification of UBIRO (Wheelchair Type)

Item	Description
Size	800(W) × 1100(D) × 1050(H) (mm)
Weight	110 kg
Battery weight	SS-SEB35-T 30 kg (two 12V 35A)
Operation system	Windows XP (Pentium III 850 MHz/512 MB)
Speed	Min. 2.5 km/h ~ max. 6.0 km/h
Range of travel	30 km

To evaluate the validity of localization and pose estimation using the proposed algorithm for mobile robot navigation, we conducted experiments under three kinds of conditions. In condition I, the initial pose of the mobile robot was 0°, which is the positive y-axis. Condition II was 90°. Finally, condition III was −90°. The experimental environment is shown in Figure 30.3. We assumed the environment to be an indoor field where the RFID tags were deployed. The 198 passive RFID tags were laid on the floor in a grid-like pattern over an area measuring 420 cm × 620 cm, with a spacing of 34 cm. The field was free of obstacles. The robot stops navigating when the antenna detects the RFID tag of the goal area (within 34 cm × 34 cm).

Figure 30.4 shows the results with an initial pose of 0°. Shown are the start location, the goal location, and the path trajectories. Here, we compared the

Figure 30.3 Experimental environment.

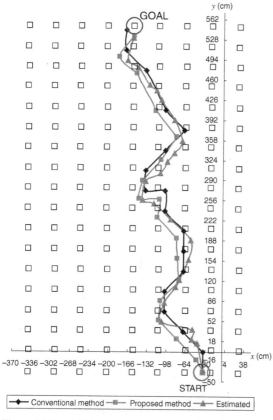

Figure 30.4 Path trajectories under condition I.

TABLE 30.2.   Comparison of Localization Methods with
Proposed Method (Regular Pattern)

Localization Error	x-Axis (cm)	y-Axis (cm)
Conventional method [2]	17.0	17.0
Proposed method	13.3	5.7

conventional method [2] with and without the proposed error reduction for each condition. The robot stops and ends navigation if the antenna detects the RFID tag representing the goal. Also, we arranged the RFID tags around the field in order to prevent the robot from exiting the RFID tag field. The conventional method [2] acquired an intermediate location of the RFID tag, ignoring the pose of robot. The moving distance of the robot relies on the direction of the front wheel, which is a free caster, and ranges from 0 cm to 60 cm, approximately. The proposed method could more accurately measure the localization and pose of the robot in real time. Table 30.2 illustrates the localization error between the conventional and proposed methods. While localization error in the proposed method was on average 13.3 cm in the x-axis and 5.7 cm in the y-axis, the conventional method has a 17-cm error for both x- and y-axes.

According to the results of these experiments, when sufficient RFID tags are deployed in the environment, the mobile robot was able to reach the goal without the aid of other sensors. In path trajectories under each condition, the proposed method showed that it was able to estimate more precisely both the location and pose of a mobile robot during navigation. However, there are problems such as wheel slippage and drift, as mentioned before.

The front wheels of the robot are free casters, so it is difficult for the robot to move as commanded. For these reasons, estimation of location and pose of robot are sometimes unstable. We realized through various conditions in navigation that the x-axis error is larger when the robot makes a wider turn from left to right, and vice versa. Due to this reason, the y-axis error is smaller than that of the x-axis. Although the mobile robot using the proposed method sometimes navigated in a zigzag toward the goal, the overall movement of the robot was smoother. On average, the robot was able to reach the goal within 85 seconds, shorter than the conventional method's 97 seconds.

30.2   REDUCTION OF LOCALIZATION ERROR: READ TIME

30.2.1   Problem with RFID-Based Localization

The localization error in navigation based on the RFID system is proportional to the sensing range of the antenna. It is difficult to obtain the precise location of a robot due to the nature of the antenna, which can only detect if an RFID tag is present or not. To cope with this problem, many researchers have used external sensors such

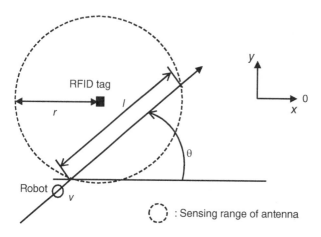

Figure 30.5   The relationship between the robot trajectory(ℓ) and angle(θ).

as the dead reckoning (DR), vision systems, and sonar in order to reduce the localization error. In using the RFID system, some research introduces methods that raise the arrangement density of RFID tags. However, the system cost is also increased because of the increased number of RFID tags. Moreover, a problem with detection collisions arises, as sometimes the antenna senses more RFID tags than it can handle at once. Consequently, the distance between RFID tags and their distribution pattern are crucial factors which determine the localization accuracy.

30.2.2   Read-Time Concept

In order to reduce the localization error mentioned in the previous sections, we can utilize the read time to reduce localization error without increasing the number of RFID tags. The read-time concept is shown in Figure 30.5. Read time (R_t) is defined as the duration of time the antenna recognizes and reads a given RFID tag, that is, the length of response time of an RFID tag that is in the scanning range. The length (ℓ) of the robot trajectory in the detection area of a tag is given as

$$\ell = 2r\sin\theta, \tag{30.9}$$

where θ and r represent the robot orientation and the radius of the sensing range of antenna. Therefore, by using the speed, v, of the robot, R_t is given as

$$R_t(\theta, r, v) = \frac{\ell}{v} = \frac{2r\sin\theta}{v}. \tag{30.10}$$

Table 30.3 summarizes the actual read time when the robot is moving at 12.2 cm/s. Although there are some deviations due to unexpected effects such as the slipping of the wheels and fluctuations in the robot speed, the read time can be used to estimate orientation. Using this read time, we are able to acquire the difference between the expected angle and the robot's actual moving angle every time the antenna detects a new RFID tag.

TABLE 30.3. Measured Read Time for Each Angle

Angle (Degrees [°]) (θ)	Read Time (s) (R_t)
90	2.6 ~ 2.8
70 ~ 89	2.2 ~ 2.6
60 ~ 69	1.8 ~ 2.2
50 ~ 59	1.1 ~ 1.8
40 ~ 49	0.8 ~ 1.1

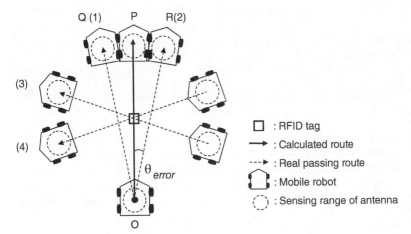

Figure 30.6   Possible routes in read time.

The estimation of the angle θ from R_t is not unique, as

$$sin\,\theta = sin(\pi - \theta). \tag{30.11}$$

We assume that the robot direction does not change very much from the intended one. Therefore, we select the closer angle based on orientation previously set as the appropriate one.

Let us assume that the center of the mobile robot passed through OP based on calculations using the RFID tag's location information to reach the goal. However, the mobile robot may appear to pass through OR (2) or OQ (1), because the prototype robot UBIRO occasionally veers on its trajectory because the front wheels are free casters that we cannot control. θ_{error} represents the difference in angle between calculated route OP and real route OR (2) as shown in Figure 30.6. We obtain θ_{error} when the mobile robot passed route OR (2) or OQ (1) by choosing the path most likely taken based on the last turn the robot executed. In order to compensate for the localization error by using an RFID system, the mobile robot gets feedback using the read time after every step during localization estimation and navigation. The θ_{error} is calculated using:

$$\theta_{error} = \theta(R_t) - \theta_{current}, \tag{30.12}$$

where θ (R_t) is a function of R_t and $\theta_{current}$ is the intended angle of the robot. An orientation error may accumulate through repeated rotation; however, this can be compensated for by maximizing the read time when moving between two RFID tags. If read time is maximized, the robot has moved through a straight line between two RFID tags. In this way, the robot can detect its orientation as the coordinates of the two RFID tags are known. As mentioned above, localization error will exist without sufficient RFID tags in the antenna area. However, simply increasing the number of RFID tags is impractical due to the increasing cost and possibility of collision. In light of this, while most research involves using the center point of a number of RFID tags read simultaneously, we can use an arrangement and density that ensure only a single RFID tag can be read at any one time. The proposed read time is applied to the navigation of the mobile robot in order to achieve a reliable localization. We assume that the robot moves straight, so that the incident angle (θ') equals the orientation of the robot; thus:

$$\theta_{previous \to current} = \arctan\left(\frac{y_{current} - y_{previous}}{x_{current} - x_{previous}}\right) + \theta_{error}, \tag{30.13}$$

where ($x_{previous}$, $y_{previous}$) and ($x_{current}$, $y_{current}$) denote the coordinates of the previously scanned location and the current one, respectively, and θ_{error} represents the difference between the theoretical and physical location of the robot. Using the proposed read time, θ_{error} is updated every step as feedback whenever the antenna detects a new RFID tag. As a result, it is possible to estimate localization and orientation accurately.

30.2.3 Randomly Distributed RFID Tags

We describe an autonomous navigation system for a mobile robot using a randomly distributed arrangement of RFID tags and the read-time method. In most RFID systems, the arrangement of RFID tags affects the localization accuracy. It is desirable for reliable localization to be achieved using a small quantity of RFID tags. The aim of our research is to realize robot navigation in a randomly distributed passive RFID tag layout that has a small quantity of RFID tags in contrast with the grid or triangle pattern. Unlike most other studies of this kind, we employ only one RFID antenna to estimate the location and orientation of a mobile robot that does not use odometry, gyro, and vision system. Furthermore, we do not assume that multiple RFID tags are present under the antenna in order to reduce localization error, as this would increase the number of required tags. Using our approach, we can navigate autonomously even in a randomly distributed passive RFID tag environment.

30.2.4 Experimental Verification: Grid-Like and Random Pattern

We envisioned the environment as an indoor field where the RFID tags (size: 76 (W) $\times$ 45(D) $\times$ 0.23(H)[mm]) were deployed at 34 cm $\times$ 34 cm intervals on the floor. As shown in Figure 30.3, the 216 passive RFID tags were laid on the floor in a grid-like pattern over an area measuring 420 cm × 620 cm, with a spacing of 34 cm. There are no obstacles on the floor. The robot stops navigating when the antenna detects the RFID tag representing the goal location (within 34 cm × 34 cm).

In the case of randomly distributed RFID tags, high localization and orientation estimation are required because there are not enough RFID tags to acquire their location information. In order to evaluate the feasibility of the proposed localization method using read time in a randomly distributed arrangement, we conducted the experiment with five different initial orientations of the mobile robot, five times each. We compared the results with the conventional method [2] without using read time under the same conditions. For condition I, the initial orientation of the mobile robot was 0°, that is, the positive y-axis. Condition II was 45°, condition III was −45°, condition IV was 90°, and condition V was −90°. The random distribution consisted of 102 passive RFID tags laid on the floor over an area measuring 420 cm × 620 cm and free of obstacles. The experimental environment and the procedure of the navigation using read time in a randomly distributed arrangement are shown in Figures 30.7 and 30.8.

The proposed and conventional method's path trajectories can be seen in Figure 30.9. The trajectories may look similar, but there is actually a significant difference. The localization and orientation in the conventional method [2] could not be specified more precisely than the RFID tags interval. On the other hand, the proposed method using the read time could make a more accurate estimation of the location and orientation of the robot in real time, even with no additional sensors to modify that error. Tables 30.4 and 30.5 illustrate the average localization and orientation errors throughout a series of experiments. The average localization error in the proposed method was 7.78 cm for the x-axis and 7.16 cm for the y-axis, while in the conventional method it was 17 cm for both axes. Although the results for localization

Figure 30.7   Experimental environment (random pattern).

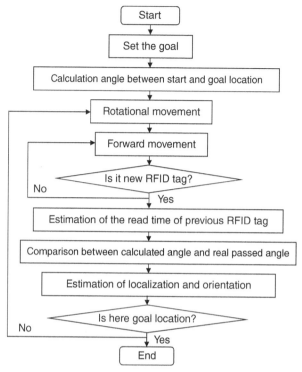

Figure 30.8   Navigation procedure using read time.

estimation are not as good as the results in a grid-like pattern, our proposed method is effective, especially considering the cost of the system.

30.2.5   Path Trajectories with Grid-Like Pattern

We conducted experiments from various initial orientations ($-90°$, $-45°$, $0°$, $45°$, and $90°$) of the mobile robot. Figure 30.10 shows the results when the initial orientation was $0°$.

Here, we compared the results with those of the conventional method described in [2], without using read time. The figure illustrates the start location, the goal location, and the path trajectories until the goal is reached. The estimated result was compared with real trajectories during navigation under the same condition. In general, conventional methods regard the robot location as the location of the RFID tag regardless of the orientation of the robot when it detected an RFID tag. Accordingly, although the robot is able to navigate successfully to the goal, the estimation error of location and orientation of the robot has persisted as much as intermediate location during navigation. Table 30.6 illustrates the localization error between the conventional method and the proposed method. Likewise, Table 30.7 illustrates orientation error. While the localization error in the proposed method is

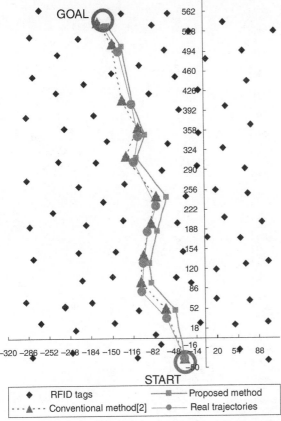

Figure 30.9   Path trajectories with randomly distributed RFID tags.

TABLE 30.4.   Comparison of Localization Methods (Random Pattern)

	x-Axis (cm)	y-Axis (cm)	Number of RFID Tags
Conventional method [2]	17.0	17.0	216
Proposed method	7.78	7.16	102

TABLE 30.5.   Comparison of Orientation Estimation Methods
(Random Pattern)

	Degrees (°)
Conventional method [2]	26.5
Proposed method	11.1

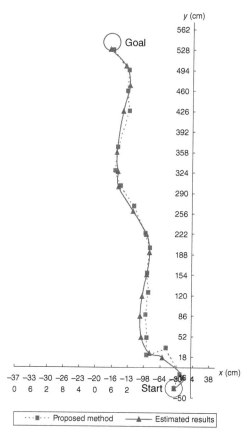

Figure 30.10   Path trajectories with initial orientation 0° (grid-like pattern).

TABLE 30.6.   Comparison of Localization Methods (Grid-Like Pattern)

	x-Axis (cm)	y-Axis (cm)
Conventional method [2]	17.0	17.0
Proposed method	6.2	5.4

TABLE 30.7.   Comparison of Orientation Estimation Methods (Grid-Like Pattern)

	Degrees (°)
Conventional method [2]	26.5
Proposed method	10.3

TABLE 30.8. RFID-Based Method Comparison

Reference	Technique	Sensor	Tag Layout	Sizes (cm)	Req. Tags (5 m × 5 m)	Accuracy (for Interval)
Han et al. [3]	Cartesian coordinates and encoder error reduction	Encoder and single RFID reader	Square and triangle pattern	Interval: 5 Tag: 3 × 3 Antenna: 10 × 10	10,000	1.6 cm (312%)
Kotaka et al. [4]	Monte Carlo localization	Encoder and four RFID readers	Grid-like pattern	Interval: 30 Tag: 26 × 26 Antenna: 1.5 cm (diameter)	277	10 cm (300%)
Lee and Lee [5]	Weighted average and Hough transform	Encoder and single RFID reader	Grid-like pattern	Interval: 5 Tag: 3 × 4 Antenna: 10 × 10	10,000	2.6 cm (192%)
Park and Hashimoto [6]	Trigonometric functions and Cartesian coordinates	Single RFID reader	Grid-like pattern	Interval: 34 Tag: 7.6 × 4.5 Antenna: 24 (diameter)	216	x-axis: 13.3 cm (255%) y-axis: 5.7 cm (596%)

6.2 cm in the x-axis and 5.4 cm in the y-axis on average, the conventional method has a 17-cm error in the x- and y-axes. Additionally, the total navigation time when using read time in a randomly distributed RFID tag arrangement was within 60 seconds.

30.2.6 Comparison with Other RFID-Based Methods

In Table 30.8, we see a comparison of these results with those of similar RFID-based research [3–6]. In [3] and [5], the tags are allocated at every 0.05 m in a row for both the square and triangular patterns. In other words, the antenna always detects more than two tags at any given location. If we compare this approach with our experimental environment, the number of tags deployed would need to be more than 10,000 in order to acquire the same localization accuracy. This is a significantly larger number of RFID tags to improve the accuracy and would create a larger computation burden on the server. In [4], although this method could obtain location accuracy about 10 cm less than the tag interval (30 cm), the robot has an encoder that monitors its rotation and is equipped with an RFID reader in four places. Such approaches usually use an inertial sensor and multiple antennas to obtain the location and orientation of the robot. In contrast to them, we developed an effective method that used trigonometric functions and the RFID tags' Cartesian coordinates in a regular grid-like pattern in order to reduce localization error [6]. However, in this method, the accuracy of the location depends on the incident angle and is unstable due to the rotation angle error that arises when the robot makes wide turns.

30.3 EXTENDING THE READ-TIME PARADIGM

For mobile robots, obstacle avoidance is a critical issue for navigating to a desired location without collision. To do this, many approaches that employ multiple sensors, such as vision, location retrieval function (LRF), and sonar, have been introduced. However, such approaches have some drawbacks, including the influence of ambient light or other environmental factors. To cope with them, we can apply a passive RFID-only system to single and multiple static obstacle avoidance. In the case of single static obstacle avoidance, like a pillar, we can deploy passive RFID tags to surround the obstacle. Each passive RFID tag has an x and y relation between obstacle and absolute coordinate. When the antenna detects their RFID tag, the direction of a robot is changed to avoid the obstacle. In the case of multiple static obstacle avoidance, like a doorway or between tables, we set an RFID tag in a specific region which is located in front of and between the obstacles. Using the read time in the specific region, we can accurately estimate the orientation of the robot, and the robot is able to pass between the obstacles without the aid of any other sensor.

30.3.1 Static Obstacle Avoidance

To avoid static obstacles, we need to add information to the RFID tags surrounding the obstacle. There are two types of RFID tags surrounding the obstacle as shown in Figure 30.11.

The squares labeled "X" represent RFID tags that run horizontally along the x-axis in relation to the obstacle. Likewise, the squares labeled "Y" represent RFID tags which run vertically along the y-axis. In the other areas, the robot calculates the localization and orientation using the previous and current RFID tag location information. If the robot detects an RFID tag surrounding the obstacle, it selects one direction from among 0°, 90°, 180°, and 270° as the direction it should take to reach the goal. To select the direction of the robot, the orientation of the robot and the goal location are considered.

30.3.2 Multiple Obstacles Avoidance

For passing through obstacles such as elevator entrances or between tables, the precise localization and orientation of the robot are indispensable. The RFID tags are laid on the floor in a grid-like pattern and are correlated to two dimensional

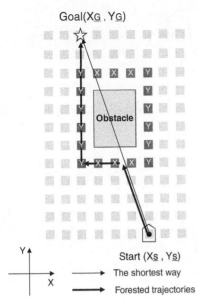

Figure 30.11   Layout of RFID tags surrounding an obstacle.

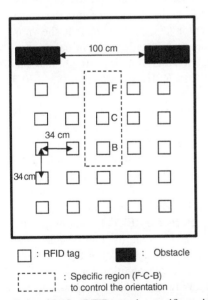

Figure 30.12   RFID tags in specific region.

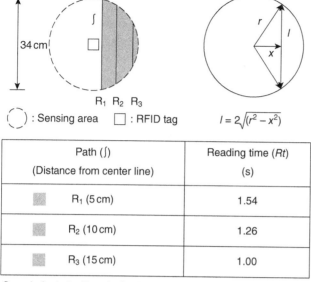

$): Sensing area$ □ : RFID tag $\qquad l = 2\sqrt{(r^2 - x^2)}$

Path (∫) (Distance from center line)	Reading time (Rt) (s)
R₁ (5 cm)	1.54
R₂ (10 cm)	1.26
R₃ (15 cm)	1.00

Speed of robot : 19 cm/s, S = V·T (Center : 1.78 s)

Figure 30.13   Read time of each path.

coordinates in the environment. However, the RFID tag in the specific region is perpendicular to the center point between the obstacles. Therefore, the key in the proposed algorithm is for RFID tags in the specific region to pass through the center of the antenna. To do this, we set the specific region (F_{orward}, C_{enter}, B_{ack}) as shown in Figure 30.12 in order to precisely control the robot's orientation. The initial orientation of the robot is unknown. Therefore, the robot starts by going straight and calculates the orientation when the antenna detects a new RFID tag using the 30.1 ~ 30.5 mentioned above.

This read-time usage, Figure 30.13, is similar to that proposed in the previous section, but the read time in the specific region is used to achieve precise orientation of the robot along a path between obstacles. Figure 30.14 shows the control of orientation using read time. The robot is able to decide the rotation direction to align the RFID tag to the center of the antenna by comparing previous and current read times. The forward movement speed of the robot was 19.0 cm/s in this experiment. Using this read time, the robot is able to calculate how far away the RFID tag is from the center of the antenna. When the robot detects the RFID tags in the specific region, it moves forward and back repeatedly until obtaining the longest read time of RFID tag C(R_C) and the shortest read time from when RFID tag B(R_B) is read until RFID tag F(R_F) is read. These two things are a precondition for the robot to pass and avoid the obstacles without collision. With the method proposed in the previous section, we are able to calculate rotation direction and angle correctly. The robot estimates the read time of RFID tag C(R_C) and between RFID tag B and F(R_{BF}). The robot rotates to align the RFID tag with the antenna center based on read time and orientation of the robot. If the robot cannot satisfy the above two preconditions, it goes back until detecting RFID tag B and repeats the process.

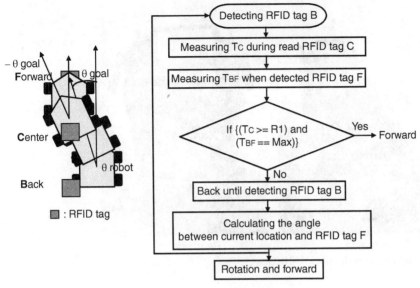

Figure 30.14   Control algorithm in specific region.

30.3.3   Experimental Verification: Static Single/Multiple Obstacles

For the navigation experiments with static obstacle avoidance, we developed the mobile robot UBIKO, which is equipped with a steering device for the front wheels. Therefore, unlike the previously utilized mobile base UBIRO, we can control the front wheels according to the desired rotation angle. UBIKO and its specifications are shown in Figure 30.15 and Table 30.9.

UBIKO has a photoelectric sensor for detecting obstacles, but it was not utilized in these experiments. Moreover, the navigation operations can also be controlled using a touch screen monitor. The antenna that acquires the location of robot was attached to the underside of the mobile base.

30.3.4   Navigation with a Single Static Obstacle

To show the validity of the proposed method, we ran the navigation experiment with a single static obstacle. The 216 passive RFID tags were deployed on the floor in a grid-like pattern over an area measuring 420 cm × 620 cm.

The experimental environment is shown in Figure 30.16. The obstacle size was 70 cm × 70 cm × 100 cm. Figure 30.17 shows the trajectories from navigation with a single static obstacle when the initial orientation of the robot was 0°. Using this method, the mobile robot was able to avoid the static obstacle while navigating and estimating localization and orientation. The navigation time to reach the goal was 50 seconds.

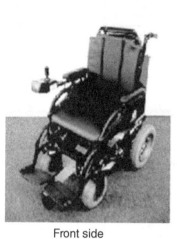 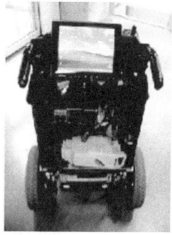

Front side   Rear side

Figure 30.15   Mobile robot: UBIKO.

TABLE 30.9.   Specification of UBIKO

Item	Description
Size	1040(L) × 600(W) × 950(H) (mm)
Weight	95 kg (includes battery)
Battery weight	Two batteries for mobile base (12V 35A)
	One portable battery for PC (26V 8A)
Operation system	Windows XP
CPU	3.0 GHz/4 GB
Speed	Min. 2.5 km/h ~ max. 6.0 km/h

Figure 30.16   Experimental environment with a single obstacle.

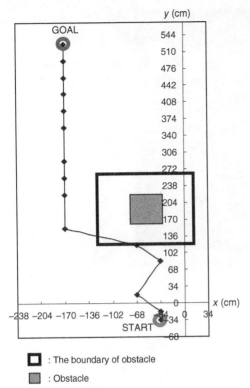

□ : The boundary of obstacle

▦ : Obstacle

Figure 30.17   Path trajectories with a single obstacle.

30.3.5   Navigation with Multiple Static Obstacles

To verify the effectiveness of the proposed method, we ran navigation experiments with multiple static obstacles. For this experiment, 35 passive RFID tags were laid on the floor in a grid-like pattern over an area measuring $200\,cm \times 300\,cm$ with two obstacles. The experimental environment is shown in Figure 30.18. The robot goes between the obstacles when the robot aligns itself based on the read time (R_C and R_{BF}) in the specific region. From the trajectories of navigation as shown in Figure 30.19, the robot was able to precisely estimate localization and orientation and pass in between the obstacles without the aid of any other sensors. This result shows that an RFID system can be utilized in navigation with static obstacles. In Figure 30.19, the initial orientation of the robot was $45°$. The initial orientation, however, was not given to the robot when navigation began. The robot calculated the orientation using the previous and current location information. Additionally, in the specific region, the robot could measure the read time accurately and pass through the center of the RFID tag utilizing the proposed method, resulting in an overall smooth movement.

According to the results of both experiments, the mobile robot could avoid obstacles without collision based on localization and orientation estimation. Although

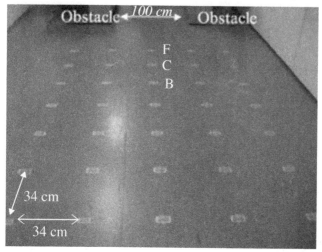

Figure 30.18   Experimental environment with multiple obstacles.

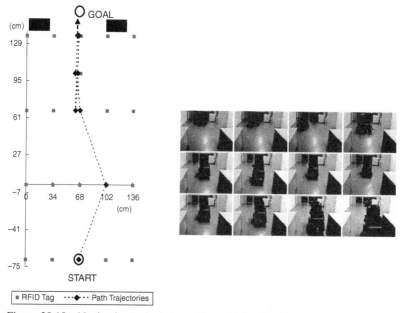

Figure 30.19   Navigation trajectories with multiple obstacle.

we need to add information to the RFID tag surrounding the obstacle to achieve this, our approach can navigate autonomously while estimating the localization and orientation of the robot. Unlike other approaches that usually use additional sensors, the proposed method utilized only RFID. By comparing the read time between the previous R_C, R_{BF}, and the current R_C, R_{BF}, we could control the orientation and estimate the localization in the specific region. With this, we could accurately measure localization while controlling the orientation of the robot to complete the pass. To obtain them, we do not employ any other sensors or deploy a large amount of RFID tags. The speed of robot was also improved from 12.2 cm/s to 19.0 cm/s.

30.4 APPLICATIONS AND EXTENSIONS

30.4.1 Application Concepts

Using the techniques discussed, there are numerous real-world applications for RFID-based mobile robot navigation systems. In general, anywhere a predominantly static, indoor navigation setting is involved, this type of system can be utilized. For example, hospitals and similar care facilities are often in need of extra support both for patient mobility and the transportation of supplies. Additionally, their interior arrangement does not often change, meaning once RFID tags have been arranged throughout the building, they will require little or no maintainance. Hotels and museums are also good examples of places where a robot could be used to perform guidance within the building or perhaps be used for automated cleaning services. A point particular to museums, however, is that often times, while the walls may be static, the exhibits in a given room may be rearranged periodically. Especially when using only RFID, it is essential for the system to update RFID information according to environmental changes. To support this, we presented a semiautomatic device, as shown in Figure 30.20, to update RFID information for object position changes. The RFID system obtains two-dimensional location information and the presence of an obstacle when detecting RFID tags. It consists of two parts: a system to transmit RFID information to the PC using Bluetooth and an RFID reader to obtain location information in an indoor environment when detecting a passive RFID tag, as shown in Figure 30.21.

At the same time, we are able to detect voltage changes when reading RFID tags. If this value is above a threshold level, three LEDs (red, green, blue) and an

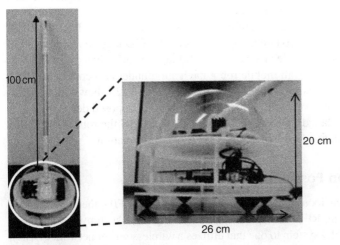

Figure 30.20   Prototype device.

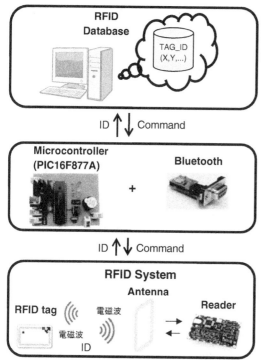

Figure 30.21   Composition of the proposed device.

alarm signify the presence of an RFID tag to the user. The RFID tag's information is sent to the PC via Bluetooth. The operations of updating, deleting, and adding information into the database (DB) are executed by DB management software on the PC. For instance, in the case of changing the object position in the environment, the user moves the device to the RFID tags around the object and the device obtains the ID of the detected RFID tag. Based on this ID, the Cartesian coordinate information of the RFID tag and the presence of the obstacle in DB are updated. Figure 30.22 shows the experimental environment, and 110 passive RFID tags were deployed in a grid-like pattern. The obstacle size is 80 cm × 80 cm. The effectiveness of the proposed device is shown through mobile robot navigation under two conditions, with object position changes, as shown in Figure 30.22. Figure 30.23 shows the trajectories from navigation with obstacle and without. Using the proposed device, the user was able to easily update data about the object's location, and the robot was able to navigate without collision based on the new RFID tag information.

30.4.2   Extension Possibilities

While this chapter has focused on approaches that only use RFID, there are obvious arguments for utilizing RFID to augment other systems. One such hybrid approach is the fusion of an RFID system to one that utilizes a wireless sensor network (WSN).

With obstacle

Without obstacle

Figure 30.22   Experimental environment.

WSN technology can give a mobile robot not only the ability to navigate but also provides access to a network that can be used to remotely actuate other systems in the environment, another example of the environment improvement paradigm. Hybrid systems are best used when the systems being combined strengthen each other's weaknesses. WSN systems typically allow for fast movement of the robot but are not as precise in localization as RFID. Conversely, RFID systems can provide high accuracy but often force the robot to move at relatively slower speeds to ensure all tags can be detected. The system presented in [7] is such a WSN/RFID fusion system. The WSN provides fast, general navigation in wide-open areas, with the RFID providing accurate navigation near obstacles.

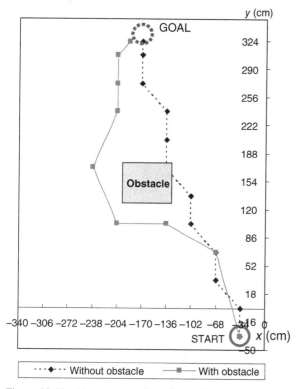

Figure 30.23   Navigation trajectories under two conditions.

The WSN navigation component works on a virtual potential field concept, as shown in Figure 30.24. First, WSN nodes are distributed around the environment, near the floor, at key points such as corners and long stretches of wall. Then, knowing the location of each of these nodes and the location of the goal, calculations are made to determine the best direction to move in from each node. In general, this direction points straight at the goal. In the event an obstacle is present between the node and the goal, the node's direction points at a sub goal. As the robot moves around the environment, the received signal strength intensity (RSSI) of each node is measured. These values are used to sort the nodes in order of strongest RSSI. Then, the RSSIs are combined with their respective node's suggested direction in a weighted sum, using

$$\vec{V} = \sum_{n=1}^{2} \frac{s_n}{s_1} \vec{d}_n, \tag{30.14}$$

$$\vec{V} = \sum_{n=1}^{3} \frac{s_n}{s_1} \vec{d}_n. \tag{30.15}$$

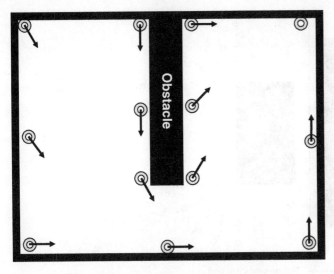

⊚ WSN Mote   � Suggested direction

Figure 30.24   WSN-based field navigation concept.

Using this approach, we can use the WSN to navigate around larger, static obstacles and for general navigation. In open areas, the WSN alone is used as described for navigation at the robot's maximum speeds. However, by using the RSSI of certain nodes, the robot can detect that it is approaching an obstacle, and can reduce its speed to improve maneuverability and reduce the risk of collision. Accordingly, RFID tags are placed at key areas such as around obstacles or doorways in one of two arrangements, strips or radials, as shown in Figure 30.25. Each tag is linked to a direction that is normal to the nearest obstacle, which is stored in a DB on the robot. When a tag is read, which implies the robot is near an obstacle, the two angles perpendicular to the RFID tag's direction are calculated, and the robot moves in the direction closest to the WSN's imperative. The resulting direction of movement of robots correspond to:

$$
\theta_{fusion} = \begin{cases} \theta_{tag} + 90° : \theta_{tag} < \theta_{wsn} \leq \theta_{tag} + 180° \\ \theta_{tag} - 90° : \theta_{tag} \geq \theta_{wsn} > \theta_{tag} - 180° \end{cases} \tag{30.16}
$$

where θ_{tag} is the currently read tag's direction from the DB, and θ_{wsn} is the WSN's directional imperative.

This system was tested under two conditions. In type A, the robot navigated a straight passage that was relatively narrow. In type B, the robot made several tight consecutive turns. For both types, the experiment was run with and without RFID fusion. For tests type A and B, 101 and 104 tags were used, respectively. For both, eight WSN nodes were used. The RFID antenna used was 20 cm in diameter with an interval of 15 cm between tags. The type A layout had all of the tags in a straight line, pointing in the same direction. For type B, a quarter circle at both the entrance and exit point between each obstacle was used. The prototype robot Chamuko, shown in Figure 30.26, was used.

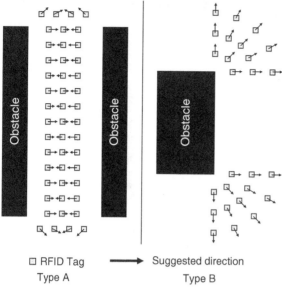

□ RFID Tag   ➝   Suggested direction

Type A   Type B

Figure 30.25   Type A and B experiment layout.

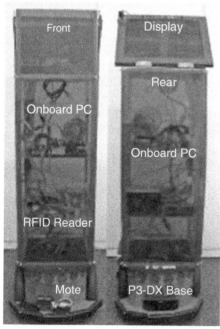

Figure 30.26   Chamuko, the robot used for testing.

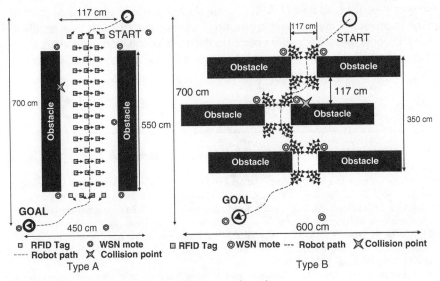

Figure 30.27   Type A and B experiment example path.

Built on a MobileRobots Pioneer 3-DX base, it was equipped with an antenna attached to the underside of the chassis, and a WSN node was mounted on the front bumper. An internal compass was utilized for pose estimation. Both types A and B experiments were conducted with and without RFID support. When using the WSN navigation system only, it was difficult for the robot to avoid obstacles due to a lack of precision. However, when tested with RFID support, the robot could reach the goal without colliding with the obstacles. Figure 30.27 shows examples of paths taken by the robot in the two types of experiments. A dotted line represents the path of the robot and the X shows where the robot would normally collide with the obstacle when not utilizing RFID support.

30.5   CONCLUSIONS

Mobile robots are becoming intelligent, autonomous systems, gradually becoming able to accomplish assigned tasks such as guidance, transportation, and human–robot cooperation. For these tasks, a robust and stable navigation system is a key requirement. In this chapter, an autonomous navigation system using only passive RFID was presented. Utilizing the location information of the previous and currently detected RFID tags, pose estimation can be performed without the aid of additional internal or external sensors. In order to further refine the localization, the use of read time, the total time a given tag is detected, was described. In the presented experimental results, the localization error was reduced to less than one-fifth of the scanning range of the antenna. The concept of read time was further extended for use in single and multiple static obstacle avoidance. For single static obstacle avoidance, directional information was added to RFID tags surrounding obstacles. In the case of multiple static obstacles, the read time of RFID tags in a specific region was used to provide precise orientation. Finally, some applications were discussed, such as a handheld device that could be used to easily update tag database information in the event static obstacles are repositioned, and a wireless sensor network fusion system, which can provide speedy and precise navigation.

REFERENCES

[1] D. C. Wyld, "RFID 101: The next big thing in management," *Manag. Res. News*, vol. 29, no. 4, pp. 154–173, 2006.

[2] S. Park, R. Saegusa, and S. Hashimoto, "Autonomous navigation of a mobile robot based on passive RFID," *IEICE Trans. Fundam. Electron. Commun. Comput. Sci.*, vol. J90-A, pp. 901–909, 2007.

[3] S. Han, H. S. Lim, and J. M. Lee, "An efficient localization scheme for a differential-driving mobile robot based on RFID system," *IEEE Trans. Ind. Electron.*, vol. 54, no. 6, pp. 3362–3369, 2007.

[4] M. Kotaka, M. Niwa, Y. Sakamoto, M. Otake, Y. Kanemori, and S. Sugano, "Pose estimation of a mobile robot on a lattice of RFID tags," in *Proc. of the 2008 IEEE/RSJ Int. Conf. on Intelligent Robots and Systems*, 2008, pp. 1385–1390.

[5] H. J. Lee and M. C. Lee, "Localization of mobile robot based on radio frequency identification devices," in *SICE-ICASE Int. Joint Conf.*, 2006, pp. 5934–5939.

[6] S. Park and S. Hashimoto, "Autonomous mobile robot navigation using passive RFID in indoor environment," *IEEE Trans. Ind. Electron.*, vol. 56, no. 67, pp. 2366–2373, 2009.

[7] E. Guillermo, S. Park, and S. Hashimoto, "Wireless sensor network-based mobile robot navigation with RFID path refinement," presented at The 27th Annual Conference of the Robotics Society of Japan, RSJ20093Q2-04, Kanagawa, Japan, Sep. 2009.

CHAPTER 31

VISIBLE LIGHT-BASED COMMUNICATION AND LOCALIZATION

Lisandro Lovisolo, Michel P. Tcheou & Flávio R. Ávila

PROSAICO - Signal Processing, Applied Intelligence and Communications Lab. Rio de Janeiro State University (UERJ)

SOLID-STATE DEVICE light-emitting diodes (LEDs) are rapidly being embedded in different commonly used devices, as are photodiodes and image sensors. Furthermore, illumination using LED reduces power consumption and motivates decorative illumination design. Because LED-based light sources are widely available, there are several options for visible light communication and positioning systems. This chapter provides an overview of the technologies that have paved the way for visible light communication and positioning systems.

31.1 INTRODUCTION

There are many real-time location-based applications. For outdoors, most of them rely on GNSSs (global navigation satellite systems). For indoors, other technologies need to be employed, because it is difficult for GNSS-born signals to penetrate walls and ceilings. In this case, techniques using different RF signals must be used, as we discuss in several chapters in this book. An example are the dead-reckoning techniques discussed in Chapter 29 that use embedded sensors (accelerometers and inertial sensors).

Positioning systems have diverse coverage and accuracy. For example, global coverage is provided by GNSSs; wide-area service is attainable using mobile network signals; Wi-Fi networks are used for small to campus-sized positioning applications; and Bluetooth, ultrawide band (UWB) and radio frequency identification (RFID) standards are used to deploy short-range positioning systems. There is an increasing demand for self-owned short-range positioning systems. For example, it may be of interest to know where an asset is located within a building. In a museum,

Handbook of Position Location: Theory, Practice, and Advances, Second Edition.
Edited by S. A. (Reza) Zekavat and R. Michael Buehrer.
© 2019 by the Institute of Electrical and Electronics Engineers, Inc.
Published 2019 by John Wiley & Sons, Inc.
Companion Website: www.wiley.com/go/zekavat/positionlocation2e

context-dependent (positioning- and user-dependent) data may be provided to enrich the visitor's experience. In shopping centers and markets, it may be used to advertise products and offer discounts to the consumer [86]. Industrial robots may also benefit from high-accuracy indoor positioning systems (IPS). Visible light positioning (VLP) is expected to be a key technology for IPS and to leverage high-accuracy location-dependent data delivery at high data rates and at great availability.

Positioning systems using light are emerging due to the large quantity of proposals and standardization activities for light-based communication systems. These in turn derive from the staggering use of the nonlicensed RF bands (such as the Wi-Fi and Bluetooth bands) and the developments in LED, PDs (photodiodes), and image sensors. The light within a room does not interfere with the light in another room, easing reuse of the VLP system and the medium for indoor applications. The dual wave-particle nature of light (very short wavelength electromagnetic wave/photon) makes possible the research and development of very high accuracy VLP systems.

Several techniques for VLP systems have already been designed and researched [6, 12, 20, 47, 50]. In this chapter, we will focus on VLP systems using information arriving at light detectors from specific light sources and also on how the arrangements of different light detectors can improve VLP system accuracy and usability. Some of these VLP systems require light to convey information—i.e., they use light as a communication channel. Many techniques to use visible light for communications have also been developed. This has been generically called visible light communications (VLC) [11, 17, 36, 60].

31.1.1 VLC Systems

The very first example of a visible light communication system is the use of flags of different colors to share information between towns or troops. Another is the use of lighthouses that help ship locate the coast (such as the famous one in Alexandria). While the use of flags requires sunlight, lighthouses can be seen in the dark. Traffic semaphores are a modern-day example using a similar approach. Yet another example of light-based communications are systems that encode information by turning on and off lights to send messages between towers or boats by different military forces. These examples can all be categorized as optical wireless communications (OWC): wireless communications by means of optical technology—regardless of the age of the technology. Currently, OWC is understood to encompass mainly infrared (IR) communications for short range and free-space optics (FSO) communications for long range.

VLC systems resemble the aforementioned ones because visible light sources are used to transmit information. The signal source is a light emitter, the transmission medium is the air, and the receiver is an appropriate detector, a PD. Different communication techniques (modulation, channel coding, equalization, etc.) have been adopted, designed and properly put to work for VLC [11, 17, 25, 60].

Besides the aforementioned historical (and now rudimentary) techniques, optical communications have been around for some time. It is said that Alexander Graham Bell experimented with a "photophone" [9, 11, 63]. Fiber optic links proliferate as heavy traffic communication infrastructure [84]. Meanwhile, IR remote controllers are employed by almost everyone on a daily basis to control different types of equipment and devices. While fiber optics guide the light from the emitter to the detector, IR control uses the air as a channel, causing some remote commands to fail if they are not correctly aligned or if conditions are foggy.

Since the 1970s, lasers, usually at IR wavelengths, have been used to send signals between buildings—an FSO communication link. FSO assumes an optical propagation path—i.e, ray-tracing, since the carrier has a visible light or near visible wavelength. The FSO channel degradations are more severe than for IR remote control, since weather and some other atmospheric aspects come into play [11, 17]. These examples prove the relevance of light-based communication technologies and their role in the electronics and communications devices market. In the development of light-based communication systems, the IEEE 802.11 has been the most common standard for WLAN (wireless local area networks). Since its birth, the IEEE 802.11 has standardized the use of IR communication links between devices. The so-called IEEE 802.11 (legacy mode) or IEEE 802.11-1997 or even the IEEE 802.11-1999 produced standards about links of 1 and 2 megabits per second (Mbps) using IR signals. However, although IR remains in the IEEE 802.11 standard scope, it is not currently implemented.

Expectations: With the increasing adoption of LED-based illumination, the possibility of using LED-based light bulbs for providing the last information hop has attracted attention. It is expected that LED lights will replace incandescent and fluorescent lights because they last longer, are mercury free, may mix different colors more easily, and switch faster, among other features. For example, fast switching makes it possible to modulate LED-based light sources to encode information, embedding VLC transmitters in illumination devices (the term "light bulb" may seem inappropriate for VLC, as it is in general associated with the incandescent lamp). This leads to a reduction of infrastructure cost, since the same infrastructure may be used for light ambiance and communication. Even if the VLC technology employed requires the light from the emitter to the detector to be free of obstacles (i.e., a free-space link), this condition is not hard to obtain in some scenarios, mainly in buildings that have a large set of illuminant sources, such as commercial stores, shopping centers, markets, and schools. Nevertheless, this two-purpose illumination requires the information to be transmitted by the illuminant. For this purpose, PLC (power line communications), using the power cables to both energize and deliver the data to the light bulb (somewhat similar to power over Ethernet technology), might be employed.

Another not-to-be-overstated push for VLC systems comes from the increasing interference among Wi-Fi networks. The ISM (instrumentation, scientific, and medical) band employed by these networks can be used without restriction, except for power limitations. Therefore, in office buildings, commerce centers, and even in houses, such systems may severely interfere with each other, hampering and even precluding communication. The polluted and overpopulated RF bandwidth is not an issue in VLC. This saves RF bandwidth, reducing its usage and congestion and saving it for other uses. Obviously, other RF bandwidth could be used instead.

However, there are not a lot of vacancies left in RF spectrum that can be used for new services. Neither governments nor communications operators are willing to free the RF spectrum for usage without a financial counterpart. In contrast, light spectrum usage is currently free of charge. Its spectrum cannot be commercialized (at least in principle) for the provision of communication systems since the visible spectrum is already associated with vision—one of our senses.

A not-to-be-underestimated feature of VLC is that light is assumed to be safe. While there is a large debate about the possible health issues of nonionizing radiation on human bodies and other life forms, light is considered to be safe (considering here the use of incoherent light, as opposed to lasers, which may harm people); when it is too strong, people are capable of feeling the danger by means of its heat. In addition, as it does not interfere with other electromagnetic equipment, it has a path to be used in environments heavily dependent on electronic equipment, such as hospitals.

LiFi: The acronym LiFi comes from li(ght) fi(delity) and is the term used for marketing hot-spot technologies for providing the last communication hop using light. It simply refers to the use of a VLC system to implement the last hop between a host and its serving network. The bet is that the illumination devices will provide this downstream link in the near future. For the upstream link, either some kind of specialized light system or the existing WLAN could be used. This is mainly motivated by the fact that in general users tend to consume a larger rate in the downstream than in the upstream. The probability of this scenario increases because wireless devices are in general used much more indoors and in transportation vehicles than outdoors. In the two first arrangements, illumination is commonly present, and its use for communication may have some advantages. Although not yet standardized (and we cannot predict if there will actually be a LiFi standard), there are already several applications designed to benefit from the depicted LiFi scenario.

Existing devices have embedded cameras that can already be used to provide low-data-rate VLC at the expenses of high energy consumption. However, for high-rate ones (the LiFi paradigm) additional photodetectors with the necessary optical setup and signal processing capabilities need to be embedded on host devices (smart-phones, computers, etc.). The opacity of walls and objects to light provides certain antipoaching capabilities. Nevertheless, instead of using one Wi-Fi hot spot for one house, one LiFi hot spot will be required in each room. Note that if the device is in a bag or in a pocket, LiFi cannot be used at all.

Figure 31.1 illustrates the LiFi scenario. For the LiFi upstream, there are several options on the table: IR, specialized VLC back-channel (also LiFi), and even Wi-Fi. Figure 31.1 illustrates some of these cases. The LiFi scenario is appropriate for broadcast. Possibly LiFi will not preclude Wi-Fi but will instead provide a large downstream bandwidth, which is, for example, well suited to the high data demands of video streaming. However, this scenario will require lights to be on during transmission, which might not be appropriate for video sessions. It is said that we are on the verge of so-called heterogeneous networks, in which multiple and different media and devices use the available communications resources opportunistically so that an acceptable performance (and ideally the best possible one) is achieved, and LiFi may contribute to this.

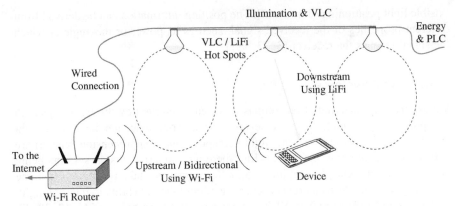

Figure 31.1   Possible LiFi scenarios. The depicted scenario considers the possibility of using different transmission channels for downstream and upstream.

The light within a room does not interfere with the light in another room. Therefore, within a room, if the wireless link between the "access point" (possibly implemented by illumination systems) and the "hosts" connected to it is light-based, this means that the interference is null among rooms. It is somewhat ironic that something treated as an advantage for RF-based communications can now be considered exactly one of its disadvantages. Besides a high reuse of the resource, the confined nature of light also brings some kind of security, as any eavesdropper will necessarily need to be located in the room.

Despite the disruptive and appealing nature of VLC systems, large-scale commercialization of VLC-capable devices will depend on the development of robust and efficient engineering solutions and applications, along with the execution of incremental commercialization strategies [29]. Without the objective of exhausting the subject, we may mention some creative and interesting proposals and applications that can be classified within VLC. For example, the authors of [70] push the vision of the Internet of Things by proposing Linux-based VLC devices. This principle can be incorporated into consumer LED light bulbs, such as in [71]. Another example is the use [8] of VLC by means of LED and smartphones for toy networking. Some bet on a large-scale use of VLC to allow vehicles to communicate, for example, to transmit relevant security information such as sudden breaks, reducing the chance of collisions, and also including the provision of the relative position of vehicles [69]. The healthcare industry may benefit from VLC to deliver bandwidth inside hospitals (light does not interfere with the equipment), although some places in hospitals cannot have illuminated lights all the time. Airlines use the same "noninterfering-over-RF-channel" argument to advocate the use of a VLC standard to provide data services without interference concerns while avoiding employing RF signals.

There is a large group advocating location-dependent data delivery for the first applications in scale. The in-building asset tracking, the shopper announcements, and museum and gallery IPS service discussed above largely benefit from using VLC links. Likewise, VLC-based positioning is a very active field, falling within the VLP (visible light positioning) domain [3]. The position information can be derived from different properties of the received signal, such as its power or the angle at which the signal reaches the receiver [3, 33, 111].

31.1.2   VLP Systems

We use the light arriving at our retinas reflected or emitted by different objects to interpret the space and to locate ourselves. Although our understanding of both light and vision has evolved over time, light perception and comprehension (vision) are still of utmost importance for daily life; there are numerous vision systems being developed to infer and abstract information from images, and navigation is just one of the applications. As a natural consequence of our inherent ability to interpret light, together with VLC techniques, VLP systems have also been proposed [6, 12, 20, 47, 50]. VLP technologies aim at localizing devices using information generated by specially designed light sources.

LEDs are rapidly replacing older lighting technologies inside buildings, and there are many light sensors embedded in portable devices such as smartphones, smart watches, and tablets. These facts jointly create a fruitful market niche for the development of pervasive, secure, and low-cost VLC systems for localization purposes. For indoor scenarios, Wi-Fi is the popular option since there are many access points, providing positioning accuracy up to a few meters with reasonable cost and complexity [102]. VLP arises as a competitive alternative since it presents higher accuracy by a few centimeters [95]. Besides, it can be safely used in areas where RF interference is undesirable, such as hospitals, mines, gas stations, and aircraft. In addition, the multipath effects are negligible in visible light propagation, thus reducing the complexity at the receiver. It is also cost-effective because it can be installed in existing illumination systems with few changes.

Using Light for Positioning: A VLP system uses light sources as beacons for positioning purposes. An indoor scenario is depicted in Figure 31.2. The beacons can be VLC access points or specially deployed LEDs or even common incandescent and fluorescent bulbs. The different colors employed for the beacons in Figure 31.2 indicate that a multiplexing technique must be incorporated in order to allow the device to detect and extract position-related data from the different beacons. The device to be located must incorporate at least a PD for detecting the light from the different beacons, although there are several proposals that involve embedding multiple PDs or using image sensors. We will focus on the use of LED-born beacons and IPS, although many of the concepts can be applied to VLP systems using other light beacons and for outdoor scenarios.

Regarding VLP techniques, we will discuss the case in which multiple references (light sources) may exist and a sensor captures information from them. This information is then employed to compute a position fix. Nevertheless, many of the discussed techniques with the proper adaptions can be used the other way around: one emitter (reference signal) and multiple receivers/detectors capturing the signal to compute the position fix.

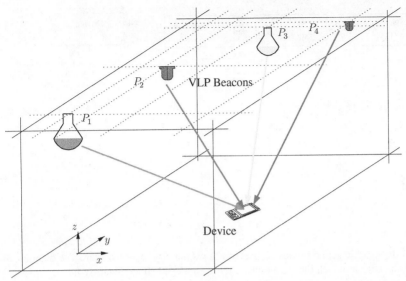

Figure 31.2   Each beacon or anchor is located at a different point and, although not shown, they might even be at different heights. In this figure, different colors are used in order to allow the detector to separate the signals from each source; in practice, other multiplexing techniques could be used instead.

31.2   VISIBLE LIGHT SYSTEMS BASICS

Figure 31.3 presents a general block diagram for visible light-based systems [39]. The transmitter is considered to act simultaneously as a lighting device and a communication interface (for communications or positioning purposes); while the color and intensity of the source are set by the block "Illumination control," the block "Modulation and Coding" provides the signal to be transmitted using the light as carrier; therefore, the driving circuit must be capable of mixing both without impacting the desired light configuration such that the possible variations in the light source (flickers) are not perceived by people in the room. Conversely, the design of the data modulation scheme may also consider this fact about the human eye.

The transmitter is assumed to be connected to a network via Ethernet, WLAN or PLC ("Physical Link") with the appropriate network protocol ("Network Interface"). The transmitter and receiver use the same modulation and coding schemes ("Modulation and Coding" and "Clock and Data Recovery"). Light emission and detection have specific issues that shall be considered: the color or, more precisely, the spectral profile of the emission device that shall be somehow matched by the detector, which may require an optical filter and "lenses" being used to diffuse or to concentrate light at the emitter and detector.

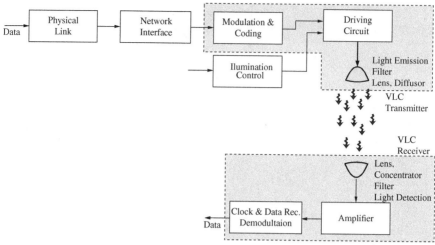

Figure 31.3 A basic general block diagram for a visible light-based system. The diagram considers the case in which the light transmitter is employed simultaneously for illumination and communications or positioning.

31.2.1 Light as Carrier

The visible light spectrum ranges from 380 nm to 750 nm, as illustrated in Figure 31.4. The visible light spectrum is hard to use in some scenarios, and we dispose of neither light sources that can be controlled to emit in this entire bandwidth (or some of its specific regions) nor detectors that work properly on some specific frequencies. Consequently, communications using visible light data rates are highly dependent on the emitter, detector, and modulation scheme employed.

Incoherent Nature of Light: Visible light communication systems use the incoherent light emitted by LEDs (FSO systems may use coherent laser-based light sources instead). The incoherent nature of light limits what can be manipulated, restricting modulation strategies. For example, while laser sources may have both phase and amplitude adjusted, incoherent light can only have its amplitude or intensity modified.

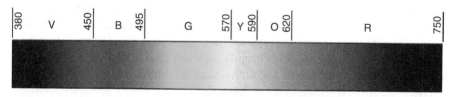

Figure 31.4 The visible light spectrum. The numbers refer to the wavelength $\lambda = c/f$ of the colors in nanometers. V stands for violet, B for blue, G for green, Y for yellow, O for orange, and R for red.

Bandwidth: Current technology provides LEDs that can be switched off and on very fast; even simple devices can offer a bandwidth of 20 MHz[*]; nevertheless, an appropriate detector is required.

Modulation Schemes: The modulation schemes should aim at high data rates while constrained by requirements of perceived light—that is, dimming support and flicker mitigation. Communication should not be affected when one arbitrarily chooses the local illuminant level considering different types of activities. For instance, office building areas such as single offices and conference rooms need higher illumination levels when compared to circulation areas like corridors and stairs [68]. Communication should also not imply perceivable fluctuations in light brightness since such flickering may provoke physiological changes that lead to headaches and dizziness[†]. All these constraints need to be considered when choosing the modulation scheme, which will be further discussed in Section 31.2.8. We follow this by briefly describing some relevant characteristics of LEDs and photodetectors in Sections 31.2.2 and 31.2.2, respectively.

31.2.2 Transmitter and Receiver for Visible Light Carriers

When using visible light for sending and receiving information, the transmitter is in general a LED lamp or a set of them, and the receiver is some kind of photodetector or a collection of them.

LEDs: An LED is a device that relies on electroluminescence to generate light in a semiconductor. In a forwarded-biased junction, minority carriers are injected into a region of the device to recombine with majority carriers and, as a result, recombination radiation is emitted. More prosaically, free electrons in the conduction band recombine with the holes in the valence band, releasing energy in the form of light. A distinction between direct and indirect gap materials has to be made in order to have a considerable light emission. Diodes made from indirect gap materials such as Si or Ge, when forwarded-biased, release energy as heat and are not suited to light emission. Conversely, direct gap materials such as GaAs, GaP, InP, and mixed alloys of these compounds emit light and are then suited to an optical communication system.

LEDs used for general illumination are made of a triple-chip RGB or phosphor-based white LED. The phosphor-based LED is cheaper and is more attractive for general illumination, while the RGB type has the potential for three different communication channels, one for each color.

Another distinction exists between homojunction and heterojunction LEDs. The major difference is that light emitted from a homojunction LED is back absorbed by the host substrate, reducing device efficiency. The active region of heterojunction LEDs is made of a narrow gap material in between two wide gap material—for example, AlGaAs/GaAs/AlGaAs and InAlAs/InGaAs/InP. Photons generated in the

---

[*]Just for comparison purposes, this corresponds to three times the bandwidth of a 3G channel, and it's comparable to that of 4G (LTE-A) and Wi-Fi channels.
[†]People suffering from photoepilepsy and migraines are more sensitive to these fluctuations in brightness [35]

narrow gap material do not have sufficient energy for being reabsorbed by the wide gap material, and light can be redirected to improve device efficiency.

White LEDs: There are several WLED (white LED) technologies that produce white light through inorganic and organic sources, as well as phosphors [62, 73]. WLEDS can be made of di-, tri-, or tetrachromatic sources in a LED-only or LED-plus-phosphor arrangement. In the first, the white light is produced by mixing the emissions of LEDs with different colors. In the latter, a phosphor material is excited by a monochromatic light from a blue or UV LED (ultraviolet LED) and consequently emits light having a broader spectrum. Thus, the technologies for WLEDs are divided into [73]:

1. LED-only:

   - Mix of emissions from blue and yellow LEDs (dichromatic);
   - Mix of emissions from red, green, and blues LEDs—RGB LEDs (trichromatic);
   - Mix of emissions from blue, cyan, green, and red LEDS (tetrachromatic);

2. LED-plus-phosphor:

   - Use of a blue LED to excite a yellow phosphor (dichromatic);
   - Use of near-UV or UV LED to excite triphosphor—red, green, and blue (trichromatic);
   - Use of blue and red LEDs to excite cyan and green phosphors (tetrachromatic).

The RGB LED is used for backlighting in displays, although it is not well suited for illumination due to gaps in spectrum. The UV LED produces narrow band light exciting an RGB phosphor, which in turn emits light having a larger bandwidth. The blue LED plus yellow phosphor approach is more efficient and also more commonly used for LED lamps when compared to those two approaches. The LED-plus-phosphor technologies use short-wavelength-emitting LEDs to excite a phosphor. The short-wavelength light is absorbed by the phosphor, which in turn emits light in a larger spectrum. Thus, many wavelengths are observed as emitted by the "encapsulated device" (a WLED). Figure 31.5 illustrates the typical spectrum of LEDs: red (wavelength around 625 nm), green (wavelength around 525 nm) and blue (wavelength around 470 nm), as well as other WLED technologies. The presented graphs are loosely based on typical LED spectra and descriptions from [40, 52, 60, 62, 75, 105]. RGB-LED-based WLEDs have the combined spectral curves for blue, yellow-green, and high-brightness red solid-state semiconductor LEDs and their FWHM (full width at half maximum)* spectral bandwidth is approximately 24–27 nm for all three colors. As alternative technologies, we have the mostly blue light directly emitted by the GaN-based LED having a peak at about 465 nm, and the more broadband Stokes-shifted light emitted by the Cerium-doped YAG phosphor, roughly in the 500–700 nm wavelength range.

---

*FWHM expresses the "width" of a function as the difference between the two extreme values of the independent variable at which the function equals half its maximum. It is the width of a spectrum curve measured as the distance between the *x*-axis points at which the function attains the half of the maximum amplitude.

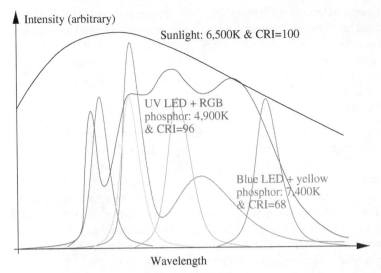

Figure 31.5 Typical characteristics of WLED sources—a tentative representation of their color/spectral distribution. In addition, green, red, cyan, and blue LED emissions are also presented. The black line shows as reference the sunlight spectral distribution. The vertical axis is to be used only as a reference.

Besides bandwidth, there are other important metrics for assessing white light sources [62]:

1. Luminous efficacy: it quantifies the energy efficiency of a white light source through the luminous flux—that is, the ratio of the produced visual sensation (in lumens) to the electrical power required to produce the light (in watts).

2. Color temperature: the white light produced by WLEDs may be classified as warm, neutral, or cold with respect to the white light emitted by an ideal blackbody radiator such as the sun.

3. Color-rendering index (CRI): it measures the ability of a light source to render the true colors of an object as if it were illuminated directly by the sun. In this case sunlight and incandescent bulbs have an ideal value of 100. Values above 80 are considered sufficient for indoor lighting, whereas lower values are acceptable for outdoor lighting.

Clearly, there is a compromise between color rendering and luminous efficacy of radiation. The blue-phosphor LED presents high efficacy but low CRI values; the UV-phosphor LED has low efficacy with high CRI values, and the RGB LED achieves medium efficacy with high CRI values [62]. In short, dichromatic sources can achieve high efficacy (as high as 425 lm/W) but poor color rendering. Tetrachromatic sources have exceptional color-rendering properties but lower efficacy than dichromatic and trichromatic sources. Trichromatic sources possesses the best compromise between luminous efficacy (as high as 320 lm/W) and color rendering capacity [73].

Lightning × Communication: The advantage of RGB LED mixing is that it is a well-known technology and it allows for a very fast switching rate. However, it may present some balancing difficulties that limit its use and cost. "Blue chip + phosphor" technology is the most popular for the lightening industry and has a lower cost, although it is slower than its predecessors (it has longer rise/fall times due to its process of absorption and re emission–a "two stage emission process," which indeed slows down the possible switching rate) and its switch may provoke color shift. OLEDs (organic LEDs) are an emerging technology that may surpass previous technologies, but they are still in development.

Power × Current: The power versus current characteristic of an LED can be understood as the amplification transfer function of the LED. It is often estimated following the Shockley model. For an asymmetric junction, the current injection efficiency is maximized—for example, with a high doping p-type substrate and a low doping n-type surface. With the definition of radiative efficiency (optical power generated by the LED with radiative and non radiative recombination divided by the optical power as if only radiative recombination were taken into account) and the transmission efficiency for normal incidence (photons are not extracted from the device in the air-semiconductor interface due to internal reflection), the output power is

$$P_{out} = E_G \eta_x I_D,$$ (31.1)

where E_G is the gap energy of the material in eV, I_D is the total diode current, and η_x is the total LED external efficiency, which is the multiplication of injection efficiency, radiative efficiency, and transmission efficiency.

The slope efficiency of the LED is

$$\frac{dP_{out}}{dI_D} = E_G \eta_x.$$ (31.2)

This relation is linear, which is an expected result for low power LEDs—i.e., the output is proportional to the current. For high-power LEDs, the power vs. current curve saturates at high current and power, and the linear relation no longer holds due to a reduction in the non radiative lifetimes—thermal lifetime (device heating) and Auger recombination (high injection current). High-brightness applications usually use AlGaInP and GaInN[*] LEDs and a more careful analysis needs to be done.

Modulation Response: Asymmetric homojunction LED bandwidths can be modeled by a charge-control approach. By estimating the excess electron charge in the P side and relating it to the current needed to replace charge recombination over time, we can evaluate the LED modulation response [1]:

$$P_{out} = \eta_r \frac{\hbar\omega}{q} \frac{\tau_n}{\tau_{n,r}} \frac{1}{1 + j\omega_m \tau_n} I_D,$$ (31.3)

where η_r is the transmission efficiency, $\hbar\omega$ is the photon energy, q is the charge, $\frac{\hbar\omega}{q}$ is the gap energy, ω_m is the modulation frequency, τ_n is the total lifetime, $\tau_{n,r}$ is the total lifetime associated with radiative recombination, and $\frac{\tau_n}{\tau_{n,r}}$ is the radiative efficiency.

---

[*]The 2014 Nobel prize in Physics was awarded to Isamu Akasaki, Hiroshi Amano, and Shuji Nakamura for the "invention of efficient blue light-emitting diodes which has enabled bright and energy-saving white light sources."

Bandwidth: Equation (31.3) is a low-pass transfer function with cutoff frequency related to the carrier total lifetime [1]:

$$f_{3dB} = \frac{\sqrt{3}}{2\pi\tau_n}. \tag{31.4}$$

The radiative lifetime is inversely proportional to the carrier injection, so the modulation bandwidth is bias dependent. The asymptotic value for the total radiative lifetime is τ_0 (spontaneous radiative lifetime). Assuming unitary radiative efficiency, we can relate τ_n to the total device current to obtain a relation for the cutoff frequency with the current

$$f_{3dB} \propto \sqrt{I_D}. \tag{31.5}$$

The cutoff frequency f_{3dB} increases with the square root of the current, with a limit value of $1/2\pi\tau_0$. The limiting value for τ_0 is material dependent. GaAs has a maximum modulation bandwidth of ~300 MHz. Heterojunction LEDs can be analyzed in a similar fashion by using charge control analysis, and its dynamic regime gives the same result for its bandwidth limit.

Phosphor-based WLEDs have their bandwidth further limited due to the time constant associated with the phosphor coating emission. Therefore, solutions that circumvent this limitation need to be explored [38]. Overall, the data rates will be highly dependent on the LED modulation bandwidth. Obviously, when the emitter/transmitter has illumination purposes besides communications and positioning, there is much more freedom to design the receiver than the transmitter.

Photodetectors: The most common choice of receivers for visible light are photodetectors or PDs; these are semiconductor devices that convert the incoming light into a current, allowing for the retrieval of illumination variations and the demodulation the data they may be carrying. The basic physics mechanism to explain a photodetector involves the absorption of photons and the generation of carriers. In a PD, an electric field separates the electron-hole pair and generates current, also called photocurrent. A PD behaves as a photocontrolled current source in parallel with a semiconductor diode and is governed by the standard diode equation. As shown in Table 31.1, current commercial PDs can be used to sample the received visible light at rates of tens of MHz.

TABLE 31.1 Typical Photodetector Characteristics

Materials and Junction	Wavelength (nm)	Responsivity (A/W)	Dark Current (nA)	Response Time (ns)
Si PN	550–850	0.41–0.7	1–5	5–10
Si PIN	850–950	0.6–0.8	10	0.070
InGaAs PIN	1310–1550	0.85	0.5–1.0	0.005–5
InGaAs APD	1310–1550	0.80	30	0.100
Ge	1000–1500	0.70	1000	1–2

Other examples of receivers are the ones based on CCD (charged coupled device) or CMOS (complementary metal oxide semiconductor) image sensors; these are used in digital cameras and are currently embedded into many devices. The large availability of such sensors motivates image sensor-based visible light communications and positioning applications. An imaging sensor consists of many photodetectors (disposed in a matrix) aiming to enable high-resolution photography, and the number of photodetectors can be very high. Unfortunately, its physical construction significantly reduces the number of times per second that each pixel is read (actually, the unit is fps [frames per second]), which in smartphones is on the order of hundreds at best. Therefore, using a camera sensor as a receiver in VLC can provide a reliable but only low data rate.

When the PD is unbiased it is in the photovoltaic mode, meaning that in a dark environment the (dark) current is very low and a low response speed is obtained. Nevertheless, this mode can be used in scenarios where low speed is acceptable as far as low light levels can be detected. When an external reverse bias voltage is applied to the PD, it is said to be in the photoconductive mode. This increases the width of the depletion region and reduces the junction capacitance, which translates into an increased response speed. However, the reverse bias also increases the dark current, and there is a high noise current (due to the reverse saturation current). This mode can be used for high-speed applications under high light levels.

Several types of photodetectors can be considered as a detector for a VLC system. The PIN PD is characterized by having a large sensor area, which translates into a limited sensitivity, but at low cost. The avalanche PD has a higher sensitivity than the PIN PD due to its smaller area. This generates a high reverse bias, which translates into a higher circuitry cost. The main issue of using a single-element PD is that it cannot be used effectively in direct sunlight. In this scenario, image sensors come to play due to their ability to spatially separate sources. The image sensors mainly used are the common CCD and the array type. The first has a low implementation cost but a slow response due to it serial readout. The array type consists of pixels that can be read out in parallel, like parallel PDs; thus it is faster but also more expensive, although its mass-market deployment may revolutionize and boost VLC systems.

Responsivity: The responsivity R_p is the ratio between the amount of current generated and incident optical power. In photoconductors, it is measured in amperes per watt (A/W) while in photovoltaic it is measured in volts per watts (V/W). Responsivity depends on the wavelength of the incident light and on the material of which the detector is made. The band-gap of the semiconductor material is responsible for the spectral response obtained in different photodetectors. For VLC, the materials used are typically Si or InGaAs. The responsivity of a photocurrent detector is expressed as

$$R_p = \frac{\eta q \lambda}{hc},$$ (31.6)

where η denotes the quantum efficiency (in the conversion from photons to electrons) at the wavelength λ, q is the electron charge, and h is Planck's constant.

Linearity: When reverse-biased, the PD output is extremely linear with respect to the illuminance applied to the PD junction, making regular photodetectors to present a linear response as a function of the incident power.

Dark Current: The dark current I_d has a straightforward interpretation; even when no light hits the photodetector, a current is still present in the device. The dark current is related to the thermal generation of electron-hole pairs in the depletion region. For that reason, detectors have an increased sensitivity when operating at low temperatures. Defects in the semiconductor crystal structure also increase the dark current of a device.

Response Time: The response time of a photodetector limits the available bandwidth for signal transmission. For high-speed photodetectors to be used in VLC, the frequency response of the photodetector has to be limited by the transit time of the charge carriers. Thus, PDs are operated in photoconductive mode (reverse bias) and the dominant factor for the response time is the drift time.

Noise Equivalent Power: The noise-equivalent power (NEP) is a measure of detector sensitivity and it is expressed in $W/\sqrt{Hz}$. Assuming a one-hertz bandwidth[*], the NEP is the signal power generating a unitary signal-to-noise ratio (SNR). NEP is the noise spectral density divided by the responsivity. The smaller the NEP, the more sensitive the detector. The noise in the PD technologies discussed bellow (PN and PIN) can be modeled as shot noise originating by the discrete nature of the carriers and its dynamic fluctuation.

Specific Detectivity: The specific detectivity of a photodetector is the figure of merit used to compare photodetectors based on different materials with different sizes [96].

PN × PIN photodetectors: We compare the properties of two types of photodetectors: PN junction PDs and PIN PDs.[†] We compare them regarding linearity, speed/bandwidth, sensibility, and efficiency, along with noise and its physical dimension compromises.

The PN PD is made of a thin, highly p-doped layer on top of a bulk n-doped substrate. By illuminating the device, carriers that contribute to the photocurrent (e-h pairs) are generated in the depletion region and in the diffusion length of the carriers (electrons on the p-doped layer and holes in the n-doped layers). The frequency analysis of input optical power indicates that carrier lifetime affects the high-frequency response (giving typical frequencies of the order of 100–200 MHz [18]). Carriers generated in the diffusion length have a smaller contribution to the photocurrent with increased frequencies, because of their short lifetime in comparison to the transit time and RC time constant of readout.

A large reverse bias voltage would increase a PN PD depletion region but would also increase its dark current. A much larger depletion region would require

---

[*]This is equivalent to an integration of .5 seconds.

[†]The I stands for insulation, while the P and N refer to semiconductor doping.

an impractically large reverse voltage in a PN PD, but this can be achieved with a PIN PD. At high frequencies, contribution to the photocurrent from carriers generated in the diffusion length is reduced in comparison to the carriers generated in the long depletion region of a PIN PD. Using semiconductor materials like AlGaAs/ GaAs, InAlAs/InP, contributions outside the depletion region can be extremely reduced to negligible values, thus increasing the cutoff frequency. The speed is now limited by transit time and RC time constant. The frequency response bandwidth of such photodetectors can reach 10 GHz.

Usually, high responsivity on a PIN PD is preferable, leading to large areas and wide intrinsic region. However, increasing the area increases the capacitance, which in turn increases the RC time constant and decreases the high-frequency bandwidth. By making the intrinsic region wider, the RC time constant decreases (the intrinsic capacitance decreases), which would be beneficial. In contrast, the transit time increases, reducing the transit time-limited bandwidth. In the external quantum efficiency of the PIN PD, there is a trade-off between responsivity and frequency response. To have a better frequency response, a PIN PD has a small area with a small intrinsic region, but the responsivity is larger for large areas and wide intrinsic region. Thus, PIN PDs used for high speed would have a limited responsivity.

The response time of a photodetector limits the available bandwidth for signal transmission. Typical bandwidths go up to 200 MHz for PN photodetectors and up to 10 GHz for PIN photodetectors. To increase responsivity, larger detector regions are required. But increasing the responsivity, one decreases the RC time constant and increases the transit time, thus reducing the transit time-limited bandwidth. There is a trade-off between responsivity and frequency response, since lar=equency response requires smaller regions and responsivity is larger for larger areas. The higher the speed, the more limited the responsivity. Synthesizing some of the above PD designs, Table 31.1 shows typical ranges of the photodetector characteristics previously discussed for different semiconductors and junctions.*

31.2.3 The Visible Light Channel

The visible light channel is expected to be optical, meaning that the ray (signal) travels in straight line segments (reflections may occur). Nevertheless, the link between emitter and detector can significantly vary over time. First of all, the emitter and detector may be in a line of sight (LOS) or not. The ray may go from the emitter to the detector through reflection on surfaces. Another relevant link classification is related to the link being directed or nondirected. While LOS classification considers the existence or absence of an obstruction, the directed one considers exactly the direction of the link [39,67]. The non directed LOS channel is of utmost relevance for VLC. General illumination is neither focused nor directed (actually, it may be the exact opposite, diffuse) although it may be in LOS with the detector.

*APD stands for avalanche photodiode.

31.2.4 LOS Channel Model

Figure 31.6 presents a simple model for an indoor, non directed LOS channel. The received optical power can be modeled as [39]:

$$P = P_T S \frac{m+1}{2\pi d^2} \cos^m(\phi) T(\phi) G(\psi) \cos(\psi), \ 0 \leq \psi \leq \Psi_C, \tag{31.7}$$

where P_T is the transmitted power and S the detector area, ϕ the angle of irradiation, measured with respect to the normal to the plane in which the transmitter is located, and ψ is the incidence angle, measured similarly but at the receiver. $T(\phi)$ is the filter transmission while $G(\psi)$ is the receiver concentration. An optical filter may exist in the transmitter and a concentrator may be employed in the receiver to improve detection. Ψ_C is the concentrator field of view (FOV), also measured at half of the maximum concentration.

Above, parameter m is the order of the Lambertian emission, which depends on the emitter semi-angle $\Phi_{1/2}$ (the one giving half-power), i.e.,

$$m = -\frac{\ln(2)}{\ln(\Phi_{1/2})}. \tag{31.8}$$

The Lambertian is used to obtain the irradiation at angle ϕ:

$$I_\phi = \frac{m+1}{2\pi} \cos^m(\phi). \tag{31.9}$$

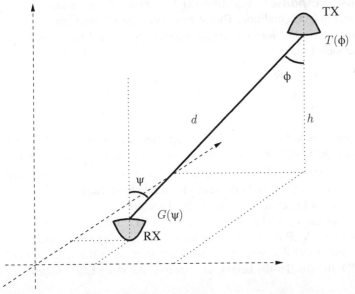

Figure 31.6 LOS non directed optical link model. ϕ is the irradiation angle, and ψ is the angle of incidence. Emitter and detector are at a distance d, and the transmitter is placed at h meters above the plane of the detector (or vice versa). If the emitting and detection planes are parallel, then these angles are equal, which further simplifies the model.

In doing that, Equation (31.7) simplifies into

$$P = \log P_T + \log I_\phi - 2\log d + \log S + \log T(\phi) + \log G(\psi) + \log \cos(\psi). \quad (31.10)$$

(Mis)alignment between the emitter (Tx) and the detector (Rx) may hamper the optical link. Fish-eye lenses can be placed on top of PDs so that the optical link may be established at a wide angle. Other alternatives include using multiple detectors in the receiver or even image sensors for detection [81].

Attenuation and Received Electrical Signal: In LOS conditions, assuming a generalized Lambertian LED with order m, the attenuation of the optical channel is given by [31]:

$$\alpha = \frac{(m+1)S}{2\pi d^2}\cos^m(\phi)\cos\psi = \frac{(m+1)S}{2\pi}\left(\frac{h^{m+1}}{d^{m+3}}\right), \quad (31.11)$$

meaning that the received signal after optical to electrical conversion will be

$$r(t) = \alpha R_p x(t-\tau) + n(t), \quad (31.12)$$

where $x(t)$ corresponds to the electrical signal to be converted for transmission using the optical link, $n(t)$ is the noise, $\tau = (d/c)$ is the time taken for light to travel from transmitter to receiver, and R_p is the PD responsivity; see Section 31.2.2. The angles ϕ and ψ required to compute the attenuation are shown in Figure 31.6.

Channel Impulse Response: Equation (31.12) shows the attenuation for a visible light channel in LOS condition. This attenuation is the channel impulse from an emitter to a detector. Therefore for the ith emitter, its channel response to the detector can be modeled using

$$h(t) = \begin{cases} \dfrac{(m+1)S}{2\pi d^2}\cos^m(\phi)\cos\psi\,\delta\left(t-\dfrac{d}{c}\right), & \psi \le \text{FOV} \\ 0, & \psi > \text{FOV} \end{cases}. \quad (31.13)$$

This impulse response makes the received signal an attenuated version of the transmitted signal with a delay of $\frac{d}{c}$ seconds. The attenuation is $\alpha = \frac{(m+1)S}{2\pi d^2}\cos^m(\phi)\cos\psi$ and depends on the distance and the relative arrangement between transmitter and receiver (see Fig. 31.6). In (31.13), FOV stands for the field of view, i.e., the angle in which the signal from the emitter arrives at the detector.

For example, assuming a VLP indoor scenario, a VLP IPS, it is required that in all regions where the VLP must operate there is light arriving from a sufficient quantity of LEDs for the VLP to work properly. Therefore, since the value of FOV depends on the PD, the smaller the height, in principle, the more LEDs required.

31.2.5   Non-Line-of-Sight Channel Model

In the case of an optical link by reflection—i.e., a non-line-of-sight (NLOS) channel—the surface and its physical properties are now of importance. Nevertheless, this can be grouped into a combined reflectance effect $\rho < 1$. The received power can be described as

$$
\begin{aligned}
P = \log P_T + \log I_\phi - 2\log(d_1 + d_2) + \log\rho + \log S \\
+ \log T(\phi) + \log G(\psi) + \log\cos(\psi), \text{ in dB,}
\end{aligned}
\tag{31.14}
$$

where d_1 is the length of the path from the transmitter to the reflective surface, and d_2 is the length from this surface to the receiver. Although NLOS visible light links may be relevant in some scenarios, most visible light systems assume LOS conditions. Therefore, in this chapter, unless otherwise stated, we assume LOS.

31.2.6 Multiple Wavelengths

Illumination devices must follow photometric criteria (in lx) instead of radiometric ones (in W). The optical power (in Watts) of a monochromatic light of wavelength λ is related to illuminance I by means of [39] (using Fig. 31.6):

$$
P_{rec} = \frac{I\cos^m(\phi)}{683d^2 V(\lambda)\cos(\psi)},
\tag{31.15}
$$

$V(\lambda)$ represents the eye sensitivity function [72],[*] the constant 683 converts from lx to Watt ($[lx] = 683\frac{lm}{[W]}V(\lambda)$, i.e., photometric to radiometric conversion).

As described in Section 31.2.2, LED illuminants have a multiple-wavelength spectra in the visible range (380–750 nm). As a result, the model for the received optical power in (31.7) must incorporate that fact. Therefore, any calculation of illuminance and received optical power must integrate over the light source wavelengths according to the eye sensitivity function. Assuming that the power spectral density of the light source is $P(\lambda)$, we obtain

$$
P_{rec} = \frac{I\cos^m(\phi)}{683d^2\cos(\psi)\int V(\lambda)P(\lambda)d\lambda},
\tag{31.16}
$$

where, for simplicity, $P(\lambda)$ accommodates the transmission filter and concentrator effects.

31.2.7 Optical Interference and Noise

The indoor channel is very well behaved. However, smoke, fog, and mist may severely hamper visible light links since they occlude and deviate light. Visible light systems are also subject to noise. Indoors, the main source of noise is the presence of other illumination devices, like incandescent and fluorescent lights, or even concurrent visible light systems. Such sources may even make it impossible to use the visible light to convey information depending on their power. In an outdoor scenario, sunlight appears as a great source of noise. In addition, thermal or shot noise is always present in any optoelectronic detector. In the detected signal, different noise contributions exist from different sources.

---

[*]Assuming that the receiver/detector is human, for example, $V(550nm) = 1$ (green) and $V(720nm) = .001$ (red).

Thermal Noise: Thermal noise exists over a resistance. In this case the Johnson thermal noise has a variance

$$\sigma_{\text{Th}}^2 = \frac{4kT_eFB}{R_L}, \tag{31.17}$$

where R_L is the load resistance, and F is the figure of merit of the system. The above expression is device specific (R_L and F); for example, assuming a PIN/FET (Field Effect Transistor) trans impedance receiver [36], we can derive a two-term expression for it composed of the feedback-resistor noise and FET channel noise given by

$$\sigma_{\text{Th}}^2 = \frac{8\pi\kappa T_e}{G} c_{\text{PD}} S\beta^2 I_2 + \frac{16\pi^2\kappa T_e\gamma}{g_m} c_{\text{PD}}^2 S^2\beta^3 I_3, \tag{31.18}$$

where κ is Boltzmann constant, T_e is the absolute temperature, G is the open-loop voltage gain, c_{PD} is the fixed capacitance of the photodetector per unit area, γ is the FET channel noise factor, β is the equivalent noise bandwidth, g_m is the FET trans conductance, and the constants are $I_2 = 0.562$ and $I_3 = 0.0868$.

Shot noise occurs due to different contributions: background radiation, dark current, and even from the received signal. Shot noise comes from the conversion of different radiation into current and the nature of electronic devices.

Shot Noise Due to Background Radiation: Considering that the ambient light (background radiation) is isotropic, its power at the detector is given by:

$$\sigma_{\text{BG}}^2 = 2qR_p S\delta\lambda p_n\beta, \tag{31.19}$$

where q is the electron charge, p_n is the background spectral irradiance [31], $\Delta\lambda$ is the bandwidth of the optical filter in front of the PD, and β is the effective bandwidth.*

Shot Noise Due to the Dark Current: The dark current also produces a shot noise, given by [49]:

$$\sigma_{\text{ID}}^2 = 2qI_D\beta. \tag{31.20}$$

Shot Noise Due to the Received Signal: The received signal also produces a shot noise, given by [49]:

$$\sigma_{\text{RS}}^2 = 2qR_p P_r\beta, \tag{31.21}$$

where P_r is the power of the received optical signal at the detector surface.

Summing all the contributions, the variance of the Gaussian noise detected by the PD is

$$\sigma_N^2 = \sigma_{\text{Th}}^2 + \sigma_{\text{BG}}^2 + \sigma_{\text{ID}}^2 + \sigma_{\text{RS}}^2. \tag{31.22}$$

---

*The effective signal bandwidth of a signal $x(t)$ is given by $\beta = \dfrac{\int_{-\infty}^{\infty} f^2|X(f)|df}{\int_{-\infty}^{\infty} f^2|X(f)|df}$, where $X(f)$ the Fourier transform of $x(t)$.

31.2.8   Modulation Techniques for Visible Light Systems

Conventional modulation techniques adopted in RF systems cannot be readily applied to visible light based systems, since for light one is not able to encode data by modifying the wave phase, frequency, and amplitude. Hence, the information has to be transmitted through intensity variation of the emitting light wave and recovered at the receiver by direct detection. This modulation framework is usually referred to as IM/DD (intensity modulated/direct detection).

Most practical visible light communication or positioning systems that are currently being considered employ IM/DD for outdoor as well as indoor applications. Atmospheric conditions considerably reduce light intensity, especially when light propagates through a dense fog. Instinctively, it appears that the best way to surpass high attenuation would be to employ more optical power or to concentrate and focus more power into smaller areas. However, eye safety regulations introduce a limitation on the amount of optical power being transmitted. For indoor applications, such regulation imposes an even more rigid limit on transmission optical power. Selecting the coding and modulation schemes is one of the fundamental decisions in the design of any communication system. The coding and modulation techniques may impact the data rates and the positioning accuracy of VLP as well.

Common Modulation Techniques: The most popular modulation methods are on–off keying (OOK), pulse position modulation (PPM), pulse width modulation (PWM), orthogonal frequency division multiplexing (OFDM), and color-shift keying (CSK). Moreover, there are myriad variants of each, as well as several combinations. A general description of these is given below.

OOK: OOK is the simplest technique, in which the intensity of an optical source is directly modulated by the information sequence. The data bits are transmitted by turning the LED on (bit 1) and off (bit 0). In fact, the off state corresponds to light intensity reduction, so the LED is not fully turned off. There are many proposed systems employing the OOK modulation scheme, especially the NRZ (non-return-to-zero) OOK, along with two types of WLEDs: the one combining a blue emitter with a yellow phosphor and the other combining RGB frequencies [15, 16, 59, 91, 92]. At the receiver, avalanche PD can be used instead of PIN PD [16]. The reported achievable data rate ranges from 10 to 614 Mbps. Although OOK is simple and ease to implement, it provides low data rates when dimming support is considered.

PWM: PWM offers an efficient way to modulate signals with dimming level variation. Information is conveyed through the width of the pulses, which are tuned according to the dimming level. Then, LED can operate with full brightness during the pulses. PWM has the advantage of providing dimming support without affecting the pulse intensity level, avoiding LED color shift. Unfortunately, PWM can only provide small data rates [80]. An alternative way to surpass this limitation relies on combining PWM along with discrete multitone (DMT), which enables simultaneous dimming control and communication. By using QAM (Quadrature Amplitude Modulation) on DMT subcarriers, 513 Mbps can be achieved [93].

Pulse Position Modulation: Pulse position modulation is an attractive modulation option for visible light LOS links, where the bandwidth constraint is not so stringent [17]. It presents better power efficiency with respect to OOK but at the expense of a worse spectral efficiency and greater complexity. In an *L*-PPM scheme, a symbol consists of a pulse of constant power occupying one time slot within $L = 2^M$ ($M \in \mathbb{N}$) possible time slots. The position of the pulse identifies the transmitted symbol. A number of modified PPM schemes have been proposed to improve bandwidth efficiency and resilience to multipath ISI effect in NLOS links [17, 55, 56].

CSK: Despite its name, CSK is an intensity modulation scheme that has attracted increased attention over recent years [13, 51, 74]. CSK is found in IEEE 802.15.7 [23], which transmits data imperceptibly through the variation of the color emitted by red, green, and blue LEDs (TriLEDs). In CSK, the power envelope of the transmitted signal is constant, reducing the potential for possible human health problems derived from fluctuations in light intensity. This modulation scheme is based on the *x-y* color coordinates, defined by the international commission on illumination, the CIE 1931 color space [7,14], shown in Figure 31.7.

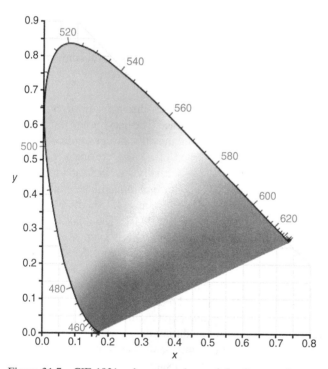

Figure 31.7   CIE 1931 color space chromaticity diagram. It represents all the colors visible to the human eye using the *x-y* chroma coordinates. The colorful region corresponds to the extent or gamut of human vision. The curved edge with wavelengths listed in nanometers is known as the monochromatic or spectral loci.

The IEEE 802.15.7 standard has proposed nine valid sets of the three potential sources for use in CSK, known as color band combinations [23]. In TriLED CSK, the constellation diagram has a triangular shape, the ABC triangle shown in Figure 31.7. The mixture of light produced from the LED sources allows CSK to regenerate various colors present in the constellation triangle and to represent these colors as data symbols. The symbols should be equally distributed in the triangle so that the combined light when transmitting different symbols does not produce human-perceivable color variation. IEEE 802.15.7 provides four-, eight-, and 16-point CSK constellations for communications at rates of up to 96 Mbps.

The use of QLEDS (blue, cyan, yellow, and red) enables quadrilateral constellation shape, allowing QAM-based constellation configuration [74]. This improves energy efficiency and ISI resilience. Approaches employing WDM (wavelength division multiplexing) along with TriLED promise to achieve data rates of 3.22 Gbps [90,98].

OFDM: As the name suggests, OFDM consists of a channelization scheme that can efficiently use the available bandwidth. It divides the bandwidth into orthogonal subcarriers, putting forward a great capacity for reducing intersymbol interference and multipath fading effect. In addition, it enables the use of low-complexity equalizers. The ease of interconnecting with other PHY layer (physical) systems (using very similar data frames) justifies using OFDM for channelization and modulation in visible light-based systems; for example, PLC, long term evolution (LTE), and digital TV standards all use OFDM for channelization and modulation.

When modulating light, unipolar real-valued signals must be used instead of complex-valued bipolar signals employed in RF communication systems. This may be accomplished by imposing Hermitian symmetry on the subcarriers, which ensures that the time domain symbol is real, and by adding a DC level to the resulting time-domain signals so that it is unipolar. Two commonly discussed OFDM variants for visible light are the asymmetrically-clipped optical (ACO-OFDM) [79,87] and the DC-biased optical OFDM (DCO-OFDM) [82, 83, 108].

Figure 31.8 illustrates a typical ACO-OFDM frame transmitting N (possibly complex) symbols $S_0, S_1, \ldots, S_{N-1}$. The frequency domain signal of length $4N$, after inserting zeroes and enforcing Hermitian symmetry, is defined as

$$\mathbf{X} = \begin{bmatrix} 0 & S_0 & 0 & \ldots & S_{N-1} & 0 & S_{N-1} & 0 & \ldots & S_0 & 0 \end{bmatrix}^T. \quad (31.23)$$

Time domain symbol $\mathbf{x}_{TD}$ of length $4N$ is generated as $\mathbf{x} = \mathbf{W}^T\mathbf{X}$, where $\mathbf{W}$ is the DFT (Discrete Fourier Transform) matrix. The clipped version of this signal $\mathbf{x}_{clip}$ is generated by simply mapping negative samples of $\mathbf{x}$ into zero. A cyclic prefix sequence is combined with this clipped signal, as is done in traditional OFDM systems, in order to make the linear channel matrix circulant, which is instrumental to simple frequency domain equalization. The inverse operations are carried out in the receiver and are not detailed here for space reasons.

Other variants of optical-OFDM (O-OFDM) exist, and all of them keep the time domain symbol real and nonnegative, as required in VLC. In DCO-OFDM, a DC bias is added to force the resulting signal to lie mostly in the positive range. In

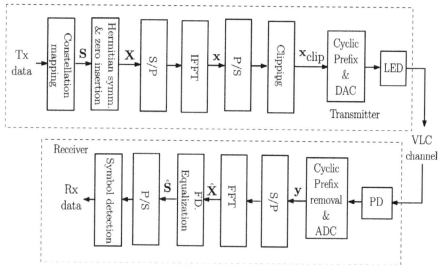

Figure 31.8 Structure of an ACO-OFDM visible light communication system. The schematic considers the use of different modulation constellations. Hermitian symmetry ensures real-valued signals, and clipping in the signal is introduced to avoid negative voltage inputing the LED. This scheme fits ACO-OFDM and DCO-OFDM visible light channelization and modulation with minor modifications.

ACO-OFDM, in contrast, the negative parts of the time domain signal are clipped at zero, and the positive ones are transmitted without modification; in order to avoid loss of information, only odd carriers are loaded with useful data and the even ones are loaded with zeroes.[*] In comparison, in ACO-OFDM, because of Hermitian symmetry and insertion of zeros, only one-fourth of the carriers transmit useful information. On the other hand, ACO-OFDM requires less power to achieve a given BER (Bit Error Rate) level than DCO-OFDM.[†]

Standardization Activities: The IEEE 802.15 WPAN (wireless and personal area network) group has formed the task group 7 (TG7) to study VLC [24]. In Japan, the Visible Light Communication Consortium (VLCC) was organized to promote technology exchange, system development, demonstration, and standardization of VLC inside Japan [89], and the JEITA (Japan Electronics and Information Technology Industries Association) has produced standards for VLC [26], for Light ID [27], and for light beacons [28]. In Europe, the working group 5 of the Wireless World Research Forum (WWRF) deals with VLC technology as one of next-generation wireless access technologies [99].

---

[*]It can be shown that the error caused by negative clipping will affect only the odd frequencies, which will be discarded in the receiver.

[†]For a more detailed comparison between ACO-OFDM and DCO-OFDM in an AWGN channel see [2].

31.3 VLP

The majority of the positioning techniques aforementioned in this book can and are already employed for designing VLP systems. VLP systems may use the identity (proximity) of light anchors, fingerprinting methods, ranging (distance estimates) for trilateration (time of arrival [TOA], time difference of arrival [TDOA], and received signal strength [RSS]) triangulation using angle measurements (angle of arrival [AOA]), and other methods based on image sensors and computer vision to provide a position fix. In the following sections, we discuss some of the pros and cons of these techniques for VLP. We present a geometrical analysis only for the image-sensor based VLP systems, since it is the only method requiring a profoundly different analysis than the others already presented and discussed in different chapters.

31.3.1 Multiple Light Sources and Beacons

As previously stated, depending on the positioning technique employed in the VLP system, measurements from more than one LED may be necessary. Let $x_i(t)$ be the signal transmitted from the ith LED; then the received signal at the detector can be modeled as

$$r(t) = R_p \sum_{i=1}^{I} h_i(t) * x_i(t) + n(t) = R_p \sum_{i=1}^{I} \alpha_i x_i(t - \tau_i) + n(t), \quad (31.24)$$

where I is the number of anchors/beacons inside the PD FOV—the beacons whose light impinge the PD. In (31.24), h_i corresponds to the channel impulse response from the ith emitter to the detector, Equation (31.13). On the path from the ith emitter to the detector, the signal undergoes an attenuation α_i and suffers a delay of τ_i seconds, given by

$$\alpha_i = \frac{(m+1)S}{2\pi d_i^2} \cos^m(\phi_i) \cos \psi_i \text{ and } \tau_i = \frac{d_i}{c}, \quad (31.25)$$

where the terms in the above equation were defined in Section 31.2.4.

One readily sees that simple multiple access techniques may be allowed for easy recovery of the RSS or the TOA of different LED emissions. In this book, the accuracy limitations of these techniques for RF positioning have been analyzed using the CRLB (Cramer–Rao Lower Bound). Nevertheless, the results cannot be directly utilized for VLP systems due to the nature of the optical link, and thus the VLP-specific CRLB is presented under certain conditions. However, before discussing those techniques, let us discuss the much more prosaic identity-/proximity-based techniques that require only one beacon to be in the FOV to provide a position fix, although with great inaccuracy.

31.3.2 Identity-Based Techniques

If a device uses any access point or transceiver station as a communication channel, then a reasonable although possibly inaccurate guess for the device position is the site of the station. This strategy resembles the so-called cell-identity methods for mobile cellular networks [5]; using visible light these techniques have been generally called visible light cell ID (VLID) [12]. Obviously, such a strategy can be used even if the station is not employed as a gateway by the device; that is, the station can be designed to act solely as a beacon carrying its ID and possibly its position for localization purposes. The ID can be conveyed by the beacon using very different VLC techniques ranging from simple OOK coded information to a color (spectral identity) beacon ID [33, 42, 48].

Although VLID techniques may provide inaccurate position estimates, they require only one beacon. Consider the scenario of providing communication access inside buildings using VLC techniques (e.g., the LiFi scenario), where the simple association to an access point already provides the room in which the device is located. The VLID principle can be scaled by placing several anchors, each sending a different identity encoded in its beacon. The identity provides a reasonable estimate of the device position, which is at most the room where the device is located.

Obviously, one may change from a device-computed position estimate to a detectors-network-computed position estimate. A tag emitting a beacon (its identity) can be placed at the asset to be located or tracked, and a network of detectors is used to capture the device identity from its emitted light and then to estimate the asset position.

Visible light ID positioning systems cannot be used to compute relative or even absolute position estimates; they provide positioning support within the coverage of the light source. For this reason, some refer to identity-based positioning methods as proximity positioning methods. Such methods have accuracy of only a few meters, but because they are very simple and may inherently provide accuracy at the room level, they have many applications [3, 10, 33, 42, 43, 48, 57]. For example, the VLID rationale has been used for tracking medical assets in a hospital [3]. Other approaches combine the VLID support with a communications network to resolve the asset position [42, 43]. Although some specialists advocate that VLP systems will replace RF-based positioning systems, the tendency is that both forms of transmission will coexist [61].

Neither LEDs nor PDs can be modeled as omnidirectional or isotropic. Therefore, the existence of a link enabling the beacon to be detected will highly depend on the physical arrangement between emitter and detector, as shown in Figure 31.6; NLOS scenarios may also be supported by reflection on walls and furniture. LEDs with larger diffusion area, larger PDs, and lenses (such as as the fish-eye) may be used to augment system coverage [107]. Conversely, specific arrangements and physical constructions of LEDs and PDs (in the transmitter and in the receiver, respectively) may be used. This will allow for beacon detection in different link angles and/or alignments, or to extract features corresponding to the physical disposition of the transmitter and receiver from the light impinging the detector [53, 54, 76, 78].

Alternatively, image sensors can also be used to recover the ID information [45,58]. This scenario is exemplified in Figure 31.9. One notes the resemblance of this to quick response (QR) codes or any other specially designed graphical tags in sites and devices. Nevertheless, they have very different usages. The VLP systems are not designed for the user having an active role, as happens when a user points

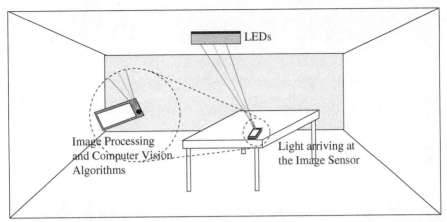

Figure 31.9   Image sensor-based positioning: the image captured at the device can be used to compute position estimates.

the camera to QR codes. The use of image sensors for capturing data from visible light sources opens new avenues. VLP systems can be designed using computer vision techniques as localization engines; let us delve into that.

31.3.3   Computer Vision Techniques

Image sensors can be used to detect beacons and obtain the emitter identity in VLID. However, the image sensors allow for image-based methods VLP systems [44, 66, 106, 110]. If a camera captures a scene or a video sequence containing different light sources, in addition to extracting the identity of the light sources the system can obtain the sources' positions in the image. Image-based methods use the light sources, positions extracted from the sensed images to compute the position estimate. Moreover, the underlying VLC can be used to convey not only the beacon's IDs but also the sources' absolute or relative positions to enhance and facilitate computing the image sensor estimated position [22, 44].

The Pinhole Camera Model: Figure 31.10 illustrates the projection of a point P on an image sensor through a pinhole camera. The projection of P in the image sensor is p'. The image sensor is at a distance f from the pinhole. The analysis of such a configuration must encompass different coordinate systems: a three-dimensional (3-D) coordinate system for the camera (the X-, Y-, and Z-axes shown in Fig. 31.10); a two-dimensional (2-D) coordinate system (the x- and y-axes in Fig. 31.10) to describe the points on the image sensor; and a 3-D coordinate system for the space in which the objects exist—the world, the X'-, Y'-, and Z'-axes in Figure 31.10.

For the coordinates of P in the image sensor, we employ triangle similarity (for simplicity, reflection on the x- and y-axes is not considered) to obtain

$$\frac{f}{Z_p} = \frac{x'}{X_p} = \frac{y'}{Y_p} \Rightarrow x' = \frac{fX_p}{Z_p} \text{ and } y' = \frac{fY_p}{Z_p}. \tag{31.26}$$

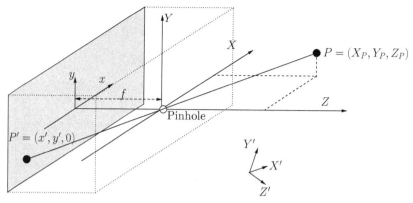

Figure 31.10   Pinhole camera model: The figure illustrates the projection of P through a pinhole, resulting in the point P' at the image sensor. The distance f between the image plane-sensor and the pinhole-center of the projection-lens is the focal length. As f gets smaller, more points can be projected onto the image plane. Conversely, as f increases, the FOV becomes smaller.

Using homogeneous coordinates* we may write

$$\begin{bmatrix} x' \\ y' \\ w' \end{bmatrix} = \begin{bmatrix} f & 0 & 0 \\ 0 & f & 0 \\ 0 & 0 & 1 \end{bmatrix} \begin{bmatrix} X_P \\ Y_P \\ Z_P \end{bmatrix}. \tag{31.27}$$

It is important to understand that the above result is only valid if the origin of the 3-D camera coordinate system and the 2-D image sensor coordinate system are aligned and the planes xy and XY are parallel. If that is not the case, further terms must be included in the above model.

Let $\mathbf{R}$ denote the rotation/alignment between the world coordinate system axes and the camera system axes, i.e., $\mathbf{R}$ is a 3×3 matrix, and $\mathbf{t}$ denotes the translation between the two 3-D systems, i.e., a 3×1 matrix—a vector. The coordinates of P in the camera coordinate system are derived from its world system coordinates (X_P', Y_P', Z_P') by means of [19]:

$$\begin{bmatrix} X_P \\ Y_P \\ Z_P \\ 1 \end{bmatrix} = \begin{bmatrix} \mathbf{R} & \mathbf{t} \\ 0 & 1 \end{bmatrix} \begin{bmatrix} X_P' \\ Y_P' \\ Z_P' \\ 1 \end{bmatrix}. \tag{31.28}$$

*This system avoids the division by Z_p which is stored in the homogenization coordinate w.

Employing four-dimensional (4-D) homogenous coordinates, we obtain

$$
\begin{bmatrix} x' \\ y' \\ z' \\ w \end{bmatrix} = \begin{bmatrix} f & 0 & 0 & 0 \\ 0 & f & 0 & 0 \\ 0 & 0 & 0 & 0 \\ 0 & 0 & 0 & 1 \end{bmatrix} \begin{bmatrix} \mathbf{R} & \mathbf{t} \\ 0 & 1 \end{bmatrix} \begin{bmatrix} X'_P \\ Y'_P \\ Z'_P \\ 1 \end{bmatrix}. \tag{31.29}
$$

If we assume that there are m_x and m_y pixels per unit length in the directions x and y, respectively, then a length of x' in the x image sensor corresponds to $m_x x'$ pixels. If we define the intrinsic camera parameters matrix

$$
\mathbf{K} = \begin{bmatrix} m_x f & 0 & 0 & 0 \\ 0 & m_y f & 0 & 0 \\ 0 & 0 & 0 & 0 \\ 0 & 0 & 0 & 1 \end{bmatrix}, \tag{31.30}
$$

then the point coordinates in pixels (c, l) (column and line) are given by

$$
\begin{bmatrix} c \\ l \\ z \\ w \end{bmatrix} = K \begin{bmatrix} \mathbf{R} & \mathbf{t} \\ 0 & 1 \end{bmatrix} \begin{bmatrix} X'_P \\ Y'_P \\ Z'_P \\ 1 \end{bmatrix}, \tag{31.31}
$$

Single View Geometry: Assume that one camera is used, meaning that a single view of the scene is available.[*] Consider a scenario in which different LEDs are available for the VLP system; this scenario is illustrated in Figure 31.11. Each LED P_i generates a set of coordinates for p_i depending on the sensor position and its orientation with respect to the LEDs. For each point $P_i = (X_i, Y_i, Z_i)$, one obtains

$$
c_i = fm_x \frac{r_{1,1}(X_i - X'_o) + (Y_i - Y'_o)r_{1,2} + (Z_i - Z'_o)r_{1,3}}{r_{3,1}(X_i - X'_o) + (Y_i - Y'_o)r_{3,2} + (Z_i - Z'_o)r_{3,3}} \text{ and} \tag{31.32}
$$

$$
l_i = fm_y \frac{r_{2,1}(X_i - X'_o) + (Y_i - Y'_o)r_{2,2} + (Z_i - Z'_o)r_{2,3}}{r_{3,1}(X_i - X'_o) + (Y_i - Y'_o)r_{3,2} + (Z_i - Z'_o)r_{3,3}}, \tag{31.33}
$$

where $(X'_o, Y'_o, Z'_o) = \mathbf{t}$ is the pinhole position, i.e., translation from the world reference frame to the camera centered one. Similarly, the entries of $\mathbf{R}$ ($r_{i,j}$) depend on the angles between the two 3-D coordinate systems. Let ω be the horizontal angle, i.e., the angle between the axes X and X', let the vertical angle between Y and Y' be ψ, and let the optical axis angle between Z and Z' be θ. Then, the resulting $\mathbf{R}$ matrix entries are

$$
r_{1,1} = \cos(\psi)\cos(\theta), \tag{31.34}
$$

$$
r_{1,2} = -\cos(\psi)\sin(\theta), \tag{31.35}
$$

---

[*]Different poses could be captured using more than one camera or one camera with accelerometers or inertial sensors used to estimate displacement and then generate multiple views from different poses of the same camera.

$$r_{1,3} = \sin(\psi), \tag{31.36}$$

$$r_{2,1} = \sin(\omega)\sin(\psi),\cos(\theta) + \cos(\omega)\cos(\theta), \tag{31.37}$$

$$r_{2,2} = -\sin(\omega)\sin(\psi)\sin(\theta) + \cos(\omega)\cos(\theta), \tag{31.38}$$

$$r_{2,3} = -\sin(\omega)\sin(\psi), \tag{31.39}$$

$$r_{3,1} = -\cos(\omega)\sin(\psi)\cos(\theta) + \sin(\omega)\sin(\theta), \tag{31.40}$$

$$r_{3,2} = \cos(\omega)\sin(\psi)\sin(\theta) + \sin(\omega)\sin(\theta), \text{ and} \tag{31.41}$$

$$r_{3,3} = -\cos(\omega)\cos(\psi). \tag{31.42}$$

Finding the Position Estimate: Consequently, from the coordinates of the pixels corresponding to several points/LEDs, we could, in principle, obtain the sensor position (X'_o, Y'_o, Z'_o) and its orientation (ω, ψ, θ). Obviously, the problem lies in inverting the nonlinear mapping composed by (31.32–31.42). Since each point provides two values (c_i and l_i) and there are six unknown variables (X'_o, Y'_o, Z'_o) and (ω, ψ, θ), at least three LEDs are required to produce a system of six nonlinear equations to be solved.

Using linearization and other numeric techniques, the above strategy was evaluated in [21] and [106], showing that in principle very high accuracy in the position estimate is possible. The smaller the pixel dimension (equivalently, the greater the quantity of pixels per unit length), the smaller the positioning error. In addition, increasing the number of transmitters improves accuracy. The accuracy of the estimated position depends also on the focal length of the lens f and exposure time of the imaging sensor; the smaller it is, the larger the uncertainty [110].

The pair of (31.32) and (31.33) requires the coordinates of the LED to be used as a reference. LEDs may encode their identification and world coordinates in their beacons/lighting [37,65] to improve the position fix (as GNSS satellites do). Other data and multiple sensors can also reduce tilt influence on the position fix [22].

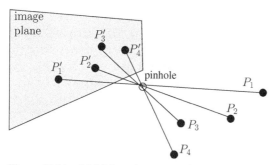

Figure 31.11 Multiple points are projected onto an image sensor. For image-based VLP systems, if the positions of the points (absolute or relative) are known, then spatial information about the sensing device can be obtained.

Cameras are increasingly being embedded in devices, and dual-camera configurations are widely available, making the use of two views another possibility to estimate the sensor position [65]. Knowledge of the two-camera arrangement may in principle reduce the quantity of references required to compute the position fix.

31.3.4 Trilateration Methods and Ranging Techniques

Identity/proximity and image sensor methods do not require estimating ranges or distances between emitter and detector. Several positioning methods already presented in this book rest on estimated ranges to compute a position fix. Methods based on trilateration use distance or estimated distance differences to provide a position fix. The estimation of such distances or ranges relies on the measurement or analysis of some received signal features—for example, strength or timing, which require techniques to measure TOA, TDOA, and RSS [20,32]. Methods for calculating position fix from TOA were presented in Part II of this book, and methods for RSS-based position fix computation were discussed in Part III. Below, we briefly discuss some aspects of using these measurements in VLP systems.

Time-Based Ranging: Time-based ranging requires some synchronization between emitter and detector, as discussed in Chapter 3. However, if we can ensure this synchronization, then the LOS condition of optical links offers the possibility to obtain very precise range estimates. Obviously, the precision will also be limited by the clock of the systems (as the clock increases, the bandwidth also increases) and the synchronization accuracy.

From Detected Light to Distance Estimates: The inaccuracy incurred in converting TOA into distance in VLP systems is studied in [95]. The authors assume (1) that the link follows the LOS attenuation model described in Section 31.2.4 (the perpendicular axis of LEDs and PDs are parallel); and (2) that emitter and that the detector are perfectly synchronized. They also consider (3) that the ith anchor modulates light using a (windowed) sinusoid of frequency f_{c_i} added to a DC level over an (4) AWGN (additive white Gaussian noise) channel. Using these assumptions, it is derived that the CRLB of the range estimate $\hat{d}_i$ from the detector to the ith beacon is given by

$$\sqrt{\text{var}(\hat{d}_i)} = \frac{c}{\sqrt{2}\pi} \frac{1}{\alpha_i} \frac{1}{R_p} \frac{1}{\sqrt{\varepsilon_i/N_0}} \frac{1}{\beta_i}, \tag{31.43}$$

where ε_i denotes the electrical energy of the signal $x_i(t)$,[*] β_i is the effective signal bandwidth of the ith signal, and in this case $\beta_i \approx 1/(3f_{c_i})$.

The larger the SNR and the bandwidth, the more accurate the ranging estimate is. The TOA-based distance estimate accuracy decreases as a function of the attenuation, meaning that the inaccuracy increases as a function of the distance d_i between

---

[*]In (31.12), $x(t)$ is used to define the received electrical signal after optical-electrical conversion and $\varepsilon_x = \int_{t-T}^{T} x(t)^2 \, dt$; T is the observation interval.

emitter and detector. Again, the larger the detector area and the responsivity, the smaller the expected error for the distance estimate. The result above also clearly stated that as the clock of the conveyed information increases (see the term $\beta_i \approx 1/(3f_{c_i})$ appearing in the denominator), the CRLB decreases—i.e., better range estimates are obtained, exactly as previously studied for RF-based TOA systems.

The Detector Influence: Let us further analyze the effects of the detector itself. The main aspect is that its physical properties contribute not only to the signal but also to noise. From the discussion in Section 31.2.7, $N_0/2 = qR_pP_a = qR_pS\Delta\lambda p_n$ and (31.43), we obtain

$$\sqrt{\text{var}(\hat{d}_i)} = \frac{\sqrt{qc}}{2\pi} \frac{S}{\alpha_i} \sqrt{\frac{\Delta\lambda}{SR_p}} \frac{\sqrt{3}}{f_{c_i}} \frac{1}{\sqrt{\varepsilon_i/p_n}}. \tag{31.44}$$

The more selective the PD, the smaller the range estimate inaccuracy; both the responsivity and the detector area improve the bound (although S appears in the numerator, it also appears in the denominator since α_i is proportional to S; see (31.11)).

Multiplexing Strategies: The schemes using TOA measurements for VLP also need to employ some multiplexing techniques. The techniques previously discussed for RSS measurement can also be employed for TOA, although specific signal features must exist so that some sort of timing information can be extracted from the received signal. Let us discuss some examples of TDOA-based VLP systems.

TDOA in VLP Systems: To loosen the synchronization requirement between the receiver to be located and the network of beacons, techniques such as TDOA are an option. Since TDOA employs the difference of arrival of signals from different anchors, only the anchors need to be synchronized. Several VLP systems have been devised using TDOA for trilateration. Examples of this approach are in [30] and [34]. The techniques to obtain the TDOAs between different beacons/anchors may employ the alignment of encoded data on the beacon burst. Another option is to use simple pulses of arbitrary positions and amplitudes within a frame (a time interval) so that from the received profile of the frame the TDOA can be extracted. Even sinusoidal waves and the detection of the phase differences between each pair of different sources have been used to compute the difference in the light travel time. Such techniques are been called PDOA—phase difference of arrival. From these, the position fix is computed using the methods described in Chapter 3.

RSS-Based Ranging: The incident power from a light source on a detector may provide an estimate of the distance between the light source and the detector. The distance between a transmitter and a receiver may also be estimated by measuring the RSS; in this case, the Lambertian transmission property (31.7) has to be inverted [111]. Several VLP systems use that feature of the optical link [33, 42, 48]. It is important to highlight that LOS reception is usually assumed for visible light systems and consequently the variability of the RSS is smaller than in RF scenarios.

From RSS to Distances: One may estimate the distance between emitter and detectors by analytically solving the Lambertian transmission channel in (31.7) [111]. Simulation results indicate the capability to provide positioning resolution in the order of 0.5 mm. The same principle was investigated adding an RF carrier modulated using phase shift keying (PSK) to a DC level and using the resulting current to emit light [33]. This scheme is very similar to the optical signal previously employed for for TOA estimation (and for which the CRLB was obtained); the RF carrier is summed to a DC level before being applied to the LED. Several beacons (LED) coexist by employing a different carrier frequency for each of them. The distances are estimated from the received RF power, although a specific method is employed to obtain the range estimates; it was observed that the received RF power can vary due to radiation directivity and angle of incidence. Then, the position fix is obtained using a compensated distance trilateration method.

The inaccuracy incurred in converting received light into distances estimates is studied in [109]. Assuming (1) a sinusoid modulating the light that is added to a DC level and that this signal is corrupted by a Gaussian white noise of variance $\sigma = N_0/2$, (2) using the LOS attenuation model described in Section 31.2.4,[*] (3) an AWGN channel, and (4) that an optical filter exists with ideal band-pass causing no loss in the pass-band and completely blocking the stop-band in the detector, it can be shown that

$$\sqrt{\text{var}(\hat{d}_i)} = \frac{d_i^{m+4}}{h_i^{m+1}} \frac{1}{(m+1)(m+3)} \frac{\sqrt{2}\pi}{S R_p} \frac{1}{P_i/N_0}, \tag{31.45}$$

where P_i denotes the optical power of the signal irradiated from the ith transmitter. This bound is inversely proportional to the SNR (as it is inversely proportional to the term P_i/N_0). That is, the larger optical power of the signal, the better the range estimate. In addition, at larger distances d_i (or greater the attenuations α_i) there is an inherent increase in the estimate inaccuracy. Likewise in the TOA ranging case, RSS-based ranging improves as detector area and responsivity increases. As the reader readily sees, the bandwidth does not play an important role anymore, as expected.

Multiplexing Strategies: These schemes use time division multiplexing [77, 109] or frequency division multiplexing [33] to share the optical medium among different light emitters and thus measure the RSS from different modulated light sources. Obviously, other medium sharing techniques, such as code division multiplexing and even color/wavelength multiplexing (using light sources of different colors), can also be used.

31.3.5 Triangulation

Triangulation methods can also be applied to build VLP systems. Such methods employ the DOA (direction of arrival)/AOA estimates from different beacons to the receiver to estimate the receiver position, as presented in Chapter 9. Each AOA

---

[*]Actually, the room diffusion component is also considered in [109], but it is shown that this component can be accounted for and does not impair the distance estimate accuracy.

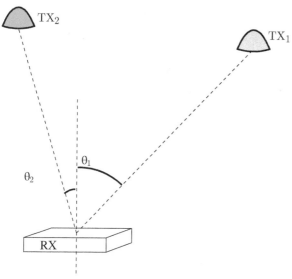

Figure 31.12   AOA VLP system principle of operation. The angles in which the light arrives at the sensor can be used to find the sensor position if the beacon positions are known. With two references a 2-D position fix can be obtained; with three references a 3-D position fix can be obtained.

between the beacon and the detector corresponds to a line where both are collocated. Two lines cross in a point as illustrated in Figure 31.12. Thus, a set of AOAs may feed a system of equations that is solved to produce the position fix. For VLP these methods are also possible [64, 101].

The use of AOA in VLP systems has the inherent advantage of the LOS nature of the optical link, a feature not granted for RF AOA-based positioning systems. Compared to the other techniques, AOA does not require any positioning-designed synchronization scheme, and AOA can provide a 2-D or a 3-D position fix using only two or three measurements, respectively, although AOA requires more complex detectors.

AOA of Light Rays: LED-radiated light follows Lambert's cosine law; as a result, changes in the AOA or angle of incidence lead to changes in the received power in a predictable manner. This is readily seen from (31.24). Consequently, we could use the difference between the received power and the received power at a known angle to estimate the angle of incidence of the light detected from a source. However, this strategy is limited by FOV and possible modeling difficulties of the Lambertian equation; to mitigate the possible problems, a PD array (multiple receivers [101]) can be used.

PD Arrangements: A half-circle PD arrangement is illustrated in Figure 31.13 [41]. The light received by the ith PD is affected by the distance d_i, the radiance angle ϕ_i, and the incidence angle ψ_i. If the distance between the PD array and the LED is sufficiently greater than the physical dimensions of the arrangement,

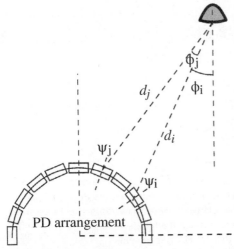

Figure 31.13 Half-circle PD arrangement that can be used to obtain AOA estimates of incident light. The different orientations of the PD force the light arriving from a source to have different angles of incidence in each PD.

then the distances d_i and the radiance angles ϕ_i can be assumed to be the same. In this case, the difference in the received powers are presumably solely due to differences in the incidence angles ψ_i. The differences between these angles can then be used to obtain the angle between the array and the LED, the AOA.

Help from Other Sensors: In order to ease the task of obtaining AOA, some propose the use of accelerometers to obtain the sensor orientation [103]. Making different measurements with the same PD at different orientations provides different AOAs. Nevertheless, the device should be static (not suffering any translation besides possible necessary rotations) between different AOA measurements, compromising device mobility. Similarly, instead of using multiple tilts of the device, we can embed some PDs with different tilts in the device and use an accelerometer to provide the device orientation [104], reducing the time for the position fix and consequently allowing more device mobility.

Aperture Array Sensors: AOA-based VLP systems may benefit from embedding several PDs. Unfortunately, traditional image sensors cannot be used since all the PDs lie in the same plane, and therefore the light impinges on all of them at the same angle, approximately. A solution is to use an aperture-based receiver [76,94]. The aperture-based receiver is a spatial distribution of PDs with a superimposed plane of holes. The relative placement of the PDs and the apertures make the receiver capable of offering angular diversity. Consequently, accurate positioning is achievable even with nondirectional LEDs. This receiver structure can be used to obtain AOA estimates from the relative values of RSSs at different PDs in the sensor, even not knowing the transmitted signal strength. Figure 31.14 illustrates a circular

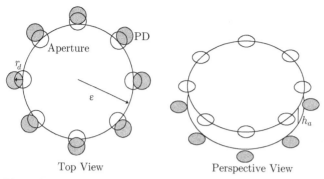

Figure 31.14   Aperture array sensor using a circular arrangement of photodetectors. The apertures are displaced with respect to the PDs underneath. Therefore, the PD areas that are covered by the light of a source depend on the relative position between the light path from source to PD and the aperture. Different angles between the light source and the array will produce different electric signal levels at the PDs in the arrangement.

arrangement of PDs and the apertures [78]; the main geometrical properties are shown. Other works present a square arrangement [76]. Depending on the relative orientation of the LED and the aperture-sensor, different areas of the PDs are illuminated, receiving more or less energy and generating different outputs. Obviously, the accuracy of the estimated AOA depends on the geometrical dimensions of the arrangement [76, 78].

31.3.6   Fingerprinting

Fingerprinting is also known as scene analysis, as it considers the pattern matching of a target reference vector against a set of reference vectors [4, 85]. Due to its similarity with the identification of persons by means of the patterns in their fingers, the name "fingerprinting" is largely employed and the target vector and reference vectors are now called target fingerprint and reference fingerprints.

   Continuing the personal fingerprint analogy, fingerprinting positioning methods also require the collection of information associated with the sources. This collection phase produces the correlation database, against which a reference will be analyzed/ compared. This phase may be labor intensive and may impede the adoption of the fingerprinting strategy. Although, for IPS, as the area is small and access to it can be controlled, it is possible for the collection phase to be simpler and easily executable. Once the database is constructed, the training phase takes place. This phase will depend on the specific pattern matching/machine intelligence algorithm employed; it uses collected fingerprints to provide an optimized localization engine that is capable of returning good estimates of the position where fingerprints are collected.

   Fingerprinting techniques are a possibility for VLP systems [3, 12, 20]. Nevertheless, we will not examine the specific fingerprinting techniques to estimate the light sensor position. In this book, many such techniques have already been discussed, and it is not hard to adapt them so a fingerprinting engine can be used in a VLP system. Also, we will not discuss the construction of the database because, with appropriate care, the concepts for database construction already presented in this book can be used for VLP fingerprinting systems. Nevertheless, using light as the medium allows or requires (depending on the case and the reader) different characteristics to compose the fingerprint.

Fingerprint Features for VLP: Fingerprinting positioning accuracy will depend not only on the algorithm of scene analysis but also on the features composing the fingerprint. The simplest approach would be to use the identities of the anchors/beacons detected to compute the position fix. The identities can be accompanied by their RSS. There are many fingerprinting examples both for RF positioning and VLP using RSS or its "quantized" counterpart, the RSSI (RSS index). Several RSS values or RSSIs from different anchors can be used as fingerprinting features. Regarding the aforementioned features, we could add TOA and POA (phase of arrival) with respect to some beacons to possibly boost performance and improve location accuracy.

AOA is also a possible fingerprint feature [101]. The detected power varies with the distance between emitter and receiver and also with the position, since the angle of incidence varies accordingly. This is boosted by using an array of multiple PDs that are tilted at different angles. Since the light arrives at the different PDs in the arrays at different angles, the detected power also varies accordingly. The resulting received power can be measured and used as a fingerprint.

We have seen that RSS, TOA, POA, and AOA can be used as fingerprinting features. Nevertheless, the capability to extract the features depends on how (modulation scheme, coding, clock/bandwidth) and which data is conveyed in the physical layer. Therefore, specific modulation and coding schemes could be designed by taking into account its usability to extract fingerprinting features. For example, OOK modulation and the construction of fingerprints using the ratio between the received amplitudes of the on and the off states has been considered [97, 100]. Such features are used to find the most probable location by means of an algorithm that uses stored values of the received on/off amplitudes ratio. Other features, such as impulse response, can also be included in the fingerprint [88].

As a final consideration, there is the possibility of using images as fingerprinting features. This is somehow similar to using QR codes but with leverage for much higher position precision and accuracy. Image sensors can be used to extract signal patterns. Beacons can be designed to emit patterns using high-frequency blinking of LEDs. The image sensor can be used to collect the patterns configured for different rates. The fingerprints are composed of light intensity plus a signal pattern [46] and by means of a probabilistic approach, the image sensor position is determined.

31.4 CONCLUSION

Visible light communication systems are in the process of being adopted. Such systems are primarily derived from recent advances in LED technology. Concomitantly, VLP systems have emerged.

LED-based illumination technologies have some environmental appeal:

- Energy efficiency: at similar brightness and illumination levels, LED lighting infrastructure saves up to 85% or 60% of energy compared to incandescent and fluorescent lamps, respectively;

- Harmless technology: Visible light does not interfere with common electronic devices, while the alternatives IR and lasers may harm humans;

- Environmentally friendly: LED lighting is green, as it requires less emission of CO_2 [62] and does not use mercury or other toxic elements;

- Longer life expectancy: LED lights are expected to last several years, more than incandescent and fluorescent lamps.

These factors motivated the research for effectively using light as carrier for both communications and positioning purposes. Such use provides the following features:

- High data rates: LEDs can be switched on and off at high frequencies, allowing for wireless data communication at high rates;

- Adoption in restricted sites: Visible light can be used even at restricted site such as hospitals, airplanes etc, as it does not produce electromagnetic interference;

- Security: As light cannot penetrate thick walls, the data conveyed by light is confined and is consequently more secure than RF systems;

- High reuse rate: Light confinement allows independent systems in adjacent rooms to use the same medium without interference.

To which some economic positives must be added:

- Cost-effectiveness: High energy efficiency and longer life expectancy: additional costs to incorporate extra features are marginally small as the benefits may last longer;

- License-exempt deployment: visible light systems (VLC/VLP) use a license-free bandwidth.

These features leverage the design of VLP systems. They are mainly intended for IPS, although initiatives for outdoor counterparts mainly using public illumination also exist. Joint VLC and VLP IPS seem to be technologies that we will see on a daily basis. VLP systems may employ very different techniques. Most of the RF methods and techniques can be applied to designing and implementing VLP systems, provided that proper adaptations considering the specificity of the VLC channel are made. However, VLP systems may use image sensor and computer vision techniques to extend their capabilities.

VLP systems, similarly to RF positioning systems, may be constructed using different features and resources. In comparison to RF positioning systems, VLP systems rely on LOS and an optical link. This feature eases analysis but also allows for a receiving sensor with smaller dimensions. These provide easier integration of sensors (the photodetectors), which allow us to go beyond the techniques used in

RF positioning systems and use image sensors to capture the light from beacons. As a result, computer vision techniques can also be used to obtain the position fix.

The localization accuracy provided by the VLP designs is very high, ranging from tens of centimeters to micrometers. However, practical implementations may not achieve such a high accuracy due to several factors:

- Light propagation does not follow the simplified Lambertian model, which provides the channel impulse response in (31.13) and (31.24) and is assumed in the VLP performance analyses;

- While LOS transmission is mostly assumed for VLP, reflections, scattering, and diffusion of light from various physical objects in indoor environments may also exist;

- Simplified models are assumed for thermal and shot noises; therefore, more accurate analysis of their effects on positioning is still required;

- In image sensor VLP systems, noise impacts different features of pixels, such as color and brightness, which may compromise the VLP performance, requiring thus further analysis.

Despite these possible drawbacks, the current drive for VLC systems and the reliability of VLP contribute to improve the design of VLP systems. Some of them are already in use.

VLP systems can be applied in several scenarios, including shopping malls, museums, hospitals, offices, and retail stores, to name a few. In shopping malls and retails stores, location-based services and advertisements to the shoppers can benefit retailers and buyers. Location-aware search is useful to find the nearby shops or products, and it can also be used for locating a tagged worker, monitoring assets, and managing inventory. In museums, location-aware services may conduct visitors through different galleries and exhibitions, indicating nearby works of art. In office buildings, VLP can track a person inside a building, detecting suspicious activities and unauthorized access to restricted areas. In hospitals, it can help orient elderly, handicapped, or visually-impaired patients, thus preventing falls or collisions with other objects. In these scenarios, the communication systems embedded in the lighting infrastructure are normally used for downlink transmission, whereas RF-based technologies are used for uplink transmission. Although some specialists advocated that VLP systems will replace RF-based positioning systems, the tendency is that both forms of transmission will coexist. The main reason is that VLP needs free LOS to work properly. Thus, when the portable device is out of sight—for example inside a purse—no location service can be provided. However, VLC can be combined with other techniques to mitigate this shortcoming.

REFERENCES

[1] G. P. Agrawal, *Fiber-Optic Communication Systems*, 4th ed. John Wiley & Sons, 2010.

[2] J. Armstrong and B. J. Schmidt, "Comparison of asymmetrically clipped optical OFDM and DC-biased optical OFDM in AWGN," *IEEE Commun. Lett.*, vol. 12, no. 5, pp. 343–345, 2008.

[3] J. Armstrong, Y. Sekercioglu, and A. Neild, "Visible light positioning: a roadmap for international standardization," *IEEE Commun. Mag.*, vol. 51, no. 12, pp. 68–73, 2013.

[4] C. M. Bishop, *Pattern Recognition and Machine Learning*. Springer, 2006.

[5] R. S. Campos, and L. Lovisolo, *RF Positioning: Fundamentals, Applications, and Tools*. Artech House, 2015.

[6] W. Chunyue, W. Lang, C. Xuefen, L. Shuangxing, S. Wenxiao, and D. Jing, "The research of indoor positioning based on visible light communication," *China Commun.*, vol. 12, no. 8, pp. 85–92, 2015.

[7] CIE. *Commission Internationale de l'Eclairage Proc.*, 1931, Cambridge University Press, 1932.

[8] G. Corbellini and K. Aks, "Connecting networks of toys and smartphones with visible light communication," *IEEE Commun. Mag.*, vol. 52, no. 7, pp. 72–78, Jul. 2014.

[9] P. Daukantas, "Optical wireless communications: The new hot spots?" *Opt. Photonics News*, vol. 25, no. 3, pp. 34–41, Mar. 2014.

[10] G. del Campo-Jimenez, J. M. Perandones, and F. J. Lopez-Hernandez, "A VLC-based beacon location system for mobile applications," in *2013 IEEE Int. Conf. on Localization and GNSS (ICL-GNSS 2013)*, Jun. 25–27, 2013, pp. 1–4.

[11] S. Dimitrov and H. Haas, *Principles of LED Light Communications: Towards Networked Li-Fi*, 1st ed. Cambridge, UK: Cambridge University Press, 2015.

[12] T.-H. Do and M. Yoo, "An in-depth survey of visible light communication based positioning systems," *Sensors*, vol. 16, no. 5, p. 678, 2016.

[13] R. J. Drost and B. M. Sadler, "Constellation design for color-shift keying using billiards algorithms," in *2010 IEEE GLOBECOM Workshops (GC Wkshps)*, Dec. 2010, pp. 980–984.

[14] M. Ebner, *Color Constancy*, 1st ed. UK: John Wiley & Sons, 2007.

[15] N. Fujimoto and H. Mochizuki, "614 Mbit/s OOK-based transmission by the duobinary technique using a single commercially available visible LED for high-speed visible light communications," in *2012 European Conf. and Exhibition on Optical Communication*, Sep. 2012, p. P4.03.

[16] N. Fujimoto and H. Mochizuki, "477 Mbit/s visible light transmission based on OOK-NRZ modulation using a single commercially available visible LED and a practical LED driver with a pre-emphasis circuit," in *2013 Optical Fiber Communication Conf. and Exposition and National Fiber Optic Engineers Conf.*, Mar. 2013, p. JTh2A.73.

[17] Z. Ghassemlooy, W. Popoola, and S. Rajbhandari, *Optical Wireless Communications: Systems and Channel Modelling with MatLab*. Taylor & Francis Group, 2013.

[18] G. Ghione, *Semiconductor Devices for High-Speed Optoelectronics*, 1st ed. Cambridge, UK: Cambridge University Press, 2009.

[19] R. Hartley and A. Zisserman, *Multiple view Geometry in Computer Vision*. Cambridge University Press, 2003.

[20] N. U. Hassan, A. Naeem, M. A. Pasha, T. Jadoon, and C. Yuen, "Indoor positioning using visible LED lights: A survey," *ACM Comput. Surv. (CSUR)*, vol. 48, no. 2, p. 20, 2015.

[21] M. S. Hossen, Y. Park, and K.-D. Kim, "Performance improvement of indoor positioning using light-emitting diodes and an image sensor for light-emitting diode communication," *Opt. Eng.*, vol. 54, no. 4, p. 045101, 2015.

[22] P. Huynh and M. Yoo, "VLC-based positioning system for an indoor environment using an image sensor and an accelerometer sensor," *Sensors*, vol. 16, no. 6, p. 783, 2016.

[23] IEEE-802-WPAN-TG7. "IEEE standard for local and metropolitan area networks–part 15.7: Short-range wireless optical communication using visible light," *IEEE Std 802.15.7-2011*, pp. 1–309, Sep. 2011.

[24] IEEE-802-WPAN-TG7, *IEEE 802.15.7 Visible Light Communication Task Group*, 2017. Available: http://www.ieee802.org/15/pub/TG7.html

[25] Y. Ikeda, K. Okuda, and W. Uemura, "The coded hierarchical modulation with amplitude for estimating the position in visible light communications," in *2014 IEEE 3rd Global Conf. on Consumer Electronics (GCCE)*, Oct. 2014, pp. 412–413.

[26] JEITA. Visible Light Communications System. *JEITA CP-1221* (March 2007).

[27] JEITA. Visible Light ID System. *JEITA CP-1222* (June 2007).

[28] JEITA. Visible Light Beacon System. *JEITA CP-1223* (May 2013).

[29] A. Jovicic, J. Li, and T. Richardson, "Visible light communication: Opportunities, challenges and the path to market," *IEEE Commun. Mag.*, vol. 51, no. 12, pp. 26–32, 2013.

[30] S.-Y. Jung, S. Hann, and C.-S. Park, "TDOA-based optical wireless indoor localization using LED ceiling lamps," *IEEE Trans. Consum. Electron.*, vol. 57, no. 4, pp. 1592–1597, 2011.

[31] J. M. Kahn and J. R. Barry, "Wireless infrared communications," in *Proc. of the IEEE*, vol. 85, no. 2, 1997, pp. 265–298.

[32] M. F. Keskin and S. Gezici, "Comparative theoretical analysis of distance estimation in visible light positioning systems," *J. Lightw. Technol.*, vol. 34, no. 3, pp. 854–865, 2016.

[33] H.-S. Kim, D.-R. Kim, S.-H. Yang, Y.-H. Son, and S.-K. Han, "An indoor visible light communication positioning system using a RF carrier allocation technique," *J. Lightw. Technol.*, vol. 31, no. 1, pp. 134–144, 2013.

[34] Y. Kim, Y. Shin, and M. Yoo, "VLC-TDOA using sinusoidal pilot signal," in *2013 Int. Conf. on IT Convergence and Security (ICITCS)*, IEEE, Dec. 16–18, 2013, pp. 1–3.

[35] S. Kitsinelis and S. Kitsinelis, *Light Sources: Basics of Lighting Technologies and Applications*, 2nd ed. Boca Raton, FL: CRC Press, 2015.

[36] T. Komine and M. Nakagawa, "Fundamental analysis for visible-light communication system using led lights," *IEEE Trans. Consum. Electron.*, vol. 50, no. 1, pp. 100–107, 2004.

[37] Y.-S. Kuo, P. Pannuto, K.-J. Hsiao, and P. Dutta, "Luxapose: Indoor positioning with mobile phones and visible light," in *Proc. of the 20th Annual Int. Conf. on Mobile Computing and Networking (ACM MobiCom '14)*, Sep. 2014, pp. 447–458.

[38] H. Le Minh, D. O'Brien, G. Faulkner, L. Zeng, K. Lee, D. Jung, and Y. Oh, "High-speed visible light communications using multiple-resonant equalization," *IEEE Photon. Technol. Lett.*, vol. 20, no. 14, pp. 1243–1245, 2008.

[39] C. G. Lee and M. Katz, "Visible light communication," in *Short-Range Wireless Communications: Emerging Technologies and Applications,* Wiley Online Library *11*, 2015.

[40] J.-H. Lee, D.-Y. Lee, B.-W. Oh, and J.-H. Lee, "Comparison of InGaN-based LEDs grown on conventional sapphire and cone-shape-patterned sapphire substrate," *IEEE Trans. Electron Devices*, vol. 57, no. 1, pp. 157–163, 2010.

[41] S. Lee and S.-Y. Jung, "Location awareness using angle-of-arrival based circular-PD-array for visible light communication," in *IEEE 18th Asia-Pacific Conf. on Communications (APCC)*, Dec. 24, 2012, pp. 480–485.

[42] Y. Lee, S. Baang, J. Park, Z. Zhou, and M. Kavehrad, "Hybrid positioning with lighting LEDs and ZigBee multihop wireless network," in *SPIE OPTO 2012*, International Society for Optics and Photonics, Jan. 2012, pp. 82820L–82820L-7.

[43] Y. U. Lee and M. Kavehrad, "Two hybrid positioning system design techniques with lighting LEDs and ad-hoc wireless network," *IEEE Trans. Consum. Electron.*, vol. 58, no. 4, pp. 1176–1184, 2012.

[44] B. Lin, Z. Ghassemlooy, C. Lin, X. Tang, Y. Li, and S. Zhang, "An indoor visible light positioning system based on optical camera communications," *IEEE Photon. Technol. Lett.*, vol. 29, no. 7, pp. 579–582, 2017.

[45] H. S. Liu and G. Pang, "Positioning beacon system using digital camera and LEDs," *IEEE Trans. Veh. Technol.*, vol. 52, no. 2, pp. 406–419, 2003.

[46] M. Liu, K. Qiu, F. Che, S. Li, B. Hussain, L. Wu, and C. P. Yue, "Towards indoor localization using visible light communication for consumer electronic devices," in *2014 IEEE/RSJ Int. Conf. on Intelligent Robots and Systems (IROS 2014)*, Sep. 2014, pp. 143–148.

[47] J. Luo, L. Fan, and H. Li, "Indoor positioning systems based on visible light communication: State of the art," *IEEE Commun. Surveys Tuts.*, vol. 19, no. 4, pp. 2871–2893, 2017.

[48] P. Luo, M. Zhang, X. Zhang, G. Cai, D. Han, and Q. Li, "An indoor visible light communication positioning system using dual-tone multi-frequency technique," in *IEEE 2013 2nd Int. Workshop on Optical Wireless Communications (IWOW)*, Oct. 21, 2013, pp. 25–29.

[49] H. Manor and S. Arnon, "Performance of an optical wireless communication system as a function of wavelength," *Appl. Opt.*, vol. 42, no. 21, pp. 4285–4294, 2003.

[50] R. Mautz and S. Tilch, "Survey of optical indoor positioning systems," in *IEEE 2011 Int. Conf. on Indoor Positioning and Indoor Navigation (IPIN)*, Sep. 2011, pp. 1–7.

[51] E. Monteiro and S. Hranilovic, "Design and implementation of color-shift keying for visible light communications," *J. Lightw. Technol.*, vol. 32, no. 10, pp. 2053–2060, 2014.

[52] S. Muthu, F. J. Schuurmans, and M. D. Pashley, "Red, green, and blue LED based white light generation: Issues and control," in *Conf. Record of the 2002 IEEE Industry Applications Conf.—37th IAS Annual Meeting*, vol. 1, 2002, pp. 327–333.

[53] Y. Nakazawa, H. Makino, K. Nishimori, D. Wakatsuki, and H. Komagata, "Indoor positioning using a high-speed, fish-eye lens-equipped camera in visible light communication," in *IEEE Int. Conf. on Indoor Positioning and Indoor Navigation (IPIN)*, Oct 28–31, 2013, pp. 1–8.

[54] Y. Nakazawa, H. Makino, K. Nishimori, D. Wakatsuki, and H. Komagata, "LED-tracking and ID-estimation for indoor positioning using visible light communication," in *IEEE 2014 Int. Conf. on Indoor Positioning and Indoor Navigation (IPIN)*, Oct 27–30, 2014, pp. 87–94.

[55] M. Noshad and M. Brandt-Pearce, "Expurgated PPM using symmetric balanced incomplete block designs," *IEEE Commun. Lett.*, vol. 16, no. 7, pp. 968–971, 2012.

[56] M. Noshad and M. Brandt-Pearce, "Multilevel pulse-position modulation based on balanced incomplete block designs," in *2012 IEEE Global Communications Conf. (GLOBECOM)*, Dec. 3–7, 2012, pp. 2930–2935.

[57] G. Pang, H. Liu, C.-H. Chan, and T. Kwan, "Vehicle location and navigation systems based on LEDs," in *Proc. of the 5th World Congress on Intelligent Transport Systems,* 1998, pp. 12–16.

[58] G. K. Pang and H. H. Liu, "Led location beacon system based on processing of digital images," *IEEE Trans. Intell. Trans. Syst.*, vol. 2, no. 3, pp. 135–150, 2001.

[59] S. Park, D. Jung, H. Shin, D. Shin, Y. Hyun, K. Lee, and Y. Oh, "Information broadcasting system based on visible light signboard," in *Proc. of the Wireless and Optical Communications*, 2007, pp. 311–313.

[60] P. H. Pathak, X. Feng, P. Hu, and P. Mohapatra, "Visible light communication, networking, and sensing: A survey, potential and challenges," *IEEE Commun. Surveys Tuts.*, vol. 17, no. 4, pp. 2047–2077, 2015.

[61] Philips Inc. Unlocking the value of retail apps with lighting. White Paper (2016).

[62] S. Pimputkar, J. S. Speck, S. P. DenBaars, and S. Nakamura, "Prospects for LED lighting," *Nat. Photon.*, vol. 3, no. 4, pp. 180–182, 2009.

[63] C. Pohlmann, "Visible light communication," in *Seminar Kommunikationsstandards in der Medizintechnik,* 2010, pp. 1–14.

[64] G. B. Prince and T. D. Little, "A two phase hybrid RSS/AOA algorithm for indoor device localization using visible light," in *2012 IEEE Global Communications Conf. (GLOBECOM)*, Dec. 3–7, 2012, pp. 3347–3352.

[65] M. S. Rahman, M. M. Haque, and K.-D. Kim, "High precision indoor positioning using lighting LED and image sensor," in *14th IEEE Int. Conf. on Computer and Information Technology (ICCIT)*, Dec. 22–24, 2011, pp. 309–314.

[66] M. S. Rahman and K.-D. Kim, "Indoor location estimation using visible light communication and image sensors," *International Journal of Smart Home*, vol. 7, no. 1, pp. 99–114, 2013.

[67] R. Ramirez-Iniguez, S. M. Idrus, and Z. Sun, *Optical Wireless Communications: IR for Wireless Connectivity*. CRC press, 2008.

[68] M. S. Rea, *The IESNA Lighting Handbook: Reference & Application.* Illuminating Engineering Society of North America, 2000.

[69] R. Roberts, P. Gopalakrishnan, and S. Rathi, "Visible light positioning: automotive use case," in *IEEE Vehicular Networking Conf. (VNC)*, 2010, pp. 309–314.

[70] S. Schmid, T. Bourchas, S. Mangold, and T. R. Gross, "Linux Light Bulbs," in *Proc. of the 2nd Int. Workshop on Visible Light Communications Systems: VLCS '15*, 2015, pp. 3–8.

[71] S. Schmid, J. Ziegler, G. Corbellini, T. R. Gross, and S. Mangold, "Using consumer LED light bulbs for low-cost visible light communication systems," in *Proc. of the 1st ACM MobiCom Workshop on Visible Light Communication Systems: VLCS '14*, 2014, pp. 9–14.

[72] E. F. Schubert, T. Gessmann, and J. K. Kim, *Light Emitting Diodes*. Wiley Online Library, 2005.

[73] E. F. Schubert and J. K. Kim, "Solid-state light sources getting smart," *Science*, vol. 308, no. 5726, pp. 1274–1278, 2005.

[74] R. Singh, T. O'Farrell, and J. P. David, "An enhanced color shift keying modulation scheme for high-speed wireless visible light communications," *J. Lightw. Technol.*, vol. 32, no. 14, pp. 2582–2592, 2014.

[75] R. Srividya, and C. P. Kurian, "White light source towards spectrum tunable lighting: A review," in *IEEE Int. Conf. on Advances in Energy Conversion Technologies (ICAECT)*, 2014, pp. 203–208.

[76] H. Steendam, T. Q. Wang, and J. Armstrong, "Cramer-Rao bound for indoor visible light positioning using an aperture-based angular-diversity receiver," in *IEEE Int. Conf. on Communications (ICC)*, 2016, pp. 1–6.

[77] H. Steendam, T. Q. Wang, and J. Armstrong, "Theoretical lower bound for indoor visible light positioning using received signal strength measurements and an aperture-based receiver," *J. Lightw. Technol.,* vol. 35, no. 2, pp. 309–319, 2016.

[78] H. Steendam, T. Q. Wang, and J. Armstrong, "Cramer-Rao bound for AOA-based VLP with an aperture-based receiver," in *IEEE Int. Conf. on Communications (ICC),* 2017, pp. 1–6.

[79] I. Stefan, H. Elgala, and H. Haas, "Study of dimming and LED nonlinearity for aco-OFDM based vlc systems," in *IEEE Wireless Communications and Networking Conf. (WCNC),* 2012, pp. 990–994.

[80] H. Sugiyama, S. Haruyama, and M. Nakagawa, "Brightness control methods for illumination and visible-light communication systems," in *Third IEEE Int. Conf. on Wireless and Mobile Communications* (ICWMC'07), 2007, pp. 78–78.

[81] I. Takai, S. Ito, K. Yasutomi, K. Kagawa, M. Andoh, and S. Kawahito, LED and EMOS image sensor based optical wireless communication system for automotive applications," *IEEE Photonics J.,* vol. 5, no. 5, pp. 6801418–6801418, 2013.

[82] J. Tan, Z. Wang, Q. Wang, and L. Dai, "BICM-ID scheme for clipped DCO-OFDM in visible light communications," *Opt. Express,* vol. 24, no. 5, pp. 4573–4581, 2016.

[83] J. Tan, Z. Wang, Q. Wang, and L. Dai, "Near-optimal low-complexity sequence detection for clipped DCO-OFDM. *IEEE Photon. Technol. Lett.,* vol. 28, no. 3, pp. 233–236, 2016.

[84] TeleGeography, *Submarine cable map 2016.* Available: http://submarine-cable-map-2016.telege-ography.com/, 2016

[85] S. Theodoridis and K. Koutroumbas, *Pattern Recognition.* Elsevier, 2009.

[86] C.-Y. Tsai, M.-H. Li, and R. Kuo, "A shopping behavior prediction system: Considering moving patterns and product characteristics," *Comput. Ind. Eng.,* vol. 106, pp. 192–204, 2017.

[87] D. Tsonev and H. Haas, "Avoiding spectral efficiency loss in unipolar OFDM for optical wireless communication," in *IEEE Int. Conf. on Communications (ICC),* 2014, pp. 3336–3341.

[88] A. M. Vegni and M. Biagi, "An indoor localization algorithm in a small-cell LED-based lighting system," in *IEEE Int. Conf. on Indoor Positioning and Indoor Navigation (IPIN),* 2012, pp. 1–7.

[89] VLCC, *Visible light communication consortium.* Available: http://www.vlcc.net, 2017

[90] Vučić J., C. Kottke, K. Habel, and K.-D. Langer, "803 Mbit/s visible light WDM link based on DMT modulation of a single RGB LED luminary," in *2011 IEEE Optical Fiber Communication Conf. and Exposition and the National Fiber Optic Engineers Conf.,* Mar. 2011, pp. 1–3.

[91] J. Vucic, C. Kottke, S. Nerreter, K. Habel, A. Buttner, K.-D. Langer, and J. Walewski, "125 Mbit/s over 5 m wireless distance by use of OOK-modulated phosphorescent white LEDs," in *IEEE 35th European Conf. on Optical Communication* (ECOC'09), 2009, pp. 1–2.

[92] Vučić J., C. Kottke, S. Nerreter, K. Habel, A. Büttner, K.-D. Langer and J. W. Walewski, "230 Mbit/s via a wireless visible-light link based on OOK modulation of phosphorescent white LEDs," in *Optical Fiber Communication Conf.,* Optical Society of America, Mar. 2010, pp. 1–3.

[93] J. Vucic, C. Kottke, S. Nerreter, K.-D. Langer, and J. W. Walewski, "513 Mbit/s visible light communications link based on DMT-modulation of a white LED." *J. Lightw. Technol.,* vol. 28, no. 24, pp. 3512–3518, 2010.

[94] T. Q. Wang, C. He, and J. Armstrong, "Angular diversity for indoor mimo optical wireless communications," in *IEEE Int. Conf. on Communications (ICC),* 2015, pp. 5066–5071.

[95] T. Q. Wang, Y. A. Sekercioglu, A. Neild, and J. Armstrong, "Position accuracy of time-of-arrival based ranging using visible light with application in indoor localization systems," *J. Lightw. Technol.,* vol. 31, no. 20, pp. 3302–3308, 2013.

[96] W. L. Wolfe, *Introduction to Radiometry,* vol. 29. SPIE Press, 1998.

[97] Y.-Y. Won, S.-H. Yang, D.-H. Kim, and S.-K. Han, "Three-dimensional optical wireless indoor positioning system using location code map based on power distribution of visible light emitting diode," *IET Optoelectron.,* vol. 7, no. 3, pp. 77–83, 2013.

[98] F.-M. Wu, C.-T. Lin, C.-C. Wei, C.-W. Chen, Z.-Y. Chen, and K. Huang, "3.22-Gb/s WDM visible light communication of a single RGB LED employing carrier-less amplitude and phase modulation," in *Optical Fiber Communication Conf.,* Optical Society of America, 2013, pp. 1–3.

[99] WWRF, *Wireless world research forum.* Available: http://www.wwrf.ch/, 2017

[100] S.-H. Yang, D.-R. Kim, H.-S. Kim, Y.-H. Son, and S.-K. Han, "Visible light based high accuracy indoor localization using the extinction ratio distributions of light signals," *Microw. Opt. Techn. Lett.,* vol. 55, no. 6, pp. 1385–1389, 2013.

[101] S.-H. Yang, H.-S. Kim, Y.-H. Son, and S.-K. Han, "Three-dimensional visible light indoor localization using aoa and rss with multiple optical receivers," *J. Lightw. Technol.*, vol. 32, no. 14, pp. 2480–2485, 2014.

[102] Z. Yang, Z. Zhou, and Y. Liu, "From RSSI to CSI: Indoor localization via channel response," *ACM Comput. Surv. (CSUR)*, vol. 46, no. 2, p. 25, 2013.

[103] M. Yasir, S.-W. Ho, and B. N. Vellambi, "Indoor positioning system using visible light and accelerometer," *J. Lightw. Technol.*, vol. 32, no. 19, pp. 3306–3316, 2014.

[104] M. Yasir, S.-W. Ho, and B. N. Vellambi, "Indoor position tracking using multiple optical receivers," *J. Lightw. Technol.*, vol. 34, no. 4, pp. 1166–1176, 2016.

[105] S. Ye, F. Xiao, Y. Pan, Y. Ma, and Q. Zhang, "Phosphors in phosphor-converted white light-emitting diodes: Recent advances in materials, techniques and properties," *Mater. Sci. Eng. R. Rep.*, vol. 71, no. 1, pp. 1–34, 2010.

[106] M. Yoshino, S. Haruyama, and M. Nakagawa, "High-accuracy positioning system using visible led lights and image sensor," in *IEEE Radio and Wireless Symp.*, 2008, pp. 439–442.

[107] G. Yun and M. Kavehrad, "Spot-diffusing and fly-eye receivers for indoor infrared wireless communications," in *1992 IEEE Int. Conf. on Selected Topics in Wireless Communications. Conference Proceedings (Cat. No.92TH0462-2)*, 1992, pp. 262–265.

[108] H. Zhang, Y. Yuan, and W. Xu, "PAPR reduction for DCO-OFDM visible light communications via semidefinite relaxation," *IEEE Photon. Technol. Lett.*, vol. 26, no. 17, pp. 1718–1721, 2014.

[109] X. Zhang, J. Duan, Y. Fu, and A. Shi, "Theoretical accuracy analysis of indoor visible light communication positioning system based on received signal strength indicator," *J. Lightw. Technol.*, vol. 32, no. 21, pp. 3578–3584, 2014.

[110] X. Zhao, and J. Lin, "Theoretical limits analysis of indoor positioning system using visible light and image sensor," *ETRI J.*, vol. 38, no. 3, pp. 560–567, 2016.

[111] Z. Zhou, M. Kavehrad, and P. Deng, "Indoor positioning algorithm using light-emitting diode visible light communications," *Opt. Eng.*, vol. 51, no. 8, p. 085009–1, 2012.

POSITIONING IN LTE

Ari Kangas,[1] Iana Siomina,[2] and Torbjörn Wigren[1]
[1]4G/5G RAN System Management,
Ericsson AB, SE-16480 Stockholm, Sweden
[2]Ericsson Research, Ericsson AB, SE-16480 Stockholm, Sweden

32.1 INTRODUCTION

Long-term evolution (LTE) [18] is standardized in the Third Generation Partnership Program (3GPP) as an evolution of the Universal Mobile Telecommunications System (UMTS). The chapter discusses the localization technology available in LTE and gives remarks on likely fifth-generation (5G) 3GPP developments.

32.1.1 System Architecture

The LTE system architecture [9], depicted in Figure 32.1, comprises two major functional parts, the *radio access network* and the *core network*. The design philosophy behind the LTE radio network, Evolved Universal Terrestrial Radio Access Network (E-UTRAN), has been to avoid the functional split between different radio network node types, which resulted in the single eNodeB. An eNodeB can support frequency-division duplex (FDD) mode, time-division duplex (TDD) mode, or dual mode operation. There may also be base stations supporting multiple radio access technologies (RATs), including LTE. In later releases positioning beacons and location measurement units (LMUs) have also been added to enhance positioning performance. The design philosophy for the LTE core network, Evolved Packet Core (EPC), has been to make the LTE core network as independent of the radio access technology as possible.

32.1.2 Radio Access Network

A *cell* is the smallest radio network entity having its own identification number, or *cell ID*, publicly visible for the user equipment (UE). Each eNodeB, and each cell has a globally unique ID, which is the eNodeB ID and the evolved cell global identity

Handbook of Position Location: Theory, Practice, and Advances, Second Edition.
Edited by S. A. (Reza) Zekavat and R. Michael Buehrer.
© 2019 by the Institute of Electrical and Electronics Engineers, Inc.
Published 2019 by John Wiley & Sons, Inc.
Companion Website: www.wiley.com/go/zekavat/positionlocation2e

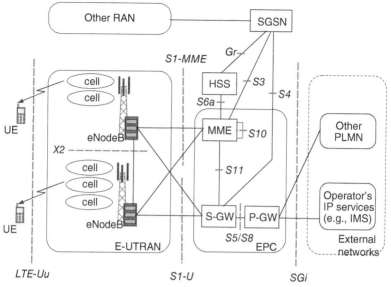

Figure 32.1 LTE system architecture.

(ECGI), respectively. An eNodeB may serve several cells, with antenna sites that are not necessarily co-located.

Radio communication between E-UTRAN and the UE is conducted over the LTE-Uu interface. Within E-UTRAN, the eNodeBs can be interconnected through the logical X2 interface, which is mainly used for mobility and some radio resource management. The E-UTRAN is connected to the EPC through a logical interface S1.

The LTE E-UTRAN interfaces and protocol structures are organized into two logically independent planes, the *control plane* and the *user plane*. The control plane involves the Application Protocol, which operates over several interfaces involving different network nodes (e.g., S1-AP is the Application Protocol transmitted over the S1 interface, and X2-AP is the Application protocol transmitted over the X2 interface), and the signalling bearer for *transporting*, i.e. delivering, the Application Protocol messages. The top protocol level in the control plane is the non-access stratum (NAS), which operates between the UE and the EPC. It uses the radio resource control (RRC) protocol over the LTE-Uu interface and S1-AP over the S1 interface as transport. The user plane includes the data bearers for the data streams transported by a user-plane tunneling protocol.

32.1.3 Core Network

The mobility management entity (MME) is the EPC node responsible for the control-plane functionality, and it is the reference node in the EPC for NAS signaling. The user-plane functionality is implemented in the serving gateway (S-GW), which is separated from the MME. The S1 interface between the eNodeB and the MME, denoted S1-MME, is used in control-plane positioning. The S1 interface between the eNodeB and the S-GW, denoted S1-U, is involved in user plane positioning. User subscription information, mobility and service data are stored in the home subscriber server (HSS) node, which also has the Authentication Center functionality.

32.1.4   Air Interface

Downlink

The transmission scheme used in the downlink (DL) of LTE is orthogonal frequency division multiplexing (OFDM) [18]. In OFDM, modulated symbols are transmitted on many parallel orthogonal subcarriers. In order to preserve orthogonality in multipath channels, a cyclic prefix (CP) is added before the OFDM symbol. The use of the CP simplifies receiver processing in multipath channels since the subcarriers are orthogonal as long as the channel delay spread is smaller than the CP. The structure of the OFDM modulation waveform makes it natural to use fast Fourier transforms (FFTs)/inverse fast Fourier transforms (IFFTs) in implementation, as outlined in Figure 32.2.

In Figure 32.2, modulated symbols a_i, $i = 0, \ldots, N-1$ are mapped on to N subcarriers, starting with negative frequencies $-N/2/N_{fft}, \ldots, -1/N_{fft}$, skipping the zero subcarrier and continuing with positive frequencies $1/N_{fft}, \ldots, N/2/N_{fft}$. Since typical FFT implementations only use positive frequencies, the symbols mapped to negative frequencies $(a_0, \ldots, a_{N/2-1})$ are placed at the end of the IFFT input sequence. Since the FFT size N_{fft} typically is larger than N, the central part of the FFT is zero-padded. The N_{fft}-point IFFT produces an N_{fft}-length time domain sequence s_i, $i = 0, \ldots, N_{fft} - 1$. The last N_{cp} samples of this sequence are copied and prepended to the sequence, resulting in a time-domain sequence of length $N_{fft} + N_{cp}$, which is then digital-to-analog converted and mixed up to the carrier frequency and sent to the antenna port. Phase and amplitude modulation schemes are both used, e.g., binary phase-shift keying (BPSK), quadrature phase-shift keying (QPSK), and quadrature amplitude modulation (QAM).

Figure 32.3 shows a simplified block diagram of an LTE receiver. In the first step, the CP is removed and the remaining N_{fft}-length sequence is processed with an FTT. The N used subcarriers are extracted, and the received signal is multiplied

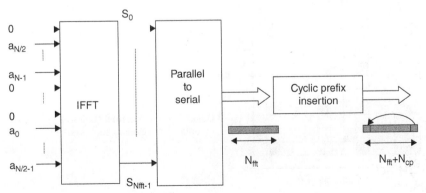

Figure 32.2   A logical block diagram for an example LTE transmitter.

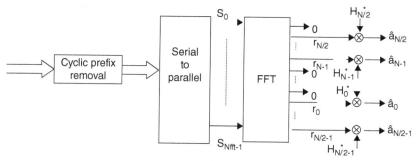

Figure 32.3 A logical block diagram for an example LTE receiver.

with the conjugate of the channel response H_i, $i = 0, \ldots , N - 1$, producing an estimate $\hat{a}_i$ of the transmitted signal a_i.

Figure 32.4 shows an example of an allocation of different physical channels and signals in a *resource block*. When a normal CP is assumed, the resource grid consists of 14 OFDM symbols in time and 12 subcarriers in frequency. Each subcarrier occupies 15 kHz and the OFDM symbol duration including the CP is 1/14 ms. The transmission bandwidth configuration of an LTE carrier is in the range from 6 to 100 resource blocks corresponding to channel bandwidths of 1.4 MHz to 20 MHz. For positioning purposes it is desirable for the UE to be able to perform measurements (e.g., time, signal, strength) on signals that are known to the UE. Examples of such signals in LTE are cell-specific reference signals (CRS) and positioning reference signals (PRS) [6]. For each of the two signal types, 504 different signals exist, and different cells can use up to six different shifts in frequency. CRS and PRS symbols are transmitted on the resource elements shown in Figure 32.4. Further details on PRS are provided in Section 32.6.4. CRS are transmitted every subframe and over the entire bandwidth, being used, for example, for channel estimation.

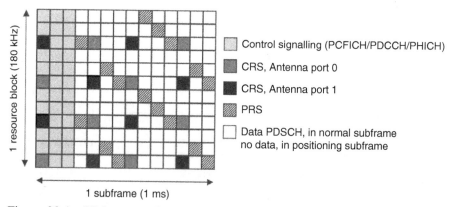

Figure 32.4 PRS pattern for normal CP with two transmit antennas.

Uplink

The access method of the LTE uplink (UL) is single carrier frequency division multiple access (SC-FDMA). Similarly to OFDM, SC-FDMA transmits data over the air interface in many subcarriers. However, it adds an additional processing step—spreading the signal over all subcarriers, which ultimately reduces the power consumption of the UE compared to OFDM. At the receiver end, the signal is demodulated, amplified, and treated by FFT processing in the same way as in OFDM; however, an equalizer is also required due to the spreading at the transmitting end, which causes intersymbol interference. The time-frequency grid of resources is the same as the one defined for the DL. In the radio frequency spectrum the DL and UL is separated when FDD is used.

In LTE Rel-9, only UL angle of arrival (AoA) and eNodeB Rx-Tx time difference positioning measurements utilized the UL radio interface. In LTE Rel-11, UL time difference of arrival (UTDOA) positioning based on sounding reference signals (SRS) was introduced, thereby extending the utilization of the UL radio interface.

SRS are modulated and power-controlled in a way similar to that used for the physical uplink control channel (PUCCH). The modulated signals on SRS are based on so-called Zadoff–Chu sequences [6] and have good auto- and cross-correlation properties. SRS can be periodic and aperiodic, the alternatives are separately configured. Some of the configuration parameters are cell specific, some are user specific, and some can be both. For most subframes, the SRS is transmitted in the last symbol of the subframe. The transmission SRS bandwidth is configurable between four and 96 resource blocks in frequency (which is possible with the 40MHz system bandwidth), and may be either cell specific or user specific. Pseudorandom frequency hopping is also possible for SRS, for example, to randomize the interference and better adapt to frequency selective fading. The SRS hopping may be configured as group hopping, which relies on a pattern of frequency-time resources on which SRS are to be transmitted, and/or as sequence hopping, which implies SRS sequence randomization. The periodicity and the offset from the first subframe in the cell may be configured by cell specific parameters, and values are selected from a predefined set of values. Thus, SRS may be configured and transmitted in each subframe, every second, every fifth or every tenth subframe, whilst the offset, measured in terms of the number of full subframes, can be any integer between zero and nine, that is, any subframe within a radio frame.

32.1.5   Advanced Antenna Systems and Beamforming

In recent releases of LTE functionality exploiting advanced antenna systems (AAS) has been introduced. This opens up new opportunities to do enhanced AoA positioning. The development is expected to accelerate with the introduction of the 5G wireless systems. In the millimeter-wave (mmw) bands considered for parts of the 5G deployment, the antenna gain provided by large antenna arrays will be of central importance to obtain a sufficient coverage [31]. In the lower frequency bands, below 6 GHz, coverage is not a major problem, and therefore the AAS systems are likely to be exploited for capacity enhancement by means of multiple-input-multiple-output (MIMO) transmission [21]. The focus in LTE is also on MIMO communication based on AAS.

Feedback-Based Beamforming

To describe how UE feedback is used to derive angular information, consider Figure 32.5. There the signal ss_{UE} received at the UE antenna is transmitted from a correlated linear antenna array arrangement with antenna element separation $k\lambda$, where λ is the wavelength. The factor k is normally $0.5 - 0.7$. Assuming that the UE is located in the far field of the antenna array, this gives the following expression for the received signal:

$$s_{UE}(t) = \sum_{n=1}^{N} s_n h_n e^{j\omega(t - nk\lambda\sin(\varphi)/c)} = e^{j\omega t} \sum_{n=1}^{N} s_n h_n e^{-j2\pi nk\sin(\varphi)}. \qquad (32.1)$$

Here s_n is the signal transmitted from the nth antenna element, h_n is the corresponding complex channel gain, $\omega = 2\pi f_c$, where f_c is the carrier frequency and c is the speed of light. Note that (32.1) means that the directional information represented by the combination of n, k, and $\sin(\varphi)$ can be added to the signal in base band before modulation. This is done by application of a selected complex baseband precoder in LTE. The precoder is taken from a codebook of beam directions given by steering vector coefficients motivated by (32.1) and given by $w_{m,n} = e^{-jf(m,n)}$. Note that the UE actually estimates $h_n e^{-j2\pi nk\sin(\varphi)}$ rather than h_n. A straightforward way to see that (32.1) represents a beam is to study it for the case $h_n = h$.

The UE finds the best code book entry by a search over the codebook and the complex channel, typically maximizing the signal-to-noise ratio. This is possible since the transmitted signals, s_n, are known reference signals. The so obtained codebook entry is fed back to the eNodeB encoded as precoder matrix index (PMI) information. This way, exploiting channel reciprocity, the eNodeB can acquire information of the AoA as represented by φ of Figure 32.5.

The channel state information reference signals (CSI-RS), which were introduced in LTE release 11, are assigned to a specific antenna port [1]. These reference signals may be transmitted to the whole cell or may be beamformed in a UE-specific manner. In LTE release 13, two classes of CSI-RS reporting modes were introduced.

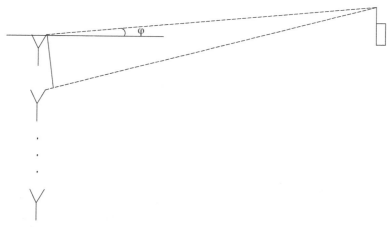

Figure 32.5   Antenna array beamforming.

Class A CSI-RS refers to the use of fixed-beam codebook-based beamforming, while a class B CSI-RS process may send beamformed CSI-RS in any manner. A CSI-RS process in a UE comprises detection of selected CSI-RS signals, measuring interference and noise on CSI-IM, and reporting of the related CSI, in terms of CQI, RI, and PMI. Here CQI denotes Channel Quality Indication, RI denotes (channel matrix) Rank Indication and PMI denotes the selected codebook entry. A UE may report more than one set of CQI, RI, and PMI, that is, information for more than one codebook entry. Up to four CSI-RS processes can be set up for each UE.

The discovery reference signal (DRS) was introduced in LTE release 12. DRS can serve many purposes, for example, supporting cell identification, coarse time/frequency synchronization, as well as intra-/interfrequency measurement of cells. The discovery signals in a DRS occasion are composed of the primary synchronization signal (PSS), secondary synchronization signal (SSS), CRS, and when configured, the CSI-RS.

As stated above, the codebook of the 3GPP standard is defined to represent certain directions. In release 13, directions in both azimuth and elevation are defined, thereby allowing two-dimensional (2-D) beamforming to be used, where 2-D refers to the existence of azimuth and elevation information. The codebooks are specified in detail in [1]. That technical report also discusses the antenna port mappings to achieve different antenna configurations. In order to illustrate that the codebooks indeed define specific directions, it can be noted that the formulas for one of the sets of azimuth and elevation codebooks are

$$w_m = \frac{1}{\sqrt{M}} e^{-j2\pi k(m-1)\cos(\theta_{etilt})}, \quad m = 1,...,M, \tag{32.2}$$

$$w_{l,i} = \frac{1}{\sqrt{L}} e^{-j2\pi k(l-1)\sin(\theta_i)}, \quad l = 1,...,L. \tag{32.3}$$

The structure is evidently as in (32.1). The 2-D beams are obtained by multiplication of the azimuth and elevation codebooks. The angles θ_{etilt} and θ_i are defined in [1].

Reciprocity-Based Beamforming

Channel reciprocity is a consequence of fundamental electrodynamics, and it means that the DL and UL channels of the same specific frequency band are the same at a specific time [21]. Therefore, knowledge of the UL channel can be exploited for DL transmission as long as the latency between the UL channel estimation and the DL transmission is below the channel decorrelation time. In LTE the TDD mode exploits this, for example, for high-capacity optimized MIMO transmission in the DL.

The UL channel estimate is normally obtained from measurements on SRSs transmitted from the UE. Based on these, the eNodeB can estimate the complex channel matrix **H** from each transmit antenna element to each receive antenna element. Given the SRS measurements and **H**, a variety of direction-of-arrival estimation techniques can be applied, which can be exploited for positioning, see, for example, [34].

32.2   REQUIREMENTS ON POSITIONING IN LTE

32.2.1   3GPP Requirements

3GPP specifies functional requirements [2] that describe the minimum set of mechanisms for supporting and developing various location services (LCS), as well as a set of the attributes to describe or characterize LCS. The 3GPP also specifies performance requirements [5].

32.2.2   Emergency Positioning

The U.S. Federal Communication Commission (FCC) enhanced 911 (E-911) emergency positioning requirements [19] mandate the positioning performance that must be met by cellular operators in the U.S. To account for the increased amount of cellular indoor traffic, the FCC tightened its requirements in February 2015 to specify:

- A 50-meter horizontal accuracy must be provided for 40%, 50%, 70%, and 80% of *all* emergency calls within 2, 3, 5, and 6 years, respectively.

- For vertical positioning performance, operators should propose an accuracy metric within 3 years to be approved by the FCC. Operators need to comply with the metric within 6 years. To facilitate an enhanced vertical accuracy, all services must make uncompensated barometric data available from any handset that has the capability to deliver barometric sensor data.

- The response time shall be less than 30 seconds after the E-911 call is placed for 90% of all calls.

Support for barometric measurement reporting was introduced in LTE release 14.

The above requirements are intended to eventually replace the earlier requirements of the FCC:

- A horizontal accuracy of 50 m and 150 m in 67% and 95% of all positioning attempts for UE-based and UE-assisted positioning methods.

- A horizontal accuracy of 100 m and 300 m in 67% and 90% of all positioning attempts for network-based positioning methods

- A response time less than 30 seconds after the E-911 call is placed for 90% of the calls.

The accuracy requirements implicitly require an availability of at least 90%. This creates a difficult situation, since much more than 50% of all mobile calls are placed indoors where the most accurate method, assisted GPS (A-GPS) [23], has a very limited availability. As a result, cellular operators are forced to use other positioning methods as a fallback or select other methods than A-GPS for high-accuracy positioning [14]. The fallback methods also need to be fast to meet the overall 30-second requirement since they are sometimes applied in a sequential manner.

The current emergency call delivery standards in the U.S. impose additional requirements by a limitation on the allowed geometrical shapes used by cellular systems for reporting of positions [3]. For LTE, the consequence is a requirement to be able to transform between the shapes outlined in [3].

32.2.3 Location-Based Services

One consequence of the growing number of location-based services (LBS) is that the LTE system needs to be prepared for service-dependent positioning to a higher degree than previously. The positioning node therefore needs to be able to select the best mix of positioning technologies to meet the requested positioning quality of service (QoS).

The large bandwidth in LTE enables real-time video transmissions from and between terminals. Today's smartphones also integrate inertial sensors [25] like accelerometers and gyrometers. This development is expected to have implications for the future positioning technology of LTE and 5G wireless systems. The tactile internet is one particularly interesting emerging technology. There a combination of transmitted audio, video, locations, and force feedback information is bound to enable new advanced virtual reality based applications; see, for example, [20] and [31]. See also Chapters 29 and 33 of this book for discussion of two additional emerging applications.

The convergence of the telecommunication and computer industries means that it will no longer be acceptables for cellular operators to mainly provide high-performance positioning services outdoors. Users will expect applications to work irrespective of where they are located, with indoor and outdoor performance not being too different. Here high-accuracy AoA positioning could prove to be particularly useful in small cells without too much non- line-of-sight (NLOS) radio propagation.

32.2.3 Positioning Architecture and Signaling in LTE

The three important network elements in any positioning architecture are the *LCS client*, the *LCS target,* and the *LCS server.* The LCS server is a physical or logical entity managing positioning for a LCS target device by obtaining measurements and other location information, providing assistance data to assist UEs in measurements, and computing or verifying the final position estimate. A LCS client is a software and/or hardware entity that interacts with a LCS server for the purpose of obtaining location information for one or more LCS targets, that is, the entities being positioned. LCS clients may reside in the LCS targets themselves. LCS clients subscribe to LCS to obtain location information, and LCS servers process and serve the received requests and send the positioning result to the LCS target.

Depending on the positioning method, a location may be obtained based on radio signals received by and/or transmitted from the LCS target. The radio signals transmitted from the LCS target are typically received by eNodeBs, and the radio signals received by the LCS target are typically transmitted from eNodeBs or satellites. However, to further enhance the positioning methods based on radio signals received by the LCS target, an additional terrestrial beacon system (TBS) may be deployed. The TBS consists of a network of ground-based transmitters, broadcasting signals only for positioning purposes. The signals comprise the non-LTE metropolitan beacon system (MBS) signals or the LTE PRS, with these signals corresponding to two different beacon types. To enhance positioning methods based on radio signals transmitted from the LCS target, additional LMUs may be deployed in a standalone manner. The LMUs may also be colocated with or integrated into eNodeBs. The purpose of the LMUs is to perform measurements on UL radio signals transmitted by the LCS target and to report the measurements to the LCS server.

The positioning result comprises estimated location coordinates, although it may also include a velocity estimate or the location failure indication in case of a failure. Seven formats for reporting location coordinates are currently supported in LTE [3]: ellipsoid point, ellipsoid point with uncertainty circle, ellipsoid point with uncertainty ellipse, polygon, ellipsoid point with altitude, ellipsoid point with altitude and uncertainty ellipsoid, and ellipsoid arc.

Example 32.1: 3GPP Encoding of Position Reporting Formats

An ellipsoid point is a point on the surface of the World geodetic system 84 (WGS84) earth ellipsoid [3, 25], described by a latitude and a longitude. The latitude is the angle between the equatorial plane and the normal to the tangent plane to the ellipsoid surface at the ellipsoid point. Positive latitudes correspond to the Northern Hemisphere. The longitude is the angle between the half-plane determined by the Greenwich meridian and the half-plane defined by the ellipsoid point and the polar axis, measured eastward. The coordinates of an ellipsoid point are coded with an uncertainty of less than 3 m. The latitude is coded with 24 bits: 1 bit of sign and a number between 0 and $2^{23} - 1$ binary coded with 23 bits. The longitude, expressed in the range $[-180°, +180°]$ is coded as a number between -2^{23} and $2^{23} - 1$, with a binary 2-complement representation.

Solution: The formal encoding of the ellipsoid point is performed as follows [3].

```
Ellipsoid-Point ::= SEQUENCE {
    latitudeSign ENUMERATED {north, south},
    degreesLatitude INTEGER (0..8388607),
    degreesLongitude INTEGER (-8388608..8388607)
}
```

The gateway mobile location center (GMLC) is the first node an external LCS client accesses in a public land mobile network (PLMN), through the Le interface. After performing registration authorization, it sends positioning requests to the MME and receives final location estimates from the corresponding entity via the SLg interface. A location retrieval function (LRF), which may or may not be colocated with the GMLC, is responsible for retrieving or validating location information, providing routing and/or correlation information of a UE that has initiated an IP multimedia subsystem (IMS) emergency session.

In LTE release 9, two new protocols were standardized specifically to support positioning, the LTE positioning protocol (LPP) [8] and the LTE positioning protocol annex (LPPa) [10]. The LPP [8] is a point-to-point protocol between a LCS server and a LCS target device used in order to position the target device. The following transactions have been specified: capability transfer procedure (request/provide

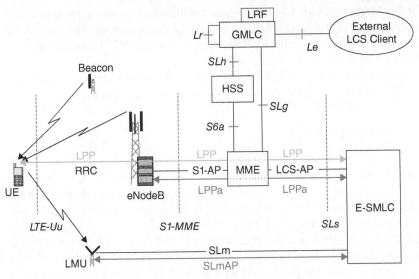

Figure 32.6   LTE positioning architecture, control plane.

messages), assistance data transfer procedure (request/provide messages), and a location information transfer procedure (request/provide messages). Multiple LPP procedures of any of the aforementioned types can be used in series and/or in *parallel*. LPP is used both by control-plane (see, e.g., Fig. 32.6) and user-plane (see, e.g., Fig. 32.7) positioning solutions.

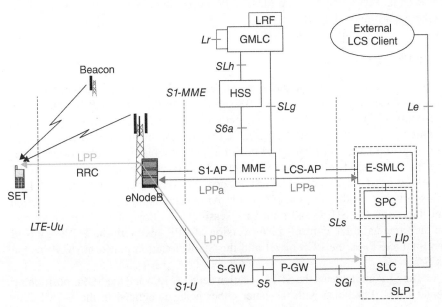

Figure 32.7   LTE positioning architecture, user plane.

LPPa is specified only for control-plane positioning procedures. However, with the user plane and the control plane interworking, LPPa can also assist the user-plane solution by querying eNodeBs for information and eNodeB measurements not related to a UE connection.

Another positioning protocol, SLmAP, over the SLm interface between the LMU and the LCS server was introduced in LTE release 11 to support UTDOA.

32.3.1   Control Plane

To support LCS, at least two functional nodes must be present in the LTE control-plane architecture: the evolved serving mobile location center (E-SMLC), which controls the coordination and scheduling of the resources required to locate the mobile device, and the GMLC, which controls the delivery of position data, user authorization, charging, and more. The LPP messages are transmitted transparently to the MME, using RRC as a transport over the LTE-Uu interface between the UE and the E-SMLC, using S1-AP over the S1-MME interface, and using LCS-AP over the SLs interface between the eNodeB and the E-SMLC. LPPa conducts the LPPa location information transfer procedures for positioning-related information. LPPa is also transparent to the MME, which routes the LPPa packets transparently over the S1-MME and SLs interfaces without knowledge of the involved LPPa transaction. The LTE positioning architecture for the control plane is shown in Figure 32.6. To describe the operation of the architecture, consider the case where the MME receives a positioning request for some LCS associated with a particular LCS target (e.g., a UE). The MME then sends a LCS request in a LCS-AP Location Request message [4] to an E-SMLC. The E-SMLC processes the location services request to perform a positioning of the target UE. The E-SMLC then returns the result of the location service back to the MME. The MME then forwards the result at least to the requesting node.

32.3.2   User Plane

In general, secure user plane location (SUPL) [29, 30] supports and complements control-plane protocols to enable LBS support with the least-possible impact on the control plane and the deployed network. The LTE user-plane positioning architecture is shown in Figure 32.7. SUPL uses established data-bearing channels (i.e., the LTE user plane) and positioning protocols (i.e., LPP) for exchanging the positioning-related data between a LCS target and a LCS server. In the general user-plane protocol stack, SUPL occupies the application layer, with LPP transported as another layer above SUPL. After establishing a TCP/IP connection and initiating the SUPL and then LPP sessions, the flow of LPP messages can be the same as in the control- plane version of LPP, just with the SUPL enabled terminal (SET) as the LCS target and the SUPL location platform (SLP) as the LCS server.

The SLP implements the SUPL location center (SLC) and the SUPL positioning center (SPC) functions, with the latter either being integrated in the E-SMLC or attached to it with a proprietary interface. The SLC system coordinates the operations of SUPL in the network and implements the following SUPL functions as it interacts with the SET over the user- plane bearer: privacy function, initiation function, security function (c.f. chapter 3), roaming support, charging function, service management, and position calculation. The SPC supports the following SUPL functions: security function, assistance delivery function, SUPL reference retrieval function (e.g., retrieving data from a GPS reference network), and SUPL position calculation function.

32.4 POSITIONING PROCEDURES IN LTE

32.4.1 Signaling of Client Type and QoS

The type of service is normally received in the E-SMLC in the form of a Client Type [4]. The Client Type may be determined in the MME, the GMLC, or some other node from the knowledge of the type of service that requests positioning. Currently there are eight Client Types available in LTE [4].

The most important QoS parameters are the response time, the horizontal accuracy and the vertical accuracy. In LTE, these can be signalled to the E-SMLC from the MME node [4]. The QoS parameters can also be requested from the UE over the LPP protocol [8], that is, after the positioning session has been established. The response time is within $[1,128]$ s. The horizontal accuracy is expressed as 128 accuracy codes, corresponding to uncertainty radii of uncertainty circles. The vertical accuracy is expressed as 128 other accuracy codes. Both the horizontal and vertical accuracy are associated with a confidence value. This confidence value expresses the probability that the UE is located within the uncertainty region associated with the positioning result. The importance of the confidence cannot be underestimated, since positioning results are statistical quantities. The accuracy alone may, for example, be enhanced by a reduction of the associated confidence. See [36, 38, 40], and [45] for further details.

32.4.2 Positioning Method selection

The E-SMLC uses the Client Type and the QoS information to determine the best possible use of the available positioning resources. One way to do this is to first let the Client Type determine a service class, mapped to a set of LBSs, each service class with its own positioning sequence.

The Positioning Sequence and Prior Performance Information

When the method for the first positioning attempt is to be determined, the positioning method selection mechanism must rely on stored, possibly preconfigured, information regarding the performance of the different available positioning methods. A more accurate solution than preconfiguration is to build up the prior information on positioning performance adaptively [45]. This may be done by registration of the obtained positioning results and the associated QoS information in the E-SMLC. This registered information can then be used for online generation of histograms describing the probability density functions of the response time, the horizontal accuracy, the vertical accuracy, and the availability for each positioning method.

To determine the positioning method for the first attempt, one approach is to check QoS parameters sequentially, which is used in some UMTS implementations where the order of precedence of the QoS information is specified [35]. More advanced approaches, suitable for LTE, include the use of performance indices that can be used to find a best fit to the requested QoS.

Once the first positioning attempt has been performed, the position method selection mechanism may proceed with one or more positioning reattempts in case the requested QoS has not been met.

QoS evaluation

In order to apply the above principles, methods for calculation of the achieved QoS need to be available. The achieved positioning time can be directly measured with timers. The vertical accuracy is a one-dimensional quantity, and there are no computations needed to assess it. The computation of the achieved horizontal accuracy is more intricate since the uncertainty region associated with each geographical reporting format [3] needs to be transformed to a radius of an uncertainty circle to comply with the reporting format for horizontal accuracy [8]. One common principle is to first compute the area of the reported geographical region. The radius of a circle with the same area is then computed and used as the achieved horizontal accuracy.

The most important geometrical shapes horizontally are the polygon, the ellipse, and the ellipsoid arc. To describe the QoS computations, the corners of the 3GPP polygon [3] are denoted $(x_i\ y_i)^T$, $i = 1,...,N_p$. It is assumed that the last corner is identical to the first. Furthermore, the inner radius of the ellipsoid arc [3] is denoted by R, the thickness is denoted by ΔR, and the opening angle in radians is denoted by α. The semimajor and semiminor axes of the 3GPP ellipse [3] are denoted by a and b, respectively.

The area of the polygon can now be computed as in [13]. The area of the ellipse is πab, while the area of the ellipsoid arc is easily computed as the difference of the areas of the circle sectors limited by the outer and inner radius. The equivalent radius of the uncertainty circle is then computed by setting these areas equal to πr^2, and solving for r. This gives the uncertainty radii of the equation below, which are then quantized as described in [3]:

$$
\begin{aligned}
r_{polygon} &= \sqrt{\frac{\sum_{i=0}^{N_p-1}(x_i y_{i+1} - x_{i+1} y_i)}{2\pi}} \\
r_{ellipsoid\ arc} &= \sqrt{\frac{\alpha}{2\pi}\Delta R(2R + \Delta R)} \\
r_{uncertainty\ ellipse} &= \sqrt{ab}.
\end{aligned}
\tag{32.4}
$$

Example 32.2: QoS of Ellipsoid Point with Uncertainty Ellipse

The QoS computations described above are necessary, for example, to determine if a positioning sequence can be terminated or if it needs to improve the QoS by applying another positioning method. One QoS computation algorithm is needed for each of the seven reporting formats of LTE. The example code Chapter_32_Example_2.m illustrates how the QoS is computed in one such case, for an ellipsoid point with uncertainty ellipse. As stated above, the uncertainty is represented by the radius of a circle that has the same area as the ellipse. The formula of (32.4) is used for the computation. The example code also contains the encoding/decoding steps according to [3], thereby providing further insight and examples of this.

Solution: The example code is run from the MATLAB workspace. When executing the code, the variables positionFormat and positionFormatDescription are expected to be present in the work space of MATLAB. positionFormat contains the name of the input reporting format, while positionFormatDescription contains a 3GPP-encoded ellipsoid point with uncertainty ellipse. The 3GPP-encoded format contains the following elements (in order): The number of the format, the encoded latitude sign, the encoded latitude, the encoded longitude, the encoded semiminor axis, the encoded semimajor axis, the encoded orientation angle (clockwise from North), and the encoded confidence. A typical MATLAB command sequence is as follows:

```
>> positionFormat='ELLIPSOID_POINT_WITH_UNCERTAINTY_ELLIPSE';
>> positionFormatDescription=[3 0 400 400 25 15 130 39];
>> Chapter_32_Example_2
>> AccuracyCode

   AccuracyCode =

     20
```

It can be noted that the encoded accuracy code is between the encoded semiminor and semimajor axes, as can be expected.

32.5 COORDINATES

To measure, compute, signal, and use cellular positioning results, it is necessary to agree on two basic things—the coordinate system and the time.

32.5.1 Time

The Universal Time Code (UTC) and GPS system time are two common time bases used for cellular positioning. UTC is derived from two time scales, the first one being defined by atomic clocks and the second one by the rotation of the Earth. The international atomic time is associated with the SI-definition of the second. The time scale related to the rotation of the Earth helps in defining the so-called Earth-centered

Earth-fixed (ECEF) coordinate frame. The two time scales are merged to provide UTC time, which is aligned with solar time [26, 27].

GPS system time is derived from atomic clocks in the GPS space vehicles (SVs) and in the GPS ground control segments [25]. GPS system time therefore drifts with respect to UTC and in case the LTE system is not synchronized to GPS system time, the relation to the time base used needs to be maintained. GPS system time is expressed in terms of the GPS week number and the GPS time of week, counted from midnight Saturday/Sunday.

32.5.2 Coordinate Systems

The A-GPS and assisted Global Navigation Satellite System (A-GNSS) methods utilize Cartesian ECEF coordinates for measurements and position calculation, whereas the other positioning methods of LTE utilize Cartesian Earth tangential (ET) coordinate systems. Assistance data for A-GPS and A-GNSS, as well as results, are signaled as latitudes, longitudes, and altitudes expressed in the WGS84 geodetic model [3]. Figure 32.8 defines these coordinate systems and the most important variables. The latitude of the location **r** is measured as the angle between the equatorial plane of the WGS84 Earth ellipsoid, and the normal of the tangent plane at **r**. The longitude is counted eastward from the Greenwich meridian.

The ECEF coordinate system has the origin in the center of the WGS84 ellipsoid, the x-axis ($\hat{x}_{ECEF}$) in the Greenwich meridian plane and the y-axis ($\hat{y}_{ECEF}$) 90° to the east of $\hat{x}_{ECEF}$. The z-axis of the ECEF coordinate frame coincides with the axis of rotation of the WGS84 ellipsoid.

The ET coordinate system has the origin on the surface of the WGS84 ellipsoid with the x-axis $\hat{x}_{ET}$ pointing East, the y-axis ($\hat{y}_{ET}$) pointing North and the z-axis ($\hat{z}_{ET}$) pointing up. The latitude at **r** is denoted θ, the longitude φ, the altitude h, and Cartesian coordinates in any coordinate frame are denoted $(x\ y\ z)^T$.

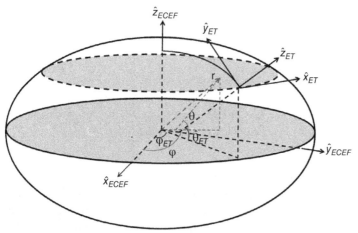

Figure 32.8 Coordinate systems and variables for LTE positioning.

32.5.3 Coordinate Transformations

It is assumed that the positions handled by the E-SMLC are relatively close. The vectors from the center of the Earth to $\mathbf{r}$ and from the center of the Earth to the center of the ET coordinate system are then almost parallel, a fact that allows the altitude to be directly added to the ET coordinates and to be neglected when transforming from WGS84 coordinates to ECEF coordinates and to ET coordinates.

Using the ellipsoidal geometry of Figure 32.8, the transformation of $\mathbf{r}$ from WGS84 latitude and longitude to Cartesian ECEF coordinates is

$$
\begin{aligned}
x_{ECEF} &= N \, \cos(\theta) \, \cos(\varphi) \\
y_{ECEF} &= N \, \cos(\theta) \, \sin(\varphi) \\
z_{ECEF} &= \left(\frac{b}{a}\right)^2 N \sin(\theta) \\
N &= \frac{a}{\sqrt{1 - e^2 \sin(\theta)^2}},
\end{aligned}
\tag{32.5}
$$

where e, a, and b denote the eccentricity, the semimajor axis and the semi-minor axis of the WGS84 ellipsoid, respectively, and N is a help variable. Similarly, the selected center of the ET coordinate system is transformed to ECEF coordinates, resulting in the position $\left(x_{ECEF}^{ET_{system}} \; y_{ECEF}^{ET_{system}} \; z_{ECEF}^{ET_{system}}\right)^T$ The transformation from ECEF to ET coordinates of the position $\mathbf{r}$ is then

$$
\begin{aligned}
x_{ET} &= -\sin\left(\varphi^{ETsystem}\right)\left(x_{ECEF} - x_{ECEF}^{ETsystem}\right) \\
&\quad + \cos\left(\varphi^{ETsystem}\right)\left(y_{ECEF} - y_{ECEF}^{ETsystem}\right) \\
y_{ET} &= -\sin\left(\theta^{ETsystem}\right)\cos\left(\varphi^{ETsystem}\right)\left(x_{ECEF} - x_{ECEF}^{ETsystem}\right) \\
&\quad - \sin\left(\theta^{ETsystem}\right)\sin\left(\varphi^{ETsystem}\right)\left(y_{ECEF} - y_{ECEF}^{ETsystem}\right) \\
&\quad + \cos\left(\theta^{ETsystem}\right)\left(z_{ECEF} - z_{ECEF}^{ETsystem}\right) \\
z_{ET} &= h.
\end{aligned}
\tag{32.6}
$$

In the other direction the ET-to-ECEF transformation becomes

$$
\begin{aligned}
x_{ECEF} &= x_{ECEF}^{ETsystem} - \sin\left(\varphi^{ETsystem}\right)x_{ET} \\
&\quad - \sin\left(\theta^{ETsystem}\right)\cos\left(\varphi^{ETsystem}\right)y_{ET} \\
y_{ECEF} &= y_{ECEF}^{ETsystem} + \cos\left(\varphi^{ETsystem}\right)x_{ET} \\
&\quad - \sin\left(\theta^{ETsystem}\right)\sin\left(\varphi^{ETsystem}\right)y_{ET} \\
z_{ECEF} &= z_{ECEF}^{ETsystem} + \cos\left(\theta^{ETsystem}\right)y_{ET}.
\end{aligned}
\tag{32.7}
$$

Finally, the transformation from ECEF to geodetic coordinates is given by

$$
\begin{aligned}
\theta &= \text{sign}\left(z_{ECEF}\right)\arccos\left(\frac{x_{ECEF}^2 + y_{ECEF}^2}{M^2}\right) \\
\varphi &= \text{atan}2\left(\frac{y_{ECEF}}{x_{ECEF}}\right) \\
h &= z_{ET} \\
M &= \sqrt{x_{ECEF}^2 + y_{ECEF}^2 + \left(\frac{a}{b}\right)^4 z_{ECEF}^2},
\end{aligned}
\tag{32.8}
$$

where atan 2(·) denotes the four-quadrant inverse of the tangent function.

Example 32.3: Coordinate Transformations

The coordinate transformations described previously are used in virtually all computations of a cellular positioning systems. The reason is that most cellular measurements are performed in an environment best described in an ET coordinate system, this is, for example, the case when the travel time of radio waves between base stations and terminals are used. It is therefore of central importance to have a good understanding of these transformations. The example code Chapter_32_Example_3.m illustrates how the transformation from WGS84 latitude/longitude to ET coordinates (and back) works in one such case, for the polygon geographical format of LTE. The example code includes the encoding/decoding steps according to [3], thereby providing further insight in format encoding/decoding. The example code makes use of (32.5–32.8), which are implemented sequentially. The code hence exploits the simplifying assumption that the altitude is handled separately in the coordinate transformations.

Solution: The example illustrates the transformation for a small four-corner polygon close to the equator. The example code is run from the MATLAB workspace. When executing the code, the variable `positionFormatDescription` is expected to be present in the work space of MATLAB. `positionFormatDescription` contains the 3GPP-encoded polygon. In this case the polygon format contains the following elements (in order): the number of the polygon format, the number of corners in the polygon, the encoded latitude sign of the first corner, the encoded latitude of the first corner, the encoded longitude of the first corner, the encoded latitude sign of the second corner, the encoded latitude of the second corner, the encoded longitude of the second corner, the encoded latitude sign of the third corner, the encoded latitude of the third corner, the encoded longitude of the third corner, the encoded latitude sign of the fourth corner, the encoded latitude of the fourth corner, and the encoded longitude of the fourth corner. After definition of the polygon, the transformations are performed by typing Chapter_32_Example_3 in the MATLAB command window. The result is then available in the variables CellLatLong (the latitude and longitude of the four corners), CellXYZ (the ECEF coordinates of the four corners), polygonXY (the ET coordinates of the four corners, with first corner selected as the ET origin). The command sequence is as follows:

```
>> positionFormatDescription=[5 4 0 418 -210 0 376 188 1 418
209 1 376 -189]; >> Chapter_32_Example_3 % Execute the code

>> CellLatLong % Display the latitudes and longitudes of the
polygon corners
CellLatLong =

1.0e-004 *

   0.7827      0.7041     -0.7827     -0.7041
  -0.7865      0.7041      0.7827     -0.7078.

>> CellXYZ % Display the ECEF coordinates

CellXYZ =

1.0e+006 *

   6.3781      6.3781      6.3781      6.3781
  -0.0005      0.0004      0.0005     -0.0005
   0.0005      0.0004     -0.0005     -0.0004,

>> polygonXY % Display    the ET coordinates

polygonXY =

1.0e+003 *

   0          0.9507      1.0008      0.0502
   0         -0.0498     -0.9918     -0.9419
```

32.6   POSITIONING METHODS IN LTE

To meet LBS demands, the LTE network will deploy a range of methods, which have different performance. Depending on where the measurements are conducted and the final position is calculated, the methods can be *UE- based*, *UE-assisted*, or *network-based*, each having specific advantages. The following methods are available in the LTE standard for both the control plane and the user plane [7]:

- Cell ID (CID);
- UE-assisted and network-based enhanced CID (E-CID), including network-based AoA;
- UE-based and UE-assisted A-GNSS,

- UE-assisted observed time difference of arrival (OTDOA),
- UTDOA.

Hybrid positioning and fingerprinting positioning do not require additional standardization and are therefore also possible [41].

32.6.1   CID

Given the CID of the serving cell, the UE position is associated with the cell coverage area which can be described, for example, by a prestored polygon. The polygon format is one of the standardized positioning reporting formats in 3GPP [3], where a polygon is defined as a list of 3–15 corners, with each corner represented by latitude and longitude encoded in the WGS84 system [3]. The cell boundary is modeled by the set of non-intersecting polygon segments connecting all the corners. The UE is assumed to be within the polygon with a certain confidence, although the polygon format as such does not contain the confidence information. This is the fastest positioning method, since no measurements are needed.

It can be noted that for high carrier frequencies like in 5G, cells are expected to be small, thereby increasing the usefulness of CID. Provided that a beam ID could be used as an enhancement to CID further accuracy improvements would result.

32.6.2   E-CID

E-CID methods exploit four sources of position information: the CID and the corresponding geographical description of the serving cell, the timing advance (TA) of the serving cell, the CIDs and the corresponding signal measurements of the cells (up to 32 cells in LTE, including the serving cell), and AoA measurements. The following measurements are available for E-CID in LTE [5, 7]:

- *UE measurements:* E-UTRAN carrier received signal strength indicator (RSSI), reference signal received power (RSRP), reference signal received quality (RSRQ), and UE receiver - transmitter (Rx-Tx) time difference;
- *E-UTRAN measurements*: TA Type 1 (eNodeB Rx-Tx time difference) + (UE Rx-Tx time difference), TA Type 2 (eNodeB Rx-Tx time difference), and UL AoA,

where the UE E-CID measurements are reported by the UE to the E-SMLC over the LPP protocol, and the E-UTRAN E-CID measurements are reported by the eNodeB to the E-SMLC over the LPPa protocol (see Section 32.3). It should be noted that time measurements are expected to become more accurate in 5G 3GPP wireless systems since the radio bandwidth increases [31]. AoA accuracy is also expected to improve significantly both in LTE and 5G 3GPP systems with the introduction of AAS [1, 31].

Next, four techniques for E-CID are described.

CID and TA

One common E-CID method combines the geographical cell description, the eNodeB position, and the distance between the eNodeB and the UE obtained from a time measurement. A previous example of this technique is round trip time (RTT) positioning in the wideband code division multiple access (WCDMA) system [12, 35, 46]. Applied to the LTE situation, the distance is obtained from the TA time measurement as

$$R_{TA} = \frac{c \cdot TA}{2}, \tag{32.9}$$

where c is the speed of light. The uncertainty of the TA distance ΔR_{TA} may be determined by field trials [35] and can then be configured in the system as a parameter. In the Global System for Communications (GSM) the granularity of the TA is of the order of 1 km. In LTE the granularity is much finer, of the order of 10 m. Since the bandwidth of LTE is normally larger than that of WCMDA, a radial accuracy at least as good as for RTT-positioning in WCDMA is expected. Note that the accuracy varies with the radio environment (c.f. [35]).

The combination of the TA distance with the geographical cell description can be performed in several ways. Figure 32.9 illustrates an algorithm for a case where the cell boundary is described by a 3GPP polygon [3]. In the figure, the resulting ellipsoid arc describes the intersection between the 3GPP polygon and the 360° arc corresponding to R_{TA} with uncertainty ΔR_{TA}. To compute the left and

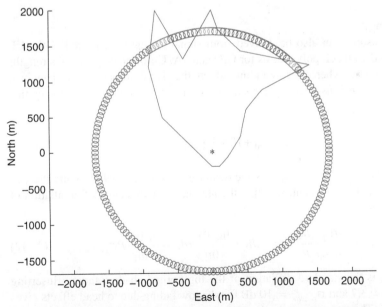

Figure 32.9 Combination of TA and the serving cell polygon. The eNodeB is marked with a blue star, the cell polygon is plotted with a blue solid line, test points are plotted as red circles, with green circles being located in the interior of the cell polygon. The resulting ellipsoid arc is plotted with a blue solid line.

opening angle of the ellipsoid arc, test points are distributed uniformly within the circular strip at a distance of $R_{TA} + \Delta R_{TA}/2$ from the eNodeB. Next, it is checked for each test point $(x_0, y_0)^T$ whether it is in the interior of the polygon. The test exploits a test ray from the test point to infinity, parallel to the East axis of the coordinate system. When a test points is in the interior, the cell polygon must be intersected an odd number of times when the test ray moves from the test point to infinity. The crossings with the polygon boundary are easily determined by checking for intersections between the test ray and the line segments between two adjacent cell polygon corner points. The intersection between the horizontal ray $y = y_0$, $x \geq x_0$, and the line segment between the polygon corners with index i, $(x_i^{ET} \quad y_i^{ET})^T$, and $i + 1$, $(x_{i+1}^{Et} \quad y_{i+1}^{ET})^T$ of the cell polygon with N_P corners, is given by a solution to

$$
\begin{pmatrix} x \\ y_0 \end{pmatrix} = \alpha \begin{pmatrix} x_{i+1}^{ET} \\ y_{i+1}^{ET} \end{pmatrix} + (1-\alpha) \begin{pmatrix} x_i^{ET} \\ y_i^{ET} \end{pmatrix}, \quad x \geq x_0, \quad 0 \leq \alpha < 1, \tag{32.10}
$$

when a unique solution exists. Repetition of this procedure for all line segments between corners allows for a count of the number of intersections for one specific test point. The complete algorithm of [35] and [46] repeats this for all test points. In order to find the angles, a search is performed for the largest set of adjacent test points that are exterior to the polygon. The complement of this set corresponds to the set of test points that defines the ellipsoid arc. Note that this procedure handles the case with more than one intersection between the circular strip and the cell polygon; see Figure 32.9.

Signal Strength

Distance measures can also be derived from signal strengths measured in the UE and combined with cell polygons as for CID and TA. Unfortunately, signal strengths become inaccurate when shadow fading affects the signal propagation of handheld UEs. This can be illustrated by consideration of the Okumura–Hata propagation model:

$$
L = L_0 + 10\eta \cdot \log_{10}(R), \tag{32.11}
$$

where L is the path loss in dB, R is the distance between the eNodeB and the UE, L_0 is the attenuation constant, and η is the attenuation exponent. Differentiation of (32.11) results in

$$
dL = \frac{10\eta}{\ln(10)} \frac{dR}{R} \quad \Leftrightarrow \quad dR = \frac{\ln(10)}{10\eta} RdL \approx \frac{\ln(10)}{10\eta} R\sigma_{shadow}, \tag{32.12}
$$

where σ_{shadow} is the standard deviation of the log-normal shadow fading. Inserting the values $\eta = 3.7$ and $\sigma_{shadow} = 10$ dB for shadow fading due to head effects gives $dR = 0.62R$. This means that the 1-sigma error of a received signal strength measurement amounts to more than 60% of the measured value. The inaccuracy associated with raw signal strength measurements is demonstrated in the following example.

Example 32.4: Shadow Fading Due to Head Effects

The example highlights the fact that the orientation of handheld UEs has a very significant effect on the accuracy of any positioning method that relies on signal strength measurements. The reason is obvious considering the direction-dependent damping of the head and body of the users when different eNodeBs are considered.

Solution

In the example, the signal strength of six eNodeBs was generated with the Okumura–Hata model (32.11) in multiple points of a cell. The signal strengths were quantized as high/low, resulting in signals strengths characterized by 1 of 64 different cases, for each point in the cell. The geographical points appear color-coded according to the signal strength case in Figure 32.10. The smearing due to shadow fading of the geographical information content in the combined signal strength measurements is very significant.

Admittedly, the large uncertainty of a single signal strength measurement is reduced somewhat when several measurements with respect to different eNodeBs are combined. Relative signal strengths, using one eNodeB as the reference, can contribute to a further reduction of the errors. The use of advanced pattern matching techniques and advanced signal processing is known to improve accuracy even more [47]. Further comments are given in Chapters 11–14 and in the discussion on fingerprinting positioning in Section 32.6.3.

AoA: early LTE releases

The AoA measurement standardized for LTE is defined as the estimated angle of a UE with respect to a reference direction, which is the geographical North, positive in a clockwise direction. AoA can improve accuracy as compared to the CID and TA method by reducing the angular uncertainty, as shown in Figure 32.11. For a given beam direction φ and beam width $\Delta\varphi$, the set of possible UE positions are all

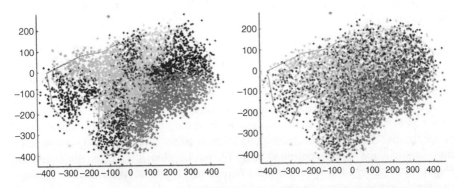

Figure 32.10   Inaccuracy with shadow fading. No shadow fading (left) and 6dB shadow fading (right). The eNodeB positions are marked with stars. The colors represent positions with the same fingerprinted signal strength.

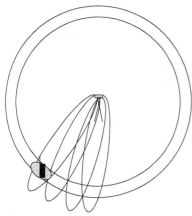

Figure 32.11 Combination of TA and AoA.

locations $(x\ y)^T$ given by (32.13), parameterized by γ and α that each vary in the range of [0,1],

$$\begin{pmatrix} x \\ y \end{pmatrix} = \left(R_{TA} + \Delta R_{TA}\left(\gamma - \frac{1}{2} \right) \right) \cdot \begin{pmatrix} \sin\left(\varphi + \Delta\varphi\left(\alpha - \frac{1}{2} \right) \right) \\ \cos\left(\varphi + \Delta\varphi\left(\alpha - \frac{1}{2} \right) \right) \end{pmatrix}. \qquad (32.13)$$

Reporting of an ellipsoid arc is anticipated for this case. Due to the geometric orthogonality of distance and AoA, this is expected to be an efficient hybrid positioning method.

AoA is traditionally measured by the eNodeB with an antenna array, (see [22] and Chapter 9), but with LTE it may also be possible to use the PMI reported by the UE [16, 24]. Each PMI corresponds to the use of an "antenna beam," as indicated in Figure 32.12. It can be seen that the reported PMI gives a useful indication of the direction to the UE.

Example 32.5: AoA Antenna Diagrams

This example illustrates the angular accuracies that can be expected when PMIs are used for positioning purposes using pre-release-13 codebooks.

Solution: The expected SNRs for an example 4-TX antenna codebook are shown in Figure 32.13. As can be seen from these diagrams, the central major lobes cover approximately 30 degrees, making PMIs a useful source of location information, in particular when combined with TA information in LTE as illustrated in Figure 32.11. The radial TA uncertainty may be about 100 m. At a distance of 1 km, the region corresponding to these measurements would hence cover an area of about 100×500 m^2.

Figure 32.12 Antenna configuration and PMI selection. The results are based on live measurements.

2-D AoA and TA

As stated in Section 32.1.5, AAS arrays can be exploited from LTE release 13. In particular, beamforming with CSI-RS feedback has potential to provide highly accurate 2-D AoA information. At the same time, an increasing radio bandwidth will provide better TA measurements that will contribute to more accurate measurements of the distance between the eNodeB and the UE.

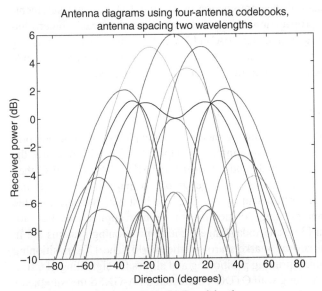

Figure 32.13   SNRs for a 4-TX PMI codebook.

In the 2D AoA case, Equation (32.13) generalizes to the following 3-D cartesian region:

$$
\begin{pmatrix} x \\ y \\ z \end{pmatrix} = \left(R_{TA} + \Delta R_{TA} \left(\gamma - \frac{1}{2} \right) \right)
$$
$$
\cdot \begin{pmatrix} \cos\left(\theta + \Delta\theta\left(\alpha - \frac{1}{2} \right) \right) \cos\left(\varphi + \Delta\varphi\left(\beta - \frac{1}{2} \right) \right) \\ \cos\left(\theta + \Delta\theta\left(\alpha - \frac{1}{2} \right) \right) \sin\left(\varphi + \Delta\varphi\left(\beta - \frac{1}{2} \right) \right) \\ \sin\left(\theta + \Delta\theta\left(\alpha - \frac{1}{2} \right) \right) \end{pmatrix},
\tag{32.14}
$$

expressed with respect to the eNodeB location. Here ΔR_{TA} is the TA range measurement inaccuracy, $\Delta\varphi$ the azimuth angle beam width and $\Delta\theta$ the elevation angle beam width. As previously, the azimuth angle is denoted by φ, while the elevation angle is denoted by θ. The location of the UE is described by the set of Cartesian coordinates generated when α, β, and γ vary independently in [0,1]. Since the beam width is inversely proportional to the number of antenna elements in azimuth and elevation, this region can become quite small when the cell size is small, as in indoor deployments. When large antenna arrays are used in small cells in future 5G systems, this technique may therefore meet the vertical E-911 requirement indoors. Depending on the size of the Cartesian region, the positioning result may be transformed to an ellipsoid point with altitude and uncertainty ellipsoid before reporting. Alternatively, proprietary reporting methods may be used [40].

The potential success of the combination of TA and 2-D AoA measurements is subject to NLOS radio propagation. In cases where this is not the case, fingerprinting may be added for mitigation, using AoA and TA as the radio properties forming the fingerprint. Other methods for NLOS mitigation appear in Chapters 16–19 while AoA localization in general is treated in Chapters 9 of this handbook.

32.6.3 Fingerprinting

Fingerprinting denotes a set of positioning methods that exploit detailed geographical maps of radio properties for positioning (c.f. Chapter 15). The UE measures the radio properties it experiences and sends it to the E-SMLC. The E-SMLC then searches for a best match between its stored geographical map of radio properties and the measured radio properties sent by the UE. The best match determines the position of the UE. Fingerprinting positioning for cellular applications has, for example, been studied in [33], and today fingerprinting positioning systems are in commercial operation in several markets around the world. The LTE positioning standard is prepared for fingerprinting positioning and allows for signalling of CIDs, signal strengths, TA, AAA along with OTDOA and A-GPS / A-GNSS measurements between the UE, the eNodeB, and the E-SMLC. The information is carried over the LPP [8] and LPPa [10] protocols.

Radio Frequency fingerprinting

The most common technique for fingerprinting is denoted *radio frequency (RF) fingerprinting* or *RF pattern matching*. In LTE, it would exploit UE measurements of received signal strength from a number of eNodeBs. The geographical RF maps can be created by advanced radio signal strength prediction software, using very detailed information of the 3-D geographical topology together with accurate information of the cell plan, tower locations, tower heights, antenna directions, antenna tilting, antenna patterns, and transmission power. To achieve a good-enough accuracy, such prediction software still may need to be complemented with surveying. Another approach would be to rely entirely on surveying, which, however, would be very costly even for normally sized cellular networks since the positioning accuracy will always be bounded by the density of the geographical grid of the RF map.

Even when accurate predictions of the signal strength can be made, another problem limits performance. This is due to the fact that the antenna diagrams of handheld UEs vary a lot with user orientation and the way the UE is held, for example, against the user's head. Such effects can easily amount to more than 10 dB of uncertainty (c.f. Fig. 32.10). There is no perfect way around this situation, and this will affect any positioning technology exploiting received signal strength measurements, while time measurements are likely to be less sensitive (c.f. (32.12) and [14]). A way to mitigate the effects of this problem is to apply averaging or to use relative signal strength measurements. Other types of signal processing can be used as well, exploiting aspects related to the signal strength [47], possibly in combination with advanced multiple-hypothesis strategies.

Adaptive E-CID

Another way to enhance fingerprinting positioning performance is to extend the number of radio properties that are used. In LTE, at least CIDs, TA, and AoA are suitable in addition to received signal strengths. Unfortunately, the geographical radio map then becomes much more difficult to generate. In addition, the problem with handheld UEs becomes even more complex. Furthermore, there is a need to address the accuracy of the measured position, using multiple radio properties. Ideally, this requires that both an inaccuracy and an associated confidence value are determined together with the position.

The adaptive E-CID (AECID) positioning method [36] addresses the above problems. It fuses geographical cell descriptions (corresponding to CIDs), received signal strengths, and TA, and can be extended to include AoA information. The method replaces the radio property prediction software and the surveying by the following self-learning mechanism:

1. Whenever an A-GPS, A-GNSS, OTDOA, or UTDOA high-precision positioning is performed, the E-SMLC orders measurements of the radio properties, this being a subset of geographical cell descriptions, TA, signal strengths, and AoA. The radio property measurements are quantized, producing the fingerprint of the obtained high-precision position.

2. All high-precision positions with the same fingerprint are collected in clusters. These clusters describe geographical regions where the same fingerprint persists.

3. A model of the cluster boundaries are computed by an algorithm that generates a 3GPP polygon that contains a pre specified fraction of the high-precision measurements of the cluster. An advantage with this is that the polygon is determined to have a confidence equal to the pre-specified fraction. This polygon, together with its fingerprint and confidence value, is stored in a database accessible by the E-SMLC. Note that other shapes than the polygon can be used.

The clustering can be extended to perform smoothing by the clustering computation algorithms of [37]. Those algorithms can also split weakly coupled parts of a cluster into subclusters that can be individually described by polygons, after which these polygons are merged into one polygon for reporting.

The polygon computation is initiated by a calculation of the center of gravity, r_{CG}, of the cluster of high-precision reference points, and by creation of an initial polygon with all points of the cluster in the interior [36].

Each step of the so-called contracting polygon algorithm performs tentative movements of the corners, $\mathbf{r}_i^P, i = 1,...,N_p$ of the polygon p inwards towards the center of gravity of the cluster. The movement is continued until one point of the cluster becomes an exterior point of the polygon. This is repeated for all corners, and the corner that results in the largest decrease of the polygon area is selected for the actual movement, for each iteration. The algorithm continues iteration until a pre specified fraction of the points remain in the interior of the polygon.

When moving a corner inwards, it needs to be checked when a specific reference measurement point (superscript m) of the cluster, $\mathbf{r}_j^{m,P}, j = 1,...,N$, becomes exterior to the polygon. This check needs to be performed for all points of the cluster that remain in the interior of the polygon at the start of an iteration. Figure 32.14. shows three adjacent polygon corners $\mathbf{r}_k^P, r_i^P$, and $\mathbf{r}_l^P$. The middle point $\mathbf{r}_i^P$ is then moved inwards. The line segments connecting $\mathbf{r}_k^P$ and $\mathbf{r}_i^P$, as well as $\mathbf{r}_i^P$ and $\mathbf{r}_l^P$, move. At some point, $\mathbf{r}_j^{m,P}$ may be intersected by either of these two line segments.

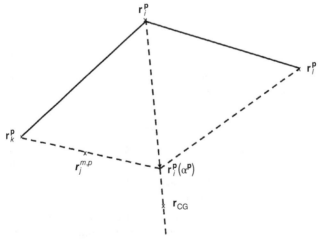

Figure 32.14   Tentative polygon corner movement geometry for AECID.

To determine the tentative points of intersection, it is noted that the movement of $\mathbf{r}_i^p$ can be expressed as

$$\mathbf{r}_i^p\left(\alpha^p\right) = \mathbf{r}_i^p + \alpha^p\left(\mathbf{r}_{CG} - \mathbf{r}_i^p\right), \tag{32.15}$$

where α^p is a scalar parameter that varies between zero and one when the point moves between the original position and the center of gravity. A necessary requirement for an intersection between the moving boundary of the polygon and $\mathbf{r}_j^{m,p}$ is that $r_i^p(\alpha^p) - \mathbf{r}_k^p$ and $r_j^{m,p} - \mathbf{r}_k^p$ become parallel, or that $r_i^p(\alpha^p) - r_l^p$ and $r_j^{m,p} - r_l^p$ become parallel. The fact that the cross-product between parallel vectors is zero allows for a computation of α^p. Straightforward algebra gives

$$
\alpha_{i,k}^{j,p} = \frac{-\left(x_i^p - x_k^p\right)\left(y_j^{m,p} - y_k^p\right) + \left(x_j^{m,p} - x_k^p\right)\left(y_i^p - y_k^p\right)}{\left(x_{CG} - x_i^p\right)\left(y_j^{m,p} - y_k^p\right) - \left(x_j^{m,p} - x_k^p\right)\left(y_{CG} - y_i^p\right)}
$$

$$
\alpha_{i,l}^{j,p} = \frac{-\left(x_i^p - x_l^p\right)\left(y_j^{m,p} - y_l^p\right) + \left(x_j^{m,p} - x_l^p\right)\left(y_i^p - y_l^p\right)}{\left(x_{CG} - x_i^p\right)\left(y_j^{m,p} - y_l^p\right) - \left(x_j^{m,p} - x_l^p\right)\left(y_{CG} - y_i^p\right)}. \tag{32.16}
$$

The subscripts indicate the polygon corner points that define the line segment under evaluation. The superscript denotes the index of the high-precision measurement point. Both the above are candidates for being an active constraint, for which it is required that $\alpha_{i,k}^{j,p} > 0$ and $\alpha_{i,l}^{j,p} > 0$.

One must now check if the intersection point falls between the corner points that limit the line segment of the polygon. This means that the following equations need to be fulfilled for $\beta_{i,k}^{j,p} > 0$ and $\beta_{i,l}^{j,p} \in [0,1]$:

$$
\mathbf{r}_i^{m,p} = \mathbf{r}_i^p\left(\alpha_{i,k}^{j,p}\right) + \beta_{i,k}^{j,p}\left(\mathbf{r}_k^p - \mathbf{r}_i^p\right)
$$

$$
\mathbf{r}_j^{m,p} = \mathbf{r}_i^p\left(\alpha_{i,l}^{j,p}\right) + \beta_{i,l}^{j,p}\left(\mathbf{r}_l^p - \mathbf{r}_i^p\right). \tag{32.17}
$$

Since the vectors leading to the above equations are parallel, it is enough to consider one of the coordinates when solving for the β parameters.

The conditions $\alpha_{i,k}^{j,p} > 0, \alpha_{i,l}^{j,p} > 0, \beta_{i,k}^{j,p} \in [0,1]$, and $\beta_{i,l}^{j,p} \in [0,1]$, are then checked for each point (for the specific polygon corner), and the point that meets the conditions and first limits the inward movement is determined as the point that first becomes exterior to the polygon. The procedure is then repeated for each corner. Finally, the corner that generates the largest decrease of the area of the polygon is selected for the current iteration of the algorithm. The mathematical details appear in [36].

In case TA is used, it may happen that the fingerprinted regions become very wide in the lateral direction and narrow in the radial direction at the same time as they become curved. This follows since cells may be very large. Two problems then occur. First, the contraction point, given by the center of gravity of the cluster, may be located outside the cluster itself. This may prevent convergence of the algorithm to an accurate description of the boundary of the cluster. Secondly, all polygon corner points tend to converge towards the lateral center of the cluster. This follows since the corners move toward the center of gravity of the cluster, and since the cluster is

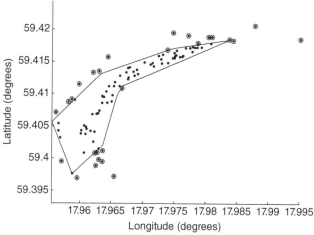

Figure 32.15   The high-precision position measurement cluster (black stars) and the computed polygon (black solid). The stars with surrounding circles represent high-precision position measurements that are exterior to the polygon after the computation using [36] and [39].

much more wide in the lateral direction than in the radial direction. This leaves few points for modeling of the lateral end points of the cluster boundary, a fact that reduces accuracy. Author in [39] discusses this further and provides solutions to these problems.

The AECID positioning method is self-learning, using positioning attempts of opportunity. Furthermore, real-world radio conditions are captured. The problem with hand held UEs is also reflected, which means that over optimistic accuracies are avoided. Also, the confidence is reflected by real user data. The availability of OTDOA and UTDOA in LTE ensures that the database of fingerprinted polygons will cover indoor locations. In fact, the OTDOA positioning method can also provide GSM and WCDMA with indoor high-precision position measurements, using the inter-RAT signal strength measurements defined in LTE [43].

When a user is to be positioned by the AECID method, the E-SMLC orders radio property measurements, constructs the fingerprint, and looks up the fingerprinted polygon and confidence in its database. The position result is then reported. An example of the output is illustrated in Figure 32.15.

That figure is computed with reference code for an AECID product, including algorithms of both [36] and [39].

Example 32.6   Polygon Computation Video Clip

The polygon computation algorithm is quite involved. For this reason, the algorithm was illustrated in MATLAB by creating a video clip, composed of all iterations applied in the computation of a cell boundary, that is, in this case only CID was used for fingerprinting.

Solution: The video clip is available in the file Chapter_32_Example_6.avi. The result can be displayed by running the file in a suitable media player. In the video the red dots represent the cluster points of the algorithm.

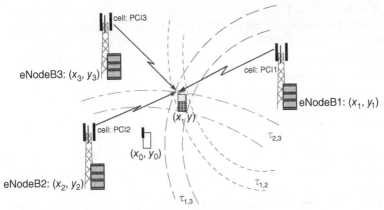

Figure 32.16   Multilateration in OTDOA positioning.

32.6.4   OTDOA

OTDOA is a DL positioning method standardized in LTE that exploits time difference measurements conducted on DL reference signals received from multiple locations. An OTDOA measurement, reference signal time difference (RSTD), is defined as the relative timing difference between two cells, the reference and a measured cell, calculated as the smallest time difference between two subframes received from the different cells. At least three timing measurements from geographically dispersed base stations with a good geometry are needed to solve for two coordinates of the UE, where the good geometry means that no two branches of the distinct hyperbolas intersect twice so that a unique solution can be found. In practice, a larger number of measurements, typically at least six to seven, is desirable. The position calculation is based on the multilateration approach by which an intersection of hyperbolas is found, where a hyperbola for a pair of cells corresponds to a set of points with the same RSTD for the two cells. The OTDOA positioning method is illustrated in Figure 32.16, where a UE is receiving DL signals from the eNodeBs and performing RSTD measurements $\tau_{1,2}$, $\tau_{1,3}$, and $\tau_{2,3}$, which form three intersecting hyperbolic strips, the widths of which correspond to the RSTD measurement uncertainties. The distance equivalents of the RSTD measurements are $\hat{d}_{1,2}$, $\hat{d}_{1,3}$, and $\hat{d}_{2,3}$, respectively. Assuming the UE is located in $(x\ y)^T$ in an ET system, and eNodeBs are located at $(x_1\ y_1)^T (x_2\ y_2)^T$, and $(x_3\ y_3)^T$, respectively, the system of equations to be solved in the least-squares sense to find 2-D coordinates $(x\ y)^T$ is as follows:

$$\sqrt{(x_1-x)^2+(y_1-y)^2}-\sqrt{(x_2-x)^2+(y_2-y)^2} = \hat{d}_{1,2}-\Delta d_{1,2}$$

$$\sqrt{(x_2-x)^2+(y_2-y)^2}-\sqrt{(x_3-x)^2+(y_3-y)^2} = \hat{d}_{2,3}-\Delta d_{2,3} \qquad (32.18)$$

$$\sqrt{(x_1-x)^2+(y_1-y)^2}-\sqrt{(x_3-x)^2+(y_3-y)^2} = \hat{d}_{1,3}-\Delta d_{1,3},$$

where $\Delta d_{i,j}$ is the distance equivalent of the transmit time difference for cells i and j ($\Delta d_{i,j} = 0$ when the two cells are perfectly synchronized and $\Delta d_{i,j} \neq 0$ for asynchronous systems, e.g. UMTS or the asynchronous mode of LTE). As follows from the system of equations (32.18), the precise knowledge of the transmitter locations and timing offsets are needed to find the UE coordinates. Determining transmitter timing is non trivial at the required accuracy level. As for two-way ranging, the advantage of OTDOA is that synchronization between the eNodeBs and the UE is not required, unlike when time of arrival measurements are used. The timing offsets may, however, need to be determined. One way is to exploit dedicated LMUs that are deployed at carefully surveyed locations. The systems of equations (32.18) can then be applied with known left-hand sides and measured $\hat{d}_{i,j}$ to obtain the $\Delta d_{i,j}$.

Many approaches exist for solving the system of equations (see, for example, [15] and Chapters 6–7 for further details). It can be noted that many of the methods involve linearization of least-squares minimization problems. With the common Taylor series based-approach, the UE coordinates $(x\ y)^T$ are found iteratively starting from an initial UE position estimate $(x_0\ y_0)^T$. It is straightforward to extend the formulation to a larger set of eNodeBs and RSTD measurements for 3-D coordinates.

In LTE, the OTDOA position calculation is performed in the network, while the UE performs and reports RSTD measurements. Given that the RSTD accuracy may vary a lot in different deployments and for different cell pairs, so-called adaptive RSTD reporting granularity has been introduced in LTE release 14.

Although RSTD measurements can be performed on a number of DL signals, for example, CRS or synchronization signals [6], it has been recognized that the existing signals suffer from poor hearability, which is crucial for OTDOA when multiple remote neighbor cells have to be detected. Therefore, to ensure the possibility of positioning measurements of a proper quality and for a sufficient number of distinct locations, PRS [6] having transmission patterns with an effective reuse of six have been introduced in LTE.

PRS are transmitted in certain *positioning subframes* grouped into *positioning occasions*. A positioning occasion is characterized by its length and periodicity. These two parameters describe the PRS configuration, which is a part of the assistance data signalled to the UE. Since LTE release 9, a positioning occasion can be 1, 2, 4, or 6 consecutive positioning subframes, which occurs periodically with a 16, 320, 640, or 1280 ms interval. Since release 14, some UEs also support new positioning occasion lengths, which can be any number between 2 and 160, and new periodicities of 5, 10, 20, 40, and 80 ms.

Within each positioning occasion, PRS are transmitted with a constant power. PRS can also be transmitted with zero power, or *muted*, which can be utilized to avoid measuring in the presence of the strongest interferers. The PRS muting configuration of the serving and neighbor cells are to be decided by the network and signalled to the UE over the LPP protocol.

To further improve hearability of PRS, positioning subframes have been designed as low-interference subframes (LIS), that is, with no or low transmission activity on data channels. As a result, ideally in synchronous networks PRS are interfered by other-cell PRS with the same PRS pattern index, that is, the same frequency shift, but not by data transmissions. To achieve good positioning performance, interference coordination for PRS is crucial in these subframes. Therefore, configuring aligned positioning subframes over cells is a justified PRS planning

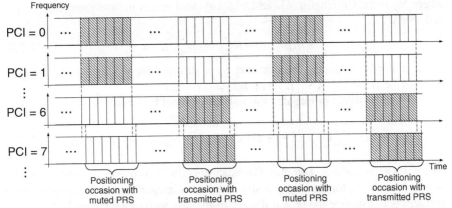

Figure 32.17 A PRS configuration example for three subframe-synchronized cells that are half-subframe-aligned with the fourth cell. The dashed rectangles are the positioning occasions with transmitted PRS that are otherwise muted in other positioning occasions. A positioning occasion comprises six consecutive PRS subframes.

strategy giving synchronized positioning occasions in subframe-synchronized LTE networks and up to half-subframe-aligned positioning subframes in asynchronous LTE networks.

The interference to PRS can be further minimized, for example, by allowing simultaneous PRS transmissions for groups of cells with either orthogonal PRS patterns or low mutual interference. A PRS configuration example for four cells is illustrated in Figure 32.17. In OTDOA, the UE receiver has to deal with PRS from neighbor cells, much weaker than those received from the serving cell. Furthermore, without the approximate knowledge of when the measured signals are expected to arrive in time and what is the exact PRS pattern, the UE would need to search the signals blindly, which would impact the time and accuracy of the measurements. To facilitate UE measurements, the network transmits assistance data to the UE [8], including, among the others, neighbor cell list with PCIs, the number of antenna ports, the number of consecutive DL subframes per positioning occasion, PRS transmission bandwidth, expected RSTD, and the estimated uncertainty.

The signalled expected RSTD and the estimated uncertainty defines the search window. Given the distance between the UE and the reference cell 1 (e.g., R_{TA} obtained from (32.9) when the reference cell is the serving cell) and the distance $d_{1,2}$ between the reference cell 1 and the measured cell 2, the search window for signals from the measured cell is the range $[-R_{TA}/c, R_{TA}/c]$ centered at $(\Delta d_{2,1} + d_{1,2} - R_{TA})/c$ with respect to the reference cell, where c is the speed of light.

The PRS signal is transmitted in N_{prs} consecutive subframes. The total number of OFDM symbols containing PRS resource elements over all the consecutive subframes is N_l. Let the time domain signal transmitted in the lth OFDM symbol be denoted

$$s_l(n), n = -N_{cp},\ldots, N-1, l = 0,\ldots, N_l - 1, \qquad (32.19)$$

where N_{cp} is the CP length, N is an OFDM symbol length in time samples, and where

$$s_l(n) = s_l(n+N), n = -N_{cp}, \ldots, -1. \quad (32.20)$$

Let a multipath channel characterized by a discrete finite length time impulse response be $(h(0) \ldots h(K-1))^T$, with $h(i) = 0, i < 0, i \geq K$.

Then the signal propagated through the channel becomes

$$y_l(n) = \sum_{i=0}^{K-1} h(i) s_l(n-i-\tau) + \epsilon_l(n), n = -N_{cp}, \ldots, N+K+\tau-2, \quad (32.21)$$

where $\epsilon_i(n)$ is additive complex Gaussian noise with variance N_0, and τ is the propagation delay, for simplicity assumed to be an integer number of samples. The problem is to estimate the arrival time τ of the first path of the channel.

For OFDM, the reference signal is specified in the frequency domain. It is well-known that the circular convolution of two sequences is equal to multiplication of the FFTs of the signals in the frequency domain. In the circular convolution, the linear convolution (32.21) is replaced with

$$y_l(n) = \sum_{i=0}^{K-1} h(i) s_l(n-i-\tau)_N + \epsilon_l(n), n = 0, \ldots, N-1, \quad (32.22)$$

where $s_l(\cdot)_N$ means that the index is taken modulo N. Due to the use of a CP of length $N_{cp}, s_l(-n) = s_l(N-n)$, for $n = 1, \ldots, N_{cp}$. Hence, as long as $K+\tau-1 \leq N_{cp}$, the circular convolution is equal to the linear convolution. To exploit this, the FFT of (32.22) is

$$Y_l(k) = \frac{1}{N} \sum_{n=0}^{N-1} \left(\sum_{i=0}^{K-1} h(i) s_l(n-i-\tau)_N + \epsilon_l(n) \right) e^{-i2\pi nk/N}$$
$$= H_l(k) S_l(k) e^{-i2\pi\tau k/N} + E_l(k), k = 0, \ldots, N-1. \quad (32.23)$$

Since $S_l(k)$ is known, multiplication in frequency domain yields an estimate of the channel in the frequency domain as

$$\widehat{H}_l(k) = Y_l(k) S_l^*(k) = H_l(k) e^{-i2\pi\tau k/N} + S_l^*(k) E_l(k). \quad (32.24)$$

The PRS signal spans several OFDM symbols, so the processing is repeated for a number of consecutive OFDM symbols containing PRS, resulting in

$$\widehat{H}(k) = \frac{1}{N_k} \sum_{l=0}^{N_k-1} \widehat{H}_l(k), k = 0, \ldots, N-1, \quad (32.25)$$

where N_k is the number of coherently accumulated PRS symbols in subcarrier k. If N_k and/or the Doppler spread are large, the terms in the complex sum (32.25) may start to cancel each other. In such a case, it is necessary to divide the coherent accumulation into M segments and perform an IFFT on each, resulting in

$$\hat{h}_m(n) = IFFT\left(\widehat{H}(k)\right), \quad (32.26)$$

which is the time-domain complex channel estimate for the mth segment. Finally the M channel estimates are non coherently added. The resulting correlator output can then be written as

$$\hat{r}(n) = \sum_{m=0}^{M-1} \left| \hat{h}_m(n) \right|^2. \tag{32.27}$$

The values of $\hat{r}(n)$ are compared to a threshold, and if any values fall above the threshold, then a fine search is performed to interpolate the position of the first peak.

Since there is no guarantee that the signal has enough power to be detected, the threshold should be selected so as to avoid false alarms. Assuming only complex-Gaussian noise as input in (32.26), the terms are complex Gaussian distributed with variance $\sum_{k=0}^{N-1} N_0 / N_k$. Hence $\hat{r}(n)$ of (32.27) is the sum of squares of Gaussian variables. Therefore it is convenient to modify the detection variable to be $\lambda = 2\hat{r}(n) / (\sum_{k=0}^{N-1} N_0 / N_k)$, which is $\mathcal{X}^2(2M)$ distributed. This can be used to determine an appropriate threshold for a desired false alarm rate.

Assuming the PRS pattern in Figure 32.4, a system using N_{RB} resource blocks, N_{prs} positioning subframes, and coherent accumulation over N_c subframes, it can be readily determined that $\sum_{k=0}^{N-1} 1 / N_k = 7 N_{RB} / N_c$. Furthermore, when the signal part dominates over the noise in (32.27), this expression becomes $\hat{r}(n) \approx |h(n-\tau)|^2 L_{pres}^2 M$, where $L_{prs} = 10 N_{RB}$ is the number of subcarriers occupied by PRS, and $M = N_{prs} / N_c$. Therefore, $\lambda \approx (200 N_{RB} N_{prs} / 7) |h(n-\tau)|^2 / N_0$. The factor

$$\gamma_{prs} = 200 N_{RB} N_{prs} / 7 \tag{32.28}$$

can be interpreted as the processing gain of the PRS detector. For the CRS pattern in Figure 32.4, it can similarly be verified that $\sum_{k=0}^{N-1} 1 / N_k = 3 N_{RB} / N_c$. Furthermore, the signal part in (32.27), becomes $\hat{r}(n) \approx |h(n-\tau)|^2 L_{crs}^2 M$, where $L_{prs} = 4 N_{RB}$ is the number of subcarriers occupied by CRS. Therefore, $\lambda \approx \gamma_{crs} |h(n-\tau)|^2 / N_0$ with

$$\gamma_{crs} = 32 N_{RB} N_{prs} / 3. \tag{32.29}$$

The probability of false alarm assuming threshold λ^* can be computed as

$$P_{fa} = 1 - chi2cdf(\lambda^*, 2M). \tag{32.30}$$

Here $chi2cdf(\cdot, \cdot)$ denotes a chi-squared cumulative distribution function. The above discussion provides the tools needed to assess the probability of detection of a certain number of base stations. This translates into predictions of the availability of the OTDOA method, which together with accuracy is the most important performance characteristic.

Example 32.7   OTDOA Coverage

The expected coverage of OTDOA, that is, the probability that at least three base stations can be measured, is evaluated for a simulated case in this example. The use of PRS and CRS is compared. It should be noted that the detection of three eNodeBs

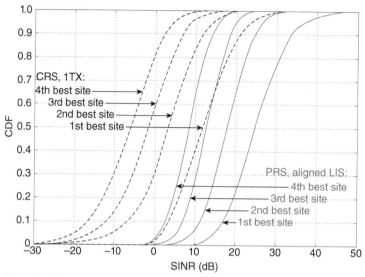

Figure 32.18   SINR CDFs with CRS and PRS.

is a bare minimum; in practice excess detections are often necessary to be able to suppress outliers in the measurements. Such outliers can, for example, arise due to NLOS propagation. For the above reason, the detection capability of a TDOA method can often be the limiting factor, which makes it important to understand how the detection performance is computed.

Solution

In Figure 32.19, the false alarm probability is plotted versus $SINR^* = \lambda^* / Y_{prs,crs}$, for $M = 1, 2, 3, 6$ using (32.30). For the calculation of $Y_{prs,crs}$ using (32.28) and

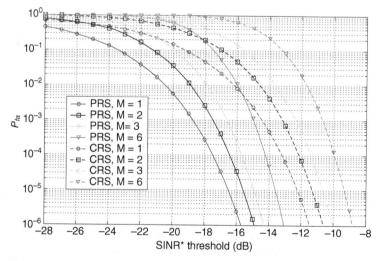

Figure 32.19   Probability of false alarm as a function of detection threshold.

(32.29), $N_{RB} = 6$ and $N_{prs} = MN_c = 6$ was assumed. It can, for example, be seen that if the desired $P_{fa} = 10^{-4}$, then for PRS and $M = 1$ the threshold $SINR^*$ is roughly -17.5 dB, whereas for $M = 6$ the threshold becomes roughly -14dB. The corresponding number for CRS are -13 dB and -10 dB, respectively. Thus the detection probability of CRS is approximately 3.5–4 dB worse than that for PRS. Furthermore, higher UE speed (larger M) requires higher $SINR^*$. In Figure 32.18, cumulative distribution functions (CDFs) of simulated signal-to-interference-plus-noise ratios (SINRs) are shown for the best first, second, third, and fourth sites, respectively, for a system using CRS and PRS. The presented results are for uniformly distributed users in a network of 19 sites with inter site distance of 500 meters, three-sector realistic 3-D antennas, and 25% data load. When interference is not specifically handled, the CRS also get interfered by data transmitted in cells that are not in the same frequency reuse group as the measured cell. By comparing Figure 32.19 with Figure 32.18 it can be concluded that the probability to be able to measure three or more base stations is close to 100% for PRS for all choices of M. For CRS, the probability is around 91% when using $M = 1$ and 88% when using $M = 6$. Note that these calculations assume a single-tap channel h(n). In reality the channel energy is usually spread over several lags, meaning that the probability to detect 3 base stations is lower.

32.6.5 UTDOA

The UTDOA positioning method was introduced as a high-accuracy method to meet E-911 requirements in GSM deployments [14]. The method was standardized for LTE in release 11. Exactly as the OTDOA method, UTDOA relies on time-of-arrival measurements from which time differences are formed. These time differences are then used together with known timing relations and eNodeB locations to compute the location of the UE. Many aspects, for example, detection performance, can therefore be treated with techniques that are similar to those described in the previous section on OTDOA. There are also differences since the functionality for signal transmission and signal detection in the UE and eNodeBs are reversed as compared to OTDOA; see [7].

A major conceptual difference between UTDOA [14] and OTDOA is thus that the latter requires multiple transmit points, whilst the former utilizes multiple receive points at different locations, although the position calculation principle is the same. Referring to Figure 32.16, applying it for signals from the UE to the eNodeBs gives the following equations for the 2-D UTDOA positioning fix

$$\sqrt{(x_1 - x)^2 + (y_1 - y)^2} - \sqrt{(x_2 - x)^2 + (y_2 - y)^2} = \hat{d}_{1,2} - \Delta d_{1,2}$$
$$\sqrt{(x_2 - x)^2 + (y_2 - y)^2} - \sqrt{(x_3 - x)^2 + (y_3 - y)^2} = \hat{d}_{2,3} - \Delta d_{2,3} \qquad (32.31)$$
$$\sqrt{(x_1 - x)^2 + (y_1 - y)^2} - \sqrt{(x_3 - x)^2 + (y_3 - y)^2} = \hat{d}_{1,3} - \Delta d_{1,3}$$

which is identical to (32.18). However, the interpretation of the involved quantities differ slightly. Here $\hat{d}_{i,j} = c(t_i - t_j)$ is the distance corresponding to the difference in travel time, with the speed of light c, from the UE to eNobeB i and j, respectively.

As for OTDOA, $\Delta d_{i,j}$ denotes the distance equivalent of the difference between the time bases of eNodeB i and j, that is, the synchronization error in case the eNodeBs would be synchronized. $(x_i \ y_i)^T$ denotes the configured coordinates of eNodeB i, while $(x \ y)^T$ denotes the unknown UE position.

Similarly as for OTDOA, the $\Delta d_{i,j}$ need to be determined in case the eNodeBs are not synchronized accuarately enough. Note that if the UE location would be known, measurements of the $d_{i,j}$ would allow the $\Delta d_{i,j}$ to be solved for using (32.31). This means that outdoor UEs may exploit A-GNSS positioning fixes to operate as LMUs of opportunity. Alternatively, LMUs transmitting UTDOA UL signals can be used in the same way.

To perform UTDOA timing measurements on user data as in GSM, one reference receiver would need to decode the UE signals and forward the sequence to cooperating non colocated receivers. This procedure is relatively complex and requires a significant amount of signalling, although it could be used in LTE if the UTDOA measurements were integrated in the eNodeB and if inter-eNodeB signaling could be made available. Therefore, SRSs [6] have been selected for UTDOA measurements. In UTDOA each UE needs to perform a separate transmission of SRSs. For UTDOA positioning, periodically transmitted SRSs are scheduled in a non dynamic way to allow a sufficiently long time for the measurements. This reduces the need for signaling of the full scheduling information to communicate the SRS configuration parameters. SRSs occupy a selected part of the last OFDM symbol in a subframe. The used bandwidth and the periodicity may be configured to meet the demands for positioning, however system performance for other services than positioning also need to be accounted for since SRSs consi- tute a shared resource. SRS measurements for positioning may either be performed and processed directly at an eNodeB, or the operation may be performed by specific LMUs that may or may not share the antennas with the radio base stations.

32.6.6 A-GNSS

GNSS is a generic name for satellite-based positioning systems with global coverage; see chapter 20. Examples of GNSS systems include the U.S. GPS [25], the European Galileo system, the Russian GLObal NAvigation Satellite System (GLONASS) system, and the Chinese Compass and BeiDou systems. In addition, there exist regional satellite-based augmentation systems (SBAS), including WAAS (U.S.), EG- NOS (EU), MSAS (Japan) and GAGAN (India). SBAS typically use a few satellites to provide improved ionospheric models and real-time integrity monitoring. The SBAS satellite signals can also be used as an additional satellite in computing the receiver position.

Of these systems, GPS has been fully operational for many years. The basic characteristics of the transmitted signals from GPS satellites is shown in Figure 32.20. The signal is transmitted on the L1 frequency (1175.28MHz), and it is a code division multiple access (CDMA) signal characterized by a coarse acquisition (C/A) pseudorandom code that is unique to each satellite. A simplified model of the re-ceived, sampled baseband signal at time sample t_k is

$$y(t_k) = \sqrt{P}c(t_k + \tau_{tr})d(t_k + \tau_{tr})e^{-j(\omega_r t_k + \varphi)} + n(t_k), \qquad (32.32)$$

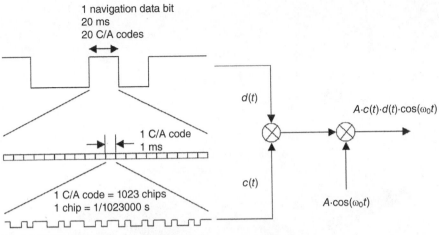

Figure 32.20   The GPS signal structure.

where P is the received SV signal power $c(t)$ is the C/A code of the satellite, a known sequence of -1 and $+1$ values that switch with a rate of $1/T_c = 1023106$ Hz. The C/A code repeats itself every 1 ms. The quantity τ_{tr} is the code phase, $d(t)$ is a sequence of -1 and $+1$ values containing navigation data, which switches at a rate of 20 ms; ω_{tr} is the unknown Doppler shift; and ϕ is an unknown phase offset. The noise $n(t)$ is assumed to be white with spectral density $N_F k_B T_0$, where N_F is the receiver noise figure, $k_B = 1.38 \times 10^{-23}$ J/K, and T_0 is the receiver noise temperature. Interference from other satellites is not modelled here. Since the multiple access interference adds of the order of 0.1 dB to the noise power, GPS is essentially a noise-limited CDMA system. The task of the GPS receiver is to find the correct code phase and doppler, detect the data bit edges, and perform navigation data decoding. When the GPS receiver has acquired the timing and decoded the navigation signal from at least four satellites, it can determine its location. The A-GPS receiver attempts to improve or eliminate some of the steps above. In order to do so, assistance data is collected from a network of GPS reference receivers. These reference receivers are located at sites with favorable signal conditions. The receivers continuously track visible SVs, decode their messages and transfer the information to the cellular network for further distribution to the GPS receivers in the UEs. In this way, the decoding may be avoided, and the GPS receiver in the UE gets access to complete navigation models and correction parameters.

Assistance data is also helpful in reducing the time to find correct code phase and Doppler [44]. By using knowledge of the approximate receiver location, the Doppler shift of each visible SV can be predicted and the search space in the Doppler domain reduced; see Figure 32.21. It is also possible to provide the GPS receiver with GPS time information, accurate to a few microseconds, by exploiting, for example, a GPS receiver in the base stations of the cellular network [44]. With such information, only coherent code and Doppler search over a small search window would remain; see Figure 32.21. Further information on A-GNSS signal processing appears in Chapters 20–23.

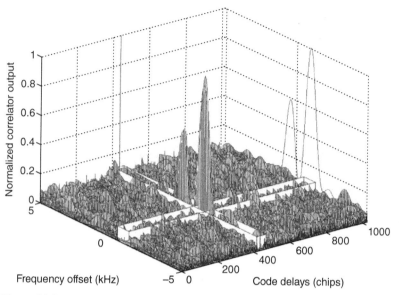

Figure 32.21   Search window reduction.

TABLE 32.1.   A-GPS Comparison (a Noise Figure of 5dB is Assumed)

A-GPS Variant	Assistance Data	Response Time (s)	Sensitivity(dBm)
Stand-alone	FR	20-40	-171
Basic	FR, NM, CT ($\approx$2s)	8-15	-178
Synchronized	FR, NM, AT ($\approx$10 μs)	1-8	-185

FR = frequency reference, NM = navigation models,
CT = location coarse time, AT = location accurate time

A comparison of the time to first fix and sensitivity performance of different A-GPS variants is shown in Table 32.1; see also [23].

32.7   SHAPE CONVERSION

The term shape conversion denotes the geometrical transformations between the 3GPP geographical shapes [3]. Such algorithms are needed for three main reasons. First, the public safety answer point (PSAP) in the U.S. expect E-911 position results in the form of an ellipsoid point with uncertainty circle. Secondly, similar restrictions may be in place for certain LBSs. Thirdly, it is common that E-UTRAN and EPC vendors only support a restricted set of positioning reporting formats and then interoperability requires shape conversions.

Although shape conversions enhance the flexibility of positioning solutions, they have a distinct drawback in that they destroy information. This actually makes some E-911 reporting standards in the U.S. one of the major bottlenecks as far as E-911 compliance is concerned.

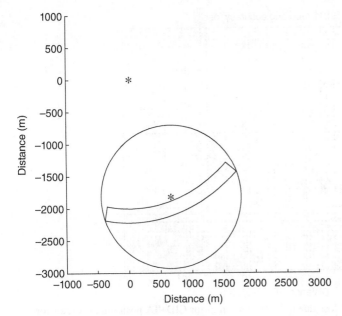

Figure 32.22   Shape conversion between the ellipsoid arc and the ellipsoid point with uncertainty. The eNodeB (star) is in the origin; the other star represents the center point of the uncertainty circle. The ellipsoid arc and the circle are shown solid.

In order to understand why, the conversion between the ellipsoid arc and the point with uncertainty circle formats is considered—this transformation is highly likely to be needed for E-911 reporting when the E-CID positioning method is used. The transformation is depicted in Figure 32.22. In this transformation it is assumed that the underlying probability density function of the ellipsoid arc is uniform [42] since the extension of this region is normally dependent on radio propagation properties rather than random measurement errors. It is also assumed that the required confidence of the transformed shape, that is, the circle, shall be the same as for the ellipsoid arc. The consequence is that the circle needs to cover the entire ellipsoid arc, a fact that is exploited by the algorithm of [42]. It is evident that the area of the circle can become much larger than the area of the original shape, in particular in rural regions where the lateral extension of the ellipsoid arc may be tens of kilometers. As discussed in [42], the loss of accuracy expressed as an area may exceed a factor of 100 for E-CID E-911 reporting in such cases. A typical E-SMLC may need more than 15 shape conversions in order to be flexible enough [35, 46].

Example 32.8   Shape Conversion Accuracy Loss

The example illustrates the severe loss of accuracy caused by the need to perform a shape conversion when doing E-911 positioning, using CID+TA E-CID positioning.

Solution

Figure 32.23 illustrates the loss of accuracy when the ellipsoid arc is transformed to an ellipsoid point with uncertainty circle. The loss of accuracy was calculated in

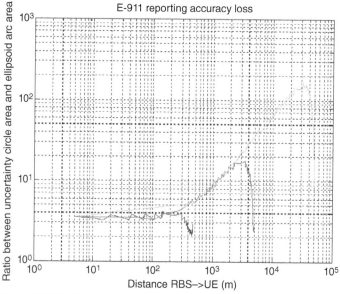

Figure 32.23   Accuracy loss due to E-911 reporting for CID+TA positioning. Values for urban (left curve), suburban (middle curve), and rural (right curve) cells with radii 500 m, 5000 m and 50,000 m, respectively.

terms of the quotient of the areas of the two reported regions. The TA uncertainty was 200 m. The calculations used cell shapes that were roughly hexagonal. Similar but less severe results occur for A-GPS positioning; see [42] and the example code file Chap- ter_32_Example_9.m for that case.

Example 32.9   Ellipse to Circle Shape Conversion

The shape conversion from ellipsoid point with uncertainty ellipse to ellipsoid point with uncertainty circle, based on a Gaussian distribution, is described in [42]. This transformation is typically used for E-911 reporting when A-GPS positioning has been performed. The file Chapter_32_Example_9.m contains a complete MATLAB code for this case. The example code also contains the 3GPP encoding/ decoding needed.

Solution

The example code is run from the MATLAB workspace. When executing the code, the variable positionFormatDescription is expected to be present in the work space of MATLAB. positionFormatDescription contains a 3GPP-encoded ellipsoid point with uncertainty ellipse. The 3GPP-encoded format contains the following elements (in order): the number of the format, the encoded latitude sign, the encoded latitude, the encoded longitude, the encoded semi major axis, the encoded semi minor axis, the encoded orientation angle (clockwise from North), and the encoded confidence. In order to execute the code the reporting confidence of 0.95 also needs to be defined in the MATLAB workspace. The encoded result is stored in encodedEllipsoidPointWithUncertaintyCircle. This variable stores the elements (in order): the number of the format, the encoded latitude sign, the encoded latitude, the encoded longitude, and the encoded accuracy code. A typical MATLAB command sequence is as follows:

```
>> positionFormatDescription=[3 0 400 400 25 15 130 39];
>> confidenceAfterTransformation=0.95; % Change the
confidence level
>> Chapter_32_Example_9 % Perform transformation
>> encodedEllipsoidPointWithUncertaintyCircle % Display the
result encodedEllipsoidPointWithUncertaintyCircle =

          1      0     400    400    32
```

It can be noted that the accuracy code is larger than both the semi major axis and the semi minor axis. This follows since the reporting confidence is larger than the input confidence.

32.8 POSITIONING PERFORMANCE IN LTE

32.8.1 Limiting Factors

A first factor that influences the availability of a positioning method is the UE support, since large parts of the 3GPP specifications are optional. Other critical parameters affecting availability include the detection performance of the UE, the link budgets, and the geometry of the cellular network. For example, it is well-known that the weak link budget for A-GNSS makes indoor positioning availability poor [23], whereas the geometry in rural areas may limit the number of detectable eNodeBs for OTDOA and the number of detecting eNodeBs for UTDOA. In large rural cells the vertical accuracy provided by AoA- and TA-based positioning is not likely to meet the vertical E-911 requirement, even for very large antenna array systems in LOS.

The response time is affected by the complexity of the positioning measurements and the prior knowledge of the UE position and time. For time- measurement-based positioning methods like A-GNSS, OTDOA, and UTDOA the two last parameters affect the size of the search window over which correlations need to be evaluated to find the pseudorange or time of arrival. It can be noted that for A-GPS, the availability of fine time assistance [23, 44] may reduce the positioning time and enhance the availability.

In the literature, positioning accuracy has often been addressed by a combination of geometrical effects and pure measurement inaccuracy, using linearization [25] and theoretical performance measures like the Cramer-Rao–lower bound. However, in particular for terrestrial time-measurement-based methods, like OTDOA and UTDOA, this is seldom true. Positioning accuracy is rather dominated by NLOS propagation and the ability to detect a large enough number of eNodeBs to handle outlier measurements.

32.8.2 Accuracy Metrics

Due to the factors that limit practical positioning performance, it is preferred to evaluate positioning performance in field trials. Both availability and response time can be directly measured, whereas it is more difficult to address the horizontal accuracy.

Since the geographical reporting shapes are geometrically very different, a common method to address the horizontal accuracy for a specific positioning method is to compute the area of the reported region, followed by a computation of the radius, $r_{uncertainty}$, of a circle that covers the same area, $A_{reported}$, as the reported geographical shape. Averaging over the results gives the performance metric

$$\bar{r}_{uncertainty} = \frac{1}{N} \sum_{i=1}^{N} \sqrt{\frac{A_{reported,i}}{\pi}}. \tag{32.33}$$

The metric (32.33) does measure the uncertainty, but the relation to the actual UE position is not captured. To define such a metric, it is assumed that a number of accurately surveyed truth points, $(x_{truth,j} \; y_{truth,j})^T$, $j = 1,\ldots, M$, are available in a Cartesian ET coordinate system. A number of measurements, $(x_{i,j} \; y_{ij})^T$, $i = 1,\ldots, N$, $j = 1,\ldots, M$, are then performed at each of these truth points. The average 2-D error, $d_{uncertainty}$, can then be computed as

$$\bar{d}_{uncertainty} = \frac{1}{MN} \sum_{j=1}^{M} \sum_{i=1}^{N} \sqrt{\left(x_{i,j} - x_{truth,j}\right)^2 + \left(y_{i,j} - y_{truth,j}\right)^2}. \tag{32.34}$$

32.8.3 Indoor Aspects

The LTE positioning standard does not make much difference between indoor and outdoor positioning. However, in practice the window-, roof- and wall-propagation losses can exceed 20 dB between the outdoor and indoor environments also at low carrier frequencies, with significantly higher losses expected for the 5G millimeter-wave bands [31]. This is the reason why A-GNSS has a very limited coverage indoors. Also terrestrial high-accuracy methods like OTDOA and UTDOA that provide good performance for LTE outdoors are likely to experience reduced accuracy indoors. The reason for this is the increased NLOS propagation and reduced number of detected/detecting (outdoor) sites. Therefore, the need for dedicated indoor positioning methods is likely to increase significantly in the future, in particular for 5G deployments. The importance cannot be underestimated given the fact that the vast majority of cellular connections are nowadays placed by indoor users.

Today, the commercially available indoor positioning techniques typically apply CID positioning. Dedicated beacon systems provide one possibility, however deployment costs would become very high on a national scale. As an alternative, Wi-Fi transmission nodes may be used as beacons. The position accuracy is quite good since Wi-Fi cells are usually quite small. However, as stated previously, CID alone is not enough to meet the new E-911 vertical accuracy requirements. The new 2-D AoA methods enabled by the codebooks of [1] may solve this issue, in particular if combined with the improved time resolution of new high bandwidth radios (c.f. Section 32.6.2). Another technology that is deployed, for example, underground in subways, uses leaky antenna cables, where the attenuation can be related to distance as measured from the power amplifier.

Indoor obstacles causing NLOS propagation is, however, a significant difficulty, in particular for AoA based techniques. Here fingerprinting methods may become useful, in which the AoA measurements are made a part of the radio fingerprint at surveyed locations.

32.8.4 Expected Performance

CID

The CID positioning method is based on preconfigured geographical cell information, ranging from a point to a complete polygon. The method thus consists of reporting the CID to the E-SMLC, followed by a database lookup and reporting to the end user. The CID method is therefore available wherever there is LTE coverage. The response time can be expected to be very short, significantly less than 1s. The drawback is the accuracy, which is determined by the size of the cell. Typical values for cell sizes, that is, the largest distance from the base station to the cell edge, are 150-400 m (urban), and 1000-3000 m (suburban). However, the cell sizes of LTE are decreasing over time, and for 5G systems at high mmw frequencies small indoor cells may dominate [31].

A disadvantage of the CID method is the new emerging vertical E- 911 requirement [19], which cannot be met since the CID method lacks altitude information. Here the 3GPP polygon format could be enhanced with altitude information in the standardization of the 3GPP 5G cellular system, a fact that would enhance CID and could enable accurate large cell vertical fingerprinting and hybrid positioning without A-GPS and A-GNSS (c.f. [40] and [41]).

E-CID

The first E-CID method augments the CID method by TA and signal strength measurements. These measurements are available almost anywhere where there is cellular coverage, resulting in almost the same availability as for CID. The response time is increased somewhat due to the need to perform measurements and reporting these, but it is expected to be below 1-2 s.

The radial accuracy can be estimated from field trials performed for the RTT positioning method in WCDMA [35]. These results indicate a radial accuracy of about 80 m (67%) and 170 m (95%), with the nominal (90%) measurement accuracy of that method being 78 m according to 3GPP specifications. The distribution of the radial error appears in Figure 32.24. This means that propagation conditions is starting to dominate the radial error at the WCDMA bandwidth of 3.84 MHz. For LTE the specified reporting quantization of the TA corresponds to 10-40 m [5], which means that the quantization is comparable to the one used for RTT positioning in WCDMA. Since the bandwidth of the LTE system is normally higher than that of WCDMA, it seems safe to conclude that E-CID in LTE is likely to perform at least slightly better than RTT positioning in WCDMA. The larger radio bandwidths of the 5G cellular systems are expected to enhance the radial accuracy even further.

The lateral inaccuracy is mainly determined by the right and left limits of the geographical extension of the cell [46]. When signal strength measurements are

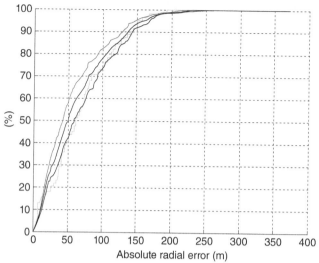

Figure 32.24 Cumulative distribution of the absolute radial distance error in all radio environments for E-CID. The middle black curverepresents the results with all measurements from all terminals included, whereas the other curves show results for each of three terminal types.

combined with the cell description and the TA, the lateral accuracy can be further improved.

A second E-CID positioning method that is becoming increasingly important follows from the combination of CID with TA and AoA measurements. The reasons include the fact that the TA inaccuracy is reduced with increasing radio bandwidth and the fact that the AoA inaccuracy is being rapidly reduced with the introduction of large antenna arrays in LTE. These antenna arrays and codebooks allow 2-D angular measurements, thereby enabling 3-D Cartesian E-CID based on TA and AoA measurements. This development is expected to accelerate in 5G 3GPP cellular systems. As an example, 800-MHz radios should easily result in distance measurements accurate to less than 1 m in LOS. In addition large antenna arrays should allow LOS AoA detection with accuracies of the order of few degrees. Together this would allow the 3-D position of an object to be pointed out relative to the antenna with an inaccuracy of the order of a meter in a small cell with a cell radius of approximately 100 m. Such a method could solve, for example, the indoor E-911 positioning problem, using a single base station site with an accurately configured antenna position. The remaining difficulty would be NLOS propagation mitigation. One way to address the NLOS problem would be to apply fingerprinting to AoA and TA measurements in a single site.

Fingerprinting

The availability and response time for fingerprinting in LTE can be expected to be comparable to the availability and response time of the E-CID method since the positioning step requires similar measurements.

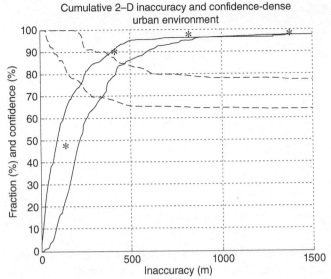

Figure 32.25   Cumulative radial inaccuracy (solid), TA-distances (stars) and confidence (dashed), for AECID (left) and CGI-TA (right). Note that CID+TA are the main sources of information and that the AECID results include effects of the shape conversion of [38].

It is not straightforward to state the accuracy for RF fingerprinting since algorithmic refinements like the use of differential signals strengths and adaptive reporting can significantly improve the accuracy beyond what is predicted by (32.12). Vendors claim that E-911 requirements can be met, at least in rural areas. What is clear is that the error will scale with the cell size, as shown by (32.12).

For the AECID positioning method, field trial results for GSM have been presented in [32]. Those results indicate that the adaptation to measured reference positions of opportunity from A-GPS improves the performance by a factor of about two (measured by (32.34)) as compared to an E-CID method using the same measurements; see Figure 32.25. This scaling is expected to carry over to LTE, and also when AoA information is included in the fingerprint. It is further expected that the sensitivity to NLOS propagation of AoA will be mitigated by the AECID adaptation.

OTDOA

The OTDOA position accuracy depends mainly on the RSTD measurement quality and the geometry. The best accuracy is achieved with a PRS transmission bandwidth of 10 MHz or larger. The minimum required measurement accuracy [5] of about 50 m is achieved after measuring over one positioning subframe at the PRS SINR of −6dB and −13dB for the reference and measured cells, respectively. The RSTD reporting granularity is about 10 m for the largest part of the feasible range. Example results for simulated OTDOA accuracy are shown in Figure 32.26 using ETU and EPA channel models for synchronous (1.4MHz) and asynchronous (10 MHz)

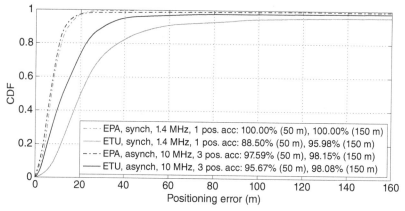

Figure 32.26   OTDOA accuracy for intersite distance of 500m, Case 1 [11].

networks. The number of positioning occasions were one and three, respectively, and the signals are coherently accumulated within $N_{prs} = 6$. The impact of real measured propagation channels on OTDOA positioning has been studied in [28], where highly accurate and well-calibrated channel measurement data from a measurement campaign in an urban macro cellular LTE scenario showed that a positioning accuracy better than 20 m for 50% and 63 m for 95% of the time may be achieved. Note that in more difficult radio environments accuracy may be further degraded as compared to Figure 32.26.

UTDOA

The UTDOA method is discussed at length in [14] primarily for GSM. Due to the larger bandwidth of the LTE system, the expected performance of UTDOA in LTE is expected to be in the same class of accuracy as OTDOA. The detailed accuracy depends on the positioning geometry, the scheduling of the SRS pilot signals, and the UE transmit power.

A-GNSS

The field performance of A-GPS and A-GNSS have been discussed in many publications, and these methods are known to be very accurate in clear-sky conditions; see Figure 32.27. Here it should be noted though, that in dense urban canyons the cell plan may in fact give a CID accuracy that is comparable to A-GNSS due to the NLOS propagation caused by very tall buildings, which deteriorates A-GNSS accuracy; see Figure 32.28.

Comparison of the Expected Performance for Different Methods

The main characteristics of the different positioning methods in LTE and their expected positioning performance are summarized in Tables 32.2 and 32.3. The

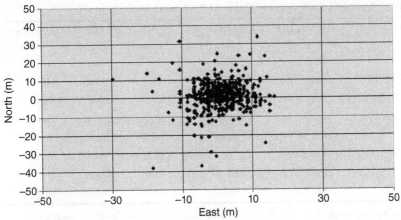

Figure 32.27   A-GPS accuracy in a rural environment.

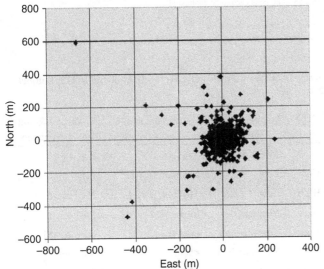

Figure 32.28   A-GPS accuracy in a dense urban environment.

footnotes in Table 32.2 give an indication of the dominating factors behind the described impacts, when there are such. Note that the impact on UEs, base stations (sites), and the positioning system (mainly the positioning node) described in Table 32.2 also include the efforts associated with the implementation of new interfaces, new protocols (e.g., LPP and LPPa) and their transport over the lower-layer protocols.

The accuracy and the response time, as indicated in Table 32.3, may vary significantly and are highly dependent on the network deployment and implementation. In general, the response time of CID is the lowest (mainly table look-up), and the response time of A-GNSS is the longest (mainly due to the long signal accumulation time needed to cope with low received powers at cold starts). For

TABLE 32.2. Typical Characteristics of Different Positioning Methods in LTE

Positioning Method	Earliest LTE Release/Optional	Environment Limitations	UE Impact	Site Impact	System Impact
CID	All/No	No	No	No	Small[‡]
E-CID, CID+TA	Rel 9/Yes	No	No	Small	Medium[‡,§]
E-CID, TA+AoA	Rel 13/Yes	Medium[†]	No	Large[†]	Small
RF Fingerprinting	Rel 9/Yes	Rural[*]	Small	Small	Large[§,§§]
AECID	Rel 9/Yes	Small	Small	Small	Medium[§]
OTDOA	Rel 9/Yes	Rural[*]	Medium	Small	Medium
UTDOA	Rel 11/Yes	Rural[*]	Small	Large	Medium
A-GNSS	Rel 9/Yes	Indoors	Large[**]	Small	Medium[§§§]

[*]No limitation if a sufficient number of sites with a good joint geometry are detectable (for DL) or can detect the signal (for UL).

[**]An A-GNSS receiver is necessary.

[†]AoA measurements may suffer from NLOS conditions and require antenna arrays.

[‡]Mainly configuration, since coordinates of all cells need to be configured.

[§]Large database maintenance.

[§§]Advanced signal processing, filtering over time and test drives.

[§§§]Assistance data build up, by the network or by external providers (A-GNSS).

TABLE 32.3. Typical Accuracy of Different Positioning Methods Available in LTE, Subject to Restrictions of Table 32.2

Positioning Method	Availability	Response Time (in RAN)	Horizontal Result Uncertainty	Vertical Result Uncertainty
CID	100%	Very Low	High, $\alpha = 1$[*]	n.a.
E-CID, CID+TA	Very High	Low	Medium, $\alpha < 1$[*]	n.a.
E-CID, TA + AoA	Very High	Low	Medium, $\alpha \ll 1$[*,†]	Medium, $\alpha \ll 1$[*,4]
RF Fingerprinting	Very High	Medium	Low, $\alpha < 1$[*]	Low $\alpha < 1$[*,**]
AECID	Very High	Low/Medium	Low, $\alpha \ll 1$[*,4]	Low, $\alpha \ll 1$[*,4]
OTDOA	High	Medium/High	<50m	Medium[**]
UTDOA	High	Medium/High	<50m	Medium[**]
A-GNSS	Medium	Medium/High	<5 m	<20m

[*]Proportional to the cell range, with a proportionality constant α.

[**]Optional support.

[†]Requiring AAS and release 13 codebooks.

OTDOA, the measurement time depends mainly on the number of cells in the neighbor cell list. UTDOA should be subject to similar constraints, albeit reversed between the involved nodes. For RF fingerprinting, advanced signal processing and the use of measurements during time intervals of several seconds probably contribute the most, which impacts both the UE and the positioning node complexity. The availability aspect covered in Table 32.3 refers to the average success rate of the corresponding method in different networks and their parts. In general a high availability is required, especially in networks where regulatory positioning requirements apply, for example, in the U.S.

A comparison of the performance of the available TDOA methods is beyond the scope of this work. The reasons include the dependence on specific deployment details, the applied signal processing techniques, as well as the differences between low- and high-frequency bands. Some specifics of each method that may affect the performance are therefore discussed instead. The transmission power level is important for both OTDOA and UTDOA since detectability with respect to a sufficient number of neighbor base stations is crucial to achieve an accurate position solution. Here OTDOA applies cell-specific CRS and/or PRS references, while UTDOA applies UE-specific SRS. The UTDOA transmission power level of the UE is power controlled, resulting in UEs close to the own base station transmitting SRSs at a reduced power level to avoid creation of unnecessarily high levels of interference—this is a general effect in cellular systems, denoted the near–far problem. A consequence is that additional signal processing techniques is typically applied to enhance the detection performance to the required level. The introduction of AAS with high beamforming gain in LTE and 5G means that also the CRS and PRS used by OTDOA are likely to require beamforming gains, a fact that would increase the transmission resources for OTDOA toward a more UE-specific resource consumption as for UTDOA. However, presently it is unclear how beamforming from/toward non serving cells would be implemented. For mmw 5G deployments, current human exposure restrictions may limit the available UE transmit power further as compared to the current situation below 6 GHz, where 23 dBm is commonly used for LTE handsets; see [17]. This effect may however have a small impact since 5G mmw cells are more likely to be UL-limited due to the handset exposure limitations. In addition, work is underway in the wireless industry aiming to reduce these restrictions.

For all the methods, the reported result contains at least the horizontal position. Note, though, that the importance of highly accurate vertical position information is increasing and is mandatory in the U.S., in terms of the E-911 requirements. Although the uncertainty of the horizontal position, that is, the estimated positioning error, cannot always be reported to the requesting node using the native shape of the positioning method, the uncertainty can always be reported after a suitable shape-converting transformation. A vertical positioning result is not always available, for example, for CID and E-CID. However, in case a large antenna array would allow accurate 2-D AoA information to be obtained, this situation could change. In combination with TA information for high bandwidth radios a highly accurate Cartesian 3-D position could then be computed and reported, possibly after a shape conversion. Vertical information and uncertainty are almost always reported when A-GPS is used. The vertical accuracy of OTDOA and UTDOA are, however, most often worse than that of A-GPS and A-GNSS. The reason is the (mostly) dominating 2-D geometry of the eNodeBs that results in a poor vertical dilution of precision (VDOP); c.f. [25]. In order to mitigate this effect, additional altitude information need to be provided as a function of the horizontal coordinates. Such surface models can be obtained by a geographical information system (GIS) providing digital maps. In the 5G wireless systems an augmentation of the polygon format could provide an alternative solution that does not need GIS information; see [41].

32.9 SUMMARY

This chapter discussed the positioning functionality standardized in 3GPP for the LTE system and outlined a few likely fifth-generation developments. The LTE architecture was reviewed, emphasizing aspects on nodes, signalling, and the air-interface properties that are important for positioning. A discussion on requirements was used to highlight the need and techniques for LBS discrimination. The backbone positioning method in LTE is the CID method, which is augmented by, for example, timing advance and AoA information in the E-CID class of methods.

The LTE system is also prepared for fingerprinting positioning. The TDOA methods performs hyperbolic trilateration on DL positioning reference signals or on UL sounding reference signals, while the satellite navigation functionality provided by A-GNSS is similar to that in other cellular systems. The LTE system uses the seven reporting formats defined by 3GPP for the GSM and WCDMA systems. Techniques for conversions between these shapes were reviewed. The predicted performance of the positioning methods were discussed, thereby indicating how they can be applied to address different positioning services.

REFERENCES

[1] 3GPP TR 36.897, "*Study on Elevation Beamforming/Full-Dimension (FD) MIMO for LTE,*" Release *13* (V1.0.0), 2015.

[2] 3GPP TS 22.071, *Technical Specification Group Services and System Aspects; Location Services (LCS); Service description; Stage 1.* Available: www.3gpp.org/specifications/specifications

[3] 3GPP TS 23.032, *Universal Geographical Area Description (GAD).* Available: www.3gpp.org/specifications/specifications

[4] 3GPP TS 29.171, *LCS Application Protocol (LCS-AP) between the MME and E-SMLC; SLs Interface.* Available: www.3gpp.org/specifications/specifications

[5] 3GPP TS 36.133, *Evolved Universal Terrestrial Radio Access (E-UTRA); Requirements for Support of Radio Resource Management.* Available: www.3gpp.org/specifications/specifications

[6] 3GPP TS 36.211, *Evolved Universal Terrestrial Radio Access (E-UTRA); Physical Channels and Modulation.* Available: www.3gpp.org/specifications/specifications

[7] 3GPP TS 36.305, *Stage 2 Functional Specification of User Equipment (UE) Positioning in E-UTRAN.* Available: www.3gpp.org/specifications/specifications

[8] 3GPP TS 36.355, *Evolved Universal Terrestrial Radio Access (E-UTRA); LTE Positioning Protocol (LPP).* Available: www.3gpp.org/specifications/specifications

[9] 3GPP TS 36.401, *Evolved Universal Terrestrial Radio Access Network (E-UTRAN); Architecture Description.* Available: www.3gpp.org/specifications/specifications

[10] 3GPP TS 36.455, *Evolved Universal Terrestrial Radio Access (E-UTRA); LTE Positioning Protocol A (LPPa).* Available: www.3gpp.org/specifications/specifications

[11] 3GPP TS 36.814, *Evolved Universal Terrestrial Radio Access (E-UTRA); Further Advancements for E-UTRA Physical Layer Aspects.* Available: www.3gpp.org/specifications/specifications

[12] J. Borkowski, J. Niemälä, and J. Lämpiäinen, "Performance of cell ID+RTT hybrid positioning method for UMTS radio networks," in *Proc. 5th European Wireless Conf.*, 2004, pp. 487–492.

[13] P. Bourke, *Calculating the Area and Centroid of a Polygon*. Available: http://local.wasp.uwa.edu .au/pbourke/geometry/polyarea, 1988

[14] J. F. Bull, "Wireless geolocation—advantages and disadvantages of the two basic approaches for E-911," *IEEE Veh. Tech. Mag.*, vol. 4, no. 4, pp. 45–53, 2009.

[15] Y. T. Chan and K. C. Ho, "A simple and efficient estimator for hyperbolic location," *IEEE Trans. Signal Process.*, vol. 42, no. 8, 1905–1915, 1994.

[16] G. W. Coleman, M. Wang, and S. Watson, Direction of arrival estimation for MIMO systems employing constellation based pre-coding. In *IEEE VTC 2012 - Fall*, Quebec City, Canada, 2012.

[17] D. Colombi, B. Thors, and C. Törnevik, "Implications of EMF exposure limits on output power levels for 5G devices above 6 GHz," *IEEE Antennas Wireless Propag. Lett.*, vol. 14, pp. 1247–1249, 2015.

[18] E. Dahlman, S. Parkvall, J. Sköld, and P. Beming, *3G Evolution: HSPA and LTE for Mobile Broadband*, 2nd ed. Oxford: Academic Press, 2008.

[19] FCC. Wireless e911 location accuracy requirements, fourth report and order. FCC 15-9, February 2015.

[20] G. P. Fettweis, "The tactile internet: Applications and challenges," *IEEE Veh. Tech. Mag.*, vol. 9, no. 1, pp. 64–70, 2014.

[21] A. Goldsmith, *Wireless Communications*. Cambridge, MA: Cambridge University Press, 2005.

[22] D. H. Johnson and D. E. Dugdeon, *Array Signal Processing: Concepts and Techniques*. Englewood Cliffs, NJ: Prentice Hall, 1996.

[23] A. Kangas and T. Wigren, "Location coverage and sensitivity with A-GPS," presented at the URSI Int. Symp. on Electromagnetic Theory, Pisa, Italy, 2004.

[24] A. Kangas and T. Wigren, "Angle of arrival localization in LTE using MIMO pre-coder index feedback," *IEEE Comm. Lett.*, vol. 17, no. 8, pp. 1584–1587, 2013.

[25] E. D. Kaplan, *Understanding GPS Principles and Applications*. Norwood, MA: Artech House, 1996.

[26] R. Langley, Time, clocks and GPS. GPS *World Magazine*, pages 38–42, 1991.

[27] W. Lewandowski and C. Thomas, GPS time transfer. *Proc. of the IEEE*, 79(7), 1993.

[28] J. Medbo, I. Siomina, A. Kangas, and J. Furuskog, "Propagation channel impact on LTE positioning accuracy: A study based on real measurements of observed time difference of arrival," in *Proc. IEEE 20th Int. Symp. on Personal, Indoor and Mobile Radio Communications (PIMRC)*, 2009, pp. 2213–2217.

[29] Open Mobile Alliance. Secure user plane location architecture. *OMA-AD-SUPL-V2-0*.

[30] Open Mobile Alliance. Secure user plane location protocol. *OMA-TS-ULP-V2-0*.

[31] T. S. Rappaport, S. Sun, R. Mayzus, H. Zhao, Y. Azar, K. Wang, G. N. Wong, J. K. Schulz, M. Samimi, and F. Gutierrez, "Millimeter wave mobile communications for 5G cellular: It will work!" *IEEE Access*, vol. 1, pp. 335–349, 2013.

[32] L. Shi and T. Wigren, "AECID fingerprinting positioning performance," presented at the Proc. Globecomm 2009, Honolulu, HI, 2009.

[33] M. I. Simic and P. V. Pejovic, "An algorithm for determining mobile station location based on space segmentation," *IEEE Comm. Lett.*, vol. 12, no. 7, pp. 499–501, 2008.

[34] P. Stoica and R. L. Moses, *Spectral Analysis of Signal*. Upper Saddle River, NJ: Prentice Hall, 2004.

[35] J. Wennervirta and T. Wigren, "RTT positioning field performance," *IEEE Trans. Veh. Tech.*, vol. 59, no. 7, pp. 3656–3661, 2010.

[36] T. Wigren, "Adaptive enhanced cell-ID fingerprinting localization by clustering of precise position measurements," *IEEE Trans. Veh. Tech.*, vol. 56, no. 5, pp. 3199–3209, 2007.

[37] T. Wigren, "Clustering and polygon merging algorithms for fingerprinting positioning in LTE," presented at the Proc. 5th ICSPCS, Honolulu, HI, 2011.

[38] T. Wigren, "A polygon to ellipse transformation enabling fingerprinting and emergency localization in GSM," *IEEE Trans. Veh. Tech.*, vol. 60, no. 4, pp. 1971–1976, 2011.

[39] T. Wigren, "Fingerprinting localization using RTT and TA," *IET Comm.*, vol. 6, no. 4, pp. 419–427, 2012.

[40] T. Wigren, "LTE fingerprinting localization with altitude," presented at the Proc. VTC-Fall 2012, Quebec City, Canada, 2012.

[41] T. Wigren, "Wireless hybrid positioning based on surface modeling with polygon support," presented at the Proc. VTC-Spring 2018, Porto, Portugal, 2018.

[42] T. Wigren, M. Anderson, and A. Kangas, "Emergency call delivery standards impair cellular positioning accuracy," presented at the *Proc. ICC 2010*, Cape Town, South Africa, 2010.

[43] T. Wigren, A. Kangas, Y. Jading, I. Siomina, and C. Tidestav, "Enhanced WCDMA fingerprinting localization using OTDOA positioning measurements from LTE," presented at the Proc. VTC-Fall 2012, Quebec City, Canada, 2012.

[44] T. Wigren and T. Palenius, "Optimized search window alignment for A-GPS," *IEEE Trans. Veh. Tech.*, vol. 58, no. 8, pp. 4670–4675, 2009.

[45] T. Wigren, I. Siomina, and M. Anderson, "Estimation of prior positioning method performance in LTE," presented at the Proc. PIMRC, Toronto, Canada, 2011.

[46] T. Wigren and J. Wennervirta, "RTT positioning in WCDMA," in *Proc. 5th Int. Conf. on Wireless and Mobile Communications, ICWMC 2009*, Cannes/La Bocca, France, 2009, pp. 303–308.

[47] J. Zhu, S. Spain, T. Bhattacharya, and G. D. Durgin, "Performance of an indoor/outdoor RSS signature cellular handset location method in Manhattan," in *Proc. IEEE Antennas and Propagation Society Int. Symp. 2006*, Albuquerque, U.S.A., 2006, pp. 3069–3072.

CHAPTER 33

AUTOMATED WILDLIFE RADIO TRACKING

Dr. Robert B. MacCurdy,[1] *Dr. Allert I. Bijleveld,*[2]
Richard M. Gabrielson,[3] *and Dr. Kathryn A. Cortopassi,*[3]

[1]*University of Colorado, Boulder*
[2]*NIOZ Royal Netherlands Institute for Sea Research Texel, the Netherlands*
[3]*Cornell University*

33.1 INTRODUCTION

Radio direction-finding techniques have been widely employed by the wildlife tracking community because they offer powerful, flexible tools for monitoring animal movements and behavior. Reductions in the size and power consumption of GPS chipsets have recently allowed GPS location-finding techniques to also be applied to wildlife monitoring. Despite their successes, these approaches still have significant shortcomings, primarily due to the energy constraints imposed by the allowable mass of the electrochemical battery that can be carried by the animal. This requirement causes tag lifetimes to be shorter than desired. Attaching a tracking collar is a risky procedure for all participants, and maximizing the tag service intervals is extremely important. This is true even for large animals, which can carry significant tag mass without behavior disruptions. Therefore, disturbances to the animal (which are primarily driven by tag energy consumption), followed closely by cost are the primary design requirements. These requirements motivate a new tracking system, based on time-of-arrival (TOA) measurements. This system is similar in many respects to GPS; the primary difference is that in this system the mobile asset to be tracked emits, rather than receives, signals. Although transmitting a radiofrequency signal is often the most power-intensive operation for a tag, this choice yields a system with average tag energy requirements that are lower than any current radio-tracking method. Though this chapter focuses on the application of this technology to monitoring animal movements, the same set of design criteria apply to generic asset tracking, and we believe that there is a universal need for a local terrestrial tracking system that offers precise positioning with tiny, cheap, long-lived tracking tags.

Handbook of Position Location: Theory, Practice, and Advances, Second Edition.
Edited by S. A. (Reza) Zekavat and R. Michael Buehrer.
© 2019 by The Institute of Electrical and Electronics Engineers, Inc.
Published 2019 by John Wiley & Sons, Inc.
Companion Website: www.wiley.com/go/zekavat/positionlocation2e

33.2   A REVIEW OF WILDLIFE TRACKING TECHNIQUES

Radio tracking has been widely used to monitor wildlife movements since the 1960s [13], [28] with countless scientific papers published using some variant of this method. In the majority of these studies, an operator in the field monitors received signal strength while manually changing the orientation of a directional receiving antenna. The direction yielding the maximal signal strength is recorded as a pointing vector to the tagged animal (Chapter 9 provides a detailed discussion of direction estimation). This simple method is adequate to guide a researcher to the location of a focal individual, and triangulation using two or more receiving stations can be used to track a few individuals simultaneously. However, this method yields relatively few position fixes per hour and fully absorbs an operator's attention and effort. Automatic or supervised tracking systems have been developed using stationary receiving towers [14], [22], [34], [48] in an attempt to increase the number of animals that can be tracked simultaneously. Most efforts involve directional antennas and rely on the beam pattern of the antennas to infer a direction of arrival. These approaches generally show error in the 1–10 degree range, depending on the implementation [6], [16], [25], [34], and the cross-bearing positional error for each receiving station increases linearly with range. An excellent, comprehensive manual for employing current radio tracking tools in the study of wildlife can be found in [24]. Several other technologies, including satellite-based transmitters or receivers, cellular communications, solar geolocation, and radar have been used to study wildlife movements; these methods are discussed later in this section.

33.2.1   Wildlife Tag Design Constraints

Electronic tags offer the possibility of monitoring animals in their native habitat with minimal disturbance. If this capability is to be realized, these devices must be unobtrusive. Wildlife tracking tags could also enable vastly larger study sample sizes than methods that use unaided observations by field personnel; however, the cost of the equipment required must not prohibit its use. We consider these two parameters, disturbance to the animal and cost, to be the primary factors in the design of wildlife tracking tags, and they motivate all design choices. Animals are disturbed by the use of tags in two primary ways: (1) they are captured in order to apply the tags, and (2) they must then accommodate carrying a foreign object on their body. The relative impacts of these two factors differ depending on the animal being studied. For example, elephants are capable of carrying a large tag on a collar around their neck; however, the process of anesthetizing the animal and applying the tag is dangerous for the animal as well as the researchers. In contrast, small birds can be easily captured in mist nets, but they are only capable of carrying a small percentage (no more than 2.5% to 5%) of their body mass as additional payload [15], [35]. While the factors differ, both cases require carefully minimizing the energy used by the tag's electronics: the elephant can carry a large battery, but long intervals between servicing the tag are desired, while the bird can carry little mass, and therefore requires a small battery.

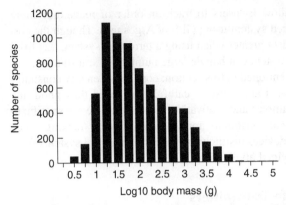

Figure 33.1   Bird body mass distributions. Data from [19].

Bird body mass varies widely, and though no one tag design will be appropriate for all birds, tag designers should strive to make their tags as widely applicable as possible, within functional limits. Figure 33.1 [19] shows a strong peak in bird body-mass distribution, and when combined with the loading heuristic mentioned previously, implies that a tag between 1 and 9 grams can be safely applied to roughly 50% of all bird species. This metric can be used as an upper limit on the mass of a general purpose bird-tracking tag. Though larger tags have utility in some specific applications, the majority of migratory bird species cannot carry them.

Limitations on tag mass apply to species other than birds. Even though larger animals are capable of carrying tags with heavier batteries that provide long run-times, there may be compelling reasons to limit tag mass. Whale tagging provides an apt example. Whale tags are often applied via ballistic darts because: (1) this method is far safer and less intrusive than capturing and anesthetizing an animal (if possible at all), and (2) the thick blubber present in many whales provides a sound anchor for the dart. The darts are delivered by a crossbow, so the overall mass of the dart/tag system must be limited. Many large land mammals could carry a small dart or tag affixed to the ear more easily than the current collar attachment; cattle have long worn plastic ear tags for identification. The primary issue is mass: the mass of these tags must be less than a few grams for this approach to succeed.

Tag lifetime, in addition to tag mass, factors strongly into the relative disturbance to the animal and the scientific utility of the method. The movements of migratory birds are strongly tied to seasonal changes, and migration behavior is becoming recognized as an important indicator of climate change [29]; however, recording migrations requires equipment with a useful lifetime of at least one year.

Electronic tags offer the potential to dramatically increase the study sample size achievable with a small research team; however, this capability is hindered by cost. Current research tools generally impose severe cost constraints on the study, either in the form of high individual tag costs (as is the case with GPS or Argos), or high labor costs to track and maintain the tags. An automatic wildlife tracking system that used $200 USD tags would allow its users to track an order of magnitude more animals than a comparably priced system using GPS or Argos tags. Though the cost of installing receivers is obviously greater when using a terrestrial system, this fixed cost becomes negligible if the system can handle large numbers of transmitters.

The design constraints mentioned in this section: cost, mass, energy consumption, and lifetime, are interrelated and must be carefully balanced throughout the design process. Though the examples and motivation that we provide are specific to wildlife tracking, nearly identical constraints apply to mobile asset tracking, since small, low-cost and long-lived devices distinguish successful practical systems from those that work in narrowly defined applications.

33.2.2 Terrestrial Wildlife Transmitters

Early wildlife tracking transmitters [28] began to achieve acceptance in the 1950s and their use accelerated in subsequent decades. The first tag designs used radio frequency (RF) tank circuits for frequency control, with a single active element to drive the system into oscillation. These devices broadcast a single frequency carrier, with rudimentary *on–off key* (OOK) modulation. Subsequent designs employed crystal resonators to achieve tighter frequency specifications and added additional output amplification stages to increase the output power (and thereby increase range). Incremental refinements in the intervening 50 years have yielded transmitters that can be extremely light (see fig. 33.2) and are generally low cost; commercial tags using this technology usually cost no more than $250. Despite these improvements, the underlying technology has remained essentially unchanged: a carrier-frequency oscillator circuit is turned on and off by a secondary timing circuit with a period between 2 and 60 seconds. The carrier is typically turned on for a short transmit pulse lasting approximately 20 milliseconds and is off for the duration of the period. The resulting transmit duty cycle determines the tag's average energy consumption. Shorter transmit pulses or longer intervals between transmissions can

Figure 33.2   A 110-milligram tag developed by Julian Kapoor at Cornell University.

reduce the average energy consumption but make the tag more difficult to identify and track. Individual tags that will be used in overlapping geographical regions must be assigned unique operating frequencies via crystal selection. These channels are typically spaced in 5 or 10 kHz intervals. Typical carrier frequencies for wildlife tracking range from 140 to 225 MHz. Several bands in this range have historically been reserved for narrowband amateur use, and a few are allocated exclusively for wildlife tracking. In addition to OOK, some transmitters modulate data using *frequency modulation* (FM). This method has been used to telemeter numerous types of analog data, including heart rate and acoustic information.

The tag designs just described have been successful for so long because they use relatively few components, and yield tags that are inexpensive and simple to use. Unfortunately, the approach also uses the energy required to send RF transmissions inefficiently, since the information content of the signal is low and because the signal is transmitted frequently but is rarely actually received. The signal is received infrequently (relative to the number of transmissions) because it would be extremely arduous for human observers to monitor a radio receiver continuously over the lifetime of a tag (tag lifetimes range from weeks to years). Though automated receivers can monitor a single tag frequency continuously, or sequentially switch between channels, no wildlife-tracking systems have been built that monitor multiple tag frequencies simultaneously. This is discussed in the next section.

33.2.3   Terrestrial Wildlife Receivers

The evolution of wildlife tracking receivers has followed a trajectory similar to that of tags: a proven design has been continually refined, with few fundamental changes. These receivers typically use a conventional narrow-band heterodyne architecture with analog components, and employ either an FM detector/decoder or a tunable *beat frequency oscillator* (BFO) in the final down conversion stage. The BFO approach mixes the signal at the intermediate frequency with a tunable oscillator that is one or two kHz away from the intermediate frequency. When a carrier is present, it is mixed into an audio frequency that is easily heard. This approach is simple, reliable, and yields impressive overall bearing sensitivity (+/− 1.4°) when paired with a trained operator [34]. However, it requires that the signal being received has no complex modulation and must last long enough for the human operator to hear. In practice, this means about 20 milliseconds, though some tags use slightly shorter transmitter pulse lengths.

33.2.3.1 Handheld Receivers   Handheld receivers constitute the vast majority of wildlife tracking systems in use. Locating and tracking a tagged animal using a conventional handheld receiver requires an operator to tune the receiver to the channel allocated to the tagged animal. The operator sweeps a directional antenna through all expected bearings while listening (typically through headphones) for a "beep" from the receiver. The audible tone is generated by the BFO circuit, and the system relies on a human listener for detection. This is why the tag's transmissions must be relatively lengthy; the human observer's ear integrates the acoustical signal, so longer tones sound louder. When a signal is detected, the operator makes fine adjustments to the bearing of the antenna while listening to the amplitude of the beep in order to determine the actual line of bearing to the animal. When the direction to the animal has been established, the operator moves to a second location along a baseline roughly perpendicular to the original bearing in order to make a second bearing measurement. The distance between the two locations should be large enough that the second bearing measurement is significantly different from the first. The animal's location is then estimated by the intersection of the two lines of bearing, a process known as triangulation (see Chapter 1). The operator can also simply follow successive lines of bearing to the animal if the goal is to approach the animal. This method though time tested, leaves much to be desired. It is labor intensive and slow, which limits the number of animals that can be studied and the amount of position data that can be gathered. The cost of achieving round-the-clock observations is prohibitive. It also requires many tag transmissions to achieve a single position estimate.

33.2.3.2 Automatic Receivers

Automatic receivers have been developed to increase the number of animals that can be simultaneously tracked and to reduce the labor cost of the effort. These devices are functionally very similar to handheld receivers; however, they typically employ a microcontroller to perform the channel scanning and signal detection operations. The simplest automatic receivers do not attempt to estimate transmitter direction at all and are used to determine the presence or absence of a tag within a given detection radius [22]. Directional antennas can be added to these receivers, and the presence/absence information is associated with a particular range of bearings (the main "lobe" of the antenna), which yields a rudimentary location. More sophisticated receivers employ multiple directional antennas whose main sensitivity lobes are uniformly distributed around 360°. A specific variant of this, known as the crossed-Adcock antenna, uses two pairs of matched dipole antennas. Each pair of antennas is connected to a phasing element that combines their individual outputs into a single output. The output of each phased pair of antennas reaches a maximum when the phase of the incident signal is equal at both antennas and a minimum when the phase differs by 180°. The two antenna pairs are arranged in a cross, and the signal strengths of the two outputs are compared in order to establish signal direction. Receivers with more antennas have been successfully used; in general, directional receivers of this type use the relative signal strength at each of the directional antennas to establish a line of bearing when a transmitter is detected. Networks of these receivers are established in a study area and when multiple receivers detect the same transmission event they can locate the transmitter by intersecting their estimated lines of bearing [14], [16], [22], [34], [48]. Digitally steered phased-array approaches have also been employed to determine tag bearing, as described in Chapter 9. These methods use multiple antennas, usually in a circular or linear array, and establish signal direction by measuring the phase difference of the signal at each receiving antenna. This method usually requires multiple synchronized receiver signal paths; modern directional receivers usually accomplish the task with high-speed synchronous analog-to-digital converters operating at the intermediate-frequency (IF) stage of the receiver.

The sampled multichannel signal is then digitally down-converted, and the bearing is established via software; the multiple signal classification (MUSIC) [42] algorithm is widely used for this purpose (Chapter 9 has a detailed description of the MUSIC algorithm). The complexity of this approach can be problematic for wildlife tracking applications, and receivers of this type are costly. Additionally, phased array receive antennas must be mounted far from other objects; experiments conducted by our group showed significant variation in signal phase due to nearby vegetation. This constraint requires phased array antennas to be mounted on tall, sturdy masts, rather than opportunistically placed in trees.

A less expensive variation of the phased array receiver, borrowed from the amateur radio community, uses multiple antennas but only a single receive channel. This receiver, known as a pseudo-Doppler direction finder, arranges the antennas around the circumference of a circle whose diameter is half the wavelength of the received signal. A many-to-one multiplexer sequentially selects each antenna around the circle as the input for a single-channel FM receiver with a phase locked loop (PLL) detector. As each antenna is switched in, different phases of the signal are presented to the PLL detector, and the phase changes produced each time the receiver is switched from one antenna to the next cause the PLL to produce output pulses that are proportional to the magnitude of the phase change. This pulse train, when low pass filtered, has a sinusoidal shape and the relative phase between this sinusoid and the antenna switching signal provides an estimate of the incident signal's bearing. Though less complex (and far less expensive) than the multichannel receiver described previously, pseudo-Doppler directional receivers generally sacrifice sensitivity for simplicity.

Though receivers that automatically detect transmitter direction have been successful in reducing the labor cost of tracking, and can increase the number of location estimates per unit time, they have drawbacks. Their detection sensitivity is lower than that for a system with a human operator, which reduces the radius of detection, and their bearing estimate is several times less accurate than when using a human-rotated antenna.

Most automated receivers operate by continuously scanning a sequence of narrow-band channels for tag transmissions. This approach is adapted from the manually tuned receivers that preceded them, and restricts the number of channels that can be monitored per unit time. The primary challenge is synchronization with the tag's transmissions, since a typical tag transmits for a few milliseconds every few seconds, while the receiver has to scan tens to hundreds of channels, depending on the number of tags that must be accommodated. The tag must transmit while the receiver is listening for its carrier frequency, an occurrence that becomes increasingly unlikely as the number of channels to be scanned increases. This situation can be partially mitigated, on average, by randomizing the channel scan sequence, or by ensuring that a particular channel's scan period is not a multiple of the tag's transmit period. Even so, the scanning approach breaks down when the channel count exceeds thirty to fifty, with many tag transmission events going unnoticed by the receiver. An alternative approach selects each channel for a long enough period to guarantee that a tag will transmit several times and then moves to the next channel. Though this method guarantees that all tags within range will eventually be heard, it also guarantees that most tag transmissions will not be detected. These undetected tag transmissions represent system-level energy inefficiency, and the energy is wasted where it can be tolerated least: in the battery-powered mobile tags.

33.2.4 Satellite Tracking Systems

Two satellite-based systems are widely used for wildlife tracking: GPS and Argos. These two systems provide location information using different techniques. GPS employs a network of orbiting satellites that broadcast signals to a terrestrial receiver that uses a TOA algorithm to estimate its position. Chapters 20–23 discuss the GPS system in detail. In addition, Misra and Enge provide a comprehensive description of the GPS system in their book [33]. Tag mass is the primary limiting factor for applying satellite-based systems to wildlife tracking. As mentioned previously, the typical maximum allowable tag-to-body mass ratio is 5%. Commercial GPS tags typically weigh between 22 to 150 grams, which limits their application to larger animals (>440 g), and cost between $1500 and $3500. One new, very low-mass (4.5g) tag is now available [44]; however, the small battery used in order to achieve low mass limits the system to no more than several hundred position fixes. Though continuous refinements have yielded ever smaller and more sensitive receivers, the positioning approach used by GPS makes it unsuitable for ultra-low-power tracking systems. The principle drawback of GPS is that it does not directly provide a means of reporting position information back to the researcher. The position information is either stored and retrieved later, or downloaded via an auxiliary RF link. The energy cost required to transmit position data from the animal to the researcher is often prohibitive. Additionally, GPS receivers require a relatively low-noise, broadband RF front-end, coupled with fast digitizers and signal processing hardware. The power consumption of these elements, integrated over the satellite signal acquisition time, imposes a significant energy demand on the tag's battery because the signal acquisition time can be long. The duration, which depends on several factors, varies from one second to one minute in modern receivers. A specialized type of GPS logger, which can yield position fixes by postprocessing recorded satellite transmissions, reduces energy consumption by limiting the signal acquisition time to approximately 60 milliseconds [49]. These tags, referred to as fast lock GPS tags, store the raw digital IF data from all satellites in view rather than attempt to acquire each satellite's signal via matched filtering. The trade-off of this approach is that many kilobytes of data must be written to nonvolatile memory. Despite a considerable improvement in energy consumption, relative to conventional GPS, this approach must transmit the recorded data if animal recapture is not possible, or if real-time operation is desired. The energy cost of transmitting data from a GPS logger is even higher than for a conventional GPS tag, since the logger must send far more data.

In contrast to GPS's TOA-based operation, the Argos system determines tag (transmitter) position by exploiting the frequency shift in the tag's signal, measured by the satellite's receiver. This frequency shift is caused by the satellite's motion relative to the transmitter and is primarily dependent on the orbital parameters and the earth's rotation. The computation of transmitter location takes place within the

Argos satellite control system, rather than on the tag, so tag positions are immediately available to the researcher. For a complete description of the Argos system, see [11]. The Argos system can achieve reasonably good accuracy; the best service class available advertises 250 m error bounds. However, this level of accuracy is often not available, and the other three accuracy classes range from 250 to 1500 m. The Argos system's link budget requires significant transmit power from the tags; typical tag power consumption varies from 150 mW to 500 mW during a 300 to 900 millisecond transmission. These transmissions must repeat with a 90- to 300-second period. These parameters set the minimum energy for operation, and necessitate relatively large tags, though some newer models have incorporated solar cells and ultracapacitors to reduce mass. Commercial Argos tags are generally smaller than GPS tags (the lowest-mass Argos tag is currently about 5 grams) but still only allow animals heavier than 100 grams to be tracked. This weight constraint excludes 75% of all bird species [19]. The cost of Argos tags is also prohibitive for large-scale studies. The complexity and low production volumes of these tags lead to typical single unit prices in the $1500 to $4000 range, with little cost reduction at larger volumes.

33.2.5 Solar Geolocation tracking

Each tracking method mentioned previously has employed a form of RF technology; however they do so at a price: energy consumption. As shown in Figure 33.1, many birds are so small that their maximum payload is less than one gram. Remarkably, some of these tiny animals perform very long migrations and researchers would like to track their movements, yet their size makes the application of continent-scale tracking via GPS or Argos tags impossible. Another approach, using measurements of the time of local dawn and dusk, can be performed at a very low energy cost. The British Antarctic Survey has taken the lead in the application of this technique, known as solar geolocation [4]. Their tiny tags use a microcontroller to keep track of time and sample the ambient light level every few minutes with a photodiode. The tags create a continuous record of the light level as a function of time (usually Universal Time Coordinated (UTC) is used as the reference). The time of sunrise and sunset is a function of longitude, while the length of the day is predominantly a function of latitude (each is also affected by the time of year). This method can produce global position estimates with accuracy better than 100 km, which is sufficient for determining migratory flyways and critical stopover points.

33.2.6 Cellular Tracking

Despite several decades of growth, and near-ubiquitous availability (with frustrating exceptions) cellular telephone operators in the U.S. have historically been reluctant to open their networks to nonvoice traffic. This has begun to change recently, as the network operators recognize the market potential for *machine-to-machine* (M2M) communications. They are beginning to offer nonvoice short message service (SMS) and à la carte data rate plans that are suitable for M2M use. At least one wildlife tracking company, Cell Track Tech of Rector, Pennsylvania, has begun to use cellular communications technology to enable real-time data downloads and software updates for deployed tags. Their lightest tag weighs about 50 grams, making it suitable only for larger birds. Nevertheless, these tags allow researchers to monitor the movements of migratory birds within the global system for mobile communications (GSM) network and can archive GPS data for future download when no communication network is available. Cellular data connectivity is now even available to hobbyists via a line of cellular communications modules manufactured by Telit Wireless, an Italian company. Their products can be easily connected to a microcontroller and abstract the specific details of the cellular network from the developer.

33.2.7   Radar Tracking

Migratory birds have long been visible on weather and transportation radar systems. At times the considerable backscatter from large flocks of birds or even insects can be a nuisance for the operators. Biologists, however, have begun to use this information to study migratory species [20]. Though individuals cannot be followed, flocks of birds can be clearly tracked over hundreds of square miles using the existing Doppler radar infrastructure. Smaller, portable radars have been used with extremely small tags, some weighing less than 12 milligrams, in order to track individual flying insects [8], [17], [40]. The tags employ passive nonlinear switching elements in the antenna that reflect a frequency-doubled version of the incident signal. This coherent signal is easily detected by the radar. This approach has been used to determine range and bearing over half-kilometer distances.

33.2.8   Summary and Motivation for Improvements

Although they have been separated into two categories in this description, wildlife receivers and tags should be viewed as integral components of the same system; one is not useful without the other, and neither can be substantially modified without impacting the design of the other. This linkage causes discrete design choices to propagate throughout the system. For example, the use of narrowband transmitters, which has historical as well as practical underpinnings, limits the application of signal processing techniques at the receiver. As a consequence, the tags must transmit long sequences relatively often in order to satisfy the link budget, with an attendant energy cost. The use of narrowband signals also reduces the options for mitigating multipath interference, which can cause errors in the estimated position. Tags are usually designed without a microcontroller in the name of simplicity; however this limits them to very simple control schemes (for example: on for 20 milliseconds, off for 2 seconds), that cannot turn off the transmitter during lengthy periods in which tracking will not occur (at night, or when the animal is in a remote segment of its migration). Channel-scanning in the receivers causes many transmissions to be missed, which wastes tag battery energy.

33.3 A NEW APPROACH TO WILDLIFE TRACKING

Although traditional narrow-band radio tracking systems work well when a single researcher is following a small number of animals, this technology does not lend itself to automation. An alternative approach, using TOA information captured by fixed terrestrial receivers, offers the potential for significant improvements when tracking wildlife within limited regions. The TOA approach, which can be imagined as "inverse GPS," uses small tags that periodically transmit a short carrier that is modulated by a *pseudorandom noise sequence* (PRN). A network of nearby receivers continuously listens for tag transmissions and records the arrival time when a transmission is detected. Each of the receivers sends its arrival time measurement to a centralized server where the transmitter position is computed. This system dramatically reduces tag energy consumption by using very short RF transmissions and sending the transmissions infrequently. TOA receivers should be capable of detecting a transmission from any tag in the system at any time, so the tags only need to transmit as frequently as a position update is desired. In addition, this requirement avoids any costly synchronization or registration process between the tag and receiver (the fixed receivers maintain synchronization with each other instead).

Despite (or perhaps because of) the long success of GPS, relatively few TOA-based terrestrial tracking systems have been built. One business, Recon Dynamics of Kirkland, Washington (which acquired the technology from S5 Wireless), is attempting to commercialize a system that uses small transmitters and TOA measurements for asset tracking. A prototype system using TOA to track flying foxes was developed by researchers in Australia [41]. A similar system, using fairly powerful transmitters, was developed to track moose in Sweden [26]. An early vehicle tracking system from Teletrac used terrestrial TOA spread spectrum techniques (though they now use GPS with terrestrial data links). Our group designed, built and installed a prototype system based on CDMA and TOA that is capable of automatically locating thousands of tags in real time. In addition, the tags utilize onboard microcontrollers that can implement sophisticated calendar functions; the tags can be put into long periods of "deep sleep" and awakened only when the animal is expected to be within range of the system. This system borrowed concepts from the GPS system in general, and from *pseudolites* in particular (pseudolites, or pseudo-satellites, are terrestrial devices which transmit GPS signals; see [12] for additional information on this topic). Finally, the recent ATLAS wildlife tracking system employs similar techniques and has been demonstrated at a regional scale in several ecologically significant locations [47]. These TOA-based tracking systems rely on several concepts, including PRN codes, CDMA, matched filters, and digital signal processing. These concepts, along with their applications in TOA tracking systems, are reviewed in the following section. For comprehensive resources, see the earlier chapters in in this book or [5], [18], and [33].

33.3.1.1 PRN Sequences, Radio-tracking equipment has traditionally employed simple modulation schemes, such as OOK or FM, because they are easy to implement with simple analog circuits; however this simplicity comes at a cost. The signals used are not readily distinguished from noise, and they interfere with each other. Two adjacent tags that use OOK and share a carrier frequency will be difficult to differentiate if they transmit simultaneously. Using PRN signals can reduce these problems. Their autocorrelation properties allow detection and precise synchronization even when they are contaminated by significantly more powerful noise. Additionally, certain PRN sequences have guaranteed cross-correlation behaviors that allow many different signals to coexist with minimal interference.

We begin the discussion of pseudorandom sequences with a description of random sequences. Consider the random sequence snippet shown in Figure 33.3, which has nine entries whose magnitudes are either 1 or −1. Each of these entries represents a bit (as indicated in the figure), has a duration, a magnitude, and is called a *chip*. The duration of each of the chips is T_C seconds, the entire sequence has N entries, and the sequence's total duration is T_N seconds. The normalized *cross-correlation*, $C(\tau)$ of two signals $x(t)$ and $y(t)$, is defined as

$$C(\tau) = \frac{1}{2T_N} \int_{-T_N}^{T_N} x^*(t) y(\tau + t) dt \qquad (33.1)$$

where the superscript * indicates the complex conjugate operator. Cross-correlation can be thought of as a measure of how similar two different signals are as they are shifted past each other in time. We define the similar *autocorrelation*, $R(\tau)$ of $x(t)$, by setting $y(t) = x(t)$ in (33.1). The autocorrelation can be thought of as the degree to which a signal is similar to time-shifted replicas of itself.

The magnitude of the time shift is usually referred to as a time *lag*, and has the same units as the domain of the signal that was shifted (in this case, seconds). The mean value of a portion of the autocorrelation function of a signal $x(t)$ composed of a repeated N-length random sequence snippet is shown in Figure 33.4 (the N-length sequence $s(t)$ is a finite-length snippet from an infinite-length random signal, and copies are concatenated to form $x(t)$). The mean is shown because the actual autocorrelation values depend on the particular random sequence chosen; however, the mean value of the autocorrelation function of *any* random sequence corresponds to [33]:

$$\bar{R}(\tau) = \begin{cases} \dfrac{\tau}{T_c} + 1 & \text{if } -T_c < \tau < 0 \\ \dfrac{-\tau}{T_c} + 1 & \text{if } 0 < \tau < T_c \\ 0 & \text{else} \end{cases} \qquad (33.2)$$

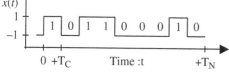

Figure 33.3 A random sequence snippet, $s(t)$ with chip duration T_C and period T_N.

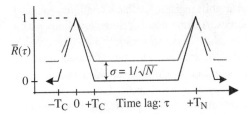

Figure 33.4   The mean value, $\bar{R}(\tau)$, of the autocorrelation of the random sequence, $x(t)$. The grey envelope is one standard deviation from the autocorrelation.

Notice that the peak in this function occurs at the zero lag. Also note that the function has a prominent peak whose width is equal to twice the chip duration. A sharp peak allows precise time alignment and aids detection, as we will see later. The standard deviation σ of $R(\tau)$ is

$$\sigma = \begin{cases} 0 & \text{if } \tau = 0 \\ \frac{1}{\sqrt{3}} & \text{else} \end{cases} \tag{33.3}$$

standard deviation is depicted as an envelope (grey line) of $\bar{R}(\tau)$ in Figure 33.4.

Although random sequences have appealing autocorrelation properties, the cross-correlation of any two distinct equal-length random snippets is not guaranteed to be small. In the worst case, two random codes could differ by a single bit and would have a maximum cross correlation value close to one, making them very difficult to differentiate in the presence of noise. Additionally, generating a truly random sequence in a simple piece of hardware is not an easy task. For these reasons PRN generators were developed.

These devices generate periodic signals that share many properties with random signals, but are easy to implement. A PRN generator can be constructed by performing modulo-2 summation of multiple taps on a shift register and feeding the result back into the shift register. Figure 33.5 shows a simple example of this approach. The output of this generator will repeat after a certain number of chips, referred to earlier as N. If the position of the taps is properly chosen, $N = 2^m - 1$. This arrangement, known as a *maximal-length* generator, yields the longest codes possible from a shift register with m cells.

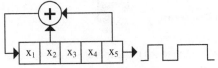

Figure 33.5   A shift register circuit for generating PRN sequences.

There are many variations on the basic shift register feedback configuration, and they yield sequences with different properties. An important architecture, known as a Gold code generator [21], offers whole families of deterministic codes that provide beneficial autocorrelation properties similar to random codes but guarantee that the cross-correlation of any two member codes will be below a threshold. If the time lag τ is constrained to an integer number of chips ($\tau = iT_C$, $i = \{0,1,2,...\}$), the cross-correlation of any two different Gold code sequences of length N, "generated from a shift register with m cells," takes on only three values. For the case of $N = 1023$, these values correspond to:

$$C(\tau) = \left\{ \frac{-1}{N}, \frac{-\beta(m)}{N}, \frac{\beta(m)-2}{N} \right\}, \beta(m) = 1 + 2^{floor\left(\frac{m+2}{2}\right)} \qquad (33.4)$$

$$C(\tau) = \left\{ \begin{array}{ll} \dfrac{63}{1023} & 10\log_{10}\left|\dfrac{63}{1023}\right|^2 = -24\text{dB} \\[2ex] \dfrac{-65}{1023} \text{ or } & 10\log_{10}\left|\dfrac{-65}{1023}\right|^2 = -24\text{dB} \\[2ex] \dfrac{-1}{1023} & 10\log_{10}\left|\dfrac{-1}{1023}\right|^2 = -60\text{dB} \end{array} \right\}, \qquad (33.5)$$

which uses a magnitude-squared metric for differentiating the side lobes from the peak. The expression for the autocorrelation of any Gold code sequence includes the three values in (33.4) and adds a fourth: 1 or 0 dB, which corresponds to $\tau = 0$ or "zero-lag." Therefore, if two different Gold codes (from the same family of $N = 2^m - 1$ length codes) are transmitted simultaneously in the same region, with equal signal power, they can be readily distinguished both from random noise, and from each other.

Finally, there are $2^m + 1$ Gold codes available from a shift register generator of length m, which allows large numbers of codes with low cross-correlation to be easily created (see [18] for helpful Gold code-generator tables). This idea forms the basis of CDMA systems, which enable multiple simultaneous transmitters to coexist by assigning each a unique code.

33.3.1.2 Chip Rate and Bandwidth

The use of CDMA signals has significant implications for the spectrum utilization of a wildlife radio tracking system, since the bandwidth of a CDMA signal is typically several orders of magnitude larger than the signals used by conventional tracking systems. The amplitude spectrum of one period of a random code sequence is given by

$$X(f) = \underbrace{T_C\sqrt{N}\left|\text{sinc}\left(\pi T_C f\right)\right|}_{\text{Envelope}} \left|\frac{1}{\sqrt{N}} \sum_{n=0}^{N-1} x_n e^{(-j2\pi f T_{cn})}\right|, \qquad (33.6)$$

where x_n is the discrete-time version of the sequence $x(t)$, sampled at intervals of T_C seconds [33].

This expression has been evaluated for a 1023-chip random noise signal with a 1 MHz chip rate and plotted in Figure 33.6. The signal amplitude spectrum $X(f)$ is plotted, as is the envelope of the signal. Notice that the 1-MHz chip rate causes the signal to have a 2-MHz-wide main lobe between the two nulls. Though it is not obvious from the figure's log-amplitude scaling, roughly 90% of the signal's power occurs in the 2-MHz-wide main lobe [18]. This fact has practical utility because although the code's spectral energy extends to +/− ∞ along the frequency axis, the signal can be band-limited to the 2-MHz main lobe with a filter and suffer minimal energy loss. Though not identical to Figure 33.6, the spectrum of a PRN sequence is very similar. As Figure 33.6 indicates, modulating the carrier with a random or PRN sequence adds significant bandwidth to the signal, reducing the signal power at any particular frequency (relative to an unmodulated carrier of equal total power). This feature of direct sequence spread spectrum systems reduces the likelihood that the transmitters will interfere with each other or with conventional narrow-band receivers outside the system. Also important to note, though not obvious from the figure, is the fact that spreading a carrier with a PRN sequence actually makes the signal easier to detect than an un-modulated carrier with the same transmitter power and duration. This result, referred to as *processing gain*, will be discussed next. This improvement comes at almost no energy cost to the tag (relative to a narrow-band tag), since the primary energy cost during transmission is in the output amplifier stage rather than in the modulation stage.

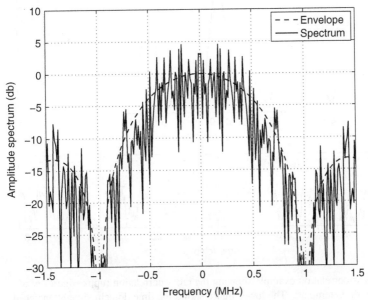

Figure 33.6   Normalized amplitude spectrum of a 1023 chip random noise signal with a 1 MHz chip rate.

33.3.1.3 Detection via Matched Filters

Matched filters exist in many forms, and the term generically refers to the linear filter whose impulse response is the time-reversed replica of the signal to be detected. It can be shown that the matched filter is the optimal linear detector when white Gaussian noise is present. A digital matched filter can be implemented by correlating the incoming signal, which may be heavily contaminated by noise, with an uncorrupted replica (template) of the expected signal. The correlator runs continuously, shifting the incoming signal past the stored template by one sample at each time step. It multiplies aligned samples of the signal and template, and accumulates the result; the output of each time step is a single number that indicates how well the signal and template agree. A detection decision is made when the correlation output exceeds a threshold. Figure 33.7 shows an example of the correlation of a noisy, time-delayed PRN signal with its template. The template in this case is a 31-chip Gold sequence. The incoming signal is delayed by half a chip, causing the strongest peak to occur at −0.5 lags. The strong peak is easily distinguished from the background noise via thresholding methods. In this case, the four visible peaks to the right and left of the strongest peak are not noise but the side lobes of the autocorrelation function. The cross-correlation plot in Figure 33.7 reveals an important feature of Gold codes: they provide a guaranteed signal-to-noise (SNR) ratio between the main peak and any side lobes. This relationship

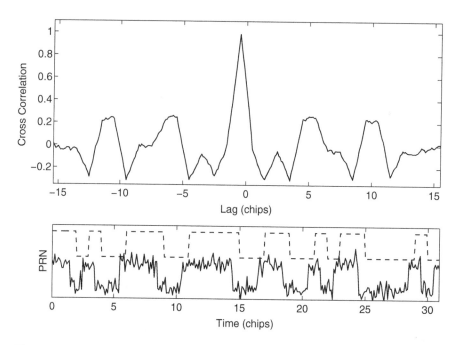

Figure 33.7 Cross-correlation example showing the cross-correlation (upper figure) of a received signal with its template. The lower figure shows the time domain signals: matched filter template (dashed line) and noisy and delayed received signal from the transmitter (solid line).

is shown exactly in (33.4), and an example of this property is shown in Figure 33.8, which compares the autocorrelation functions of a 31-chip length Gold code and a random code of the same length. The autocorrelation magnitudes are displayed on a log scale, showing the nearly 6-dB advantage in sidelobe magnitude that the Gold code has against this particular random code. This difference has a practical significance in signal detection, since sharper and more prominent autocorrelation peaks yield better SNR.

As stated previously, in addition to providing a sharp correlation peak, which aids precise time synchronization, matched filters offer excellent detection sensitivity. They achieve this sensitivity because the received signal is strongly correlated with the template, while the noise corrupting the signal is not. A system using matched filters for communication effectively replaces each data bit with a chip sequence that is coherently matched at the receiver. Because each bit is represented by a longer sequence of chips, the system is said to yield processing gain. As shown in Figure 33.6, the spectrum of the new sequence is broad, relative to the data that the system is trying to send. The processing gain PG is related to the data rate B_D and the chip rate B_C or the data bit duration T_D and the chip duration T_C by

$$PG \approx \frac{B_C}{B_D} = \frac{T_D}{T_C} \tag{33.7}$$

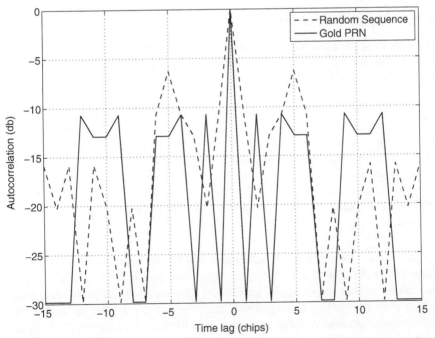

Figure 33.8   Comparison of the autocorrelation functions of two 31-chip sequences: Gold code and random code.

Note that the approximate symbol is used because this is a close approximation, but it is not exact for PRN codes like Gold codes. See [33] for specific details. Since each chip sequence has a total duration $T_D = NT_C$, we can write

$$PG \approx N, \text{ or } PG \approx 10log_{10}(N)dB \qquad (33.8)$$

This convenient result provides an estimate of the improvement in SNR that can be achieved by using a digital matched filter detector of length N chips.

At this point, we have demonstrated the ability of matched filters to provide processing gain: however we have not said anything about signal detection. Ideally, we desire a method that provides an unambiguous result and detects all true signal transmissions with no false positives. This is a tall order, and in the end the performance of the system nearly always boils down to SNR, in this case at the output of the matched filter. The wildlife tracking system will need to be able to make detection decisions based solely on the output of the matched filter (no contextual information is available). One approach is to accumulate statistics for the current and several previous cross-correlation buffers, and use that information to set an adaptive threshold for detections. The GPS system, in contrast, operates under the assumption that a GPS signal is always present to detect, though it may be below the detection threshold. Because of this, GPS receivers can take the signal's history into account during the tracking or acquisition process by tracking with a very narrow loop filter or applying noncoherent correlation over multiple adjacent data bits.

33.3.2 Signal Processing

The previous section introduced matched filter detectors and described a few core concepts that we have employed in the tracking system's implementation. This section describes a few additional details, and addresses some practical design considerations when implementing a matched filter. The first subsection discusses the performance of a matched filter when the received carrier frequency differs from the expected value. The next subsection addresses the computational requirements of the detector algorithm and provides methods to reduce them. The final subsection describes the problems caused by the asynchronous arrival of tag transmissions, and invokes the time-shifting property of the Fourier transform to efficiently handle this issue. See Chapter 7 for additional information on the methods used below as well as alternative detection and TOA estimation approaches.

33.3.2.1 Code Phase Search, Doppler Shift and Frequency Error The matched filter detector in a TOA receiver runs continuously, shifting new samples into a buffer as they arrive, and searching for a tag signal by cross-correlating the samples with a template. This operation occurs at *baseband*, after the carrier has been removed completely. At this stage, in theory, the frequency content of the signal is due solely to the PRN sequence, as shown in Figure 33.6.

Unfortunately, the situation is not this simple in reality. Two different sources of carrier frequency uncertainty, clock error and Doppler shift, prevent the carrier from being completely removed. The result corresponds to:

$$E\{S\} = Ae^{j(\Delta\theta + \pi\Delta f T_D)}\overline{R}(\Delta\tau)\,sinc\left(\pi\Delta f T_D\right) \tag{33.9}$$

which shows the expected value of the correlator output signal, S as a function of $\Delta\theta, \Delta f, T_D$, and $\Delta\tau$.

These parameters are: $\Delta\theta$, the difference between the phase of the tag's carrier and the receiver's local oscillator; Δf, the difference between the frequency of the tag's carrier and the receiver's local oscillator; T_D, the duration of the PRN sequence; $\Delta\tau$, the difference in phase between the received PRN sequence and the local template; and A, the carrier amplitude. Assuming that the matched filter finds the correct code phase, which causes $\Delta\tau = 0$ and maximizes $\overline{R}$, $E\{S\}$ might still have a small magnitude if the *sinc* term is small. This situation is illustrated in Figure 33.9, which shows that although the noise level of the received signal is low and the incoming signal is PRN code-phase-matched to the template, the correlation between these signals is low. This error is caused by a small residual frequency mismatch between

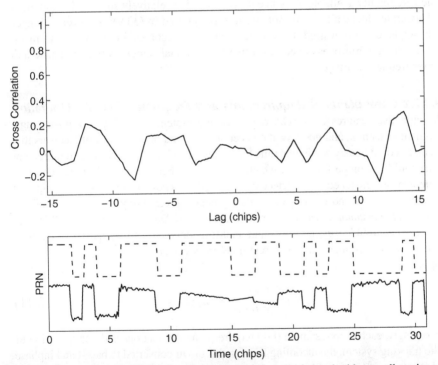

Figure 33.9   Cross-correlation between template and received signal with a small carrier and local oscillator (LO) frequency mismatch; the cross-correlation peak is eliminated, despite high SNR. The lower figure shows the time domain template (dashed line) and the received signal from the transmitter (solid line). Note the low noise level in the received signal.

the incoming signal and its template. Notice that the sign of the incoming signal is inverted halfway through its length. This causes the correlation disagreements in the second half of the signal to exactly cancel the agreements in the first half. Equation (33.9) provides a simple criterion for ensuring frequency uncertainty does not adversely impact the matched filter detection. The half-power-point of the *sinc* function in (33.9) is reached when $\Delta f T_D \approx 2/5$, so choosing $\Delta f T_D < 2/5$ ensures that a reasonable amount of the signal will always be available to detect. For example, if the PRN duration $T_D = 1.5$ milliseconds, we require that the difference between the transmitter's carrier and the receiver's local oscillator be no more than 266 Hz. This difference could be caused by Doppler shift or by oscillator error. The Doppler shift is equal to:

$$\Delta f_D = \frac{\Delta v}{c} f_0 \qquad (33.10)$$

The velocity difference Δv between the receiver and the animal carrying the tag is assumed to be less than 50 m/s, which yields a maximum Δf_D of about 23 Hz, with a carrier frequency $f_0 = 140$ MHz.

Therefore, Doppler shift is not a significant concern for the animal tracking system. Oscillator error can be bounded by choosing high-precision oscillators, which are readily available. As you can see, choosing relatively short code sequences and accurate clocks allows the *sinc* term to be ignored in (33.9); however, if longer code sequences are desired, the criterion on $\Delta f T_D$ becomes more difficult to meet, and the detection process becomes a two-dimensional search over code phase and carrier frequency offset.

33.3.2.2 Computational Requirements and Frequency Domain Operation

The attractive features of a TOA radio-tracking system, including low power tags, automatic detection, and good location accuracy, depend on a network of receivers that can listen continuously and in real time for tag transmissions. The real-time requirement sets a hard limit on performance, which impacts all other design choices. We chose to implement the matched filter detector on a *digital signal processor* (DSP) chip that operates at 1 GHz and offers parallel data processing capability, with up to 8 billion *multiply and accumulate* (MAC) operations per second. Despite the powerful DSP chip, the receivers would struggle to maintain real-time operation if they used a straightforward, time-domain algorithm. Discrete-time cross correlation version of (33.11) corresponds to:

$$C(k) \equiv \frac{1}{N} \sum_{n=0}^{N} x^*(n) y(k+n) \qquad (33.11)$$

Accordingly, each output sample (lag) requires N multiplications and additions. In our radio tracking system, the incoming RF signal is down-converted to base-band inphase and quadrature (I & Q) channels and then sampled at 2.8125 MHz.

An 11-bit PRN sequence at the tag's 1MHz chip rate would occupy 5760 samples, and processing the I and Q channels with a straightforward, time-domain matched filter implementation would require approximately 32G MAC per second in order to guarantee real-time operation. This requirement clearly cannot be met by the DSP. The options available are to either reduce the bandwidth of the transmitted PRN signal, which reduces the ranging accuracy, reduce the PN sequence length, which reduces the processing gain, or to use a frequency domain algorithm to implement the matched filter.

Ordinary time-domain correlation is an $O(N^2)$ operation, where N is the number of elements to be cross-correlated. Operation in the frequency domain in contrast is an $O(Nlog_2 N)$ operation, thanks to the remarkable efficiency of the FFT. An early application of this technique to GPS was demonstrated in [46]. We chose this alternative, and used available FFT routines along with the established practice of computing a cross-correlation by conjugate multiplication (element by element) in the frequency domain to meet the real-time requirements of the system. In this approach, C corresponds to:

$$C = IFFT\left\{X(f)^* Y(f)\right\}$$ (33.12)

where * indicates the complex conjugate, IFFT refers to the inverse fast Fourier Transform operation, and $X(f)$, $Y(f)$ refer to the complex spectra of the time series $x(t)$ and $y(t)$ that result from applying the Fourier transform. This simple expression masks two important caveats: the signals are finite in duration, since they are stored in random access memory (RAM), and the signals are not periodic. Real-time operation requires that the FFT lengths be as short as possible, since the processing load scales faster than the length of the buffer to be transformed. As an implementation of the discrete Fourier transform, the FFT assumes that its input data are periodic in N samples, where N is the length of the input buffer. The cross-correlation technique based on the FFT exhibits a circular behavior, and will "wrap around" data from the end of one buffer onto its beginning for any lag other than zero (where the two buffers are exactly aligned). The solution to this problem is to zero-pad the data buffers at their ends. If +/− k lags are desired, then the data buffers must both be padded with k trailing zeros. The Numerical Recipes book [39] explains this technique in greater detail.

33.3.2.3 Time-Shifting and Windowing

Operation in the frequency domain offers significant reductions in computational load but also adds complications. Unlike a time-domain correlator, in which incoming samples are continuously shifted and accumulated as data arrive at each time step, frequency domain operation involves processing complete, contiguous blocks of samples, then gathering another whole block and repeating the process. Each block of data represents a "snapshot" or window of the data stream arriving at the receiver. As the previous section demonstrated, we wish to keep the data windows as short as possible in order to reduce the computational load. Of course, the windows must be long enough to at least contain the data from one complete tag transmission, in order to maintain the full autocorrelation of the PRN code. Tag transmissions occur asynchronous to any processing that occurs at the receiver, so the arrival of the first chip from a transmission may fall anywhere in the receiver's window. In the worst case, the incoming signal is misaligned with the buffer boundary by N/2 samples, so that the first half the signal is in one buffer and the second half is in the next buffer. In this case, the matched filter detector will still register a maximum at the N/2 lag, but the maximum will be ½ its autocorrelation value, since only half of the PRN signal is in the buffer. This reduction in signal strength becomes problematic in low signal to noise situations. A common solution to this problem is to overlap the buffer by 50% so that the second half of the last buffer becomes the first half of the next buffer to be cross-correlated with the template (in the next iteration of the matched filter). This approach requires twice the processing effort of 0% overlap, and it also computes redundant information, since half of the sample data from the previous cross-correlation are present in the next correlation.

An alternative approach exploits the time shifting property of the Fourier transform, equals:

$$x(n - n_0) \overset{F}{\Leftrightarrow} e^{-j\frac{2\pi}{A} f n_0} X(f) \tag{33.13}$$

where x is the sampled time series data, n_0 is the number of samples to shift, A is the length of x, $X(f)$ is the complex spectra of $x[n]$, and F is the Fourier transform operator, shown as a bidirectional transform to indicate the equivalence of the representations these signals in the time and frequency domains.

This property becomes particularly useful when the time shift is $A/2$, since the complex exponential reduces to the sequence $[1, -1, 1, \ldots]$. This sequence can be stored in memory, rather than computed at run time. Multiplying $X(f)$ by this simple sequence yields the same spectra as would time-shifting $x(n)$ by n_0 seconds, and recomputing the Fourier transform. In fact, no actual multiplications need take place at all, since this is merely a sign change on every other data entry.

The complete frequency domain cross-correlation algorithm diagram is shown in Figure 33.10; each conceptual step is identified by a Roman numeral, and time-domain data are represented by lowercase letters, while frequency-domain data are represented by uppercase. Data arrive from the analog-to-digital

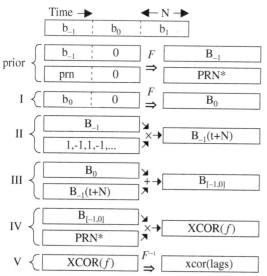

Figure 33.10 Efficient frequency domain cross-correlation algorithm.

converter (ADC) in a time-domain buffer, shown at the top of the figure. This buffer is arranged as a circular buffer and is subdivided into three N-sample segments. The segments are synchronized with the ADC in a way that ensures no samples are changed by the converter while the cross-correlation algorithm is operating on that particular buffer segment. For the sake of clarity, Figure 33.10 was drawn with the assumption that the algorithm always begins at buffer b_0, which has recently been filled with new data. The first step (I) is to compute the discrete Fourier transform of the $2N$-sample buffer that is formed by copying and zero-padding buffer b_0 to length $2N$. The second step (II) time shifts the frequency spectra of the previous buffer, b_{-1}, by multiplication with the simple time-shift sequence $[1,-1,1,...]$. Note that the frequency spectra of this buffer were retained from the previous iteration and need not be recomputed. The third step (III) adds the current frequency spectra to the time-shifted spectra from the previous buffer. The fourth step (IV) multiplies the result of (III) by the complex conjugate of the PRN sequence's spectra.

Note that PRN* does not change, and can be precomputed. The final step is to convert the cross-spectra back into the time domain, yielding the cross-correlation. Note that although this buffer contains $2N$ lag entries, only the first N are valid. This makes intuitive sense because we have effectively shifted an N-long template inside a $2N$-long data window. Moving the template beyond N lags would cause the template to fall outside the data window, reducing the overlap to below 100%. This block processing algorithm provides an efficient, 100% overlap, frequency-domain, digital matched filter.

33.3.3 System Description

The following section delves into the specific implementation of the TOA system that we built. Though certain details are unique to our system, the following ideas can be adapted to any TOA system. We begin with a basic description of the structure of a CDMA-capable transmitter and how we implemented one. The receiver's analog signal chain is discussed next, including the radio frequency and signal conversion components. The final stage in the signal path involves the baseband components, which perform signal detection, timing, and communication. See [30] and [31] for additional descriptions of the system.

33.3.3.1 Transmitters

The transmitter is based on an inexpensive, very-low-power microcontroller, along with a precision reference clock, frequency synthesizer, modulator, and amplifier. Our design, shown in Figure 33.11, integrates off-the-shelf components in order to avoid the high cost and long development time of a custom application-specific integrated circuit (ASIC). This choice results in an implementation

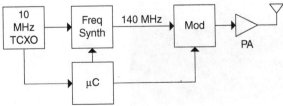

Figure 33.11   Block diagram for the CDMA tracking tags.

that is larger than it could otherwise be, but this trade-off allows rapid development. The blocks shown in Figure 33.11 are: a 10-MHz crystal oscillator (10-MHz Xtal), a sinusoidal frequency synthesizer that creates a carrier signal with a programmable frequency (Freq Synth), a microcontroller (µC) that programs the Freq Synth and controls the modulator, a modulator (Mod) that performs binary phase shift keying by inverting the sign of the carrier signal when commanded by the µC, and a power amplifier (PA) that drives the antenna. The tag uses a *binary phase shift keyed* (BPSK) modulation scheme to directly modulate and spread the carrier power. Unlike the shift register examples shown earlier, the complete Gold code sequences used in our system are simply stored in the microcontroller's flash memory and are used to toggle a digital output line that is connected to the modulator. The modulation rate (chip rate) is 1 MHz, resulting in a 2-MHz-wide main band. The tag is programmable for center frequency, transmission interval, pseudonoise code, chip-rate, RF output power, and operating schedule. This programmability allows tailoring the tag parameters to the application, which maximizes lifetime for a particular tag mass. Typical settings call for operation during the early morning and evening, when birds are most active. During these periods the tag sends one two-millisecond-long signal once every minute. This signal is actually the concatenation of two different Gold codes. The first code is common to all tags and allows the receiver to achieve phase synchronization. The second code immediately follows the first, is phase synchronous to the first, and is unique to the tag that sends it. Although we could, in principle, transmit only the unique tag identifier code, this scheme dramatically reduces the processing load on the receivers since they only need to perform a code-phase search on a single synchronization code.

The current tag, shown in Figure 33.12, weighs 1.4 grams without the battery and epoxy encapsulation. The tag's 140-MHz center frequency implies a ¼-wave antenna length of approximately ½ meter. This is too long for most small birds to

Figure 33.12   BPSK tracking tag, shown with inhibit magnet.

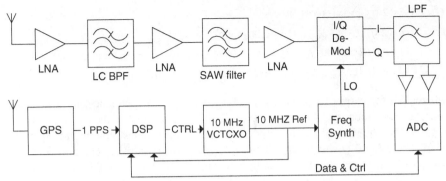

Figure 33.13   Automatic tracking receiver block diagram.

manage, so the actual antenna used is often between 15 and 25 cm. Despite the efficiency penalty that these electrically short whip antennas impose, they are very common in animal tracking applications because they are relatively unobtrusive and mechanically robust. At the maximum setting, the tag's total output power in the 2-MHz main lobe is 12 dBm (measured into a fifty-Ohm load). The actual power broadcast into free space is significantly lower than this since the efficiency of the short antenna is low.

33.3.3.2 Receiver Architecture

The block diagram for an individual receiver in the automatic tracking system is shown in Figure 33.13. Tag transmissions are received at a 2-7/8 λ phased element monopole antenna that yields approximately 6dB of gain and is omnidirectional in azimuth. The antenna is mounted atop a four-meter portable mast. The signal is then immediately amplified by a *low-noise amplifier* (LNA) and then passed through an 8-MHz-wide, six-pole inductor-capacitor (LC) passive band-pass filter with a center frequency of 140 MHz. This filter blocks strong, nearby signals before they can overload subsequent gain stages. The signal then passes through an LNA, a 2-MHz-wide *surface acoustic wave* (SAW) band-pass filter, and another LNA. These additional gain stages are necessary because although SAW filters offer very high selectivity, they are usually fairly lossy. Note that this receiver architecture is not frequency agile; the SAW filter has one set pass-band and cannot be tuned. This approach is simple, and works well if the local RF environment is free from interference at the 140-MHz operating frequency. A more traditional heterodyne architecture would afford the receiver greater flexibility in operating frequency. Next, the signal is down-converted from 140 MHz directly to 0 Hz, or the so-called baseband, by the demodulator. The demodulator uses an internal 90°phase shift circuit to derive sine and cosine signals from the local oscillator (LO) and it multiplies the input with each of these signals. The two outputs, inphase (*I*) and quadrature (*Q*), correspond to:

$$I(t) = cos(2\pi f_{LO}t) Ax(t) cos(2\pi f_C t + \varphi)$$
$$= 1/2 Ax(t)\{cos(2\pi t(f_C - f_{LO}) + \varphi) + cos(2\pi t(f_C + f_{LO}) + \varphi)\} \qquad (33.14)$$
$$= 1/2 Ax(t)\{cos(\varphi) + cos(2\pi t(2f_C) + \varphi)\}, \quad \text{if } f_C = f_{LO}$$

$$Q(t) = sin(2\pi f_{LO}t)Ax(t)cos(2\pi f_c t + \varphi)$$
$$= cos(2\pi f_{LO}t - \pi/2)Ax(t)cos(2\pi f_c t + \varphi)$$
$$= 1/2Ax(t)\{cos(2\pi t(f_c - f_{LO}) + \varphi - \pi/2) + cos(2\pi t(f_c + f_{LO}) + \varphi - \pi/2)\}$$
$$= 1/2Ax(t)\{cos(\varphi - \pi/2) + cos(2\pi t(2fc) + \varphi - \pi/2)\}, \quad if\ fc = f_{LO} \quad (33.15)$$

where f_C is the carrier frequency, f_{LO} is the local oscillator frequency, φ is the phase difference between the LO and the carrier, A is the amplitude of the carrier, and $x(t)$ is the PRN sequence. Since the I and Q signals contain undesirable high-frequency content, the I and Q signals are passed through a low pass filter. The output of the low pass filter is the chip sequence that the tag used to modulate the carrier.

After the low-pass filter, we have:

$$I(t) = 1/2Ax(t)cos(\varphi)$$
$$Q(t) = 1/2Ax(t)cos(\varphi - \pi/2) = 1/2Ax(t)sin(\varphi) \quad (33.16)$$

If we use complex notation, we can write the complex input to the ADC as:

$$S(t) = I(t) + jQ(t) = 1/2Ax(t)e^{j\varphi} \quad (33.17)$$

Note that (33.14–33.17) neglect the case when f_C and f_{LO} differ by a small amount. This situation was discussed in section 33.3.2.1 and is addressed in the example at the end of the section.

The complex signal $S(t)$ is buffered and sampled by a two-channel high-speed ADC (one channel for each of the I and Q signals). The sample rate used by the ADC should be greater than the Nyqyist frequency, 2 MHz in this case; our receiver uses a sample clock of 2.8125 MHz for each of the channels. This sample rate, which is a noninteger multiple of the chip rate, ensures that the sampling operation is not synchronous with the chip sequence. This improves the timing resolution by better aligning the samples with the chip edges, on average. Figure 33.14 illustrates the problem with an extreme case. It shows two misaligned PRN signal snippets that are sampled at regular intervals (vertical dashed lines). The intervals are synchronous with the chips, which causes the digital samples to take identical values, even though the two signals are misaligned by nearly half of a chip. Figure 33.14 is an example of aliasing, and in practice the signals to be sampled are usually band-limited to avoid this problem. There are cases, however, when it is convenient, from a system design standpoint, to violate the sampling theorem. DSP performance limitations

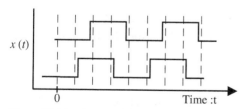

Figure 33.14 A potential problem with chip-synchronous sampling: two misaligned signals yield identical digital samples.

may make a lower than desired sampling rate necessary, or require the samples to be low resolution (many GPS receivers use only two bits per sample). Noninteger rate sampling can sometimes improve the performance of systems operating under these conditions.

The complex samples from the ADC are placed into a circular buffer in the DSP's memory, and the matched filter detector algorithm described in Section 33.3.2.3 is used to find the cross-correlation peaks within each buffer. Once a peak has been found by the cross-correlation algorithm, its arrival time must be measured. The cross-correlation output includes the sample number (lag) of the peak, so its position within the buffer is known and its time can therefore be calculated, provided that the DSP notes the time when the first sample in the buffer was acquired by the ADC. The timing resolution provided by this method is limited by the ADC sample rate, in this case 1/2.8125 MHz or about 356 nanoseconds, which corresponds to roughly 106 meters. Fortunately, we can do better, provided that the SNR is high enough. Recall that the autocorrelation peak from Figure 33.4 is $2T_C$ chips wide, or 5.625 samples wide, and is triangular. The cross-correlation peak and several samples from either side of it can be used to curve-fit the autocorrelation peak to the cross-correlation data. This method significantly improves the timing resolution, provided that the cross-correlation peak is substantial enough to provide a good fit.

The final step, after the DSP has computed a precise arrival-time estimate, is for each receiver to share that information with the central server that will calculate the position estimate. Each receiver must be connected to the server via a data network (if the positions are required in real time) so that they can submit the arrival information. Our system uses an internet protocol network, and submits TCP data packets via the network to the server where they were placed into a database. The server then groups the arrival events by tag ID, and computes positions with the stochastic search method (K. A. Cortopassi, unpublished data) described in Section 33.3.4.4.

Portability and low cost were significant considerations during the design process. Each receiver, including all antenna components, weighs 30 pounds and can be easily transported by a single person. The total power consumption of each receiver, including wireless networking equipment, is 16 watts. Power is typically supplied from two 12-volt car batteries that are charged by solar panels.

Example 33.1   Numerical Simulation of a TOA receiver

Many of the design issues mentioned in this chapter can be explored with a numerical simulation in MATLAB. Implement a software simulation of the tag/receiver system that allows parameters such as chip rate, carrier offset frequency, SNR, sample rate, and receiver bandwidth to be modified.

Solution

The example code provided ("Chapter _33_Example_1.m") uses MATLAB to generate a carrier signal, modulates that carrier signal with a Gold code, adds

Gaussian noise, down converts the signal into I/Q baseband signals, cross-correlates the signals with the Gold code template, and uses a threshold detector to indicate signal detection. The script illustrates the importance of I/Q baseband processing in a communications system that is not phase synchronous and provides a way of investigating the impact of clock frequency offsets in the tag and receiver. Also provided is a Simulink model that illustrates the use of the Gold code sequence generator block.

The example begins by invoking the Gold code generator Simulink block to create the PRN that the tag will send. This block is programmable for a particular sequence; see the appendix in [18] for appropriate generator polynomial coefficients. The script next generates an appropriate carrier sinusoid and multiplies the PRN with the carrier to create the tag output signal. The following step makes in-phase and quadrature local oscillator sinusoids and multiplies the tag output signal with each in order to downconvert the signal. These signals are then low pass filtered using the butter function to generate filter coefficients for filtfilt. The output of the filter is decimated to simulate sampling with an ADC (the resulting sample rate is the ADC's sample rate). Finally, noise is added using randn; this noise accounts for all noise that the signal would encounter. The cross-correlation of the individual I and Q channels with the PRN is calculated via xcorr, and individual plots illustrate the "fading" in each channel as the phase between the carrier and LO changes as the IQ vector rotates. A full, complex cross-correlation solves this issue, and the magnitude of this cross-correlation is shown in another plot. Finally, a simple threshold detector based on a median magnitude measure of the cross-correlation is used to determine if a tag transmission is present.

33.3.3.3 Time Base

33.3.3.3 Time Base Several components in a TOA receiver require very precise frequency or time information. These include the ADC sampling, the LO generator, and the buffer timestamp. Although very precise quartz frequency references are available, even these devices (which advertise frequency tolerances as low as 0.1 parts per million) do not offer sufficient stability to maintain precise synchronization between the receivers over a long period of time. The distance between the receivers, which is typically several kilometers, precludes a cabled or even a point-to-point radio link for synchronization. Fortunately, GPS receivers are capable of providing very precise 1 pulse per second (1 pps) and 10 MHz signals, which supply the reference signals for the rest of the receiver. Each TOA receiver uses an independent GPS receiver to maintain very tight synchronization with UTC. GPS receivers that are specifically designed for time-keeping purposes are now available for embedded applications. These devices assume a fixed location in order to overdetermine a solution that yields very accurate 1 pps edges. These edges are used to discipline a voltage controlled, temperature compensated crystal oscillator (VCTCXO), or in some cases an oven-compensated crystal oscillator (OCXO). Each of these devices provides excellent short term stability, and the GPS synchronization maintains their long-term accuracy. See [2] for an excellent overview of modern timekeeping technology.

33.3.4 Arrival-Time Location Finding Algorithms

Several methods exist for computing location estimates from arrival-time measurements in a TOA system. We briefly present four of them below. Two of the methods (stochastic search and the Newton–Raphson method) rely on iterative searching within an assumed solution space; the other two (hyperbolic and spherical intersection) are closed-form solutions based on some simplification of the problem. It should be noted that the scale and requirements of most wildlife tracking systems permit calculations in two dimensions (easting and northing); altitude is ignored. This assumption, which simplifies the system design somewhat, can be made because the primary application of a TOA system is in tracking animals over medium ranges (5 to 50 km). Any birds being tracked will be near the ground rather than migrating at altitude, since migrating animals would pass through the relatively small array too quickly for the system to be of use (apart from presence/absence detection). This is an important constraint, because tags that are substantially out-of-plane will add a significant source of error if a two-dimensional solution is assumed. Most environments lend themselves to this planar assumption, since the variation in elevation over a typical 5 km x 5 km array cell is small, relative to the 5 km receiver spacing (this spacing is set by the maximum range of detection, which depends on various factors). In terrain that violates this condition, some receivers must be placed on hilltops so they are significantly out-of-plane, and a full three-dimensional solution must be found. Note that more detailed treatments of TOA localization computations are available in Chapters 2, 6, and 7.

33.3.4.1 Hyperbolic Positioning The hyperbolic positioning method, described by Ho and Chan [10], is a popular approach for determining transmitter position from arrival times. The method works by observing that the locus of points satisfying a signed time difference of arrival (TDOA) between two receivers is one branch of a hyperbola (TDOA pairs can be computed from TOAs). The dashed lines in Figure 33.15 show the hyperbolae induced by three noise-free TDOAs received at the locations represented by orange dots. The transmitter is presumed to lie at the intersection of those three (ideal) hyperbolae.

The squared distance between the source at (x,y) and sensor i at (x_i, y_i) is

$$r_i^2 = (x_i - x)^2 + (y_i - y)^2. \tag{33.18}$$

One of the receivers is chosen as the origin, and the computation finds the solution to

$$\begin{bmatrix} x \\ y \end{bmatrix} = -\begin{bmatrix} x_2 & y_2 \\ x_3 & y_3 \end{bmatrix}^{-1} \times \left\{ \begin{bmatrix} r_2 \\ r_3 \end{bmatrix} r + \frac{1}{2} \begin{bmatrix} r_2^2 - K_2 \\ r_3^2 - K_3 \end{bmatrix} \right\}, \tag{33.19}$$

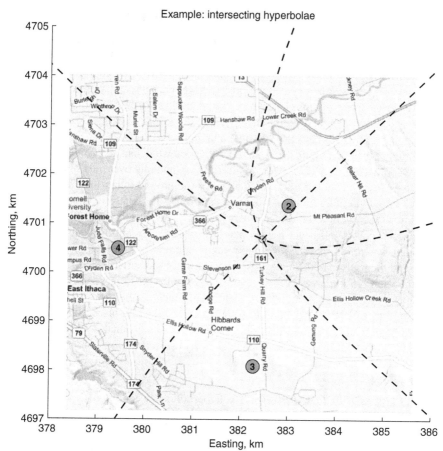

Figure 33.15 Hyperbolic location example.

subject to the constraint,

$$r^2 = x^2 + y^2 \tag{33.20}$$

where the range difference (TDOA x propagation speed) between receiver 1 and receiver i is

$$r_{i,1} = r_i - r_1, \tag{33.21}$$

and

$$K_i = x_i^2 + y_i^2, \tag{33.22}$$

and the unsubscripted variables are the x and y coordinates of the source and its distance (r) from the receiver at the origin. Where TOAs are available from more than three receivers, replacing the matrix inverse with the pseudoinverse of the receiver coordinates yields a least-squares solution.

33.3.4.2 Spherical Positioning Another formulation, spherical interpolation (SI), is due to Smith and Abel [1]. The solution is presumed to lie on the surface of a sphere (or a circle in the two-dimensional case) whose radius is the distance to one of the receiver towers chosen as a reference. The perpendicular distance between the surface of the sphere (the circumference of the circle) and any other receiver tower is the range difference between that tower and the reference tower. The SI method inserts an equation error term (corresponding to measurement noise) into the distance formula described above and minimizes the error term in a least-squares sense to yield the actual solution. One disadvantage of this approach is that it requires one more TDOA than the other methods described here.

33.3.4.3 Iterative Root Finding (Newton–Raphson Method) The Newton–Raphson method starts with an arbitrary initial guess of the transmitter's location and time of transmission and proceeds by comparing the measured TOAs against the TOAs computed from the initial guess. A correction to the guessed position and transmission time is estimated by linearizing the problem at the current transmitter position estimate, and the corrected position is used as the new guess. The process is repeated until a specified convergence criterion is met. Because the method's error term is a nonconvex function of position, this method is sensitive to the quality of the initial guess. In our analysis, because the area of interest was only slightly larger than the bounds of the receiver array, the centroid of the receiver array was used as the initial guess. In more general applications a suboptimal closed-form solution, such as one of those described above, may yield a better initial guess. This method, which is similar to the approach used in the GPS system, merits an example.

Example 33.2   Estimating Position from TOA Measurements with a Newton–Raphson Method.

Use the Newton–Raphson method to estimate the location of a transmitter, given arrival-time measurements at several nearby receivers. Additionally, show the impact of the receiver geometry on the location error.

Solution

We begin by looking at the range, $r^{(k)}$ that equals:

$$r^{(k)} = \left(t^{(k)}_{Rx(R)} - t^{(k)}_{Tx(R)}\right)c = \left\|x^{(k)} - x\right\| \tag{33.23}$$

between a transmitter and the kth receiver, where $t^{(k)}_{Rx(R)}$ is the signal's receive time, measured by the kth receiver's clock, $t^{(k)}_{Tx(R)}$ is the signal's transmit time, measured by the kth receiver's clock, $x^{(k)}$ is the position vector of the kth receiver, x is the position vector of the transmitter, and c is the signal's propagation velocity.

Unfortunately, we do not know $t^{(k)}_{Tx(R)}$, since the transmissions happen at the transmitter, which is asynchronous to the receiver. Instead of ranges, we can express the distance from the transmitter to the kth receiver as a *pseudorange* $\rho^{(k)}$, which is a combination of the true range and some offset; in this case the pseudorange is measured between the receiver's clock and the tag's clock (these clocks are assumed to have an unknown, constant offset). Through substitution of a new variable, we can express $\rho^{(k)}$ as the true range, plus an offset b that corresponds to:

$$
\begin{aligned}
\rho^{(k)} &= \left(t_{Rx(R)}^{(k)} - t_{Tx(T)}\right)c = \left(t_{Rx(R)}^{(k)} - t_{Tx(R)}^{(k)} + \left(t_{Tx(R)}^{(k)} - t_{Tx(T)}\right)\right)c \\
&= \left(t_{Rx(R)}^{(k)} - t_{Tx(R)}^{(k)}\right)c + b = \left\| x^{(k)} - x \right\| + b
\end{aligned}
\tag{33.24}
$$

Notice that b is the clock offset between the tag and the receivers, expressed in meters. Although $t_{Tx(R)}$ is a column vector with k entries, b is scalar since all receivers have synchronized clocks and therefore will all have the same offset from the tag.

Let $\rho_0^{(k)}$ be an approximation of $\rho^{(k)}$ with initial guesses for tag position x_0, and clock offset b_0:

$$
\rho_0^{(k)} = \left\| x^{(k)} - x_0 \right\| + b_0.
\tag{33.25}
$$

The difference between the measured pseudorange $\rho^{(k)}$ and the initial guess $\rho_0^{(k)}$ is $\delta\rho^{(k)}$, that equals:

$$
\delta\rho^{(k)} = \rho^{(k)} - \rho_0^{(k)}, \; x = x_0 + \delta x, \; b = b_0 + \delta b
\tag{33.26}
$$

This difference starts out large, since our initial guess is poor, and eventually goes to zero as our guess for the tag position and time offset improves. Additionally, we introduce the variables δx and δb, that represent the changes to our initial guesses in order to move closer to the actual position and time. With these relationships, and a Taylor series approximation of the vector norm, the relationship is equal to:

$$
\delta\rho^{(k)} = \left\| x^{(k)} - x_0 - \delta x \right\| - \left\| x^{(k)} - x_0 \right\| + (b - b_0) \approx -\frac{(x^{(k)} - x_0)}{\left\| x^{(k)} - x_{(0)} \right\|} \cdot \delta x + \delta b
\tag{33.27}
$$

Equation (33.27) can be rewritten in matrix form as

$$
\delta\rho = \begin{bmatrix} \delta\rho^{(1)} \\ \vdots \\ \delta\rho^{(k)} \end{bmatrix} = \begin{bmatrix} -\dfrac{(x^{(1)} - x_0)}{\left\| x^{(1)} - x_0 \right\|} & 1 \\ \vdots & \vdots \\ -\dfrac{(x^{(k)} - x_0)}{\left\| x^{(k)} - x_0 \right\|} & 1 \end{bmatrix} \begin{bmatrix} \delta x \\ \delta b \end{bmatrix} = G \begin{bmatrix} \delta x \\ \delta b \end{bmatrix}
\tag{33.28}
$$

where $\delta\rho$ is a column vector with as many entries as receivers that participated in this particular arrival time measurement.

Thus far we have not made use of the actual arrival time measurements, $t_{Rx(R)}^{(k)}$. We can say that $\rho^{(k)} = t_{Rx(R)}^{(k)}c$ if we choose $t_{Tx(T)} = 0$. This is an arbitrary but allowable choice since we are solving for b, the offset between the receiver and transmitter clocks. The linear system in (33.28) is easily solved numerically for $\begin{bmatrix} \delta x \\ \delta b \end{bmatrix}$; the next estimates for tag position and time, x_1 and b_1, are given by $x_1 = x_0 + \delta x$, and $b_1 = b_0 + \delta b$.

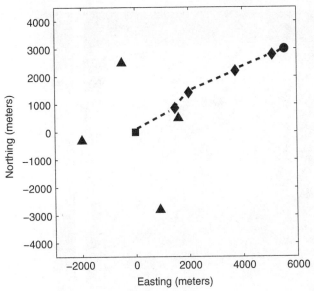

Figure 33.16   Example of NR positioning, showing receivers (triangles), initial guess (square), subsequent guesses (diamonds), and final position (circle).

These updated guesses are used to compute new values for $\delta\rho$ and $\mathbf{G}$, and (33.28) is solved again. This process continues with successively better estimates for position and time offset until the error is below a termination threshold. The algorithm converges quickly, usually requiring only a few iterations. The example code provided illustrates this algorithm by creating synthetic TOA measurements from a tag at a known location, and then uses only those measurements to find the location. It plots the array geometry, the true tag location, and shows the positions of the guesses as they converge to the true position, shown in Figure 33.16.

The precision of TOA location estimates in the presence of noise depends on the location of the transmitter, relative to the receivers, with some locations yielding much higher error than others. This phenomenon, known as position dilution of precision (PDOP), is a function of the array geometry, and the position of the tag within the array. PDOP can be thought of as a scaling factor that makes our measurement errors (from timing resolution, RF noise, etc.) more pronounced in some locations of the array than in others that can be highlighted via:

$$\text{RMS position error} = \sigma \cdot \text{PDOP} \qquad (33.29)$$

σ is a lumped error term that represents all sources of timing error, and is expressed in meters. The Newton–Raphson method provides a convenient means of estimating PDOP, that corresponds to:

$$\mathbf{H} = (\mathbf{G}^{\mathrm{T}}\mathbf{G})^{-1}, \quad \text{PDOP} = \sqrt{H_{11} + H_{22} + H_{33}} \qquad (33.30)$$

The example code provided ("Chapter_33_Example_2.m") estimates position from TOA information and plots PDOP (Fig. 33.17) for any desired array geometry.

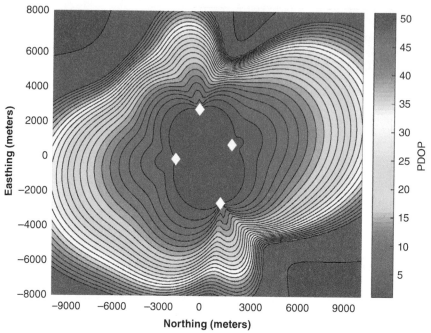

Figure 33.17 A heat map of PDOP for a 4 receiver array (receivers are white triangles). Cooler colors represent lower PDOP values.

33.3.4.4 Stochastic Search In contrast to the previous example, which started with a single position guess, the stochastic search algorithm begins with a number of initial guesses for the transmitter location, spaced within specified search bounds, and the corresponding theoretical demeaned TOA vectors are compared with the measured demeaned TOAs. A fraction n of the initial guesses with the smallest squared error are retained and duplicated, and small random perturbations are added to the duplicate points. The best n of these new guesses are expected to be closer to the actual solution, and so are retained, duplicated, perturbed, and passed to the next iteration. This process is continued until the current guesses converge to within a specified radius. The random perturbations are drawn from a uniform distribution over an interval that decreases with each iteration (K. A. Cortopassi, unpublished data). Because this process uses multiple guesses throughout, its initial condition is not limited to a single guess. The random perturbations are relatively large at first and decrease in size as the search proceeds, making the search unlikely to converge on a local minimum rather than the best solution.

33.4 PERFORMANCE OF A DEMONSTRATION WILDLIFE TRACKING SYSTEM

33.4.1 Testing a TOA System

Although the ultimate purpose of a TOA tracking network is to determine geographical position, the role of each individual receiver is to make accurate signal

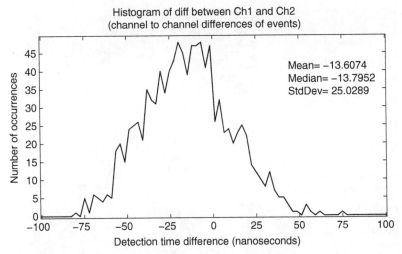

Figure 33.18   Detection time test of two colocated receivers.

arrival-time measurements, since any errors in the arrival-time estimate translate into errors in the position estimate. We measured the timing error of our system by setting up two receivers side by side and injecting a test signal into each receiver via a splitter and equal cable lengths. The arrival times of each test transmission were compared, and the differences between the two receiver's measurements were computed. The results of this test are shown in Figure 33.18 for a test with high SNR (30 dB). The receivers are able to achieve tight synchronization; however, a slight time offset was evident in this particular test. We later attributed this to cabling differences in each receiver's GPS antenna. The standard deviation of this test was 25 nanoseconds, or roughly seven meters.

We employed a similar test to determine the receiver's minimum signal detection. We injected a test signal into the receiver via a variable attenuator and increased the attenuation until the signal was not detected. The minimum signal that can typically be detected (with no added in-band noise) is −124 dBm.

Range of detection is the system parameter that is most often requested and is also the most difficult to estimate because it is so dependent on the application's location. Wildlife tracking occurs in a wide range of environments, and tag signals can encounter everything from foliage to free-space. Many resources exist for estimating the likely attenuation in forests ([27], [32], [45]), though we found a wide variation in practice. We performed numerous field tests in flat and rolling terrain, as well as transmissions through clear areas and transects obstructed by foliage. We also tested the free-space range by placing a receiver on a tall hill or building and moving the transmitter to another suitably prominent location. As described earlier, the radiation efficiency and radiation pattern of the tags are difficult to measure and change in response to how and where the tag is mounted to an animal. In general, the range of a tag 1 m above the ground is 3 to 5 km. A tag in free-space can be detected up to 10 km away. These estimates lead to a recommended receiver spacing of 5 km, with the receivers arrayed in a square or hexagonal grid to provide coverage over a large region.

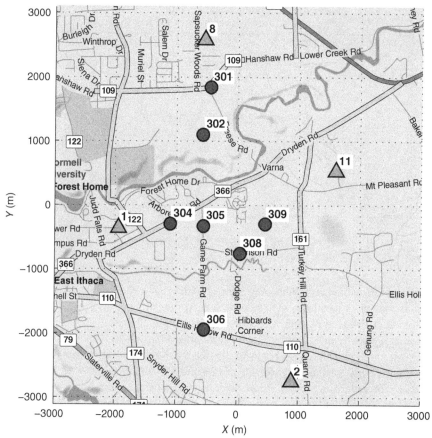

Figure 33.19 Setup for field testing of localization system. Sites are red circles, receiving towers are blue triangles.

We compared the performance of four location estimation methods—Hyperbolic localization (HL), SI, Newton-Raphson iteration (NR), and Stochastic Search (SS)—discussed in Section 33.3.4 using actual field data, acquired on August 23, 2007, in Ithaca, NY, between approximately 12:30–14:30. An RF tag transmitter was stationed in each of the seven sites (red circles) shown in Figure 33.19; several hundred repeated transmissions were made from each site, and the arrival times were measured by each receiver (blue triangles). The arrival-time data from our test site were used to evaluate the performance of the four localization techniques mentioned. The field test results are shown in Figure 33.20, which compares the median location error (distance from actual transmission location) for the four different methods with different transmitter locations. Similar experiments were carried out with simulated arrival-time data, which allowed the effect of different noise models to be investigated. In these tests the two iterative methods (NR, SS) outperformed the closed-form solutions (SI, HL). The iterative methods perform particularly well, relative to other methods, in the presence of radiofrequency noise, timing measurement error, and receiver survey position error.

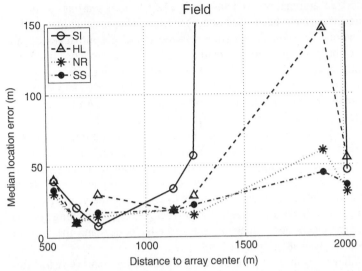

Figure 33.20   Median location error for each of the four estimation methods (SI, open circle; HL, open triangle; NR, asterisk; SS, closed circle) applied to the *field-derived* TOA vectors plotted against source distance from the array center. The median error for the SI method at site 301 was 3401 m.

33.4.2 TOA Systems Enable New Science*

Movement is fundamental to life and crucial to most ecological and evolutionary processes [37]. Moreover, as climate change and human encroachment on critical habitat place ever-increasing pressure on wildlife [9], data on movement will help to illuminate whether and how species cope with these rapid environmental changes and possibly allow mitigating its effects. Understanding why, how, when, and where animals move starts by describing movement patterns at fine spatiotemporal resolution. When coupled with an ecological understanding of the species and large-scale resource sampling, studying animal movement could provide ecologists an unprecedented mechanistic view into how animals utilize their habitat and available resources. For these reasons, we predict that regional-scale automatic location-finding systems will make a significant impact on the science of Movement Ecology; high-resolution tracks of many small animals over long time-scales will reveal the mechanisms underlying animal movement, which will have implications for the study of behavior, animal ecology, population dynamics, biodiversity, and ecosystems [23].

Realizing this potential will require scaling the capabilities of conventional tracking technologies across multiple dimensions. Tracking systems must be applicable to a wide variety of animal species, be cheap enough to deploy in large

---

*Alternative position-finding approaches might also yield new scientific insights; however, the unique combination of high-accuracy location estimates using small, lightweight, low-cost tags makes the TOA approach particularly appealing for long-duration tracking applications involving small wildlife.

numbers, provide true automation, and last over one or more seasons of animal behavior. Although they have been used successfully in certain situations with a relatively small number of animals, existing wildlife tracking approaches do not scale well. Simple "beeper tags," though small and inexpensive, are labor-intensive, requiring field staff to track each individual animal. Automated tracking tags based on satellites are larger, and their size has restricted their application to larger species only, which are far less prevalent than smaller ones (see Fig. 33.1). Additionally, tag costs for automated systems like GPS or Argos preclude large study sizes. Regional TOA tracking systems offer a means to resolve these issues and can track very large numbers of tiny, low-cost tags automatically. When installed in regions of particular scientific importance, such as biodiversity hotspots or migration corridors, they will create a "tracking and monitoring hotspot," allowing all tagged organisms within the region to be monitored. For these reasons it seems likely that in the years ahead we will see an increasing number of local- and regional-scale terrestrial tracking systems deployed, as land managers and scientists seek to better understand the ways in which organisms use physical space over time.

Our groups have begun to exploit the capabilities of regional TOA tracking systems at study sites around the world. One example that highlights the advantages of this approach is given by a study on shorebirds (Red Knots, *Calidris canutus*) in the Netherlands. With a weight of roughly 120 grams, Red Knots had been too small to be tracked with the high spatiotemporal accuracy necessary for studying individual variations in resource selection. In 2011, we setup an array of nine TOA receiver stations on the tidal mudflats of the Dutch Wadden Sea (Fig. 33.21; [7][38]). We equipped 47 Red Knots with a 7-gram prototype tag (less than 5% of body mass), and simultaneously—across 50 km^2—measured the spatial distribution of their prey (Edible Cockles, *Cerastoderma edule*). In approximately one month, this study resulted in over 2 million position fixes. Contrary to the common understanding that predators should select habitats with the most abundant resources, we found that Red Knots trade-off prey quantity with quality and select habitat with intermediate prey densities [7]. Moreover, by measuring relevant characteristics of individual foragers (their physiology), we were able to show that individuals differed consistently in modulating this trade-off, that is, some selected prey quality over quantity and vice versa. This detailed knowledge on animal movement, resource selection, and spatial distributions has led to a better understanding of how animals interact with the environment and other species and how these processes determine survival and fitness [6].

An important benefit of TOA systems is their approximately hundred-fold cheaper tags compared to conventional GPS. Cheap tags will enable tracking all individuals of a population simultaneously. To date, such studies have only been possible in the laboratory or with larger animals over a constrained timeline [43]. TOA could bring the study of group dynamics, social information use, and collective decision-making into the field under natural conditions. Cheap tags will also allow simultaneously monitoring all species in a particular region, which we predict will provide incredible insights into the complex inner workings of ecological systems, for example, predators–prey interaction, trophic cascades, and emergent facilitation. Additionally, simultaneous tracking of an entire food web may also—through animal

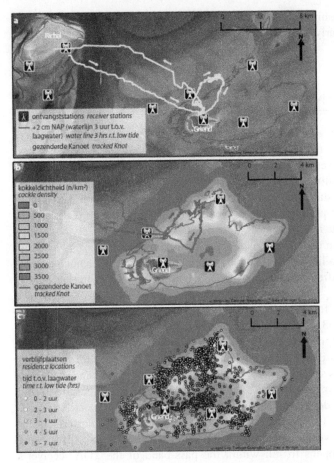

Figure 33.21   TOA observations during August–September 2011 of Red Knots roosting on the islets of Richel and Griend in the western Dutch Wadden Sea and foraging on the mudflats around Griend. (a) Study area with the nine TOA receiver stations and the movements of a Red Knot during the low-tide period on 31 August 2011 (yellow line). As the tide recedes this bird first flies to the beach north of Griend, and as the water recedes further it moves to the mudflats north and east of Griend, and eventually returns to the Richel high-tide roost as the water rises again. The blue line is the depth contour of +2 cm above new Amsterdam level (NAP). (b) The Griend mudflats with interpolated densities of Cockles small enough for Red Knots to ingest, together with the route taken by a Knot during the low tide period on 15 August 2011. After roosting on Richel, it arrives on the mudflats north of Griend the moment the receding tide has reached +2 cm NAP and suitable foraging grounds are exposed. The bird carries on towards the northeast, and with the incoming tide (again when this reaches +2cm NAP) moves back to Richel via the elevated mudflats northeast of Griend. "Beads" on the bird's track show areas where it stayed for longer periods of time. (c) For the mudflats around Griend, we show such "residence locations" for all tagged Red Knots during the study period, superimposed on the density distribution of ingestible Cockles. The colors of the dots indicate the time that the tagged knots were present relative to the tide. Figure reproduced from [7]. Used with permission.

movement—allow monitoring our rapidly changing environment and perhaps provide us with an early warning system for the impacts of regional and global climate changes [23].

33.5   CAVEATS AND LIMITATIONS

The use of terrestrial receivers with limited range means that unlike GPS and Argos, fixed TOA systems are not appropriate for tracking over very large spatial scales. The typical TOA receiver spacing is 5 km, so a grid of 16 receivers could cover an area of 400 km^2 depending on the terrain. Therefore, these systems are appropriate for covering a limited study site, but establishing coverage over a large geographic region would require a prohibitive number of receivers. Also, though these system can in principle provide real-time position updates, the tags would likely be configured to provide position updates relatively infrequently in order to conserve energy. This feature becomes a liability if field personnel must make an unplanned capture of an animal, since infrequent position updates could make following an animal difficult.

Signals suitable for TOA estimation occupy a fairly wide bandwidth, as illustrated in Figure 33.6. This can be problematic for TOA wildlife-tracking systems which often opportunistically share spectrum with narrowband transmitters, including other wildlife tags, amateur radio operators, and licensed high-power communications systems in the very-high-frequency (VHF) band. Although the choice of low transmitter power, the use of short transmission durations, and the low power spectral density afforded by direct sequence spread spectrum essentially ensure that the TOA system's transmitters will not cause interference for other systems, the receivers are susceptible to narrow-band interference. If a strong nearby transmitter happens to fall within the pass-band of the TOA receiver's front-end filters, the interfering transmitter can overwhelm the receiver's dynamic range, or cause the receiver's automatic gain control circuits to adjust, dropping the much weaker desired tag signals below the detection threshold. Though these issues can be partially addressed through careful design of the RF front-end, and through adaptive digital notch filters, strong narrow-band interference is a persistent design issue for CDMA systems.

Conventional handheld direction-finding equipment is far simpler than a complete TOA system. This leads to the two primary advantages of conventional RF tracking tools: up-front cost and reliability. Though no TOA-based wildlife tracking system is currently available for sale, the cost of a small, four-receiver TOA system is likely to be substantially higher than four handheld directional receivers. Individual TOA receivers are also components of a complicated, networked system, and individual component failures in any of the receivers could cause substantial portions of the tracking system to fail, since the coordination of multiple receivers is required for proper operation.

33.6 CONCLUSION

Wildlife tracking tools have undergone evolutionary improvements since their introduction over sixty years ago, and the last 20 years have seen a proliferation of
complementary technologies brought to bear on the problem. Improvements in
technology, including the application of microcircuits to wildlife tracking equipment, have enabled small tags, which in turn permits their use with a much greater
diversity of animals. As tag sizes have dropped, their energy consumption has
become a critical design parameter. Existing terrestrial wildlife tag technology,
though simple and reliable, uses precious transmitter energy poorly; updated communications techniques, including improved modulation and the application of
signal processing in the receiver will enable further tag mass reductions. Additionally,
appropriate tag signals will enable future wildlife tracking systems to provide accurate, automated localization via TOA measurements. We have demonstrated the
feasibility of this approach with small, inexpensive, portable receivers that can be
combined to form a tracking network capable of high fidelity localizations. This
system is appropriate for a wide range of animals and animal tracking studies, and
its small, low-cost, long-lived, spread-spectrum transmitters also make it appropriate
for generic mobile asset tracking applications.

REFERENCES

[1] J. Abel and J. Smith, "The spherical interpolation method for closed-form passive source localization using range difference measurements," in *Proc. IEEE Intl. Conf. on Acoustics Speech and Signal
Processing (ICASSP)* '87, vol. 12, Apr. 1987, pp. 471–474.

[2] D. Allan, N. Ashby, and C. Hodge, "The science of timekeeping," Hewlett Packard Corporation,
Application note 1289, 1997.

[3] International Telecommunications Union, "Attenuation in vegetation," Rec. ITU-R P.833-4, 2001.
Available: www.itu.int.

[4] E. Bächler, S. Hahn, M. Schaub, R. Arlettaz, L. Jenni, J. W. Fox, V. Afanasyev, and F. Liechti,
"Year-round tracking of small trans-Saharan migrants using light-level geolocators," *PLoS One*,
vol. 5, p. e9566, 2010.

[5] A. Bensky, *Wireless Positioning Technologies and Applications*. Norwood, MA: Artech House,
2008.

[6] A. I. Bijleveld, G. Massourakis, A. van der Marel, A. Dekinga, B. Spaans, J. A. van Gils, and T.
Piersma, "Personality drives physiological adjustments and is not related to survival," *Proc. Royal
Soc. B*, vol. 281, p. 20133135, 2014.

[7] A. I. Bijleveld, R. B. MacCurdy, Y. C. Chan, E. Penning, R. M. Gabrielson, J. Cluderay, E. L.
Spaulding, A. Dekinga, S. Holthuijsen, J. ten Horn, M. Brugge, J. A. van Gils, D. W. Winkler, and
T. Piersma, "Understanding spatial distributions: Negative density-dependence in prey causes preda-
tors to trade-off prey quantity with quality," *Proc. R. Soc. B*, vol. 283, p. 20151557, 2016.

[8] E. T. Cant, A. D. Smith, D. R. Reynolds, and J. L. Osborne, "Tracking butterfly flight paths across
the landscape with harmonic radar," *Proc. R. Soc. B*, vol. 272, pp. 785–790, Apr. 2005.

[9] G. Ceballos, P. R. Ehrlich, A. D. Barnosky, A. García, R. M. Pringle, and T. M. Palmer, "Accelerated
modern human–induced species losses: Entering the sixth mass extinction," *Sci. Adv.*, vol. 1, no. 5,
p. e1400253, 2015.

[10] Y. T. Chan and K. C. Ho, "A simple and efficient estimator for hyperbolic location," *IEEE Trans.
Signal Process.*, vol. 42, no. 8, pp. 1905–1915, Aug. 1994.

[11] CLS, Inc., *A User manual for the Argos System*, 2007. Available: http://www.argos-system.org/
documents /userarea/argos_manual_en.pdf

[12] S. Cobb. "GPS pseudolites: Theory, design and applications," Ph.D. thesis, Stanford University, 1997.

[13] W. W. Cochran and R. D. Lord, "A radio tracking system for wild animals," *J. Wildl. Mgmt.*, vol. 27, no. 1, pp. 9–24, Jan. 1963.

[14] W. W. Cochran, D. W. Warner, J. R. Tester, and V. B. Kuechle, "Automatic radio-tracking system for monitoring animal movements," *BioScience*, vol. 15, no. 2, pp. 98–100, Feb. 1965.

[15] W. W. Cochran, "Long distance tracking of birds," *Animal Orientation and Navigation*, NASA SP-262, pp. 39–59, 1972.

[16] W. W. Cochran, G. Swenson, Jr., and L. Pater, "Radio direction-finding for wildlife research" 2002. Available: http://userweb.springnet1.com/sparrow/ Direction-finding.html

[17] B. Colpitts and G. Boiteau, "Harmonic radar transceiver design: Miniature tags for insect tracking," *IEEE Trans. Antennas Propag.*, vol. 52, no. 11, pp. 2825–2832, Nov. 2004.

[18] R. Dixon, *Spread Spectrum Systems with Commercial Applications,* 3rd ed. New York: John Wiley & Sons, 1994

[19] K. J. Gaston and T. M. Blackburn, "The frequency distribution of bird body weights: Aquatic and terrestrial species," *Ibis*, vol. 137, pp. 237–240, 1995.

[20] S. Gauthreaux, J. Livingston, and C. Belser, "Detection and discrimination of fauna in the aerosphere using Doppler weather surveillance radar," *Integr. Comp. Biol.*, vol. 48, no. 1, pp. 12–23, Jan. 2008.

[21] R. Gold, "Optimal binary sequences for spread spectrum multiplexing," *IEEE Trans. Inf. Theory*, vol. 13, no. 4, pp. 619–621, Oct. 1967.

[22] M. Green, T. Piersma, J. Jukema, P. De Goeij, B. Spaans, and J. Van Gils, "Radio-telemetry observations of the first 650 km of the migration of Bar-tailed Godwits *Limosa lapponica* from the Wadden Sea to the Russian Arctic," *Ardea*, vol. 90, no. 1, pp. 71–80, 2002.

[23] R. Kays, M. C. Crofoot, W. Jetz, and M. Wikelski, "Terrestrial animal tracking as an eye on life and planet," *Science*, vol. 348, no. 6240, p. aaa2478, 2015.

[24] R. Kenward, *A Manual for Wildlife Tagging.* London: Academic Press, 2001.

[25] J. E. Lee, "Accessing accuracy of a radiotelemetry system for estimating animal locations," *J. Wildlife Manage.*, vol. 49, pp. 658–663, Jul. 1985.

[26] P. Lemnell, C. Johnsson, H. Helmersson, O. Holmstrand, and L. Norling, "An automatic radiotelemetry system for position determination," in *Proc. of the Int. Conf. on Biotelemetry*, vol. 4, 1983, pp. 76–93.

[27] M. Le Palud, T. Dupaquier, and L. Bertel, "Experimental study of VHF propagation in forested environment and modeling techniques," in *Proc. of the IEEE Int. Radar Conf.*, Alexandria, Virginia, May 2000, pp. 539–544.

[28] C. D. LeMunyan, W. White, and E. Nybert, "Design of a miniature radio transmitter for use in animal studies," *J. Wildl. Mgmt.*, vol. 23, no. 1, pp. 107–110, 1959.

[29] B. Lyon, A. Chaine, and D. Winkler, "A matter of timing," *Science*, vol. 321, pp. 1051–1052, 2008.

[30] R. MacCurdy, R. Gabrielson, E. Spaulding, A. Purgue, K. Cortopassi, and K. Fristrup, "Automatic animal tracking using matched filters and time difference of arrival," *J. Commun.*, vol. 4, no. 7, pp. 487–495, Aug. 2009.

[31] R. MacCurdy, R. Gabrielson, E. Spaulding, A. Purgue, K. Cortopassi, and K. Fristrup, "Real-time, automatic animal tracking using direct sequence spread spectrum," in *Proc. European Conf. on Wireless Technology (EuWiT)*, Amsterdam, Oct. 2008, pp. 53–56.

[32] International Telecommunications Union, "Method for point-to-area predictions for terrestrial services in the frequency range 30 MHz to 3000 MHz," Rec. ITU-R P.1546-1, 2003. Available: http://www.itu.int

[33] P. Misra and P. Enge, *Global Positioning System: Signals, Measurements, and Performance,* 2nd ed. Lincoln, MA: Ganga-Jamuna Press, 2006.

[34] B. Naef-Daenzer, "A new transmitter for small animals and enhanced methods of home-range analysis," *J. Wildl. Mgmt.*, vol. 57, no. 4, pp. 680–689, 1993.

[35] B. Naef-Daenzer, F. Widmer, and M. Nuber, "A test for effects of radio-tagging on survival and movements of small birds," *Avian Sci.*, vol. 1, no. 1, pp. 15–23, 2001.

[36] B. Naef-Daenzer, D. Früh, M. Stalder, P. Wetli, and E. Weise, "Miniaturization (0.2 g) and evaluation of attachment techniques of telemetry transmitters," *J. Exp. Biology*, vol. 208, pp. 4063–4068, 2005.

[37] R. Nathan, W. M. Getz, E. Revilla, M. Holyoak, R. Kadmon, D. Saltz, and P. E. Smouse, "A movement ecology paradigm for unifying organismal movement research," *Proc. Natl. Acad. Sci.*, vol. 105, pp. 19052–19059, 2008.

[38] T. Piersma, R. B. MacCurdy, R. M. Gabrielson, J. Cluderay, A. Dekinga, E. L. Spaulding, T. Oudman, J. Onrust, J. A. van Gils, D. W. Winkler, and A. I. Bijleveld, "Fine-scale measurements of individual movements within bird flocks: The principles and three applications of TOA tracking," *Limosa*, vol. 87, pp. 156–167, 2014.

[39] W. Press, S. Teukolsky, W. Vetterling, and B. Flannery, *Numerical Recipes in C: The Art of Scientific Computing,* 2nd ed. Cambridge: Cambridge University Press, 1992.

[40] J. Riley and A. Smith, "Design considerations for an harmonic radar to investigate the flight of insects at low altitude," *Comput. Electron. Agr.,* vol. 35, pp. 151–169, 2002.

[41] F. Savaglio, D. Maskell, and H. Spencer, "Direct sequence spread spectrum burst transmissions in a hyperbolic automatic radio tracking system," in *Int. Conf. on Telecommunications,* Melbourne, Australia, Apr. 1997, pp. 903–908.

[42] R. O. Schmidt, "Multiple emitter location and signal parameter estimation," *IEEE Trans. Antennas Propag.,* vol. 34, no. 3, pp. 276–280, Mar. 1986.

[43] A. Strandburg-Peshkin, D. R. Farine, M. C. Crofoot, and I. D. Couzin, "Habitat and social factors shape individual decisions and emergent group structure during baboon collective movement," *eLife,* vol. 6, p. e19505, 2017.

[44] TECHNOSMART, Montecelio, Italy. http://www.technosmart.eu/

[45] R. Tewari and S. Swarup, "Radio wave propagation through rain forests of India," *IEEE Trans. Antennas Propag.,* vol. 38, no. 4, pp. 433–449, Apr. 1990.

[46] D. Van Nee and A. Coenen, "New fast GPS code-acquisition technique using FFT," *IEEE Electron. Lett.,* vol. 27, no. 2, pp. 158–160, Jan. 1991.

[47] A. W. Weiser, Y. Orchan, R. Nathan, M. Charter, A. J. Weiss, and S. Toledo, "Characterizing the accuracy of a self-synchronized reverse-GPS wildlife localization system," in *Proc. of the 15th Int. Conf. on Information Processing in Sensor Networks,* 2016, 1:1–1:12.

[48] R. Kays, S. Tilak, M. Crofoot, T. Fountain, D. Obando, A. Ortega, F. Kuemmeth, J. Mandel, G. Swenson, T. Lambert, B. Hirsch, and M. Wikelski, "Tracking animal location and activity with an automated radio telemetry system in a tropical rainforest," *Comput. J.,* vol. 54, pp. 1931–1948, 2011.

[49] Wildtrack Telemetry Systems Ltd., Leeds, UK. http://www.wildtracker.com/fastloc.htm

CHAPTER 34

WIRELESS LOCAL POSITIONING SYSTEMS*

S. A. (Reza) Zekavat,
Michigan Technological University

Aₛ MENTIONED in Chapter 1, a wireless local positioning system (WLPS) is an active remote positioning system. The details and the implementation of WLPSs are discussed in this chapter. WLPS is a positioning system that functions in GPS-denied environments and/or when the GPS is jammed. It allows single-node localization through roundtrip time-of-arrival (TOA) estimation and direction-of-arrival (DOA) estimation via antenna arrays. It is a critical localization technique for urban and indoor areas. The implementation of WLPS via direct-sequence code division multiple access system (DS-CDMA) is explained. DS-CDMA systems allow high-performance detection and localization in urban areas by exploiting path diversity. In addition, the implementation of beamforming and DOA techniques via WLPS antenna arrays is discussed. Finally, we will discuss the design stages of WLPS. This discussion provides designers with a good example of the design stages of a radio prototype.

34.1 INTRODUCTION

As shown in Figure 34.1, WLPS consists of two main components [1–11]: (1) a base station that can be carried by security personnel (*dynamic base station* [DBS]), and (2) a transponder (TRX) that is mounted on all mobiles, thus introducing them as *active targets*. In this system, each DBS is capable of remotely localizing the TRXs located in its coverage area. The DBS and TRX each can be integrated with a node in an ad hoc or sensor network. This enables the process of resource allocation in sensor networks.

Handbook of Position Location: Theory, Practice, and Advances, Second Edition.
Edited by S. A. (Reza) Zekavat and R. Michael Buehrer.
© 2019 by the Institute of Electrical and Electronics Engineers, Inc.
Published 2019 by John Wiley & Sons, Inc.
Companion Website: www.wiley.com/go/zekavat/positionlocation2e

* The materials in this chapter have been partially presented in [1–12].

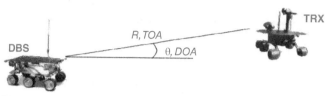

Figure 34.1   WLPS equipped with DBS and TRX.

Because targets in this system are active and contribute to the process of identification by specific ID codes, the achieved probability of detection is high, while the probability of false alarm is minimal. In this system, each DBS finds the position of all TRXs in its coverage area via a combination of DOA and TOA estimation. The DBS receiver estimates the round-trip time, and thus the distance between TRX and DBS, which is established by the TOA of a TRX's response signal (ID code) with reference to the starting time of a DBS's ID request signal. Installation of antenna array at the DBS receiver is required for DOA estimation. In addition, as will be explained later in this chapter, antenna arrays will support higher performance for the proposed system.

In WLPS, depending on applications, each mobile may be equipped with a DBS, TRX, or both. For example, for security monitoring of airports or buildings, security personnel should be equipped with a DBS, while passengers or employees only need to carry the TRX. In battlefields, commanders carry a DBS to monitor and command soldiers holding a simple TRX. For road safety, all vehicles should carry both the DBS and the TRX to prevent vehicle-to-vehicle accidents. By special ID code assignments, the DBS installed in police vehicles can identify and track a specific group of cars for law enforcement. Such a paradigm of DBS/TRX deployment in wireless devices is increasingly attractive, considering that the market of wireless handhelds is growing at an amazing speed, already passing a billion users worldwide.

Therefore, the complexity of the overall WLPS system is not high. Note that the main complexity of WLPS is in its DBS because this system is equipped with antenna elements. In addition, note that each antenna element in an antenna array should be connected to an RF component. This also increases the power consumption of the DBS system. However, the complexity of a TRX is minimal, as it is a simple single antenna transceiver. A TRX can be easily integrated with the current phones and handheld devices. It should be noted that in many applications, only a small number of nodes need to be equipped with DBSs and majority nodes need to be equipped with TRXs.

For example, as shown in Figure 34.2, using WLPS in battlefield or law enforcement operations, each commander might be equipped with a DBS and all soldiers can be equipped with a simple TRX. Therefore, the commander would be able to localize the position of all officers or soldiers under his command. Accordingly, the overall price of this system across all nodes is minimal. The position of these soldiers can be next directed to a center to allow central command and control of battlefield. This system can avoid problems such, as bombardment of friendly soldiers.

As shown in Figure 34.2, similar to GPS, WLPS can also be integrated with other positioning systems such as inertial navigation systems (INS), as detailed in

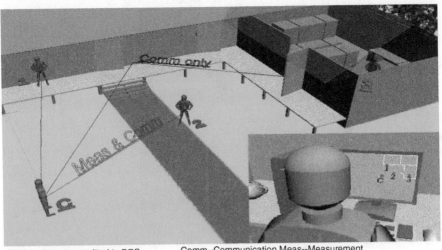

TRX package transmitted to DBS   Comm--Communication Meas--Measurement
IDS   INS data   Other sensor data   C--Commander: Equiped with DBS/INS/GPS/Comm
                                        1,2,3--Officer: Equiped with TRX/INS/other sensor

Figure 34.2   NLOS localization is the core of emergency services and law enforcement applications. The information obtained by WLPS can also be integrated with an INS for NLOS localization.

Chapter 1. This enables localization in indoor areas, and in non-line-of-sight (NLOS) situations. In this case, the signal package transmitted by WLPS would include the ID signal as well as INS.

WLPS enables many other indoor and outdoor applications. For example, using this system, security guards would be capable of localizing the position of all of (desired) targets independently. DBSs can also be installed on walls to facilitate the process of NLOS or multifloor positioning in large buildings. Each TRX is allocated a unique ID code. By allocating different categories of codes to different category of targets (mobiles, robots, security personnel, airplanes, etc.), the proposed system can be used for central command and control. By applying different codes to airplanes, the proposed system can also be used for aircraft ground traffic control in airports.

WLPS can also be installed on spacecraft orbiting around the earth in order to maintain localization across them and support collaborative tasks such as imaging. The relative and absolute position estimation of spacecraft is a fundamental task in many space missions. Nowadays, relative position estimation plays an important role in spacecraft formation flying (SFF) missions (see Fig. 34.3). Specifically, the author has already shown that incorporating a high-precision WLPS-based system improves the orbit control process for many space-based applications such as space-based solar power (SBSP) [13–16]. In the process of multisatellite synchronization for the purpose of power transfer, precise localization plays a vital role.

WLPS can also be installed on aircrafts (probably as a substitute for the current airborne traffic alert and collision avoidance [TCAS] systems) [16]. In that case, each aircraft forms a node while it is in space or on the ground. Hence, as shown in

Figure 34.3 Satellite Formation via WLPS.

Figure 34.4, aircrafts form a huge ad hoc network, in which each node would be capable of independently finding the position of other nodes in its coverage area. Note that at any time thousands of aircraft fly over the U.S. alone. The proposed ad hoc network can support the process of a distributed air traffic management (ATM) system for both air and ground traffic control. WLPS is a strong candidate for the future communication navigation surveillance ATM (CNS-ATM). Figure 34.4 represents an imaginary network created by aircrafts in the space.

CNS-ATM needs a set of navigation, communication, and surveillance equipment and protocols to support the concept of free flight. Free flight refers to direct flight from any point to other points. Currently, to maintain flight safety, each flight should follow through specific points between the source point and the destination point. Flights going through these nodes are controlled by flight control centers. Flight control centers maintain the separation (horizontal and vertical) between the aircrafts to avoid collisions. Installation of WLPS reduces the need for central controllers and/or flying through middle nodes to reach a destination. In addition, it can be

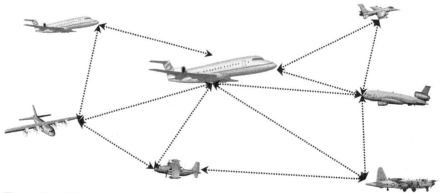

Figure 34.4 Forming a huge ad hoc network in the space supports the process of ATM.

considered an additional safety measure on the top of current aircraft safety systems. Hence, this system has the potential of reducing the flight length and the required fuel for flight. Accordingly, it reduces the flight cost and supports cleaner air [17].

Other applications of WLPS are in imaging capsule localization [27, 28] and also simultaneous vehicle-to-vehicle (V2V) communication and localization. Indeed, the same packet used by the WLPS transmitter for localization can be used communication. In addition, while based on the category of localization methods discussed in Chapter 1, WLPS is basically considered an active localization; the reflected signal from the vehicles (if sufficiently strong) can also be used for passive localization. Thus, WLPS can be used for V2V communication and passive and active localization.

It should be highlighted that WLPS incorporates beamforming and DOA estimation. DOA estimation incorporates antenna arrays, and its process usually needs eigen value decomposition of the observed covariance matrix to exclude noise subspace from signal subspace. Eigen value decomposition increases the signal processing complexity; specifically, when the number of antenna elements increases. A higher number of antenna elements enable high-resolution DOA estimation and accordingly high-precision localization. The number of antenna elements, specifically, is very high for millimeter (mm)-wave communication that incorporates massive multi-input multi-output (MIMO). In these scenarios, specific algorithms should be used to enable high-speed eigen value decomposition [29].

WLPS has also applications in nonhomogeneous media such as capsule endoscopy and underground localization. As discussed in Chapter 10, TOA estimation in nonhomogenous environments is subject to considerable error, and specific techniques should be incorporated to address the error imposed by nonhomogeneous media [30, 31].

34.2   WLPS STRUCTURE

The two WLPS main parts include: A DBS and a (TRX). The DBS transmitter generates an ID code request (IDR) signal every *IRT* (ID request repetition time) to all TRXs in the coverage area; then, it waits to receive a response back from the TRX within IRT. TRX transmits a unique ID code as soon as it detects the IDR signal transmitted by the DBS. The ID code may be selected from simple pseudorandom codes which consist of +1 and −1 (see Fig. 34.5). Hence, the number of bits in the

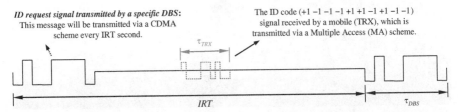

Figure 34.5   Transmission of ID request and reception from TRX in DBS. Assuming pseudorandom ID codes, the number of the bits in the code represents the maximum capacity of the WLPS (see [2]).

code is a measure of the maximum capacity of the WLPS. Depending on the application, the ID code can be assigned permanently or can be assigned by the DBS.

As shown in Figure 34.6a, in WLPS, each DBS communicates with a number of TRX in its coverage area simultaneously. This is the same as usual cellular communication systems. However, in contrast to cellular systems, in WLPS, each TRX communicates with a number of DBSs simultaneously as well (see Fig. 34.6b). In addition, as seen in Figure 34.5, at the DBS a specific time frame is allocated to transmission and another time frame is allocated to reception. Moreover, DBSs and TRXs use different transmission frequencies. Thus, the overall system is considered a time division duplex (TDD)–frequency division duplex (FDD), that is, a hybrid TDD/FDD communication system (differ from cellular systems that are either TDD or FDD). This allows WLPS to reduce the interference effects via a proper selection of IRT.

The minimum allowable value for IRT (IRT_{min}) is calculated to avoid range ambiguity or second-time-around echo [18]. That is, if the response to each IDR signal is received in DBS within IDT, the mobile range is calculated correctly;

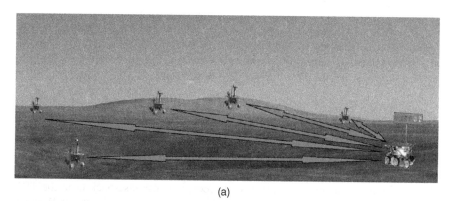

(a)

(b)

Figure 34.6 (a) Each DBS communicates with a number of TRXs in its coverage area. (b) Each TRX communicates with a large number of DBSs in its coverage area (see [2]).

however, if it is received after the next IDR transmission, the range is not correctly calculated. Hence, IRT is a function of the maximum coverage or the maximum range R_{max}. As mentioned earlier, the DBS receiver is equipped with antenna arrays to support DOA estimation.

Considering the maximum uplink antenna array half power beam widths (HPBW) β to be less than $90°$, the minimum allowable IRT corresponds to:

$$IRT_{min} = 2T_{max} + T_d + T_G, \tag{34.1}$$

where T_{max} denotes the maximum possible time delay between the TRX transmission and the DBS reception, T_d is the TRX time delay in responding to the ID request signal, and T_G is the guard band time, which corresponds to

$$T_G = 5T_m + \tau_{DBS} + \tau_{TRX}. \tag{34.2}$$

Here, T_m is the wireless channel delay spread, and τ_{DBS} and τ_{TRX} are the durations of DBS and TRX transmitted signals, respectively. Using simple geometry, T_{max} is determined by R_{max} and β via:

$$T_{max} = \left(\frac{R_{max}}{2c}\right) \cdot \left(1 + \cos^{-1}\beta\right), \tag{34.3}$$

where c denotes the speed of light. Equation (34.1) defines a lower limit for IRT (i.e., IRT_{min}). The upper limit for IRT (IRT_{max}) is a function of the speed of moving TRX and DBS, and the required speed of processing, which varies with applications. In general, IRT is selected to belarge enough to reduce the interference effects at the TRX.

Because more than one DBS may transmit IDR signals in the coverage area of a TRX, the TRX receiver is subject to inter-DBS interference (IBI). Large selection of IRT reduces the probability of overlap or collision of the DBSs' transmitted signals at the TRX receiver. In addition, a number of TRXs in the DBS coverage area re,spond to the IDR signal of one DBS simultaneously, causing inter-TRX-interference (IXI) at the DBS receiver. Figure 34.6a and 34.6b represent IBI and IXI effects at the TRX and DBS receivers, respectively. Both IXI and IBI are a function of the probability of overlap, p_{ovl}, of the received signals from TRXs and DBSs respectively.

Probability of overlap has a profound effect on the performance of the receiver, is a function of the number of mobiles or transmitters (DBS or TRX), K, in their coverage area, and corresponds to

$$p_{ovl} = 1 - (1 - d_c)^{K-1}, \tag{34.4}$$

where, $d_c = \tau / T$, $\tau = \tau_{DBS}$ (τ_{TRX}) is the duration of DBS (TRX) transmitted signal (see Fig. 34.5), and $T = T_{DBS}$ ($T = T_{TRX}$), where $T_{DBS} = IRT_{min}$, IRT_{min} refers to minimum allowable IRT ($T_{TRX} = IRT_{max}$, IRT_{max} denotes the maximum allowable IRT,

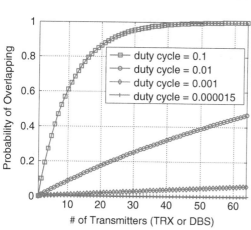

Figure 34.7   The probability of overlapping.

or the selected IRT). Figure 34.7 represents this probability as a function of the number of transmitters (TRX or DBS) for different values of duty cycle.

In general, large selection of IRT reduces IBI effects at the TRX receiver and highly enhances its probability of detection performance. However, the large selection of ID request repetition time does not affect the IXI: all of the signals are received by DBS receiver within the T_{max} time frame, which is mainly a function of the maximum coverage range (see Fig. 34.8a). It is also worth mentioning that each TRX located in the coverage area of more than one DBS may generate ID codes in response to more than one DBS within each IRT. This leads to both IXI as well as range ambiguity (see Fig. 34.8b). Range ambiguity can be resolved via changing code assignments (multiple access codes, ID codes, or both) for different DBS. In

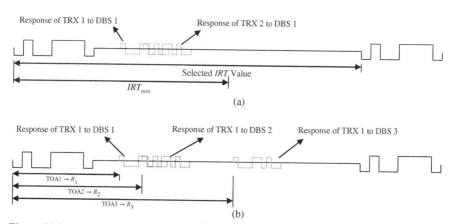

Figure 34.8   (a) At the DBS1 receiver all signals are received within IRT_{min} independent of selection of IRT; this leads to high probability of overlap at the DBS receiver. (b) DBS1 receives the TRX1 response to its IDR as well as DBS2's and DBS3's ID requests. This leads to range ambiguity if TRX uses the same ID code and/or multiple access code for all DBSs.

addition, in (34.4), the parameter d_c is a function of τ_{TRX} and τ_{DBS}, and accordingly the probability of overlap and the probability of detection performance would be functions of d_c.

In general, selection of τ_{DBS} and τ_{TRX} depends on the probability of detection, desired system capacity (in terms of the number of TRX/DBS accommodated), bandwidth, positioning accuracy, and maximum coverage range, and may vary with WLPS application. The duration of the transmitted signal by the DBS (τ_{DBS}) and TRX (τ_{TRX}) should be much smaller than the ID repetition time to reduce probability of overlap among signals received by receivers of TRX and DBS, respectively. A smaller probability of overlap decreases both the IBI (at TRX) and IXI effects (at DBS), which in turn enhances the probability of detection performance, positioning accuracy and user capacity of the WLPS system.

On the other hand, the system maximum capacity expressed by the maximum number of TRX (DBS) determines the number of bits within each ID code, which is to be transmitted over a period of τ_{TRX} (τ_{DBS}). The required bandwidth is inversely proportional to τ_{DBS} and τ_{TRX} for a given capacity. A large selection of ID repetition time allows τ_{DBS} to be selected much larger than τ_{TRX} without sacrificing probability of overlap at the TRX receiver. Hence, WLPS bandwidth is mainly determined by the value of τ_{TRX}.

Large ID repetition time (IRT) values lead to low probability of overlap, and to reduce IBI effects at the TRX receiver, a simple structure consists of an omni-directional antenna and a standard or code division multiple access (CDMA) receiver suffices. However, to reduce IXI effects, DS-CDMA along with beamforming are employed for DBS receivers.

In general, the purpose of multiple-access (MA) schemes is to maintain orthogonality across user signals. This orthogonality reduces the interference effects and improves the performance of receivers. In CDMA systems this orthogonality is maintained via allocating unique and orthogonal codes to each user. In DS-CDMA systems this codes are applied in time domain [18]. In spatial division multiple access (SDMA) schemes this orthogonality is maintained in space domain. Antenna arrays are implemented to create beam patterns in the direction of the desired users and create nulls in the direction of interfering users [20].

Thus, DS-CDMA suppresses the interference via orthogonal codes, and beam-forming reduces the IXI via SDMA. The TRX needs a simple demodulator and a DS-CDMA transmitter (which leads to a very simple structure): Antenna arrays and beamforming techniques are not required at the TRX, and its complexity, cost, and size are minimal. For some applications with a larger selection of ID repetition time, a DS-CDMA receiver is suggested for TRX. However, the DBS would have a more complex structure at its receiver.

Figure 34.9 represents the structure of a DBS. A DBS receiver calculates the position of active targets (TRX) via both the TOA) and DOA information. A DBS receiver estimates the signal TOA with respect to the time of the transmission of the ID repetition time signal in order to find the distance of the mobile. It uses antenna arrays to find the DOA via various schemes [20]. The DBS downlink signal is transmitted via omnidirectional antennas, and the uplink signal is received via

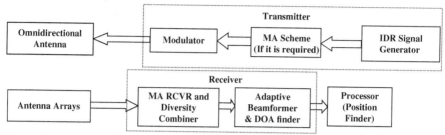

Figure 34.9   DBS Structure.

directional antennas that direct power toward desired users (see Fig. 34.10). Figure 34.10 also represents that IXI, which is more severe than IBI can be reduced by beamforming.

The proposed positioning system should possess the following features:

a. *Contiguous positioning* requires a high probability-of-detection performance. To achieve this goal, we: (a) employ both CDMA and SDMA to mitigate the interference effects; and (b) develop and examine various robust beamforming (BF) techniques to support SDMA for different indoor and outdoor environments.

b. *Precise positioning* requires accurate DOA and TOA estimation. DOA techniques are critical for the development of many beamforming schemes as well. Many DOA estimation techniques have already been developed by WLPS research group [7–10].

c. *NLOS coverage* (i.e., beyond single-hop transmissions) needs the implementation of mobility estimation and routing protocols at DBSs. NLOS coverage

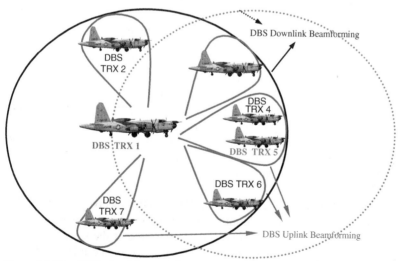

Figure 34.10   DBS signal is omnidirectionally transmitted but directionally received.

can also be addressed via an integration of the proposed WLPS system with other local positioning systems such as INS. An important issue with INS is its reduced performance with time. WLPS is used to update the positioning information of INS and avoid error propagation.

34.3   WLPS PERFORMANCE INVESTIGATION

Interference effects at the receiver can be mitigated via reducing the probability of overlap and incorporating multiple access schemes. Per our previous discussions, a large selection of *ID repetition time* reduces the duty cycle d_c and consequently the interference at the TRX receiver; however, the selection of ID repetition time does not have any effect on the interference at the DBS receiver. Hence, while a simple receiver may ensure a high TRX probability-of-error performance, the DBS performance is improved just via multiple access schemes. Here, we combine of two multiple access schemes, DS-CDMA and SDMA. This combination leads to a very high DBS probability-of-detection performance. In addition, DS-CDMA systems exploit path diversity via RAKE receivers (see Fig. 34.11) that mitigates fading effects and enhances the bit-error-rate performance. The fading effects are mitigated after combining the signals across different paths. Note that each path might be received through different directions. Therefore, beamforming toward specific directions helps to increase signal-to-noise ratio and, accordingly, the performance. The theoretical performance has been evaluated in [3]. Here, we introduce the details of DBS receiver structure and discuss a summary of simulation results.

34.3.1   The DS-CDMA Receiver

The transmitted DS-CDMA signal by the kth TRX (DBS) corresponds to (see [5] for details):

$$S^k(t) = g_\tau(t) \sum_{n=0}^{N-1} b^k[n] \cdot g_{T_b}(t - nT_b) \cdot a^k(t - nT_b) \cdot \cos(2\pi f_c t), \qquad (34.5)$$

where $a^k(t) = \sum_{i=0}^{G-1} C_i^k g_{T_c}(t - iT_c)$, $C_i^k \in \{-1,1\}$, denotes the spreading code, G is the processing gain (code length), N is the number of bits, $T_C = \tau / N \cdot G$ represents TRX chip duration, and $g_{T_C}(t)$ is a rectangular pulse with the duration of T_C. Other $g(t)$ pulse shapes in this equation are defined similarly. The structure of the DBS receiver has been shown in Figure 34.11. In the figure, $r(t)$ represents the received signal.

As depicted in Figure 34.11, a RAKE receiver is required for each antenna array. A RAKE receiver includes multiple receivers, each allocated to different time delays. Usually, the number of receivers per RAKE does not exceed three receivers. Exceeding the number of receivers beyond three highly increases the complexity of

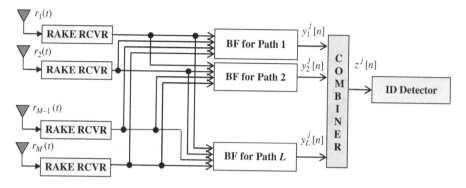

Figure 34.11 j^{th} TRX, DBS receiver structure assuming a frequency selective channel (see [26]).

the receiver. For a typical airport indoor environment, given huge and long walkways, up to two receivers per RAKE is suggested.

After dispreading, the nth bit output for the jth user's lth path corresponds to $y_l^j[n]$. The received signals across different paths are then combined to generate $Z^j[n]$, that is, the input to the detector. Typically, we apply maximal ratio combining (MRC) across the path diversity components, i.e., $z^j[n] = \sum_{l=1}^{L} \alpha_l^j y_l^j[n]$. Here, α_l^j refers to the fading amplitude of the lth path.

34.3.2   Simulation Results

In this subsection, we evaluate the probability-of-detection performance and capacity (defined in terms of number of users) of the proposed WLPS system. Here, we consider multiple TRXs and a multipath environment. Simulation assumptions have been summarized in Table 34.1.

Using the parameters defined in Table 34.1, the minimum *ID request Repetition Time*, IRT_{min}, is 9.83 μ sec (see (34.1)). We select a much larger value $IRT = 24$ ms

TABLE 34.1.   Simulation Assumptions

8 bits per ID code ($N = 8$)
DS-CDMA code with 64 chips ($G = 64$)
Channel delay spread for typical airport = 60 n Sec [22]
Carrier frequency = 3 GHz, $\tau_{TRX} = 1.2\mu$ sec, $\tau_{DBS} = 24\mu$ sec
Uniform multi-path intensity profile, $SNR = 20$dB
4-element linear antenna array, $d = 0.05$m (HPBW = 27°)
Maximum coverage range, $R_{max} = 1000$m
Four multi-paths, lead to 4 fold path diversity
TRX position U[0 1]km, TRX angle U[0 π]

in order to reduce the IBI effects, as discussed in the previous section. With the assumed τ_{DBS} and τ_{DBS}, the required bandwidth of a DS-CDMA transmitter is 320 MHz for a TRX (with CDMA), 16 MHz for a DBS (with CDMA), which is much smaller than the TRX bandwidth. In addition, the required bandwidth of a standard transmitter would be 5 MHz for a TRX and 250 MHz for a DBS. Accordingly, the TRX transmission bandwidth determines the WLPS bandwidth.

Using the parameters of Table 34.1, the duty cycle for DBS and TRX receivers correspond to $d_{c,DBS} \cong 0.32$ and $d_{c,TRX} = 0.002$, respectively. Figure 34.7 depicts the probability of overlap as the number of transmitters (TRX or DBS) changes for various values of the duty cycle that is a function of *ID request Repetition Time*. Here, we show that as the probability of overlap increases, the detection performance decreases. It should be noted that when the probability of overlap is small, the nature of the received signals would not be stationary. This nonstationarity behavior of the received signals leads to poor performance of adaptive beamforming techniques. The cyclostationary behavior of WLPS is exploited to address this issue [6].

In general, IBI at the TRX receiver can be considerably reduced by selecting the *ID request repetition time* large enough. This selection will not affect IXI at the DBS receiver. Hence, a TRX receiver can just be implemented by a simple transceiver (or DS-CDMA) systems without the employment of beamforming techniques, while a DBS receiver needs a combination of DS-CDMA and beamforming. A small $d_{c,TRX} = 0.002$ at the TRX receiver leads to a small probability of overlap, which leads to small IBI and high probability of detection. In contrast, a large $d_{c,DBS} \cong 0.32$ at the DBS receiver leads to a high probability of overlap that results in high IXI. Both BF and CDMA techniques help to reduce the IXI effects at DBS.

The probability of detection, p_d, of the DBS receiver is depicted in Figure 34.12a. This figure compares p_d with respect to the number of TRX for a standard transceiver and a DS-CDMA transceiver, with or without antenna arrays and BF. It shows that in general the p_d decreases as the number of TRX increases, which is a direct result of inter-TRX-interference. As to the impact of beam forming (BF), the use of BF does not affect much the capacity (in terms of number of TRX) for a standard receiver (the lower two curves), but it considerably enhances the capacity of the DS-CDMA system (the upper two curves). Merging DS-CDMA with beamforming is thus highly promising for enhancing the probability-of-detection performance of WLPS systems.

The probability-of-detection results for a TRX receiver using a standard receiver is shown in Figure 34.12b. Although simple, a standard TRX receiver typically achieves good p_d performance. For example, occupying the same bandwidth as DS-CDMA, a standard receiver should choose τ_{DBS} (τ_{TRX}) to be 1/64th of that of a DS-CDMA system. (In this case, the same number of path diversity as the DS-CDMA receiver (i.e., four-fold diversity is achievable.) This corresponds to $d_{c,TRX} \cong 0.000032$ ($d_{c,DBS} \cong 0.005$), which leads to a very small *probability of overlap* at the TRX (DBS) receiver and very high p_d. Further improvement is possible by selecting a larger *ID repetition time* value, or a smaller τ_{DBS} value.

Figure 34.12a shows that, with similar bandwidths, it seems that a DS-CDMA system with duty cycle 64 times higher than a standard receiver leads to (almost)

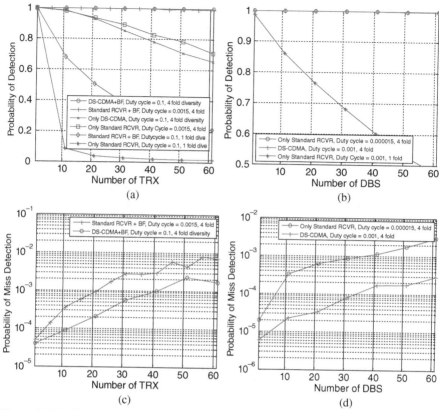

Figure 34.12 Performance simulation results for (a) DBS and (b) TRX receivers. Figures (c) and (d) represent the two top curves in (a) and (b), respectively.

the same performance curve. To have a finer observation, the two top curves in Figure 34.9a for standard and DS-CDMA receivers with beamforming are sketched in Figure 34.12c. It is seen that a DS-CDMA receiver outperforms a standard receiver for DS-CDMA with high duty cycles ($dc \cong 0.12$ leads to a high probability of overlap, as shown in Figure 34.7). In this case, the DS-CDMA receiver leads to a capacity about two times that of a standard receiver at the $p_d = 0.999$.

For the TRX receiver, probability-of-overlap statistics with $d_c \cong 0.000032$ is depicted in Figure 34.12d. This low probability leads to the high TRX standard receiver probability-of-detection performance shown in Figure 34.12b. However, it is seen that a DS-CDMA receiver outperforms a standard receiver for DS-CDMA with low duty cycles ($d_c \cong 0.001$ leads to a low probability of overlap as shown in Figure 34.7). For airport security, where the data rate is not critical, the *ID request repetition time* can be chosen larger than 24 ms for even better TRX standard receiver performance (in terms of probability of detection).

Based on the above observation, it can be concluded the DS-CDMA receiver always outperforms the standard receiver, no matter if it is combined with BF or not. The reason is that standard receivers experience "bursty" interference, that is, the interference power is strong for only a certain short duration. During the other time duration, the interference power is small or zero. On the other hand, the DS-CDMA technique "spreads" the interference over a long time period. Thus the DS-CDMA receiver (almost) always experiences interference, but the interference power is always small. Although the total interference powers are the same for the two techniques, the averaging effects of DS-CDMA causes a much better performance than a standard receiver [23].

In addition to airport security, another possible application of the WLPS is aircraft navigation. In these applications, WLPS allows each node (aircraft) to independently localize and track other aircraft in its coverage area. The system structure is extremely similar to the airport security system, except the following parameters change:

i) Aircrafts are uniformly distributed in [0 20] kilometers.

ii) Each ID code contains 32 bits.

iii) No multipath effect for aircrafts at high altitude, and two paths for aircrafts at low altitude.

iv) Eight-element antenna array (because aircrafts have more space to install antenna compared to vehicles). The simulation results are illustrated in Figure 34.13.

In Figure 34.13a, comparing the BF and CDMA performance, it is observed that BF alone outperforms CDMA alone when the number of aircraft is small. The reason is that the system is mainly noise-limited when the number of aircraft is small, while BF can effectively reduce noise power by eight (the number of antenna) times, but CDMA does not. In contrast, when the number of aircraft is high, the system is mainly interference limited. In this situation, CDMA outperforms BF. The reason is in this simulation CDMA processing gain is 64, while the number of antenna is high, hence CDMA has a better interference suppression capability than

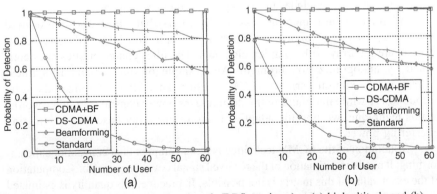

Figure 34.13   Simulation results for aircraft (DBS receivers) at (a) high altitude and (b) low altitude (see [6]).

BF. Finally, it is shown that a combination of CDMA and BF can obtain a satisfactory performance.

Figure 34.13b represents the same p_d results for low altitude aircraft, assuming two paths are available. It should be further noted that low altitude not only introduces a multipath effect, but also introduces faster power decay with respect to the distance between TRX and DBS, because the two paths cancel each other when they have similar single trip time [24]. Correspondingly, can we assume the signal-to-noise ratio (SNR) is 10 dB in Figure 34.13b, while SNR is 20 dB in Figure 34.13a. Again, a simple standard receiver cannot achieve satisfactory performance. Moreover, BF or CDMA improves performance greatly. It is noted that the crossover point of BF and CDMA performance is shifted to the right side. The reason is that noise power in Figure 34.13b is higher; hence, noise power dominates system performance for a larger number of aircrafts compared with Figure 34.13a. Finally, a BF and CDMA merger also achieves satisfactory performance.

34.4   ADAPTIVE BEAMFORMING TECHNIQUES

The simulation results depicted in Figures 34.12 and 34.13 are based on the assumption that a conventional beamformer is installed at the receiver of Figure 34.11. A conventional beamformer estimates the DOA of the desired user, θ, and creates a beam in that direction by applying constant complex weights to each antenna element, m, that correspond to

$$\vec{V}(\theta_l^k) = \left[1 \quad \exp\left(\frac{-2\pi d \cos\left(\theta_l^k\right)}{\lambda} \right) \quad \cdots \quad \exp\left(\frac{-2(M-1)\pi d \cos\left(\theta_l^k\right)}{\lambda} \right) \right]. \quad (34.6)$$

$\vec{V}(\theta_l^k)$ is the array vector corresponding to the lth path of kth user. Apparently, $\vec{V}(\theta_l^k)$ is an array that includes unity amplitude complex elements.

Adaptive beamformers are essentially filters that can extract the desired signal from interfering signals via applying an optimum filter. Adaptive beamformers are an array with complex elements. The filter characteristics are determined by an optimization problem. An example of those filters is a linear constrained maximum variance (LCMV) filter. This filter minimizes the total output power while maintaining the desired user power constant. In other words, in this technique, we find the adaptive beamformer array based on minimizing interfering power.

Installation of the LCMV technique is more complex than conventional beamforming: It needs an estimation of the received signal covariance matrix. Computation of the exact value of this matrix is not possible. In practice, this quantity is estimated by the sampled covariance matrix. To do so, the received signal should be a stationary process. However, due to the periodic and discontinuous nature of WLPS signals,

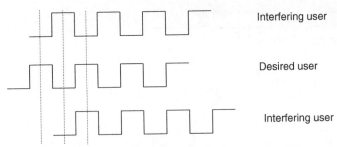

Interfering user

Desired user

Interfering user

Figure 34.14   Different bits of different users may suffer different interferences: Thus, the received signal is not stationary (see [6]).

the received signal would not be stationary. This point has been discussed in detail in Chapter 9 and has been depicted in Figure 34.14. This figure shows that the interference effects on different arrived bits are not similar. Therefore, the received signal over different time sequences (bits) would be different, and this leads to the nonstationary nature of the received signals.

Accordingly, the proposed system suffers from poor estimation of the sampled covariance matrix, which leads to a poor performance of the developed receiver. Figure 34.15 confirms that in spite of its higher complexity, the performance of the LCMV beamforming technique is similar to the conventional beamforming (a low complex beamforming system). Thus, we have increased the complexity without achieving higher performance.

Researchers have proposed a solution to this problem [6]. The proposed technique uses the cyclostationarity property of the received signal. As shown in Figure 34.16, the same bit over consecutive reception of signals at the DBS receiver will experience the same interference. Accordingly, a sample covariance matrix can

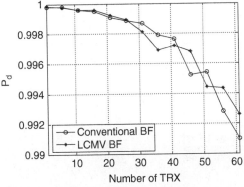

Figure 34.15   LCMV beamforming performance does not exceed conventional due to covariance matrix estimation error created because of nonstationarity.

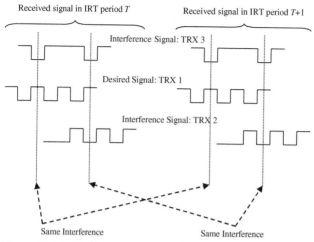

Figure 34.16   The bits over consecutive transmissions experience the same fading.

be applied to the same bits of consecutive signals. Accordingly, we generate a beam-former for every unique bit in our system.

This solution highly improves the performance, as depicted in Figure 34.17. As shown in this figure, the probability of misdetection of LCMV beamforming becomes less than that of conventional beamforming. This shows that indeed the LCMV beamforming generated based on the cyclostationary property improves the performance of the DBS receiver. The number of periods over which the cylcosta-tionary of the received signal remains stable depends on the cyclostationary time and the IRT. Cyclostationary time duration has been investigated in detail in [6].

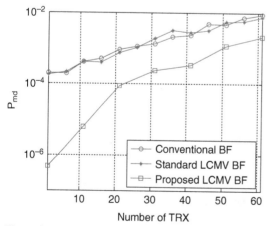

Figure 34.17   LCMV cyclostationary merger improves DBS receiver performance (see [6]).

34.5 NOVEL DOA AND TOA ESTIMATION ALGORITHMS

An important component of the proposed WLPS system is DOA and TOA estimation. DOA estimation is required for both beamforming and positioning. TOA estimation is required for the implementation of RAKE receivers and for positioning. Similar to beamforming, the cyclostationary nature of the proposed system can be incorporated to improve the DOA and TOA performance of the proposed system. The DOA estimated by the same bit based on multiple receptions (Fig. 34.17) can be combined in order to improve the DOA estimation performance [8]. The same technique can be applied to improve TOA estimation performance. Incorporation of Kalman filtering or Wiener filtering across multiple receptions is another technique for improving the performance of this system. Chapter 9 details incorporating cyclostationary techniques in order to improve DOA estimation.

An important issue for the implementation of WLPS is complexity. To support the process of beamforming and channel estimation, online DOA and TOA estimation techniques should be implemented. Online estimation techniques need higher speed, and thus, lower complexity. As detailed in Chapter 9, authors in [25] and [26] have shown that the implementation of high-resolution techniques (such as MUSIC) needs considerable complexity. In [10], authors have depicted that a fusion of a low-complexity/low-performance DOA estimation scheme such as delay and sum and high-complexity/high-performance techniques such as MUSIC may fulfill the needs of two worlds: We can achieve high performance with a relatively low complexity. The same technique can be applied to TOA estimation to ensure higher performance and lower complexity.

34.6 WLPS DESIGN AND STRUCTURE

In Figure 34.9, we present the block diagram of the WLPS, and in Figure 34.11, we present the block diagram of the RAKE receiver and beamforming system. In this section, we present the design approach and block diagram details of the WLPS system. The WLPS is composed of six-element antenna arrays. As discussed in Chapter 9, in order to apply DOA estimation full synchronization across all antenna elements is required. Therefore, all antenna elements should be fed by the same oscillator in the RF front-end. Next, the whole system was designed based on the structure shown in Figure 34.18. In this system, each antenna element is connected to an RF front-end which, brings the frequency down from 2.4 GHz to 70 MHz.

After passing through an analog-to-digital converter (ADC), the signal is applied to a memory buffer; then it is applied to a digital down converter (DDC), which reduces the frequency to 40 MHz. A DDC has two functions: (1) digital down conversion and coarse frequency tracking and (2) to act as a part of the fine frequency tracking loop. Next, signals across all antenna elements are phase, frequency, and amplitude synchronized before being applied to DOA estimator and beamformer. Equalization and detection is applied after beamforming. The block

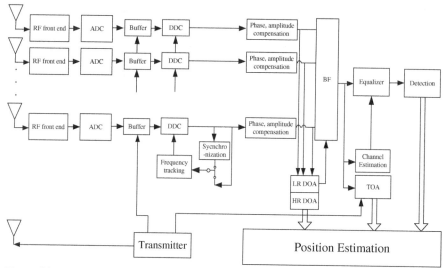

Figure 34.18　The detailed block diagram of the DBS system.

diagram of Figure 34.18 can serve as the block diagram of one RAKE receiver in Figure 34.11.

Figure 34.19 represents the details of the RF front-end of Figure 34.18. This figure also represents the synchronization across all antenna elements. In addition, the clock pulse of all ADC should be well-synchronized. The figure also shows that each antenna element receiver is decomposed to I and Q components, within which the process of phase and frequency synchronization is applied prior to the application of beamforming and localization.

The whole system is then developed over a six-element antenna software-defined radio (SDR) that can be programmed to function as a DBS. The system was designed as shown in Figure 34.20a, and ordered from Sundance DSP Co.

As shown in Figure 34.20a, the system is composed of:

(1) three RF front-end boards, which each can be connected to two antenna elements, and called SMT 349;

(2) three ADC cards called SMT 350;

(3) a system clock generator called SMT 399; and

(4) a processor called SMT 374.

Each of these SMT cards include many Field Programmable Gate Array (FPGA) chips that can be programmed to function as synchronizer, detector, TOA and DOA estimator, positioner, and tracker. Figure 34.20b represents the picture of the real radio system. The whole DBS system was designed on the SDR of Figure 34.20.

The flowchart of Figure 34.21 represents the approach to system design. Many components of the flowchart of Figure 34.21 are not limited to the design of WLPS

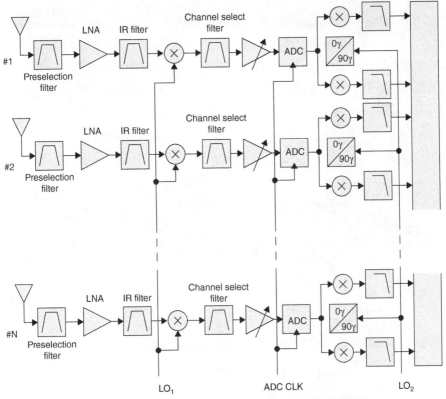

Figure 34.19   The details of RF front-end of the design of Figure 34.18.

and can be applied to the design of any radio system. As shown in this figure, a radio can be first developed on MATLAB simulink. The performance of a developed system is tested for different channel models. Next, we design and test a spread spectrum transmitter, which can operate as the transmitter of the TRX and can be used for real channel test and data collection. Figure 34.22 represents the design block diagram of the spread spectrum transmitter.

Next, a suitable six-element antenna array was designed for this system. Each element of this antenna is selected to be a patch antenna. A patch antenna is very light, easy to construct, and its material price is very low. It is constructed on printed circuit boards. The dimension of each antenna element is in the order of 1.2 inches. The bandwidth of each antenna element is in the order of 50 MHz, its beam width is in the order of 85°, and its gain is 7.2 dBi. Chapter 9 discussed the antenna features. Figure 34.23 represents the antenna array, and Figure 34.24 represents its pattern created by ANSOFT.

The next step in the process design is real data acquisition. The collected data is used to test the performance of the designed system on the MATLAB simulink using real channel models. After completing the design of the system over simulink,

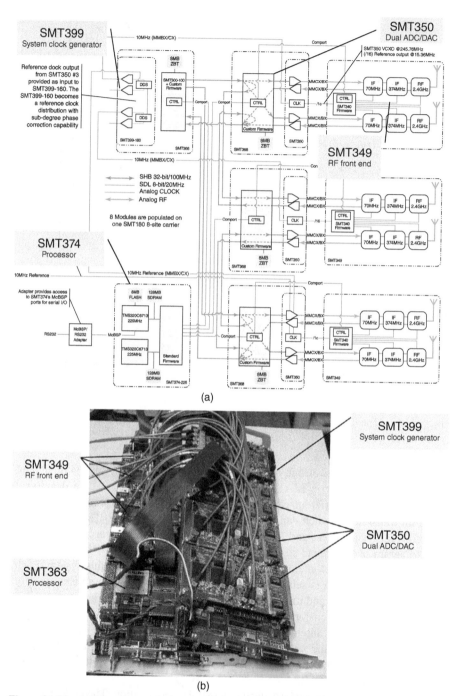

(a)

(b)

Figure 34.20 (a) The system design details prepared by Sundance DSP co. (b) A picture of the developed software radio system by Sundance DSP co.

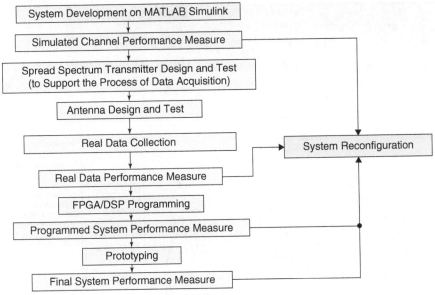

Figure 34.21   The WLPS prototype development process.

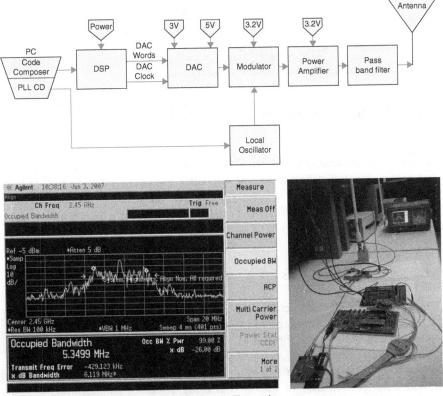

Figure 34.22   The developed Spread Spectrum Transmitter.

Figure 34.23 Patch antenna array prototype.

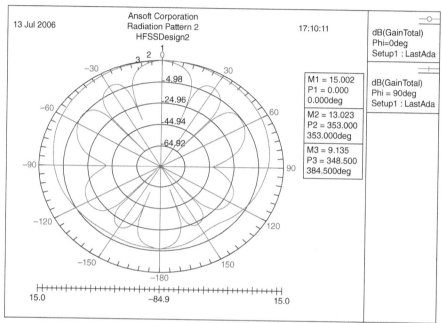

Figure 34.24 The antenna array beam pattern created by ANSOFT for the six-element antenna of Figure 34.23.

the system is implemented on the FPGA of our system (Fig. 34.20). The programmed system is next tested, and the test data is used to in order to adjust further the system design. Next, the system is prototyped and used for real environment tests and final adjustments. After this step, the whole system is submitted for mass production.

34.7 CONCLUSIONS

This chapter presented the details of a novel local positioning system with active targets called WLPS. This system has the potential to be used for many applications, such as emergency services, fire fighter applications, indoor (e.g., airport) security, central command and control, V2V collision avoidance and multirobot coordination and collaboration. The detection performance of this system was investigated. In addition, the performance of this system using conventional beamforming and optimal beamforming techniques was studied. It was depicted how the cyclostation-ary property inherent in the transmission signal of WLPS can be used to improve the detection and the localization performance. Moreover, the design of this system using a software defined radio was detailed.

ACKNOWLEDGEMENTS

Many former students in the WLPS Lab have contributed to preparation of pictures and materials in this chapter. Specifically, I would like to thank Dr. Hui Tong, Dr. Zhonghai Wang, Dr. Wenjie Xu, and Dr. Mohsen Pourkhaatoun.

REFRENCES

[1] S. A. Zekavat, "Mobile base station (MBS) wireless with applications in vehicle early warning systems," in *Proc. of University of Texas-Austin Symp. on Wireless Networking*, Oct. 2003, pp. 527–531.

[2] S. A. Zekavat, H. Tong, and J. Tan, "A novel wireless local positioning system for airport (indoor) security," *Sensors, and Command, Control, Communications, and Intelligence (C3I) Technologies for Homeland Security and Homeland Defense IV, Proc. of SPIE*, vol. 5403, pp. 522–533, San Diego, CA, Feb. 2004.

[3] H. Tong and S. A. Zekavat, "LCMV beamforming for a novel wireless local positioning system: a stationarity analysis," in *Sensors, and Command, Control, Communications, and Intelligence (C3I) Technologies for Homeland Security and Homeland Defense IV, Proceedings of SPIE*, vol. 5778, Orlando, FL, Mar.–Apr. 2005, pp. 851–862.

[4] H. Tong and S. A. Zekavat, "Wireless local positioning system via DS-CDMA and beamforming: A perturbation analysis," presented at the Proc. of the IEEE Wireless and Communication Networking Conf. (WCNC) 2005, New Orleans, LA, Apr. 2005.

[5] H. Tong and S. A. Zekavat, "A novel wireless local positioning system via a merger of DS-CDMA and beamforming: Probability-of-detection performance analysis under array perturbations," *IEEE Trans. on Veh. Technol.*, vol. 56, no. 3, pp. 1307–1320, 2007.

[6] H. Tong, J. Pourrostam, and S. A. Zekavat, "LCMV beamforming for a novel wireless local posi-tioning system: Nonstationarity and cyclostationarity analysis," *EURASIP J. Adv. Signal Process.*, vol. 2007, Jun. 2007, Art. ID 98243.

[7] Z. Wang and S. A. Zekavat, "MANET localization via multi-node TOA-DOA optimal fusion," in *Proc. of the Military Communications Conf. (MILCOM '06)*, Washington, DC, Oct. 2006, pp. 1–7.

[8] J. Pourrostam, S. A. Zekavat, and H. Tong, "Novel DOA estimation techniques for periodic-sense local positioning systems," presented at the Proc. of the IEEE Radar Conf. (RADAR '07), Waltham, MA, Apr. 2007.

[9] M. Pourkhaatoun, S. A. Zekavat, and J. Pourrostam, "A high resolution ICA based TOA estimation technique," presented at the Proc. of the IEEE Radar Conf. (RADAR '07), Waltham, MA, Apr. 2007.

[10] S. A. Zekavat, "Wireless Local Positioning System (WLPS)," U.S. Patent 7,489,935.

[11] Z. Wang and S. A. Zekavat, "A novel semi-distributed cooperative localization technique for MANET: Achieving high performance," presented at the Proc. of the IEEE Wireless and Communication Networking Conf. (WCNC), Las Vegas, NV, Mar. 31 – Apr. 3, 2008.

[12] S. A. Zekavat and H. Tong, "Single node multi-antenna positioning systems for airport security," in *Protecting Airline Passengers in the Age of Terrorism*, P. Seidenstat and, F. X. Splane, Eds. Praeger Publishers, 2009.

[13] S. G. Ting, O. Abdelkhalik, and S. A. Zekavat, "Spacecraft constellation orbit estimation via a novel wireless positioning system," presented at the Proc. AAS/AIAA Space Flight Mechanics Meeting, Savannah, GA, Feb. 8–12, 2009.

[14] S. A. Zekavat, O. Abdelkhalek, and H. Tong, "Aircraft navigation and spacecraft coordination via wireless local positioning systems," presented at the Proc. USA 2008 Integrated Comm. Navigation and Surveillance (ICNS) Conf., Bethesda, MD, May 5–7, 2008.

[15] S. G. Ting, O. Abdelkhalik, and S. A. Zekavat, "Differential geometric estimation for spacecraft formations orbits via a novel wireless positioning," presented at the Proc. IEEE Aerospace Conf., Big Sky, MT, Mar. 6–13, 2010.

[16] S. A. Zekavat, O. Abdelkhalik, S. G. Ting, and D. Fuhrmann, "A novel space-based solar power collection via LEO satellite networks: Orbital via a novel wireless local positioning system," presented at the Proc. IEEE Aerospace Conf., Big Sky, MT, Mar. 6–13, 2010.

[17] M. Kojima, "Cleaner transport fuels for cleaner air in Central Asia and the Caucasus," *World Bank Publications*, 2000.

[18] M. I. Skolnik, *Introduction to Radar Systems*. Mc-Graw Hill , 1981.

[19] S. Verdú, *Multiuser Detection*. New York: Cambridge University Press, 1998.

[20] M. Kojima, "Cleaner transport fuels for cleaner air in Central Asia and the Caucasus," *World Bank Publications*, 2000.

[21] L. Ni, Y. Liu, Y. C. Lau, and A. P. Patil, "LANDMARC: Indoor location sensing using active RFID," in *Proc. of the First IEEE Int. Conf. on Personal Communications*, Mar. 2003, pp. 407–415.

[22] A. A. Arowojolu, A. M. D. Turkman, and J. D. Parsons, "Time dispersion in urban microcellular enviroments," in *IEEE 44th Vehicular Technical Conf.*, vol. 1, Jun. 8–10, 1994, pp. 150–154.

[23] D. Tse and P. Viswanath, *Fundamentals of Wireless Communication*. New York: Cambridge University Press, 2005.

[24] T. S. Rappaport, *Wireless Communications: Principles and Practice*, 2nd ed. Prentice Hall, 2002.

[25] R. Schmidt, "Multiple emitter location and signal parameter estimation," *IEEE Trans. on Antennas and Propag.*, vol. 34, no. 3, pp. 276–280, Mar. 1986.

[26] P. Stoica and A. Nehorai, "MUSIC, maximum likelihood, and Cramer-Rao bound," *IEEE Trans. on Acoustics, Speech, and Signal Process.*, vol. 37, no. 5, pp. 720–741, 1989.

[27] S. G. Ting, S. A. Zekavat, and K. Pahlavan, "DOA-based endoscopy capsule localization and orientation estimation via unscented Kalman filter," *IEEE Sensors J.*, vol. 14, no. 11, pp. 3819–3829, Sept. 16, 2014.

[28] A. Reisi, S. G. Ting, and S. A. Zekavat, "Circular arrays and inertial measurement unit for DOA/TOA-based endoscopy capsule localization: Performance and complexity investigation," *IEEE Sensors J.*, vol. 14, no. 11, pp. 150–154, Sept. 16, 2014.

[29] M. Athi, S. A. Zekavat, and A. A. Struthers, "Real time signal processing of massive sensor arrays via a novel fast converging SVD algorithm: Latency, throughput and resource analysis," *IEEE Sensors J.*, vol. 16, no. 8, pp. 2519–2526, 2016.

[30] M. Jamalabdollahi and S. A. Zekavat, "High resolution ToA estimation via optimal waveform design," *IEEE Trans. Commun.*, vol. 65, no, 3, pp. 1207–1218, 2017.

[31] M. Jamalabdollahi and S. Zekavat, "Range measurements in non-homogenous, time and frequency dispersive channels via time and direction of arrival merger," *IEEE Trans. Geosci. Remote Sens.*, vol. 55, no. 2, pp. 742–752, 2017.

NEAR-GROUND CHANNEL MODELING WITH APPLICATIONS IN WIRELESS SENSOR NETWORKS AND AUTONOMOUS DRIVING

Amir Torabi and S. A. (Reza) Zekavat

Michigan Technological University, Houghton, MI

WIRELESS SENSOR networks (WSNs) have found diverse applications in environmental, security, and infrastructure monitoring as well as in location-based services. In most of WSNs' emerging applications, sensor nodes work at or slightly above the ground level [1]. A key example is autonomous driving. As discussed in Chapter 29, many sensors and devices are emerging to enable reliable autonomous driving. Most of these devices such as radars or wireless local positioning systems (detailed in Chapter 34) are planned for installation on car bumpers that are close to the ground. There is a lack of accurate and computationally efficient radio models tailored for near-ground communications. In most available channel models antennas are assumed to be far above the ground [2, 3]. There are a few near-ground models proposed in the literature that are mainly based on measurement campaigns. These models offer limited simulation scalability and are only accurate for certain environments [3–13]. In Chapter 4, we briefly touched upon near-ground channel modeling. In this chapter, we discuss the details and impact of a versatile theoretical model that is developed to predict the feasible transceiver (TRX) range and node connectivity for WSNs deployed for diverse applications. The applicability of this model is verified by comparing the results with the near-ground measurements carried out by independent researchers in rural, forested, and urban settings.

Handbook of Position Location: Theory, Practice, and Advances, Second Edition.
Edited by S. A. (Reza) Zekavat and R. Michael Buehrer.
© 2019 by The Institute of Electrical and Electronics Engineers, Inc.
Published 2019 by John Wiley & Sons, Inc.
Companion Website: www.wiley.com/go/zekavat/positionlocation2e

35.1 INTRODUCTION

Promising applications of WSNs and mobile ad hoc networks (MANETs) have stimulated growing interest in modeling and optimizing their performance in various environments [5]. Based on the measurement results reported by several researchers, it is known that lowering the antennas' altitude significantly decreases the signal strength, hence reducing the system range. This effect is addressed in [5] by proposing a two-slope log-normal path loss model for a WSN at 868 MHz in an open area. In [6], the impact of foliage on near-ground radio-wave propagation is studied for battlefield sensor networks operating at 300 MHz and 1900 MHz. Measurement results for ground-based ultra-high frequency (UHF) band communicators in urban terrain are reported in [7] for both line-of-sight (LOS) and non-line-of-sight (NLOS) links. Numerical solvers are prescribed in [8] and [9] to characterize near-ground long-range propagation, but their computational complexity limits the number of nodes in the simulated network. In [10], a simple mathematical path loss model for near-ground links is introduced. Nonetheless, a flat, perfectly conducting ground is assumed in the derivation of the model, which overlooks the significant impact of terrain roughness and electrical properties on the channel transfer characteristics. In addition, the break-point distance after which, according to the two-ray propagation model, the path loss increases at the rate of 40 dB per decade is set too far, which results in underestimation of the path loss at larger distances.

According to the plane-earth (two-ray) model, at small range, strong oscillations take place around the direct ray level [14]. However, the median power falloff rate obtained through regression fits is roughly the same as in free space and the total loss can be approximated by the free space loss [15, 16]. The distance where the last maximum in the received wave pattern occurs is called the break distance, d_B, which is a function of antenna heights and operating frequency. At this distance, the first Fresnel zone touches the ground and the direct and ground-reflected waves, collectively called the space waves, only combine destructively beyond this range. Owing to destructive interference between the space waves, the power falloff rate increases from 20 dB/decade before d_B to 40 dB/decade after it. At almost three times the break distance, we will reach the critical distance, d_C. This is the distance where approximately 57% of the first Fresnel zone is still clear of obstruction. If we move farther than the critical distance, ground turns into a significant obstruction for the transmitted energy, and diffraction loss should also be included in the total loss. As explained in [17], using the two-ray analysis, we can make out another distinct region. At distances that are smaller than the transmitter height $(d < h_t)$, space waves only combine constructively and the received signal strength increases slowly. However, this region does not have any practical importance for near-ground WSNs and will be neglected in this chapter.

The plane-earth propagation model offers a simple but useful path loss model that properly predicts the rise of the falloff rate at the break distance. However, in order to arrive at a more accurate model suitable for sensor network design, we shall also consider the geometrical and electrical properties of terrain. For propagation above an irregular terrain, the physical statistical properties of the ground surface have a considerable impact on the statistics of the received signal by decreasing the ground reflectivity and generating local surface waves [18]. For rough surfaces, an equivalent reflection coefficient can be derived by multiplying the plane surface reflection coefficient by a scattering loss factor to account for the reduction in the reflected signal amplitude. Two commonly used approximations for the scattering loss factor are derived by Ament and Boithias [19, 20].

In WSNs, due to low heights of the sensor nodes, propagation often approaches the grazing condition. In this scenario, according to the Rayleigh criterion, the surface appears electrically smooth and the space waves cancel each other, leaving only the higher order surface waves. It is shown in [21] that as long as the TRX altitude is low in terms of the wavelength, these surface waves are dominant regardless of frequency of operation. Nevertheless, the traditional loss factors found by Ament and Boithias disregard the effects of terrain self-shadowing and surface waves, which render them inadequate for near-ground propagation.

In this chapter, a new computationally tractable path loss model is introduced for WSNs working above a dielectric rough terrain. Principles of the Fresnel zones are exploited to split the proposed path loss model into three segments. The distances that define the edges of each segment are derived theoretically. In the first region, the LOS ray dominates the signal transmission, while in the second region both the direct and ground-reflected signals impact the received energy. In the third region, diffraction loss caused by insufficient path clearance is also added to the reference loss. The effective reflection coefficients used in the proposed model include the effect of higher-order surface waves and are founded on the perturbation approach applied to a volumetric integral equation and are applicable to grazing propagation when the surface roughness is less than a wavelength [22, 23]. Path loss predictions offered by the proposed model are consistent with the measurement results in rural and urban areas reported by independent researchers. Moreover, it is verified that by adding an empirically modeled foliage loss to the proposed model, it is possible to accurately evaluate the near-ground propagation in a foliage environment. Finally, the proposed model is used to examine the influence of communication and link parameters on coverage range and network connectivity.

The rest of the chapter is organized as follows. Section 35.2 derives the break point distances. In Section 35.3, the near-ground path loss model is developed. In Section 35.4, the proposed model is verified and used in WSN connectivity analysis. Section 35.5 concludes the chapter.

35.2   DERIVATION OF THE BREAK POINTS

The locus of all points having a constant value of excess path length, Δd, as compared to the direct path, forms an ellipsoid of revolution with the two terminals at the foci and the LOS path as the axis of revolution. A family of such ellipsoids in which Δd varies in integer multiples of half-wavelengths, $n\lambda / 2$ with n an integer and λ the wavelength, is called the Fresnel zones [15]. As shown in Figure 35.1, the intersection of these Fresnel ellipsoids with an imaginary

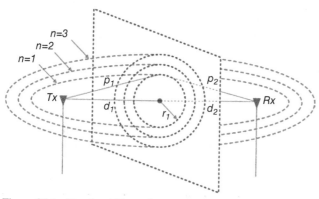

Figure 35.1   Family of Fresnel zones. An imaginary plane perpendicular to the LOS cuts the successive Fresnel ellipsoids in concentric circles.

plane perpendicular to the LOS path constructs a family of concentric circles with radii of:

$$r_n = \sqrt{\frac{n\lambda d_1 d_2}{d_1 + d_2}} \; , \tag{35.1}$$

where d_1 and d_2 are the distances of the plane from the transmitter and the receiver, respectively [15]. Equation 35.1 is valid if $d_1, d_2 \gg r_n$. The radii of the circles depend on the location of the plane and reach their maximum of $r_{n,max} = \sqrt{n\lambda d}\,/\,2$ when the plane is midway between the terminals where $d = d_1 + d_2$.

Most of the radio energy is concentrated in the first Fresnel zone; hence, to prevent the blockage of energy, we try to site the antennas such that the first Fresnel zone is clear of obstacles. As sketched in Figure 35.2, when one terminal moves away from the other, the radius of the first Fresnel zone increases until it touches the earth surface at the break distance, d_B. d_B divides the LOS propagation path into two distinct near and far regions. In the near region, mean signal attenuation is equivalent to free space wavefront spreading loss, whereas beyond d_B obstruction of the first Fresnel zone also contributes to attenuation loss which results in a steeper falloff rate of the signal strength [24].

Figure 35.3 illustrates the concept of the near/far regions using the two-ray model. In the near region, regression fits about the oscillatory pattern of the received signal result in a path loss exponent of two, which corresponds to free space loss. However, in the far region, destructive interference of the direct and reflected signals leads to a path loss exponent of four. Break distance indicates the position where the gradual transition from square law to fourth law begins. This phenomenon serves as the basis for the well-known dual-slope piecewise linear path loss model for microcellular LOS topographies.

Specular reflected ray travels the shortest path among all the rays scattered by the ground for any given transmitter and receiver heights and separation. Therefore, when the range increases, any ellipsoid with the transmitter and receiver at the foci first touches

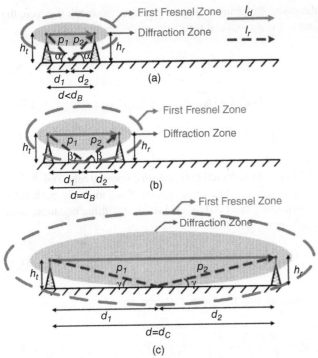

Figure 35.2 Schematic representation of the first Fresnel zone and the diffraction zone in a radio link as the distance between two terminals increases. (a) First Fresnel zone is clear of obstructions; (b) first Fresnel zone touches the ground surface; (c) diffraction zone is tangent to the ground surface.

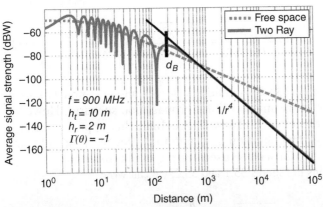

Figure 35.3 Path loss versus terminal separation for free space model and two-ray model. Depiction of the break distance and dual-slope piecewise linear regression fit in a microcellular propagation scenario assuming a flat, perfectly conducting ground plane.

the ground on the specular point. Referring to Figure 35.2b, at the break distance, the first Fresnel zone is tangent to the ground and the excess path length equals

$$\Delta d = l_r - l_d = \frac{\lambda}{2} \tag{35.2}$$

$$l_d = \sqrt{d_B^2 + (h_t - h_r)^2} \tag{35.3}$$

$$l_r = \sqrt{d_B^2 + (h_t + h_r)^2}, \tag{35.4}$$

in which l_d and l_r are the direct and reflected path lengths, respectively, and h_t and h_r are the transmitter and receiver heights, respectively. If we multiply both sides of (35.2) by $(l_d + l_r)$ and substitute (35.3) and (35.4) into the resulting equation, after simple algebraic manipulations we will find:

$$l_d + l_r = \frac{8 h_t h_r}{\lambda}. \tag{35.5}$$

Adding (35.2) and (35.5), we can find the break distance as

$$d_B = \sqrt{\left(\frac{4 h_t h_r}{\lambda} + \frac{\lambda}{4}\right)^2 - (h_t + h_r)^2}. \tag{35.6}$$

Alternatively, subtracting (35.2) from (35.5), we can write

$$d_B = \sqrt{\left(\frac{4 h_t h_r}{\lambda} - \frac{\lambda}{4}\right)^2 - (h_t - h_r)^2}. \tag{35.7}$$

We can rewrite (35.7) as

$$d_B = \frac{4 h_t h_r}{\lambda} \sqrt{1 - \frac{\lambda^2 (h_t^2 + h_r^2)}{(4)^2 h_t^2 h_r^2} + \left(\frac{\lambda^2}{(4)^2 h_t h_r}\right)^2} \approx \frac{4 h_t h_r}{\lambda}. \tag{35.8}$$

The approximate value in (35.8) is valid if $h_t, h_r \gg \lambda$ and is famously known as the break distance in the microcellular propagation scenario. However, in near-ground sensor network channel modeling, this approximation is no longer accurate, as the condition does not hold. Hence, in this chapter, the exact formula in (35.7) will be used for the break distance.

Using the notations in Figures 35.1 and 35.2, we can write

$$p_1 = \sqrt{d_1^2 + h^2} \approx d_1 + \frac{h^2}{2 d_1}, \quad h \ll d_1 \tag{35.9}$$

$$p_2 = \sqrt{d_2^2 + h^2} \approx d_2 + \frac{h^2}{2 d_2}, \quad h \ll d_2, \tag{35.10}$$

in which d_1 and d_2 are the horizontal distances between the transmitter-obstacle and obstacle-receiver, respectively. h is the obstruction height, which is always a negative value in LOS topographies and can be expressed as:

$$h = -\left[h_r + (h_t - h_r)\frac{d_2}{d} \right], 0 \leq d_2 \leq d, \tag{35.11}$$

where $d = d_1 + d_2$ is the horizontal distance between the transmitter and receiver. As mentioned earlier, the point where the ellipsoid touches the surface of the earth is the specular reflection point, which can be used to find d_2 and d in a more convenient form given by

$$d_2 = \left(\frac{h_r}{h_t + h_r} \right) d \tag{35.12}$$

$$h = \frac{-2h_t h_r}{h_t + h_r}. \tag{35.13}$$

Now, we can write

$$\Delta d = l_r - l_d = (p_1 + p_2) - (d_1 + d_2) \approx \frac{h^2}{2}\left(\frac{d}{d_1 d_2} \right). \tag{35.14}$$

We can express the phase difference between the LOS and the reflected path as

$$\Delta\varphi = \frac{2\pi\Delta d}{\lambda} = \frac{\pi}{2}v^2. \tag{35.15}$$

v is the dimensionless Fresnel–Kirchhoff diffraction parameter that corresponds to [17]:

$$v = h\sqrt{\frac{2d}{\lambda d_1 d_2}} = \frac{h}{r_n}\sqrt{2n}. \tag{35.16}$$

h / r_n is called the Fresnel zone clearance. According to the Fresnel knife-edge diffraction model, when v is approximately -0.8 or less, there is sufficient LOS path clearance and the diffraction loss is minimal. Substituting $v = -0.8$ in (35.16), it is found that $h = -0.566$, which implies that as long as almost 57% of the first Fresnel zone is kept clear of obstacles, diffraction loss can be neglected. To find the critical distance, d_C, we first substitute $v = -0.8$ in (35.15) and find $\Delta d = 0.16\lambda$ is the path lengths difference at the critical distance. Therefore, using the same approach that was employed to derive the break distance,

$$l_r - l_d = \Delta d = 0.16\lambda \tag{35.17}$$

$$l_r + l_d = \frac{25 h_t h_r}{\lambda}. \tag{35.18}$$

Adding (35.17) and (35.18), we can find the critical distance as

$$d_C = \sqrt{\left(\frac{12.5 h_t h_r}{\lambda} + \frac{\lambda}{12.5} \right)^2 - (h_t + h_r)^2}. \tag{35.19}$$

Alternatively, subtracting (35.17) from (35.18), we can write

$$d_C = \sqrt{\left(\frac{12.5 h_t h_r}{\lambda} - \frac{\lambda}{12.5}\right)^2 - (h_t - h_r)^2}. \tag{35.20}$$

We can rewrite (35.20) as

$$d_C = \frac{12.5 h_t h_r}{\lambda} \sqrt{1 - \frac{\lambda^2 (h_t^2 + h_r^2)}{(12.5)^2 h_t^2 h_r^2} + \left(\frac{\lambda^2}{(12.5)^2 h_t h_r}\right)^2} \approx \frac{12.5 h_t h_r}{\lambda}. \tag{35.21}$$

However, the approximate value in (35.21) is valid as long as $h_t, h_r \gg \lambda$, which does not apply to near-ground WSN scenario. Therefore, in this chapter, the exact formula in (35.20) will be used for the critical distance.

As witnessed in [25], dual-slope piecewise linear path loss model based on Fresnel zone clearance can predict the LOS microcellular propagation loss as accurately as a minimum mean square error (MMSE) regression fit on the measured data in open, urban and suburban areas. In the remainder of this chapter, we will inspect the accuracy and adequacy of such models for near-ground sensor networks.

35.3 NEAR-GROUND PATH LOSS MODELING

35.3.1 Short-Range Communication

As shown in Figure 35.2a, when $d < d_B$ the first Fresnel zone is clear of obstacles. In this region, reflected field from the ground interferes with the direct signal causing the signal level to oscillate widely. However, the attenuation of the median received signal with distance is almost the same as that of free space. Hence, attenuation is entirely due to the spreading of wavefront, which corresponds to the free-space path loss,

$$L_{fs} = (2kd)^2, \tag{35.22}$$

in which $k = \omega\sqrt{\mu\varepsilon_0} = 2\pi/\lambda$ is the free space wavenumber, ω is the angular frequency, ε_0 is the free space permittivity, μ is the permeability, and λ is the wavelength in free space.

35.3.2 Medium-Range Communication

As shown in Figure 35.2b, when $d > d_B$ part of the energy in the first Fresnel zone is intercepted by the ground. Therefore, attenuation results from both spherical wavefront spreading and obstruction of the first Fresnel zone, which leads to a more pronounced decay rate. The plane-earth propagation model offers a good prediction of signal attenuation in this region; however, assuming a flat, perfectly conducting ground leads to an overestimation of the destructive interference between the direct and the geometrical-optics (GO) reflection fields, which gives an imprecisely high power falloff rate. As found in [26], disregarding the electrical properties of the terrain can lead to grave errors as large as 10 dB, which is unacceptable in designing energy-constrained WSNs. Moreover, according to [23], including the physical statistics of the terrain roughness in the model may attenuate the ground wave by up to 6 dB for realistic ground surfaces.

To investigate the impact of roughness statistics on the signal transmission, the Fresnel reflection coefficient of a flat surface is modified using a correction factor to create an effective reflection coefficient. Traditional correction factors such as those proposed by Ament and Boithias are simple to implement and are sufficiently accurate for remote sensing and microcellular applications. However, for sensor network design, they have fundamental shortcomings, which stem from the near-grazing propagation condition. At low grazing angles, in conformity with the Rayleigh criterion, since the incident waves are almost parallel to the surface of the plane the direct and ground reflected components cancel each other even for larger roughness heights. Therefore, the surface waves that are smaller in magnitude and are highly localized to the ground boundary become the dominant propagation mechanism. However, surface waves are not considered in the conventional correction factors. They also overlook the impacts of polarization of the incident wave and roughness correlation length.

The effective reflection coefficients used in our model address the impacts of surface waves, incident polarization, and surface correlation distance and are founded on the perturbation theory applied to a volumetric integral equation originally prescribed in [22, 23]. They are valid for low grazing propagation as far as the surface roughness height is less than a wavelength. The two-ray model over rough dielectric terrain is given by

$$L_{tr} = L_{fs}\Delta L_{ex}, \tag{35.23}$$

in which L_{fs} is the free-space attenuation and L_{ex} is the excess loss factor due to obstruction of the first Fresnel zone, which corresponds to

$$L_{ex} = \left|1 + \frac{l_d}{l_r} R_\alpha^{eff} e^{(-j\Delta\varphi)}\right|^2, \tag{35.24}$$

where l_d, l_r, and $\Delta\varphi$ can be calculated using (35.3), (35.4), and (35.15), respectively. R_α^{eff} is the effective reflection coefficient where $\alpha = v, h$ denotes the vertical or horizontal incident polarization, respectively [22, 23]. We have

$$R_v^{eff} = R_v + \frac{\sigma^2}{2}(K_1^2 - k^2)(R_v^2 - 1) - \sigma^2 \frac{k_{ix}^2(k_1^2 - k^2)}{k^2(\mu_1 k_{iz} + k_{1zi})}[(R_v - 1)k_{iz} - (R_v + 1)k_{1zi}]$$

$$- \sigma^2 \frac{(k_1^2 - k^2)^2}{2\pi k^4(\mu_1 k_{iz} + k_{1zi})}[(R_v - 1)k_{iz}A_x(k_x) + (R_v + 1)k_{ix}A_z(k_x)] \tag{35.25}$$

$$R_h^{eff} = R_h - \frac{\sigma^2}{2}(k_1^2 - k^2)(R_h^2 - 1) - \sigma^2 \frac{(k_1^2 - k^2)^2}{2\pi(k_{iz} + k_{1zi})}(R_h + 1)A_y(k_x), \tag{35.26}$$

where

$$A_x(k_x) = \int_{-\infty}^{\infty} \frac{k_{1z}(k_{1zi}k_z' + k_{ix}k_x')}{(\varepsilon_1 k_z' + k_{1z})} W(k_x' - k_{ix})dk_x' \tag{35.27}$$

$$A_y(k_x) = \int_{-\infty}^{\infty} \frac{1}{k_z' + k_{1z}'} W(k_x' - k_{ix}) dk_x' \qquad (35.28)$$

$$A_z(k_x) = \int_{-\infty}^{\infty} \frac{k_x' (k_{1zi} k_z' + k_{ix} k_x')}{(\varepsilon_1 k_z' k_{1z}')} W(k_x' - k_{ix}) dk_x'. \qquad (35.29)$$

k is the wavenumber in the air and k_1 is the wavenumber in the dielectric. k_{iu} / k_u stands for the wavenumber components in the incident/scattered direction, where $u = x, z$. We have $k_x = k \sin\theta$ and $k_z = k \cos\theta$, in which θ is the elevation angle. The rough surface is characterized by a random height function $z = f(x)$, in which $f(x)$ is a random function with zero mean, that is, $\langle f(x) \rangle = 0$. Statistics of the roughness are included in the model using surface root mean square (RMS) height, σ, and roughness correlation length, L, which is incorporated in the Gaussian spectral density function $W(k_x) = \sqrt{\pi} L e^{(-k_x^2 L^2 / 4)}$. Electrical properties of the terrain are taken into account using the effective permittivity, ε_1, of the underlying dielectric layer. Integrals in (35.27–35.29) are fast converging and are evaluated numerically. Most of their contribution comes from a narrow angular range centered on the specular direction ($k_x = k_{ix}$), which expands by increasing the roughness height. R_v and R_h are the Fresnel reflection coefficients of the flat surface for vertical and horizontal polarizations, respectively,

$$R_v = \frac{k_1^2 k_{iz} - k^2 k_{1zi}}{k_1^2 k_{iz} + k^2 k_{1zi}} \qquad (35.30)$$

$$R_h = \frac{k_{iz} - k_{1zi}}{k_{iz} + k_{1zi}}. \qquad (35.31)$$

At this point, it is worth noting that a one dimensional (1-D) surface model is adopted to derive the effective reflection coefficients. A 1-D rough surface refers to a surface with protuberance along one horizontal coordinate and constant profile along the other, whereas a two-dimensional (2-D) rough surface has variations along both horizontal coordinates. As verified throughout our simulations, assuming a 1-D rough surface in the construction of a path loss model leads to tremendous computational savings while being perfectly adequate for the following reasons: (1) Surfaces under study are assumed to be geometrically isotropic. Examples of anisotropic surfaces are some wind-driven and cultivated lands; (2) in a path loss model, we are only interested in scattering effects in the plane of incidence. Scattering outside the plane of incidence cannot be accurately predicted using a 1-D model unless the roughness under consideration is truly 1-D [27]. On the other hand, off the plane scattering becomes critical when dealing with three-dimensional (3-D) spatial channel characterization; and (3) it is well-known that cross-polarized fields are only created by 2-D surfaces. However, the amplitude of the cross-polarized fields generated by realistic rough surfaces is generally several orders of magnitude smaller than that of the copolarized fields. Hence, cross-polarized components do not serve an important role in the transmission of energy and can be ignored in the path loss model.

35.3.3 Long-Range Communication

In cellular communications, cell radii are commonly much smaller than d_B to reduce the transmit power and increase the capacity. However, in WSNs, antenna heights are very low, and a significant part of the first Fresnel zone is always occupied by the ground; therefore, in most applications, the link range by far exceeds d_C. For example, at the frequency of 915 MHZ, if we consider the antenna heights are 7 cm, according to (35.7) and (35.20), the break distance, d_B, is only 2.2 cm and the critical distance, d_C, is roughly 16 cm. Hence, computing the diffraction loss, which can attenuate the transmitted signal by up to 6 dB, is integral to WSN channel characterization.

Here, total path attenuation is the sum of the loss due to an ideal knife-edge diffraction and an additional reference loss that takes account of the diffractor (rough terrain) parameters such as permittivity and roughness statistics that are discussed in the previous subsection. The path loss associated with knife-edge diffraction is calculated by assuming an asymptotically thin diffracting object halfway between the transmitter and receiver, which corresponds to [15]:

$$L_{ke} = \left(0.5 + \frac{0.877(h_t + h_r)}{\sqrt{\lambda d}} \right)^2. \tag{35.32}$$

Finally, the proposed near-ground WSN path loss model in decibel is summarized as

$$L_{NG}(dB) = \begin{cases} L_{fs}, & d \le d_B \\ L_{fs} + L_{ex}, & d_B \le d \le d_C \\ L_{fs} + L_{ex} + L_{ke}, & d \ge d_C \end{cases}, \tag{35.33}$$

where

$$L_{fs}(dB) = -27.56 + 20\log_{10}(f) + 20\log_{10}(d) \tag{35.34}$$

$$L_{ex}(dB) = 20\log_{10}\left| 1 + \frac{l_d}{l_r} R_\alpha^{eff} e^{(-j\Delta\varphi)} \right| \tag{35.35}$$

$$L_{ke}(dB) = 20\log_{10}\left(0.5 + \frac{0.877(h_t + h_r)}{\sqrt{\lambda d}} \right) \tag{35.36}$$

and d_B and d_C have been introduced in (35.6) and (35.19), respectively. Here, f is in MHz and d, h_t, h_r, and λ are all in meters.

35.3.4 Communication in Urban Setting

LOS urban propagation, for instance in an urban canyon, can be well predicted using the proposed model because the canyon only rises the spatial fading but the average trend does not change noticeably [7]. However, in NLOS scenarios, an additional term in the model is required to account for the steeper falloff rate. Here, to include the excess NLOS loss, we exploit an empirical formula originally derived for ground-based communicators in an L-shaped urban path at UHF [7],

$$L_{NLOS}(dB) = 8 + 10n_{NLOS}\log_{10}\frac{d}{d_L}, \tag{35.37}$$

in which n_{NLOS} is the NLOS path loss factor given by

$$n_{NLOS} = \left[4.27 \times 10^{-5} (f - 225) - 8.22 \times 10^{-4} (W - 17) + 0.033 \right] d_L + 2.7. \quad (35.38)$$

Here, f is the frequency in MHz, W is the width of the NLOS route in meters, and d_L is the length of the LOS route in meters prior to turning into the NLOS route. As we turn from the LOS street to the NLOS street, a discontinuity in the received power level occurs, which is referred to as the corner loss in [7] and is set to 8 dB based on averaging the measurement results which is included in (35.37).

35.3.5 Communication in Foliage Environments

WSNs deployed in wild environments often undergo an excess attenuation due to signal transmission through a depth of foliage. We shall add an empirically modeled foliage loss to our model in order to assess its prediction accuracy in forested environments. Foliage-induced excess loss is generally represented in mathematical form $L_{fo} (dB) = u f^v d_{fo}^w$, where f is the frequency of operation typically in MHz or GHz and d_{fo} is the propagation distance through foliage in meters. u, v, and w are numerical values evaluated using least squared error fitting on the measured data. Table 35.1 summarizes some of the well-known foliage loss models in the literature [13, 28–31].

TABLE 35.1. Summary of Foliage Loss Models

Model	Expression	Notes
Weissberger's Model [24]	$L_{fo} (dB) = \begin{cases} 0.45 f^{0.284} d_{fo}^1, & 0 < d_{fo} \leq 14\,m \\ 1.33 f^{0.284} d_{fo}^{0.588}, & 14\,m < d_{fo} \leq 1400\,m \end{cases}$	f in GHZ; d_{fo} in meters $0.230\,GHz \leq f \leq 95\,GHz$ for dense, dry, in leaf trees
COST 235 Model [25]	$L_{fo} (dB) = \begin{cases} 26.6 f^{-0.2} d_{fo}^{0.5}, & \text{out of leaf} \\ 15.6 f^{-0.009} d_{fo}^{0.26}, & \text{in leaf} \end{cases}$	f in MHz; d_{fo} in meters $9.6\,GHz \leq f \leq 57.6\,GHz$ $d_{fo} \leq 200\,m$
ITU-R Model [26]	$L_{fo} (dB) = 0.2 f^{0.3} d_{fo}^{0.6}$	f in MHz; d_{fo} in meters for UHF $d_{fo} \leq 400$ m
Fitted ITU-R Model [27]	$L_{fo} (dB) = \begin{cases} 0.37 f^{0.18} d_{fo}^{0.59}, & \text{out of leaf} \\ 0.39 f^{0.39} d_{fo}^{0.25}, & \text{in leaf} \end{cases}$	f in MHz; d_{fo} in meters for SHF
Lateral ITU-R Model [12]	$L_{fo} (dB) = 0.48 f^{0.43} d_{fo}^{0.13}$	f in MHz; d_{fo} in meters for VHF $d_{fo} \leq 5$ km for UHF $d_{fo} \leq 1.2$ km

35.4 MODEL VALIDATION AND WSN CONNECTIVITY ANALYSIS

In this section, prediction ability of the proposed model is verified by comparing it to the near-ground measurements in open, urban, and forested areas reported by independent researchers. Next, the influence of terrain electrical and geometrical properties on the range and connectivity of low-altitude WSNs are discussed.

35.4.1 Model Validation and Discussion

Here, the prediction results of our model are validated against available near-earth measurement campaigns. Many empirical models in the literature are provided without stating the dielectric and roughness properties of the terrain over which the measurements are performed. In these cases, we consider $\varepsilon = \varepsilon_r - j60\kappa\lambda$, where for average ground the dielectric constant is $\varepsilon_r = 15$, conductivity is $\kappa = 0.005 mhos / m$, $\sigma = 1.13 cm$ is the roughness RMS height, and surface correlation length is $L = 7.39 cm$ [15, 32]. For standard asphalt $\varepsilon_r = 6$, $\kappa = 0.0001 mhos / m$, $\sigma = 0.34 cm$, and $L = 4.2 cm$ [33, 34]. Moreover, transceiver antenna gains are removed from the received power to find the path loss whenever need be.

Figure 35.4a compares the proposed model predictions with the LOS stationary measurements at distance of 75 m using monopole antennas working at 300 MHz [6]. Model predictions match the measurements closely. It is noted that path loss depends largely on the terminal heights. Therefore, any model that does not explicitly include the impact of antennas altitude is ineffective in WSN design. In Figure 35.4b, measurements at 868 MHz are carried out in three different quasi-flat open environments, namely, a ground plain, a yard, and a grass park. $\lambda / 4$ monopole antennas are employed at height of 13 cm [5]. There is a vast disparity between the measurements, especially at long range, which is attributed to uneven terrain and presence of large scatterers in proximity of the measurement setup. Nevertheless, it is observed that the proposed model follows the slope of the measured curves along their entire range.

WSN nodes deployed in urban and suburban settings are generally much closer to the ground (usually asphalt) than to the building facades, which implies that the break distances formulated in this study apply to LOS urban topographies as well. Moreover, this model also works for a NLOS scenario provided it is supplemented by the excess loss term in (35.37). Figure 35.5 shows the path loss along an L-shaped urban route measured using omnidirectional wideband discone antennas at 225 MHz along with the model predictions [7]. Antennas height is 2 m and the width of the NLOS route is 17 m. Each result is piecewise linear with three distinct segments that correspond to a

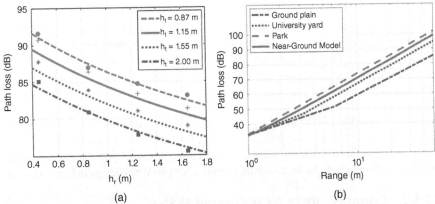

Figure 35.4 Comparison of the near-ground models predictions with measured data in open area. (a) Stationary measurements at distance of 75 m at 300 MHz ([5], fig. 12); (b) Measurements at 868 MHz recorded in three diverse open environments ([4], table 1).

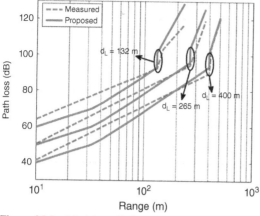

Figure 35.5 Model predictions versus measured path loss along an L-shaped urban route at 225 MHz ([6], Figures 5-7). Plots for d_L = 265 m and d_L = 132 m are offset by 10 dB and 20 dB, respectively, for illustration purposes.

LOS segment, a corner loss segment, and a NLOS segment. Since the measured LOS data are fitted into a single slope line, it appears that the measured and predicted results are diverging up to the break distance, which does not have a factual basis. It is observed that the LOS path loss, the discontinuity in the signal level due to corner loss, and the NLOS steeper falloff rate are all predicted with a reasonable accuracy.

It is very common for wireless sensors to communicate through a depth of foliage. Several empirical models are proposed to account for the excess loss caused by the foliage. These models are generally based on experimental data acquired over a shallow depth of a specific type of vegetation. This suggests that they do not address the impact of density and texture of foliage, as well as the propagation of lateral waves on the canopy-air interface. However, the latter phenomenon is only dominant at very large foliage depth, which does not apply to WSN usage.

In Figure 35.6, some well-known empirical foliage loss models are added to near-ground path loss model to evaluate their performance in low-altitude deployments. In Figure 35.6a, measurements are recorded using discone antennas in a deciduous forest at the frequency of 300 MHz. Antennas height is 0.75 m [6]. Another measurement campaign is conducted in a tropical palm plantation using omnidirectional antennas at height of 2.15 m [13]. Measurements are taken at the frequencies of 240 MHz and 700 MHz and are depicted in Figures 35.6b and Figure 35.6c, respectively. It is found that at a short range, predicted values of all the models are in good agreement with measured data. Nevertheless, Weissburger and International Telecommunication Union Radiocommunication Sector (ITU-R) models predictions diverge sharply from the measured data as the range in the foliage increases. Predicted path loss using the FITU-R and LITU-R models match the measured data closely, especially at the lower antenna height of 0.75 m, which demonstrates their applicability to WSN design in forested areas.

35.4.2 Connectivity in Near-Ground WSNs

In this subsection, the near-ground path loss model is used in WSN design by evaluating the maximum transmission range of nodes and studying network connectivity via Monte Carlo simulations.

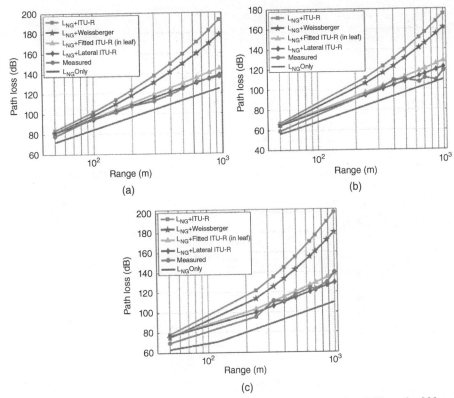

Figure 35.6 Measured and predicted path loss in forested areas. (a) $h_a = 0.75$ m, $f = 300$ MHz ([5], fi.g 10a); (b) $h_a = 2.15$ m, $f = 240$ MHz ([12], fig. 4a); (c) $h_a = 2.15$ m, $f = 700$ MHz ([12], fig. 4b).

Example 35.1 Evaluation of Mote Transmission Range Using the Near-Ground Path Loss Model

Compare the maximum link range found using the free-space model, two-ray model, and near-ground model and examine the impact of polarization, frequency, and dielectric properties of the ground on the link range. In this example, sensor motes have a maximum transmission power of 10 dBm and receiver sensitivity of −101 dBm for a maximum dynamic range of 111 dB, which are typical values for popular Mica2 motes [35]. Center frequency is 915 MHz unless otherwise specified.

Solution

The MATLAB programs for this example are provided as Figure35_7a.m, Figure35_7b.m, and Figure35_7c.m. Maximum link range is found by equating the path loss and the dynamic range of motes. When path loss exceeds the dynamic range, the communication between nodes is lost. Evaluation of the coverage range is critical in sensor deployment to find the optimum node density for which the quality of service (QoS), scalability, and reliability requirements are met while the cost is minimized.

In Figure 35.7a, coverage range is evaluated using different models introduced in Section 35.3, namely, the free-space model, the two-ray model and the

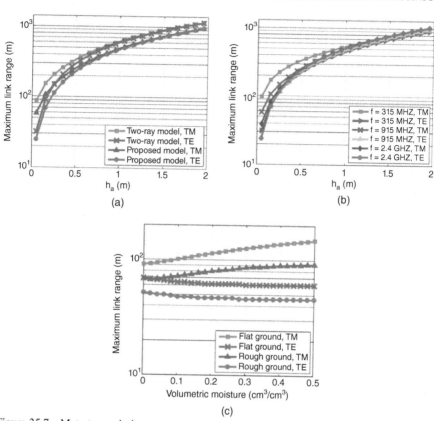

Figure 35.7 Mote transmission range: (a) versus antennas height using different models; (b) versus antennas height at different frequencies; (c) versus VWC over flat and rough ground.

near-ground model. The free-space model gives the unrealistically long coverage range of 4640 m independent of the terminal's height and is not shown on the plot. The antenna height, h_a, is observed to have a prominent role in confining the coverage area, and the maximum link range increases drastically when the antenna is elevated from the ground level. Comparing the results from the two-ray and the near-ground models, it is realized that incorporating the diffraction loss can significantly impact the link range, especially at lower altitudes. Closer to the ground level, antenna polarization also becomes important, and it is seen that vertically-polarized antennas can communicate over longer distances.

According to the plane-earth model, loss has no frequency dependence when $d \gg h_t, h_r$ [36]. However, it is shown in Figure 35.7b that lowering the antenna height undermines the frequency independence. To explain this observation, we note that in (35.25) and (35.26), with all the geometrical parameters fixed, the amplitude of higher-order waves rise as the wavelength increases; hence, in order to improve network connectivity, we should lower the operating frequency which increases the communication range.

Figure 35.7c depicts the influence of soil volumetric water content (VWC) on the communication range. Water content has a major impact on the electrical properties of soil [37]. Silt loam with 17.16% sand, 63.84% silt, and 19% clay represents the terrain dielectric in this simulation and the antenna height is 10 cm. Using the well-known semi-empirical dielectric mixing model in [38] and [39], we find that as the moisture content changes from 0 to 0.5 cm^3/cm^3, the dielectric constant of soil changes from $\varepsilon = 2.39$ to $\varepsilon = 33.83 - 3.24i$. This steep increase in the dielectric constant, however, only slightly alters the Fresnel reflection coefficient at grazing incidence. The amplitude of the reflection coefficient will decrease (increase) slowly for vertical (horizontal) polarization, and as VWC increases it will remain very close to −1, which warrants the near cancellation of the direct and specularly reflected waves. It is perceived that even slight variation in the reflection coefficient causes a noticeable change in the coverage radius. Moreover, the plane-earth model tends to overestimate the coverage range compared to the near-ground model that takes surface irregularities into consideration.

In high-node-count applications such as environmental or security monitoring, a large number of low-cost autonomous sensors are spatially distributed to cooperatively monitor certain physical or environmental conditions. Here, mesh connectivity and the number of neighbors in range per node have a crucial influence on network performance, reliability, and power conservation.

Example 35.2 Estimation of Average Number of Single-Hop Neighbors in the Network

Using Monte Carlo simulations, find the average degree of a node in a network consisting of 100 nodes randomly distributed in the 2-D plane of 1000 ft × 1000 ft (304.8 m × 304.8 m).

Solution

The MATLAB programs for this example are provided as Figure35_8a.m, Figure35_8b.m, and Figure35_8c.m. In order to assess the average number of single-hop neighbors in range (average degree of a node) in the network, Monte Carlo simulations are used. In each iteration, 100 nodes are randomly distributed in the 2-D plane of 1000 ft × 1000 ft (304.8 m × 304.8 m) and average number of neighbors is calculated as the ratio of the total number of links within range to the number of nodes in the network.

In Figure 35.8a, different loss models are used to predict the average number of neighbors. The free-space model predicts full mesh connectivity; that is, each node has 99 single-hop neighbors independent of terminal height. Vertical polarization yields far better connectivity at lower altitudes in comparison to horizontal polarization. It is also found that connectivity has high sensitivity to terminal height and almost spans full range as height increases from the ground level to 0.8 m; particularly, closer to the ground level, the height effect magnifies. Figure 35.8b shows the impact of frequency on average number of neighbors. At lower frequencies, network connectivity is improved. In Figure 35.8c, we study the effect of soil VWC on network connectivity. It is noticed that a fluctuating terrain degrades network connectivity in contrast with a flat one. An interesting observation is that as soil moisture content increases, for example, as a result of precipitation, connectivity enhances if the antennas are vertically polarized but deteriorates if the antennas are horizontally polarized. However, horizontally polarized antennas are minimally affected by change in roughness and dielectric properties of the ground.

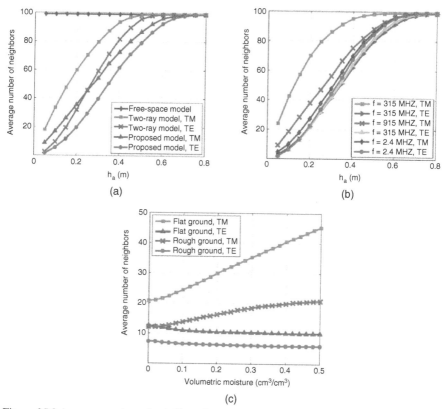

Figure 35.8 Average number of neighbors for a 100-node network with uniform topology: (a) versus antennas height using different models; (b) versus antennas height at different frequencies; (c) versus VWC over flat and rough ground.

35.5 CONCLUSION

In this chapter, an improved computationally feasible near-ground field prediction model was presented to facilitate highly accurate WSN simulations. The model was validated against published measured data in open, urban, and forested areas. The accuracy of this model is due to careful assessment of the impact of first Fresnel zone obstruction, terrain irregularities, and dielectric properties of the ground on the LOS, specular reflection, and higher-order waves. It is realized that the near-ground model applies to LOS urban topographies as well. Moreover, this model also works for NLOS urban scenario, provided it is supplemented by a proper NLOS excess loss term. Various empirical foliage loss models were added to near-ground path loss model and compared to measured data in near-ground foliage environments to determine which one is more suitable for low-altitude applications.

The near-ground model was also used to evaluate the effects of radio link and terrain parameters on the transmission range and network connectivity of WSNs. Some practical implications learned from this model include: (a) the critical distance is very small in WSN applications and, therefore, the diffraction loss is integral to WSN channel characterization; (b) at grazing angles, Fresnel reflection coefficient displays a very low sensitivity to terrain dielectric constant; (c) provided the geometrical parameters are fixed, higher-order waves intensify as the wavelength increases; (d) antenna height is by far the most influential geometrical parameter to link range and network connectivity; (e) coverage radius and connectivity are fairly sensitive to the reflection coefficient when antennas are placed near the ground; (f) terrain roughness degrades the accessible range and connectivity; (g) lowering the frequency of operation, enhances the reachable communication distance and network connectivity; (h) close to the ground level, vertically polarized antennas outperform their horizontally polarized counterparts in terms of coverage range and connectivity; (i) precipitation boosts/reduces the maximum link range and network connectivity when motes are equipped with vertically/horizontally polarized antennas.

REFERENCES

[1] I. F. Akyildiz, S. Weilian, Y. Sankarasubramaniam, and E. Cayirci, "A survey on sensor networks," *IEEE Commun. Mag.*, vol. 40, pp. 102–114, 2002.

[2] F. P. Fontán and P. M. Espiñeira, *Modelling the Wireless Propagation Channel: A Simulation Approach with MATLAB*, vol. 5. Hoboken, NJ, John Wiley & Sons, 2008.

[3] M. Rodriguez, R. Feick, H. Carrasco, R. Valenzuela, M. Derpich, and L. Ahumada, "Wireless access channels with near-ground level antennas," *IEEE Trans. Wireless Commun.*, vol. 11, pp. 2204–2211, 2012.

[4] S. L. Javali, A. Torabi, and S. A. Zekavat, "Snow covered forest channel modeling for near-ground wireless sensor networks," in *2017 IEEE Int. Conf. on Wireless for Space and Extreme Environments (WiSEE)*, Montreal, QC, 2017, pp. 69–74.

[5] A. Martinez-Sala, J.-M. Molina-Garcia-Pardo, E. Egea-Lopez, J. Vales-Alonso, L. Juan-Llacer, and J. Garcia-Haro, "An accurate radio channel model for wireless sensor networks simulation," *J. Commun. Netw.*, vol. 7, pp. 401–407, 2005.

[6] G. G. Joshi, C. B. Dietrich, C. R. Anderson, W. G. Newhall, W. A. Davis, J. Isaacs, and G. Barnett, "Near-ground channel measurements over line-of-sight and forested paths," *IEEE Proc. on Microwaves, Antennas and Propag.*, vol. 152, pp. 589–596, 2005.

[7] J. R. Hampton, N. M. Merheb, W. L. Lain, D. E. Paunil, R. M. Shuford, and W. Kasch, "Urban propagation measurements for ground based communication in the military UHF band," *IEEE Trans. Antennas Propag.*, vol. 54, pp. 644–654, 2006.

[8] L. DaHan and K. Sarabandi, "Terminal-to-terminal hybrid full-wave simulation of low-profile, electrically-small, near-ground antennas," *IEEE Trans. Antennas Propag.*, vol. 56, pp. 806–814, 2008.

[9] L. DaHan and K. Sarabandi, "Simulation of near-ground long-distance radiowave propagation over terrain using Nystrom method with phase extraction technique and FMM-acceleration," *IEEE Trans. Antennas Propag.*, vol. 57, pp. 3882–3890, 2009.

[10] M. I. Aslam and S. A. Zekavat, "New channel path loss model for near-ground antenna sensor networks," *Wireless Sensor Systems, IET*, vol. 2, pp. 103–107, 2012.

[11] R. A. Foran, T. B. Welch, and M. J. Walker, "Very near ground radio frequency propagation measurements and analysis for military applications," in *Military Communications Conf. Proc., MILCOM 1999*, Atlantic City, NJ, vol. 1, 1999, pp. 336–340.

[12] L. DaHan and K. Sarabandi, "Near-earth wave propagation characteristics of electric dipole in presence of vegetation or snow layer," *IEEE Trans. Antennas Propag.*, vol. 53, pp. 3747–3756, 2005.

[13] M. Yu Song, L. Yee Hui, and N. Boon Chong, "Empirical near ground path loss modeling in a forest at VHF and UHF bands," *IEEE Trans. Antennas Propag.*, vol. 57, pp. 1461–1468, 2009.

[14] A. Torabi and S. A. R. Zekavat, "MIMO channel characterization over random rough dielectric terrain," in *2014 IEEE 25th Annual Int. Symp. on Personal, Indoor, and Mobile Radio Communication (PIMRC)*, Washington, DC, 2014, pp. 161–165.

[15] T. S. Rappaport, *Wireless Communications: Principles and Practice*, vol. 2. Englewood Cliffs, NJ: Prentice Hall, 1996.

[16] A. Torabi and S. A. Zekavat, "Near-ground channel modeling for distributed cooperative communications," *IEEE Trans. Antennas Propag.*, vol. 64, pp. 2494–2502, 2016.

[17] A. Goldsmith, *Wireless Communications*. Cambridge University Press, 2005.

[18] A. Torabi, S. A. Zekavat, and A. A. Rasheed, "Millimeter wave directional channel modeling," in *2015 IEEE Int. Conf. on Wireless for Space and Extreme Environments (WiSEE)*, Orlando, FL, 2015, pp. 1–6.

[19] W. S. Ament, "Toward a theory of reflection by a rough surface," *Proc. of the IRE*, vol. 41, 1953, pp. 142–146.

[20] L. Boithias and L.-J. Libols, *Radio Wave Propagation*. London: North Oxford Academy, 1987.

[21] F. T. Dagefu and K. Sarabandi, "Analysis and modeling of near-ground wave propagation in the presence of building walls," *IEEE Trans. Antennas Propag.*, vol. 59, pp. 2368–2378, 2011.

[22] K. Sarabandi and C. Tsenchieh, "Electromagnetic scattering from slightly rough surfaces with inhomogeneous dielectric profiles," *IEEE Trans. Antennas Propag.*, vol. 45, pp. 1419–1430, 1997.

[23] L. DaHan and K. Sarabandi, "On the effective low-grazing reflection coefficient of random terrain roughness for modeling near-earth radiowave propagation," *IEEE Trans. Antennas Propag.*, vol. 58, pp. 1315–1324, 2010.

[24] A. Torabi and S. A. Zekavat, "A rigorous model for predicting the path loss in near-ground wireless sensor networks," in *2015 IEEE 82nd Vehicular Technology Conf. (VTC2015-Fall)*, Boston, MA, 2015, pp. 1–5.

[25] M. J. Feuerstein, K. L. Blackard, T. S. Rappaport, S. Y. Seidel, and H. Xia, "Path loss, delay spread, and outage models as functions of antenna height for microcellular system design," *IEEE Trans. Veh. Technol.* vol. 43, pp. 487–498, 1994.

[26] A. Torabi, S. A. Zekavat, and K. Sarabandi, "Wideband Directional Channel Characterization for Multiuser MIMO Systems Over a Random Rough Dielectric Ground," in *IEEE Transactions on Wireless Communications*, vol. 15, no. 5, pp. 3103–3113, May 2016.

[27] J. T. Johnson, "Computer simulations of rough surface scattering," in *Light Scattering and Nanoscale Surface Roughness*, Ed. Springer, 2007, pp. 181–210.

[28] M. A. Weissberger, "An initial critical summary of models for predicting the attenuation of radio waves by trees," DTIC Document 1982.

[29] COST 235, "Radio propagation effects on next-generation fixed-service terrestrial telecommunication systems," Luxembourg, Final Rep., 1996.

[30] CCIR, "Influences of terrain irregularities and vegetation on troposphere propagation," Geneva, pp. 235–236, CCIR Rep, 1986.

[31] M. O. Al-Nuaimi and R. B. L. Stephens, "Measurements and prediction model optimisation for signal attenuation in vegetation media at centimetre wave frequencies," *IEEE Proc. Microwaves, Antennas Propag.* vol. 145, pp. 201–206, 1998.

[32] M. Rahman, M. Moran, D. Thoma, R. Bryant, E. Sano, C. Holifield Collins, S. Skirvin, C. Kershner, and B. Orr, "A derivation of roughness correlation length for parameterizing radar backscatter models," *Int. J. Remote Sens.*, vol. 28, pp. 3995–4012, 2007.

[33] J. Shang, J. Umana, F. Bartlett, and J. Rossiter, "Measurement of complex permittivity of asphalt pavement materials," *J. Transp. Eng.*, vol. 125, pp. 347–356, 1999.

[34] K. Sarabandi, E. S. Li, and A. Nashashibi, "Modeling and measurements of scattering from road surfaces at millimeter-wave frequencies," *IEEE Trans. Antennas Propag.*, vol. 45, pp. 1679–1688, 1997.

[35] Mica2 MOTES, http://www.moog-crossbow.com

[36] A. Torabi, "Channel modeling for fifth generation cellular networks and wireless sensor networks," Ph.D. dissertation, Michigan Technological University, 2016.

[37] F. T. Ulaby, D. G. Long, W. J. Blackwell, C. Elachi, A. K. Fung, C. Ruf, K. Sarabandi, H. A. Zebker, and J. Van Zyl, *Microwave Radar and Radiometric Remote Sensing*. Ann Arbor: University of Michigan Press, 2014.

[38] N. R. Peplinski, F. T. Ulaby, and M. C. Dobson, "Dielectric properties of soils in the 0.3-1.3-GHz range," *IEEE Trans. Geosci. Remote Sens.* vol. 33, pp. 803–807, 1995.

[39] N. R. Peplinski, F. T. Ulaby, and M. C. Dobson, "Corrections to dielectric properties of soils in the 0.3-1.3-GHz range," *IEEE Trans. Geosci. Remote Sens.* vol. 33, p. 1340, 1995.

INDEX

Note: Page numbers in *italics* refer to figures, those in **bold** to tables.

Handbook of Position Location: Theory, Practice, and Advances, Second Edition.
Edited by S. A. (Reza) Zekavat and R. Michael Buehrer.
© 2019 by the Institute of Electrical and Electronics Engineers, Inc.
Published 2019 by John Wiley & Sons, Inc.
Companion Website: www.wiley.com/go/zekavat/positionlocation2e

IEEE PRESS SERIES ON DIGITAL AND MOBILE COMMUNICATION

John B. Anderson, *Series Editor*
University of Lund

The manufacturer's authorised representative in the EU for product safety is Oxford University Press España S.A. of El Parque Empresarial San Fernando de Henares, Avenida de Castilla, 2 – 28830 Madrid (www.oup.es/en or product.safety@oup.com). OUP España S.A. also acts as importer into Spain of products made by the manufacturer.

Printed in the USA/Agawam, MA
January 27, 2025

880951.012